More information about this series at http://www.springer.com/series/7409

Leif Azzopardi · Benno Stein ·
Norbert Fuhr · Philipp Mayr ·
Claudia Hauff · Djoerd Hiemstra (Eds.)

Advances in Information Retrieval

41st European Conference on IR Research, ECIR 2019
Cologne, Germany, April 14–18, 2019
Proceedings, Part I

Editors
Leif Azzopardi
University of Strathclyde
Glasgow, UK

Norbert Fuhr
Universität Duisburg-Essen
Duisburg, Germany

Claudia Hauff
Delft University of Technology
Delft, The Netherlands

Benno Stein
Bauhaus Universität Weimar
Weimar, Germany

Philipp Mayr
GESIS - Leibniz Institute
for the Social Sciences
Cologne, Germany

Djoerd Hiemstra
University of Twente
Enschede, The Netherlands

ISSN 0302-9743 ISSN 1611-3349 (electronic)
Lecture Notes in Computer Science
ISBN 978-3-030-15711-1 ISBN 978-3-030-15712-8 (eBook)
https://doi.org/10.1007/978-3-030-15712-8

Library of Congress Control Number: 2019934339

LNCS Sublibrary: SL3 – Information Systems and Applications, incl. Internet/Web, and HCI

This Springer imprint is published by the registered company Springer Nature Switzerland AG
The registered company address is: Gewerbestrasse 11, 6330 Cham, Switzerland

Preface

The 41st European Conference on Information Retrieval (ECIR) was held in Cologne, Germany, during April 14–18, 2019, and brought together hundreds of researchers from Europe and abroad. The conference was organized by GESIS – Leibniz Institute for the Social Sciences and the University of Duisburg-Essen—in cooperation with the British Computer Society's Information Retrieval Specialist Group (BCS-IRSG).

These proceedings contain the papers, presentations, workshops, and tutorials given during the conference. This year the ECIR 2019 program boasted a variety of novel work from contributors from all around the world and provided new platforms for promoting information retrieval-related (IR) activities from the CLEF Initiative. In total, 365 submissions were fielded across the tracks from 50 different countries.

The final program included 39 full papers (23% acceptance rate), 44 short papers (29% acceptance rate), eight demonstration papers (67% acceptance rate), nine reproducibility full papers (75% acceptance rate), and eight invited CLEF papers. All submissions were peer reviewed by at least three international Program Committee members to ensure that only submissions of the highest quality were included in the final program. As part of the reviewing process we also provided more detailed review forms and guidelines to help reviewers identify common errors in IR experimentation as a way to help ensure consistency and quality across the reviews.

The accepted papers cover the state of the art in IR: evaluation, deep learning, dialogue and conversational approaches, diversity, knowledge graphs, recommender systems, retrieval methods, user behavior, topic modelling, etc., and also included novel application areas beyond traditional text and Web documents such as the processing and retrieval of narrative histories, images, jobs, biodiversity, medical text, and math. The program boasted a high proportion of papers with students as first authors, as well as papers from a variety of universities, research institutes, and commercial organizations.

In addition to the papers, the program also included two keynotes, four tutorials, four workshops, a doctoral consortium, and an industry day. The first keynote was presented by this year's BCS IRSG Karen Sparck Jones Award winner, Prof. Krisztian Balog, On Entities and Evaluation, and the second keynote was presented by Prof. Markus Strohmaier, On Ranking People. The tutorials covered a range of topics from conducting lab-based experiments and statistical analysis to categorization and deep learning, while the workshops brought together participants to discuss algorithm selection (AMIR), narrative extraction (Text2Story), Bibliometrics (BIR), as well as social media personalization and search (SoMePeAS). As part of this year's ECIR we also introduced a new CLEF session to enable CLEF organizers to report on and promote their upcoming tracks. In sum, this added to the success and diversity of ECIR and helped build bridges between communities.

The success of ECIR 2019 would not have been possible without all the help from the team of volunteers and reviewers. We wish to thank all our track chairs for coordinating the different tracks along with the teams of meta-reviewers and reviewers who helped ensure the high quality of the program. We also wish to thank the demo chairs: Christina Lioma and Dagmar Kern; student mentorship chairs: Ahmet Aker and Laura Dietz; doctoral consortium chairs: Ahmet Aker, Dimitar Dimitrov and Zeljko Carevic; workshop chairs: Diane Kelly and Andreas Rauber; tutorial chairs: Guillaume Cabanac and Suzan Verberne; industry chair: Udo Kruschwitz; publicity chair: Ingo Frommholz; and sponsorship chairs: Jochen L. Leidner and Karam Abdulahhad. We would like to thank our webmaster, Sascha Schüller and our local chair, Nina Dietzel along with all the student volunteers who helped to create an excellent online and offline experience for participants and attendees.

ECIR 2019 was sponsored by: DFG (Deutsche Forschungsgemeinschaft), BCS (British Computer Society), SIGIR (Special Interest Group on Information Retrieval), City of Cologne, Signal Media Ltd, Bloomberg, Knowledge Spaces, Polygon Analytics Ltd., Google, Textkernel, MDPI Open Access Journals, and Springer. We thank them all for their support and contributions to the conference.

Finally, we wish to thank all the authors, reviewers, and contributors to the conference.

April 2019

Leif Azzopardi
Benno Stein
Norbert Fuhr
Philipp Mayr
Claudia Hauff
Djoerd Hiemstra

Organization

General Chairs

Norbert Fuhr	Universität Duisburg-Essen, Germany
Philipp Mayr	GESIS – Leibniz Institute for the Social Sciences, Germany

Program Chairs

Leif Azzopardi	University of Glasgow, UK
Benno Stein	Bauhaus-Universität Weimar, Germany

Short Papers and Poster Chairs

Claudia Hauff	Delft University of Technology, The Netherlands
Djoerd Hiemstra	University of Twente, The Netherlands

Workshop Chairs

Diane Kelly	University of Tennessee, USA
Andreas Rauber	Vienna University of Technology, Austria

Tutorial Chairs

Guillaume Cabanac	University of Toulouse, France
Suzan Verberne	Leiden University, The Netherlands

Demo Chairs

Christina Lioma	University of Copenhagen, Denmark
Dagmar Kern	GESIS – Leibniz Institute for the Social Sciences, Germany

Industry Day Chair

Udo Kruschwitz	University of Essex, UK

Proceedings Chair

Philipp Mayr	GESIS – Leibniz Institute for the Social Sciences, Germany

Publicity Chair

Ingo Frommholz	University of Bedfordshire, UK

Sponsor Chairs

Jochen L. Leidner	Thomson Reuters/University of Sheffield, UK
Karam Abdulahhad	GESIS – Leibniz Institute for the Social Sciences, Germany

Student Mentoring Chairs

Ahmet Aker	Universität Duisburg-Essen, Germany
Laura Dietz	University of New Hampshire, USA

Website Infrastructure

Sascha Schüller	GESIS – Leibniz Institute for the Social Sciences, Germany

Local Chair

Nina Dietzel	GESIS – Leibniz Institute for the Social Sciences, Germany

Program Committee

Mohamed Abdel Maksoud	Codoma.tech Advanced Technologies, Egypt
Ahmed Abdelali	Research Administration
Karam Abdulahhad	GESIS - Leibniz institute for the Social Sciences, Germany
Dirk Ahlers	Norwegian University of Science and Technology, Norway
Qingyao Ai	University of Massachusetts Amherst, USA
Ahmet Aker	University of Duisburg Essen, Germany
Elif Aktolga	Apple Inc., USA
Dyaa Albakour	Signal Media, UK
Giambattista Amati	Fondazione Ugo Bordoni, Italy
Linda Andersson	Vienna University of Technology, Austria
Avi Arampatzis	Democritus University of Thrace, Greece
Ioannis Arapakis	Telefonica Research, Spain
Jaime Arguello	The University of North Carolina at Chapel Hill, USA
Leif Azzopardi	University of Strathclyde, UK
Ebrahim Bagheri	Ryerson University, USA
Krisztian Balog	University of Stavanger, Norway
Alvaro Barreiro	University of A Coruña, Spain

Carsten Eickhoff	Brown University, USA
Tamer Elsayed	Qatar University, Qatar
Liana Ermakova	UBO, France
Cristina España-Bonet	UdS and DFKI, Germany
Jose Alberto Esquivel	Signal Media, UK
Hui Fang	University of Delaware, USA
Hossein Fani	University of New Brunswick, Canada
Paulo Fernandes	PUCRS, Brazil
Nicola Ferro	University of Padua, Italy
Ingo Frommholz	University of Bedfordshire, UK
Norbert Fuhr	University of Duisburg-Essen, Germany
Michael Färber	University of Freiburg, Germany
Patrick Gallinari	LIP6 - University of Paris 6, France
Shreyansh Gandhi	Walmart Labs, USA
Debasis Ganguly	IBM Ireland Research Lab, Ireland
Wei Gao	Victoria University of Wellington, New Zealand
Dario Garigliotti	University of Stavanger, Norway
Anastasia Giachanou	University of Lugano, Switzerland
Giorgos Giannopoulos	Imis Institute, Athena R.C., Greece
Lorraine Goeuriot	University Grenoble Alpes, CNRS, France
Julio Gonzalo	UNED, Spain
Pawan Goyal	IIT Kharagpur, India
Michael Granitzer	University of Passau, Germany
Guillaume Gravier	CNRS, IRISA, France
Shashank Gupta	International Institute of Information Technology, Hyderabad, India
Cathal Gurrin	Dublin City University, Ireland
Matthias Hagen	Martin-Luther-Universität Halle-Wittenberg, Germany
Shuguang Han	Google, USA
Allan Hanbury	Vienna University of Technology, Austria
Preben Hansen	Stockholm University, Sweden
Donna Harman	NIST, USA
Morgan Harvey	Northumbria University, UK
Faegheh Hasibi	Norwegian University of Science and Technology, Norway
Claudia Hauff	Delft University of Technology, The Netherlands
Jer Hayes	Accenture, Ireland
Ben He	University of Chinese Academy of Sciences, China
Nathalie Hernandez	IRIT, France
Djoerd Hiemstra	University of Twente, The Netherlands
Frank Hopfgartner	The University of Sheffield, UK
Andreas Hotho	University of Würzburg, Germany
Gilles Hubert	IRIT, France
Ali Hürriyetoğlu	Koc University, Turkey
Dmitry Ignatov	National Research University Higher School of Economics, Russia

Bogdan Ionescu	University Politehnica of Bucharest, Romania
Radu Tudor Ionescu	University of Bucharest, Romania
Shoaib Jameel	Kent University, UK
Adam Jatowt	Kyoto University, Japan
Shen Jialie	Queen's University, Belfast, UK
Jiepu Jiang	Virginia Tech, USA
Xiaorui Jiang	Aston University, UK
Alexis Joly	Inria, France
Gareth Jones	Dublin City University, Ireland
Jaap Kamps	University of Amsterdam, The Netherlands
Nattiya Kanhabua	NTENT, Spain
Jaana Kekäläinen	University of Tampere, Finland
Diane Kelly	University of Tennessee, USA
Liadh Kelly	Maynooth University, Ireland
Dagmar Kern	GESIS, Germany
Roman Kern	Graz University of Technology, Austria
Dhruv Khattar	International Institute of Information Technology Hyderabad, India
Julia Kiseleva	Microsoft Research AI, USA
Dietrich Klakow	Saarland University, Germany
Yiannis Kompatsiaris	CERTH - ITI, Greece
Kriste Krstovski	University of Massachusetts Amherst, USA
Udo Kruschwitz	University of Essex, UK
Vaibhav Kumar	Carnegie Mellon University, USA
Oren Kurland	Technion, Israel
Chiraz Latiri	University of Tunis, Tunisia
Wang-Chien Lee	The Pennsylvania State University, USA
Teerapong Leelanupab	King Mongkut's Institute of Technology Ladkrabang
Mark Levene	Birkbeck, University of London, UK
Liz Liddy	Center for Natural Language Processing, Syracuse University, USA
Nut Limsopatham	Amazon
Chunbin Lin	Amazon AWS, USA
Christina Lioma	University of Copenhagen, Denmark
Aldo Lipani	University College London, UK
Nedim Lipka	Adobe, USA
Elena Lloret	University of Alicante, Spain
Fernando Loizides	Cardiff University, UK
David Losada	University of Santiago de Compostela, Spain
Natalia Loukachevitch	Research Computing Center of Moscow State University, Russia
Bernd Ludwig	University of Regensburg, Germany
Mihai Lupu	Research Studios, Austria
Craig Macdonald	University of Glasgow, UK
Andrew Macfarlane	City University London, UK
Joao Magalhaes	Universidade NOVA de Lisboa, Portugal

Walid Magdy	The University of Edinburgh, UK
Marco Maggini	University of Siena, Italy
Maria Maistro	University of Padua, Italy
Thomas Mandl	University of Hildesheim, Germany
Behrooz Mansouri	Rochester Institute of Technology, USA
Ilya Markov	University of Amsterdam, The Netherlands
Stefania Marrara	Consorzio C2T, Italy
Miguel Martinez-Alvarez	Signal, UK
Bruno Martins	IST and INESC-ID - Instituto Superior Técnico, University of Lisbon, Portugal
Luis Marujo	Snap Inc., USA
Yosi Mass	IBM Haifa Research Lab, Israel
Philipp Mayr	GESIS, Germany
Edgar Meij	Bloomberg L.P., UK
Massimo Melucci	University of Padua, Italy
Marcelo Mendoza	Universidad Técnica Federico Santa María, Brazil
Alessandro Micarelli	Roma Tre University, Italy
Dmitrijs Milajevs	Queen Mary University of London, UK
Seyed Abolghasem Mirroshandel	Sharif University of Technology, Iran
Stefano Mizzaro	University of Udine, Italy
Boughanem Mohand	IRIT University Paul Sabatier Toulouse, France
Felipe Moraes	Delft University of Technology, The Netherlands
Ajinkya More	Netflix, USA
Jose Moreno	IRIT/UPS, France
Yashar Moshfeghi	University of Strathclyde, UK
Josiane Mothe	Institut de Recherche en Informatique de Toulouse, France
André Mourão	Universidade NOVA de Lisboa, Portugal
Henning Müller	HES-SO, Switzerland
Franco Maria Nardini	ISTI-CNR, Italy
Wolfgang Nejdl	L3S and University of Hannover, Germany
Mahmood Neshati	Sharif University of Technology, Iran
Dong Nguyen	University of Twente, The Netherlands
Massimo Nicosia	University of Trento, Italy
Jian-Yun Nie	University of Montreal, Canada
Qiang Ning	University of Illinois at Urbana-Champaign, USA
Boris Novikov	National Research University Higher School of Economics, Russia
Andreas Nuernberger	Otto-von-Guericke University of Magdeburg, Germany
Aurélie Névéol	LIMSI, CNRS, Université Paris-Saclay, France
Neil O'Hare	Yahoo Research, USA
Salvatore Orlando	Università Ca' Foscari Venezia, Italy
Iadh Ounis	University of Glasgow, UK
Mourad Oussalah	University of Oulu, Finland
Deepak P.	Queen's University Belfast, Ireland

Jiaul Paik	IIT Kharagpur, India
Joao Palotti	Vienna University of Technology/Qatar Computing Research Institute, Austria/Qatar
Girish Palshikar	Tata Research Development and Design Centre, India
Javier Parapar	University of A Coruña, Spain
Gabriella Pasi	Università degli Studi di Milano Bicocca, Italy
Arian Pasquali	University of Porto, Portugal
Bidyut Kr. Patra	VTT Technical Research Centre, Finland
Pavel Pecina	Charles University in Prague, Czech Republic
Gustavo Penha	UFMG
Avar Pentel	University of Tallinn, Estonia
Raffaele Perego	ISTI-CNR, Italy
Vivien Petras	Humboldt-Universität zu Berlin, Germany
Jeremy Pickens	Catalyst Repository Systems, USA
Karen Pinel-Sauvagnat	IRIT, France
Florina Piroi	Vienna University of Technology, Austria
Benjamin Piwowarski	CNRS, Pierre et Marie Curie University, France
Vassilis Plachouras	Facebook, UK
Bob Planque	Vrije Universiteit Amsterdam, The Netherlands
Senja Pollak	University of Ljubljana, Slovenia
Martin Potthast	Leipzig University, Germany
Georges Quénot	Laboratoire d'Informatique de Grenoble, CNRS, France
Razieh Rahimi	University of Tehran, Iran
Nitin Ramrakhiyani	TCS Research, Tata Consultancy Services Ltd., India
Jinfeng Rao	University of Maryland, USA
Andreas Rauber	Vienna University of Technology, Austria
Traian Rebedea	University Politehnica of Bucharest, Romania
Navid Rekabsaz	Idiap Research Institute, Switzerland
Steffen Remus	University of Hamburg, Germany
Paolo Rosso	Universitat Politècnica de València, Spain
Dmitri Roussinov	University of Strathclyde, UK
Stefan Rueger	Knowledge Media Institute, UK
Tony Russell-Rose	UXLabs, UK
Alan Said	University of Skövde, Sweden
Mark Sanderson	RMIT University, Australia
Eric Sanjuan	Laboratoire Informatique d'Avignon- Université d'Avignon, France
Rodrygo Santos	Universidade Federal de Minas Gerais, Brazil
Kamal Sarkar	Jadavpur University, Kolkata, India
Fabrizio Sebastiani	Italian National Council of Research, Italy
Florence Sedes	IRIT P. Sabatier University, France
Giovanni Semeraro	University of Bari, Italy
Procheta Sen	Indian Statistical Institute, India
Armin Seyeditabari	UNC Charlotte, USA
Mahsa Shahshahani	University of Amsterdam, The Netherlands
Azadeh Shakery	University of Tehran, Iran

Manish Shrivastava	International Institute of Information Technology, Hyderabad, India
Ritvik Shrivastava	Columbia University, USA
Rajat Singh	International Institute of Information Technology, Hyderabad, India
Eero Sormunen	University of Tampere, Finland
Laure Soulier	Sorbonne Universités UPMC-LIP6, France
Rene Spijker	Cochran, The Netherlands
Efstathios Stamatatos	University of the Aegean, Greece
Benno Stein	Bauhaus-Universität Weimar, Germany
L. Venkata Subramaniam	IBM Research, India
Hanna Suominen	The ANU, Australia
Pascale Sébillot	IRISA, France
Lynda Tamine	IRIT, France
Thibaut Thonet	University of Grenoble Alpes, France
Marko Tkalcic	Free University of Bozen-Bolzano, Italy
Nicola Tonellotto	ISTI-CNR, Italy
Michael Tschuggnall	Institute for computer science, DBIS, Innsbruck, Austria
Theodora Tsikrika	Information Technologies Institute, CERTH, Greece
Denis Turdakov	Institute for System Programming RAS
Ferhan Ture	Comcast Labs, USA
Yannis Tzitzikas	University of Crete and FORTH-ICS, Greece
Sumithra Velupillai	KTH Royal Institute of Technology, Sweden
Suzan Verberne	Leiden University, The Netherlands
Vishwa Vinay	Adobe Research Bangalore, India
Marco Viviani	Università degli Studi di Milano-Bicocca - DISCo, Italy
Stefanos Vrochidis	Information Technologies Institute, Greece
Shuohang Wang	Singapore Management University, Singapore
Christa Womser-Hacker	Universität Hildesheim, Germany
Chenyan Xiong	Carnegie Mellon University; Microsoft, USA
Grace Hui Yang	Georgetown University, USA
Peilin Yang	Twitter Inc., USA
Tao Yang	University of California at Santa Barbara, USA
Andrew Yates	Max Planck Institute for Informatics, Germany
Hai-Tao Yu	University of Tsukuba, Japan
Hamed Zamani	University of Massachusetts Amherst, USA
Eva Zangerle	University of Innsbruck, Austria
Fattane Zarrinkalam	Ferdowsi University, Iran
Dan Zhang	Facebook, USA
Duo Zhang	Kunlun Inc.
Shuo Zhang	University of Stavanger, Norway
Sicong Zhang	Georgetown University, USA
Guoqing Zheng	Carnegie Mellon University, USA
Leonardo Zilio	Université catholique de Louvain, Belgium
Guido Zuccon	The University of Queensland, Australia

Sponsors

Deutsche Forschungsgemeinschaft

German Research Foundation

Information Retrieval Specialist Group

SIGIR

Special Interest Group on Information Retrieval

SIGNAL

Bloomberg

Engineering

POLYGON ANALYTICS™

Open Access Journals by MDPI

textkernel

Machine Intelligence for People and Jobs

Google

Keynote Papers

On Entities and Evaluation

Krisztian Balog

University of Stavanger, Stavanger, Norway
krisztian.balog@uis.no

This talk addresses two broad topics, entities and evaluation, which have been the main focus of my research for over ten years. Over the past decade, we have witnessed entities becoming first-class citizens in many information access scenarios [1]. With this has also come an increased reliance on knowledge bases (a.k.a. knowledge graphs), which organize information about entities in a structured and semantically meaningful way. Knowledge bases have enabled significant advancements on specific retrieval tasks, such as entity retrieval and entity linking [2], as well as have contributed to the grand challenge effort of building intelligent personal assistants. The talk provides a brief synthesis of progress thus far, then highlights some open challenges that remain in this space. In particular, the concept of a *personal knowledge graph* is introduced, which is a resource of structured information about entities personally relevant to a given user. A range of tasks associated to personal knowledge graphs are also discussed.

The second part of the talk concentrates on evaluation, which has been a central theme in information retrieval since the inception of the field. For a long time, system-oriented evaluation has primarily been performed using offline test collections, following the Cranfield paradigm. While this rigorous methodology ensures the repeatability and reproducibility of experiments, it is inherently limited by abstracting the actual user, to a large extent, away. In this talk, an argument is made for the (complementary) need of online evaluation. Specifically, the "living labs" evaluation methodology is presented, along with past and current efforts to implement it as a collaborative research and development scheme [3, 4].

References

1. Balog, K.: Entity-Oriented Search. The Information Retrieval Series. Springer, Cham (2018). https://eos-book.org
2. Hasibi, F., Balog, K. Bratsberg, S.E.: Exploiting entity linking in queries for entity retrieval. In: Proceedings of the 2nd ACM International Conference on the Theory of Information Retrieval, ICTIR 2016, pp. 209–218 (2016)
3. Jagerman, R., Balog, K., Rijke, M.D.: Opensearch: Lessons learned from an online evaluation campaign. J. Data Inf. Qual. **10**(3), 13:1–13:15 (2018)
4. Schuth, A., Balog, K., Kelly, L.: Overview of the living labs for information retrieval evaluation (LL4IR) CLEF lab 2015. In: Mothe, J., et al. (eds.) CLEF 2015. LNCS, vol. 9283, pp. 484–496. Springer, Cham (2015). https://doi.org/10.1007/978-3-319-24027-5_47

Ranking People

Markus Strohmaier

RWTH Aachen University and GESIS - Leibniz Institute for the Social Sciences
markus.strohmaier@cssh.rwth-aachen.de
http://cssh.rwth-aachen.de

Abstract. The popularity of search engines on the World Wide Web is a testament to the broad impact of the work done by the information retrieval community over the last decades. The advances achieved by this community have not only made the World Wide Web more accessible, they have also made it appealing to consider the application of ranking algorithms to other domains, beyond the ranking of documents. One of the most interesting examples is the domain of *ranking people*. In this talk, I will first highlight some of the many challenges that come with deploying ranking algorithms to individuals. I will then show how mechanisms that are perfectly fine to utilize when ranking documents can have undesired or even detrimental effects when ranking people. This talk intends to stimulate a discussion on the manifold, interdisciplinary challenges around the increasing adoption of ranking algorithms in computational social systems.

Keywords: Information retrieval · Ranking · Computational social science

Contents – Part I

Modeling Relations

Classification and Search

Recommender Systems I

Graphs

Recommender Systems II

Query Analytics

Representation

Reproducibility (Systems)

Reproducibility (Application)

Neural IR

Cross Lingual IR

QA and Conversational Search

Topic Modeling

Metrics

Image IR

Short Papers

Contents – Part II

Demonstration Papers

CLEF Organizers Lab Track

Doctoral Consortium Papers

Workshops

Tutorials

Modeling Relations

Learning Lexical-Semantic Relations Using Intuitive Cognitive Links

Georgios Balikas[1], Gaël Dias[2(✉)], Rumen Moraliyski[3], Houssam Akhmouch[2,4], and Massih-Reza Amini[5]

[1] Kelkoo Group, Grenoble, France
[2] Normandy University, CNRS GREYC, Caen, France
gael.dias@unicaen.fr
[3] Kodar Ltd., Plovdiv, Bulgaria
[4] Credit Agricole Brie Picardie, Amiens, France
[5] University of Grenoble Alps, CNRS LIG, Grenoble, France

Abstract. Identifying the specific semantic relations between words is crucial for IR and NLP systems. Our goal in this paper is twofold. First, we want to understand whether learning a classifier for one semantic relation (e.g. hypernymy) can gain from concurrently learning another classifier for a cognitively-linked semantic relation (e.g. co-hyponymy). Second, we evaluate how these systems perform where only few labeled examples exist. To answer the first question, we rely on a multi-task neural network architecture, while for the second we use self-learning to evaluate whether semi-supervision improves performance. Our results on two popular datasets as well as a novel dataset proposed in this paper show that concurrent learning of semantic relations consistently benefits performance. On the other hand, we find that semi-supervised learning can be useful depending on the semantic relation. The code and the datasets are available at https://bit.ly/2Qitasd.

1 Introduction

The ability to automatically identify lexical-semantic relations is an important issue for Information Retrieval (IR) and Natural Language Processing (NLP) applications such as question answering [13], query expansion [19], or text summarization [14]. Lexical-semantic relations embody a large number of symmetric and asymmetric linguistic phenomena such as synonymy (bike ↔ bicycle), co-hyponymy (bike ↔ scooter), hypernymy (bike → tandem) or meronymy (bike → chain), but more exist [38].

Most approaches focus on a single semantic relation and consist in deciding whether a given relation r holds between a pair of words (x, y). Within this binary classification framework, the vast majority of efforts [26,30,36,37] concentrate on hypernymy, as it is the key organization principle of semantic memory. Other studies can be found on antonymy [27], meronymy [15] and co-hyponymy [39]. Another research direction consists in dealing with several semantic relations simultaneously. This is defined as deciding which semantic relation r_i

L. Azzopardi et al. (Eds.): ECIR 2019, LNCS 11437, pp. 3–18, 2019.
https://doi.org/10.1007/978-3-030-15712-8_1

(if any) holds between a pair of words (x, y). This multi-class problem is challenging as it is known that distinguishing between different semantic relations (e.g. synonymy and hypernymy) is difficult [36].

Recently, [33] showed that symmetric similarity measures that capture synonymy [20] are important features in hypernymy detection. Second, [40] showed that learning term embeddings that take into account co-hyponymy similarity improves hypernymy identification. Such observations imply that learning features that encode one lexical relation can benefit the task of identifying another lexical relation. In this work, we evaluate to what extent this hypothesis holds using four semantic relations: synonymy, co-hyponymy, hypernymy and meronymy. For this purpose, we use multi-task learning where the associated tasks that are learned concurrently are the binary classification problems, which determine the semantic relations between word pairs. Our hypothesis is that if the tasks are cognitively linked, multi-task learning approaches should improve the performance on the tasks as the decision functions are learned concurrently.

In this paper, we also explore the effect of relying on a small amount of labeled data and a larger number of unlabeled data when learning classification models. Indeed, previous works use several (rather small) gold standard datasets of word pairs ignoring the potential of weakly labeled word pairs that can be obtained through selected lexico-syntactic patterns [18] or paraphrase alignments [12]. We argue that such gold-standard datasets may not be available for specific languages or domains. Moreover, human cognition and its generalization capacity is unlikely to rely on the equivalent number of positive examples. Therefore, we propose to use semi-supervised learning methods, both with and without multi-task learning, and evaluate whether they can benefit overall performance amongst all experimented tasks.

Our contributions in this paper are as follows: (1) we show that **multi-task learning** consistently improves the classification performance of semantic relations, (2) we build a **novel dataset** for this specific task that is larger than previously published datasets and will serve the community when developing and evaluating classification methods for semantic relations, and (3) we show that **semi-supervised learning** can benefit performance depending on the used dataset and semantic relation.

2 Related Work

Whether semantic relation identification has been tackled as a binary or a multi-class problem, two main families of approaches have been proposed to capture the semantic links between two words (x, y): pattern-based and distributional. Pattern-based (also called path-based) methods base their decisions on the analysis of the lexico-syntactic patterns (e.g. X *such as* Y) that connect the joint occurrences of x and y. Within this context, earlier works proposed unsupervised [18] and supervised [37] methods to detect hypernymy. However, path-based approaches suffer from sparse coverage and benefit precision over recall. To overcome these limitations, recent two-class studies on hypernymy [36] and antonymy [27], as well as multi-class approaches [35] have been focusing on representing dependency patterns as continuous vectors using long short-term memory networks. Within this context, successful results have been evidenced but [27,36] also show that the combination of pattern-based methods with the distributional approach greatly improves performance.

In distributional methods, the decision whether x is within a semantic relation with y is based on the distributional representation of these words following the distributional hypothesis [17], i.e. on the separate contexts of x and y. Earlier works developed symmetric [12] and asymmetric [21] similarity measures based on discrete representation vectors, followed by numerous supervised learning strategies for a wide range of semantic relations [4,30,39], where word pairs are encoded as the concatenation of the constituent words representations ($\overrightarrow{x} \oplus \overrightarrow{y}$) or their vector difference ($\overrightarrow{x} - \overrightarrow{y}$). More recently, attention has been focusing on identifying semantic relations using neural language embeddings, as such semantic spaces encode linguistic regularities [24]. Within this context, [38] proposed an exhaustive study for a wide range of semantic relations and showed that under suitable supervised training, high performance can be obtained. However, [38] also showed that some relations such as hypernymy are more difficult to model than others. As a consequence, new proposals have appeared that tune word embeddings for this specific task, where hypernyms and hyponyms should be closed to each other in the semantic space [26,40].

In this paper, we propose an attempt to deal with semantic relation identification based on a multi-task strategy, as opposed to previous two-class and multi-class approaches. Our main scope is to analyze whether a link exists between the learning process of related semantic relations. The closest approach to ours is proposed by [3], which develops a multi-task convolutional neural network for multi-class semantic relation classification supported by relatedness classification. As such, it can be seen as a domain adaptation problem. Within the scope of our paper, we aim at studying semantic inter-relationships at a much finer grain and understanding the cognitive links that may exist between synonymy, co-hyponymy, hypernymy and meronymy, that represent a large proportion of any taxonomic structure. For this first attempt, we follow the distributional approach as in [3], although we are aware that improvements may be obtained by the inclusion of pattern-based representations[1]. Moreover, to the best of our knowledge, we propose the first attempt to deal with semantic relation identification based on a semi-supervised approach, thus avoiding the existence of a large number of training examples. As a consequence, we aim at providing a more natural learning framework where only a few labeled examples are initially provided and massively-gathered related word pairs iteratively improve learning.

[1] This issue is out of the scope of this paper.

3 Methodology

3.1 Multi-task with Hard Parameter Sharing

As discussed in [7], not every task combination is beneficial. But, concurrent learning of tasks that have cognitive similarities is often beneficial. We may hypothesize that recognizing the different semantic relations that hold between words can benefit classification models across similar tasks. For instance, learning that *bike* is the hypernym of *mountain bike* should help while classifying *mountain bike* and *tandem bicycle* as co-hyponyms, as it is likely that *tandem bicycle* shares some relation with *bike*. To test this hypothesis, we propose to use a multi-task learning approach. Multi-task learning [9] has been empirically validated and has shown to be effective in a variety of NLP tasks ranging from sentiment analysis to part-of-speech tagging and text parsing [7,8]. The hope is that by jointly learning the decision functions for related tasks, one can achieve better performance. It may be first due to knowledge transfer across tasks that is achieved either in the form of learning more robust representations or due to the use of more data. Second, it has been argued that multi-task learning can act as a regularization process thus preventing from overfitting by requiring competitive performance across different tasks [9].

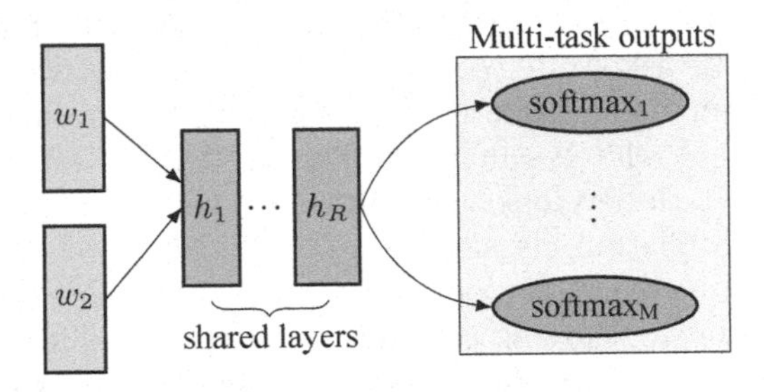

Fig. 1. Feed-forward neural network, where the layers $h_1 \cdots h_R$ are shared across tasks while the output layers softmax$_1 \cdots$ softmax$_M$ are task-dependent.

In this paper, we propose to use a multi-task learning algorithm that relies on hard parameter sharing. Using a simple neural network architecture, our primary objective is to validate our initial hypotheses limiting the effect that choices of architectures and free parameters may have, to the extent of possible. The idea is that the shared parameters (e.g. word representations or weights of some hidden layers) can benefit the performance of all tasks learned concurrently if the tasks are related. In particular, we propose a hard parameter sharing architecture based on a feed-forward neural network (NN) to perform the classification task. The NN architecture is illustrated in Fig. 1, based on the overall idea is that there exists a common representation of the input features that can serve to solve all tasks at hand.

In this model, all tasks base their decision on the same shared representation[2]. In particular, the input of the network is the concatenation of the word embeddings of the word pairs followed by a series of non-linear hidden layers. Then, a number of softmax layers gives the network predictions. Here, a softmax layer corresponds to a task, and concurrently learning M tasks requires M separate output softmax layers. For example, if the network tries to solve two problems concurrently, like *hypernymy* vs. *random* and *hyponymy* vs. *random*, there will be two independent outputs with separate loss functions, each one solving a dedicated problem. The efficiency of hard parameter sharing architectures relies on the fact that the first layers that are shared are tuned by back-propagating the classification errors of every task. That way, the architecture uses the datasets of all tasks (the two dataset of the two problems in the above example), instead of just one at a time. In Algorithm 1, we detail the training protocol. Note that the different tasks learned by the NN share the same weights as batches are randomly sampled from their corresponding datasets. Notice that the architecture can be used with different weights for the tasks or can even be tuned with in order to achieve better results on one of the tasks. Automatically learning these weights is an interesting future research direction.

Algorithm 1. Multi-task Training Process

```
Data: Labeled words pairs L^i for each of the M tasks, batch size b, epochs
epoch = 1 ;
while epoch < epochs do
    for i = 0; i < M; i = i + 1 do
        Randomly select a batch of size b for task i ;
        Update the parameters of the neural network architecture according to
        the errors observed for the batch;
        Calculate the performance on the validation set of task i.
    end
end
```

3.2 Semi-supervision via Self-learning

Semi-supervised learning, also referred as learning with partially labeled data, concerns the case where a prediction function is learned on both labeled and unlabeled training examples [2,10]. As in the supervised learning framework, we assume that we are given access to a set $\mathcal{L} = \{(w_i, w'_i, rel)\}_{i=1}^{i=K}$ that consists of K pairs of words labeled according to the relationship rel. Complementary to that, we also assume to have access to a set of K' words pairs $\mathcal{U} = \{(w_i, w'_i)\}_{i=1}^{i=K'}$ distinct from those of $\mathcal{L}$, and totally unlabeled. The challenge in this setting is

[2] We are aware that this architecture can further be improved by additional task-specific inputs, but as a great deal of possible models can be proposed, which deserve intensive research, this issue remains out of the scope of this paper.

to surpass the performance of classification models trained exclusively on $\mathcal{L}$ by using the available data in $\mathcal{U}$. To do so we use self-learning, boosting the training set with confident predictions of an initial classifier. Formally, the underlying idea of self-learning is to train a learner on the set $\mathcal{L}$, and then progressively expand $\mathcal{L}$, by pseudo-labeling N pairs within $\mathcal{U}$, for which the current prediction function is the most confident and adding them to $\mathcal{L}$. This process is repeated until no more pairs are available in $\mathcal{U}$ or, that the performance on a validation set degrades due to the newly-added possibly noisy examples. Algorithm 2 details this process. One point illustrated in Algorithm 2 to be highlighted is that the training set $\mathcal{L}$ is augmented after each iteration of self-learning in a stratified way. In this case, the class distribution of the N pseudo-labeled examples that are added to $\mathcal{L}$ is the same as the class distribution of $\mathcal{L}$. This constraint follows from the independent and identically distributed (i.i.d.) assumption between the $\mathcal{L}$ and $\mathcal{U}$ sets and ensures that the distribution on the classes in the training set does not change as training proceeds. Another point to be mentioned is that the examples that are added to $\mathcal{L}$ may be noisy. Despite the confident predictions of the classifier C, one should expect that some of the instances added are wrongly classified [1]. To reduce the impact of the noise to the training set, we monitor the performance of the classifier using the validation set $\mathcal{V}$ and if the performance degrades the self-learning iteration stops.

Algorithm 2. Self-learning

Data: Word pairs: labeled $\mathcal{L}$, unlabeled $\mathcal{U}$, validation $\mathcal{V}$; integer N
$\mathcal{L}_0 = \mathcal{L}, \mathcal{U}_0 = \mathcal{U}$;
Train classifier C using $\mathcal{L}_0$, V_0 : Performance of C on $\mathcal{V}$;
Set $t = 0$;
while *Size*$(\mathcal{U}_t) > 0$ *and* $V_t \geqslant V_0$ **do**
  Get probability scores p of C on $\mathcal{U}_t$;
  pseudo_labeled$(N) = \arg\max(p)$, **stratified wrt** $\mathcal{L}_0$;
  t = t + 1;
  $\mathcal{L}_t = \mathcal{L}_{t-1} +$ pseudo_labeled ;
  $\mathcal{U}_t = \mathcal{U}_{t-1} -$ pseudo_labeled;
  Retrain C using $\mathcal{L}_t$, V_t : Performance of C on $\mathcal{V}$;
end

4 Experimental Setups

4.1 Datasets

In order to perform our experiments, we use the ROOT9 dataset[3] [32] that contains 9,600 word pairs, randomly extracted from three well-known datasets: EVALution [34], Lenci/Benotto [6] and BLESS [5]. The word pairs are equally

[3] https://github.com/esantus/ROOT9.

distributed among three classes (hypernymy, co-hyponymy and random) and involve several part-of-speech tags (adjectives, nouns and verbs). Here, we exclusively focus on nouns and keep 1,212 hypernyms, 1,604 co-hyponyms and 549 random pairs that can be represented by GloVe embeddings [29].

In order to include synonymy as a third studied semantic relation, we build the RUMEN dataset[4] that contains 18,978 noun pairs automatically gathered from WordNet 3.0[5] [25] and equally organized amongst three classes (hypernymy, synonymy and random). Note that the words in the pairs are single words and do not contain multi-word expressions. In particular, the RUMEN dataset contains 9,125 word types (i.e. unique nouns) distributed as follows for each semantic relation: 5,054 for hypernymy, 5,201 for synonymy and 6,042 for random. In order to evidence the ambiguity level of the dataset, Table 1 presents the distribution of the types by the number of senses they cover in WordNet. It can be evidenced that while the random category is mainly composed (by construction) of weakly ambiguous nouns, synonymy embodies a large set of polysemous words, while hypernymy contains both weakly polysemous words (usually the hyponym) and more polysemous words (usually the hypernym).

Note that with respect to hypernyms, all noun pairs are randomly selected such that they are not necessarily in direct relation. Exactly 17.2% of the hypernyms are in direct relation, 19.9% have a path length of 2, 20.2% of 3, 16.2% of 4, and 26.5% have a path length superior or equal to 5. Note also that random pairs have as lowest common ancestor the root of the hierarchy with a minimum path distance equals to 7[6] between both words so to ensure semantic separateness. On average, each random pair is separated by a path length of 13.2.

Table 1. Distribution of types by number of senses discriminated by semantic category.

# of senses	1	2	3	4	5	6	≥7
RUMEN Hypernymy	24.69%	19.67%	14.21%	9.34%	7.40%	4.82%	19.87%
RUMEN Synonymy	18.46%	18.80%	14.52%	11.04%	8.52%	5.77%	22.89%
RUMEN Random	39.71%	23.28%	12.89%	7.54%	5.05%	3.08%	8.45%

In order to better understand the particularities of the RUMEN dataset, we present the portions of the hierarchy, which are covered by the word pairs in Table 2. To do so, we compute the path length that holds between the root and the highest word of the noun pair in the hierarchy, for all pairs, and calculate the respective distribution.

[4] Available at https://bit.ly/2Qitasd.
[5] http://wordnetcode.princeton.edu/3.0/.
[6] This value was set experimentally.

Table 2. Distribution of pairs by length path from the root discriminated by semantic category.

Path length	0	1	2	3	4	5	6	≥7
RUMEN Hypernymy	5.72%	2.99%	10.22%	27.91%	23.58%	14.86%	9.21%	5.51%
RUMEN Synonymy	0.00%	0.00%	0.08%	1.72%	10.92%	26.14%	26.88%	34.26%
RUMEN Random	0.00%	0.03%	0.69%	5.56%	20.34%	30.07%	22.63%	20.68%

Note that for the synonymy relation, mostly the bottom part (i.e. near the leaves) is covered, while for the hypernymy relation, most pairs have their hypernym in the middle of the hierarchy (levels 3 and 4)[7]. As for the random relation, the pairs are rather uniformly distributed from level 4 to bottom.

Finally, note that for our experiment, we keep 3,375 hypernym, 3,213 synonym and 3,192 random word pairs encoded by GloVe embeddings as many pairs contain unknown words.

4.2 Lexical Splits

Following a classical learning procedure, the datasets must be split into different subsets: train, validation, test and unlabeled in the case of semi-supervision. The standard procedure is random splitting where word pairs are randomly selected without other constraint to form the subsets. However, the authors of [22] point out that using distributional representations in the context of supervised learning tends to perform lexical memorization. In this case, the model mostly learns independent properties of single terms in pairs. For instance, if the training set contains word pairs like (*bike*, *tandem*), (*bike*, *off-roader*) and (*bike*, *velocipede*) tagged as hypernyms, the algorithm may learn that *bike* is a prototypical hypernym and all new pairs (*bike*, y) may be classified as hypernyms, regardless of the relation that holds between *bike* and y. To overcome this situation and prevent the model from overfitting by lexical memorization, [22] suggested to split the train and test sets such that each one contains a distinct vocabulary. This procedure is called lexical split. Within the scope of this study, we propose to apply lexical split as defined in [22]. So, lexical repetition exists in the train, validation and the unlabeled subsets, but the test set is exclusive in terms of vocabulary. Table 3 shows the vocabulary and the pairs before and after the lexical splits.

For the specific case of semi-supervised learning, we have further split the pairs dubbed as train so that 60% of them are unlabeled examples. From the remaining 40%, we have randomly selected 30% for validation, resulting in few training examples, which resembles more to a realistic learning scenario where

[7] A large number of hypernym pairs contain the root synset "entity", i.e. path length equals to 0.

Table 3. Statistics on the datasets and the lexical splits we performed to obtain the train and test subsets. V is the vocabulary size in the original dataset; V_{train} (resp. V_{test}) corresponds to the vocabulary size in the train (resp. test) dataset for the lexical split after removing all words that do not belong to GloVe dictionary. Then, for each lexical relation, we provide the number of word pairs in the train/test datasets.

Dataset	ROOT9	RUMEN	ROOT9+RUMEN	BLESS
Co-hyponyms	939/665	-	1,193/350	1,361/502
Hypernyms	806/486	2,638/737	3,330/1,238	525/218
Meronyms	-	-	-	559/256
Synonyms	-	2,256/957	2,297/1,002	-
Random	339/210	2,227/965	2,630/1,160	2,343/971
V	2,373	9,125	9,779	3,582
$V_{\text{train}}/V_{\text{test}}$	1,423/950	5,475/3,650	5,867/3,912	3,181/2,121

only few positive examples are known. This process is illustrated in Fig. 2, with the percentages of the overall dataset. Note that lexical split is not performed between the train, validation and unlabeled subsets[8]. So, while lexical split ensures that the network generalizes to unseen words, it also results in significantly smaller datasets due to the way that these datasets are produced.

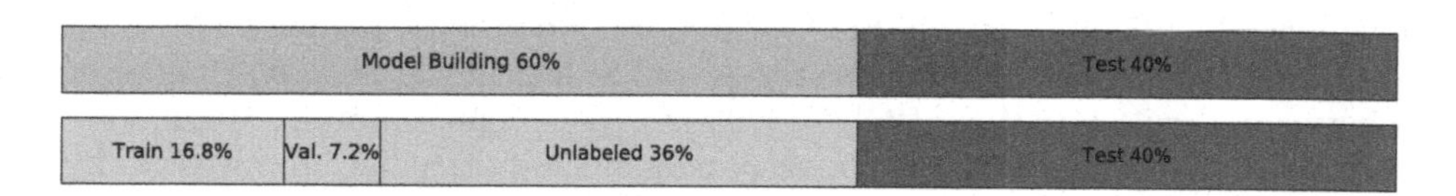

Fig. 2. Illustration of the lexical split of the datasets. The percentages in parentheses correspond to the portions of the original data, used for each purpose.

4.3 Learning Frameworks

In order to evaluate the effects of our learning strategy, we implement the following baseline systems: (1) Multi-class Logistic Regression using a one-vs-rest approach[9], (2) Logistic Regression that has shown positive results in [33] for hypernymy (i.e. a binary problem), and (3) Feed-forward neural network with two hidden layers of 50 neurons each, which is the direct binary counterpart of our multi-task NN.

For the multi-task learning algorithm, we implemented the architecture shown in Fig. 1 using Keras [11]. In particular, we define 2 fully-connected hidden layers (i.e. h_1, h_2, $R = 2$) of 50 neurons each. While the number of hidden

[8] All datasets are available at https://bit.ly/2Qitasd.

[9] A multi-class model learns to separate between several classes and direct comparison with binary models is not fair. Nevertheless, we report its performance as it highlights the potential of multi-class learning for problems that are cognitively similar.

layers is a free parameter to tune, we select two hidden layers in advance so that the complexity of the multi-task models are comparable to the neural network baseline. The activation function of the hidden layers is the sigmoid function and the weights of the layers are initialized with a uniform distribution scaled as described in [16]. As for the learning process, we use the Root Mean Square Propagation optimization method with learning rate set to 0.001 and the default value for $\rho = 0.9$. For every task, we use the binary cross-entropy loss function. The network is trained with batches of 32 examples[10]. The word embeddings are initialized with the 300-dimensional representations of GloVe [29].

For the Logistic Regression, we used the implementation of scikit-learn [28]. In particular, a grid search with stratified 3-fold cross validation was used to select the C value in $[0.001, 0.01, 0.1, 1, 10]$.

5 Results

In the following experiments, we report two evaluation measures: Accuracy and Macro-average F_1 measure (MaF_1). Accuracy captures the number of correct predictions over the total predictions, while MaF_1 evaluates how the model performs across the different relations as it uniformly averages the F_1 measures of each relation. In the remaining paragraphs, we comment on three experiments.

In the first experiment, we propose to study the impact of the concurrent learning of co-hyponymy (bike $\leftrightarrow$ scooter) and hypernymy (bike $\rightarrow$ tandem) following the first findings of [40]. For that purpose, we propose to apply our (semi-supervised) multi-task learning strategy over the lexically split ROOT9 dataset using vector concatenation of GloVe [29] as feature representation. Results are illustrated in Table 4. The multi-task paradigm shows that an improved MaF_1 score can be achieved by concurrent learning without semi-supervision achieving a value of 77.3% (maximum value overall). In this case, a 1.1% improvement is obtained over the best baseline (i.e. logistic regression) for hypernymy classification, indeed suggesting that there exists a learning link between hypernymy and co-hyponymy. However, the results for co-hyponymy classification can not compete with a classical supervised strategy using logistic regression. In this case, a 2.1% decrease in MaF_1 is evidenced suggesting that the gains for hypernymy classification are not positively balanced by the performance of co-hyponymy. So, we can expect an improvement for hypernymy classification but not for co-hyponymy, suggesting a positive influence of co-hyponymy learning towards hypernymy but not the opposite. Interestingly, the results of the semi-supervised strategy reach comparable figures compared to the multi-task proposal (even superior in some cases), but do not complement each other for the semi-supervised multi-task experiment. In this case, worst results are obtained for both classification tasks suggesting that the multi-task model is not able to correctly generalize from a large number of unlabeled examples, while this is the case for the one-task architecture.

[10] The code is available at https://bit.ly/2Qitasd.

Table 4. Accuracy and MaF_1 scores on ROOT9 and RUMEN datasets using GloVe.

Algorithm		Co-hypo. vs Random		Hyper. vs Random		Average results	
		Accuracy	MaF_1	Accuracy	MaF_1	Accuracy	MaF_1
ROOT9	Multi-class Logistic Regression	0.740	0.500	0.781	0.507	0.760	0.500
	Logistic Regression	**0.893**	0.854	0.814	0.762	**0.854**	0.808
	NN Baseline	0.890	0.851	0.803	0.748	0.847	0.800
	Self-learning	0.869	**0.859**	0.816	0.772	0.843	**0.815**
	Multi-task learning	0.882	0.833	**0.818**	**0.773**	0.850	0.803
	Multi-task learning + Self-learning	0.854	0.811	0.810	0.767	0.832	0.789
Algorithm		Syn. vs Random		Hyper. vs Random		Average results	
		Accuracy	MaF_1	Accuracy	MaF_1	Accuracy	MaF_1
RUMEN	Multi-class Logistic Regression	0.600	0.430	0.620	0.467	0.610	0.448
	Logistic Regression	0.628	0.628	0.711	0.706	0.670	0.667
	NN Baseline	0.679	0.678	0.752	0.748	0.716	0.713
	Self-learning	0.686	0.685	0.757	0.754	0.722	0.720
	Multi-task learning	0.706	0.700	0.755	0.750	0.731	0.725
	Multi-task learning + Self-learning	**0.708**	**0.708**	**0.760**	**0.755**	**0.734**	**0.732**

In the second experiment, we propose to study the impact of the concurrent learning of synonymy (bike ↔ bicycle) and hypernymy following the experiments of [33] which suggest that symmetric similarity measures (usually tuned to detect synonymy [20]) improve hypernymy classification. For that purpose, we propose to apply the same models over the lexically split RUMEN dataset. Results are illustrated in Table 4. The best configuration is the combination of multi-task learning with self-learning achieving maximum accuracy and MaF_1 scores for both tasks. The improvement equals to 0.7% in terms of MaF_1 for hypernymy and reaches 3% in terms of MaF_1 for synonymy when compared to the best baseline (i.e. neural network). The overall average improvement (i.e. both tasks combined[11]) reaches 1.8% for accuracy and 1.9% for MaF_1 over the best baseline. So, these results tend to suggest that synonymy identification may positively be impacted by the concurrent learning of hypernymy and vice versa (although to a less extent). In fact, these results consistently build upon the positive results of the multi-task strategy without semi-supervision and the self-learning approach alone that both improve over the best baseline results. Nevertheless, the improvement obtained by combining multi-task learning and semi-supervision is negligible compared to multi-task alone. Note also that the results obtained over the RUMEN dataset by the baseline classifiers are lower than the ones reached over ROOT9 for hypernymy, certainly due to the complexity of the datasets themselves. So, we may hypothesize that the multi-task strategy plays an important role by acting as a regularization process and helping in solving learning ambiguities, and reaches improved results over the two-task classifiers.

[11] Column 3 of Table 4.

In the **third experiment**, we propose to study the impact of the concurrent learning of co-hyponymy, synonymy and hypernymy together. The idea is to understand the inter-relation between these three semantic relations that form the backbone of any taxonomic structure. For that purpose, we propose to apply the models proposed in this paper over the lexically split ROOT9+RUMEN dataset[12]. Results are illustrated in Table 5. The best configuration for all the tasks combined (i.e. co-hyponymy, synonymy and hypernymy) is multi-task learning without semi-supervision. Overall, improvements up to 1.4% in terms of accuracy and 2% in terms of $\mathrm{MaF_1}$ can be reached over the best baseline (i.e. neural network). In particular, the $\mathrm{MaF_1}$ score increases 4.4% with the multi-task strategy without self-learning for co-hyponymy, while the best result for synonymy is obtained by the semi-supervised multi-task strategy with an improvement of 1.1% $\mathrm{MaF_1}$ score. The best configuration for hypernymy is evidenced by self-learning alone, closely followed by the multi-task model, reaching improvements in $\mathrm{MaF_1}$ scores of 1.7% (resp. 1%) for self-learning (resp. multi-task learning). Comparatively to the first experiment, both learning paradigms (i.e. semi-supervision and multi-task) tend to produce competitive results alone, both exceeding results of the best baseline. However, the multi-task model hardly generalizes from the set of unlabeled examples, being synonymy the only exception. Finally, note that co-hyponymy seems to be the simplest task to solve, while synonymy is the most difficult one, over all experiments.

In the **fourth experiment**, We now study the meronymy relation (bike → chain) into a multi-task environment, as it has traditionally been studied together with hypernymy [15]. The overall idea is to verify whether meronymy can benefit from the concurrent learning of the backbone semantic relations that form knowledge bases. For that purpose, we apply our learning models over the lexically split BLESS dataset [5] that includes three semantic relations: co-hyponymy, hypernymy and meronymy. The details of the lexical split is presented in Table 3 and note that the BLESS dataset has been processed in the exact same way as ROOT9 and RUMEN, i.e. retaining only noun categories and word pairs that can be represented by the GloVe semantic space. Results are presented in Table 5. The best configuration over the three tasks combined is obtained by the semi-supervised multi-task strategy with a $\mathrm{MaF_1}$ score equals to 80.3%, thus improving 1.2% over the best baseline (i.e. neural network). In particular, we can notice that the most important improvement is obtained for the meronymy relation that reaches 73.3% for $\mathrm{MaF_1}$ and 76.4% for accuracy with the multi-task model without semi-supervision. In this particular case, the improvement is up to 2.6% in accuracy and 2.4% in $\mathrm{MaF_1}$ over the neural network baseline. For co-hyponymy (resp. hypernymy), best results are obtained by multi-task with semi-supervision (resp. without semi-supervision), but show limited improvements over the best baseline, suggesting that meronymy gains more in performance from the concurrent learning of co-hyponymy and hypernymy than the contrary, although improvements are obtained in all cases.

[12] Note that due to the lexical split process, results can not directly be compared to the ones obtained over ROOT9 or RUMEN.

Table 5. Accuracy and MaF_1 scores on ROOT9+RUMEN and BLESS datasets using GloVe.

System		Co-hypo. vs Random		Hyper. vs Random		Syn. vs Random		Average results	
		Accuracy	MaF_1	Accuracy	MaF_1	Accuracy	MaF_1	Accuracy	MaF_1
ROOT+ RUMEN	Multi-class Log. Reg.	0.606	0.370	0.560	0.320	0.500	0.280	0.555	0.323
	Logistic Regression	0.909	0.872	0.669	0.669	0.634	0.632	0.737	0.724
	NN Baseline	0.914	0.875	0.712	0.712	0.663	0.659	0.763	0.748
	Self-learning	0.928	0.900	**0.729**	**0.729**	0.668	0.665	0.775	0.765
	Multi-task learning	**0.943**	**0.919**	0.723	0.722	0.666	0.664	**0.777**	**0.768**
	Multi-task learning + Self.	0.939	0.911	0.711	0.711	**0.672**	**0.670**	0.774	0.764
System		Co-hypo. vs Random		Hyper. vs Random		Mero. vs Random		Average Results	
		Accuracy	MaF_1	Accuracy	MaF_1	Accuracy	MaF_1	Accuracy	MaF_1
BLESS	Multi-class Log. Reg.	0.760	0.408	0.720	0.355	0.722	0.362	0.734	0.375
	Logistic Regression	0.845	0.830	0.888	0.794	0.748	0.723	0.827	0.782
	NN Baseline	0.870	0.855	0.892	0.809	0.738	0.709	0.833	0.791
	Self-learning	0.877	**0.863**	0.900	0.807	0.749	0.723	0.842	0.798
	Multi-task learning	0.866	0.847	**0.903**	**0.816**	**0.764**	**0.733**	**0.844**	0.799
	Multi-task learning + Self.	**0.878**	**0.863**	0.900	0.813	0.754	**0.733**	**0.844**	**0.803**

Comparatively to the other experiments, we also notice that although the self-learning algorithm and the multi-task framework without semi-supervision perform well alone, the combination of both strategies does not necessary lead to the best results overall, suggesting that the present architecture can be improved by the massive extraction of unlabeled examples.

6 Conclusions

In this paper, we proposed to study the concurrent learning of cognitively-linked semantic relations (co-hyponymy, hypernymy, synonymy and meronymy) using **semi-supervised** and **multi-task learning**. Our results show that concurrent learning leads to improvements in most tested situations and datasets, including the **newly-built dataset** called RUMEN. In particular, results show that hypernymy can gain from co-hyponymy, synonymy from hypernymy, co-hyponymy from both hypernymy and synonymy, and meronymy from both co-hyponymy and hypernymy. Moreover, it is interesting to notice that in three cases out of four, the improvement achieved by the multi-task strategy is obtained for the most difficult task to handle. Nevertheless, there still exists a great margin for improvement. First, we intend to propose new multi-task architectures that include task-specific features similarly to [23] as well as LSTM path-based features as in [36]. Second, we expect to build on new semi-supervised multi-task architectures such as Tri-training [31] to positively combine semi-supervision and multi-task learning as their combination is currently not beneficial in a vast majority of cases. Third, we intend to massively gather unlabeled examples by lexico-syntactic patterns [18] or by paraphrase alignment [12] instead of simulating such a behaviour, as we do currently. Finally, we plan to test all our configurations in "noisy" situations as proposed in [38].

References

1. Amini, M., Laviolette, F., Usunier, N.: A transductive bound for the voted classifier with an application to semi-supervised learning. In: 22nd Annual Conference on Neural Information Processing Systems (NIPS), pp. 65–72 (2008)
2. Amini, M.R., Usunier, N.: Learning with Partially Labeled and Interdependent Data. Springer, Cham (2015). https://doi.org/10.1007/978-3-319-15726-9
3. Attia, M., Maharjan, S., Samih, Y., Kallmeyer, L., Solorio, T.: Cogalex-V shared task: GHHH - detecting semantic relations via word embeddings. In: Workshop on Cognitive Aspects of the Lexicon, pp. 86–91 (2016)
4. Baroni, M., Bernardi, R., Do, N.Q., Shan, C.-C: Entailment above the word level in distributional semantics. In: 13th Conference of the European Chapter of the Association for Computational Linguistics (EACL), pp. 23–32 (2012)
5. Baroni, M., Lenci, A.: How we blessed distributional semantic evaluation. In: Workshop on Geometrical Models of Natural Language Semantics (GEMS) Associated to Conference on Empirical Methods on Natural Language Processing (EMNLP), pp. 1–10 (2011)
6. Benotto, G.: Distributional Models for Semantic Relations: A Study on Hyponymy and Antonymy. Ph.D. thesis, University of Pisa (2015)
7. Bingel, J., Søgaard, A.: Identifying beneficial task relations for multi-task learning in deep neural networks. arXiv preprint arXiv:1702.08303 (2017)
8. Braud, C., Plank, B., Søgaard, A.: Multi-view and multi-task training of RST discourse parsers. In: 26th International Conference on Computational Linguistics (COLING), pp. 1903–1913 (2016)
9. Caruana, R.: Multitask learning. In: Thrun, S., Pratt, L. (eds.) Learning to Learn, pp. 95–133. Springer, Boston (1998). https://doi.org/10.1007/978-1-4615-5529-2_5
10. Chapelle, O., Scholkopf, B., Zien, A.: Semi-supervised learning. IEEE Trans. Neural Networks **20**(3), 542 (2009)
11. Chollet, F.: Keras. https://keras.io (2015)
12. Dias, G., Moraliyski, R., Cordeiro, J., Doucet, A., Ahonen-Myka, H.: Automatic discovery of word semantic relations using paraphrase alignment and distributional lexical semantics analysis. Nat. Lang. Eng. **16**(4), 439–467 (2010)
13. Dong, L., Mallinson, J., Reddy, S., Lapata, M.: Learning to paraphrase for question answering. In: Conference on Empirical Methods in Natural Language Processing (EMNLP), pp. 886–897 (2017)
14. Gambhir, M., Gupta, V.: Recent automatic text summarization techniques: a survey. Artif. Intell. Rev. **47**(1), 1–66 (2017)
15. Glavas, G., Ponzetto, S.P.: Dual tensor model for detecting asymmetric lexico-semantic relations. In: Conference on Empirical Methods in Natural Language Processing (EMNLP), pp. 1758–1768 (2017)
16. Glorot, X., Bengio, Y.: Understanding the difficulty of training deep feedforward neural networks. In: 13th International Conference on Artificial Intelligence and Statistics, pp. 249–256 (2010)

17. Harris, Z.S.: Distributional structure. Word **10**(2–3), 146–162 (1954)
18. Hearst, M.: Automatic acquisition of hyponyms from large text corpora. In: 14th Conference on Computational Linguistics (COLING), pp. 539–545 (1992)
19. Kathuria, N., Mittal, K., Chhabra, A.: A comprehensive survey on query expansion techniques, their issues and challenges. Int. J. Comput. Appl. **168**(12), (2017)
20. Kiela, D., Hill, F., Clark, S.: Specializing word embeddings for similarity or relatedness. In: Conference on Empirical Methods in Natural Language Processing (EMNLP), pp. 2044–2048 (2015)
21. Kotlerman, L., Dagan, I., Szpektor, I., Zhitomirsky-Geffet, M.: Directional distributional similarity for lexical inference. Nat. Lang. Eng. **16**(4), 359–389 (2010)
22. Levy, O., Remus, S., Biemann, C., Dagan, I.: Do supervised distributional methods really learn lexical inference relations? In: Conference of the North American Chapter of the Association for Computational Linguistics: Human Language Technologies, pp. 970–976 (2015)
23. Liu, P., Qiu, X., Huang, X.: Adversarial multi-task learning for text classification. In: 55th Annual Meeting of the Association for Computational Linguistics (ACL) (2017)
24. Mikolov, T., Chen, K., Corrado, G., Dean, J.: Efficient estimation of word representations in vector space. CoRR abs/1301.3781 (2013)
25. Miller, G.A., Beckwith, R., Fellbaum, C., Gross, D., Miller, K.J.: Introduction to wordnet: an on-line lexical database. Int. J. Lexicogr. **3**(4), 235–244 (1990)
26. Nguyen, K.A., Köper, M., Schulte im Walde, S., Vu, N.T.: Hierarchical embeddings for hypernymy detection and directionality. In: Conference on Empirical Methods in Natural Language Processing, pp. 233–243 (2017)
27. Nguyen, K.A., Schulte im Walde, S., Vu, N.T.: Distinguishing antonyms and synonyms in a pattern-based neural network. In: 15th Conference of the European Chapter of the Association for Computational Linguistics (EACL), pp. 76–85 (2017)
28. Pedregosa, F., et al.: Scikit-learn: machine learning in python. J. Mach. Learn. Res. **12**(Oct), 2825–2830 (2011)
29. Pennington, J., Socher, R., Manning, C.D.: Glove: global vectors for word representation. In: Conference on Empirical Methods on Natural Language Processing (EMNLP), pp. 1532–1543 (2014)
30. Roller, S., Erk, K., Boleda, G.: Inclusive yet selective: Supervised distributional hypernymy detection. In: 25th International Conference on Computational Linguistics (COLING), pp. 1025–1036 (2014)
31. Ruder, S., Plank, B.: Strong baselines for neural semi-supervised learning under domain shift. In: 56th Annual Meeting of the Association for Computational Linguistics (ACL) (2018)
32. Santus, E., Lenci, A., Chiu, T., Lu, Q., Huang, C.: Nine features in a random forest to learn taxonomical semantic relations. In: 10th International Conference on Language Resources and Evaluation, pp. 4557–4564 (2016)
33. Santus, E., Shwartz, V., Schlechtweg, D.: Hypernyms under siege: linguistically-motivated artillery for hypernymy detection. In: 15th Conference of the European Chapter of the Association for Computational Linguistics, pp. 65–75 (2017)
34. Santus, E., Yung, F., Lenci, A., Huang, C.R.: Evalution 1.0: an evolving semantic dataset for training and evaluation of distributional semantic models. In: 4th Workshop on Linked Data in Linguistics (LDL) Associated to Association for Computational Linguistics and Asian Federation of Natural Language Processing (ACL-IJCNLP), pp. 64–69 (2015)

35. Shwartz, V., Dagan, I.: Cogalex-V shared task: lexnet - integrated path-based and distributional method for the identification of semantic relations. CoRR abs/1610.08694 (2016)
36. Shwartz, V., Goldberg, Y., Dagan, I.: Improving hypernymy detection with an integrated path-based and distributional method. In: 54th Annual Meeting of the Association for Computational Linguistics, pp. 2389–2398 (2016)
37. Snow, R., Jurafsky, D., Ng, A.Y.: Learning syntactic patterns for automatic hypernym discovery. In: 17th International Conference on Neural Information Processing Systems (NIPS), pp. 1297–1304 (2004)
38. Vylomova, E., Rimell, L., Cohn, T., Baldwin, T.: Take and took, gaggle and goose, book and read: evaluating the utility of vector differences for lexical relation learning. In: 54th Annual Meeting of the Association for Computational Linguistics, pp. 1671–1682 (2016)
39. Weeds, J., Clarke, D., Reffin, J., Weir, D.J., Keller, B.: Learning to distinguish hypernyms and co-hyponyms. In: 5th International Conference on Computational Linguistics (COLING), pp. 2249–2259 (2014)
40. Yu, Z., Wang, H., Lin, X., Wang, M.: Learning term embeddings for hypernymy identification. In: 24th International Joint Conference on Artificial Intelligence, pp. 1390–1397 (2015)

Relationship Prediction in Dynamic Heterogeneous Information Networks

Amin Milani Fard[1(✉)], Ebrahim Bagheri[2], and Ke Wang[3]

[1] New York Institute of Technology, Vancouver, Canada
amilanif@nyit.edu
[2] Ryerson University, Toronto, Canada
bagheri@ee.ryerson.ca
[3] Simon Fraser University, Burnaby, Canada
wangk@cs.sfu.ca

Abstract. Most real-world information networks, such as social networks, are heterogeneous and as such, relationships in these networks can be of different types and hence carry differing semantics. Therefore techniques for link prediction in homogeneous networks cannot be directly applied on heterogeneous ones. On the other hand, works that investigate link prediction in heterogeneous networks do not necessarily consider network dynamism in sequential time intervals. In this work we propose a technique that leverages a combination of latent and topological features to predict a target relationship between two nodes in a dynamic heterogeneous information network. Our technique, called MetaDynaMix, effectively combines meta path-based topology features and inferred latent features that incorporate temporal network changes in order to capture network (1) heterogeneity and (2) temporal evolution, when making link predictions. Our experiment results on two real-world datasets show statistically significant improvement over AUCROC and prediction accuracy compared to the state of the art techniques.

1 Introduction

The goal of link prediction [18] is to estimate the likelihood of a future relationship between two nodes based on the observed network graph. Predicting such relationships in a network can be applied in different contexts such as recommendation systems [4,13,17,20,29], network reconstruction [12], node classification [11], or biomedical applications such as predicting protein-protein interactions [15]. Traditional link prediction techniques, such as [18], consider networks to be homogeneous, i.e., graphs with only one type of nodes and edges. However, most real-world networks, such as social networks, scholar networks, patient networks [6] and knowledge graphs [35] are heterogeneous information networks (HINs) [28] and have multiple node and relation types. For example, in a bibliographic network, there are nodes of types authors, papers, and venues, and edges of types writes, cites and publishes.

L. Azzopardi et al. (Eds.): ECIR 2019, LNCS 11437, pp. 19–34, 2019.
https://doi.org/10.1007/978-3-030-15712-8_2

In a HIN, relations between different entities carry different semantics. For instance, the relationship between two authors is different in meaning when they are co-authors compared to the case when one cites another's paper. Thus techniques for homogeneous networks [1,16,18,19,34] cannot be directly applied on heterogeneous ones. A few works such as [30,31] investigated the problem of link prediction in HINs, however, they do not consider the dynamism of networks and overlook the potential benefits of analyzing a heterogeneous graph as a sequence of network snapshots. Previous work on temporal link prediction scarcely studied HINs and to the best of our knowledge, the problem of predicting relationships in dynamic heterogeneous information networks (DHINs) has not been studied before. In this work we study the problem of relationship prediction in a DHIN, which can be stated as: *Given a DHIN graph G at t consecutive time intervals, the objective is to predict the existence of a particular relationship between two given nodes at time* $t+1$. In the context of this problem, the main contributions of our work can be enumerated as follows:

- We propose the problem of relationship prediction in a DHIN, and draw contrast between this problem and existing link prediction techniques that have been proposed for dynamic and/or heterogeneous networks;
- We present a simple yet effective technique, called *MetaDynaMix*, that leverages topological meta path-based and latent features to predict a target relationship between two nodes in a DHIN;
- We empirically evaluate the performance of our work on two real-world datasets, and the results show statistically significant improvement over AUCROC and prediction accuracy compared to the state of the art techniques.

2 Problem Statement

Our work is focused on heterogeneous information networks (graphs) that can change and evolve over time. As such, we first formally define the concept of *Dynamic Heterogeneous Information Networks*, as follows:

Definition 1 (Dynamic heterogeneous information network). *A dynamic heterogeneous information network (DHIN) is a directed graph $G = (V, E)$ with a node type mapping function $\phi : V \to \mathcal{A}$ and a link type mapping function $\psi : E \to \mathcal{R}$, where V, E, $\mathcal{A}$, and $\mathcal{R}$ denote sets of nodes, links, node types, and relation types, respectively. Each node $v \in V$ belongs to a node type $\phi(v) \in \mathcal{A}$, each link $e \in E$ belongs to a relation $\psi(e) \in \mathcal{R}$, and $|\mathcal{A}| > 1$ and $|\mathcal{R}| > 1$. Also each edge $e = (u, v, t)$ connects two vertices u and v with a timestamp t.* □

The DBLP bibliographic network is an example of a DHIN, containing different types of nodes such as papers, authors, topics, and publication venues, with publication links associated with a date. In the context of a heterogeneous network, a *relation* can be in the form of a *direct link* or an *indirect link*, where

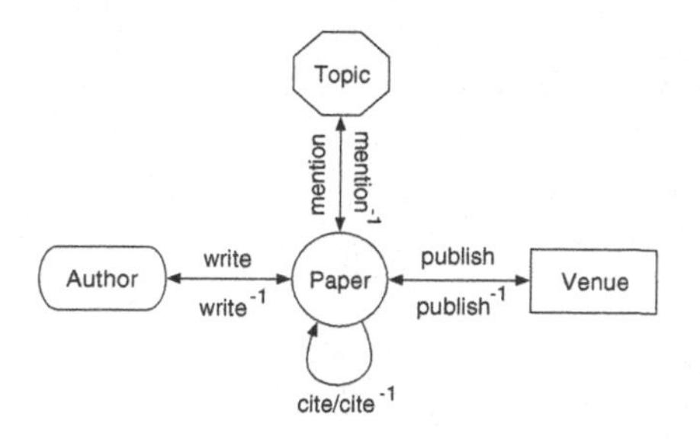

Fig. 1. Network schema for DBLP network.

an indirect link is a sequence of direct links in the network. Thus, two nodes might not be directly connected, however they might be considered to be indirectly connected through a set of intermediary links. In this work, we use the terms *relationship prediction* and *link prediction* interchangeably referring to predicting whether two nodes will be connected in the future via a *sequence of relations* in the graph, where the *length* of a sequence is greater than or equal to one. For instance in a bibliographic network, a direct link exists between an author and a paper she wrote, and an indirect link exists between her and her co-authors through the paper, which they wrote together. In order to better capture different types of nodes and their relation in a network, the concept of *network schema* [32] is used. A network schema is a meta graph structure that summarizes a HIN and is formally defined as follows:

Definition 2 (Network schema). *For a heterogeneous network $G = (V, E)$, the network schema $S_G = (\mathcal{A}, \mathcal{R})$ is a directed meta graph where $\mathcal{A}$ is the set of node types in V and $\mathcal{R}$ is the set of relation types in E.* □

Figure 1 shows the network schema for the DBLP bibliographic network with $\mathcal{A} = \{Author, Paper, Venue, Topic\}$. In this paper, we refer to different types of nodes in the DBLP bibliographic network with abbreviations P for paper, A for author, T for topic, and V for venue.

Similar to the notion of network schema that provides a meta structure for the network, a *meta path* [32] provides a meta structure for paths between different node types in the network.

Definition 3 (Meta path). *A meta path $\mathcal{P}$ is a path in a network schema graph $S_G = (\mathcal{A}, \mathcal{R})$, denoted by $\mathcal{P}(A_1, A_{n+1}) = A_1 \xrightarrow{R_1} A_2 ... \xrightarrow{R_n} A_{n+1}$, as a sequence of links between node types defining a composite relationship between a node of type A_1 and one of type A_{n+1}, where $A_i \in \mathcal{A}$ and $R_i \in \mathcal{R}$.* □

The *length* of a meta path is the number of relations in it. Note that given two node types A_i and A_j, there may exist multiple meta paths of different lengths between them. We call a path $p = (a_1 a_2 ... a_{n+1})$ a *path instance* of a meta path $\mathcal{P} = A_1 - A_2 ... - A_{n+1}$ if p follows $\mathcal{P}$ in the corresponding HIN, i.e., for each node a_i in p, we have $\phi(a_i) = A_i$. The co-author relationship in DBLP

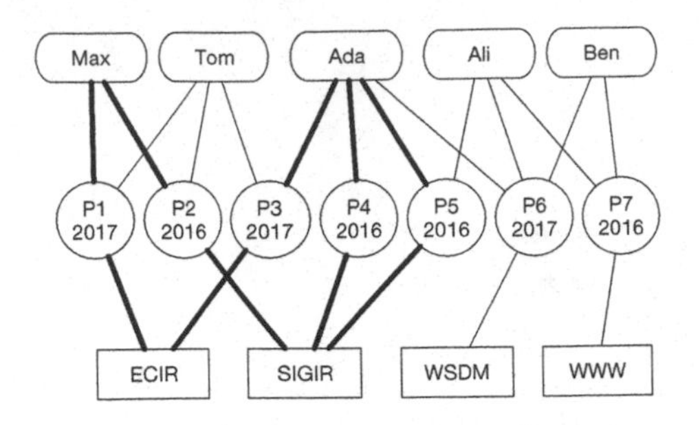

(a) An example of A–P–V–P–A meta paths between two authors Max and Ada.

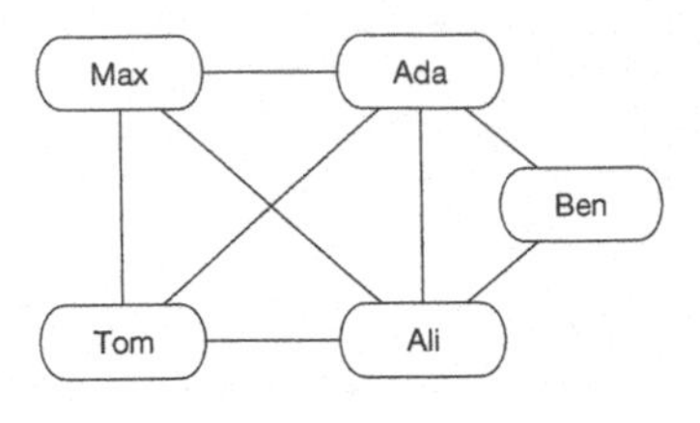

(b) The augmented reduced graph based on $\mathcal{P}(A, A)$=A–P–V–P–A

Fig. 2. An example of a publications network. Link formation time is shown below the paper ID.

can be described with the meta path $A \xrightarrow{write} P \xrightarrow{write^{-1}} A$ or in short A–P–A. Paths in thick solid lines in Fig. 2(a) correspond to A–P–V–P–A meta paths between *Max* and *Ada*, indicating they published in the same venue, such as *Max–P1–ECIR–P3–Ada*. Each meta path carries different semantics and defines a unique topology representing a special relation.

Meta Path-Based Similarity Measures. Given a meta path $\mathcal{P} = (A_i, A_j)$ and a pair of nodes a and b such that $\phi(a) = A_i$ and $\phi(b) = A_j$, several *similarity measures* can be defined between a and b based on the path instances of $\mathcal{P}$. Examples of such similarity or proximity measures in a HIN are *path count* [30, 32], *PathSim* [32] or *normalized path count* [30], *random walk* [30], *HeteSim* [27], and *KnowSim* [36]. Without loss of generality, in this work, we use Path Count (PC) as the default similarity measure. For example, given the meta path A–P–V–P–A and the HIN in Fig. 2(a), $PC(Max, Ada) = 3$ and $PC(Tom, Ada) = 4$. We now formally define the problem that we target in this work as follows:

Definition 4 (Relationship prediction problem). *Given a DHIN graph G at time t, and a target relation meta path $\mathcal{P}(A_i, A_j)$ between nodes of type A_i and A_j, we aim to predict the existence of a path instance of $\mathcal{P}$ between two given nodes of types A_i and A_j at time $t + 1$.* □

3 Proposed Relationship Prediction Approach

Given a DHIN graph $G = (V, E)$, we decompose G into a sequence of t HIN graphs $G_1, .., G_t$ based on links with associated timestamps and then predict relationships in G_{t+1}. As mentioned in Definition 4, we intend to predict existence of a given type of relationship (target meta path) between two given nodes. Thus we define a new type of graph, called *augmented reduced graph* that is generated according to an input heterogeneous network and a target relation meta path.

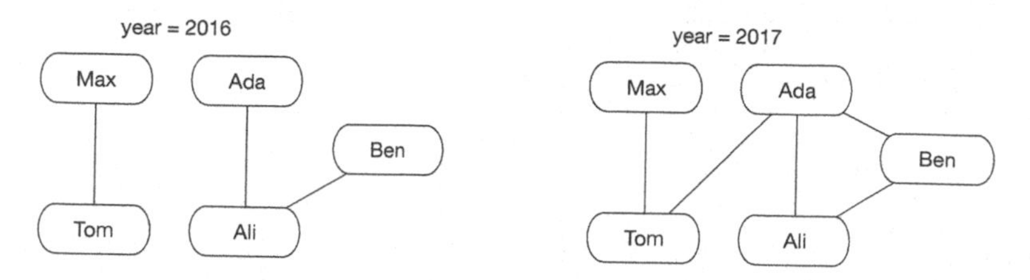

Fig. 3. Augmented reduced graphs for the network in Fig. 2(a) with respect to the target meta path A–P–A (co-authorship) in 2016 and 2017.

Definition 5 (Augmented reduced graph). *Given a HIN graph $G = (V, E)$ and a target meta path $\mathcal{P}(A_i, A_j)$ between nodes of type A_i and A_j, an augmented reduced graph $G^{\mathcal{P}} = (V^{\mathcal{P}}, E^{\mathcal{P}})$ is a graph, where $V^{\mathcal{P}} \subseteq V$ and nodes in $V^{\mathcal{P}}$ are of type A_i and A_j, and edges in $E^{\mathcal{P}}$ indicate relationships of type $\mathcal{P}$ in G.* □

For example, an augmented reduced graph for the network in Fig. 2(a) and target meta path $\mathcal{P}(A, A) = A$–P–V–P–A is a graph shown in Fig. 2(b) whose nodes are of type *Author* and whose edges represent *publishing in the same venue*.

3.1 Homogenized Link Prediction

Once the given DHIN graph $G = (V, E)$ is decomposed into t HIN graphs $G_1, .., G_t$, one solution to the relationship prediction problem (Definition 4) is to build an augmented reduced graph $G_i^{\mathcal{P}}$ for each G_i with respect to the given target meta path $\mathcal{P}$ and then predict a link in $G_i^{\mathcal{P}}$ instead of a path in G_i. In other words, we generate a homogenized version of a graph snapshot and apply a link prediction method. Figure 3 shows examples of such graphs at different time intervals. The intuition behind considering different snapshots, i.e., a dynamic network, rather than a single snapshot for link prediction is that we can incorporate network evolution patterns to increase prediction accuracy. Our hypothesis is that the estimated graph $\hat{G}_{i+1}^{\mathcal{P}}$ is dependent on $\hat{G}_i^{\mathcal{P}}$.

Recent research in link prediction has focused on network latent space inference [7,22,25,37,41] with the assumption that the probability of a link between two nodes depends on their positions in the latent space. Each dimension of the latent space characterizes an attribute, and the more two nodes share such attributes, the more likely they are to connect (also known as homophily). Amongst such graph embedding methods, a few [7,41] considered dynamic networks. Inspired by Zhu et al. [41], we formulate our problem as follows: Given a sequence of augmented reduced graphs $G_1^{\mathcal{P}}, .., G_t^{\mathcal{P}}$, we aim to infer a low rank k-dimensional latent space matrix Z_i for each adjacency matrix $G_i^{\mathcal{P}}$ at time i by minimizing

$$\begin{aligned} &\underset{Z_1,..,Z_t}{\operatorname{argmin}} \sum_{i=1}^{t} \left(\|G_i^{\mathcal{P}} - Z_i Z_i^T\|_F^2 + \lambda \sum_{x \in V^{\mathcal{P}}} (1 - Z_i(x) Z_{i-1}(x)^T) \right) \\ &\text{subject to: } \forall x \in V^{\mathcal{P}}, i, Z_i \geq 0, Z_i(x) Z_i(x)^T = 1 \end{aligned} \quad (1)$$

Algorithm 1. Homogenized Link Prediction

Input: A DHIN graph G, the number of snapshots t, a target meta path $\mathcal{P}(A, B)$, the latent space dimension k, the link to predict (a, b) at $t + 1$
Output: The probability of existence of link (a, b) in $G^{\mathcal{P}}_{t+1}$
1: $\{G_1, .., G_t\} \leftarrow DecomposeGraph(G, t)$
2: **for** each graph $G_i = (V_i, E_i)$ **do**
3: **for** each node $x \in V_i$ that $\phi(x) = A$ **do**
4: Follow $\mathcal{P}$ to reach a node $y \in V_i$ that $\phi(y) = B$
5: Add nodes x and y, and edge (x, y) to the augmented reduced graph $G^{\mathcal{P}}_i$
6: **end for**
7: **end for**
8: $\{Z_1, .., Z_t\} \leftarrow MatrixFactorization(G^{\mathcal{P}}_1, .., G^{\mathcal{P}}_t, k)$
9: Return $Pr((a, b) \in E^{\mathcal{P}}_{t+1}) \leftarrow \sum_{i=1}^{k} Z_t(a, i)Z_t(b, i)$

where $Z_i(x)$ is a temporal latent vector for node x at time i, λ is a regularization parameter, and $1 - Z_i(x)Z_{i-1}(x)^T$ penalizes sudden changes for x in the latent space. This optimization problem can be solved using gradient descent. The intuition behind the above formulation is two fold: (1) nodes with similar latent space representation are more likely to connect with each other, and (2) nodes typically evolve slowly over time and abrupt changes in their connection network are less likely to happen [39]. The matrix $G^{\mathcal{P}}_{t+1}$ can be estimated by $\Phi(f(Z_1, ...Z_t))$, where Φ and f are link and temporal functions, or simply by $Z_t Z_t^T$. Note that Z_i depends on Z_{i-1} as used in the temporal regularization term in Eq. (1).

Algorithm 1 presents a concrete implementation of Eq. 1 for relation prediction. It takes as input a DHIN graph G, the number of graph snapshots t, a target relation meta path $\mathcal{P}(A, B)$, the latent space dimension k, and the link to predict (a, b) at $t+1$. It first decomposes G into a sequence of t graphs $G_1, .., G_t$ by considering the associated timestamps on edges (line 1). Next from each graph G_i, a corresponding augmented reduced graph $G^{\mathcal{P}}_i$ is generated (lines 2–7) for which nodes are of type a and b (beginning and end of target meta path $\mathcal{P}$). For example given $\mathcal{P}(A, A) = A–P–A$, each $G^{\mathcal{P}}_i$ represents the co-authorship graph at time i. Finally by optimizing Eq. (1), it infers latent spaces $Z_1, ..., Z_t$ (line 8) and estimates $G^{\mathcal{P}}_{t+1}$ using $Z_t Z_t^T$ (line 9).

3.2 Dynamic Meta Path-Based Relationship Prediction

The above homogenized approach does not consider different semantics of meta paths between the source and destination nodes and assumes that the probability of a link between nodes depends only on their latent features. For instance, as depicted in Fig. 3, *Tom* and *Ada* became co-authors in 2017 that can be due to publishing at the same venue in 2016, i.e., having two paths between them that passes through *SIGIR*, as shown in Fig. 2. Similarly *Ben* and *Ada* who published with a same author, *Ali* in 2016, became co-authors in 2017.

We would like to further hypothesize that combining latent and topological features can increase prediction accuracy as we can learn latent features that fit the residual of meta path-based features. One way to combine these features is to incorporate meta path measures in Eq. (1) by changing the loss function and regularization term as:

$$\begin{aligned}\underset{\theta_i, Z_i}{\operatorname{argmin}} \sum_{i=1}^{t} \left\| G_i^{\mathcal{P}} - (Z_i Z_i^T + \sum_{i=1}^{n} \theta_{i_{i-1}} \mathcal{F}_{i-1}^{\mathcal{P}_i}) \right\|_F^2 \\ +\lambda \sum_{i=1}^{t} \left(\sum_{x \in V^{\mathcal{P}}} (1 - Z_i(x) Z_{i-1}(x)^T) + \sum_{i=1}^{n} \theta_{i_i}^2 \right)\end{aligned} \quad (2)$$

where n is the number of meta path-based features, $\mathcal{F}^{\mathcal{P}_i}$ is the ith meta path-based feature matrix defined on G_i, and θ_i is the weight for feature f_i. Although we can use a fast block-coordinate gradient descent [41] to infer Z_is, it cannot be efficiently applied to the above changed loss function. This is because it requires computing meta paths for all possible pairs of nodes in $\mathcal{F}^{\mathcal{P}_i}$ for all snapshots, which is not scalable, as calculating similarity measures, such as Path Count or PathSim, can be very costly. For example computing path counts for the A–P–V–P–A meta path can be done by multiply adjacency matrices $AP \times PV \times VP \times PA$.

As an alternative solution, we build a predictive model that considers a linear interpolation of topological and latent features. Given the training pairs of nodes and their corresponding meta path-based and latent features, we apply logistic regression to learn the weights associated with these features. We define the probability of forming a *new link* in time $t+1$ from node a to b as $Pr(label = 1|a, b; \boldsymbol{\theta}) = \frac{1}{e^{-z}+1}$, where $z = \sum_{i=1}^{n} \theta_i f_t^{\mathcal{P}_i}(a, b) + \sum_{j=1}^{k} \theta_{n+j} Z_t(a, j) Z_t(b, j)$, and $\theta_1, \theta_2, ..., \theta_n$ and $\theta_{n+1}, \theta_{n+2}, ..., \theta_{n+k}$ are associated weights for meta path-based features and latent features at time t between a and b. Given a training dataset with l instance-label pairs, we use logistic regression with L_2 regularization to estimate the optimal $\boldsymbol{\theta}$ as:

$$\hat{\boldsymbol{\theta}} = \underset{\boldsymbol{\theta}}{\operatorname{argmin}} \sum_{i=1}^{l} -log Pr(label|a_i, b_i; \boldsymbol{\theta}) + \lambda \sum_{j=1}^{n+k} \theta_j^2 \quad (3)$$

We prefer to combine features in this learning framework since G_i is very sparse and thus the number of newly formed links are much less compared to all possible links. Consequently calculating meta path-based features for the training dataset is scalable compared to the matrix factorization technique. Moreover, similar to [30], in order to avoid excessive computation of meta path-based measures between nodes that might not be related, we confine samples to pairs that are located in a nearby neighborhood. More specifically, for each source node x in $G_i^{\mathcal{P}}$, we choose target nodes that are within two hops of x but not in 1-hop, i.e, are not connected to x in $G_i^{\mathcal{P}}$. We first find all target nodes that make a new relationship with x in $G_{i+1}^{\mathcal{P}}$ and label respective samples as positive. Next we sample an equivalent number of negative pairs, i.e., those targets that do not make new connection, in order to balance our training set. Once the dataset is built, we perform logistic regression to learn the model and then apply the predictive model to the feature vector for the target link. The output probability can be later interpreted as a binary value based on a cut-off threshold.

Algorithm 2. Dynamic Meta path-based Relationship Prediction

Input: A DHIN graph G, the number of snapshots t, a network schema S, a target meta path $\mathcal{P}(A, B)$, the maximum length of a meta path l, the latent space dimension k, the link to predict (a, b) at $t+1$

Output: The probability of existence of link (a, b) in $G_{t+1}^{\mathcal{P}}$

1: $\{G_1, .., G_t\} \leftarrow DecomposeGraph(G, t)$
2: Generate target augmented reduced graphs $G_1^{\mathcal{P}}, .., G_t^{\mathcal{P}}$ following Algorithm 1 lines 2-7
3: $\{\mathcal{P}_1, .., \mathcal{P}_n\} \leftarrow GenerateMetaPaths(S, \mathcal{P}(A, B), l)$
4: $\{Z_1, .., Z_t\} \leftarrow MatrixFactorization(G_1^{\mathcal{P}}, .., G_t^{\mathcal{P}}, k)$
5: **for** each pair (x, y), where $x \in V_{t-1}^{\mathcal{P}}$ and $y \in N(x)$ is a nearby neighbor of x in $G_{t-1}^{\mathcal{P}}$ **do**
6: Add feature vector $\langle f_{t-1}^{\mathcal{P}_i}(x, y)$ for $i = 1..n$, $Z_{t-1}(x, j)Z_{t-1}(y, j)$ for $j = 1..k\rangle\rangle$ to the training set T with $label = 1$ if (x, y) is a new link in $E_t^{\mathcal{P}}$ otherwise $label = 0$.
7: **end for**
8: $model \leftarrow Train(T)$
9: Return $Pr((a, b) \in E_{t+1}^{\mathcal{P}}) \leftarrow Test(model, \langle f_t^{\mathcal{P}_i}(a, b)$ for $i = 1..n$, $Z_t(a, j)Z_t(b, j)$ for $j = 1..k\rangle\rangle)$

We describe steps for building and applying our predictive model, called *MetaDynaMix*, in Algorithm 2. The algorithm takes as input a DHIN graph G, the number of graph snapshots t, a network schema S, a target relation meta path $\mathcal{P}(A, B)$, the maximum length of a meta path l, the latent space dimension k, and the link to predict (a, b) at $t+1$. Similar to Algorithm 1, it decomposes G into a sequence of graphs (line 1). Next it generates augmented reduced graphs $G_i^{\mathcal{P}}$s from G_is based on $\mathcal{P}$ for nodes which are of type A and B (beginning and end of meta path $\mathcal{P}$) (line 2) as explained in Algorithm 1. It then produces the set of all meta paths between nodes of type A and type B defined in $\mathcal{P}(A, B)$ (line 3). This is done by traversing the network schema S (for instance through BFS traversal) and generating meta paths with the maximum length of l. It then applies matrix factorization to find latent space matrices Z_i (line 4). Next it creates a training dataset for sample pairs (x, y) with feature set containing meta path-based measures $f_t^{\mathcal{P}_i}(x, y)$ for each meta path $\mathcal{P}_i$, and latent features $Z_t(a, j)Z_t(b, j)$ for $j = 1..k$ at time t, and *label*=1 if (x, y) is a new link in $G_{t+1}^{\mathcal{P}}$ otherwise *label*=0 (lines 5–7). Subsequently the algorithm trains the predictive model (line 8), generates features for the given pair (a, b), and tests it using the trained model (line 9).

4 Experiments

4.1 Experiment Setup

Dataset. We conduct our experiments on two real-world network datasets that have different characteristics and evolution behaviour.

Publications Dataset: The AMiner citation dataset [33] version 8 (2016-07-14) is extracted from DBLP, ACM, and other sources. It contains 3,272,991 papers and 8,466,859 citation relationships for 1,752,443 authors, who published in 10,436 venues, from 1936 to 2016. Each paper is associated with an abstract, authors, year, venue, and title. We confined our experiments to papers published since

1996, which includes 2,935,679 papers. Similar to [30], we considered only authors with at least 5 papers.

Movies Dataset: The RecSys HetRec movie dataset [3] is an extension of MovieLens10M published by the GroupLens research group that links the movies of MovieLens dataset with their corresponding web pages on IMDB and Rotten Tomatoes. It contains information of 2,113 users, 10,197 movies, 20 movie genres (avg. 2.04 genres per movie), 4,060 directors, 95,321 actors (avg. 22.78 actors per movie), 72 countries, and 855,598 ratings (avg. 404.92 ratings per user, and avg. 84.64 ratings per movie).

Experiment Settings. Here, we describe meta paths and target relationships, baseline methods, and different parameter settings that have been used in our experiments.

Meta Paths and Target Relationships. Figure 4 depicts network schemas for the two datasets. Note that we consider a simplified version and ignore nodes such as topic for papers or tag for movies. Table 1 presents a number of meta paths that we employed in our experiments where target meta path relations are *co-authorship* and *watching*. Note that in the publications network, each paper is published only once and authorship relationships are formed at the time of publication whereas in the movies network, users can watch/rate a movie at any given point in time and hence user-movie relations are not as rigid as the authorship relations in the publication dataset.

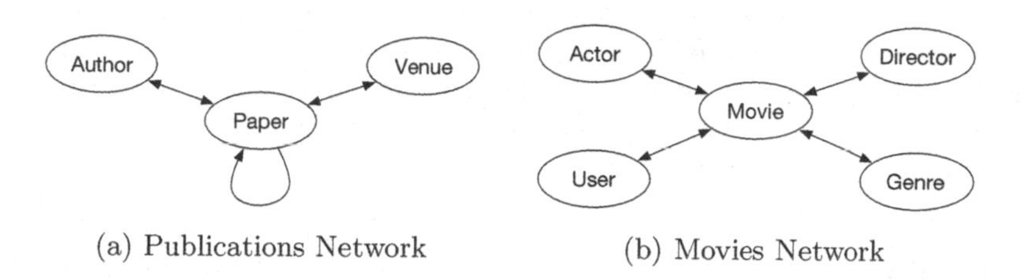

(a) Publications Network (b) Movies Network

Fig. 4. The simplified network schema used for our experiments.

Baseline Methods. Sun et al. [30] proposed a supervised learning framework for link prediction in HINs, called PathPredict, that learns coefficients associated with meta path-based features by maximizing the likelihood of new relationship formation. Their model is learned based on one past interval and does not consider temporal changes in different intervals. Since to our knowledge there is no baseline for relationship prediction in DHINs, we perform comparative analysis of our work, denoted as MetaDynaMix, with four techniques: (1) The original PathPredict [30] that considers only 3 intervals, (2) PathPredict applied on different time intervals, denoted as PathPredict+, (3) homogenized link prediction (Sect. 3.1) by applying [41], denoted as HLP, and (4) logistic regression on HLP

Table 1. Meta paths for publications dataset with $V = \{\text{Author, Paper, Venue}\}$ and movies dataset with $V = \{\text{User, Movie, Actor, Director, Genre}\}$.

Network	Meta path	Meaning
Publications	*A–P–A*	[*The target relation*] Authors are coauthors
	A–P–V–P–A	Authors publish in the same venue
	A–P–A–P–A	Authors have the same co-author
	A–P–P–P–A	Authors cite the same papers
Movies	*U–M*	[*The target relation*] A user watches a movie
	U–M–A–M	A user watches a movie with the same actor
	U–M–D–M	A user watches a movie with the same director
	U–M–G–M	A user watches a movie of the same genre
	U–M–U–M	A user watches a movie that another user

latent features, denoted as LRHLP. Note that PathPredict [30] was shown to outperform traditional link prediction approaches that use topological features defined in homogeneous networks such as common neighbors or Katzβ, and thus we do not include these techniques in our experiments.

Parameters. We set the number of snapshots $t = 3, 5$, and 7 to evaluate the effect of dynamic analysis of different time intervals. Note that $t = 3$ refers to the default case for many link prediction algorithms that learn based on one interval and test based on another. More specifically in the training phase, features are extracted based on T1 and labels are determined based on T2, and for the testing phase, features are calculated based on T2 and labels are derived from T3. In our experiments we did not observe a considerable change in prediction performance by setting the number of latent features k to 5, 10, and 20, and thus all presented results are based on setting k to 20.

Implementation. We use the implementation of matrix factorization for inferring temporal latent spaces of a sequence of graph snapshots presented in [41]. We use all the default settings such as the number of latent features k to be 20, and the optimization algorithm to be the local block-coordinate gradient descent. For the classification part, we use the efficient LIBLINEAR [8] package and set the type of solver to L2-regularized logistic regression (primal).

Evaluation Metrics. To assess link prediction performance, we use Area Under Curves (AUC) for Receiver Operating Characteristic (ROC) [5] and accuracy (ACC). We also perform the McNemar's test [21] to assess the statistical significance of the difference between classification techniques.

4.2 Results and Findings

Link Prediction Accuracy. We now compare the prediction accuracy of different methods. The results shown in Fig. 5 are based on setting the number of time intervals t to 7 for dynamic methods and 3 intervals for PathPredict. Table 2 shows more details considering different intervals. These results show the statistically significant improvement provided by the proposed MetaDynaMix prediction method compared to the baselines. The authors in [22,41] showed that latent features are more predictive compared to unsupervised scoring techniques such as Katz, or Adamic. In our experiments we observed that combining latent features with meta path-based features (MetaDynaMix) can increase prediction accuracy. However, if latent features learn similar structure as topological features do, then mixing them may not be beneficial. In such cases feature engineering techniques could be applied.

We also observe that PathPredict+ performs better than LRHLP in predicting links for the publications network but LRHLP offers more accurate predictions on the movies network. This implies that unlike the publications network, our meta path-based features for the movies network are not as predictive as latent features. However, in both cases combining the two set of features gives better performance than either model individually.

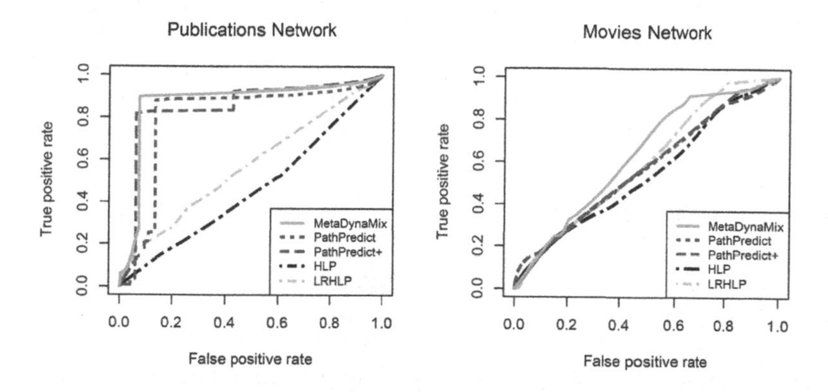

Fig. 5. The ROC curves for different methods and datasets.

Significance of Improvement. McNemar's test, also called within-subjects χ^2 test, is used to compare statistically significant difference between the accuracy of two predictive models based on the contingency table of their predictions. The null hypothesis assumes that the performances of the two models are equal. We compare MetaDynaMix with the other four baselines and the test results show a p-value < 0.0001 for all cases and hence we reject the null hypothesis.

The Effect of Time Intervals. We set the number of time intervals t to 3, 5, and 7 and assess its impact on prediction performance. As presented in Table 2, accuracy increases with the number of snapshots. The intuition is that shorter

Table 2. Relationship prediction accuracy comparison. Bold values are determined to be statistically significant compared to the baselines based on McNemar's test.

Method	Metric	Publications network			Movies network		
		$t = 3$	$t = 5$	$t = 7$	$t = 3$	$t = 5$	$t = 7$
PathPredict	ROC	0.78	–	–	0.56	–	–
	ACC	0.55	–	–	0.54	–	–
PathPredict+	ROC	0.78	0.80	0.83	0.56	0.57	0.57
	ACC	0.55	0.58	0.60	0.54	0.54	0.55
HLP	ROC	0.42	0.43	0.46	0.51	0.53	0.54
	ACC	0.50	0.50	0.50	0.51	0.52	0.53
LRHLP	ROC	0.49	0.50	0.52	0.52	0.56	0.59
	ACC	0.47	0.50	0.51	0.52	0.56	0.58
MetaDynaMix	ROC	0.85	0.87	**0.87**	0.57	0.59	**0.63**
	ACC	0.78	0.80	**0.82**	0.56	0.60	**0.62**

time intervals result in less changes in the graph and thus leads to more reliable predictions. For example considering a meta path A–P–V–P–A, with smaller number of intervals, i.e., longer time intervals, we have more distinct authors who have published in a venue in different years and thus more similar path count values. However, by considering more intervals fewer authors will have such relations and more diverse path counts can contribute to a more accurate prediction for the next time interval.

5 Related Work

The problem of link prediction in static and homogeneous networks has been extensively studied in the past [1,2,16,18,19,34], for which the probability of forming a link between two nodes is generally considered as a function of their topological similarity. However, such techniques cannot be directly applied to heterogeneous networks. A few works such as [30,31] investigated the problem of link prediction in HINs. Sun et al. [30] showed that *PathPredict* outperforms traditional link prediction approaches that use topological features defined on homogeneous networks such as common neighbors, preferential attachment, and Katzβ. Different from the original link prediction problem, Sun et al. [31] studied the problem of predicting the time of relationship building in HINs. These works, however, do not consider the dynamism of networks and overlook the potential benefits of analyzing a HIN as a sequence of network snapshots.

Research works on static latent space inference of networks [22,25,26,37,38] have assumed that the latent positions of nodes are fixed, and only few graph embedding methods [7,10,41] have considered dynamic networks. Dunlavy et al. [7] developed a tensor-based latent space modeling technique to predict temporal links. Zhu et al. [41] added a temporal-smoothing regularization term to a non-negative matrix factorization objective to penalize abrupt large changes in the latent positions. These works do not consider heterogeneity of network structure.

6 Conclusions and Future Work

We have studied the problem of relationship prediction in DHINs and proposed a supervised learning framework based on a combined set of latent and topological meta path-based features. Our results show that the proposed technique significantly improves prediction accuracy compared to the baseline methods. As a part of future work and given the major computational bottleneck of methods that rely on meta-paths, such as our approach, is calculating meta path-based measures, we would like to investigate approximation techniques to make the prediction process scalable. Furthermore, we are interested in enhancing the matrix factorization technique based on a loss function that does not require the full topological features matrix. Another interesting direction to investigate is the effectiveness of our proposed approach in other application domains such as predicting user interests in a social network that is both temporally dynamic and heterogeneous by nature. Link prediction techniques may also increase the risk of link disclosure, such as through link reconstruction and re-identification attacks [9,40], and thus increase privacy concern. It is interesting to study the effect of our technique in performance of link privacy preserving methods, such as [14,23,24,40], and propose suggestions for improvement.

References

1. Al Hasan, M., Chaoji, V., Salem, S., Zaki, M.: Link prediction using supervised learning. In: SDM06: Workshop on Link Analysis, Counter-Terrorism and Security (2006)
2. Al Hasan, M., Zaki, M.J.: A survey of link prediction in social networks. In: Aggarwal, C. (ed.) Social Network Data Analytics, pp. 243–275. Springer, Boston (2011). https://doi.org/10.1007/978-1-4419-8462-3_9
3. Cantador, I., Brusilovsky, P., Kuflik, T.: 2nd workshop on information heterogeneity and fusion in recommender systems (HetRec 2011). In: Proceedings of the 5th ACM Conference on Recommender Systems, RecSys 2011, ACM. New York (2011). http://www.grouplens.org
4. Chen, H., Li, X., Huang, Z.: Link prediction approach to collaborative filtering. In: Proceedings of the 5th ACM/IEEE-CS Joint Conference on Digital Libraries, JCDL 2005, pp. 141–142. IEEE (2005)
5. Davis, J., Goadrich, M.: The relationship between precision-recall and ROC curves. In: Proceedings of the 23rd International Conference on Machine Learning, pp. 233–240. ACM (2006)

6. Denny, J.C.: Mining electronic health records in the genomics era. PLoS Comput. Biol. **8**(12), e1002823 (2012)
7. Dunlavy, D.M., Kolda, T.G., Acar, E.: Temporal link prediction using matrix and tensor factorizations. ACM Trans. Knowl. Discovery Data (TKDD) **5**(2), 10 (2011)
8. Fan, R.E., Chang, K.W., Hsieh, C.J., Wang, X.R., Lin, C.J.: Liblinear: a library for large linear classification. J. Mach. Learn. Res. **9**, 1871–1874 (2008). https://github.com/cjlin1/liblinear
9. Fire, M., Katz, G., Rokach, L., Elovici, Y.: Links reconstruction attack. In: Altshuler, Y., Elovici, Y., Cremers, A., Aharony, N., Pentland, A. (eds.) Security and Privacy in Social Networks, pp. 181–196. Springer, New York (2013). https://doi.org/10.1007/978-1-4614-4139-7_9
10. Fu, W., Song, L., Xing, E.P.: Dynamic mixed membership blockmodel for evolving networks. In: Proceedings of the 26th Annual International Conference on Machine Learning, pp. 329–336. ACM (2009)
11. Gallagher, B., Tong, H., Eliassi-Rad, T., Faloutsos, C.: Using ghost edges for classification in sparsely labeled networks. In: Proceedings of the 14th ACM SIGKDD International Conference on Knowledge Discovery and Data Mining, pp. 256–264. ACM (2008)
12. Guimerà, R., Sales-Pardo, M.: Missing and spurious interactions and the reconstruction of complex networks. Proc. Nat. Acad. Sci. **106**(52), 22073–22078 (2009)
13. Guy, I.: Social recommender systems. In: Ricci, F., Rokach, L., Shapira, B. (eds.) Recommender Systems Handbook, pp. 511–543. Springer, Boston (2015). https://doi.org/10.1007/978-1-4899-7637-6_15
14. Hay, M., Miklau, G., Jensen, D., Towsley, D., Weis, P.: Resisting structural re-identification in anonymized social networks. Proc. VLDB Endowment **1**(1), 102–114 (2008)
15. Lei, C., Ruan, J.: A novel link prediction algorithm for reconstructing protein-protein interaction networks by topological similarity. Bioinformatics **29**(3), 355–364 (2012)
16. Leroy, V., Cambazoglu, B.B., Bonchi, F.: Cold start link prediction. In: Proceedings of the 16th ACM SIGKDD International Conference on Knowledge Discovery and Data Mining, pp. 393–402. ACM (2010)
17. Li, X., Chen, H.: Recommendation as link prediction in bipartite graphs: a graph kernel-based machine learning approach. Decis. Support Syst. **54**(2), 880–890 (2013)
18. Liben-Nowell, D., Kleinberg, J.: The link-prediction problem for social networks. J. Am. Soc. Inform. Sci. Technol. **58**(7), 1019–1031 (2007)
19. Lichtenwalter, R.N., Lussier, J.T., Chawla, N.V.: New perspectives and methods in link prediction. In: Proceedings of the 16th ACM SIGKDD International Conference on Knowledge Discovery and Data Mining, pp. 243–252. ACM (2010)
20. Lü, L., Medo, M., Yeung, C.H., Zhang, Y.C., Zhang, Z.K., Zhou, T.: Recommender systems. Phys. Rep. **519**(1), 1–49 (2012)
21. McNemar, Q.: Note on the sampling error of the difference between correlated proportions or percentages. Psychometrika **12**(2), 153–157 (1947)

22. Menon, A.K., Elkan, C.: Link prediction via matrix factorization. In: Gunopulos, D., Hofmann, T., Malerba, D., Vazirgiannis, M. (eds.) ECML PKDD 2011. LNCS (LNAI), vol. 6912, pp. 437–452. Springer, Heidelberg (2011). https://doi.org/10.1007/978-3-642-23783-6_28
23. Milani Fard, A., Wang, K.: Neighborhood randomization for link privacy in social network analysis. World Wide Web **18**(1), 9–32 (2015)
24. Milani Fard, A., Wang, K., Yu, P.S.: Limiting link disclosure in social network analysis through subgraph-wise perturbation. In: Proceedings of the International Conference on Extending Database Technology (EDBT), pp. 109–119. ACM (2012)
25. Qi, G.J., Aggarwal, C.C., Huang, T.: Link prediction across networks by biased cross-network sampling. In: IEEE 29th International Conference on Data Engineering (ICDE), pp. 793–804. IEEE (2013)
26. Sarkar, P., Moore, A.W.: Dynamic social network analysis using latent space models. ACM SIGKDD Explor. Newsl. **7**(2), 31–40 (2005)
27. Shi, C., Kong, X., Huang, Y., Philip, S.Y., Wu, B.: HeteSim: a general framework for relevance measure in heterogeneous networks. IEEE Trans. Knowl. Data Eng. **26**(10), 2479–2492 (2014)
28. Shi, C., Li, Y., Zhang, J., Sun, Y., Philip, S.Y.: A survey of heterogeneous information network analysis. IEEE Trans. Knowl. Data Eng. **29**(1), 17–37 (2017)
29. Song, H.H., Cho, T.W., Dave, V., Zhang, Y., Qiu, L.: Scalable proximity estimation and link prediction in online social networks. In: Proceedings of the 9th ACM SIGCOMM Conference on Internet measurement, pp. 322–335. ACM (2009)
30. Sun, Y., Barber, R., Gupta, M., Aggarwal, C.C., Han, J.: Co-author relationship prediction in heterogeneous bibliographic networks. In: Proceedings of the 2011 International Conference on Advances in Social Networks Analysis and Mining, ASONAM 2011, pp. 121–128. IEEE Computer Society (2011)
31. Sun, Y., Han, J., Aggarwal, C.C., Chawla, N.V.: When will it happen? Relationship prediction in heterogeneous information networks. In: Proceedings of the Fifth ACM International Conference on Web Search and Data Mining, WSDM 2012, pp. 663–672. ACM, New York (2012)
32. Sun, Y., Han, J., Yan, X., Yu, P.S., Wu, T.: PathSim: meta path-based top-k similarity search in heterogeneous information networks. In: Proceedings of the VLDB Endowment, vol. 4, no. 11, pp. 992–1003. VLDB Endowment (2011)
33. Tang, J., Zhang, J., Yao, L., Li, J., Zhang, L., Su, Z.: ArnetMiner: extraction and mining of academic social networks. In: KDD 2008, pp. 990–998 (2008). https://aminer.org/citation
34. Wang, C., Satuluri, V., Parthasarathy, S.: Local probabilistic models for link prediction. In: ICDM, pp. 322–331. IEEE (2007)
35. Wang, C., Song, Y., El-Kishky, A., Roth, D., Zhang, M., Han, J.: Incorporating world knowledge to document clustering via heterogeneous information networks. In: Proceedings of the 21th ACM SIGKDD International Conference on Knowledge Discovery and Data Mining, pp. 1215–1224. ACM (2015)
36. Wang, C., Song, Y., Li, H., Zhang, M., Han, J.: Text classification with heterogeneous information network kernels. In: AAAI, pp. 2130–2136 (2016)
37. Ye, J., Cheng, H., Zhu, Z., Chen, M.: Predicting positive and negative links in signed social networks by transfer learning. In: Proceedings of the 22nd International Conference on World Wide Web, pp. 1477–1488. ACM (2013)

38. Yin, J., Ho, Q., Xing, E.P.: A scalable approach to probabilistic latent space inference of large-scale networks. In: Advances in Neural Information Processing Systems, pp. 422–430 (2013)
39. Zhang, J., Wang, C., Wang, J., Yu, J.X.: Inferring continuous dynamic social influence and personal preference for temporal behavior prediction. Proc. VLDB Endowment **8**(3), 269–280 (2014)
40. Zheleva, E., Getoor, L.: Preserving the privacy of sensitive relationships in graph data. In: Bonchi, F., Ferrari, E., Malin, B., Saygin, Y. (eds.) PInKDD 2007. LNCS, vol. 4890, pp. 153–171. Springer, Heidelberg (2008). https://doi.org/10.1007/978-3-540-78478-4_9
41. Zhu, L., Guo, D., Yin, J., Steeg, G.V., Galstyan, A.: Scalable temporal latent space inference for link prediction in dynamic social networks. IEEE Trans. Knowl. Data Eng. (TKDE) **28**(10), 2765–2777 (2016). https://github.com/linhongseba/Temporal-Network-Embedding

Retrieving Relationships from a Knowledge Graph for Question Answering

Puneet Agarwal[1,2(✉)], Maya Ramanath[1], and Gautam Shroff[2]

[1] Indian Institute of Technology, Hauz Khas, New Delhi, India
ramanath@cse.iitd.ac.in
[2] TCS Research, New Delhi, India
{puneet.a,gautam.shroff}@tcs.com

Abstract. Answering natural language questions posed on a knowledge graph requires traversing an appropriate sequence of relationships starting from the mentioned entities. To answer complex queries, we often need to traverse more than two relationships. Traditional approaches traverse at most two relationships, as well as typically first retrieve candidate sets of relationships using indexing etc., which are then compared via machine-learning. Such approaches rely on the textual labels of the relationships, rather than the structure of the knowledge graph. In this paper, we present a novel approach KG-REP that directly predicts the embeddings of the target relationships against a natural language query, avoiding the candidate retrieval step, using a sequence to sequence neural network. Our model takes into account the knowledge graph structure via novel entity and relationship embeddings. We release a new dataset containing complex queries on a public knowledge graph that typically require traversal of as many as four relationships to answer. We also present a new benchmark result on a public dataset for this problem.

Keywords: Relationship retrieval · Knowledge graph · Seq2Seq model

1 Introduction

Knowledge graphs can be queried in many different ways, via structured queries using SPARQL, relationship queries using keyword search [1,27], or natural language (NL) queries [47], etc. Natural language queries are especially challenging as these require sense disambiguation and mapping of words mentioned in the query to the appropriate entity and/or relationships in the knowledge graph. Such issues arise when the words mentioned in the query and the text associated with the corresponding entities and relationships (of knowledge graph) are different, as explained in detail below. With recent advances in deep learning [23,30] it has become possible to predict such mappings using supervised learning. Such advances have given rise to mainstream usage of digital assistants, and conversational systems [36]. However, users expect that such assistants should also be

L. Azzopardi et al. (Eds.): ECIR 2019, LNCS 11437, pp. 35–50, 2019.
https://doi.org/10.1007/978-3-030-15712-8_3

able to answer more complex NL questions that perform a form of reasoning: Answering natural language questions requiring multi-hop traversal on a knowledge graph can be seen as a form of compositional reasoning [21].

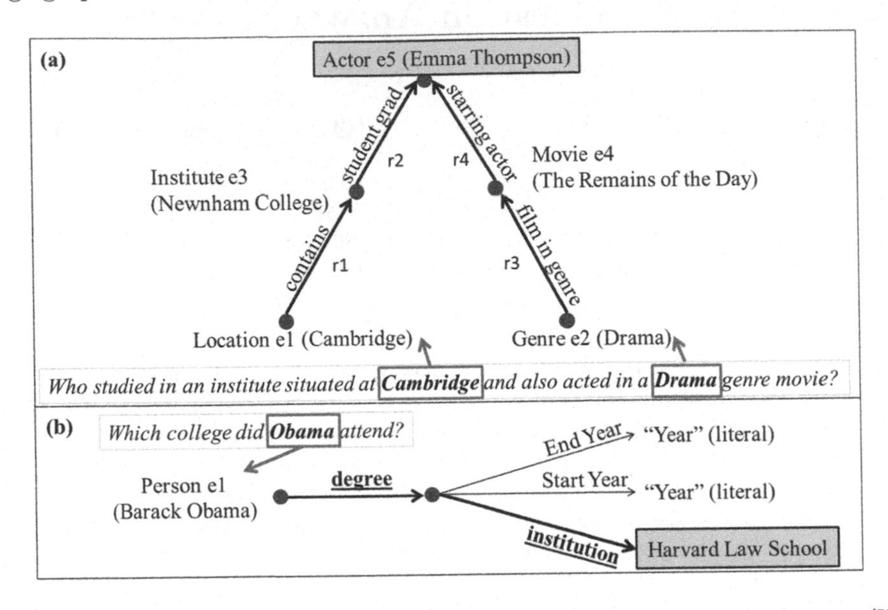

Fig. 1. (a) Freebase subgraph, for a sample query taken from QSMRQ dataset. (We release this dataset through this paper. See Sect. 6 for details.) (b) Freebase subgraph for a query, taken from WebQSP dataset [51]

In this paper, we are concerned with natural language queries such as, "*Which college did Obama attend?*" that mention one or more entities, and expect another entity (one or more) as answer: Such queries are hereafter referred to as *factoid queries*; a similarly worded query can be asked about another person (e.g., Bill Clinton). We assume that knowledge required for answering queries is stored in a knowledge graph such as Freebase knowledge graph [7]. Note that to answer the above query in Freebase requires traversal of two relationships (knowledge graph predicates) {'degree', 'institution'} to retrieve the answer, see Fig. 1(b). Here, the query words 'which college' and 'attend' should help us identify the required relationships out of many relationships of the mentioned entity (Barack Obama). Similarly, another query "*Who studied in an institute situated at Cambridge and also acted in a Drama genre movie?*", requires four relationships {'contains', 'student graduates', 'films in genre', 'starring actor'} to be traversed to retrieve the answer, as shown in Fig. 1(a). Here, again query word 'situated' should point to 'contains' relationship. We can therefore say that the words mentioned in the query and the names of the relationships required to answer the queries are different. In this paper we formulate and address this problem, i.e., the retrieval of *multiple* relationships required for answering a factoid query. To the best of our knowledge prior work does not address this problem for complex queries requiring more than two relationships.

To answer factoid queries against a knowledge graph, two components are required: (a) Entity Document Linking [48], which refers to identification and association of entities mentioned in the query to appropriate entity in the knowledge graph, e.g., 'Barack Obama' in Fig. 1(b); and (b) Relationship Retrieval, which is the focus of our work in this paper. For the problem of relationship retrieval, prior works [13,51], first retrieve candidate sets of relationships using variety of heuristic approaches, such as inverted index search [22]. After that they use supervised learning to predict whether a pair comprising the query and a candidate relationship set, match with each other or not, using binary classification. The number of candidate sets against a query can vary from 10 to 100+ in WebQSP dataset [24]. Such approaches do not scale when a query requires traversing of more than two relationships, since the number of candidate sets to evaluate can increase in multiples of node degree with every increase in number of relationships. Prior works [13,51] also does not take into account the structure of the underlying knowledge graph; while there has been significant research on representation learning of knowledge graphs [10], where they learn vector representations (also referred to as embeddings) of entities and relationships. These embeddings have been used for knowledge graph completion, link prediction, community detection, etc.

We present a novel and efficient approach KG-REP (Knowledge Graph Relationship Embedding Prediction) that uses a sequence to sequence neural network [39] to predict the relationship embeddings of the target set of relationships required for answering factoid queries. We then use beam search [18], against the predicted embeddings to retrieve the target set of relationships. Therefore, we completely avoid the candidate retrieval step, leading to better efficiency. While predicting the relationship embeddings, our model takes into account the structure of knowledge graph, using pre-learned knowledge graph entity and relationship embeddings. We also present a novel method of learning vector representation of the knowledge graph, which is inspired by DeepWalk [33]. Through empirical experiments on two datasets, we demonstrate that by using entity and relationship embeddings in the neural network, we can predict the set of relationships with better accuracy. We also demonstrate that our approach outperforms a prior benchmark for the relationship prediction task on WebQSP dataset [51].

Novelty of Our Approach: Classification models normally do not predict characteristics of a target class; rather they predict a discrete probability distribution in the domain of classes, e.g., the softmax layer in neural networks [26]; the most probable class is taken as the predicted class. In our approach the model is trained on vector representations of the target classes, and also predicts vector representation of the target class. Retrieval proceeds via comparison of the predicted embedding with target embeddings using `cosine` similarity, and we show how to do this efficiently via beam search. Very recently, a similar approach has been used in the domain of image processing [28], in which embeddings for many constituent image patches are predicted and then filtered using a conditional random field to choose the best set of predictions. In summary, the principal contributions of our work are:

- We present a novel approach KG-REP for retrieval of multiple relationships that need to be traversed in a knowledge graph for answering natural language factoid queries; further, we directly predict these relationships in a scalable manner rather than relying on candidate retrieval using indexing.
- We present a novel method of learning the vector representations for entities and relationships of a knowledge graph - *ere-embedding*. We demonstrate that, using these representations, significantly improves the accuracy of relationship prediction.
- We present a novel deep-learning approach based on a sequence to sequence model that takes a factoid query and entity embeddings as input and predicts the required relationship embeddings.
- We demonstrate the capability to predict multiple relationships to answer queries and also release a dataset of such complex queries on public knowledge graph. We also present a new benchmark for relationship prediction on a prior dataset, using KG-REP.

Organization of the Paper: We begin with a description of the problem of relationship prediction in Sect. 2 and then provide a brief background of the technique used to solve this problem in Sect. 3. We present a perspective of prior work done in the related area in Sect. 4. After that in Sect. 5 we describe our novel approach KG-REP and then conclude in Sect. 7 after a brief analysis of the experimental results in Sect. 6.

2 Problem Description

We assume that a knowledge graph G, comprises a set of entities $e_i \in E$ and relationships $r_k \in R$, i.e., $G = (E, R)$. A relationship $r_i \in R$ (also referred to as *predicates*) is defined against a pair of entities $(e_i, e_j) \in E$, and the knowledge graph G is represented as a set of triples, i.e., $t_i = (e_i^1, r_i, e_i^2)$. The first entity e_i^1 is usually referred to as subject and the second entity e_i^2 is referred to as object. The entities $e_i \in E$ as well as the relationships r_i have associated textual label. Many open domain knowledge graphs are publicly available: Yago [38], Freebase [7], NELL [12] DBpedia [2], etc.

We are interested in a special type of natural language query $q = \{w_1^q, \ldots, w_n^q\}$, which is a sequence of words w_i^q. Such queries can be answered by retrieval of appropriate information from a knowledge graph. The query q should mention at-least one entity e_i^q of the knowledge graph, as shown in the examples given in Fig. 1. Set of all mentioned entities, in a query q, is denoted as $E_q = \{e_1^q, ...\}$. We assume that a mapping between words of the query w_i^q, which mention an entity, and corresponding knowledge graph entity e_i^q is available, i.e., $\forall e_i^q \in E_q, map(e_i^q) = \{w_j^q, ...\}, w_j^q \in q$. Through such a query, users intend to retrieve a set of entities $A_q = \{a_1^q, a_2^q, \ldots\}, a_i^q \in E$. We refer to such natural language queries as *factoid queries*.

The objective of multiple relationship prediction problem, against a factoid query q, is to detect the set of relationships $R_q = \{r_1^q, r_2^q, ...\}$, such that if we traverse along these relationships in the knowledge graph G starting from the mentioned entities $e_i^q \in E_q$, we can arrive at the answer entities A_q, which are then retrieved. *Note:* We assume that the mapping between all mentioned entities E_q and corresponding query words w_i^q is available a priori.

3 Brief Background

Recurrent Neural Network and LSTM. Recurrent neural network (RNN), are normally used to encode a sequence $x_i = (v_1, v_2, ..., v_n)$. At every timestamp in the sequence, the input unit takes the sequence value along with the activation of previous unit as input, i.e., $h_t = \sigma(\theta(h_{t-1} + v_t) + b)$. When learning such a neural network using back propagation we will need to use the chain rule of differentiation as many times as the sequence length, which can also be referred to as back propagation through time. Which has been shown [23] to either vanish or explode, making it hard to train the network. This problem has later solved by Hochreiter and Schmidhuber in their famous work popularly known as LSTM [23]. Here, for a sequence input to RNN x_i, the output is controlled by a set of gates in $\mathbb{R}^d$ as a function of the previous hidden state h_{t-1} and the input at the current time step v_t as defined below.

$$\begin{aligned}
\text{input gate}, i_t &= \sigma(\theta_{vi}v_t + \theta_{hi}h_{t-1} + b_i)\\
\text{forget gate}, f_t &= \sigma(\theta_{vf}v_t + \theta_{hf}h_{t-1} + b_f)\\
\text{output gate}, o_t &= \sigma(\theta_{vo}v_t + \theta_{ho}h_{t-1} + b_o)\\
\text{candidate hidden st.}, g_t &= tanh(\theta_{vg}v_t + \theta_{hg}h_{t-1} + b_q)\\
\text{internal memory}, c_t &= f_t \oplus c_{t-1} + i_t \oplus g_t\\
\text{hidden state}, h_t &= o_t \oplus tanh(c_t)
\end{aligned}$$

Here, σ is the logistic sigmoid function with output in $[0, 1]$, `tanh` denotes the hyperbolic tangent function with output in $[-1, 1]$, and $\oplus$ denotes the element wise multiplication. We can view f_t as a function that decides how much information from the old memory cell should be forgotten, i_t controls how much new information should be stored in the current memory cell, and o_t controls output based on the memory cell c_t.

Sequence to Sequence Model. In a sequence to sequence model [39] we first arrive at a lower dimension representations of input sequence of symbols $\mathbf{x_i} = (v_1, v_2, ..., v_T)$, referred to as latent representation ($\mathbf{h_n}$). For example, using RNN, referred to as *encoder network*. A sequence of latent representation ($\mathbf{h_n}$) repeated many times is input to another neural network (also RNN), which is referred to as *decoder network*. The decoder network outputs a sequence of symbols, typically using a softmax layer [26]. The vocabulary as well as length of the input sequence and the output sequence can be different. The *encoder network* and *decoder network* are jointly trained using a common loss function via backward propagation of the error against target output sequence.

4 Related Work

Knowledge Graph Representation Learning. Initial approaches for learning the latent representation of graphs were Isomap [40], LLE [35] and Laplacian Eigenmap [3]. For example, Laplacian Eigenmap [3] uses eigen vector optimization to minimize distance between a node and its neighbors using Gaussian kernel. Lately, three types of approaches have emerged, (a) Based on walk of the graph DeepWalk [33], Node2Vec [20], etc. (b) Based on transformation of an entity into its neighboring entities, e.g., TransE [10], TransH [42], TransG [45], TransM [17], TransR [29], TransD [25], etc. Similarly, many other approaches such as RESCAL [31], Structural Embeddings [11], etc. attempt to learn knowledge graph representations encoding structural properties of the graph (c) Joint encoding of knowledge graph structure and entity descriptions, such as NTN [37], TEKE [43], DKRL [46], etc. In almost all these approaches they have attempted knowledge graph completion/link prediction task, or community detection task, etc., but not the task of factoid question answering.

Factoid Queries. In order to run factoid queries, we first need to identify and map the mentioned entities to corresponding entities in the knowledge graph, this step is usually referred to as entity document linking (EDL) [34]. We rely on prior approaches for entity linking, and assume that it is already available. We then need to identify the relationship(s) required to traverse to the target entities. There has been significant work in the area of *factoid queries*. Based on functional aspects of the prior works we can divide the prior work in following categories: (a) When the factoid queries can be answered using single knowledge graph triple, i.e., single relationship. For example, Large Scale QA [9], or Character Level Attention based [19] (b) When the factoid queries require one or two relationships STAGG [13], HR-BiLSTM [51], SEMPRE [4], etc.; (c) When the task of EDL and single relationship extraction both are performed together EARL [16], AMPCNN [50], CFO-QA [14], etc. Similar to Bordes et al. in Large Scale QA [9], we also assume that the entity linking is available. However, we attempt the queries that require traversal of more than two relationships.

Based on the technique used to answer the factoid queries we can group the prior works in following categories: (a) Using Semantic Parsing, SEMPRE [4–6, 41] (b) Deep neural network (CNN) based approach, e.g., [13] (c) Deep neural network (LSTM) based approach, e.g., HR-BiLSTM [51]. Most of these approaches and many other [24], first generate a set of candidates and then evaluate them, while our approach predicts the relationship embeddings without the need to retrieve the candidates first. Unlike prior work, we take into account structure of the knowledge graph rather than textual descriptions of entities and relationships.

Multiple Relationship Extraction. Guu et al. [21] traverse the knowledge graph up to five hops. Here, they traverse to a knowledge graph triple using a transformation operation and calculate a score for it as $score(s/r,t) = \mathbf{e_i^1 W_r e_i^2}$. They progressively apply $\{W_{r_1}, W_{r_2}, ...\}$ to finally get an embedding which represents all the entities that are on the path $\{r_1, r_2, ...\}$ starting from x_s. This operation is called as *compositionalization*. Here the assumption is that sequence of relations to be traversed is given a priori, which is very different from our work. Similarly, Wang et al. [44], extract relationships with respect to images.

Based on this analysis, we find that our work is most similar to HR-BiLSTM [51], while they extract at the most two relationships against a factoid query. We therefore consider this as a baseline approach.

Embedding Prediction. Predicting embedding from a neural network, and then using a search in embedding space has been attempted [28]. Li and Ping [28] first split the image into different patches, then predict an embedding against every patch and then use Conditional Random Fields to identify most correlated embeddings for every patch. Similarly, for a talk of similar image retrieval [15] they generate the embedding of the query image and then search for the similar images in embedding space. However, to the best our knowledge we are the first ones to use embedding prediction followed by knowledge graph driven beam search in embedding space.

5 Knowledge Graph Relationship Retrieval

We train a neural network (which can be thought of as a function approximation) that maps the query and the vector representations of the mentioned entities into relationship embeddings, i.e., f: $(E_q, q) \rightarrow R_q$. These relationships are required to be traversed from the mentioned entities to reach the answer entities in the knowledge graph. Since the output R_q is a set (of relationships), we use a sequence to sequence model [39] for this function approximation. Here, to encode the words of a query, we use pre-trained Glove [32] embeddings. We demonstrate that using this approach, we can retrieve the target relationship set with better accuracy. In this section, we first present our method of learning entity and relationship representation followed by a description of our approach KG-REP.

5.1 Knowledge Graph Representation Learning

Recent approaches of Knowledge Graph representation learning [10] learn the entity and relationship representation based on one fact at a time. While, Deepwalk [33] attempts multiple hops in a graph, via word2Vec [30], but they focus on graph node representation only. We therefore present an approach similar to Deepwalk [33], for knowledge graphs, which also learns relationship embeddings and performs better than TransE [10], and TransD [25] as shown later.

In Deepwalk [33] they perform a random walk on the knowledge graph and various entities encountered on the walk are recorded as a sequence of symbols. They perform such a walk starting from every node of the knowledge graph, and generate many such sequences. Entity-ids are assumed to be words and sequences are assumed to be sentences and they learn vector representation of the entities using Word2Vec [30]. Word2Vec [30] attempts to predict words in window rather than next words only. Later Grover and Leskovec [20] showed that if we choose the next node to walk from a biased probability distribution, we can encode the structure of the graph better, with respect to distant nodes.

Having been inspired by the node2vec [20] and DeepWalk [33] algorithms, we learned vector representations of the entities and relationships of a knowledge graph in a similar manner. We generated random walks from all entities of the knowledge graph to their neighboring entities and relationships involved. We perform k iterations of the walk on the knowledge graph, using a maximum walk length as l. In every iteration of the walk, we start the walk from every entity relationship pair present in the knowledge graph, to ensure that all relationships are covered in the random walk. As a result, we obtained a sequence of entities and relationships as traversed during random walk (therefore the name *ere*). These sequences were considered as sentences, and nodes and relationships as words. We then use Word2Vec [30] to learn vector representation for the nodes and relationships of the graph. We call these vector representations as *ere embeddings*. As a result, we obtain embeddings for all entities $\boldsymbol{e_i} \in E$ and relationships $\boldsymbol{r_k} \in R$ of the knowledge graph G.

5.2 KG-REP: KG Relationship Embedding Prediction

We use a supervised learning approach to retrieve the sequence of relationships which is also referred to as *path*, required for answering factoid queries against a knowledge graph. This model can also be seen as an approximation to a mapping function $f : (q, E_q) \rightarrow R_q$, i.e., it predicts the set of relationships directly, instead of evaluating the candidate paths. The neural network architecture for this model is shown in Fig. 2. We use bidirectional LSTM layers (BiLTSM) in the sequence to sequence model, i.e., the sequence is given as input in forward and reverse order, as a result with respect to every word in the query it retains the context of words on left and right hand side both.

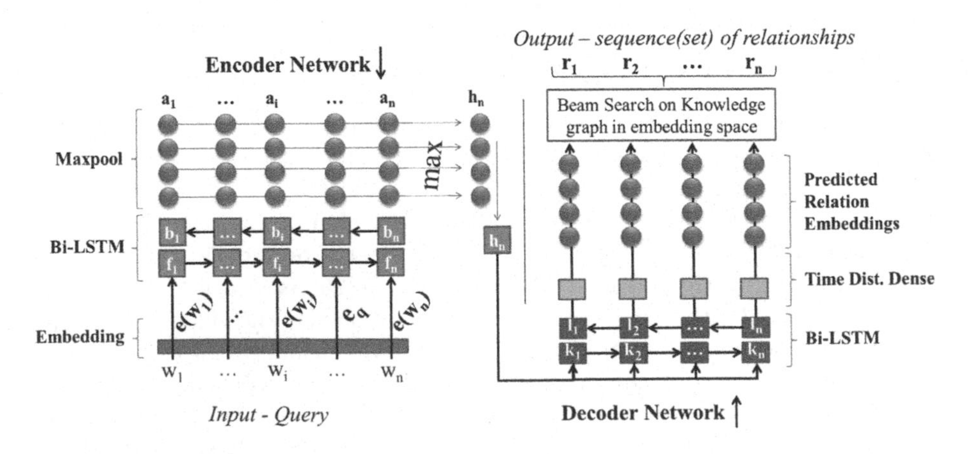

Fig. 2. Deep Neural Network architecture for Sequence to Sequence Model

Encoder Network: Before passing the sentence into the neural network, using the entities linked to the query, we replace the words that mention the entity with the corresponding entity id, and get a modified query q'. We also merge the entity and relationship embeddings (i.e., *ere embedding*) with the Glove embeddings [32] pre-trained on Wikipedia data, which forms the first layer of our neural network, i.e., *embeddings layer* as shown in Fig. 2. The next layer in our sequence to sequence model is a *Bi-directional LSTM* network. At every timestamp this network has two activation states (one for forward direction (f_i) and another one for backward (b_i)), which are concatenated to obtain one vector per timestamp a_i. The dimension of these vectors (a_i) depends on the number of units (l_1) in the *Bidirectional LSTM* layer, which is a hyper-parameter. The output vectors of all the timestamps of the Bi-LSTM layer $(\{\mathbf{a_1}, ..., \mathbf{a_n}\})$ are passed to *max-pool* layer, which gives us final activation vector or latent representation (h_n) of the input query q'. In the *maxpool* layer, we take the maximum value of every dimension of input vectors $(\mathbf{a_i})$, to obtain the latent representation h_n of the query q', which is also considered as the output of the *encoder network*.

Decoder Network: The latent representation h_n is then passed to a *decoder network*, which comprises a *Bi-directional LSTM layer* (with $l2$ units) and a *time distributed dense* layer with `tanh` as activation function. In the *Bi-directional LSTM layer* the encoded state $(\mathbf{h_n})$ is passed repeatedly as many times as the length of the output sequence. The number of units in the time distributed dense layer is kept same as the number of dimensions in the relationship embedding. As a result, the decoder network gives us as many embeddings as the length of the output sequence. We call these output embeddings as predicted embeddings, i.e., $\{\hat{r}^q_1, \hat{r}^q_2, ...\}$. This model directly outputs the embeddings of the target set of relationships, and we do not need to identify the candidates first. As the predicted embeddings do not directly map to the embedding of any relationship, we can identify the target relationship using a nearest neighbor, i.e., the relationship embedding which has highest `cosine` similarity with the predicted embedding. In fact this would be the simplest form of our beam search based approach, which we describe next.

Beam Search: In order to find our final predicted set of relationships we use a beam search algorithm, with a beam width of b, on the predicted embedding of the above mentioned neural network. Here, the relationships to be considered are those connected with the mentioned entity, e.g., 'Barack Obama' in Fig. 1(b). In beam search, we identify the b relationships out of all relationships of the mentioned entity, which have the highest cosine similarity to the first predicted relationship embedding $\mathbf{r^q_1}$. We then arrive at a set of all target entities e_m of these b relationships starting from the mentioned entity. We then identify the set of all relationships connected to all the e_m entities. From this set also we choose the best b relationships. Here, the relationships are chosen based on product of cosine similarity of previous relationship with its predicted embedding (i.e., $cs_1 = \mathtt{cos}(\mathbf{r^q_1}, \hat{r}^q_1)$) and the current relationship with its corresponding predicted embedding (i.e., $cs_2 = \mathtt{cos}(\mathbf{r^q_2}, \hat{r}^q_2)$). The product of the `cosine` similarities $p = (cs_1 \times cs_2 \times ...)$ is also referred to as beam score. As a result, at every timestamp of the output sequence we maintain only the b best (with maximum beam score) paths so far. Finally, we identify the best set of relationships at the last timestamp based on highest beam score.

Training the Network: In order to train the neural network shown in Fig. 2, we use *cosine distance* between the predicted vector representation and true vector representation of the relationships, i.e., $L = 1 - \cos(\hat{\boldsymbol{r}}_i^q, \boldsymbol{r}_i^q)$, as a loss function for back propagation. Here, $\hat{\boldsymbol{r}}_i^q$ and $\boldsymbol{r}_i^q$ are predicted relationship embedding and the true relationship embeddings respectively.

6 Experimental Results

In this section analyze the results of experiments performed on two datasets WebQSP and QSMRQ and find that KG-REP performs better than baseline.

6.1 Description of Datasets Used

WebQSP: Berant et al. in SEMPRE [4] released a dataset, called *WebQuestions*, for researchers to use for the knowledge graph question answering task. This dataset was created by crawling the Google Suggest API and the answers were obtained using Amazon mechanical turk, the crowd sourcing platform. The training set of this data contained 3,778 questions and the test set contained 2,032 questions. Later, Yih et al. [49] found that about 18.5% of the questions in this dataset are not answerable, so they created a subset of this dataset and called it as WebQSP dataset. In this dataset, mentioned entities are marked against Freebase, which is about 490 GB in size.

Knowledge Graph for WebQSP: Similar to [8] we also created a subset of Freebase to make it feasible to learn the KG embeddings fast. For this we took the Freebase data from SEMPRE [4] distribution, which contains about 79 million entities and about 15 million relationships, and is about 40GB in size. To make a subset of this dataset we took suspected set of mentioned entities based on S-Mart linking [48] and their corresponding knowledge graph facts. We also included additional entities that were identified as mentioned entities in the WebQSP dataset. Similar to Large Scale QA [9] we also removed the triples for which entity-relationship pair count was more than a threshold τ. While removing the extra triples, care was taken to not remove the triples required for answering any of the queries present in training or test data. The value of threshold τ was taken as 3. We also included all the facts required for answering the questions. As a result, we obtained a subset of Freebase comprising about 400 thousand facts, having about 234 thousand entities and 3,528 relationships. Here the literal values are also being considered as entities.

FB15K and QSMRQ: Bordes et al. [10], released a subgraph of Freebase knowledge graph named FB15K. This dataset has been extensively used by researchers to demonstrate the efficacy of their proposed approaches. We also took this dataset and created quiz style natural language queries ourselves, which require four relations to answer a question. We call this dataset as QSMRQ (Quiz Style Multi Relation Queries) dataset. This dataset comprises of three sets of 79 template queries involving 4 relations, e.g., "*Who studied in an institute at* $< e1 >$, *and also acted in a* $< e2 >$ *genre movie*", as also shown in Fig. 1. We filled valid entity names in these template questions from the knowledge graph to generate three sets of 755 questions, giving us the training, validation and test datasets. Here, three different natural language variation of every template were used for creation of the training, validation and test dataset. We release (https://github.com/apuneet/KG-REP) this dataset, the code used for generation of the questions, as well as code for KG-REP, for future research.

ERE Embeddings: We ran $k = 10$ iterations of the random walk, with maximum walk length as $l = 10$ on both the knowledge graphs as described above. When running Word2Vec [30], we used skip-graph method of learning with 100 epochs, keeping the window size as 5. Number of dimensions for these embeddings were taken as 300.

6.2 Baseline Comparison - WebQSP

We compare the performance of our algorithm with HR-BiLSTM [51], on WebQSP dataset. When creating the knowledge graph against the WebQSP queries from Freebase (SEMPRE [4] distribution) we could not find corresponding triples for some of the queries. We therefore also took a subsets of the WebQSP dataset to report the results. We report the % of queries against which our approach could retrieve correct set of relationships in Table 1. We report the results with respect to the best of ten different initializations of the neural network parameters chosen before training. Here, we used beam width of 20 and $l_1 = 384$ and $l_2 = 320$ unit in the encoder and decoder Bi-LSTM networks respectively. These were chosen after a round of hyper-parameter tuning on a set $[384, 448, 512]$ and $[128, 192, 256, 320, 384]$ for l_1 and l_2 respectively.

We could not find knowledge graph triples for 35 and 64 queries for test and training sets respectively (either the mentioned entity of the relationship was not present, in the SEMPRE [4] distribution of Freebase). *Note:* We are the first one to use a knowledge graph representation with WebQSP. We created a subset of WebQSP, by removing such queries. (Recall that in order to run KG-REP we need to have embeddings of the mentioned entities and target relationships). When the embeddings of the mentioned entity was not available, similar to HR-BiLSTM [51], we use a token '<e>'. When embedding of a relationship present in the ground truth of a query was not available, we used random embedding.

Through various experiments performed on the WebQSP dataset, we observed that our algorithm KG-REP performs better when we feed the embedding of the mentioned entity in the neural network as compared to using a standard token '<e>' for mentioned entities in the queries. This is evident from the second and third row of the Table 1. This finding corroborates our hypothesis that the neural network attempts to learn a function that maps the mentioned entities into the relationships required for answering the factoid query. Our approach of using *ere embedding* with a sequence to sequence model is also better

Table 1. Results on WebQSP Dataset, comparing the accuracies for two situations, when using the *ere-embedding* of the mentioned entities in input query (ERE), and without using them (no-ERE), and baseline approach.

S.N.	Algorithm	% Accuracy (All queries)	% Accuracy (Queries having data in KG)
1	HR-BiLSTM [51]	82.53 % (all)	-
2	S2S Model (no-ERE)	81.19 % (all)	81.97 % (1581)
3	S2S Model + ERE	**83.23 %** (all)	**84.12 %** (1581)

than the state of the art approach [51], overall. It is important to observe that our approach of using the sequence to sequence model is much easier to train since it works with at-least 10 times lesser data (assuming there are 10 candidates per query on an average).

6.3 Experiments on Quiz-Style Multi-Relation Queries (QSMRQ)

If the candidate path is longer than 2 hops, number of candidates can increase to a large number, which can become in-efficient when using approaches such as [51]. However, our approach KG-REP that uses a sequence to sequence model can scale to longer sequences also. In order to demonstrate this, we report the performance of our approach KG-REP with and without using the *ere embeddings* in the input queries in Table 2, on the QSMRQ dataset. We report the results based on best of ten different neural network parameter initializations. The best model was chosen based on validation split of the data, and the results are being reported on hold out set (test data). For no-ERE situation, we did not input the entity embedding in the input query, keeping rest of the approach same. We observe that not only can our approach be efficiently used for relationship retrieval for larger number of relationships but it also performs better by using the ere embeddings. When using TransE [10], and TransD [25] approaches, we get 94.97% and 93.78% accuracy, respectively. We cannot compare with DeepWalk [33], as they don't learn relationship embeddings.

Table 2. Results on QSMRQ Dataset, comparing the accuracies for: (1) when using the *ere embedding* in both input and output, (2) without using them in the input queries.

S.N.	Algorithm	% Accuracy
1	S2S Model + ERE	**97.35 %**
2	S2S Model (no-ERE)	92.05 %

7 Conclusion

We have described a novel approach, KG-REP, for identifying the multiple relationships that need to be traversed for answering complex natural language queries on a knowledge graph. KG-REP uses a sequence to sequence model for prediction of relationships embeddings, followed by beam search to efficiently retrieve the required relationships. We have demonstrated that KG-REP outperforms a prior benchmark for relationship prediction while answering simple questions on a knowledge graph. We have shown that our approach also works for complex queries that require more than two relationships to retrieve corresponding answer. We have prepared a new dataset of such complex queries that we have released for public use. We have also used a novel approach *ere-embedding* for learning vector representation of nodes and relationships in a knowledge graph, and have shown that using this embedding can significantly improve accuracy of the relationship prediction task as compared to commonly used graph embeddings.

References

1. Agarwal, P., Ramanath, M., Shroff, G.: Relationship queries on large graphs. In: Proceedings of International Conference on Management of Data. COMAD (2017)
2. Auer, S., Bizer, C., Kobilarov, G., Lehmann, J., Cyganiak, R., Ives, Z.: DBpedia: a nucleus for a web of open data. In: Aberer, K., et al. (eds.) ASWC/ISWC-2007. LNCS, vol. 4825, pp. 722–735. Springer, Heidelberg (2007). https://doi.org/10.1007/978-3-540-76298-0_52
3. Belkin, M., Niyogi, P.: Laplacian eigenmaps and spectral techniques for embedding and clustering. In: Advances in Neural Information Processing Systems, pp. 585–591. NIPS (2002)
4. Berant, J., Chou, A., Frostig, R., Liang, P.: Semantic parsing on freebase from question-answer pairs. In: Proceedings of the 2013 Conference on Empirical Methods in Natural Language Processing, EMNLP (2013)
5. Berant, J., Liang, P.: Semantic parsing via paraphrasing. In: Proceedings of the 52nd Annual Meeting of the Association for Computational Linguistics (Volume 1: Long Papers). vol. 1, pp. 1415–1425 (2014)
6. Berant, J., Liang, P.: Imitation learning of agenda-based semantic parsers. Trans. Assoc. Comput. Linguist. **3**, 545–558 (2015)
7. Bollacker, K., Evans, C., Paritosh, P., Sturge, T., Taylor, J.: Freebase: a collaboratively created graph database for structuring human knowledge. In: Proceedings of the 2008 ACM SIGMOD international conference on Management of data, pp. 1247–1250. SIGMOD. ACM (2008)
8. Bordes, A., Chopra, S., Weston, J.: Question answering with subgraph embeddings. arXiv preprint arXiv:1406.3676 (2014)
9. Bordes, A., Usunier, N., Chopra, S., Weston, J.: Large-scale simple question answering with memory networks. arXiv preprint arXiv:1506.02075 (2015)
10. Bordes, A., Usunier, N., Garcia-Duran, A., Weston, J., Yakhnenko, O.: Translating embeddings for modeling multi-relational data. In: Advances in neural information processing systems. NIPS (2013)

11. Bordes, A., Weston, J., Collobert, R., Bengio, Y.: Learning structured embeddings of knowledge bases. In: Conference on artificial intelligence, No. EPFL-CONF-192344 (2011)
12. Carlson, A., Betteridge, J., Kisiel, B., Settles, B., Hruschka Jr., E.R., Mitchell, T.M.: Toward an architecture for never-ending language learning. In: Proceedings of AAAI Conference on Artificial Intelligence. AAAI, vol. 5, p. 3. Atlanta (2010)
13. Chang, W.t.Y.M.W., Gao, X.H.J.: Semantic parsing via staged query graph generation: question answering with knowledge base. In: Proceedings of the 53rd Annual Meeting on Association for Computational Linguistics. ACL (2015)
14. Dai, Z., Li, L., Xu, W.: Cfo: conditional focused neural question answering with large-scale knowledge bases (2016)
15. Do, T.T., Cheung, N.M.: Embedding based on function approximation for large scale image search. IEEE Trans. Pattern Anal. Mach. Intell. **40**(3), 626–638 (2018)
16. Dubey, M., Banerjee, D., Chaudhuri, D., Lehmann, J.: EARL: joint entity and relation linking for question answering over knowledge graphs. In: Vrandečić, D., et al. (eds.) ISWC 2018, Part I. LNCS, vol. 11136, pp. 108–126. Springer, Cham (2018). https://doi.org/10.1007/978-3-030-00671-6_7
17. Fan, M., Zhou, Q., Chang, E., Zheng, T.F.: Transition-based knowledge graph embedding with relational mapping properties. In: Proceedings of the 28th Pacific Asia Conference on Language, Information and Computing (2014)
18. Furcy, D., Koenig, S.: Limited discrepancy beam search. In: Proceedings of the International Joint Conferences on Artificial Intelligence, pp. 125–131 (2005)
19. Golub, D., He, X.: Character-level question answering with attention. In: Proceedings of the 2016 Conference on Empirical Methods in Natural Language Processing. EMNLP (2016)
20. Grover, A., Leskovec, J.: node2vec: scalable feature learning for networks. In: Proceedings of the 22nd ACM SIGKDD International Conference on Knowledge Discovery and Data Mining. KDD, ACM (2016)
21. Guu, K., Miller, J., Liang, P.: Traversing Knowledge Graphs in Vector Space (2015)
22. Hatcher, E., Gospodnetic, O., McCandless, M.: Lucene in Action. Manning Publications, Greenwich (2004)
23. Hochreiter, S., Schmidhuber, J.: Long short-term memory. Neural comput. **9**(8), 1735–1780 (1997)
24. Jain, S.: Question answering over knowledge base using factual memory networks. In: Proceedings of the NAACL Student Research Workshop (2016)
25. Ji, G., He, S., Xu, L., Liu, K., Zhao, J.: Knowledge graph embedding via dynamic mapping matrix. In: Proceedings of the 53rd Annual Meeting of the Association for Computational Linguistics and the 7th International Joint Conference on Natural Language Processing (Volume 1: Long Papers), vol. 1, pp. 687–696 (2015)
26. Lai, S., Xu, L., Liu, K., Zhao, J.: Recurrent convolutional neural networks for text classification. In: Proceedings of AAAI Conference on Artificial Intelligence. AAAI (2015)
27. Li, R.H., Qin, L., Xu Yu, J., Mao, R.: Efficient and progressive group steiner tree search. In: International Conference on Database Theory. SIGMOD (2016)
28. Li, Y., Ping, W.: Cancer metastasis detection with neural conditional random field. In: Medical Imaging with Deep Learning (2018)
29. Lin, Y., Liu, Z., Sun, M., Liu, Y., Zhu, X.: Learning entity and relation embeddings for knowledge graph completion. In: Proceedings of AAAI Conference on Artificial Intelligence. AAAI, vol. 15, pp. 2181–2187 (2015)

30. Mikolov, T., Sutskever, I., Chen, K., Corrado, G.S., Dean, J.: Distributed representations of words and phrases and their compositionality. In: Advances in neural information processing systems. NIPS (2013)
31. Nickel, M., Tresp, V., Kriegel, H.P.: A three-way model for collective learning on multi-relational data. ICML **11**, 809–816 (2011)
32. Pennington, J., Socher, R., Manning, C.: Glove: Global vectors for word representation. In: Proceedings of the 2014 Conference on Empirical Methods in Natural Language Processing. EMNLP (2014)
33. Perozzi, B., Al-Rfou, R., Skiena, S.: Deepwalk: online learning of social representations. In: Proceedings of the 20th ACM SIGKDD International Conference on Knowledge Discovery and Data Mining. KDD (2014)
34. Radhakrishnan, P., Talukdar, P., Varma, V.: Elden: Improved entity linking using densified knowledge graphs. In: Proceedings of the 2018 Conference of the North American Chapter of the Association for Computational Linguistics: Human Language Technologies, Volume 1 (Long Papers), vol. 1, pp. 1844–1853 (2018)
35. Roweis, S.T., Saul, L.K.: Nonlinear dimensionality reduction by locally linear embedding. Science **290**(5500), 2323–2326 (2000)
36. Singh, M.P., et al.: Knadia: enterprise knowledge assisted dialogue systems using deep learning. In: Proceedings of International Conference on Data Engineering. ICDE (2018)
37. Socher, R., Chen, D., Manning, C.D., Ng, A.: Reasoning with neural tensor networks for knowledge base completion. In: Advances in neural information processing systems. NIPS (2013)
38. Suchanek, F.M., Kasneci, G., Weikum, G.: Yago: a core of semantic knowledge. In: Proceedings of the 16th International Conference on World Wide Web, WWW. pp. 697–706. ACM (2007)
39. Sutskever, I., Vinyals, O., Le, Q.V.: Sequence to sequence learning with neural networks. In: Advances in Neural Information Processing Systems. NIPS (2014)
40. Tenenbaum, J.B., De Silva, V., Langford, J.C.: A global geometric framework for nonlinear dimensionality reduction. Science **290**(5500), 2319–2323 (2000)
41. Wang, Y., Berant, J., Liang, P.: Building a semantic parser overnight. In: Proceedings of the 53rd Annual Meeting of the Association for Computational Linguistics and the 7th International Joint Conference on Natural Language Processing (Volume 1: Long Papers), vol. 1, pp. 1332–1342 (2015)
42. Wang, Z., Zhang, J., Feng, J., Chen, Z.: Knowledge graph embedding by translating on hyperplanes. In: Proceedings of AAAI Conference on Artificial Intelligence. AAAI, vol. 14, pp. 1112–1119 (2014)
43. Wang, Z., Li, J.Z.: Text-enhanced representation learning for knowledge graph. In: Proceedings of the International Joint Conferences on Artificial Intelligence, pp. 1293–1299. IJCAI (2016)
44. Wang, Z., Chen, T., Ren, J.S.J., Yu, W., Cheng, H., Lin, L.: Deep reasoning with knowledge graph for social relationship understanding. In: Proceedings of the International Joint Conferences on Artificial Intelligence. IJCAI (2018)
45. Xiao, H., Huang, M., Zhu, X.: Transg: A generative model for knowledge graph embedding. In: Proceedings of the 54th Annual Meeting of the Association for Computational Linguistics (Volume 1: Long Papers), ACL, vol. 1, pp. 2316–2325 (2016)
46. Xie, R., Liu, Z., Jia, J., Luan, H., Sun, M.: Representation learning of knowledge graphs with entity descriptions. In: Proceedings of AAAI Conference on Artificial Intelligence, pp. 2659–2665. AAAI (2016)

47. Yahya, M., Berberich, K., Elbassuoni, S., Ramanath, M., Tresp, V., Weikum, G.: Natural language questions for the web of data. In: Proceedings of the 2012 Joint Conference on Empirical Methods in Natural Language Processing and Computational Natural Language Learning. EMNLP-CoNLL, pp. 379–390 (2012)
48. Yang, Y., Chang, M.W.: S-mart: novel tree-based structured learning algorithms applied to tweet entity linking. In: Proceedings of the 53rd Annual Meeting on Association for Computational Linguistics. ACL (2015)
49. Yih, W.t., Richardson, M., Meek, C., Chang, M.W., Suh, J.: The value of semantic parse labeling for knowledge base question answering. In: Proceedings of the 54th Annual Meeting of the Association for Computational Linguistics (Volume 2: Short Papers), vol. 2, pp. 201–206 (2016)
50. Yin, W., Yu, M., Xiang, B., Zhou, B., Schütze, H.: Simple question answering by attentive convolutional neural network. In: Proceedings of the 26th International Conference on Computational Linguistics. COLING (2016)
51. Yu, M., Yin, W., Hasan, K.S., Santos, C.d., Xiang, B., Zhou, B.: Improved neural relation detection for knowledge base question answering. In: Proceedings of the 55th Annual Meeting on Association for Computational Linguistics. ACL (2017)

Embedding Geographic Locations for Modelling the Natural Environment Using Flickr Tags and Structured Data

Shelan S. Jeawak[1,2(✉)], Christopher B. Jones[1], and Steven Schockaert[1]

[1] School of Computer Science and Informatics, Cardiff University, Cardiff, UK
{JeawakSS,JonesCB2,SchockaertS1}@cardiff.ac.uk

[2] Department of Computer Science, Al-Nahrain University, Baghdad, Iraq

Abstract. Meta-data from photo-sharing websites such as Flickr can be used to obtain rich bag-of-words descriptions of geographic locations, which have proven valuable, among others, for modelling and predicting ecological features. One important insight from previous work is that the descriptions obtained from Flickr tend to be complementary to the structured information that is available from traditional scientific resources. To better integrate these two diverse sources of information, in this paper we consider a method for learning vector space embeddings of geographic locations. We show experimentally that this method improves on existing approaches, especially in cases where structured information is available.

Keywords: Social media · Text mining · Vector space embeddings · Volunteered geographic information · Ecology

1 Introduction

Users of photo-sharing websites such as Flickr[1] often provide short textual descriptions in the form of tags to help others find the images. Besides, for a large number of Flickr photos, the latitude and longitude coordinates have been recorded as meta-data. The tags associated with such georeferenced photos often describe the location where these photos were taken, and Flickr can thus be regarded as a source of environmental information. The use of Flickr for modelling urban environments has already received considerable attention. For instance, various approaches have been proposed for modelling urban regions [5], and for identifying points-of-interest [45] and itineraries [7,36]. However, the usefulness of Flickr for characterizing the natural environment, which is the focus of this paper, is less well-understood.

Many recent studies have highlighted that Flickr tags capture valuable ecological information, which can be used as a complementary source to more traditional sources. To date, however, ecologists have mostly used social media to

[1] http://www.flickr.com.

L. Azzopardi et al. (Eds.): ECIR 2019, LNCS 11437, pp. 51–66, 2019.
https://doi.org/10.1007/978-3-030-15712-8_4

conduct manual evaluations of image content with little automated exploitation of the associated tags [6,10,39]. One recent exception is [21], where bag-of-words representations derived from Flickr tags were found to give promising result for predicting a range of different environmental phenomena.

Our main hypothesis in this paper is that by using vector space embeddings instead of bag-of-words representations, the ecological information which is implicitly captured by Flickr tags can be utilized in a more effective way. Vector space embeddings are representations in which the objects from a given domain are encoded using relatively low-dimensional vectors. They have proven useful in natural language processing, especially for encoding word meaning [30,34], and in machine learning more generally. In this paper, we are interested in the use of such representations for modelling geographic locations. Our main motivation for using vector space embeddings is that they allow us to integrate the textual information we get from Flickr with available structured information in a very natural way. To this end, we rely on an adaptation of the GloVe word embedding model [34], but rather than learning word vectors, we learn vectors representing locations. Similar to how the representation of a word in GloVe is determined by the context words surrounding it, the representation of a location in our model is determined by the tags of the photos that have been taken near that location. To incorporate numerical features from structured environmental datasets (e.g. average temperature), we associate with each such feature a linear mapping that can be used to predict that feature from a given location vector. This is inspired by the fact that salient properties of a given domain can often be modelled as directions in vector space embeddings [8,17,40]. Finally, evidence from categorical datasets (e.g. land cover types) is taken into account by requiring that locations belonging to the same category are represented using similar vectors, similar to how semantic types are sometimes modelled in the context of knowledge graph embedding [16].

While our point-of-departure is a standard word embedding model, we found that the off-the-shelf GloVe model performed surprisingly poorly, meaning that a number of modifications are needed to achieve good results. Our main findings are as follows. First, given that the number of tags associated with a given location can be quite small, it is important to apply some kind of spatial smoothing, i.e. the importance of a given tag for a given location should not only depend on the occurrences of the tag at that location, but also on its occurrences at nearby locations. To this end, we use a formulation which is based on spatially smoothed version of pointwise mutual information. Second, given the wide diversity in the kind of information that is covered by Flickr tags, we find that term selection is in some cases critical to obtain vector spaces that capture the relevant aspects of geographic locations. For instance, many tags on Flickr refer to photography related terms, which we would normally not want to affect the vector representation of a given location[2]. Finally, even with these modifications, vector

[2] One exception is perhaps when we want to predict the scenicness of a given location, where e.g. terms that are related to professional landscape photography might be a strong indicator of scenicness.

space embeddings learned from Flickr tags alone are sometimes outperformed by bag-of-words representations. However, our vector space embeddings lead to substantially better predictions in cases where structured (scientific) information is also taken into account. In this sense, the main value of using vector space embeddings in this context is not so much about abstracting away from specific tag usages, but rather about the fact that such representations allow us to integrate textual, numerical and categorical features in a much more natural way than is possible with bag-of-words representations.

The remainder of this paper is organized as follows. In the next section, we provide a discussion of existing work. Section 3 then presents our model for embedding geographic locations from Flickr tags and structured data. Next, in Sect. 4 we provide a detailed discussion about the experimental results. Finally, Sect. 5 summarizes our conclusions.

2 Related Work

2.1 Vector Space Embeddings

The use of low-dimensional vector space embeddings for representing objects has already proven effective in a large number of applications, including natural language processing (NLP), image processing, and pattern recognition. In the context of NLP, the most prominent example is that of word embeddings, which represent word meaning using vectors of typically around 300 dimensions. A large number of different methods for learning such word embeddings have already been proposed, including Skip-gram and the Continuous Bag-of-Words (CBOW) model [30], GloVe [34], and fastText [13]. They have been applied effectively in many downstream NLP tasks such as sentiment analysis [43], part of speech tagging [28,35], and text classification [12,26]. The model we consider in this paper builds on GloVe, which was designed to capture linear regularities of word-word co-occurrence. In GloVe, there are two word vectors w_i and $\tilde{w_j}$ for each word in the vocabulary, which are learned by minimizing the following objective:

$$J = \sum_{i,j=1}^{V} f(x_{ij})(w_i.\tilde{w_j} + b_i + \tilde{b}_j - \log x_{ij})^2$$

where x_{ij} is the number of times that word i appears in the context of word j, V is the vocabulary size, b_i is the target word bias, $\tilde{b}_j$ is the context word bias. The weighting function f is used to limit the impact of rare terms. It is defined as 1 if $x > x_{max}$ and as $(\frac{x}{x_{max}})^{\alpha}$ otherwise, where x_{max} is usually fixed to 100 and α to 0.75. Intuitively, the target word vectors w_i correspond to the actual word representations which we would like to find, while the context word vectors $\tilde{w_j}$ model how occurrences of j in the context of a given word i affect the representation of this latter word. In this paper we will use a similar model, which will however be aimed at learning location vectors instead of the target word vectors.

Beyond word embeddings, various methods have been proposed for learning vector space representations from structured data such as knowledge graphs [2,44,52], social networks [15,48] and taxonomies [31,47]. The idea of combining a word embedding model with structured information has also been explored by several authors, for example to improve the word embeddings based on information coming from knowledge graphs [42,50]. Along similar lines, various lexicons have been used to obtain word embeddings that are better suited at modelling sentiment [43] and antonymy [33], among others. The method proposed by [27] imposes the condition that words that belong to the same semantic category are closer together than words from different categories, which is somewhat similar in spirit to how we will model categorical datasets in our model.

2.2 Embeddings for Geographic Information

The problem of representing geographic locations using embeddings has also attracted some attention. An early example is [41], which used principal component analysis and stacked autoencoders to learn low-dimensional vector representations of city neighbourhoods based on census data. They use these representations to predict attributes such as crime, which is not included in the given census data, and find that in most of the considered evaluation tasks, the low-dimensional vector representations lead to more faithful predictions than the original high-dimensional census data.

Some existing works combine word embedding models with geographic coordinates. For example, in [4] an approach is proposed to learn word embeddings based on the assumption that words which tend to be used in the same geographic locations are likely to be similar. Note that their aim is dual to our aim in this paper: while they use geographic location to learn word vectors, we use textual descriptions to learn vectors representing geographic locations.

Several methods also use word embedding models to learn representations of Points-of-Interest (POIs) that can be used for predicting user visits [11,29,55]. These works use the machinery of existing word embedding models to learn POI representations, intuitively by letting sequences of POI visits by a user play the role of sequences of words in a sentence. In other words, despite the use of word embedding models, many of these approaches do not actually consider any textual information. For example, in [29] the Skip-gram model is utilized to create a global pattern of users' POIs. Each location was treated as a word and the other locations visited before or after were treated as context words. They then use a pair-wise ranking loss [49] which takes into account the user's location visit frequency to personalize the location recommendations. The methods of [29] were extended in [55] to use a temporal embedding and to take more account of geographic context, in particular the distances between preferred and non-preferred neighboring POIs, to create a "geographically hierarchical pairwise preference ranking model". Similarly, in [53] the CBOW model was trained with POI data. They ordered POIs spatially within the traffic-based zones of urban areas. The ordering was used to generate characteristic vectors of POI types. Zone vectors represented by averaging the vectors of the POIs contained in them, were then used as features to predict land use types. In the CrossMap method [54] they learned embeddings for spatio-temporal hotspots obtained from social media data of locations, times and text. In one form of embedding, intended to enable reconstruction of records, neighbourhood relations in space and time were encoded by averaging hotspots in a target location's spatial and temporal neighborhoods. They also proposed a graph-based embedding method with nodes of location, time and text. The concatenation of the location, time and text vectors were then used as features to predict peoples' activities in urban environments. Finally, in [51], a method is proposed that uses the Skip-gram model to represent POI types, based on the intuition that the vector representing a given POI type should be predictive of the POI types that found near places of that type.

Our work is different from these studies, as our focus is on representing locations based on a given text description of that location (in the form of Flickr tags), along with numerical and categorical features from scientific datasets.

2.3 Analyzing Flickr Tags

Many studies have focused on analyzing Flickr tags to extract useful information in domains such as linguistics [9], geography [5,14], and ecology [1,21,22]. Most closely related to our work, [21] found that the tags of georeferenced Flickr photos can effectively supplement traditional scientific environmental data in tasks such as predicting climate features, land cover, species occurrence, and human assessments of scenicness. To encode locations, they simply combine a bag-of-words representation of geographically nearby tags with a feature vector that encodes associated structured scientific data. They found that the predictive value of Flickr tags is roughly on a par with that of the scientific datasets, and that combining both types of information leads to significantly better results than using either of them alone. As we show in this paper, however, their straightforward way of combining both information sources, by concatenating the two types of feature vectors, is far from optimal.

Despite the proven importance of Flickr tags, the problem of embedding Flickr tags has so far received very limited attention. To the best of our knowledge, [18] is the only work that generated embeddings for Flickr tags. However, their focus was on learning embeddings that capture word meaning (being evaluated on word similarity tasks), whereas we use such embeddings as part of our method for representing locations.

3 Model Description

In this section, we introduce our embedding model, which combines Flickr tags and structured scientific information to represent a set of locations L. The proposed model uses Adagrad to minimize the following objective:

$$J = \alpha J_{tags} + (1-\alpha) J_{nf} + \beta J_{cat} \tag{1}$$

where $\alpha \in [0, 1]$ and $\beta \in [0, +\infty]$ are parameters to control the importance of each component in the model. Component J_{tags} will be used to constrain the representation of the locations based on their textual description (i.e. Flickr tags), J_{nf} will be used to constrain the representation of the locations based on their numerical features, and J_{cat} will impose the constraint that locations belonging to the same category should be close together in the space. We will discuss each of these components in more detail in the following sections.

3.1 Tag Based Location Embedding

Many of the tags associated with Flickr photos describe characteristics of the places where these photos were taken [19,37,38]. For example, tags may correspond to place names (e.g. Brussels, England, Scandinavia), landmarks (e.g. Eiffel Tower, Empire State Building) or land cover types (e.g. mountain, forest, beach). To allow us to build location models using such tags, we collected the tags and meta-data of 70 million Flickr photos with coordinates in Europe (which is the region our experiments will focus on), all of which were uploaded to Flickr before the end of September 2015. In this section we first explain how tags can be weighted to obtain bag-of-words representations of locations from Flickr. Subsequently we describe a tag selection method, which will allow us to specialize the embedding depending on which aspects of the considered locations are of interest, after which we discuss the actual embedding model.

Tag Weighting. Let $L = \{l_1, ..., l_m\}$ be a set of geographic locations, each characterized by latitude and longitude coordinates. To generate a bag-of-words representation of a given location, we have to weight the relevance of each tag to that location. To this end, we have followed the weighting scheme from [21], which combines a Gaussian kernel (to model spatial proximity) with Positive Pointwise Mutual Information (PPMI) [3,32].

Let us write $U_{t,l}$ for the set of users who have assigned tag t to a photo with coordinates near l. To assess how relevant t is to the location l, the number of times t occurs in photos near l is clearly an important criterion. However, rather than simply counting the number of occurrences within some fixed radius, we use a Gaussian kernel to weight the tag occurrences according to their distance from that location:

$$w(t, l) = \sum_{d(l,r) \leq D} |U_{t,l}| \cdot \exp\left(-\frac{d^2(l, r)}{2\sigma^2}\right)$$

where the threshold $D > 0$ is assumed to be fixed, r is the location of a Flickr photo, d is the Haversine distance, and we will assume that the bandwidth parameter σ is set to $D/3$. A tag occurrence is counted only once for all photos by the same user at the same location, which is important to reduce the impact of bulk uploading. The value $w(t, l)$ reflects how frequent tag t is near location l, but it does not yet take into account the total number of tag occurrences near l, nor how popular the tag t is overall. To measure how strongly tag t is associated with location l, we use PPMI, which is a commonly used measure of association in natural language processing. However, rather than estimating PPMI scores from term frequencies, we will use the $w(t,l)$ values instead:

$$PPMI(t,l) = \max\left(0, \log\left(\frac{p_{t,l}}{p_t p_l}\right)\right)$$

where:

$$p_{t,l} = \frac{w(t,l)}{N} \quad p_t = \frac{\sum_{l' \in L} w(t,l')}{N} \quad N = \sum_{t' \in T} \sum_{l' \in L} w(t',l') \quad p_l = \frac{\sum_{t' \in T} w(t',l)}{N}$$

with T the set of all tags, and L the set of locations.

Tag Selection. Inspired by [25], we use a term selection method in order to focus on the tags that are most important for the tasks that we want to consider and reduce the impact of tags that might relate only to a given individual or a group of users. In particular, we obtained good results with a method based on Kullback-Leibler (KL) divergence, which is based on [46]. Let $C_1, ..., C_n$ be a set of (mutually exclusive) properties of locations in which we are interested (e.g. land cover categories). For the ease of presentation, we will identify C_i with the set of locations that have the corresponding property. Then, we select tags from T that maximize the following score:

$$KL(t) = \sum_{i=1}^{n} P(C_i|t) \log \frac{P(C_i|t)}{Q(C_i)}$$

where $P(C_i|t)$ is the probability that a photo with tag t has a location near C_i and $Q(C_i)$ is the probability that an arbitrary tag occurrence is assigned to a photo near a location in C_i. Since $P(C_i|t)$ often has to be estimated from a small number of tag occurrences, it is estimated using Bayesian smoothing:

$$P(C_i|t) = \frac{\left(\sum_{l \in C_i} w(t,l)\right) + \gamma \cdot Q(C_i)}{N + \gamma}$$

where γ is a parameter controlling the amount of smoothing, which will be tuned in the experiments. On the other hand, for $Q(C_i)$ we can simply use a maximum likelihood estimation:

$$Q(C_i) = \frac{\sum_{l \in C_i} \sum_{t \in T} w(t,l)}{\sum_{j=1}^{n} \sum_{l \in C_j} \sum_{t \in T} w(t,l)}$$

Location Embedding. We now want to find a vector $v_{l_i} \in V$ for each location l_i such that similar locations are represented using similar vectors. To achieve this, we use a close variant of the GloVe model, where tag occurrences are treated as context words of geographic locations. In particular, with each location l we associate a vector v_l and with each tag t we associate a vector $\tilde{w_t}$ and a bias term $\tilde{b_{t_j}}$, and consider the following objective (which in our full model (1) will be combined with components that are derived from the structured information):

$$J_{tags} = \sum_{l_i \in L} \sum_{t_j \in T} (v_{l_i} \tilde{w_{t_j}} + \tilde{b_{t_j}} - PPMI(t_j, l_i))^2$$

Note how tags play the role of the context words in the GloVe model, but instead of learning target word vectors we now learn location vectors. In contrast to GloVe, our objective does not directly refer to co-occurrence statistics, but instead uses the *PPMI* scores. One important consequence is that we can also consider pairs (l_i, t_j) for which t_j does not occur in l_i at all; such pairs are usually called *negative examples*. While they cannot be used in the standard GloVe model, some authors have already reported that introducing negative examples in variants of GloVe can lead to improvements [20]. In practice, evaluating the full objective above would not be computationally feasible, as we may need to consider millions of locations and tags. Therefore, rather than considering all tags in T for the inner summation, we only consider those tags that appear at least once near location l_i together with a sample of negative examples.

3.2 Structured Environmental Data

There is a wide variety of structured data that can be used to describe locations. In this work, we have restricted ourselves to the same datasets as [21]. These include nine (real-valued) numerical features, which are latitude, longitude, elevation[3], population[4], and five climate[5] related features (avg. temperature, avg. precipitation, avg. solar radiation, avg. wind speed, and avg. water vapor pressure). In addition, 180 categorical features were used, which are CORINE[6] land cover classes at level 1 (5 classes), level 2 (15 classes) and level 3 (44 classes) and 116 soil types (SoilGrids[7]). Note that each location should belong to exactly 4 categories: one CORINE class at each of the three levels and a soil type.

Numerical Features. Numerical features can be treated similarly to the tag occurrences, i.e. we will assume that the value of a given numerical feature can be predicted from the location vectors using a linear mapping. In particular, for each numerical feature f_k we consider a vector $\tilde{w_{f_k}}$ and a bias term $\tilde{b_{f_k}}$, and the following objective:

$$J_{nf} = \sum_{l_i \in L} \sum_{f_k \in NF} (v_{l_i} . \tilde{w_{f_k}} + \tilde{b_{f_k}} - score(f_k, l_i))^2$$

where we write *NF* for set of all numerical features and $score(f_k, l_i)$ is the value of feature f_k for location l_i, after z-score normalization.

[3] http://www.eea.europa.eu/data-and-maps/data/eu-dem.
[4] http://data.europa.eu/89h/jrc-luisa-europopmap06.
[5] http://worldclim.org.
[6] http://www.eea.europa.eu/data-and-maps/data/corine-land-cover-2006-raster-2.
[7] https://www.soilgrids.org.

Categorical Features. To take into account the categorical features, we impose the constraint that locations belonging to the same category should be close together in the space. To formalize this, we represent each category type cat_l as a vector w_{cat_l}, and consider the following objective:

$$J_{cat} = \sum_{l_i \in R} \sum_{cat_l \in C} (v_{l_i} - w_{cat_l})^2$$

4 Experimental Results

Evaluation Tasks. We will use the method from [21] as our main baseline. This will allow us to directly evaluate the effectiveness of embeddings for the considered problem, since we have used the same structured datasets and same tag weighting scheme. For this reason, we will also follow their evaluation methodology. In particular, we will consider three classification tasks:

1. Predicting the distribution of 100 species across Europe, using the European network of nature protected sites Natura 2000[8] dataset as ground truth. For each of these species, a binary classification problem is considered. The set of locations L is defined as the 26,425 distinct sites occurring in the dataset.
2. Predicting soil type, again each time treating the task as a binary classification problem, using the same set of locations L as in the species distribution experiments. For these experiments, none of the soil type features are used for generating the embeddings.
3. Predicting CORINE land cover classes at levels 1, 2 and level 3, each time treating the task as a binary classification problem, using the same set of locations L as in the species distribution experiments. For these experiments, none of the CORINE features are used for generating the embeddings.

In addition, we will also consider the following regression tasks:

1. Predicting 5 climate related features: the average precipitation, temperature, solar radiation, water vapor pressure, and wind speed. We again use the same set of locations L as for species distribution in this experiment. None of the climate features is used for constructing the embeddings for this experiment.
2. Predicting people's subjective opinions of landscape beauty in Britain, using the crowdsourced dataset from the ScenicOrNot website[9] as ground truth. The set L is chosen as the set of locations of 191 605 rated locations from the ScenicOrNot dataset for which at least one georeferenced Flickr photo exists within a 1 km radius.

Experimental Setup. In all experiments, we use Support Vector Machines (SVMs) for classification problems and Support Vector Regression (SVR) for regression problems to make predictions from our representations of geographic

[8] http://ec.europa.eu/environment/nature/natura2000/index_en.htm.
[9] http://scenic.mysociety.org/.

locations. In both cases, we used the SVM^{light} implementation[10] [23]. For each experiment, the set of locations L was split into two-thirds for training, one-sixth for testing, and one-sixth for tuning the parameters. All embedding models are learned with Adagrad using 30 iterations. The number of dimensions is chosen for each experiment from $\{10, 50, 300\}$ based on the tuning data. For the parameters of our model in Eq. 1, we considered values of α from $\{0.1, 0.01, 0.001, 0.0001\}$ and values of β from $\{1, 10, 100, 1000\}$.

Table 1. Results for predicting species.

	Prec	Rec	F1
BOW-Tags	0.57	0.11	0.18
BOW-KL(Tags)	0.11	0.86	0.19
GloVe	0.10	0.88	0.17
EGEL-Tags	0.10	0.88	0.18
EGEL-Tags+NS	0.12	0.82	0.21
EGEL-KL(Tags+NS)	0.15	0.64	0.25
BOW-All	0.65	0.50	0.56
EGEL-All	0.56	0.60	**0.58**

Table 2. Results for predicting soil type.

	Prec	Rec	F1
BOW-Tags	0.17	0.44	0.24
BOW-KL(Tags)	0.30	0.43	0.36
GloVe	0.32	0.39	0.35
EGEL-Tags	0.32	0.40	0.36
EGEL-Tags+NS	0.30	0.44	0.36
EGEL-KL(Tags+NS)	0.32	0.44	0.37
BOW-All	0.39	0.43	0.41
EGEL-All	0.33	0.67	**0.44**

Table 3. Results for predicting CORINE land cover classes, at levels 1, 2 and 3.

	CORINE level 1			CORINE level 2			CORINE level 3		
	Prec	Rec	F1	Prec	Rec	F1	Prec	Rec	F1
BOW-Tags	0.49	0.45	0.47	0.20	0.13	0.16	0.14	0.08	0.10
BOW-KL(Tags)	0.40	0.47	0.43	0.39	0.12	0.18	0.24	0.13	0.17
GloVe	0.20	0.90	0.33	0.12	0.53	0.19	0.12	0.25	0.17
EGEL-Tags	0.20	0.89	0.33	0.12	0.56	0.20	0.16	0.21	0.18
EGEL-Tags+NS	0.23	0.73	0.35	0.12	0.52	0.20	0.18	0.22	0.19
EGEL-KL(Tags+NS)	0.26	0.62	0.37	0.14	0.58	0.23	0.19	0.25	0.22
BOW-All	0.52	0.51	0.51	0.27	0.19	0.22	0.18	0.11	0.13
EGEL-All	0.45	0.66	**0.54**	0.27	0.48	**0.35**	0.23	0.33	**0.27**

To compute KL divergence, we need to determine a set of classes $C_1, ..., C_n$ for each experiment. For classification problems, we can simply consider the given categories, but for the regression problems we need to define such classes by discretizing the numerical values. For the scenicness experiments, we considered scores 3 and 7 as cut-off points, leading to three classes (i.e. less than 3, between 3 and 7, and above 7). Similarly, for each climate related features, we consider two

[10] http://www.cs.cornell.edu/people/tj/svm_light/.

cut-off values for discretization: 5 and 15 for average temperature, 50 and 100 for average precipitation, 10 000 and 17 000 for average solar radiation, 0.7 and 1 for average water vapor pressure, and 3 and 5 for wind speed. The smoothing parameter γ was selected among $\{10, 100, 1000\}$ based on the tuning data. In all experiments where term selection is used, we select the top 100 000 tags. We fixed the radius D at 1 km when counting the number of tag occurrences. Finally, we set the number of negative examples as 10 times the number of positive examples for each location, but with a cap at 1000 negative examples in each region for computational reasons. We tune all parameters with respect to the F1 score for the classification tasks, and Spearman ρ for the regression tasks.

Variants and Baseline Methods. We will refer to our model as EGEL[11] (Embedding GEographic Locations), and will consider the following variants. **EGEL-Tags** only uses the information from the Flickr tags (i.e. component J_{tags}), without using any negative examples and without feature selection. **EGEL-Tags+NS** is similar to **EGEL-Tags** but with the addition of negative examples. **EGEL-KL(Tags+NS)** additionally considers term selection. **EGEL-All** is our full method, i.e. it additionally uses the structured information. We also consider the following baselines. **BOW-Tags** represents locations using a bag-of-words representation, using the same tag weighting as the embedding model. **BOW-KL(Tags)** uses the same representation but after term selection, using the same KL-based method as the embedding model. **BOW-All** combines the bag-of-words representation with the structured information, encoded as proposed in [21]. **GloVe** uses the objective from the original GloVe model for learning location vectors, i.e. this variant differs from **EGEL-Tags** in that instead of $PPMI(t_j, l_i)$ we use the number of co-occurrences of tag t_j near location l_i, measured as $|U_{t_j l_i}|$.

Results and Discussion. We present our results for the binary classification tasks in Tables 1, 2 and 3 in terms of average precision, average recall and macro average F1 score. The results of the regression tasks are reported in Tables 4 and 5 in terms of the mean absolute error between the predicted and actual scores, as well as the Spearman ρ correlation between the rankings induced by both sets of scores. It can be clearly seen from the results that our proposed method (EGEL-All) can effectively integrate Flickr tags with the available structured information. It outperforms the baselines for all the considered tasks. Furthermore, note that the PPMI-based weighting in EGEL-Tags consistently outperforms GloVe and that both the addition of negative examples and term selection lead to further improvements. The use of term selection leads to particularly substantial improvements for the regression problems. While our experimental results confirm the usefulness of embeddings for predicting environmental features, this is only consistently the case for the variants that use both the tags and the structured datasets. In particular, comparing BOW-Tags with EGEL-Tags, we sometimes see that the former achieves the best results. While this might seem surprising, it is in accordance with the findings in [24,54], among others, where it was also found that bag-of-words

[11] The EGEL source code is available online at https://github.com/shsabah84/EGEL-Model.git.

Table 4. Results for predicting average climate data.

	Temp		Precip		Solar rad		Water vap		Wind speed	
	MAE	ρ	MAE	ρ	MAE	ρ	MAE	ρ	MAE	ρ
BOW-Tags	1.62	0.84	11.66	0.68	926	0.83	0.08	0.71	0.54	0.75
BOW-KL(Tags)	1.69	0.81	12.85	0.65	1057	0.75	0.08	0.71	0.53	0.73
GloVe	1.96	0.44	15.37	0.31	1507	0.36	0.11	0.47	0.74	0.28
EGEL-Tags	1.95	0.47	15.03	0.31	1426	0.41	0.10	0.46	0.73	0.32
EGEL-Tags+NS	1.97	0.44	14.93	0.32	1330	0.44	0.10	0.46	0.72	0.36
EGEL-KL(Tags+NS)	1.48	0.73	13.55	0.52	1008	0.77	0.08	0.66	0.65	0.59
BOW-All	0.72	0.94	10.52	0.75	484	0.93	**0.05**	0.91	**0.43**	0.84
EGEL-All	**0.71**	**0.95**	**10.03**	**0.79**	**436**	**0.95**	**0.05**	**0.92**	**0.43**	**0.88**

representations can sometimes lead to surprisingly effective baselines. Interestingly, we note that in all cases where EGEL-KL(Tags+NS) performs worse than BOW-Tags, we also find that BOW-KL(Tags) performs worse than BOW-Tags. This suggests that for these tasks there is a very large variation in the kind of tags that can inform the prediction model, possibly including e.g. user-specific tags. Some of the information captured by such highly specific but rare tags is likely to be lost in the embedding.

Table 5. Results for predicting scenicness.

	MAE	ρ
BOW-Tags	1.01	0.57
BOW-KL(Tags)	1.09	0.51
GloVe	1.27	0.19
EGEL-Tags	1.12	0.37
EGEL-Tags+NS	1.14	0.40
EGEL-KL(Tags+NS)	1.05	0.53
BOW-All	1.00	0.58
EGEL-All	**0.94**	**0.64**

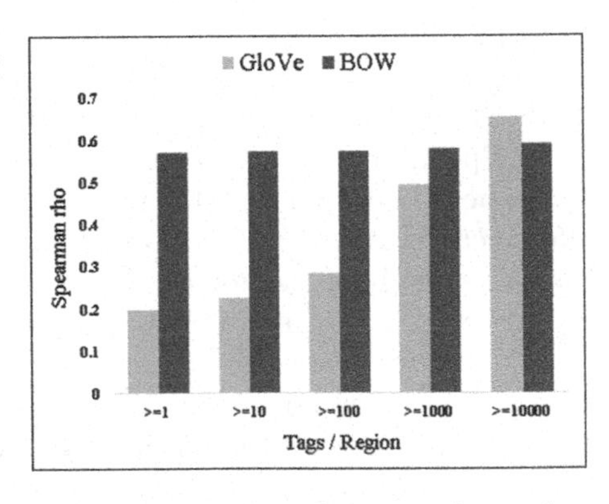

Fig. 1. Comparison between the performance of the GloVe and bag-of-words models for predicting scenicness, as a function of the number of tag occurrences at the considered locations.

To further analyze the difference in performance between BoW representations and embeddings, Fig. 1 compares the performance of the GloVe model with the bag-of-words model for predicting place scenicness, as a function of the number of tag occurrences at the considered locations. What is clearly noticeable in Fig. 1 is that GloVe performs better than the bag-of-words model for large corpora and worse for smaller corpora. This issue has been alleviated in our embedding method by the addition of negative examples.

5 Conclusions

In this paper, we have proposed a model to learn geographic location embeddings using Flickr tags, numerical environmental features, and categorical information. The experimental results show that our model can integrate Flickr tags with structured information in a more effective way than existing methods, leading to substantial improvements over baseline methods on various prediction tasks about the natural environment.

Acknowledgments. Shelan Jeawak has been sponsored by HCED Iraq. Steven Schockaert has been supported by ERC Starting Grant 637277.

References

1. Barve, V.V.: Discovering and developing primary biodiversity data from social networking sites. Ph.D. thesis, University of Kansas (2015)
2. Bordes, A., Usunier, N., Garcia-Duran, A., Weston, J., Yakhnenko, O.: Translating embeddings for modeling multi-relational data. In: Advances in Neural Information Processing Systems, pp. 2787–2795 (2013)
3. Church, K.W., Hanks, P.: Word association norms, mutual information, and lexicography. Comput. Linguist. **16**(1), 22–29 (1990)
4. Cocos, A., Callison-Burch, C.: The language of place: semantic value from geospatial context. In: Proceedings of the 15th Conference of the European Chapter of the Association for Computational Linguistics: Volume 2, Short Papers, vol. 2, pp. 99–104 (2017)
5. Cunha, E., Martins, B.: Using one-class classifiers and multiple kernel learning for defining imprecise geographic regions. Int. J. Geogr. Inf. Sci. **28**(11), 2220–2241 (2014)
6. Daume, S.: Mining Twitter to monitor invasive alien species - an analytical framework and sample information topologies. Ecol. Inform. **31**, 70–82 (2016)
7. De Choudhury, M., Feldman, M., Amer-Yahia, S., Golbandi, N., Lempel, R., Yu, C.: Constructing travel itineraries from tagged geo-temporal breadcrumbs. In: Proceedings of the 19th International Conference on World Wide Web, pp. 1083–1084 (2010)
8. Derrac, J., Schockaert, S.: Inducing semantic relations from conceptual spaces: a data-driven approach to plausible reasoning. Artif. Intell. **228**, 74–105 (2015)
9. Eisenstein, J., O'Connor, B., Smith, N.A., Xing, E.P.: A latent variable model for geographic lexical variation. In: Proceedings of the 2010 Conference on Empirical Methods in Natural Language Processing, pp. 1277–1287 (2010)

10. ElQadi, M.M., Dorin, A., Dyer, A., Burd, M., Bukovac, Z., Shrestha, M.: Mapping species distributions with social media geo-tagged images: case studies of bees and flowering plants in Australia. Ecol. Inform. **39**, 23–31 (2017)
11. Feng, S., Cong, G., An, B., Chee, Y.M.: Poi2vec: geographical latent representation for predicting future visitors. In: Proceedings of the Thirty-First AAAI Conference on Artificial Intelligence, pp. 102–108 (2017)
12. Ge, L., Moh, T.S.: Improving text classification with word embedding. In: IEEE International Conference on Big Data, pp. 1796–1805 (2017)
13. Grave, E., Mikolov, T., Joulin, A., Bojanowski, P.: Bag of tricks for efficient text classification. In: Proceedings of the 15th Conference of the European Chapter of the Association for Computational Linguistics, pp. 427–431 (2017)
14. Grothe, C., Schaab, J.: Automated footprint generation from geotags with kernel density estimation and support vector machines. Spat. Cogn. Comput. **9**(3), 195–211 (2009)
15. Grover, A., Leskovec, J.: node2vec: scalable feature learning for networks. In: Proceedings of the 22nd ACM SIGKDD International Conference on Knowledge Discovery and Data Mining, pp. 855–864 (2016)
16. Guo, S., Wang, Q., Wang, B., Wang, L., Guo, L.: Semantically smooth knowledge graph embedding. In: Proceedings of the 53rd Annual Meeting of the Association for Computational Linguistics, pp. 84–94 (2015)
17. Gupta, A., Boleda, G., Baroni, M., Padó, S.: Distributional vectors encode referential attributes. In: Proceedings of the 2015 Conference on Empirical Methods in Natural Language Processing, pp. 12–21 (2015)
18. Hasegawa, M., Kobayashi, T., Hayashi, Y.: Social image tags as a source of word embeddings: a task-oriented evaluation. In: LREC, pp. 969–973 (2018)
19. Hollenstein, L., Purves, R.: Exploring place through user-generated content: using Flickr tags to describe city cores. J. Spat. Inf. Sci. **1**, 21–48 (2010)
20. Jameel, S., Schockaert, S.: D-glove: a feasible least squares model for estimating word embedding densities. In: Proceedings of the 26th International Conference on Computational Linguistics: Technical Papers, pp. 1849–1860 (2016)
21. Jeawak, S., Jones, C., Schockaert, S.: Using Flickr for characterizing the environment: an exploratory analysis. In: 13th International Conference on Spatial Information Theory, vol. 86, pp. 21:1–21:13 (2017)
22. Jeawak, S., Jones, C., Schockaert, S.: Mapping wildlife species distribution with social media: augmenting text classification with species names. In: Proceedings of the 10th International Conference on Geographic Information Science, pp. 34:1–34:6 (2018)
23. Joachims, T.: Making large-scale SVM learning practical. Technical report, SFB 475: Komplexitätsreduktion in Multivariaten Datenstrukturen, Universität Dortmund (1998)
24. Joulin, A., Grave, E., Bojanowski, P., Nickel, M., Mikolov, T.: Fast linear model for knowledge graph embeddings. arXiv preprint arXiv:1710.10881 (2017)
25. Kuang, S., Davison, B.D.: Learning word embeddings with chi-square weights for healthcare tweet classification. Appl.Sci. **7**(8), 846 (2017)
26. Lilleberg, J., Zhu, Y., Zhang, Y.: Support vector machines and word2vec for text classification with semantic features. In: IEEE 14th International Conference on Cognitive Informatics & Cognitive Computing (ICCI*CC), pp. 136–140 (2015)
27. Liu, Q., Jiang, H., Wei, S., Ling, Z.H., Hu, Y.: Learning semantic word embeddings based on ordinal knowledge constraints. In: Proceedings of the 53rd Annual Meeting of the Association for Computational Linguistics, pp. 1501–1511 (2015)

28. Liu, Q., Ling, Z.H., Jiang, H., Hu, Y.: Part-of-speech relevance weights for learning word embeddings. arXiv preprint arXiv:1603.07695 (2016)
29. Liu, X., Liu, Y., Li, X.: Exploring the context of locations for personalized location recommendations. In: Proceedings of the Twenty-Fifth International Joint Conference on Artificial Intelligence, pp. 1188–1194 (2016)
30. Mikolov, T., Sutskever, I., Chen, K., Corrado, G.S., Dean, J.: Distributed representations of words and phrases and their compositionality. In: Advances in Neural Information Processing Systems, pp. 3111–3119 (2013)
31. Nickel, M., Kiela, D.: Poincaré embeddings for learning hierarchical representations. In: Advances in Neural Information Processing Systems, pp. 6341–6350 (2017)
32. Niwa, Y., Nitta, Y.: Co-occurrence vectors from corpora vs. distance vectors from dictionaries. In: Proceedings of the 15th Conference on Computational Linguistics-Volume 1, pp. 304–309 (1994)
33. Ono, M., Miwa, M., Sasaki, Y.: Word embedding-based antonym detection using thesauri and distributional information. In: Proceedings of the 2015 Conference of the North American Chapter of the Association for Computational Linguistics: Human Language Technologies, pp. 984–989 (2015)
34. Pennington, J., Socher, R., Manning, C.: Glove: global vectors for word representation. In: Proceedings of the 2014 Conference on Empirical Methods in Natural Language Processing, pp. 1532–1543 (2014)
35. Qiu, L., Cao, Y., Nie, Z., Yu, Y., Rui, Y.: Learning word representation considering proximity and ambiguity. In: AAAI, pp. 1572–1578 (2014)
36. Quercia, D., Schifanella, R., Aiello, L.M.: The shortest path to happiness: recommending beautiful, quiet, and happy routes in the city. In: Proceedings of the 25th ACM Conference on Hypertext and Social Media, pp. 116–125 (2014)
37. Rattenbury, T., Good, N., Naaman, M.: Towards automatic extraction of event and place semantics from Flickr tags. In: Proceedings of the 30th Annual International ACM SIGIR Conference on Research and Development in Information Retrieval, pp. 103–110 (2007)
38. Rattenbury, T., Naaman, M.: Methods for extracting place semantics from Flickr tags. ACM Trans. Web **3**(1), 1 (2009)
39. Richards, D.R., Friess, D.A.: A rapid indicator of cultural ecosystem service usage at a fine spatial scale: content analysis of social media photographs. Ecol. Ind. **53**, 187–195 (2015)
40. Rothe, S., Schütze, H.: Word embedding calculus in meaningful ultradense subspaces. In: Proceedings of the 54th Annual Meeting of the Association for Computational Linguistics (Volume 2: Short Papers), pp. 512–517 (2016)
41. Saeidi, M., Riedel, S., Capra, L.: Lower dimensional representations of city neighbourhoods. In: AAAI Workshop: AI for Cities (2015)
42. Speer, R., Chin, J., Havasi, C.: Conceptnet 5.5: an open multilingual graph of general knowledge. In: Proceedings of the Thirty-First AAAI Conference on Artificial Intelligence, pp. 4444–4451 (2017)
43. Tang, D., Wei, F., Yang, N., Zhou, M., Liu, T., Qin, B.: Learning sentiment-specific word embedding for twitter sentiment classification. In: Proceedings of the 52nd Annual Meeting of the Association for Computational Linguistics (Volume 1: Long Papers), vol. 1, pp. 1555–1565 (2014)
44. Trouillon, T., Welbl, J., Riedel, S., Gaussier, É., Bouchard, G.: Complex embeddings for simple link prediction. In: International Conference on Machine Learning, pp. 2071–2080 (2016)

45. Van Canneyt, S., Schockaert, S., Dhoedt, B.: Discovering and characterizing places of interest using Flickr and Twitter. Int. J. Semant. Web Inf. Syst. (IJSWIS) **9**(3), 77–104 (2013)
46. Van Laere, O., Quinn, J.A., Schockaert, S., Dhoedt, B.: Spatially aware term selection for geotagging. IEEE Trans. Knowl. Data Eng. **26**, 221–234 (2014)
47. Vendrov, I., Kiros, R., Fidler, S., Urtasun, R.: Order-embeddings of images and language. arXiv preprint arXiv:1511.06361 (2015)
48. Wang, X., Cui, P., Wang, J., Pei, J., Zhu, W., Yang, S.: Community preserving network embedding. In: Proceedings of the Thirty-First AAAI Conference on Artificial Intelligence, pp. 203–209 (2017)
49. Weston, J., Bengio, S., Usunier, N.: Large scale image annotation: learning to rank with joint word-image embeddings. Mach. Learn. **81**(1), 21–35 (2010)
50. Xu, C., et al.: RC-NET: a general framework for incorporating knowledge into word representations. In: Proceedings of the 23rd ACM International Conference on Conference on Information and Knowledge Management, pp. 1219–1228 (2014)
51. Yan, B., Janowicz, K., Mai, G., Gao, S.: From ITDL to Place2Vec: reasoning about place type similarity and relatedness by learning embeddings from augmented spatial contexts. In: Proceedings of the 25th ACM SIGSPATIAL International Conference on Advances in Geographic Information Systems, pp. 35:1–35:10 (2017)
52. Yang, B., Yih, W., He, X., Gao, J., Deng, L.: Embedding entities and relations for learning and inference in knowledge bases. In: Proceeding of ICLR 2015 (2015)
53. Yao, Y., et al.: Sensing spatial distribution of urban land use by integrating points-of-interest and Google word2vec model. Int. J. Geogr. Inf. Sci. **31**(4), 825–848 (2017)
54. Zhang, C., et al.: Regions, periods, activities: uncovering urban dynamics via cross-modal representation learning. In: Proceedings of the 26th International Conference on World Wide Web, pp. 361–370 (2017)
55. Zhao, S., Zhao, T., King, I., Lyu, M.R.: Geo-teaser: geo-temporal sequential embedding rank for point-of-interest recommendation. In: Proceedings of the 26th International Conference on World Wide Web Companion, pp. 153–162 (2017)

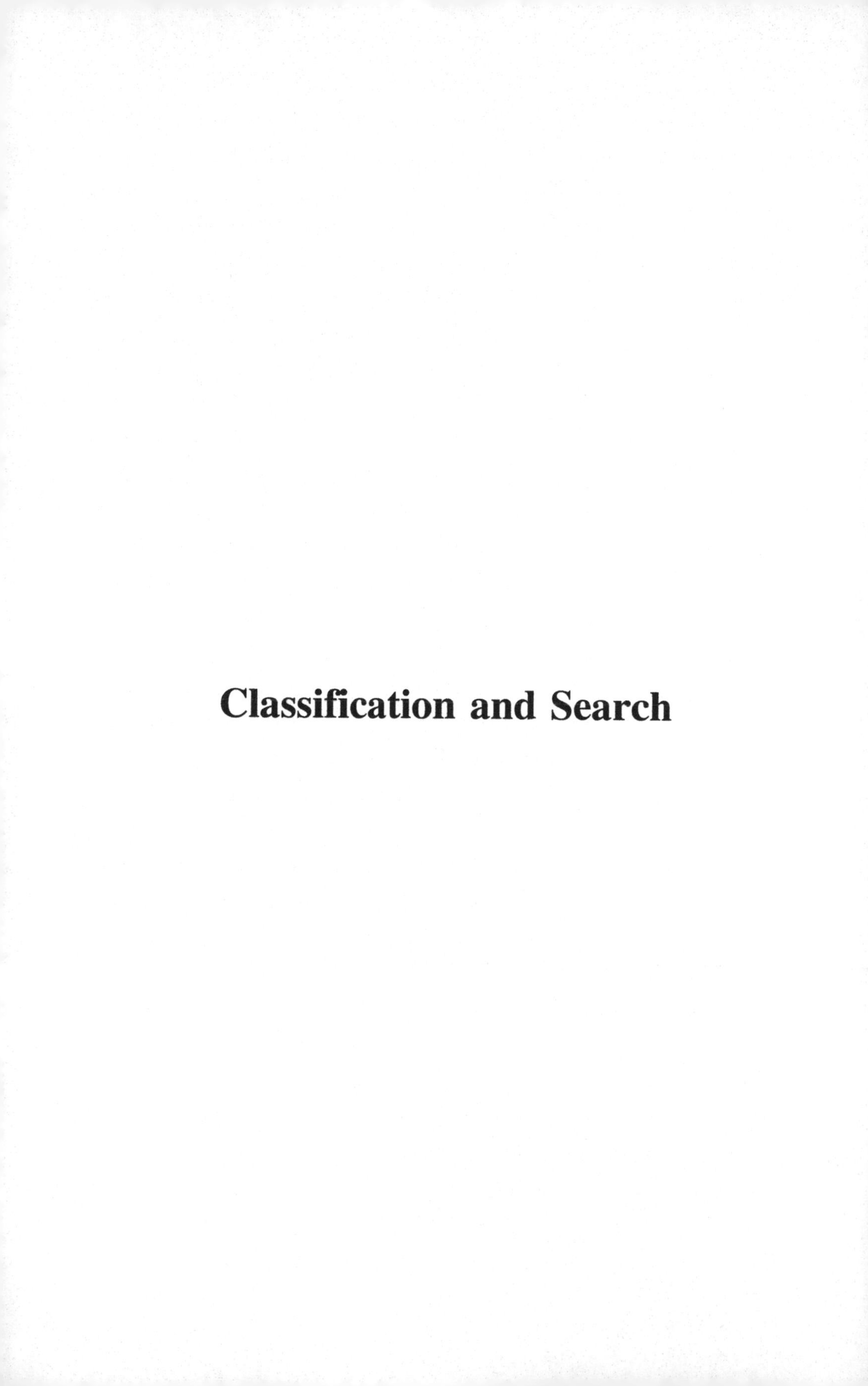

Classification and Search

Recognising Summary Articles

Mark Fisher[1,2], Dyaa Albakour[2(✉)], Udo Kruschwitz[1], and Miguel Martinez[2]

[1] School of Computer Science and Electronic Engineering, University of Essex, Colchester, UK
fishbpm@googlemail.com, udo@essex.ac.uk
[2] Signal, 145 City Road, London EC1V 1AZ, UK
{dyaa.albakour,miguel.martinez}@signal-ai.com

Abstract. Online content providers process massive streams of texts to supply topics and entities of interest to their customers. In this process, they face several information overload problems. Apart from identifying topically relevant articles, this includes identifying duplicates as well as filtering *summary* articles that comprise of disparate topical sections. Such summary articles would be treated as noise from a media monitoring perspective, an end user might however be interested in just those articles. In this paper, we introduce the recognition of summary articles as a novel task and present theoretical and experimental work towards addressing the problem. Rather than treating this as a single-step binary classification task, we propose a framework to tackle it as a two-step approach of *boundary detection* followed by classification. Boundary detection is achieved with a bi-directional LSTM sequence learner. Structural features are then extracted using the *boundaries* and *clusters* devised with the output of this LSTM. A range of classifiers are applied for ensuing summary recognition including a convolutional neural network (CNN) where we treat articles as 1-dimensional structural 'images'. A corpus of natural summary articles is collected for evaluation using the Signal 1M news dataset. To assess the *generalisation* properties of our framework, we also investigate its performance on synthetic summaries. We show that our structural features sustain their performance on generalisation in comparison to baseline bag-of-words and word2vec classifiers.

1 Introduction

As the news domain becomes increasingly digitalised, individuals and companies are now ever more reliant on monitoring tools to filter streams of online news articles. In particular, individuals seek to find news relevant to their interests, while companies proactively monitor online content to manage their own brand image, and position themselves to react quickly to changes in the industry and market [1].

Figure 1 depicts a typical processing pipeline for such monitoring tools. The key steps of this pipeline are the topic classification of articles, and the identification of relevant entities within each article. This pipeline must cope with massive streams of documents from various online sources, which can be noisy. Hence, the pre-processing step plays an important role in removing undesirable content (noise) that may affect the output of the latter steps.

L. Azzopardi et al. (Eds.): ECIR 2019, LNCS 11437, pp. 69–85, 2019.
https://doi.org/10.1007/978-3-030-15712-8_5

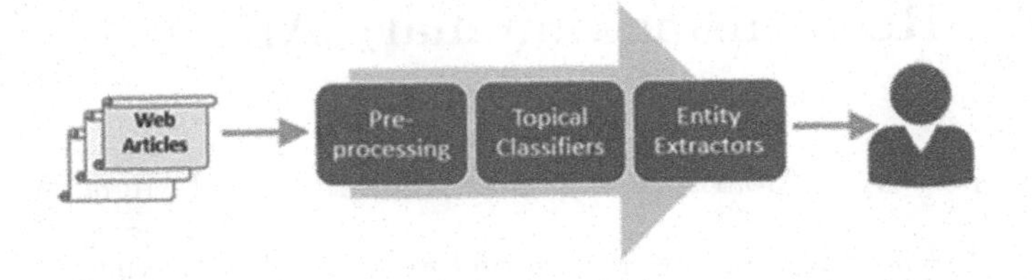

Fig. 1. A typical media monitoring pipeline.

One distinct example of such noise is what we define as 'summary' articles. Our definition is as follows: a summary article aggregates several otherwise disparate topical sections. If one writes a summary of another topical article (an article discussing one topic), the resulting article is still clearly topical. By contrast, in our definition, a summary article encompasses a collection of topics that do not bear any manifest relation. Such articles are often created by web aggregators[1], but are also published by other more traditional news sources, for example when reporting on today's current affairs. An example of a summary article is provided in Fig. 2 (right). If these articles are passed over to the topical classifier of the pipeline in Fig. 1, they might become classified under any of their constituent topics, rather than being discarded. Therefore, it is important to automatically identify these articles within a media monitoring context. In this paper, we introduce the new task of 'summary article recognition', which involves the binary classification of news articles into summaries or not (i.e. topical).

```
Poland - Factors to Watch Sept 9

Following are news stories, press reports and
events to watch that may affect Poland's financial
markets on Wednesday. ALL TIMES GMT (Poland: GMT +
2 hours):

SWISS FRANC MORTGAGES
Poland's junior coalition partner the Polish
Peasant Party (PSL), still wants banks to cover the
bulk of costs of converting Swiss-franc mortgages
into zlotys, a PSL parliamentarian said on Tuesday.

MIGRANTS
Polish Prime Minister Ewa Kopacz said the European
Council which is to deal with Europe's migrant
crisis is likely to take place earlier than
planned.
Poland could accept more migrants than the 2,000 it
declared earlier, but under certain conditions,
Kopacz also said.

KGHM
The chief executive of Poland's KGHM, Europe's No.2
copper producer, said on Tuesday he expected copper
prices to stabilise at $5,000 dollars per tonne.
```

```
10 Things to Know for Thursday

Your daily look at late-breaking news, upcoming
events and the stories that will be talked about
Thursday:

GOP OPPONENTS TARGET TRUMP
Rand Paul, calling The Donald too brash to lead, is
among candidates ganging up on the front-runner at
the second Republican debate.

CHAOS ERUPTS ON SERBIA-HUNGARY BORDER
Baton-wielding Hungarian police unleash tear gas
and water cannons against hundreds of migrants
trying to cross the border from Serbia.

WHY MUSLIM TEEN HAS BECOME SOCIAL MEDIA CAUSE
Sympathy spreads for 14-year-old Ahmed Mohamed
after he was placed in handcuffs and suspended for
taking a homemade clock to his Texas school that
teachers thought resembled a bomb.

POWERFUL MAGNITUDE-8.3 EARTHQUAKE HITS CHILE
The quake causes buildings to sway in Santiago and
other cities and sends people running into the
streets.
```

Fig. 2. Examples of topical (left) and summary (right) articles.

[1] Examples are https://news360.com and https://www.bloomberg.com/series/top-headlines.

One may assume that summary articles will be presented in a characteristic form making them easy to recognise. Taking the two examples in Fig. 2, the first (left) article is topical, because each paragraph discusses some aspect of one topic "Poland's financial markets". The article on the right consists of paragraphs which do not share any connective topic, and the article is hence a summary. In terms of their visual form, however, the two articles cannot be distinguished. Therefore, it is the underlying flow of topics and the entities, 'the linear structure' of an article, that is the principal determinant of its class, irrespective of its apparent visual form. Although the first article in Fig. 2 also comprises of linearly segmented topics, these topics are each connected under the article's principal theme, thus rendering it topical.

There exists a range of established tools that can help in modelling the article's content in terms of topics and entities. These include generative methods with topic modelling as proposed by [2,3], as well as entity-based approaches, the most effective of which leverage a knowledge graph [4]. Although powerful, these methods have certain limitations. They either rely on corpus specific parameterisation that impairs generalisation, or require elaborate processing that limits extension to large datasets or document streams.

To this end, in this paper, we propose a framework for summary recognition that does not suffer from the aforementioned limitations. Our framework consists of two steps: structure extraction followed by classification. For structure extraction, we employ *boundary detection* to characterise the 'linear structure' of an article. In particular, boundary detection quantifies whether there is a topic shift at the end of every sentence[2] in the article. The output of boundary detection is then used to devise 'structural' features used for the classification step in a supervised manner. To illustrate the intuition behind our framework, in Fig. 3 we visualise the output of boundary detection for each sentence in both

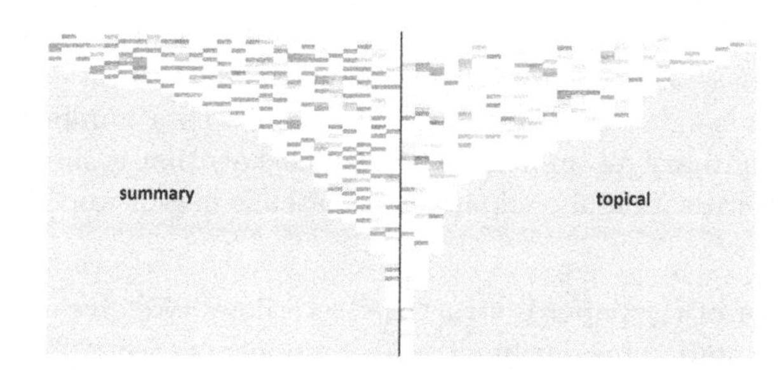

Fig. 3. Each column represents an article and rows the output of boundary detection applied subsequently on each sentence. Darker colours represents higher probability of boundaries. (Color figure online)

[2] Paragraph delimiters are not consistently available, especially in the realm of digital web content. For robustness we thus perform all boundary detection at the sentence level.

topical and summary articles. One can visually discriminate between summary and topical articles, as summary articles exhibit blunt topic shifts in the text. It is therefore reasonable to rely solely on structural features devised from the boundary detection step to classify summary articles.

For boundary detection, we use neural networks and word embeddings for a supervised strategy. Building on the work of Koshorek *et al.* [5], we perform boundary detection with a pre-trained LSTM (Long Short-Term Memory) neural network. Unlike [5] who train on Wikipedia content headers for detection of 'narrative' boundaries, the distinctive aspect of our approach is a synthetic training set of 1 Million summary articles which we tailor to detect 'topical' boundaries.

As summary recognition has not, as far as we are aware, been previously attempted, we trial a variety of structural features. This enables us to comparatively assess their generalisation performance and identify potential candidates for future research. Our proposed features are as follows: (i) *boundary* features directly derived from the output of the boundary detector (example in Fig. 3), (ii) *cluster* features derived by applying a linear clustering step on top of the boundary detection output. Finally, for the *classification* component of our framework, we evaluate a number of binary classification models, but foremost we propose boundaries be treated as 'structural images'; this enables us to capture the overall aggregate structure of an article regardless of its length, and to leverage the power of a convolutional neural network (CNN).

To evaluate our framework, we build a *natural* dataset of summary and topical articles by annotating a sample of the Signal 1M news dataset [6]. To further gauge generalisation performance, we also construct an additional *composite* training set comprising of synthetic summaries. Our contributions can be summarised as follows:

- We introduce the new task of summary recognition and devise a dataset to foster further research on this task. This dataset is built on top of the public Signal 1M dataset and we make it publicly available[3].
- Using this dataset, we evaluate our framework with a number of structural features for summary recognition. The results show that it sustains its performance on generalisation in comparison to baseline bag-of-words and word2vec classifiers.

The remainder of the paper is structured as follows. We give a brief review of related work that underpins our functional components (Sect. 2), before presenting our framework thoroughly (Sect. 3). In Sect. 4, we present the three datasets we employ for experimentation. These experiments and their results are then detailed in Sects. 5 and 6, followed by our conclusions (Sect. 7).

[3] https://research.signal-ai.com/datasets/signal1m-summaries.html.

2 Related Work

Although our goal is recognising summary articles, the principal facet of our framework is boundary detection. Boundary detection is most prominently employed in the field of text segmentation; also known equivalently as 'linear clustering'. This involves the detection of boundaries within a text that together form an optimally cohesive sequence of contiguous segments.

2.1 Text Segmentation

Text segmentation is inherently an unsupervised problem as there are rarely true objective boundaries. Hence, supervised methods are usually domain specific relying on supplementary sources to mark out these boundaries. One area where such 'multi-source' approaches have proven effective is the transcript and newswire domains [7,8], where content breaks are more explicit, aligned with natural prosodic features such as pauses, speaker change and cue phrases.

Unusually, Koshorek *et al.* [5] overcome this by leveraging Wikipedia headers to label sentence boundaries in Wikipedia articles, producing a somewhat 'narrative' segmentation. We take a similar approach, but instead synthesise our training data to produce a more blunt 'topical' segmentation suited to our summary recognition task.

Aside from these few supervised methods, the bulk of segmentation research is in the *un*supervised field where a wide variety of algorithms [9–11] have arisen. This includes statistical and hierarchical methods that involve dynamic programming [12] and more elaborate probabilistic modelling [13] to infer optimal clustering of a text. These generative methods all require parameterisation, for example in Misra *et al.*'s LDA method [3] the number of topics and Dirichlet priors must first be specified for the sample corpus. In this paper, we propose a framework that is generic and extensible, so that our model can be deployed dynamically to new document streams.

2.2 Neural Methods

The rise of neural networks has provided new mechanisms for representing words and sentences, which as demonstrated by [5] can also be employed for our boundary detection component. Although we opt to employ pre-trained word2vec embeddings, as first introduced by [14], more elaborate pre-trained embeddings are also available. This includes CPHRASE [15], which use syntactic relations to selectively determine which context words are used for training. Garten *et al.* [16] also show that combining multiple embeddings, via aggregation or maxima, further improves performance. There is thus plenty of scope for trialling different trained embeddings.

Sentence embeddings can also be trained using similar methods. Implementations such as FastSent [17] and Skip Thought [18] have adopted equivalent (Continuous Bag of Words) CBOW and Skip Gram training mechanisms, but employing contextual sentences rather than words. More recently, Sent2Vec [19] augmented the CBOW approach by including n-grams within the context window. Hill *et al.* [17] observed that simple aggregation of word and n-gram embeddings such as neural bag-of-words can still achieve equivalent performance to the aforementioned sentence embedding approaches on unsupervised tasks.

As shown by [5], LSTM neural networks can also be used in an equivalent unsupervised capacity to encode sentences. Here, aggregation is performed in alignment with the LSTM's bi-directional context window to capture a more sequential embedding. These recurrent LSTMs, as first proposed by Hochreiter and Schmidhuber [20], are well known for their performance in sequence learning such as machine translation (MT). Sutskever *et al.* [21] outperformed a statistical MT system, despite their LSTM being restricted to a more limited vocabulary. Moreover, Xu *et al.* [22] developed an elaborate bi-directional LSTM architecture incorporating Viterbi decoding to model prosodic and lexical features for sentence boundary detection in broadcast news. But as far as we know from our research, Koshorek *et al.*'s [5] is the first attempt at text segmentation using neural methods. We employ their model directly, but extract the softmax boundary layer for input into our feature-based approaches.

3 Framework for Structural Summary Recognition

We propose a framework comprising two generic functional components: the structure extractor, followed by a binary classifier (Fig. 4). The structure extractor aims to characterise the linear structure of the article. It outputs structural features which are then used by a binary classifier for summary recognition. The output of the framework is a binary label for the article: positive (summary) or negative (topical)[4].

For the **structure extractor** component of the framework, we propose to employ a boundary detection approach as shown in Fig. 5. The boundary detector produces a probability for each sentence in the article, denoting the likelihood of a topic shift in the following sentence. These probabilities are used to engineer structural features. In our implementation of the framework, we propose two families of structural features; sentence *boundaries* and word *clusters* (Fig. 5).

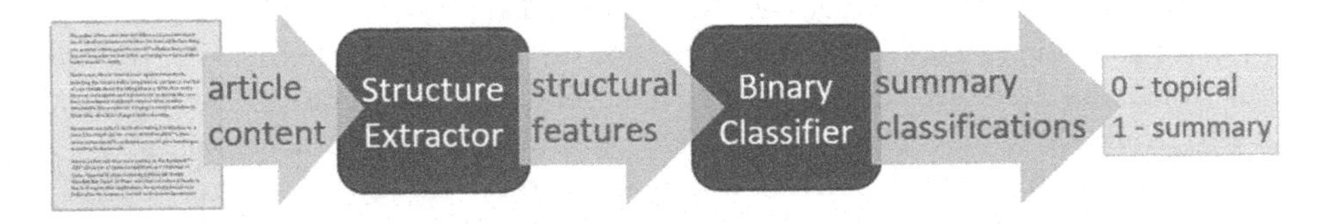

Fig. 4. Framework for structural summary recognition.

[4] We refer to negative articles by our classification as 'topical', as the vast majority of non-summary articles are typically topical.

For the **binary classifier** component of the framework, we trial a number of models as suited to each class of structural features. As choice of classifier is intrinsically tied to the features, we present these classifier components alongside the respective features in Sects. 3.2 and 3.3. Before this, we present our boundary detection approach in Sect. 3.1.

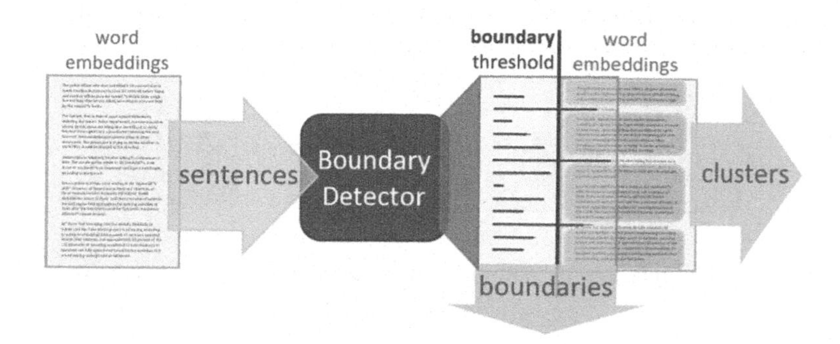

Fig. 5. Our structure extractor model. We employ a boundary detector component to extract *boundary* and *cluster* features. Sections 3.2 and 3.3 describe how these features are extracted.

3.1 Boundary Detector

For boundary detection, we employ the LSTM model proposed by Koshorek *et al.* [5]. This model has a dual architecture commencing with a first a *sentence encoder* followed by a *sequence labeller*. The sentence encoder is an *un*supervised LSTM network. It acts as an aggregator, encoding the set of word embeddings within each sentence. Rather than a flat aggregation of words, it aggregates the word sequences in the sentence, making it well suited to the boundary detection objective. The encoded sentences are then supplied to the sequence labeller, which is a supervised bi-directional LSTM trained to label a sequence of sentences as boundaries or not. The final softmax output layer of the network thus provides a boundary probability for each sentence in the article, that are used in constructing the boundary features in Sect. 3.2.

It should be noted that there are other options to implement the boundary detector component of our framework. This includes, for example, state-of-the-art unsupervised models such as GraphSeg [23]. We leave this for future work.

3.2 Boundary Features

Here, we use the boundary probabilities directly to characterise an article's linear structure. We extract these structural features in two forms.

In the first form, we apply the full sequence of boundary probabilities to feed a CNN classifier. We refer to this set as an '*image*', as it captures the complete sequential structure of the article. Just as a CNN convolves over

2-dimensional visual images, here it convolves over our 1-dimensional structural image. This would enable it to recognise any intrinsic elements or artefacts that might typify the style of summary articles, whilst maintaining invariance to the specific position of these elements. We start with the foundation architecture optimised for image recognition by Lecun *et al.* [24], but make its convolutional layers 1-dimensional. The number of filters and dense layers are also adjusted through general experimentation to maximise performance. Drop out layers in particular were found to be beneficial, as shown in Fig. 6.

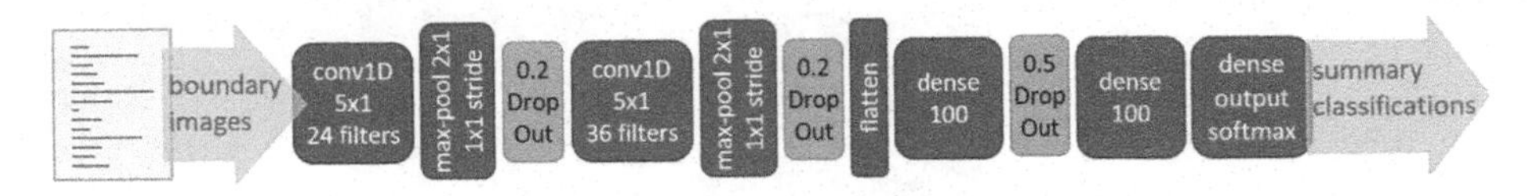

Fig. 6. Binary classification of summary articles using a CNN with image-based boundaries.

In the second form, we 'average' the boundary probabilities. Here, as an additional feature, we also include the quantity of detected boundaries by applying a threshold to the boundary probabilities as visualised in Fig. 5. These two features provide a more generic representation of the article, which we hypothesise may improve its scope in summary classification. We trial the SVM model for binary classification of summary articles using these features.

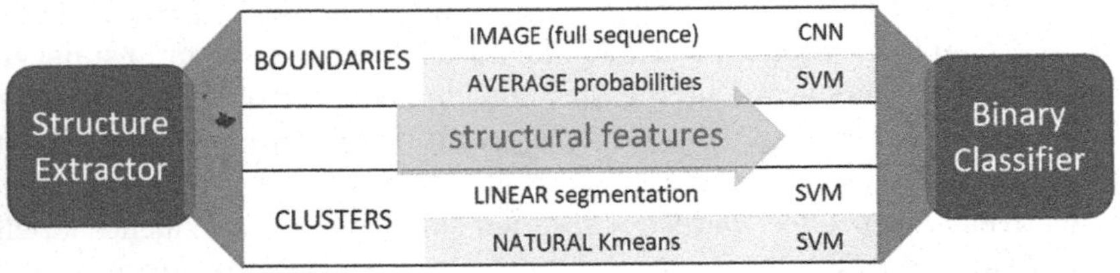

Fig. 7. Our two structural features. For cluster features we employ Silhouette [25] and Calinski Harabaz scores [26]. Apart from the image features, all features additionally include the normalised quantity of detected boundaries.

3.3 Clustering Features

Here, we aim to capture richer structural features of the articles. The hypothesis is that summary articles will consist of distant clusters, while topical ones will have close and cohesive clusters. We employ two forms of clustering; 'linear' and 'natural'.

Linear clusters are formed by segmenting the text in compliance with the boundary threshold as visualised in Fig. 5. We also relax this linear constraint to perform *natural* K-Means clustering of the article's word embeddings. Here, to encourage a clustering that may still correlate with the article's potential topics, the *quantity* of detected boundaries is used to seed the K-Means algorithm. Therefore, these clusters are still indirectly dependent upon the boundary detection. While linear clusters are fully supervised by the boundary detector, natural clusters are semi-supervised by virtue of this dependency. Relaxing the linear constraint should allow K-Means to yield more cohesive clusters, but it is unclear whether these will still sufficiently correlate with the contiguous structure of the article, hence why we opt to trial both forms of clustering.

After clustering, we compute clustering metrics using Silhouette [25] and Calinski Harabaz scores [26], which characterise the cohesion and distribution of the resulting clusters. The clustering features comprise of both these metrics and the quantity of clusters (normalised by number of sentences). These features are then used for binary summary classification where, as before, we trial the SVM classifier.

4 Datasets

Three datasets are assembled for training and evaluation purposes. These are sampled from two distinct data sources as shown in Table 1 which outlines the composition of each dataset. We devise a *natural* dataset for evaluating summary classification performance. We make this dataset available for public use.[5] We also construct a much larger set of *synthetic* summaries for training of the LSTM boundary detector (described in Sect. 3.1). Additional summaries are then synthesised, alongside a randomised selection of topical articles to build a *composite* dataset to evaluate the generalisation performance of our summary classifiers. Next, we describe our datasets in more detail.

Table 1. The sizes and the sources of our three datasets.

		Boundary training	Summary classification	
		Synthetic	Natural	Composite
Data source	Size	1,000,000	892	892
Signal 1M	1,000,000		892	
Topical news articles	31,000			446
[a]Aggregated (2–10 topics)	1,000,000	1,000,000		446

[a] synthesised from the 'Topical News Articles' source, as set out in Sect. 4.2

4.1 Natural Summaries Dataset

To evaluate summary article recognition, we collect summary articles using the Signal 1M news article dataset [6]. This dataset covers a typical stream of news articles processed for media monitoring purposes. It includes 1 million articles

[5] https://research.signal-ai.com/datasets/signal1m-summaries.html.

during September 2015 from 93 K sources ranging from web aggregators to premium publications.

To label articles from Signal 1M, we obtained a biased sample of 2900 articles. To create this sample, we use a Lucene index and apply search terms such as 'month', 'week', 'review', 'report' and 'roundup', using vector-based tf-idf ranking to retrieve the highest scoring articles. In conducting this search, certain sources were found to produce a larger proportion of summary-style articles. As summaries are relatively uncommon, queries were also further tailored to promote articles of these sources. This biased sample (rather than a random one) somewhat limits the variance of the resulting dataset, but was necessary in order to obtain a sufficient quantity of summary articles.

From the biased sample (2900 articles), a subset of 400 articles was first labelled by 4 independent annotators, in order to gauge labelling accuracy, yielding a pair-wise agreement of 85%. This reflects the ambiguity inherent in the recognition of summary articles, as demonstrated by the examples in Fig. 2. Due to resource limitations, the remaining 2500 articles were then labelled by one of our 4 annotators. Using the labels of this annotator, the biased sample had 446 summary articles and 2454 topical articles. To create a *balanced* natural dataset, we selected all the 446 summary articles and an equal quantity of topical articles drawn randomly, providing a total of 892 articles. Although surplus articles remain, balancing is very important for effective training of our binary classifiers.

4.2 Synthetic Summaries Dataset

Effective sequence training of an LSTM network requires a very large volume of labelled articles. As a large labelled dataset of summary articles is not readily available, we opt to instead *synthesise* our boundary training set, using topical articles.

To construct this training set, we first obtained a private set of 31,000 topical news articles, from similar sources of those covered by Signal 1M, but during a larger time frame (September 2015 till July 2018). Each article is manually labelled by independent commercial annotators with one of 50 different topical classes. To synthesise summary articles from these topical articles, we follow the protocols employed by Choi on his 'Choi dataset' [10], which has become recognised as the reference baseline for segmentation evaluation. In particular, these protocols are designed to mirror the variability of natural articles. With this protocol, to synthesise a summary article consisting of subsequent segments, a distinct topical article is selected (from the 31,000 topical articles) for each segment, ensuring no topic repetition. Then, a random position within this topical article is selected to extract the requisite quantity of contiguous sentences. Applying these protocols we synthesise 1 Million articles forming our synthetic summary dataset.

4.3 Composite Summaries Dataset

In order to assess the 'generalisation' capabilities of our framework, we assemble an additional *composite* dataset of summary articles. This set is for training purposes only, to evaluate its impact on natural summary classification.

We call this set 'composite' as its summary portion is synthetic (synthesised using the protocol described in Sect. 4.2), while its topical portion is natural (see Table 1). The basis for its use in generalisation is 2-fold, upon both content and structure:

(i) *Content*: The entire dataset is sourced from a much wider variety of articles, encompassing three years (Sept-2015 through July-2018), unlike our natural dataset which is sampled from the more restrictive 1 month (Sept-2015) Signal 1M dataset.
(ii) *Structure*: Having been algorithmically synthesised, its summary samples are each structurally much more uniform; They do not exhibit the same stochastic variation as natural occurring summaries.

To aid direct evaluation against our natural corpus, we also make this composite set the same size (892 articles), again balancing summary and topical. Topical articles are drawn *randomly* from the 31,000 dataset to maximise its variance.

5 Experiments

Our experiments seek to evaluate the main objectives we set out in Sect. 1. In particular, our experiments aim to answer the following research questions:

RQ1: How does our framework compare to existing content-feature and semantic-feature approaches, such as, bag-of-words or pairwise sentence similarity?
RQ2: Can our framework generalise effectively to new settings and content shift?
RQ3: Within our framework, which structural features are most effective for summary article recognition?

Next, we describe the baselines we use. Then, we detail our experimental setup and implementation details.

5.1 Baselines

As far as we are aware, the task of summary article recognition has not yet been attempted. Therefore, there is no obvious state-of-the-art baseline to compare our framework. We therefore experimented with three supervised baselines, encompassing both conventional and word embedding approaches. For all the baselines, we use standard binary classification applied on different sets of features. As an initial content-based baseline, we apply conventional **bag-of-word** features. We then leverage word embeddings for two further semantic baselines. First, we use the *relatedness score* [23] for every adjacent pair of sentences to calculate the **average pairwise sentence similarity** of each article. This exploits the same semantic relatedness properties as used to train our LSTM, but in an unsupervised setting, so is more directly comparable to our structural framework. Finally, we apply **aggregated embeddings** [27] where for each article, a single aggregated vector is produced by averaging the word2vec embeddings for each unique word in the article.

5.2 Experimental Setup

The experiments assess the performance of the baselines and our framework using all the combination of our structural features and summary classifier described in Sect. 3 and summarised in Fig. 7.

We ran two distinct experiments. In the first experiment, *natural training*, we use our balanced natural dataset of 892 articles (Sect. 4.1). We employ 5-fold stratified cross validation (CV), reporting average independent performance on each fold. In the second experiment, *composite training*, we aim to evaluate the generalisation properties of our models. Here, we train separately on our identical size composite dataset (see Sect. 4.3). The identical 5-fold CV strategy is applied but we test on the corresponding natural dataset folds, as previously stratified, allowing direct comparison with natural training.

5.3 Implementation Details

Pre-processing: As sentences are the base unit for boundary detection, pre-processing is aimed at yielding a coherent and contiguous set of candidate sentences, each of adequate length; Length is important to avoid data sparsity issues that might arise due to a lack of matched word2vec embeddings. First, employing a SpaCy syntactic parser, we clean all non-content words, numerics and punctuation. Thereafter, pre-processing involved discarding small paragraphs, which typically constitute headers (we opt for a 50-character limit on paragraph retention), then concatenating short sentences.

Word Embeddings: For all our approaches, features are built upon foundation word embeddings. As we source our articles from the Signal 1M collection [6], there is no particular relevant domain that would offer the potential to train tailored embeddings. We therefore employ word2vec pre-trained Google News embeddings [14], which are also well suited to the general news domain of our corpus. To maximise semantic interpretation, we allow Google News to enforce its own limit on stop words.

Neural Networks: Our LSTM two-layer network for boundary detection was trained using our synthetic datasets (See Sect. 4.2). It was trained in 40 h using an Nvidia Tesla GPU. When performing CV, the CNN network weights are re-initialised on each fold to ensure a new model is fitted.

Feature Normalisation: As articles vary in size, structural features must also be normalised to enable effective use in classification. For most of our approaches this is achieved in straightforward fashion, normalising by the total quantity of clusters or boundaries as respective to the class of feature. For our *image* features, we perform normalisation using a bicubic image filter. Here, to best preserve the structural representation of the article we opt for a target size of 40 boundaries, which approximates the average size in our natural test dataset.

6 Results

The cross validation results for our framework (using boundaries and clustering features) and the baselines in both experiments are reported in Table 2.

In the *natural* training experiment, our bag-of-words and aggregated embeddings baselines show strongest classification performance in terms of both precision and recall (and thus F1 accuracy), exceeding our best performing linear segmentation features (0.8067 vs. 0.6834). This points to the degree of content consistency in the natural dataset, likely contributed by some publishers we have selected in our biased sampling of Signal 1M (see Sect. 4.1) having a consistent style.

In the *composite* training experiment, the performance of all of the baselines drop, up by 12% points for aggregated embeddings, from 0.8067 to 0.6801. In other words, their performance is not resilient to content shift. By contrast, most variants of our framework maintain a similar classification performance when tested on a different setting. They exhibit a marginal drop in F1 when comparing their performance between natural training and composite training. With composite training, one variant of our framework 'linear segmentation/SVM' is significantly better than both the bag-of-words and pairwise sentence similarity baselines using McNemar Test ($p < 0.01$).

To summarise, as an answer to our research question RQ1, we can conclude that conventional content-features approaches may be adequate for summary recognition, and they outperform our framework. The aggregated embeddings in particular show strongest performance. This is only true, however, when re-training a model is feasible, and a budget is available to collect labelled data. For RQ2, we conclude our framework has a strong generalisation performance. It has shows to be resilient to content shift (the composite training). This suggests that it has the potential to sustain its performance if deployed on dynamic content streams that continuously change.

Next, we analyse the differences between our proposed combinations of *structural features* and binary classifiers within the framework. Table 2 shows that the 'image/CNN' achieves the strongest performance on *natural* training. On generalisation, however, this performance drops noticeably, unlike 'average probabilities' which sustains its performance. This suggests that the CNN is better equipped to learn the distinctive aspects of summaries in the training domain. From this viewpoint, the CNN's performance on natural training is perhaps still muted. We suggest this is due to the small size of our dataset; with

Table 2. Average CV classification performance for summary articles. All test results encompass the full 892 articles of our *natural* summaries dataset. For baselines, we report the best performing classifier (with composite training) from logistic regression, Naïve-Bayes (NB) and SVM.

			Natural training			Composite training		
	Features	Class.	P	R	F1	P	R	F1
Boundaries	Image	CNN	0.7382	0.6143	0.6703	0.6621	0.5492	0.6001
	Average probabilities	SVM	0.6459	0.6728	0.6585	0.6107	0.6032	0.6416
Clusters	Linear segmentation	SVM	0.6630	0.7062	0.6834	0.6581	0.7084	**0.6820**[a]
	Natural K-Means	SVM	0.6628	0.6524	0.6573	**0.7210**	0.5828	0.6425
Baselines	Aggregated embeddings	log. reg.	**0.7647**	0.8542	**0.8067**	0.6152	0.7622	0.6801
	Avg. pairwise sent. sim.	SVM	0.5853	0.7534	0.6588	0.5839	0.7265	0.6469
	Bag-of-words	NB	0.6429	**0.9685**	0.7728	0.5023	**0.9955**	0.6677

[a] Denote statistically significant differences of classification decisions when compared to bag-of-words and average pairwise sentence similarity respectively using McNemar Test ($p < 0.01$).

more training samples, performance of this neural method can reasonably be expected to improve. The 'average probabilities/SVM' achieves a lower F1 than 'image/CNN', but appear to more resilient; with more balance between precision and recall. Also, it is more capable of sustaining performance on generalisation (dropping F1 only marginally 0.6585 to 0.6416), we thus suggest these averaging boundary probabilities provides a better foundation for improvement.

Our clustering feature variants 'linear segmentation/SVM' and 'natural K-Means/SVM' show the strongest overall performance on 'composite training' within our framework. In practical use, however, these clustering features may not be the optimal choice. As a gauge, our trained LSTM generates boundary predictions at the rate of 4 articles per second on an Nvidia Tesla GPU; thereafter, averaging these features is trivial. On the other hand, clustering costs additional CPU time for cluster assembly.

In summary, and as an answer to RQ3, we conclude that the clustering features have strong generalisation capabilities, and are more resilient to content changes than sophisticated CNN approaches trained on sequence of boundary probabilities. The caveat however is their computational complexity.

7 Conclusion

We present the new task of 'summary recognition', that is relevant in a media monitoring context in particular but is also applicable in many other scenarios.

To address this task, we propose a structural framework for summary article prediction aimed principally at achieving generalised performance, which is resilient to variations and shifts in content. The salient component of the framework is a structure extractor that identifies the linear (or semantic) structure of the article to aid summary classification. Building on the work of boundary detection and text segmentation, we show that we can effectively devise structural features that are robust for summary recognition. In particular, we show that our structural features sustain their performance upon generalisation to new content distributions, compared to established aggregated embeddings and bag-of-words baselines that both markedly degrade in performance.

Central to our experiments in this paper, is the construction of new datasets (natural and synthetic) to evaluate the effectiveness of our framework and its generalisation behaviour. The natural is made public to foster further research in this area.

As we are the first to experiment with summary recognition, an important aspect of our work is to provide a foundation for further research. Based on performance of our boundary structural features, for future work, we suggest methods for tailoring the word embeddings to produce more purpose-specific boundaries that may enhance their performance. As entities are a principal topical indicator, we suggest incorporating knowledge graph concept vectors [28] or training embeddings on their entity usage contexts only. Following the findings of Garten *et al.* [16], such approaches may also be combined if beneficial to augment pre-trained generalised embeddings.

References

1. Martinez, M., et al.: Report on the 1st International Workshop on Recent Trends in News Information Retrieval (NewsIR 2016). SIGIR Forum, volo. 50, no. 1, pp. 58–67 (2016)
2. Blei, D.M., Ng, A.Y., Jordan, M.I.: Latent dirichlet allocation. J. Mach. Learn. Res. **3**, 993–1022 (2003)
3. Misra, H., Yvon, F., Jose, J.M., Cappe, O.: Text segmentation via topic modeling: an analytical study. In: Proceedings of the 18th ACM Conference on Information and Knowledge Management, CIKM 2009, New York, NY, USA, pp. 1553–1556. ACM (2009)
4. Schuhmacher, M., Ponzetto, S.P.: Knowledge-based graph document modeling. In: Proceedings of the 7th ACM International Conference on Web Search and Data Mining (WSDM), pp. 543–552 (2014)
5. Koshorek, O., Cohen, A., Mor, N., Rotman, M., Berant, J.: Text segmentation as a supervised learning task. In: Proceedings of the 2018 Conference of the North American Chapter of the Association for Computational Linguistics: Human Language Technologies, Volume 2 (Short Papers), pp. 469–473. Association for Computational Linguistics (2018)
6. Corney, D., Albakour, D., Martinez-Alvarez, M., Moussa, S.: What do a million news articles look like? In: Proceedings of the First International Workshop on Recent Trends in News Information Retrieval Co-located with 38th European Conference on Information Retrieval (ECIR 2016), Padua, Italy, 20 March 2016, pp. 42–47 (2016)

7. Pillai, R.R., Idicula, S.M.: Linear text segmentation using classification techniques. In: Proceedings of the 1st Amrita ACM-W Celebration on Women in Computing in India. A2CWiC 2010, New York, NY, USA, pp. 58:1–58:4. ACM (2010)
8. Galley, M., McKeown, K., Fosler-Lussier, E., Jing, H.: Discourse segmentation of multi-party conversation. In: Proceedings of the 41st Annual Meeting on Association for Computational Linguistics - Volume 1, ACL 2003, Stroudsburg, PA, USA, pp. 562–569. Association for Computational Linguistics (2003)
9. Hearst, M.A.: TextTiling: segmenting text into multi-paragraph subtopic passages. Comput. Lingust. **23**(1), 33–64 (1997)
10. Choi, F.Y.Y.: Advances in domain independent linear text segmentation. In: Proceedings of the 1st North American Chapter of the Association for Computational Linguistics Conference. NAACL 2000, Stroudsburg, PA, USA, pp. 26–33. Association for Computational Linguistics (2000)
11. Dadachev, B., Balinsky, A., Balinsky, H.: On automatic text segmentation. In: Proceedings of the 2014 ACM Symposium on Document Engineering, DocEng 2014, New York, NY, USA, pp. 73–80. ACM (2014)
12. Utiyama, M., Isahara, H.: A statistical model for domain-independent text segmentation. In: Proceedings of the 39th Annual Meeting on Association for Computational Linguistics. ACL 2001, Stroudsburg, PA, USA, pp. 499–506. Association for Computational Linguistics (2001)
13. Riedl, M., Biemann, C.: TopicTiling: a text segmentation algorithm based on LDA. In: Proceedings of ACL 2012 Student Research Workshop, pp. 37–42. Association for Computational Linguistics (2012)
14. Mikolov, T., Chen, K., Corrado, G.S., Dean, J.: Efficient estimation of word representations in vector space. CoRR abs/1301.3781 (2013)
15. Pham, N.T., Kruszewski, G., Lazaridou, A., Baroni, M.: Jointly optimizing word representations for lexical and sentential tasks with the c-phrase model. In: Proceedings of the 53rd Annual Meeting of the Association for Computational Linguistics and the 7th International Joint Conference on Natural Language Processing (Volume 1: Long Papers), pp. 971–981. Association for Computational Linguistics (2015)
16. Garten, J., Sagae, K., Ustun, V., Dehghani, M.: Combining distributed vector representations for words. In: Proceedings of the 1st Workshop on Vector Space Modeling for Natural Language Processing, pp. 95–101. Association for Computational Linguistics (2015)
17. Hill, F., Cho, K., Korhonen, A.: Learning distributed representations of sentences from unlabelled data. In: Proceedings of the 2016 Conference of the North American Chapter of the Association for Computational Linguistics: Human Language Technologies, pp. 1367–1377. Association for Computational Linguistics (2016)
18. Kiros, R., et al.: Skip-thought vectors. In: Cortes, C., Lawrence, N.D., Lee, D.D., Sugiyama, M., Garnett, R. (eds.) Advances in Neural Information Processing Systems 28, pp. 3294–3302. Curran Associates Inc., New York (2015)
19. Pagliardini, M., Gupta, P., Jaggi, M.: Unsupervised learning of sentence embeddings using compositional n-gram features. In: Proceedings of the 2018 Conference of the North American Chapter of the Association for Computational Linguistics: Human Language Technologies, Volume 1 (Long Papers), pp. 528–540. Association for Computational Linguistics (2018)
20. Hochreiter, S., Schmidhuber, J.: Long short-term memory. Neural Comput. **9**(8), 1735–1780 (1997)

21. Sutskever, I., Vinyals, O., Le, Q.V.: Sequence to sequence learning with neural networks. In: Ghahramani, Z., Welling, M., Cortes, C., Lawrence, N.D., Weinberger, K.Q. (eds.) Advances in Neural Information Processing Systems 27, pp. 3104–3112. Curran Associates Inc, New York (2014)
22. Xu, C., Xie, L., Xiao, X.: A bidirectional lstm approach with word embeddings for sentence boundary detection. J. Signal Process. Syst. **90**(7), 1063–1075 (2018)
23. Glavaš, G., Nanni, F., Ponzetto, S.P.: Unsupervised text segmentation using semantic relatedness graphs. In: Proceedings of the Fifth Joint Conference on Lexical and Computational Semantics, pp. 125–130. Association for Computational Linguistics (2016)
24. LeCun, Y., Bottou, L., Bengio, Y., Haffner, P.: Gradientbased learning applied to document recognition. Proc. IEEE **86**(11), 2278–2324 (1998)
25. Rousseeuw, P.J.: Silhouettes: a graphical aid to the interpretation and validation of cluster analysis. J. Comput. Appl. Math. **20**, 53–65 (1987)
26. Calinksi, T., Harabasz, J.: A dendrite method for cluster analysis. Commun. Stat. **3**(1), 1–27 (1974)
27. Balikas, G., Amini, M.R.: An empirical study on large scale text classification with skip-gram embeddings. arXiv preprint arXiv:1606.06623 (2016)
28. Schuhmacher, M., Ponzetto, S.P.: Exploiting dbpedia for web search results clustering. In: Proceedings of the 2013 Workshop on Automated Knowledge Base Construction, AKBC@CIKM 13, San Francisco, California, USA, 27–28 October 2013, pp. 91–96 (2013)

Towards *Content Expiry Date* Determination: Predicting Validity Periods of Sentences

Axel Almquist[1] and Adam Jatowt[2(✉)]

[1] SentiSum, London, UK
axel.almquist@sentisum.com
[2] Kyoto University, Kyoto 606-8501, Japan
adam@dl.kuis.kyoto-u.ac.jp

Abstract. Knowing how long text content will remain valid can be useful in many cases such as supporting the creation of documents to prolong their usefulness, improving document retrieval or enhancing credibility estimation. In this paper we introduce a novel research task of forecasting *content's validity period*. Given an input sentence the task is to approximately determine until when the information stated in the content will remain valid. We propose machine learning approaches equipped with NLP and statistical features that can successfully work on a relatively small number of annotated data.

Keywords: Content validity scope estimation · Text classification · Natural language processing · Machine learning

1 Introduction

Estimating validity and outdatedness of information is paramount in the pursuit of knowledge, something that we humans, luckily, are good at. If we stumble upon a month-old news article stating *"Trump is visiting Sweden"* we would be fairly certain that this information would no longer be true. Yet, facing a sentence such as *"Stefan Löfven is the prime minister of Sweden"* we would most likely think the contrary. It is knowledge of the world that permits us to make such judgments: we know that a presidential visit only lasts for a couple of days and that if someone is a prime minister they will probably remain so for a few months or years. Unfortunately, computers generally lack knowledge of this kind and are thus still incapable of making such judgments. In a world where the amount of information is perpetually increasing at a fast rate and the need for correct and valid information is at its peak, this is a problem to be solved.

In this paper, we introduce a novel research task of predicting how long information expressed in natural language content remains valid, a notion that we will refer to as *the validity period of content*. In analogy to product's expiry

A. Almquist—This work was mainly done at the University of Gothenburg.

L. Azzopardi et al. (Eds.): ECIR 2019, LNCS 11437, pp. 86–101, 2019.
https://doi.org/10.1007/978-3-030-15712-8_6

date, the validity period of content can be used for assessing content's *expiry date*. This would define the approximate time point until which the content can be *"safely consumed"* (i.e., used or published), meaning it should retain its validity until that date. The applications of the proposed task are multiple. Few examples are listed below:

Support for Document Editing: Methods that will flag content with short expected scopes of validity could help with document creation and editing, especially, with documents that are meant to be used over longer time frames.

Fact Checking: Fact checking has become recently increasingly important [6, 15,16,22,25,34]. A model for identifying validity periods of sentences could be useful to help recognizing outdated or unreliable facts. Given the current time, the creation time and the predicted validity period of a sentence, one could conclude that the sentence is at risk of being outdated if the current time is outside the predicted validity period.

Enhancing Document Retrieval: Search engines are in a constant state of improvement. Many approaches have tried to make use of temporal information to improve content rankings, often in the form of prioritizing recency [8,12,24, 26,28,33]. Validity period could be used to flag outdated content while making the still reliable content rise in ranks. By taking the aggregated validity period of sentences in a document, one could filter or flag documents whose validity scope does not cover the current time.

Maintenance of Collaborative Spaces: A part of the information in large knowledge spaces such as Wikipedia, and sites such as Stack Overflow and Quora, will sooner or later go out-of-date. To keep track of outdatedness, validity period estimation could be used to help flag outdated or soon-to-be outdated content. This content could later be removed or changed appropriately. Validity period estimation could also be used to enhance existing approaches for maintenance of such spaces [14].

Besides introducing a novel research task our goal is to create a model that only uses linguistic and statistical features, and is independent from any domain or knowledge graph. We train machine learning models that given a sentence provide an estimate in the form of a selection over fixed validity periods representing how long that sentence will remain valid. We set up the task to be challenging by accepting content of short length (i.e., a sentence) as input, which means there is often limited or no context available. A sentence-level approach can be especially useful for social network services where the length of messages is typically constrained. The experiments are done on an annotated dataset of sentences extracted from blog posts, Wikipedia and news articles.

In summary, we make the following contributions in this paper:

1. We propose a novel task of predicting the validity period of arbitrary textual content and we discuss its applications as well as future extensions.
2. We train machine learning models to predict the validity periods of sentences given a range of linguistic and statistical features, and analyze their impact.
3. We release a dataset which has a high level of annotator agreement for fostering further research.

2 Related Work

Temporal Information Extraction (T-IE) is concerned with extracting temporal information from text [3]. A large portion of T-IE have been focused on extracting and normalizing temporal expressions, such as "*today*" or "*1995*", a task referred to as temporal tagging—to mention a few temporal taggers: GUTime[1], SUTime[2] [4] and HeidelTime[3]. As temporal expressions are not always present, methods and resources for finding *implicit* temporal cues have been used and developed, including language modeling [19,20], word occurrences statistics [17], word embeddings [9] or TempoWordNet [11].

The importance and usefulness of the T-IE has become increasingly recognized and related tasks have been developed including *focus time estimation* [9,17,23], which is the task of identifying what time the texts refer to, *future-related content summarization* [1,10,18] — the task of collecting or summarizing future related information expressed in text, *text date estimation* [5,19,20] which is about detecting the creation time of text, and *temporal scoping of facts*. As for the last one, systems such as T-Yago [31], CoTS [30], PRAVDA [32], TIE [21], and approaches developed by Gupta and Berberich [2] and Sil and Cucerzan [27] have been developed to give facts temporal scopes. Most of these works rely on the existence of temporal expressions in the context of the facts, i.e. that the fact is expressed along with temporal information (T-Yago, PRAVDA, IE, Gupta and Berberich, Sil and Cucerzan). Other approaches rely on occurrence-based statistics of facts to identify temporal scopes (CoTS). For example, if *"Trump is president of the USA"* starts to occur more often than *"Obama is President of the USA"*, this would be an indication of the end of the temporal scope of the fact in the latter sentence. Some approaches, e.g. T-Yago [31], are mainly aimed on Wikipedia infoboxes and lists instead of on free text, and some only focus on a certain type of facts such as relational facts, e.g. *"X was married to Y"* [27].

Due to the above-mentioned limitations, the previous methods are incapable of dealing with either information that is stated in the absence of any temporal expression or with non-factual information, such as *"I am leaving the office now and I will soon be home."* These limitations may not be of serious concern for the above-listed approaches as they mainly focus on major facts regarding past or scheduled events and states. Hence, these approaches are not applicable on many other types of information (i.e., future and often minor actions and events).

Lastly, while our approach would not extract facts and define exact scopes, we should keep in mind that determining exact scopes is generally impossible for many ongoing or future actions and events, especially, if such events or actions lack any predefined period (cf. eating dinner vs. presidential term).

The closest work to ours is research by Takemura and Tajima [29], who classify tweets into lifetime durations, which are used to decide the urgency of Twitter messages. Their objective is to develop an approach for improving the flow

[1] http://www.timeml.org/tarsqi/modules/gutime/index.html.
[2] https://nlp.stanford.edu/software/sutime.html.
[3] https://github.com/HeidelTime/heideltime.

of tweets by taking into account when messages go out of date, hence prioritizing tweets with short life-lengths and ignoring outdated messages. Although Takemura and Tajima try to predict message's lifetime duration, they focus on Twitter messages rather than arbitrary texts. The authors also only use classes of rather short scope, from minutes to weeks, as they want their classification of urgency to be useful for Twitter. Furthermore, and most importantly, Takemura and Tajima's method relies on non-linguistic features, of which many are rather specific to Twitter (e.g., presence of URLs and a user type which is based on the user's previous messages, frequency of their replies, follow relationships and such), rendering their approach less useful on data outside the platform. Our approach does not share this limitation and is meant to be applicable on any text.

3 Method

3.1 Problem Definition and Setting

The validity period of a sentence in our task is a measure of how long the information in that sentence remains valid after it has been expressed. More formally, we define it as follows:

Definition 1. *Given a sentence s created at time t_s, its validity period is the maximum length of time after which the information expressed in s still remains valid.*[4]

While the above definition is general, we use the following validity periods in this work: *few hours, few days, few weeks, few months* and *few years or more.* The granularity of these scopes is unequal ranging from fine-grained (hours) to more coarse (years), which resembles forward-looking logarithmic timeline representation[5]. This is a more natural way for humans to refer to the future, where the uncertainty increases along with the time span extension. Also note that while we could try to pose the problem as a regression task, the simplification of the prediction to a multi-class problem reduces the complexity of the prediction. Besides, given the cost of data annotation and inherent difficulty even for humans to pinpoint the exact validity range, relying on few fixed classes is a more natural choice.

Formally, our model takes as the input a sentence $s_i = \langle w_1, w_2, ..., w_{|s_i|} \rangle$ where w_j denotes a word and $|s_i|$ is the sentence's length, and outputs validity period y_i of a sentence, $y_i = \{$ *few hours, few days, few weeks, few months, few years or more* $\}$[6]. To simplify the computation, we assume that the sentence is created during the assessment time[7].

---

4 We assume that content is valid at its creation time.

5 https://en.wikipedia.org/wiki/Logarithmic_timeline.

6 In Experiments in Sec 5, we also test the case with the reduced set of three classes.

7 Determining the approximate expiry date requires then extending the actual creation time of a sentence with its predicted validity period.

3.2 Feature Engineering

In this section we motivate and explain the modeling of feature groups we use.

LSA: Certain words are bound to be more related to some temporal spans than others. It is fairly intuitive that words such as *"election"*, *"economy"* or *"investigate"* would occur more often in sentences with rather long validity periods than with short ones, and vice versa for words such as *"moment"*, *"walk"* or *"dinner"*. We use Gensim[8] to build a TF-IDF model based on Wikipedia[9]. Based on a vocabulary with over 600k words we create an LSA model using T-SVD (Truncated Singular Value Decomposition) to identify such lexical trends as mentioned above. We reduce the dimensions down to 200 which means that each sentence is represented by the top 200 trends identified using T-SVD.

Average Word Length: The intuition here is that more complicated sentences with longer words might tend to have a longer period of validity. For example, sentences about species, statistics, economics or science in general, in contrast to sentences referring to day-to-day things. While it might not always be the case, this feature might still capture some useful shallow patterns.

Sentence Length: Similar to the average word length, the sentence length may be a sign of the validity period. Longer sentences could be characterized by a longer validity or the vice versa.

POS-Tags: This feature is meant to elicit grammatical patterns throughout the classes. Each sentence in represented by a vector of counts for each POS-tag.

Temporal Expressions: Temporal expressions, if present in text, may serve as explicit markers for when the information expressed in a sentence ceases to be valid. CoreNLP's Name Entity Recognition parser identifies four different types of temporal expressions: DATE, TIME, DURATION and SET. We discard the SET type due to the ambiguous nature of SET expressions[10], their less frequent occurrence and difficulty to be mapped into time granularities. DATE, TIME and DURATION expressions are converted by CoreNLP into Timex expressions, which are normalizations of temporal expressions. These are then converted into one of the eight following time granularities we have chosen to use which give us a generalized representation of temporal expressions in the sense of which time granularity they are related to:

[year, month, week, day, hour, minute, second, now]

The conversion, as exemplified in Table 1, is done by using Regex to find the time measure of finest granularity mentioned in the Timex expression. Looking at the first row in Table 1 we can see that the finest granularity that is mentioned in the Timex expression is referring to a day, thus the time granularity for that expression will be *day*. The sentences are represented by a vector of counts for each granularity for each time type.

---

[8] https://radimrehurek.com/gensim/.

[9] Text dump from 2018-05-01.

[10] TIME, DATE and DURATION expressions often point to a specific point (or duration) in time which means that they can be used as explicit markers for when information ceases to be valid. However, SET expressions, such as "*every day*", does not.

Table 1. Conversion of example temporal expressions (in bold) to time granularities.

Sentence	Time type	Timex	Time granularity
*"Today is **the 9th of July**"*	DATE	2018-07-09	Day
*"It is **12.15** and she is still not here"*	TIME	2018-07-09T12.15	Minute
*"I am going away for **a few months**"*	DURATION	PXM	Month

Sentence Embedding: The meaning of a sentence is naturally an important marker for its validity duration. A sentence about the geographical location of a town has a widely different meaning than a person reporting that they will soon be going to bed. To catch this difference in meaning throughout the classes we created sentence embeddings from the average word embeddings[11] of a sentence.

TempoWordNet: TempoWordNet is an extension to WordNet[12]. TempoWordNet gives information about how WordNet senses are related to the past, present, future or if they are a-temporal. To retrieve this information, we need to know the senses of the words in a sentence. We use a naive disambiguation approach and pick the most frequent sense for each word. Each sentence is represented by the probability for past, present, future and a-temporal which is the average probability across all the words, e.g., the past probability for the sentence is the average probability for past for all words in the sentence.

Lexical Categories: In addition to LSA we use a set of manually picked and validated lexical categories which were extracted from modern fiction. The objective is to capture more informal themes which are representative for day-to-day life-related situations. These lexical categories are provided by Empath [13][13]. Empath generates lexical categories by creating category specific term lexicons from a vector space model using cosine similarity. These lexicons then are manually pruned through crowd validation. Empath identifies categories in a piece of text by looking at which lexicons the occurring terms belong to. We use Empath's 194 pre-validated categories, of which 10 are exemplified below.

[help, office, dance, money, wedding, domestic_work, sleep, ...]

Each sentence is represented by a vector containing normalized scores for all categories.

Global Temporal Associations (GTA): Next we propose a method for finding and representing temporal properties of words and combinations of words. The intuition is that certain words and their combinations such as *("build", "house")* or *("kick", "ball")* presuppose temporal aspects. For example, building a house is something long-lasting while kicking a ball is not. The idea

[11] We used pre-trained word embeddings by Google created based on news: https://code.google.com/archive/p/word2vec/.

[12] https://wordnet.princeton.edu/.

[13] https://github.com/Ejhfast/empath-client.

is that by looking at the time granularities of temporal expressions associated with a word or a combination of words one might find their underlying temporal properties. For example, the combination *("build", "house")* should have a stronger association with temporal expression of a granularity of *year* rather than *hour*, unlike, *("kick", "ball")* or *("kick")*.

Table 2. Statistics about sentences extracted from Common Crawl.

Sentences	Sentences with temp exp	Temp exp	Temp exp as verb modifiers
8,800,000	1,811,608	2,249,309	823,944

We calculate Global Temporal Associations (GTA) of content elements based on statistical approach over large scale data to discover the global co-occurrences of time granularities with words and their combinations. For this, we use Common Crawl dataset[14] which is a web dump composed of billions of websites with plain text versions available. To handle noise, a few filtering conditions are used, such as allowing only Latin characters and punctuation-ended lines. After filtering, slightly less than 9 million sentences were parsed.

For each sentence found in the Common Crawl dataset we identify DATE, TIME and DURATION expressions. We only use temporal expressions that are modifiers of a verb. These are identified by looking at the sentence dependency relations. The choice of only using temporal expressions that are verb modifiers is because the meaning of the sentence heavily relies on verbs which often dictate the temporal properties of the sentence. For each extracted verb we find the related subject and object. This results in obtaining SVO combinations related to each temporal expression. The temporal expressions are then converted into one-hot-vectors representing time granularities and added to the count vector for DATE, TIME or DURATION for the verbs, nouns and SVO combinations related to the temporal expressions. These count vectors are stored and updated throughout the extraction process.

It is important to note that we store information about a noun in a subject position separate from information about the same noun in an object position. This is meant to catch potential differences in temporal properties of when a noun occurs as an object or as a subject. Other statistics are also collected to exemplify the dataset and to calculate the GTA, such as word counts, corpus size, numbers of temporal expressions that are of type DURATION, DATE and TIME, and how many of these are verb modifiers, and so on. Looking at Table 2 we can see that out of the almost 9 million sentences extracted, only 823k sentences included a temporal expression that is a verb modifier.

Having collected the co-occurrence statistics of words and their combinations with temporal granularites as described above, we can proceed to estimating temporal associations in target sentences. For a given sentence from our dataset

[14] http://commoncrawl.org/.

we extract the subject (S), object (O) and root verb (V) along with the SVO combinations (i.e., SV, VO, SVO). Every word and word combination found in the analyzed sentence is queried for in the count data which contain all the time granularity count vectors. If we can not find a word or combinations we use the count vectors of the most similar word or combination, with the assumption that they share similar temporal properties. To find the most similar word or combination of words we use the concatenation of word embeddings and cosine similarity measure. When we find the time granularity count vectors for a word or a combination of words we calculate the PMI (Pointwise Mutual Information) of that word or word combination with each level of time granularity. For example, PMI(*('build', 'house')*, DAY_DURATION) gives the association strength between the term combination *('build', 'house')* and the duration of *day* granularity.

The GTA features for a target sentence are the PMI vectors for all the six possible words and combinations related to the sentence's root verb (S, V, O, SV, VO, SVO) along with the similarity scores for each. If a certain grammatic position or combination is not found in a sentence, the vectors for that position or combination are filled with zeros.

3.3 Feature Normalization and Selection

After all the feature groups are prepared, each sentence is represented by 907 features. We normalize all features using L2-norm and scale them to fit to [0,1].

To avoid overfitting and to improve efficiency, a feature selection method - Recursive Feature Elimination (RFE) - is used. RFE recursively removes features to obtain the best possible model relative to a machine learning algorithm. After initial testing we found it beneficial to reduce feature count down to 100.

4 Evaluation

In this section, we report experimental results starting with dataset preparation and experiment settings.

4.1 Dataset Construction

As this is a novel task, we needed to create a dataset[15]. This comprised of two challenges; selecting sentences and manually annotating them into the five classes of validity periods. To cover a broad linguistic variations and build a model reflecting the real world, sentences were randomly extracted from three different datasets: blog posts[16], news articles[17] and Wikipedia articles[18]. These different datasets use different types of language, cover different type information, and can be more related to certain time spans than others. Blogs tend to be about

[15] https://github.com/AxlAlm/ValidityPeriods-dataset.
[16] http://u.cs.biu.ac.il/~koppel/BlogCorpus.htm.
[17] https://www.kaggle.com/snapcrack/all-the-news/data.
[18] http://kopiwiki.dsd.sztaki.hu/.

day-to-day things, while Wikipedia articles tend to be about general and objective concepts. News are somewhat in between and are often more formal and objective than blogs.

During the extraction, a set of conditions were used to improve the quality of sentences. The conditions removed sentences starting with certain words such as "*and*" and "*this*", ones that were too short or too long based on the character count, ones containing past or future tense verbs, and non-English sentences. Furthermore, we tried to equalize the numbers of sentences with and without temporal expression to maintain balance between sentences with explicit and implicit temporal properties.

The annotation consisted of categorizing sentences into one of the five temporal classes. Sentences that could not be understood or for which a validity period could not be estimated were discarded. Annotators were asked to assume that every sentence was true and created at the annotation time and to estimate when the information stated in each sentence ceases to be valid. All sentences that did not have 100% agreement between two annotators were removed.

Table 3. Example sentences for each class taken from the dataset.

Few hours	"So Michi, Audrey, Joel and myself are all hanging out in Linda's basement."
Few days	"School starts at a later time on Wednesday but that's no big deal."
Few weeks	"I am taking a course on learning how to use the program 3d studio max."
Few months	"I am also playing a gig with the new millennium string orchestra at the beginning of next month."
Few years or more	"The middle eastern nation of Israel is planning to expand its settlements, its housing areas in the west bank."

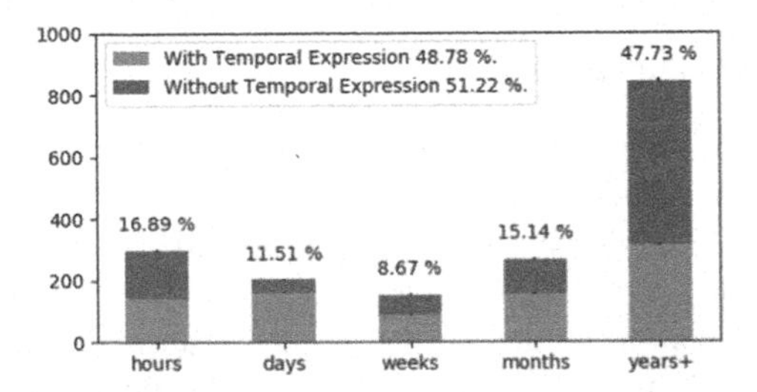

Fig. 1. Distribution of classes and sentences with and without temporal expressions for the original five classes.

Table 4. F1-micro using all sentences with original classes. Statistical significance in relation to baselines in each previous box is marked with **.

Models	F1-micro
Random	19.61
Majority Class	47.76
RNN	59.49
MLP (LSA)	39.17**
KNN (LSA)	56.95**
RandomForest (LSA)	60.01**
SVC_RBF (LSA)	61.77**
LinearSVC (LSA)	62.39**
MLP (all features)	53.76**
KNN (all features)	60.07**
RandomForest (all features)	62.75**
SVC_RBF (all features)	67.44**
LinearSVC (all features)	**68.69****

Table 5. F1-micro using all sentences with reduced classes. Statistical significance in relation to baselines in each previous box is marked with **.

Models	F1-micro
Random	34.94
Majority Class	50.14
RNN	70.51
MLP (LSA)	63.61**
KNN (LSA)	61.77**
RandomForest (LSA)	66.15**
SVC_RBF (LSA)	69.48**
LinearSVC (LSA)	70.11**
MLP (all features)	72.50**
KNN (all features)	68.75**
RandomForest (all features)	70.90**
SVC_RBF (all features)	77.37**
LinearSVC (all features)	**78.11****

The final dataset consists of 1,762 sentences. Table 3 contains example sentences for each class, while Fig. 1 describes the distribution of classes along with the portion of sentences with and without temporal expressions.

4.2 Experimental Setting

When extracting sentences for the dataset, Tokenization and POS-Tagging were performed using NLTK[19] and Temporal Tagging by CoreNLP's[20] Name Entity Recognition parser. For feature modeling all previously mentioned NLP methods plus dependency parsing were done with CoreNLP. Normalization was done in the form of lowercasing and abbreviation resolving[21]. For certain features[22] stopwords were also removed.

As the main metric, we use F1-micro to account for the class imbalance. To get robust results and reduce overfitting we use 5-fold cross-validation. T-test was performed on each model using all features in references to all the baselines to mark any significant differences. We contrast the results of LinearSVC[23],

[19] https://www.nltk.org/.
[20] https://stanfordnlp.github.io/CoreNLP/.
[21] e.g. "i'm to "I" and "am".
[22] LSA, average word length, sentence length, POS-tags, Sentence embeddings, TempoWordNet.
[23] For LinearSVC we use $C = 0.7$.

SVC_RBF[24], RandomForest[25], KNN[26] and MLP (Multi-Layered Perceptron) to baseline models that only use LSA as well as to two naive baselines, one that classifies everything randomly and one that classifies every sample with the most common class. MLP uses two dense hidden layers with 500 cells each. Each layer has a 75% dropout and uses Relu activation. The last layer is a dense layer with a Softmax activation and with cells equal to the amount of classes.

We also compare the previously introduced models to RNN where the input is a sentence represented as a sequence of word embeddings. The RNN in constructed by two stacked LSTM layers with 40 cells each preceded by a dense layer with 120 cells. All hidden layers were set to have 75% dropout after tuning and use Tanh activation. The last layer is identical to the last layer of the MLP.

The avoid overfitting the NN's, we stop training if there are no imporvements after 5 epochs for the RNN and 7 for the MLP. The neural networks are implemented in Keras[27] and for other methods we use the implementations provided by Scikit-learn[28].

Table 6. The impact on the F1-micro when each feature when using LinearSVC is removed for both sets of classes.

Features removed	Original classes	Reduced classes
None	**68.69**	**78.11**
LSA	67.38 ↓	78.33 ↑
avrg word len	67.73 ↓	79.18 ↑
pos-tags	69.14 ↑	79.47 ↑
sent emb	65.29 ↓	74.3 ↓
temp exp	67.44 ↓	77.82 ↓
TempoWordNet	68.63 ↓	79.01 ↑
Lexical categories	69.65 ↑	79.24 ↑
GTA	**70.22** ↑	**79.98** ↑

Table 7. F1 score of two different models that only use the best features from Table 6.

Features	F1 score
LSA, avrg word len, sent len, sent emb, temp exp, TempoWordNet	69.65
sent emb, temp exp	69.03

[24] For RBF we use $C = 60$.
[25] We use 150 trees.
[26] Five neighbors are used together with distance weighting.
[27] https://keras.io/.
[28] http://scikit-learn.org/stable/.

5 Experimental Results

Table 4 shows the main results. We can see that classifiers with all the features outperform the baseline models. Also we see that the NN-based approaches, the RNN and MLP, did not perform well, likely due to the small amount of data. LinearSVC is the most prominent model achieving 68.69% F1-micro.

Next, in Table 5 we test how the models perform with reduced classes. The reduction to obtain the three new classes is done in the following way: the *short-term* class is obtained by merging the classes *few hours* and *few days*, *middle-term* is obtained after combining *few weeks* and *few months*, and *long-term* being equal to *few years or more.* The results display similar tendencies observed in the main task: LinearSVC using all features outperforms all other models.

To understand the influence of each feature we create a model using LinearSVC where each feature was omitted. The results are displayed in Table 6. Features with a downward-pointing arrow signify loss in F1-micro when being removed. We can see that LinearSVC achieved a F1-micro of 70.22% when GTA was removed. We can also observe that sentence embeddings is the most influential feature.

Based on the information in Table 6, additional tests are done with combinations of positively influential features. The results are shown in Table 7. Given these results we can conclude that the best observed model uses all features without GTA and achieves a F1-micro of 70.22%.

In the light of the poor performance of GTA, different representations of GTA are tested as shown in Table 8 with mixed results. The possible reason for under-performance of GTA is the low co-occurrence of words and their combinations with temporal expressions. For SVO's (Fig. 2) we can see that they mainly co-occurred with zero temporal expressions, and only a minority co-occurred with more than five temporal expressions, a trend that is similar for subject, verbs, objects, VO's and SV's.

Finally, taking a closer look at the classification confusion of our best model (linearSVC with all features except GTA) in Table 9, we see that although the model predicts hours and years+ with reasonably good accuracy, it tends to confuse weeks with months and years+. Moreover, we see that months are often confused as years+. This might be due to class imbalance or difficulties annotators had in judging the temporal spans of sentences of these classes.

Table 8. F1 score using different representations of GTA along with other features. First row uses average DURATION PMI vectors, the second uses the weighted average by the cosine similarity score, the third uses the average TIME PMI vectors and the last uses the average DATE PMI vectors.

GTA version	LinearSVC	SVC_RBF	RandomForest	KNN
Average DURATION	69.26	68.52	63.58	61.37
Weighted DURATION	70.05	68.23	64.04	62.96
Average TIME	69.54	69.09	64.26	63.3
Average DATE	69.77	68.52	64.95	62.9

Table 9. Confusion probabilities: true class in row and predicted class in column.

Classes	Hours	Days	Weeks	Months	Years+
Hours	**75.17**	6.04	1.68	3.36	13.76
Days	19.21	**57.14**	3.45	4.93	15.27
Weeks	17.65	9.15	13.07	**30.72**	29.41
Months	5.24	6.37	3.0	36.7	**48.69**
Years+	1.78	1.31	0.71	3.56	**92.64**

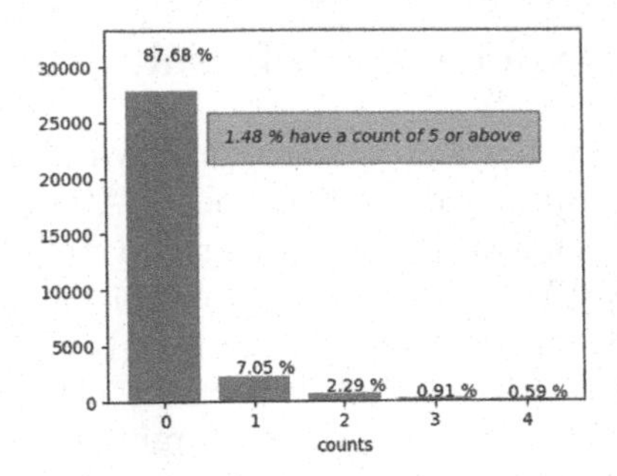

Fig. 2. Five most common co-occurrence counts for SVO and DURATION temporal expressions.

6 Conclusions and Future Work

The goal of our work is to introduce the task of forecasting the validity period of sentences and to create the first model using only linguistic features. This was motivated by the fact that humans can make such judgments and giving such an ability to computers would help in preventing outdatedness and misinformation.

To be able to design the model, several challenges had to be overcome. The first was to define and represent the task. The second was to create a dataset, which included understanding how sentences should be selected and then annotated to obtain quality data. The second challenge was creating and representing effective features. While sentence embeddings proved to be the most influential features, the core feature, GTA was unfortunately under-performing. Nevertheless, its analysis can provide valuable insights for further improvements. We emphasize that the task is challenging also due to short length of input.

Several further improvements can be proposed. First, as sentence embeddings proved to be the best features, a lot more could be done with their construction and representation. For example, sentence embedding that could reflect the grammatical roles and structure of the sentences [7] might be utilized to further boost the results. Furthermore, including more fine grained validity classes, e.g., classes like "*few minutes*" would make the model more applicable on social media platforms such as Twitter where one might stumble upon sentences such as "*Brad Pitt is standing in front of me in the line at Starbucks!*".

Finally, we propose another task of *extending content's validity period*. It would mean proposing a set of minimal updates of a sentence to move it to a validity class of a longer duration, or to make an invalid sentence valid again while maintaning its semantics. The example applications of this kind of task would be make content of obsolete documents to be useful again or automatically maintaining online content.

Acknowledgements. We thank Nina Tahmasebi for valuable comments and encouragement. This research has been supported by JSPS KAKENHI Grants (#17H01828, #18K19841) and by Microsoft Research Asia 2018 Collaborative Research Grant.

References

1. Baeza-Yates, R.: Searching the future. In: ACM SIGIR Workshop MF/IR (2005)
2. Berberich, K., Gupta, D.: Identifying time intervals for knowledge graph facts. In: Proceeding WWW 2018 Companion. In: Proceedings of the Web Conference 2018, Lyon, France, 23–27 April 2018, pp. 37–38 (2018)
3. Campos, R., Dias, G., Jorge, A.M., Jatowt, A.: Survey of temporal information retrieval and related applications. ACM Comput. Surv. (CSUR) **47**(2), 15 (2015)
4. Chang, A., Manning, C.: SUTIME: a library for recognizing and normalizing time expressions. In: Proceedings of the LREC 2012, Istanbul, Turkey, 23–25 May 2012
5. Chambers, N.: Labeling documents with timestamps: learning from their time expressions. In: ACL 2012 Proceedings of the 50th Annual Meeting of the Association for Computational Linguistics: Long Papers - Volume 1, Jeju Island, Korea, 08–14 July 2012, pp. 98–106 (2012)
6. Ciampaglia, G.L., Shiralkar, P., Rocha, L.M., Bollen, J., Menczer, F., Flammini, A.: Computational fact checking from knowledge networks. PLoS One **10**(6), e0128193 (2015)
7. Clark, S.: Vector space models of lexical meaning. In: Handbook of Contemporary Semantics (2015). https://doi.org/10.1002/9781118882139.ch16
8. Dai, N., Shokouhi, M., Davison, B.D.: Learning to rank for freshness and relevance. In: Proceedings of the SIGIR 2011. Beijing, China, 24–28 July, pp. 95–104. ACM Press (2011)
9. Das, S., Mishra, A., Berberich, K., Setty, V.: Estimating event focus time using neural word embeddings. In: Proceedings of the 2017 ACM on Conference on Information and Knowledge Management, CIKM 2017, Singapore, 06–10 November 2017, pp. 2039–2042 (2017)
10. Dias, G., Campos, R., Jorge, A.: Future retrieval: what does the future talk about? In: Workshop on Enriching Information Retrieval of the 34th ACM Annual SIGIR Conference (SIGIR 2011), July 2011, Pekin, China, p. 3 (2011)
11. Dias, G., Hasanuzzaman, M., Ferrari, S., Mathet, Y.: TempoWordNet for sentence time tagging. In: 23rd International Conference on World Wide Web Companion, April 2014, Seoul, South Korea. WWW Companion 2014. In: Proceedings of the Companion Publication of the 23rd International Conference on World Wide Web Companion, pp. 833–838 (2014)
12. Efron, M., Golovchinsky, G.: Estimation methods for ranking recent information. In: Proceedings of the SIGIR 2011. Beijing, China, 24–28 July, pp. 495–504. ACM Press (2011)
13. Fast, E., Chen, B., Bernstein, M.S.: Empath: understanding topic signals in large-scale text. In: Proceedings of the 2016 CHI Conference on Human Factors in Computing Systems, CHI 2016, pp. 4647–4657 (2016)
14. Grosser, Z., Schmidt, A.P., Bachl, M., Kunzmann, C.: Determining the outdatedness level of knowledge in collaboration spaces using a machine learning-based approach. Professionelles Wissensmanagement. Tagungsband der 9. Konferenz Professionelles Wissensmanagement (Professional Knowledge Management) Karlsruhe, Germany, 5–7 April 2017

15. Hassan, N., et al.: The quest to automate fact-checking. In: Proceedings of the 2015 Computation + Journalism Symposium (2015)
16. Hassan, N., Arslan, F., Li, C., Tremayne, M.: Towards automated fact-checking: detecting check-worthy factual claims by claimbuster. In: Proceedings of the 23rd ACM SIGKDD International Conference on Knowledge Discovery and Data Mining, KDD 2017, Halifax, NS, Canada, 13–17 August, pp. 1803–1812 (2017)
17. Jatowt, A., Yeung, C.-M.A., Tanaka, K.: Estimating document focus time. In: Proceedings of the 22nd ACM international conference on Information & Knowledge Management, CIKM 2013, San Francisco, California, USA, October 27 – November 01, pp. 2273–2278 (2013)
18. Jatowt, A., Yeung, C.-M.A.: Extracting collective expectations about the future from large text collections. In: Proceedings of the 20th ACM International Conference on Information and Knowledge Management, CIKM 2011, Glasgow, Scotland, UK, October 24–28, pp. 1259–1264 (2011)
19. Kanhabua, N., Nørvåg, K.: Improving temporal language models for determining time of non-timestamped documents. In: Proceedings of the 12th European conference on Research and Advanced Technology for Digital Libraries, ECDL 2008, Aarhus, Denmark, 14–19 September, pp. 358–370 (2008)
20. Kumar, A., Baldridge, J., Lease, M., Ghosh, J.: Dating Texts without Explicit Temporal Cues. CoRR, abs/1211.2290 (2012)
21. Ling, X., Weld, D.: Temporal information extraction. In: Twenty-Fourth AAAI Conference on Artificial Intelligence (2010)
22. Graves, L.: Understanding the Promise and Limits of Automated Fact-Checking. University of Oxford, Reuters institute, 28 February 2018. https://reutersinstitute.politics.ox.ac.uk/our-research/understanding-promise-and-limits-automated-fact-checking
23. Morbidoni, C., Cucchiarelli, A., Ursino, D.: Leveraging linked entities to estimate focus time of short texts. In: IDEAS 2018 Proceedings of the 22nd International Database Engineering & Applications Symposium, Villa San Giovanni, Italy, 18–20, pp. 282–286, June 2018
24. Perkiö, J., Buntine, W., Tirri, H.: A temporally adaptative content-based relevance ranking algorithm. In Proceedings of the SIGIR 2005. Salvador, Brazil. 15–16 August, pp. 647–648. ACM Press (2005)
25. Popat, K., Mukherjee, S., Weikum, G.: Credibility assessment of textual claims on the web. In: CIKM 2016 Proceedings of the 25th ACM International on Conference on Information and Knowledge Management, Indianapolis, Indiana, USA, 24–28 October, pp. 2173–2178 (2016)
26. Sato, S., Uehar, M., Sakai, Y.: Temporal ranking for fresh information retrieval. In: Proceeding AsianIR 2003 Proceedings of the Sixth International Workshop on Information Retrieval with Asian Languages, Sapporo, Japan, vol. 11, pp. 116–123, 07 July 2003
27. Sil, A., Cucerzan, S.: Temporal scoping of relational facts based on Wikipedia data. In: Proceedings of the Eighteenth Conference on Computational Language Learning, Baltimore, Maryland USA, 26–27 June 2014, pp. 109–118. Association for Computational Linguistics (2014)
28. Styskin, A., Romanenko, F., Vorobyev, F., Serdyukov, P.: Recency ranking by diversification of result set. In: Proceedings of the CIKM 2011, Glasgow, Scotland, UK, 24–28 October, pp. 1949–1952. ACM Press (2011)
29. Takemura, H., Tajima, K.: Tweet classification based on their lifetime duration published. In: Proceedings of CIKM 2012, Maui, pp. 2367–2370, October 2012

30. Talukdar, P.P., Wijaya, D., Mitchell, T.: Coupled temporal scoping of relational facts. In: Proceedings of the WSDM 2012, Seattle, USA, 8–12 February, pp. 73–82. ACM Press (2012)
31. Wang, Y., Zhu, M., Qu, L., Spaniol, M., Weikum, G.: Timely YAGO: harvesting, querying, and visualizing temporal knowledge from Wikipedia. In: EDBT 2010 Proceedings of the 13th International Conference on Extending Database Technology, Lausanne, Switzerland, 22–26 March, pp. 697–700 (2010)
32. Wang, Y., Yang, B., Qu, L., Spaniol, M., Weikum, G.: Harvesting facts from textual web sources by constrained label propagation. In: CIKM 2011 Proceedings of the 20th ACM International Conference on Information and Knowledge Management, Glasgow, Scotland, UK, 24–28 October, pp. 837–846 (2011)
33. Yamamoto, Y., Tezuka, T., Jatowt, A., Tanaka, K.: Honto? search: estimating trustworthiness of web information by search results aggregation and temporal analysis. In: Dong, G., Lin, X., Wang, W., Yang, Y., Yu, J.X. (eds.) APWeb/WAIM -2007. LNCS, vol. 4505, pp. 253–264. Springer, Heidelberg (2007). https://doi.org/10.1007/978-3-540-72524-4_28
34. You, W., Agarwal, P.K., Li, C., Yang, J., Cong, Y.: Toward computational fact-checking. J. Proc. VLDB Endowment **7**(7), 589–600 (2014)

Dynamic Ensemble Selection for Author Verification

Nektaria Potha and Efstathios Stamatatos(✉)

University of the Aegean, 83200 Karlovassi, Greece
{nekpotha,stamatatos}@aegean.gr

Abstract. Author verification is a fundamental task in authorship analysis and associated with significant applications in humanities, cyber-security, and social media analytics. In some of the relevant studies, there is evidence that heterogeneous ensembles can provide very reliable solutions, better than any individual verification model. However, there is no systematic study of examining the application of ensemble methods in this task. In this paper, we start from a large set of base verification models covering the main paradigms in this area and study how they can be combined to build an accurate ensemble. We propose a simple stacking ensemble as well as a dynamic ensemble selection approach that can use the most reliable base models for each verification case separately. The experimental results in ten benchmark corpora covering multiple languages and genres verify the suitability of ensembles for this task and demonstrate the effectiveness of our method, in some cases improving the best reported results by more than 10%.

Keywords: Author verification · Authorship analysis · Ensemble learning · Dynamic ensemble selection

1 Introduction

Authorship analysis is a research area in text mining that attempts to reveal information about the authors of electronic documents. Among authorship analysis tasks, author verification is considered to be fundamental [19] since it focuses on the most basic question: whether two documents are written by the same author. More complex tasks like authorship attribution (i.e., identifying the most likely author given a closed-set or open-set of suspects) [31] or authorship clustering (grouping a collection of documents by authorship) can be decomposed into a series of author verification cases [21].

Technology in author verification is strongly associated with applications in several fields. In digital humanities, author verification can be used to reveal the identity of authors of documents of high historical and literary importance [35,36]. In cyber-security, it can be used to detect compromised accounts [3] or spearphishing attacks [8] and enable continuous authentication of users [5]. Author verification can also be used to detect multiple accounts controlled by the same user [1] and facilitate deception detection [22] in social media.

L. Azzopardi et al. (Eds.): ECIR 2019, LNCS 11437, pp. 102–115, 2019.
https://doi.org/10.1007/978-3-030-15712-8_7

A typical author verification case (or instance) is a tuple $(D_{known}, d_{unknown})$ where D_{known} is a set of documents of known authorship, all by the same author, and $d_{unknown}$ is another document of questioned authorship. An author verification method should be able to decide whether or not the author of D_{known} is also the author of $d_{unknown}$. Apart from a binary (yes/no) answer, author verification methods usually produce a verification score in [0,1] that can be viewed as a confidence estimation [33,34]. Essentially, author verification is a one-class classification task since only labelled samples from the positive class are available [10]. However, there are approaches that attempt to transform it to a binary classification task by sampling the negative class (i.e., all documents by all other authors) [21].

Recently, several methods have been proposed in the relevant literature [32], largely motivated by the corresponding PAN shared tasks organized from 2013 to 2015 [14,33,34]. In some previous works, there is evidence that an ensemble of verifiers could be better than any single model. The organizers of PAN used a simple heterogeneous ensemble by averaging all verification scores produced by the submitted methods and found that this simple meta-model was far better than any individual verifier in PAN-2014 [34]. A similar attempt in PAN-2015 shared task did not provide equally impressive results, mainly due to the very low performance of many submissions in that case [33]. However, another heterogeneous ensemble combining five verifiers won the second-best overall rank in PAN-2015 [23]. So far, there is lack of more systematic studies examining a large pool of verifiers and more sophisticated ensemble learning approaches.

In the current paper, we attempt to fill that gap by starting from a wide range of base verification models covering the most important paradigms in the relevant literature. Then, we propose two ensemble learning approaches. First a simple stacking method using a meta-learner to combine the outputs of base models. Second, a dynamic ensemble selection method that can focus on the most effective models for each verification case separately. Experimental results on several benchmark datasets covering different languages and genres demonstrate the effectiveness of both approaches especially in challenging cases where D_{known} is of limited size and in cross-domain conditions.

The rest of this paper is organized as follows: Sect. 2 presents previous work in author verification while Sect. 3 describes our proposed methods. The performed experiments are analytically presented in Sect. 4 while Sect. 5 discusses the main conclusions and suggests future work directions.

2 Previous Work

Early work in this field is marked by the *unmasking* approach [20] that builds a classifier to distinguish between two documents and examines how fast the accuracy drops when the most important features are gradually removed. This method was found to be very effective in long literary documents but not reliable when only short text samples are available [29] or when cross-genre conditions are met [15]. Research in author verification has been strongly influenced by

the recent PAN shared tasks [14,33,34] where multiple submitted methods [illegible]
evaluated in several benchmark datasets covering different languages and g[illegible]

In general, author verification methods follow specific paradigms [32]. [illegible]
intrinsic methods attempt to handle a one-class classification task by in[illegible]
estimating the similarity of $d_{unknown}$ to D_{known} and then deciding wh[illegible]
this similarity is significant [10,13,18,25]. Usually, such approaches are fas[illegible]
robust across different domains and languages.

On the other hand, *extrinsic* methods attempt to transform author[illegible]
fication to a binary classification task by sampling the negative class [illegible]
is huge and extremely heterogeneous (it comprises all other possible aut[illegible]
[16,21,26,30]. Then, extrinsic methods attempt to decide if the similari[illegible]
$d_{unknown}$ to D_{known} is higher than the similarity of $d_{unknown}$ to $D_{external}$ [illegible]
collected samples of the negative class). All top-ranked submissions to
shared tasks from 2013 to 2015 follow this paradigm [2,16,30] demonstrati[illegible]
effectiveness. However, the performance of such methods heavily depends o[illegible]
quality of the collected external documents [21].

From another point of view, author verification methods can be distingu[illegible]
according to the way they handle the members of D_{known}. The *instance-*[illegible]
approaches [31] treat each known document separately and then combin[illegible]
corresponding decisions [6,16,30]. If there is only one known document, [illegible]
instance-based methods segment it into parts to enable the estimation of var[illegible]
of similarity within D_{known} [13]. On the contrary, *profile-based* techniques [illegible]
concatenate all known documents attempting to better represent the prop[illegible]
of the style of author, rather than the style of each document [10,18,25]. [illegible]
approach is better able to handle short texts in comparison to instance-b[illegible]
methods. However, it disregards any useful information about the variatio[illegible]
might exist within the set of known documents.

Another category of methods focus on the representation of verific[illegible]
instances (i.e., the tuple $(D_{known}, d_{unknown})$) rather than the individual [illegible]
uments they contain. Each verification instance may be positive (same aut[illegible]
or negative (different author) and given a training dataset of such instan[illegible]
classifier is build to learn to distinguish between these two classes [4,9,12]. [illegible]
eager approaches [32] attempt to learn a general verification model and its [illegible]
tiveness strongly depends on the volume, representativeness and distributi[illegible]
the training dataset [33].

Regarding the stylometric information extracted from documents, [illegible]
author verification approaches are based on simple but effective features [illegible]
word and character n-grams [13,18,21,25]. There are also language-agno[illegible]
approaches using information extracted from text compression [10]. The [illegible]
of NLP tools to extract syntactic-related information is limited [4,23]. A re[illegible]
study based on representation learning uses a neural network to jointly [illegible]
character, lexical, topical, and syntactic modalities and achieved very [illegible]
results [7]. Another deep learning method, the winning approach in PAN-2[illegible]
uses a character-level recurrent neural network language model [2]. In addi[illegible]
recently, the use of topic modeling techniques provided promising results [11[illegible]

A typical author verification case (or instance) is a tuple $(D_{known}, d_{unknown})$ where D_{known} is a set of documents of known authorship, all by the same author, and $d_{unknown}$ is another document of questioned authorship. An author verification method should be able to decide whether or not the author of D_{known} is also the author of $d_{unknown}$. Apart from a binary (yes/no) answer, author verification methods usually produce a verification score in [0,1] that can be viewed as a confidence estimation [33,34]. Essentially, author verification is a one-class classification task since only labelled samples from the positive class are available [10]. However, there are approaches that attempt to transform it to a binary classification task by sampling the negative class (i.e., all documents by all other authors) [21].

Recently, several methods have been proposed in the relevant literature [32], largely motivated by the corresponding PAN shared tasks organized from 2013 to 2015 [14,33,34]. In some previous works, there is evidence that an ensemble of verifiers could be better than any single model. The organizers of PAN used a simple heterogeneous ensemble by averaging all verification scores produced by the submitted methods and found that this simple meta-model was far better than any individual verifier in PAN-2014 [34]. A similar attempt in PAN-2015 shared task did not provide equally impressive results, mainly due to the very low performance of many submissions in that case [33]. However, another heterogeneous ensemble combining five verifiers won the second-best overall rank in PAN-2015 [23]. So far, there is lack of more systematic studies examining a large pool of verifiers and more sophisticated ensemble learning approaches.

In the current paper, we attempt to fill that gap by starting from a wide range of base verification models covering the most important paradigms in the relevant literature. Then, we propose two ensemble learning approaches. First a simple stacking method using a meta-learner to combine the outputs of base models. Second, a dynamic ensemble selection method that can focus on the most effective models for each verification case separately. Experimental results on several benchmark datasets covering different languages and genres demonstrate the effectiveness of both approaches especially in challenging cases where D_{known} is of limited size and in cross-domain conditions.

The rest of this paper is organized as follows: Sect. 2 presents previous work in author verification while Sect. 3 describes our proposed methods. The performed experiments are analytically presented in Sect. 4 while Sect. 5 discusses the main conclusions and suggests future work directions.

2 Previous Work

Early work in this field is marked by the *unmasking* approach [20] that builds a classifier to distinguish between two documents and examines how fast the accuracy drops when the most important features are gradually removed. This method was found to be very effective in long literary documents but not reliable when only short text samples are available [29] or when cross-genre conditions are met [15]. Research in author verification has been strongly influenced by the recent PAN shared tasks [14,33,34] where multiple submitted methods were evaluated in several benchmark datasets covering different languages and genres.

In general, author verification methods follow specific paradigms [32]. First, *intrinsic* methods attempt to handle a one-class classification task by initially estimating the similarity of $d_{unknown}$ to D_{known} and then deciding whether this similarity is significant [10,13,18,25]. Usually, such approaches are fast and robust across different domains and languages.

On the other hand, *extrinsic* methods attempt to transform author verification to a binary classification task by sampling the negative class which is huge and extremely heterogeneous (it comprises all other possible authors) [16,21,26,30]. Then, extrinsic methods attempt to decide if the similarity of $d_{unknown}$ to D_{known} is higher than the similarity of $d_{unknown}$ to $D_{external}$ (the collected samples of the negative class). All top-ranked submissions to PAN shared tasks from 2013 to 2015 follow this paradigm [2,16,30] demonstrating its effectiveness. However, the performance of such methods heavily depends on the quality of the collected external documents [21].

From another point of view, author verification methods can be distinguished according to the way they handle the members of D_{known}. The *instance-based* approaches [31] treat each known document separately and then combine the corresponding decisions [6,16,30]. If there is only one known document, some instance-based methods segment it into parts to enable the estimation of variance of similarity within D_{known} [13]. On the contrary, *profile-based* techniques [31] concatenate all known documents attempting to better represent the properties of the style of author, rather than the style of each document [10,18,25]. This approach is better able to handle short texts in comparison to instance-based methods. However, it disregards any useful information about the variation it might exist within the set of known documents.

Another category of methods focus on the representation of verification instances (i.e., the tuple (D_{known},$d_{unknown}$)) rather than the individual documents they contain. Each verification instance may be positive (same author) or negative (different author) and given a training dataset of such instances a classifier is build to learn to distinguish between these two classes [4,9,12]. Such *eager* approaches [32] attempt to learn a general verification model and its effectiveness strongly depends on the volume, representativeness and distribution of the training dataset [33].

Regarding the stylometric information extracted from documents, most author verification approaches are based on simple but effective features like word and character n-grams [13,18,21,25]. There are also language-agnostic approaches using information extracted from text compression [10]. The use of NLP tools to extract syntactic-related information is limited [4,23]. A recent study based on representation learning uses a neural network to jointly learn character, lexical, topical, and syntactic modalities and achieved very good results [7]. Another deep learning method, the winning approach in PAN-2015, uses a character-level recurrent neural network language model [2]. In addition, recently, the use of topic modeling techniques provided promising results [11,27].

Table 1. Distribution of base verification models over the different paradigms.

	Intrinsic	Extrinsic
Instance-based	16	10
Profile-based	16	5

3 The Proposed Methods

3.1 Base Verification Models

In this paper, we use an extended list of author verification models that cover the main paradigms in this area. More, specifically, we implemented 47 base models that belong to the following categories:

- *Instance-based Intrinsic Models*: These are inspired from [13], a very robust method that was used as baseline in PAN-2014 and PAN-2015 shared tasks [33,34].
- *Profile-based Intrinsic Models*: We adopt a simple but effective method described in [27].
- *Instance-based Extrinsic Models*: The well-known General Impostors (GI) method [30] as well as a recently-proposed modification called ranking-based impostors [26] are used.
- *Profile-based Extrinsic Models*: We use another modification of GI that follows the profile-based paradigm [28]. For each verification instance, all known documents (D_{known}) are first concatenated. Then, multiple artificial impostor documents of similar properties are formed by concatenating an equal number ($|D_{known}|$) of external documents.

The variation of models in each category consists of using different text representation schemes. Several feature types (word unigrams, character 3-grams, 4-grams, or 5-grams) are used and two topic modeling techniques (Latent Semantic Indexing (LSI) or Latent Dirichlet Allocation (LDA)) are applied to build various version of a certain verifier. We also examine two different corpora (a small and a larger one) to extract the topic models. More details on how the base models are used and tuned in the performed experiments are given in Sect. 4.2. The distribution of base verification models is shown in Table 1. Extrinsic models are fewer than intrinsic ones because less variation in feature types is used in that case. That way the distribution of our base verifiers is similar to those of methods submitted to PAN shared tasks [32], where intrinsic methods were more popular than extrinsic ones while the majority of submissions followed the instance-based paradigm.

3.2 Stacking Ensemble

First, we focus on the use of a simple method to construct heterogeneous ensembles. A meta-learner (binary classifier) can be trained based on the output of

```
input  : v, T, M, a, b
output: fusedScore
foreach t ∈ T do
    if cosine(vector(v), vector(t)) ≥ a then
        T_similar = T_similar ∪ t
    end
end
foreach m ∈ M do
    weight(m) = accuracy(m, T_similar)
    if weight(m) ≥ b then
        M_suitable = M_suitable ∪ m
    end
end
fusedScore = Σ_{m∈M_suitable} weight(m) · score(m, v)
```

Algorithm 1. The proposed DES author verification method.

the 47 base verifiers following a well-known *stacked generalization* approach [37]. Such a model can learn the correlations between the input features and the correctness of base models. After performing some preliminary experiments examining several alternative classifiers (e.g. multi-layer perceptron, k-nn), we finally use a Support Vector Machine (SVM) meta-learner in our stacking ensemble since in most of the cases it provides the most competitive results.

It has to be noted that the performance of such an approach heavily depends on the distribution of verification instances over the two classes (same author or different author). As it is explained in Sect. 4.1, all datasets we use in this study are balanced. However, in case the distribution of instances over the classes is not balanced, or not known, then a more carefully selected meta-learner (able to handle the class imbalance problem) should probably be used.

3.3 Dynamic Ensemble Selection

In this paper, we also propose a more sophisticated ensemble that is based on Dynamic Ensemble Selection (DES) [17] that focuses on suitable base models for each verification instance separately. Our method requires a training dataset, i.e. a collection of verification instances in the same language and genre with respect to the test dataset. In particular, given a test verification instance v, a collection of training instances T, and a collection of base verification models M (each model can produce a score in [0,1] when it gets as input a verification instance), our DES method performs the following steps (see also Algorithm 1):

1. Represent the characteristics of each (training or test) verification instance $(D_{known}, d_{unknown})$ as a numerical vector reflecting how homogeneous known documents are and how distant they are from the unknown document. First, each (known or unknown) document is represented based on a certain feature type (word or character n-grams) and then similarity between documents

Table 2. The similarity features used to represent each (training or test) verification instance ($D_{known}, d_{unknown}$). An additional feature is the size of D_{known}, so the total number of features is 73.

Similarity	Function	Fusion	Representation	#Features
D_{known} vs. $d_{unknown}$, or within D_{known}	Cosine, Minmax, or Euclidean	min, max, or avg	word unigrams, char 3-grams, char 4-grams, or char 5-grams	$2 \times 3 \times 3 \times 4 = 72$

is calculated. We focus on two types of similarity. First, within members of D_{known} that shows the degree of homogeneity in the known document set. Second, we compare all members of D_{known} to the unknown document to estimate how close they are. Since D_{known} usually includes multiple documents, a fusion method is needed to combine the obtained similarity values for each known document. In more detail, we use 3 similarity functions, 3 fusion methods, and 4 text representation types (see Table 2) to calculate 72 similarity features for each verification instance. The final vector contains one more feature that corresponds to the size of D_{known}.
2. Calculate the similarity of the test instance vector to each of the training instance vectors using cosine similarity.
3. Filter out all training instances with similarity to the test instance lower than a threshold a. Let $T_{similar} \subset T$ be the set of the remaining training instances highly similar to the test instance.
4. Calculate the effectiveness of each base verification model (47 in total) on $T_{similar}$.
5. Filter out all base verification models with effectiveness on $T_{similar}$ lower than a threshold b. Let $M_{suitable} \subset M$ be the set of the selected verification models.
6. Apply the remaining base verification models ($M_{suitable}$) to the test instance.
7. Fuse the scores of $M_{suitable}$ on v according to a weighted average where the weight of each model is determined by its effectiveness on $T_{similar}$.

The proposed method has two important parameters, thresholds a and b. The former determines the size of the set of selected training instances. If it is set too high (e.g., 0.9), very few similar training instances would be found. The latter affects the number of selected base verification models. If it is set too high, very few base verification models will be considered to provide the final answer. It should also be noted that some of the features used to represent a verification instance become useless when there is only one known document. In more detail, when there is exactly one known document then the features that calculate the similarity within the set of known documents are all equal.

Table 3. The PAN benchmark datasets used in this study ($|d|$ denotes text length in words).

	Dataset	Training instances	Test instances	$avg(\|D_{known}\|)$	$avg(\|d\|)$
PAN-2014	DE (Dutch Essays)	96	96	1.89	405
	DR (Dutch Reviews)	100	100	1.02	114
	EE (English Essays)	200	200	2.62	841
	EN (English Novels)	100	200	1.00	5115
	GR (Greek Articles)	100	100	2.77	1470
	SP (Spanish Articles)	100	100	5.00	1129
PAN-2015	DU (Dutch Cross-genre)	100	165	1.75	357
	EN (English Cross-topic)	100	500	1.00	508
	GR (Greek Cross-topic)	100	100	2.87	717
	SP (Spanish Mixed)	100	100	4.00	950

4 Experiments

4.1 Description of Data

We consider benchmark corpora built in the relevant PAN evaluation campaigns on authorship verification in 2014 and 2015 (Table 3). These corpora cover four languages (Dutch, English, Greek, and Spanish) and several genres (newspaper articles, essays, reviews, literary texts etc.) [33,34]. Each corpus is divided into a training and a test part and in each case multiple verification instances are provided. Each instance includes a small number (up to 10) of known documents, all by the same author, and exactly one questioned document(unknown document). It is noticeable each dataset, either training or test, is balanced with respect to the distribution of positive (same-author) and negative (different-author) instances.

In PAN-2014 datasets, all (known and unknown) documents within a verification instance share the same language, genre, and thematic area. On the other hand, PAN-2015 datasets are more challenging since they include cross-domain cases, i.e., all documents within a verification instance are in the same language but they may belong to distinct thematic areas or genres.

4.2 Setup

We follow the same evaluation procedure with PAN shared tasks to achieve compatibility of evaluation results with PAN participants [33,34]. We use the training part of the each dataset to tune the parameters and calibrate the verification score of each model and then we apply the tuned models to the test part of the dataset. The parameter tuning is performed by grid search trying to optimize the Area Under the Receiver-Operating Characteristic Curve (AUC), an evaluation measure also used in PAN shared tasks. In addition, we perform five runs for the non-deterministic models (i.e., all variants of GI) and consider their average verification score. The output of each base model for the training instances of a dataset is used to train a logistic regression classifier that can provide binary (same author or different author) answers.

Apart from the tuned models, we also examine base models with fixed parameter settings. The idea behind this is that DES is based on the performance of the (selected) base models on the (selected) training instances. Thus, if the base models are tuned based on the same dataset, their output on that specific dataset could be biased. Taking into account results of previous work [13,26,27,30], we set each parameter of the base model to a default value (e.g., 250 latent topics when LSI or LDA is applied, 150 impostors per repetition in all variations of GI).

The set of external documents ($D_{external}$) used in the framework of extrinsic verification models is collected from the world wide web for each dataset separately following the procedure described in [26]. When topic modeling is applied, the latent topic models are extracted either from the documents in the training dataset exclusively or from a larger collection consisting of the training documents and the set of impostors.

As baselines, we use the top-ranked submissions and the meta-models combining all submissions of PAN-2014 and PAN-2015 shared tasks, as well as other recent studies which report AUC results in the same datasets. More specifically:

- Khonji and Iraqi [16]: This is a modification of GI [30] and the winning submission of PAN-2014 [34].
- Fréry et al. [9]: The second-best submission in PAN-2014, it is an eager verification approach using a decision tree classifier.
- META-PAN14 [34]: This is a simple heterogeneous ensemble reported by PAN organizers. It is based on the average of all 13 PAN-2014 submissions.
- Bagnall [2]: The winning approach of PAN-2015 [33]. It uses a multi-headed recurrent neural network language model.
- Moreau et al. [23]: The second-best submission of PAN-2015. It is an heterogeneous ensemble of 5 verification models.
- META-PAN15 [33]: A simple heterogeneous ensemble based on the average of all 18 PAN-2015 submissions.
- Potha and Stamatatos [26]: This is another modification of GI with improved results.
- Potha and Stamatatos [27]: This is a profile-based and intrinsic method using topic modeling.
- Ding et al. [7]: A neural network approach that jointly learns distributed word representations together with topical and lexical biases achieving improved results in PAN-2014 datasets.

4.3 Results

First, we compare the baselines with the stacking ensemble (based on a SVM meta-learner with its hyper-parameters tuned based on the training dataset[1])

[1] This is done for each PAN dataset separately. In all cases, an RBF kernel is selected.

as well as the proposed DES method. For the latter, thresholds are set to $a = 0.75$ and $b = 0.6$. Tables 4 and 5 present the evaluation results (AUC) per test dataset and the average performance over PAN-2014 and PAN-2015 datasets, respectively. Moreover, two versions of stacking and DES are reported: one using base verification models that have been tuned using the training dataset and another one using base models with fixed parameter settings.

Table 4. Evaluation results (AUC) on PAN-2014 datasets.

	DE	DR	EE	EN	GR	SP	Avg
Baselines							
Khonji and Iraqi (2014)	0.913	0.736	0.590	0.750	0.889	0.898	0.797
Fréry et al. (2014)	0.906	0.601	0.723	0.612	0.679	0.774	0.716
META-PAN14 (2014)	0.957	0.737	0.781	0.732	0.836	0.898	0.824
Potha and Stamatatos (2017)	0.976	0.685	0.762	0.767	0.929	0.878	0.833
Potha and Stamatatos (2018)	0.982	0.646	0.781	0.761	0.919	0.902	0.832
Ding et al. (2019)	**0.998**	0.658	0.887	0.767	0.924	0.934	0.876
Proposed ensembles							
Stacking_{tuned}	0.988	**0.890**	0.861	0.832	0.969	0.922	0.910
Stacking_{fixed}	0.986	0.854	0.818	0.828	0.965	0.940	0.898
DES_{tuned}	0.983	0.879	0.876	0.843	0.973	0.945	0.916
DES_{fixed}	0.985	0.885	**0.901**	**0.857**	**0.977**	**0.963**	**0.928**

As can be seen, the proposed DES method is the most effective one in most of the cases improving the best reported results for the specific datasets. Its performance is higher when fixed parameter settings are used in comparison to tuned models. This sounds reasonable since fixed models are less biased in the training dataset and the weight of each model is more reliably estimated. Nevertheless, DES based on tuned models also provides very good results. The stacking methods are also very effective surpassing in terms of average performance all baselines. In this case, the tuned models seem to be the best option. Again, this can be explained since the meta-learner needs as accurate base models as possible and tuned models are more likely to be more accurate than models with fixed settings. The improvement in average performance of the best ensemble models with respect to that of the best baselines is higher than 5% in the PAN-2014 datasets and more than 10% in the PAN-2015 datasets. It is also remarkable that the biggest improvement is achieved in datasets with very limited D_{known} size (PAN-2014-DR, PAN-2014-EE, PAN-2014-EN). All these indicate that the ensemble approach is much more reliable and effective in difficult verification cases where there are few known documents or documents belong to different domains.

We also examine the statistical significance of pairwise differences of all tested (both the proposed and baseline) methods using an approximate randomization

Table 5. Evaluation results (AUC) on PAN-2015 datasets.

	DU	EN	GR	SP	Avg
Baselines					
Bagnall (2015)	0.700	0.811	0.882	0.886	0.820
Moreau et al. (2015)	0.825	0.709	0.887	0.853	0.819
META-PAN15 (2015)	0.696	0.786	0.779	0.894	0.754
Potha and Stamatatos (2017)	0.709	0.798	0.844	0.851	0.801
Potha and Stamatatos (2018)	0.572	0.764	0.859	0.946	0.785
Proposed Ensembles					
Stacking_{tuned}	0.858	0.864	0.955	0.976	0.913
Stacking_{fixed}	0.814	0.867	0.968	0.977	0.907
DES_{tuned}	0.849	**0.898**	0.962	0.971	0.920
DES_{fixed}	**0.866**	0.879	**0.988**	**0.990**	**0.930**

test [24]. The null hypothesis assumes there is no difference between a pair of tested methods when each PAN dataset is considered separately. The baseline approach of Ding et al. (2019) [7] is not included in these tests since we did not have access to the original output of this method for each individual verification instance. In most of the cases, the proposed ensembles are significantly ($p < 0.05$) better than the baselines. Notable exceptions are PAN-2014-DE, where the proposed ensembles are not significantly better than the baselines of Potha and Stamatatos (2017) and Potha and Stamatatos (2018), and PAN-2015-DU, where the difference with Moreau et al. (2015) is not significant. On the other hand, the differences between the stacking and DES ensembles in most of the cases are not statistically significant. The full results of this analysis are not included here due to lack of space.

Next, we focus on the effect of thresholds a, b in the average performance of DES. Figure 1 depicts the average AUC (for all PAN-2014 and PAN-2015 datasets) of DES using either tuned or fixed base models for a range of threshold a values while we fix threshold $b = 0.6$. Recall that the higher threshold a is, the less similar training instances are retrieved. In case of very high values of a it is possible that the retrieved set of training instances is empty. In such cases, the test instance is left unanswered by getting a fix verification score = 0.5. This is in accordance with the evaluation setup of PAN shared tasks [33,34]. From the obtained results, it is clear that the fixed models are better than the tuned models in almost all examined cases. In addition, DES is clearly better than the best baseline for the whole range of threshold a values. With respect to the best stacking ensemble, it seems that DES is better in both datasets when $0.7 \leq a \leq 0.9$. This means that a should be set to a relatively large value to filter out most dissimilar training instances.

Figure 2 depicts the corresponding average performance of DES method, based either on tuned or fixed models, on PAN-2014 and PAN-2015 datasets varying

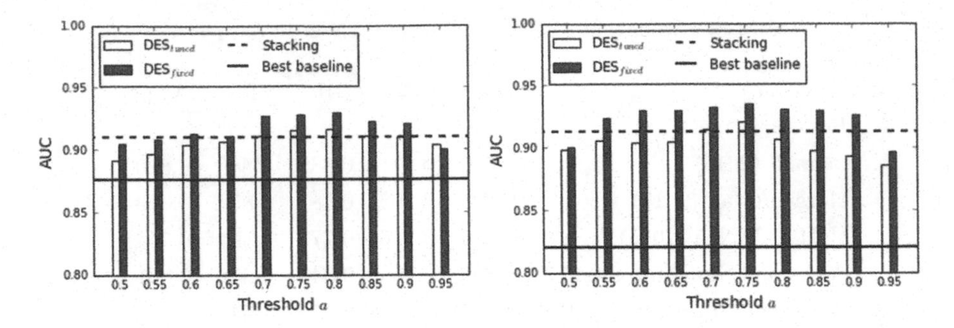

Fig. 1. Average AUC of DES method, using either tuned or fixed base models, on PAN-2014 (left) and PAN-2015 (right) datasets for varying threshold a. The performances of the best stacking ensemble (based on tuned models) and the best baseline are also shown.

threshold b while we fix threshold $a = 0.75$. Similar to the previous case, for high values of b, if none of the verifiers is selected for a test instance, then it is assigned a fix verification score $= 0.5$, namely it is left unanswered. Again, DES based on tuned models is outperformed by the DES using fixed models in almost all b values. In this case, this difference is higher in comparison to that of Fig. 1. This clearly shows that the tuned models are more biased in the training dataset and, therefore, less useful in DES. It is also important that the performance of DES remains better than the best baseline for the whole range of examined b values. As concerns the comparison to the stacking ensemble, we see that DES is better when $b \leq 0.6$. It seems that the performance of DES remains robust when b decreases, even in case it is set to zero. In that extreme case, all base models are taken into account. However, some of them (the ones with poor performance in the selected training instances) will be considered with very small weight, so practically they are filtered out. Actually, when $b = 0$ DES achieves comparatively good results. This means that it is possible to reduce the parameters of DES by setting $b = 0$ (use all base models) and still getting respectable performance.

5 Discussion

In this paper, we present author verification approaches based on ensemble learning. We collect a relatively large pool of 47 base verifiers covering the basic paradigms in this area, namely, both intrinsic and extrinsic methods as well as both instance-based and profile-based methods. This collection of verifiers provides a pluralism of verification scores and we attempt to take advantage of their correlations by building two ensembles. The first one is based on stacked generalization and learns patterns of agreement/disagreement among verifiers. In other words, it learns when to trust a verifier. The second, more sophisticated approach, is based on dynamic ensemble selection and attempts to find relevant training instances with respect to each verification case separately and then filters out verifiers that are not too specialized for the selected subset of instances.

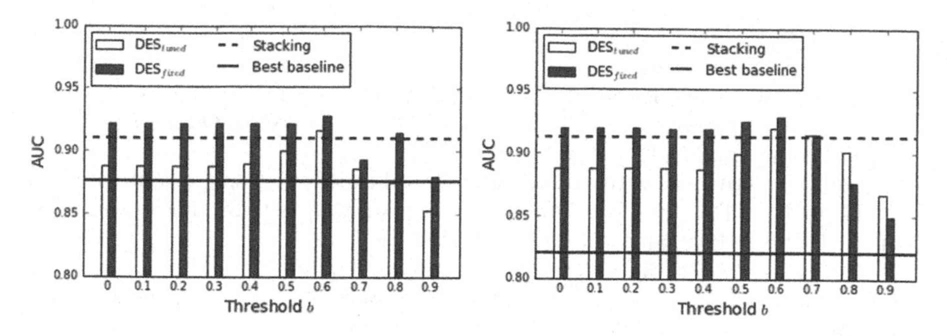

Fig. 2. Average AUC of DES method, using either tuned or fixed base models, on PAN-2014 (left) and PAN-2015 (right) datasets for varying threshold b. The performances of the best stacking ensemble (based on tuned models) and the best baseline are also shown.

Both ensemble approaches outperform a set of strong baselines according to experiments using ten PAN benchmark datasets. The performance of DES is (in average) more than 5% better than the best baseline in PAN-2014 datasets and more than 10% better in PAN-2015 datasets. Recall that PAN-2015 datasets consist of difficult cross-domain cases (where the known and unknown documents are about distant topics or belong to different genres). In addition, the performance of the proposed ensembles is much better than the strongest baseline in datasets where only one known document is provided (PAN-2014-DR, PAN-2014-EN, PAN-2015-EN). This indicates that our ensembles are able to handle challenging verification scenarios and are more robust than individual models.

DES has two parameters that control how many similar training instances will be retrieved and how many base classifiers will be considered. It has been shown that a relatively high threshold a value is required to filter out most irrelevant training instances. In addition, relatively good results are obtained when b takes low values including the case where $b = 0$. This means it does not harm to consider all possible base models given that their weight (determined by their performance on the similar training instances) will be quite low.

Our experiments demonstrate that the stacking ensemble works better with base models that are tuned to maximize performance in the training dataset. On the other hand, the DES method is more effective when fixed parameter settings are used in the base models. Tuned verifiers are biased in the training dataset and the estimation of their weight within DES becomes less reliable. This could be used to further enrich the pool of our base verifiers considering several versions of the same approach with different fixed parameter settings. Another possible future work direction is to try to combine the stacking and DES ensembles in a more complex approach.

References

1. Almishari, M., Oguz, E., Tsudik, G.: Fighting authorship linkability with crowdsourcing. In: Proceedings of the Second ACM Conference on Online Social Networks, COSN, pp. 69–82 (2014)
2. Bagnall, D.: Author identification using multi-headed recurrent neural networks. In: Cappellato, L., Ferro, N., Gareth, J., San Juan, E. (eds.) Working Notes Papers of the CLEF 2015 Evaluation Labs (2015)
3. Barbon, S., Igawa, R., Bogaz Zarpelão, B.: Authorship verification applied to detection of compromised accounts on online social networks: a continuous approach. Multimed. Tools Appl. **76**(3), 3213–3233 (2017)
4. Bartoli, A., Dagri, A., Lorenzo, A.D., Medvet, E., Tarlao, F.: An author verification approach based on differential features. In: Cappellato, L., Ferro, N., Gareth, J., San Juan, E. (eds.) Working Notes Papers of the CLEF 2015 Evaluation Labs (2015)
5. Brocardo, M., Traore, I., Woungang, I., Obaidat, M.: Authorship verification using deep belief network systems. Int. J. Commun. Syst. **30**(12) (2017). Article no. e3259
6. Castro-Castro, D., Arcia, Y.A., Brioso, M.P., Guillena, R.M.: Authorship verification, average similarity analysis. In: Recent Advances in Natural Language Processing, pp. 84–90 (2015)
7. Ding, S., Fung, B., Iqbal, F., Cheung, W.: Learning stylometric representations for authorship analysis. IEEE Trans. Cybern. **49**(1), 107–121 (2019)
8. Duman, S., Kalkan-Cakmakci, K., Egele, M., Robertson, W., Kirda, E.: Emailprofiler: Spearphishing filtering with header and stylometric features of emails. In: Proceedings - International Computer Software and Applications Conference, vol. 1, pp. 408–416 (2016)
9. Fréry, J., Largeron, C., Juganaru-Mathieu, M.: UJM at CLEF in author identification. In: Proceedings CLEF-2014, Working Notes, pp. 1042–1048 (2014)
10. Halvani, O., Graner, L., Vogel, I.: Authorship verification in the absence of explicit features and thresholds. In: Pasi, G., Piwowarski, B., Azzopardi, L., Hanbury, A. (eds.) ECIR 2018. LNCS, vol. 10772, pp. 454–465. Springer, Cham (2018). https://doi.org/10.1007/978-3-319-76941-7_34
11. Hernández, C.A., Calvo, H.: Author verification using a semantic space model. Computación y Sistemas **21**(2) (2017)
12. Hürlimann, M., Weck, B., van den Berg, E., Šuster, S., Nissim, M.: GLAD: groningen lightweight authorship detection. In: Cappellato, L., Ferro, N., Jones, G., San Juan, E. (eds.) CLEF 2015 Evaluation Labs and Workshop - Working Notes Papers. CEUR-WS.org (2015)
13. Jankowska, M., Milios, E., Keselj, V.: Author verification using common n-gram profiles of text documents. In: Proceedings of COLING 2014, the 25th International Conference on Computational Linguistics: Technical Papers, pp. 387–397 (2014)
14. Juola, P., Stamatatos, E.: Overview of the author identification task at PAN 2013. In: Working Notes for CLEF 2013 Conference (2013)
15. Kestemont, M., Luyckx, K., Daelemans, W.T.C.: Cross-genre authorship verification using unmasking. Engl. Stud. **93**(3), 340–356 (2012)
16. Khonji, M., Iraqi, Y.: A slightly-modified GI-based author-verifier with lots of features (ASGALF). In: CLEF 2014 Labs and Workshops, Notebook Papers. CLEF and CEUR-WS.org (2014)
17. Ko, A.H., Sabourin, R., de Souza Britto Jr., A.: From dynamic classifier selection to dynamic ensemble selection. Pattern Recogn. **41**(5), 1718–1731 (2008)

18. Kocher, M., Savoy, J.: A simple and efficient algorithm for authorship verification. J. Assoc. Inf. Sci. Technol. **68**(1), 259–269 (2017)
19. Koppel, M., Schler, J., Argamon, S., Winter, Y.: The fundamental problem of authorship attribution. Engl. Stud. **93**(3), 284–291 (2012)
20. Koppel, M., Schler, J., Bonchek-Dokow, E.: Measuring differentiability: unmasking pseudonymous authors. J. Mach. Learn. Res. **8**, 1261–1276 (2007)
21. Koppel, M., Winter, Y.: Determining if two documents are written by the same author. J. Am. Soc. Inf. Sci. Technol. **65**(1), 178–187 (2014)
22. Layton, R., Watters, P., Ureche, O.: Identifying faked hotel reviews using authorship analysis. In: Proceedings - 4th Cybercrime and Trustworthy Computing Workshop, CTC 2013, pp. 1–6 (2013)
23. Moreau, E., Jayapal, A., Lynch, G., Vogel, C.: Author verification: basic stacked generalization applied to predictions from a set of heterogeneous learners-notebook for PAN at CLEF 2015. In: CLEF 2015-Conference and Labs of the Evaluation forum. CEUR (2015)
24. Noreen, E.: Computer-Intensive Methods for Testing Hypotheses: An Introduction. Wiley, New York (1989)
25. Potha, N., Stamatatos, E.: A profile-based method for authorship verification. In: Likas, A., Blekas, K., Kalles, D. (eds.) SETN 2014. LNCS (LNAI), vol. 8445, pp. 313–326. Springer, Cham (2014). https://doi.org/10.1007/978-3-319-07064-3_25
26. Potha, N., Stamatatos, E.: An improved *impostors* method for authorship verification. In: Jones, G.J.F., et al. (eds.) CLEF 2017. LNCS, vol. 10456, pp. 138–144. Springer, Cham (2017). https://doi.org/10.1007/978-3-319-65813-1_14
27. Potha, N., Stamatatos, E.: Intrinsic author verification using topic modeling. In: Artificial Intelligence: Methods and Applications - Proceedings of the 10th Hellenic Conference on AI, SETN (2018)
28. Potha, N., Stamatatos, E.: Improving author verification based on topic modeling. J. Assoc. Inf. Sci. Technol. (2019)
29. Sanderson, C., Guenter, S.: Short text authorship attribution via sequence kernels, markov chains and author unmasking: an investigation. In: Proceedings of the International Conference on Empirical Methods in Natural Language Engineering, pp. 482–491 (2006)
30. Seidman, S.: Authorship verification using the impostors method. In: Forner, P., Navigli, R., Tufis, D. (eds.) CLEF 2013 Evaluation Labs and Workshop - Working Notes Papers (2013)
31. Stamatatos, E.: A survey of modern authorship attribution methods. J. Am. Soc. Inf. Sci. Technol. **60**, 538–556 (2009)
32. Stamatatos, E.: Authorship verification: a review of recent advances. Res. Comput. Sci. **123**, 9–25 (2016)
33. Stamatatos, E., et al.: Overview of the author identification task at PAN 2015. In: Working Notes of CLEF 2015 - Conference and Labs of the Evaluation Forum (2015)
34. Stamatatos, E., et al.: Overview of the author identification task at PAN 2014. In: CLEF Working Notes, pp. 877–897 (2014)
35. Stover, J.A., Winter, Y., Koppel, M., Kestemont, M.: Computational authorship verification method attributes a new work to a major 2nd century African author. J. Am. Soc. Inf. Sci. Technol. **67**(1), 239–242 (2016)
36. Tuccinardi, E.: An application of a profile-based method for authorship verification: investigating the authenticity of Pliny the Younger's letter to Trajan concerning the Christians. Digit. Scholarsh. Humanit. **32**(2), 435–447 (2017)
37. Wolpert, D.H.: Stacked generalization. Neural Netw. **5**, 241–259 (1992)

Structural Similarity Search for Formulas Using Leaf-Root Paths in Operator Subtrees

Wei Zhong(✉) and Richard Zanibbi

Rochester Institute of Technology, Rochester, USA
wxz8033@rit.edu, rlaz@cs.rit.edu

Abstract. We present a new search method for mathematical formulas based on Operator Trees (OPTs) representing the application of operators to operands. Our method provides (1) a simple indexing scheme using OPT leaf-root paths, (2) practical matching of the K largest common subexpressions, and (3) scoring matched OPT subtrees by counting nodes corresponding to visible symbols, weighting operators lower than operands. Using the largest common subexpression (K = 1), we outperform existing formula search engines for non-wildcard queries on the NTCIR-12 Wikipedia Formula Browsing Task. Stronger results are obtained when using additional subexpressions for scoring. Without parallelization or pruning, our system has practical execution times with low variance when compared to other state-of-the-art formula search engines.

Keywords: Mathematical Information Retrieval · Formula search · Similarity search · Subexpression matching

1 Introduction

Mathematical Information Retrieval (MIR [5,21]) requires specialized tasks including detecting and recognizing math in documents, math computation and knowledge search (e.g., in Wolfram Alpha), and similarity search for math expressions. Formula search engines are useful for looking up unfamiliar notation and math question answering.

Traditional text search engines are unaware of many basic characteristics of math formulas. Key problems in math formula similarity search include:

- How do we represent math formulas for search?
- How do we measure math formula similarity?
  - Structural similarity: Common subexpression(s), operator commutativity and operator associativity.
  - Symbol set similarity: Being aware of unifiable/interchangeable elements (e.g., $(1+1/n)^n$ and $(1+1/x)^x$), while still distinguishing $e=mc^2$ from $y=ax^2$; weighting identical symbols appropriately.

L. Azzopardi et al. (Eds.): ECIR 2019, LNCS 11437, pp. 116–129, 2019.
https://doi.org/10.1007/978-3-030-15712-8_8

  - Semantic similarity of mathematical formulas, including equivalent formulas (e.g., x^{-1} and $1/x$).
- What is a good trade-off between feature-based matching and costly structure matching, to identify similar formulas efficiently and effectively?

We present a new formula search engine based on Operator Trees (OPTs). OPTs represent the semantics of a formula, in terms of the application of operators to operands in an expression. We adapt the leaf-root path indexing of all subtrees used in MCAT [8], where these paths act as the retrieval units or "keywords" for formulas. MCAT uses additional encodings (e.g., Presentation MathML) that we do not consider in this work. Our scoring function generalizes subtree scoring methods which only consider single best matched tree such as the Maximum Subtree Similarity (MSS) of Tangent [22]. To the best of our knowledge, our model is the first using multiple common subexpressions to score formula hits. Our approach has achieved usable execution times using a single process without any dynamic pruning applied so far, and produces state-of-the-art results for non-wildcard queries in the NTCIR-12 Wikipedia Formula Browsing Task [20]. Our system is available for download.[1]

2 Related Work

There are two major approaches to math representation and indexing, *Text-based* and *Tree-based* [21]. Text-based approaches apply traditional text search engines, converting formulas to canonically ordered text strings with index augmentation [11,12,14] to deal with operator commutativity, operator associativity, and subexpression matching. Tree-based approaches index formulas directly from hierarchical representations of appearance and semantics. In the recent NTCIR-12 MIR tasks [20], tree-based MIR systems achieve the best accuracy.

Tree representations are primarily divided into SLTs (Symbol Layout Trees) and OPTs (Operator Trees). SLTs capture appearance based on the arrangement of symbols on writing lines (i.e., topology). OPTs represent semantics: internal nodes represent operators, and leaves represent operands. SLTs capture appearance with few ambiguities, and require few spatial relationships to represent structure. However, they cannot capture semantic equivalences, operator commutativity, or operator associativity. By representing operations explicitly, visually distinct but mathematically equivalent formulas have identical OPTs (e.g., $\frac{1}{x}$ and $1/x$) and operator commutativity is captured explicitly (e.g., allowing us to determine that $1 + x^2$ and $x^2 + 1$ are equivalent). OPT construction requires an expression grammar, which for real-world data needs to accommodate ambiguous and malformed expressions (e.g., unpaired parentheses). In our work, we parse LaTeX formulas into OPTs.

Measuring similarity using both SLTs and OPTs, Gao et al. [9] uses *sibling patterns* extracted from semi-OPTs (which do not identify implicit multiplication). They extract "level content" from OPTs, identifying the depth at which a

[1] Source code: https://github.com/approach0/search-engine/tree/ecir2019.

sibling pattern appears. Arguments are represented as wildcards. For example, at level one, $(x + y)z$ is represented by $(*) \times *$, at level two by $(*)$, and then recursively $* + *$. Extracted *(pattern, level)* tuples are used for search.

Some systems use leaf-root paths extracted from formula trees (*vertical paths*). Hijikata et al. [6] use leaf-root paths in Content MathML (a form of OPT), but to be considered as a match, a candidate path and query path must be identical. Similarly, OPMES [23,24] requires complete leaf-root query paths to match some *prefix* of candidate leaf-root paths. This allows for retrieving partially matched candidate expressions, but still requires a complete match in the query expression. Yokoi and Aizawa [19] adopt a more flexible approach, using all possible *subpaths* of leaf-root paths, and do not require all leaf-root paths in the query to be matched. The MCAT system of Kristianto et al. [8] combines path features (both ordered paths and unordered paths) generated from leaf-root paths, and also uses sibling patterns for search.

Stalnaker et al. [15] use *symbol pairs* extracted from SLTs, where node to ancestor symbol pairs along with their relative position in an SLT are used for search. Later Davila et al. [4,22] use labeled paths between symbols and generate symbol pairs falling within a given maximum path length (window size). In their approach, expressions are a candidate as long as they share one symbol pair with the query. This method has high recall due to low granularity in the search unit; however, it produces a large candidate set with few structural constraints, thus Davila et al. [22] introduce a second stage to rerank by structural similarity. They find an alignment between query and candidate formulas maximizing a similarity score with $O(|T_d||T_q|^2 \log T_q)$ time complexity, where T_q and T_d are query and candidate trees. Later they apply similar techniques in both SLTs and OPTs, and combine results to obtain better results in the *Tangent-S* system [4].

There are also techniques that capture structural similarity more precisely, e.g., Kamali et al. [7] use tree edit distance to measure differences between MathML DOM trees for formula similarity (in SLTs), however, the computation has non-linear time complexity in terms of expression size. How to determine the costs of edit operations to reflect similarity remains an open problem. There have been studies on similarity distance metrics that do not depend on edit operations, and subgraph-based graph similarity metrics have been explored for a long time in the pattern recognition literature [2].

3 Methodology

In our context, matching subexpressions means finding subtrees that are structurally identical and the matched nodes have the same tokens (we will use uppercase words to indicate tokens, e.g. variables x, y will both be VAR tokens after tokenization). To formally define our structure matching approach, we incorporate the graph/subtree isomorphism definition [13] and add a few definitions based on the *formula subtree* [23]. In addition to general subtree isomorphism, a formula subtree (indicated by $\preceq_l$) requires leaves in a subtree to be also mapped to leaves in the other tree.

Definition 1. *A common formula subtree of two formula trees T_q and T_d consists of two corresponding formula subtrees $\hat{T}_q$ of T_q and $\hat{T}_d$ of T_d where they are isomorphic and they are subgraphs of T_q and T_d respectively. Let $CFS(T_q, T_d)$ denote the set of all such common formula subtrees of T_q, T_d, i.e.,*
$CFS(T_q, T_d) = \{\hat{T}_q, \hat{T}_d : \hat{T}_q \preceq_l T_q, \hat{T}_d \preceq_l T_d, \hat{T}_q \cong \hat{T}_d, \hat{T}_q \subseteq T_q, \hat{T}_d \subseteq T_d\}$ where "$\cong$" and "$\subseteq$" indicate graph isomorphism and subgraph relation respectively.

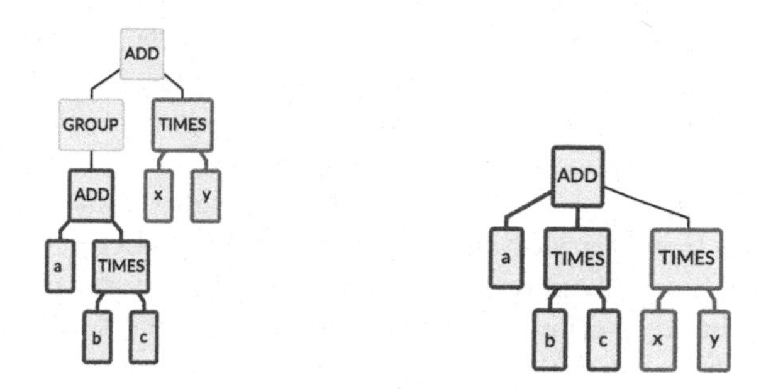

Fig. 1. Common formula forest in OPT. Left to right: $(a + bc) + xy$ and $a + bc + xy$.

Similar to common forest definitions [18], we define a form of disjoint common subtrees to describe multiple subexpression matches. Figure 1 illustrates two matching common subexpressions ($a + bc$ and xy), with the matches highlighted in blue and green. We call these matches a *common formula forest.* It consists of common formula subtree(s) identified by $(\hat{T}_q^i, \hat{T}_d^i)$ as defined below.

Definition 2. *A set of common formula subtrees π is called a common formula forest of two formula trees T_q and T_d,*

$$\pi = \{(\hat{T}_q^1, \hat{T}_d^1), (\hat{T}_q^2, \hat{T}_d^2), ...(\hat{T}_q^n, \hat{T}_d^n)\} \in \Pi(T_q, T_d) \tag{1}$$

iff for $i = 1, 2, ...n$:
(1) $\hat{T}_q^i, \hat{T}_d^i \in CFS(T_q, T_d)$
(2) $\hat{T}_q^1, \hat{T}_q^2, ...\hat{T}_q^n$ are disconnected, and $\hat{T}_d^1, \hat{T}_d^2, ...\hat{T}_d^n$ are disconnected.
where $\Pi(T_q, T_d)$ denote all possible common formula forests of T_q and T_d.

For our structural similarity metric, we want to find the "largest" common formula forest to represent the most similar parts of two math expressions. In order to define "large" generally, our similarity scoring formula between two formula trees is parameterized by some scoring function γ of $\pi \in \Pi(T_q, T_d)$.

Definition 3 (General multi-tree structure similarity). *The formula tree similarity of T_q and T_d given scoring function γ is*

$$\Gamma_\gamma(T_q, T_d) = \max_{\pi \in \Pi(T_q, T_d)} \gamma(\pi) \tag{2}$$

Intuitively, we choose the number of matched tree nodes to measure matched "size". Since the similarity contribution of different nodes (i.e. operands and operators) may be non-uniform, we propose using the similarity scoring function γ defined by

$$\gamma(\pi) = \sum_{(\hat{T}_q^i, \hat{T}_d^i) \in \pi} \beta_i \cdot \left(\alpha \cdot \text{internals}(\hat{T}_d^i) + (1-\alpha) \cdot \text{leaves}(\hat{T}_d^i)\right) \tag{3}$$

where internals(T) is the number of internal nodes/operators in T, leaves(T) is the number of leaves/operands in T, and $\alpha \in [0, 1]$ defines the contribution weight of operators. $\beta_i \geq 0$ are the contribution weights for different matched subexpressions. For the convenience of later discussion, we refer to trees in Eq. (3) indexed by i, e.g. $\hat{T}_d^i$, as the *i-th widest* match (in terms of number of matched leaves) in π. We set $\beta_1 \geq \beta_2 \geq ... \geq \beta_n$ in order to weight "wider" subexpressions higher. And in practice, it is wasteful to compute all terms in Eq. (3), if we assume the largest K matched subexpressions cover most of the total matched size, we obtain an approximate scoring function where only a subset of terms in Eq. (3) are computed by fixing $\beta_i = 0$ for $i \geq \min(n, K)$.

3.1 Subexpression Matching

Valiente [17] has shown $O(m+n)$ time complexity for computing similar multiple-tree similarity, but this requires matching vertex out degree (i.e., complete subtrees), which is too strict for retrieval, e.g., $a+b$ will not match $a+b+c$ because the operand number does not agree. To practically compute formula tree similarity, we propose an greedy algorithm.

Using paths as units, it is easier to count matched operands than operators (each matched operand is identified by a matched path), so we first greedily find a common formula forest π^* that consists of the widest common formula subtree (in terms of matched operands), and then calculate the corresponding number of matched operators. In order for π^* to be the optimizer in Eq. (2), it requires the following assumption.

Assumption 1. *If* $\pi^* = \{(\hat{T}_q^{1*}, \hat{T}_d^{1*}), (\hat{T}_q^{2*}, \hat{T}_d^{2*}) ... (\hat{T}_q^{n*}, \hat{T}_d^{n*})\} \in \Pi(T_q, T_d)$ *is the maximizer in Eq. (2) for* $\alpha = 0$ *and* $\beta_1 \gg \beta_2 \gg ... \gg \beta_n$*, then we assume* π^* *is also maximizer in Eq. (2) for all* $\alpha \neq 0$ *and all* $\beta_1 \geq \beta_2 \geq ... \geq \beta_n$.

Under this assumption, finding the widest matched subtrees in order will yield our defined formula tree similarity, while in reality, greedily finding widest matched subtrees may not maximize Eq. (3).

We also want to use paths to test identical structures efficiently. Let $\mathcal{P}(T)$ be all leaf-root paths from rooted tree T, and a matching between path sets S_1, S_2 is defined as bipartite graph $M(S_1, S_2, E)$ where E is edges representing assigned matches. In our context, two paths match if they are identical after tokenization (e.g. The OPTs represent $a+b$ and $x+y$ have the same set of tokenized leaf-root path "VAR/ADD"). To compare structure efficiently, we also assume that two subtrees are structurally identical if only their leaf-root paths match:

Assumption 2. *For any tree* T_q, T_d*, let* $S_q = \mathcal{P}(T_q), S_d = \mathcal{P}(T_d)$*, if there exists perfect matching* $M(S_q, S_d, E)$*, then we assume* $T_q \cong T_d$*.*

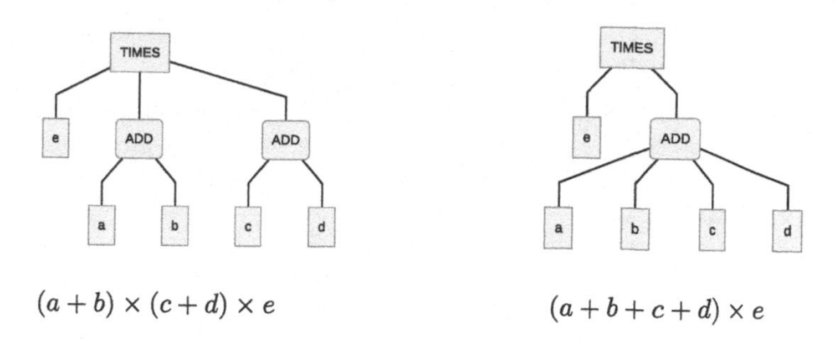

$(a + b) \times (c + d) \times e$ $\qquad$ $(a + b + c + d) \times e$

Fig. 2. Formulas with identical leaf-root paths, but different structure

This assumption does not always hold true (see Fig. 2), nevertheless, we expect this to be relatively rare in practice, and it allows us to design a practical algorithm for computing formula tree similarity.

Under Assumptions 1 and 2, it can be shown that if $\text{CFS}(T_q, T_d) \neq \emptyset$, and $S_q^m, S_d^m = \arg\max |E|$ for any matching $M(S_q^m \subseteq S_q, S_d^m \subseteq S_d, E)$, where $S_q = \mathcal{P}(\hat{T}_q), S_d = \mathcal{P}(\hat{T}_d), \hat{T}_q, \hat{T}_d \in \text{CFS}(T_q, T_d)$ then $\text{leaves}(\hat{T}_q^{1*}) = \text{leaves}(\hat{T}_d^{1*}) = |S_q^m| = |S_d^m|$. In other words, we can use leaf-root paths from query and document OPT subtrees to get the number of leaves of the widest matched tree in a common formula forest π^* that maximizes scoring function γ in Eq. (2). After $\text{leaves}(\hat{T}_d^{1*})$ is obtained, we can exclude already matched paths and similarly compute other $\text{leaves}(\hat{T}_d^{i*}), i = 2, 3...k$. The process of matching, i.e., finding S_q^m, S_d^m in any $M(S_q, S_d, E)$, can be implemented using bit masks and the output value $|S_q^m|$ does not depend on input order (matching order).

In scoring function (3), we also want the number of operators associated with matched leaves. Adding the number of matched operators in Eq. (3) helps better assess similarity when Assumption 2 fails. Consider the example in Fig. 2: only one of the two "ADD" operators on the left tree can match the "ADD" operator on the right. If we count the matched operators correctly, we can differentiate the two expressions in Fig. 2. To calculate the number of operators, assume we have found a common formula forest π^* that maximizes function γ, then we go through all subtree pairs (T_q^x, T_d^y) rooted at $x \in T_q, y \in T_d$, and examine if it joins with any pair of matched trees in π^* by looking at whether their leaves intersect. If true, we will count x, y as matched operators if both of them are not marked as matched yet.

Algorithm 1 describes our matching procedure in detail. In experiments, we found that counting only visible operators improves results in most cases. This is because some internal OPT nodes do not appear in the rendered expression, so counting them will bias the similarity measurement in our model. In particular, we do not count SUBSCRIPT and SUPERSCRIPT operator nodes.

Algorithm 1. Formula tree matching algorithm

```
Let (S_q^m, S_d^m) be the maximum matching path set of given path set (S_q, S_d).
Define ℓ(S) to be all the leaf nodes (equivalently, path IDs) for path set S.
function OPERANDMATCH(Q^m, D^m, L, k, leavesCounter)
    Q^X := { }, D^X := { }                                  ▷ Excluded path set
    for i < k do
        Q^max, D^max, max := 0                              ▷ Best matched tree records
        for (S_q, S_d) from L do
            if Q^X ∩ ℓ(S_q) = ∅ and D^X ∩ ℓ(S_d) = ∅ then   ▷ Disjoint tree pairs
                if |S_q^m| > max then                       ▷ Greedily find widest matches
                    max := |S_q^m|
                    Q^max, D^max := ℓ(S_q^m), ℓ(S_d^m)
        if max > 0 then
            Q^X := Q^X ∪ Q^max
            D^X := D^X ∪ D^max
            Q_i^m, D_i^m := Q^max, D^max
            leavesCounter[i] = max
        else                                                ▷ No more possible operand matchings
            break
    return Q^m, D^m, leavesCounter
function OPERATORMATCH(Q^m, D^m, L, k, operatorsCounter)
    Let Q^map, D^map be maps of matched internal nodes, initially empty.
    for (S_q, S_d) from L do
        for i < k do
            if Q_i^m ∩ ℓ(S_q) ≠ ∅ and D_i^m ∩ ℓ(S_d) ≠ ∅ then   ▷ Joint tree pairs
                Let n_q, n_d be the root-end nodes of S_q, S_d respectively.
                if Q^map[n_q], D^map[n_d] are both empty then
                    Q^map[n_q], D^map[n_d] := n_d, n_q
                    if visible(n_q) then
                        operatorsCounter[i] := operatorsCounter[i] + 1
                break
    return operatorsCounter
function FORMULATREEMATCH(T_q, T_d, k)
    for i < k do
        Q_i^m := { }, D_i^m := { }          ▷ Matched path set for i-th largest matched tree
        leavesCounter[i] := 0
        operatorsCounter[i] := 0
    L := List of (S_q, S_d) where S_q, S_d ∈ P(T_q^x), P(T_d^y) for each node x ∈ T_q, y ∈ T_d.
    Q^m, D^m, leavesCounter := OPERANDMATCH(Q^m, D^m, L, k, leavesCounter)
    operatorsCounter := OPERATORMATCH(Q^m, D^m, L, k, operatorsCounter)
    return leavesCounter[i], operatorsCounter[i] for i = 1, 2, ...k
```

Algorithm 1 avoids counting those nodes by consulting a pre-built "visibility" mapping for operators (i.e., the VISIBLE function).

3.2 Indexing and Retrieval

At the indexing stage, every math expression in the corpus is parsed into an OPT T_d. For all internal (operator) nodes n in T_d, we extract all leaf-root paths of T_d^n rooted at n. This path set $S = \bigcup_{n \in T_d} \mathcal{P}(T_d^n)$ is *tokenized* (e.g. operand symbols a, b, c are tokenized into VAR, operators fraction and division ($\div$) are tokenized into FRAC) by pre-defined OPT parser rules[2] to allow results from unification/substitution and boost recall. Each unique tokenized path is associated with a posting list, where the IDs of expressions containing the path are

[2] Our expression grammar has roughly 100 grammar rules and 50 token types.

stored. The IDs of endpoint nodes (leaf and operator) of each path are also stored in the posting lists. This allows the structure of matched subexpressions to be recovered from hit paths at the posting list merge stage.

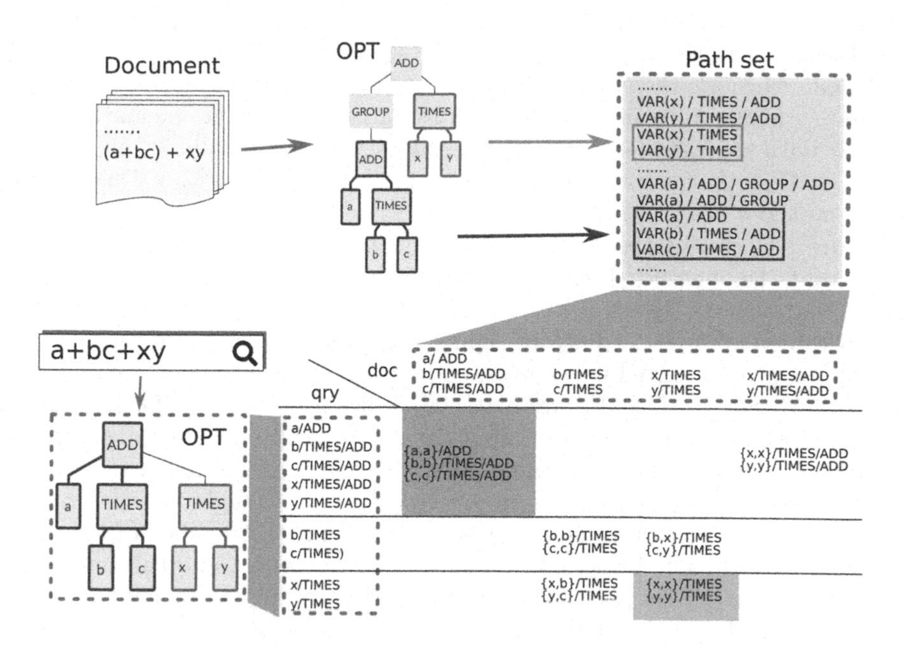

Fig. 3. Illustration of path retrieval and subexpression matching. After matching the largest common subexpression, i.e., $a + bc$ (in blue), the remaining largest disjoint common subexpression is xy (in green). (Color figure online)

During query processing, a query expression tree T_q is decomposed in the same way. Posting lists associated to its tokenized path set are retrieved and merged. During merging, we examine the matched paths from a document expression one at a time, input as list L in Algorithm 1 and compute the structural matching. Then we compute the overall similarity score (considering both structural and symbolic similarity) as follows:

$$\frac{S_{\text{st}} S_{\text{sy}}}{S_{\text{st}} + S_{\text{sy}}} \left[(1-\theta) + \theta \frac{1}{\log(1 + \text{leaves}(T_d))} \right], \qquad \theta \in [0, 1] \tag{4}$$

where structure similarity S_{st} is normalized formula tree similarity

$$S_{\text{st}} = \begin{cases} \frac{\Gamma_\gamma(T_q, T_d)}{\text{leaves}(T_q)} & \text{if } \alpha = 0 \\ \\ \frac{\Gamma_\gamma(T_q, T_d)}{\text{leaves}(T_q) + \text{internals}(T_q)} & \text{if } \alpha \neq 0 \end{cases} \tag{5}$$

and S_{sy} is the normalized *operand* symbol set similarity score $y \in [0, 1]$ produced from the Mark-and-Cross algorithm [23] which scores exact symbol matches higher than unified symbol matches:

$$S_{\text{sy}} = \frac{1}{1 + (1 - y)^2} \quad (6)$$

The final scoring function (4) is a F-measure form of structure similarity and symbol set similarity combination, partially (θ) penalized by document math formula size measured by total number of its operands, i.e. leaves(T_d).

We can calculate the maximum matchings using bit operations if we assume the number of operands and the number of subexpressions that one math expression can have are less than a constant. And because the number of elements in L is $|T_q| \times |T_d|$, after maximum matchings are obtained, Algorithm 1 has overall time complexity $O(k|T_q||T_d|)$.

Figure 3 illustrates the path retrieval and subexpression matching process. Notice that the operands/leaves are not shown as tokenized in some places, so that we can identify which paths are matched. Algorithm 1 can be visualized using a table (as shown at bottom right), where pairs of matched {query, document} paths are inserted into corresponding cells when we merge posting lists (e.g. b/TIMES and x/TIMES are matched, indicated by {b, x}/TIMES). Each table cell represents an element of input list L in Algorithm 1. At the end of the algorithm, we obtain the highlighted cells with the largest number of matched leaves. Then the matched operators are counted for highlighted cells. Finally, we calculate the structural similarity score S_{st} from the number of operators and operands associated with each matched subexpression, and the symbol set similarity score S_{sy} from matched operands symbolic differences, then plug these into Eq. (4) to obtain the final similarity score for ranking.

4 Evaluation

We evaluate our system using the NTCIR-12 MathIR Wikipedia Formula Browsing Task (in the following, use *NTCIR-12* for short), which is the most current benchmark for isolated formula retrieval. The dataset contains over 590,000 math expressions taken from English Wikipedia. We consider all the 20 non-wildcards queries in NTCIR-12. During the task, pooled hits from participating systems were each evaluated by two human assessors. Assessors score a hit from highly relevant to irrelevant using 2, 1, or 0. The final hit relevance rating is the sum of the two assessor scores (between 0 and 4), with scores of 3 or higher considered *fully relevant* and other scores of 1 and higher considered *partially relevant.* We use *bpref* [1] on top-1000 results as our primary effectiveness metric because our system does not contribute to pooling, and bpref is computed over only judged hits. In addition to *standard* Precision@K values, we compute Precision@K metrics using only judged hits (*condensed*), and provide *upper bound* values by treating unjudged hits as relevant [10].

First, we explored the impact of different parameter values using up to 3-tree matching ($K = 3$). The θ parameter for penalizing overly large formulas is fixed at 0.05. Figure 4 shows representative parameter values that we have tried. We started with a single tree match (first four rows in table at left), finding that weighting operator symbol matches slightly lower than operands ($\alpha = 0.4$) have

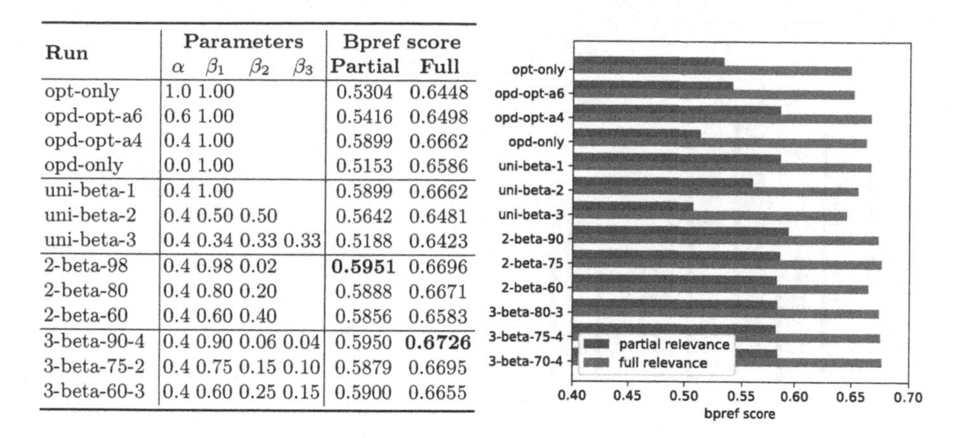

Run	Parameters				Bpref score	
	α	β_1	β_2	β_3	Partial	Full
opt-only	1.0	1.00			0.5304	0.6448
opd-opt-a6	0.6	1.00			0.5416	0.6498
opd-opt-a4	0.4	1.00			0.5899	0.6662
opd-only	0.0	1.00			0.5153	0.6586
uni-beta-1	0.4	1.00			0.5899	0.6662
uni-beta-2	0.4	0.50	0.50		0.5642	0.6481
uni-beta-3	0.4	0.34	0.33	0.33	0.5188	0.6423
2-beta-98	0.4	0.98	0.02		**0.5951**	0.6696
2-beta-80	0.4	0.80	0.20		0.5888	0.6671
2-beta-60	0.4	0.60	0.40		0.5856	0.6583
3-beta-90-4	0.4	0.90	0.06	0.04	0.5950	**0.6726**
3-beta-75-2	0.4	0.75	0.15	0.10	0.5879	0.6695
3-beta-60-3	0.4	0.60	0.25	0.15	0.5900	0.6655

Fig. 4. Relevance results from representative parameter values (table and bar graph).

produced the best results (we tried α in $[0, 1]$ using an increment of 0.1). We then fixed $\alpha = 0.4$, $\sum_i^n \beta_i = 1$ and tried uniform weights (rows 5–7) and non-uniform weights for two trees (rows 8–10) and three trees (rows 11–13). We examined uniform β weights for multiple matches from K = 1 to 3. For non-uniform weights in two-trees, we consider β_1 in $[0.5, 0.99]$ using increments of 0.05 or 0.01; for three-tree matching, we considered β_1 in $[0.5, 0.95]$ and β_2 in $[0.05, 0.45]$ using increments of 0.05. Figure 4 shows uniform weights generally yield worse results than non-uniform ones. And two runs from non-uniform weights when K = 2 and 3 obtain the best partial and full relevance scores respectively. This observation is intuitive because our setting non-uniform weights emphasizes larger subexpressions, which arguably have more visual impact.

Second, to illustrate the effect of matching multiple subexpressions, Fig. 5 shows changes in fully relevant bpref scores for different queries, when changing the maximum number of matched trees (K) with uniform weights. Figure 5 omits queries whose score remains unchanged or differs negligibly across values of K. We can observe that different queries have different behaviours as K increases, e.g., introducing secondary matching into queries 4 and 6 improves results, while multi-tree matching hurts performance noticeably in queries 16, 18 and 20. Looking at the queries in Fig. 6, due to the differences in their structural complexity, extracting partial components in queries 4 and 6 produces better similarity than matching partial components in more complex queries (e.g., 16 and 18). This makes Queries 4 and 6 benefit from multiple-tree scoring while queries 16 and 18 perform better using a single tree.

Table 1 compares our system with two other state-of-the-art formula search engines. Our model is able to outperform both of them in bpref full relevance and partial relevance. We compare our best runs for $K = 1, 2, 3$ (uni-beta-1,

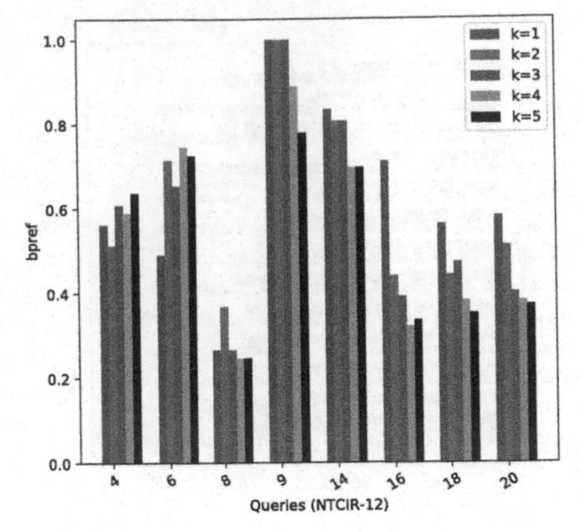

Fig. 5. NTCIR-12 Full relevance scores for matching uniformly-weighted subtrees (1 to 5 trees).

No.	Query Formula
4)	$\nabla \times \mathbf{B} = \mu_0 \mathbf{J} + \underbrace{\mu_0 \epsilon_0 \frac{\partial}{\partial t} \mathbf{E}}_{\text{Maxwell's term}}$
6)	${}^{238}_{92}\mathrm{U} + {}^{64}_{28}\mathrm{Ni} \rightarrow {}^{302}_{120}\mathrm{Ubn}^* \rightarrow \ldots$
16)	$\tau_{\mathrm{rms}} = \sqrt{\frac{\int_0^\infty (\tau - \overline{\tau})^2 A_c(\tau) d\tau}{\int_0^\infty A_c(\tau) d\tau}}$
18)	$P_i^x = \frac{N!}{n_x!(N-n_x)!} p_x^{n_x} (1 - p_x)^{N-n_x}$

Fig. 6. A few example queries in Fig. 5.

2-beta-98, 3-beta-90-4) with the *Tangent-S* system[3] and the best performing system at NTCIR-12, *MCAT* [20]. Using only one subexpression match (uni-beta-1), we outperform the other systems in bpref score. Although lower bound Precision@k values are lower for our system and Tangent-S partly due to some relevant hits being unjudged (all MCAT results are judged), we can achieve equal or better condensed scores in all the full relevance evaluations, and potentially can have higher precision than the other two systems according to upper bound values.

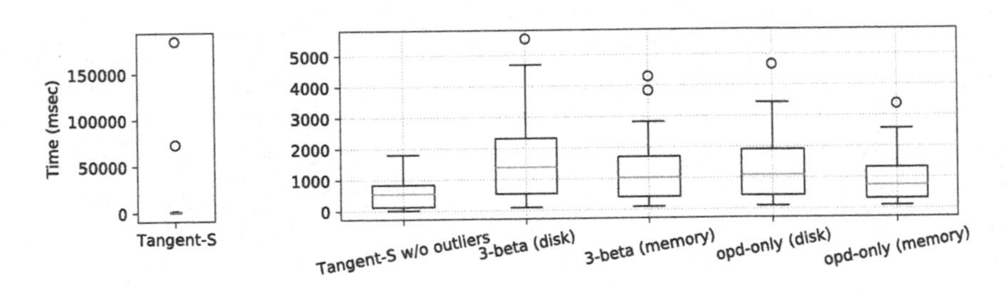

Fig. 7. Query processing times in milliseconds for the 20 NTCIR-12 queries

In terms of efficiency, MCAT reportedly has a median query execution time of 25 s, using a server machine and multi-threading [8]. Figure 7 shows query run times for our system and Tangent-S in the same environment using a single thread (Intel Core i5 CPU @ 3.8 GHz each core, DDR4 32 GB RAM, 256 GB SSD

[3] Tangent-S is an improved version of the Tangent system [3] that participated in NTCIR-12.

Table 1. NTCIR-12 Wikpiedia Formula Browsing Task Results (top-1000 hits). $k - \beta$ represents our best run using k matched subtrees in scoring.

Metrics		Fully relevant					Partially relevant				
		$1-\beta$	$2-\beta$	$3-\beta$	MCAT	Tangent-S	$1-\beta$	$2-\beta$	$3-\beta$	MCAT	Tangent-S
Bpref		0.6662	0.6696	**0.6726**	0.5678	0.6361	0.5899	**0.5951**	0.5950	0.5698	0.5872
P@5	Standard	0.4000	0.4000	0.4000	**0.4800**	**0.4800**	0.5300	0.5300	0.5300	**0.9500**	0.7900
	Condensed	0.5400	0.5400	**0.5400**	0.4800	0.5200	0.8900	0.9000	0.9000	**0.9500**	0.9300
	Upper bound	**0.8400**	**0.8400**	**0.8400**	0.4800	0.6500	**0.9700**	**0.9700**	**0.9700**	0.9500	0.9600
P@10	Standard	0.2850	0.2800	0.2900	**0.3550**	0.3500	0.4600	0.4650	0.4650	**0.8650**	0.7000
	Condensed	0.4050	0.4050	**0.4150**	0.3550	**0.4150**	0.8600	0.8650	0.8600	0.8650	**0.9200**
	Upper bound	**0.7850**	0.7750	**0.7850**	0.3550	0.5850	**0.9600**	**0.9600**	**0.9600**	0.8650	0.9350
P@15	Standard	0.2200	0.2233	0.2233	0.2867	**0.2900**	0.3967	0.4067	0.4100	**0.8333**	0.6433
	Condensed	0.3367	0.3433	**0.3467**	0.2867	0.3233	0.8233	0.8333	0.8333	0.8333	**0.8633**
	Upper bound	**0.7833**	0.7800	0.7767	0.2867	0.5600	0.9600	**0.9633**	**0.9633**	0.8333	0.9133
P@20	Standard	0.1950	0.1950	0.1900	**0.2450**	0.2300	0.3775	0.3850	0.3800	**0.8100**	0.6050
	Condensed	0.3125	**0.3175**	**0.3175**	0.2450	0.2825	0.8000	0.7950	0.7925	0.8100	**0.8350**
	Upper bound	**0.7800**	**0.7800**	**0.7800**	0.2450	0.5350	0.9625	**0.9700**	**0.9700**	0.8100	0.9100

drive). We compare our most effective run for full-relevance bpref scores (3-beta-90-4), and the most efficient run opd-only which only matches the single largest subtree, counting only leaves for structure scoring. Both of our runs have two versions, one with posting lists read from disk, and another where posting lists are cached in memory. Tangent has two substantial outlier queries (due to the non-linear complexity of its structure alignment algorithm), although it is faster in general. However, our execution times are more consistent, with a median time of about 1.5 s or less. Our higher typical run time is likely caused by the large number of query "keywords", as the query path set contains all leaf-root paths in all subtrees. Our in-memory posting lists are compressed by Frame-Of-Reference variances [25]. The in-memory version reduces the variance in run times, but the relatively small shift in median times suggests our system is more computation-bound than IO-bound. Our on-disk path index is stored as a naive file-system directory hierarchy where each posting list is a single uncompressed file. The on-disk index takes about 0.8 GB in a reiserFS partition.

5 Conclusion and Future Work

We have introduced a math formula search engine that obtains state-of-the-art results using simple and consistent path-based indexing. Our system uses a novel structural matching scheme that incorporates multiple subtree matches. It achieves better results when considering only visible symbols, and giving greater weight to operands than operators. Our algorithm allows trading-off between effectiveness and efficiency by changing the maximum number of matched subexpressions, or choosing to count matched operators or not. Because the current system examines *all* hits and merges posting lists without any skipping, and our query path set is typically large, there may be great potential in single-process efficiency if we can skip documents and avoid unnecessary computations (e.g. by applying dynamic pruning techniques such as MaxScore [16]). In the future we will extend our retrieval model to support query expansion of math synonyms to improve recall (e.g. expand $1/x$ for x^{-1}), and provide support for wildcard symbols in queries.

References

1. Buckley, C., Voorhees, E.M.: Retrieval evaluation with incomplete information. In: Proceedings of the 27th Annual International ACM SIGIR Conference on Research and Development in Information Retrieval, pp. 25–32. ACM (2004)
2. Bunke, H., Shearer, K.: A graph distance metric based on the maximal common subgraph. Pattern Recogn. Lett. **19**(3–4), 255–259 (1998)
3. Davila, K.: Tangent-3 at the NTCIR-12 MathIR Task (2016). http://research.nii.ac.jp/ntcir/workshop/OnlineProceedings12/pdf/ntcir/MathIR/06-NTCIR12-MathIR-DavilaK.pdf
4. Davila, K., Zanibbi, R.: Layout and semantics: combining representations for mathematical formula search. In: Proceedings of the 40th International ACM SIGIR Conference on Research and Development in Information Retrieval, pp. 1165–1168. ACM (2017)
5. Guidi, F., Sacerdoti Coen, C.: A survey on retrieval of mathematical knowledge. In: Kerber, M., Carette, J., Kaliszyk, C., Rabe, F., Sorge, V. (eds.) CICM 2015. LNCS (LNAI), vol. 9150, pp. 296–315. Springer, Cham (2015). https://doi.org/10.1007/978-3-319-20615-8_20
6. Hijikata, Y., Hashimoto, H., Nishida, S.: An investigation of index formats for the search of MathML objects. In: 2007 IEEE/WIC/ACM International Conferences on Web Intelligence and Intelligent Agent Technology, pp. 244–248, November 2007
7. Kamali, S., Tompa, F.W.: Structural similarity search for mathematics retrieval. In: Carette, J., Aspinall, D., Lange, C., Sojka, P., Windsteiger, W. (eds.) CICM 2013. LNCS (LNAI), vol. 7961, pp. 246–262. Springer, Heidelberg (2013). https://doi.org/10.1007/978-3-642-39320-4_16
8. Kristianto, G., Topic, G., Aizawa, A.: MCAT Math Retrieval System for NTCIR-12 MathIR Task, June 2016
9. Lin, X., Gao, L., Hu, X., Tang, Z., Xiao, Y., Liu, X.: A mathematics retrieval system for formulae in layout presentations. In: Proceedings of the 37th International ACM SIGIR Conference on Research & Development in Information Retrieval, SIGIR 2014. ACM, New York (2014)
10. Lu, X., Moffat, A., Culpepper, J.S.: The effect of pooling and evaluation depth on IR metrics. Inf. Retr. **19**(4), 416–445 (2016). https://doi.org/10.1007/s10791-016-9282-6
11. Miller, B.R., Youssef, A.: Technical aspects of the digital library of mathematical functions. Ann. Math. Artif. Intell. **38**(1–3), 121–136 (2003). https://link.springer.com/article/10.1023/A:1022967814992
12. Misutka, J., Galambos, L.: Extending Full Text Search Engine for Mathematical Content, pp. 55–67, January 2008
13. Shamir, R., Tsur, D.: Faster subtree isomorphism. J. Algorithms **33**(2), 267–280 (1999)

14. Sojka, P., Líška, M.: Indexing and searching mathematics in digital libraries. In: Davenport, J.H., Farmer, W.M., Urban, J., Rabe, F. (eds.) CICM 2011. LNCS (LNAI), vol. 6824, pp. 228–243. Springer, Heidelberg (2011). https://doi.org/10.1007/978-3-642-22673-1_16
15. Stalnaker, D., Zanibbi, R.: Math expression retrieval using an inverted index over symbol pairs. In: Document recognition and retrieval XXII, vol. 9402, p. 940207. International Society for Optics and Photonics (2015)
16. Turtle, H., Flood, J.: Query evaluation: strategies and optimizations. Inf. Process. Manage. **31**(6), 831–850 (1995). https://doi.org/10.1016/0306-4573(95)00020-H
17. Valiente, G.: An efficient bottom-up distance between trees. In: Proceedings of Eighth Symposium on String Processing and Information Retrieval, pp. 212–219, November 2001
18. Valiente Feruglio, G.A.: Simple and Efficient Tree Comparison (2001)
19. Yokoi, K., Aizawa, A.: An approach to similarity search for mathematical expressions using MathML. In: Towards a Digital Mathematics Library, Grand Bend, Ontario, Canada, 8–9th July 2009, pp. 27–35 (2009)
20. Zanibbi, R., Aizawa, A., Kohlhase, M., Ounis, I., Topic, G., Davila, K.: NTCIR-12 MathIR task overview. In: NTCIR (2016)
21. Zanibbi, R., Blostein, D.: Recognition and retrieval of mathematical expressions. Int. J. Doc. Anal. Recognit. **15**(4), 331–357 (2012)
22. Zanibbi, R., Davila, K., Kane, A., Tompa, F.W.: Multi-stage math formula search: using appearance-based similarity metrics at scale. In: Proceedings of the 39th International ACM SIGIR Conference on Research & Development in Information Retrieval. SIGIR 2016. ACM, New York (2016)
23. Zhong, W., Fang, H.: A novel similarity-search method for mathematical content in LaTeX markup and its implementation. Master's thesis, University of Delaware (2015)
24. Zhong, W., Fang, H.: OPMES: a similarity search engine for mathematical content. In: Ferro, N., et al. (eds.) ECIR 2016. LNCS, vol. 9626, pp. 849–852. Springer, Cham (2016). https://doi.org/10.1007/978-3-319-30671-1_79
25. Zukowski, M., Heman, S., Nes, N., Boncz, P.: Super-scalar RAM-CPU cache compression. In: Proceedings of the 22nd International Conference on Data Engineering, 2006. ICDE 2006, p. 59. IEEE (2006)

Recommender Systems I

PRIN: A Probabilistic Recommender with Item Priors and Neural Models

Alfonso Landin(✉), Daniel Valcarce, Javier Parapar, and Álvaro Barreiro

Information Retrieval Lab, Department of Computer Science, University of A Coruña, A Coruña, Spain
{alfonso.landin,daniel.valcarce,javierparapar,barreiro}@udc.es

Abstract. In this paper, we present PRIN, a probabilistic collaborative filtering approach for top-N recommendation. Our proposal relies on continuous bag-of-words (CBOW) neural model. This fully connected feedforward network takes as input the item profile and produces as output the conditional probabilities of the users given the item. With that information, our model produces item recommendations through Bayesian inversion. The inversion requires the estimation of item priors. We propose different estimates based on centrality measures on a graph that models user-item interactions. An exhaustive evaluation of this proposal shows that our technique outperforms popular state-of-the-art baselines regarding ranking accuracy while showing good values of diversity and novelty.

Keywords: Collaborative filtering · Neural models · Centrality measures

1 Introduction

In recent years, the way users interact with different services has shifted from a proactive approach, where users were actively looking for content, to one where users play a more passive role receiving content suggestions. This transformation has been possible thanks to the advances in the field of Recommender Systems (RS). These models produce personalized item recommendations based on user-item past interactions.

Approaches to item recommendation are usually classified in three families [2]. The first algorithms, content-based systems, use item metadata to produce tailored recommendations [13]. The second family, collaborative filtering (CF), exploits the past interactions of the users with the items to compute recommendations [21,26]. These interactions can be ratings, clicks, purchases, reproductions, etc. The third family, hybrid systems, combines techniques from the other two approaches to generate recommendations.

Collaborative filtering algorithms, which is the focus of this paper, can, in turn, be divided into two types. Model-based techniques, which build predictive models from the interaction data, and neighborhood-based techniques [26]

L. Azzopardi et al. (Eds.): ECIR 2019, LNCS 11437, pp. 133–147, 2019.
https://doi.org/10.1007/978-3-030-15712-8_9

(also called memory-based methods), which exploit past interactions directly. Neighborhood-based techniques rely on similar users or items, the neighborhoods, to compute the recommendations.

In this paper, we address the item recommendation task by proposing a model-based collaborative filtering technique. Our method is inspired by a word embedding model recently developed in the field of Natural Language Processing: the continuous bag-of-words (CBOW) model. Word embedding models are capable of learning word representations that take the form of dense vectors, called embeddings. Embeddings have much lower dimensionality than traditional sparse one-hot and bag-of-words representations and, moreover, are effective state-of-the-art methods in several tasks [22,24,25]. In particular, `word2vec` [24,25] has attracted great attention because of their efficiency and effectiveness. This tool provides two different models for generating word embeddings: the continuous bag-of-words model, which is designed to predict a word given its context, and the skip-gram model, which aims to predict the context of a word. When working on textual data, the context of a word in a document is composed of the surrounding words inside a fixed window.

In this paper, we propose PRIN, Probabilistic Recommender with Item Priors and Neural Models, a novel probabilistic model for the top-N recommendation task. PRIN uses the neural network topology of the CBOW model. However, instead of generating embeddings, we use the network to compute the conditional probabilities of the users given an item. Our probabilistic model requires the estimation of item prior probabilities to perform the Bayesian inversion of the conditional probabilities. To compute these item priors, we develop a graph-based interpretation of the user-item interactions. Over that graphs, we propose several estimates of item priors based on well-known graph centrality measures.

One additional advantage of the PRIN model is that it can be computed by leveraging current `word2vec` CBOW implementations. Moreover, our adaptation is even able to incorporate graded preference values (such as ratings) into the neural model.

Experiments are conducted on three datasets from different domains, with distinct sizes and sparsity figures. We show that our model can outperform several state-of-the-art collaborative baselines in ranking accuracy while maintaining good values of novelty and diversity. Moreover, PRIN inherits the efficiency and scalability of the CBOW model. For the sake of reproducibility, we also make our software publicly available[1].

2 Related Work

In this section, we present previous work on embeddings models, initially proposed in Natural Language Processing and nowadays commonly used in Recommender Systems. After that, we introduce graph centrality measures, whose aim is to reveal the importance of a node in a graph.

[1] https://gitlab.irlab.org/alfonso.landin/prin.

2.1 Embeddings Models

Word and documents were traditionally represented using sparse high-dimensional vectors based on one-hot and bags-of-words (BOW [16]) models. However, nowadays, neural embedding methods provide more effective fixed-length dense vector representations [24,25,28]. In particular, the `word2vec` tool [24,25] implements efficient estimations of the continuous bag-of-words (CBOW) and the skip-gram (SG) word embedding models. While the SG model aims to predict the surrounding words within a fixed window, the CBOW model predicts the actual word given the surrounding words [24]. The neural network architecture of these models is the same: a fully connected feedforward network with a single hidden layer. The size of the input and output layers is the size of the vocabulary, and the size of the hidden layer size is given by the desired number of dimensions of the embeddings.

Our proposal approaches the recommendation task from a different perspective than previous efforts that used embedding models in collaborative filtering. In particular, the SG model has been previously adapted in [3,14] for the generation of item embeddings. In both cases, the methods discard the model once trained, and the embeddings are merely used with some memory-based techniques in the case of [14] and for category classification in [3]. In our proposal, we use the output of the neural model in combination with the item priors to produce the ranking of recommended items for a user in a model-based approach. Moreover, previous approaches do not tackle graded preference into the training process but use the data in a binarized form.

2.2 Centrality Measures

The importance of a node in a graph has been a subject of study for a long time. Researchers started exploring the dynamics in social groups from a mathematical perspective [9,34]. With this objective in mind, graphs were proposed to model the groups and the relations between their members. Finding the influence of a user has been reduced to the problem of measuring the importance of a node inside the social graph. Research from Bavelas [4] or Katz [19] in the early 50s showed the first attempts of defining centrality measures that can capture this property of the nodes of a graph. With the emergence of the World Wide Web, centrality measures were once again bought to the forefront as a way to analyze the graph formed by the pages contained within it. In this context, PageRank [27] and HITS [20] were defined.

Graph representations of collaborative filtering data has been previously used in tasks such as neighborhood selection for memory-based recommenders [7]. Centrality measures have also been used in the recommendation field, especially in social-based recommender systems since they exploit the social relationships between users [5,15]. In contrast, in our work, we use centrality measures to compute prior probabilities over items, taking advantage of their ability to capture the importance of the items in the whole graph.

3 Proposal

In this section we present our probabilistic recommender, PRIN, explaining how we train it and how the model computes the recommendations. A brief introduction to the notation used to present the model precedes this description. Lastly, we include some comments on the implementation details.

3.1 Notation

We denote the set of users of the systems as $\mathcal{U}$ and the set of items as $\mathcal{I}$. For a user $u \in \mathcal{U}$ and an item $i \in \mathcal{I}$, we use $r_{u,i}$ to indicate the rating given by u to i, having a value of zero in case the user did not rate the item. The set of items rated by a user u is represented by $\mathcal{I}_u$ and the set of users that have rated an item i is denoted by $\mathcal{U}_i$.

3.2 Probabilistic Recommender with Neural Model

The idea behind word embedding models is that both words occurring close to each other, inside a window of fixed length, or words that appear in different sentences surrounded by the same words are similar. We postulate that this also applies to collaborative filtering data.

We propose a probabilistic model based on the adaptation of the continuous bag-of-words (CBOW) model for the task of top-N recommendation. The CBOW model predicts a word given its context, defined by the surrounding words inside a fixed-length window [24]. In our scenario, users play the role of words and item profiles that of documents, defining an item profile as the set of users that have rated it. We choose the CBOW model for two main reasons. On the one hand, its efficiency is superior to the skip-gram model [24]. On the other hand, we think the task of finding if a user fits inside an item profile is more natural for the recommendation task than the skip-gram objective of finding the context that fits a user.

Figure 1 presents the architecture of the model, inspired by the CBOW neural model. The output of the network is the target user, and the input is its context, i.e. the item profile without the user. For a particular item i and target user u the input consists of the input context vectors $\{\mathbf{x}_1, \ldots, \mathbf{x}_{u-1}, \mathbf{x}_{u+1}, \ldots, \mathbf{x}_{|\mathcal{U}_i|}\}$, that are all the users that rated the item except user u, with $|\mathcal{U}_i|$ being the number of users that have rated item i. These vectors are encoded using a one-hot representation. For user v, $\mathbf{x}_v$ is a vector of the form $\{x_{v1}, \ldots, x_{v|\mathcal{U}|}\}$, where all components are zero except the v-th component which is one and $|\mathcal{U}|$ is the number of users in the dataset. This way the training examples are created from the item profiles, being able to generate $|\mathcal{U}_i|$ training examples for each item profile, one example for each user in the item profile.

The amount of units in the hidden layer, d, is a hyperparameter of the model that determines the dimension of the embeddings. These units have a linear activation function. We use the matrix $\mathbf{W} \in \mathbb{R}^{|\mathcal{U}| \times d}$ to denote the weights of the connections between the input layer and the hidden layer. Each row of the

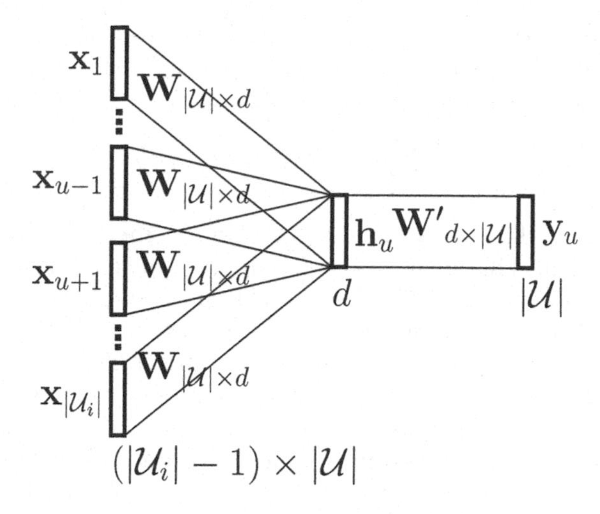

Fig. 1. Architecture of the model. The input layer consists of the aggregation of $|\mathcal{U}_i|-1$ one-hot encoded vectors, each vector of dimension $|\mathcal{U}|$, the hidden layer has d units and the output layer has $|\mathcal{U}|$ units.

matrix, $\mathbf{v}_v$, corresponds to the input embedding vector of dimension d of the user v. The output of the hidden layer for the target user u, $\mathbf{h}_u$, is computed by averaging the embeddings of the input users corresponding to the context, weighted by the rating given by the users to the item i:

$$\mathbf{h}_u = \frac{\mathbf{W}}{\sum\limits_{v\in\mathcal{U}_i\setminus\{u\}} r_{v,i}} \sum_{v\in\mathcal{U}_i\setminus\{u\}} r_{v,i}\,\mathbf{x}_v = \frac{\sum\limits_{v\in\mathcal{U}_i\setminus\{u\}} r_{v,i}\,\mathbf{v}_v}{\sum\limits_{v\in\mathcal{U}_i\setminus\{u\}} r_{v,i}} \tag{1}$$

By weighting the average by the ratings given by the users, we can incorporate these values into the training process. Although we evaluated our proposal with explicit feedback dataset, one can incorporate information from implicit feedback, such as clicks or play counts, by substituting the ratings in Eq. 1.

The output layer is composed of $|\mathcal{U}|$ units with a softmax activation function. Similar to what we did before, we use the matrix $\mathbf{W}' \in \mathbb{R}^{d\times|\mathcal{U}|}$ to denote the weights of the connection between the hidden and the output layer. Each column of this matrix, $\mathbf{v}'_u$, is the d-dimensional output embedding vector of user u. This way the input of the output layer is given by $\mathbf{v}'^T_u\mathbf{h}_u$. The output of the network is the posterior probability distribution of users for the context, i.e. the item profile without the target user. These probabilities are calculated using the softmax function. The component u of the output vector for the target user, $\mathbf{y}_u$, is calculated as:

$$p(u\,|\,\mathcal{U}_i\setminus\{u\}) = (\mathbf{y}_u)_u = \frac{\exp\left(\mathbf{v}'^T_u\mathbf{h}_u\right)}{\sum\limits_{v\in\mathcal{U}}\exp\left(\mathbf{v}'^T_v\mathbf{h}_u\right)} \tag{2}$$

Each example of the training set consists of the profile of an item i and a target user u from the profile. Maximizing the likelihood of the data is equivalent to minimizing the negative log likelihood. Therefore, the objective function is:

$$\mathcal{L} = -\sum_{i \in \mathcal{I}} \sum_{u \in \mathcal{U}_i} \log p(u \,|\, \mathcal{U}_i \setminus \{u\}) \tag{3}$$

The training of the model consists in learning the matrices $\mathbf{W}$ and $\mathbf{W}'$ by backpropagation. This model becomes impractical in large-scale scenarios because the cost of computing the gradient of $\log p(u \,|\, \mathcal{U}_i \setminus \{u\})$ for each training example is proportional to the number of users, due to the softmax function (see Eq. 2). Mikolov et al. already noted this problem in [24,25], where they propose to solve it by using one of two approximations to the softmax function: hierarchical softmax and negative sampling. For this work, we choose negative sampling as it provides faster training than hierarchical softmax and similar effectiveness [24,25].

It can be observed that the objective function in Eq. 3 does not include any regularization term. Early experiments with the model showed that it was overfitting the data when training with too many iterations. To solve this problem we choose to use dropout regularization in the input layer [32]. We decided this over other forms of regularization, such as ℓ_2 regularization, because it provided better effectiveness and improved training time [29]. At the same time, we can leverage existing `word2vec` implementations when using dropout as we will see later on.

Once the parameters have been trained, it is possible to use the model to compute the posterior probability distribution of a user for each item. This calculation is done by applying Eqs. 1 and 2 to the whole item profile, without removing any user. It should be noted that after applying dropout regularization, during evaluation, the activations are reduced to account for the missing activations during training [32]. This process is not necessary in our case because the inputs to the hidden layer are averaged, as we can see in Eq. 1.

The output for each unit of the output layer is the probability of the corresponding user u given the item i, $p(u|i) = p(u|\mathcal{U}_i)$. It is not possible to use these probabilities to make a ranking of items for a user as $p(u|i)$ and $p(u|j)$ are not comparable $(i, j \in \mathcal{I}, i \neq j)$. It is possible to apply Bayes' rule to transform the probabilities and make them comparable:

$$p(i|u) = \frac{p(u|i)\, p(i)}{p(u)} \overset{\text{rank}}{=} p(u|i)\, p(i) \tag{4}$$

We describe in the next subsection the options we explored for the computation of the prior distribution of items, $p(i)$.

3.3 Item Priors with Centrality Measures

The objective of the centrality measures is to capture the importance of a node inside a graph. Several measures have been defined over the years, and their suitability for a task depends on the flow inside the graph [9]. For this reason, we examined different types of measures [8]. The first category, geometric measures, is comprised of measures that assume that importance is a function of distances. In this group, we examined the indegree measure (the number of incoming edges of a node) and closeness [4]. The measures of the second category, spectral measures, compute the left dominant eigenvector of some matrix derived from the graph. We studied Katz's index [19], PageRank [27] and HITS [20] from this category. Lastly, we analyzed betweenness centrality, a path-based measure that takes into examination all the shortest path coming into a node [1,12].

To apply these measures to the computation of the prior distribution of items, we need to construct a graph-based model of the interactions between users and items in the system. We propose to construct a bipartite graph, where users and items play the role of the nodes, and the user-item interactions define the edges between users and the items. The weight of these edges is the rating assigned by the user to the item. We built two variants of the graph, one directed with the orientation of the edges going from users to the items and an undirected version. We do this because the direction of the edges can be meaningful when computing some centrality measures, but other measures are not useful when applied on a graph whose paths have a maximum length of one edge, as is the case of the directed graph.

3.4 Implementation Details

One of the advantages of the popularity of `word2vec` is that there are several publicly available implementations of the CBOW model. It is possible to leverage these implementations to build our model. We explain how to do that, also making possible to introduce the ratings of the items in the process and simulating the dropout in the input layer.

The original `word2vec` model is trained with a text corpus as the input. A corpus is composed of ordered sequences of words which we call documents, but can also be any other grouping of words such as sentences or paragraphs. The model has a hyperparameter for the window size, w, that controls how large is the context of a word, i.e. how many words before and after it are part of the context. For example, for $w = 1$, the context would be the preceding and the following words of the target word.

To train our model using collaborative filtering data, we build the analogous of a document in the format expected by the tool for each item profile. This pseudo-document contains all the identifiers of the users that have rated the item. To consider the whole item profile as the context, we set the window hyperparameter to the size of the larger item profile. The order of the items in the profile does not matter because the input of the hidden layer is the average of the input embeddings. To introduce the preference values into the model, we repeat each user identifier as many times as the rating given by the user to the item. Computing the average of the input constructed in this way is equivalent to the weighted average of Eq. 1.

Finally, we can introduce the dropout effect by modifying the hyperparameter w. If we set this parameter to a value smaller than the size of the profile, the

Table 1. Datasets statistics.

Dataset	Users	Items	Ratings	Density
MovieLens 20M	138,493	26,744	20,000,263	0.540%
R3-Yahoo	15,400	1,000	365,703	2.375%
LibraryThing	7,279	37,232	749,401	0.277%

context will be comprised of only some of the users of the item profile, *dropping out* the rest. We can add randomness to this procedure by shuffling the input each iteration. The combination of setting the w to a suitable value with the shuffling of the item profiles each iteration produces a similar effect to the dropout. This approach is a variant of the original technique that drops units randomly with a probability p [32]. Using this variant allows us to reuse existing `word2vec` CBOW implementations.

Training the PRIN model leads to a complexity for each training step of $\mathcal{O}(d \times (w + n))$, when using d dimensions, window size w and n negative samples. At each training step, there are w input embeddings, each corresponding to each input, of dimension d, that are averaged. It should be noted that w is bounded by the size of the larger item profile. The use of dropout in the form of a window produces notable improvements in the average computational cost of training the model. Moreover, using negative sampling allow approximating the softmax function with only n samples, instead of the whole user set. Finally, the number of training examples scales linearly with the number of user-item interactions in the collection. With all these facts, we can see that scalability of PRIN is well-suited for large-scale scenarios.

4 Experiments

In this section, we introduce the datasets, the evaluation protocol and the metrics used in our experiments. We finish the section by presenting the results of the experiments, confronting them with representative baselines.

4.1 Datasets

To evaluate the effectiveness of our proposal we conducted experiments on several collections, from different domains: the MovieLens 20M movie dataset[2], the R3-Yahoo! music dataset[3] and the LibraryThing book dataset. Details from each collection can be seen in Table 1. The datasets where partitioned randomly in two sets, one containing 80% of the ratings of each user, used for training, and a second split, with the remaining 20%, used for evaluation purposes.

[2] http://grouplens.org/datasets/movielens.
[3] http://webscope.sandbox.yahoo.com.

4.2 Evaluation Protocol

To evaluate the algorithms for the top-N recommendation task, we use the TestItems evaluation approach as described in [6]. For each user u, we rank all the items that have a rating by any user in the test set and were not rated by user u in the training set. This protocol provides a reliable assessment of the quality of the recommendation because it measures how well a recommender discerns relevant items in the collection [6].

To assess the accuracy of the recommendation rankings we use the Normalized Discounted Cumulative Gain (nDCG), using the *standard formulation* as described in [33], with the ratings in the test set as graded relevance judgments. We also measured diversity using the complement of the Gini index [11]. Last, we assess the novelty of the recommendations using the mean self-information (MSI) [35]. All the metrics are evaluated at a cut-off of 10 because we want to study the quality of the top recommendations, the ones the user usually consumes. To penalize a recommender not being able to provide recommendations to every user, the score in all metrics for those users is assigned a value of zero.

We study the statistical significance of the improvements regarding nDCG@10 and MSI@10 using a permutation test ($p < 0.01$) [31]. We cannot apply this procedure to the Gini index because we are using a paired test and Gini is a global metric. The statistical significance of the results is annotated in Table 3.

4.3 Baselines

We compare our proposed model to a representative set of state-of-the-art baselines. First, from the memory-based category of recommenders, we use NNCosNgbr [10], an item-based neighborhood approach. We also employ several techniques based in matrix factorization: PureSVD [10], BPRMF [30] and WRMF [18]. We compared with CoFactor [23], a variant of WRMF that jointly factorizes the user-item matrix and an embedding model, and NeuMF, a novel neural collaborative filtering approach [17]. Finally, we also include the results of the item-based counterpart of our model. We called this probabilistic recommender with neural models PRN.

The networks architecture of PRN is analogous to the architecture of PRIN (shown in Fig. 1), with an input of $|\mathcal{I}_u| - 1$ one-hot encoded vectors of dimension $|\mathcal{I}|$, where $\mathcal{I}$ is the set of items and $\mathcal{I}_u$ is the set of items rated by user u. The output is calculated with the dual equations of Eqs. 1 and 2. PRN takes as input a user profile and the output is the posterior probability distribution of the items for that user. This probability is usable as the basis of ranking, obviating the need for a prior distribution as in the case of PRIN.

We performed a grid search to tune all the hyperparameters of the baselines to maximize nDCG@10. Table 2 reports the optimal values of the hyperparameters (using the notation from the original papers) for all the techniques to favour reproducibility.

Table 2. Optimal values of the hyperparameters for nDCG@10 for NNCosNgbr, PureSVD, BPRMF, WRMF, CoFactor, NeuMF and our proposal PRIN and its item-based counterpart PRN.

Model	MovieLens 20M	R3-Yahoo!	LibraryThing
NNCosNgbr	$k = 50$	$k = 25$	$k = 25$
PureSVD	$d = 30, \kappa = 10^{-6}$	$d = 15, \kappa = 10^{-6}$	$d = 700, \kappa = 10^{-6}$
BPRMF	$d = 50$, $\lambda = 0.01$, $\alpha = 0.01$, $i = 10^5$	$d = 175$, $\lambda = 0.01$, $\alpha = 0.01$, $i = 10^5$	$d = 600$, $\lambda = 0.001$, $\alpha = 0.01$, $i = 10^6$
WRMF	$d = 50$, $\lambda = 0.01$, $\alpha = 1$, $i = 50$	$d = 50$, $\lambda = 1$, $\alpha = 2$, $i = 50$	$d = 400$, $\lambda = 0.1$, $\alpha = 1$, $i = 50$
CoFactor	$d = 100$, $c_0 = 0.3$, $c_1 = 3$, $\lambda_\theta = \lambda_\beta = \lambda_\gamma = 10^{-5}$, $k = 1$	$d = 30$, $c_0 = 1$, $c_1 = 10$, $\lambda_\theta = \lambda_\beta = \lambda_\gamma = 10^{-5}$, $k = 1$	$d = 500$, $c_0 = 1$, $c_1 = 10$, $\lambda_\theta = \lambda_\beta = \lambda_\gamma = 10^{-5}$, $k = 1$
NeuMF	$d = 64$, $i = 20$, $n = 5$	$d = 12$, $i = 5$, $n = 5$	$d = 1024$, $i = 20$, $n = 5$
PRIN	$d = 1000$, $w = 50$, $it = 1000$, indegree	$d = 50$, $w = 10$, $it = 200$, Katz	$d = 200$, $w = 10$, $it = 200$, PageRank
PRN	$d = 500$, $w = 100$, $it = 300$	$d = 200$, $w = 2$, $it = 100$	$d = 500$, $w = 1$, $it = 1000$

4.4 Results and Discussion

To tune our model, we perform a grid search over the hyperparameters, the same way we did with the baselines, to maximize nDCG@10. Although our implementation is based on the CBOW model, to keep things simple we only tune the parameters relevant to our model: the dimension of the hidden layer d, the window size w for the regularization effect and the number of training iterations it. The parameter for the negative sampling training is fixed, with a value of 10 negative samples. We also report the centrality measure that yields the best results. Table 2 reports the optimal values for each collection with the values of the hyperparameters of the baselines.

Table 3 shows the values for nDCG@10, Gini@10 and MSI@10 for all the recommenders. The results show that PRIN outperforms all the baselines concerning nDCG@10. In the MovieLens dataset, it surpasses the best baseline, WRMF, while also obtaining a better result in diversity but a lower score in novelty. When comparing to the next best result in R3-Yahoo!, BPRMF, our model is also able to perform better in novelty and diversity. In the case of the LibraryThing dataset, the improvement in nDCG@10 is statistically significant over all the baselines except CoFactor. When it comes to novelty and diversity in this dataset, the results are not as good as other baselines. This fact is not unexpected, diversity and accuracy are frequently considered as two irreconcilable goals in the field of Recommender Systems. Usually, systems with good figures

Table 3. Values of nDGC@10, Gini@10, MSI@10 on MovieLens 20M, R3-Yahoo! and LibraryThing datasets. Statistical significant improvements (according to permutation test with $p < 0.01$) in nDCG@10 and MSI@10 with respect to NNCosNgbr, PureSVD, BPRMF, WRMF, CoFactor, NeuMF and our proposal PRIN and its dual model PRN are superscripted with a, b, c, d, e, f, g and h, respectively.

Model	Metric	ML 20M	R3-Yahoo!	LibraryThing
NNCosNgbr	nDCG@10	0.1037	0.0172	0.1438
	Gini@10	0.0209	0.1356	0.1067
	MSI@10	29.3332^{bcdefg}	$\mathbf{36.8264}^{bcdefgh}$	$\mathbf{47.0790}^{bcdefgh}$
PureSVD	nDCG@10	0.3477^{acfh}	0.0233^{a}	0.2283^{af}
	Gini@10	0.0079	0.0587	0.0535
	MSI@10	15.4201	21.9703^{c}	40.7276^{cdefg}
BPRMF	nDCG@10	0.2671^{ah}	0.0278^{abdf}	0.2479^{abf}
	Gini@10	0.0103	0.1071	0.0474
	MSI@10	15.9674^{b}	21.4253	34.5252
WRMF	nDCG@10	0.3682^{abcefh}	0.0266^{a}	0.2532^{abcfh}
	Gini@10	0.0138	0.1191	0.0512
	MSI@10	17.3695^{bcg}	24.7479^{bcefg}	38.2290^{cfg}
CoFactor	nDCG@10	0.3555^{abcfh}	0.0258^{ab}	0.2568^{abcdfh}
	Gini@10	0.0215	0.1407	0.0690
	MSI@10	19.5491^{bcdg}	25.7688^{bcfg}	39.7497^{cdfg}
NeuMF	nDCG@10	0.3185^{ach}	0.0258^{ab}	0.1835^{a}
	Gini@10	0.0328	0.0993	0.0613
	MSI@10	21.2605^{bcdeg}	22.2208^{bc}	36.5621
PRIN	nDCG@10	$\mathbf{0.3751}^{abcdefh}$	$\mathbf{0.0299}^{abcdefh}$	$\mathbf{0.2578}^{abcdfh}$
	Gini@10	0.0155	0.1966	0.0482
	MSI@10	16.5353^{bc}	24.0921^{bcf}	34.4458^{cg}
PRN	nDCG@10	0.1909^{a}	0.0276^{abd}	0.2423^{abf}
	Gini@10	**0.2175**	**0.3221**	**0.1208**
	MSI@10	$\mathbf{49.4532}^{abcdefg}$	29.4596^{bcdefg}	42.1890^{bcdefg}

of accuracy tend to degrade de diversity of the recommendation, and systems with bad performance in accuracy show better diversity, in the extreme case a random recommender would produce very diverse recommendations.

Another significant result is that PRIN is consistently the best method regarding accuracy across collections. This property is essential in order to select an algorithm for use in a commercial solution. This property does not appear with the other methods. For instance, when observing the other neural/embedding-based models we can observe that CoFactor ranks third in the ML dataset, fifth in the R3 collection and sixth with the LibraryThing data, in turn, NeuMF ranks fifth, sixth and seventh respectively.

The optimal values for the hyperparameters vary for each collection. This fact indicates the need, shared with all the baselines, to tune these hyperparameters to the particular data. In the case of the size of the hidden layer, there is a trend for the need of larger hidden layers the larger the dataset. This fact supports the intuition that with more data there is a need for more features to be able to capture the properties of the data.

Regarding the centrality measures, each dataset performs better with a different one. The best results with MovieLens are obtained using indegree, whose value for the items is independent of whether the directed or the undirected graph is used. For R3-Yahoo!, using Katz's index on the directed graph yields the best results. In this dataset, using the indegree measure leads to similar results in nDCG@10 but worse on novelty and diversity. When it comes to LibraryThing, it is PageRank, computed on the undirected graph, that gives the best performance. Therefore, we can conclude that the centrality measures have to be adapted to the nature of the graph. For instance, dataset sparsity affects the connectivity of the graph, and the existence of the connected components reflects user communities. Therefore before selecting an item prior, we have to analyse the connectivity, edge meaning and size of the graph.

5 Conclusions

In this paper, we presented PRIN, a novel probabilistic collaborative filtering technique. Our probabilistic model exploits the output of a neural user embedding model. This embedding model can be computed by leveraging existing `word2vec` CBOW implementations. The probabilistic formulation of PRIN also requires an item prior estimate. We evaluated several centrality measures of two graph-based interpretations of the user-item interactions as item prior probability estimates.

Our experiments showed that PRIN outperforms all the baselines on three datasets regarding ranking accuracy. PRIN is also able to provide good figures of diversity and novelty.

As future work, we envision to study other no graph-based prior estimates to further improve PRIN. Additionally, we think that it would be interesting to analyze the adaptation of the skip-gram model and also explore deeper or more complex network topologies for the neural model. Another prospect is the evaluation of the model when using an implicit feedback dataset.

Acknowledgments. This work has received financial support from project TIN2015-64282-R (MINECO/ERDF) and accreditation ED431G/01 (Xunta de Galicia/ERDF). The first author acknowledges the support of grant FPU17/03210 (MICIU) and the second author acknowledges the support of grant FPU014/01724 (MICIU).

References

1. Anthonisse, J.: The rush in a directed graph. Stichting Mathematisch Centrum. Mathematische Besliskunde (BN 9/71), January 1971
2. Balabanović, M., Shoham, Y.: Fab: content-based, collaborative recommendation. Commun. ACM **40**(3), 66–72 (1997). https://doi.org/10.1145/245108.245124
3. Barkan, O., Koenigstein, N.: Item2vec: neural item embedding for collaborative filtering. In: 2016 IEEE 26th International Workshop on Machine Learning for Signal Processing (MLSP), pp. 1–6, September 2016. https://doi.org/10.1109/MLSP.2016.7738886
4. Bavelas, A., Barrett, D.: An Experimental Approach to Organizational Communication. American Management Association (1951)
5. Bellogín, A., Cantador, I., Díez, F., Castells, P., Chavarriaga, E.: An empirical comparison of social, collaborative filtering, and hybrid recommenders. ACM Trans. Intell. Syst. Technol. **4**(1), 1–29 (2013). https://doi.org/10.1145/2414425.2414439
6. Bellogín, A., Castells, P., Cantador, I.: Precision-oriented evaluation of recommender systems. In: Proceedings of the 5th ACM Conference on Recommender systems, RecSys 2011, pp. 333–336. ACM, New York (2011). https://doi.org/10.1145/2043932.2043996
7. Bellogin, A., Parapar, J.: Using graph partitioning techniques for neighbour selection in user-based collaborative filtering. In: Proceedings of the Sixth ACM Conference on Recommender Systems, RecSys 2012, pp. 213–216. ACM, New York (2012). https://doi.org/10.1145/2365952.2365997
8. Boldi, P., Vigna, S.: Axioms for centrality. Internet Math. **10**(3–4), 222–262 (2014). https://doi.org/10.1080/15427951.2013.865686
9. Borgatti, S.P.: Centrality and network flow. Soc. Netw. **27**(1), 55–71 (2005). https://doi.org/10.1016/j.socnet.2004.11.008
10. Cremonesi, P., Koren, Y., Turrin, R.: Performance of recommender algorithms on Top-N recommendation tasks. In: Proceedings of the 4th ACM Conference on Recommender Systems, RecSys 2010, pp. 39–46. ACM, New York (2010). https://doi.org/10.1145/1864708.1864721
11. Fleder, D., Hosanagar, K.: Blockbuster culture's next rise or fall: the impact of recommender systems on sales diversity. Manage. Sci. **55**(5), 697–712 (2009). https://doi.org/10.1287/mnsc.1080.0974
12. Freeman, L.C.: A set of measures of centrality based on betweenness. Sociometry **40**(1), 35–41 (1977). https://doi.org/10.2307/3033543
13. de Gemmis, M., Lops, P., Musto, C., Narducci, F., Semeraro, G.: Semantics-aware content-based recommender systems. In: Ricci, F., Rokach, L., Shapira, B. (eds.) Recommender Systems Handbook, pp. 119–159. Springer, Boston (2015). https://doi.org/10.1007/978-1-4899-7637-6_4
14. Grbovic, M., et al.: E-commerce in your inbox: Product recommendations at scale. In: Proceedings of the 21th ACM SIGKDD International Conference on Knowledge Discovery and Data Mining, KDD 2015, pp. 1809–1818. ACM, New York (2015). https://doi.org/10.1145/2783258.2788627
15. Guy, I.: Social recommender systems. In: Ricci, F., Rokach, L., Shapira, B. (eds.) Recommender Systems Handbook, pp. 511–543. Springer, Boston, MA (2015). https://doi.org/10.1007/978-1-4899-7637-6_15
16. Harris, Z.S.: Distributional structure. WORD **10**(2–3), 146–162 (1954). https://doi.org/10.1080/00437956.1954.11659520

17. He, X., Liao, L., Zhang, H., Nie, L., Hu, X., Chua, T.S.: Neural collaborative filtering. In: Proceedings of the 26th International Conference on World Wide Web, WWW 2017, pp. 173–182. International World Wide Web Conferences Steering Committee, Republic and Canton of Geneva, Switzerland (2017). https://doi.org/10.1145/3038912.3052569
18. Hu, Y., Koren, Y., Volinsky, C.: Collaborative filtering for implicit feedback datasets. In: Proceedings of the 2008 Eighth IEEE International Conference on Data Mining, ICDM 2008, pp. 263–272. IEEE, Washington (2008). https://doi.org/10.1109/ICDM.2008.22
19. Katz, L.: A new status index derived from sociometric analysis. Psychometrika **18**(1), 39–43 (1953). https://doi.org/10.1007/BF02289026
20. Kleinberg, J.M.: Authoritative sources in a hyperlinked environment. J. ACM **46**(5), 604–632 (1999). https://doi.org/10.1145/324133.324140
21. Koren, Y., Bell, R.: Advances in collaborative filtering. In: Ricci, F., Rokach, L., Shapira, B. (eds.) Recommender Systems Handbook, pp. 77–118. Springer, Boston (2015). https://doi.org/10.1007/978-1-4899-7637-6_3
22. Le, Q., Mikolov, T.: Distributed representations of sentences and documents. In: Xing, E.P., Jebara, T. (eds.) Proceedings of the 31st International Conference on Machine Learning, pp. 1188–1196. Proceedings of Machine Learning Research, PMLR, Bejing, China, 22–24 June 2014
23. Liang, D., Altosaar, J., Charlin, L., Blei, D.M.: Factorization meets the item embedding: regularizing matrix factorization with item co-occurrence. In: Proceedings of the 10th ACM Conference on Recommender Systems, RecSys 2016, pp. 59–66. ACM, New York (2016). https://doi.org/10.1145/2959100.2959182
24. Mikolov, T., Chen, K., Corrado, G., Dean, J.: Efficient Estimation of Word Representations in Vector Space. CoRR abs/1301.3, January 2013
25. Mikolov, T., Sutskever, I., Chen, K., Corrado, G.S., Dean, J.: Distributed representations of words and phrases and their compositionality. In: Advances in Neural Information Processing Systems 26, NIPS 2013, pp. 3111–3119. Curran Associates, Inc. (2013)
26. Ning, X., Desrosiers, C., Karypis, G.: A comprehensive survey of neighborhood-based recommendation methods. In: Ricci, F., Rokach, L., Shapira, B. (eds.) Recommender Systems Handbook, pp. 37–76. Springer, Boston (2015). https://doi.org/10.1007/978-1-4899-7637-6_2
27. Page, L., Brin, S., Motwani, R., Winograd, T.: The pagerank citation ranking: Bringing order to the web. Technical report, Stanford InfoLab, November 1999
28. Pennington, J., Socher, R., Manning, C.: Glove: global vectors for word representation. In: Proceedings of the 2014 Conference on Empirical Methods in Natural Language Processing, EMNLP 2014, pp. 1532–1543. ACL, Stroudsburg (2014). https://doi.org/10.3115/v1/D14-1162
29. Phaisangittisagul, E.: An Analysis of the regularization between L2 and dropout in single hidden layer neural network. In: Proceedings of the 7th International Conference on Intelligent Systems, Modelling and Simulation, ISMS 2016, pp. 174–179. IEEE (2016). https://doi.org/10.1109/ISMS.2016.14
30. Rendle, S., Freudenthaler, C., Gantner, Z., Schmidt-Thieme, L.: BPR: Bayesian personalized ranking from implicit feedback. In: Proceedings of the Twenty-Fifth Conference on Uncertainty in Artificial Intelligence, UAI 2009, pp. 452–461. AUAI Press, Arlington (2009)

31. Smucker, M.D., Allan, J., Carterette, B.: A comparison of statistical significance tests for information retrieval evaluation. In: Proceedings of the Sixteenth ACM Conference on Conference on Information and Knowledge Management, CIKM 2007, p. 623. ACM, New York (2007). https://doi.org/10.1145/1321440.1321528
32. Srivastava, N., Hinton, G., Krizhevsky, A., Sutskever, I., Salakhutdinov, R.: Dropout: a simple way to prevent neural networks from overfitting. J. Mach. Learn. Res. **15**(1), 1929–1958 (2014). https://doi.org/10.1214/12-AOS1000
33. Wang, Y., Wang, L., Li, Y., He, D., Chen, W., Liu, T.Y.: A theoretical analysis of NDCG ranking measures. In: Proceedings of the 26th Annual Conference on Learning Theory, COLT 2013, pp. 1–30. JMLR.org (2013)
34. Wasserman, S., Faust, K.: Social Network Analysis: Methods and Applications. Structural Analysis in the Social Sciences. Cambridge University Press, Cambridge (1994)
35. Zhou, T., Kuscsik, Z., Liu, J.G., Medo, M., Wakeling, J.R., Zhang, Y.C.: Solving the apparent diversity-accuracy dilemma of recommender systems. Proc. Nat. Acad. Sci. **107**(10), 4511–4515 (2010). https://doi.org/10.1073/pnas.1000488107

Information Retrieval Models for Contact Recommendation in Social Networks

Javier Sanz-Cruzado(✉) and Pablo Castells(✉)

Universidad Autónoma de Madrid, Escuela Politécnica Superior, Madrid, Spain
{javier.sanz-cruzado,pablo.castells}@uam.es

Abstract. The fast growth and development of online social networks has posed new challenges for information retrieval and, as a particular case, recommender systems. A particularly compelling problem in this context is recommending network edges, that is, automatically predicting people that a given user may wish or benefit from connecting to in the network. This task has interesting particularities compared to more traditional recommendation domains, a salient one being that recommended items belong to the same space as the users they are recommended to. In this paper, we investigate the connection between the contact recommendation and the text retrieval tasks. Specifically, we research the adaptation of IR models for recommending contacts in social networks. We report experiments over data downloaded from Twitter where we observe that IR models, particularly BM25, are competitive compared to state-of-the art contact recommendation methods. We further find that IR models have additional advantages in computational efficiency, and allow for fast incremental updates of recommendations as the network grows.

Keywords: Social networks · Contact recommendation · Text information retrieval

1 Introduction

The creation of online social network applications such as Twitter, Facebook and LinkedIn, and their subsequent expansion along the 2,000s has given rise to new perspectives and challenges in the information retrieval (IR) field, and, as a particular case, recommender systems. One of the most compelling problems in this area is recommending people with whom users might want to engage in an online network. The social nature of these networks, and the massive amount of users accessing them every day has raised the interest for contact recommendation of both industry [10, 11] and several research communities [4, 12, 13]. The most prominent social platforms offer user recommendation services since the end of the past decade, with systems such as 'Who-to-follow' on Twitter [10, 11] or 'People you may know' on Facebook and LinkedIn.

Contact recommendation represents a very particular perspective of the recommendation task. On the one hand, the recommendation domain lays connections to social network analysis and network science, with rich potential implications [12, 28]. On the other, while in most domains users and items are different objects, this one has

L. Azzopardi et al. (Eds.): ECIR 2019, LNCS 11437, pp. 148–163, 2019.
https://doi.org/10.1007/978-3-030-15712-8_10

the peculiar and interesting characteristic that users and items are the same set. These particularities have motivated the creation of a wide variety of people recommendation algorithms from diverse fields, such as network science [18, 19], machine learning [14], recommender systems [13] or, to a lesser extent, information retrieval [13].

In our present work we focus on this last line of research: we investigate the relation between contact recommendation in social networks, and text retrieval. For this purpose, we establish associations between the fundamental elements involved both tasks, in order to adapt classic IR models to the task of suggesting people in a social network. We explore the adaptation of well-known models: the vector space model [26] (VSM), BM25 [25] and query likelihood [22]. We empirically compare the effectiveness of the resulting algorithms to state-of-the-art contact recommendation methods over data samples extracted from Twitter, and we find the adapted IR models, particularly BM25, to be competitive with the best alternatives. Moreover, we find important additional advantages in terms of computational efficiency, both in producing recommendations from scratch, and in incrementally updating them as the network grows with new links and users.

2 Related Work

In the context of online social networks, contact recommendation aims at identifying people in a social network that a given user would benefit from relating to [30]. The problem is in many aspects equivalent to the link prediction task [18, 19], which aims to identify unobserved links that exist or will form in the future in a real network. Link prediction and recommendation is an established topic at the confluence of social network analysis and recommender systems for which many methods have been proposed in the literature, based on the network topology [18], random walks across the network graph [4, 10], or user-generated content [13].

In this paper, we investigate the adaptation of classic text IR models to the contact recommendation task. The connections between recommendation and text IR date back to the earliest recommender systems and their relation to the information filtering task [6]. Even though most of this connection has focused on content-based methods [2], it has also developed into collaborative filtering algorithms [7, 31, 32].

A particularly representative and relevant approach for our present work was developed by Bellogín et al. [7], allowing the adaptation of any IR term weighting scheme to create a collaborative filtering algorithm. To this end, the approach represents users and items in a common space, where users are the equivalent of queries, and items play the role of the documents to be retrieved. Our work pursues a similar goal, but taking a step further: if Bellogín et al. folded three spaces (terms, documents, queries) into two (users, items), we fold them into just one, as we shall explain.

Some authors have likewise connected IR techniques to the specific task of recommending users in social networks. For example, some link prediction approaches, such as the ones based on the Jaccard index [16, 18, 27] have their roots in IR. More recently, Hannon et al. [13] adapted the vector-space model [26] to recommend users on

Twitter, based on both content-based and collaborative filtering algorithms. Our work seeks to extend, generalize and systematize this point of view to adapt any state-of-the-art IR model to contact recommendation.

3 Preliminaries

We start by formally stating the contact recommendation task, and introducing the notation we shall use in our formulation. We can represent the structure of a social network as a graph $\mathcal{G} = \langle \mathcal{U}, E \rangle$, where $\mathcal{U}$ is the set of network users, and $E \in \mathcal{U}_*^2$ is the set of relations between users (friendship, interactions, whatever the network is representing), where $\mathcal{U}_*^2 = \{(u,v) \in \mathcal{U}^2 | u \neq v\}$ is the set of pairs formed by different users.

For each user $u \in \mathcal{U}$, we denote her neighborhood as $\Gamma(u)$ (the set of users that u has established relations with). In directed networks, three different neighborhoods can be considered: the incoming neighborhood $\Gamma_{\text{in}}(u)$ (users who create links towards u), the outgoing neighborhood $\Gamma_{\text{out}}(u)$ (users towards whom u creates links), and the union of both neighborhoods $\Gamma_{\text{und}}(u)$. In weighted graphs we have additionally a weight function $w : \mathcal{U}_*^2 \rightarrow \mathbb{R}$, which returns the weight of an edge if $(u,v) \in E$, and 0 otherwise. In unweighted graphs, we can consider that $w(u,v) = 1$ if the link exists, and 0 otherwise.

Now given a target user u, the contact recommendation task consists in finding a subset of users $\tilde{\Gamma}_{\text{out}}(u) \subset \mathcal{U} \setminus \Gamma_{\text{out}}(u)$ towards whom u has no links but who might be of interest for her. We address the recommendation task as a ranking problem, in which we find a fixed number of users $n = \left|\tilde{\Gamma}_{\text{out}}(u)\right|$ sorted by decreasing value of a ranking function $f_u : \mathcal{U} \setminus \Gamma_{\text{out}}(u) \rightarrow \mathbb{R}$.

4 IR Model Adaptation Framework for Contact Recommendation

Even though recommendation and text retrieval have been traditionally addressed as separate problems, it is possible to establish analogies and equivalences between both tasks. Recommender systems are indeed often described as retrieval systems where the query is absent, and records of user activity are available instead [7], and the approaches we develop follow this perspective.

4.1 Task Unification

In order to adapt text IR models to the recommendation task, we need to establish equivalences between the elements in the contact recommendation task (users and interactions between them) and the spaces involved in text search (queries, documents and terms). In previous adaptations of IR models for recommendation, the three IR spaces commonly folded into two: the set of users and the set of items [7]. However, when we seek to recommend people in social networks, the latter two spaces are the same. Therefore, to adapt the IR models to our task, we fold the three IR spaces into a single dimension: the set of users in the social network, playing the three different roles, as we illustrate in Fig. 1. We explain next in more detail how we carry this mapping through.

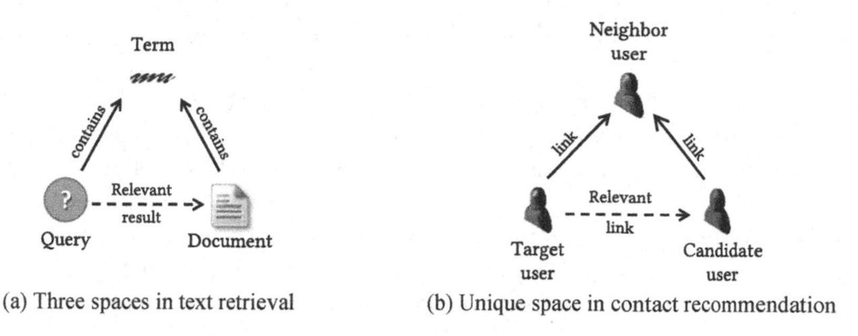

(a) Three spaces in text retrieval (b) Unique space in contact recommendation

Fig. 1. Text IR elements (a) vs. contact recommendation elements (b).

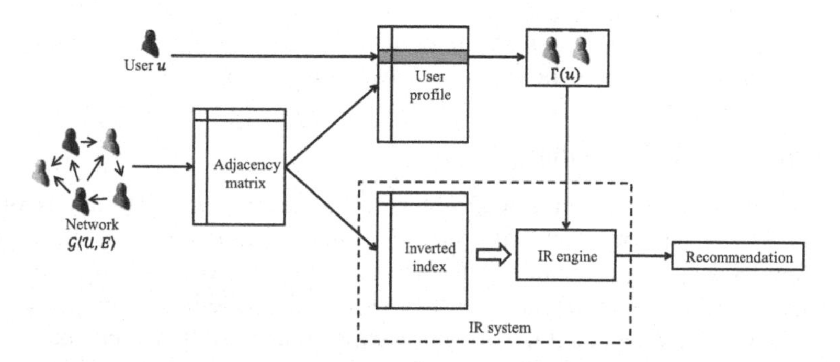

Fig. 2. Adaptation of IR models to recommend users in social networks.

First, the natural equivalent of documents in the search space are candidate users (to be recommended as contacts), as they play the same role: they are the elements to be retrieved in order to fulfil a user need. The need is explicit in the search task, expressed by a query; and it is implicit in contact recommendation: the need for creating new bonds. This social need is to be predicted based on records of past user activity, which therefore play an equivalent role to the query keywords in text IR. In a social network, past user activity is encoded in existing links to and/or from the target user.

Finally, we need an equivalent to the term representation of documents. In prior adaptations of IR models for recommendation, this was the main difficulty: users and items were different objects, so a representation that suits one might not work for the other [7]. In contact recommendation this becomes in fact easier: users and items are the same thing, so any term representation for target users is automatically valid for the "items" (the candidate users). The possibilities for defining an equivalent to terms are manifold, and result in very different algorithms. For instance, we can define content-based recommendation methods by using texts associated to users, such as messages or documents posted or liked by the users [13]. On the other hand, if we take users as the term space, and we equate the term-document relationship to interactions between users, we obtain collaborative filtering algorithms. We shall focus on the latter approach in this paper.

Figure 2 illustrates the collaborative filtering adaptation approach. A social network is encoded as a weighted adjacency matrix A, where $A_{uv} = w(u, v)$. Using link data, we build two elements: on one hand, an inverted index that allows for fast retrieval of candidate users and, on the other, a structure that provides direct access to the neighborhood of the target users, i.e. the query term representation. The inverted index uses network users as keys (playing the role of terms), and postings lists store the set of candidate users to whose neighborhood representation (as "documents") the "key" users belong to.

Using this index and the "query" structure, any text IR engine can be used as a contact recommendation algorithm. Additional details and options remain open however when developing a specific instance of this framework in full detail, as we will describe in the following sections. An important one concerns the direction of social links in the reinterpretation of IR models, to which we shall pay specific attention.

4.2 Neighborhood Orientation

In directed social networks such as Twitter or Instagram, three definitions of user neighborhood can be considered: the incoming neighborhood $\Gamma_{\text{in}}(u)$, the outgoing neighborhood $\Gamma_{\text{out}}(u)$ and the union of both, $\Gamma_{\text{und}}(u) = \Gamma_{\text{in}}(u) \cup \Gamma_{\text{out}}(u)$. Any of the three options is valid in our adaptation of IR models. Since the inverted index and user profiles are created independently, it is even possible to take a different choice for target and candidate users: since we still use the same elements to represent (the equivalent of) both queries and documents, it is possible to work just smoothly with different neighborhood orientation choices for targets and candidates.

Identifying which neighborhood best characterizes the candidate and target users in the social network is an interesting problem by itself [13]. It concerns many state-of the-art contact recommendation algorithms –besides IR adaptations– such as Adamic-Adar [1] or Jaccard similarity [18, 27] which use neighborhoods in their ranking functions. We shall therefore explore this issue in our experiments in Sect. 6.

5 Adaptation of Specific IR Models

As an example of the general unification framework, we now show in some detail the adaptation of two particular IR models: BIR and BM25 [25]. In the formulations in this section, we shall denote the neighborhood representation of the target user as $\Gamma^q(u)$, and the neighborhood representation of the candidate users as $\Gamma^d(v)$.

5.1 Binary Independence Retrieval

The model known as BIR (binary independence retrieval) [25] is the simplest representative of IR models building on the probability ranking principle [24]. Under the assumption that term occurrence follows a (multiple) Bernoulli distribution, this model estimates the probability of relevance of a document d for a query q as:

$$P(r|d,q) \propto \sum\nolimits_{t \in d \cap q} \mathrm{RSJ}(t) \tag{1}$$

where r denotes the event that the document is relevant, and RSJ is the Robertson-Spärck-Jones formula [25], which is defined as:

$$\mathrm{RSJ}(t) = \log \frac{|R_t|(|D| - |D_t| - |R| - |R_t|)}{(|R| - |R_t|)(|D_t| - |R_t|)} \tag{2}$$

In the above equation R is the set of relevant documents for the query, R_t is the set of relevant documents containing the term t, D is the document collection, and D_t is the set of documents containing t. Since the set R of relevant documents is not known, the following approximation can be taken, considering that typically only a tiny fraction of documents are relevant:

$$\mathrm{RSJ}(t) = \log \frac{|D| - |D_t| + 0.5}{|D_t| + 0.5} \tag{3}$$

As described in Sect. 4, to adapt this model for contact recommendation, we equate queries and documents to target and candidate users respectively, and the term-document relationship to social network edges. Under this equivalence, $|D|$ is the number of users in the network, and $|D_t|$ is the number of users that t is a neighbor of (i.e. her neighbor size in the transposed network). Denoting inverse neighborhoods as $\Gamma^d_{\mathrm{inv}}(t)$, the adapted BIR equation becomes:

$$f_u(v) = \sum_{t \in \Gamma^q(u) \cap \Gamma^d(v)} \mathrm{RSJ}(w) = \sum_{t \in \Gamma^q(u) \cap \Gamma^d(v)} \log \frac{|\mathcal{U}| - |\Gamma^d_{\mathrm{inv}}(t)| + 0.5}{|\Gamma^d_{\mathrm{inv}}(t)| + 0.5} \tag{4}$$

5.2 BM25

BM25 is one of the best-known and most effective probabilistic IR models [25]. It starts from similar principles as BIR, but modeling term occurrence in documents as a Poisson instead of a Bernoulli distribution. Its ranking function is defined as:

$$P(r|d,q) \propto \sum_{t \in d \cap q} \frac{(k+1)\mathrm{freq}(t,d)}{k(1 - b + b|d|/\mathrm{avg}_{d'}(|d'|)) + \mathrm{freq}(t,d)} \mathrm{RSJ}(t) \tag{5}$$

where $\mathrm{freq}(t, d)$ denotes the frequency of t in d, $|d|$ is the document length, $\mathrm{RSJ}(w)$ is defined in Eq. 3, and $k = [0, \infty)$ and $b \in [0, 1]$ are free parameters controlling the effect of term frequencies and the influence of the document length, respectively.

The text retrieval space can be mapped to a social network just as before, now taking, additionally, edge weights as the equivalent of term frequency. In directed networks, we will need to make a choice between the weight of incoming or outgoing links as the equivalent of frequency. We shall link this decision to the edge orientation selected for candidate users (as pointed out earlier in Sect. 4.2 and beginning of Sect. 5), as follows:

$$\mathrm{freq}(t, v) = w^d(v, t) = \begin{cases} w(t, v) & \text{if } \Gamma^d \equiv \Gamma_{\text{in}} \\ w(v, t) & \text{if } \Gamma^d \equiv \Gamma_{\text{out}} \\ w(v, t) + w(t, v) & \text{otherwise} \end{cases} \tag{6}$$

Finally, document length can be now defined as the sum of edge weights of the candidate user. In unweighted graphs this is simply equivalent to the degree of the node; in directed networks we have again different choices. The BM25 formulation for text retrieval considers different options in defining document length (number of unique terms, sum of frequencies, etc.) [25]. We have found similarly worthwhile to decouple the orientation choice for document length from the one for the term representation of candidate users. We reflect this by defining length as:

$$\mathrm{len}^l(v) = \sum_{t \in \Gamma^l(v)} w^l(v, t) \tag{7}$$

where $\Gamma^l(v)$ represents the candidate's neighborhood in a specific orientation choice for document length. Based on all this, the adaptation of BM25 becomes:

$$f_u(v) = \sum_{t \in \Gamma^q(u) \cap \Gamma^d(v)} \frac{(k+1) w^d(v, t)}{k\left(1 - b + b\,\mathrm{len}^l(v) / \mathrm{avg}_x \mathrm{len}^l(x)\right) + w^d(v, t)} \mathrm{RSJ}(t) \tag{8}$$

5.3 Other IR Models

Analogous adaptations can be defined for virtually any other IR model, such as the vector space model [26] or query likelihood [22], which we summarize in Table 1, including Jelinek-Mercer [17] (QLJM), Dirichlet [20] (QLD), and Laplace smoothing [32] (QLL) for query likelihood, which were adapted in prior work for general recommendation [7, 31, 32] –we now adapt them to the specific contact recommendation task.

Table 1. Adaptation of IR models to contact recommendation.

Model	Ranking function
VSM	$f_u(v) = \sum_{t \in \Gamma^q(u) \cap \Gamma^d(v)} u_t v_t / \sqrt{\sum_{t \in \Gamma^d(v)} v_t^2}$ $u_t = \mathrm{tf} - \mathrm{idf}^q(u,t) = w^q(u,t) \log\left(1 + \lvert\mathcal{U}\rvert / \left(1 + \lvert\Gamma^q_{\mathrm{inv}}(t)\rvert\right)\right)$ $v_t = \mathrm{tf} - \mathrm{idf}^d(v,t)$
BIR	$f_u(v) = \sum_{t \in \Gamma^q(u) \cap \Gamma^d(v)} \mathrm{RSJ}(t) \quad \mathrm{RSJ}(t) = \log \frac{\lvert\mathcal{U}\rvert - \lvert\Gamma^d_{\mathrm{inv}}(t)\rvert - 0.5}{\lvert\Gamma^d_{\mathrm{inv}}(t)\rvert - 0.5}$
BM25	$f_u(v) = \sum_{t \in \Gamma^q(u) \cap \Gamma^d(v)} \frac{(k+1) w^d(v,t) \cdot \mathrm{RSJ}(t)}{k\left(1 - b + b\,\mathrm{len}^l(v) / \mathrm{avg}_x \mathrm{len}^l(x)\right) + w^d(v,t)}$
QLJM	$f_u(v) = \sum_{t \in \Gamma^q} w^q(u,t) \log\left((1-\lambda) \frac{w^d(v,t)}{\mathrm{len}^d(v)} + \lambda \frac{\mathrm{len}^d_{\mathrm{inv}}(t)}{\sum_{x \in \mathcal{U}} \mathrm{len}^d(x)}\right)$
QLD	$f_u(v) = \sum_{t \in \Gamma^q} w^q(u,t) \log\left(\frac{w^d(v,t) + \mu\,\mathrm{len}^d_{\mathrm{inv}}(t) / \sum_{x \in \mathcal{U}} \mathrm{len}^d(x)}{\mathrm{len}^d(v) + \mu}\right)$
QLL	$f_u(v) = \sum_{t \in \Gamma^q} w^q(u,t) \log\left(\frac{w^d(v,t) + \gamma}{\mathrm{len}^d(v) + \gamma\lvert\mathcal{U}\rvert}\right)$

6 Experiments

In order to analyze the performance of the adaptation of IR methods to contact recommendation and compare them to baseline alternatives, we conduct several offline experiments using social network data extracted from Twitter. We describe the experimental approach, setup and results in the paragraphs that follow.

6.1 Data and Experimental Setup

We run our experiments over dynamic, implicit networks induced by the interactions between users (i.e. $(u,v) \in E$ if u retweeted, mentioned or replied v). We built two datasets: one containing all tweets posted by a set of around 10,000 users from June 19th to July 19th 2015, and one containing the last 200 tweets posted by 10,000 users as of August 2nd 2015. Users are sampled in a snowball graph crawling approach starting with a single seed user, and taking the interaction tweets (retweets, mentions, replies) by each user as outgoing network edges to be traversed. User sampling stops when 10,000 users are reached in the traversal; at that point, any outgoing edges from remaining users in the crawl frontier pointing to sampled users are added to the network.

For evaluation purposes, we partition the network into a training graph that is supplied as input to the recommendation algorithms, and a test graph that is held out from them for evaluation. IR metrics such as precision, recall or nDCG [5] can be computed on the output of a recommendation algorithm by considering test edges as binary relevance judgments: a user v is relevant to a user u if –and only if– the edge (u,v) appears in the test graph. In our experiments we apply a temporal split, which

Table 2. Twitter network dataset details.

Network	Users	Training edges	Test edges
1 Month	9,528	170,425	54,355
200 Tweets	9,985	137,850	21,598

better represents a real setting: the training data includes edges created before a given time point, and the test set includes the links created afterwards. The split point for the "1 month" dataset is July 12th (thus taking three weeks for training and one for test); and in "200 tweets" the split date is July 29th in order to have 80% of edges in the training graph. Edges appearing in both sides of the split are removed from the test network, and the frequency of training interaction between every pair of users is available to the evaluated systems as part of the training information. We show the resulting dataset statistics in Table 2.

Finally, to avoid trivializing the recommendation task, reciprocating links are excluded from both the test network and the systems' output. Given the high reciprocation ratio on Twitter, recommending reciprocal links would be a trivial hard to beat baseline. Moreover, users already notice when someone retweets or mentions them since Twitter sends notifications every time, whereby an additional recommendation would be redundant and would barely add any value.

6.2 Recommendation Algorithms

We assess the IR model adaptations by comparing them to a selection of the most effective and representative algorithms in the link prediction and contact recommendation literature. These include Adamic-Adar [1], most common neighbors (MCN) [18], personalized PageRank [1], and collaborative filtering (item-based and user-based kNN [21], and implicit matrix factorization (iMF) [15], as implemented in the RankSys library [23]). In addition, we implement the Money algorithm [10, 11] developed at Twitter, in which, for simplicity, we include all users in the circle of trust. We also include random and most-popular recommendation as sanity-check baselines.

We optimize all algorithms (edge orientation and parameter settings) by grid search targeting P@10. For those that can take advantage of edge weights (IR models and collaborative filtering algorithms), we select the best option. The resulting optimal settings are detailed in Table 3.

6.3 Experimental Results

We show in Table 4 the results for both datasets. We observe that only four of the algorithms in our comparison achieve good results in both datasets: the implicit matrix factorization approach, BM25 and, to a lesser extent, Adamic-Adar and BIR. Indeed, iMF is the best algorithm in terms of precision and recall for the "1 month" dataset, whereas BM25 achieves the maximum accuracy in terms of P@10 for the "200 tweets" dataset, with a technical tie (non-significant difference) in R@10. For the rest of algorithms, we see three different trends: Jaccard and VSM are far from the best

Table 3. Parameter settings for each algorithm and dataset. We take $\Gamma^q \equiv \Gamma_{\text{und}}$ and $\Gamma^d \equiv \Gamma_{\text{in}}$ for all algorithms, except $\Gamma^d \equiv \Gamma_{\text{und}}$ for VSM on 200 tweets. For BM25 we take $\Gamma^l \equiv \Gamma_{\text{out}}$. All algorithms perform best without weights, except BM25 on both datasets, and VSM on 1 month. In Adamic-Adar, Γ^l represents the direction on the selection of common neighbors between the target and candidate users (see [30]).

Algorithm	1 Month	200 Tweets
BM25	$b = 0.1, k = 1$	$b = 0.5, k = 1$
QLD	$\mu = 1000$	$\mu = 1000$
QLJM	$\lambda = 0.1$	$\lambda = 0.1$
QLL	$\gamma = 100$	$\gamma = 100$
Money	Authorities, $\alpha = 0.99$	Authorities, $\alpha = 0.99$
Adamic-Adar	$\Gamma^l \equiv \Gamma_{\text{out}}$	$\Gamma^l \equiv \Gamma_{\text{und}}$
Personalized PageRank	$r = 0.4$	$r = 0.4$
iMF	$k = 250, \alpha = 40, \lambda = 150$	$k = 290, \alpha = 40, \lambda = 150$
User-based kNN	$k = 120$	$k = 90$
Item-based kNN	$k = 300$	$k = 290$

Table 4. Effectiveness of the IR model adaptations and baselines. Cell color goes from red (lower) to blue (higher values) for each metric/dataset, with the top value highlighted in bold. The differences between BM25 (the best IR model) and iMF (the best baseline) are always statistically significant (two-tailed paired t-test at $p = 0.05$) except in R@10 on 200 tweets.

	1 month			200 tweets		
	P@10	R@10	nDCG@10	P@10	R@10	nDCG@10
BM25	0.0691	0.1010	0.1030	**0.0572**	0.1313	**0.1102**
BIR	0.0675	0.0943	0.0995	0.0534	0.1234	0.1016
QLL	0.0609	0.0798	0.0869	0.0490	0.1108	0.0929
QLJM	0.0580	0.0758	0.0823	0.0492	0.1124	0.0943
QLD	0.0441	0.0644	0.0682	0.0482	0.1112	0.0931
VSM	0.0191	0.0287	0.0292	0.0268	0.0597	0.0498
Money	0.0772	0.1325	0.1315	0.0476	0.1180	0.0932
Adamic-Adar	0.0676	0.0936	0.0991	0.0532	0.1236	0.1006
MCN	0.0631	0.0847	0.0920	0.0501	0.1141	0.0948
PageRank Pers.	0.0598	0.1076	0.0996	0.0336	0.0855	0.0635
Jaccard	0.0226	0.0281	0.0320	0.0304	0.0700	0.0586
iMF	**0.0834**	**0.1414**	**0.1384**	0.0541	**0.1351**	0.1045
User-based kNN	0.0805	0.1308	0.1360	0.0479	0.1211	0.0955
Item-based kNN	0.0739	0.1119	0.1174	0.0360	0.0859	0.0724
Popularity	0.0255	0.0368	0.0376	0.0225	0.0505	0.0422
Random	0.0009	0.0017	0.0013	0.0003	0.0006	0.0003

approaches, and near to the popularity baseline. Query likelihood, personalized PageRank and MCN stand as mid-packers in both datasets. Finally, classic collaborative filtering and Money show very different behaviors in both datasets: on 1 month they are among the top 5 algorithms, while on 200 tweets they are far from the best, leveled with query likelihood.

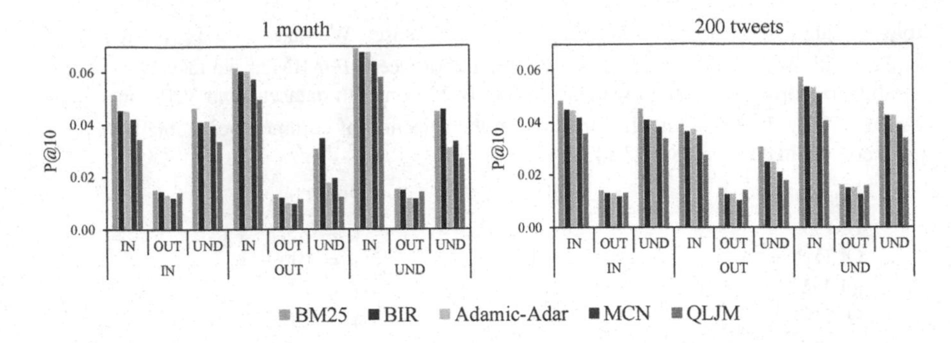

Fig. 3. P@10 values for the different possible choices for Γ^d and Γ^q on a selection of the most effective algorithms in the comparative included in Table 4.

We can also examine which neighbor orientation works best in the neighborhood-based algorithms –whether users are better represented by their followers, their followees, or both. Figure 3 shows a detailed comparison of all combinations for this setting. The outer labels on the *x* axis show the neighborhood orientation for the target user, and the inner ones for the candidate user. We can see that the undirected neighborhood Γ_{und} is consistently the most effective representation for target users, whereas the incoming neighborhood Γ_{in} works best for candidate users.

All in all, we find that BM25 makes for a highly competitive contact recommendation approach. One of the reasons for this is likely its ability to take advantage of interaction frequency (edge weights) better than any other algorithm –in fact, all other algorithms except VSM produce worse results when using a non-binary edge representation. BM25 is however not the top algorithm, since iMF overall has a slight advantage in effectiveness. Money and kNN get decent results in one dataset, but quite suboptimal in the other. We may therefore say BM25 is a decent second best in recommendation accuracy after matrix factorization. We find however important advantages to BM25 in terms of computational cost and simplicity, as we examine in the next section.

7 Complexity Analysis: BM25 Vs. Matrix Factorization

Computational cost and simplicity are critical in a commercial deployment of recommendation algorithms, which have to provide recommendations in real time. We focus on two aspects in our analysis: (a) generating recommendations from scratch, and (b) updating or retraining the algorithms each time a new user or a new link is added to the network. We first examine the cost analytically, and then we run a small test to observe the empirical difference.

7.1 Theoretical Analysis

The complexity analysis for generating recommendations for scratch is shown in Table 5, for the algorithms tested in the previous section. We can see that, in general, IR models are the fastest, along with MCN, Jaccard and Adamic-Adar, whereas implicit MF is among the costliest algorithms.

The reason why IR models (and, similarly, MCN, Jaccard and Adamic-Adar) are so fast is that we can take advantage of IR index-based optimizations, such as the "term-at-a-time" or "document-at-a-time" algorithms for fast query response-time [8]. If we store the network as an inverted index, as shown in Fig. 2, it suffices to run over the "posting lists" of target user neighbors (the "query terms") in linear time to generate a recommendation. The resulting average complexity of this is the square of the average network degree. The training time $O(|E|)$ in the table for these algorithms just corresponds to the straightforward computation of certain values such as the length of the neighborhoods.

Table 5. Running time complexity of the different algorithms, grouped by families. We show the complexity for both the full training, and the recommendation score computation (excluding the additional $\log N$ for final rank sorting). The variable m denotes the average network degree; c is the number of iterations for personalized PageRank, Money and iMF; and k represents the number of latent factors in iMF, and the number of neighbors in kNN.

Algorithms	Training	Recommendation
IR models	$O(\|E\|)$	$O(\|\mathcal{U}\|m^2)$
Jaccard, Adamic-Adar	$O(\|E\|)$	$O(\|\mathcal{U}\|m^2)$
MCN	-	$O(\|\mathcal{U}\|m^2)$
Random walks	$O\left(c\|\mathcal{U}\|^2 + c\|\mathcal{U}\|\|E\|\right)$	$O(\|\mathcal{U}\|)$
User/Item-based kNN	$O(\|E\| + \|\mathcal{U}\|m^2 \log k)$	$O\left(\|\mathcal{U}\|^2 k\right)$
iMF	$O(c(k^2\|E\| + k^3\|\mathcal{U}\|))$	$O\left(\|\mathcal{U}\|^2 k\right)$
Popularity	$O(\|E\|)$	$O(\|\mathcal{U}\|)$
Random	-	$O\left(\|\mathcal{U}\|^2\right)$

Implicit MF, on its side, is quadratic on the number of users, linearly multiplied by the number of latent factors. Yet worse, the same cost is incurred to produce recommendations after the training phase. Adding to this, iMF has three parameters to configure while BM25 has only two, which implies additional savings on the side of BM25 in parameter tuning cost. In terms of memory spending, assuming an all-in-memory implementation, iMF uses $2k^2|\mathcal{U}|$ decimal values, whereas BM25 only needs $3|\mathcal{U}|$ values (neighborhood length, size, and RSJ), which can make a considerable difference.

Matrix factorization is moreover not particularly flexible to incremental updates for incoming data. Update approaches have been proposed [34] by which a new link can be added in $O(mk^2 + k^3)$ time –though this does not work as an exact retraining, as it comes at the expense of incremental accuracy losses in the updated model. In contrast, BM25 can be updated in $O(1)$ for a single new link, by storing neighborhood lengths and RSJ values in the index. When a new user comes in, all values of RSJ need updating, involving an additional $O(|\mathcal{U}|)$. BM25 therefore enables fast updates, and better yet, equivalent to a full retraining. User-based kNN also enables lossless updates, but these take $O((|\mathcal{U}| + m)\log|\mathcal{U}|)$ time, which is even significantly heavier than the iMF update.

7.2 Empirical Observation

In order to observe what the theoretical analysis translates to in quantitative terms, we carry out an incremental update experiment where we test the running times for BM25, implicit MF, and –as a third-best algorithm– user-based kNN. For the 1 month network, we randomly sample 10% of the users, along with all links between them, and take this small graph as the starting point for a growing network. Over that reduced network, we train and run both recommendation algorithms. Then, we randomly sample and add one user at a time from the remaining 90% of users. For each new user, we add all its edges pointing to or from the users in the growing network. Then, we generate recommendations for all users in the subset. We continue the process until all users have been added to the growing network. We compute separately the time taken to update the recommender, and the time spent in generating the corresponding recommendations.

Figure 4 shows the time cost for both tasks: the advantage of BM25 over iMF and kNN is apparent. In incremental update (Fig. 4 right), the difference is in fact overwhelming –notice the logarithmic scale in the y axis, which means that updating BM25 is indeed orders of magnitudes faster than its two counterparts. It should moreover be noted that iMF and kNN are configured here with $k = 10$ (factors and neighbors, respectively). If we increased this parameter –as in the optimal configurations shown in Table 3– the cost would increase even further and faster.

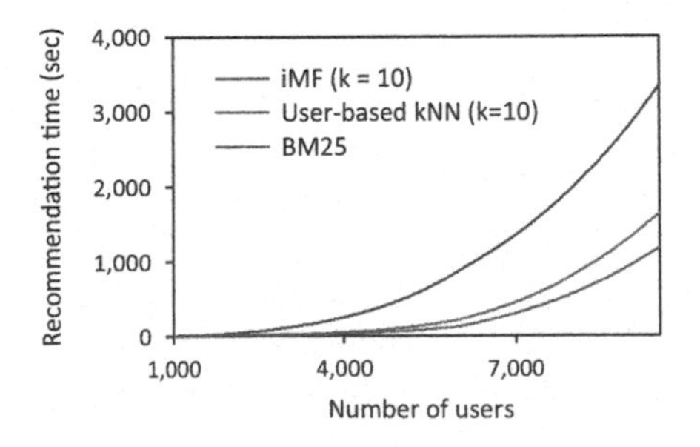

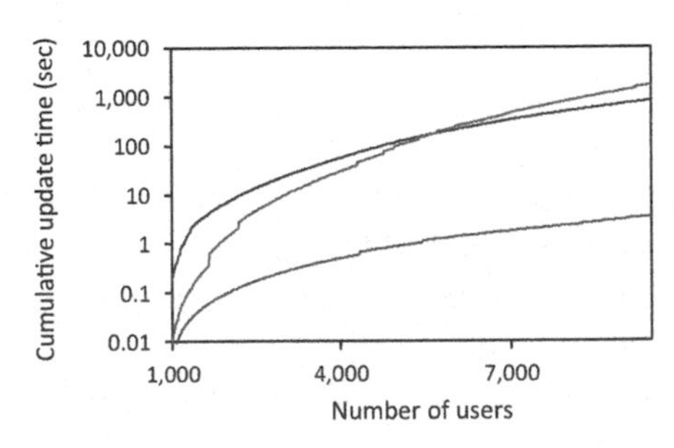

Fig. 4. Time comparison between BM25, user-based kNN and implicit matrix factorization.

8 Conclusions and Future Work

Though separately developed to much extent by different communities, text-based search and recommendation are very related tasks. This relation has been explored in prior research on the general perspective of adapting IR techniques to item recommendation [7, 31]. In our present work, we particularize this step to the recommendation of contacts in social networks. Our research has found that adapting IR models leads to empirically effective solutions, and to some extent simpler than previously developed adaptations for the general item recommendation task. We find that BM25 in particular is competitive with the best state-of-the-art approaches in terms of effectiveness over Twitter interaction networks. At the same time, IR models are orders of magnitude faster to run and update than the most effective recommendation algorithms.

Compared with alternative heuristic solutions, translating new and principled IR models to contact recommendation can add new and deeper insights to our understanding of the task and how we solve it, by importing the theory and foundations upon which the IR models were developed. Reciprocally, this can contribute to a broader perspective on IR models, their meaning, interpretation, and usefulness in different tasks, bringing higher levels of abstraction. We have thus found for instance that IR models tend to take better advantage of user-user interaction frequency than heuristic algorithms. We have likewise observed that followers seem to describe the social value of candidate recommendations better than followees, whereas the union of both consistently appears to best represent the social needs of target users.

We envision continuing this line of research to deeper levels in future work. We also plan to extend our current research by considering further evaluation dimensions beyond accuracy, such as recommendation novelty and diversity [9, 29], or the effects that recommendation can have on the evolution of the network structure [3, 28].

Acknowledgements. This work was funded by the Spanish Government (grant nr. TIN2016-80630-P).

References

1. Adamic, L.A., Adar, E.: Friends and neighbors on the web. Soc. Netw. **25**(3), 211–230 (2003)
2. Adomavicius, G., Tuzhilin, A.: Toward the next generation of recommender systems: a survey of the state-of-the-art and possible extensions. IEEE Trans. Knowl. Data Eng. **17**(6), 734–749 (2005)
3. Aiello, L., Barbieri, N.: Evolution of ego-networks in social media with link recommendations. In: 10th ACM International Conference on Web Search and Data Mining (WSDM 2017), pp. 111–120. ACM, New York (2017)
4. Backstrom, L., Leskovec, J.: Supervised random walks: predicting and recommending links in social networks. In: 4th ACM International Conference on Web Search and Data Mining (WSDM 2011), pp. 635–644. ACM, New York (2011)
5. Baeza-Yates, R., Ribeiro-Neto, B.: Modern information retrieval: the concepts and technology behind search, 2nd edn. Addison-Wesley, Harlow (2011)

6. Belkin, N.J., Croft, W.B.: Information filtering and information retrieval: two sides of the same coin? Commun. ACM **35**(12), 29–38 (1992)
7. Bellogín, A., Wang, J., Castells, P.: Bridging memory-based collaborative filtering and text retrieval. Inf. Retrieval **16**(6), 697–724 (2013)
8. Büttcher, S., Clarke, C.L.A., Cormack, G.V.: Information retrieval: implementing and evaluating search engines. MIT Press, Cambridge (2010)
9. Castells, P., Hurley, N.J., Vargas, S.: Novelty and Diversity in Recommender Systems. In: Ricci, F., Rokach, L., Shapira, B. (eds.) Recommender Systems Handbook, pp. 881–918. Springer, Boston, MA (2015). https://doi.org/10.1007/978-1-4899-7637-6_26
10. Goel, A., Gupta, P., Sirois, J., Wang, D., Sharma, A., Gurumurthy, S.: The who-to-follow system at twitter: strategy, algorithms and revenue impact. Interfaces **45**(1), 98–107 (2015)
11. Goel, A., Gupta, P., Sirois, J., Wang, D., Sharma, A., Gurumurthy, S.: WTF: the who to follow service at Twitter. In: 22nd International Conference on World Wide Web (WWW 2013), pp. 505–514. ACM, New York (2013)
12. Guy, I.: Social Recommender Systems. In: Ricci, F., Rokach, L., Shapira, B. (eds.) Recommender Systems Handbook, 2nd edn, pp. 511–543. Springer, New York (2015)
13. Hannon, J., Bennet, M., Smyth, B.: Recommending Twitter users to follow using content and collaborative filtering approaches. In: 4th ACM conference on Recommender Systems (RecSys 2010), pp. 199–206. ACM, New York (2010)
14. Al Hasan, M., Chaoji, V., Salem, S., Zaki, M.: Link prediction using supervised learning. In: SDM 06 Workshop on Link Analysis, Counterterrorism and Security at the 5th SIAM International Conference of Data Mining (2006)
15. Hu, Y., Koren, Y., Volinsky, C.: Collaborative filtering for implicit feedback datasets. In: 8th IEEE International Conference on Data Mining (ICDM 2008), pp. 263–272, IEEE Press, New York (2008)
16. Jaccard, P.: Étude de la distribution florale dans une portion des Alpes et du Jura. Bulletin de la Société Vaudoise des Sciences Naturelles **37**(142), 547–579 (1901)
17. Jelinek, F., Mercer, R.L.: Interpolated estimation of Markov source parameters from sparse data. In: Gelsema, E.S., Kanal, L.N. (eds.) Pattern Recognition in Practice, pp. 381–397. North-Holland, Amsterdam (1980)
18. Liben-Nowell, D., Kleinberg, J.: The link-prediction problem for social networks. J. Am. Soc. Inform. Sci. Technol. **58**(7), 1019–1031 (2007)
19. Lü, L., Zhou, T.: Link prediction in social networks: a survey. Phys. A **390**(6), 1150–1170 (2010)
20. Mackay, D., Bauman, L.: A hierarchical Dirichlet language model. Nat. Lang. Eng. **1**(3), 289–307 (1995)
21. Ning, X., Desrosiers, C., Karypis, G.: A comprehensive survey of neighborhood-based recommendation methods. In: Ricci, F., Rokach, L., Shapira, B. (eds.) Recommender Systems Handbook, pp. 37–76. Springer, Boston, MA (2015). https://doi.org/10.1007/978-1-4899-7637-6_2
22. Ponte, J.M., Croft, W.B.: A language modelling approach to information retrieval. In: 21st Annual International ACM SIGIR Conference on Research and Development in Information Retrieval (SIGIR 1998), pp. 275–281. ACM Press, New York (1998)
23. RankSys library. http://ranksys.org. Accessed 11 Oct 2018
24. Robertson, S.E.: The probability ranking principle in IR. J. Doc. **33**(4), 294–304 (1977)
25. Robertson, S.E., Zaragoza, H.: The probabilistic relevance framework: BM25 and beyond. Found. Trends Inform. Retrieval **3**(4), 333–389 (2009)
26. Salton, G., Wong, A., Yang, C.: A vector space model for automatic indexing. Commun. ACM **18**(11), 613–620 (1975)

27. Salton, G., McGill, M.: Introduction to Modern Information Retrieval. McGraw-Hill, New York (1983)
28. Sanz-Cruzado, J., Castells, P.: Enhancing structural diversity in social networks by recommending weak ties. In: 12th ACM Conference on Recommender Systems (RecSys 2018), pp. 233–241. ACM, New York (2018)
29. Sanz-Cruzado, J., Pepa, S.M., Castells, P.: Structural novelty and diversity in link prediction. In: 9th International Workshop on Modelling Social Media (MSM 2018) at The Web Conference (WWW Companion 2018), pp. 1347–1351. International World Wide Web Conferences Steering Committee, Republic and Canton of Geneva, Switzerland (2018)
30. Sanz-Cruzado, J., Castells, P.: Contact recommendations in social networks. In: Berkovsky, S., Cantador, I., Tikk, D. (eds.) Collaborative recommendations: algorithms, Practical Challenges and Applications, pp. 525–577. World Scientific Publishing, Singapore (2018)
31. Valcarce, D.: Exploring statistical language models for recommender systems. In: 9th ACM Conference on Recommender Systems (RecSys 2015), pp. 375–378. ACM Press, New York (2015)
32. Valcarce, D., Parapar, J., Barreiro, A.: Axiomatic analysis of language modelling of recommender systems. Int. J. Uncertainty, Fuzziness Knowl.-based Syst. **25**(Suppl. 2), 113–127 (2017)
33. White, S., Smyth, P.: Algorithms for estimating relative importance in networks. In: Proceedings of the 9th ACM SIGKDD International Conference on Knowledge Discovery and Data Mining (KDD 2003), pp. 266–275. ACM, New York (2003)
34. Yu, T., Mengshoel, O., Jude, A., Feller, E., Forgeat, J., Radia, N.: Incremental learning for matrix factorization in recommender systems. In: 2016 IEEE Conference on Big Data (BigData 2016), pp. 1056–1063. IEEE Press, New York (2016)

Conformative Filtering for Implicit Feedback Data

Farhan Khawar(✉) and Nevin L. Zhang

Department of Computer Science and Engineering,
The Hong Kong University of Science and Technology, Clear Water Bay, Hong Kong
{fkhawar,lzhang}@cse.ust.hk

Abstract. Implicit feedback is the simplest form of user feedback that can be used for item recommendation. It is easy to collect and is domain independent. However, there is a lack of negative examples. Previous work tackles this problem by assuming that users are not interested or not as much interested in the unconsumed items. Those assumptions are often severely violated since non-consumption can be due to factors like unawareness or lack of resources. Therefore, non-consumption by a user does not always mean disinterest or irrelevance. In this paper, we propose a novel method called Conformative Filtering (CoF) to address the issue. The motivating observation is that if there is a large group of users who share the same taste and none of them have consumed an item before, then it is likely that the item is not of interest to the group. We perform multidimensional clustering on implicit feedback data using hierarchical latent tree analysis (HLTA) to identify user "taste" groups and make recommendations for a user based on her memberships in the groups and on the past behavior of the groups. Experiments on two real-world datasets from different domains show that CoF has superior performance compared to several common baselines.

Keywords: Implicit feedback · One class collaborative filtering · Recommender systems

1 Introduction

With the advent of the online marketplace, an average user is presented with an un-ending choice of items to consume. Those could be books to buy, web-pages to click, songs to listen, movies to watch, and so on. Online stores and content providers no longer have to worry about shelf space to display their items. However, too much choice is not always a luxury. It can also be an unwanted distraction and makes it difficult for a user to find the items she desires. It is necessary to automatically filter a vast amount of items and identify those that are of interest to a user.

Collaborative filtering (CF) [7] is one commonly used technique to deal with the problem. Most research work on CF focuses on explicit feedback data, where ratings on items have been previously provided by users [13]. Items with high

L. Azzopardi et al. (Eds.): ECIR 2019, LNCS 11437, pp. 164–178, 2019.
https://doi.org/10.1007/978-3-030-15712-8_11

ratings are preferred over those with low ratings. In other words, items with high ratings are positive examples, while those with low ratings are negative examples. Unrated items are missing data.

In practice, one often encounters implicit feedback data, where users did not explicitly rate items [18]. Recommendations need to be made based on user activities such as clicks, page views, and purchase actions. Those are positive-only data and contain information regarding which items were consumed. There is no information about the unconsumed items. In other words, there are no negative examples. The problem is hence called *one class collaborative filtering (OCCF)* [20].

In previous works, the lack of negative examples in OCCF is addressed by adopting one of the following four strategies with respect to each user: (1) Treat unconsumed items as negative examples [19]; (2) Treat unconsumed items as negative examples with low confidence [11]; (3) Identify some unconsumed items as negative examples using heuristics [20]; (4) Assume the user prefers consumed items over unconsumed items [25]. We refer to the strategies as the *unconsumed as negative (UAN)*, *UAN-with-low-confidence*, *UAN-with-chance* and *consumed preferred over unconsumed (CPU)* assumptions respectively.

All the assumptions are problematic. The UAN assumption is in contradiction with the very objective of collaborative filtering—to identify items that might be of interest to a user among those she did not consume before. Moreover, if we assume a user does not like two items to exactly the same degree, then theoretically there is 50% chance that she would prefer the next item she chooses to consume to the last item she consumed.

In this paper, we adopt a new assumption: If there is a large group of users who share the same taste and none of them have consumed an item before, then the item is not of interest to the group. By a *taste* we mean the tendency to consume a certain collection of items such as comedy movies, pop songs, or spicy food. We call our assumption the *group UAN* assumption because it is with respect to a user group. In contrast, we refer to the first assumption mentioned above as the *individual UAN* because it is with respect to an individual user. Group UAN is more reasonable than individual UAN because there is less chance of treating unawareness as disinterest.

We identify user taste groups by performing multidimensional clustering using hierarchical latent class analysis (HLTA) [2]. HLTA can detect sets of items that tend to be *co-consumed* in the sense users who consumed some of the items in a set often also consumed others items in the set, albeit not necessarily at the same time. HLTA can also determine the users who showed the tendency to consume the items in a co-consumption set. Those users make up a taste group. To make recommendation for a user, we consider her memberships in the taste groups and past behaviors of those groups. We call this method *Conformative Filtering (CoF)* because a user is expected to conform to the behaviors of the groups she belongs to.

The main contributions of this paper include:

1. Proposing an intuitively appealing strategy, namely group UAN, to deal with the lack of negative examples;
2. Proposing a novel framework for OCCF, i.e., CoF, that is based on this assumption;
3. Using HLTA, an algorithm proposed for text analysis, to solve a fundamental problem in collaborative filtering;

The empirical results show that CoF significantly outperforms the state-of-the-art OCCF recommenders in predicting the items that users want to consume in the future. In addition, the latent factors in CoF are more interpretable than those in matrix factorization methods.

2 Related Work

In the model-based approach to collaborative filtering, the goal is to find a feature vector $\mathbf{f_u}$ for each user u and a feature vector $\mathbf{f_i}$ for each item i, and predict the rating of user u for item i using the inner products of the two vectors, i.e., $\hat{r}_{ui} = <\mathbf{f_u}, \mathbf{f_i}>$. The dimension of $\mathbf{f_u}$ and $\mathbf{f_i}$ is usually much smaller than the number of users and the number of items.

Let $\mathcal{C}$ be the set all *consumption pairs*, i.e., user-item pairs (u, i) such that u consumed i before. The complement $\mathcal{U}$ of $\mathcal{C}$ consists of *non-consumption pairs*. In the case of explicit feedback data, we have a rating r_{ui} for each pair $(u, i) \in \mathcal{C}$. It is the rating for item i given by user u and its possible values are usually the integers between 1 and 5. The feature vectors can obtained by minimizing the following loss function:

$$\sum_{(u,i)\in\mathcal{C}} (r_{ui} - \hat{r}_{ui})^2 + \text{regularization terms.}$$

In the literature, this is known as the matrix factorization (MF) method [14] because $[\mathbf{f_u}]^\top[\mathbf{f_i}]$ is an approximate low-rank factorization of the user-item matrix $[r_{ui}]$.

For implicit feedback data, researchers usually set $r_{ui} = 1$ for consumption pairs $(u, i) \in \mathcal{C}$. There is no information about r_{uj} for non-consumption pairs $(u, j) \in \mathcal{U}$. In this case, minimizing Eq. (1) would lead to non-sensible solutions. Several methods have been proposed to solve the problem. We briefly review them below. Regularization terms and constraints are ignored for simplicity.

The sparse linear method (SLIM) [19] makes the individual UAN assumption and sets $r_{uj} = 0$ for all $(u, j) \in \mathcal{U}$. It minimizes:

$$\sum_{(u,i)\in\mathcal{C}} (1 - \hat{r}_{ui})^2 + \sum_{(u,j)\in\mathcal{U}} (0 - \hat{r}_{uj})^2 + \text{regularization terms.}$$

In addition, it lets $\mathbf{f_u}$ be the binary vector over items that represents past consumptions of user u, and it only finds $\mathbf{f_i}$.

The weighted regularized MF (WRMF) [11,20] algorithm makes the UAN-with-low-confidence assumption and minimizes:

$$\sum_{(u,i)\in\mathcal{C}} (1-\hat{r}_{ui})^2 + \sum_{(u,j)\in\mathcal{U}} c_{uj}(0-\hat{r}_{uj})^2 + \text{regularization terms},$$

where $0 \leq c_{uj} \leq 1$ for all $(u,j) \in \mathcal{U}$. The values of the weights c_{uj} indicate the confidence in treating the non-consumption pairs as negative examples.

The negative sampling method [20] makes the UAN-with-chance assumption and minimizes:

$$\sum_{(u,i)\in\mathcal{C}} (1-\hat{r}_{ui})^2 + \sum_{(u,j)\in\mathcal{U}'} (0-\hat{r}_{uj})^2 + \text{regularization terms},$$

where $\mathcal{U}'$ is a randomly sampled subset of $\mathcal{U}$.

The overlapping co-cluster recommendation (Ocular) algorithm [9] minimizes:

$$-\sum_{(u,i)\in\mathcal{C}} \log(|1-e^{-\hat{r}_{ui}}|) - \sum_{(u,j)\in\mathcal{U}} \log(|0-e^{-\hat{r}_{uj}}|) + \text{regularization terms}.$$

This loss functions gives large penalty if $\hat{r}_{ui}$ is close to 0 for consumption pairs (u,i) and small penalty if $\hat{r}_{uj}$ is close to 1 for non-consumption pairs (u,j). There is stronger "force" pushing $\hat{r}_{ui}$ toward 1 and weaker "force" pushing $\hat{r}_{uj}$ toward 0. So, ocular is implicitly making the UAN-with-low-confidence assumption.

The Bayesian personalized ranking MF (BPRMF) [25] algorithm makes the CPU assumption and minimizes:

$$\sum_{u}\sum_{i\in\mathcal{C}_u}\sum_{j\in\mathcal{U}_u} -\log\sigma(\hat{r}_{ui}-\hat{r}_{uj}) + \text{regularization terms},$$

where $\mathcal{C}_u = \{i|(u,i) \in \mathcal{C}\}$, $\mathcal{U}_u = \{j|(u,j) \in \mathcal{U}\}$, and σ is the sigmoid function. The penalty for a user u is small if the predicted scores $\hat{r}_{ui}$ for the consumed items are large relative to the predicted scores r_{uj} for the unconsumed items.

Various extensions of the aforementioned methods have been proposed. SLIM has been extended by [4,5,15], WRMF has been extended by [8,24,27], and BPRMF has been extended by [10,21,22,26].

Clustering algorithms have also been applied to CF. The k-means algorithm and hierarchical clustering have been used to group either users or items to reduce the time complexity of CF methods such as user-kNN and item-kNN [1,13]. Co-clustering has been used to identify user-item co-clusters so as to model user group heterogeneity [9,29,30]. In [29,30], the authors find multiple sub-matrices (co-clusters) of the user-item matrix, apply another CF method on each sub-matrix, and then aggregate the results. In ocular [9], the authors first obtain feature vectors f_u and f_i, and then use those vectors to produce user-item co-clusters for the sake of interpretability of the results. The notion of user groups is used in an extension of BPRMF called GBPR [22]. In GBPR

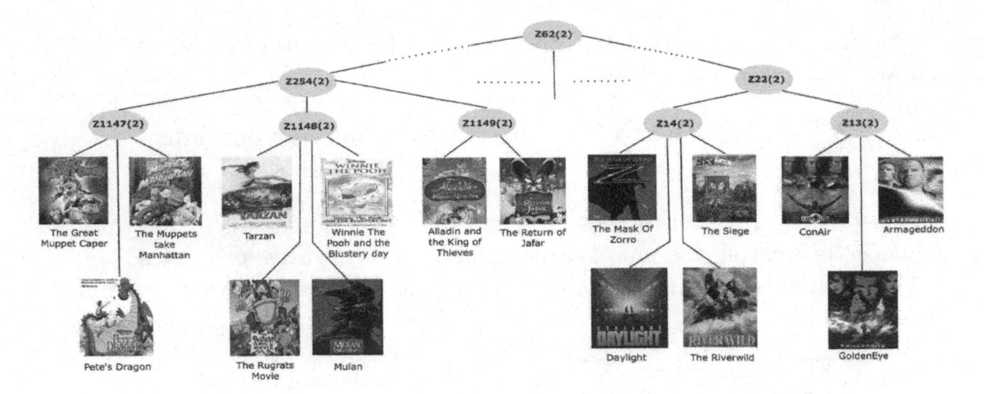

Fig. 1. A part of the hierarchical latent tree model learned from Movielens dataset. The level-1 latent variables reveal co-consumption of items by users and identify user tastes for various subsets of items. Latent variables at higher levels reveal co-occurrence of the tastes at the level below and identify more broad tastes.

Table 1. User clusters identified by the latent variables Z_{13} and Z_{1147}. High percentages of the users in the cluster $Z_{13} = s_1$ have watched the three movies **Armageddon**, **Golden Eye** and **Con Air**. Hence, the cluster is regarded as a user group with a taste for the three movies. Similarly, the cluster $Z_{1147} = s_1$ is regarded as a user group with a taste for the three movies **The Great Muppet Caper**, **Pete's Dragon** and **The Muppets take Manhattan**.

Action-Adventure-Thriller	$Z_{13}=s_1$ (0.21)	$Z_{13}=s_0$ (0.79)
Armageddon	0.610	0.055
Golden_Eye	0.588	0.013
Con_Air	0.635	0.014

Children-Comedy	$Z_{1147}=s_1$ (0.09)	$Z_{1147}=s_0$ (0.91)
Great_Muppet_Caper_The	0.456	0.009
Petes_Dragon	0.450	0.004
Muppets_Take_Manhattan_The	0.457	0.005

a group of users is formed for each consumption pair (u, i), and it consists of a few randomly selected other users who also consumed the item i before. In CoF a user taste group is determined based on a set of items that tend to be co-consumed, and it is typically quite large. However, none of the aforementioned methods use the (user, item, or user-item) clusters obtained to deal with the lack of negative preference in implicit feedback data.

3 User Taste Group Detection Using HLTA

When applied to implicit feedback data, HLTA[1] learns models such as the one shown in Fig. 1, which was obtained from the Movielens dataset. Movielens is an explicit feedback dataset. It was turned into an implicit feedback dataset by ignoring the item ratings[2].

[1] https://github.com/kmpoon/hlta.

[2] Movielens is used for illustration since movies genres are easier to interpret.

The model is a tree-structured Bayesian network, where there is a layer of observed variables at the bottom, and multiple layers of latent variables on top. It is called a *hierarchical latent tree (HLTM)* model [2,17]. The model parameters include a marginal distribution for the root[3] and a conditional distribution for each of the other nodes given its parent. The product of the distributions defines a joint distribution over all the variables.

In this paper, all the variables are assumed to be binary. The observed variables indicate whether the items were consumed by a user. For example, the value of the variable `Mulan` for a user is 1 if she watched the movie before, and 0 otherwise. Note that here the value 0 means non-consumption, not disinterest.

The latent variables are introduced during data analysis to explain co-consumption patterns detected in data. For example, the fact that the variables `Armageddon`, `Golden Eye` and `Con Air` are grouped under Z_{13} indicates that the three movies tend to be co-consumed, in the sense users who watched one of them often also watched the other two. The pattern is explained by assuming that there is a taste, denoted by Z_{13}, such that users with the taste tend to watch the movies and users without it do not tend to watch the movies. Similarly, Z_{14} explains the co-consumption of `The Seige`, `Mask of Zorro`, `Daylight` and `The River Wild`. Z_{22} indicates that the patterns represented by Z_{13} and Z_{14} tend to co-occur.

HLTMs are a generalization of latent class models (LCMs) [12], which is a type of finite mixture models for discrete data. In a finite mixture model, there is one latent variable and it is used to partition objects into soft clusters. Similarly, in an HLTM, each latent variable partitions all the users into two clusters. Since there are multiple latent variables, multiple partitions are obtained. In this sense, HLTMs are a tool for *multidimensional clustering* [3,16,31].

Information about the partition given by Z_{13} is shown in Table 1. The first cluster $Z_{13} = s_1$ consists of 21% of the users. High percentages of the users in the cluster have watched the three movies `Armageddon`, `Golden Eye` and `Con Air`. So, they have a taste for them. In contrast, few users in the second cluster $Z_{13} = s_0$ have watched these movies and hence they do not possess the taste.

Similarly, Z_{14} identifies another group of users with a taste for the movies `The Seige`, `Mask of Zorro`, `Daylight` and `The River Wild`. Z_{14} and Z_{13} are grouped under Z_{22} in the model structure, which indicates that the two tastes tend to be co-possessed, and Z_{22} identifies the users who tend to have both tastes.

4 Conformative Filtering

Suppose we have learned an HLTM m from an implicit feedback dataset and suppose there are K latent variables on the l-th level of the model, each with two states s_0 and s_1. Denote the latent variables as $Z_{l1}, \ldots, Z_{lK}$. They give us

[3] When there are multiple latent variables at the top level, arbitrarily pick one of them as the root.

K user taste groups $Z_{l1} = s_1, \ldots, Z_{lK} = s_1$, which will sometimes be denoted as $G_1, \ldots, G_K$ for simplicity. In this section, we explain how these user taste groups can be used for item recommendation.

4.1 User Group Characterization

A natural way to characterize the preferences of a user group for items is to aggregate past behaviors of the group members. The issue is somewhat complex for us because our user groups are soft clusters. Let $\mathbb{I}(i|u, \mathcal{D})$ be the indicator function which takes value 1 if user u has consumed item i before, and 0 otherwise. We determine the preference of a taste group G_k (i.e., $Z_{lk} = s_1$) for an item i using the *relative frequency that the item was consumed by users in the group*, i.e.:

$$\phi(i|G_k, \mathcal{D}) = \frac{\sum_u \mathbb{I}(i|u, \mathcal{D}) P(G_k|u, m)}{\sum_u P(G_k|u, m)}, \tag{1}$$

where $P(G_k|u, m)$ is the probability of user u belonging to group G_k, and the summations are over all the users who consumed item i before.

Note that $\phi(i|G_k, \mathcal{D}) = 0$ if no users in G_k have consumed the item i before. In other words, we assume that a group is not interested in an item if none of the group members have consumed the item before.

There is an important remark to make. The reason we determine the preferences of a user group G_k is that we want to predict future behavior of the group. As such, we might want to base the prediction on recent behaviors of the group members instead of their entire consumption histories. For example, we might want to choose to use a subset $\mathcal{D}_H$ of the data that consists of only the latest H consumptions for each user. We will empirically investigate this strategy and will show that the choice of H has an impact on the quality of item recommendations.

4.2 Item Recommendation

Having characterized the user taste groups, we now give feature vectors for items and users. We characterize item i using a vector where the k-th component is the relative frequency that it was consumed by members of group G_k, i.e.,

$$\mathbf{f}_i = (\phi(i|G_1, \mathcal{D}_H), \ldots, \phi(i|G_K, \mathcal{D}_H)). \tag{2}$$

Note that $\mathcal{D}_H$ is used instead of $\mathcal{D}$, which means that the latent representation is obtained from the H most recent consumptions of users.

We characterize user u using a vector where the k-th component is the probability that user u belongs to the group G_k, i.e.,

$$\mathbf{f}_u = (P(G_1|u, m), \ldots, P(G_k|u, m)). \tag{3}$$

The latent representations require the computation the posterior probabilities $P(G_k|u,m) = P(Z_{lk} = s_1|u,m)$ for $k = 1,\ldots,K$. Because m is a tree-structured model, all the posterior probabilities can be computed by propagating messages over the tree twice [23]. It takes time linear in the number of variables in the model, and hence linear in the number of items.

We use the inner product of the two vectors $\mathbf{f_i}$ and $\mathbf{f_u}$ as the predicted score $\hat{r}_{ui}$ for the user-item pair (u,i), i.e.,

$$\hat{r}_{ui} = \sum_{k=1}^{K} \phi(i|G_k, \mathcal{D}_H) P(G_k|u,m). \tag{4}$$

To make recommendations for a user u, we sort all the items i in descending order of the predicted scores $\hat{r}_{ui}$, and recommend the items with the highest scores.

4.3 Discussions

Matrix factorization (MF) is often used in collaborative filtering to map items and users to feature vectors in the same Euclidean space. The components of the vectors are called *latent factors*, which are not be confused with latent variables in latent tree models.

CoF differs fundamentally from MF. The latent factors in MF are obtained by factorizing the user-item matrix and they are not interpretable. In contrast, the latent factors in CoF are characteristics of user taste groups and they have clear semantics.

CoF naturally incorporates the group UAN assumption, which is the most reasonable assumption to deal with the lack of negative examples to date. In contrast, MF has only been extended to incorporate the individual UAN assumption and its variants. It cannot incorporate the group UAN assumption because there is no notion of user groups.

In addition, CoF has a desirable characteristic that is not shared by MF. It considers the entire consumption histories of the users when grouping them and uses only recent user behaviors when predicting what the groups would like to consume in the future. Consumption behaviors long ago are useful when identifying similar users. However, they might not be very useful when predicting what the users would like to consume in the future. Actually, they might be misleading.

5 Experiments

We performed experiments on two real-world datasets to compare CoF with five baselines using two evaluation metrics.

5.1 Datasets

The datasets used in our experiments are Movielens20M[4] and Ta-feng[5]. Movielens20M contains ratings given by users to the movies they watched. It is an explicit feedback dataset and was converted to implicit feedback data by keeping all the rating events and ignoring the rating values. Ta-feng contains purchase events at a supermarket, where each event is a customer-item pair and a checking-out action by a customer involves multiple events.

Statistics about the datasets are as follows:

Movielens20M	Users	Items	Sparsity
Train	118,526	15,046	99.047%
Validation	22,684	14,888	99.112%
Test	25,561	25,843	99.546%
Ta-feng			
Train	27,574	22,226	99.907%
Validation	12,261	15,206	99.934%
Test	13,191	14,561	99.936%

Each dataset is comprised of *(user, item, time-stamp)* tuples. Following [28], we split each dataset into training, validation and test sets by time. This is so that all the training instances came before all the testing instances, which matches real-world scenarios better than splits that do not consider time. We tested on several splits and the results were similar. In the following, we will only report the results on the split with 70% of the data for training, 15% for validation and 15% for test.

5.2 Baselines

In the Related Work section, we discussed five representative OCCF methods, namely WRMF, BPRMF, ocular (co-clustering), GBPR (group based) and SLIM (model based neighborhood method). They were all included in our experiments.[6] The implementation of the original authors was used for ocular, and the LibRec implementations[7] and the MyMediaLite implementations [6] were used for all other baselines.

All the algorithms require some input parameters. For the baselines, we tuned their parameters on the validation set through grid search, as is commonly done in the literature. The best parameters were chosen based on recall@R. The details

[4] https://grouplens.org/datasets/movielens/20m/.

[5] www.bigdatalab.ac.cn/benchmark/bm/dd?data=Ta-Feng.

[6] SLIM failed to finish a single run in one week on the Movielens20M dataset during validation, therefore its performance is not reported on this dataset.

[7] https://www.librec.net/index.html.

of the parameters chosen can be found in Table 2. The key parameters are: F—the number of latent factors; λ, $\beta/2$—weights for regularization terms; k—the size of neighborhood; $|G|$—group size; ρ—tuning parameter. We refer the reader to the original papers for the meanings of other parameters.

Table 2. Parameters selected by validation.

	Movielens20M	Ta-feng
CoF	$l=1, H=5$	$l=1$, $H=40$
WRMF	$F=40$, $\lambda=10^{-2}$	$F=10$, $\lambda=10$
BPRMF	$F=80$, $\lambda=10^{-2}$	$F=80$, $\lambda=10^{-4}$
Ocular	$F=120$, $\lambda=80$	$F=60$, $\lambda=120$
SLIM	N.A	$k=500$, $\lambda=10^{-2}$, $\beta/2=10$
GBPR	$F=160$, $\lambda = 10^{-4}$, $\|G\| = 10$, $\rho = 0.2$	$F=20$, $\lambda=0.1$, $\|G\| = 4$, $\rho = 0.4$

For CoF, we searched for the value of H over the set $\{2, 3, 4, 5, 10, 20, \ldots, 100\}$, and for l we considered all levels of the hierarchical model obtained by HLTA. The number K of latent variables on a level of the model was automatically determined by HLTA.

5.3 Evaluation Metrics

Two standard evaluation metrics are used in our experiments, namely recall@R, and area under the curve (AUC). The evaluation metrics are briefly described below:

- Recall@R: It is the fraction, among all items consumed by a user in the test set, of those that are placed at the top R positions in a recommended list. Formally, recall is defined as: $\frac{TP}{TP+FN}$, where TP denotes true positive and FN denotes false negative.
- AUC: Is the area under the Receiver Operating Characteristic curve. It is the probability that two items randomly picked from the recommendation list are in the correct order, i.e., the first is a consumed item and the second is an unconsumed item.

5.4 Results

The recall@R results, at each cutoff position R, are shown in Fig. 3. We see that CoF achieved the best performance on both datasets. Since CoF attempts to model each taste of a user individually, a higher recall suggests that these tastes are catered for while making recommendations. The improvements are comparatively larger on the Movielens20M dataset. This indicates that when the data is relatively less sparse, CoF is able to extract meaningful information much more effectively than other methods.

Ocular performed competitively with BPR and WRMF in terms of recall@R. However, WRMF and BPRMF performed better at larger cutoff values. Interestingly, we found that the performance of GBPR was better than all baselines on the ta-feng dataset but was unimpressive on Movielens20M despite extensive parameter tuning. A possible reason for this could be the tendency of GBPR to focus on popular items[8]. Ta-feng is grocery dataset and certain common items are in every customer's basket (e.g. bread), focusing on these may lead to a higher recall. On the other hand, the movie domain is comparatively more personalized and focusing on popular items might not cater to different tastes of a user.

Table 3 shows the performance in terms of AUC. CoF outperforms the baselines on both datasets. This suggests that CoF is able to identify the "true" negatives and it puts them lower than the items of interest in the ranked list. We see that BPRMF performs the second best w.r.t. AUC on both datasets. This is expected since BPRMF optimizes for AUC. Moreover, we note that WRMF performed better than BPRMF in terms of recall@R, however, it's performance in terms of AUC is lower. This provides further evidence that the score of the top-R metrics and those which evaluate over the whole list might not correlate and depending on the target of the recommendation an appropriate metric should be chosen. SLIM gave the lowest performance over all metrics in our experiments. It is worth noting that our experimental setup (splitting the data by time) and the metrics used differ from the experimental conditions under which SLIM is normally evaluated.

Table 3. The AUC for each recommender is shown. CoF outperforms other methods.

	BPRMF	WRMF	CoF	Ocular	SLIM	GBPR
Ta-feng	0.74977	0.71316	**0.7793**	0.63653	0.68321	0.71117
ML-20M	0.87289	0.85258	**0.88816**	0.84879	N.A	0.80367

5.5 Impact of Parameters

CoF begins by running HLTA to learn a hierarchical model and then uses the model for item recommendation. There are two parameters. The first parameter l determines which level of the hierarchy to use. The larger the l, the fewer the number of user taste groups. The second parameter H determines the amount of consumption history to use when characterizing user groups. Although both parameters are selected via validation, it would be interesting to gain some insights about how they impact the performance of CoF.

[8] During our experiments we found that GBPR has low global diversity. These results are not reported in the interest of space.

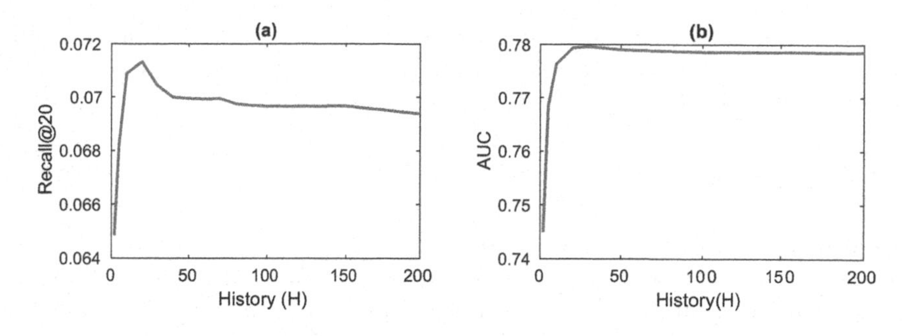

Fig. 2. Impact of parameters on the performance of CoF on Ta-feng dataset.

Figure 2(a) shows the recall@20 scores on ta-feng as a function of H when $l = 1$. We see that the curve first increases with H and then decreases with H. It reaches the maximum value when $H = 20$. The reason is that, when H is too small, the data used for user taste group characterization contain too little information. When H is too large, on the other hand, too much history is included and the data do not reflect the current interests of the user groups.

Figure 2(b) shows the AUC scores on ta-feng as a function of H when $l = 1$. We observe a similar trend, but the impacts of H are not as pronounced on AUC. CoF is more or less robust to the choice of l and the performance is almost the

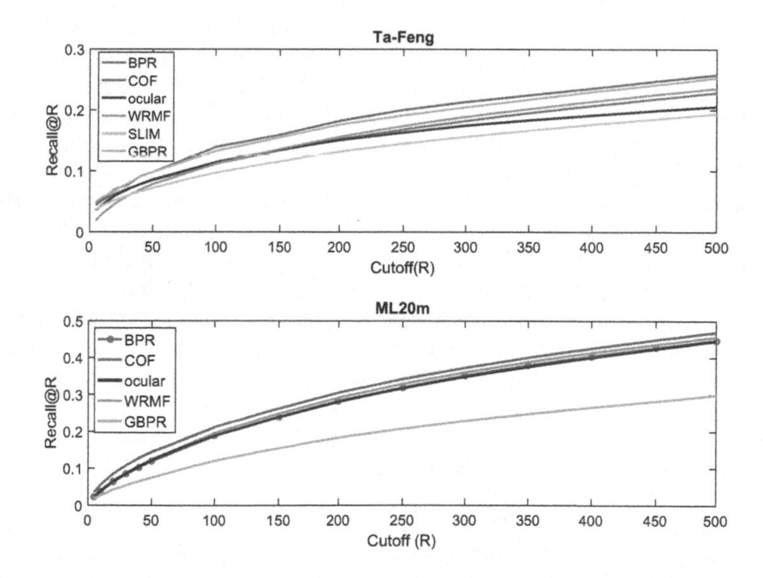

Fig. 3. The recall@R performance of different recommenders on the ta-feng and Movielens20M dataset. CoF exhibits the best performance on both datasets due to better representation of individual tastes.

same regardless of the level chosen. The results are not shown for brevity. This was somewhat unexpected as when l increases the taste become more general and one would expect the performance to deteriorate[9].

6 Conclusion

A novel method called CoF is proposed for collaborative filtering with implicit feedback data. It deals with the lack of negative examples which does not perform negative sampling, rather it uses the group UAN assumption, which is more reasonable than assumptions made by previous works. Extensive experiments were performed to compare CoF with a variety of baselines on two real-world datasets. CoF achieved the best performance over both recall@R and AUC signifying that various taste of an individual is captured and the true negatives are placed at the bottom of the ranked list.

Acknowledgment. Research on this article was supported by Hong Kong Research Grants Council under grant 16202118.

References

1. Aggarwal, C.C.: Recommender Systems: The Textbook, 1st edn. Springer, Cham (2016). https://doi.org/10.1007/978-3-319-29659-3
2. Chen, P., Zhang, N.L., Liu, T., Poon, L.K., Chen, Z., Khawar, F.: Latent tree models for hierarchical topic detection. Artif. Intell. **250**, 105–124 (2017)
3. Chen, T., Zhang, N.L., Liu, T., Poon, K.M., Wang, Y.: Model-based multidimensional clustering of categorical data. Artif. Intell. **176**(1), 2246–2269 (2012)
4. Cheng, Y., Yin, L., Yu, Y.: Lorslim: low rank sparse linear methods for top-n recommendations. In: 2014 IEEE International Conference on Data Mining (ICDM), pp. 90–99. IEEE (2014)
5. Christakopoulou, E., Karypis, G.: HOSLIM: higher-order sparse linear method for top-*n* recommender systems. In: Tseng, V.S., Ho, T.B., Zhou, Z.-H., Chen, A.L.P., Kao, H.-Y. (eds.) PAKDD 2014, Part II. LNCS (LNAI), vol. 8444, pp. 38–49. Springer, Cham (2014). https://doi.org/10.1007/978-3-319-06605-9_4
6. Gantner, Z., Rendle, S., Freudenthaler, C., Schmidt-Thieme, L.: Mymedialite: a free recommender system library. In: Proceedings of the Fifth ACM Conference on Recommender Systems, RecSys 2011, pp. 305–308. ACM, New York, NY, USA (2011). https://doi.org/10.1145/2043932.2043989
7. Goldberg, D., Nichols, D., Oki, B.M., Terry, D.: Using collaborative filtering to weave an information tapestry. Commun. ACM **35**(12), 61–70 (1992). https://doi.org/10.1145/138859.138867
8. He, X., Zhang, H., Kan, M.Y., Chua, T.S.: Fast matrix factorization for online recommendation with implicit feedback. In: Proceedings of the 39th International ACM SIGIR Conference on Research and Development in Information Retrieval, SIGIR 2016, pp. 549–558. ACM, New York, NY, USA (2016). https://doi.org/10.1145/2911451.2911489

[9] We did observe slight deterioration but the magnitude was too small to draw conclusions from.

9. Heckel, R., Vlachos, M., Parnell, T., Dünner, C.: Scalable and interpretable product recommendations via overlapping co-clustering. In: 2017 IEEE 33rd International Conference on Data Engineering (ICDE), pp. 1033–1044. IEEE (2017)
10. Hong, L., Bekkerman, R., Adler, J., Davison, B.D.: Learning to rank social update streams. In: Proceedings of the 35th International ACM SIGIR Conference on Research and Development in Information Retrieval, pp. 651–660. ACM (2012)
11. Hu, Y., Koren, Y., Volinsky, C.: Collaborative filtering for implicit feedback datasets. In: Proceedings of the 2008 Eighth IEEE International Conference on Data Mining, ICDM 2008, pp. 263–272. IEEE Computer Society, Washington, DC, USA (2008). https://doi.org/10.1109/ICDM.2008.22
12. Knott, M., Bartholomew, D.J.: Latent Variable Models and Factor Analysis. Edward Arnold, London (1999)
13. Koren, Y., Bell, R.: Advances in collaborative filtering. In: Ricci, F., Rokach, L., Shapira, B., Kantor, P. (eds.) Recommender Systems Handbook, pp. 77–118. Springer, Boston (2015). https://doi.org/10.1007/978-0-387-85820-3_5
14. Koren, Y., Bell, R., Volinsky, C.: Matrix factorization techniques for recommender systems. Computer **42**(8), 30–37 (2009)
15. Levy, M., Jack, K.: Efficient top-n recommendation by linear regression. In: RecSys Large Scale Recommender Systems Workshop (2013)
16. Liu, T.F., Zhang, N.L., Chen, P., Liu, A.H., Poon, L.K., Wang, Y.: Greedy learning of latent tree models for multidimensional clustering. Mach. Learn. **98**(1–2), 301–330 (2015)
17. Liu, T., Zhang, N.L., Chen, P.: Hierarchical latent tree analysis for topic detection. In: Calders, T., Esposito, F., Hüllermeier, E., Meo, R. (eds.) ECML PKDD 2014, Part II. LNCS (LNAI), vol. 8725, pp. 256–272. Springer, Heidelberg (2014). https://doi.org/10.1007/978-3-662-44851-9_17
18. Nichols, D.M.: Implicit ratings and filtering. In: Proceedings of the 5th DELOS Workshop on Filtering and Collaborative Filtering, vol. 12. ERCIM, Budapaest (1997)
19. Ning, X., Karypis, G.: Slim: Sparse linear methods for top-n recommender systems. In: 2011 IEEE 11th International Conference on Data Mining (ICDM), pp. 497–506. IEEE (2011)
20. Pan, R., et al.: One-class collaborative filtering. In: Proceedings of the 2008 Eighth IEEE International Conference on Data Mining, ICDM 2008, pp. 502–511. IEEE Computer Society, Washington, DC, USA (2008). https://doi.org/10.1109/ICDM.2008.16
21. Pan, W., Chen, L.: Cofiset: collaborative filtering via learning pairwise preferences over item-sets. In: Proceedings of the 2013 SIAM International Conference on Data Mining, pp. 180–188. SIAM (2013)
22. Pan, W., Chen, L.: GBPR: group preference based bayesian personalized ranking for one-class collaborative filtering. IJCAI **13**, 2691–2697 (2013)
23. Pearl, J.: Probabilistic Reasoning in Intelligent Systems: Networks of Plausible Inference. Morgan Kaufmann Publishers Inc., San Francisco (1988)
24. Pilászy, I., Zibriczky, D., Tikk, D.: Fast als-based matrix factorization for explicit and implicit feedback datasets. In: Proceedings of the Fourth ACM Conference on Recommender Systems, RecSys 2010, pp. 71–78. ACM, New York, NY, USA (2010). https://doi.org/10.1145/1864708.1864726

25. Rendle, S., Freudenthaler, C., Gantner, Z., Schmidt-Thieme, L.: BPR: Bayesian personalized ranking from implicit feedback. In: Proceedings of the Twenty-Fifth Conference on Uncertainty in Artificial Intelligence, UAI 2009, pp. 452–461. AUAI Press, Arlington, Virginia, United States (2009). http://dl.acm.org/citation.cfm?id=1795114.1795167
26. Takács, G., Tikk, D.: Alternating least squares for personalized ranking. In: Proceedings of the Sixth ACM Conference on Recommender Systems, pp. 83–90. ACM (2012)
27. Wang, K., Peng, H., Jin, Y., Sha, C., Wang, X.: Local weighted matrix factorization for top-n recommendation with implicit feedback. Data Sci. Eng. **1**(4), 252–264 (2016). https://doi.org/10.1007/s41019-017-0032-6
28. Wu, C.Y., Ahmed, A., Beutel, A., Smola, A.J., Jing, H.: Recurrent recommender networks. In: Proceedings of the Tenth ACM International Conference on Web Search and Data Mining, WSDM 2017, pp. 495–503. ACM, New York, NY, USA (2017). https://doi.org/10.1145/3018661.3018689
29. Wu, Y., Liu, X., Xie, M., Ester, M., Yang, Q.: Cccf: Improving collaborative filtering via scalable user-item co-clustering. In: Proceedings of the Ninth ACM International Conference on Web Search and Data Mining, WSDM 2016, pp. 73–82. ACM, New York, NY, USA (2016). https://doi.org/10.1145/2835776.2835836
30. Xu, B., Bu, J., Chen, C., Cai, D.: An exploration of improving collaborative recommender systems via user-item subgroups. In: Proceedings of the 21st International Conference on World Wide Web, WWW 2012, pp. 21–30. ACM, New York, NY, USA (2012). https://doi.org/10.1145/2187836.2187840
31. Zhang, N.L.: Hierarchical latent class models for cluster analysis. J. Mach. Learn. Res. **5**, 697–723 (2004). http://dl.acm.org/citation.cfm?id=1005332.1016782

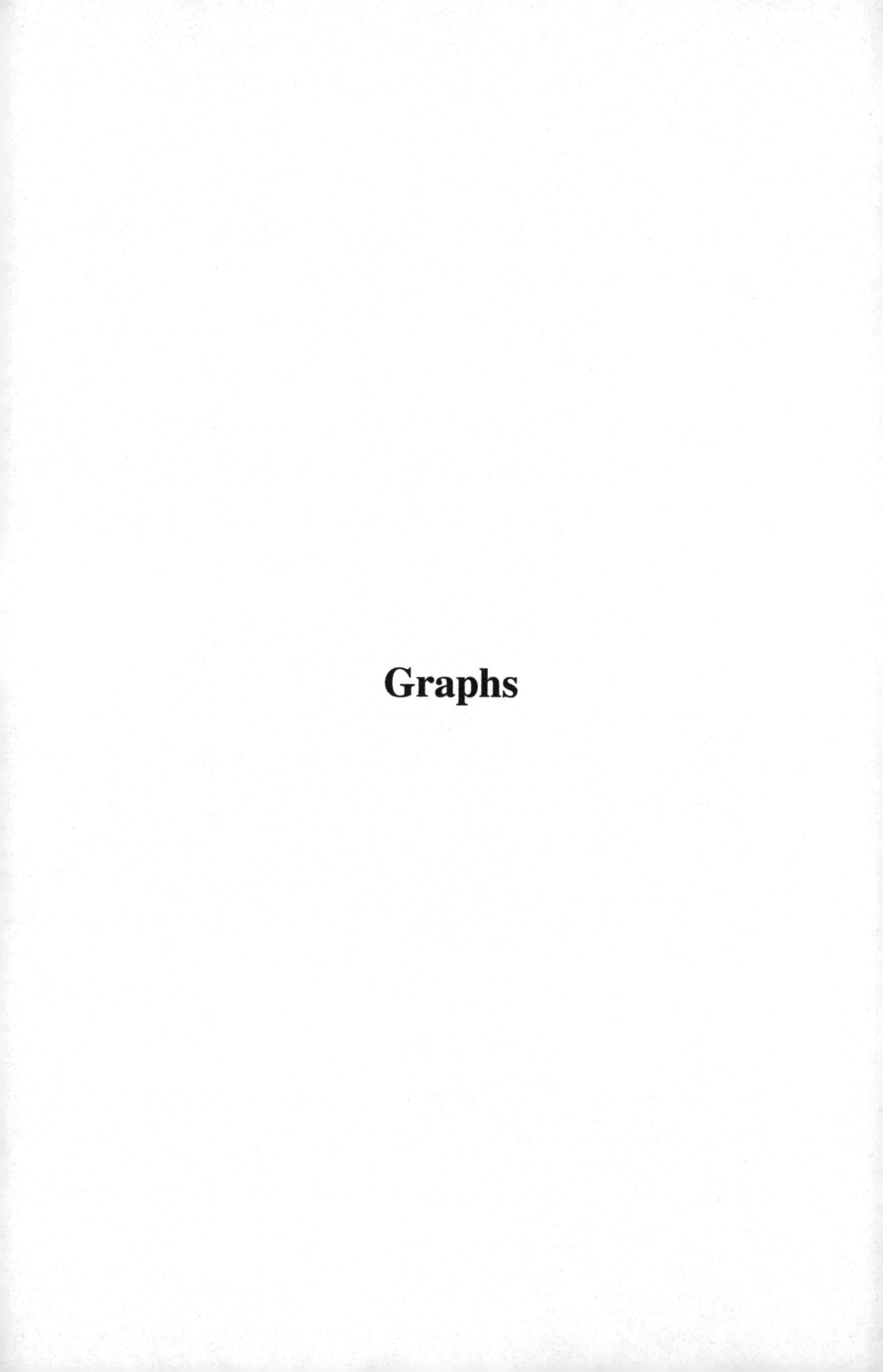

Graphs

Binarized Knowledge Graph Embeddings

Koki Kishimoto[1(✉)], Katsuhiko Hayashi[1,3], Genki Akai[1], Masashi Shimbo[2,3], and Kazunori Komatani[1]

[1] Osaka University, Osaka, Japan
kishimoto@ei.sanken.osaka-u.ac.jp
[2] Nara Institute of Science and Technology, Nara, Japan
[3] RIKEN Center for Advanced Intelligence Project, Tokyo, Japan

Abstract. Tensor factorization has become an increasingly popular approach to knowledge graph completion (KGC), which is the task of automatically predicting missing facts in a knowledge graph. However, even with a simple model like CANDECOMP/PARAFAC (CP) tensor decomposition, KGC on existing knowledge graphs is impractical in resource-limited environments, as a large amount of memory is required to store parameters represented as 32-bit or 64-bit floating point numbers. This limitation is expected to become more stringent as existing knowledge graphs, which are already huge, keep steadily growing in scale. To reduce the memory requirement, we present a method for binarizing the parameters of the CP tensor decomposition by introducing a quantization function to the optimization problem. This method replaces floating point–valued parameters with binary ones after training, which drastically reduces the model size at run time. We investigate the trade-off between the quality and size of tensor factorization models for several KGC benchmark datasets. In our experiments, the proposed method successfully reduced the model size by more than an order of magnitude while maintaining the task performance. Moreover, a fast score computation technique can be developed with bitwise operations.

Keywords: Knowledge graph completion · Tensor factorization · Model compression

1 Introduction

Knowledge graphs, such as YAGO [24] and Freebase [2], have proven useful in many applications such as question answering [3], dialog [17] and recommender [21] systems. A knowledge graph consists of triples (e_i, e_j, r_k), each of which represents a fact that relation r_k holds between subject entity e_i and object entity e_j. Although a typical knowledge graph may have billions of triples, it is still far from complete. Filling in the missing triples is of importance in carrying out various inference over knowledge graphs. *Knowledge graph completion* (KGC) aims to perform this task automatically.

L. Azzopardi et al. (Eds.): ECIR 2019, LNCS 11437, pp. 181–196, 2019.
https://doi.org/10.1007/978-3-030-15712-8_12

In recent years, *knowledge graph embedding* (KGE) has been actively pursued as a promising approach to KGC. In KGE, entities and relations are embedded in vector space, and operations in this space are used for defining a confidence score (or simply score) function θ_{ijk} that approximates the truth value of a given triple (e_i, e_j, r_k). Although a variety of KGE methods [4,6,20,23,25] have been proposed, Kazemi and Poole [11] and Lacroix et al. [14] found that a classical tensor factorization algorithm, known as the CANDECOMP/PARAFAC (CP) decomposition [10], achieves the state-of-art performances on several benchmark datasets for KGC.

In CP decomposition of a knowledge graph, the confidence score θ_{ijk} for a triple (e_i, e_j, r_k) is calculated simply by $\boldsymbol{a}_{i:}(\boldsymbol{b}_{j:} \circ \boldsymbol{c}_{k:})^{\mathrm{T}}$ where $\boldsymbol{a}_{i:}$, $\boldsymbol{b}_{j:}$, and $\boldsymbol{c}_{k:}$ denote the D-dimensional row vectors representing e_i, e_j, and r_k, respectively, and $\circ$ is the Hadamard (element-wise) product. In spite of the model's simplicity, it needs to maintain $(2N_e + N_r)$ D-dimensional 32-bit or 64-bit valued vectors, where N_e, and N_r denote the number of entities and relations, respectively. Because typical knowledge graphs contain enormous number of entities and relations, this leads to a significant memory requirement. As mentioned in [6], CP with $D = 200$ applied to Freebase will require about 66 GB of memory to store parameters. This large memory consumption poses issues especially when KGC is conducted on resource-limited devices. Moreover, the size of existing knowledge graphs is still growing rapidly, and a method for shrinking the embedding vectors is in strong demand.

To address the problem, this paper presents a new CP decomposition algorithm to learn compact knowledge graph embeddings. The basic idea is to introduce a quantization function built into the optimization problem. This function forces the embedding vectors to be binary, and optimization is done with respect to the binarized vectors. After training, the binarized embeddings can be used in place of the original vectors of floating-point numbers, which drastically reduces the memory footprint of the resulting model.

In addition, the binary vector representation contributes to efficiently computing the dot product by using bitwise operations. This fast computation allows the proposed model to substantially reduce the amount of time required to compute the confidence scores of triples.

Note that our method only improves the run-time (i.e., predicting missing triples) memory footprint and speed but not those for training a prediction model. However, the reduced memory footprint of the produced model enables KGC to be run on many affordable resource-limited devices (e.g., personal computers). Unlike research-level benchmarks in which one is required to compute the scores of a small set of test triples, completion of an entire knowledge graph requires computing the scores of all missing triples in a knowledge graph, whose number is enormous because knowledge graphs are sparse. Thus, improved memory footprints and reduced score computation time are of practical importance, and these are what our proposed model provides.

The quantization technique has been commonly used in the community of deep neural networks to shrink network components [5,8]. To the best of our knowledge, this technique has not been studied in the field of tensor factorization. The main contribution of this paper is that we introduce the quantization function to a tensor factorization model for the first time. This is also the first study to investigate the benefits of the quantization for KGC. Our experimental results on several KGC benchmark datasets showed that the proposed method reduced the model size nearly 10- to 20-fold compared to the standard CP decomposition without a decrease in the task performance. Besides, with bitwise operations, B-CP got a bonus of speed-up in score computation time.

2 Related Work

Approaches to knowledge graph embedding (KGE) can be classified into three types: models based on bilinear mapping, translation, and neural network-based transformation.

RESCAL [20] is a bilinear-based KGE method whose score function is formulated as $\theta_{ijk} = \boldsymbol{a}_{e_i}^{\mathrm{T}} \boldsymbol{B}_{r_k} \boldsymbol{a}_{e_j}$, where $\boldsymbol{a}_{e_i}, \boldsymbol{a}_{e_j} \in \mathbb{R}^D$ are the vector representations of entities e_i and e_j, respectively, and matrix $\boldsymbol{B}_{r_k} \in \mathbb{R}^{D \times D}$ represents a relation r_k. Although RESCAL is able to output non-symmetric score functions, each relation matrix $\boldsymbol{B}_{r_k}$ holds D^2 parameters. This can be problematic both in terms of overfitting and computational cost. To avoid this problem, several methods have been proposed recently. DistMult [26] restricts the relation matrix to be diagonal, $\boldsymbol{B}_{r_k} = \mathrm{diag}(\boldsymbol{b}_{r_k})$. However, this form of function is necessarily symmetric in i and j; i.e., $\theta_{ijk} = \theta_{jik}$. To reconcile efficiency and expressiveness, Trouillon et al. [25] proposed ComplEx, using the complex-valued representations and Hermitian inner product to define the score function, which unlike DistMult, can be non-symmetric in i and j. Hayashi and Shimbo [9] found that ComplEx is equivalent to another state-of-the-art KGE method, holographic embeddings (HolE) [18]. ANALOGY [16] is a model that can be view as a hybrid of ComplEx and DistMult. Lacroix et al. [14] and Kazemi and Pool [11] independently showed that CP decomposition (called SimplE in [11]) achieves a comparable KGC performance to other bilinear methods such as ComplEx and ANALOGY. To achieve this performance, they introduced an "inverse" triple (e_j, e_i, r_k^{-1}) to the training data for each existing triple (e_i, e_j, r_k), where r_k^{-1} denotes the inverse relation of r_k.

TransE [4] is the first KGE model based on vector translation. It employs the principle $\boldsymbol{a}_{e_i} + \boldsymbol{b}_{r_k} \approx \boldsymbol{a}_{e_j}$ to define a distance-based score function $\theta_{ijk} = -\|\boldsymbol{a}_{e_i} + \boldsymbol{b}_{r_k} - \boldsymbol{a}_{e_j}\|^2$. Since TransE was recognized as too limited to model complex properties (e.g., symmetric/reflexive/one-to-many/many-to-one relations) in knowledge graphs, many extended versions of TransE have been proposed.

Neural-based models, such as NTN [23] and ConvE [6], employ non-linear functions to define score function, and thus they have a better expressiveness. Compared to bilinear and translation approaches, however, neural-based models require more complex operations to compute interactions between a relation and two entities in vector space.

It should be noted that the binarization technique proposed in this paper can be applied to other KGE models besides CP decomposition, such as those mentioned above. Our choice of CP as the implementation platform only reflects the fact that it is one of the strongest baseline KGE methods.

Numerous recent publications have studied methods for training quantized neural networks to compact the models without performance degradation. The BinaryConnect algorithm [5] is the first study to show that binarized neural networks can achieve almost the state-of-the-art results on datasets such as MNIST and CIFAR-10 [8]. BinaryConnect uses the binarization function $Q_1(x)$ to replace floating point weights of deep neural networks with binary weights during the forward and backward propagation. Lam [15] used the same quantization method as BinaryConnect to learn compact word embeddings. To binarize knowledge graph embeddings, this paper also applied the quantization method to the CP decomposition algorithm. To the best of our knowledge, this paper is the first study to examine the benefits of the quantization for KGC.

3 Notation and Preliminaries

We follow the notation and terminology established in [12] for the most part. These are summarized below mainly for third-order tensors, by which a knowledge graph is represented (see Sect. 4.1).

Vectors are represented by boldface lowercase letters, e.g., $\boldsymbol{a}$. Matrices are represented by boldface capital letters, e.g., $\boldsymbol{A}$. Third-order tensors are represented by boldface calligraphic letters, e.g., $\boldsymbol{\mathcal{X}}$.

The ith row of a matrix $\boldsymbol{A}$ is represented by $\boldsymbol{a}_{i:}$, and the jth column of $\boldsymbol{A}$ is represented by $\boldsymbol{a}_{:j}$, or simply as $\boldsymbol{a}_j$. The symbol $\circ$ represents the Hadamard product for matrices and also for vectors, and $\otimes$ represents the outer product.

A third-order tensor $\boldsymbol{\mathcal{X}} \in \mathbb{R}^{I_1 \times I_2 \times I_3}$ is rank one if it can be written as the outer product of three vectors, i.e., $\boldsymbol{\mathcal{X}} = \boldsymbol{a} \otimes \boldsymbol{b} \otimes \boldsymbol{c}$. This means that each element $x_{i_1 i_2 i_3}$ of $\boldsymbol{\mathcal{X}}$ is the product of the corresponding vector elements:

$$x_{i_1 i_2 i_3} = a_{i_1} b_{i_2} c_{i_3} \quad \text{for } i_1 \in [I_1],\ i_2 \in [I_2],\ i_3 \in [I_3],$$

where $[I_n]$ denotes the set of natural numbers $1, 2, \cdots, I_n$.

The norm of a tensor $\boldsymbol{\mathcal{X}} \in \mathbb{R}^{I_1 \times I_2 \times \cdots \times I_k}$ is the square root of the sum of the squares of all its elements, i.e.,

$$\|\boldsymbol{\mathcal{X}}\| = \sqrt{\sum_{i_1 \in [I_1]} \sum_{i_2 \in [I_2]} \cdots \sum_{i_k \in [I_k]} x_{i_1 i_2 \cdots i_k}^2}.$$

For a matrix (or a second-order tensor), this norm is called the Frobenius norm and is represented by $\| \cdot \|_F$.

4 Tensor Factorization for Knowledge Graphs

4.1 Knowledge Graph Representation

A knowledge graph $\mathcal{G}$ is a labeled multigraph $(\mathcal{E}, \mathcal{R}, \mathcal{F})$, where $\mathcal{E} = \{e_1, \ldots, e_{N_e}\}$ is the set of entities (vertices), $\mathcal{R} = \{r_1, \ldots, r_{N_r}\}$ is the set of all relation types (edge labels), and $\mathcal{F} \subset \mathcal{E} \times \mathcal{E} \times \mathcal{R}$ denotes the observed instances of relations over entities (edges). The presence of an edge, or a triple, $(e_i, e_j, r_k) \in \mathcal{F}$ represents the fact that relation r_k holds between subject entity e_i and object entity e_j.

A knowledge graph can be represented as a boolean third order tensor $\boldsymbol{\mathcal{X}} \in \{0,1\}^{N_e \times N_e \times N_r}$ whose elements are set such as

$$x_{ijk} = \begin{cases} 1 & \text{if } (e_i, e_j, r_k) \in \mathcal{F} \\ 0 & \text{otherwise} \end{cases}.$$

KGC is concerned with incomplete knowledge graphs, i.e., $\mathcal{F} \subsetneq \mathcal{F}^*$, where $\mathcal{F}^* \subset \mathcal{E} \times \mathcal{E} \times \mathcal{R}$ is the unobservable set of ground truth facts (and a superset of $\mathcal{F}$). KGE has been recognized as a promising approach to predicting the truth value of unknown triples in $\mathcal{F}^* \setminus \mathcal{F}$. KGE can be generally formulated as the tensor factorization problem and defines a score function θ_{ijk} using the latent vectors of entities and relations.

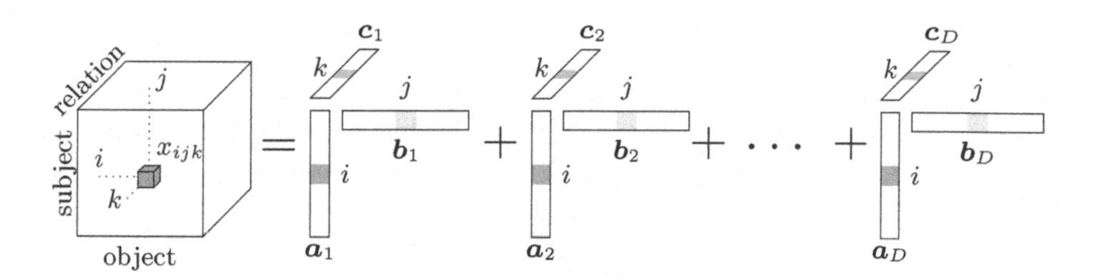

Fig. 1. Illustration of a D-component CP model for a third-order tensor $\boldsymbol{\mathcal{X}}$.

4.2 CP Decomposition

CP decomposition [10] factorizes a given tensor as a linear combination of D rank-one tensors. For a third-order tensor $\boldsymbol{\mathcal{X}} \in \mathbb{R}^{N_e \times N_e \times N_r}$, its CP decomposition is

$$\boldsymbol{\mathcal{X}} \approx \sum_{d \in [D]} \boldsymbol{a}_d \otimes \boldsymbol{b}_d \otimes \boldsymbol{c}_d,$$

where $\boldsymbol{a}_d \in \mathbb{R}^{N_e}$, $\boldsymbol{b}_d \in \mathbb{R}^{N_e}$ and $\boldsymbol{c}_d \in \mathbb{R}^{N_r}$. Figure 1 illustrates CP for third-order tensors, which demonstrates how we can formulate knowledge graphs. The elements x_{ijk} of $\boldsymbol{\mathcal{X}}$ can be written as

$$x_{ijk} \approx \boldsymbol{a}_{i:}(\boldsymbol{b}_{j:} \circ \boldsymbol{c}_{k:})^{\mathrm{T}} = \sum_{d \in [D]} a_{id} b_{jd} c_{kd} \quad \text{for } i, j \in [N_e],\ k \in [N_r].$$

A *factor matrix* refers to a matrix composed of vectors from the rank one components. We use $\boldsymbol{A} = [\boldsymbol{a}_1\, \boldsymbol{a}_2\, \cdots\, \boldsymbol{a}_D]$ to denote the factor matrix, and likewise $\boldsymbol{B}$, $\boldsymbol{C}$. Note that $\boldsymbol{a}_{i:}$, $\boldsymbol{b}_{j:}$ and $\boldsymbol{c}_{k:}$ represent the D-dimensional embedding vectors of subject e_i, object e_j, and relation r_k, respectively.

4.3 Logistic Regression

Following literature [19], we formulate a logistic regression model for solving the CP decomposition problem. This model considers CP decomposition from a probabilistic viewpoint. We regard x_{ijk} as a random variable and compute the maximum a posteriori (MAP) estimates of $\boldsymbol{A}$, $\boldsymbol{B}$, and $\boldsymbol{C}$ for the joint distribution

$$p(\boldsymbol{\mathcal{X}}|\boldsymbol{A}, \boldsymbol{B}, \boldsymbol{C}) = \prod_{i\in[N_e]} \prod_{j\in[N_e]} \prod_{k\in[N_r]} p(x_{ijk}|\theta_{ijk}).$$

We define the score function $\theta_{ijk} = \boldsymbol{a}_{i:}(\boldsymbol{b}_{j:} \circ \boldsymbol{c}_{k:})^{\mathrm{T}}$ that represents the CP decomposition model's confidence that a triple (e_i, e_j, r_k) is a fact; i.e., that it must be present in the knowledge graph. By assuming that x_{ijk} follows the Bernoulli distribution, $x_{ijk} \sim \text{Bernoulli}(\sigma(\theta_{ijk}))$, the posterior probability is defined as the following equation

$$p(x_{ijk}|\theta_{ijk}) = \begin{cases} \sigma(\theta_{ijk}) & \text{if } x_{ijk} = 1 \\ 1 - \sigma(\theta_{ijk}) & \text{if } x_{ijk} = 0 \end{cases},$$

where $\sigma(x) = 1/(1 + \exp(-x))$ is the sigmoid function.

Furthermore, we minimize the negative log-likelihood of the MAP estimates, such that the general form of the objective function to optimize is

$$E = \sum_{i\in[N_e]} \sum_{j\in[N_e]} \sum_{k\in[N_r]} E_{ijk},$$

where

$$E_{ijk} = \underbrace{-x_{ijk} \log \sigma(\theta_{ijk}) + (x_{ijk} - 1) \log(1 - \sigma(\theta_{ijk}))}_{\ell_{ijk}} + \underbrace{\lambda_A \|\boldsymbol{a}_{i:}\|^2 + \lambda_B \|\boldsymbol{b}_{j:}\|^2 + \lambda_C \|\boldsymbol{c}_{k:}\|^2}_{\text{L2 regularizer}}.$$

ℓ_{ijk} represents the logistic loss function for a triple (e_i, e_j, r_k). While most knowledge graphs contain only positive examples, negative examples (false facts) are needed to optimize the objective function. However, if all unknown triples are treated as negative samples, calculating the loss function requires a prohibitive amount of time. To approximately minimize the objective function, following previous studies, we used negative sampling in our experiments.

The objective function is minimized with an online learning method based on stochastic gradient descent (SGD). For each training example, SGD iteratively updates parameters by $\boldsymbol{a}_{i:} \leftarrow \boldsymbol{a}_{i:} - \eta\frac{\partial E_{ijk}}{\partial \boldsymbol{a}_{i:}}$, $\boldsymbol{b}_{j:} \leftarrow \boldsymbol{b}_{j:} - \eta\frac{\partial E_{ijk}}{\partial \boldsymbol{b}_{j:}}$, and $\boldsymbol{c}_{k:} \leftarrow \boldsymbol{c}_{k:} - \eta\frac{\partial E_{ijk}}{\partial \boldsymbol{c}_{k:}}$ with a learning rate η. The partial gradient of the objective function with respect to $\boldsymbol{a}_{i:}$ is

$$\frac{\partial E_{ijk}}{\partial \boldsymbol{a}_{i:}} = -x_{ijk}\exp\left(-\theta_{ijk}\right)\sigma(\theta_{ijk})\boldsymbol{b}_{j:} \circ \boldsymbol{c}_{k:} + (1 - x_{ijk})\,\sigma(\theta_{ijk})\boldsymbol{b}_{j:} \circ \boldsymbol{c}_{k:} + 2\lambda_A \boldsymbol{a}_{i:}.$$

Those with respect to $\boldsymbol{b}_{j:}$ and $\boldsymbol{c}_{k:}$ can be calculated in the same manner.

5 Binarized CP Decomposition

We propose a binarized CP decomposition algorithm to make CP factor matrices $\boldsymbol{A}$, $\boldsymbol{B}$ and $\boldsymbol{C}$ binary, i.e., the elements of these matrices are constrained to only two possible values.

In this algorithm, we formulate the score function $\theta_{ijk}^{(b)} = \sum_{d \in [D]} a_{id}^{(b)} b_{jd}^{(b)} c_{kd}^{(b)}$, where $a_{id}^{(b)} = Q_\Delta(a_{id})$, $b_{jd}^{(b)} = Q_\Delta(b_{jd})$, $c_{kd}^{(b)} = Q_\Delta(c_{kd})$ are obtained by binarizing a_{id}, b_{jd}, c_{kd} through the following quantization function

$$x^{(b)} = Q_\Delta(x) = \begin{cases} \Delta & \text{if } x \geq 0 \\ -\Delta & \text{if } x < 0 \end{cases},$$

where Δ is a positive constant value. We extend the binarization function to vectors in a natural way: $\boldsymbol{x}^{(b)} = Q_\Delta(\boldsymbol{x})$ whose ith element $x_i^{(b)}$ is $Q_\Delta(x_i)$.

Using the new score function, we reformulate the loss function defined in Sect. 4.3 as follows

$$\ell_{ijk}^{(b)} = -x_{ijk}\log\sigma(\theta_{ijk}^{(b)}) + (x_{ijk} - 1)\log(1 - \sigma(\theta_{ijk}^{(b)})).$$

To train the binarized CP decomposition model, we optimize the same objective function E as in Sect. 4.3 except using the binarized loss function given above. We also employ the SGD algorithm to minimize the objective function. One issue here is that the parameters cannot be updated properly since the gradients of Q_Δ are zero almost everywhere. To solve the issue, we simply use an identity matrix as the surrogate for the derivative of Q_Δ:

$$\frac{\partial Q_\Delta(\boldsymbol{x})}{\partial \boldsymbol{x}} \approx \boldsymbol{I}.$$

The simple trick enables us to calculate the partial gradient of the objective function with respect to $\boldsymbol{a}_{i:}$ through the chain rule:

$$\frac{\partial \ell_{ijk}^{(b)}}{\partial \boldsymbol{a}_{i:}} = \frac{\partial Q_\Delta(\boldsymbol{a}_{i:})}{\partial \boldsymbol{a}_{i:}} \frac{\partial \ell_{ijk}^{(b)}}{\partial Q_\Delta(\boldsymbol{a}_{i:})} \approx \boldsymbol{I}\frac{\partial \ell_{ijk}^{(b)}}{\partial Q_\Delta(\boldsymbol{a}_{i:})} = \frac{\partial \ell_{ijk}^{(b)}}{\partial \boldsymbol{a}_{i:}^{(b)}}.$$

This strategy is known as Hinton's straight-through estimator [1] and has been developed in the community of deep neural networks to quantize network components [5,8]. Using this trick, we finally obtain the partial gradient as follows:

$$\frac{\partial E_{ijk}}{\partial \boldsymbol{a}_{i:}} = -x_{ijk} \exp\left(-\theta_{ijk}^{(b)}\right) \sigma(\theta_{ijk}^{(b)}) \boldsymbol{b}_{j:}^{(b)} \circ \boldsymbol{c}_{k:}^{(b)} + (1 - x_{ijk})\, \sigma(\theta_{ijk}^{(b)}) \boldsymbol{b}_{j:}^{(b)} \circ \boldsymbol{c}_{k:}^{(b)} + 2\lambda_A \boldsymbol{a}_{i:}.$$

The partial gradients with respect to $\boldsymbol{b}_{j:}$ and $\boldsymbol{c}_{k:}$ can be computed similarly.

Binary vector representations bring benefits in faster computation of scores $\theta_{ijk}^{(b)}$, because the inner product between binary vectors can be implemented by bitwise operations: To compute $\theta_{ijk}^{(b)}$, we can use XNOR and Bitcount operations:

$$\theta_{ijk}^{(b)} = \boldsymbol{a}_{i:}^{(b)} (\boldsymbol{b}_{j:}^{(b)} \circ \boldsymbol{c}_{k:}^{(b)})^{\mathrm{T}} = \Delta^3 \{2BitC - D\}$$

Table 1. Benchmark datasets for KGC.

	WN18	FB15k	WN18RR	FB15k-237
N_e	40,943	14,951	40,559	14,505
N_r	18	1,345	11	237
#Training triples	141,442	483,142	86,835	272,115
#Validation triples	5,000	50,000	3,034	17,535
#Test triples	5,000	59,071	3,134	20,466

Table 2. KGC results on WN18 and FB15k: Filtered MRR and Hits@{1, 3, 10} (%). Letters in boldface signify the best performers in individual evaluation metrics.

Models	WN18				FB15k			
	MRR	Hits@			MRR	Hits@		
		1	3	10		1	3	10
TransE*	45.4	8.9	82.3	93.4	38.0	23.1	47.2	64.1
DistMult*	82.2	72.8	91.4	93.6	65.4	54.6	73.3	82.4
HolE*	93.8	93.0	94.5	94.9	52.4	40.2	61.3	73.9
ComplEx*	94.1	93.6	94.5	94.7	69.2	59.9	75.9	84.0
ANALOGY**	94.2	93.9	94.4	94.7	72.5	64.6	**78.5**	**85.4**
CP***	94.2	93.9	94.4	94.7	72.7	66.0	77.3	83.9
ConvE**	94.3	93.5	94.6	**95.6**	65.7	55.8	72.3	83.1
CP ($D = 200$)	94.2	93.9	94.5	94.7	71.9	66.2	75.2	82.0
B-CP ($D = 200$)	90.1	88.1	91.8	93.3	69.5	61.1	76.0	83.5
B-CP ($D = 400$)	94.5	94.1	94.8	95.0	72.2	66.3	77.5	84.2
B-CP ($D = 300 \times 3$)	**94.6**	**94.2**	**95.0**	95.3	**72.9**	**66.5**	77.7	84.9

*, ** and *** indicate the results transcribed from [6,11,25], respectively.

where $BitC = \text{Bitcount}(\text{XNOR}(\text{XNOR}(\overline{\boldsymbol{a}}_{i:}^{(b)}, \overline{\boldsymbol{b}}_{j:}^{(b)}), \overline{\boldsymbol{c}}_{k:}^{(b)}))$. $\overline{\boldsymbol{x}}^{(b)}$ denotes the boolean vector whose ith element $\overline{x}_i^{(b)}$ is set to 1 if $x_i^{(b)} = \Delta$, otherwise to 0. Bitcount returns the number of one-bits in a binary vector and XNOR represents the logical complement of the exclusive OR operation.

6 Experiments

6.1 Datasets and Evaluation Protocol

We evaluated the performance of our proposal in the standard knowledge graph completion (KGC) task. We used four standard datasets, WN18, FB15k [4], WN18RR, and FB15k-237 [6]. Table 1 shows the data statistics[1].

We followed the standard evaluation procedure to evaluate the KGC performance: Given a test triple (e_i, e_j, r_k), we corrupted it by replacing e_i or e_j with every entity e_ℓ in $\mathcal{E}$ and calculated $\theta_{i,\ell,k}$ or $\theta_{\ell,j,k}$. We then ranked all these triples by their scores in decreasing order. To measure the quality of the ranking, we used the mean reciprocal rank (MRR) and Hits at N (Hits@N). We here report only results in the filtered setting [4], which provides a more reliable performance metric in the presence of multiple correct triples.

Table 3. KGC results on WN18RR and FB15k-237: Filtered MRR and Hits@$\{1, 3, 10\}$ (%).

Models	WN18RR				FB15k-237			
	MRR	Hits@			MRR	Hits@		
		1	3	10		1	3	10
DistMult*	43.0	39.0	44.0	49.0	24.1	15.5	26.3	41.9
ComplEx*	44.0	41.0	46.0	51.0	24.7	15.8	27.5	42.8
R-GCN*	–	–	–	–	24.8	15.3	25.8	41.7
ConvE*	43.0	40.0	44.0	52.0	**32.5**	**23.7**	**35.6**	**50.1**
CP ($D = 200$)	44.0	42.0	46.0	51.0	29.0	19.8	32.2	47.9
B-CP ($D = 200$)	45.0	43.0	46.0	50.0	27.8	19.4	30.4	44.6
B-CP ($D = 400$)	45.0	43.0	46.0	52.0	29.2	20.8	31.8	46.1
B-CP ($D = 300 \times 3$)	**48.0**	**45.0**	**49.0**	**53.0**	30.3	21.4	33.3	48.2

* indicates the results transcribed from [6].

[1] Following [11,14], for each triple (e_i, e_j, r_k) observed in the training set, we added its inverse triple (e_j, e_i, r_k^{-1}) also in the training set.

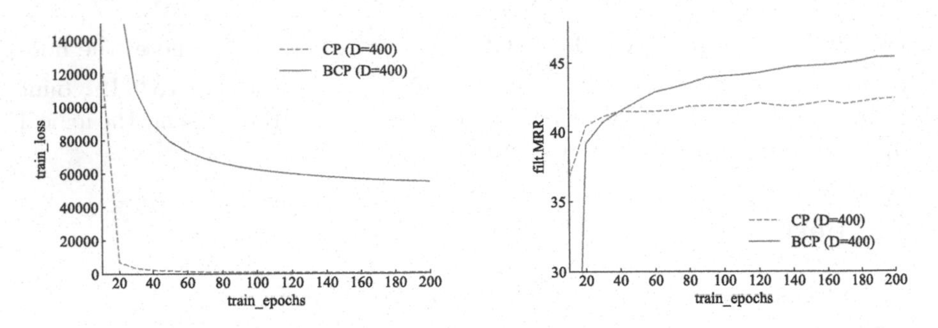

Fig. 2. Training loss and filtered MRR vs. epochs trained on WN18RR.

6.2 Experiment Setup

To train CP models, we selected the hyperparameters via grid search such that the filtered MRR is maximized on the validation set. For standard CP model, we tried all combinations of $\lambda_A, \lambda_B, \lambda_C \in \{0.0001, 0\}$, learning rate $\eta \in \{0.025, 0.05\}$, and the embedding dimension $D \in \{15, 25, 50, 100, 150, 200, 300, 400, 500\}$ during grid search. For our binarized CP (B-CP) model, all combinations of $\lambda_A, \lambda_B, \lambda_C \in \{0.0001, 0\}$, $\eta \in \{0.025, 0.05\}$, $\Delta \in \{0.3, 0.5\}$ and $D \in \{100, 200, 300, 400, 500\}$ were tried. We randomly generated the initial values of the representation vectors from the uniform distribution $U[-\frac{\sqrt{6}}{\sqrt{2D}}, \frac{\sqrt{6}}{\sqrt{2D}}]$ [7]. The maximum number of training epochs was set to 1000. For SGD training, negative samples were generated using the local closed-world assumption [4]. The number of negative samples generated per positive sample was 5 for WN18/WN18RR and 10 for FB15k/FB15k-237. In addition, to further take advantage of the benign run-time memory footprint of B-CP, we also tested the *model ensemble* of three independently trained B-CP models[2], in which the final ranking is computed by the sum of the scores of the three models. For this ensemble, the embedding dimension of each model was set to $D = 300$, yet the total required run-time memory is still smaller than CP with $D = 200$.

We implemented our CP decomposition systems in C++ and conducted all experiments on a 64-bit 16-Core AMD Ryzen Threadripper 1950x with 3.4 GHz CPUs. The program codes were compiled using GCC 7.3 with -O3 option.

[2] As the original CP model has much larger memory consumption than B-CP, we did not test model ensemble with the CP model in our experiments.

Table 4. Results on WN18RR and FB15k-237 with varying embedding dimensions.

Model	Bits per entity	Bits per relation	MRR	
			WN18RR	FB15k-237
DistMult* ($D = 200$)	6,400	6,400	43.0	24.1
ComplEx* ($D = 200$)	12,800	12,800	44.0	24.7
ConvE* ($D = 200$)	6,400	6,400	43.0	**32.5**
CP ($D = 15$)	960	480	40.0	22.0
CP ($D = 50$)	3,200	1,600	43.0	24.8
CP ($D = 200$)	12,800	6,400	44.0	29.0
CP ($D = 500$)	32,000	16,000	43.0	29.2
VQ-CP ($D = 200$)	400	200	36.0	8.7
VQ-CP ($D = 500$)	1,000	500	36.0	8.3
B-CP ($D = 100$)	200	100	38.0	23.2
B-CP ($D = 200$)	400	200	45.0	27.8
B-CP ($D = 300$)	600	300	46.0	29.0
B-CP ($D = 400$)	800	400	45.0	29.2
B-CP ($D = 500$)	1,000	500	45.0	29.1
B-CP ($D = 300 \times 3$)	1,800	900	**48.0**	30.3

6.3 Results

Main Results. We compared standard CP and B-CP models with other state-of-the-art KGE models. Table 2 shows the results on WN18 and FB15k, and Table 3 displays the results on WN18RR and FB15k-237. For most of the evaluation metrics, our B-CP model ($D = 400$) outperformed or was competitive to the best baseline, although with a small vector dimension ($D = 200$), B-CP showed tendency to degrade in its performance. In the table, B-CP ($D = 300 \times 3$) indicates an ensemble of three different B-CP models (each with $D = 300$). This ensemble approach outperformed the baseline B-CP constantly on all datasets. Figure 2 shows training loss and accuracy versus epochs of training for CP ($D = 400$) and B-CP ($D = 400$) on WN18RR. The results indicate that CP is prone to overfitting with increased epochs of training. By contrast, B-CP appears less susceptible to overfitting than CP.

KGC Performance Vs. Model Size. We also investigated how our B-CP method can maintain the KGC performance while reducing the model size. For a fair evaluation, we also examined a naive vector quantization method (VQ) [22] that can reduce the model size. Given a real valued matrix $\boldsymbol{X} \in \mathbb{R}^{D_1 \times D_2}$, the VQ method solves the following optimization problem:

$$\hat{\boldsymbol{X}}^{(b)}, \hat{\alpha} = \underset{\boldsymbol{X}^{(b)}, \alpha}{\operatorname{argmin}} \|\boldsymbol{X} - \alpha \boldsymbol{X}^{(b)}\|_F^2$$

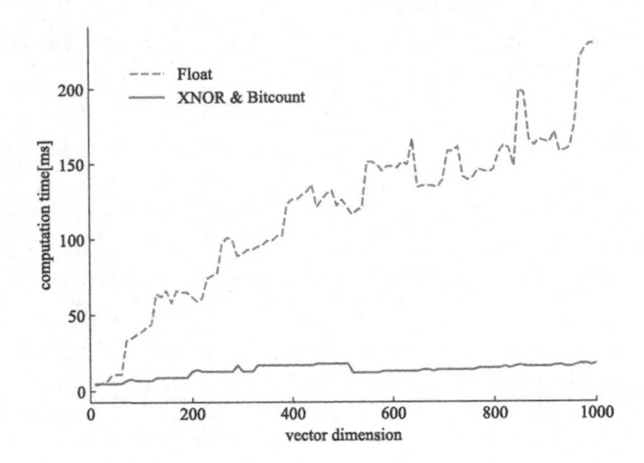

Fig. 3. CPU run time per 100,000-times score computations with single CPU thread.

where $\boldsymbol{X}^{(b)} \in \{+1, -1\}^{D_1 \times D_2}$ is a binary matrix and α is a positive real value. The optimal solutions $\hat{\boldsymbol{X}}^{(b)}$ and $\hat{\alpha}$ are given by $Q_1(\boldsymbol{X})$ and $\frac{1}{D_1 \times D_2}\|\boldsymbol{X}\|_1$, respectively, where $\|\cdot\|_1$ denotes l_1-norm, and $Q_1(\boldsymbol{X})$ is a sign function whose behavior in each element x of $\boldsymbol{X}$ is as per the sign function $Q_1(x)$. After obtaining factor matrices $\boldsymbol{A}$, $\boldsymbol{B}$ and $\boldsymbol{C}$ via CP decomposition, we solved the above optimization problem independently for each matrix. We call this method VQ-CP.

Table 4 shows the results when the dimension size of the embeddings was varied. While CP requires $64 \times D$ and $32 \times D$ bits per entity and relation, respectively, both B-CP and VQ-CP have only to take one thirty-second of them. Obviously, the task performance dropped significantly after vector quantization (VQ-CP). The performance of CP also degraded when reducing the vector dimension from 200 to 15 or 50. While simply reducing the number of dimensions degraded the accuracy, B-CP successfully reduced the model size nearly 10- to 20-fold compared to CP and other KGE models without performance degradation. Even in the case of B-CP ($D = 300 \times 3$), the model size was 6 times smaller than that of CP ($D = 200$).

Computation Time. As described in Sect. 5, the B-CP model can accelerate the computation of confidence scores by using the bitwise operations (XNOR and Bitcount). To compare the score computation speed between CP (Float) and B-CP (XNOR and Bitcount), we calculated the confidence scores 100,000 times for both CP and B-CP, varying the vector size D from 10 to 1000 at 10 increments. Figure 3 clearly shows that bitwise operations provide significant speed-up compared to standard multiply-accumulate operations.

Table 5. Results on the Freebase-music dataset.

	Accuracy	Model size
CP ($D = 15$)	50.3	0.4 GB
CP ($D = 200$)	89.2	4.8 GB
B-CP ($D = 400$)	**92.8**	**0.3 GB**

6.4 Evaluation on Large-Scale Freebase

To verify the effectiveness of B-CP over larger datasets, we also conducted experiments on the Freebase-music data[3]. To reduce noises, we removed triples from the data whose relation and entities occur less than 10 times. The number of the remaining triples were 18,482,832 which consist of 138 relations and 3,025,684 entities. We split them randomly into three subsets: 18,462,832 training, 10,000 validation, and 10,000 test triples. We randomly generated 20,000 triples that are not in the knowledge graph, and used them as negative samples; half of them were placed in the validation set, and the other half in the test set. Experiments were conducted under the same hyperparameters and negative samples setting we achieved the best results on the FB15k dataset. We here report the triple classification accuracy. Table 5 gives the results. As it was with the small datasets, the performance of CP ($D = 15$) was again poor. Meanwhile, B-CP successfully reduced the model size while achieving better performance than CP ($D = 200$). These results show that B-CP is robust to the data size.

7 Conclusion

In this paper, we showed that it is possible to obtain binary vectors of relations and entities in knowledge graphs that take 10–20 times less storage/memory than the original representations with floating point numbers. Additionally, with the help of bitwise operations, the time required for score computation was considerably reduced. Tensor factorization arises in many machine learning applications such as item recommendation [21] and web link analysis [13]. Applying our B-CP algorithm to the analysis of other relational datasets is an interesting avenue for future work.

The program codes for the binarized CP decomposition algorithm proposed here will be provided on the first author's GitHub page[4].

[3] https://datalab.snu.ac.kr/haten2/.

[4] https://github.com/KokiKishimoto/cp_decomposition.git.

References

1. Bengio, Y., Léonard, N., Courville, A.C.: Estimating or propagating gradients through stochastic neurons for conditional computation. CoRR abs/1308.3432 (2013). http://arxiv.org/abs/1308.3432
2. Bollacker, K.D., Evans, C., Paritosh, P., Sturge, T., Taylor, J.: Freebase: a collaboratively created graph database for structuring human knowledge. In: Proceedings of the ACM SIGMOD International Conference on Management of Data, SIGMOD 2008, Vancouver, BC, Canada, 10–12 June 2008, pp. 1247–1250 (2008). https://doi.org/10.1145/1376616.1376746
3. Bordes, A., Chopra, S., Weston, J.: Question answering with subgraph embeddings. In: Proceedings of the 2014 Conference on Empirical Methods in Natural Language Processing, EMNLP 2014, Doha, Qatar, 25–29 October 2014, A meeting of SIGDAT, a Special Interest Group of the ACL, pp. 615–620 (2014). http://aclweb.org/anthology/D/D14/D14-1067.pdf
4. Bordes, A., Usunier, N., García-Durán, A., Weston, J., Yakhnenko, O.: Translating embeddings for modeling multi-relational data. In: Advances in Neural Information Processing Systems 26: 27th Annual Conference on Neural Information Processing Systems 2013. Proceedings of A meeting held 5–8 December 2013, Lake Tahoe, Nevada, United States, pp. 2787–2795 (2013). http://papers.nips.cc/paper/5071-translating-embeddings-for-modeling-multi-relational-data
5. Courbariaux, M., Bengio, Y., David, J.: Binaryconnect: training deep neural networks with binary weights during propagations. In: Advances in Neural Information Processing Systems 28: Annual Conference on Neural Information Processing Systems 2015, Montreal, Quebec, Canada, 7–12 December 2015, pp. 3123–3131 (2015). http://papers.nips.cc/paper/5647-binaryconnect-training-deep-neural-networks-with-binary-weights-during-propagations
6. Dettmers, T., Minervini, P., Stenetorp, P., Riedel, S.: Convolutional 2D knowledge graph embeddings. In: Proceedings of the Thirty-Second AAAI Conference on Artificial Intelligence, 2–7 February 2018, New Orleans, Louisiana, USA (2018). https://www.aaai.org/ocs/index.php/AAAI/AAAI18/paper/view/17366
7. Glorot, X., Bengio, Y.: Understanding the difficulty of training deep feedforward neural networks. In: Proceedings of the Thirteenth International Conference on Artificial Intelligence and Statistics, AISTATS 2010, Chia Laguna Resort, Sardinia, Italy, 13–15 May 2010, pp. 249–256 (2010). http://www.jmlr.org/proceedings/papers/v9/glorot10a.html
8. Guo, Y.: A survey on methods and theories of quantized neural networks. CoRR abs/1808.04752 (2018). http://arxiv.org/abs/1808.04752
9. Hayashi, K., Shimbo, M.: On the equivalence of holographic and complex embeddings for link prediction. In: Proceedings of the 55th Annual Meeting of the Association for Computational Linguistics, ACL 2017, Vancouver, Canada, 30 July–4 August, Volume 2: Short Papers, pp. 554–559 (2017). https://doi.org/10.18653/v1/P17-2088
10. Hitchcock, F.L.: The expression of a tensor or a polyadic as a sum of products. J. Math. Phys **6**(1), 164–189 (1927)
11. Kazemi, S.M., Poole, D.: Simple embedding for link prediction in knowledge graphs. In: Advances in Neural Information Processing Systems 31: Annual Conference on Neural Information Processing Systems 2018, NeurIPS 2018, Montréal, Canada, 3–8 December 2018. pp. 4289–4300 (2018). http://papers.nips.cc/paper/7682-simple-embedding-for-link-prediction-in-knowledge-graphs

12. Kolda, T.G., Bader, B.W.: Tensor decompositions and applications. SIAM Rev. **51**(3), 455–500 (2009). https://doi.org/10.1137/07070111X
13. Kolda, T.G., Bader, B.W., Kenny, J.P.: Higher-order web link analysis using multilinear algebra. In: Proceedings of the 5th IEEE International Conference on Data Mining (ICDM 2005), Houston, Texas, USA, 27–30 November 2005, pp. 242–249 (2005). https://doi.org/10.1109/ICDM.2005.77
14. Lacroix, T., Usunier, N., Obozinski, G.: Canonical tensor decomposition for knowledge base completion. In: Proceedings of the 35th International Conference on Machine Learning, ICML 2018, Stockholmsmässan, Stockholm, Sweden, 10–15 July 2018, pp. 2869–2878 (2018). http://proceedings.mlr.press/v80/lacroix18a.html
15. Lam, M.: Word2bits - quantized word vectors. CoRR abs/1803.05651 (2018). http://arxiv.org/abs/1803.05651
16. Liu, H., Wu, Y., Yang, Y.: Analogical inference for multi-relational embeddings. In: Proceedings of the 34th International Conference on Machine Learning, ICML 2017, Sydney, NSW, Australia, 6–11 August 2017, pp. 2168–2178 (2017). http://proceedings.mlr.press/v70/liu17d.html
17. Ma, Y., Crook, P.A., Sarikaya, R., Fosler-Lussier, E.: Knowledge graph inference for spoken dialog systems. In: 2015 IEEE International Conference on Acoustics, Speech and Signal Processing, ICASSP 2015, South Brisbane, Queensland, Australia, 19–24 April 2015, pp. 5346–5350 (2015). https://doi.org/10.1109/ICASSP.2015.7178992
18. Nickel, M., Rosasco, L., Poggio, T.A.: Holographic embeddings of knowledge graphs. In: Proceedings of the Thirtieth AAAI Conference on Artificial Intelligence, Phoenix, Arizona, USA, 12–17 February 2016, pp. 1955–1961 (2016). http://www.aaai.org/ocs/index.php/AAAI/AAAI16/paper/view/12484
19. Nickel, M., Tresp, V.: Logistic tensor factorization for multi-relational data. CoRR abs/1306.2084 (2013). http://arxiv.org/abs/1306.2084
20. Nickel, M., Tresp, V., Kriegel, H.: A three-way model for collective learning on multi-relational data. In: Proceedings of the 28th International Conference on Machine Learning, ICML 2011, Bellevue, Washington, USA, 28 June–2 July 2011, pp. 809–816 (2011)
21. Palumbo, E., Rizzo, G., Troncy, R., Baralis, E., Osella, M., Ferro, E.: An empirical comparison of knowledge graph embeddings for item recommendation. In: Proceedings of the First Workshop on Deep Learning for Knowledge Graphs and Semantic Technologies (DL4KGS) Co-located with the 15th Extended Semantic Web Conerence (ESWC 2018), Heraklion, Crete, Greece, 4 June 2018, pp. 14–20 (2018). http://ceur-ws.org/Vol-2106/paper2.pdf
22. Rastegari, M., Ordonez, V., Redmon, J., Farhadi, A.: XNOR-net: imagenet classification using binary convolutional neural networks. In: Leibe, B., Matas, J., Sebe, N., Welling, M. (eds.) ECCV 2016, Part IV. LNCS, vol. 9908, pp. 525–542. Springer, Cham (2016). https://doi.org/10.1007/978-3-319-46493-0_32
23. Socher, R., Chen, D., Manning, C.D., Ng, A.Y.: Reasoning with neural tensor networks for knowledge base completion. In: Advances in Neural Information Processing Systems 26: 27th Annual Conference on Neural Information Processing Systems 2013, Proceedings of a meeting held 5–8 December 2013, Lake Tahoe, Nevada, United States, pp. 926–934 (2013). http://papers.nips.cc/paper/5028-reasoning-with-neural-tensor-networks-for-knowledge-base-completion
24. Suchanek, F.M., Kasneci, G., Weikum, G.: Yago: a core of semantic knowledge. In: Proceedings of the 16th International Conference on World Wide Web, WWW 2007, Banff, Alberta, Canada, 8–12 May 2007, pp. 697–706 (2007). https://doi.org/10.1145/1242572.1242667

25. Trouillon, T., Welbl, J., Riedel, S., Gaussier, É., Bouchard, G.: Complex embeddings for simple link prediction. In: Proceedings of the 33rd International Conference on Machine Learning, ICML 2016, New York City, NY, USA, 19–24 June 2016, pp. 2071–2080 (2016). http://jmlr.org/proceedings/papers/v48/trouillon16.html
26. Yang, B., Yih, W., He, X., Gao, J., Deng, L.: Embedding entities and relations for learning and inference in knowledge bases. CoRR abs/1412.6575 (2014). http://arxiv.org/abs/1412.6575

A Supervised Keyphrase Extraction System Based on Graph Representation Learning

Corina Florescu(✉) and Wei Jin

Computer Science and Engineering, University of North Texas, Denton, TX, USA
corinaflorescu@my.unt.edu, wei.jin@unt.edu

Abstract. Current supervised approaches for keyphrase extraction represent each candidate phrase with a set of hand-crafted features and machine learning algorithms are trained to discriminate keyphrases from non-keyphrases. Although the manually-designed features have shown to work well in practice, feature engineering is a labor-intensive process that requires expert knowledge and normally does not generalize well. To address this, we present SurfKE, an approach that represents the document as a word graph and exploits its structure in order to reveal underlying explanatory factors hidden in the data that may distinguish keyphrases from non-keyphrases. Experimental results show that SurfKE, which uses its self-discovered features in a supervised probabilistic framework, obtains remarkable improvements in performance over previous supervised and unsupervised keyphrase extraction systems.

Keywords: Keyphrase extraction · Feature learning · Phrase embeddings · Graph representation learning

1 Introduction

Keyphrases associated with a document typically provide a high-level topic description of the document and can allow for efficient information processing. In addition, keyphrases are shown to be particularly useful in many applications ranging from information search and retrieval to summarization, clustering, recommendation or simply to contextual advertisement [14,34,39]. Due to their importance, many approaches to automatic keyphrase extraction (KE) have been proposed in the literature along two lines of research: supervised and unsupervised [15,35].

In the supervised line of research, KE is formulated as a classification problem, where candidate phrases are classified as either positive (i.e., keyphrases) or negative (i.e., non-keyphrases). Specifically, each candidate phrase is encoded with a set of features such as its *tf-idf*, position in the document or part-of-speech tag, and annotated documents with "correct" keyphrases are used to train classifiers for discriminating keyphrases from non-keyphrases. Although these features have shown to work well in practice, many of them are computed based on observations and statistical information collected from the training documents which

L. Azzopardi et al. (Eds.): ECIR 2019, LNCS 11437, pp. 197–212, 2019.
https://doi.org/10.1007/978-3-030-15712-8_13

may be less suited outside of that domain. For example, position of a phrase and its presence in different sections of a document have shown to facilitate KE from scientific papers [10,22]. Other types of documents such as news articles or web pages do not follow standard formats which may give less useful location or structural information. Thus, it is essential to automatically discover such features or representations without relying on feature engineering which requires expert knowledge and normally does not generalize well.

Feature learning or representation learning is a set of techniques that allows a system to automatically discover characteristics that explain some structure underlying the data. Word embeddings, a representation learning technique which allows words with similar meaning to have similar representations has been shown to provide a fresh perspective to many NLP tasks. Recently, the word embedding models have been adopted by network science as a new learning paradigm to embed network vertices into a low-dimensional vector space, while preserving the network structure [13,33].

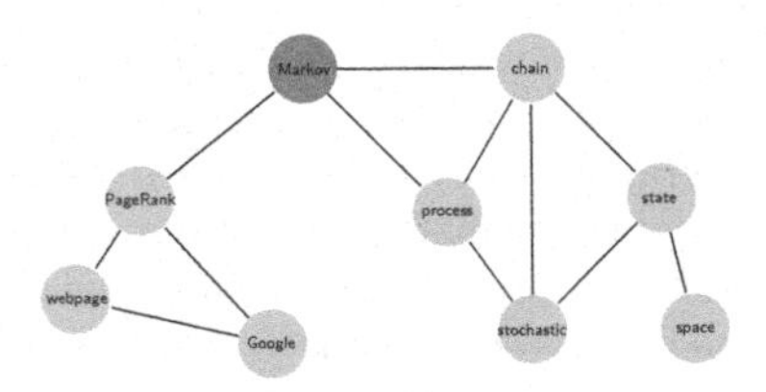

Fig. 1. An example of a bridge node (word).

Networks are becoming increasingly popular to capture complex relationships across various fields (e.g., social networks, biological networks). Additionally, it has been shown that the NLP tasks can benefit from a graph representation of text which connects words with meaningful relations [27]. For example, existing graph-based approaches for KE build a word graph according to word co-occurrences within the document, and then use graph centrality measures such as PageRank to measure word importance. These methods compute a single importance score for each candidate word by considering the number and quality of its associated words in the graph. However, the nodes in a network may exhibit diverse connectivity patterns which are not captured by the graph-based ranking methods. For example, nodes in a network may play different structural roles (e.g., hubs, bridge nodes, peripherals) and they can belong to the same or different communities. In a document graph representation, communities may represent topics, hub words might indicate central words in their topics while bridge nodes serve as transition words between topics. Figure 1 shows an example illustrating this behavior using a portion of a word graph built with text from Markov Chain Wikipedia page. Notice in this example that the word "Markov" acts as a bridge between the concepts concerning "PageRank" (e.g. webpage, Google) and the various terms commonly associated with the word "Markov" (e.g., state, process). Hence, one question that can be raised is the following: *Can we design an effective approach to automatically discover informative features for the keyphrase extraction tasks by exploiting the document graph representation?*

The rest of the paper is organized as follows. We summarize related work in the next section. Our proposed approach is described in Sect. 3. We present the data, our experiments and results in Sect. 4. We conclude the paper in Sect. 5.

2 Related Work

Many supervised and unsupervised approaches to KE have been proposed which target different types of documents such as research papers [10], news articles [20,36], meeting transcripts [24] or web pages [39].

In unsupervised approaches, various measures such as *tf-idf* and topic proportions are used to score words, which are later aggregated to obtain scores for phrases [1,41]. The ranking based on *tf-idf* has shown to work well in practice [8], despite its simplicity. Graph-based ranking methods and centrality measures are considered state-of-the-art for unsupervised KE. These methods build a word graph from each target document, such that nodes correspond to words and edges correspond to word association patterns. Nodes are then ranked using graph centrality measures such as PageRank [20,27] or HITS [19], and the top ranked phrases are returned as keyphrases. Since their introduction, many graph-based extensions have been proposed, which aim at modeling various types of information [9,11,23,36,38]. For example, textually-similar documents, information from the WordNet lexical database as well as the word embedding models are leveraged to both compute more accurate word co-occurrences and measure the relatedness between words in graph [23,36,38]. Several unsupervised approaches leverage word clustering techniques to group candidate words into topics, and then extract one representative keyphrase from each topic [4,21].

In supervised approaches, KE is commonly formulated as a binary classification problem where various feature sets and classification algorithms generally produce different models. For example, KEA [10], a representative supervised approach for KE, extracts two features for each candidate phrase, i.e., the *tf-idf* of a phrase and its distance from the beginning of the document, and uses them as input to Naïve Bayes. Hulth [16] used a combination of lexical and syntactic features such as the collection frequency, the relative position of the first occurrence and the part-of-speech tag of a phrase in conjunction with a bagging technique. Medelyan [25] extended KEA to incorporate various information from Wikipedia. Structural features (e.g., the presence of a phrase in particular sections of a document) have been extensively used for extracting keyphrases from research papers [22,30,31]. Chuang et al. [7] used a set of statistical and linguistic features (e.g., *tf-idf*, BM25, part-of-speech filters) for identifying descriptive terms in text. Jiang et al. [17] used a set of traditional features in conjunction with the pairwise learning-to-rank technique in order to design a ranking approach to KE. Tagging approaches such as Conditional Random Fields (CRFs) were also proposed for the KE task where a set of features such as, a word occurrences in different sections of a document, part-of-speech information, or named-entities are used to train CRFs for identifying keyphrases [2,12]. Recently, Meng et al. [26] proposed a generative model for predicting keyphrases which attempts to capture the deep semantic meaning of the content.

In contrast to the above approaches that focus on designing domain-tailored features to improve the task of KE, we aim at replacing the feature engineering process with a feature learning framework. Precisely, we propose a model that automatically harnesses the document structure in order to capture those patterns that keyphrases express. The learned representations are used to build a supervised classification model which predicts if a candidate phrase is a keyphrase for a document.

3 Proposed Model

Word graph representations are powerful structures that go beyond the linear nature of text to capture richer relationships between words. Regardless of a document type or domain, its text normally establishes a natural structural flow that is "encoded" in its word graph. For instance, a word that appears multiple times in a document may have different co-occurring neighbors within a short distance. As a result, the word graph may form densely connected regions for those terms. We conjecture that keyphrases, which are "important" terms in a document, display certain patterns that are captured by the wealth of relationships existing in the word graph. These clues give rise to the novel design of our model, called SurfKE.

SurfKE represents the document as a word graph and uses a biased random walk model to explore the proximity of nodes. The random walks are then used to produce node representations which reflect the text's intrinsic properties. The latent representations of words are then used as features to train a machine learning algorithm to discriminate the keyphrases from non-keyphrases.

SurfKE involves four main steps: (1) the graph construction at word level; (2) the process of learning continuous feature representations; (3) the formation of candidate phrases; and (4) the process of training the model. These steps are detailed below.

3.1 Graph Construction

Let d_t be a target document for extracting keyphrases. We first apply part-of-speech filters using the spaCy Python toolkit and then build an undirected weighted word graph $G = (V, E)$, where each unique word that passes certain part-of-speech tags corresponds to a vertex $v_i \in V$. Since keyphrases are usually noun phrases [16,20], our filter allows only for nouns and adjectives to be added to the graph. Two vertices v_i and v_j are linked by an edge $(v_i, v_j) \in E$ if the words corresponding to these vertices co-occur within a fixed window of w contiguous tokens in d_t. The weight of an edge is computed based on the co-occurrence count of the two words within a window of w successive tokens in d_t.

3.2 Feature Learning Framework

Let $G = (V, E)$ be a word graph constructed as above. Our goal is to learn latent representations of vertices which capture the connectivity patterns observed in the graph. Several network representation learning (NRL) algorithms have been proposed that aim to encode the graph structure by exploiting matrix factorization [5], random walks [33] and deep learning approaches [37]. Although the NRL algorithms are designed to be task independent, it has been shown that their quality may vary across different applications [40]. Recent NRL methods successfully applied for node classification use a searching strategy to explore the neighborhood of nodes and then leverage the Skip-Gram model [28] to learn the node representations [13,33]. To generate node proximity information, we use a random walk strategy which has been shown to work well for approximating many properties in the graph including node centrality and local structure information. The nodes in our word graph present various levels of connectivity which are reflected in the weight of an edge. To incorporate the strength of the association between two words into the model, we sample biased random walks with respect to the edge weights. For instance, let us assume that the word graph presented in Fig. 2 was built from a target document as explained in Sect. 3.1. We used bold black lines to mark those edges that reflect strong connections between words (words that co-occur frequently) and dashed blue lines to mark weak connection between words. As can be noticed from the figure, the graph expresses strong associations between words such as “random”, “walk”, “markov” or “chain” which may indicate that the topic of the document resides around these concepts. Next, let us assume that we want to explore the neighborhood of word “random”. Our search procedure should explore the local information of word “random” while focusing on its strong connections since these links may lead to the main topics of the document. Building on the these observations, we formulate our search strategy as a biased random walk model which is described next.

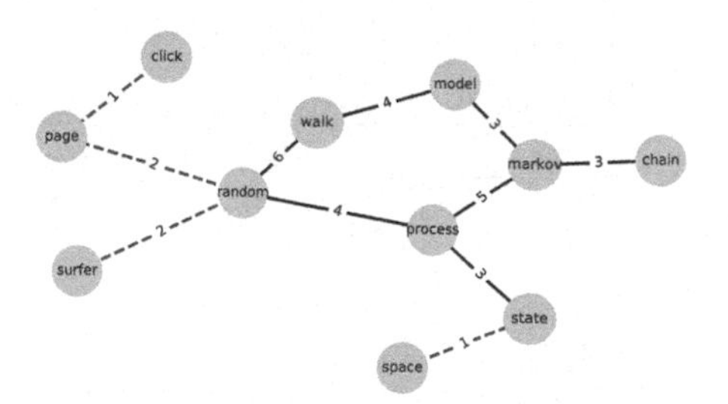

Fig. 2. A word graph example.

Biased Random Walks. Let $G = (V, E)$ be an undirected weighted graph built as described in Sect. 3.1 and let A be its adjacency matrix. An element

$a_{ij} \in A$ is set to the weight of edge (v_i, v_j) if there exists an edge between v_i and v_j, and is set to 0 otherwise. Let us denote by $\widetilde{A}$ the normalized form of matrix A with $\widetilde{a_{ij}} \in \widetilde{A}$ defined as:

$$\widetilde{a_{ij}} = \begin{cases} a_{ij} / \sum_{k=1}^{|V|} a_{ik} & \text{if } \sum_{k=1}^{|V|} a_{ik} \neq 0 \\ 0 & \text{otherwise} \end{cases}$$

Given a vertex $v_i \in V$, we then define a biased random walk of length L starting at vertex v_i as a stochastic process with random variables $X_1, X_2, ..., X_L$ such that $X_1 = v_i$ and X_{i+1} is a vertex chosen from the adjacent nodes of X_i based on the transition probabilities described in the stochastic matrix $\widetilde{A}$. Specifically, each random variable X_i, $i = 2, ..., L$ is generated by the following transition distribution:

$$P(X_{i+1} = v_j | X_i = v_i) = \begin{cases} \widetilde{a_{ij}} & (v_i, v_j) \in E \\ 0 & \text{otherwise} \end{cases}$$

Feature Learning Algorithm. Given a word graph $G = (V, E)$, we want a mapping function that associates with each node $v_i \in V$ a vector representation that preserves some graph properties. Let $f : V \rightarrow \mathbb{R}^d, d \leq |V|$ be such a mapping function, where d is a parameter specifying the number of dimensions for the embeddings. In particular, we want to learn vertex representations that maximize the likelihood of conserving neighborhood information of nodes. To capture node proximity information, we sample for each node $v_i \in V$, a fixed number (λ) of truncated vertex sequences according to our search strategy (i.e., biased random walks). The set of all generated biased random walks, denoted as C, ($|C| = \lambda * |V|$) forms our corpus. Then, given the set of vertex sequences (C), we leverage the Skip-Gram language model to train the vertex representations. Specifically, let $s = \{v_1, v_2, ..., v_T\}, s \in C$ be a biased random walk sampled from the word graph. We regard the vertices $v \in \{v_{i-t}, ..., v_{i+t}\} \setminus \{v_i\}$ as the context of the center word v_i, where t is the context window size. The objective of the Skip-gram model is to maximize the following function:

$$O = \sum_{s \in C} [\frac{1}{T} \sum_{i=1}^{T} \sum_{\substack{-t \leq j \leq t \\ j \neq 0}} \log P(v_{i+j} | v_i)]$$

That is, the objective function aims to maximize the average log probability of all vertex-context pairs in the random walk s for all $s \in C$. The probability of predicting the context node v_j given node v_i is computed using the softmax function:

$$P(v_j | f(v_i)) = \frac{exp(f'(v_j) \cdot f(v_i))}{\sum_{v_k \in V} exp(f'(v_k) \cdot f(v_i))}$$

We have denoted by f and f' the vector representation of the target node v_i and the vector representation of its context (neighbor) node v_j, respectively.

Namely, in the Skip-gram model, every word is associated with two learnable parameter vectors f when it plays the role of input vector and f' when it is the context vector. Calculating the denominator is computationally expensive and two strategies, hierarchical softmax [29] and negative sampling [28] have been widely used for speeding up the model optimization. We tried both methods and found that hierarchical softmax performs slightly better, although the difference was not statistically significant. The results presented in this paper are obtained using hierarchical softmax.

3.3 Forming Candidate Phrases

Candidate words that have contiguous positions in a document are concatenated into phrases. We extract noun phrases with pattern (adjective)*(noun)+ of length up to five tokens. We regard these noun phrases as candidate keyphrases. The feature vector for a multi-word phrase (e.g., *keyphrase extraction*) is obtained by taking the component-wise mean of the vectors of words constituting the phrase.

3.4 Building the Model

To build our supervised model, we parse each training document and create its corresponding word graph (Sect. 3.1), then we extract the candidate phrases (Sect. 3.3) and compute their feature vectors (Sect. 3.2). Each candidate phrase is then marked as a positive (i.e. keyphrase) or negative (i.e. non-keyphrase) example, based on the gold standard keyphrases of that document. The positive and negative phrase examples collected from the entire training set are inputted into a machine learning algorithm which is trained to distinguish keyphrases from non-keyphrases. The Näive Bayes algorithm has shown to perform well over a wide range of classification problems including keyphrase extraction [6, 10]. However, since our features are all continuous, we use Gaussian Näive Bayes to train our model.

To assign keyphrases to a new document, SurfKE determines its candidate phrases and their feature vectors as described above, and then applies the model built from the training documents. The model determines the probability of a candidate to be a keyphrase and uses it to output the keyphrases of that document. This probability can be used to output either the top k keyphrases or all keyphrases whose predicted probabilities are above a threshold (e.g., 50% confidence).

4 Experiments and Results

4.1 Datasets

To evaluate the performance of our model, we carried out experiments on a heterogeneous dataset of news articles and research papers from several domains.

The synthesized dataset contains 2,808 documents collected from two benchmark collections for KE, DUC [36] and Inspec [16], and a set of documents collected from the MEDLINE/PubMed database[1]. DUC, previously used by [36], is the DUC 2001 collection [32] which contains 308 news articles. The collection was manually annotated, each article being labeled with a set of at most 10 keyphrases. The Inspec dataset contains 2000 abstracts of journal papers from Computer Science and Information Technology, and 19,254 manually assigned keywords. To increase the diversity of our collection, we further collected 500 medical research papers from the NLM website. For each paper, we collected its title, abstract and the author-assigned keyphrases.

4.2 Experimental Design

To evaluate our model, we randomly split our heterogeneous dataset into 75% training (82,864 phrase instances) and 25% testing (27,713 phrase instances) such that each type of document is proportionally distributed among the two sets (e.g., news articles were proportionally divided between train and test sets). The model parameters such as the vector dimension or the number of walks per node were estimated on the training set using a 10-fold cross-validation setting where folds are created at the document level, i.e., the keyphrases of one document are not found in two different folds.

We measure the performance of our model by computing Precision, Recall and F1-score. Similar to our comparative models, we evaluate the top 10 predicted keyphrases returned by the model, where candidates are ranked based on the confidence scores as output by the classifier.

4.3 Results and Discussions

Our experiments are organized around several questions, which are discussed below.

How Sensitive is SurfKE to its Parameters? There are four major parameters that can influence the performance of our proposed model: (1) the window size (w), which determines how edges are added to the word graph; (2) the dimension of the vector embedding (d); (3) the maximum length of a sampled walk (L); and (4) the number of walks (λ) that we sample for each node. To answer this question, we analyze the influence of these parameters on our model. The parameters are set to the following values: $w = 6$, $\lambda = 80$, $d = 24$, $L = 10$, except for the parameter under investigation. We use Precision, Recall, and F1-score curves to illustrate our parameter sensitivity findings.

Window Size w. The first parameter that can influence the performance of SurfKE is the window size used to add edges between candidate words in the graph. We experimented with values of w ranging from 2 to 12 in steps of 2 and plot

[1] https://www.nlm.nih.gov/databases/download/pubmed_medline.html.

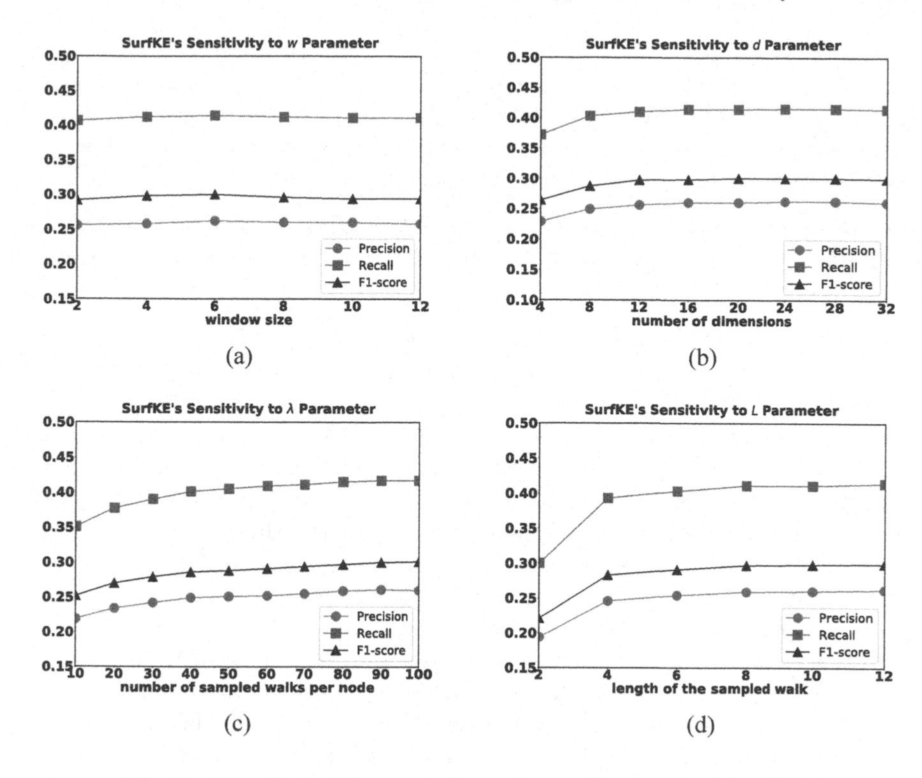

Fig. 3. The effect of parameters on the performance of SurfKE.

the performance of the model in Fig. 3(a) for illustration. We can observe in this figure that the performance of our model is stable over different values of w.

Number of Latent Dimensions d. To illustrate the effect of embeddings' dimensionality on our model, we span the value of d from 8 to 32 in steps of 4 and plot the achieved performance in Fig. 3(b). As can be seen from the figure, the performance of our model does not change significantly as we increase the dimension of the vector. For example, the model is stable for 12 to 24 latent dimensions and it has no benefit beyond a size of 24. This shows that we are able to store meaningful information for the keyphrase extraction task in quite small dimensional vectors.

Number of Walks λ. The biased random walks that we sample for each node allow for the model to surf nodes' neighborhoods in order to learn their structure. A small number of walks may not be sufficient to observe the patterns in the data while a large number of sampled walks might not bring additional information. To investigate the influence of λ on our model, we plot the performance of SurfKE when the number of walks sampled for each node ranges from 10 to 100 in steps of 10. As can be observed in Fig. 3(c), our model performs consistently well for $\lambda > 30$ walks per node. However, when we use less than 30 walks per node our model cannot learn good representations and slightly decreases its performance.

Length of the Sampled Walk L. The length of the walk offers flexibility in exploring smaller or larger portions of the graph. The influence of L on our model is illustrated in Fig. 3(d). As can be observed in the plot, the length of the walk affects the model's performance for very small values of L ($L < 6$) but its impact quickly slows and the model achieves consistent performance.

How does SurfKE Compare with Other Existing State-of-the-art Methods? We compare our model against several supervised and unsupervised approaches for KE.

Supervised Approaches. Many supervised approaches for KE target specific corpora and rely on structural features which are not commonly available for all types of documents. Since our goal is to extract keyphrases from all sorts of documents, we compare SurfKE with two supervised models, KEA [10] and Maui [25], which have shown to perform well on various domains and documents. KEA is a representative system for KE with the most important features being the frequency and the position of a phrase in a document. Maui is an extension of KEA which includes features such as keyphraseness, the spread of a phrase and statistics gathered from Wikipedia such as the likelihood of a term in being a link in Wikipedia or the number of pages that link to a Wikipedia page.

Table 1 shows the results of the comparison of SurfKE with the supervised models on the heterogeneous dataset, in terms of our evaluation metrics. As we can see in the table, SurfKE substantially outperforms both supervised approaches on the synthesized dataset for all evaluation metrics. For instance, SurfKE obtains an F1-score of 0.300 compared to 0.168 and 0.245 achieved by the two systems, KEA and Maui, respectively.

Table 1. The performance of SurfKE in comparison with the supervised models.

Approach	Precision	Recall	F1-score
KEA	0.146	0.247	0.168
Maui	0.207	0.362	0.245
SurfKE	**0.260**	**0.414**	**0.300**

Unsupervised Approaches. Unsupervised approaches to KE have started to attract significant attention, since they typically do not require linguistic knowledge, nor domain specific annotated corpora, which makes them easily transferable to other domains. Hence, we also compare SurfKE with three unsupervised models, KPMiner [8], TopicRank [4] and PositionRank [9]. KPMiner was the best performing unsupervised system in the SemEval 2010 competition [18]. It first uses statistical observations such as term frequencies to filter out phrases that are unlikely to be keyphrases. Then, candidate phrases are ranked using the *tf-idf* model in conjunction with a boosting factor which aims at reducing the bias towards single word terms. TopicRank groups candidate phrases into topics and uses them as vertices in a complete graph. The importance of each topic is computed using graph-based ranking functions, and keyphrases are selected from the highest ranked topics. PositionRank is a graph-based model, which exploits the positions of words in the target document to compute a weight for each word. This weight is then incorporated into a biased PageRank algorithm to score words that are later used to rank candidate phrases.

Table 2. The performance of SurfKE in comparison with the unsupervised models.

Approach	Precision	Recall	F1-score
KpMiner	0.182	0.335	0.219
TopicRank	0.230	0.356	0.261
PositionRank	0.231	0.389	0.270
SurfKE	**0.260**	**0.414**	**0.300**

Table 2 shows the comparison of SurfKE with the three unsupervised baselines described above. As we can see in the table, SurfKE substantially outperforms all unsupervised systems on the synthesized dataset. For instance, with relative improvements of 36.98%, 14.94% and 11.11% over KPMiner, TopicRank and PositionRank respectively, the F1-score of our model is significantly superior to that of the other models.

With a paired t-test, we found that the improvements in Precision, Recall, F1-score achieved by SurfKE are all statistically significant with $p \leq 0.05$. The results presented in this paper are obtained using the authors' publicly-available implementations for KEA, Maui and, PositionRank and the implementation from the `pke` package [3] for TopicRank and KPMiner. Our code and data will be made freely available.

What is the Impact of the Training Size on the Performance of SurfKE? Human labeling is a time consuming process hence, we investigate the effect that the training size has on the performance of our proposed model. For this experiment, we randomly selected 11 samples of n documents from the training set, where n ranges from 1 to 101 in steps of 10. We used these samples one at a time to train SurfKE and recorded its performance in terms of F1-score on the test set. We show the results of this experiment in Fig. 4 together with the performance of Maui, the best performing supervised approach. Maui was trained and tested accordingly to this experiment. The blue and red dotted lines indicate the F1 values obtained by SurfKE respectively Maui when all training documents are used to train the model. As can be observed from the figure, SurfKE reaches its best performance (i.e., performance achieved with all training documents) with a quite small set of training documents. Specifically, a number of about 20 documents is enough for our model to obtain good performance. Furthermore, we can notice that our model trained using only one document archives a higher F1-score than the Maui model trained on all training documents ($\approx 2,100$). In the same figure, we can observed that Maui highly depends on the size of training data.

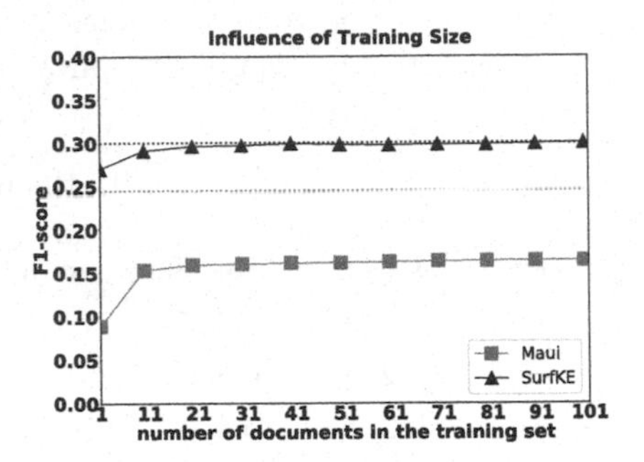

Fig. 4. Performance of SurfKE when smaller sets of documents are used for training.

What is the Performance of SurfKE in a Cross-domain Setting? In previous experiments, we ensured that train and test sets contained each type of document present in the synthesized collection. Such a requirement is not necessary and such an environment may not be always possible in reality. To investigate the transferable nature of our proposed model, we carried out an additional experiment. Specifically, we used an external collection to train the model, and evaluated its performance on our test set. The data that we used to train the model was made available by Nguyen and Kan [30]. It consists of 211 research papers from various disciplines and the author-input keyphrases as standard for evaluation. We show the results of this experiment in Table 3 together with the performance of our supervised models. Note that the two baselines were trained and tested accordingly to this experiment. We denote by SurfKE (*) the model trained on our training set (see Sect. 4.2) and copied its performance here for comparison purpose.

As can be seen from the Table 3, the type of documents in the training set does not affect the performance of SurfKE. For example, our model achieves an F1-score of 0.299 in a cross-domain setting (SurfKE) compared with 0.300 when train and test sets have a similar distribution of documents (SurfKE (*)). Our model performs well and it outperforms the comparative approaches. For instance, in a cross-domain setting, SurfKE gets an F1-score of 0.299 compared with 0.168 and 0.157 achieved by KEA and Maui, respectively.

What type of Errors Does our Model Output? Finally, we performed an error analysis by manually comparing predicted keyphrases with gold standard for evaluation of 30 randomly selected documents from our datasets. We found that the most common errors occur when (1) a system incorrectly predicts several candidates as keyphrases because they contain a word that represents the main topic of the target document. For example, document "90.txt" (Inspec)

Table 3. Performance of SurfKE and supervised models in a cross-domain setting.

Approach	Precision	Recall	F1-score
KEA	0.147	0.248	0.168
Maui	0.128	0.249	0.157
SurfKE	**0.260**	**0.412**	**0.299**
SurfKE (*)	0.260	0.414	0.300

has main topic "surveillance robot". Models built on word embeddings succeed in identifying "surveillance robot" as a keyphrase but they incorrectly output "robot system" and "robot prototype" as keyphrases; (2) a model predicts a candidate as a keyphrase (e.g. sprinter Ben Johnson (la030889-0163.txt (DUC))), but it also outputs an alternative form as a keyphrase (e.g., Ben Johnson); (3) a model correctly predicts a keyphrase but the gold standard is a alternative form of that concept. For instance, a systems predicts the phrase "link prediction" as a keyphrase but the gold standard says "link prediction model". Some of these errors can be reduced by clustering semantically related candidates [4] or applying filters on the list of predicted keyphrases. We believe that a better understanding of errors has the potential to improve feature learning models for KE.

5 Conclusion and Future Work

In this paper, we designed a novel feature learning framework for the keyphrase extraction task. Specifically, we proposed to represent the document as a word graph and exploit its structure in order to capture various connectivity patterns that keyphrases might express. The learned phrase representations are then used to build a supervised model that predicts the keyphrases of a document.

Our experiments show that (1) our supervised model (SurfKE) obtains remarkable improvements in performance over 5 previous proposed models for KE; (2) SurfKE requires a small amount of annotated documents (about 20) to achieve good performance, which makes it an effective supervised approach; (3) the performance of SurfKE is consistent across domains, which makes it easily transferable to other domains.

In future work, it would be interesting to explore other network representation learning techniques to jointly learn the network structure and the node attribute information (e.g., to incorporate text features of vertices into network representation learning).

Acknowledgments. This research is supported by the National Science Foundation award IIS-1739095.

References

1. Barker, K., Cornacchia, N.: Using noun phrase heads to extract document keyphrases. In: Hamilton, H.J. (ed.) AI 2000. LNCS (LNAI), vol. 1822, pp. 40–52. Springer, Heidelberg (2000). https://doi.org/10.1007/3-540-45486-1_4
2. Bhaskar, P., Nongmeikapam, K., Bandyopadhyay, S.: Keyphrase extraction in scientific articles: a supervised approach. In: COLING (Demos), pp. 17–24 (2012)
3. Boudin, F.: pke: an open source python-based keyphrase extraction toolkit. In: COLING 2016, pp. 69–73 (2016)
4. Bougouin, A., Boudin, F., Daille, B.: Topicrank: graph-based topic ranking for keyphrase extraction. In: International Joint Conference on Natural Language Processing (IJCNLP), pp. 543–551 (2013)
5. Cao, S., Lu, W., Xu, Q.: GraRep: learning graph representations with global structural information. In: Proceedings of the 24th ACM International on Conference on Information and Knowledge Management, pp. 891–900. ACM (2015)
6. Caragea, C., Bulgarov, F.A., Godea, A., Gollapalli, S.D.: Citation-enhanced keyphrase extraction from research papers: a supervised approach. In: Proceedings of the Conference on Empirical Methods in Natural Language Processing, pp. 1435–1446 (2014)
7. Chuang, J., Manning, C.D., Heer, J.: Without the clutter of unimportant words: descriptive keyphrases for text visualization. ACM Trans. Comput. Hum. Interact. **19**(3), 19 (2012)
8. El-Beltagy, S.R., Rafea, A.: KP-miner: participation in semeval-2. In: Proceedings of the 5th International Workshop on Semantic Evaluation, pp. 190–193. Association for Computational Linguistics (2010)
9. Florescu, C., Caragea, C.: Positionrank: an unsupervised approach to keyphrase extraction from scholarly documents. In: Proceedings of the 55th Annual Meeting of the Association for Computational Linguistics (Volume 1: Long Papers), vol. 1, pp. 1105–1115 (2017)
10. Frank, E., Paynter, G.W., Witten, I.H., Gutwin, C., Nevill-Manning, C.G.: Domain-specific keyphrase extraction. In: Proceedings of the 16th International Joint Conference on Artificial Intelligence, pp. 668–673 (1999)
11. Gollapalli, S.D., Caragea, C.: Extracting keyphrases from research papers using citation networks. In: Proceedings of the 28th American Association for Artificial Intelligence, pp. 1629–1635 (2014)
12. Gollapalli, S.D., Li, X.L., Yang, P.: Incorporating expert knowledge into keyphrase extraction. In: American Association for Artificial Intelligence, pp. 3180–3187 (2017)
13. Grover, A., Leskovec, J.: node2vec: scalable feature learning for networks. In: Proceedings of the 22nd ACM SIGKDD International Conference on Knowledge Discovery and Data Mining, pp. 855–864. ACM (2016)
14. Hammouda, K.M., Matute, D.N., Kamel, M.S.: CorePhrase: keyphrase extraction for document clustering. In: Perner, P., Imiya, A. (eds.) MLDM 2005. LNCS (LNAI), vol. 3587, pp. 265–274. Springer, Heidelberg (2005). https://doi.org/10.1007/11510888_26
15. Hasan, K.S., Ng, V.: Automatic keyphrase extraction: a survey of the state of the art. In: Proceedings of the 27th International Conference on Computational Linguistics, pp. 1262–1273 (2014)
16. Hulth, A.: Improved automatic keyword extraction given more linguistic knowledge. In: Proceedings of the Conference on Empirical Methods in Natural Language Processing, pp. 216–223 (2003)

17. Jiang, X., Hu, Y., Li, H.: A ranking approach to keyphrase extraction. In: Proceedings of the 32nd International ACM SIGIR Conference on Research and Development in Information Retrieval, pp. 756–757. ACM (2009)
18. Kim, S.N., Medelyan, O., Kan, M.Y., Baldwin, T.: SemEval-2010 task 5: automatic keyphrase extraction from scientific articles. In: Proceedings of the 5th International Workshop on Semantic Evaluation, SemEval 2010, pp. 21–26 (2010)
19. Litvak, M., Last, M.: Graph-based keyword extraction for single-document summarization. In: Proceedings of the Workshop on Multi-source Multilingual Information Extraction and Summarization, pp. 17–24 (2008)
20. Liu, Z., Huang, W., Zheng, Y., Sun, M.: Automatic keyphrase extraction via topic decomposition. In: Proceedings of the Conference on Empirical Methods in Natural Language Processing, pp. 366–376 (2010)
21. Liu, Z., Li, P., Zheng, Y., Sun, M.: Clustering to find exemplar terms for keyphrase extraction. In: Proceedings of the 2009 Conference on Empirical Methods in Natural Language Processing, pp. 257–266. ACL (2009)
22. Lopez, P., Romary, L.: HUMB: automatic key term extraction from scientific articles in GROBID. In: Proceedings of the 5th International Workshop on Semantic Evaluation, pp. 248–251. Association for Computational Linguistics (2010)
23. Martinez-Romo, J., Araujo, L., Duque Fernandez, A.: Semgraph: extracting keyphrases following a novel semantic graph-based approach. J. Assoc. Inf. Sci. Technol. **67**(1), 71–82 (2016)
24. Marujo, L., Ribeiro, R., de Matos, D.M., Neto, J.P., Gershman, A., Carbonell, J.: Key phrase extraction of lightly filtered broadcast news. In: Sojka, P., Horák, A., Kopeček, I., Pala, K. (eds.) TSD 2012. LNCS (LNAI), vol. 7499, pp. 290–297. Springer, Heidelberg (2012). https://doi.org/10.1007/978-3-642-32790-2_35
25. Medelyan, O., Frank, E., Witten, I.H.: Human-competitive tagging using automatic keyphrase extraction. In: Proceedings of the 2009 Conference on Empirical Methods in Natural Language Processing, pp. 1318–1327. ACL (2009)
26. Meng, R., Zhao, S., Han, S., He, D., Brusilovsky, P., Chi, Y.: Deep keyphrase generation. arXiv preprint arXiv:1704.06879 (2017)
27. Mihalcea, R., Tarau, P.: TextRank: bringing order into text. In: Proceedings of the 2004 Conference on Empirical Methods in Natural Language Processing, pp. 404–411 (2004)
28. Mikolov, T., Sutskever, I., Chen, K., Corrado, G.S., Dean, J.: Distributed representations of words and phrases and their compositionality. In: Advances in Neural Information Processing Systems, pp. 3111–3119 (2013)
29. Morin, F., Bengio, Y.: Hierarchical probabilistic neural network language model. In: AISTATS, vol. 5, pp. 246–252 (2005)
30. Nguyen, T.D., Kan, M.-Y.: Keyphrase extraction in scientific publications. In: Goh, D.H.-L., Cao, T.H., Sølvberg, I.T., Rasmussen, E. (eds.) ICADL 2007. LNCS, vol. 4822, pp. 317–326. Springer, Heidelberg (2007). https://doi.org/10.1007/978-3-540-77094-7_41
31. Nguyen, T.D., Luong, M.T.: Wingnus: keyphrase extraction utilizing document logical structure. In: Proceedings of the 5th International Workshop on Semantic Evaluation, pp. 166–169. Association for Computational Linguistics (2010)
32. Over, P.: Introduction to DUC-2001: an intrinsic evaluation of generic news text summarization systems. In: Proceedings of DUC 2001 (2001)
33. Perozzi, B., Al-Rfou, R., Skiena, S.: Deepwalk: online learning of social representations. In: KDD, pp. 701–710 (2014)

34. Pudota, N., Dattolo, A., Baruzzo, A., Ferrara, F., Tasso, C.: Automatic keyphrase extraction and ontology mining for content-based tag recommendation. Int. J. Intell. Syst. **25**(12), 1158–1186 (2010)
35. Verberne, S., Sappelli, M., Hiemstra, D., Kraaij, W.: Evaluation and analysis of term scoring methods for term extraction. Inform. Retrieval J. **19**(5), 510–545 (2016)
36. Wan, X., Xiao, J.: Single document keyphrase extraction using neighborhood knowledge. In: Proceedings of the 2008 American Association for Artificial Intelligence, pp. 855–860 (2008)
37. Wang, D., Cui, P., Zhu, W.: Structural deep network embedding. In: Proceedings of the 22nd ACM SIGKDD International Conference on Knowledge Discovery and Data Mining, pp. 1225–1234. ACM (2016)
38. Wang, R., Liu, W., McDonald, C.: Corpus-independent generic keyphrase extraction using word embedding vectors. In: Software Engineering Research Conference, p. 39 (2014)
39. Yih, W.t., Goodman, J., Carvalho, V.R.: Finding advertising keywords on web pages. In: WWW 2006, pp. 213–222 (2006)
40. Zhang, D., Yin, J., Zhu, X., Zhang, C.: Network representation learning: a survey. arXiv preprint arXiv:1801.05852 (2017)
41. Zhang, Y., Milios, E., Zincir-Heywood, N.: A comparative study on key phrase extraction methods in automatic web site summarization. J. Digit. Inf. Manage. **5**(5), 323 (2007)

Recommender Systems II

Comparison of Sentiment Analysis and User Ratings in Venue Recommendation

Xi Wang(✉), Iadh Ounis, and Craig Macdonald

University of Glasgow, Glasgow, UK
x.wang.6@research.gla.ac.uk,
{iadh.ounis,craig.macdonald}@glasgow.gla.ac.uk

Abstract. Venue recommendation aims to provide users with venues to visit, taking into account historical visits to venues. Many venue recommendation approaches make use of the provided users' ratings to elicit the users' preferences on the venues when making recommendations. In fact, many also consider the users' ratings as the ground truth for assessing their recommendation performance. However, users are often reported to exhibit inconsistent rating behaviour, leading to less accurate preferences information being collected for the recommendation task. To alleviate this problem, we consider instead the use of the sentiment information collected from comments posted by the users on the venues as a surrogate to the users' ratings. We experiment with various sentiment analysis classifiers, including the recent neural networks-based sentiment analysers, to examine the effectiveness of replacing users' ratings with sentiment information. We integrate the sentiment information into the widely used matrix factorization and GeoSoCa multi feature-based venue recommendation models, thereby replacing the users' ratings with the obtained sentiment scores. Our results, using three Yelp Challenge-based datasets, show that it is indeed possible to effectively replace users' ratings with sentiment scores when state-of-the-art sentiment classifiers are used. Our findings show that the sentiment scores can provide accurate user preferences information, thereby increasing the prediction accuracy. In addition, our results suggest that a simple binary rating with 'like' and 'dislike' is a sufficient substitute of the current used multi-rating scales for venue recommendation in location-based social networks.

1 Introduction

Location-Based Social Networks (LBSNs), such as Yelp, are increasingly used by users to discover new venues and share information about such venues. These networks are nowadays collecting a large volume of user information such as ratings, check-ins, tips, user comments, and so on. This large volume of interaction data makes it more difficult for users to select venues to visit without the help of a recommendation engine. Indeed, many systems [1–3] have been proposed to address the data overload problem on LBSNs by automatically

L. Azzopardi et al. (Eds.): ECIR 2019, LNCS 11437, pp. 215–228, 2019.
https://doi.org/10.1007/978-3-030-15712-8_14

suggesting venues for users to visit based on their profile and visiting history. In particular, the explicit ratings of venues by users are widely used in various recommendation systems to elicit users' preferences, including in collaborative filtering systems [4,5], matrix factorization (MF) approaches [6] and more recent advanced venue recommendation approaches [7–9].

However, in practice, user ratings are not always effective in representing the users' preferences. For example, it has been reported that users have distinct and inconsistent rating behaviour [10] and find it difficult to provide accurate feedback on the venues when faced with selecting among multi-rating values [11]. Several previous studies aimed to assess the impact of users' location [10], their personal and situational characteristics [12], and the nature of rating scales available to users on the quality of the users' ratings [11].

On the other hand, sentiment analysis is a widely used technique to gauge users' opinions and attitudes, for instance towards products and venues, from textual user reviews [13]. Sentiment analysis not only predicts the polarity of user opinions (e.g. positive vs. negative) but can also provide a summary of the users' opinions of a product or a venue from their reviews [13,14]. In fact, sentiment analysis has been adopted by many studies to adjust user ratings or provide extra features to enhance the performance of recommendation systems [15–17].

The integration of sentiment analysis into recommendation systems is still limited to adjusting users' ratings to overcome their inconsistency. Instead, Lak et al. [18] substituted user' ratings with sentiment analysis, but concluded that sentiment analysis was insufficient to replace ratings in their experiments. Since then, sentiment analysis has seen a lot of attention in the literature cumulating in the development of advanced effective neural networks-based sentiment analysis of long and short texts [19,20]. Indeed, previous sentiment analysis approaches mainly relied on human-crafted sentiment dictionaries [21,22], which are not necessarily sufficiently effective on the wide variety of words used in LBSNs [20].

Therefore, in this paper, we hypothesise that it is possible to replace the users' explicit ratings by leveraging state-of-the-art sentiment analysers on the users' comments, thereby increasing the consistency of the user's preferences when making venue recommendations. We integrate the obtained users' preference scores through sentiment analysis into the widely used MF and GeoSoCa multi feature-based venue recommendation models [8]. Our results, using three different Yelp Challenge-based datasets, show that it is indeed possible to effectively replace users' ratings with sentiment scores when state-of-the-art sentiment analysers are used and still produce accurate venue recommendations. Our findings also suggest that it is possible to alleviate the users' ratings inconsistency by substituting the popular five-star rating scale used by LBSNs with a binary rating scale (i.e. 'like' or 'dislike') based on the sentiment analysis of their comments. In particular, the main contributions of our study are as follows:

- We explore the sentiment polarity classification accuracy of various recent sentiment analysis approaches based on neural-networks such as convolutional neural networks (CNN) and long short-term memory (LSTM) along the more conventional SentiWordNet and support vector machine (SVM)-based approaches.

- We replace the user ratings with sentiment scores in the MF model, as well as within the popular multi-feature GeoSoCa venue recommendation model [8]. To the best of our knowledge, this is the first study to examine the performance of solely using sentiment scores to substitute explicit user ratings in venue recommendation.
- We conduct thorough sentiment classification and venue recommendation experiments on datasets from the Yelp Dataset Challenge[1]. First, we use part of the dataset to conduct sentiment classification experiments and then conduct venue recommendation experiments on two other different types of datasets, which are extracted from Yelp. These two types of datasets include two city-based datasets (i.e. Phoenix and Las Vegas) and one cross-city dataset (multiple cities are covered).

The rest of our paper is structured as follows. We review the literature on the effectiveness of rating, sentiment analysis development and the application of sentiment analysis to venue recommendation in Sect. 2. In Sect. 3, we describe the GeoSoCa model and the rating substitution strategy. Section 4 describes the sentiment analysis techniques that we deploy. After that, we detail the setup for our experiments (Sect. 5). Then, in Sect. 6, we present our obtained results for evaluating the effectiveness of substituting ratings with sentiment scores. Finally, Sect. 7 provides concluding remarks.

2 Related Work

Ratings and Rating Scale. Many venue recommendation systems use explicit user ratings as the ground truth, both when learning user preferences and to evaluate the performance [6,23,24]. Ratings are a simple way for users to express opinions. However, the effectiveness of rating and rating scale have been well studied in the literature – for instance, Cosley et al. [11] argued that ratings are not sufficient for recommendation systems to effectively model the complex user preferences and opinions, while also biasing recommendation evaluation with inaccurate user opinion information. Moreover, as argued by Amoo et al. [25], the users' rating distributions are affected by different rating scales. Pennock et al. [26] recognised that the ratings of a user for the same item may not be consistent at different times. Therefore, in this paper, we similarly argue that using explicit ratings may not well represent users' opinions or attitudes. In particular, with the development of further refined techniques for sentiment analysis (discussed next), we propose to use sentiment scores to replace ratings for representing users opinions.

Sentiment Analysis. Sentiment analysis has been developed over many years, to automatically estimate users' opinions and attitudes from review texts. In the early period, sentiment analysis mainly relied on manually collected sentiment

[1] Yelp Dataset Challenge: https://www.yelp.co.uk/dataset/challenge.

words. Ohana et al. [27] used SentiWordNet[2] [28] to identify word features, and constructed a learned model using SVM on such features. Mohammad et al. [21] leveraged tweet-specific sentiment lexicons to construct features and also included a collection of negated words. However, with the increasing applications of deep neural networks (NN), NN-based sentiment analysis techniques have achieved excellent accuracies [20]. We note the work of Kim [19], who exploited a convolutional NN (CNN) to run on a pre-trained word embedding vector and obtained much improvement in the sentiment analysis performance. Moreover, Baziotis et al. [20] leveraged long short-term memory (LSTM) to capture word order information when conducting sentiment analysis on tweets. Their approach outperformed other approaches in the 2017 SemEval competition [29]. Therefore, in this paper, with these comparably recent improvements in sentiment analysis performances, we aim to measure their usefulness in leveraging sentiment scores expressed in user reviews for the purposes of venue recommendation.

Venue Recommendation with Sentiment Analysis. Various studies [15,30,31] have been concerned with integrating sentiment analysis in venue recommendation. However, to the best of our knowledge, the used sentiment analysis in venue recommendation models have not encompassed the most recent state-of-the-art sentiment analysis approaches. For instance, Yang et al. adopted SentiWordNet3.0 [28] and the NTLK toolkit [32] for sentiment analysis. Gao et al. [30] used unsupervised sentiment classification on the sentiment polarity of words to generate a user sentiment indication matrix. Zhao et al. [31] constructed a probabilistic inference model to predict the user sentiment based on a limited number of sentiment seed words. Wang et al. [17] extracted latent semantic topics from the user reviews using a Latent Dirichlet allocation (LDA) model and inferred a user preference distribution. According to the recent sentiment analysis competition in SemEval [29], these sentiment analysis approaches are not competitive with the current state-of-the-art approaches for sentiment analysis. Therefore, we postulate that applying the state-of-the-art sentiment analysis approaches in venue recommendation could improve the performance of the sentiment-based venue recommendation approaches. Moreover, we argue that the sentiment scores could effectively replace the users' ratings to represent users' preferences.

3 Venue Recommendation Model and Rating Substitution Strategy

In this section, we first state the venue recommendation problem and the notations used in this paper (Sect. 3.1). Next, we introduce the two models (i.e. MF and GeoSoCa) that we use to learn the ability of sentiment analysis in capturing user preferences instead of user ratings. Finally, in Sect. 3.3, we describe the rating substitution strategy which we apply to MF and GeoSoCa.

[2] SentiWordNet is an opinion lexicon, where the sentiment and polarity of each term is quantified.

3.1 Problem Statement

The venue recommendation task aims to rank highly venues that users would like to visit, based on users' previous venue visits and other sources of information. For instance, considering sets of U users and V venues (of size m and n, respectively), the previous ratings of users can be encoded in $R \in \mathbb{R}^{m \times n}$, where entries $r_{u,v} \in R$ can represent the previous venue ratings (1..5) or the check-ins (0..1) of user $u \in U$ to venue $v \in V$. Venue recommendation can then be described to accurately estimate the value $r_{u,v}$ for a venue that the user has not previously visited, or to rank highly venues that they would highly likely visit.

3.2 Venue Recommendation Approaches

In this work, we examine the behaviour of two venue recommendation approaches, namely MF and GeoSoCa, and how they perform when we change the definition of $r_{u,v}$.

Matrix Factorization (MF). MF is a classic recommendation approach, which has been adopted by many recommendation model studies as a baseline [33–35]. MF adopts singular value decomposition to learn latent semantic vectors q_v and p_u for user u and item v, respectively on known ratings $r_{u,v} \in R$.

GeoSoCa. GeoSoCa [8] is a popular venue recommendation approach, proposed by Zhang et al. in 2015. Compared to MF, it encompasses three additional important sources of information in making improved venue recommendation, namely geography, social and category information [23,30]. Since then, it has been frequently used and discussed in various studies [1,36,37]. GeoSoCa estimates the probability of users visiting an unvisited venue according to the influence of three additional sources of information, namely the geography, social and category features. The geographical and social influence features use the geographical distance and users' social connections to measure the influence of different venues on users, respectively. The categorical influence estimates users' preferences distribution over categories of venues (restaurants, bars, etc.). In particular, $p_{c,v}$ indicates the popularity of venue $v \in V$, which belongs to category $c \in C$, where C denotes all venue categories[3]. In computing all these three additional features, GeoSoCa makes use of the users' ratings to estimate the probability of user u visiting venue v [8]. Note that, following Zhang et al., in our experiments, we also deploy both GeoSoCa and its components individually, i.e. Geo, So and Ca.

3.3 Rating Substitution Strategy

As argued earlier, the advent of accurate sentiment analysis approaches offers new opportunities for more refined venue recommendation. In particular, since

[3] A venue might belong to more than one category in the Yelp dataset. For such venues, we use the category that is uppermost in the hierarchy.

the resulting sentiment classifiers can be formulated in a probabilistic manner, we assume that the users' preferences are indicated by the classifier's confidence, denoted as sentiment score $s_{u,v}$. This score captures the classifier confidence in user's u comment on venue v is positive. Indeed, our work examines if the sentiment score, $s_{u,v}$, can effectively replace the rating $r_{u,v}$ as an indicator of users' preferences. We now describe our adaptations of MF and GeoSoCa.

In MF, the sentiment-based MF approach replaces user ratings on venues, $r_{u,v} \in R$, with sentiment scores $s_{u,v}$. In contrast, for GeoSoCa, we consider the substitution strategy on each component: In the geographical and social influence features, we replace users' ratings $r_{u,v} \in R$ with $s_{u,v} \in R$. Moreover, different from the previous two features, in the categorical influence feature, we not only replace users' ratings $r_{u,v} \in R$ with $s_{u,v}$, but we also modify the venue category popularity, $p_{c,v}$, as follows:

$$p_{c,v} = \sum_{u \in U} s_{u,v} \tag{1}$$

Therefore, we evaluate the ability of the sentiment scores to accurately capture the overall venue popularity. In the next section, we discuss the sentiment classification approaches that we apply to calculate $s_{u,v}$.

4 Sentiment Classification Approaches

As discussed in Sect. 2, sentiment analysis approaches can be broadly classified into dictionary-based, learned, and deep-learned. We apply four approaches that represent all of the categories, as well as a **Random** (Rand) classifier that matches the class distribution in its predictions, as a weak baseline.

1. **SentiWordNet-Based Classifier (SWN).** The SentiWordNet-based classification approach is constructed following the approach proposed by [38], which used the updated SentiWordNet3.0 dictionary [28]. In addition, we use the 'geometric' weighting strategy that considers the word frequency to compute the prior polarity of each sentiment lexicon. The sentiment score is obtained by averaging the sentiment score of words in each user's comments.
2. **SVM-Based Classifier (SVM).** Following the experimental setup of Pang et al. [39], we implement an SVM-based classifier, using the labelled word frequency vector for each review, trained using a linear kernel.
3. **CNN-Based Classifier (CNN).** We use a CNN-based classifier [19] for sentiment classification. In addition, we also follow the 'CNN-Static' model setup in [19], which reported a good performance without the need for tuning the word embedding vectors.
4. **LSTM-Based Classifier (LSTM).** We deploy an LSTM-based classifier [20], which obtained the top performance in the sentiment classification competition in SemEval 2017. We follow the experimental model construction process and configuration described by Baziotis et al. [20].

5 Experimental Setup

In the following, we evaluate the sentiment classification approaches compared to the corresponding users' ratings of these venues. Thereafter, we examine the difference in venue recommendation effectiveness between models that leverage user ratings and those that use sentiment scores instead. Therefore, our experiments aim at answering the following research questions:

- **RQ1.** Which sentiment analysis approaches exhibit the highest performance for user review classification?
- **RQ2.** Can sentiment scores sufficiently capture the users' preferences so as to replace ratings for the purposes of effective venue recommendation? Does increased sentiment classification accuracy results in improved venue recommendation effectiveness?

To address these research questions, we perform two experiments using the Yelp Challenge dataset Round 11. We use the Yelp dataset as it is the only available public dataset that fulfils our experimental requirements, i.e. to include geographical, social and category information, as well as user reviews.

Sentiment Classification: The statistics of the dataset extracted from Yelp for the sentiment classification experiments are shown in Table 1. In Yelp, all ratings are given in a 5-star rating scale (1 is poor, 5 is great). Following Koppel et al. [40], we label the polarity of each review according to the user's rating of the venue, which we regard positive if the rating $\geq$4, and negative if rating $\leq$2. Then we randomly select equal numbers of positive and negative reviews to construct the training and testing datasets, which also avoids the class bias phenomena. Moreover, as the CNN and LSTM approaches rely on trained word embedding models, we use the remaining reviews (minus the reviews found in the Phoenix and Las Vegas city-based Yelp datasets, discussed below) in the Yelp dataset to train a word embedding model using the GenSim tool. For out-of-vocabulary words, we randomly initialise the embedding vectors, as suggested by Yang et al. [41].

We vary the size of the training dataset, from 10,000 to 600,000 reviews, to examine the stability and accuracy of the sentiment classification approaches. We use a 5-fold cross-validation setup on the training dataset before reporting the accuracy on the test datasets.

Venue Recommendation: We use three subsets of the Yelp dataset to evaluate the performance differences between the rating-based and sentiment-based venue recommendation models. Unless otherwise stated, the sentiment scores are generated after the classifiers have been trained on 600,000 comments[4]. Table 2 shows the statistics of the three Yelp-based datasets we use to evaluate the venue recommendation effectiveness. In particular, for generalisation purposes, we include

[4] As will be shown in Sect. 6, this is the best training setup in terms of sentiment classification accuracy.

two city-based datasets (namely Phoenix and Las Vegas) following other recent works [42–44], and one cross-city dataset. Indeed, we use these different Yelp subsets to obtain an overall understanding of the venue recommendation models' performances in different settings. To alleviate extreme sparseness, following Yuan et al. [43], for each dataset, we remove users with less than 20 reviews and venues with less than 5 visits. Figure 1 shows the ratings distribution of the three datasets. It is of note that for all three datasets, the number of positive reviews (ratings 4 & 5) outweighs the number of negative reviews (ratings 1 & 2) by quite a margin. Finally, experiments are conducted using a 5-fold cross-validation on each dataset, and evaluated for recommendation quality using Precision@5 & @10 and mean average precision (MAP)[5].

Table 1. Dataset summary for sentiment classification usage

Dataset name	Number of reviews
Training	600,000
Testing	200,000

Table 2. Datasets summaries for the venue recommendation task

Dataset	#Users	#Venues	#Reviews	Density
Phoenix	2,781	9,678	124,425	0.46%
Las Vegas	8,315	17,791	386,486	0.26%
Cross city	11,536	54,922	564,216	0.089%

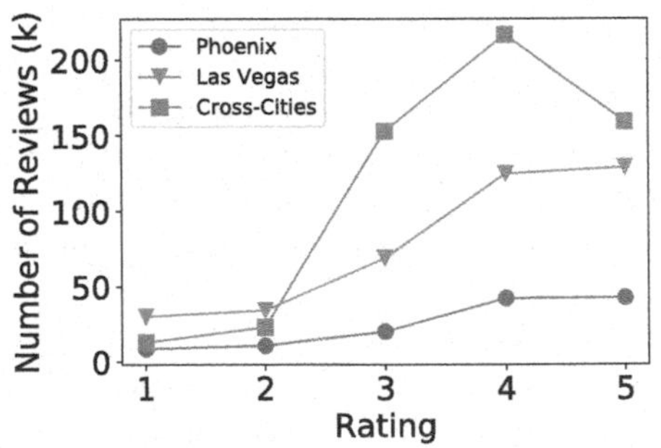

Fig. 1. Ratings distribution within the datasets

6 Results Analysis

We now present the experiments that address our two research questions, concerning the sentiment classification accuracy (Sect. 6.1) and the usefulness of sentiment classification as an effective proxy for ratings in venue recommendation (Sect. 6.2).

6.1 RQ1: Opinion Classification

Figure 2 presents the classification accuracy of our sentiment classification approaches described in Sect. 4, while varying the amount of training review data. In particular, we show the overall accuracy (Fig. 2(a)) as well as the accuracy on the positive and negative comments alone ((b) & (c), respectively).

Figure 2(a) shows that our selected sentiment analysis approaches are divided into three groups with different classification performances – SVM, CNN and LSTM all exhibit similar top performances, followed by SWN with medium

[5] The NDCG metric is not used since not all users will consistently use the rating scale (1-5), as discussed in this paper.

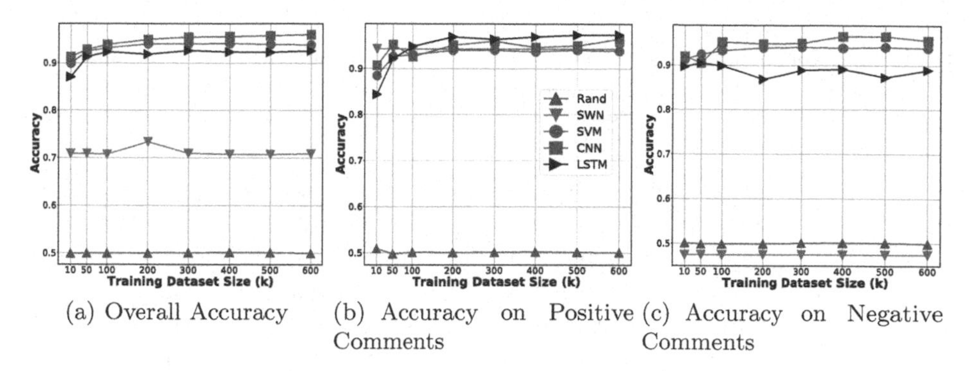

(a) Overall Accuracy (b) Accuracy on Positive Comments (c) Accuracy on Negative Comments

Fig. 2. Sentiment classification accuracy of different approaches.

accuracy, and Rand with (expected) low accuracy. Among the highest performing group (SVM, CNN and LSTM), CNN is the highest performer.

Next, we consider the accuracy of the classifiers separately on the positive and negative classes. From Figs. 2(b) & (c), we find that SVM, CNN and LSTM still provide high accuracy ($\geq$0.9) for both classes, while LSTM varies in accuracy across the classes. Indeed, LSTM surpasses CNN on the positive comments yet underperforms on the negative comments (indicating a higher false positive rate). Finally, since SWN exhibits a high accuracy on the positive comments but a low accuracy on the negative comments, it is mostly identifying comments as having a positive polarity.

Overall, in answer to research question RQ1, we find that SVM, CNN and LSTM exhibit high sentiment classification accuracy, with CNN outperforming all other techniques in terms of overall accuracy. In particular, LSTM performs better than SVM and CNN for positive comments and CNN is more accurate than the other classifiers for negative comments.

6.2 RQ2: Sentiment Classification in the MF and GeoSoCa Models

We now consider if the sentiment scores generated from the classifiers evaluated in Sect. 6.1 can be used for effective venue recommendation by MF and GeoSoCa. All results are presented in Table 3. Each column denotes the evaluation metric on the corresponding datasets. Each group of rows defines a particular venue recommendation approach: MF, GeoSoCa, or the latter's respective components (Geo, So, Ca). Each row in a group specifies the rating-based performance or the sentiment scores-based performances from the corresponding applied sentiment classification approaches. Finally, the rightmost column indicates the number of significant increases and decreases compared to the rating-based (baseline) model in that group of rows.

On analysing the general trends between MF, Geo|So|Ca and GeoSoCa, we find that the MF approach exhibits a weak effectiveness for this ranking-based recommendation task. Indeed, the observed performances for the

Table 3. Recommendation performances of rating and sentiment-based approaches on three datasets (reported evaluation measures are *100). Using the t-test, statistically significant increases (resp. decreases) with respect to the corresponding rating-based baseline are indicated by ↑ (resp. ↓).

		Phoenix (*100)			Las Vegas (*100)			Cross-City (*100)			Signf. #
		P@5	P@10	MAP	P@5	P@10	MAP	P@5	P@10	MAP	(↑ / ↓)
MF	Rating	0.12	**0.11**	**0.29**	0.01	0.02	0.12	0.02	0.01	0.05	—
	Rand	0.06	0.06	0.23	0.01	0.01	0.10	**0.05**	0.02	0.02	0 / 0
	SWN	**0.13**	**0.11**	0.28	**0.05**	**0.05**	**0.17**	0.02	0.02	0.05	0 / 0
	SVM	0.06	0.06	**0.29**	0.01	0.01	0.12	0.02	0.01	0.04	0 / 0
	CNN	0.08	0.08	0.27	0.01	0.01	0.11	0.02	**0.03**	**0.06**	0 / 0
	LSTM	0.04	0.03	0.25	0.03	0.02	0.11	0.04	0.02	**0.06**	0 / 0
Geo	Rating	0.67	0.73	0.83	**0.47**	0.43	**0.55**	0.78	0.74	1.02	—
	Rand	0.64	0.61	0.78	0.35	0.39	0.51	0.73	0.75	1.01	0 / 0
	SWN	**0.69**	0.69	0.86	0.44	**0.45**	**0.55**	0.73	0.77	1.03	0 / 0
	SVM	0.61	0.73	0.81	**0.47**	0.44	0.53	**0.76**	**0.79**	1.03	0 / 0
	CNN	0.66	0.73	0.82	0.45	0.44	**0.55**	0.75	**0.79**	**1.04**	0 / 0
	LSTM	0.67	**0.75**	0.84	0.45	**0.45**	**0.55**	0.75	0.78	**1.04**	0 / 0
So	Rating	3.38	2.88	1.95	**2.81**	**2.36**	1.76	2.97	**2.61**	1.53	—
	Rand	2.52↓	2.10↓	1.14↓	1.98↓	1.76↓	0.98↓	2.07↓	1.80↓	0.77↓	0 / 9
	SWN	2.73↓	2.36↓	1.74↓	2.15↓	1.82↓	1.36↓	2.15↓	1.89↓	1.28↓	0 / 9
	SVM	3.34	2.27↓	2.02	2.69	2.33	1.67	2.56↓	2.15↓	1.49	0/ 3
	CNN	3.39	2.87	2.06	2.70	2.33	1.70	2.84	2.49	1.57	0/ 0
	LSTM	**3.43**	**2.89**	**2.16**↑	2.71	2.34	**1.75**	**2.98**	2.54	**1.71**	1/ 0
Ca	Rating	3.51	**3.17**	**2.79**	2.54	2.27	**2.11**	**0.79**	**0.72**	0.54	—
	Rand	3.03↓	2.83	2.57	2.35	2.16	1.98↓	0.72	0.68	0.50	0/ 1
	SWN	1.88↓	2.14↓	2.72	0.72↓	0.85↓	2.01↓	0.49↓	0.55↓	0.54↓	0/ 9
	SVM	3.35	3.15	2.69	2.53	2.25	2.07	0.74	0.70	0.53	0/ 0
	CNN	**3.52**	**3.17**	2.77	2.51	2.26	2.07	0.78	0.71	0.54	0/ 0
	LSTM	3.50	3.15	2.78	**2.56**	2.26	2.10	0.78	**0.72**	**0.55**	0/ 0
GeoSoCa	Rating	3.68	**3.04**	2.23	2.64	2.09	1.51	3.92	3.17	**2.22**	—
	Rand	3.08↓	2.57↓	1.79↓	2.44	2.03	1.47	3.59	2.97	2.06	0/ 3
	SWN	1.32↓	1.35↓	1.39↓	0.52↓	0.58↓	1.06↓	1.83↓	1.75↓	1.44↓	0/ 9
	SVM	3.52	2.93	2.17	2.84	2.21	1.78↑	3.68↓	2.98↓	2.06	1/ 2
	CNN	3.62	2.90	2.18	2.86↑	2.29	1.79↑	3.71↓	3.02↓	2.09	2/ 2
	LSTM	**3.73**	2.96	**2.28**	**2.97**↑	**2.38**↑	**1.87**↑	**3.96**	**3.21**	2.15	3/ 0

combined GeoSoCa approach are markedly higher (0.0029 vs. 0.0223 MAP)[6]. Overall, the lower performance of MF is expected, as MF is intended as a rating prediction approach, rather than a ranking approach, where the objective is to rank highly the actual venues that the user visited. Using sentiment information

[6] The low absolute MAP values on this dataset are inline with other papers, e.g. [45].

shows some minor improvements, but none of the sentiment classifiers causes significant enhancements to this weaker rating-based MF baseline.

Next, we consider GeoSoCa and its components Geo|So|Ca for each dataset. For the geographical information, the rating and sentiment-based models provide statistically indistinguishable results (according to a paired t-test; p-value <0.05), regardless of the sentiment classification approach used. Next, for the social influence model (i.e. So), the distinction among the approaches is clear: SWN and Rand significantly degrade effectiveness compared to the rating-based baseline in 9 cases; the learned approach (SVM) significantly degrades effectiveness in 3 cases (P@10 for Phoenix and P@5 & P@10 for Cross-City); on the other hand, the deep-learned sentiment approaches (CNN and LSTM) are at least statistically indistinguishable from the corresponding rating-based baseline (only one significant increase: 1.95 → 2.16). Indeed, it is promising that the latest approaches (CNN and LSTM), which were shown to be the most effective sentiment classifiers in Sect. 6.1, also result in the recommendation models with the highest effectiveness, suggesting that they could be a suitable proxy for user ratings.

For the categorical information (i.e. Ca), recall that our substitution strategy replaces not only the users' preferences but also the aggregated popularity of the category for that user, as per Eq. (1). On examining Table 3, the learned and deep learning sentiment approaches are able to provide comparable performances to the corresponding rating-based baseline. The same observation also holds with the social information-based model. Moreover, similar to the social information-based model, the recommendation effectiveness also aligns with the performances of sentiment classifications, with CNN and LSTM providing the most effective results.

Finally, we consider the combined GeoSoCa model - where we observe that the product of the geographical, social and category influence scores, when using the sentiment scores from CNN or LSTM, could still provide performances that cannot be statistically distinguished from those based on ratings (only 1 significant decrease). Moreover, in 5 cases there were actually significant increases in effectiveness by deploying CNN or LSTM. Therefore, in answer to research question RQ2, we find that only the sentiment-based user preference scores from the state-of-the-art deep-learning-based sentiment classification approaches (i.e. CNN and LSTM) can provide similar effectiveness to the rating-based models. It is also of note that in these experiments, given that we regard a user rating ≥4 as positive, and a user rating ≤2 as negative, our results in Table 3 suggest that the sentiment scores can simply be binary (i.e. 'like' and 'dislike'). Such a binary rating scale (as might be determined by a sentiment polarity classifier) is a sufficient substitute for the currently used multi-rating scales to effectively capture the users' preferences in venue recommendation.

7 Conclusions

In this paper, we explored the performances of various sentiment analysis approaches at identifying the polarity of comments about venues, while also considering their use as a replacement for the users' explicit ratings in venue recommendation. For the sentiment classification approaches, we found that CNN outperforms other approaches in terms of overall accuracy, while LSTM performs better in classifying positively labelled reviews. Next, when substituting users' ratings with sentiment scores from state-of-the-art sentiment classification approaches (i.e. CNN and LSTM), we found that the resulting GeoSoCa-models were rarely significantly degraded in effectiveness, and were actually seen to be significantly enhanced in several cases. Overall, our results suggest that, for venue recommendation, a simple binary rating with 'like' and 'dislike' (as might be determined by a sentiment polarity classifier) is an effective substitute for the currently used multi-rating scales in location-based social networks. As future work, we plan to apply our rating substitution strategy in additional venue recommendation approaches. We will also investigate how to improve the performances of venue recommendation models by exploiting user reviews posted on other platforms (e.g. Twitter or Facebook) where no multi-scaling rating is used.

References

1. Manotumruksa, J., Macdonald, C., Ounis, I.: Modelling user preferences using word embeddings for context-aware venue recommendation. In: Neu-IR: The SIGIR 2016 Workshop on Neural Information Retrieval (2016)
2. Noulas, A., Scellato, S., Lathia, N., Mascolo, C.: A random walk around the city: new venue recommendation in location-based social networks. In: Proceedings of SocialCom-PASSAT (2012)
3. Lian, D., et al.: Scalable content-aware collaborative filtering for location recommendation. IEEE Trans. Knowl. Data Eng. **30**(6), 1122–1135 (2018)
4. Frankowski, D., Herlocker, J., Sen, S., et al.: Collaborative filtering recommender systems. Adapt. Web **4321**, 291–324 (2007)
5. Hu, L., Sun, A., Liu, Y.: Your neighbors affect your ratings: on geographical neighborhood influence to rating prediction. In: Proceedings of SIGIR (2014)
6. Koren, Y., Bell, R., Volinsky, C.: Matrix factorization techniques for recommender systems. Computer **42**(8), 30–37 (2009)
7. Manotumruksa, J., Macdonald, C., Ounis, I.: A contextual attention recurrent architecture for context-aware venue recommendation. In: Proceedings of SIGIR (2018)
8. Zhang, J.D., Chow, C.Y.: GeoSoSa: exploiting geographical, social and categorical correlations for point-of-interest recommendations. In: Proceedings of SIGIR (2015)
9. Zhu, Q., Wang, S., Cheng, B., Sun, Q., Yang, F., Chang, R.N.: Context-aware group recommendation for point-of-interests. IEEE Access **6**, 12129–12144 (2018)
10. Hu, R., Pu, P.: Exploring relations between personality and user rating behaviors. In: Proceedings of Workshop on Emotions and Personality in Personalized Services (EMPIRE at UMAP) (2013)
11. Cosley, D., Lam, S.K., Albert, I., Konstan, J.A., Riedl, J.: Is seeing believing?: how recommender system interfaces affect users' opinions. In: Proceedings of SIGCHI (2003)
12. Knijnenburg, B.P., Willemsen, M.C., Gantner, Z., Soncu, H., Newell, C.: Explaining the user experience of recommender systems. User Model. User-Adap. Inter. **22**(4–5), 441–504 (2012)

13. Han, J., Pei, J., Kamber, M.: Data Mining: Concepts and Techniques. Elsevier, New York (2011)
14. López Barbosa, R.R., Sánchez-Alonso, S., Sicilia-Urban, M.A.: Evaluating hotels rating prediction based on sentiment analysis services. Aslib J. Inf. Manage. **67**(4), 392–407 (2015)
15. Yang, D., Zhang, D., Yu, Z., Wang, Z.: A sentiment-enhanced personalized location recommendation system. In: Proceedings of ACM Conference on Hypertext and Social Media (2013)
16. Gurini, D.F., Gasparetti, F., Micarelli, A., Sansonetti, G.: Temporal people-to-people recommendation on social networks with sentiment-based matrix factorization. Future Gener. Comput. Syst. **78**(P1), 430–439 (2018)
17. Wang, H., Fu, Y., Wang, Q., Yin, H., Du, C., Xiong, H.: A location-sentiment-aware recommender system for both home-town and out-of-town users. In: Proceedings of SIGKDD (2017)
18. Lak, P., Turetken, O.: Star ratings versus sentiment analysis-a comparison of explicit and implicit measures of opinions. In: Proceedings of HICSS (2014)
19. Kim, Y.: Convolutional neural networks for sentence classification. arXiv preprint arXiv:1408.5882 (2014)
20. Baziotis, C., Pelekis, N., Doulkeridis, C.: Datastories at SemEval-2017 task 4: deep LSTM with attention for message-level and topic-based sentiment analysis. In: Proceedings of SemEval (2017)
21. Mohammad, S., Kiritchenko, S., Zhu, X.D.: NRC-Canada: building the state-of-the-art in sentiment analysis of tweets. In: Proceedings of SemEval (2013)
22. Kiritchenko, S., Zhu, X., Mohammad, S.M.: Sentiment analysis of short informal texts. J. Artif. Intell. Res. **50**, 723–762 (2014)
23. Cheng, C., Yang, H., King, I., Lyu, M.R.: Fused matrix factorization with geographical and social influence in location-based social networks. In: Proceedings of AAAI (2012)
24. Manotumruksa, J., Macdonald, C., Ounis, I.: Regularising factorised models for venue recommendation using friends and their comments. In: Proceedings of CIKM (2016)
25. Amoo, T., Friedman, H.: Do numeric values influence subjects' responses to rating scales. Int. Mark. Mark. Res. **26**(1), 41–46 (2001)
26. Pennock, D.M., Horvitz, E., Lawrence, S., Giles, C.L.: Collaborative filtering by personality diagnosis: a hybrid memory-and model-based approach. In: Proceedings of UAI (2000)
27. Ohana, B., Tierney, B.: Sentiment classification of reviews using SentiWordNet. In: Proceedings of Information Technology & Telecommunication (2009)
28. Baccianella, S., Esuli, A., Sebastiani, F.: SentiWordNet 3.0: an enhanced lexical resource for sentiment analysis and opinion mining. In: Proceedings of LREC (2010)
29. Rosenthal, S., Farra, N., Nakov, P.: SemEval-2017 task 4: sentiment analysis in Twitter. In: Proceedings of SemEval (2017)
30. Gao, H., Tang, J., Hu, X., Liu, H.: Content-aware point of interest recommendation on location-based social networks. In: Proceedings of AAAI (2015)
31. Zhao, K., Cong, G., Yuan, Q., Zhu, K.Q.: SAR: a sentiment-aspect-region model for user preference analysis in geo-tagged reviews. In: Proceedings of ICDE (2015)
32. Bird, S., Loper, E.: NLTK: the natural language toolkit. In: Proceedings of ACL (2004)

33. Lian, D., Zhao, C., Xie, X., Sun, G., Chen, E., Rui, Y.: GeoMF: joint geographical modeling and matrix factorization for point-of-interest recommendation. In: Proceedings of SIGKDD (2014)
34. Zhao, G., Qian, X., Xie, X.: User-service rating prediction by exploring social users' rating behaviors. Trans. Multimedia **18**(3), 496–506 (2016)
35. He, J., Li, X., Liao, L., Song, D., Cheung, W.K.: Inferring a personalized next point-of-interest recommendation model with latent behavior patterns. In: Proceedings of AAAI (2016)
36. Zhao, S., Zhao, T., King, I., Lyu, M.R.: Geo-teaser: geo-temporal sequential embedding rank for point-of-interest recommendation. In: Proceedings of WWW (2017)
37. Zhao, S., Zhao, T., Yang, H., Lyu, M.R., King, I.: Stellar: sapatial-temporal latent ranking for successive point-of-interest recommendation. In: Proceedings of AAAI (2016)
38. Guerini, M., Gatti, L., Turchi, M.: Sentiment analysis: how to derive prior polarities from sentiwordnet. In: Proceedings of EMNLP (2013)
39. Pang, B., Lee, L., Vaithyanathan, S.: Thumbs up?: sentiment classification using machine learning techniques. In: Proceedings of ACL (2002)
40. Koppel, M., Schler, J.: The importance of neutral examples for learning sentiment. Comput. Intell. **22**(2), 100–109 (2006)
41. Yang, X., Macdonald, C., Ounis, I.: Using word embeddings in Twitter election classification. Inf. Retrieval J. **21**(2–3), 183–207 (2018)
42. Zimba, B., Chibuta, S., Chisanga, D., Banda, F., Phiri, J.: Point of interest recommendation methods in location based social networks: traveling to a new geographical region. arXiv preprint arXiv:1711.09471 (2017)
43. Yuan, F., Jose, J.M., Guo, G., Chen, L., Yu, H., Alkhawaldeh, R.S.: Joint geo-spatial preference and pairwise ranking for point-of-interest recommendation. In: Proceedings of ICTAI (2016)
44. Guo, Q., Sun, Z., Zhang, J., Chen, Q., Theng, Y.L.: Aspect-aware point-of-interest recommendation with geo-social influence. In: Proceedings of UMAP (2017)
45. Liu, Y., Pham, T.A.N., Cong, G., Yuan, Q.: An experimental evaluation of point-of-interest recommendation in location-based social networks. In: Proceedings of VLDB (2017)

AntRS: Recommending Lists Through a Multi-objective Ant Colony System

Pierre-Edouard Osche(✉), Sylvain Castagnos, and Anne Boyer

Univ. of Lorraine - CNRS - LORIA, Campus Scientifique, B.P. 239, Nancy, France
{pierre-edouard.osche,sylvain.castagnos,anne.boyer}@loria.fr

Abstract. When people use recommender systems, they generally expect coherent lists of items. Depending on the application domain, it can be a playlist of songs they are likely to enjoy in their favorite online music service, a set of educational resources to acquire new competencies through an intelligent tutoring system, or a sequence of exhibits to discover from an adaptive mobile museum guide. To make these lists coherent from the users' perspective, recommendations must find the best compromise between multiple objectives (best possible precision, need for diversity and novelty). We propose to achieve that goal through a multi-agent recommender system, called AntRS. We evaluated our approach with a music dataset with about 500 users and more than 13,000 sessions. The experiments show that we obtain good results as regards to precision, novelty and coverage in comparison with typical state-of-the-art single and multi-objective algorithms.

Keywords: Recommender systems · Multi-agent systems · Multi-agent reinforcement learning

1 Introduction

Recommending an appropriate list or sequence of items to a specific user can be seen as a multi-objective problem. Let us illustrate this with a use case: Imagine a user who enjoys listening to music while doing sport through a mobile app. Such an online service should generate a playlist that is adapted to her preferences (*precision*). The tempo and the energy of the proposed songs should fit the context (*similarity*). The playlist should offer an appropriate level of *diversity* to avoid boredom. It could also bring *novelty* and *serendipity* according to her desires. The scientific challenge thus consists in taking into account different constraints that are contextualized and potentially not compatible.

In this paper, we propose a new multi-objective recommender system, called AntRS. Our model relies on a Multi-Agent System. The environment is a graph whose nodes are the items in the item set and whose edges connect items that have been co-consulted by several users. An Ant Colony Optimization algorithm allows to explore this environment until the target state is reached for each objective. Our model is generic since it is possible to add as many colonies as

L. Azzopardi et al. (Eds.): ECIR 2019, LNCS 11437, pp. 229–243, 2019.
https://doi.org/10.1007/978-3-030-15712-8_15

the domain context requires. The paths generated by the different colonies are then merged to offer a good compromise between the objectives. We have validated our approach by choosing 4 objectives which can be antagonist (similarity vs. diversity, preferences vs. novelty). We relied on a music dataset made of 180,000 songs and 500 users, and compared our approach to 4 state-of-the-art algorithms. We measured the performances using several metrics (accuracy, novelty, diversity, coverage...). Results show that AntRS achieves a better accuracy than others, while offering a better compromise to users on other objectives.

This paper is organized as follows: Sect. 2 presents the related work on multi-objective recommenders and the principle of the Ant Colony Systems from which our system took inspiration. Section 3 describes our AntRS model. Sections 4 and 5 respectively describe the experiments carried out and the results obtained. Finally Sect. 6 concludes this paper and presents our perspectives.

2 Related Work

2.1 Multi-objective Recommender Systems

A recommender system can either propose a list of independent items at each time step, or it can propose a sequence of items [20]. In Sequence-Aware Recommender Systems, one can both consider the importance of the order of the past events (by looking for co-occurrence patterns [4] or for sequential patterns [12] in past sessions) and the expected order of the future recommendations (e.g. continuation in playlists [13] or transitions between items [18]). In this paper, we based our experiment on a music dataset. As recent research has found little evidence that the exact order of songs actually matters to users [25], we limited our state-of-the-art to the recommendations of lists.

Transversely a recommender system can be mono-objective or multi-objective. Most recommenders solely focus on the accuracy (precision and recall) [23]. Others attempt to find a compromise between precision, serendipity and novelty [15], or between precision and diversity [17,32] for example. There are several ways to address a multi-objective optimization problem. One can either look for a set of Pareto solutions, considering that a solution is optimal if it is not possible to make any objective better off, without making at least one objective worse off [33]. In that case, recommender systems aim at producing as many solutions as possible in order to cover as much as possible of the problem's Pareto front. Or one can rank the items in a single list by aggregating or reordering the results of each single objective [21]. This list can be produced in one stage [10], or come out of a 2-step process consisting in generating several lists of candidate items for the active user and in merging them [8,9,22,27–29]. Recommending several Pareto solutions offers the advantage to leave the choice to the active user. It can be interesting in some application domains such as e-commerce where an explicit validation process from the user is mandatory. However, in the context of online music services, it is not conceivable to request a user decision at each timestep. The songs must come one after another without disturbing the user in his/her main current task. For this reason, we focused this paper on recommenders which produce only one solution (i.e. only one list of recommendations).

The existing multi-objective single-list recommenders suffer from several limitations: they are dependent from the application domain (any change in the set of objectives has a drastic impact on the implementation of the model) and they are very time-consuming. To bypass these difficulties, we propose a new approach relying on a Multi-Agent System (MAS), and more precisely on an Ant Colony System explained below. MAS have multiple advantages in our context:

- they have a relatively low computational complexity;
- they are efficient at tackling multi-objective problems [1,3];
- they can easily be adapted to new configurations and are resilient to changes.

2.2 Ant Colony Systems

The Ant system algorithm (AS) is inspired by the foraging behavior of ants, specifically the pheromone communication between ants, to find shortest paths in an environment between a starting node and a target node. Dorigo proposed a few different versions of this AS model [7]. Our model took inspiration from one of those variants, the Ant Colony System (ACS) algorithm. In comparison to the classic AS model, ACS proposes a different way for the ants to deposit pheromones. Instead of having all the ants deposit their pheromones at the end of one iteration (i.e. after all the ants have finished their tour), only the ant that found the best path can deposit pheromones. Furthermore, ants perform a so called local pheromone update where, after each construction step, they deposit some pheromone on the last edge they visited. In other words, each time an ant takes an edge, it deposits some pheromones along its way, regardless of the quality of the path. As explained by Dorigo, this version of the ACO algorithms is known to "*diversify the search performed by subsequent ants during an iteration: by decreasing the pheromone concentration on the traversed edges, ants encourage subsequent ants to choose other edges and, hence, to produce different solutions. This makes it less likely that several ants produce identical solutions during one iteration*". As our search space is large (there are millions of songs on an online music service) and as we promote not only the precision but other characteristics in the recommended lists, we chose the ACS algorithm. In the rest of this subsection, we explain the main formulas of the ACS.

State Transition Rule - The state transition rule uses the pseudo-random proportional rule where a random variable $q \in [0;1]$ is compared to a parameter q_0 to decide if the ant will explore the graph or if it will exploit the knowledge collected by previous ants. $q_0 = 0$ is equivalent to the AS model where ants only explore the graph while $q_0 = 1$ refers to a pure reinforcement behavior with no exploration. The Eq. 1 let the algorithm decide between knowledge exploitation and biased exploration of the graph.

$$\begin{cases} \text{Exploitation (Equation 2) if } q \leq q_0 \\ \text{Biased exploration (Equation 3) if } q \geq q_0 \end{cases} \quad (1)$$

The Eq. 2 represents the direct exploitation of the knowledge in the graph where the best edge is always chosen.

$$\arg\max_{l \in V_i} \{\tau_{il}^{\alpha} \cdot \eta_{il}^{\beta}\}, \tag{2}$$

where V_i is the set of available nodes from the node i, $\tau_{il} \in [0;1]$ is the amount of pheromones left on an edge (i,l) by previous ants, η_{il} is the heuristic information on an edge (i,l), α and β are two parameters representing respectively the weight of the pheromones and the weight of the heuristic.

The Eq. 3 represents the biased exploration where best edges have more chances to be picked and p_{ij} is the probability for an ant at the node i to choose the edge (i,j).

$$p_{ij} = \begin{cases} \dfrac{\tau_{ij}^{\alpha} \cdot \eta_{ij}^{\beta}}{\sum\limits_{l \in V_i} \tau_{il}^{\alpha} \cdot \eta_{il}^{\beta}}, & \text{if } j \in V_i, \\ 0, & \text{otherwise,} \end{cases} \tag{3}$$

Global Pheromone Update - After each iteration, only the ant who found the best tour is allowed to update the pheromone level τ_{ij}:

$$\tau_{ij} = \begin{cases} (1-\rho) \cdot \tau_{ij} + \rho \cdot \Delta\tau_{ij} & \text{if } (i,j) \text{ belongs to best tour,} \\ \tau_{ij} & \text{otherwise,} \end{cases} \tag{4}$$

where ρ is the evaporation rate of the pheromones and $\Delta\tau_{ij} = 1/L_{best}$ where L_{best} is the length of the best tour.

Local Pheromone Update - Another addition of the ACS model over the AS model is the local pheromone update performed after each step by each ant described in Eq. 5.

$$\tau_{ij} = (1-\rho) \cdot \tau_{ij} + \rho \cdot \tau_0, \tag{5}$$

where τ_0 is the pheromone level set on every edge at the initialization.

Heuristic Information - the heuristic information η_{ij} represents the information that ants possess *a priori* on an edge (i,j). In ACS, the heuristic information is computed based on the distance between the two nodes of the edge: the farther both nodes are from each other and the lower η_{ij} will be.

$$\eta_{ij} = \frac{1}{d_{ij}}, \text{ where } d_{ij} \text{ is the distance between nodes } i \text{ and } j. \tag{6}$$

3 Our Model: AntRS

As previously stated, AntRS has been built with several goals in mind: (1) be as generic as possible; (2) be able to include several competing objectives in a single list; (3) be resilient to changes in the environment (new items, new preferences, ...). Our model takes inspiration from the ACS algorithm because the latter gathers all the quality needed to satisfy those objectives. However, we want to point out the differences between the classic ACS as described in Sect. 2.2 and our model AntRS. First of all, we had to develop our own method to create a graph to model as best as possible the large environment we were working in without hindering the execution time of the system. In Subsect. 3.1 we present our graph creation method. Secondly, we wanted to optimize many objectives while the ACS optimizes only one attribute which is usually the distance. In Subsect. 3.2 we introduce more formally the objectives used in our model. Thirdly, as we are generating several paths during the algorithm execution, a merging procedure has to be executed at one point to be able to propose the best possible recommendation list for each user. The Subsect. 3.3 explains two tactics we used to do so.

3.1 Graph Creation

The first step of our model is the creation of the graph. This is often an overlooked part in the literature as the datasets used are usually small and/or the links between nodes of the graph are manually picked by a field expert. One of the main differences between ACO simulations and real ants is the definition of the search space. Real ants are evolving in a continuous search space without any landmarks (or vertices) and are free to go everywhere whereas agents are released in a discrete environment and have to follow predetermined paths (or edges) between set landmarks. One of the ways to be as close as possible to real ants' behavior would be to compute distances between each and every node of the graph to build a complete graph. It is nonetheless an unpractical solution for more than a few thousands vertices as the number of edges depends on the number of vertices n with $|E| = \frac{n(n-1)}{2}$. As our goal is to use our model in a realistic situation with many potential items represented by vertices, we decided to find a workaround without sacrificing the quality in the solutions found. To do so, we needed to select a few "best" edges between each vertex. At this point, we formulated two hypotheses to help us construct the graph: (1) past sequences created by previous users represent useful domain knowledge which should be exploited; (2) past sequences done by previous users are not always the best possible ones and could have been improved with clever recommendations.

To take into account those two hypotheses, we first computed the number of transitions (i.e. co-consultations) between each pair of items in our dataset and we added (1) all the transitions above a specific threshold, and (2) only some of those below this threshold as edges. Finally, if a given connectivity degree was not reached, we added new edges between items who were not connected in our dataset to allow our model to discover new potential interesting paths not known by users. The process of creating an edge is shown in Eq. 7.

$$e_{ij} = \begin{cases} \text{if } t_{ij} \geq m \\ \text{or if } t_{ij} < m \text{ and } q < t_{ij} \text{ where } q \in [min\ t_{il}; \log(max\ t_{il})] \\ \text{or if } t_{ij} = 0 \text{ and if } deg(i) < d \text{ then pick a random} \\ \text{transition until } deg(i) = d \end{cases} \tag{7}$$

where e_{ij} is the presence of an edge between vertices i and j, t_{ij} is the number of transitions performed from item i to item j by the users in the dataset, m is the threshold where transitions are not directly added to the graph as edges, $q \in [min\ t_{il}; \log(max\ t_{il})]$ is a random variable uniformly distributed, $min\ t_{il}$ is the minimal number of transitions between the item i and all the others items $l \in V_i$ where at least one transition has been found, $deg(i)$ is the current degree of the node i in the graph and d is a parameter specifying the minimal degree each vertex must have in the final graph.

3.2 Objectives

It is now widely admitted that the sole precision is not sufficient to produce good recommendations to users. We thus propose to define a set of 4 concurrent objectives that have to often be considered in the literature while recommending a list of items. The objectives we considered are all transposable in different application domains, guaranteeing the genericity of our approach.

Furthermore, the ability to add, to modulate the importance or to remove objectives on the fly was essential for having an adaptive model. To address this issue, we chose to integer as many colonies as objectives in our model, and each colony is specialized in maximizing its own objective. To do so, we modified the way the ACS model computes the distance d between two nodes of the graph while the calculation of the heuristic η_{ij} was left untouched. The rest of this Section describes the equations used to compute the distance for each colony.

Similarity - This is one of the main factor considered by nearly all the recommender systems. The main goal of a recommender system is to propose items similar to what the user liked before. Even if similarity is a well-known and widely used characteristic, we think that a good recommender system cannot overlook it. We also do consider that similarity should not be the cornerstone of each and every recommender system anymore. The goal of this colony is to find a list with items as similar as possible of what the user previously viewed or is currently viewing. A lot of methods exist to compute the similarity of two vectors and, based on our dataset and on the metadata available, we decided to use a cosine similarity measure [24]. To compute the distance value on the edges of the graph, we simply computed the cosine similarity between the two items represented by the vertices. More formally, for an edge (i, j), its associated distance d_{ij} is computed with the cosine similarity between the vectors of the descriptive characteristics of the items i and j.

$$d_{ij} = \frac{1}{sim(C_i, C_j)} \tag{8}$$

where C_i are the characteristics of the item i. The item characteristics depend obviously on the dataset and on the meta-information available but, we can formalize that each item of the dataset is described by n characteristics as follow $C_i = \{c_1, c_2, \ldots, c_n\}$. We used the multiplicative inverse to transform the similarity metric $sim \in [0; 1]$ into a distance $d \in [1; +\infty]$. Therefore, a distance value near 1 on an edge (i, j) means that the two items i and j are similar.

Diversity - This characteristic and the similarity are often described together as they are both related to the distance/correlation between the items liked by the user and his/her recommendations. But unlike similarity, diversity depicts how dissimilar two items are relatively to each other. Similarity and diversity are complementing each other in the sense that they are both needed to adapt the system to the needs of different users [14]. To compute this objective, we chose to apply one of the classic diversity metric which is obtained by computing the inverse of the similarity between two items, as shown in Eq. 9. As for the similarity, we used the multiplicative inverse of the diversity to obtain a distance $d \in [1; +\infty]$.

$$d_{ij} = \frac{1}{1 - sim(C_i, C_j)} \tag{9}$$

Novelty - This characteristic represents the items that are not yet known by the user. It could be new items recently added to the system or old but not so popular items that the user missed. Novelty should not be confused with diversity, since novel items could be either similar or dissimilar to what the user usually likes. Novelty is an important characteristic of a recommender system to avoid a potential lack of interest of users due to too much foreseeability in the recommended items [26].

To determine if an item is novel or not relatively to a specific user, we used the work of Zhang [31] who defined the novelty as a notion composed of three characteristics: (1) Unknown: the item is unknown to the user; (2) Satisfactory: the item is liked by the user; (3) Dissimilarity: the item is dissimilar to the other items known by the user. The author proposed to evaluate the novelty of the item i for the user u as follow:

$$novelty(i, u) = p(i|unknown, u) \cdot dis(i, pref_u) \cdot p(i|like, u) \tag{10}$$

where $p(i|unknown, u)$ is the probability that the user u does not know the item i, $dis(i, pref_u)$ is the dissimilarity between i and the set of items in the users' profile and $p(i|like, u)$ is the probability that u will like i. However, the dissimilarity and the satisfaction of the user relatively to i are closely related to other objectives in our model, respectively maximized by the diversity colony and by both the preferences and the similarity colonies. Hence we decided to trim down the Eq. 10 to the probability $p(i|unknown, u)$ only (see Eq. 11).

$$p(i|unknown, u) = -log(1-pop_i), \text{ where } pop_i \text{ is the popularity of item } i. \tag{11}$$

Preferences - The preferences characteristic corresponds to what the user really likes. It intersects with the similarity notion but, again as with diversity and novelty, we think that preferences express another aspect of a good recommendation for a user. The similarity characteristic allows the recommender to propose items that are similar to the preferences of the user, but it is not guaranteed that he will like those items. It is for example perfectly common to both like and dislike some songs coming from the same album and artist, yet those songs will probably be treated as very similar relative to each other. The preferences characteristic favors items that are known to be liked by the user.

The goal for this colony is to find a sequence in the graph prioritizing items that are already known to be liked by the user. Thereby, items must have criteria conveying how the user like an item or not. This can be done either with explicit feedback (*e.g.* item rating, ...), with implicit feedback (*e.g.* number of times the user viewed an item, ...) or with a combination of both. The nature of the feedback will heavily depends on the domain, but we can formalize that each collected information concerning the behavior of a user on an item must be taken into account. Let C_u be the set of criteria representing all the actions that a user u may perform on the items of the system, thus $c_{u,i}$ is the sum of all interactions specific to a single criterion c that a user u performed on an item i (*e.g.* the number of times a user u viewed i). To aggregate all the different interactions possible in a single value, we use the presumed interest formula proposed by Castagnos et al. in [6] and described in the Eq. 12.

$$presumed\ interest_{u,i} = v_{min} + \frac{\sum\limits_{c \in C} (w(c) \cdot c(u,i))}{\sum\limits_{c \in C} w(c)} \cdot \frac{(v_{max} - v_{min})}{c_{max}} \tag{12}$$

where $c(u,i)$ corresponds to normalized values given to the item i by the user u to each criterion c, $w(c)$ is the weight of the criterion c, v_{min} and v_{max} are the minimal and maximal expected values for the presumed interest and c_{max} is the maximal value that $c(u,i)$ can take regardless of the criteria. In our case, we considered the following criteria for each song: number of consultations, number of skips, number of bans (when the user do not want to listen to the song ever again) and number of likes.

3.3 Merging Tactics

In the previous section, we described four objectives that could provide suitable recommendations for users. Each of these four objectives is associated to a specific ant colony in our model. Thus, after this step, we are left with as many lists of recommendations as colonies, where each one should represent a part of the final recommended list. In order to build it, we needed a tactic to merge all the colonies' lists into one. To do so, we propose two techniques described below.

Merging colony - The first merging tactic relies once more on the ACS algorithm but with one additional colony that we called "merging colony". Starting from the set of items found by the other colonies (**step 1**), the merging colony considers all the objectives at once with a weighted sum to calculate the distances on the graph's edges (**step 2**), as shown in Eq. 13.

$$d_{ij} = \sum_{col \in colonies} w(col) \cdot d_{ij}(col) \tag{13}$$

where $w(col)$ is the weight representing the expected importance of the colony's objective in the final recommendation. To estimate those weights, we calculated the average values of each objective (similarity, diversity, novelty and preferences) on the last n sessions processed of the user. This gave us the general importance of each objective while taking into account contextual information and recent tendencies in the user's behavior. We also built a new graph for this path of the algorithm. To construct it, we used all the items in the lists found by the other colonies as vertices, we added edges to each consecutive pair of items in the lists and finally we added random edges in the same way that is described in the last part of the Eq. 7 to give the possibility of new paths to be found and chosen by the merging colony's agents.

Lists merging - For the second merging tactic, we calculated the weight $w(col)$ of each objective in the same way that for the merging colony (see above). We then built the list step by step by considering all the items found by the different colonies. We iterated through all the available items for each step of the list construction and we added to the final recommended list the item which yield the best amelioration towards the expected values. This process was stopped either when the remaining items degraded the list's metrics, when there was no items left or when the last item of the initial listened session was found.

4 Experimentation

4.1 Comparison Algorithms

We compared the performances of AntRS with four state-of-the-art algorithms capable of producing lists of items in the same conditions than our model. The first three are classical techniques spanning most of the work in the recommender systems domain: (1) UserKNN [5]; (2) TrustMF [30] and (3) SVD++ [16]. Those three algorithms were implemented using the Java library librec [11]. We also implemented a fourth hybrid multi-objective model named PEH described in [22] to be able to compare AntRS to a state-of-the-art multi-objective recommender system. We used the three algorithms described above in the hybridization process. Furthermore, we also ran several version of AntRS to assess the strenghts and the weaknesses of our model.

4.2 Dataset

For our experimentations, we decided to use a dataset from Deezer as they offer the possibility to get metadata and information on listened tracks with their API. Our dataset spans one month of listenings starting from 5th Dec. 2016.

We split the dataset in listening sessions which corresponded to a listening with a break not longer than 900+$track\_duration$ seconds. Among all 1,871,919 consultations, we were able to determine 91,468 unique sessions with a mean length of 18.3 tracks each. The full dataset contains 3,561 unique users, 178,910 unique songs, and 1,871,919 listenings. However, as the PEH algorithm is not highly scalable (similarly to most of multi-objective algorithms) and so as to compare all algorithms in the same conditions, we limited the experiment to 500 randomly chosen users.

Each track is described by a number of metadata provided by Deezer, including song name, artist name, album name, music genre, related artists and some numerical values like acousticness, danceability, energy, instrumentalness, liveness, loudness, speechiness, tempo and valence. We used those characteristics to implement the metrics of the four different colonies described in Sect. 3.2.

Finally, so as to transform consultations (more precisely metadata such as duration, frequency or recency of consultations) into ratings usable by collaborative filtering algorithms (UserKNN, TrustMF, SVD++), we used the Formula 12 proposed by [6]. We recall that this same formula was also used by our model for the objective of preferences, taking no advantage on other algorithms.

4.3 Experimental Protocol

We performed several evaluations of our model in diverse configurations to measure its performance relatively to itself and to other models. To guarantee a fair chance for each model and configuration, we set the same starting and stopping conditions and we used the same data for all the experiments we did. For each listening session of the test base users, each algorithm produced one recommended list and its performance was measured in comparison to the initial listening session. The first item of each session was given to the algorithm as starting point for the recommended list. The last item was not given but, if the algorithm reached it during the recommendation process then it was stopped. A minimum size was set for the recommended list which was half of the size of the initial listened session. After the recommendation, the initial listened session was added to the training base to simulate a real-case scenario where a system first has no information on a new user and then gather more and more data on him as he interacts with the system. Finally, all the tests were performed on a cross-validation dataset with a training base of 400 users and a test base of 100 users each time. For each listening session composed of 5 items or more in the test base, a recommended session was produced. The users of the training base had listened to 10,621 sessions while there were 2,569 sessions in the test base.

We used different metrics capturing all the aspects of what we consider a good recommendation: Precision, Recall and F-measure [2], Similarity and Intra-list Similarity [32], Diversity and Relative diversity [19,26]. We also used the preferences and the novelty metrics of Eqs. 11 and 12 as well as in [26]. We empirically fixed the meta-parameters values of the baseline algorithm described in Sect. 2.2 as follows: $q_0 = 0.3$, $\alpha = 0.1$, $\beta = 0.9$, $\rho_0 = 0.2$, $\tau_0 = 0.1$.

5 Results

5.1 Single-Objective AntRS

We first wanted to see how each of our objective performed alone. As our model allows us to change the number of objectives on the fly, we performed four different tests for the four different objectives without any merging step and we measured how each of the tests performed considering the metrics described above. The Table 1 presents those results. We measured the statistical significance of the score of the colony that was supposed to perform best on each metric in comparison with the results obtained by the other colonies. Thus, the similarity and the intra-list similarity of the similarity colony were compared to the similarity and intra-list similarity of the 3 others colonies and so on. As the Shapiro-Wilk test revealed that our data did not follow a normal distribution, we used the non-parametric Wilcoxon test which allowed us to compare the means of two related samples (same users with different algorithms). As expected, we can see that each of our colonies produce lists that are specialized in a single objective. Thus, the similarity colony produces lists that are the most similar relatively to all the other colonies; the novelty colony produces lists that are the most novel, and so on. Those experiments showed that our model was working as intended and that we could combine the four objectives together.

Table 1. Experimentations with AntRS as a single-objective model

	Precision	Recall	F-measure	Similarity	ILS	Diversity	RD	Novelty	Preferences
Similarity colony	0.317	0.126	0.165	**0.952**	**0.923**	0.048***	0.049***	0.72***	0.447***
Diversity colony	0.211	0.104	0.127	0.862***	0.77***	**0.138**	**0.19**	0.806***	0.393***
Novelty colony	0.132	0.112	0.114	0.895***	0.837***	0.105***	0.144***	**0.909**	0.379***
Preferences colony	0.296	0.159	0.191	0.946***	0.91***	0.054***	0.08***	0.695***	**0.804**

Significance codes: 0 '***' 0.001 '**' 0.01 '*' 0.05 '.' 0.1 '' 1

5.2 Multi-objective AntRS

In Table 2, we present the summary of the results obtained for all the models tested and their variations. We also measured the statistical significance of the results of our best performing model, AntRS with the lists merging, in comparison with the results of all the other models. As for the previous subsection, the statistical test used was Wilcoxon signed-rank test.

As explained in Sect. 3.3, we proposed and tested two merging tactics to combine the results of our four objectives. We also tested to run the merging colony alone, without the **step 1** in the first merging tactic of Sect. 3.3: in that scenario, the merging colony operates on the whole graph, rather than on the subgraph of items recommended by the four colonies. We hypothesize that running directly the merging colony without the **step 1** will degrade the quality of the final solutions found. This first step with the four colonies gave our model the ability to find very specialized lists of items, which are associated to optimal solutions in a Pareto front, and this process was supposed to help the merging colony to find a better compromise between those solutions. This hypothesis has been confirmed in Table 2 since the two variants of AntRS with the 4 colonies got better results than the sole merging colony on each metric, except for the similarity, thus offering a better deal. The first merging tactic of our model outperforms the second one in terms of precision, but the latter obtains the best relative diversity of all the AntRS variants.

We can also see that both AntRS variations obtain the best precision (up to +78.23%), recall (up to +29.05%) and F-measure (up to +31.28%) of all the models tested, which means that our model was the best to capture the preferences from the lists initially listened by the users in the training set. AntRS also outperformed the other models for the preferences and the novelty metrics, while still managing to maintain a correct level of similarity and diversity. Let us remind that we deliberately chose non-compatible objectives (Similarity vs. Diversity, Novelty vs. Preferences), which makes the task harder for the multi-objective algorithms (AntRS and PEH) compared to others. Despite a lower diversity, AntRS offers a better compromise between all the objectives than PEH, and in a much shorter execution time. Within the frame of this experiment, we considered that all the objectives had an equal importance. However, it would be easy to weight the different objectives in AntRS according to user expectations, like they did in [8]. Finally, we can notice that AntRS and PEH got a much better coverage compared to other algorithms.

Table 2. Experimentations with AntRS as a multi-objective model

	Precision	Recall	F-measure	Similarity	ILS	Diversity	RD	Novelty	Preferences	Coverage
AntRS (4 colonies + lists merging)	**0.344**	0.131	**0.1838**	0.947	0.908	0.053	0.069	**0.741**	**0.61**	**96.92%**
AntRS (4 colonies + merging colony)	**0.288*****	**0.151*****	**0.1836** **	0.945	0.905***	0.055	**0.082*****	0.702***	**0.612** **	**96.92%**
AntRS (merging colony alone)	0.197***	0.118***	0.141***	0.953***	0.93***	0.047***	0.068	0.324***	0.389***	96.92%
UserKNN	0.224***	0.138	0.162***	0.95***	0.905***	0.05***	0.081***	0.68***	0.541	61.39%
TrustMF	0.195***	0.117***	0.14***	0.95***	0.894***	0.058***	0.09***	0.697***	0.458***	68.17%
SVD++	0.195***	0.12***	0.14***	0.941***	0.892***	0.06***	0.093***	0.70***	0.455***	68.17%
PEH	0.193***	0.118***	0.14***	0.941***	0.892***	0.059***	0.096***	0.697***	0.452***	97,52%

Significance codes (compared to AntRS with 4 colonies + lists merging): 0 '***' 0.001 '**' 0.01 '*' 0.05 '.' 0.1 '' 1

6 Conclusion and Perspectives

In this paper, we showed that our model, AntRS, is able to generate lists with a higher precision than other methods, while still offering a good compromise between similarity, diversity, novelty and preferences. AntRS offers many advantages compared to the state-of-the-art models: (1) the multi-agents part of our model guarantees that it is highly adaptable to changes in the environment, (2) the objective-oriented colonies can be added or removed on the fly, (3) it is generic enough to be adapted in all the domains where a list recommender is relevant, and (4) it is highly parallelizable and resilient to the cold-start problem.

We have some interesting ideas on how to pursue our work in the future. First, we would like to improve the quality of the recommended lists by personalizing the construction of the graph. At the moment, a unique graph is created for all the users while the distances on the edges are recalculated for each user. We would like to improve the personalization by creating a unique graph for each user, and even a unique graph for each colony.

Secondly, we would like to work on the notion of sequence. Instead of recommending simple lists to user, we think that offering a coherent sequence of items with a start, an end and a good progressivity could be beneficial in other domains (a path in a museum, a sequence of courses for a student...). This could be achieved by adding a colony dedicated to the progressivity.

References

1. Ariyasingha, I., Fernando, T.: Performance analysis of the multi-objective ant colony optimization algorithms for the traveling salesman problem. Swarm Evol. Comput. **23**, 11–26 (2015)
2. Baeza-Yates, R., Ribeiro-Neto, B., et al.: Modern Information Retrieval, vol. 463. ACM Press, New York (1999)
3. Barán, B., Schaerer, M.: A multiobjective ant colony system for vehicle routing problem with time windows. In: Applied Informatics, pp. 97–102 (2003)
4. Bonnin, G., Jannach, D.: Automated generation of music playlists: survey and experiments. ACM Comput. Surv. **47**(2), 26:1–26:35 (2014)
5. Breese, J.S., Heckerman, D., Kadie, C.: Empirical analysis of predictive algorithms for collaborative filtering. In: Proceedings of the Fourteenth conference on Uncertainty in artificial intelligence, pp. 43–52. Morgan Kaufmann Publishers Inc. (1998)
6. Castagnos, S., Boyer, A.: A client/server user-based collaborative filtering algorithm: model and implementation. In: 17th European Conference on Artificial Intelligence (ECAI 2006), pp. 617–621 (2006)
7. Dorigo, M., Birattari, M.: Ant colony optimization. In: Encyclopedia of Machine Learning, pp. 36–39. Springer (2011)
8. Fortes, R.S., Lacerda, A., Freitas, A., Bruckner, C., Coelho, D., Gonçalves, M.: User-oriented objective prioritization for meta-featured multi-objective recommender systems. In: Adjunct Publication of the 26th Conference on User Modeling, Adaptation and Personalization, pp. 311–316. ACM (2018)
9. Geng, B., Li, L., Jiao, L., Gong, M., Cai, Q., Wu, Y.: NNIA-RS: a multi-objective optimization based recommender system. Physica A **424**, 383–397 (2015)
10. Guimarães, A., Costa, T.F., Lacerda, A., Pappa, G.L., Ziviani, N.: GUARD: a genetic unified approach for recommendation. J. Inf. Data Manage. **4**(3), 295 (2013)
11. Guo, G., Zhang, J., Sun, Z., Yorke-Smith, N.: LibRec: a Java library for recommender systems. In: UMAP Workshops (2015)
12. Hariri, N., Mobasher, B., Burke, R.: Context-aware music recommendation based on latenttopic sequential patterns. In: Proceedings of the Sixth ACM Conference on Recommender Systems, RecSys 2012, pp. 131–138 (2012)
13. Jannach, D., Lerche, L., Kamehkhosh, I.: Beyond hitting the hits: generating coherent music playlist continuations with the right tracks. In: Proceedings of the 9th ACM Conference on Recommender Systems, pp. 187–194. ACM (2015)

14. Jones, N.: User perceived qualities and acceptance of recommender systems: the role of diversity. Ph.D. thesis, EPFL (2010)
15. Kaminskas, M., Bridge, D.: Diversity, serendipity, novelty, and coverage: a survey and empirical analysis of beyond-accuracy objectives in recommender systems. ACM Trans. Interact. Intell. Syst. **7**(1), 2:1–2:42 (2016)
16. Koren, Y.: Factorization meets the neighborhood: a multifaceted collaborative filtering model. In: Proceedings of the 14th ACM SIGKDD International Conference on Knowledge Discovery and Data Mining, pp. 426–434. ACM (2008)
17. L'Huillier, A., Castagnos, S., Boyer, A.: Understanding usages by modeling diversity over time. In: 22nd Conference on User Modeling, Adaptation, and Personalization, vol. 1181 (2014)
18. Maillet, F., Eck, D., Desjardins, G., Lamere, P.: Steerable playlist generation by learning song similarity from radio station playlists. In: In Proceedings of the 10th International Conference on Music Information Retrieval (2009)
19. McGinty, L., Smyth, B.: On the role of diversity in conversational recommender systems. In: Ashley, K.D., Bridge, D.G. (eds.) ICCBR 2003. LNCS (LNAI), vol. 2689, pp. 276–290. Springer, Heidelberg (2003). https://doi.org/10.1007/3-540-45006-8_23
20. Quadrana, M., Cremonesi, P., Jannach, D.: Sequence-aware recommender systems. CoRR abs/1802.08452 (2018)
21. Ribeiro, M.T., Lacerda, A., Veloso, A., Ziviani, N.: Pareto-efficient hybridization for multi-objective recommender systems. In: Proceedings of the Sixth ACM Conference on Recommender Systems, RecSys 2012, pp. 19–26 (2012)
22. Ribeiro, M.T., Ziviani, N., Moura, E.S.D., Hata, I., Lacerda, A., Veloso, A.: Multi-objective pareto-efficient approaches for recommender systems. ACM Trans. Intell. Syst. Technol. **5**(4), 53:1–53:20 (2014)
23. Ricci, F., Rokach, L., Shapira, B.: Recommender Systems Handbook. Springer, Boston (2015). https://doi.org/10.1007/978-0-387-85820-3
24. Su, X., Khoshgoftaar, T.M.: A survey of collaborative filtering techniques. Adv. Artif. Intell. **2009**, 4 (2009)
25. Tintarev, N., Lofi, C., Liem, C.C.: Sequences of diverse song recommendations: an exploratory study in a commercial system. In: Proceedings of the 25th Conference on User Modeling, Adaptation and Personalization, UMAP 2017, pp. 391–392 (2017)
26. Vargas, S., Castells, P.: Rank and relevance in novelty and diversity metrics for recommender systems. In: Proceedings of the Fifth ACM Conference on Recommender Systems, pp. 109–116. ACM (2011)
27. Wang, S., Gong, M., Ma, L., Cai, Q., Jiao, L.: Decomposition based multiobjective evolutionary algorithm for collaborative filtering recommender systems. In: IEEE Congress on Evolutionary Computation (CEC), pp. 672–679 (2014)
28. Wang, S., Gong, M., Li, H., Yang, J.: Multi-objective optimization for long tail recommendation. Knowl. Based Syst. **104**, 145–155 (2016)
29. Xia, X., Wang, X., Li, J., Zhou, X.: Multi-objective mobile app recommendation: a system-level collaboration approach. Comput. Electr. Eng. **40**(1), 203–215 (2014)

30. Yang, B., Lei, Y., Liu, J., Li, W.: Social collaborative filtering by trust. IEEE Trans. Pattern Anal. Mach. Intell. **39**(8), 1633–1647 (2017)
31. Zhang, L.: The definition of novelty in recommendation system. Int. J. Eng. Sci. Technol. Rev. **6**(3), 141–145 (2013)
32. Ziegler, C.N., McNee, S.M., Konstan, J.A., Lausen, G.: Improving recommendation lists through topic diversification. In: Proceedings of the 14th International Conference on World Wide Web, pp. 22–32. ACM (2005)
33. Zuo, Y., Gong, M., Zeng, J., Ma, L., Jiao, L.: Personalized recommendation based on evolutionary multi-objective optimization [research frontier]. IEEE Comput. Intell. Mag. **10**(1), 52–62 (2015)

Automated Early Leaderboard Generation from Comparative Tables

Mayank Singh[1](✉), Rajdeep Sarkar[2], Atharva Vyas[2], Pawan Goyal[2], Animesh Mukherjee[2], and Soumen Chakrabarti[3]

[1] IIT Gandhinagar, Gandhinagar, India
singh.mayank@iitgn.ac.in
[2] IIT Kharagpur, Kharagpur, India
rajdeep.sarkar@iitkgp.ac.in,atharvavyas139@gmail.com,
{pawang,animeshm}@cse.iitkgp.ac.in
[3] IIT Bombay, Mumbai, India
soumen@cse.iitb.ac.in

Abstract. A *leaderboard* is a tabular presentation of performance scores of the best competing techniques that address a specific scientific problem. Manually maintained leaderboards take time to emerge, which induces a latency in performance discovery and meaningful comparison. This can delay dissemination of best practices to non-experts and practitioners. Regarding papers as proxies for techniques, we present a new system to automatically discover and maintain leaderboards in the form of partial orders between papers, based on performance reported therein. In principle, a leaderboard depends on the task, data set, other experimental settings, and the choice of performance metrics. Often there are also tradeoffs between different metrics. Thus, leaderboard discovery is not just a matter of accurately extracting performance numbers and comparing them. In fact, the levels of noise and uncertainty around performance comparisons are so large that reliable traditional extraction is infeasible. We mitigate these challenges by using relatively cleaner, structured parts of the papers, e.g., performance tables. We propose a novel *performance improvement graph* with papers as nodes, where edges encode noisy performance comparison information extracted from tables. Every individual performance edge is extracted from a table with citations to other papers. These extractions resemble (noisy) outcomes of 'matches' in an incomplete tournament. We propose several approaches to rank papers from these noisy 'match' outcomes. We show that our ranking scheme can reproduce various manually curated leaderboards very well. Using widely-used lists of state-of-the-art papers in 27 areas of Computer Science, we demonstrate that our system produces very reliable rankings. We also show that commercial scholarly search systems cannot be used for leaderboard discovery, because of their emphasis on citations, which favors classic papers over recent performance breakthroughs. Our code and data sets will be placed in the public domain.

Electronic supplementary material The online version of this chapter (https://doi.org/10.1007/978-3-030-15712-8_16) contains supplementary material, which is available to authorized users.

L. Azzopardi et al. (Eds.): ECIR 2019, LNCS 11437, pp. 244–257, 2019.
https://doi.org/10.1007/978-3-030-15712-8_16

1 Introduction

Comparison against best prior art is critical for publishing experimental research. With the explosion of online research paper repositories like arXiv, and the frenetic level of activity in some research areas, keeping track of the best techniques and their reported performance on benchmark tasks has become increasingly challenging. *Leaderboards*, a tabular representation of the performance scores of some of the most competitive techniques to solve a scientific task, are now commonplace. However, most of these leaderboards are manually curated and therefore take time to emerge. The resulting latency presents a barrier to entry of new researchers and ideas, trapping "wisdom" about winning techniques to small coteries, disseminated by word of mouth. Thus, automatic leaderboard generation is an interesting research challenge. Recent work [4] has focused on automatic synthesis of reviews from multiple scientific documents. However, to the best of our knowledge, no existing system incorporates *comparative experimental performance* reported in papers into the process of leaderboard generation.

Limitations of Conventional Information Extraction: The ordering of competing techniques in a leaderboard depends on a large number of factors, including the task being solved, the data set(s) used, sampling protocols, experimental conditions such as hyperparameters, and the choice of performance metrics. Further, there are often tradeoffs between various competing metrics, such as recall vs. precision, or space vs. time. In fact, an accurate extraction, in conjunction with all the contextual details listed above, is almost impossible. We argue that conventional table and quantity extraction [2,10] is neither practical, nor sufficient, for leaderboard induction. In fact, numeric data is often presented as combinations of comparative charts and tables embedded together in a single figure [11]. These may even use subplots with multicolor bars representing baseline and proposed approaches.

Table Citations: A practical way to work around the difficult extraction problem is to focus on the relatively cleaner and more structured parts of a paper, viz., tables. Performance numbers are very commonly presented in tables. A prototypical performance table is shown in Singh et al. [11, Figure 3]. Each row shows the name of a competing system or algorithm, along with a citation. (A transposed table style is easily identified with simple rules.) Each subsequent column is dedicated to some performance **metric**. The rows make it simple to associate performance numbers with specific papers. In recent years, tables with citations (here, named **table citations**) and performance summaries have become extremely popular in arXiv.

Performance Improvement Graphs: We digest a multitude of tables in different papers into a novel **performance improvement graph**. Each edge represents an instance of comparison between two papers, labeled with the ID of the paper where the comparison is reported, the metric (e.g., recall, precision, F1 score, etc.) used for the comparison, and the numeric values of the metric in the two papers. Note that every individual performance edge is extracted from a table with citations to other papers. Each such extracted edge is noisy. Apart from the challenge of extracting quantities from tables and recognizing their numeric types [2,10], there is no control on the metric names, as they come from an open vocabulary (i.e., the column headers are arbitrary strings). Processing one table is a form of 'micro' reading; we must aggregate these 'micro' readings into a satisfactory 'macro' reading comparing two papers. We propose several reasonable edge aggregation strategies to simplify and featurize the performance improvement graph, in preparation for ranking papers.

Ranking Papers Using Table Citation Tournaments: Ranking sports teams into total orders, on the basis of the win/loss outcomes of a limited number of matches played between them, has a long history [3,5,9]. We adapt two widely-used tournament solvers and find that they are better than some simple baselines. However, we can further improve on tournament solvers using simple variations of PageRank [8,12] on a graph suitably derived from the tournament. Overall, our best ranking algorithms are able to produce high-quality leaderboards that agree very well with various manually curated leaderboards. In addition, using a popular list of papers spanning 27 different areas of Computer Science, we show that our system is able to produce reliable rankings of the state-of-the-art papers. We also demonstrate that commercial academic search systems like Google Scholar (GS)[1] and Semantic Scholar (SS)[2] cannot be used (and, in fact, are not intended to be used) for discovering leaderboards, because of their emphasis on aggregate citations, which typically favors classic papers over latest performance leaders.

2 Emergence of Leaderboards

Experts in an area are usually familiar with latest approaches and their performance. In contrast, new members of the community and practitioners need guidance to identify the best-performing techniques. This gap is currently bridged by "organically emerging" leaderboards that organize and publish the names and the performance scores of the best algorithms in a tabular form. Such leaderboards are commonplace in Computer Science, and in many other applied sciences.

The prime limitation of manually curated leaderboards is the natural latency until the performance numbers in a freshly-published paper are noticed, verified, and assimilated. This can induce delays in the dissemination of the best techniques to non-experts. In this paper, we build an end-to-end system to automate the process of leaderboard generation. The system is able to mine table citations, extract noisy performance comparisons from these table citations, aggregate the micro readings to a smooth macro reading and finally obtain rankings of papers.

In Table 1, we show an example leaderboard generated by our system (details of the system to be discussed later in the subsequent sections) for the PASCAL VOC Challenge (which involves semantic segmentation of images).

[1] https://scholar.google.com/.
[2] https://semanticscholar.org/.

Table 1. Ability of GS, SS, and our system to recall prominent leaderboard papers for the PASCAL VOC Challenge.

Paper	GS	SS	Our
Encoder-Decoder with Atrous Separable Convolution for Semantic Image Segmentation	×	×	×
Rethinking Atrous Convolution for Semantic Image Segmentation	×	×	✓
Pyramid Scene Parsing Network	×	×	✓
Wider or Deeper: Revisiting the ResNet Model for Visual Recognition	×	×	×
RefineNet: Multi-Path Refinement Networks for High-Resolution Semantic Segmentation	✓	×	×
Understanding Convolution for Semantic Segmentation	×	×	✓
Not All Pixels Are Equal: Difficulty-aware Semantic Segmentation via Deep Layer Cascade	×	×	✓
Identifying Most Walkable Direction for Navigation in an Outdoor Environment	×	×	×
Fast, Exact and Multi-Scale Inference for Semantic Image Segmentation with Deep ...	×	×	×
DeepLab: Semantic Image Segmentation with Deep Convolutional Nets, Atrous Convolution, ...	×	✓	✓
Laplacian Pyramid Reconstruction and Refinement for Semantic Segmentation	✓	×	✓
High-performance Semantic Segmentation Using Very Deep Fully Convolutional Networks	×	×	✓
Higher Order Conditional Random Fields in Deep Neural Networks	×	×	×
Efficient piecewise training of deep structured models for semantic segmentation	✓	✓	✓
Semantic Image Segmentation via Deep Parsing Network	×	✓	✓
Semantic Image Segmentation with Task-Specific Edge Detection Using CNNs ...	✓	×	✓
Pushing the Boundaries of Boundary Detection using Deep Learning	×	×	✓
Attention to Scale: Scale-aware Semantic Image Segmentation	✓	✓	✓
BoxSup: Exploiting Bounding Boxes to Supervise Convolutional Networks ...	✓	✓	×
Learning Deconvolution Network for Semantic Segmentation	✓	✓	✓
Conditional Random Fields as Recurrent Neural Networks	×	×	×
Weakly- and Semi-Supervised Learning of a DCNN for Semantic Image Segmentation	×	×	✓
Bayesian SegNet: Model Uncertainty in Deep Convolutional Encoder-Decoder Architectures ...	×	×	×
Semantic Image Segmentation with Deep Convolutional Nets and Fully Connected CRFs	×	✓	✓
Global Deconvolutional Networks for Semantic Segmentation	×	×	×
Convolutional Feature Masking for Joint Object and Stuff Segmentation	×	×	×

Similar results reproducing other leaderboards are presented by Singh et al. [11]. We observe that our system is able to find many of the papers present in this human-curated leaderboard. Traditional academic search systems like GS and SS do not fare well in finding leaderboard entries; each returned only seven papers (see Table 1) in their top 50 results retrieved for the query 'semantic segmentation'. Systems that emphasize cumulative citations rather than performance scores cannot be used for leaderboard discovery. Citations to a paper that make incremental improvements, resulting in the best experimental performance, may never catch up with the seminal paper that introduced a general problem or technique.

3 Limits of Conventional Table Information Extraction

Performance displays are implicitly connected to a complex context developed in the paper, including the task, the data set, choice of training and test folds, hyperparameters and other experimental settings, performance metrics etc. Millions of reviewer hours are spent each year weighing experimental evidence based on the totality of the experimental context. "Micro-reading" one table at a time is not likely to replace that intellectual process. Beyond contextual ambiguities, there are often trade-offs between different metrics like space vs. time, recall vs. precision, etc. In summary, leaderboard induction is not merely a matter of accurately extracting performance numbers and numerically comparing them.

One way to partly mitigate the above challenges is to use relatively cleaner, structured parts of the papers, e.g., single tables or single charts. We focus on tables in our first-generation system. However, with advanced visual chart mining and OCR [1,6,7], we can conceivably extend the system to charts as well.

We concentrate on (the increasing number of) tables that also cite papers, which are surrogates for techniques. Table 2 shows the average number of citations in a paper p that occur in tables, against the year of publication of p. Clearly, there is a huge surge in the use of table citations in the last five years, which further motivates us to exploit them for building our system.

Table 2. Average number of table citations made by an arXiv paper between 2005 and 2017.

Year	2005	2006	2007	2008	2009	2010	2011	2012	2013	2014	2015	2016	2017
Average	0.0	0.0	0.12	0.17	0.082	0.18	0.40	0.46	0.57	1.04	3.22	3.61	4.06

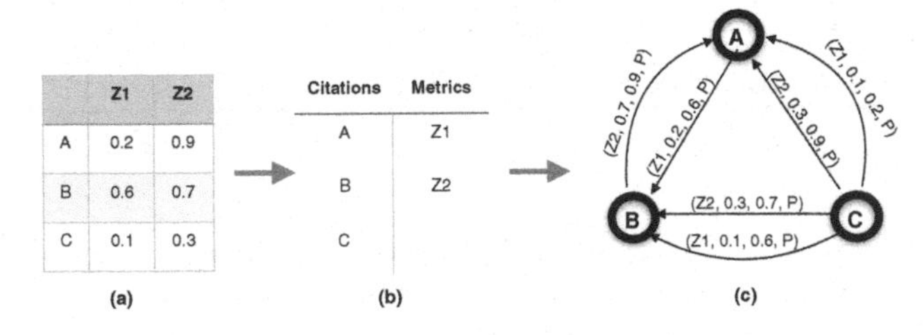

Fig. 1. First table extraction step toward performance tournament graph construction: (a) An example table present in paper P comparing three methods, A, B and C, for two evaluation metrics, $Z1$ and $Z2$. (b) Unique citations to the methods as well the evaluation metrics used are extracted, and (c) an abstract performance tournament graph is constructed.

4 Performance Improvement Graph

4.1 Raw Performance Improvement Graph

The performance improvement graph $G(V, E, Z)$ is a directed graph among a set of research papers V that are compared against each other. Here, Z represents the set of all the evaluation metrics. An edge between two papers (A, B) (see Fig. 1) is annotated with four-tuple (z, v_1, v_2, P), where $z \in Z$, v_1 and v_2 represent the metric value ('recall', 'F1', 'time') and lower and higher performing papers respectively. P denotes the paper that compared A and B. The directionality of an edge e $(e \in E)$ is determined by the performance comparison between two endpoints. The paper with lower performance points toward better performing paper. Simple heuristic rules are used to orient the edges. E.g., large F1 but small running times[3] are preferred. Figure 1 shows a toy example of the construction of a raw performance improvement graph from an extracted table.

One table provides just one noisy comparison signal between two papers or techniques. Although table citations allow us to make numerical comparisons, there is no guarantee of the same data set or experimental conditions across different tables, leave alone different papers. Therefore, we process the raw performance improvement graph in two steps:

Local sanitization: All directed edges connecting a pair of papers in the raw performance improvement graph are replaced with one directed edge in the sanitized performance improvement graph. This is partly a denoising step, described through the rest of this Sect. 4.2.

Global aggregation: In Sect. 5, we present and propose various methods of analyzing the sanitized performance improvement graph to arrive at a total order for the nodes (papers) to present in a synthetic leaderboard.

4.2 Sanitized Performance Improvement Graph

Relative Edge Improvement (REI) Distribution: One unavoidable characteristic of the raw performance improvement graph is the existence of noisy edges from incomparable or botched extractions. We define

$$\mathrm{REI}_z(u, v) = \frac{v_z - u_z}{u_z} \tag{1}$$

where (u, v) represents a directed edge from paper u to v; u_z and v_z denote performance scores of paper u and v respectively against a metric z. As described in previous section, u_z is lower than v_z.

We computed REIs from four leaderboards described in Sect. 6.1. These improvement scores are computed by considering all pairs of papers present in the respective leaderboards. We note that less than 0.5% of the edges have REI above 100%. In contrast, manual inspection of various erroneously extracted edges revealed that their REI was much larger than 100%. Therefore, we sanitize the raw performance improvement graph by pruning edges having improvement scores larger than 100%. This simple thresholding yielded graphs as clean as by using supervised learning (details omitted) to remove noisy edges.

Sanitizing Multi-edges: Every comparison creates a directed edge with different tuple value. A pair of papers can be compared in multiple tables, resulting in (anti-) parallel edges or multi-edges. Two directed edges are termed as *anti-parallel* if they are between the same pair of papers, but in opposite directions. Whereas, two directed edges are said to be *parallel* if they are between the same

[3] 'Time' is ambiguous by itself: a long time on battery but short training time are preferred. Our system is meant to take such errors it might make in stride.

pair of papers and in the same direction. In Fig. 1c, two parallel edges exist between papers B and C and two anti-parallel edges exist between papers A and B.

Multiple strategies can be utilized to summarize and aggregate multi-edges into a condensed tournament graph. We consider the following variations. Note that all of these are directed graphs. In each case, we discuss if and how a directed edge (i, j) is assigned a summarized weight.

UNW — Unweighted Graph: The simplest variant preserves the directed edges without any weights. This is equivalent to giving a weight of 1 for each of these directed edge (i, j), if there is any comparison.
ALL — Weighted graph (total number of comparisons): This variation uses the total number of comparisons between two papers p_i and p_j as the weights of the directed edge. Thus, each time an improvement is reported, it is used as an additional vote to obtain the edge weight.
SIG — Sigmoid of actual improvements on edges: This variation takes into account the sigmoid value of the actual improvement score. If paper u having a score of u_z on a specific metric z, improves upon paper v which has a score of v_z in the same table and same metric, we compute the improvement score using Eq. (1). We then pass this score through a sigmoid function of the form:

$$\sigma_z(u, v) = \frac{1}{1 + e^{-\mathrm{REI}_z(u,v)}} \tag{2}$$

To combine the multiple improvement scores of u over v on different metrics and, thereby, obtain the edge weights, we use the following two techniques.

Max: We set the weight of the edge pointing from v to u as the maximum of all the sigmoid values of the improvement scores across the different metrics.
Average: We set the weight of the edge pointing from v to u as the average of all the sigmoid values of the improvement scores across the different metrics.

Dummy Winner and Loser Nodes: In the tournament ranking literature that we shall discuss in the next section, the most prominent factor that guarantees convergence is that the tournament must be connected. However, performance tournament graphs are mostly disconnected due to extraction inaccuracies, incomplete article collection, etc. Therefore, we introduce a dummy node that either wins or loses over all other nodes in the graph. A dummy node has a suitably directed edge to every other node.

5 Mining Sanitized Performance Improvement Graphs

In this section, we explore several ranking schemes to select the most competitive papers by analyzing the sanitized performance graph. We begin with basic baselines, then explore and adapt the tournament literature, and finally present adaptations of PageRank-style algorithms. Solving an incomplete tournament over n teams means to assign each team a score or rank inducing a total order over them, and presents a natural analogy with incomplete pairwise observations. The literature on tournaments seeks to extrapolate the anticipated outcome of a match between teams i and j (which was never played, say) in terms of the statistics of known outcomes, e.g., i defeated k and k defeated j.

Sink Nodes: We can ignore the numeric values in table cells and regard each table as comparing some papers, a pair at a time, and inserting an edge from paper p_1 to paper p_2 if the table lists a better (greater or smaller depending on metric) number against p_2 than p_1. In such a directed graph, sink nodes that have no out-links are locally maximal. Thus, the hunt for leaders may be characterized as a hunt for sink nodes. We do not expect this to work well, because our graphs contain many biconnected components, thanks to papers being compared on multiple metrics.

Cocitation: An indirect indication that a paper has pushed the envelope of performance on a task is that it is later compared with many papers. We can capture this signal in a graph where nodes are papers, and an edge and its reverse edge (both unweighted) are added between papers p_1 and p_2 if they are cited by any paper. Edges in both directions are added without considering the numbers extracted from the tables.

Linear Tournament: As described earlier, incomplete tournament presents a natural analogy to performance comparisons. [9] started with an incomplete tournament matrix M where $m_{ij} = m_{ji}$ is the number of matches played between teams i and j. $\boldsymbol{m} = (m_i)$ where $m_i = \sum_j m_{ij}$ is the number of matches played by team i. Abusing the division operator, let $\bar{M} = M/\boldsymbol{m}$ denote M after normalizing rows to add up to 1.

Of the m_{ij} matches between teams i and j, suppose i won r_{ij} times and j won $r_{ji} = m_{ij} - r_{ij}$ times. Then the *dominance* of i over j is $d_{ij} = r_{ij} - r_{ji}$ and the dominance of j over i is $d_{ji} = r_{ji} - r_{ij} = -d_{ij}$. Setting the dominance of a team over itself as zero in one dummy match, we can calculate the average dominance of a team i as $\bar{d}_i = \left[\sum_j d_{ij}\right] \Big/ \left[\sum_j m_{ij}\right]$, and this produces a reasonable ranking of the teams to a first approximation, i.e., up to "first generation" or direct matches. To extrapolate to"second generation" matches, we consider all (i,k) and (k,j) matches, which is given by the matrix M^2. Third generation matches are likewise counted in M^3, and so on. [3] showed that a meaningful scoring of teams can be obtained as the limit $\lim_{T\to\infty} \sum_{t=0}^{T} \bar{M}^t \cdot \bar{\boldsymbol{d}}$, where $\bar{\boldsymbol{d}} = (\bar{d}_i)$.

Exponential Tournament: The exponential tournament model [5] is somewhat different, and based on a probabilistic model. Given $R = (r_{ij})$ as above, it computes row sums $\rho_i = \sum_j r_{ij}$. Let $\boldsymbol{\rho} = (\rho_i)$ be the empirically observed team scores. Again, we can sort teams by decreasing ρ_i as an initial estimate, but this is based on an incomplete and noisy tournament. Between teams i and j there are (latent/unknown) probabilities $p_{ij} + p_{ji} = 1$ such that the probability that

i defeats j in a match is p_{ij}. Then the MLE estimate is $p_{ij} = r_{ij}/m_{ij}$. [5] shows that there exist team 'values' $\boldsymbol{v} = (v_i)$ such that $\sum_i v_i = 0$ and

$$\rho_i = \sum_j m_{ij} p_{ij} = \sum_j \frac{m_{ij}}{1 + \exp(v_j - v_i)}. \tag{3}$$

Here M and $\boldsymbol{\rho}$ are observed and fixed, and $\boldsymbol{v}$ are variables. Values $\boldsymbol{v}$ can be fitted using gradient descent. Once the matrix $\boldsymbol{P} = (p_{ij})$ is thus built, it gives a consistent probability for all possible permutations of the teams. In particular, $\prod_j p_{ij}$ gives the probability that i defeats all other teams (marginalized over all orders within the other teams j). Sorting teams i by decreasing $\prod_j p_{ij}$ is thus a reasonable rating scheme.

PageRank: PageRank computes a ranking of the competitive papers in the (suitably aggregated) tournament graph based on the structure of the incoming links. We utilize standard PageRank implementation[4] to rank nodes in the directed weighted tournament graph. We found best results (see Table 5) when damping factor (α) is set at 0.90. We run this weighted variant of PageRank on each induced tournament graph corresponding to each query. The induced tournament graph consists of papers (P) relevant to the query along with the papers compared with P. These candidate response papers are ordered using PR values. These scores can also be used for tie-breaking sink nodes.

6 Experiments

6.1 Datasets

***ArXiv* Dataset:** We downloaded (in June 2017) the entire arXiv document source dump but restricted this study to the field of Computer Science. Table 3 shows statistics of arXiv's Computer Science papers. ArXiv mandates uploading the source of DVI, PS, or PDF articles generated from LaTeX code resulting in a large volume of papers (1,181,349 out of 1,297,992 papers) with source code.

Table 3. Salient statistics about the arXiv and Computer Science data sets.

Full	Year range	1991–2017
	papers	1,297,992
	papers with LaTeX code	1,181,349
	Total fields	9

Comp. Science	Number of papers	107,795
	Year range	1993–2017
	Total references	2,841,554
	Total indexed papers	1,145,083
	Total tables	204,264
	Total table citations	98,943
	Unique extracted metrics	14,947

Preprocessing and Extracting Table Citations: The curation process involves several sub-tasks such as reference extraction, reference mapping, table extraction, collecting table citations, performance metrics extraction and edge orientation. Due to space constraints, we present detailed description and evaluation of each sub-tasks elsewhere [11].

[4] https://networkx.github.io.

State-of-the-Art Deep Learning Papers: A representative example from the rapidly growing and evolving area of deep learning is https://github.com/sbrugman/deep-learning-papers. The website contains state-of-the-art (SOTA) papers on malware detection/security, code generation, NLP tasks like summarization, classification, sentiment analysis etc., as well as computer vision tasks like style transfer, image segmentation, and self-driving cars. This Github repository is very popular and has more than 2,600 stargazers and has been forked 330 times. The repository notes 27 different popular topics shown in Table 4. The table also shows that the SOTA papers curated by knowledgeable experts rarely find a place in the top results returned by the two popular academic search systems — GS and SS. To be fair, these systems were not tuned to find SOTA papers, but we argue that this is an important missing search feature. As fields saturate and stabilize, citations to "the last of the SOTA papers" may eclipse citations to older ones, rendering citation-biased ranking satisfactory. But we again argue that recognizing SOTA papers quickly is critical to researchers, especially new comers and practitioners.

Organic Leaderboards: We identify manually curated leaderboards that compare competitive papers on specific tasks. The four popular leaderboards that we choose for our subsequent experiments are (i) The Stanford Question Answering Dataset (SQuAD)[5], (ii) Pixel-Level Semantic Labeling Task (Cityscapes)[6], (iii) VOC Challenge (PASCAL)[7], and (iv) MIT Saliency (MIT-300)[8]. Each leaderboard consists of several competitive papers compared against multiple metrics. For example, the SQuAD leaderboard consists of 117 competitive papers compared against two metrics 'Exact Match' and 'F1 score'. The tasks mostly include topics from natural language processing (e.g., question answering) and image processing (e.g., semantic labeling, image segmentation and saliency prediction).

6.2 Ranking State-of-the-Art Papers

Table 5 shows comparisons between Google Scholar (GS), Semantic Scholar (SS), and several ranking variations implemented in our testbed. Recall@10, Recall@20, NDCG@10, and NDCG@20 are used as the evaluation measures, averaged over the 27 topics shown in Table 4. Since our primary objective is to find competitive prior art, recall is more important in case of Web search, where precision at the top (NDCG) is paid more attention.

[5] https://rajpurkar.github.io/SQuAD-explorer/.
[6] https://www.cityscapes-dataset.com/benchmarks/.
[7] https://goo.gl/6xTWxB.
[8] http://saliency.mit.edu/results_mit300.html.

Table 4. Recall of human-curated state-of-the-art (SOTA) deep learning papers within top-10 and top-20 responses from two popular academic search engines (Google Scholar and Semantic Scholar). Both systems show low visibility of SOTA papers.

		Code Generation	Malware Detection	Summarization	Taskbots	Text Classification	Question Answering	Sentiment Analysis	Machine Translation	Chatbots	Reasoning	Gaming	Style Transfer	Object Tracking	Visual Q&A	Image Segmentation	Text Recognition	Brain Comp. Interfacing	Self Driving Cars	Object Recognition	Logo Recognition	Super Resolution	Pose Estimation	Image Captioning	Image Compression	Image Synthesis	Face Recognition	Audio Synthesis	**Total**
#SOTA		7	3	3	2	15	1	2	6	2	1	14	6	1	1	15	6	3	2	30	4	5	4	9	1	9	8	6	166
GS	Top-10	0	0	0	0	0	0	0	1	0	0	0	1	0	1	0	0	0	1	1	0	0	0	1	0	0	0	0	6(3.6%)
	Top-20	0	0	0	0	1	0	0	1	0	0	0	3	0	1	1	1	0	1	1	0	0	0	1	0	0	0	1	12 (7.2%)
SS	Top-10	0	0	0	0	0	0	0	0	0	0	0	2	0	1	0	0	0	1	1	0	0	0	1	0	0	1	0	7 (4.2%)
	Top-20	0	0	0	0	0	0	0	1	0	0	0	2	0	1	1	1	0	1	1	0	1	0	1	0	0	1	0	11 (6.6%)

Given the complex nature of performance tournament ranking, our absolute recall and NDCG are modest. Among naive baselines, sink node search led to generally worst performance, which was expected. The numeric comparison is slightly better, but not much.

GS and SS are mediocre as well. Despite the obvious fit between our problem and tournament algorithms, they are surprisingly lackluster. In fact, many of the tournament variants lose to simple cocitation. PageRank on unweighted improvement graphs performs beyond cocitation. However, the "sigmoid" versions of PageRank improve upon the unweighted case, almost doubling the gains beyond GS and SS, and are clearly the best choice.

Table 5. Comparison between several ranking schemes. Recall@10, Recall@20, NDCG@10, NDCG@20 measures are averaged over the 27 tasks (queries). OS: Online Systems; LT: Linear Tournament; ET: Exponential Tournament; ALL: Weighted graph (total number of comparisons); UNW: Unweighted directed performance graph; SIG: Sigmoid of the actual performance improvement; DW: Dummy Winner; DL: Dummy Loser, DCC: Dense co-citation, NC: Numeric comparison.

		OS		LT				ET				PageRank				Sink	BS	
		GS	SS	DW	DL	DW	DL	DW	DL	DW	DL	UNW	ALL	Avg.	Max.	ALL	DCC	NC
				ALL		SIG		ALL		SIG			SIG					
T-10	Recall %	7.38	7.84	4.63	4.63	1.8	1.93	1.7	2.31	1.7	1.7	19.35	16.86	19.35	19.35	0.62	12.91	6.73
	NDCG	0.073	0.065	0.029	0.029	0.016	0.019	0.027	0.024	0.02	0.02	0.151	0.131	0.154	0.149	0.009	0.142	0.036
T-20	Recall %	10.48	10.08	5.86	5.86	6.5	6.63	4.17	2.93	2.93	2.93	21.74	21.95	22.36	22.09	0.62	19.25	7.35
	NDCG	0.086	0.074	0.034	0.034	0.028	0.03	0.036	0.026	0.025	0.025	0.159	0.151	0.164	0.159	0.009	0.152	0.037

Table 6. Recall@50 and NDCG@50 measures for four leaderboards. Green cells indicate best scores and red cells indicate worst scores.

Leaderboard name	GS		SS		PageRank UNW		PageRank SIG (Avg)		PageRank SIG (Max)	
	Recall (%)	NDCG	Recall (%)	NDCG	Recall (%)	NDCG	Recall (%)	NDCG	Recall (%)	NDCG
SQuAD	0	0	7.14	0.014	21.42	0.206	21.42	0.205	14.29	0.177
Cityscapes	25	0.067	37.5	0.159	62.5	0.303	62.5	0.310	62.5	0.295
PASCAL	26.92	0.12	26.92	0.179	57.69	0.497	57.69	0.500	57.69	0.502
MIT-300	42.86	0.115	14.28	0.036	50.00	0.465	50.00	0.437	50.00	0.438

6.3 Leaderboard Generation

In this section, we demonstrate our system's capability to automatically generate task-specific leaderboards. We utilize four manually curated leaderboards for this study. Automatic leaderboard generation procedure is divided into two phases:

Obtaining List of Candidate Papers Relevant to a Task: We, first, obtain a list of candidate papers relevant to a given task. We utilize textual information such as title and abstract to find relevant candidate papers. These candidate papers are further ranked by utilizing best performing PageRank schemes (described in Sect. 6.2). We consider top-50 ranked results and show comparisons between Google Scholar (GS), Semantic Scholar (SS), and top-3 high performing PageRank variations against two evaluation measures — Recall@50 and NDCG@50 — in Table 6. As expected, GS and SS performed poorly for all of the four leaderboards. PageRank variations have almost double the gains beyond GS and SS and are clearly the best choice. Some generated leaderboards are listed in Singh et al. [11].

Ranking Candidate Papers to Generate Leaderboard: Next, we compute the correlation between ranks in generated leaderboards with the ground-truth ranks obtained from the organic leaderboards. Table 7 presents the Spearman's rank correlation of rankings produced by PageRank variations, UNW, SIG (Avg) and SIG (Max), with the corresponding ground-truth rankings for the four leaderboards. SQuAD shows the highest correlation (0.94 for F1 and 0.89 for EM) for all of the three PageRank variations. CityScapes and PASCAL also exhibit impressive correlation coefficients for all the PageRank variants. For the MIT-300 leaderboard, while the correlation coefficient is decent for the SIM metric it is somewhat low for the AUC metric. The reason for the low correlation is existence of multiple weakly connected components. A local winner in one component is affecting the global ranks across all components.

6.4 Effect of Graph Sanitization

As described in Sect. 4.2, graph sanitization is a necessary preprocessing step. In this section, we present several real examples that resulted in greater visibility of state-of-the-art after sanitization. As representative examples, we consider two tasks, "image segmentation" and "gaming", to show how graph sanitization results in noise reduction in the performance improvement graphs. We find

Table 7. Spearman's rank correlation of rankings produced by UNW, SIG (Avg) and SIG (Max) with the corresponding ground-truth rankings for the four leaderboards.

Name	Nodes	Metric	UNW	SIG (AVG)	SIG (MAX)
SQuAD	9	F1	0.94	0.94	0.94
		EM	0.89	0.89	0.89
CityScapes	7	iIoU	0.7	0.7	0.7
PASCAL	26	AP	0.57	0.57	0.57
MIT-300	9	AUC	0.23	0.23	0.23
		SIM	0.53	0.45	0.45

several state-of-the-art papers that performed poorer than a competitive paper with high improvement score (>700%). This anomaly resulted in the poorer visibility of the state-of-the-art papers in top ranks. However, after sanitization, the visibility gets improved. For example, Table 8 shows four examples of high improvement edges whose removal resulted in the higher recall of the state-of-the-art papers.

Table 8. Effect of graph sanitization. The first two edges correspond to the task of "image segmentation" and the last two to the task of "gaming". Removal of these edges resulted in higher visibility of SOTA papers.

Source	Destination	Improvement %	Back-edge (Y/N)
1511.07122	1504.01013	775	Y
1511.07122	1511.00561	6597	Y
1611.02205	1207.4708	4012.3	N
1412.6564	1511.06410	928.8	N

6.5 Why Is PageRank Better Than Tournaments?

PageRank variations performed significantly better than tournament variations. Several assumptions of tournament literature do not hold true for scientific performance graphs; for instance, existence of disconnected components is a common characteristic of performance graphs. Unequal number of comparisons between a pair of papers in performance graphs is another characteristic that demarcates it from the tournament settings. We observe that in a majority of task-specific performance graphs, tournament-based ranking scheme is biased toward papers with zero out-degrees. Therefore the tournaments mostly converge to the global sinks; in fact, we observe more than half of the tournament based top-ranked papers are sink nodes. This is why recall and NDCG in Table 5 for these two methods are close.

7 Conclusion and Future Scope

We introduce performance improvement graphs that encode information about performance comparisons between scientific papers. The process of extracting tournaments is designed to be robust, flexible, and domain-independent, but this makes our labeled tournament graphs rather noisy. We present a number of ways to aggregate the tournament edges and a number of ways to score and rank nodes on the basis of this incomplete and noisy information. In ongoing work, we are extending beyond LaTeX tables to line, bar and pie charts [1,7].

Acknowledgment. Partly supported by grants from IBM and Amazon.

References

1. Al-Zaidy, R.A., Giles, C.L.: Automatic extraction of data from bar charts. In: Proceedings of the 8th International Conference on Knowledge Capture, p. 30. ACM (2015)
2. Cafarella, M.J., Halevy, A., Wang, D.Z., Wu, E., Zhang, Y.: WebTables: exploring the power of tables on the web. PVLDB **1**(1), 538–549 (2008). ISSN 2150-8097. http://doi.acm.org/10.1145/1453856.1453916, http://www.eecs.umich.edu/~michjc/papers/webtables_vldb08.pdf
3. David, H.A.: Ranking from unbalanced paired-comparison data. Biometrika **74**(2), 432–436 (1987). https://academic.oup.com/biomet/article-pdf/74/2/432/659083/74-2-432.pdf
4. Hashimoto, H., Shinoda, K., Yokono, H., Aizawa, A.: Automatic generation of review matrices as multi-document summarization of scientific papers. In: Workshop on Bibliometric-Enhanced Information Retrieval and Natural Language Processing for Digital Libraries (BIRNDL), vol. 7, pp. 850–865 (2017)
5. Jech, T.: The ranking of incomplete tournaments: a mathematician's guide to popular sports. Am. Math. Monthly **90**(4), 246–266 (1983). http://www.jstor.org/stable/2975756
6. Jung, D., et al.: Chartsense: interactive data extraction from chart images. In: Proceedings of the 2017 CHI Conference on Human Factors in Computing Systems CHI 2017, pp. 6706–6717 (2017). ISBN 978-1-4503-4655-9
7. Mitra, P., Giles, C.L., Wang, J.Z., Lu, X.: Automatic categorization of figures in scientific documents. In: Proceedings of the 6th ACM/IEEE-CS Joint Conference on Digital Libraries JCDL2006, pp. 129–138. IEEE (2006)
8. Page, L., Brin, S., Motwani, R., Winograd, T.: The PageRank citation ranking: bringing order to the Web. Manuscript, Stanford University (1998)
9. Redmond, C.: A natural generalization of the win-loss rating system. Math. Mag. **76**(2), 119–126 (2003). http://www.jstor.org/stable/3219304
10. Sarawagi, S., Chakrabarti, S.: Open-domain quantity queries on web tables: annotation, response, and consensus models. In: SIGKDD Conference (2014)
11. Singh, M., Sarkar, R., Vyas, A., Goyal, P., Mukherjee, A., Chakrabarti, s.: Automated early leaderboard generation from comparative tables. In: ECIR 2019 arXiv:1802.04538 (2018). https://arxiv.org/abs/1802.04538
12. Xing, W., Ghorbani, A.: Weighted pagerank algorithm. In: Proceedings of Second Annual Conference on Communication Networks and Services Research, pp. 305–314. IEEE (2004)

Query Analytics

Predicting the Topic of Your Next Query for Just-In-Time IR

Seyed Ali Bahrainian[1](✉), Fattane Zarrinkalam[2], Ida Mele[3], and Fabio Crestani[1]

[1] Faculty of Informatics, University of Lugano (USI), Lugano, Switzerland
{bahres,fabio.cerstani}@usi.ch
[2] Laboratory for Systems, Software and Semantics (LS3), Ryerson University, Toronto, Canada
fzarrinkalam@ryerson.ca
[3] ISTI-CNR, Pisa, Italy
ida.mele@isti.cnr.it

Abstract. Proactive search technologies aim at modeling the users' information seeking behaviors for a *just-in-time information retrieval* (JITIR) and to address the information needs of users even before they ask. Modern virtual personal assistants, such as Microsoft Cortana and Google Now, are moving towards utilizing various signals from users' search history to model the users and to identify their short-term as well as long-term future searches. As a result, they are able to recommend relevant pieces of information to the users at just the right time and even before they explicitly ask (e.g., before submitting a query). In this paper, we propose a novel neural model for JITIR which tracks the users' search behavior over time in order to anticipate the future search topics. Such technology can be employed as part of a personal assistant for enabling the *proactive* retrieval of information. Our experimental results on real-world data from a commercial search engine indicate that our model outperforms several important baselines in terms of predictive power, measuring those topics that will be of interest in the near-future. Moreover, our proposed model is capable of not only predicting the near-future topics of interest but also predicting an approximate time of the day when a user would be interested in a given search topic.

Keywords: Topic prediction · Topic modeling · Just-In-Time Information Retrieval · Neural IR

1 Introduction

With the rapid proliferation of web search as the primary mean for addressing the users' information needs, search engines are becoming more sophisticated with the purpose of improving the user experience and of assisting users in their search tasks more effectively. As an example, with the increasing and ubiquitous usage of mobile devices, it has become more important for search engines to offer

L. Azzopardi et al. (Eds.): ECIR 2019, LNCS 11437, pp. 261–275, 2019.
https://doi.org/10.1007/978-3-030-15712-8_17

also *Just-In-Time Information Retrieval* (JITIR) [18] experiences. This means retrieving the right information at just the right time [10] to save users from the hassle of typing queries on mobile devices [2,3].

The notion of "personalized search" [25] has shown to be effective in improving the ranking of search results. However, such personalization comes at the cost of lower speed, which in some cases might even cause the retrieval of the results only after the user search session has ended. Moreover, possible discovery of newly available content related to a previous search is another application of JITIR models for presenting results to a user at a future time.

As a result, researchers have focused on improving search personalization with respect to not only the retrieved content but also the user's habits (e.g., *when and what* information the users consume). While such models can benefit desktop users in better addressing their information needs at just the right time, they are essential on mobile platforms. Indeed, Microsoft Cortana and Google Now aim at offering a proactive experience to the users showing *the right information before they ask* [24].

As pointed out by Agichtein et al. [1], knowing the user's information needs at a particular time of the day allows to improve the search results ranking. For example, the search results can be personalized based on the specific search task (of a given user at a given time) rather than based on the more general information of user interests which have been inferred by the entire user's profile. This would also support users in resuming unfinished search tasks (e.g., if a search is likely to be continued one can save the results already found for a faster or more convenient access once the task is resumed).

Figure 1(a) shows the behavior of a randomly selected user from our dataset in issuing search queries related to a topic about `movies` over different week days. For example, the user might have searched the word "imdb" along with the title of a movie. As we can see, the user exhibits a higher tendency to search for movies in the afternoons and evenings as well as on Saturdays. Hence, we can infer that the user is interested in watching movies on Saturday evenings

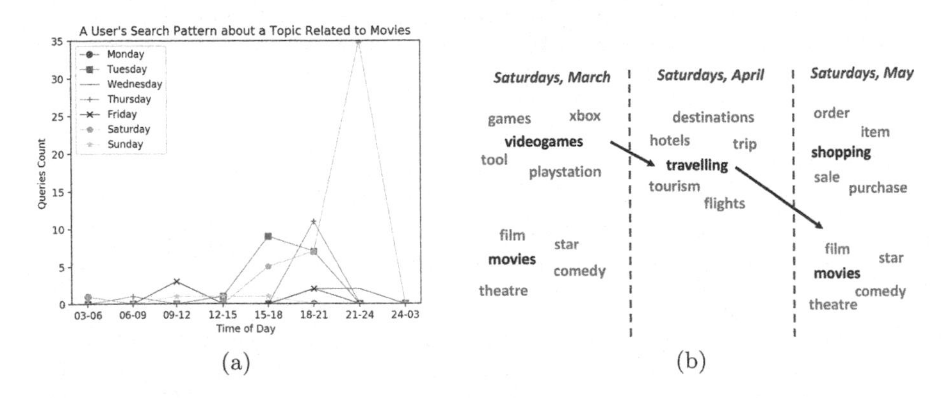

Fig. 1. (a) The number of queries about `movies` submitted by a randomly selected user. (b) Evolution of user-search patterns on different Saturdays.

and thus it is likely that her queries are related to movies. Moreover, as shown in Fig. 1(b) a user changes search behavior over time. To address such changes in search behavior we propose a dynamic memory system.

Addressing the near-future information needs of the users has been also studied in the context of personal assistants, such as Google Now, Microsoft Cortana, or Apple's Siri and in the context of memory augmentation in meetings [6]. These systems offer proactive experiences [22] that aim to recommend useful information to a user at just the right time.

In this paper, we focus on predicting the topics of the users' future search queries. Specifically, we propose a model which predicts the topic of the search queries submitted by the users in the next 24 h. Moreover, our model leverages the user's behavior patterns over time in order to predict the topic of the user's query on a specific weekday (e.g., Mondays, Tuesdays) and at an approximate time of the day. The main contributions of this paper are:

C1: we propose a time-series model based on neural networks to predict the topic of near-future queries of users.
C2: our model is equipped with a dynamic memory learning users' behavior over time. This memory evolves over time when the search patterns change. We demonstrate that our dynamic memory architecture is beneficial as it increases the prediction performance. Further, we believe that this model could be useful in other domains that involve temporal data.

The organization of this paper is as follows: Sect. 2 presents the related work, Sect. 3 describes our research goals and Sect. 4 presents our model for predicting the topics of the users? future search queries. In Sect. 5, we evaluate our method against the baseline methods based on their predictive performance. Finally, Sect. 6 concludes this paper and gives insight into future work.

2 Related Work

2.1 Just-In-Time Information Retrieval

Addressing the users' near-future information needs has been studied in the context of personal assistants [4] such as Google Now, Microsoft Cortana, or Apple's Siri. These systems offer proactive experiences and aim to recommend the right information at just the right time [22,24]. As an example, Sun et al. [24] proposed an approach for tracking the context and intent of a user leveraging smartphone data [5] in order to discover complex co-occurring and sequential correlations between different signals. Then, they utilized the discovered patterns to predict the users' intents and to address their information needs.

In the context of proactive search and recommendation, Song and Guo [23] aimed at predicting task repetition for offering a proactive search experience. They focused on predicting when and what type of tasks will be repeated by the users in the future. Their model was based on time series and classification. They tested the effectiveness of their approach for future query and future app predictions. Our work differs from their work since we take a collaborative time-series approach for predicting the topics of future user queries. Moreover, our goal is to predict the topic of one's next query and not only predicting the repetition of a search task.

Agichtein et al. [1] tried to predict the continuation of a previously started task within the next few days. Similarly to [23], they defined the prediction of the continuation of a task as a classification problem. They used an extensive set of features for training the classifiers. Such features include query topics, level of user engagement and focus, user profile features such as total number of unique topics in prior history, and repeating behavior among others. Our work differs from this work as we do not simply try to predict the search task continuation in the future but we also aim at predicting the day of the week and the approximate time of the day when a query topic will occur. Moreover, unlike their model which is a classifier based on a number of hand-engineered features, our model has a time-series structure and it evolves over time by learning from the data and correcting itself over time.

Furthermore, another interesting but different work consists in the identification of recurrent event queries and was presented by Zhang et al. [28]. In this work, the authors aimed at identifying search queries that occur at regular and predictable time intervals. To accomplish this, they train various classifiers such as Support Vector Machines and Naïve Bayes, on a number of proposed features such as query frequency, click information, and auto correlation. They conclude that a combination of all features leads to the highest performance.

2.2 Topic Models and Word Embeddings

Topic models are defined as hierarchical Bayesian models of discrete data, where each topic is a set of words, drawn from a fixed vocabulary, which together represent a high level concept [26]. According to this definition, Blei et al. introduced Latent Dirichlet Allocation (LDA) [8]. In our work, we use LDA for discovering the topics of the users' search queries. In particular, as we will see in Sect. 4, we created a collection of documents consisting of some of the query results, then we run LDA to extract the topics of the various search queries.

Another form of word vectors is represented by word embeddings which map semantically related words to nearby positions in a vector space. Topic modeling approach is also unsupervised. Some well-known approaches are the word2vec model [15] and the Glove model [17]. As explained in Sect. 4, we needed to use word embeddings to model the dependencies between different attributes.

3 Research Goals

We can summarize the goals of our work as follows: (1) predicting the topics of future search queries of a user, and (2) predicting the day of the week and the approximate time of the day when the topic will be queried by a user.

Given the search history of each user u in the last n consecutive time slices, as well as a set of corresponding query topics $\mathbb{Z}$, we aim to predict the topic $z \in \mathbb{Z}$ of the query of user u in the $(n+1)_{th}$ time slice.

For achieving this, we first model the search tasks as topics using LDA. Then, leveraging a time-series model we discover the latent patterns in search tasks and predict the continuation of a search task in the near-future. In other words, we aim at predicting the topics of the user's future queries. Such technology will enable the proactive retrieval of relevant information in a JITIR setting. However, estimating the time of the day when a user would access a particular content is the second piece of the puzzle in order to recommend content more precisely and more effectively. Thus, our second goal consists in correctly predicting when (day and time of the day) the users will consume what content (topic) knowing the users' habits in requesting the various topics at the different times.

4 Query Topic Prediction Model

We now present our novel time-series evolutionary model for predicting the topic of a user's near-future queries. The model is based on the notion of reinforcement learning so that it adapts itself to the data over time and corrects itself. We formally define our model as a function f which takes as input the search history of users and predicts which topics occur in the near future.

The model consists of a dynamic memory in the form of a word embedding connected with a *Bi-directional Long Short Term Memory* (BiLSTM) [21] used to capture the behaviour of a user over time. The dynamic memory implements two different effects of persistence and recency. At each point in time, based on the possible changes in the input data, it updates the word vectors to provide as input to the BiLSTM network.

In the following, we first describe the dynamic memory system in Sects. 4.1 and 4.2. Then, we present the BiLSTM network in Sect. 4.3.

4.1 A Dynamic Memory Based on Word Embeddings

Our intuition behind the design of such memory model is that people often show similar behavior over time (i.e., persistence) but they also have a tendency to explore new things (i.e., recency). As a result, over a timeline people may show very different behaviors and the model should be capable of capturing them in order to accurately anticipate the users' future behaviors [27]. Therefore, we believe that dividing the temporal input data into a number of time slices and weighting them based on identified patterns in the data is important.

The dynamic memory is based on the word2vec word embeddings. Throughout this paper whenever we use the term word2vec, we refer to a Skip-Gram with Negative Sampling (SGNS) word embedding model. Levy et al. [14] showed that the SGNS method is implicitly factorizing a word-context matrix, whose cells are the Pointwise Mutual Information (PMI) of the respective word and context pairs, shifted by a global constant. They further elaborate that word2vec decomposes the data very similar to Singular Value Decomposition (SVD) and that under certain conditions an SVD can achieve solutions very similar to SGNS when computing word similarity. Apart from scalability and speed, SGNS is capable of removing bias towards rare words using negative sampling. Other than the few differences, at the concept level both SVD and SGNS are very similar. They both build a word-context matrix for finding similarities between words.

Based on these principles we propose a novel and effective method for integrating multiple word2vec memory components where each is trained with data from a different time slice of the input data. Let $m_t \in \boldsymbol{M}$ where m_t is a word2vec memory trained on data form time slice t and $\boldsymbol{M}$ is a vector of all word2vec models. Instead of using only one single memory to capture the global patterns in the dataset, we propose to use a different word vector from model m_t to represent time slice t where $t \in 0, 1, \ldots, n$. Then, we integrate all these word vectors into one final vector. Therefore, a temporal dataset of web search queries can be divided into n different time slices, and one word2vec memory m_t is trained for each time slice. We assume that all the vectors have the same embedding dimensions, so given two vectors m_t and m_{t+1} we can combine them using the *orthogonal Procrustes* matrix approximation. Let W^t be the matrix of word embeddings from m_t which is trained on data at the time slice t. We align across m_t and m_{t+1} which are derived from consecutive time slices while preserving the cosine similarities by optimizing:

$$argmin_{Q^TQ=I} ||W^tQ - W^{t+1}|| \quad (1)$$

Matrix Q is described in the following. We note that this process only uses orthogonal transformations like rotations and reflections. We have solved this optimization problem by applying SVD [20].
We can summarize the steps of the approach as follows:

1. The vocabulary of the resulting word vectors from the two time slices are intersected and the ones in common are kept. We note that due to our definition of an active user as well as the way we map queries to unique user, topic and time identifiers, vocabulary remains the same over all time slices (see Sect. 5.1).
2. We compute the dot product of the two matrices (for doing so, we first transpose one of the matrices).
3. The SVD of the matrix resulting from the dot product is computed. The result consists of three factorized matrices commonly known as U, the diagonal matrix S, and the matrix V.
4. We compute the dot product of U (left singular matrix) and V (right singular matrix) to have as resulting matrix Q. Since S contains information on the strength of concepts representing word-dimension relations which are not needed here as they are not modeled in word2vec, we discard the matrix S. The existence of the S matrix is also one important difference between SVD and word2vec, which word2vec does not compute.

5. Finally, we compute the dot product of Q and the embedding matrix W^t. For further detailed information we refer to [20] where the *orthogonal Procrustes* approximation using SVD is described.

We repeat the process of model alignment for all n word2vec models spread over the entire timeline.

4.2 Modifying the Dynamic Memory Using Recency and Persistence Effects

Now that we have explained the process of combining different word2vec models, in this section we explain how our proposed model takes into account the recency and persistence of the searching behaviors of users over time. Before combining W^t and W^{t+1} which are word-dimension matrices from two word2vec models (as described in Sect. 4.1), we modify each matrix based on the following effects:

Recency Effect. It modifies the strength of word embeddings by assigning higher weights to the word vectors observed in the most recent time slice. We formally define the recency effect as follows: given the query topics of the last n consecutive time slices of a sequential dataset, we would like to predict which query topics continue in the $(n+1)_{th}$ time slice. By assuming a vocabulary v of all the words occurring in the first n time slices, we construct a word vector containing the probability scores corresponding to each word in v. The assigned probability scores are higher for the words appearing in the most recent time slices. After modifying the word vectors, we then perform alignment of models as described in Sect. 4.1. According to the recency effect presented in the following equation we modify the word embedding matrices W^ts by $P_{Rec} = \sum_{n=1}^{N} \sum_{w_i \in W^t} P(w_i) * 2^n$, where n is the time slice number, P indicates probability, and w_i is a word from the word embedding matrix W^t. The 2^n is the rate with which recent word vectors are assigned higher weights. The resulting constructed word vector is an average representation of the probability of all the words present in all the n time slices.

Therefore, this effect assigns higher weight to a word which has occurred in the most recent time slice of a sequential corpus. We refer to the word vectors which are computed by the recency effect as the *recency matrix*.

Persistence Effect. Given the word embeddings of the last n consecutive time slices, we would like to predict which query topic continues in the $(n+1)_{th}$ time slice. Given a vocabulary v of all the words occurring in the first n time slices, we construct a word vector containing the probability scores corresponding to each word in v. The assigned probability scores are higher for the words which have persisted over time. We compute the updated probability of each word according to the persistence effect using $P_{Pers} = \sum_{n=1}^{N} \sum_{w_i \in W^t} P(w_i) * 2^{-n}$, where n is the time slice number, P indicates probability, and w_i is a word from W^t. The 2^{-n} is the rate with which the higher weights are assigned to persistent words.

Therefore, the more persistent words (i.e., persisting in occurrence) have higher weights, and we refer to the word vector computed with the persistent effect as the *persistence vector.*

Combining Recency and Persistence: Recency and persistence scores are combined in a linear interpolation to modify the original word embedding matrix. The linear interpolation at the time t is defined as:

$$EmbeddingMatrix_{w,t} = w_{P,t} * Score_{Pers.} + w_{R,t} * Score_{Rec.} \quad (2)$$

where $Score_{Pers.}$ and $Score_{Rec.}$ are computed by the persistence and the recency effects, respectively. Furthermore, $w_{P,t}$ and $w_{R,t}$ are persistence weights and recency weights computed at each time slice. They have the following relation and are learned from the data: $w_{P,t} + w_{R,t} = 1$. This means that at each time slice t each of the two effects corresponds to a weight. The weights can be equal (i.e. when the effects have the same intensity) or different, but their sum would be always 1.

The weight $w_{R,t}$ is then computed as square root of the sum of the difference in the number of occurrences of each query topic compared with the previous time slice divided by the number of all the queries at the same time slice. Subsequently, $w_{P,t}$ is computed based on $w_{P,t} + w_{R,t} = 1$. As a result, the dynamic memory evolves over time and updates itself proportionally to the rate of the changes in the data.

Finally, we map each query to a topic using LDA. Further details of this process are explained in Sect. 5.1. We specify each query with the ID of the user who submitted it along with the given week day and time bucket (i.e. which is an approximate time of day) of the query. Then, we train n word2vec SGNS models on this data in order to train the dynamic memory. For word2vec, we use embedding size of 300, without discarding any of the input words. The result will be n word embedding matrices derived from n time slices which are aligned and combined into one word embedding matrix which is given as input to the BiLSTM.

4.3 Bi-directional Long Short Term Memory (BiLSTM)

We train the BiLSTM network using the word embeddings of the dynamic memory. We use the BiLSTM neural network as function for generating a sequence of events given an input query. In other words, we aim at modeling the sequence of observations (i.e., the searches about certain topics) in a time-series fashion. Thus, given the user ID, the future week day and the time bucket, the model will predict the topics of the near-future queries.

As shown in Fig. 2, the architecture of our model consists of word embeddings from the dynamic memory provided as input to the BiLSTM network. We model each query in 4 recurrent time steps in order to predict the topic of near-future queries along with their weekday, and approximate time of the day (i.e., we refer to it as the time bucket in dataset description). On the other hand, when we want to only predict the near-future topics without specifying their approximate

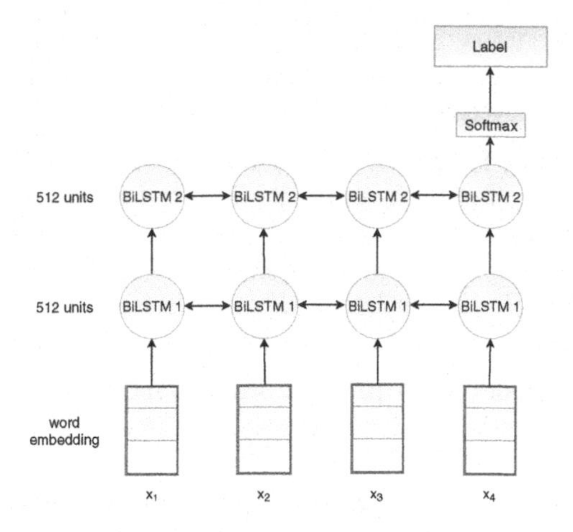

Fig. 2. The architecture of our proposed model. x_i stands for input at time step i

time of the day, we train the same network with 2 recurrent time steps (i.e., one for the user ID and the other for the topic). In both cases we set number of word2vec models to six (n = 6) to model almost every two weeks with one word2vec model. Furthermore, our model includes two fully connected BiLSTM layers, with each layer containing 512 cells or units. We applied a SoftMax layer to the final output from the BiLSTM networks.

Our intuition behind this architecture is to first find a collaborative generalization of patterns of users in issuing queries about certain topics at particular points in time by using the dynamic memory based on the word2vec model. Then, using the BiLSTM neural network we leverage the local dependencies between certain behaviors in a temporal manner. The BiLSTM network serves as a time-series model that determines the occurrence of a future event (i.e., a future query's topic) by modeling the sequences of events (i.e., sequences of topics).

5 Experimental Setup

5.1 Dataset Description

In the experiments, we use the publicly available AOL query log [16] which has been used in other research works on query caching and search result personalization. It consists of about 36M query records and spans a period of three months (from March to May 2006). Each record consists of the anonymous ID of the user, query keywords, timestamp, and rank and URL of the clicked result.

Our goal is to predict the topics of the future queries issues by a user, hence we selected those users who have a high number of queries. Formally, we define *active users* those who have searched at least one query every week and over a span of three months have issued at least 1,000 queries. From this set which contains 1,197 active users, we randomly selected 500 users to train and test our proposed model as well as the baselines. The query log made of the queries issued by these 500 users consists of 755, 966 queries.

Training and Testing Data. Our experiments aim at predicting the topics of the future queries searched by a user, so we sorted the query log by time and split it into training and test sets. The training set is used for learning the topics of interests of a user, while the test set to check the prediction performance. For our experiments, the test set consists of the queries issued in the last 24 h (which results of 10, 848 queries) while the rest of the queries is used for training.

Modeling Search Tasks as Topics. In order to model the topics of the search tasks we used the Latent Dirichlet Allocation (LDA) topic model [8].

Since the search queries are short and lack context, we decided to enrich them with the content of clicked pages. More in detail, given the queries from the training set and the URL of their clicked results, we gathered the content of 351, 301 unique web pages. We treat each query and the text of its corresponding clicked result as a document, and we run LDA over the collection made of these documents. LDA returns K list of keywords representing the latent topics discussed in the collection. Since the number of topics (K) is an unknown variable in the LDA model, it is important to estimate it. For this purpose, similar to the method proposed in [7,9], we went through a model selection process. It consists in keeping the LDA parameters (commonly known as α and η) fixed, while assigning several values to K and run the LDA model each time. We picked the model that minimize $logP(W|K)$, where W contains all the words in the vocabulary. This process is repeated until we have an optimal number of topics. The training of each LDA model takes nearly a day, so we could only repeat it for a limited number of K values. In particular, we trained the LDA model with K equals to 50 up to 500 at steps of 50, and the optimal value was 150.

Labels for Predicting the Approximate Time. The search queries have timestamps, so we could extract the day of the week and the time of the day when they were issued. We divide the 24 h into 8 time buckets of 3 h each. Each time bucket represents an approximate time of the day and we can use this for predicting the approximate time of the day when a query topic will appear. Hence, given a user, our ultimate purpose is to predict the right query topic and when it will be requested (i.e., the week day and the time bucket).

5.2 Evaluation Metrics and Baselines

Evaluation Metrics. We performed a rigorous testing of our proposed method and compared it against several baseline methods. For our evaluation, we used the standard information retrieval evaluation metrics: precision, recall, and F_1.

Baseline Methods. Since our proposed method is based on a collaborative filtering principle, we chose as baselines the following top-performing techniques:

1. ***Probabilistic Matrix Factorization (PMF)*** is a model for collaborative filtering that has achieved robust and strong results in rating prediction [19].

2. ***Non-negative Matrix Factorization (NMF)*** can analyze data with a high number of attributes [13]. It reduces the dimensions of data by converting the original matrix into two matrices with non-negative values.
3. ***User-based K Nearest Neighbours (userKNN)*** is another popular method which uses similarities among the users' behaviours to recommend items to users.
4. ***SVD++*** is a collaborating filtering method, where previously seen patterns are analyzed in order to establish connections between users and items [11]. The approach merges latent factor models that profile both the users and the items with the neighborhood models that analyze the similarities between the items or between the users.
5. ***TimeSVD++*** is one of the most successful models for capturing the temporal dynamics and has shown strong results in various prediction and ranking tasks which seek to model a generalized pattern over time [12]. The regularization parameter and the factor size are selected using a grid search over $\lambda \in \{10^0, \ldots, 10^{-5}\}$ and $k \in \{20, 40, 80, 160\}$.
6. ***BiLSTM+w2v*** we also add as a baseline our own model with only one word2vec model trained as input (i.e., see n=1 in Table 2).

5.3 Experimental Results

First Experiment. The aim of our first experiment is predicting the topics of the queries issued by a user in the next 24 h. Table 1 reports the results of our approach compared to the baselines. We observe that our method outperforms all the baseline models in terms of predicting the topics of one's queries in the future 24 h with statistically significant improvement. We averaged the prediction results over the 500 users of our sampled data. As a result of this experiment, we could observe that our model is superior in predicting the topics of future queries compared to the other collaborative-filtering baselines.

Our proposed model features incorporating some principles that we believe have caused the superiority of our model. First, the dynamic memory not only learns users' search behavior but also considers the temporal dimension when

Table 1. A comparison of our proposed method against the baselines.

	Precision	Recall	F_1
PMF (%)	12.53	31.03	19.78
NMF (%)	13.65	35.11	21.23
UserKNN (%)	14.20	38.06	22.09
SVD++ (%)	12.60	30.43	20.20
TimeSVD++ (%)	28.46	14.62	20.00
BiLSTM+w2v (%)	34.28	36.04	35.12
Our Model (%)	48.19	38.44	42.77
Our Model+time prediction (%)	26.23	34.41	29.77

Table 2. A comparison of the prediction performance varying the number of word2vec models in terms of F_1 (n=1 means one model trained over the whole dataset, etc.).

	n = 1	n = 2	n = 3	n = 4	n = 5	n = 6	n = 7	n = 8	n = 9	n = 10
24 h prediction (%)	35.12	38.63	39.14	41.56	42.93	42.77	41.46	40.34	40.52	40.12
24 h + time-bucket prediction (%)	28.64	29.32	29.57	29.83	30.21	29.77	29.40	29.43	29.21	29.06

modeling data. Furthermore, our model uses the recency and persistence effects and adjust itself to the data by measuring the behavior of the data and subsequently updating itself when needed. None of the baseline models, despite being powerful models, can model such complexities.

Second Experiment. In the second experiment, we would like to investigate whether or not running and combining different word2vec models can improve the performance compared to one trained word2vec model. In particular, we divided our input data into chunks (e.g., weeks) and trained several word2vec models over them. Then, we compared the performance of one word2vec model trained over the whole timeline against the performances achieved with different numbers of word2vec models. We started by only having one word2vec model up to 12 models (one for each week of our dataset).

The results of this experiment are reported in Table 2 and show that training the word2vec model over different time slices performs better than having only one word2vec model trained over the entire dataset. Moreover, we could observe that increasing the number of models allows to gain higher performance, however, after some point the performance plunges. We can conclude that training several models is better than one, but the number of models should be chosen depending on the application. We could observe that the best results can be achieved training the models with roughly two weeks of data.

Our research goal was to design an intuitive time-series method for modeling the user behavior, specifically regarding search queries. The broader vision and strategy that we tried to incorporate into the model was that the users have the tendency to repeat the behavior (e.g., searching about the same topic in a sequence), but they also have consistent behaviors (e.g., searching for the same topic every Saturday night). Hence, incorporating these two dimensions into our model helped to improve the prediction performance. The concept behind our model may also be used in a personal assistant environment for modeling other types of data, tracking the user behavior over time and providing the user with the right information just-in-time the user might need it. Envisaging that a system can correctly predict the topic of your near-future query more than 40% of times among all possible options (i.e., in this case 150 topics) while also predicting the time bucket when you will show interest in that topic and presenting relevant information or targeted ads to you even before you have started searching on that topic is a very interesting result.

6 Conclusions

In this paper, we addressed the problem of predicting topics of future queries for just-in-time IR. For this purpose, we proposed a novel method and compared it against six baseline methods which have been extensively used in the literature for temporal and non-temporal collaborative filtering. We showed through experimental results that our method, generalizing the users' behavior and modeling the temporal recurrent patterns, outperforms all the baselines. The developed method could be implemented as a part of a proactive search system that aids people in their every day lives.

One interesting future work would be adapting our method to other domains. For example, analyzing various data modalities gathered by current personal assistant tools such as Microsoft Cortana could be an interesting direction.

References

1. Agichtein, E., White, R.W., Dumais, S.T., Bennet, P.N.: Search, interrupted: understanding and predicting search task continuation. In: Proceedings of the 35th International ACM SIGIR Conference on Research and Development in Information Retrieval, SIGIR 2012, pp. 315–324 (2012)
2. Aliannejadi, M., Harvey, M., Costa, L., Pointon, M., Crestani, F.: Understanding Mobile Search Task Relevance and User Behaviour in Context. In: CHIIR 2019 (2018)
3. Aliannejadi, M., Zamani, H., Crestani, F., Croft, W.B.: In situ and context-aware target apps selection for unified mobile search. In: Proceedings of the 27th ACM International Conference on Information and Knowledge Management, pp. 1383–1392 (2018)
4. Bahrainian, S.A., Crestani, F.: Towards the next generation of personal assistants: systems that know when you forget. In: Proceedings of the ACM SIGIR International Conference on Theory of Information Retrieval, ICTIR 2017, pp. 169–176 (2017)
5. Bahrainian, S.A., Crestani, F.: Tracking smartphone app usage for time-aware recommendation. In: Choemprayong, S., Crestani, F., Cunningham, S.J. (eds.) ICADL 2017. LNCS, vol. 10647, pp. 161–172. Springer, Cham (2017). https://doi.org/10.1007/978-3-319-70232-2_14
6. Bahrainian, S.A., Crestani, F.: Augmentation of human memory: anticipating topics that continue in the next meeting. In: Proceedings of the 2018 Conference on Human Information Interaction & Retrieval, CHIIR 2018, pp. 150–159 (2018)
7. Bahrainian, S.A., Mele, I., Crestani, F.: Predicting topics in scholarly papers. In: Pasi, G., Piwowarski, B., Azzopardi, L., Hanbury, A. (eds.) ECIR 2018. LNCS, vol. 10772, pp. 16–28. Springer, Cham (2018). https://doi.org/10.1007/978-3-319-76941-7_2

8. Blei, D.M., Ng, A.Y., Jordan, M.I.: Latent dirichlet allocation. J. Mach. Learn. Res. **3**, 993–1022 (2003)
9. Griffiths, T.L., Steyvers, M.: Finding scientific topics. Proc. Natl. Acad. Sci. **101**(suppl. 1), 5228–5235 (2004)
10. Guha, R., Gupta, V., Raghunathan, V., Srikant, R.: User modeling for a personal assistant. In: Proceedings of the Eighth ACM International Conference on Web Search and Data Mining, WSDM 2015, pp. 275–284 (2015)
11. Koren, Y.: Factorization meets the neighborhood: a multifaceted collaborative filtering model. In: Proceedings of the 14th ACM SIGKDD International Conference on Knowledge Discovery and Data Mining, pp. 426–434 (2008)
12. Y. Koren. Collaborative filtering with temporal dynamics. In: Proceedings of the 15th ACM SIGKDD International Conference on Knowledge Discovery and Data Mining, KDD 2009, pp. 447–456 (2009)
13. Lee, D.D., Seung, H.S.: Algorithms for non-negative matrix factorization. In: Advances in neural information processing systems, pp. 556–562 (2001)
14. Levy, O., Goldberg, Y.: Neural word embedding as implicit matrix factorization. In: Advances in Neural Information Processing Systems, pp. 2177–2185 (2014)
15. Mikolov, T., Sutskever, I., Chen, K., Corrado, G.S., Dean, J.: Distributed representations of words and phrases and their compositionality. In: Advances in Neural Information Processing Systems, pp. 3111–3119 (2013)
16. G. Pass, A. Chowdhury, and C. Torgeson. A picture of search. In: Proceedings of the 1st International Conference on Scalable Information Systems (InfoScale 2006). ACM, New York (2006)
17. Pennington, J., Socher, R., Manning, C.D.: Glove: global vectors for word representation. In: Proceedings of the 2014 Conference on Empirical Methods in Natural Language Processing (EMNLP), pp. 1532-1543 (2014)
18. Rhodes, B.J.: Just-in-time information retrieval. PhD thesis, Massachusetts Institute of Technology (2000)
19. Salakhutdinov, R., Mnih, A.: Probabilistic matrix factorization. In: Proceedings of the 20th International Conference on Neural Information Processing Systems, NIPS 2007, pp. 1257–1264 (2007)
20. Schönemann, P.H.: A generalized solution of the orthogonal procrustes problem. Psychometrika **31**(1), 1–10 (1966)
21. Schuster, M., Paliwal, K.K.: Bidirectional recurrent neural networks. IEEE Trans. Signal Process. **45**(11), 2673–2681 (1997)
22. Shokouhi, M., Guo, Q.: From queries to cards: Re-ranking proactive card recommendations based on reactive search history. In: Proceedings of the 38th International ACM SIGIR Conference on Research and Development in Information Retrieval, SIGIR 2015, pp. 695–704 (2015)
23. Song, Y., Guo, Q.: Query-less: predicting task repetition for nextgen proactive search and recommendation engines. In: Proceedings of the 25th International Conference on World Wide Web, WWW 2016, pp. 543–553 (2016)
24. Sun, Y., Yuan, N.J., Wang, Y., Xie, X., McDonald, K., Zhang, R.: Contextual intent tracking for personal assistants. In: Proceedings of the 22nd ACM SIGKDD International Conference on Knowledge Discovery and Data Mining, KDD 2016, pp. 273–282 (2016)

25. Teevan, J., Dumais, S.T., Horvitz, E.: Personalizing search via automated analysis of interests and activities. In: Proceedings of the 28th Annual International ACM SIGIR Conference on Research and Development in Information Retrieval, SIGIR 2005, pp. 449–456 (2005)
26. Wang, C., Blei, D., Heckerman, D.: Continuous time dynamic topic models. In: Proceedings of UAI (2008)
27. Zarrinkalam, F., Kahani, M., Bagheri, E.: Mining user interests over active topics on social networks. Inf. Process. Manage. **54**, 339–357 (2018)
28. Zhang, R., Konda, Y., Dong, A., Kolari, P., Chang, Y., Zheng, Z.: Learning recurrent event queries for web search. In: Proceedings of the 2010 Conference on Empirical Methods in Natural Language Processing, EMNLP 2010, pp. 1129–1139 (2010)

Identifying Unclear Questions in Community Question Answering Websites

Jan Trienes[1(✉)] and Krisztian Balog[2]

[1] University of Twente, Enschede, Netherlands
jan.trienes@gmail.com
[2] University of Stavanger, Stavanger, Norway
krisztian.balog@uis.no

Abstract. Thousands of complex natural language questions are submitted to community question answering websites on a daily basis, rendering them as one of the most important information sources these days. However, oftentimes submitted questions are unclear and cannot be answered without further clarification questions by expert community members. This study is the first to investigate the complex task of classifying a question as clear or unclear, i.e., if it requires further clarification. We construct a novel dataset and propose a classification approach that is based on the notion of similar questions. This approach is compared to state-of-the-art text classification baselines. Our main finding is that the similar questions approach is a viable alternative that can be used as a stepping stone towards the development of supportive user interfaces for question formulation.

1 Introduction

The emergence of community question answering (CQA) forums has transformed the way in which people search for information on the web. As opposed to web search engines that require an information seeker to formulate their information need as a typically short keyword query, CQA systems allow users to ask questions in natural language, with an arbitrary level of detail and complexity. Once a question has been asked, community members set out to provide an answer based on their knowledge and understanding of the question. Stack Overflow, Yahoo! Answers and Quora depict popular examples of such CQA websites.

Despite their growing popularity and well established expert communities, increasing amounts of questions remain ignored and unanswered because they are too short, unclear, too specific, hard to follow or they fail to attract an expert member [2]. To prevent such questions from being asked, the prediction of question quality has been extensively studied in the past [1,13,15]. Increasing the question quality has a strong incentive as it directly affects answer quality [9], which ultimately drives the popularity and traffic of a CQA website. However, such attempts ignore the fact that even a high-quality question may

L. Azzopardi et al. (Eds.): ECIR 2019, LNCS 11437, pp. 276–289, 2019.
https://doi.org/10.1007/978-3-030-15712-8_18

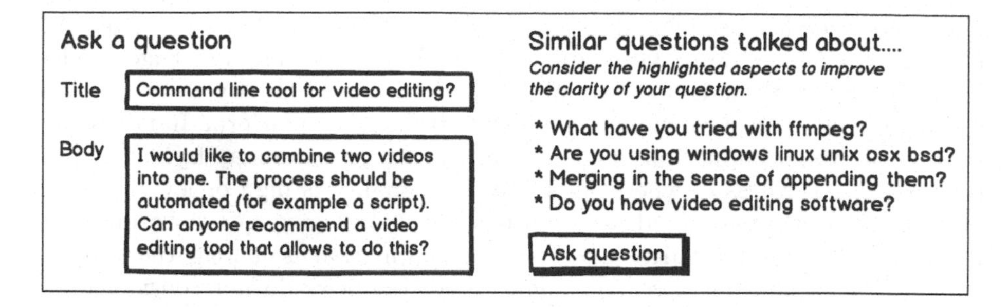

Fig. 1. Envisioned question formulation interface. If a question is found to be unclear, a list of clarification questions (obtained from similar questions) is presented to encourage the user to include information that may be required to provide an answer.

lack an important detail that requires clarification. On that note, previous work attempts to identify what aspects of a question requires editing. While the need for editing can be reliably detected, the prediction of whether or not a question lacks important details has been shown to be difficult [19]. In order to support an information seeker in the formulation of her question and to increase question quality, we envision the following two-step system: (1) determine whether a question requires clarification (i.e., it is unclear), and (2) automatically generate and ask clarifying questions that elicit the missing information. This paper addresses the first step. When successful, the automated identification of unclear questions is believed to have a strong impact on CQA websites, their efforts to increase question quality and the overall user experience; see Fig. 1 for an illustration.

We phrase the unclear question detection as a supervised, binary classification problem and introduce the Similar Questions Model (SQM), which takes characteristics of similar questions into account. This model is compared to state-of-the-art text classification baselines, including a bag-of-words model and a convolutional neural network. Our experimental results show that this is a difficult task that can be solved to a limited extent using traditional text classification models. SQM provides a sound and extendable framework that has both comparable performance and promising options for future extensions. Specifically, the model can be used to find keyphrases for question clarification that may be utilized in a question formulation interface as shown in Fig. 1. Experiments are conducted on a novel dataset including more than 6 million labeled Stack Exchange questions, which we release for future research on this task.[1]

2 Related Work

Previous work on question modeling of CQA forums can be roughly grouped into three categories: *question quality prediction*, *answerability prediction* and *question review prediction* [17]. With respect to the prediction of question quality,

[1] The dataset and sources can be found at https://github.com/jantrienes/ecir2019-qac.

user reputation has been found to be a good indicator [18]. Also, several machine learning techniques have been applied including topic and language models [1,15]. However, there is no single objective definition of quality, as such a definition depends on the community standards of the given platform. In this paper, we do not consider question quality itself, since a question may lack an important detail regardless of whether its perceived quality is high or low. Question answerability has been studied by inspecting unanswered questions on Stack Overflow [2]. Lack of clarity and missing information is among the top five reasons for a question to remain unanswered. Here, we do not consider other problems such as question duplication and too specific or off-topic questions [5]. Finally, question review prediction specifically attempts to identify questions that require future editing. Most notably, Yang et al. [19] determine if a question lacks a code example, context information or failed solution attempts based on its contents. However, they disregard the task of predicting whether detail (e.g., a software version identifier) is missing and limit their experiments to the programming domain.

Clarification questions have been studied in the context of synchronous Q&A dialog systems. Kato et al. [6] analyzed how clarification requests influence overall dialog outcomes. In contrast to them, we consider asynchronous CQA systems. With respect to asynchronous systems, Braslavski et al. [3] categorized clarification questions from two Stack Exchange domains. They point out that the detection of unclear questions is a vital step towards a system that automatically generates clarification questions. To the best of our knowledge, we are the first study to address exactly this novel unclear question detection task. Finally, our study builds on recent work by Rao and Daumé III [14]. We extend their dataset creation heuristic to obtain both clear and unclear questions.

3 Unclear Question Detection

The unclear question detection task can be seen as a binary classification problem. Given a dataset of N questions, $Q = \{q_1, ..., q_N\}$, where each question belongs to either the *clear* or *unclear* class, predict the class label for a new (unseen) question q. In this section, we propose a model that utilizes the characteristics of similar questions as classification features. This model is compared to state-of-the-art text classification models described in Sect. 4.2.

We define a question to be *unclear* if it received a clarification question, and as *clear* if an answer has been provided without such clarification requests. This information is only utilized to obtain the ground truth labels. Furthermore, it is to be emphasized that it is most useful to detect unclear questions during their creation-time in order to provide user feedback and prevent unclear questions from being asked (see the envisioned use case in Fig. 1). Consequently, the classification method should not utilize input signals available only after question creation, such as upvotes, downvotes or conversations in form of comments. Finally, we do not make any assumptions about the specific representation of a question as it depends on the CQA platform at hand. The representation we employ for our experiments on Stack Exchange is given in Sect. 4.1.

3.1 Similar Questions Model

The Similar Questions Model is centered around the idea that similar existing questions may provide useful indicators about the presence or absence of information. For example, consider the two questions in Table 1. It can be observed that the existing question specifies additional information after a clarification question has been raised. A classification system may extract keyphrases (e.g., *operating system*) from the clarification question and check whether this information is present in the given question (see Fig. 1 and Table 7 for examples). In other words, the system checks if a new question lacks information that was also missing from similar previous questions. It has been shown that this general approach can be successfully employed to find and rank suitable clarification questions [14].

Table 1. Example of a new unclear question (Left) and a similar existing question (Right). The left question fails to specify the operating system. The text in italics has been added in response to the shown comment.

Field	New question	Similar (Unclear) question
Title	Simplest XML editor	XML editing/Viewing software
Body	I need the simplest editor with utf8 support for editing xml files; It's for a non programmer (so no atom or the like), to edit existing files. Any suggestion?	What software is recommended for working with and editing large XML schemas? *I'm looking for both Windows and Linux software (doesn't have to be cross platform, just want suggestions for both)*
Tags	xml, utf8, editors	Windows, xml, linux
Comments	–	What operating system?

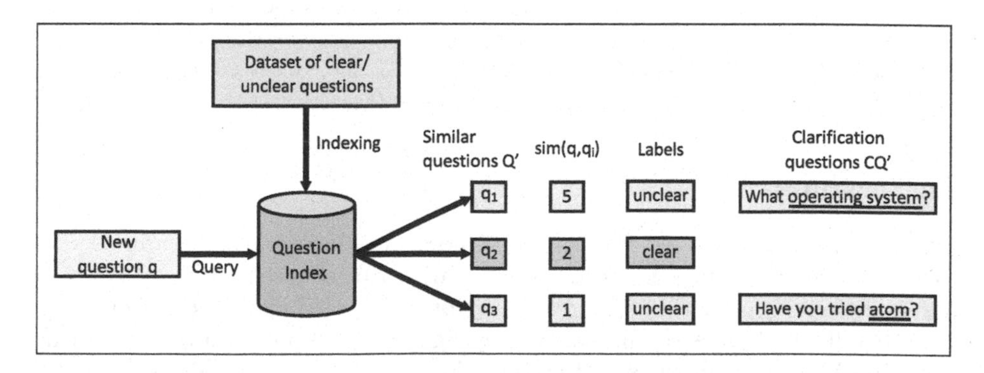

Fig. 2. Illustration of Similar Questions Model. The underlined text of a clarification question indicates a keyphrase.

The Similar Questions Model can be formalized as follows. Given a new question q, we first seek a set of k similar questions $Q' = \{q'_1, q'_2, ..., q'_k\}$ with their clear and unclear labels. As per the definition of unclear that we employ, the subset of unclear questions $Q'_{unclear}$ has a set of M corresponding clarification questions $CQ' = \{cq'_1, cq'_2, ..., cq'_M\}$. Within this framework, we design a number of indicative features that are then used to train a classifier to discriminate between the two classes. An illustration of the model can be found in Fig. 2.

Table 2. Features employed by the Similar Questions Model. The example values in the last column are based on the scenario presented in Table 1 and Fig. 2.

(i) Features based on q		Ex
$Len(q)$	Question length in the number of tokens: $\|q\|$	41
$ContainsPre(q)$	Indicator if question contains preformatted elements	0
$ContainsQuote(q)$	Indicator if question contains a quote	0
$ContainsQuest(q)$	Indicator if question contains question mark "?"	1
$Readability(q)$	Coleman-Liau Index (CLI) [4]	16.7
(ii) Features based on Q'		
$SimSum(q, Q')$	Sum of similarity scores: $\sum_{q' \in Q'} sim_{BM25}(q, q')$	8
$SimMax(q, Q')$	Maximum similarity: $\max_{q' \in Q'} sim_{BM25}(q, q')$	5
$SimAvg(q, Q')$	Average similarity: $\frac{1}{\|Q'\|} \sum_{q' \in Q'} sim_{BM25}(q, q')$	2.7
$LenSim(Q')^a$	Number of similar questions retrieved: $\|Q'\|$	3
$LenUnclear(Q')^a$	Number of similar questions that are unclear: $\|Q'_{unclear}\|$	2
$LenClear(Q')^a$	Number of similar questions that are clear: $\|Q'_{clear}\|$	1
$Majority(Q')^a$	Majority vote of labels in Q'	1
$Ratio(Q')^a$	Ratio between clear/unclear questions: $\|Q'_{clear}\|/\|Q'_{unclear}\|$	0.5
$Fraction(Q')^a$	Proportion of clear questions among similar: $\|Q'_{clear}\|/\|Q'\|$	0.3
(iii) Features based on CQ'^b		
$CQGlobal(q, CQ')$	Cosine similarity between all keyphrases in CQ' and q	0.6
$CQIndividual(q, CQ')$	Sum of cosine similarities between each keyphrase and q	1
$CQWeighted(q, CQ')$	Like above, but weighted by $sim_{BM25}(q, q')$, see Eq. 1	1

[a] These features are computed for the top-k similar questions in Q' where $k = \{10, 20, 50\}$.
[b] CQ' is obtained from the top $k = 10$ similar questions in $Q'_{unclear}$.

3.2 Features

The features employed by the Similar Questions Model can be grouped into three classes: (i) features based on q only, (ii) features based on the set of similar questions Q' and (iii) features based on the set of clarification questions CQ'. See Table 2 for a summary. We highlight the computation of the scoring features obtained from the set of clarification questions (group (iii) in Table 2). For each clarification question in CQ', one or more keyphrases are extracted. These keyphrases are the central objects of a clarification question and refer to an aspect of the original question that is unclear (see Table 7 for examples).

Afterwards, we define $f(a) = (p_1, ..., p_i, ..., p_L)$ to represent a question or clarification question as a vector, where each element indicates the number of times a keyphrase p_i occurs in q and cq', respectively. Then, a *question clarity score* is obtained by computing the cosine similarity between these vectors. The scoring features differ in the way the keyphrase vectors are created. The global model constructs a single vector consisting of all keyphrases present in CQ', whereas the individual model computes the sum of the scores considering each $cq' \in CQ'$ separately. For the individual weighted feature, the final score is given by:

$$CQWeighted(q, CQ') = \sum_{cq' \in CQ'} sim_{cos}(f(q), f(cq')) sim_{BM25}(q, q'), \quad (1)$$

where sim_{cos} is the cosine similarity between the keyphrase vectors and sim_{BM25} is the similarity between q and q'. This gives higher importance to keyphrases belonging to more similar questions.

3.3 Learning Method

We operationalize the Similar Questions Model in a variety of ways:

SimQ Majority. We obtain a simple baseline that classifies q according to the most common label of the similar questions in Q'.

SimQ Threshold. We test the scoring features in group (iii) using a threshold classifier where a threshold γ is learned on a held-out dataset. The label is then obtained as follows:

$$\hat{y} = \begin{cases} 0, & \text{if } feat(q, CQ') \geq \gamma \\ 1, & \text{otherwise,} \end{cases}$$

where $feat(q, CQ')$ is the value of the corresponding feature, 0 refers the clear class and 1 refers to the unclear class.

SimQ ML. All features of the Similar Questions Model are combined and provided as input data to a machine learning classifier.

4 Experimental Setup

This section describes our experimental setup including the dataset and methods.

4.1 Dataset Creation

The Stack Exchange CQA platform depicts a suitable data source for our experiments. It is a network of specialized communities with topics varying from programming to Unix administration, mathematics and cooking. A frequent data dump is published consisting of all questions, answers and comments submitted

Table 3. Dataset statistics and class distribution. N is the number of samples, L the median sample length in tokens, and $|V|$ the vocabulary length after tokenization. $|V^*|$ is the vocabulary length with an imposed minimum term-document frequency of 3.

Community	N	L	$\|V\|$	$\|V^*\|$	Clear	Unclear
Stack Overflow	5,859,667	159	8,939,498	1,319,587	35%	65%
Super User	121,998	121	206,249	45,432	33%	67%
Ask Ubuntu	77,712	114	188,476	40,309	27%	73%
Unix & Linux	44,936	133	162,805	31,852	27%	73%
Cross Validated	38,488	157	130,691	24,229	18%	82%

to the site. For any post, a time-stamped revision history is included. We use this dump[2] to create a labeled dataset consisting of clear and unclear questions.

To obtain unclear questions, we apply a heuristic that has been used in previous research to find clarification questions [3,14]. A question is considered to be unclear when there is a comment by a different user than the original asker and that comment contains a sentence ending with a question mark. This heuristic is not perfect as it will inevitably miss clarification requests not formulated as a question (e.g., *"Please post your code."*), while it retains rhetorical questions (e.g., *"Is this a real question?"*). We only keep those questions where the original asker has provided a clarification in form of a comment or question edit.

In order to gather clear questions, we extend the described heuristic as follows. A question is considered to be clear if it has neither edits, nor comments, but it has an accepted answer. An answer can be manually accepted by the question asker if they consider it to adequately answer their question. Again, this heuristic may introduce noise: an answer can make certain assumptions that would have ideally been asked as a clarification question instead of included in the answer itself (e.g., *"Provided you are on system X, the solution is Y"*).

We apply this heuristic to five Stack Exchange communities, each of a different size and with a different domain. The communities considered are Stack Overflow, Ask Ubuntu, Cross Validated, Unix & Linux and Super User, thus covering a broad range of topics. Table 3 summarizes the statistics of each dataset. The text has been preprocessed by combining the question title, body and tags into a single field, replacing URLs with a special token, converting every character to lowercase and removing special characters except for punctuation. Token boundaries are denoted by the remaining punctuation and whitespace. Furthermore, a minimum term-document frequency of 3 is imposed to prevent overfitting.

4.2 System Components

Obtaining Similar Questions. A general purpose search engine, Elasticsearch, is used with the BM25 retrieval model in order to obtain similar questions. The retrieval score is used as $sim_{BM25}(q, q')$ during feature computation.

[2] Available at https://archive.org/details/stackexchange.

We only index the training set of each community but retrieve similar questions for the entire dataset. Queries are constructed by combining the title and tags of a question. These queries are generally short (averaging 13 tokens).[3] To ensure efficient querying, we remove stopwords from all queries. Finally, BM25 parameters are set to common defaults ($k_1 = 1.2$, $b = 0.75$) [10].

Extracting Keyphrases. Keyphrases are extracted from clarification questions using the RAKE algorithm [16], which is an efficient way to find noun phrases. This algorithm has been used in a similar setting where CQA comments should be matched to related questions [12]. We tokenize the keyphrases and consider each token individually.

Similar Questions Classifier. Besides applying a threshold-based classifier on a selected set of features presented in Table 2, all features are combined to train a logistic regression classifier with L2 regularization (referred to as SimQ ML). The regularization strength is set to $C = 1$ which has been found to work well for all communities. All features are standardized by removing the mean and scaling to unit variance.

4.3 Baseline Models

The Similar Questions Model is compared with a number of baselines and state-of-the-art text classification approaches:

- Random: produce predictions uniformly at random.
- Majority: always predict the majority class (here: *unclear*).
- Bag-of-words logistic regression (BoW LR).
- Convolutional neural network (CNN) [7].

Within the BoW LR model, a question is represented as a vector of TF-IDF weighted n-gram frequencies. Intuitively, this approach captures question clarity on a phrase and topic level. We report model performances for unigrams ($n = 1$) and unigrams combined with phrases of length up to $n = 3$. Using 5-fold cross-validation on the training data, we find that an L2 regularization strength of $C = 1$ works best for all communities. With respect to the CNN model, we use the static architecture variant presented in [7] consisting of a single convolutional layer, followed by a fully connected layer with dropout. Model hyperparameters (number of filters, their size, learning rate and dropout) are optimized per community using a development set.[4] The network is trained with the Adam optimizer [8], a mini-batch size of 64 and early stopping. We train 300-dimensional word embeddings for each community using word2vec [11] and limit a question to its first 400 tokens (with optional padding). Out-of-vocabulary words are replaced by a special token. There are several other possible neural architectures, but an exploration of those is outside the scope of this paper.

[3] We experimented with longer queries that include 100 question body tokens. While computationally more expensive, model performance remained largely unaffected.

[4] Optimal CNN parameter settings can be found in the online appendix of this paper.

4.4 Evaluation

As the data is imbalanced, we evaluate according to the F1 score of the unclear (positive) class and the ROC AUC score. We argue that it is most important to optimize these metrics based on the envisioned use case. When the classification outcome is used as a quality guard in a user interface, it is less sever to consider a supposedly clear question as unclear as opposed to entirely missing an unclear question. We randomly divide the data for each community into 80% training and 20% testing splits. Of the training set, we use 20% of the instances for hyperparameter tuning and optimize for ROC AUC. We experimented with several class balancing methods, but the classification models were not impacted negatively by the (slight) imbalance. Statistical significance is tested using an approximate randomization test. We mark improvements with $^{\vartriangle}(p < 0.05)$ or $^{\blacktriangle}(p < 0.01)$, deteriorations with $^{\triangledown}(p < 0.05)$ or $^{\blacktriangledown}(p < 0.01)$, and no significance by $^{\circ}$.

5 Results and Analysis

This section presents and discusses our experimental results.

5.1 Results

The traditional BoW LR model provides a strong baseline across all communities that outperforms both the random and majority baselines (see Table 5). The generic CNN architecture proposed in [7] does not provide any significant improvements over the BoW LR model. This suggests that a more task-specific architecture may be needed to capture the underlying problem.

Table 4. Results for unclear question detection. The metrics are summarized over the five datasets using both micro-averaging and macro-averaging. F1, precision and recall are reported for the unclear class. Best scores for each metric are in boldface.

	Micro-average				Macro-average			
Method	Acc.	F1	Prec.	Rec.	Acc.	F1	Prec.	Rec.
Random	0.499	0.564	0.649	0.499	0.497	0.586	0.714	0.499
Majority	0.649	0.787	0.649	**1.000**	0.719	0.835	0.719	**1.000**
BoW LR ($n = 1$)	0.687	0.786	0.706	0.886	0.736	0.833	**0.752**	0.933
BoW LR ($n = 3$)	**0.699**	0.791	**0.720**	0.877	**0.741**	**0.837**	**0.752**	0.944
CNN	**0.699**	**0.794**	0.715	0.893	0.739	0.836	0.749	0.947
SimQ Models								
SimQ Majority	0.566	0.673	0.659	0.688	0.676	0.780	0.739	0.826
CQ Global	0.594	0.727	0.645	0.833	0.626	0.753	0.713	0.803
CQ Individual	0.586	0.721	0.640	0.824	0.632	0.761	0.710	0.824
CQ Weighted	0.604	0.737	0.648	0.855	0.642	0.770	0.716	0.838
SimQ ML	0.673	0.781	0.690	0.902	0.728	0.833	0.736	0.960

Table 5. Model performance for a selected set of communities. F1 scores are reported for the unclear class. Significance for model in line $i > 1$ is tested against line $i - 1$. Additionally, significance of each SimQ model is tested against the BoW LR ($n = 3$) model (second marker).

	Cross validated			Super user			Stack overflow		
Method	Acc.	AUC	F1	Acc.	AUC	F1	Acc.	AUC	F1
Random	0.493	0.500	0.618	0.502	0.500	0.575	0.499	0.500	0.563
Majority	0.818▲	0.500○	**0.900**▲	0.669▲	0.500○	0.802▲	0.646▲	0.500○	0.785▲
BoW LR ($n = 1$)	**0.819**○	0.647▲	**0.900**○	0.702▲	0.720▲	0.798○	0.685▲	0.693▲	0.784▼
BoW LR ($n = 3$)	0.818○	**0.659**▲	**0.900**○	**0.709**▲	**0.731**○	**0.807**▲	**0.697**▲	0.718▲	0.788▲
CNN	0.817○	0.626○	0.899○	0.704▽	0.715○	0.803▼	**0.697**○	**0.720**▲	**0.792**▲
SimQ Models									
SimQ Majority	0.796▼	0.584▼	0.883▼	0.639▼	0.616▼	0.738▼	0.561▼	0.515▼	0.667▼
CQ Global	0.718▼▼	0.515▼○	0.830▼▼	0.598▼▼	0.536▼▼	0.733○▼	0.592▲▼	0.520▲▼	0.725▲▼
CQ Individual	0.713○▼	0.513○○	0.827○▼	0.591▼▼	0.549▲▼	0.728▼▼	0.584▼▼	0.528▲▼	0.719▼▼
CQ Weighted	0.696▼▼	0.496▼○	0.812▼▼	0.602▲▼	0.534▼▼	0.739▲▼	0.603▲▼	0.503▼▼	0.736▲▼
SimQ ML	**0.819**▲○	0.631○○	**0.900**▲○	0.687▲▼	0.671▲▼	0.798▲▼	0.670▲▼	0.666▲▼	0.779▲▼

We make several observations with respect to the Similar Questions Model. First, a majority vote among the labels of the top $k = 10$ similar questions (SimQ Majority) consistently provides a significant improvement over the random baseline for all datasets (see Table 5). This simplistic model shows that the underlying concept of the Similar Questions Model is promising. Second, the scoring features that take clarification questions into consideration do not work well in isolation (see models prefixed with CQ in Tables 4 and 5). The assumption that one can test for the presence of keyphrases without considering spelling variations or synonyms seems too strong. For example, the phrase *"operating system"* does not match sentences such as *"my OS is X"* and thus results in false positives. Finally, the SimQ ML model outperforms both the random and majority baselines, and has comparable performance with the BoW LR model. It is to be emphasized that the SimQ ML model, in addition to classifying a question as clear or unclear, generates several valuable hints about the aspects of a question that may be unclear or missing (see demonstration in Table 7). This information is essential when realizing the envisioned user interface presented in Fig. 1, and cannot be deducted from the BoW LR or CNN models.

5.2 Feature Analysis

To gain further insights about the performance of the Similar Questions Model, we analyze the features and their predictive power. Features considering the stylistic properties of a question itself such as the length, readability and whether or not the question contains a question mark, are among the top scoring features (see Table 6). Other important features include the distribution of labels among the similar questions and their retrieval scores (*LenUnclear*, *LenClear*, *SimSum*, *Fraction*). With respect to the bag-of-words classifier, we observe that certain question topics have attracted more unclear questions. For example, a question about *windows 10* is more likely to be unclear than a question about *emacs*. Interestingly, also stylistic features are captured (e.g., a "?" token

Table 6. Subsets of learned coefficients for the Similar Questions Model (Left) and BoW LR classifier (Right). Both are trained on the Super User community. Positive numbers are indicative for the unclear class and negative values for the clear class.

SimQ ML		BoW LR ($n = 3$)	
Coef	Feature	Coef	Feature
+0.269	$Len(q)$	+4.321	windows 10
+0.166	$CQIndividual(q, CQ')$	+3.956	<URL>
+0.146	$LenUnclear(Q')$ $(k = 50)$	+3.229	problem
+0.120	$LenSim(Q')$ $(k = 20)$	+2.413	nothing
+0.094	$SimSum(q, Q')$	+2.150	help me
+0.081	$Readability(q)$	+1.760	unable to
–0.091	$ContainsPre(q)$	–1.488	documentation
–0.114	$CQIndividual(q, CQ')$	–1.742	difference between
–0.150	$LenClear(Q')$ $(k = 50)$	–1.822	can i
–0.150	$Fraction(Q')$ $(k = 50)$	–1.841	emacs
–0.171	$ContainsQuest(q)$	–3.026	vista
–0.196	$SimMax(q, Q')$	–5.306	?

and the special URL token). Finally, this model reveals characteristics of well-written, clear questions. For example, if a user articulates their problem in the form of *"difference between X and Y,"* such a question is more likely to belong to the clear class. This suggests that it may be beneficial to include phrase-level features in the Similar Questions Model to improve performance.

5.3 Error Analysis and Limitations

The feature analysis above reveals a problem which is common to both the Similar Questions Model and the traditional BoW LR model. Both models suffer from a topic bias. For example, a question about *emacs* is more likely to be classified as clear because the majority of *emacs* questions are clear. Furthermore, stylistic features can be misleading. Consider a post on Stack Overflow that contains an error message. This post does not require an explicit use of a question mark as the implied question most likely is *"How can this error message be fixed?"*. It is conceivable to design features that take such issues into account.

A potential limitation of the Similar Questions Model is its reliance on the existence of similar questions within a CQA website. It is unclear how the model would perform in the absence of such questions. It would make an interesting experiment to process a CQA dataset in chronological order, and measure how the model's effectiveness changes as more similar questions become available over time. However, we leave the exploration of this idea to future work.

Table 7. Example questions and their clarification questions retrieved by Similar Questions Model. The numbers in parenthesis indicate retrieval score $sim_{BM25}(p, q_i)$. Highlighted text corresponds to the keyphrases extracted by RAKE.

Field	Text
Title	Laptop randomly going in hibernate
Body	I have an Asus ROG G751JT laptop, and a few days ago my battery has died. The problem that I am encountering is that my laptop randomly goes to sleep after a few minutes of use even when plugged in [...].
ClarQ	(20.01) Does this happen if you boot instead from an ubuntu liveusb ?
	(17.92) Did you enable allow wake timers in power options sleep ?
	(16.88) Can you pop the battery out of the mouse ?
	(16.64) Which OS are you using?
	(16.02) Have you scanned your system for malwares ?
Title	Does ZFS make sense as local storage?
Body	I was reading about ZFS and for a moment thought of using it in my computer, but than reading about its memory requirements I thought twice. Does it make sense to use ZFS as local or only for servers used as storage?
ClarQ	(36.11) What's wrong with more redundancy ?
	(31.41) What kind of data are you trying to protect ?
	(30.77) How are you planning on doing backups and or disaster recovery ?
	(29.70) Is SSD large enough?

6 Conclusion

The paper represents the first study on the challenging task of detecting unclear questions on CQA websites. We have constructed a novel dataset and proposed a classification method that takes the characteristics of similar questions into account. This approach encodes the intuition that question aspects which have been missing or found to be unclear in previous questions, may also be unclear in a given new question. We have performed a comparison against traditional text classification methods. Our main finding is that the Similar Questions Model provides a viable alternative to these models, with the added benefit of generating cues as to why a question may be unclear; information that is hard to extract form traditional methods but that is crucial for supportive question formulation interfaces.

Future work on this task may combine traditional text classification approaches with the Similar Questions Model to unify the benefits of both. Furthermore, one may start integrating the outputs of the Similar Questions Model into a clarification question generation system, which at a later stage is embedded in the user interface of a CQA site. As an intermediate step, it would be important to evaluate the usefulness of the generated cues as to why a question is unclear. Finally, the work by Rao and Daumé III [14] provides a natural extension, by ranking the generated clarification questions in terms of their expected utility.

Acknowledgments. We would like to thank Dolf Trieschnigg and Djoerd Hiemstra for their insightful comments on this paper. This work was partially funded by the University of Twente Tech4People Datagrant project.

References

1. Arora, P., Ganguly, D., Jones, G.J.F.: The good, the bad and their kins: identifying questions with negative scores in stackoverflow. In: Proceedings of the 2015 IEEE/ACM International Conference on Advances in Social Networks Analysis and Mining 2015, ASONAM 2015, pp. 1232–1239. ACM, New York (2015)
2. Asaduzzaman, M., Mashiyat, A.S., Roy, C.K., Schneider, K.A.: Answering questions about unanswered questions of stack overflow. In: Proceedings of the 10th Working Conference on Mining Software Repositories, MSR 2013, pp. 97–100. IEEE Press, Piscataway (2013)
3. Braslavski, P., Savenkov, D., Agichtein, E., Dubatovka, A.: What do you mean exactly?: analyzing clarification questions in cqa. In: Proceedings of the 2017 Conference on Conference Human Information Interaction and Retrieval, CHIIR 2017, pp. 345–348. ACM, New York (2017)
4. Coleman, M., Liau, T.L.: A computer readability formula designed for machine scoring. J. Appl. Psychol. **60**(2), 283 (1975)
5. Correa, D., Sureka, A.: Chaff from the wheat: characterization and modeling of deleted questions on stack overflow. In: Proceedings of the 23rd International Conference on World Wide Web, WWW 2014, pp. 631–642. ACM, New York (2014)
6. Kato, M.P., White, R.W., Teevan, J., Dumais, S.T.: Clarifications and question specificity in synchronous social q&a. In: CHI 2013 Extended Abstracts on Human Factors in Computing Systems, CHI EA 2013, pp. 913–918. ACM, New York (2013)
7. Kim, Y.: Convolutional neural networks for sentence classification. In: Proceedings of the 2014 Conference on Empirical Methods in Natural Language Processing, EMNLP 2014, 25–29 October 2014, Doha, Qatar, A meeting of SIGDAT, a Special Interest Group of the ACL, pp. 1746–1751 (2014)
8. Kingma, D.P., Ba, J.L.: Adam: a method for stochastic optimization. In: Proceedings of the 3rd International Conference on Learning Representations (ICLR) (2015)
9. Li, B., Jin, T., Lyu, M.R., King, I., Mak, B.: Analyzing and predicting question quality in community question answering services. In: Proceedings of the 21st International Conference on World Wide Web, WWW 2012 Companion, pp. 775–782. ACM, New York (2012)
10. Manning, C.D., Raghavan, P., Schütze, H.: Introduction to Information Retrieval. Cambridge University Press, New York (2008)
11. Mikolov, T., Sutskever, I., Chen, K., Corrado, G., Dean, J.: Distributed representations of words and phrases and their compositionality. In: Proceedings of the 26th International Conference on Neural Information Processing Systems, NIPS 2013, vol. 2, pp. 3111–3119. Curran Associates Inc., USA (2013)
12. Nandi, T., et al.: IIT-UHH at SemEval-2017 Task 3: exploring multiple features for community question answering and implicit dialogue identification. In: Proceedings of the 11th International Workshop on Semantic Evaluation (SemEval-2017), pp. 90–97. Association for Computational Linguistics (2017)

13. Ponzanelli, L., Mocci, A., Bacchelli, A., Lanza, M.: Understanding and classifying the quality of technical forum questions. In: Proceedings of the 2014 14th International Conference on Quality Software, QSIC 2014, pp. 343–352. IEEE Computer Society, Washington (2014)
14. Rao, S., Daumé III, H.: Learning to ask good questions: ranking clarification questions using neural expected value of perfect information. In: Proceedings of the 56th Annual Meeting of the Association for Computational Linguistics (Volume 1: Long Papers), pp. 2737–2746. Association for Computational Linguistics (2018)
15. Ravi, S., Pang, B., Rastagori, V., Kumar, R.: Great question! question quality in community q&a. Int. AAAI Conf. Weblogs Social Media **1**, 426–435 (2014)
16. Rose, S., Engel, D., Cramer, N., Cowley, W.: Automatic keyword extraction from individual documents. Text Min. Appl. Theory, 1–20 (2010)
17. Srba, I., Bielikova, M.: A comprehensive survey and classification of approaches for community question answering. ACM Trans. Web **10**(3), 18:1–18:63 (2016). ISSN 1559–1131
18. Tausczik, Y.R., Pennebaker, J.W.: Predicting the perceived quality of online mathematics contributions from users' reputations. In: Proceedings of the SIGCHI Conference on Human Factors in Computing Systems, CHI 2011, pp. 1885–1888. ACM, New York (2011)
19. Yang, J., Hauff, C., Bozzon, A., Houben, G.-J.: Asking the right question in collaborative q&a systems. In: Proceedings of the 25th ACM Conference on Hypertext and Social Media, HT 2014, pp. 179–189. ACM, New York (2014)

Local and Global Query Expansion for Hierarchical Complex Topics

Jeffrey Dalton[1(✉)], Shahrzad Naseri[2], Laura Dietz[3], and James Allan[2]

[1] School of Computing Science, University of Glasgow, Glasgow, UK
jeff.dalton@glasgow.ac.uk

[2] Center for Intelligent Information Retrieval, College of Information and Computer Sciences, University of Massachusetts Amherst, Amherst, USA
{shnaseri,allan}@cs.umass.edu

[3] Department of Computer Science, University of New Hampshire, Durham, NH, USA
dietz@cs.unh.edu

Abstract. In this work we study local and global methods for query expansion for multifaceted complex topics. We study word-based and entity-based expansion methods and extend these approaches to complex topics using fine-grained expansion on different elements of the hierarchical query structure. For a source of hierarchical complex topics we use the TREC Complex Answer Retrieval (CAR) benchmark data collection. We find that leveraging the hierarchical topic structure is needed for both local and global expansion methods to be effective. Further, the results demonstrate that entity-based expansion methods show significant gains over word-based models alone, with local feedback providing the largest improvement. The results on the CAR paragraph retrieval task demonstrate that expansion models that incorporate both the hierarchical query structure and entity-based expansion result in a greater than 20% improvement over word-based expansion approaches.

1 Introduction

Current web search engines incorporate question answer (QA) results for a significant fraction of queries. These QA results are a mixture of factoid questions ["Who won the James Beard Award for best new chef 2018?"] that can be answered from web results or from entity-based knowledge bases. However, many questions require more than short fact-like responses. In particular, topics like ["What are the causes of the Civil War?"] require multifaceted essay-like responses that span a rich variety of subtopics with hierarchical structure: Geography and demographics, States' rights, The rise of abolitionism, Historical tensions and compromises, and others. These 'complex' and multifaceted topics differ significantly from simple factoid QA information needs.

Hierarchical complex topics have rich structure that could (and should) be leveraged for effective retrieval. The first type of structure is the hierarchical nature of the topics. They start with a root topic and contain more specific

L. Azzopardi et al. (Eds.): ECIR 2019, LNCS 11437, pp. 290–303, 2019.
https://doi.org/10.1007/978-3-030-15712-8_19

subtopics in a hierarchy. We propose fine-grained methods that perform expansion both at the overarching level as well as for each of the subtopics individually.

One of the fundamental issues is the well-studied problem of vocabulary mismatch. Retrieval from a paragraph collection exacerbates this problem because the content retrieved is short passages that may be taken out of context. In particular, the discourse nature of language in related series of paragraphs on the same topic makes extensive use of language variety and abbreviations. Neural ranking models address the vocabulary mismatch problem by learning dense word embedding representations [11,14,15]. In this work we use learned 'global' embedding models that use joint embeddings of both words and entities. We also experiment with 'local' models derived from pseudo-relevance feedback expansion approaches.

Our proposed expansion methods incorporate expansion features that can be used in a first-pass retrieval model. The state-of-the-art neural network models [11,15] for complex topics re-rank a candidate set of paragraphs retrieved from a simple and fast first pass baseline model, most commonly BM25. However, these simple models often fail to return relevant paragraphs in the candidate pool. The consequence is that multi-pass reranking models are limited by the (poor) effectiveness of the underlying first-pass retrieval. As a result, Nanni et al. [15] find that the improvement over the non-neural baseline models is marginal for retrieving paragraphs for complex topics. Our proposed expansion methods address this fundamental problem. In contrast to neural approaches, we perform expansion natively in the search system and combine the features using linear Learning-to-Rank (LTR) methods.

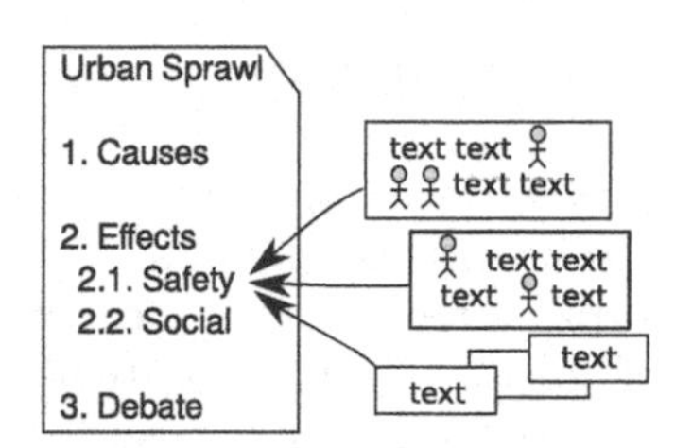

Fig. 1. Example of a complex topic from the TREC Complex Answer Retrieval track

Complex multifaceted queries (topics) of this nature are currently being studied in the TREC Complex Answer Retrieval (CAR) track. In TREC CAR, the task is **given a complex topic composed of different subtopics - a skeleton of a Wikipedia article, the system should retrieve: (1) relevant paragraphs and (2) relevant entities for each subtopic.** The first CAR task is passage (paragraph) retrieval to find relevant paragraphs for each subtopic. The second task is entity retrieval, to identify important people, locations, and other concepts that should be mentioned in the synthesis. An example of a complex topic is given in Fig. 1.

For TREC CAR, the nature of complex topics is particularly entity-centric because the subtopics are based around entities from a knowledge base. Further, recent test collections based on Wikipedia paragraphs include rich text and entity representations [7]. As a result, this topic collection is an interesting domain for models that incorporate both text and entity representations of queries and documents [4,23,24].

In this work we make several contributions to methods and understanding for expansion in complex, multifaceted, and hierarchical queries:

- We develop entity-aware query expansion methods for passage retrieval. We use probabilistic retrieval approaches and entity embedding vectors with entity-aware indicators including entity identifiers, entity aliases, and words. Entity-aware models for different levels of the topics are combined with a LTR approach.
- The experimental evaluation demonstrates that our entity-aware approach outperforms a learned combination of probabilistic word-based models by 20%. It further outperforms the best performing approach from the TREC CAR year one evaluation.

The remainder of the paper is structured as follows. First, we provide background and related work on TREC CAR as well as the broader area of entity-focused query expansion. Next, we introduce the existing and newly proposed expansion model for complex hierarchical topics. Finally, we perform an empirical study evaluation on the TREC CAR paragraph retrieval task evaluating the effectiveness of a variety of local and global expansion methods.

2 Background and Related Work

Question Answering. Although they may not be not strictly formulated as questions, retrieval for complex topics is related to approaches that answer questions from web content. Retrieval techniques for effective question answering is undergoing a resurgence of research, with a particular interest in non-factoid QA [2,3]. These works are similar to complex answer retrieval in that they perform question answering by retrieving relevant passages, in particular paragraphs from Wikipedia. The key difference with the current work is that their topics are a single question with one answer. In contrast, the complex topic retrieval addressed in this work focuses on comprehensive complex answers with topics that have explicit multifaceted hierarchical relationships.

TREC CAR. The TREC Complex Answer Retrieval track was introduced in 2017 to address retrieval for complex topics. For a survey of approaches, see Nanni et al. [15] as well as the overview [7]. Nanni et al. evaluate a variety of models, including a leading neural ranker (Duet model) [14]. They find that while the neural network gives the best performance, the gains over leading retrieval approaches are only modest. Another neural model, PACRR, by MacAvaney et al. [11] is consistently shown to improve effectiveness on CAR, we use this as one of our baseline methods. In all cases in the 2017 evaluation, BM25 is used to create candidate sets for reranking. However, BM25 fundamentally limits the effectiveness of reranking runs by constraining the candidate pool. In this work, we study methods for expansion that may be used for feature-based reranking and that incorporate entity-based representations.

Structured Queries. Complex queries in retrieval is not a new problem. In fact, some of the earliest uses of retrieval focused on boolean retrieval. Users constructed complex boolean expressions with complex subqueries [20]. This was later followed up with more complex query capability [21]. Follow-up query languages that support rich query expressions include: INQUERY, Lucene, Terrier, and Galago. However, these languages are usually only used internally to rewrite simple keyword queries, possibly using some inferred structure from natural language processing. In contrast, CAR query topics contain explicit multifaceted hierarchical structure. We test various ways of using this structure in expansion models.

Relevance Feedback Expansion Models. One of the fundamental challenges in retrieval is vocabulary mismatch and one of the primary mechanisms to address this problem is relevance feedback that takes a user judgment of a document and uses this to build an updated query model. Pseudo-relevance feedback (PRF) [1,9] approaches perform this task automatically, assuming the top documents are relevant. We build on previous work that uses mixtures of relevance models [12], but apply it to creating fine-grained expansions from complex hierarchical topic headings. Further, as the results in this work demonstrate, PRF is most effective when there is a high density of relevant documents in top ranked results. In contrast for CAR, there are few relevant documents that are often not retrieved in first-pass retrieval. To overcome this issue we propose using a fine-grained score-based fusion approach and we utilize entity-based expansion features. The results demonstrate that our approach using external entity-based features is more robust than word-based approaches.

Embedding-based Expansion Models. Another approach to overcome the word mismatch problem is using global collection word embeddings. Word embedding techniques learn a low-dimensional vector (compared to the vocabulary size) for each vocabulary term in which the similarity between the word vectors captures the semantic as well as the syntactic similarities between the corresponding words. Word embeddings are unsupervised learning methods since they only need raw text data without other explicit labels. Xiong et al. propose a model for ad-hoc document retrieval that represents documents and queries in both text and entity spaces, leveraging entity embeddings in their approach [23]. In this work we use joint entity-word embedding models to perform global term expansion.

Knowledge-base Expansion Models. Recent previous work demonstrates that query expansion using external knowledge sources and entity annotations can lead to significant improvements to a variety of retrieval tasks [4], including entity linking of queries [8], and using entity-derived language models for document representation [18]. There is also recent work on determining the salience of entities in documents [24] for ranking. Beyond salience, research focused on identifying latent entities [10,22] and connecting the query-document vocabularies in a latent space. We build on these entity-centric representations and utilize entity query annotations, explicit entity links, and related entities from entity feedback and entity embedding models. We study the differences between these different elements for the CAR task. For an overview of work in this area we refer the reader to [6].

3 Methodology

3.1 Complex Hierarchical Topics

A complex topic T consists of heading nodes constructed in a hierarchical topic tree, an example is shown in Fig. 1. Each heading node, h, represents the subtopic elements. For example, a complex topic with subtopics delimited by a slash would be: "Urban sprawl/Effects/Increased infrastructure and transportation cost". This consists of three heading nodes - the leaf heading is "Increased infrastructure and transportation cost" with the root heading "Urban sprawl" and intermediate heading "Effects". The tree structure provides information about the hierarchical relationship between subtopics. In particular, the most important relationship is that the root heading is the main focus of the overall topic.

Given a complex topic tree T, the outline consists of a representation for each of the subtopic heading nodes $h \in H$. At the basic level, each heading contains its word representation from text, $W : \{w_1, ..., w_k\}$, a sequence of words in the subtopic. Beyond words, each heading can also be represented by features extracted by information extraction and natural language processing techniques, for example part of speech tags and simple dependence relationships.

In particular, we hypothesize that another key element of effective retrieval for complex topics requires going beyond words to include entities and entity relationships. Therefore, we propose representing the topic as well as documents with entity mentions, T_M and D_M respectively, where each has $M : \{m_1, ..., m_k\}$ with m_k a mention of an entity e in a knowledge base. Given an entity-centric corpus and task along with rich structure, the mix of word and entity representation offers significant potential for retrieval with complex topics. The result is sequence of ordered entities within a heading with provenance connecting the entity annotations to free text. In TREC CAR as well as adhoc document retrieval, this representation is (partially) latent - it must inferred from the topic text.

3.2 Topic Expansion Model

In this work, we study use of different expansion methods over diverse types of representations, based on words and entities. To specify the representations we use different term vocabularies, $v \in V$, for example:

- Words, $W : \{w_1, ..., w_k\}$ are unigram words from the collection vocabulary.
- Entities, $E : \{e_1, ..., e_k\}$ are entities from a knowledge base, matched based on their entity identifiers.

Note that entities may have multiple vocabularies that interact with one another. We can match entities to word representations using the entity names and aliases $A : \{a_1, ..., a_k\}$ derived from their Wikipedia name, anchor text, redirects, and disambiguation pages.

We study two expansion approaches: (1) expansion based on local query-specific relevance feedback and (2) expansion based on global word-entity embedding similarity. We elaborate on these approaches in Sects. 3.3 and 3.4, respectively.

To perform effective expansion, our goal is to estimate the probability of relevance for an entry in the vocabulary with respect to the complex topic, T. In other words, regardless of the underlying expansion method, the overarching goal is to identify the latent representation of the topic across all vocabulary dimensions: $p(V|T)$. However, a single expansion model for an entire complex topic is unlikely to be effective. For both expansion methods we also build a mixture of fine-grained expansions for each subtopic node that are combined. For every type in the vocabulary V, and for every heading node $h \in H$, we create a feature, $f(h, D)$.

In Table 1 we illustrate different approaches for expansion that include three dimensions of the expansion: the expansion method, the representation type, and which subtopic to expand. An example is, [Antibiotic use in livestock/Use in different livestock/In swine production]. In this case, $R =$ [Antibiotic use in livestock] is the root, $I =$ [Use in different livestock] is an intermediate node, and $H =$ [In swine production] is the leaf heading. We vary the topic representation using differing combinations of these three elements. The most common approach by participants in TREC CAR is to simply concatenate the RIH context into one query and to ignore the heading relationships or boundaries. In contrast, our fine-grained method preserves these elements and handles them separately.

We use a simple and effective method for combining heading evidence up to the topic-level. Features are combined using a log-linear model with parameters, θ. The number of these features is limited to approximately 10. This scale allows it to be learned efficiently using coordinate ascent to directly optimize the target retrieval metric. All of the score-level features, both heading derived and feedback, correspond to queries that can be expressed natively in the first pass matching phase of a search system.

Table 1. Examples of topic expansion features across word and entity vocabularies. All features are for R, I, and H nodes separately. The example topic is: [Antibiotic use in livestock/Use in different livestock/In swine production]. The entities identified in the topic are: [Antibiotics, Livestock/ Livestock/ Domestic pig, Pig farming

Name	Description	Feature example
RIH-QL	Representing words from the root, intermediate, and leaf subtopics	(antibiotic use livestock different swine production)
RIH-IDs-Embed	Representing expanded entities from global embeddings from the root, intermediate, and leaf subtopics using their IDs	Antibiotics → Tetracycline.id Livestock → Cattle.id Pig farming → (Animal husbandry).id
H-Names-Embed	Expansion of entity names within the leaf subtopic using global embeddings	Pig farming → (animal husbandry dairy farming poultry ubre blanca)
R-Aliases-Embed	Expansion of aliases of entity within the root subtopic using global embeddings	Tetracycline → (tetracyn sumycin hydrochloride) Cattle → (cow bull calf bovine heifer steer moo)

3.3 Relevance Model Expansion

Lavrenko and Croft introduce relevance modeling, an approach to query expansion that derives a probabilistic model of term importance from documents that receive high scores, given the initial query [9]. In our model, we derive a distribution over all types of the vocabulary. In this case, $p(D = d|T)$ is the relevance of the document to the topic, derived from score for the document under the query model. The $p(V|d)$ is the probability of the vocabulary under the language model of the document using that representation.

3.4 Embedding-Based Expansion

In this section, we first elaborate how we learn the global embeddings. We then explain how we use the learned model for expanding complex queries.

Joint Entity-Word Embeddings. Motivated by the vocabulary mismatch problem, we learn a joint entity-word embedding following the approach presented by Ni et al. [16]. We learn a low dimensional vector representation for entities and words based on the Mikolov Skip-gram model [13] using term co-occurrence information within a text. Each entity mention is considered as a single "term". The Skip-gram model aims to maximize the probability of surrounding context terms based on the current term using a neural network. We thus model entities using their word context (and vice versa).

The following excerpt shows the transformation of text with entity mentions using special placeholders for each entity mention:

> `The_World_Health_Organization_(WHO)` is a `specialized_agency_of_the_United_Nations` that is concerned with international `public_health`. It was established on 7 April 1948, and is headquartered in `Geneva, Switzerland`.

We build a mixture of fine-grained expansions for each subtopic in a complex topic. We compute embedding-based similarity for both explicit entity mentions as well as words, two types from the vocabulary. For the global similarity between dense embedding vectors we use the cosine similarity. In addition to expanding each subtopic node individually, we also perform expansion of the complete topic tree as a whole. The embedding vector of a node (or entire query tree) is represented as the average (mean) of the embedding vector of each element within it.

4 Experimental Setup

4.1 Data

The primary dataset used for experiments is from the TREC Complex Answer Retrieval (CAR) track, v2.1 [5], released for the 2018 TREC evaluation. The CAR data is derived from a dump of Wikipedia from December 2016. There are 29,678,367 paragraphs in the V2 paragraph collection.

Each outline consists of the hierarchical skeleton of a Wikipedia article and its subtopics. Each individual heading is a complex topic for which relevant content (paragraphs) needs to be retrieved. In 2017, the test topics are chosen from articles on open information needs, i.e., not people, not organizations, not events, etc. The benchmark consists of 250 topics, split equally (roughly) into train and test sets.

The TREC CAR setup includes two types of heading-level judgments, *automatic* and *manual*. The automatic (binary) judgments are derived directly from Wikipedia and the manual judgments are created by NIST assessors. A key outcome from 2017 was that the automatic benchmark data is useful for differentiating between systems and not subject to the pooling bias in manual judgments (it's also much larger) [7]. In this work, we use the automatic judgment to evaluate our methods because the original retrieval methods were not in the pool and we found that the manual judgments had a high degree of unjudged results even for the baselines.

Knowledge Base. For the experiments here we use the non-benchmark articles from Wikipedia as a knowledge base. These include the full article text, including the heading structure. It does not include the infobox and other data that was excluded in the CAR pre-processing. In addition to the text, we use anchor text, redirects, and disambiguation metadata derived from the article collection and provided in the data.

Evaluation Measures. We use the standard measures reported in TREC CAR evaluations. The primary evaluation measure is Mean Average Precision (MAP). We report R-Precision, because the number of relevant documents in TREC CAR varies widely across topics. The NDCG@1000 metric is included following standard practice in the track. For statistical significance, we use a paired t-test and report significance at the 95% confidence interval.

4.2 System Details

In this section we provide additional details of the systems used for the implementation. The TREC CAR paragraph collection is indexed using the Galago[1] retrieval system, an open-source research system. The query models and feedback expansion models are all implemented using the Galago query language. The paragraphs are indexed with the link fields to allow exact and partial matches of entity links in the paragraphs. Stopword removal is performed on the heading queries using the 418 INQUERY stop word list. Stemming is performed using the built-in Krovetz stemmer.

In our score fusion model we use a log-linear model combination of different features for ranking. The model parameters, θ are optimized using coordinate ascent to directly optimize the target retrieval measure, Mean Average precision (MAP). The implementation of the model is available in the open-source RankLib learning-to-rank library.

Parameter Settings. For the experiments we use the provided train/test topic splits. We tune the retrieval hyper-parameters on the training data using grid search. For the Sequential Dependence Model (SDM) baseline parameters are $mu = 1200$, $uww = 0.02$, $odw = 0.10$, and $uniw = 0.82$. We observe that these parameters differ from the default settings which are optimized for short adhoc TREC queries and longer newswire documents. In contrast, the paragraph content in the CAR collection are much shorter. For relevance feedback, we use the SDM model as the baseline retrieval. The expansion parameters are tuned similarly and we find that 10 expansion documents with 20 feedback terms and an interpolation weight of 0.8 is most effective.

Query Entity Annotation. The topics in TREC CAR do not have explicit entity links. To support matching paragraph entity documents, we annotate the complex topic headings with entities. Entity linking is performed on each heading for both the train and test benchmark collections. We use the open-source state-of-the-art SMAPH entity linker[2]. Although not the main focus of the paper, we observe that the entity linker suffers from significant recall issues, missing a large fraction of the entities in the complex topic headings, which are directly derived from Wikipedia entity titles. As a result, the utility of explicit query entity links is lower than we expected.

---

[1] http://www.lemurproject.org/galago.php.
[2] https://github.com/marcocor/smaph.

Table 2. Text-based baselines and expansion methods. * indicates significance over the RH-SDM run.

Model	MAP	R-Prec	NDCG
RIH-QL	0.110	0.088	0.228
RH-SDM	0.132	0.109	0.248
RH-SDM-RM3	0.127	0.102	0.243
L2R-SDM-RM3	0.142*	0.107	0.257*
Embedding-Term	0.143*	0.119*	0.261*
GUIR (neural)	0.137	0.112	0.237
GUIR-Exp (neural)	0.142*	0.117	0.242

Table 3. Baseline: combinations of SDM and RM3 over different outline levels combined with L2R. Learned feature combination weights displayed.

Model	Weight
RIH-QL	0.288
R-SDM	0.153
H-SDM	0.340
RH-SDM	0.108
RH-SDM-RM3	0.110

Document Entity Annotations. For entity mentions in documents we use the existing entity links provided in Wikipedia. We note that the entity links in Wikipedia are sparse and biased. By convention only the first mention of an entity in an article is annotated with a link. This biases retrieval based on entity identifiers towards paragraphs that occur early in a Wikipedia article. An area for future work is to perform entity annotation on the documents to improve mention recall. For example one known issue in the current setup is that many mentions that use abbreviations are not currently linked, thereby limiting the effectiveness of entity link approaches.

Learning Embeddings. The joint entity-word embeddings are learned from the DBpedia 2016-10 full article dump. To learn the entity embeddings we use the Word2Vec implementation in gensim [19] version 3.4.0 with parameters as follow: window-size = 10, sub-sampling = 1e–3, and cutoff min-count = 0. The learned embedding dimension is equal to 200 and we learned embeddings of 3.0M entities out of 4.8M entities available in Wikipedia.

5 Results

In this section we present our main experimental results. We start with proven word-based retrieval and expansion methods. This includes state-of-the-art neural baselines. We then build on these methods and experiment with local and global entity-based expansion.

5.1 Word-Based Retrieval and Expansion

We first evaluate standard text retrieval methods for heading retrieval. The results are shown in Table 2. The baseline model, RIH-QL, is a standard bag-of-words query-likelihood model [17] on all terms in the topic. All other runs show statistically significant gains over this simple baseline. The table also shows results for the Sequential Dependence Model (SDM) that uses the root and leaf subtopics of the heading. We also experimented with other variations (H-QL, RIH-SDM, RH-QL, etc...), but these are all outperformed by RH-SDM. RH-SDM was the best performing unsupervised model for this collection in TREC 2018. We also evaluate using a relevance model term-based expansion on top of the best SDM run. We find that the RM3 performance is insignificantly worse than the SDM baseline, demonstrating the PRF based on words is challenging in this environment. We attribute this to the sparseness of relevant paragraphs to the topics, an average of 4.3 paragraphs per topic, with baselines retrieving on average about half of those, 2.2.

We experimented with combining the baseline systems with additional fine-grained SDM components from each part of the query (subtopic) separately and weighting and combining them into a linear model, the L2R-SDM-RM3 method. The features and learned weights are given in Table 3. We observe that the H-SDM feature is the most important, putting greater emphasis on the leaf subtopic (approximately 2x the root topic). Combining these baseline retrieval and subtopic heading components results in significant gains over all the models individually, including RH-SDM. The Embedding-Term method is L2R-SDM-RM3 with addition of global word expansion. The results show a small, but insignificant improvement to the model effectiveness.

The bottom of Table 2 shows a comparison with one of the leading neural ranking models from the Georgetown University IR group (GUIR). It uses the PACRR neural ranking architecture modified with heading independence and heading frequency context vectors [11]. The second row (Exp) adds expansion words of the topic's query terms. Interestingly, the learned GUIR neural run does do not improve significantly over the RH-SDM baseline, the SDM model even slightly outperforms it on NDCG. The learned word-based expansion methods L2R-SDM-RM3 and Embedding-Term are both statistically significant over the GUIR base run for MAP, but not statistically significantly different from the Exp run. This indicates that our methods are comparable to state-of-the-art word-based expansion models using deep learning for this collection.

Table 4. Entity-based expansion with varying latent entity models. * indicates significance over the L2R-SDM-RM3 Baseline.

Model	MAP	R-Prec	NDCG
L2R-SDM-RM3 Baseline	0.142	0.107	0.257
Entity_Embedding	0.154	0.127	0.277
Entity_Retrieval	0.160*	0.133	0.284*
Entity_Collection_PRF	0.172*	0.146*	0.297*

5.2 Entity Expansion

In this section we study combining the previous word-based representations with entity representations. We use entities annotated in the query as well as inferred entities from local and global sources: global embeddings, local entity retrieval, and local pseudo-relevance feedback on the paragraph collection. Each of the entity expansion models is a learned combination of subtopic expansions across the different entity vocabularies (identifiers, names, aliases, and unigram entity language models).

The results are shown in Table 4. The baseline method is L2R-SDM-RM3, the learning to rank combination of all word-based expansion features. Each entity model adds additional entity features to this baseline. The results show that adding entity-based features improves effectiveness consistently across all entity inference methods. There are benefits to using global entity embeddings, but they are not significant over the baseline. The local retrieval and collection PRF expansion models both result in significant improvements over the baseline. In particular, the collection entity representation shows the largest effectiveness gains. Additionally, all of the entity-based expansion methods show statistically significant improvements over the GUIR-Exp word-based expansion run.

We find that all entity-expansion methods consistently improve the results. When compared with the baseline word model they have a win-loss ratio varying from 2.6 up to 4.6. The best method based on collection feedback has 281 losses, 1300 wins, with a win-loss ratio of 4.6. In contrast, the win-loss ratio for the GUIR-Exp model is 1.1, hurting almost as many queries as it helps. Consequently, we conclude that entity-based expansion methods more consistently improve effectiveness for complex topics when compared with word-based expansion methods.

6 Conclusion

In this work we study local and global expansion methods that utilize word-based and entity-based features for retrieval with hierarchical semi-structured queries. We propose a method that performs a mixture of fine-grained (subtopic level) feedback models for each element of the structured query and combines them using score-based fusion. On the TREC CAR paragraph ranking task, we demonstrate that entity-centric subtopic-level expansion models constitute the most effective methods - even outperforming established neural ranking methods. Further, the entity-based expansion results show significant and consistent effectiveness gains over the word-based expansion methods, resulting in a greater than 20% improvement in mean average precision.

The new proposed expansion methods build on proven probabilistic expansion methods and combine multiple feature representations to create more robust retrieval for complex topics. As search evolves to support more complex tasks the nature of complex topics will continue to develop. We envision more complex topic structures that will grow in size. This work presents an important first step in leveraging structure effectively. We anticipate that additional modeling of the complex hierarchical relationships across diverse vocabularies (words, entities, etc.) will lead to further improvements in the future.

Acknowledgments. This work was supported in part by the Center for Intelligent Information Retrieval and in part by NSF grant #IIS-1617408. Any opinions, findings and conclusions or recommendations expressed in this material are those of the authors and do not necessarily reflect those of the sponsors.

References

1. Cao, G., Nie, J.Y., Gao, J., Robertson, S.: Selecting good expansion terms for pseudo-relevance feedback. In: Proceedings of the 31st Annual International ACM SIGIR Conference on Research and Development in Information Retrieval SIGIR 2008, pp. 243–250. ACM, New York (2008). https://doi.org/10.1145/1390334.1390377
2. Cohen, D., Croft, W.B.: End to end long short term memory networks for non-factoid question answering. In: Proceedings of the 2016 ACM International Conference on the Theory of Information Retrieval, pp. 143–146. ACM, September 2016
3. Cohen, D., Yang, L., Croft, W.B.: WikiPassageQA: A benchmark collection for research on non-factoid answer passage retrieval. In: The 41st International ACM SIGIR Conference on Research & Development in Information Retrieval SIGIR 2018, Ann Arbor, MI, USA, pp. 1165–1168, 08–12 July 2018. https://doi.org/10.1145/3209978.3210118
4. Dalton, J., Dietz, L., Allan, J.: Entity query feature expansion using knowledge base links. In: Proceedings of the 37th International ACM SIGIR Conference on Research & Development in Information Retrieval SIGIR 2014, pp. 365–374. ACM, New York (2014)
5. Dietz, L., Gamari, B., Dalton, J.: TREC CAR 2.1: A data set for complex answer retrieval (2018). http://trec-car.cs.unh.edu
6. Dietz, L., Kotov, A., Meij, E.: Utilizing knowledge graphs for text-centric information retrieval. In: The 41st International ACM SIGIR Conference on Research & Development in Information Retrieval, pp. 1387–1390. ACM (2018)
7. Dietz, L., Verma, M., Radlinski, F., Craswell, N.: TREC complex answer retrieval overview. In: Proceedings of The Twenty-Sixth Text Retrieval Conference TREC 2017, Gaithersburg, Maryland, USA, 15–17 November 2017. https://trec.nist.gov/pubs/trec26/papers/Overview-CAR.pdf
8. Hasibi, F., Balog, K., Bratsberg, S.E.: Exploiting entity linking in queries for entity retrieval. In: Proceedings of the 2016 ACM International Conference on the Theory of Information Retrieval ICTIR 2016, pp. 209–218. ACM, New York (2016)
9. Lavrenko, V., Croft, W.B.: Relevance based language models. In: Proceedings of the 24th Annual International ACM SIGIR Conference on Research and Development in Information Retrieval SIGIR 2001, pp. 120–127. ACM, New York (2001). https://doi.org/10.1145/383952.383972
10. Liu, X., Fang, H.: Latent entity space: A novel retrieval approach for entity-bearing queries. Inf. Retr. J. **18**(6), 473–503 (2015). https://doi.org/10.1007/s10791-015-9267-x

11. MacAvaney, S., et al.: Characterizing question facets for complex answer retrieval. In: The 41st International ACM SIGIR Conference on Research & Development in Information Retrieval, pp. 1205–1208. ACM, June 2018
12. Metzler, D., Diaz, F., Strohman, T., Croft, W.B.: UMass robust 2005: Using mixtures of relevance models for query expansion. In: Proceedings of the Fourteenth Text Retrieval Conference TREC 2005, Gaithersburg, Maryland, USA, 15–18 November 2005
13. Mikolov, T., Sutskever, I., Chen, K., Corrado, G.S., Dean, J.: Distributed representations of words and phrases and their compositionality. In: Advances in Neural Information Processing Systems, pp. 3111–3119 (2013)
14. Mitra, B., Diaz, F., Craswell, N.: Learning to match using local and distributed representations of text for web search. In: Proceedings of the 26th International Conference on World Wide Web, WWW 2017 International World Wide Web Conferences Steering Committee, Republic and Canton of Geneva, Switzerland, pp. 1291–1299 (2017). https://doi.org/10.1145/3038912.3052579
15. Nanni, F., Mitra, B., Magnusson, M., Dietz, L.: Benchmark for complex answer retrieval. In: Proceedings of the ACM SIGIR International Conference on Theory of Information Retrieval ICTIR 2017, pp. 293–296. ACM, New York (2017)
16. Ni, Y., et al.: Semantic documents relatedness using concept graph representation. In: Proceedings of the Ninth ACM International Conference on Web Search and Data Mining, pp. 635–644. ACM (2016)
17. Ponte, J.M., Croft, W.B.: A language modeling approach to information retrieval. In: Proceedings of the 21st Annual International ACM SIGIR Conference on Research and Development in Information Retrieval SIGIR 1998, pp. 293–296. ACM, New York (1998)
18. Raviv, H., Kurland, O., Carmel, D.: Document retrieval using entity-based language models. In: Proceedings of the 39th International ACM SIGIR Conference on Research and Development in Information Retrieval SIGIR 2016, pp. 65–74. ACM, New York (2016)
19. Řehůřek, R., Sojka, P.: Software framework for topic modelling with large corpora. In: Proceedings of the LREC 2010 Workshop on New Challenges for NLP Frameworks, pp. 45–50. ELRA, Valletta, Malta, May 2010. http://is.muni.cz/publication/884893/en
20. Salton, G., Fox, E.A., Wu, H.: Extended boolean information retrieval. Commun. ACM **26**(11), 1022–1036 (1983)
21. Turtle, H., Croft, W.B.: Evaluation of an inference network-based retrieval model. ACM Trans. Inf. Syst. Secur. **9**(3), 187–222 (1991)
22. Xiong, C., Callan, J.: EsdRank: Connecting query and documents through external semi-structured data. In: Proceedings of the 24th ACM International Conference on Information and Knowledge Management CIKM 2015, pp. 951–960. ACM, New York (2015)
23. Xiong, C., Callan, J., Liu, T.Y.: Word-entity duet representations for document ranking. In: Proceedings of the 40th International ACM SIGIR Conference on Research and Development in Information Retrieval, pp. 763–772. ACM (2017)
24. Xiong, C., Liu, Z., Callan, J., Liu, T.Y.: Towards better text understanding and retrieval through kernel entity salience modeling. In: The 41st International ACM SIGIR Conference on Research & Development in Information Retrieval, pp. 575–584. ACM, June 2018

Representation

Word Embeddings for Entity-Annotated Texts

Satya Almasian(✉), Andreas Spitz, and Michael Gertz

Heidelberg University, Heidelberg, Germany
{almasian,spitz,gertz}@informatik.uni-heidelberg.de

Abstract. Learned vector representations of words are useful tools for many information retrieval and natural language processing tasks due to their ability to capture lexical semantics. However, while many such tasks involve or even rely on named entities as central components, popular word embedding models have so far failed to include entities as first-class citizens. While it seems intuitive that annotating named entities in the training corpus should result in more intelligent word features for downstream tasks, performance issues arise when popular embedding approaches are naïvely applied to entity annotated corpora. Not only are the resulting entity embeddings less useful than expected, but one also finds that the performance of the non-entity word embeddings degrades in comparison to those trained on the raw, unannotated corpus. In this paper, we investigate approaches to jointly train word and entity embeddings on a large corpus with automatically annotated and linked entities. We discuss two distinct approaches to the generation of such embeddings, namely the training of state-of-the-art embeddings on raw-text and annotated versions of the corpus, as well as node embeddings of a co-occurrence graph representation of the annotated corpus. We compare the performance of annotated embeddings and classical word embeddings on a variety of word similarity, analogy, and clustering evaluation tasks, and investigate their performance in entity-specific tasks. Our findings show that it takes more than training popular word embedding models on an annotated corpus to create entity embeddings with acceptable performance on common test cases. Based on these results, we discuss how and when node embeddings of the co-occurrence graph representation of the text can restore the performance.

Keywords: Word embeddings · Entity embeddings · Entity graph

1 Introduction

Word embeddings are methods that represent words in a continuous vector space by mapping semantically similar or related words to nearby points. These vectors can be used as features in NLP or information retrieval tasks, such as query expansion [11,20], named entity recognition [9], or document classification [21]. The current style of word embeddings dates back to the neural probabilistic

L. Azzopardi et al. (Eds.): ECIR 2019, LNCS 11437, pp. 307–322, 2019.
https://doi.org/10.1007/978-3-030-15712-8_20

model published by Bengio et al. [6], prior to which embeddings were predominantly generated by latent semantic analysis methods [10]. However, most developments are more recent. The two most popular methods are word2vec proposed by Mikolov et al. [27], and GloVe by Pennington et al. [32]. Since then, numerous alternatives to these models have been proposed, often for specific tasks.

Common to all the above approaches is an equal treatment of words, without word type discrimination. Some effort has been directed towards embedding entire phrases [17,46] or combining compound words after training [12,29], but entities are typically disregarded, which entails muddied embeddings with ambiguous entity semantics as output. For example, an embedding that is trained on ambiguous input is unable to distinguish between instances of *Paris*, which might refer to the French capital, the American heiress, or even the Trojan prince. Even worse, entities can be conflated with homographic words, e.g., the former U.S. president *Bush*, who not only shares ambiguity with his family members, but also with shrubbery. Moreover, word embeddings are ill-equipped to handle synonymous mentions of distinct entity labels without an extensive local clustering of the neighbors around known entity labels in the embedding space.

Joint word and entity embeddings have been studied only for task-specific applications, such as entity linkage or knowledge graph completion. Yamada et al. [45] and Moreno et al. [30] propose entity and word embedding models specific to named entity linkage by using knowledge bases. Embedding entities in a knowledge graph has also been studied for relational fact extraction and knowledge base completion [41,44]. All of these methods depend on knowledge bases as an external source of information, and often train entity and word embeddings separately and combine them afterwards. However, it seems reasonable to avoid this separation and learn embeddings directly from annotated text to create general-purpose entity embeddings, jointly with word embeddings.

Typically, state-of-the-art entity recognition and linking is dependent on an extensive NLP-stack that includes sentence splitting, tokenization, part-of-speech tagging, and entity recognition, with all of their accrued cumulative errors. Thus, while embeddings stand to benefit from the annotation and resolution of entity mentions, an analysis of the drawbacks and potential applications is required. In this paper, we address this question by using popular word embedding methods to jointly learn word and entity embeddings from an automatically annotated corpus of news articles. We also use cooccurrence graph embeddings as an alternative, and rigorously evaluate these for a comprehensive set of evaluation tasks. Furthermore, we explore the properties of our models in comparison to plain word embeddings to estimate their usefulness for entity-centric tasks.

Contributions. In the following, we make five contributions. *(i)* We investigate the performance of popular word embedding methods when trained on an entity-annotated corpus, and *(ii)* introduce graph-based node embeddings as an alternative that is trained on a cooccurrence graph representations of the annotated text[1]. *(iii)* We compare all entity-based models to traditional word embeddings on a comprehensive set of word-centric intrinsic evaluation tasks, and introduce

[1] Source code available at: https://github.com/satya77/Entity_Embedding.

entity-centric intrinsic tasks. *(iv)* We explore the underlying semantics of the embeddings and implications for entity-centric downstream applications, and *(v)* discuss the advantages and drawbacks of the different models.

2 Related Work

Related work covers word and graph embeddings, as well as cooccurrence graphs.

Word Embeddings. A word embedding, defined as a mapping $V \rightarrow \mathbb{R}^d$, maps a word w from a vocabulary V to a vector θ in a d-dimensional embedding space [36]. To learn such embeddings, window-based models employ supervised learning, where the objective is to predict a word's context when given a center word in a fixed window. Mikolov et al. introduced the continuous bag-of-words (CBOW) and the skip-gram architecture as window-based models that are often referred to as word2vec [27]. The CBOW architecture predicts the current word based on the context, while skip-gram predicts surrounding words given the current word. This model was later improved by Bojanowski et al. to take character level information into account [7]. In contrast to window-based models, matrix factorization methods operate directly on the word cooccurrence matrix. Levy and Goldberg showed that implicitly factorizing a word-context matrix, whose cells contain the point-wise mutual information of the respective word and context pairs, can generate embeddings close to word2vec [22]. Finally, the global vector model (GloVe) combines the two approaches and learns word embeddings by minimizing the distance between the number of cooccurrences of words and the dot product of their vector representations [32].

Graph Node Embeddings. A square word cooccurrence matrix can be interpreted as a graph whose nodes correspond the rows and columns, while the matrix entries indicate edges between pairs of nodes. The edges can have weights, which usually reflect some distance measure between the words, such as the number of tokens between them. These networks are widely used in natural language processing, for example in summarization [26] or word sense discrimination [13]. More recent approaches have included entities in graphs to support information retrieval tasks, such as topic modeling [38]. In a graph representation of the text, the neighbors of a node can be treated as the node's context. Thus, embedding the nodes of a graph also results in embeddings of words.

Numerous node embedding techniques for graph nodes exist, which differ primarily in the similarity measure that is used to define node similarity. DeepWalk was the first model to learn latent representations of graph nodes by using sequences of fixed-length random walks around each node [33]. Node2vec improved the DeepWalk model by proposing a flexible neighborhood sampling strategy that interpolates between depth-first and breadth-first search [16]. The LINE model learns a two-part embedding, where the first part corresponds to the first-order proximity (i.e., the local pairwise proximity between two vertices) and the second part represents the second-order proximity (i.e., the similarity between their neighborhood structures) [40]. More recently, Tsitsulin et al. proposed VERSE, which supports multiple similarity functions that can be tailored to individual graph structures [43]. With VERSE, the user can choose to emphasize the structural similarity or focus on an adjacency matrix, thus emulating the first-order proximity of LINE. Due to this versatility, we thus focus on DeepWalk and VERSE as representative node embedding methods to generate joint entity and word embeddings from cooccurrence graphs.

3 Embedding Models

To jointly embed words and named entities, we tweak existing word and graph node embedding techniques. To naïvely include entities in the embeddings, we train the state-of-the-art word embedding methods on an entity annotated corpus. As an alternative, we transform the text into a cooccurrence graph and use graph-based models to train node embeddings. We compare both models against models trained on the raw (unannotated) corpus. In this section, we first give an overview of GloVe and word2vec for raw text input. Second, we describe how to include entity annotations for these models. Finally, we show how DeepWalk and VERSE can be used to obtain entity embeddings from a cooccurrence graph.

3.1 Word Embeddings on Raw Text

State-of-the-art word embedding models are typically trained on the raw text that is cleaned by removing punctuation and stop words. Since entities are not annotated, all words are considered as terms. We use skip-gram from word2vec (rW2V), and the GloVe model (rGLV), where r denotes raw text input.

Skip-gram aims to optimize the embeddings θ, which maximize the corpus probability over all words w and their contexts c in documents D [14] as

$$\arg\max_{\theta} \prod_{(w,c)\in D} p(c \mid w, \theta) \tag{1}$$

To find the optimal value of θ, the conditional probability is modelled using softmax and solved with negative sampling or hierarchical softmax.

GloVe learns word vectors such that their dot product equals the logarithm of the words' cooccurrence probability [32]. If $X \in \mathbb{R}^{W\times W}$ is a matrix of word cooccurrence counts X_{ij}, then GloVe optimizes embeddings θ_i and $\tilde{\theta_j}$ for center words i and context words j, and biases b and $\tilde{b}$ to minimize the cost function

$$J = \sum_{i,j=1}^{W} f(X_{ij})(\theta_i^T \tilde{\theta_j} + b_i + \tilde{b_j} - logX_{ij})^2, \quad f = \begin{cases} (\frac{x}{x_{max}})^{\alpha} & \text{if } x < x_{max} \\ 1 & \text{otherwise} \end{cases} \tag{2}$$

The function f serves as an upper bound on the maximum number of allowed word cooccurrences x_{max}, with $\alpha \in [0, 1]$ as an exponential dampening factor.

3.2 Word Embeddings on Annotated Text

Named entities are typically mentions of person, organization, or location names, and numeric expressions, such as dates or monetary values in a text [31]. Formally, if T denotes the set of terms in the vocabulary (i.e, words and multi-word expressions), let $N \subseteq T$ be the subset of named entities. Identifying these mentions is a central problem in natural language processing that involves part-of-speech tagging, named entity recognition, and disambiguation. Note that T is technically a multi-set since multiple entities may share ambiguous labels, but entities can be represented by unique identifiers in practice. Since annotated texts contain more information and are less ambiguous, embeddings trained on such texts thus stand to perform better in downstream applications. To generate these embeddings directly, we use word2vec and GloVe on a corpus with named entity annotations and refer to them as aW2V and aGLV, where a denotes the use of annotated text. Since entity annotation requires part-of-speech tagging, we use POS tags to remove punctuation and stop word classes. Named entity mentions are identified and replaced with unique entity identifiers. The remaining words constitute the set of terms $T \backslash N$ and are used to generate term cooccurrence counts for the word embedding methods described above.

3.3 Node Embeddings of Cooccurrence Graphs

A cooccurrence graph $G = (T, E)$ consists of a set of terms T as nodes and a set of edges E that connect cooccurring terms. Edges can be weighted, where the weights typically encode some form of textual distance or similarity between the terms. If the graph is extracted from an annotated corpus, some nodes represent named entities. For entity annotations in particular, implicit networks can serve as graph representations that use similarity-based weights derived from larger cross-sentence cooccurrences of entity mentions [37]. By embedding nodes in these networks, we also obtain embeddings of both entities and terms.

From the available node embedding methods, we select a representative subset. While it is popular, we omit node2vec since cooccurrence graphs are both large and dense, and node2vec tends to be quite inefficient for such graphs [47]. Similarly, we do not use LINE since the weighted cooccurrence graphs tend to have an unbalanced distribution of frequent and rare words, meaning that the second-order proximity of LINE becomes ill-defined. Since the adjacency similarity of VERSE correlates with the first-order proximity in LINE, we use VERSE (VRS) as a representative of the first-order proximity and DeepWalk (DW) as a representative of random walk-based models. Conceptually, graph node embeddings primarily differ from word embeddings in the sampling of the context.

DeepWalk performs a series of fixed-length random walks on the graph to learn a set of parameters $\Theta_E \in R^{T \times d}$, where d is a small number of latent dimensions. The nodes visited in a random walk are considered as context and are used to train a skip-gram model. DeepWalk thus maximizes the probability of observing the k previous and next nodes in a random walk starting at node t_i by minimizing the negative logarithmic probability to learn the node embedding θ [15]:

$$J = -\log P(t_{i-k}, ..., t_{i-1}, t_{i+1}, ..., t_{i+k} \mid \theta) \tag{3}$$

Since cooccurrence graphs are weighted, we introduce weighted random walks that employ a transition probability to replace the uniform random walks. The probability of visiting node j from node i is then proportional to the edge weight $e_{i,j}$, where E_i denotes the set of all edges starting at node t_i

$$P_{i,j} = \frac{f(e_{i,j})}{\sum_{e_{ik} \in E_i} f(e_{ik})} \tag{4}$$

and f is a normalization function. To create a more balanced weight distribution, we consider no normalization, i.e., $f = \text{id}$, and a logarithmic normalization, i.e., $f = \log$. We refer to these as (DW_{id}) and (DW_{log}), respectively. The performance of $f = \text{sqr}$ is similar to a logarithmic normalization and is omitted.

VERSE is designed to accept any node similarity measure for context selection [43]. Three measures are part of the original implementation, namely Personalized PageRank, adjacency similarity, and SimRank. SimRank is a measure of structural relatedness and thus ill-suited for word relations. Personalized PageRank is based on the stationary distribution of a random walk with restart, and essentially replicates DeepWalk. Thus, we focus on adjacency similarity, which derives node similarities from the outgoing degree $out(t_i)$ of node t_i:

$$sim_G^{ADJ}(t_i, t_j) = \begin{cases} \frac{1}{out(t_i)} & if \quad (t_i, t_j) \in E \\ 0 & \quad otherwise \end{cases} \tag{5}$$

The model then minimizes the Kullback-Leibler divergence between the similarity measure of two nodes and the dot product of their embeddings θ_i and θ_j, and thus works conceptually similar to GloVe. In the following, we use this model as our second node embedding approach and refer to it as VRS.

4 Evaluation of Embeddings

In the following, we look at the datasets used for training and evaluation, before comparing the learned models on typical tasks and discussing the results.

4.1 Evaluation Tasks

The main benefit of word embeddings is found in downstream applications (extrinsic evaluation). However, since these evaluations are task-specific, an embedding that works well for one task may fail for another. The more common test scenario is thus intrinsic and analyzes how well the embeddings capture syntactic or semantic relations [36]. The problem with such tests is that the notion of semantics is not universal [4]. Some datasets reflect semantic relatedness and some semantic similarity [19]. Since few intrinsic datasets include entities, we focus on the performance of term-based intrinsic tasks. Following the approach by Schnabel et al. [36], we use three kinds of intrinsic evaluations.

Relatedness uses datasets with relatedness scores for pairs of words annotated by humans. The cosine similarity or Euclidean distance between the embeddings of two words should have a high correlation with scores assigned by humans.

(i) *Similarity353:* 203 instances of similar word pairs from WordSim353 [3] classified as synonyms, antonyms, identical, and unrelated pairs [2].
(ii) *Relatedness353:* 252 instances of word pairs from WordSim353 [3] that are not similar but still considered related by humans, and unrelated pairs [2].
(iii) *MEN:* 3,000 word pairs with human-assigned similarity judgements [8].
(iv) *RG65:* 65 pairs with annotated similarity, scaling from 0 to 4 [35].
(v) *RareWord:* 2,034 rare word pairs with human-assigned similarity scores [23].
(vi) *SimLex-999:* 999 pairs of human-labeled examples of semantic relatedness [18].
(vii) *MTurk:* 771 words pairs with semantic relatedness scores from 0 to 5 [34].

Analogy. In the analogy task, the objective is to find a word y for a given word x, such that $x : y$ best resembles a sample relationship $a : b$. Given the triple (a, b, x) and a target word y, the nearest neighbour of $\hat{\theta} := \theta_a - \theta_b + \theta_x$ is computed and compared to y. If y is the word with the highest cosine similarity to $\hat{\theta}$, the task is solved correctly. For entity embeddings, we can also consider an easier, type-specific variation of this task, which only considers neighbors that match a given entity class, such as locations or persons.

(i) *GA*: The *Google Analogy* data consists of 19,544 morphological and semantic questions used in the original word2vec publication [27]. Beyond terms, it contains some location entities that support term to city relations.
(ii) *MSR:* The Microsoft Research Syntactic analogies dataset contains 8,000 morphological questions [28]. All word pairs are terms.

Categorization. When projecting the embeddings to a 2- or 3-dimensional space with t-SNE [24] or principle component analysis [1], we expect similar words to form meaningful clusters, which we can evaluate by computing the purity of clusters [25]. We use two datasets from the Lexical Semantics Workshop, which do not contain entities. Additionally, we create three datasets by using Wikidata to find entities of type person, location, and organization.

(i) *ESSLLI_1a:* 44 concrete nouns that belong to six semantic categories [5].
(ii) *ESSLLI_2c:* 45 verbs that belong to nine semantic classes [5].
(iii) *Cities:* 150 major cities in the U.S., the U.K., and Germany.
(iv) *Politicians:* 150 politicians from the U.S., the U.K., and Germany.
(v) *Companies:* 110 software companies, Web services, and car manufacturers.

4.2 Training Data

For training, we use 209,023 news articles from English-speaking news outlets, collected from June to November 2016 by Spitz and Gertz [38]. The data contains a total of 5,427,383 sentences. To train the regular word embeddings, we use the raw article texts, from which we remove stop words and punctuation. For the annotated embeddings, we extract named entities with Ambiverse[2], a state-of-the-art annotator that links entity mentions of persons, locations, and organizations to Wikidata identifiers. Temporal expressions of type date are annotated and normalized with HeidelTime [39], and part-of-speech annotations are obtained from the Stanford POS tagger [42]. We use POS tags to remove punctuation and stop words (wh-determiner, pronouns, auxiliary verbs, predeterminers, possessive endings, and prepositions). To generate input for the graph-based embeddings, we use the extraction code of the LOAD model [37] that generates implicit weighted graphs of locations, organizations, persons, dates, and terms, where the weights encode the textual distance between terms and entities that cooccur in the text. We include term cooccurrences only inside sentences and entity-entity cooccurrences up to a default window size of five sentences. The final graph has $T = 93,390$ nodes (terms and entities) and $E = 9,584,191$ edges. Since the evaluation datasets contain words that are not present in the training vocabulary, each data set is filtered accordingly.

4.3 Parameter Tuning

We perform extensive parameter tuning for each model and only report the settings that result in the best performance. Since the embedding dimensions have no effect on the relative difference in performance between models, all embeddings have 100 dimensions. Due to the random initialization at the beginning of the training, all models are trained 10 times and the performance is averaged.

Word2vec-based models are trained with a learning rate of 0.015 and a window size of 10. We use 8 negative samples on the raw data, and 16 on the annotated data. Words with a frequency of less than 3 are removed from the vocabulary as there is not enough data to learn a meaningful representation.

GloVe-based models are trained with a learning rate of 0.06. For the weighting function, a scaling factor of 0.5 is used with a maximum cut-off of 1000. Words that occur less than 5 times are removed from the input.

DeepWalk models use 100 random walks of length 4 from each node. Since the cooccurrence graph has a relatively small diameter, longer walks would introduce unrelated words into contexts. We use a learning rate of 0.015 and 64 negative samples for the skip-gram model that is trained on the random walk results.

VERSE models use a learning rate of 0.025 and 16 negative samples.

A central challenge in the comparison of the models is the fact that the training process of graph-based and textual methods is incomparable. On the one

[2] https://github.com/ambiverse-nlu.

hand, the textual models consider one pass through the corpus as one iteration. On the other hand, an increase in the number of random walks in DeepWalk increases both the performance and the runtime of the model, as it provides more data for the skip-gram model. In contrast, the VERSE model has no notion of iteration and samples nodes for positive and negative observations. To approach a fair evaluation, we thus use similar training times for all models (roughly 10 hours per model on a 100 core machine). We fix the number of iterations of the textual models and DeepWalk's skip-gram at 100. For VERSE, we use 50,000 sampling steps to obtain a comparable runtime.

4.4 Evaluation Results

Unsurprisingly, we find that no single model performs best for all tasks. The results of the relatedness task are shown in Table 1, which shows that word2vec performs better than GloVe with this training data. The performance of both methods degrades slightly but consistently when they are trained on the annotated data in comparison to the raw data. The DeepWalk-based models perform better than GloVe but do poorly overall. VERSE performs very well for some of the tasks, but is worse than word2vec trained on the raw data for rare words

Table 1. Word similarity results. Shown are the Pearson correlations between the cosine similarity of the embeddings and the human score on the word similarity datasets. The two best values per task are highlighted.

	rW2V	rGLV	aW2V	aGLV	DW_{id}	DW_{log}	VRS
Similarity353	**0.700**	0.497	**0.697**	0.450	0.571	0.572	0.641
Relatedness353	**0.509**	0.430	0.507	0.428	0.502	0.506	**0.608**
MEN	**0.619**	0.471	**0.619**	0.469	0.539	0.546	**0.640**
RG65	**0.477**	0.399	0.476	0.386	0.312	0.344	**0.484**
RareWord	**0.409**	0.276	**0.409**	0.274	**0.279**	0.276	0.205
SimLex-999	**0.319**	0.211	**0.319**	0.211	**0.279**	0.201	0.236
MTurk	**0.647**	0.493	0.644	0.502	0.592	0.591	**0.687**
Average	**0.526**	0.400	**0.524**	0.389	0.439	0.433	0.500

Table 2. Word analogy results. Shown are the prediction accuracy for the normal analogy tasks and the variation in which predictions are limited to the correct entity type. The best two values per task and variation are highlighted.

	Normal analogy							Typed analogy				
	rW2V	rGLV	aW2V	aGLV	DW_{id}	DW_{log}	VRS	aW2V	aGLV	DW_{id}	DW_{log}	VRS
GA	0.013	**0.019**	0.003	0.015	0.009	0.009	**0.035**	0.003	**0.016**	0.110	0.110	**0.047**
MSR	**0.014**	**0.019**	0.001	**0.014**	0.002	0.002	0.012	0.001	**0.014**	0.002	0.002	**0.012**
Avg	0.013	**0.019**	0.002	0.014	0.005	0.005	**0.023**	0.002	**0.015**	0.006	0.006	**0.030**

and the SimLex data. This is likely caused by the conceptual structure of the cooccurrence graph on which VERSE is trained, which captures relatedness and not similarity as tested by SimLex. For the purely term-based tasks in this evaluation that do not contain entity relations, word2vec is thus clearly the best choice for similarity tasks, while VERSE does well on relatedness tasks.

Table 2 shows the accuracy achieved by all models in the word analogy task, which is overall very poor. We attribute this to the size of the data set that contains less than the billions of tokens that are typically used to train for this task. GloVe performs better than word2vec for this task on both raw and annotated data, while VERSE does best overall. The typed task, in which we also provide the entity type of the target word, is easier and results in better scores. If we consider only the subset of 6,892 location targets for the GA task, we find that the graph-based models perform much better, with VERSE being able to predict up to 1,662 (24.1%) of location targets on its best run, while aW2V and aGLV are only able to predict 14 (0.20%) and 16 (0.23%), respectively. For this entity-centric subtask, VERSE is clearly better suited. For the MSR task, which does not contain entities, we do not observe such an advantage.

The purity of clusters created with agglomerative clustering and mini-batch k-means for the categorization tasks are shown in Table 3, where the number of clusters were chosen based on the ground truth data. For the raw embeddings, we represent multi-word entities by the mean of the vectors of individual words in the entity's name. In most tasks, rW2V and rGLV create clusters with the best purity, even for the entity-based datasets of cities, politicians, and companies. However, most purity values lie in the range from 0.45 to 0.65 and no method performs exceptionally poorly. Since only the words in the evaluation datasets

Table 3. Categorization results. Shown is the purity of clusters obtained with k-means and agglomerative clustering (AC). The best two values are highlighted. For the raw text models, multi-word entity names are the mean of word vectors.

		rW2V	rGLV	aW2V	aGLV	DW_{id}	DW_{log}	VRS
k-means	ESSLLI_1a	**0.575**	0.545	**0.593**	0.454	0.570	0.520	0.534
	ESSLLI_2c	0.455	0.462	**0.522**	0.464	0.471	0.480	**0.584**
	Cities	**0.638**	**0.576**	0.467	0.491	0.560	0.549	0.468
	Politicians	**0.635**	0.509	0.402	0.482	0.470	0.439	**0.540**
	Companies	**0.697**	**0.566**	0.505	0.487	0.504	0.534	0.540
	Average	**0.600**	0.532	0.498	0.476	0.515	0.504	**0.533**
AC	ESSLLI_1a	0.493	**0.518**	0.493	0.440	0.486	0.502	**0.584**
	ESSLLI_2c	0.455	0.398	0.382	0.349	**0.560**	**0.408**	0.442
	Cities	0.447	**0.580**	0.440	0.515	0.364	**0.549**	0.359
	Politicians	0.477	**0.510**	**0.482**	0.480	0.355	0.360	0.355
	Companies	**0.511**	**0.519**	0.475	0.504	0.474	0.469	0.473
	Average	**0.477**	**0.505**	0.454	0.458	0.448	0.458	0.443

are clustered, the results do not give us insight into the spatial mixing of terms and entities. We consider this property in our visual exploration in Sect. 5.

In summary, the results of the predominantly term-based intrinsic evaluation tasks indicate that a trivial embedding of words in an annotated corpus with state-of-the-art methods is possible and has acceptable performance, yet degrades the performance in comparison to a training on the raw corpus, and is thus not necessarily the best option. For tasks that include entities in general and require a measure of relatedness in particular, such as analogy task for entities or relatedness datasets, we find that the graph-based embeddings of VERSE often have a better performance. In the following, we thus explore the usefulness of the different embeddings for entity-centric tasks.

5 Experimental Exploration of Entity Embeddings

Since there are no extrinsic evaluation tasks for entity embeddings, we cannot evaluate their performance on downstream tasks. We thus consider entity clustering and the neighborhood of entities to obtain an impression of the benefits that entity embeddings can offer over word embeddings for entity-centric tasks.

Entity Clustering. To investigate how well the different methods support the clustering of similar entities, we consider 2-dimensional t-SNE projections of the embeddings of cities in Fig. 1. Since the training data is taken from news, we expect cities within a country to be spatially correlated. For the raw text embeddings, we represent cities with multi-component names as the average of the embeddings of their components. For this task, the word embeddings perform

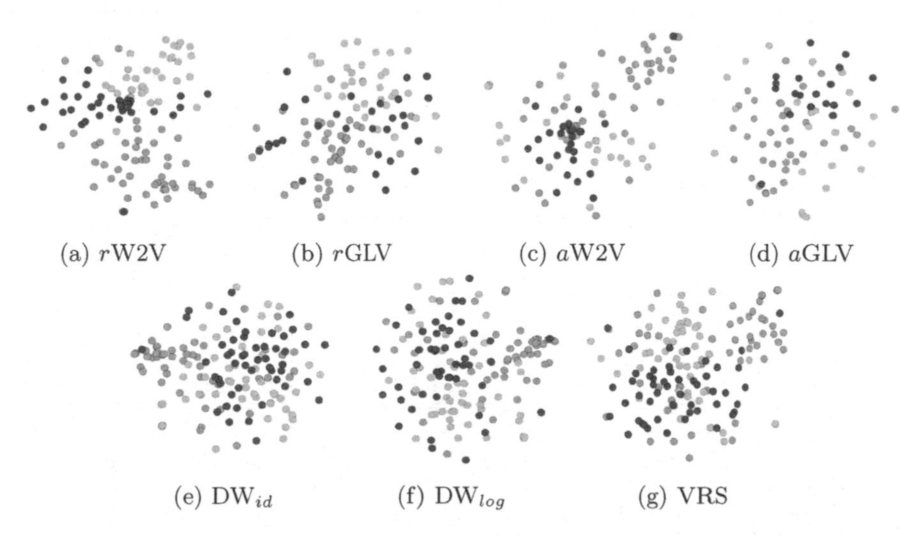

Fig. 1. t-SNE projections of the embeddings for U.S. (purple), British (orange), and German (green) cities. For the raw text models, multi-word entity names are represented as the mean of word vectors. (Color figure online)

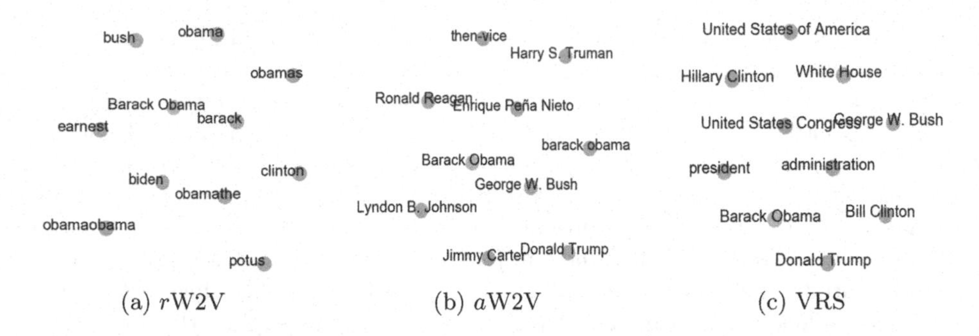

(a) rW2V (b) aW2V (c) VRS

Fig. 2. t-SNE projections of the nearest neighbours of entity *Barack Obama*.

much better on the raw text than they do on the annotated text. However, the underlying assumption for the applicability of composite embeddings for multi-word entity names is the knowledge (or perfect recognition) of such entity names, which may not be available in practice. The graph-based methods can recover some of the performance, but as long as entity labels are known, e.g., from a gazetteer, raw text embeddings are clearly preferable.

Entity Neighborhoods. To better understand the proximity of embeddings, we consider the most similar neighbors by cosine similarity on the example of the entity *Barack Obama.* Table 4 contains a list of the four nearest neighbors of Barack Obama for each embedding method. For the raw text models, we average the embeddings of the words *barack* and *obama.* Here, we find that the entity-centric models are more focused on entities, while the models that are trained on raw text put a stronger emphasis on terms in the neighborhood. In particular, rW2V performs very poorly and predominantly retrieves misspelled versions of the entity name. In contrast, even aW2V and aGLV retrieve more related entities, although the results for the graph-based embeddings are more

Table 4. Four nearest neighbours of entity *Barack Obama* with cosine similarity scores. Entity types include terms T, persons P, and locations L.

rW2V		rGLV		aW2V		aGLV	
T obama	0.90	T obama	0.99	P George W. Bush	0.76	T president	0.78
T barack	0.74	T barack	0.98	P Jimmy Carter	0.73	T administration	0.76
T obamaobama	0.68	T president	0.77	T barack obama	0.73	P George W. Bush	0.72
T obamathe	0.60	T administration	0.74	P Enrique Peña Nieto	0.67	T mr.	0.68

DW_{id}		DW_{log}		VRS	
L White House	0.88	L White House	0.88	L White House	0.87
T president	0.79	T president	0.82	T president	0.79
T presidency	0.76	P George W. Bush	0.78	L United States of America	0.76
T administration	0.75	T administration	0.78	P Donald Trump	0.75

informative of the input entity. Furthermore, we again observe a distinction in models between those that favor similarity and those that favor relatedness.

The same trend is visible in the t-SNE projections of the nearest neighbours in Fig. 2, where word2vec primarily identifies synonymously used words on the raw corpus (i.e., variations of the entity name), and entities with an identical or similar role on the annotated corpus (i.e., other presidents). In contrast, VERSE identifies related entities with different roles, such as administrative locations, or the presidential candidates and the president-elect in the 2016 U.S. election.

6 Conclusion and Ongoing Work

We investigated the usefulness of vector embeddings of words in entity-annotated news texts. We considered the naïve application of the popular models word2vec and GloVe to annotated texts, as well as node embeddings of cooccurrence graphs, and compared them to traditional word embeddings on a comprehensive set of term-focused evaluation tasks. Furthermore, we performed an entity-centric exploration of all embeddings to identify the strengths and weaknesses of each approach. While we found that word embeddings can be trained directly on annotated texts, they suffer from a degrading performance in traditional term-centric tasks, and often do poorly on tasks that require relatedness. In contrast, graph-based embeddings performed better for entity- and relatedness-centric tasks, but did worse for similarity-based tasks, and should thus not be used blindly in place of word embeddings. Instead, we see potential applications in entity-centric tasks that benefit from relatedness relations, such as improved query expansion or learning to disambiguate, which we consider to be the most promising future research directions and downstream tasks.

References

1. Abdi, H., Williams, L.J.: Principal component analysis. Wiley Interdisciplinary Reviews: Comput. Stat. **2**(4), 433–459 (2010)
2. Agirre, E., Alfonseca, E., Hall, K., Kravalova, J., Paşca, M., Soroa, A.: A study on similarity and relatedness using distributional and WordNet-based approaches. In: Proceedings of Human Language Technologies: The 2009 Annual Conference of the North American Chapter of the Association for Computational Linguistics (NAACL-HLT) (2009)
3. Agirre, E., Alfonseca, E., Hall, K.B., Kravalova, J., Pasca, M., Soroa, A.: A study on similarity and relatedness using distributional and WordNet-based approaches. In: Human Language Technologies: Conference of the North American Chapter of the Association of Computational Linguistics (NAACL-HLT) (2009)
4. Bakarov, A.: A survey of word embeddings evaluation methods. arxiv:1801.09536 (2018)
5. Baroni, M., Evert, S., Lenci, A. (eds.): Proceedings of the ESSLLI Workshop on Distributional Lexical Semantics Bridging the Gap Between Semantic Theory and Computational Simulations (2008)

6. Bengio, Y., Ducharme, R., Vincent, P.: A neural probabilistic language model. In: Advances in Neural Information Processing Systems (NIPS) (2000)
7. Bojanowski, P., Grave, E., Joulin, A., Mikolov, T.: Enriching word vectors with subword information. TACL **5**, 135–146 (2017)
8. Bruni, E., Tran, N.K., Baroni, M.: Multimodal distributional semantics. J. Artif. Int. Res. **49**(1), 1–47 (2014)
9. Das, A., Ganguly, D., Garain, U.: Named entity recognition with word embeddings and wikipedia categories for a low-resource language. ACM Trans. Asian Low-Resour. Lang. Inf. Process. **16**(3), 18 (2017)
10. Deerwester, S., Dumais, S.T., Furnas, G.W., Landauer, T.K., Harshman, R.: Indexing by latent semantic analysis. J. Am. Soc. Inform. Sci. **41**(6), 391–407 (1990)
11. Diaz, F., Mitra, B., Craswell, N.: Query expansion with locally-trained word embeddings. In: Proceedings of the 54th Annual Meeting of the Association for Computational Linguistics (ACL), Volume 1: Long Papers (2016)
12. Durme, B.V., Rastogi, P., Poliak, A., Martin, M.P.: Efficient, compositional, order-sensitive n-gram embeddings. In: Proceedings of the 15th Conference of the European Chapter of the Association for Computational Linguistics (EACL), Volume 2: Short Papers (2017)
13. Ferret, O.: Discovering word senses from a network of lexical cooccurrences. In: Proceedings of the 20th International Conference on Computational Linguistics (COLING) (2004)
14. Goldberg, Y., Levy, O.: Word2vec explained: deriving Mikolov et al.'s negative-sampling word-embedding method. CoRR abs/1402.3722 (2014)
15. Goyal, P., Ferrara, E.: Graph embedding techniques, applications, and performance: a survey. Knowl. Based Syst. **151**, 78–94 (2018)
16. Grover, A., Leskovec, J.: node2vec: scalable feature learning for networks. In: Proceedings of the 22nd ACM SIGKDD International Conference on Knowledge Discovery and Data Mining (KDD) (2016)
17. Hill, F., Cho, K., Korhonen, A., Bengio, Y.: Learning to understand phrases by embedding the dictionary. TACL **4**, 17–30 (2016)
18. Hill, F., Reichart, R., Korhonen, A.: SimLex-999: evaluating semantic models with (genuine) similarity estimation. Comput. Linguist. **41**(4), 665–695 (2015)
19. Kolb, P.: Experiments on the difference between semantic similarity and relatedness. In: Proceedings of the 17th Nordic Conference of Computational Linguistics, (NODALIDA) (2009)
20. Kuzi, S., Shtok, A., Kurland, O.: Query expansion using word embeddings. In: Proceedings of the 25th ACM International Conference on Information and Knowledge Management (CIKM) (2016)
21. Lenc, L., Král, P.: Word embeddings for multi-label document classification. In: Proceedings of the International Conference Recent Advances in Natural Language Processing (RANLP) (2017)
22. Levy, O., Goldberg, Y.: Neural word embedding as implicit matrix factorization. In: Advances in Neural Information Processing Systems 27: Annual Conference on Neural Information Processing Systems (NIPS) (2014)
23. Luong, T., Socher, R., Manning, C.D.: Better word representations with recursive neural networks for morphology. In: Proceedings of the Seventeenth Conference on Computational Natural Language Learning (CoNLL) (2013)
24. Maaten, L., Hinton, G.: Visualizing data using t-SNE. J. Mach. Learn. Res. **9**, 2579–2605 (2008)
25. Manning, C.D., Raghavan, P., Schütze, H.: Introduction to Information Retrieval. Cambridge University Press, Cambridge (2008)

26. Mihalcea, R., Tarau, P.: TextRank: bringing order into text. In: Proceedings of the 2004 Conference on Empirical Methods in Natural Language Processing (EMNLP) (2004)
27. Mikolov, T., Chen, K., Corrado, G., Dean, J.: Efficient estimation of word representations in vector space. arXiv:1301.3781 (2013)
28. Mikolov, T., Yih, W., Zweig, G.: Linguistic regularities in continuous space word representations. In: Human Language Technologies: Conference of the North American Chapter of the Association of Computational Linguistics (NAACL-HLT) (2013)
29. Mitchell, J., Lapata, M.: Composition in distributional models of semantics. Cogn. Sci. **34**(8), 1388–1429 (2010)
30. Moreno, J.G., et al.: Combining word and entity embeddings for entity linking. In: Blomqvist, E., Maynard, D., Gangemi, A., Hoekstra, R., Hitzler, P., Hartig, O. (eds.) ESWC 2017, Part I. LNCS, vol. 10249, pp. 337–352. Springer, Cham (2017). https://doi.org/10.1007/978-3-319-58068-5_21
31. Nadeau, D., Sekine, S.: A survey of named entity recognition and classification. Lingvisticae Investigationes **30**(1), 3–26 (2007)
32. Pennington, J., Socher, R., Manning, C.D.: Glove: global vectors for word representation. In: Proceedings of the 2014 Conference on Empirical Methods in Natural Language Processing (EMNLP) (2014)
33. Perozzi, B., Al-Rfou, R., Skiena, S.: DeepWalk: online learning of social representations. In: The 20th ACM SIGKDD International Conference on Knowledge Discovery and Data Mining (KDD) (2014)
34. Radinsky, K., Agichtein, E., Gabrilovich, E., Markovitch, S.: A word at a time: computing word relatedness using temporal semantic analysis. In: Proceedings of the 20th International Conference on World Wide Web (WWW) (2011)
35. Rubenstein, H., Goodenough, J.B.: Contextual correlates of synonymy. Commun. ACM **8**(10), 627–633 (1965)
36. Schnabel, T., Labutov, I., Mimno, D.M., Joachims, T.: Evaluation methods for unsupervised word embeddings. In: Proceedings of the 2015 Conference on Empirical Methods in Natural Language Processing (EMNLP) (2015)
37. Spitz, A., Gertz, M.: Terms over LOAD: leveraging named entities for cross-document extraction and summarization of events. In: Proceedings of the 39th International ACM SIGIR Conference on Research and Development in Information Retrieval (SIGIR) (2016)
38. Spitz, A., Gertz, M.: Entity-centric topic extraction and exploration: a network-based approach. In: Pasi, G., Piwowarski, B., Azzopardi, L., Hanbury, A. (eds.) ECIR 2018. LNCS, vol. 10772, pp. 3–15. Springer, Cham (2018). https://doi.org/10.1007/978-3-319-76941-7_1
39. Strötgen, J., Gertz, M.: Multilingual and cross-domain temporal tagging. Lang. Resour. Eval. **47**(2), 269–298 (2013)
40. Tang, J., Qu, M., Wang, M., Zhang, M., Yan, J., Mei, Q.: LINE: large-scale information network embedding. In: Proceedings of the 24th International Conference on World Wide Web (WWW) (2015)
41. Toutanova, K., Chen, D., Pantel, P., Poon, H., Choudhury, P., Gamon, M.: Representing text for joint embedding of text and knowledge bases. In: Proceedings of the 2015 Conference on Empirical Methods in Natural Language Processing, EMNLP, pp. 1499–1509 (2015)

42. Toutanova, K., Klein, D., Manning, C.D., Singer, Y.: Feature-rich part-of-speech tagging with a cyclic dependency network. In: Human Language Technology Conference of the North American Chapter of the Association for Computational Linguistics (HLT-NAACL) (2003)
43. Tsitsulin, A., Mottin, D., Karras, P., Müller, E.: VERSE: versatile graph embeddings from similarity measures. In: Proceedings of the 2018 World Wide Web Conference on World Wide Web (WWW) (2018)
44. Wang, Z., Zhang, J., Feng, J., Chen, Z.: Knowledge graph and text jointly embedding. In: Proceedings of the 2014 Conference on Empirical Methods in Natural Language Processing, EMNLP, A meeting of SIGDAT, a Special Interest Group of the ACL, pp. 1591–1601 (2014)
45. Yamada, I., Shindo, H., Takeda, H., Takefuji, Y.: Joint learning of the embedding of words and entities for named entity disambiguation. In: Proceedings of the 20th SIGNLL Conference on Computational Natural Language Learning, CoNLL, pp. 250–259 (2016)
46. Yin, W., Schütze, H.: An exploration of embeddings for generalized phrases. In: Proceedings of the 52nd Annual Meeting of the Association for Computational Linguistics (ACL) (2014)
47. Zhou, D., Niu, S., Chen, S.: Efficient graph computation for Node2Vec. CoRR abs/1805.00280 (2018)

Vectors of Pairwise Item Preferences

Gaurav Pandey[1(✉)], Shuaiqiang Wang[2], Zhaochun Ren[2], and Yi Chang[3]

[1] University of Jyvaskyla, Jyväskylä, Finland
gaurav.g.pandey@jyu.fi
[2] JD.com, Beijing, China
{wangshuaiqiang1,renzhaochun}@jd.com
[3] Jilin University, Changchun, China
yichang@ieee.org

Abstract. Neural embedding has been widely applied as an effective category of vectorization methods in real-world recommender systems. However, its exploration of users' explicit feedback on items, to create good quality user and item vectors is still limited. Existing neural embedding methods only consider the items that are accessed by the users, but neglect the scenario when a user gives high or low rating to a particular item. In this paper, we propose *Pref2Vec*, a method to generate vector representations of pairwise item preferences, users and items, which can be directly utilized for machine learning tasks. Specifically, *Pref2Vec* considers users' pairwise item preferences as elementary units. It vectorizes users' pairwise preferences by maximizing the likelihood estimation of the conditional probability of each pairwise item preference given another one. With the pairwise preference matrix and the generated preference vectors, the vectors of users are yielded by minimizing the difference between users' observed preferences and the product of the user and preference vectors. Similarly, the vectorization of items can be achieved with the user-item rating matrix and the users vectors. We conducted extensive experiments on three benchmark datasets to assess the quality of item vectors and the initialization independence of the user and item vectors. The utility of our vectorization results is shown by the recommendation performance achieved using them. Our experimental results show significant improvement over state-of-the-art baselines.

Keywords: Vectorization · Neural embedding · Recommender systems

1 Introduction

Based on neural networks, neural embedding has emerged as a successful category of vectorization techniques in recommender systems [2,8], among which *word2vec* [22,23] is a fundamental and effective algorithm. It was initially proposed for natural language processing problems and considers two states 1 or 0 for each word, representing either appearance or absence of the word in documents. It assumes that the words appearing closer to each other would have higher statistical dependence. Given its effectiveness, many variants have been

L. Azzopardi et al. (Eds.): ECIR 2019, LNCS 11437, pp. 323–336, 2019.
https://doi.org/10.1007/978-3-030-15712-8_21

proposed for machine learning problems, such as name speech recognition [25], entity resolution [19], machine translation [30], social embedding [12,24] and recommender systems [1,11]. Several pioneering efforts have been applied to real-world recommendation scenarios with neural embedding like *prod2vec* [11] and *item2vec* [1], that have been proposed by straightforwardly employing *word2vec*, where each user is considered as a document, and each item is simply regarded as a word. Consequently each item can only have two possible states 1 or 0, representing whether the user has performed a particular action (e.g. purchase, click, etc.) on the item or not. Using sets and sequences of items for each user, they learn the vector representations of the items.

Though such representations create good quality item vectors for some tasks, they lack the functionality to capture higher levels of granularities of users' feedback for vectorization. This could lead to incorrect interpretations, as the top-ranked item and low-ranked items would be treated equally. Thus it is expected to severely limit the vectorization quality for many tasks like calculating item similarities for single item recommendations, clustering user or items, etc. Currently, the efforts are limited for neural embedding-based methods, especially for datasets involving ratings. Therefore, we investigate the neural item embedding problem, to create quality vectorization for items using users' historical rating information with higher granularities (e.g. ratings in range 1 to 5).

To solve this problem, we propose *Pref2Vec* which involves three components: (1) The first step transforms the given user-item rating matrix into a users' pairwise preference matrix. On doing this, each pairwise preference of items has one of the two statuses i.e. occurrence or absence, which is similar to the situation of words in *word2vec*. (2) Then we employ neural embedding to create vector representations for pairwise item preferences by maximizing the likelihood estimation of the conditional probability of each pairwise item preference given another one. Using these preference vectors, the vectors of users can be generated by minimizing the difference between users' observed preferences and the product of the user and preference vectors in the second step of *Pref2Vec*. (3) In the last step, using the user vectors, the item vectors are generated similarly by minimizing the difference between items' observed ratings and the product of user and item vectors.

We evaluate the effectiveness of our *Pref2Vec* method in three experimental tasks on movie recommendation datasets to demonstrate its promising performance, where items are the movies for which user ratings are provided. (1) In the first task, we assess the quality of item vectors, by considering the movie genres as ground-truths. We find the similarities between each pair of items, using the generated item vectors and then using the ground truth (genres). The difference between these two similarities for item pairs are considered as the errors, using which we are able to compute RMSE (root mean squared error) and MAE (mean absolute error), as a quality measures for comparison. We contrast the quality of our item vectors with the quality of item vectors of other standard techniques, like: (a) item vectors generated using matrix factorization and (b) neural embedding item vectorization by using the sets of items that are rated by users as words. (2) In the second task, we run the vectorizations of the user and item vectors multiple times. We calculate the average variance of the generated values and the mean average covariance of the generated vectors, to establish that our vectorization process is highly independent of initialization. We contrast this with the vectorizations generated by matrix factorization. (3) Moreover, we compare the recommendation ranking generated using *Pref2Vec* with the standard collaborative filtering algorithms using the NDCG measure. Our results for these experimental tasks show performance gains over the comparison partners.

2 Related Work

Vectorization techniques are of great importance in machine learning. Specially in the area of natural language processing, neural embedding techniques for vectorization of words have been used in many applications [2,8,25,27–30]. Neural embedding techniques assume that the words that occur close to each other in the text are more dependent than the words that are far off. However, vectorization techniques using neural networks were inefficient to train, especially when the size and vocabulary of the dataset increased. But, the widely used word embedding technique *word2vec* that was introduced a few years ago, made creation of vector representations of words very efficient. It employs highly scalable skip-gram language model, that is fast to train and preserves the semantic relationships of the words in their vector representations. This technique for word embedding has recently shown considerable improvement in applications like name entity resolution [19] and word sense detection [3].

The success of *word2vec* has probably lead to the adoption of the neural embedding techniques in domains other than word representations. Djuric et al. [9] used vectorization of paragraphs as well as vectorization of words contained in each paragraph to create a hierarchical neural embedding framework. Also, Le et al. [20] created an algorithm that learns vector representations of sentences and text documents. They represent each document as dense vector that is utilized to predict words in the document. Moreover, Bordes et al. [4] have introduced the approach that embeds entities and relationships of multi-relational data in low-dimensional vector spaces, to be used for text classification and sentiment analysis tasks. Socher et al. [26] attempted to improve this approach by representing entities as an average of their constituting word vectors. Also, there have been recent efforts to learn the vector representations of nodes in graphs [12,24].

Moreover, several recent recommendation applications have employed neural word embedding. of *prod2vec* and *user2vec* by Grbovic et al. [11]. The *prod2vec* model creates vector representations of products by employing neural embedding on sequences of product purchases, where each product purchase is considered as a word. Whereas, the *user2vec* model considers a user as a global context in order to learn the vector representations of user and products. Similarly, *item2vec* [1] employs neural embedding on sets of items on which the user has taken action (e.g. songs played or products purchased), while ignoring the sequential information. The experimental results for these techniques show their effectiveness.

Although there have been many applications of neural embeddings in various areas including collaborative filtering, to the best of our knowledge, among the available neural embedding techniques on the rating information, there is no straightforward way to incorporate different levels of item ratings. He et al. [13] have utilized deep neural network frameworks for recommendation, but they also consider items in 1 and 0 state. Besides, there has not been an attempt to generate and utilize preference vectors. Hence, in this paper we attempt to generate preference vectors as an intermediate step, which can be utilized to generate good quality user and item vectors for various data mining tasks.

3 Problem Formulation

In this section, we formulate the neural rating vectorization problem, aiming to create vector representations for users and items by considering users' historical rating preference on items. Since matrix factorization can be actually regarded as traditional preference vectorization technique, let's firstly review its definition.

Consider a set of users U with m users, a set of items I with n items and a rating matrix R of dimension $m \times n$ containing ratings on n items given by m users. Each element $r_{u,i}$ of the uth row and ith column of R is the rating given by a particular user $u \in U$ for the item $i \in I$, where most of the elements in R are unknown as users generally can provide ratings only for a very small number of items. The objective of the rating vectorization problem is to generate a vector $\boldsymbol{u}$ for each user $u \in U$ and a vector $\boldsymbol{i}$ for each item $i \in I$, where the dot product of each user u and item i is close to the corresponding rating $r_{u,i}$ of i by u. Formally, the problem can be defined as follows:

Definition 1 (Matrix Factorization). *Given a set of users U with m users, a set of items I with n items, a rating matrix R of dimension $m \times n$ containing ratings on n items given by m users, the matrix factorization problem aims to create two low-rank dimensional matrices $\boldsymbol{U}$ of dimension $k \times m$ and $\boldsymbol{V}$ of dimension $k \times n$ for users and items respectively by minimizing the following objective function:*

$$\arg\min_{U,V} \sum_{u \in U, i \in I} \phi_{u,i} \left(r_{u,i} - \boldsymbol{u}^{\top} \boldsymbol{i} \right),$$

where $\phi_{u,i} = 1$, if u has rated i; otherwise 0, we define a novel neural rating vectorization problem. It treats the possible ratings on each item i as an intrinsic property of the item, which indicates the quality of i and thus are independent from users. The neural rating vectorization problem aims to generate rating vectors on items by maximizing the likelihood estimation of the conditional probability of each score on item given another one. Formally, the neural item embedding problem can be defined as:

Definition 2 (Neural Item Embedding). *Let U, I and R be a set of users U with m users, a set of items I with n items, and a rating matrix R of dimension $m \times n$ containing ratings on n items given by m users, respectively.*

The neural item embedding problem aims to create low rank vector representations of dimension $k \times m$ for items I by minimizing the following objective function:

$$\arg\min_{I} - \sum_{u \in U} \sum_{i,j \in I, i \neq j} \phi_{u,i} \phi_{u,j} \log Prob(r_i = r_{u,i} \mid r_j = r_{u,j}), \quad (1)$$

where $Prob(r_i = r_{u,i} \mid r_j = r_{u,j})$ is the probability that user u provides a score of $r_{u,i}$ to item i given that the same user u assigns a score of $r_{u,j}$ to another item j. Once we obtain the item vectors by solving the above problem, user vectors can be generated directly by minimizing the difference between items' observed ratings and the product of user and item vectors: $\arg\min_{U} \sum_{u \in U, i \in I} \phi_{u,i} \left(r_{u,i} - \boldsymbol{u}^{\top} \boldsymbol{i} \right)$.

Note that the probability $Prob(r_i = r_{u,i} \mid r_j = r_{u,j})$ in Eq. (1) actually involves two aspects of information: (1) the co-occurrence of ratings on each pair of items by same users, and (2) the rating scores or relative preferences of users holds on items. Thus it is extremely hard to be formulated by straightforwardly adapting that in *word2vec* [22,23] with hierarchical softmax of the vectors.

4 The *Pref2Vec* Algorithm

Pref2Vec solves the neural item embedding problem in Definition 2 in three steps. Firstly, we generate vectors of pairwise item preference. We use these preference vectors in the second step to generate user vectors, that are in turn used to create item vectors in the third step.

4.1 Pairwise Preference Vectorization

To create vectors of pairwise item preferences, we create the pairwise preference matrix and use it to create the sets of positive pairwise preferences for each user. Then, we utilize neural language models to learn representations of positive preferences in lower dimensional space using available positive preference pairs.

Consider a set of users $U = \{u_1, u_2, \ldots, u_m\}$, a set of items $I = \{I_1, I_2, \ldots, I_n\}$ and their corresponding rating matrix R of dimension $m \times n$. Each row of R contains ratings $R_u = \{r_1, r_2, \ldots, r_n\}$ given by a user u for the n items, where most of the elements in R_u are unknown as users generally can provide ratings only for a very small number of items. This allows us to build a set of pairwise preference for each user by using a preference function: $p(i,j) \in \{+1, -1\}$, where $i = 1 \ldots n$, $j = 1 \ldots n$, $i \neq j$ and both r_i and r_j are known. The preference function $p(i,j)$ has a value of $+1$ if $r_i > r_j$ and -1 otherwise.

Now, we create the sets of positive preferences P_u for each user u. Without losing generality, here we only consider the conditions of positive preference pairs, as all of the negative preferences can be straightforwardly transformed into positive ones by reversing the positions of the two items. With n items, we should consider a total of $N = n(n-1)$ unique preference pairs, denoted as $P = \{p_1, p_2, \ldots, p_N\}$. Each users' preferences P_u is a subset of P, formally $P_u \subseteq P$ for any user u. Now, *Pref2Vec* proceeds with learning the vector representations of the preferences on the collection of preference sets $\mathbb{P} = \{P_1, P_2, \ldots, P_m\}$ for all of the users.

We consider the *word2vec* framework [22,23] that generates vector representations of words. They presented the continuous *skip-gram* model, which assumed that for each target word the sequence of its surrounding words are trivial and can be ignored. This is achieved by maximizing the cumulative logarithm of the conditional probability for the surrounding words given each target word in the corpus with neural networks. Our approach is very similar, since we consider our collection of preference sets: $\mathbb{P}$ as the corpus, the preference sets $P_1, P_2, \ldots, P_m$ by the users as the sentences and the preferences $p_1, p_2, \ldots, p_N$ as the words.

However, the key difference in our approach is that we completely ignore the spatial information within the preference sets. This is because unlike words in sentences, the order of the preferences for a user (in a non-temporal setup) is inconsequential. This is the reason why we have a set representation of preferences for a user, as opposed to a sequence representation. Actually this property makes our scenario even better fit the *skip-gram* model than natural language processing, where the preferences have no sequence information and thus the sequence of the "surrounding preferences" can be ignored without any accuracy loss. Therefore, in the *Pref2Vec* framework, we learn the vector representations of the products by minimizing the following objective function over the entire collection $\mathbb{P}$ of preference sets:

$$\underset{i_1, i_2, \ldots, i_n}{\arg\min} - \sum_{P_k \in \mathbb{P}} \sum_{(p_i, p_j) \in P_k, i \neq j} \log Prob(p_j \mid p_i), \tag{2}$$

where $Prob(p_j \mid p_i)$ is the hierarchical softmax of the respective vectors of the preference p_j and p_i. In particular, $Prob(p_j \mid p_i) = \frac{\exp(\boldsymbol{i}_o^\top \boldsymbol{j}_t)}{\sum_{p_l \in P_k} \exp(\boldsymbol{i}_o^\top \boldsymbol{l}_t)}$,where $\boldsymbol{i}_o$ and $\boldsymbol{j}_t$ are the initial and target vector representations respectively of preferences p_i and p_j. $\boldsymbol{l}_t$ is the target vector representations of any preference p_l in P_k. From Eq. (2), we see that *Pref2Vec* model ignores the sequence of preferences within a user preferences set. The context is set to the level of preference sets, where the preference vectors that fall in the same preference sets will have similar vector representations.

Remarks: Our approach is also inspired by *item2vec* [1], that uses a straightforward application of *word2vec* by considering a set of items (accessed by a user) as a sentence and the individual items as words. Similar to *Pref2Vec*, *item2vec* also ignores the sequential information of items in a set. *item2vec* has been efficiently used in scenarios where we have a simple sequence of items, e.g. products purchased, videos watched, etc. In such cases, for each user the items are in 0 or 1 state. However, if the user feedback is provided in higher granularities (e.g. user ratings), then simply considering the sequence of items rated by the user and treating them equally, is expected to severely limit the quality of vectors. On the other hand, *Pref2Vec* enables the utilization of rating information by incorporating pairwise item preferences in the vectorization process.

4.2 User Vector Generation

However, the preference vectors generated in the previous section cannot be utilized directly for recommendation tasks, that often require good quality user and item vectors as an input. In this section we describe the second step of *Pref2Vec* and aim to find vectors corresponding to the m users, given the preference vectors for each pair of items and known ground truths for the preferences.

For a particular user let $\boldsymbol{p}_1, \boldsymbol{p}_2 \dots \boldsymbol{p}_r$ be the preference vectors, each of length k, for which the respective values of preference function are $p_1, p_2, \dots, p_r \in \{+1, -1\}$. The corresponding user vector can be achieved by minimizing the cumulative difference between users' each observed preference p_i and the product of the user and preference vectors $\boldsymbol{u}^\top \boldsymbol{p}_i$. Thus we can formulate this problem as linear classification, where $\boldsymbol{p}_1, \boldsymbol{p}_2 \dots \boldsymbol{p}_r$ are training instances, the values of the preference functions $p_1, p_2, \dots, p_r$ are ground truth. With consideration of a bias b, we aim to predict the coefficients of a linear classification model, which is the user vector $\boldsymbol{u}$. In this study, we use Logistic regression [16] to solve the problem. The loss function with L2 norm is: $\underset{\boldsymbol{u},b}{\arg\min} \sum_{i=1}^{r} \log(1 + \exp(-p_i(\boldsymbol{u}^T \boldsymbol{p}_i + b))) + \frac{\lambda}{2}||\boldsymbol{u}||^2$, where $\boldsymbol{u}$ is a vector of length k, b is a number and λ is the tuning parameter for L2 norm. We use the gradient descent method for optimization. Given a learning rate α, the update formulas are derived as follows:

$$\begin{aligned} \boldsymbol{u} &\leftarrow \boldsymbol{u} - \alpha\left(\sum_{i=1}^{r} \frac{-p_i}{1 + \exp(p_i(\boldsymbol{u}^T \boldsymbol{p}_i + b))} \boldsymbol{p}_i + \lambda \boldsymbol{u}\right) \\ b &\leftarrow b - \alpha\left(\sum_{i=1}^{r} \frac{-p_i}{1 + \exp(p_i(\boldsymbol{u}^T \boldsymbol{p}_i + b))}\right) \end{aligned} \tag{3}$$

The generated user vectors $\boldsymbol{u}$ corresponding to each of the m users, form a user matrix U of dimension $m \times k$.

4.3 Item Vectors Generation

The last step of *Pref2Vec* is to find item vectors given the rating matrix $R_{m \times n}$ and the user matrix $U_{m \times k}$ generated in the previous section. For this we optimize matrix $I_{n \times k}$, by minimizing the difference between items' observed ratings and the product of user and item vectors, i.e. $UI^\top \approx R$. The n rows of I would be the item vectors. We minimize the loss function: $\underset{I}{\arg\min} ||R - UI^\top||^2 + \frac{\lambda}{2}||I||^2$, where λ is the tuning parameter for L2 normalization. We use the gradient descent method for optimization. Given a learning rate η, the update formula is:

$$I \leftarrow I - \eta(-2(R - UI^T)^\top U + \lambda I) \tag{4}$$

5 Experiments

The following three research questions guide the remainder of the paper.
RQ1. Is the quality of item vectors generated using the *Pref2Vec* approach better than state-of-the-art vectorization algorithms? (See Sect. 5.1)
RQ2. Are the outputs of the proposed *Pref2Vec* algorithms independent from their initialization? (See Sect. 5.2)
RQ3. Can the vectorization results be utilized to improve the performance of recommender systems? (See Sect. 5.3)

Datasets. We use three MovieLens[1] data sets in our experiments: MovieLens-100K, MovieLens-1M and MovieLens-10M. MovieLens-100K dataset contains 100,000 ratings given by 943 users on 1682 movies. MovieLens-1M dataset is larger with 1,000,000 ratings given by 6040 users on 3952 movies. Movielens-10M is the largest dataset used, with 10 million ratings given by 69878 users on 10681 movies. In MovieLens-100K as well as MovieLens-1M the ratings are given as integers from 1 to 5. In MovieLens-10M, the ratings are given in the range 0.5 to 5 with an increment of 0.5. In these datatsets there are 18 movie genres, a movie can belong to one or more of them. For all the three datasets we randomly assign 10 ratings for each user for testing and the rest for training. We have used the vector length of 10 for all the vectorization methods.

5.1 Evaluation of Item Quality

Ground-Truth. Since the datasets provide genre information for all of the items (movies), we use the genre similarity as the ground truth. In particular, the genres of each movie are provided (or can be transformed) in the form of binary values. A value of 1 signifies that the movie belongs to a particular genre and 0 signifies the contrary. A movie can belong to more than one genre. So, let us consider that genre vectors derived from the meta-data are: $(\boldsymbol{G}_i \ldots \boldsymbol{G}_j)$, which correspond to our item vectors $\boldsymbol{I}_i \ldots \boldsymbol{I}_j$. Since, the genre vectors are binary vectors, to find similarity between them we use: Jaccard similarity [6], an efficient and popular measure for binary similarity. Jaccard similarity between two binary vectors $\boldsymbol{v}_a$ and $\boldsymbol{v}_b$ is simply calculated as: $jacSim(\boldsymbol{v}_a, \boldsymbol{v}_b) = \frac{F_{11}}{F_{01}+F_{10}+F_{11}}$,where F_{11} is the number of features for which both $\boldsymbol{v}_a$ and $\boldsymbol{v}_b$ have value 1. F_{01} is the number of features for which $\boldsymbol{v}_a$ has value 0 and $\boldsymbol{v}_b$ has 1. And, F_{10} is the number of features where $\boldsymbol{v}_a$ had the value 1 and $\boldsymbol{v}_b$ has 0.

For the item vectors $\boldsymbol{I}_1, \boldsymbol{I}_2, \ldots, \boldsymbol{I}_n$ (calculated in Sect. 4.3), the similarity can be calculated for each pair of item vectors $(\boldsymbol{I}_i, \boldsymbol{I}_j)$ as: $cosSim(\boldsymbol{I}_i, \boldsymbol{I}_j) = \frac{\boldsymbol{I}_i^\top \boldsymbol{I}_j}{|\boldsymbol{I}_i| \times |\boldsymbol{I}_j|}$, where $|\boldsymbol{I}_i|$ and $|\boldsymbol{I}_i|$ are the length of the vectors $\boldsymbol{I}_i$ and $\boldsymbol{I}_j$.

Evaluation Metrics. In order to evaluate the quality of item vectors we use the RMSE (root mean squared error) and MAE (mean absolute error) measures. To calculate these, we calculate the similarities between each pair of item vectors and the similarities between their corresponding pairs of ground truths. Since the item in our experiments are movies, the genre information about the

[1] http://grouplens.org/datasets/movielens/.

movies (available from metadata) is considered as ground truth. The differences between the two similarities for each item are considered as errors, that are in turn used to calculate RMSE and MAE. We use these measures because for good quality item vectors, the vectors that are similar should also have similarity based on their relevant meta data information. Therefore, the lower the values of RMSE and MAE, the better is the quality of vectors.

To calculate the errors we need: the difference between the similarities of two item vectors and the similarities between the corresponding two genre vectors. The errors are calculated for all pairs of items: $e_{i,j} = cosSim(\boldsymbol{I}_i, \boldsymbol{I}_j) - jacSim(\boldsymbol{G}_i, \boldsymbol{G}_j)$. Though cosine similarity and Jaccard similarity are different measurements, their difference used here is expected to be highly indicative of the error. There would be $n(n-1)/2$ such errors. Now, $RMSE = \sqrt{\frac{\sum_{i=1}^{n}\sum_{j=i+1}^{n} e_{i,j}^2}{n(n-1)/2}}$ and $MAE = \frac{\sum_{i=1}^{n}\sum_{j=i+1}^{n} |e_{i,j}|}{n(n-1)/2}$, where the function $|\cdot|$ gives the absolute value of the parameter.

Baselines. We choose the following methods to evaluate the quality of the item vectors that are generated by the *Pref2Vec* framework, i.e. P2V-Vectors.

- **RM-Vectors:** Rating matrix $R_{m \times n}$ contains ratings by m users for n items, and its columns are the simplest (and readily available) form of item vectors.
- **IS-Vectors:** Neural embeddings of items are created by considering the set of items rated by users as sentences and items as words (similar to *item2vec* [1] approach). Comparison with this method would validate the importance of using preference information for vectorization in *Pref2Vec*.
- **MF-Vectors:** In matrix factorization [18] user and item vectors are created by randomly initializing matrices $U_{m \times k}$ and $I_{n \times k}$ and then minimizing the difference between their product and the rating matrix (i.e. $R - UI^{\top}$).

Results. In Table 1, we compare the item vector qualities using RMSE and MAE. For MovieLens-100K dataset, for both RMSE and MAE, P2V-Vectors perform the best, followed by MF-Vectors. IS-Vectors are the third and the RM-Vectors are the worst performing ones. The trend is same for the dataset MovieLens-1M for RMSE. For MovieLens-1M in terms of MAE as well as for MovieLens-10M (both RMSE and MAE), though P2V-Vectors are still the best performing ones, the second best are IS-Vectors, followed by MF-Vectors and then RM-Vectors. The improvement shown by P2V-Vectors is significant.

Table 1. Quality of generated item vectors against baselines

Algorithm	ML-100K		ML-1M		ML-10M		Ref.
	RMSE	MAE	RMSE	MAE	RMSE	MAE	
RM-Vectors	0.8674	0.8163	0.8790	0.8292	0.8563	0.8151	–
IS-Vectors	0.8238	0.7508	0.7018	0.5886	0.5864	0.4758	[1]
MF-Vectors	0.6844	0.6431	0.6904	0.6478	0.6478	0.6071	[18]
P2V-Vectors	**0.4770**	**0.3846**	**0.5165**	**0.4305**	**0.4456**	**0.3695**	This paper

Pref2Vec firstly generates preference vectors, and then creates user vectors with the generated preference vectors, and finally produces item vectors with the generated user vectors. Since each step is an approximation process with certain accuracy loss, the preference and user vectors should be more accurate than the item vectors. Thus although we cannot assess the quality of user and preference vectors resulting from lack of corresponding ground-truth information, we can still claim that the quality of the preference, user and item vectors generated by our *Pref2Vec* method can significantly outperform our baselines.

5.2 Evaluation of Initialization Independence of Generated Vectors

Firstly, we describe the measurements to evaluate the independence of generated vectors from their initialization. Let us consider that x different runs (resulting from different initializations) of a vector generation method generate: user matrices $U^{(1)} \dots U^{(x)}$ and the corresponding item matrices $I^{(1)} \dots I^{(x)}$. Each user matrix is of dimension $m \times k$ with the rows corresponding to m user vectors, each of length k. Similarly each item matrix is of dimension $n \times k$ with the rows corresponding to n item vectors, each of length k. Since the features of the vectorization results might be in a different order by different runs of algorithms, we sort the generated features according to their cumulative values among all of the users. The independence of these vectors from the initialization can be measured using (a) variance of the elements of the U and I matrices and (b) correlations between the user and item vectors generated in different runs. These measures are explained in detail as follows.

Variance Calculation. Let $U_{i,j}^{(1)}, U_{i,j}^{(2)} \dots U_{i,j}^{(x)}$ be the x values in the user matrices at ith row and jth column from x different runs of a vectorization algorithm. Their variances can be calculated as: $var_U(i,j) = \frac{1}{x}\sum_{l=1}^{x}\left(U_{i,j}^{(l)} - \overline{U}_{i,j}\right)^2$, where $\overline{U}_{i,j}$ is the average of $U_{i,j}^{(1)}, U_{i,j}^{(2)} \dots U_{i,j}^{(x)}$. With $m \times k$ dimensions of the user matrix U, we can get $m \times k$ variance, and the mean variance would be: $MVU = \frac{1}{m \times k}\sum_{i=1}^{m}\sum_{j=1}^{k} var_U(i,j)$.

Similarly, the variance of the item matrices at the ith row and jth column $I_{i,j}^{(1)}, I_{i,j}^{(2)} \dots I_{i,j}^{(x)}$ from x different runs of a vectorization algorithm can be calculated as: $var_I(i,j) = \frac{1}{x}\sum_{l=1}^{x}\left(I_{i,j}^{(l)} - \overline{I}_{i,j}\right)^2$, where $\overline{I}_{i,j}$ is the average of $I_{i,j}^{(1)}, I_{i,j}^{(2)}, \dots, I_{i,j}^{(x)}$. The mean variance of the generated item vectors can be calculated as: $MVI = \frac{1}{n \times k}\sum_{i=1}^{n}\sum_{j=1}^{k} var_I(i,j)$

A lower value of the mean variance is indicative that the generated values that comprise the user or item vectors do not vary much with different initializations.

Correlation of Vectors. The independence of the vectors from the initialization of the generation technique can also be estimated by the correlation between the vectors generated in different runs. We use Pearson correlation coefficient [15] to calculate correlation $\rho(x, y)$ between variables x and y.

A user matrix $U^{(j)}$ generated in the jth run, contains m user vectors: $\boldsymbol{u}_1^{(j)} \ldots \boldsymbol{u}_m^{(j)}$. For a particular user, the average of pairwise correlations between the vectors generated in the x runs would be: $AC(i) = \frac{\sum_{j=1}^{x}\sum_{l=j+1}^{x}\rho(\boldsymbol{u}_i^{(j)},\boldsymbol{u}_i^{(l)})}{x(x-1)/2}$. And, the mean of these average correlation for all the m user vectors can simply be calculated as: $MAC = \frac{\sum_{i=1}^{m}\sum_{j=1}^{x}\sum_{l=j+1}^{x}\rho(\boldsymbol{u}_i^{(j)},\boldsymbol{u}_i^{(l)})}{m\times x(x-1)/2}$.

Similarly, the mean average correlation for the item vectors, $\boldsymbol{i}_1^{(j)} \ldots \boldsymbol{i}_n^{(j)}$ generated in x runs ($j = 1 \ldots x$), can be calculated as: $MAC = \frac{\sum_{i=1}^{n}\sum_{j=1}^{x}\sum_{l=j+1}^{x}\rho(\boldsymbol{i}_i^{(j)},\boldsymbol{i}_i^{(l)})}{n\times x(x-1)/2}$. A high value if MAC mean that the vectors generated during different runs are close to each other and hence have high level of independence to initialization.

Results. In Table 2, we show the results evaluating the initialization independence of user and item vectors generated using *Pref2Vec* (shown as P2V) and comparing them with the vectors generated by matrix factorization (shown as MF). On the dataset MovieLens-100K we run both the methods 5 times, resulting in creation of 5 different pairs of user and item vectors for both of them. We calculate MVU, MVI and MAC (for user and item vectors) for the vectorization results generated by P2V and matrix factorization (MF).

The values of MVU and MVI of our algorithm are merely 0.0015 and 0 for user and item vectors, which are sharply lower than that of the matrix factorization method. Note that although the values of matrix factorization are smaller than 0.1, they are still large because the values in the user and item matrices are very small, and most of them are less than 1. Also, the values of MAC are very high for P2V for both item and user vectors, especially in comparison with the respective values for MF. This again shows that the user and item vectors generated by P2V in different runs are highly correlated to each other.

Table 2. Initialization independence of generated vectors

Algorithm	User vectors		Item vectors	
	MVU	MAC	MVI	MAC
MF	0.0730	0.0015	0.0773	0.0315
P2V	**0.0015**	**0.8592**	0	**0.7881**

5.3 Ranking Prediction Based on Generated Vectors

Ranking Model Using User Vectors. Here we describe the method to generate rankings for items with unknown ratings for user using the available *Pref2Vec* preference and user vectors. This is done by firstly predicting the preference values $\hat{p} \in \{+1, -1\}$ for the preference vectors corresponding to the items with unknown ratings. Then we employ a greedy order algorithm to derive approximately optimal ranking of the unrated items.

In Sect. 4.2 we showed the process that generates the user vector $\boldsymbol{u}$ and the value b after optimization. Since the optimization process directly employs the Logistic regression loss function, this allows us to also directly use Logistic regression classification to predict pairwise preferences for a user. More specifically, for a user with user vector $\boldsymbol{u}$ and accompanying value b, the user's preference $\hat{p}_u$ can be predicted as: $\hat{p}_u = +1$, if $\boldsymbol{u}^\top \boldsymbol{p} + b > 0$; -1 otherwise.

Hence, for a particular user, if there are q items with unknown rankings $I_1, I_2 \ldots I_q$, the values for the preference function $\hat{p}(I_i, I_j) \in \{+1, -1\}$, can be predicted. Since the values for pairwise preference function are not a direct format to get the rankings, we use the greedy order algorithm proposed by Cohen et al. [7,21], that efficiently finds an approximately optimal ranking for the target user u. It is showed that based on reduction of cyclic ordering problem [10], the determination of optimal ranking is a NP-complete problem and the algorithm can be proved to have an approximation ratio of 2 [10].

Remarks. Alternatively, we could have directly used the user matrix U (Sect. 4.2) and the item matrix I (Sect. 4.3) to generate the ratings matrix ($R = UI^T$), that could be used to generate ranking of unrated items. However, since we follow sequential steps by first generating U from preference vectors, then using U to create I and thereafter using U and I to create R; there is accuracy loss at each step. On the other hand, our ranking model avoids such additional inaccuracies by directly using preference vectors and U to generate rankings.

Baselines. We use the following baselines to access the performance of our simple recommendation method P2VRank:

- **CF:** CF [5] is a memory-based collaborative filtering algorithm that uses the Pearson correlation coefficient to calculate the similarity between users.
- **MF:** Given a raking matrix R, in matrix factorization [18] the user matrix U and the item matrix I are optimized in order to minimize the difference: $R - UI^\top$.
- **EigenRank:** EigenRank [21] uses greedy aggregation method to aggregate the predicted pairwise preferences of items into total ranking.

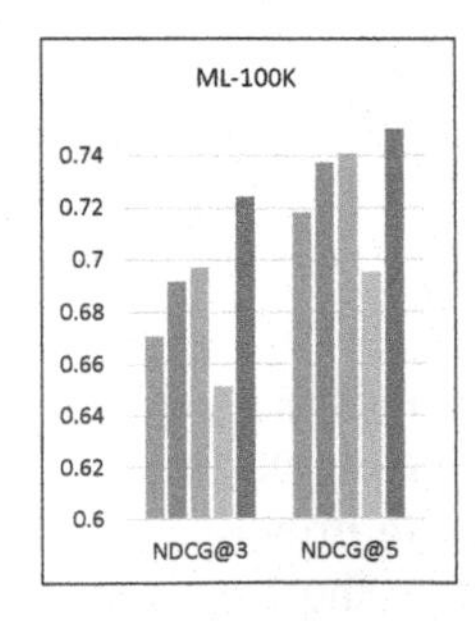

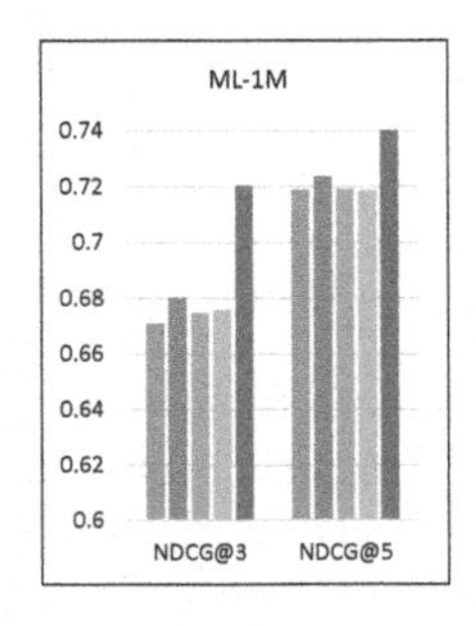

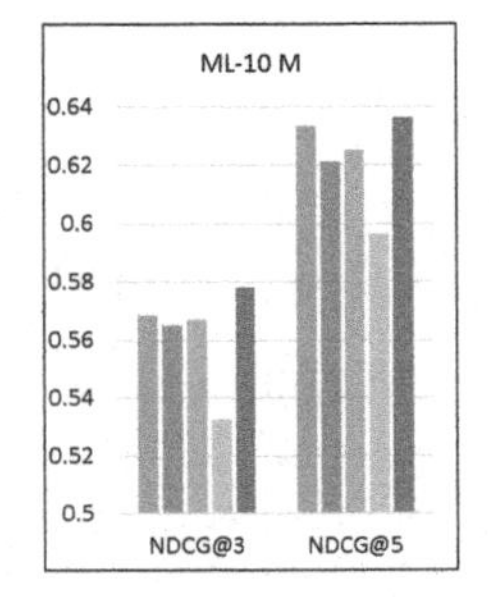

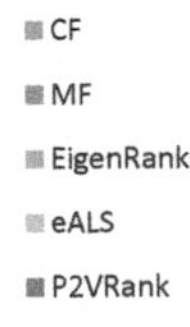

Fig. 1. Ranking performance of *Pref2Vec* against baselines

- **eALS:** Element-wise Alternating Least Squares (eALS) [14] efficiently optimizes a MF model with variably-weighted missing data. As eALS is an implicit feedback algorithm, we consider only higher ratings ($\geq$4) as positive feedback.

Results. The performance is evaluated using the standard ranking accuracy metric NDCG [17] @3 and @5. In Fig. 1, we see that P2VRank outperforms all comparison partners. Also, we also observed strong statistical significance ($\alpha = 0.05$) on comparing P2VRank against MF for all the three datasets.

6 Conclusion

We proposed *Pref2Vec* to generate vector representations of pairwise item preferences. We also presented the method to generate user and item vectors using preference vectors. Also, our experimental results demonstrated that the quality of item vectors generated by *Pref2Vec* is better than that of the standard techniques. We also verified that the generated user and item vectors are highly independent of the initializations. In addition, we presented the technique to generate rankings of items, using the generated user vectors, and showed that it outperforms the standard recommendation techniques. Currently we only consider the preference of one item over another for the creation of *Pref2Vec* and in future we would like to consider the magnitudes of these preferences.

References

1. Barkan, O., Koenigstein, N.: Item2Vec: neural item embedding for collaborative filtering. In: MLSP, pp. 1–6 (2016)
2. Bengio, Y., Ducharme, R., Vincent, P., Janvin, C.: A neural probabilistic language model. J. Mach. Learn. Res. **3**, 1137–1155 (2003)
3. Bhingardive, S., Singh, D., V, R., Redkar, H.H., Bhattacharyya, P.: Unsupervised most frequent sense detection using word embeddings. In: HLT-NAACL, pp. 1238–1243 (2015)
4. Bordes, A., Usunier, N., Garcia-Durán, A., Weston, J., Yakhnenko, O.: Translating embeddings for modeling multi-relational data. In: NIPS, pp. 2787–2795 (2013)
5. Breese, J.S., Heckerman, D., Kadie, C.: Empirical analysis of predictive algorithms for collaborative filtering. In: UAI, pp. 43–52 (1998)
6. Choi, S., Cha, S., Tapper, C.C.: A survey of binary similarity and distance measures. J. Systemics Cybern. Inform. **8**(1), 43–48 (2010)
7. Cohen, W.W., Schapire, R.E., Singer, Y.: Learning to order things. J. Art. Int. Res. **10**(1), 243–270 (1999)
8. Collobert, R., Weston, J., Bottou, L., Karlen, M., Kavukcuoglu, K., Kuksa, P.: Natural language processing (almost) from scratch. J. Mach. Learn. Res. **12**, 2493–2537 (2011)
9. Djuric, N., Wu, H., Radosavljevic, V., Grbovic, M., Bhamidipati, N.: Hierarchical neural language models for joint representation of streaming documents and their content. In: WWW, pp. 248–255 (2015)

10. Garey, M.R., Johnson, D.S.: Computers and Intractability; A Guide to the Theory of NP-Completeness. W. H. Freeman & Co., New York (1990)
11. Grbovic, M., et al.: E-commerce in your inbox: Product recommendations at scale. In: SIGKDD, pp. 1809–1818 (2015)
12. Grover, A., Leskovec, J.: Node2vec: scalable feature learning for networks. In: SIGKDD, pp. 855–864 (2016)
13. He, X., Liao, L., Zhang, H., Nie, L., Hu, X., Chua, T.S.: Neural collaborative filtering. In: WWW, pp. 173–182 (2017)
14. He, X., Zhang, H., Kan, M.Y., Chua, T.S.: Fast matrix factorization for online recommendation with implicit feedback. In: SIGIR, pp. 549–558 (2016)
15. Herlocker, J., Konstan, J.A., Riedl, J.: An empirical analysis of design choices in neighborhood-based collaborative filtering algorithms. ACM Trans. Inf. Syst. **5**, 287–310 (2002)
16. Hosmer Jr., D.W., Lemeshow, S., Sturdivant, R.X.: Applied Logistic Regression, vol. 398. Wiley, Hoboken (2013)
17. Järvelin, K., Kekäläinen, J.: Cumulated gain-based evaluation of IR techniques. ACM Trans. Inf. Syst. **20**(4), 422–446 (2002)
18. Koren, Y., Bell, R., Volinsky, C.: Matrix factorization techniques for recommender systems. Computer **42**(8), 30–37 (2009)
19. Lample, G., Ballesteros, M., Subramanian, S., Kawakami, K., Dyer, C.: Neural architectures for named entity recognition. In: HLT-NAACL, pp. 260–270 (2016)
20. Le, Q., Mikolov, T.: Distributed representations of sentences and documents. In: ICML, pp. 1188–1196 (2014)
21. Liu, N.N., Yang, Q.: Eigenrank: a ranking-oriented approach to collaborative filtering. In: SIGIR, pp. 83–90 (2008)
22. Mikolov, T., Chen, K., Corrado, G., Dean, J.: Efficient estimation of word representations in vector space. CoRR abs/1301.3781 (2013)
23. Mikolov, T., Sutskever, I., Chen, K., Corrado, G.S., Dean, J.: Distributed representations of words and phrases and their compositionality. In: NIPS, pp. 3111–3119 (2013)
24. Perozzi, B., Al-Rfou, R., Skiena, S.: Deepwalk: online learning of social representations. In: SIGKDD, pp. 701–710 (2014)
25. Schwenk, H.: Continuous space language models. Comput. Speech Lang. **21**(3), 492–518 (2007)
26. Socher, R., Chen, D., Manning, C.D., Ng, A.: Reasoning with neural tensor networks for knowledge base completion. In: NIPS, pp. 926–934 (2013)
27. Socher, R., Lin, C.C., Ng, A.Y., Manning, C.D.: Parsing natural scenes and natural language with recursive neural networks. In: ICML (2011)
28. Turian, J., Ratinov, L., Bengio, Y.: Word representations: a simple and general method for semi-supervised learning. In: ACL, pp. 384–394 (2010)
29. Turney, P.D.: Distributional semantics beyond words: supervised learning of analogy and paraphrase. TACL **1**, 353–366 (2013)
30. Zou, W.Y., Socher, R., Cer, D.M., Manning, C.D.: Bilingual word embeddings for phrase-based machine translation. In: EMNLP, pp. 1393–1398 (2013)

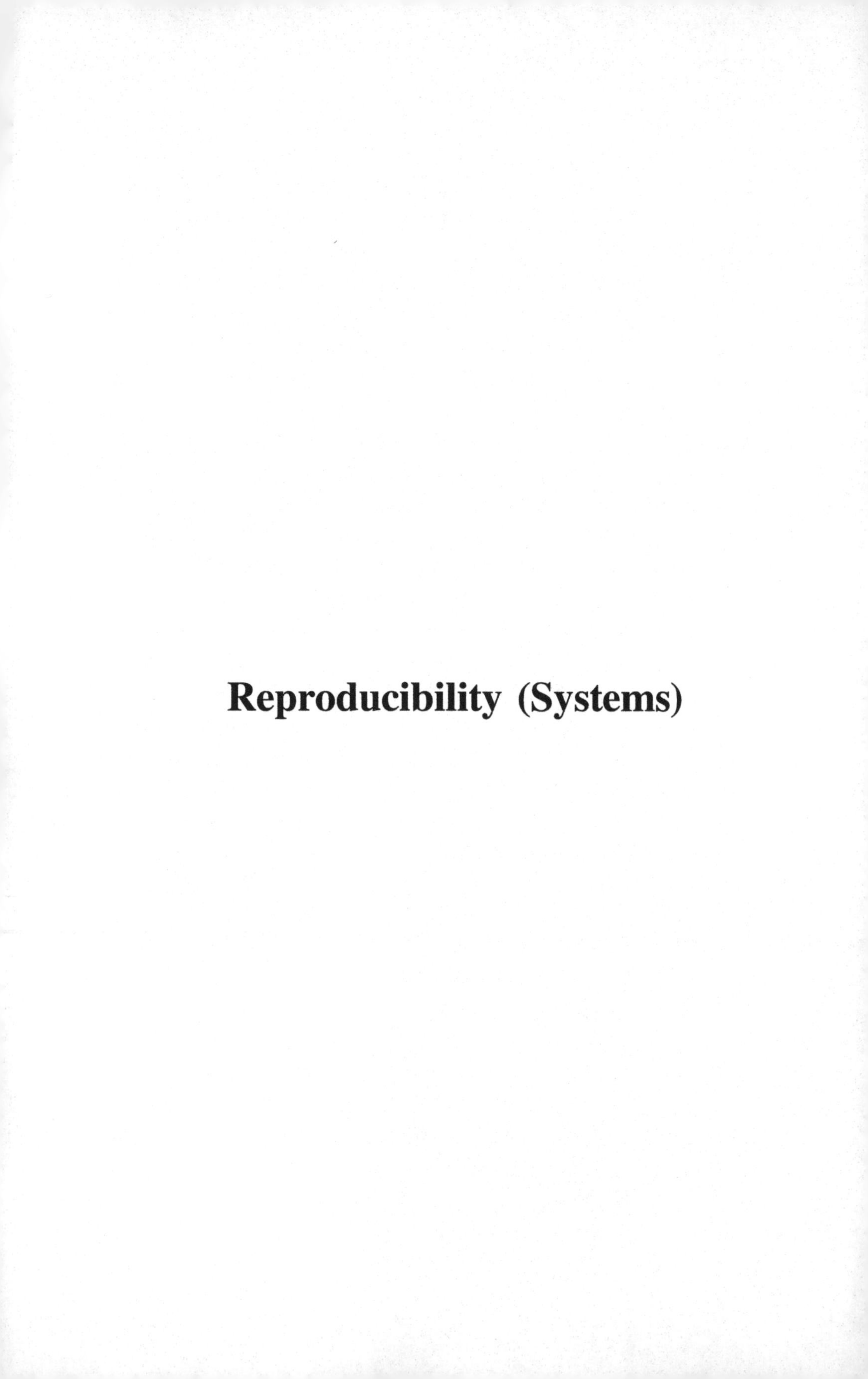

Reproducibility (Systems)

Compressing Inverted Indexes with Recursive Graph Bisection: A Reproducibility Study

Joel Mackenzie[1](✉), Antonio Mallia[2], Matthias Petri[3], J. Shane Culpepper[1], and Torsten Suel[2]

[1] RMIT University, Melbourne, Australia
{joel.mackenzie,shane.culpepper}@rmit.edu.au
[2] New York University, New York, USA
{antonio.mallia,torsten.suel}@nyu.edu
[3] The University of Melbourne, Melbourne, Australia
matthias.petri@unimelb.edu.au

Abstract. Document reordering is an important but often overlooked preprocessing stage in index construction. Reordering document identifiers in graphs and inverted indexes has been shown to reduce storage costs and improve processing efficiency in the resulting indexes. However, surprisingly few document reordering algorithms are publicly available despite their importance. A new reordering algorithm derived from recursive graph bisection was recently proposed by Dhulipala et al., and shown to be highly effective and efficient when compared against other state-of-the-art reordering strategies. In this work, we present a reproducibility study of this new algorithm. We describe the implementation challenges encountered, and explore the performance characteristics of our clean-room reimplementation. We show that we are able to successfully reproduce the core results of the original paper, and show that the algorithm generalizes to other collections and indexing frameworks. Furthermore, we make our implementation publicly available to help promote further research in this space.

Keywords: Reordering · Compression · Efficiency · Reproducibility

1 Introduction

Scalable processing and storage of large data collections has been a longstanding problem in Information Retrieval (IR). The volume of data being indexed and retrieved continues to grow, and a wealth of academic research has focused on managing this new data. A key area of focus is how to better compress the data structures used in these storage applications; better compression results in lower storage costs, and improves the efficiency of accessing data. *Document reordering* is a widely used technique that improves the compression rate of many coding schemes at the cost of additional computation at indexing time.

L. Azzopardi et al. (Eds.): ECIR 2019, LNCS 11437, pp. 339–352, 2019.
https://doi.org/10.1007/978-3-030-15712-8_22

However, finding a favorable reordering is a challenging problem for IR-scale data collections. The problem has been extensively studied in academia, and making significant improvements that are both effective and practical is quite difficult.

Documents identifiers can be assigned in many ways, such as random ordering, based on document similarity or page relevance [21], or by just following a sorted URL ordering [23]. Further, it has been noted that document reordering can also result in improved query processing efficiency [11,15,16], although a thorough understanding of this effect is still missing. Considering the advantages that reordering can yield, it is critical that tools for reordering are made publicly available, and that researchers describe the order of their index when conducting large scale efficiency studies (such as [8,11,16,18,20]).

Recently, Dhulipala et al. [10] proposed a new algorithm which aims to minimize an objective function directly related to the number of bits needed to store a graph or an index using a delta-encoding scheme. The authors experimented on both graphs and inverted indexes, obtaining notable improvements when compared to previous approaches. Their algorithm, based on *recursive graph bisection*, is currently the state-of-the-art algorithm for minimizing the compressed space used by an inverted index (or graph) through document/vertex reordering. An unfortunate aspect of this work is that the implementation was unable to be released "due to corporate restrictions," most likely because the work was done primarily at Facebook. In this paper, we perform a "clean-room" reimplementation of this algorithm, reproduce the results obtained by the original authors, and extend their original experiments to additional collections in order to confirm the effectiveness of the approach.

Our Contributions. The key contributions of this paper are:

1. We implement and validate the algorithm originally presented by Dhulipala et al. [10]. We confirm both effectiveness in compression due to reordering and efficiency in terms of execution time and memory usage.
2. We extend the experimental analysis to other large collections. The original work focused primarily on reordering graphs, with experiments shown for just two standard text collections: Gov2 and ClueWeb09. With an extensive experimental analysis over four additional text collections, we strengthen the evidence of the generalizability of the approach.
3. We evaluate an additional compression technique with the reordered index, to further examine how well the approach generalizes.
4. We make our implementation publicly available in order to motivate future analysis and experimentation on the topic.

2 Overview of Document Reordering

Several previous studies have looked at the document reordering problem. In this section, we outline the problem of document identifier assignment, review the key techniques that have been proposed in the literature, and describe the recursive bisection algorithm that is the focus of this work.

2.1 Document Identifier Assignment

The document identifier assignment problem can be described informally as finding a function that maps document identifiers to new values with the aim of minimizing the cost of coding the document gaps. More formally, different approaches exist to reduce this problem to several classical NP-Hard problems such as TSP [22], and versions of the optimal linear arrangement problem [2,7,10].

The most intuitive formalization is the *bipartite minimum logarithmic arrangement* (BiMLogA) problem [10] which models an inverted index as a bipartite graph $G = (V, E)$ with $|E| = m$ and the vertex set V consisting of a disjoint set of terms, T, and documents, D, $V = (T \cup D)$. Each edge $e \in E$ corresponds to an arc (t, d) with $t \in T$ and $d \in D$ that implies that document d contains term t. The BiMLogA problem seeks to find an ordering π of the vertices in D which minimizes the *LogGap* cost of storing the edges for each $t \in T$:

$$LogGap = \frac{1}{m} \sum_{t \in T} \sum_{i=0}^{\mathrm{d}_t} \log_2(\pi(u_{i+1}) - \pi(u_i))$$

where d_t is the degree of vertex $t \in T$, and t has neighbors $\{u_1, \ldots, u_{\mathrm{d}_q}\}$ with $\pi(u_1) < \cdots < \pi(u_{\mathrm{d}_t})$ and $u_0 = 0$. Intuitively, *LogGap* corresponds to minimizing the average logarithmic difference between adjacent entries in postings lists of an inverted index and can generally be considered a lower bound on the storage cost (in bits per integer) of a posting in an inverted index.

2.2 Document Ordering Techniques

Although there are a wide range of document ordering techniques that have been proposed [2–5,12,22,24], we focus on a few that can be efficiently run on web-scale data sets while also offering significant compression benefits. In particular, we focus on a subset of techniques that were examined in the work that we are reproducing [7,10,23].

Random Ordering. A `Random` document ordering corresponds to the case where identifiers are assigned randomly to documents with no notion of clustering or temporality. We reorder the document identifiers based on an arbitrary ordering specified by a pseudorandom number generator. This ordering represents the worst-case scenario (short of an adversarial case), and is used as a point-of-reference when comparing approaches.

Natural Ordering. Text collections usually have some notion of a `Natural` ordering. Two common orderings that can be considered as natural are either the *crawl* order of the documents, which assigns identifiers in a monotonically increasing order as they are found by the crawler, or the *URL* ordering, which is based on lexicographically sorting the URLs of the indexed documents [23].

Minhash Ordering. The Minhash (or shingle) ordering is a heuristic ordering that approximates the *Jaccard Similarity* of documents in order to cluster similar documents together [6,7]. The main idea is to obtain a fingerprint of each document (or neighbors of a graph vertex) through hashing, and to position similar documents (or vertices) close to each other. The key observation is that this improves *clustering*, which aids compression.

2.3 Recursive Graph Bisection

The BP ordering is the primary focus for this reproducibility study [10]. BP was proven to run in $\mathcal{O}(m \log n + n \log^2 n)$ time, and shown experimentally to yield excellent arrangements in practice. Unlike the aforementioned approaches, which implicitly try to cluster similar documents together (thus reducing the overall size of the delta encoding), the BP algorithm explicitly optimizes for an ordering by approximating the solution to the BiMLogA problem. The algorithm is described in Algorithm 1. On a high level, the algorithm recursively splits (bisects) the ordered set of document identifiers D into smaller ordered subsets D_1 and D_2. At each level of the recursion documents are swapped between the two subsets if a swap improves the *LogGap* objective.

At each level an initial document ordering (such as Random or Minhash) of all document identifiers in the current subset D is used to create two equally sized partitions D_1 and D_2 (line 2). Next, for a fixed set of iterations (*MaxIter*) the algorithm computes the *MoveGain* (described below) which results from moving documents in D between partitions (lines $4 - 6$). In each iteration, the documents with the highest *MoveGain*, i.e., the documents for which swapping partitions reduces the *LogGap* the most, are exchanged as long as the overall gain of the swap is beneficial (lines $8 - 10$). The current level of the recursion finishes once *MaxIter* iterations have been performed, or no document identifier swaps have occurred in the current iteration (lines $11 - 12$). Next, the same procedure is recursively applied to D_1 and D_2 until the maximum recursion depth (*MaxDepth*) is reached (lines $13 - 15$). As the recursion unwinds, the ordered partitions are 'glued' back together to form the final ordering.

Computing the *MoveGain* for a specific document/node v is shown in lines $17 - 24$. The function computes the average logarithmic gap length for all $t \in T$, for the parts of the adjacency lists (or postings lists) corresponding to documents in D_a and D_b. Specifically, the *MoveGain* of a document in D_a is defined as the *difference* in average logarithmic gap length between v remaining in D_a and v moving to D_b (and vice versa for a v in D_b). This gain can be positive (i.e., it is beneficial to move v to the other partition) or negative.

3 Reproducibility

Following other recent reproducibility studies in the field of IR [14], we adapt the following definition of reproducibility from the 2015 SIGIR Workshop on Reproducibility, Inexplicability, and Generalizability of Results [1]: *"Repeating*

Algorithm 1. Graph Reordering via Recursive Graph Bisection

```
1  Function RecursiveBisection(T, D, d)
       In: Bipartite graph (T, D) with |D| = n vertices and recursion depth d
2      D1, D2 = SortAndSplitGraph(D)
3      for iter = 0 to MaxIter do
4          forall v in D1 and u in D2 do
5              gains_D1[v] = ComputeMoveGain(T, v, D1, D2)
6              gains_D2[u] = ComputeMoveGain(T, u, D2, D1)
7          SortDecreasing(gains_D1, gains_D2)
8          forall v in gains_D1 and u in gains_D2 do
9              if gains_D1[v] + gains_D2[u] > 0 then
10                 SwapNodes(v, u)
11         if No Swaps Occurred then
12             iter = MaxIter
13     if d < MaxDepth then
14         D1 = RecursiveBisection(T, D1, ⌊n/2⌋, d + 1)
15         D2 = RecursiveBisection(T, D2, ⌈n/2⌉, d + 1)
16     return Concat(D1, D2)

17 Function ComputeMoveGain(T, v, Da, Db)
       In: Bipartite Graphs (T, Da), (T, Db) with |Da| = na, |Db| = nb and v ∈ Da
18     gain = 0
19     forall t in T do
20         if t connected to v then
21             da, db = number of edges in from t to Da and Db
22             gain = gain + da log2(na/(da+1)) + db log2(nb/(db+1))
23             gain = gain − (da − 1) log2(na/da) + (db + 1) log2(nb/(db+2))
24     return gain
```

a previous result under different but comparable conditions." To this end, we are interested in reproducing improvements in the compression of textual indexes to the same degree as the improvements reported by Dhulipala et al. [10], using a full reimplementation of the methods as described in the paper.

3.1 Implementation Details

In this section we are going to present the choices we made in our implementation. Even though the basic algorithm is conceptually simple to understand and implement, important details on implementing the algorithm so that it is scalable and efficient were omitted in the original work.

The first step is to build a forward index, which, in our case, is compressed using VarintGB [9] to optimize memory consumption. This forward index can be considered a bipartite graph, and is the input used by the BP algorithm.

To minimize the number of memory moves required by the algorithm, we create a list of references to the documents in the collection so that only pointer swaps are required when exchanging two documents between partitions. Where possible, references are used to avoid expensive memory copy or move operations.

As described by Dhulipala et al. [10], the two different recursive calls of Algorithm 1 are independent and can be executed in parallel (Algorithm 1, line 14/15). For this reason, we also employ a fork-join computation model using the Intel TBB library[1]. A pool of threads is started and every recursion call is added into a pool of tasks, so that threads can assign jobs according to the TBB scheduling policy.

After splitting the document vector into two partitions, the algorithm computes the *term degree* for every partition. In order to do so, we precompute the degree for all the terms in each partition. Since every document contains distinct terms, we can again exploit parallelism though parallel_for loops. Given the size of the collections used and the fact that the degree computations can run simultaneously on different partitions, we use a custom array implementation, which requires a one-off initialization. In contrast to constant-time initialization arrays [13], our implementation uses a global counter, C, and two arrays of the same size, V and G. The former of the two is used to store the actual values of the degrees and the latter to keep track of the validity of the data present in V at the corresponding position. The following invariant is maintained:

$$V[i] \text{ is valid} \iff G[i] = C.$$

Once the vector is allocated, which happens only once since it is marked as thread_local in our implementation, the vector G is initialized to 0. The counter C is set to 1, which indicates that none of the variables in the array are valid. Thus, an increment of the counter corresponds to a clear operation of the values in the array. If a position of the array contains a non-valid value, a default value is returned. The intuition behind this arrangement is to allow easy parallelism, and to avoid reinitialization of large vectors.

Next, for a fixed number of iterations, *MaxIter*, the gain computation, sorting, and swapping of documents is repeated (Algorithm 1, line 3–12). Gain computation is particularly interesting because it can be very expensive, so we address this operation as follows. Since terms are typically shared among multiple documents, we adopt a cache to compute every term gain at most once. However, checking if a term cost has already been cached introduces a branch that is hard to predict by the CPU; intuitively, fewer documents are processed deeper in the recursion, which implies fewer terms shared and a lower probability for each of them to be in the cache. To avoid this misprediction, we develop two functions which only differ for the check performed in the cache, where both provide branch prediction information for whether the value is likely to be in the cache or not. Furthermore, to compute a single term cost, we use SIMD instructions, allowing four values to be processed in a single CPU instruction. Sorting is, again, done in parallel and, as for the parallel_for, we use the Intel Parallel

[1] https://www.threadingbuildingblocks.org/.

STL[2] implementation. The swap function also updates the degrees of the two partitions, so recomputation is not needed for every iteration.

Finally, when the recursion reaches a segment of the document vectors which is considered small enough to terminate, a final sorting is applied to sort the otherwise unsorted leaves by their document identifiers.

4 Experiments

Testing details. All of the algorithms are implemented in C++17 and compiled with GCC 7.2.0 using the highest optimization settings. Experiments are performed on a machine with two Intel Xeon Gold 6144 CPUs (3.50 GHz), 512 GiB of RAM, running Linux 4.13.0. The CPU is based on the Skylake microarchitecture, which supports the AVX-512 instruction set, though we did not optimize for such instructions. Each CPU has L1, L2, and L3 cache sizes of 32KiB, 1024KiB, and 24.75MiB, respectively. We make use of SIMD processor intrinsics in order to speed up computation. When multithreading is used, we allow our programs to utilize all 32 threads, and our experiments assume an otherwise idle system. The source code is available[3,4] for the reader interested in further implementation details or in replicating the experiments.

Datasets. We performed our experiments using mostly standard datasets as summarized in Table 1. While most of these collections are readily available, we do note that the Wikipedia and CC-News collections are exceptions: both of these collections are temporal (and thus subject to change). To this end, we will make the raw collections available by request to any groups interested in repeating our experiments, following the best practices from Hasibi et al. [14].

- NYT corresponds to the New York Times news collection, which contains news articles between 1987 and 2007,
- Wikipedia is a crawl of English Wikipedia articles from the 22nd of May, 2018, using the Wikimedia dumps and the Wikiextractor tool[5],
- Gov2 is a crawl of `.gov` domains from 2004,
- ClueWeb09 and ClueWeb12 both correspond to the 'B' portion of the 2009 and 2012 ClueWeb crawls of the world wide web, respectively, and
- CC-News contains English news documents from the Common Crawl News[6] collection, from August 2016 to April 2018.

Postings lists were generated using Indri 5.11, with no stopword removal applied, and with Krovetz stemming. Furthermore, our results were tested for correctness by ensuring that the output of the reordered indexes matched the output of the original index for a set of test queries.

[2] https://software.intel.com/en-us/get-started-with-pstl.
[3] https://github.com/pisa-engine/pisa.
[4] https://github.com/pisa-engine/ecir19-bisection.
[5] https://github.com/attardi/wikiextractor.
[6] https://github.com/commoncrawl/news-crawl.

Table 1. Properties of our datasets. We consider only postings lists with ≥4,096 elements when conducting the reordering, but apply compression on the entire index. For completeness, we show both groups of statistics here.

		≥ 4,096		Full index	
Graph	$\|D\|$	$\|T\|$	$\|E\|$	$\|T\|$	$\|E\|$
NYT	1,855,658	10,191	457,883,999	2,970,013	501,568,918
Wikipedia	5,652,893	14,038	749,069,767	5,604,981	837,439,129
Gov2	25,205,179	42,842	5,406,607,172	39,180,840	5,880,709,591
ClueWeb09	50,220,423	101,676	15,237,650,447	90,471,982	16,253,057,031
ClueWeb12	52,343,021	88,741	14,130,264,013	165,309,501	15,319,871,265
CC-News	43,530,315	76,488	19,691,656,440	43,844,574	20,150,335,440

Reordering parameters. For most of our collections, we apply the URL ordering on the index and consider this the `Natural` ordering. Two exceptions are the NYT and Wikipedia collections. For NYT, we apply crawl ordering, as all indexed sites have the same URL prefix. For Wikipedia, we use the ordering of the crawl as specified by the Wikipedia `curid`, which is a proxy for URL ordering (as these identifiers are monotonically increasing on the page titles, which are usually the same or similar to the long URLs). For the `Minhash` scheme, we follow Dhulipala et al. [10], and sort documents lexicographically based on 10 minwise hashes of the documents (or, adjacency sets). When running `BP`, we only consider posting lists of lengths ≥ 4,096 for computing the reordering, we run 20 iterations per recursion, and we run our algorithm to $MaxDepth = \log(n) - 5$ unless otherwise specified [10]. The file orderings are available for each collection for repeatability.

4.1 Compression Ratio

Our first experiment investigates whether we are able to reproduce the *relative compression improvements* that were reported in the work of Dhulipala et al. [10] for the `BP` algorithm, while also reproducing the baselines [7,23]. Table 2 shows the effectiveness (in average bits per posting) of the various reordering techniques across each collection for a variety of state-of-the-art compression methods, including ϵ-optimal Partitioned Elias-Fano (PEF) [20], Binary Interpolative Coding (BIC) [19], and Stream Variable Byte (SVByte) [17]. We also report the *LogGap* as described in Sect. 2.1. We find that the `BP` algorithm outperforms the baselines for every collection, with improvements over the closest competitor between 2 and 15% and up to around 50% against the `Random` permutation for PEF encoding. Similar improvements are observed for the other tested compression schemes. Our findings confirm that these reordering strategies generalize to collections outside of those tested experimentally in the original work, including Newswire data such as NYT and CC-News.

Table 2. Reordered compression results for all six collections. We report the bits per edge for representing the DocIDs and the frequencies (DocID/Freq). Note that we omit the frequency data for SVByte as reordering does not impact the compression rate for this codec.

Index	Algorithm	LogGap	PEF	BIC	SVByte
NYT	Random	3.79	6.36/2.22	6.48/2.16	11.67
	Natural	3.50	6.31/2.20	6.23/2.13	11.62
	Minhash	3.18	5.91/2.19	5.79/2.11	11.51
	BP	**2.61**	**5.24/2.13**	**5.06/2.04**	**11.33**
Wikipedia	Random	5.12	8.03/2.20	8.01/1.98	12.45
	Natural	4.76	7.83/2.17	7.65/1.93	12.31
	Minhash	3.94	7.08/2.11	6.71/1.85	12.02
	BP	**3.13**	**6.17/2.03**	**5.74/1.77**	**11.69**
Gov2	Random	5.05	7.96/2.97	7.93/2.53	12.47
	Natural	1.91	4.37/2.31	4.01/2.07	11.44
	Minhash	1.99	4.57/2.34	4.17/2.10	11.51
	BP	**1.54**	**3.67/2.20**	**3.41/2.01**	**11.30**
ClueWeb09	Random	4.88	7.69/2.39	7.68/2.08	12.47
	Natural	2.71	6.12/2.20	5.36/1.84	11.71
	Minhash	3.00	6.46/2.23	5.77/1.87	11.79
	BP	**2.38**	**5.49/2.12**	**4.84/1.79**	**11.52**
ClueWeb12	Random	5.08	7.99/2.39	7.95/2.09	12.91
	Natural	2.51	6.07/2.20	5.11/1.81	12.07
	Minhash	2.89	6.08/2.17	5.49/1.86	12.06
	BP	**2.32**	**5.20/2.07**	**4.64/1.77**	**11.90**
CC-News	Random	3.56	6.06/2.19	6.16/2.06	11.48
	Natural	1.49	3.38/1.91	3.26/1.73	**10.92**
	Minhash	1.95	4.49/2.02	4.12/1.82	11.08
	BP	**1.39**	**3.31/1.90**	**3.11/1.72**	**10.92**

4.2 Efficiency

Next, we focus on the efficiency of our implementation. In the original work, the authors experimented with two implementations. One approach utilized a distributed implementation written in Java, which conducted the reordering across a cluster of "a few tens of machines." The other approach was a single-machine implementation, which used parallel processing across many cores. We opted to follow the single-machine approach as discussed in Sect. 3.1. We report the running time of `BP` for each dataset in Table 3.

Note that for these experiments, we used the `Natural` ordered index as input to `BP` (discussed further in Sect. 4.3). Clearly, our implementation of `BP` is very

Table 3. Time taken to process each dataset with recursive graph bisection, in minutes.

NYT	Wikipedia	Gov2	ClueWeb09	ClueWeb12	CC-News
2	5	28	90	86	97

efficient, completing Gov2 in 28 min, and ClueWeb09 in 90 min. This is comparable to the timings reported in the original work, which reported Gov2 and ClueWeb09 taking 29 and 129 min, respectively. We must remark that our timings are not directly comparable to those from Dhulipala et al. for a few reasons. Firstly, our indexes were built using Indri, whereas they opted to use Apache Tika for their indexing, resulting in a different number of postings to process. Furthermore, subtle differences in servers such as clock speed and cache size can impact timings. In any case, we are confident that the BP algorithm can run efficiently over large collections, and can use whatever processing pipeline adopters have available. Another aspect of efficiency is memory consumption. Dhulipala et al. report that their implementation "utilizes less than twice the space required to store the graph edges." While this is hard to interpret (how was the graph edge space consumption calculated?), we provide some intuition on our memory consumption as follows. On our largest collection, CC-News, the BP algorithm has a peak space consumption of 110 GiB, which includes the graph representation of the dataset. Given that the compressed forward index for CC-News consumes 25 GiB, it seems that we have a higher memory footprint than the original implementation (which would use up to 75 GiB). It is important to note that this is due to our caching approach, which incurs a higher memory footprint for faster execution time. Of course, alternative caching strategies may allow for lower memory consumption at the cost of a slower run time.

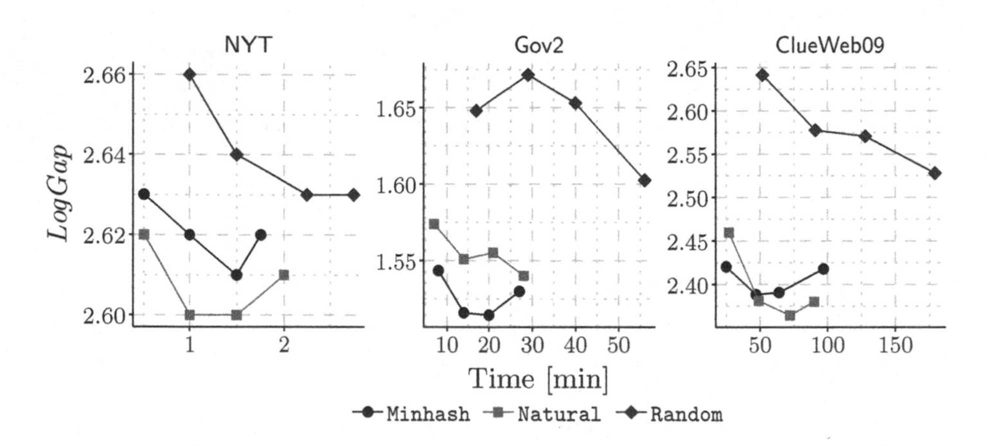

Fig. 1. *LogGap* cost of the three different input orderings after running BP as the number of iterations increases.

4.3 Parameters and Initialization

First, we investigate the impact that the number of iterations has on the algorithm. Dhulipala et al. showed that while on the higher levels of the recursion we approach convergence after just a few iterations, around $MaxIter = 20$ iterations are required at deeper levels. We are interested in understanding how effective the BP algorithm is at producing a good ordering when we do less iterations, as less iterations results in improved run-time efficiency. To measure this, we ran the algorithm across the collections setting the maximum number of iterations $MaxIter = \{5, 10, 15, 20\}$. Figure 1 shows the resulting trade-off in terms of *LogGap* and execution time. We can make a few observations from this figure. Firstly, the more optimal the input graph, the less iterations required to reach a reasonable ordering. This is intuitive, as a better input ordering implies that document clustering is already somewhat reasonable, meaning less work is required to further improve the clustering. Secondly, the quality of the input graph also seems to impact the run time of the algorithm. For example, examine Fig. 1 (right), which shows the ClueWeb09 collection. Using either the `Natural` or `Minhash` inputs achieves competitive compression levels using only 10 iterations, which takes around 50 min. On the other hand, the `Random` input takes longer to process in each iteration, and results in a less effective final ordering. Similar results were found on all tested collections. `Random` orderings are slower to compute for two main reasons. Firstly, on each iteration, more vertices are moved, which takes longer to process. Secondly, it is less likely that the convergence property will be met using a `Random` ordering, which results in more iterations in total (Algorithm 1, Line 11 and 12). Therefore, we recommend using a `Natural` or `Minhash` ordered index as input to the BP algorithm where possible, and setting $MaxIter = 20$ for to ensure a good level of compression. If the run time is critical, using smaller values of *MaxIter* allows trading off some compression effectiveness for a faster total processing time.

Our next experiment investigates the effect of initialization on the performance of the BP algorithm. Recall that in Algorithm 1, D must be partitioned into sets D_1 and D_2. As discussed by Dhulipala et al., the initialization of these sets may impact the quality of the final ordering of the vertices. Our implementation uses a generic sort-by-identifier approach to do this partitioning, so the partition is made by first sorting the documents in D by their identifiers, and then splitting it into two equal sized subgraphs. Therefore, the initialization approach used is the same as the ordering of the input collection. Based on the original work, we expect the initialization to have little impact on the final compression ratio, with no clear best practice (and negligible differences between the resultant compression levels). To test this, we ran BP using three different initialization orderings: `Random`, `Natural`, and `Minhash`. Our results confirm that the initialization order used to initialize D_1 and D_2 only has a very moderate impact on the efficacy of the bisection procedure. In particular, the largest differences in the *LogGap* for the resulting orderings was always within $\pm 5\%$. We found that there was no consistently better approach, with each initialization yielding the best compression on at least one of the tested collections. This effect can be

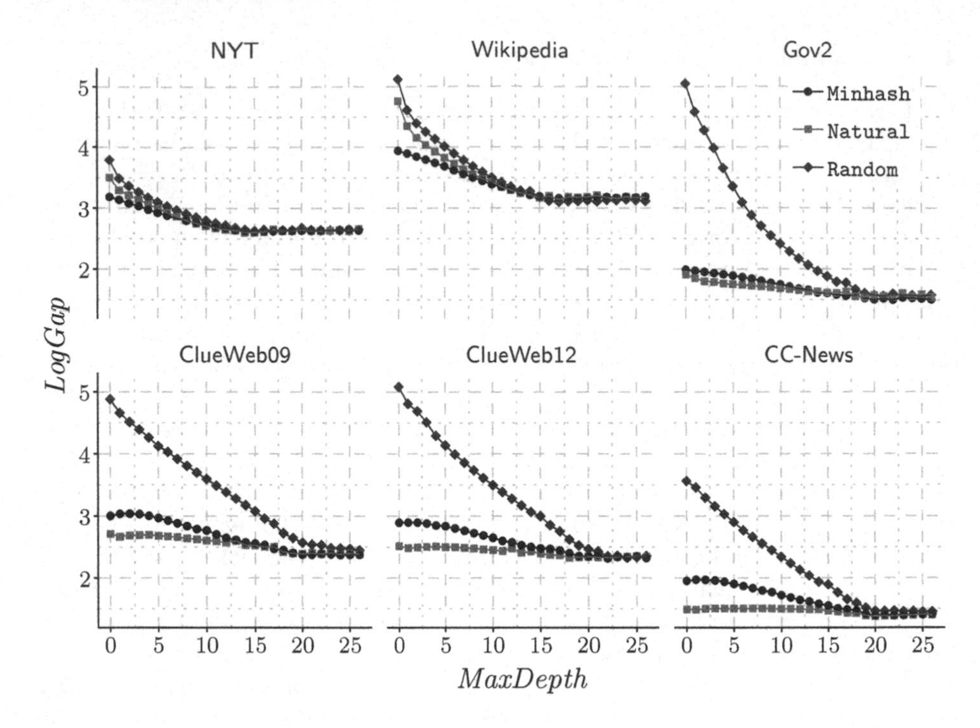

Fig. 2. *LogGap* cost of the three different input orderings as *MaxDepth* is increased. *LogGap* cost is reported up to a depth of 26 for all collections.

observed in Fig. 2, by comparing each line at depth $\log(n)-5$. For example, running BP with Random initialization results in the best ordering on the Wikipedia dataset, whereas Minhash initialization is the best on ClueWeb09.

Our final experiment examines both the depth of the recursion and the potential impact of the initialization of D_1 and D_2 on the convergence of the algorithm. For each collection, we run the BP algorithm for each recursion depth between 1 and 26 since the collection with the largest number of documents, ClueWeb12, has $\lceil \log(n) \rceil = 26$. Again, we use three varying approaches of initializing the sets D_1 and D_2, as the initialization may impact the depth at which the algorithm converges. Figure 2 shows the results for this experiment. We reiterate that the input ordering is the same as the initialization approach applied, hence the different starting points of each line. Interestingly, the ordering of the input/initialization approach does not impact the convergence level of the BP algorithm. We confirm that the recommended heuristic of using $MaxDepth = \log(n) - 5$ does indeed fit with our implementation and the additional collections we tested, with marginal (if any) gains following at further depths. Another interesting observation is that even in cases where the input ordering is very close to the compression level of the output ordering from BP (primarily in the case of the CC-News collection where the input is the Natural index), the BP algorithm still takes around 14 levels of recursion to begin improving upon the input ordering, reaching convergence at around level 19 or 20 as expected.

5 Conclusion and Discussion

During this reproducibility study, a lot of effort was spent on optimizing the implementation. In order to achieve the efficiency that is described in our experiments, considerable thought was put into various prototype algorithms and varying approaches before a final version was produced. While the original paper primarily focuses on the theoretical reasoning behind the BP algorithm, this leaves less room for explaining specific implementation details which are important in practice. Since the source code was not released in the original paper, this contributed to the difficulty of reproducibility. However, we are confident that the algorithm presented in the original work and the findings based on this algorithm are valid, as shown in this reproducibility study. By making our implementation available, we hope to stimulate further research in this interesting area of efficiency. Finally, we believe that future research should make the applied index ordering known, as is already done for other experimental factors such as stemming or stopping. This is of course important for the reproducibility of efficiency experiments conducted across inverted indexes, for both query processing speeds and compression numbers.

Acknowledgments. This work was supported by the National Science Foundation (IIS-1718680), the Australian Research Council (DP170102231), and the Australian Government (RTP Scholarship).

References

1. Arguello, J., Diaz, F., Lin, J., Trotman, A.: SIGIR 2015 workshop on reproducibility, inexplicability, and generalizability of results (RIGOR). In: Proceedings of SIGIR, pp. 1147–1148 (2015)
2. Blanco, R., Barreiro, Á.: Document identifier reassignment through dimensionality reduction. In: Losada, D.E., Fernández-Luna, J.M. (eds.) ECIR 2005. LNCS, vol. 3408, pp. 375–387. Springer, Heidelberg (2005). https://doi.org/10.1007/978-3-540-31865-1_27
3. Blanco, R., Barreiro, Á.: Characterization of a simple case of the reassignment of document identifiers as a pattern sequencing problem. In: Proceedings of SIGIR, pp. 587–588 (2005)
4. Blanco, R., Barreiro, Á.: TSP and cluster-based solutions to the reassignment of document identifiers. Inf. Retr. **9**(4), 499–517 (2006)
5. Blandford, D., Blelloch, G.: Index compression through document reordering. In: Proceedings DCC 2002, Data Compression Conference, pp. 342–352 (2002)
6. Broder, A.Z., Charikar, M., Frieze, A.M., Mitzenmacher, M.: Min-wise independent permutations. J. Comput. Syst. Sci. **60**(3), 630–659 (2000)
7. Chierichetti, F., Kumar, R., Lattanzi, S., Mitzenmacher, M., Panconesi, A., Raghavan, P.: On compressing social networks. In: Proceedings of SIGKDD, pp. 219–228 (2009)

8. Crane, M., Culpepper, J.S., Lin, J., Mackenzie, J., Trotman, A.: A comparison of Document-at-a-Time and Score-at-a-Time query evaluation. In: Proceedings of WSDM, pp. 201–210 (2017)
9. Dean, J.: Challenges in building large-scale information retrieval systems: invited talk. In: Proceedings of WSDM, pp. 1–1 (2009)
10. Dhulipala, L., Kabiljo, I., Karrer, B., Ottaviano, G., Pupyrev, S., Shalita, A.: Compressing graphs and indexes with recursive graph bisection. In: Proceedings of SIGKDD, pp. 1535–1544 (2016)
11. Ding, S., Suel, T.: Faster top-k document retrieval using block-max indexes. In: Proceedings of SIGIR, pp. 993–1002 (2011)
12. Ding, S., Attenberg, J., Suel, T.: Scalable techniques for document identifier assignment in inverted indexes. In: Proceedings of the WWW, pp. 311–320 (2010)
13. Fredriksson, K., Kilpeläinen, P.: Practically efficient array initialization. Soft. Prac. Exp. **46**(4), 435–467 (2016)
14. Hasibi, F., Balog, K., Bratsberg, S.E.: On the reproducibility of the TAGME entity linking system. In: Ferro, N., Crestani, F., Moens, M.-F., Mothe, J., Silvestri, F., Di Nunzio, G.M., Hauff, C., Silvello, G. (eds.) ECIR 2016. LNCS, vol. 9626, pp. 436–449. Springer, Cham (2016). https://doi.org/10.1007/978-3-319-30671-1_32
15. Hawking, D., Jones, T.: Reordering an index to speed query processing without loss of effectiveness. In: Proceedings of ADCS, pp. 17–24 (2012)
16. Kane, A., Tompa, F.W.: Split-lists and initial thresholds for WAND-based search. In: Proceedings of SIGIR, pp. 877–880 (2018)
17. Lemire, D., Kurz, N., Rupp, C.: Stream vbyte: faster byte-oriented integer compression. Inf. Proc. Lett. **130**, 1–6 (2018)
18. Mallia, A., Ottaviano, G., Porciani, E., Tonellotto, N., Venturini, R.: Faster Block-Max WAND with variable-sized blocks. In: Proceedings of SIGIR, pp. 625–634 (2017)
19. Moffat, A., Stuiver, L.: Binary interpolative coding for effective index compression. Inf. Retr. **3**(1), 25–47 (2000)
20. Ottaviano, G., Venturini, R.: Partitioned Elias-Fano indexes. In: Proceedings of SIGIR, pp. 273–282 (2014)
21. Richardson, M., Prakash, A., Brill, E.: Beyond pagerank: machine learning for static ranking. In: Proceedings of WWW, pp. 707–715 (2006)
22. Shieh, W.-Y., Chen, T.-F., Shann, J.J.-J., Chung, C.-P.: Inverted file compression through document identifier reassignment. Inf. Proc. Man. **39**(1), 117–131 (2003)
23. Silvestri, F.: Sorting out the document identifier assignment problem. In: Amati, G., Carpineto, C., Romano, G. (eds.) ECIR 2007. LNCS, vol. 4425, pp. 101–112. Springer, Heidelberg (2007). https://doi.org/10.1007/978-3-540-71496-5_12
24. Yan, H., Ding, S., Suel, T.: Inverted index compression and query processing with optimized document ordering. In: Proceedings of WWW, pp. 401–410 (2009)

An Experimental Study of Index Compression and DAAT Query Processing Methods

Antonio Mallia(✉), Michał Siedlaczek(✉), and Torsten Suel(✉)

Computer Science and Engineering, New York University, New York, USA
{antonio.mallia,michal.siedlaczek,torsten.suel}@nyu.edu

Abstract. In the last two decades, the IR community has seen numerous advances in top-k query processing and inverted index compression techniques. While newly proposed methods are typically compared against several baselines, these evaluations are often very limited, and we feel that there is no clear overall picture on the best choices of algorithms and compression methods. In this paper, we attempt to address this issue by evaluating a number of state-of-the-art index compression methods and safe disjunctive DAAT query processing algorithms. Our goal is to understand how much index compression performance impacts overall query processing speed, how the choice of query processing algorithm depends on the compression method used, and how performance is impacted by document reordering techniques and the number of results returned, keeping in mind that current search engines typically use sets of hundreds or thousands of candidates for further reranking.

Keywords: Compression · Query processing · Inverted indexes

1 Introduction

Over the past few decades, the IR community have been making a continuous effort to improve the efficiency of search in large collections of documents. Advances have been made in virtually all aspects of text retrieval, including index compression and top-k query processing. Although a multitude of authors have reported experimental results, comparing them across different publications poses a challenge due to varying data sets, parameters, evaluation metrics, and experimental setups. We aim to address this issue by providing an extensive experimental comparison across many index compression techniques and several query processing algorithms. Our comparison includes many recent methods, and thus provides a useful snapshot of the current state of the art in this area.

The most common structure used for text retrieval is an *inverted index*. For each term in a parsed collection, it stores a list of numerical IDs of documents containing this term, typically along with additional data, such as term frequencies or precomputed quantized impact scores. We call all values associated

L. Azzopardi et al. (Eds.): ECIR 2019, LNCS 11437, pp. 353–368, 2019.
https://doi.org/10.1007/978-3-030-15712-8_23

with a (term, document)-pair a *posting*. Postings are typically sorted in order of increasing document IDs, although there are other index organizations. We assume document-sorted posting lists throughout this paper.

The first problem we encounter is efficient index representation. In particular, compression of posting lists is of utmost importance, since they account for much of the data size and access costs. In practice, the problem we must solve is efficient encoding of non-negative integers, such as document IDs or their gaps, frequencies, positions, or quantized scores. Some encoding schemes, such as Golomb [23] or Binary Interpolative [34], can be very space-efficient but slow to decode. Other methods achieve very fast decoding while sacrificing compression ratio. In recent years, a significant boost in encoding efficiency has been achieved due to application of SIMD (Single Instruction, Multiple Data) instructions available on many modern CPUs [26,27,36,44].

Likewise, the choice of a retrieval algorithm is crucial to query efficiency. Due to large sizes of most search collections, retrieving all potential matches is infeasible and undesirable. In practice, only the top k highest ranked documents are returned. Ranking methods can be grouped into fast and simple term-wise scoring methods [39,51], and more complex rankers such as the Sequential Dependence Model (SDM) [31] or learning to rank methods [28,49]. For term-wise techniques, the score of a document with respect to a query is simply the sum of the partial scores, also called impact scores, of the document with respect to each term. Complex rankers give up this term independence assumption. They are more accurate, but also much more expensive, as they require the evaluation of fairly complex ranking functions on up to hundreds of features that need to be fetched or generated from index and document data. Thus, it would be infeasible to evaluate such complex rankers on large numbers of documents.

To combine the speed of term-wise scoring with the accuracy of the complex rankers, a cascade ranking architecture is commonly deployed: First, a fast ranker is used to obtain $k_c > k$ candidate results for a query; then the k_c candidates are reranked by a slower complex ranker to retrieve the final top k documents. In this paper, we address the first problem, also known as candidate generation, as reranking can be considered a separate, largely independent, problem. In particular, we focus on the performance of different index compression and query processing algorithms for candidate generation. We limit ourselves to safe Document-At-A-Time (DAAT) algorithms for disjunctive top-k queries.

Following the RIGOR *Generalizability* property [3], we focus on assessing how well technology performs in new contexts. There are four dimensions to our comparison study: index compression method, query processing algorithm, document ordering, and the number k of retrieved candidates. Published work proposing new compression methods or query processing algorithms typically only looks at a small slice of possible configurations, say, a new query processing algorithm compared against others using only one or two compression methods and document orderings, and only on a very limited range of k.

Catena et al. [8] showed the impact of compressing different types of posting information on the space and time efficiency of a search engine. Although they investigate several compression methods, their focus is mostly on different variations of the FOR family. They also limit their query processing comparison to exhaustive DAAT retrieval strategy, while we consider dynamic pruning techniques. On the other hand, they study several aspects not considered here, such as compression of different types of index data. Thus, we feel that our study answers different research questions by exploring many combinations of techniques that have never been reported. It can serve as a guide for choosing the best combinations of techniques in a given setup.

Contributions. We make the following major contributions:

1. We provide experimental results for an extensive range of configurations. We include almost all of the state-of-the-art compression techniques, the most commonly used DAAT query processing approaches, and several document reorderings over a wide range of k. To our knowledge, this is the most extensive recent experimental study of this space of design choices.
2. We combine already established open-source libraries and our own implementations to create a code base that provides means to reproduce the results, and that can also serve as a starting point for future research. We release this code for free and open use by the research community.

2 Outline of Our Methods

We now describe the various methods and settings we explore, organized according to the four dimensions of our study: compression method, query processing algorithm, document ordering, and number of candidates k. We decided not to include impact score quantization, another possible dimension, in this paper. Score quantization raises additional issues and trade-offs that we plan to study in a future extension of this work.

2.1 Index Compression Methods

We include in our study a total of 11 index compression methods that we consider to be a good representation of the current state of the art. For each method, we integrated what we believe to be the fastest available open-source implementation. We now briefly outline these methods.

Variable Byte Encoding. These methods encode each integer as a sequence of bytes. The simplest one is Varint (also known as Varbyte or VByte), which uses 7 bits of each byte to encode the number (or a part of it), and 1 remaining bit to state if the number continues in the next byte. Although compression of Varint is worse than that of older bit-aligned algorithms such as Elias [21], Rice [38], or Golomb [23] coding, it is much faster in memory-resident scenarios [40].

Varint basically stores a unary code for the size (number of bytes used) of each integer, distributed over several bytes. To improve decoding speed, Dean [14] proposed Group Varint, which groups the unary code bits together. One byte is used to store 4 2-bit numbers defining the lengths (in bytes) of the next 4 integers. The following bytes encode these integers.

Recently, several SIMD-based implementations of variable-byte encodings have been shown to be extremely efficient [14,36,44]. Stepanov et al. [44] analyzed a family of SIMD-based algorithms, including a SIMD version of Group Varint, and found the fastest to be *VarintG8IU*: Consecutive numbers are grouped in 8-byte blocks, preceded by a 1-byte descriptor containing unary-encoded lengths (in bytes) of the integers in the block. If the next integer cannot fit in a block, the remaining bytes are unused.

Stream VByte [27] combines the benefits of VarintG8IU and Group Varint. Like Group Varint, it stores four integers per block with a 1-byte descriptor. Thus, blocks have variable lengths, which for Group Varint means that the locations of these descriptors cannot be easily predicted by the CPU. Stream VByte avoids this issue by storing all descriptors sequentially in a different location.

PForDelta. PForDelta [54] encodes a large number of integers (say, 64 or 128) at a time by choosing a k such that most (say, 90%) of the integers can be encoded in k bits. The remaining values, called exceptions, are encoded separately using another method. More precisely, we select b and k such that most values are in the range $[b, b + 2^k - 1]$, and thus can be encoded in k bits by shifting them to the range $[0, 2^k - 1]$. Several methods have been proposed for encoding the exceptions and their locations. One variant, *OptPForDelta* [50], selects b and k to optimize for space or decoding cost, with most implementations focusing on space. A fast SIMD implementation was proposed in [26].

Elias-Fano. Given a monotonically increasing integer sequence S of size n, such that $S_{n-1} < u$, we can encode it in binary using $\lceil \log u \rceil$ bits. Instead of writing them directly, Elias-Fano coding [20,22] splits each number into two parts, a low part consisting of $l = \lceil \log \frac{u}{n} \rceil$ right-most bits, and a high part consisting of the remaining $\lceil \log u \rceil - l$ left-most bits. The low parts are explicitly written in binary for all numbers, in a single stream of bits. The high parts are compressed by writing, in negative-unary form (i.e., with the roles of 0 and 1 reversed), the gaps between the high parts of consecutive numbers. Sequential decoding is done by simultaneously retrieving low and high parts, and concatenating them. Random access requires finding the locations of the i-th 0- or 1-bit within the unary part of the data using an auxiliary succinct data structure. Furthermore, a $\mathsf{NextGEQ}(x)$ operation, which returns the smallest element that is greater than or equal to x, can be implemented efficiently. Observe that h_x, the higher bits of x, is used to find the number of elements having higher bits smaller than h_x, denoted as p. Then, a linear scan of l_x, the lower bits of x, can be employed starting from posting p of the lower bits array of the encoded list.

The above version of Elias-Fano coding cannot exploit clustered data distributions for better compression. This is achieved by a modification called *Partitioned Elias-Fano* [35] that splits the sequence into b blocks, and then uses an optimal choice of the number of bits in the high and low parts for each block. We use this version, which appears to outperform Elias-Fano in most situations.

Binary Interpolative. Similar to Elias-Fano, Binary Interpolative Coding (BIC) [34] directly encodes a monotonically increasing sequence rather than a sequence of gaps. At each step of this recursive algorithm, the middle element m is encoded by a number $m - l - p$, where l is the lowest value and p is the position of m in the currently encoded sequence. Then we recursively encode the values to the left and right of m. BIC encodings are very space-efficient, particularly on clustered data; however, decoding is relatively slow.

Word-Aligned Methods. Several algorithms including *Simple-9* [1], *Simple-16* [52], and *Simple-8b* [2] try to pack as many numbers as possible into one machine word to achieve fast decoding. For instance, Simple-9 divides each 32-bit word into a 4-bit selector and a 28-bit payload. The selector stores one of 9 possible values, indicating how the payload is partitioned into equal-sized bit fields (e.g., 7 4-bit values, or 9 3-bit values). Some of the partitionings leave up to three of the 28 payload bits unused. Later enhancements in [2,52] optimize the usage of these wasted bits or increase the word size to 64 bits.

Lemire and Boytsov [26] proposed a bit-packing method that uses SIMD instructions. The algorithm, called *SIMD-BP128*, packs 128 consecutive integers into as few 128-bit words as possible. The 16-byte selectors are stored in groups of 16 to fully utilize 128-bit SIMD reads and writes.

Very recently, Trotman [46] proposed *QMX* encoding (for Quantities, Multipliers, and eXtractor), later extended by Trotman and Lin [47]. It combines the Simple family and SIMD-BP128 by packing as many integers as possible into one or two 128-bit words. Furthermore, the descriptors are run-length encoded, allowing one selector to describe up to 16 consecutive numbers.

Asymmetric Numeral Systems. Asymmetric Numeral Systems (ANS) [19] are a recent advance in entropy coding that combines the good compression rate of arithmetic coding with a speed comparable to Huffman coding. ANS represents a sequence of symbols as a positive integer x, and depends on the frequencies of symbols from a given alphabet Σ. Each string over Σ is assigned a state, which is an integer value, with 0 for the empty string. The state of a string wa is computed recursively using the state of w and the frequency of symbol a. A more detailed description of ANS is beyond the scope of this paper, and we refer to [19,33]. Very recently, ANS was successfully used, in combination with integer encodings such as VByte and the Simple family, to encode documents and frequencies in inverted indexes [32,33].

2.2 Query Processing

Next, we describe the top-k disjunctive DAAT query processing algorithms that we study. We limit ourselves to safe methods, guaranteed to return the correct top-k, and select methods that have been extensively studied in recent years.

MaxScore. MaxScore is a family of algorithms first proposed by Turtle and Flood [48], which rely on the maximum impact scores of each term t (denoted as $\max_t$). Given a list of query terms $q = \{t_1, t_2, \ldots, t_m\}$ such that $\max_{t_i} \geq \max_{t_{i+1}}$, at any point of the algorithm, query terms (and associated posting lists) are partitioned into *essential* $q_+ = \{t_1, t_2, \ldots, t_p\}$ and *nonessential* $q_- = \{t_{p+1}, t_{p+2}, \ldots, t_m\}$. This partition depends on the current threshold T for a document to enter the top k results, and is defined by the smallest p such that $\sum_{t \in q_-} \max_t < T$. Thus, no document containing only nonessential terms can make it into the top-k results. We can now perform disjunctive processing over only the essential terms, with lookups into the nonessential lists. More precisely, for a document found in the essential lists, we can compute a score upper bound by adding its current score (from the essential lists and any non-essential lists already accessed) and the $\max_t$ of those lists not yet accessed; if this bound is lower than T, then no further lookups on this document are needed. There are TAAT and DAAT versions of MaxScore; we only consider the DAAT variant. In this case, we attempt to update p whenever T changes.

WAND. Similar to MaxScore, WAND [6] (Weighted or Weak AND) also utilizes the $\max_t$ values. During DAAT traversal, query terms and their associated posting lists are kept sorted by their current document IDs. We denote this list of sorted terms at step s of the algorithm as $q_s = \langle t_1, t_2, \ldots, t_m \rangle$. At each step, we first find the *pivot* term t_p, where p is the lowest value such that $\sum_{i=1}^{p} \max_{t_i} \geq T$. Let d_p be the current document of t_p. If all t_i with $i < p$ point to d_p, then the document is scored and accumulated, and all pointers to d_p are moved to the next document. Otherwise, no document $d < d_p$ can make it into the top-k; thus, all posting lists up to t_p are advanced to the next document $\geq d_p$, and we resort and again find the pivot. This is repeated until all documents are processed.

Block-Max WAND. One shortcoming of WAND is that it uses maximum impact scores over the entire lists. Thus, if $\max_t$ is much larger than the other scores in the list, then the impact score upper bound will usually be a significant overestimate. Block-Max WAND (BMW) [18] addresses this by introducing block-wise maximum impact scores.

First, regular WAND pivot selection is used to determine a pivot candidate. The candidate is then verified by *shallow pointer movements*: The idea is to search for the block in a posting list where the pivot might exist. This operation is fast, as it involves no block decompression. Shallow pointer movement is performed on each $t_{i<p}$, and the block-wise maximum score is computed. If it is greater than T, then t_p is the pivot. In this case, if all $t_{i \leq p}$ point at d_p, we perform the required lookups, following by advancing the pointers by one; otherwise, we pick the $t_{i<p}$ with the largest IDF, and advance its pointer to a document ID $\geq d_p + 1$. If the block-size maximum score is less than T, we must find another candidate. We consider the documents that are at the current block boundaries for $t_{i \leq p}$, and all the current documents for $t_{i>p}$. We select the minimum document ID among them and denote it as d'. Finally, we select the $t_{i<p}$ with the largest IDF, and move its pointer to d'. We repeat the entire process until all terms are processed.

Variable Block-Max WAND. [30] generalizes BMW by allowing variable lengths of blocks. More precisely, it uses a block partitioning such that the sum of differences between maximum scores and individual scores is minimized. This results in better upper bound estimation and more frequent document skipping.

Block-Max MaxScore. The idea of using per-block maximum impact scores can also be applied to MaxScore, leading to the Block-Max MaxScore [9] (BMM) algorithm. Before performing look-ups to nonessential lists, we further refine our maximum score estimate using maximum impacts of the current blocks in nonessential lists, which might lead to fewer fully evaluated documents.

2.3 Document Ordering

It is well known that a good assignment of IDs to documents can significantly improve index compression. Many query processing algorithms are also sensitive to this assignment, with speed-ups of 2 to 3 over random assignment observed for some orderings and algorithms.

The problem of finding a document ordering that minimizes compressed index size has been extensively studied [4,5,15,17,41–43]. Shieh et al. [41] propose an approach based on an approximate maximum travelling salesman problem; they build a similarity graph where documents are vertices, and edges indicate common terms. Blandford and Blelloch [5] use a similarity graph with edges weighted with cosine similarity, and run a recursive partitioning to find the ordering. A considerable downside of such algorithms are their time and space complexity. Silvestri [42] shows that a simple URL-based ordering works as well as more complex methods on many data sets. The simplicity and efficiency of this approach makes it a very attractive choice in practice. Recently, Dhulipala et al. [15] proposed the Recursive Graph Bisection algorithm for graph and index compression, and experiments show their algorithm to exhibit the best compression ratio across all tested indices. We consider three orderings in our study, random assignment, URL-based, and Recursive Graph Bisection.

While most work on document ID assignment focuses on index compression, reordering also impacts query efficiency. Yan et al. [50] found that document reordering can significantly speed up conjunctive queries. Subsequent experiments show similar results for several disjunctive top-k algorithms, and in particular for all the algorithms introduced in the previous subsection [16,18,24,25,30,45]. Thus, query processing speeds depend on both compression method and document ordering, though the trade-offs as not yet well understood.

2.4 Choice of k

The final dimension of our study is the choice of the number of results k. Much previous work has focused on smaller values of k, such as 10 or 100. However, when query processing algorithms are used for subsequent reranking by a complex ranker, more results are needed, though the optimal value of k varies according to several factors [29]. Suggested values include 20 [10], 200 [53], 1000 [37],

5000 [13], and up to tens of thousands [11], suggesting that the optimal k is context-dependent. Also, recent work in [12] indicates that many top-k algorithms slow down significantly as k becomes larger, but not always at the same rate. Given this situation, we decided to perform our evaluation over a large range of values, from $k = 10$ up to 10000.

3 Experimental Evaluation

Testing Environment. All methods are implemented in C++17 and compiled with GCC 7.3.0 using the highest optimization settings. The tests are performed on a machine with an Intel Core i7-4770 quad-core 3.40 GHz CPU, with 32 GiB RAM, running Linux 4.15.0. The CPU is based on the Haswell micro architecture which supports the AVX2 instruction set. The CPUs L1, L2, and L3 cache sizes are 32 KB, 256 KB, and 8 MB, respectively. Indexes are saved to disk after construction, and memory-mapped to be queried, so that there are no hidden space costs due to loading of additional data structures in memory. Before timing the queries, we ensure that any required lists are fully loaded in memory. All timings are measured taking the results with minimum value of five independent runs, and reported in milliseconds. We compute BM25 [39] scores during retrieval.

Data Sets. We performed experiments on two standard datasets: Gov2 and ClueWeb09 [7], summarized in Table 1. For each document in the collection the body text was extracted using Apache Tika, the words lowercased and stemmed using the Porter2 stemmer; no stopwords were removed.

Table 1. Basic statistics for the test collections

	Documents	Terms	Postings
Gov2	24,622,347	35,636,425	5,742,630,292
ClueWeb09	50,131,015	92,094,694	15,857,983,641

Document Ordering. We experimented with three document orderings: Random, URL, and BP. The first, with IDs randomly assigned to documents, serves as the baseline. URL assigns IDs in lexicographic order of URLs [42]. BP is based on the Recursive Graph Bisection algorithm [15].

Implementation Details. Our codebase is a fork of the ds2i[1] library, extended with many additional encoding, query processing, and reordering implementations used in this study. The source code is available at https://github.com/pisa-engine/pisa for readers interested in further implementation details or in replicating the experiments. We integrated what we believe are the currently best open-source implementations of the various compression algorithms. We use the

[1] https://github.com/ot/ds2i.

FastPFor[2] library for implementation of VarintGB, VarintG8IU, OptPFD, Simple16, Simple8b, SIMD-BP128, and StreamVByte. PEF and Interpolative are based on the code of the ds2i library. The QMX implementation comes from JASSv2[3]. We used the reference implementation of ANS[4] with 2d max:med contexts mechanism. All block-wise encodings use blocks of 128 postings per block.

We implemented the original Recursive Graph Bisection algorithm by Dhulipala et al. [15] and validated the results obtained against those reported in their paper. Our implementations of BMW and BMM store maximum impact scores for blocks of size 128, while VBMW uses blocks of average length 40. Both these values were also used in previous work.

Queries. To evaluate query processing speed, we use TREC 2005 and TREC 2006 Terabyte Track Efficiency Task, drawing only queries whose terms are all in the collection dictionary. This leaves us with 90% and 96% of the total TREC 2005 and TREC 2006 queries for Gov2, and 96% and 98% of the total TREC 2005 and TREC 2006 queries for ClueWeb09. From each sets of queries, we randomly

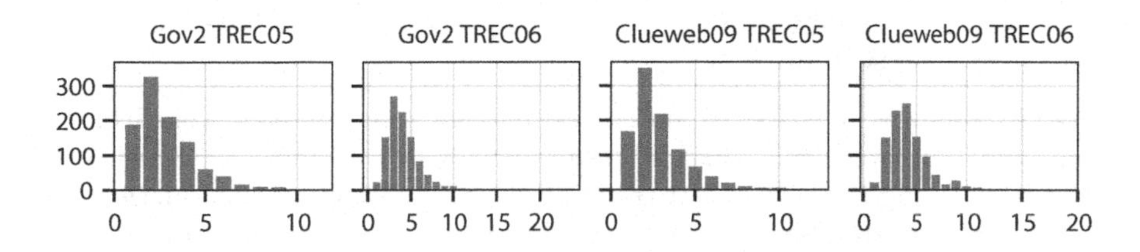

Fig. 1. Query length distributions.

Table 2. Overall space in GB, and average bits per document ID and frequency.

	Random						URL						BP					
	Gov2			ClueWeb09			Gov2			ClueWeb09			Gov2			ClueWeb09		
	space GB	doc bpi	freq bpi	space GB	doc bpi	freq bpi	space GB	doc bpi	freq bpi	space GB	doc bpi	freq bpi	space GB	doc bpi	freq bpi	space GB	doc bpi	freq bpi
Packed+ANS2	7.36	7.71	2.54	19.40	7.65	2.14	4.17	3.96	1.85	14.47	5.36	1.94	3.57	3.25	1.72	13.21	4.80	1.87
Interpolative	7.32	7.58	2.62	19.12	7.52	2.12	4.26	3.80	2.14	13.90	5.15	1.87	3.71	3.11	2.06	12.81	4.65	1.81
PEF	7.65	7.60	3.05	19.68	7.53	2.39	4.65	4.11	2.37	15.95	5.85	2.20	3.97	3.30	2.23	14.66	5.29	2.11
OptPFD	8.09	8.13	3.14	21.47	8.07	2.76	4.92	4.48	2.38	17.04	6.18	2.41	4.28	3.74	2.23	15.57	5.56	2.30
Simple16	9.53	9.43	3.85	25.30	9.40	3.35	5.96	5.34	2.90	19.36	6.92	2.79	5.28	4.62	2.73	17.82	6.34	2.65
Simple8b	9.96	9.24	4.63	26.41	9.18	4.14	6.32	5.53	3.27	21.46	7.36	3.47	5.60	4.77	3.03	20.09	6.87	3.27
QMX	10.21	9.16	5.07	27.10	9.14	4.53	6.71	5.98	3.36	23.27	7.98	3.75	5.92	5.19	3.06	21.64	7.43	3.49
SIMD-BP128	10.35	8.82	5.60	27.05	8.76	4.89	7.00	6.35	3.41	25.08	8.68	3.97	6.03	5.42	2.98	23.02	8.00	3.61
Varint-G8IU	14.51	11.38	8.83	40.51	11.60	8.84	13.75	10.35	8.81	38.82	10.75	8.83	13.57	10.09	8.81	38.35	10.51	8.83
VarintGB	15.64	12.01	9.77	43.58	12.18	9.81	15.02	11.15	9.77	42.10	11.43	9.80	14.87	10.94	9.77	41.77	11.27	9.80
StreamVByte	16.03	12.30	10.04	44.55	12.44	10.03	15.37	11.37	10.04	42.97	11.65	10.03	15.25	11.21	10.04	42.66	11.49	10.03

[2] https://github.com/lemire/FastPFor.
[3] https://github.com/andrewtrotman/JASSv2.
[4] https://github.com/mpetri/partitioned_ef_ans.

selected 1000 queries for each query length from 2 to 6+. Figure 1 shows the query length distributions. For Figs. 2 and 3, for each data point we sample 500 queries from each query set. We call this set Trec05-06.

4 Results and Discussion

Compressed Index Size. All compression results are summarized in Table 2, sorted by increasing space for the Gov2 collection under the URL ordering. We find that this order is mostly preserved across all tested scenarios. The exceptions are Packed+ANS2 and Interpolative, which compete for the top spot. However, relative differences change quite significantly. For instance, variable byte methods benefit little from URL or BP reordering. This is expected, since virtually all gain comes from decreased document gaps, and no improvement is seen for frequency encodings. On the other hand, packing methods are highly sensitive to ordering, and achieve significantly better compression with URL or BP. For instance, when using Packed+ANS2, the sizes of Gov2 and ClueWeb09 with URL ordering decrease by 43% and 27%, respectively, compared to Random. Further improvements are seen for BP.

Query Efficiency. We first executed the five early termination algorithms in all configurations with $k = 10$. The results are shown in Table 3. As expected, for a fixed ordering there is a clear trade-off between index size and query speed. Variable byte encoding is extremely fast, but also gains the least from reordering. Interestingly, SIMD-BP128 basically matches the performance of Varint-G8IU and VarintGB while achieving significantly better compression. Other packing methods—QMX, Simple8b, and Simple16—along with OptPFD, fall slightly yet noticeably behind. PEF is on average less efficient. However, it almost matches the performance of the top four encodings for algorithms utilizing many skips or lookups: BMW and especially VBMW. This is significant, as PEF decreases the index size by more than 50% over the variable byte techniques. Finally, the entropy-based encodings perform the worst across all settings, with Interpolative being by far the slowest. Based on these results, we select a set of four encoding techniques for further analysis: OptPFD, PEF, SIMD-BP128, and Varint-G8IU, each representing a different group of fast algorithms.

Overall, the fastest retrieval algorithm is VBMW. Moreover, it facilitates efficient query processing on a space-efficient PEF-encoded index. Unsurprisingly, it improves upon BMW, which in turn improves upon WAND. These two also fall short of MaxScore, which is the fastest when testing Random ordering but does not benefit as much from reordering. We find BMM to provide no improvement over MaxScore. Given these facts, and due to space constraints, we focus further experiments on the MaxScore and VBMW algorithms.

We also notice a significant difference between different document orderings. Queries on randomly ordered indexes can be almost 3 times slower than on URL ordered ones. Quite interesting, although limited, is the improvement obtained by BP over URL ordering. Even though there is some variability of the gain for Gov2, which depends on the algorithm and encoding used, it is quite evident and

Table 3. Query times (in ms) of different query processing strategies on indexes encoded using different encoding techniques.

			Gov2					ClueWeb09				
			MaxScore	WAND	BMM	BMW	VBMW	MaxScore	WAND	BMM	BMW	VBMW
RANDOM	TREC05	Packed+ANS2	13.91	41.60	14.36	22.23	13.73	36.10	117.16	36.34	64.57	46.98
		Interpolative	16.53	55.80	16.93	30.71	18.65	43.32	160.25	43.12	87.72	63.13
		PEF	11.77	16.65	12.55	10.14	6.23	29.62	47.98	31.82	33.06	23.99
		OptPFD	7.43	15.99	8.59	10.91	7.22	19.74	47.28	22.50	34.63	26.61
		Simple16	8.33	19.39	9.38	12.83	8.15	22.28	57.02	24.54	39.10	29.72
		Simple8b	7.05	16.08	8.51	10.98	7.29	18.67	47.54	22.36	34.15	26.17
		QMX	7.50	15.77	8.72	11.50	7.28	20.11	47.77	22.73	35.25	26.47
		SIMD-BP128	6.27	10.68	7.62	8.81	5.75	16.76	32.17	19.79	28.48	21.63
		Varint-G8IU	5.86	10.95	7.43	8.69	5.75	15.46	33.03	19.56	27.74	21.71
		VarintGB	5.96	11.33	7.52	9.00	5.90	15.75	34.26	19.79	28.61	21.86
		StreamVByte	6.44	11.36	7.74	9.23	6.09	17.15	34.32	20.48	29.53	22.70
	TREC06	Packed+ANS2	25.81	95.41	26.14	60.93	41.18	55.19	196.34	68.78	137.39	96.30
		Interpolative	30.84	127.23	30.90	83.35	55.84	67.06	270.54	66.90	185.83	129.58
		PEF	19.08	32.49	20.51	30.56	20.38	41.13	67.24	44.89	71.40	49.53
		OptPFD	12.93	34.24	15.21	32.73	23.04	28.84	73.09	33.78	75.62	55.61
		Simple16	14.72	42.59	16.74	37.72	25.85	32.76	91.10	37.10	84.75	61.31
		Simple8b	12.48	34.86	15.32	32.31	22.83	27.73	74.67	34.20	73.75	55.21
		QMX	13.06	34.17	15.39	33.68	23.10	29.09	73.99	34.49	78.40	55.94
		SIMD-BP128	10.56	21.80	13.20	27.04	18.88	23.95	48.20	29.73	63.25	45.76
		Varint-G8IU	9.96	22.51	13.14	26.32	18.66	22.44	49.62	29.44	61.44	45.46
		VarintGB	10.19	23.36	13.27	26.96	18.92	23.02	51.33	29.78	63.00	45.96
		StreamVByte	10.83	23.40	13.37	27.96	19.69	24.70	51.64	30.32	65.60	48.22
URL	TREC05	Packed+ANS2	9.00	19.91	9.53	9.06	5.14	29.54	79.06	29.42	33.82	20.44
		Interpolative	9.89	23.44	10.31	10.63	5.73	31.14	96.21	31.13	38.75	22.55
		PEF	8.62	9.45	8.90	3.78	2.21	25.76	31.62	26.57	14.19	8.93
		OptPFD	5.08	8.29	5.99	4.22	2.53	16.54	31.15	18.78	16.33	10.45
		Simple16	5.60	9.27	6.31	4.62	2.70	17.32	35.21	19.43	17.26	11.01
		Simple8b	4.60	7.87	5.64	4.01	2.46	14.97	29.64	17.79	15.41	9.94
		QMX	5.16	8.07	6.05	4.23	2.40	16.15	30.42	18.40	15.63	9.82
		SIMD-BP128	4.27	5.68	5.29	3.12	1.91	13.50	19.72	16.01	11.97	7.76
		Varint-G8IU	3.96	5.72	5.02	3.07	1.96	12.47	20.00	15.76	11.88	7.90
		VarintGB	4.02	5.89	5.10	3.17	1.94	12.85	20.87	16.09	12.23	8.00
		StreamVByte	4.48	5.96	5.31	3.22	1.99	13.71	20.89	16.31	12.52	8.11
	TREC06	Packed+ANS2	14.83	39.05	15.46	20.64	11.14	46.20	128.12	45.27	73.88	40.96
		Interpolative	16.23	45.66	16.65	24.26	12.51	49.40	158.23	48.14	84.89	45.85
		PEF	14.32	15.52	14.41	9.71	5.38	39.41	45.35	39.19	33.11	18.43
		OptPFD	8.22	14.12	9.85	10.49	5.95	24.98	47.76	29.42	37.37	21.61
		Simple16	8.81	16.67	10.68	11.18	6.36	26.34	55.21	29.64	39.73	22.42
		Simple8b	7.48	13.78	9.42	10.09	5.80	22.70	45.66	27.51	35.75	20.89
		QMX	8.40	13.88	9.90	10.67	5.69	24.28	45.80	28.03	36.88	20.56
		SIMD-BP128	6.80	8.98	8.64	8.27	4.63	19.96	29.18	24.49	28.56	16.50
		Varint-G8IU	6.31	9.12	8.45	8.02	4.70	18.67	29.81	24.37	28.10	16.62
		VarintGB	6.48	9.57	8.54	8.20	4.75	19.14	31.05	24.54	28.84	16.94
		StreamVByte	6.95	9.51	8.74	8.50	4.86	20.29	31.22	24.90	29.79	17.42
BP	TREC05	Packed+ANS2	8.53	14.91	9.15	6.86	4.85	27.52	59.41	27.98	25.23	17.53
		Interpolative	9.67	18.87	10.18	8.52	5.67	30.14	77.01	30.43	29.96	19.92
		PEF	8.69	8.28	8.81	2.99	2.25	24.92	26.88	25.86	10.98	7.80
		OptPFD	4.97	6.97	5.94	3.43	2.61	15.57	25.47	17.92	12.57	9.15
		Simple16	5.38	7.83	6.23	3.69	2.77	16.57	29.33	18.72	13.44	9.71
		Simple8b	4.45	6.69	5.57	3.24	2.50	14.08	24.25	17.02	11.93	8.74
		QMX	5.01	7.04	6.03	3.45	2.51	15.19	25.38	17.79	11.97	8.56
		SIMD-BP128	4.19	5.21	5.24	2.55	2.00	12.75	17.19	15.43	9.34	6.97
		Varint-G8IU	3.86	5.22	5.02	2.54	2.03	11.87	17.71	14.95	9.24	6.98
		VarintGB	3.93	5.34	5.10	2.63	2.05	12.08	18.37	15.32	9.55	7.06
		StreamVByte	4.26	5.45	5.29	2.65	2.10	12.97	18.12	15.74	9.75	7.20
	TREC06	Packed+ANS2	14.61	26.66	15.67	15.64	10.36	41.24	85.86	41.53	50.79	32.81
		Interpolative	16.50	34.54	17.49	19.30	12.29	45.86	113.44	44.90	60.75	38.32
		PEF	14.88	12.44	15.18	7.65	5.44	37.60	34.73	36.46	24.24	16.21
		OptPFD	8.31	11.15	10.26	8.61	6.03	22.62	35.20	26.18	26.95	18.40
		Simple16	8.93	12.87	10.74	9.23	6.38	24.03	40.47	27.37	28.93	19.62
		Simple8b	7.52	10.77	9.78	8.20	5.82	20.65	33.79	25.28	25.94	18.11
		QMX	8.37	11.28	10.32	8.52	5.78	21.82	34.30	25.94	26.66	17.61
		SIMD-BP128	6.97	7.73	9.06	6.80	4.74	18.09	23.00	23.41	21.21	14.47
		Varint-G8IU	6.39	7.83	8.89	6.68	4.80	16.96	23.56	22.36	20.75	14.64
		VarintGB	6.56	8.15	9.18	6.83	4.79	17.44	24.44	22.54	21.39	14.67
		StreamVByte	7.04	8.25	9.15	7.07	4.96	18.53	24.59	23.24	21.97	15.22

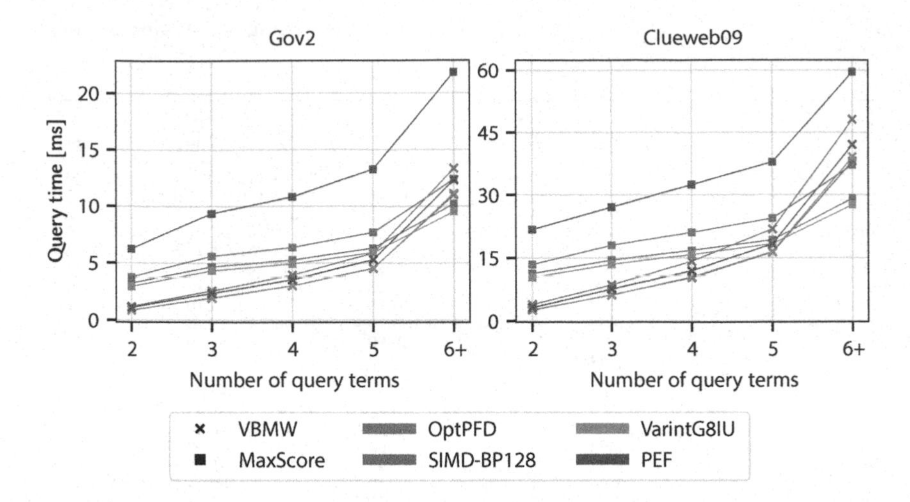

Fig. 2. Query times for different query lengths under URL ordering.

constant for ClueWeb09. To the best of our knowledge, this result for BP has not been discussed in the literature, and it would be interesting to further investigate the reasons for the improvement, with the aim of further improvements.

Query Length. Figure 2 shows average query times for different query term counts using Trec05-06 queries, under the URL ordering. Interestingly, MaxScore performs better for long queries (except when PEF encoding is used). For ClueWeb09 the difference is significant: about 10 ms. This could justify a hybrid retrieval method that switches between algorithms based on a query length.

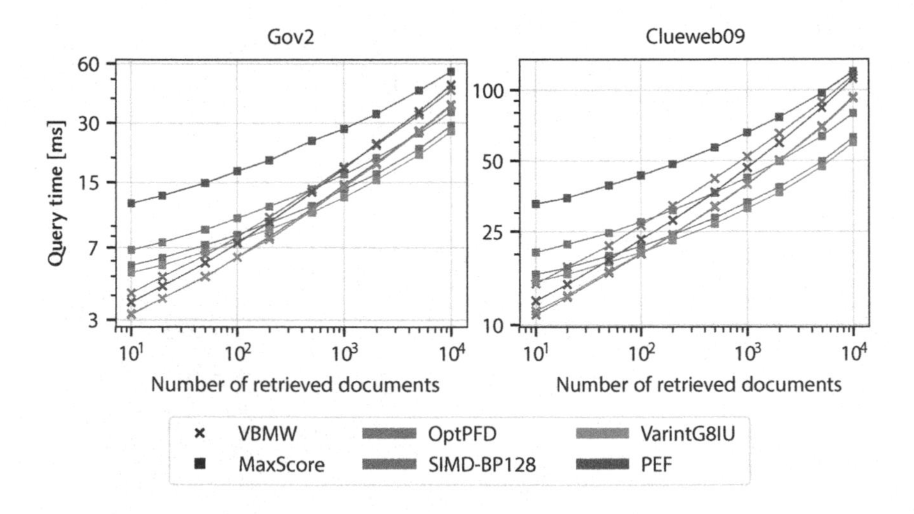

Fig. 3. Query times for different k under URL ordering.

Varying k. The results for a range of values of k (using URL ordering) are shown in Fig. 3 on a log-log scale for better readability. First, we notice a significant time increase with larger k across all encoding techniques. We find this increase to be faster for VBMW. Both algorithms are roughly equally fast for $k = 100$ using Varint-G8IU or SIMD-BP128, while for larger k MaxScore becomes faster. We note that at $k = 10,000$ even the performance of PEF, which previously performed well only for VBMW, is similar for both algorithms. This suggests that MaxScore might be better suited for some cases of candidate generation.

5 Conclusions

In this paper, we performed an experimental evaluation of a whole range of previously proposed schemes for index compression, query processing, index reordering, and the number of results k. Our experiments reproduce many previous results while filling some remaining gaps and painting a more detailed picture. We confirm known correlation between index size and query speed, and provide comprehensive data that may help to find the right trade-off for a specific application. In particular, we find SIMD-BP128 to perform on a par with the variable byte techniques while providing a significantly higher compression ratio. Moreover, PEF is both space and time-efficient when using the VBMW algorithm. VBMW is the fastest query processing method for small k while MaxScore surpasses it for $k > 100$. Query cost increases significantly with k for DAAT methods, justifying TAAT and SAAT approaches for candidate generation such as [12]. The good performance of MaxScore on long queries motivates hybrid methods that select algorithms based on query length.

Acknowledgments. This research was supported by NSF Grant IIS-1718680 and a grant from Amazon.

References

1. Anh, V.N., Moffat, A.: Inverted index compression using word-aligned binary codes. Inf. Retrieval **8**(1), 151–166 (2005)
2. Anh, V.N., Moffat, A.: Index compression using 64-bit words. Softw. Pract. Exp. **40**(2), 131–147 (2010)
3. Arguello, J., Diaz, F., Lin, J., Trotman, A.: SIGIR 2015 workshop on reproducibility, inexplicability, and generalizability of results. In: 38th International ACM SIGIR Conference on Research and Development in Information Retrieval, pp. 1147–1148. ACM (2015)
4. Blanco, R., Barreiro, Á.: Document identifier reassignment through dimensionality reduction. In: Losada, D.E., Fernández-Luna, J.M. (eds.) ECIR 2005. LNCS, vol. 3408, pp. 375–387. Springer, Heidelberg (2005). https://doi.org/10.1007/978-3-540-31865-1_27
5. Blandford, D., Blelloch, G.: Index compression through document reordering. In: 2002 Data Compression Conference, pp. 342–351 (2002)

6. Broder, A.Z., Carmel, D., Herscovici, M., Soffer, A., Zien, J.: Efficient query evaluation using a two-level retrieval process. In: Proceedings of the Twelfth International Conference on Information and Knowledge Management, pp. 426–434. ACM (2003)
7. Callan, J., Hoy, M., Yoo, C., Zhao, L.: Clueweb09 data set (2009). http://lemurproject.org/clueweb09/
8. Catena, M., Macdonald, C., Ounis, I.: On inverted index compression for search engine efficiency. In: de Rijke, M., et al. (eds.) ECIR 2014. LNCS, vol. 8416, pp. 359–371. Springer, Cham (2014). https://doi.org/10.1007/978-3-319-06028-6_30
9. Chakrabarti, K., Chaudhuri, S., Ganti, V.: Interval-based pruning for top-k processing over compressed lists. In: Proceedings of the 2011 IEEE 27th International Conference on Data Engineering, pp. 709–720 (2011)
10. Chapelle, O., Chang, Y.: Yahoo! learning to rank challenge overview. In: Proceedings of the Learning to Rank Challenge, pp. 1–24 (2011)
11. Chapelle, O., Chang, Y., Liu, T.Y.: Future directions in learning to rank. In: Proceedings of the Learning to Rank Challenge, pp. 91–100 (2011)
12. Crane, M., Culpepper, J.S., Lin, J., Mackenzie, J., Trotman, A.: A comparison of document-at-a-time and score-at-a-time query evaluation. In: Proceedings of the Tenth ACM International Conference on Web Search and Data Mining, pp. 201–210. ACM (2017)
13. Craswell, N., Fetterly, D., Najork, M., Robertson, S., Yilmaz, E.: Microsoft research at TREC 2009 web and relevance feedback tracks. Technical report, Microsoft Research (2009)
14. Dean, J.: Challenges in building large-scale information retrieval systems: invited talk. In: Proceedings of the Second ACM International Conference on Web Search and Data Mining, pp. 1–1. ACM (2009)
15. Dhulipala, L., Kabiljo, I., Karrer, B., Ottaviano, G., Pupyrev, S., Shalita, A.: Compressing graphs and indexes with recursive graph bisection. In: Proceedings of the 22nd ACM SIGKDD International Conference on Knowledge Discovery and Data Mining, pp. 1535–1544 (2016)
16. Dimopoulos, C., Nepomnyachiy, S., Suel, T.: Optimizing top-k document retrieval strategies for block-max indexes. In: Proceedings of the sixth ACM International Conference on Web Search and Data Mining, pp. 113–122. ACM (2013)
17. Ding, S., Attenberg, J., Suel, T.: Scalable techniques for document identifier assignment in inverted indexes. In: Proceedings of the 19th international conference on World wide web, pp. 311–320. ACM (2010)
18. Ding, S., Suel, T.: Faster top-k document retrieval using block-max indexes. In: Proceedings of the 34th International ACM SIGIR Conference on Research and Development in Information Retrieval, pp. 993-1002. ACM (2011)
19. Duda, J.: Asymmetric numeral systems as close to capacity low state entropy coders. CoRR abs/1311.2540 (2013)
20. Elias, P.: Efficient storage and retrieval by content and address of static files. J. ACM **21**(2), 246–260 (1974)
21. Elias, P.: Universal codeword sets and representations of the integers. IEEE Trans. Inf. Theory **21**(2), 194–203 (1975)
22. Fano, R.M.: On the number of bits required to implement an associative memory. Massachusetts Institute of Technology, Project MAC (1971)
23. Golomb, S.W.: Run-length encodings (corresp.). IEEE Trans. Inf. Theory **12**(3), 399–401 (1966)

24. Hawking, D., Jones, T.: Reordering an index to speed query processing without loss of effectiveness. In: Proceedings of the Seventeenth Australasian Document Computing Symposium, pp. 17-24. ACM (2012)
25. Kane, A., Tompa, F.W.: Split-lists and initial thresholds for wand-based search. In: The 41st International ACM SIGIR Conference on Research & Development in Information Retrieval, pp. 877-880. ACM (2018)
26. Lemire, D., Boytsov, L.: Decoding billions of integers per second through vectorization. Softw. Pract. Exper. **45**(1), 1–29 (2015)
27. Lemire, D., Kurz, N., Rupp, C.: Stream vbyte: faster byte-oriented integer compression. Inf. Process. Lett. **130**, 1–6 (2018)
28. Liu, T.Y.: Learning to rank for information retrieval. Found. Trends Inf. Retrieval **3**(3), 225–331 (2009)
29. Macdonald, C., Santos, R.L., Ounis, I.: The whens and hows of learning to rank for web search. Inf. Retr. **16**(5), 584–628 (2013)
30. Mallia, A., Ottaviano, G., Porciani, E., Tonellotto, N., Venturini, R.: Faster blockmax WAND with variable-sized blocks. In: Proceedings of the 40th International ACM SIGIR Conference on Research and Development in Information Retrieval, pp. 625–634. ACM (2017)
31. Metzler, D., Croft, W.B.: A Markov random field model for term dependencies. In: Proceedings of the 28th Annual International ACM SIGIR Conference on Research and Development in Information Retrieval, pp. 472–479 (2005)
32. Moffat, A., Petri, M.: ANS-based index compression. In: Proceedings of the 2017 ACM on Conference on Information and Knowledge Management, pp. 677-686. ACM (2017)
33. Moffat, A., Petri, M.: Index compression using byte-aligned ANS coding and two-dimensional contexts. In: Proceedings of the Eleventh ACM International Conference on Web Search and Data Mining, pp. 405-413. ACM (2018)
34. Moffat, A., Stuiver, L.: Binary interpolative coding for effective index compression. Inf. Retr. **3**(1), 25–47 (2000)
35. Ottaviano, G., Venturini, R.: Partitioned elias-fano indexes. In: Proceedings of the 37th international ACM SIGIR conference on Research & Development in Information Retrieval, pp. 273–282. ACM (2014)
36. Plaisance, J., Kurz, N., Lemire, D.: Vectorized VByte decoding. CoRR abs/1503.07387 (2015)
37. Qin, T., Liu, T.Y., Xu, J., Li, H.: LETOR: a benchmark collection for research on learning to rank for information retrieval. Inf. Retr. **13**(4), 346–374 (2010)
38. Rice, R., Plaunt, J.: Adaptive variable-length coding for efficient compression of spacecraft television data. IEEE Trans. Commun. Technol. **19**(6), 889–897 (1971)
39. Robertson, S.E., Jones, K.S.: Relevance weighting of search terms. J. Am. Soc. Inf. Sci. **27**(3), 129–146 (1976)
40. Scholer, F., Williams, H.E., Yiannis, J., Zobel, J.: Compression of inverted indexes for fast query evaluation. In: Proceedings of the 25th Annual International ACM SIGIR Conference on Research and Development in Information Retrieval, pp. 222-229. ACM (2002)
41. Shieh, W.Y., Chen, T.F., Shann, J.J.J., Chung, C.P.: Inverted file compression through document identifier reassignment. Inf. Process. Manage. **39**(1), 117–131 (2003)
42. Silvestri, F.: Sorting out the document identifier assignment problem. In: Proceedings of the 29th European Conference on IR Research, pp. 101–112 (2007)

43. Silvestri, F., Orlando, S., Perego, R.: Assigning identifiers to documents to enhance the clustering property of fulltext indexes. In: Proceedings of the 27th Annual International ACM SIGIR Conference on Research and Development in Information Retrieval, pp. 305-312. ACM (2004)
44. Stepanov, A.A., Gangolli, A.R., Rose, D.E., Ernst, R.J., Oberoi, P.S.: SIMD-based decoding of posting lists. In: Proceedings of the 20th ACM International Conference on Information and Knowledge Management, pp. 317–326 (2011)
45. Tonellotto, N., Macdonald, C., Ounis, I.: Effect of different docid orderings on dynamic pruning retrieval strategies. In: Proceedings of the 34th International ACM SIGIR Conference on Research and Development in Information Retrieval, pp. 1179–1180. ACM (2011)
46. Trotman, A.: Compression, SIMD, and postings lists. In: Proceedings of the 2014 Australasian Document Computing Symposium, pp. 50:50–50:57. ACM (2014)
47. Trotman, A., Lin, J.: In vacuo and in situ evaluation of SIMD codecs. In: Proceedings of the 21st Australasian Document Computing Symposium, pp. 1–8. ACM (2016)
48. Turtle, H., Flood, J.: Query evaluation: strategies and optimizations. Inf. Process. Manage. **31**(6), 831–850 (1995)
49. Wang, L., Lin, J., Metzler, D.: Learning to efficiently rank. In: Proceedings of the 33rd International ACM SIGIR Conference on Research and Development in Information Retrieval, pp. 138–145. ACM (2010)
50. Yan, H., Ding, S., Suel, T.: Inverted index compression and query processing with optimized document ordering. In: Proceedings of the 18th International Conference on World Wide Web, pp. 401–410. ACM (2009)
51. Zhai, C., Lafferty, J.: A study of smoothing methods for language models applied to information retrieval. ACM Trans. Inf. Syst. **22**(2), 179–214 (2004)
52. Zhang, J., Long, X., Suel, T.: Performance of compressed inverted list caching in search engines. In: Proceedings of the 17th International Conference on World Wide Web, pp. 387–396. ACM (2008)
53. Zhang, M., Kuang, D., Hua, G., Liu, Y., Ma, S.: Is learning to rank effective for web search? In: SIGIR 2009 Workshop: Learning to Rank for Information Retrieval, pp. 641–647 (2009)
54. Zukowski, M., Heman, S., Nes, N., Boncz, P.: Super-scalar RAM-CPU cache compression. In: Proceedings of the 22nd International Conference on Data Engineering (2006)

Reproducing and Generalizing Semantic Term Matching in Axiomatic Information Retrieval

Peilin Yang[1] and Jimmy Lin[2](✉)

[1] Ontario, Canada
[2] David R. Cheriton School of Computer Science, University of Waterloo, Ontario, Canada
jimmylin@uwaterloo.ca

Abstract. In the framework of axiomatic information retrieval, the semantic term matching technique proposed by Fang and Zhai in SIGIR 2006 has been shown to be effective in addressing the vocabulary mismatch problem, with experimental evidence provided from newswire collections. This paper reproduces and generalizes these results in Anserini, an open-source IR toolkit built on Lucene. In addition to making an implementation of axiomatic semantic term matching available on a widely-used open-source platform, we describe a series of experiments that help researchers and practitioners better understand its behavior across a number of test collections spanning newswire, web, and microblogs. Results show that axiomatic semantic term matching can be applied on top of different base retrieval models, and that its effectiveness varies across different document genres, each requiring different parameter settings for optimal effectiveness.

Keywords: Axiomatic retrieval · Query expansion

1 Introduction

The *vocabulary mismatch* problem is one of the most fundamental challenges in information retrieval. Frequently, query terms expressing an information need differ from those used by authors of relevant documents. Retrieval models based on exact term matches, which include instances from the probabilistic retrieval family, language modeling framework, and many others, have difficulty with this problem. "Classic" approaches to tackling this challenge include relevance feedback [7], query expansion [8,9], and modeling term relationships using statistical translation [1], while a new generation of neural ranking models offer solutions based on continuous word representations [6]. In this paper, we focus on reproducing and generalizing an alternative approach to addressing the vocabulary mismatch problem in the axiomatic retrieval framework [2]—specifically, the SIGIR 2006 paper of Fang and Zhai [3] (henceforth, FZ for short). The paper showed that semantic term matching can be incorporated into the axiomatic

L. Azzopardi et al. (Eds.): ECIR 2019, LNCS 11437, pp. 369–381, 2019.
https://doi.org/10.1007/978-3-030-15712-8_24

retrieval framework via a weighting function derived from mutual information with respect to a working set of documents. The ranking model can be formulated in terms of query expansion, and thus its implementation is well understood in the broader context of the IR literature.

The work of FZ is worthy of detailed exploration for several reasons: First, axiomatic retrieval is under-explored from a reproducibility perspective, compared to say, BM25 and language modeling approaches. For example, the large-scale study of Lin et al. [4] examined a number of different retrieval models across a number of systems, but did not include any techniques based on axiomatic retrieval. Second, axiomatic semantic term matching provides a strong non-neural baseline, since one of the purported advantages of continuous word representations (on which most neural ranking models depend) is the ability to capture word similarity based on distributional statistics. The importance of FZ has also been recognized by the recent CENTRE reproducibility initiative that cross-cuts CLEF, TREC, and NTCIR. A follow-on paper applying axiomatic semantic term matching to web collections [10] was selected as one of the targets for participants to reproduce. The organizers selected these target papers based on many different factors, including the popularity of the task that the technique tackles, as well as the impact of the work. Although the specific effort we describe here is orthogonal to the CENTRE initiative, the selection of FZ provides independent confirmation that axiomatic semantic term matching represents an important contribution that should be studied in greater detail.

We are able to successfully reproduce the work of FZ using the open-source Anserini information retrieval toolkit built on Lucene. Reproducibility here is used in a precise manner in the sense articulated in recent ACM guidelines,[1] which means "that an independent group can obtain the same result using artifacts which they develop completely independently." Whereas the original FZ paper used Indri, our reimplementation from scratch uses Anserini, sharing no common code. Our implementation, along with detailed documentation and associated run scripts, yields experimental results that are both repeatable (i.e., "a researcher can reliably repeat her own computation") and replicable (i.e., "an independent group can obtain the same result using the author's own artifacts"), both in the sense that ACM defines them (quoted from the ACM guideline referenced above). Given the widespread deployment of Lucene by a large number of organizations in production settings, our implementation increases the options that builders of real-world search applications can explore.

Having reproduced FZ, we conducted additional experiments to generalize the results in several respects: First, we applied the technique to a large number of test collections spanning many different document genres, including newswire, web, and microblogs. Axiomatic semantic term matching is effective for newswire and microblogs, but less so for web collections. Second, we examined a number of parameters that impact effectiveness. In particular, the parameter that determines the weight of semantic matches behaves quite differently across document genres. Also, the technique introduces randomness in the sampling

[1] https://www.acm.org/publications/policies/artifact-review-badging.

of non-relevant documents to construct a working document set–we characterize the impact of this non-determinism. Finally, we demonstrate that although axiomatic semantic term matching was originally developed within the axiomatic retrieval framework, the core ideas can be adapted to other ranking models as well. Specifically, axiomatic semantic term matching also works well on a base ranking model that uses BM25 or query likelihood.

2 Approach

Axiomatic semantic term matching relates document terms that do not match query terms at the lexical level, thus potentially overcoming the vocabulary mismatch problem. In this section, we provide an overview of the technique, borrowing heavily from previous papers [3,10], but refer the reader to those sources for more detailed derivations.

The matching score of term t in a document with respect to query Q comprised of terms $\{q_1, q_2, ..., q_n\}$ is computed as $S(Q,t) = \sum_{q \in Q} s(q,t)/|Q|$, where

$$s(q,t) = \begin{cases} \omega(q) & \text{if } t = q \\ \omega(q) \times \beta \times \frac{\text{MI}(q,t)}{\text{MI}(q,q)} & \text{if } t \neq q \end{cases} \tag{1}$$

For matching terms (i.e., $t = q$), $\omega(q)$ is simply the *idf* of q. In the case of lexical mismatch (i.e., $t \neq q$), the semantic distance between two terms is captured using mutual information (MI) with respect to a working set W (more details below), modulated by β, a parameter that controls how much we "trust" the semantically-related term:

$$\begin{aligned} \text{MI}(q,t) &= I(X_q, X_t|W) \\ &= \sum_{X_q, X_t \in \{0,1\}} p(X_q, X_t|W) \cdot \log \frac{p(X_q, X_t|W)}{p(X_q|W)p(X_t|W)} \end{aligned} \tag{2}$$

Here, X_q and X_t are two binary random variables that denote the presence or absence of term q and term t in the document.

The working set is assembled as follows: First, we take the R top ranked documents from an initial retrieval run, treating them as pseudo-relevant documents. We add to these $(N-1) \times R$ documents (assumed non-relevant) randomly sampled from the collection, excluding the first R documents. This yields a working set comprised of $N \times R$ total documents. Although FZ discuss sampling from external collections, particularly in the web context [10], we do not consider this variation in our study due to limited space.

Considered end to end, the steps involved in axiomatic semantic term matching are as follows:

1. Perform an initial retrieval and construct a working set for computing semantic similarity in the manner described above.

2. For each query term, select the K most similar terms using Eq. (1). From this pool of candidate terms, select the M most similar terms based on $S(Q, t)$.
3. These M terms form the weighted, expanded query. Search the collection with this expanded query and return the final ranked list.

In summary, the parameters for axiomatic semantic term matching are as follows: R, the number of pseudo-relevant documents in the working set; N, which determines the number of additional non-relevant documents to sample, $(N-1) \times R$; K, the cutoff to be considered as a potential expansion term for a query term; M, the total number of expansion terms to add; β, the weight of the expansion terms in Eq. (2).

In our effort, we decided to reproduce axiomatic semantic term matching using Anserini, an open-source information retrieval toolkit built on Lucene [11, 12]. The goal of the Anserini project is to bridge the gap between information retrieval research and real-world search applications, where Lucene has become the *de facto* platform for production deployments. We hope that a Lucene implementation will enable a broader audience (i.e., the open-source community and the long list of companies that run Lucene in production) to try out innovations from academic researchers. The source code of the implementation of axiomatic semantic term matching by Yang and Fang [10] is available online,[2] which provided us with a reference implementation to consult. This implementation is also based on Indri, but it differs from the original implementation in the FZ paper. Due to the availability of this resource, we encountered no difficulties in our implementation efforts.

Beyond reproducing the work of FZ, we explored three research questions to generalize axiomatic semantic term matching:

(RQ1) *Does axiomatic semantic term matching generalize to different types of collections?* The original FZ paper only examined newswire collections, but we experimented with many more test collections spanning three different genres: newswire, web, and microblogs. Many of these collections were not available when the original paper was published.

(RQ2) *How does axiomatic semantic term matching behave with different base ranking models?* Although the formal derivations are couched within the framework of axiomatic retrieval, the operationalization of the model in terms of query expansion means that the technique can be applied to any base ranking model. That is, we can use any number of ranking functions to construct the working set, and use the same ranking function for the expanded query. Our implementation in Anserini makes such explorations easy.

(RQ3) *What is the effect of non-determinism in sampling non-relevant documents?* Semantic term matching weights are computed with respect to a working set populated by sampling (assumed) non-relevant documents from the collection. We examine the impact of this non-determinism on effectiveness.

[2] https://github.com/Peilin-Yang/axiomatic_query_expansion.

3 Experimental Setup

Our experiments used TREC test collections spanning three different genres: newswire, web, and microblogs. The newswire collections are as follows:

- TREC Disks 1 & 2, with topics and relevance judgments from the *ad hoc* task at TREC-1 through TREC-3 (topics 51–200).
- TREC Disks 4 & 5, excluding Congressional Record, with topics and relevance judgments from the *ad hoc* task at TREC-6 through TREC-8 as well as the Robust Tracks from TREC 2003 and 2004.
- The AQUAINT Corpus of English News Text, with topics and relevance judgments from the TREC 2005 Robust Track.
- The New York Times Annotated Corpus, with topics and relevance judgments from the TREC 2017 Common Core Track.

For web collections:

- The WT10g and Gov2 collections from CSIRO (Commonwealth Scientific and Industrial Research Organisation), distributed by the University of Glasgow, with topics and relevance judgments from the web task at TREC-9 for the former, and the Terabyte Tracks at TREC 2004–2006 for the latter.
- The ClueWeb09b and ClueWeb12-B13 web crawls from Carnegie Mellon University, with topics and relevance judgments from the Web Tracks at TREC 2010–2012 for the former and the Web Tracks at TREC 2013 and 2014 for the latter. We did not run experiments on the complete ClueWeb09 and ClueWeb12 collections for two reasons: first, they are too large for running query expansion in practice (i.e., the experiments take too much time), and second, relevance judgments are too sparse to draw firm conclusions (more details later).

And finally, microblog collections:

- The Tweets 2011 collection, with topics and relevance judgments from the TREC 2011 and 2012 Microblog Tracks.
- The Tweets 2013 collection, with topics and relevance judgments from the TREC 2013 and 2014 Microblog Tracks.

All source code for replicating results reported in this paper is available in the Anserini code repository[3] (post v0.3.0 release, based on Lucene 7.6) at commit `08434ad` (dated Jan. 15, 2019).

4 Results

We begin with results from our attempts to directly reproduce the original FZ paper for those collections that overlap with our experimental settings. The original FZ paper, published in SIGIR 2006, predated many of the collections we

[3] http://anserini.io/.

Table 1. Comparisons to the original FZ results (average precision).

	SIGIR 2006		Anserini	
Run	F2EXP	+Ax	F2EXP	+Ax
Robust04	0.2480	0.2850	0.2492	0.2839
Robust05	0.1920	0.2580	0.1985	0.2481

use, and even though FZ report results on other collections, they are generally regarded as either non-standard or too small to support drawing reliable conclusions. Results in terms of average precision are shown in Table 1. Here, F2EXP is used as the base ranking model (the implementation in Anserini, not the Lucene default), with axiomatic semantic term matching denoted by "Ax". In these experiments we used the same parameter settings as in the original paper.

We see that the effectiveness metrics are quite close, despite completely different implementations. The original work of FZ was implemented in Indri, whereas our results are based on Lucene. Differences can be easily be attributed to the document processing pipeline (tokenization, stemming, stopwords, etc.) as well as the inherent non-determinism in constructing the working set (more details below). At a high level, it appears that axiomatic semantic term matching "works as advertised" in terms of effectiveness. Our narrative continues by examining the additional research questions posed in Sect. 2.

As expected, (RQ2) was straightforward to address–our implementation adopts a modular architecture that enabled us to apply different base ranking models for the construction of the working set as well as for the second stage retrieval using the expanded query. In our experiments, in addition to using F2EXP, as the original FZ paper does, we also report results with query likelihood using Dirichlet-smoothed language models (QL) and BM25 (both default Lucene implementations).

In generalizing the results of FZ, our most interesting findings centered around applications to different document genres (RQ1). Furthermore, the parameter of greatest interest is β, which determines the weight of semantically-related terms: we discovered that there are systematic variations across different genres. For these experiments, the remaining parameters were fixed as follows: $N = 30$, $R = 20$, $K = 1000$, $M = 20$. These values represent default settings recommended by FZ. As the original paper already performed a number of parameter explorations, we focused on supplementing those results, since we do not have space for exhaustive examination of all parameters. For these experiments, sampling non-relevant documents was accomplished by setting the random seed to 42, which makes our experiments repeatable.

Results on the newswire collections are shown in Fig. 1: the y axis shows average precision of the top 1000 hits, and the x axis shows the β setting. Each curve denotes a different base ranking model (in a different color): BM25, query likelihood (QL), and F2EXP. The respective baselines without axiomatic semantic term matching are shown as horizontal lines in matching colors. The same plots for the web collections are shown in Fig. 2 (note that we report average

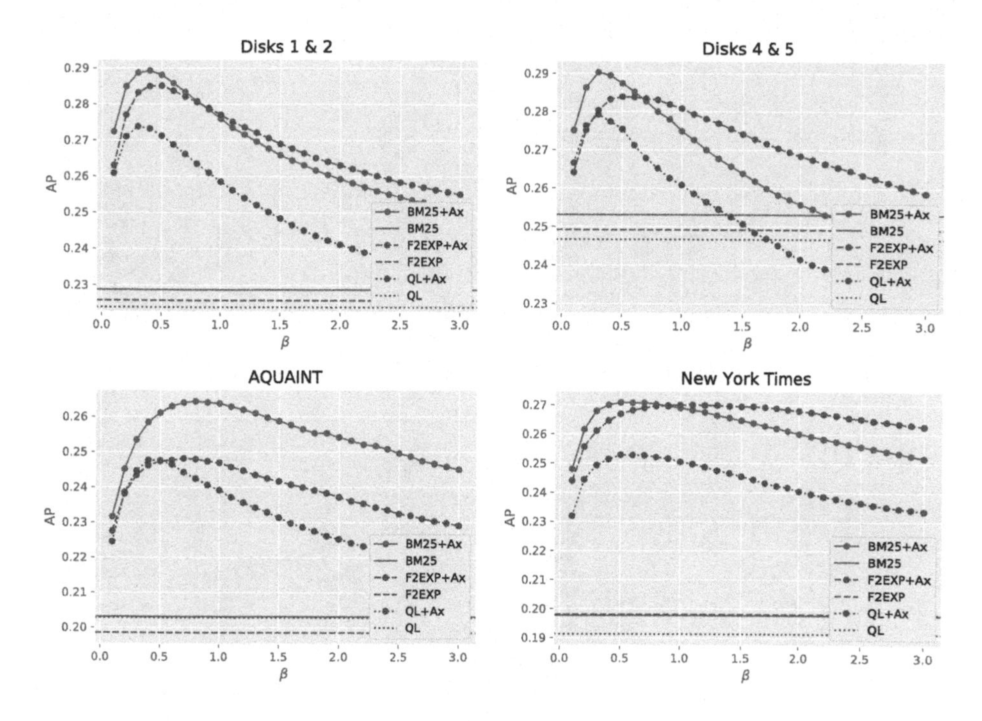

Fig. 1. Results of β tuning experiments on newswire collections.

precision for WT10g and Gov2 but NDCG@20 for the ClueWeb collections, since the shallow pool depths make AP unreliable), and the microblog collections, in Fig. 3. To aid interpretation: $\beta = 1$ places equal importance on both the original query terms and the expansion terms, while $\beta < 1$ means we "trust" expansion terms less (and the opposite for $\beta > 1$).

The newswire collections behave as we would expect—the plots in Fig. 1 are consistent with Fig. 3 in the FZ paper. However, results on the web collections are unexpected: for WT10g and Gov2, axiomatic semantic term matching yields only small improvements in average precision, and only with small values of β. For ClueWeb12-B13, no setting of β improves effectiveness. For the microblog collections, we also observe qualitatively different behavior: First, optimal effectiveness is reached at a larger value of β, which means that the ranking model places more importance on expansion terms. Second, effectiveness does not appear to be very sensitive to β at all. Whereas average precision decays sharply with larger values of β on newswire collections, effectiveness decays much more slowly for microblogs.

Before drawing any firm conclusions from these results, we need to rule out evaluation artifacts. One obvious culprit is unjudged documents—query expansion has the possibility of retrieving documents that are not part of the original evaluation pool. Figure 4 shows the results of this analysis for BM25. For each collection, we plot the fraction of unjudged documents in the top 20, 50, and 100 hits. The first row shows results for the newswire collections, the second row for

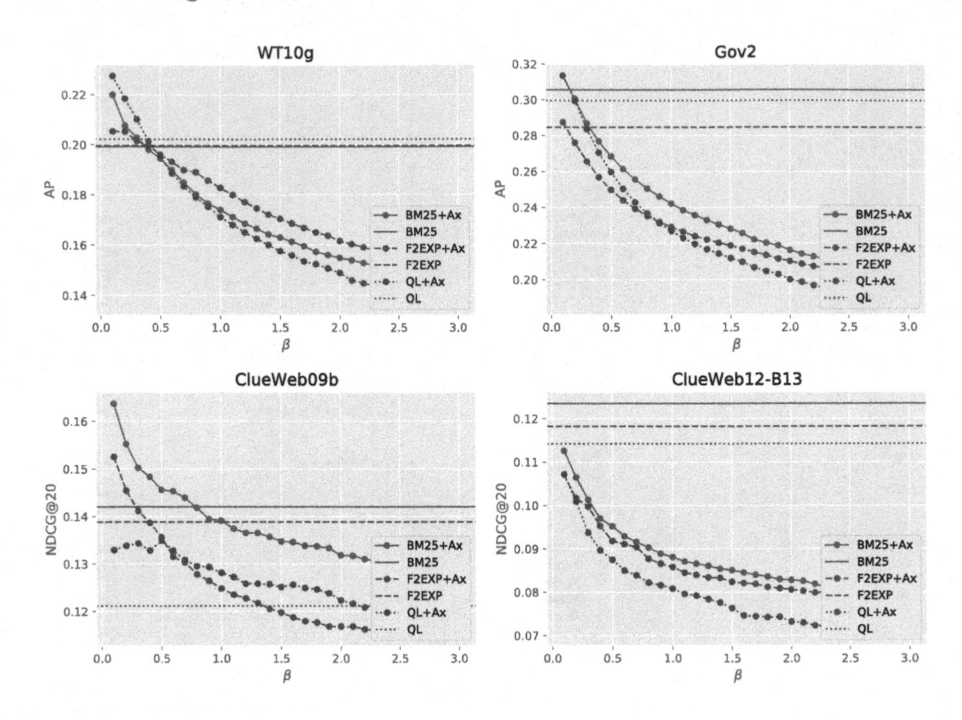

Fig. 2. Results of β tuning experiments on web collections.

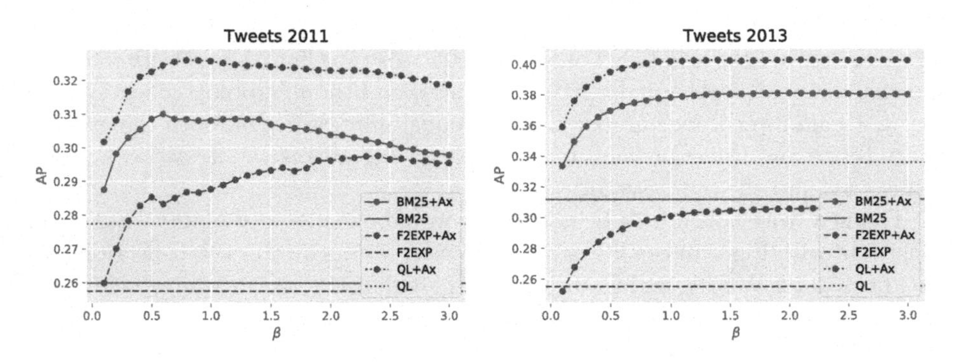

Fig. 3. Results of β tuning experiments on microblog collections.

the web collections,[4] and the third row for the microblog collections. Ideally, the fraction of unjudged documents should be constant across different β settings; that is, no setting should be penalized by retrieving more unjudged documents. The absolute value of missing judgments is less important, since judgments will always be incomplete in any pooling-based test collection. Instead, we are more interested in whether different settings of β are unfairly penalized.

[4] For the ClueWeb collections, we measured effectiveness in terms of NDCG@20, so the analysis for the top 50 and 100 documents are not applicable; nevertheless, we have included those results in the graphs for completeness.

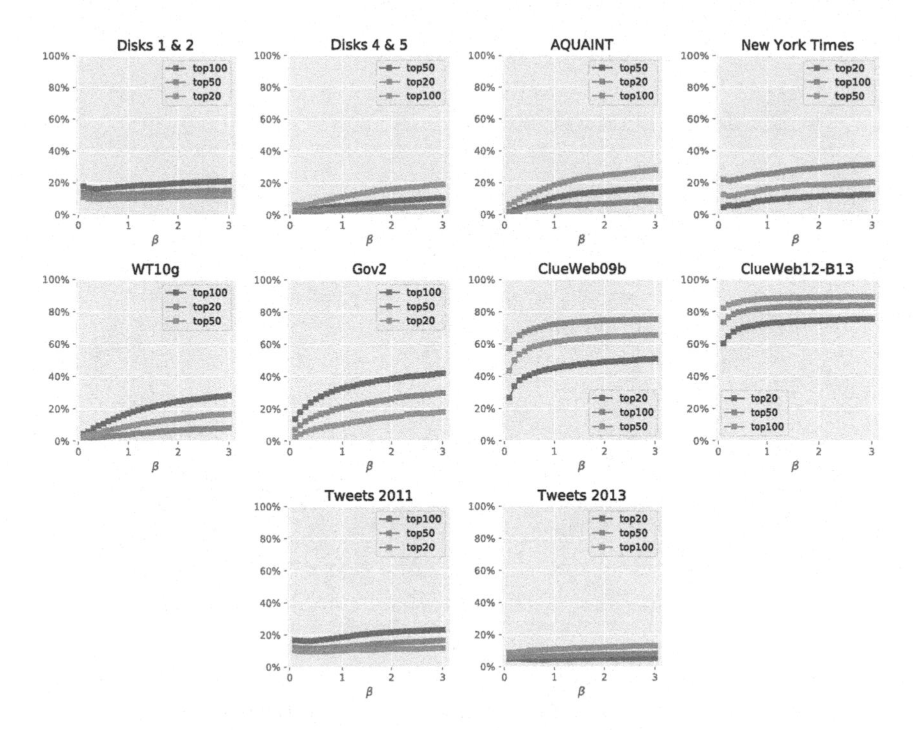

Fig. 4. Analysis of missing judgments using BM25 as the base ranking model.

Results from Disks 1 & 2 are closest to our ideal—the fraction of missing judgments does not vary much across β settings (and furthermore, the absolute values are quite low). For the newswire collections, the results on AQUAINT (Robust05) deviate the most from our ideal—for example, a setting of $\beta = 1$ yields around 10% *more* unjudged documents vs. $\beta = 0.5$ at rank 100. For web collections (second row in Fig. 4), we observe even more missing judgments. For ClueWeb12-B13, with any setting of β, over 60% of the documents are unjudged. The microblog test collections are reasonably well behaved, where the fraction of missing judgments is comparable to newswire collections.

Given the evidence presented above, the following conclusions are supported with respect to (RQ1): axiomatic semantic term matching appears to be effective across a range of newswire collections with $\beta = 0.5$; the technique also appears to be effective for microblog collections, with a setting of $\beta = 1.0$. These β values should be taken as rough, coarse-grained guides. In fact, we argue that fine-grained tuning is essentially meaningless due to missing judgments and the fact that effectiveness differences are not very large in a broad range around the above-proposed settings. For web collections, a setting of $\beta = 0.1$ yields slightly better effectiveness in some cases, but however, there is insufficient evidence to decide between two competing hypotheses: That axiomatic semantic term matching is not effective for web collections, or that current evaluation resources are unable to accurately determine its effectiveness. If the former turns out to be true, *why* would be an interesting follow-on question.

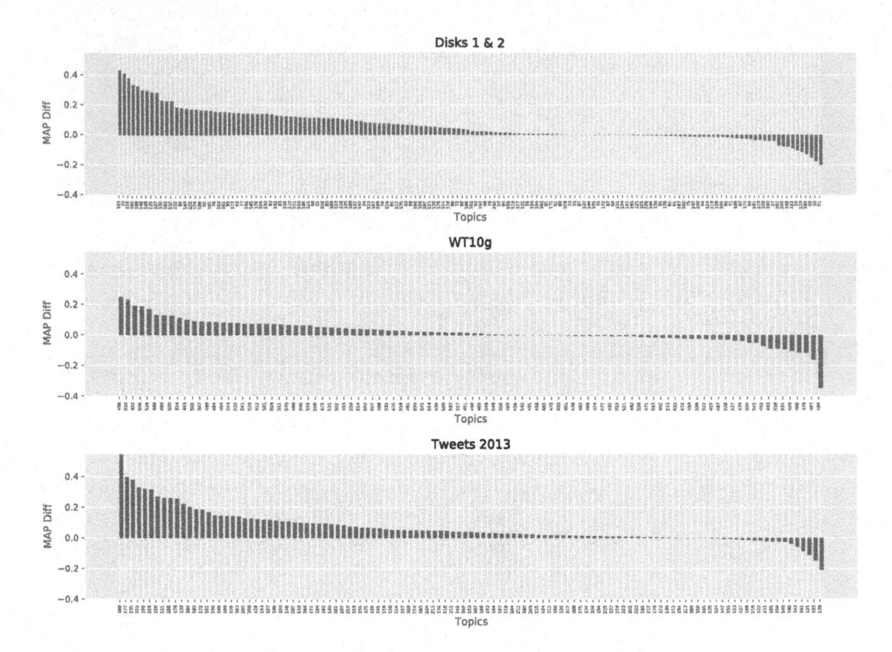

Fig. 5. Per-topic analysis for Disks 1 & 2 (top), WT10g (middle), and Tweets 2013 (bottom) comparing axiomatic semantic term matching with baseline BM25 ranking.

We attempted to dig a bit deeper into understanding the behavior of axiomatic semantic term matching across different document genres by analyzing per-topic effectiveness differences. Figure 5 shows results for a representative collection from each genre: Disks 1 & 2, WT10g, and Tweets 2013. These collections were selected because they contained the fewest unjudged documents according to the analysis in Fig. 4, thus affording us the greatest confidence in the effectiveness measurements. Each bar represents a topic and its height captures the average precision difference between baseline BM25 and axiomatic semantic term matching with BM25 as the base ranking model. Bars are sorted in descending order of effectiveness differences, from left to right, where negative bars represent topics where axiomatic semantic term matching hurts effectiveness.

As is typical of many query expansion techniques, axiomatic semantic term matching helps some topics but hurts other topics. The relative proportion of the beneficial vs. detrimental cases does not seem markedly different across genres, but it appears that even for the best topics in WT10g, the technique does not help as much as in the other two collections. Also, for WT10g, the worst-performing topics have decreases in AP that are greater than in the other collections. We followed up with manual analysis of the worst-performing topics across all three collections, comparing the original queries with the expanded queries. Unfortunately, this did not reveal any obvious insights. For example, we hypothesized that since web collections contain more noisy text, the quality of the expansion terms might be worse. However, this was not the case—the expansion terms all appeared reasonable and their quality was not markedly different from query expansion terms in the other two collections.

In answering (RQ2), looking across newswire, web, and microblog collections, it seems clear that axiomatic semantic term matching can be applied to a variety of base ranking models. For the newswire collections, effectiveness appears to be highest using BM25, with F2EXP slightly better than QL in most cases. For the web collections, the effectiveness of all three ranking models is quite similar. For the microblog collections, we observe large differences in average precision, but these results are consistent with known characteristics of the collections: BM25 does not work well for ranking microblogs because posts do not differ much in length, and thus the length normalization factor in the scoring function has little impact. For the TREC Microblog Tracks, QL is the preferred baseline [5].

Our final set of experiments tackled (RQ3) and examined the inherent non-determinism involved in the construction of the working set when sampling non-relevant documents. These experiments used the β recommendations above with the same settings of the other parameters. For each test collection, using the BM25 base ranking model, we repeated the ranking experiments 100 times with different random seeds.[5] The results are summarized in box-and-whiskers plots in Fig. 6, which report average precision except for the ClueWeb collections, which show NDCG@20. The blue dotted line in each case represents the effectiveness

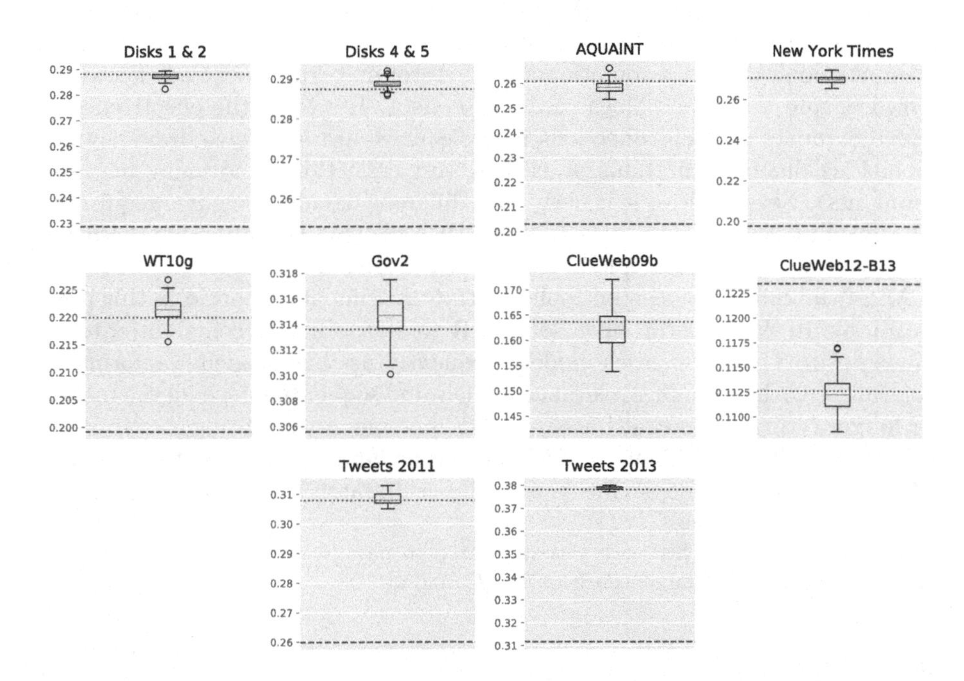

Fig. 6. Box-and-whiskers plots showing the distribution of scores across 100 random seeds when sampling non-relevant documents to construct the working set, with BM25 as the base ranking model. The BM25 baselines are shown as blue dotted lines, while the single-point measurements are shown as green dotted lines.

[5] This was accomplished by using 42 as the "meta seed" to generate a pseudo-random sequence of random seeds for each experimental run.

of the BM25 baseline, and the green dotted line represents the single-point effectiveness measurement from the comparable experiments above.

To specifically answer (RQ3): We observe that the variations in effectiveness that can be attributed to random seed selection is quite small, and that even the low effectiveness outliers are well above the BM25 baselines for both newswire and microblog collections. For both document genres, the single-point effectiveness measurement is within the range predicted by the box-and-whiskers distributions. We can conclude that axiomatic semantic term matching is robust with respect to document sampling for the working set. The results for the web collections are consistent with the findings above, and suggest that axiomatic semantic term matching helps for three of the four collections. For ClueWeb12-B13, the large fraction of unjudged documents prevents us from drawing any meaningful conclusions, as discussed above.

5 Conclusions

We have successfully reproduced the axiomatic semantic term matching work of Fang and Zhai in Anserini, based on the popular open-source Lucene search engine. The work is over a decade old, and this paper generalizes the techniques to web and microblog collections, beyond the newswire collections in the original paper. We confirm that axiomatic semantic term matching is indeed effective on newswire, and that microblogs similarly benefit. However, the effectiveness of these techniques on web collections is unclear; we are unable to draw any firm conclusions due to limitations of existing test collections (too many unjudged documents). Nevertheless, it is clear that different document genres require different weights on the importance of semantic term matches, although there does not appear to be any principled rationale for those settings.

All of the code necessary to replicate the experiments reported in this paper is available in the Anserini open-source IR toolkit. Already contributed to our code repository are numerous models frequently used in academic information retrieval research, including relevance models and sequential dependence models. Our longer term hope is that Lucene-based implementations bring academia and industry into better alignment, allowing researchers an easier path to achieve real-world impact via deployments of real-world search applications.

Acknowledgments. This work was supported in part by the Natural Sciences and Engineering Research Council (NSERC) of Canada.

References

1. Berger, A., Lafferty, J.: Information retrieval as statistical translation. In: Proceedings of the 22nd Annual International ACM SIGIR Conference on Research and Development in Information Retrieval, pp. 222–229. SIGIR 1999. ACM, New York (1999). https://doi.org/10.1145/312624.312681

2. Fang, H., Zhai, C.: An exploration of axiomatic approaches to information retrieval. In: Proceedings of the 28th Annual International ACM SIGIR Conference on Research and Development in Information Retrieval, pp. 480–487. SIGIR 2005. ACM, New York (2005). https://doi.org/10.1145/1076034.1076116
3. Fang, H., Zhai, C.: Semantic term matching in axiomatic approaches to information retrieval. In: Proceedings of the 29th Annual International ACM SIGIR Conference on Research and Development in Information Retrieval, pp. 115–122. SIGIR 2006. ACM, New York (2006). https://doi.org/10.1145/1148170.1148193
4. Lin, J., et al.: Toward reproducible baselines: the open-source IR reproducibility challenge. In: Ferro, N., et al. (eds.) ECIR 2016. LNCS, vol. 9626, pp. 408–420. Springer, Cham (2016). https://doi.org/10.1007/978-3-319-30671-1_30
5. Lin, J., Efron, M.: Overview of the TREC-2013 Microblog Track. In: Proceedings of the Twenty-Second Text REtrieval Conference (TREC 2013), Gaithersburg, Maryland (2013)
6. Onal, K.D., et al.: Neural information retrieval: at the end of the early years. Inf. Retrieval **21**(2–3), 111–182 (2018). https://doi.org/10.1007/s10791-017-9321-y
7. Rocchio, J.J.: Relevance feedback in information retrieval. In: Salton, G. (ed.) The SMART Retrieval System-Experiments in Automatic Document Processing, pp. 313–323. Prentice-Hall, Englewood Cliffs (1971)
8. Voorhees, E.M.: Query expansion using lexical-semantic relations. In: Proceedings of the 17th Annual International ACM SIGIR Conference on Research and Development in Information Retrieval. pp. 61–69. SIGIR 1994. ACM, New York (1994). http://dl.acm.org/citation.cfm?id=188490.188508
9. Xu, J., Croft, W.B.: Improving the effectiveness of information retrieval with local context analysis. ACM Trans. Inf. Syst. **18**(1), 79–112 (2000)
10. Yang, P., Fang, H.: Evaluating the effectiveness of axiomatic approaches in web track. In: Proceedings of the Twenty-Second Text REtrieval Conference (TREC 2013), Gaithersburg, Maryland (2013)
11. Yang, P., Fang, H., Lin, J.: Anserini: enabling the use of Lucene for information retrieval research. In: Proceedings of the 40th International ACM SIGIR Conference on Research and Development in Information Retrieval, pp. 1253–1256. SIGIR 2017. ACM, New York (2017). https://doi.org/10.1145/3077136.3080721
12. Yang, P., Fang, H., Lin, J.: Anserini: reproducible ranking baselines using Lucene. J. Data Inf. Qual. **10**(4) (2018). Article 16

Optimizing Ranking Models in an Online Setting

Harrie Oosterhuis(✉) and Maarten de Rijke

University of Amsterdam, Amsterdam, The Netherlands
{oosterhuis,derijke}@uva.nl

Abstract. Online Learning to Rank (OLTR) methods optimize ranking models by directly interacting with users, which allows them to be very efficient and responsive. All OLTR methods introduced during the past decade have extended on the original OLTR method: Dueling Bandit Gradient Descent (DBGD). Recently, a fundamentally different approach was introduced with the Pairwise Differentiable Gradient Descent (PDGD) algorithm. To date the only comparisons of the two approaches are limited to simulations with cascading click models and low levels of noise. The main outcome so far is that PDGD converges at higher levels of performance and learns considerably faster than DBGD-based methods. However, the PDGD algorithm assumes cascading user behavior, potentially giving it an unfair advantage. Furthermore, the robustness of both methods to high levels of noise has not been investigated. Therefore, it is unclear whether the reported advantages of PDGD over DBGD generalize to different experimental conditions. In this paper, we investigate whether the previous conclusions about the PDGD and DBGD comparison generalize from ideal to worst-case circumstances. We do so in two ways. First, we compare the theoretical properties of PDGD and DBGD, by taking a critical look at previously proven properties in the context of ranking. Second, we estimate an upper and lower bound on the performance of methods by simulating both *ideal* user behavior and extremely *difficult* behavior, i.e., almost-random non-cascading user models. Our findings show that the theoretical bounds of DBGD do not apply to any common ranking model and, furthermore, that the performance of DBGD is substantially worse than PDGD in both ideal and worst-case circumstances. These results reproduce previously published findings about the relative performance of PDGD vs. DBGD and generalize them to extremely noisy and non-cascading circumstances.

Keywords: Learning to rank · Online learning · Gradient descent

1 Introduction

Learning to Rank (LTR) plays a vital role in information retrieval. It allows us to optimize models that combine hundreds of signals to produce rankings, thereby making large collections of documents accessible to users through effective search and recommendation. Traditionally, LTR has been approached as a

L. Azzopardi et al. (Eds.): ECIR 2019, LNCS 11437, pp. 382–396, 2019.
https://doi.org/10.1007/978-3-030-15712-8_25

supervised learning problem, where annotated datasets provide human judgements indicating relevance. Over the years, many limitations of such datasets have become apparent: they are costly to produce [3,21] and actual users often disagree with the relevance annotations [23]. As an alternative, research into LTR approaches that learn from user behavior has increased. By learning from the implicit feedback in user behavior, users' true preferences can potentially be learned. However, such methods must deal with the noise and biases that are abundant in user interactions [31]. Roughly speaking, there are two approaches to LTR from user interactions: learning from historical interactions and Online Learning to Rank (OLTR). Learning from historical data allows for optimization without gathering new data [14], though it does require good models of the biases in logged user interactions [4]. In contrast, OLTR methods learn by interacting with the user, thus they gather their own learning data. As a result, these methods can adapt instantly and are potentially much more responsive than methods that use historical data.

Dueling Bandit Gradient Descent. (DBGD) [30] is the most prevalent OLTR method; it has served as the basis of the field for the past decade. DBGD samples variants of its ranking model, and compares them using interleaving to find improvements [12,22]. Subsequent work in OLTR has extended on this approach [10,25,28]. Recently, the first alternative approach to DBGD was introduced with *Pairwise Differentiable Gradient Descent* (PDGD) [19]. PDGD estimates a pairwise gradient that is reweighed to be unbiased w.r.t. users' document pair preferences. The original paper that introduced PDGD showed considerable improvements over DBGD under simulated user behavior [19]: a substantially higher point of performance at convergence and a much faster learning speed. The results in [19] are based on simulations using low-noise cascading click models. The pairwise assumption that PDGD makes, namely, that all documents preceding a clicked document were observed by the user, is always correct in these circumstances, thus potentially giving it an unfair advantage over DBGD. Furthermore, the low level of noise presents a close-to-ideal situation, and it is unclear whether the findings in [19] generalize to less perfect circumstances.

In this paper, we contrast PDGD over DBGD. Prior to an experimental comparison, we determine whether there is a theoretical advantage of DBGD over PDGD and examine the regret bounds of DBGD for ranking problems. We then investigate whether the benefits of PDGD over DBGD reported in [19] generalize to circumstances ranging from ideal to worst-case. We simulate circumstances that are perfect for both methods – behavior without noise or position-bias – and circumstances that are the worst possible scenario – almost-random, extremely-biased, non-cascading behavior. These settings provide estimates of upper and lower bounds on performance, and indicate how well previous comparisons generalize to different circumstances. Additionally, we introduce a version of DBGD that is provided with an oracle interleaving method; its performance shows us the maximum performance DBGD could reach from hypothetical extensions.

In summary, the following research questions are addressed in this paper:

RQ1. Do the regret bounds of DBGD provide a benefit over PDGD?
RQ2. Do the advantages of PDGD over DBGD observed in prior work generalize to extreme levels of noise and bias?
RQ3. Is the performance of PDGD reproducible under non-cascading user behavior?

2 Related Work

This section provides a brief overview of traditional LTR (Sect. 2.1), of LTR from historical interactions (Sect. 2.2), and OLTR (Sect. 2.3).

2.1 Learning to Rank from Annotated Datasets

Traditionally, LTR has been approached as a supervised problem; in the context of OLTR this approach is often referred to as *offline* LTR. It requires a dataset containing relevance annotations of query-document pairs, after which a variety of methods can be applied [16]. The limitations of offline LTR mainly come from obtaining such annotations. The costs of gathering annotations are high as it is both time-consuming and expensive [3,21]. Furthermore, annotators cannot judge for very specific users, i.e., gathering data for personalization problems is infeasible. Moreover, for certain applications it would be unethical to annotate items, e.g., for search in personal emails or documents [29]. Additionally, annotations are stationary and cannot account for (perceived) relevance changes [6,15,27]. Most importantly, though, annotations are not necessarily aligned with user preferences; judges often interpret queries differently from actual users [23]. As a result, there has been a shift of interest towards LTR approaches that do not require annotated data.

2.2 Learning to Rank from Historical Interactions

The idea of LTR from user interactions is long-established; one of the earliest examples is the original pairwise LTR approach [13]. This approach uses historical click-through interactions from a search engine and considers clicks as indications of relevance. Though very influential and quite effective, this approach ignores the *noise* and *biases* inherent in user interactions. Noise, i.e., any user interaction that does not reflect the user's true preference, occurs frequently, since many clicks happen for unexpected reasons [23]. Biases are systematic forms of noise that occur due to factors other than relevance. For instance, interactions will only involve displayed documents resulting in selection bias [29]. Another important form of bias in LTR is position bias, which occurs because users are less likely to consider documents that are ranked lower [31]. Thus, to learn true preferences from user interactions effectively, a LTR method should be robust to noise and handle biases correctly.

In recent years counter-factual LTR methods have been introduced that correct for some of the bias in user interactions. Such methods uses inverse propensity scoring to account for the probability that a user observed a ranking position [14]. Thus, clicks on positions that are observed less often due to position bias will have greater weight to account for that difference. However, the position bias must be learned and estimated somewhat accurately [1]. On the other side of the spectrum are click models, which attempt to model user behavior completely [4]. By predicting behavior accurately, the effect of relevance on user behavior can also be estimated [2,29].

An advantage of these approaches over OLTR is that they only require historical data and thus no new data has to be gathered. However, unlike OLTR, they do require a fairly accurate user model, and thus they cannot be applied in cold-start situations.

2.3 Online Learning to Rank

OLTR differs from the approaches listed above because its methods intervene in the search experience. They have control over what results are displayed, and can learn from their interactions instantly. Thus, the online approach performs LTR by interacting with users directly [30]. Similar to LTR methods that learn from historical interaction data, OLTR methods have the potential to learn the true user preferences. However, they also have to deal with the noise and biases that come with user interactions. Another advantage of OLTR is that the methods are very responsive, as they can apply their learned behavior instantly. Conversely, this also brings a danger as an online method that learns incorrect preferences can also worsen the experience immediately. Thus, it is important that OLTR methods are able to learn reliably in spite of noise and biases. Thus, OLTR methods have a two-fold task: they have to simultaneously present rankings that provide a good user experience *and* learn from user interactions with the presented rankings.

The original OLTR method is Dueling Bandit Gradient Descent (DBGD); it approaches optimization as a dueling bandit problem [30]. This approach requires an online comparison method that can compare two rankers w.r.t. user preferences; traditionally, DBGD methods use interleaving. Interleaving methods take the rankings produced by two rankers and combine them in a single result list, which is then displayed to users. From a large number of clicks on the presented list the interleaving methods can reliably infer a preference between the two rankers [12,22]. At each timestep, DBGD samples a candidate model, i.e., a slight variation of its current model, and compares the current and candidate models using interleaving. If a preference for the candidate is inferred, the current model is updated towards the candidate slightly. By doing so, DBGD will update its model continuously and should oscillate towards an inferred optimum. Section 3 provides a complete description of the DBGD algorithm.

Virtually all work in OLTR in the decade since the introduction of DBGD has used DBGD as a basis. A straightforward extension comes in the form of Multileave Gradient Descent [25] which compares a large number of candidates per interaction [18,24,26]. This leads to a much faster learning process, though in the long term this method does not seem to improve the point of convergence.

One of the earliest extensions of DBGD proposed a method for reusing historical interactions to guide exploration for faster learning [10]. While the initial results showed great improvements [10], later work showed performance drastically decreasing in the long term due to bias introduced by the historical data [20]. Unfortunately, OLTR work that continued this historical approach [28] also only considered short term results; moreover, the results of some work [32] are not based on held-out data. As a result, we do not know whether these extensions provide decent long-term performance and it is unclear whether the findings of these studies generalize to more realistic settings.

Recently, an inherently different approach to OLTR was introduced with PDGD [19]. PDGD interprets its ranking model as a distribution over documents; it estimates a pairwise gradient from user interactions with sampled rankings. This gradient is differentiable, allowing for non-linear models like neural networks to be optimized, something DBGD is ineffective at [17,19]. Section 4 provides a detailed description of PDGD. In the paper in which we introduced PDGD, claim that it provides substantial improvements over DBGD. However, those claims are based on cascading click models with low levels of noise. This is problematic because PDGD assumes a cascading user, and could thus have an unfair advantage in this setting. Furthermore, it is unclear whether DBGD with a perfect interleaving method could still improve over PDGD. Lastly, DBGD has proven regret bounds while PDGD has no such guarantees.

In this study, we clear up these questions about the relative strengths of DBGD and PDGD by comparing the two methods under non-cascading, high-noise click models. Additionally, by providing DBGD with an oracle comparison method, its hypothetical maximum performance can be measured; thus, we can study whether an improvement over PDGD is hypothetically possible. Finally, a brief analysis of the theoretical regret bounds of DBGD shows that they do not apply to any common ranking model, therefore hardly providing a guaranteed advantage over PDGD.

3 Dueling Bandit Gradient Descent

This section describes the DBGD algorithm in detail, before discussing the regret bounds of the algorithm.

3.1 The Dueling Bandit Gradient Descent Method

The DBGD algorithm [30] describes an indefinite loop that aims to improve a ranking model at each step; Algorithm 1 provides a formal description. The algorithm starts a given model with weights θ_1 (Line 1); then it waits for a user-submitted query (Line 3). At this point a candidate ranker is sampled from the unit sphere around the current model (Line 4), and the current and candidate

Algorithm 1. Dueling Bandit Gradient Descent (DBGD).

1: **Input**: initial weights: θ_1; unit: u; learning rate η.
2: **for** $t \leftarrow 1 \ldots \infty$ **do**
3: $\quad q_t \leftarrow receive\_query(t)$ *obtain a query from a user*
4: $\quad \theta_t^c \leftarrow \theta_t + sample\_from\_unit\_sphere(u)$ *create candidate ranker*
5: $\quad R_t \leftarrow get\_ranking(\theta_t, D_{q_t})$ *get current ranker ranking*
6: $\quad R_t^c \leftarrow get\_ranking(\theta_t^c, D_{q_t})$ *get candidate ranker ranking*
7: $\quad I_t \leftarrow interleave(R_t, R_t^c)$ *interleave both rankings*
8: $\quad \mathbf{c}_t \leftarrow display\_to\_user(I_t)$ *displayed interleaved list, record clicks*
9: $\quad$ **if** $preference\_for\_candidate(I_t, \mathbf{c}_t, R_t, R_t^c)$ **then**
10: $\quad\quad \theta_{t+1} \leftarrow \theta_t + \eta(\theta_t^c - \theta_t)$ *update model towards candidate*
11: $\quad$ **else**
12: $\quad\quad \theta_{t+1} \leftarrow \theta_t$ *no update*

model both produce a ranking for the current query (Line 5 and 6). These rankings are interleaved (Line 7) and displayed to the user (Line 8). If the interleaving method infers a preference for the candidate ranker from subsequent user interactions the current model is updated towards the candidate (Line 10), otherwise no update is performed (Line 12). Thus, the model optimized by DBGD should converge and oscillate towards an optimum.

3.2 Regret Bounds of Dueling Bandit Gradient Descent

Unlike PDGD, DBGD has proven regret bounds [30], potentially providing an advantage in the form of theoretical guarantees. In this section we answer **RQ1** by critically looking at the assumptions which form the basis of DBGD's proven regret bounds.

The original DBGD paper [30] proved a sublinear regret under several assumptions. DBGD works with the parameterized space of ranking functions $\mathcal{W}$, that is, every $\theta \in \mathcal{W}$ is a different set of parameters for a ranking function. For this study we will only consider linear models because all existing OLTR work has dealt with them [10,11,19,20,25,28,30,32]. But we note that the proof is easily extendable to neural networks where the output is a monotonic function applied to a linear combination of the last layer. Then there is assumed to be a concave utility function $u : \mathcal{W} \rightarrow \mathbb{R}$; since this function is concave, there should only be a single instance of weights that are optimal θ^*. Furthermore, this utility function is assumed to be L-Lipschitz smooth:

$$\exists L \in \mathbb{R}, \quad \forall(\theta_a, \theta_b) \in \mathcal{W}, \quad |u(\theta_a) - u(\theta_b)| < L\|\theta_a - \theta_b\|. \tag{1}$$

We will show that these assumptions are *incorrect*: there is an infinite number of optimal weights, and the utility function u cannot be L-Lipschitz smooth. Our proof relies on two assumptions that avoid cases where the ranking problem is trivial. First, the zero ranker is not the optimal model:

$$\theta^* \neq \mathbf{0}. \tag{2}$$

Second, there should be at least two models with different utility values:

$$\exists(\theta, \theta') \in \mathcal{W}, \quad u(\theta) \neq u(\theta'). \tag{3}$$

We will start by defining the set of rankings a model $f(\cdot, \theta)$ will produce as:

$$\mathcal{R}_D(f(\cdot, \theta)) = \{R \mid \forall(d, d') \in D, [f(d, \theta) > f(d', \theta) \rightarrow d \succ_R d']\}. \tag{4}$$

It is easy to see that multiplying a model with a positive scalar $\alpha > 0$ will not affect this set:

$$\forall \alpha \in \mathbb{R}_{>0}, \quad \mathcal{R}_D(f(\cdot, \theta)) = \mathcal{R}_D(\alpha f(\cdot, \theta)). \tag{5}$$

Consequently, the utility of both functions will be equal:

$$\forall \alpha \in \mathbb{R}_{>0}, \quad u(f(\cdot, \theta)) = u(\alpha f(\cdot, \theta)). \tag{6}$$

For linear models scaling weights has the same effect: $\alpha f(\cdot, \theta) = f(\cdot, \alpha\theta)$. Thus, the first assumption cannot be true since for any optimal model $f(\cdot, \theta^*)$ there is an infinite set of equally optimal models: $\{f(\cdot, \alpha\theta^*) \mid \alpha \in \mathbb{R}_{>0}\}$.

Then, regarding L-Lipschitz smoothness, using any positive scaling factor:

$$\forall \alpha \in \mathbb{R}_{>0}, \quad |u(\theta_a) - u(\theta_b)| = |u(\alpha\theta_a) - u(\alpha\theta_b)|, \tag{7}$$

$$\forall \alpha \in \mathbb{R}_{>0}, \quad \|\alpha\theta_a - \alpha\theta_b\| = \alpha\|\theta_a - \theta_b\|. \tag{8}$$

Thus the smoothness assumption can be rewritten as:

$$\exists L \in \mathbb{R}, \quad \forall \alpha \in \mathbb{R}_{>0}, \quad \forall(\theta_a, \theta_b) \in \mathcal{W}, \quad |u(\theta_a) - u(\theta_b)| < \alpha L\|\theta_a - \theta_b\|. \tag{9}$$

However, there is always an infinite number of values for α small enough to break the assumption. Therefore, we conclude that a concave L-Lipschitz smooth utility function can never exist for a linear ranking model, thus the proof for the regret bounds is not applicable when using linear models.

Consequently, the regret bounds of DBGD do not apply to the ranking problems in previous work. One may consider other models (e.g., spherical coordinate based models), however this still means that for the simplest and most common ranking problems there are no proven regret bounds. As a result, we answer **RQ1** negatively, the regret bounds of DBGD do not provide a benefit over PDGD for the ranking problems in LTR.

4 Pairwise Differentiable Gradient Descent

The Pairwise Differentiable Gradient Descent (PDGD) [19] algorithm is formally described in Algorithm 2. PDGD interprets a ranking function $f(\cdot, \theta)$ as a probability distribution over documents by applying a Plackett-Luce model:

$$P(d|D, \theta) = \frac{e^{f(d,\theta)}}{\sum_{d' \in D} e^{f(d',\theta)}}. \tag{10}$$

Algorithm 2. Pairwise Differentiable Gradient Descent (PDGD).

1: **Input**: initial weights: θ_1; scoring function: f; learning rate η.
2: **for** $t \leftarrow 1 \ldots \infty$ **do**
3: $\quad q_t \leftarrow \mathit{receive\_query}(t)$ // *obtain a query from a user*
4: $\quad \mathbf{R}_t \leftarrow \mathit{sample\_list}(f_{\theta_t}, D_{q_t})$ // *sample list according to Eq. 10*
5: $\quad \mathbf{c}_t \leftarrow \mathit{receive\_clicks}(\mathbf{R}_t)$ // *show result list to the user*
6: $\quad \nabla f(\cdot, \theta_t) \leftarrow \mathbf{0}$ // *initialize gradient*
7: $\quad$ **for** $d_i \succ_{\mathbf{c}} d_j \in \mathbf{c}_t$ **do**
8: $\quad\quad w \leftarrow \rho(d_i, d_j, R, D)$ // *initialize pair weight (Eq. 13)*
9: $\quad\quad w \leftarrow w \times P(d_i \succ d_j \mid \theta_t) P(d_j \succ d_i \mid \theta_t)$ // *pair gradient (Eq. 12)*
10: $\quad\quad \nabla f(\cdot, \theta_t) \leftarrow \nabla f_{\theta_t} + w \times (f'(d_i, \theta_t) - f'(d_j, \theta_t))$ // *model gradient (Eq. 12)*
11: $\quad \theta_{t+1} \leftarrow \theta_t + \eta \nabla f(\cdot, \theta_t)$ // *update the ranking model*

First, the algorithm waits for a user query (Line 3), then a ranking R is created by sampling documents without replacement (Line 4). Then PDGD observes clicks from the user and infers pairwise document preferences from them. All documents preceding a clicked document and the first succeeding one are assumed to be observed by the user. Preferences between clicked and unclicked observed documents are inferred by PDGD; this is a long-standing assumption in pairwise LTR [13]. We denote an *inferred* preference between documents as $d_i \succ_{\mathbf{c}} d_j$, and the probability of the model placing d_i earlier than d_j is denoted and calculated by:

$$P(d_i \succ d_j \mid \theta) = \frac{e^{f(d_i, \theta)}}{e^{f(d_i, \theta)} + e^{f(d_j, \theta)}}. \tag{11}$$

The gradient is estimated as a sum over inferred preferences with a weight ρ per pair:

$$\begin{aligned} &\Delta f(\cdot, \theta) \\ &\quad \approx \sum_{d_i \succ_{\mathbf{c}} d_j} \rho(d_i, d_j, R, D)[\Delta P(d_i \succ d_j \mid \theta)] \\ &\quad = \sum_{d_i \succ_{\mathbf{c}} d_j} \rho(d_i, d_j, R, D) P(d_i \succ d_j \mid \theta) P(d_j \succ d_i \mid \theta)(f'(d_i, \theta) - f'(d_j, \theta)). \end{aligned} \tag{12}$$

After computing the gradient (Line 10), the model is updated accordingly (Line 11). This will change the distribution (Eq. 10) towards the inferred preferences. This distribution models the confidence over which documents should be placed first; the exploration of PDGD is naturally guided by this confidence and can vary per query.

The weighting function ρ is used to make the gradient of PDGD unbiased w.r.t. document pair preferences. It uses the reverse pair ranking: $R^*(d_i, d_j, R)$, which is the same ranking as R but with the document positions of d_i and d_j swapped. Then ρ is the ratio between the probability of R and R^*:

$$\rho(d_i, d_j, R, D) = \frac{P(R^*(d_i, d_j, R) \mid D)}{P(R \mid D) + P(R^*(d_i, d_j, R) \mid D)}. \tag{13}$$

In the original PDGD paper [19], the weighted gradient is proven to be unbiased w.r.t. document pair preferences under certain assumptions about the user. Here, this unbiasedness is defined by being able to rewrite the gradient as:

$$E[\Delta f(\cdot, \theta)] = \sum_{(d_i, d_j) \in D} \alpha_{ij}(f'(\mathbf{d}_i, \theta) - f'(\mathbf{d}_j, \theta)), \tag{14}$$

and the sign of α_{ij} agreeing with the preference of the user:

$$sign(\alpha_{ij}) = sign(relevance(d_i) - relevance(d_j)). \tag{15}$$

The proof in [19] only relies on the difference in the probabilities of inferring a preference: $d_i \succ_{\mathbf{c}} d_j$ in R and the opposite preference $d_j \succ_{\mathbf{c}} d_i$ in $R^*(d_i, d_j, R)$. The proof relies on the sign of this difference to match the user's preference:

$$\begin{aligned} sign(P(d_i \succ_{\mathbf{c}} d_j \mid R) - P(d_j \succ_{\mathbf{c}} d_i \mid R^*)) = \\ sign(relevance(d_i) - relevance(d_j)). \end{aligned} \tag{16}$$

As long as Eq. 16 is true, Eqs. 14 and 15 hold as well. Interestingly, this means that other assumptions about the user can be made than in [19], and other variations of PDGD are possible, e.g., the algorithm could assume that all documents are observed and the proof still holds.

The original paper on PDGD reports large improvements over DBGD, however these improvements were observed under simulated cascading user models. This means that the assumption that PDGD makes about which documents are observed are always true. As a result, it is currently unclear whether the method is really better in cases where the assumption does not hold.

5 Experiments

In this section we detail the experiments that were performed to answer the research questions in Sect. 1.[1]

5.1 Datasets

Our experiments are performed over three large labelled datasets from commercial search engines, the largest publicly available LTR datasets. These datasets are the *MLSR-WEB10K* [21], *Yahoo! Webscope* [3], and *Istella* [5] datasets. Each contains a set of queries with corresponding preselected document sets. Query-document pairs are represented by feature vectors and five-grade relevance annotations ranging from *not relevant* (0) to *perfectly relevant* (4). Together, the datasets contain over 29,900 queries and between 136 and 700 features per representation.

[1] The resources for reproducing the experiments in this paper are available at https://github.com/HarrieO/OnlineLearningToRank.

Table 1. Click probabilities for simulated *perfect* or *almost random* behavior.

	$P(click(d) \mid relevance(d), observed(d))$				
relevance(d)	0	1	2	3	4
perfect	0.00	0.20	0.40	0.80	1.00
almost random	0.40	0.45	0.50	0.55	0.60

5.2 Simulating User Behavior

In order to simulate user behavior we partly follow the standard setup for OLTR [8,11,20,25,33]. At each step a user issued query is simulated by uniformly sampling from the datasets. The algorithm then decides what result list to display to the user, the result list is limited to $k = 10$ documents. Then user interactions are simulated using click models [4]. Past OLTR work has only considered *cascading click models* [7]; in contrast, we also use *non-cascading click models.* The probability of a click is conditioned on relevance and observance:

$$P(click(d) \mid relevance(d), observed(d)). \tag{17}$$

We use two levels of noise to simulate *perfect* user behavior and *almost random* behavior [9], Table 1 lists the probabilities of both. The *perfect* user observes all documents, never clicks on anything non-relevant, and always clicks on the most relevant documents. Two variants of *almost random* behavior are used. The first is based on cascading behavior, here the user first observes the top document, then decides to click according to Table 1. If a click occurs, then, with probability $P(stop \mid click) = 0.5$ the user stops looking at more documents, otherwise the process continues on the next document. The second *almost random* behavior is simulated in a non-cascading way; here we follow [14] and model the observing probabilities as:

$$P(observed(d) \mid rank(d)) = \frac{1}{rank(d)}. \tag{18}$$

The important distinction is that it is safe to assume that the cascading user has observed all documents ranked before a click, while this is not necessarily true for the non-cascading user. Since PDGD makes this assumption, testing under both models can show us how much of its performance relies on this assumption. Furthermore, the *almost random* model has an extreme level of noise and position bias compared to the click models used in previous OLTR work [11,20,25], and we argue it simulates an (almost) worst-case scenario.

5.3 Experimental Runs

In our experiments we simulate runs consisting of 1,000,000 impressions; each run was repeated 125 times under each of the three click models. PDGD was run with $\eta = 0.1$ and zero initialization, DBGD was run using Probabilistic Interleaving [20] with zero initialization, $\eta = 0.001$, and the unit sphere with $\delta = 1$.

Other variants like Multileave Gradient Descent [25] are not included; previous work has shown that their performance matches that of regular DBGD after around 30,000 impressions [19,20,25]. The initial boost in performance comes at a large computational cost, though, as the fastest approaches keep track of at least 50 ranking models [20], which makes running long experiments extremely impractical. Instead, we introduce a novel oracle version of DBGD, where, instead of interleaving, the NDCG values on the current query are calculated and the highest scoring model is selected. This simulates a hypothetical perfect interleaving method, and we argue that the performance of this oracle run indicates what the upper bound on DBGD performance is.

Performance is measured by NDCG@10 on a held-out test set, a two-sided t-test is performed for significance testing. We do not consider the user experience during training, because past work has already investigated this aspect thoroughly [19].

6 Experimental Results and Analysis

Recall that in Sect. 3.2 we have already provided a negative answer to **RQ1**: the regret bounds of DBGD do not provide a benefit over PDGD for the ranking problems in LTR. In this section we present our experimental results and answer

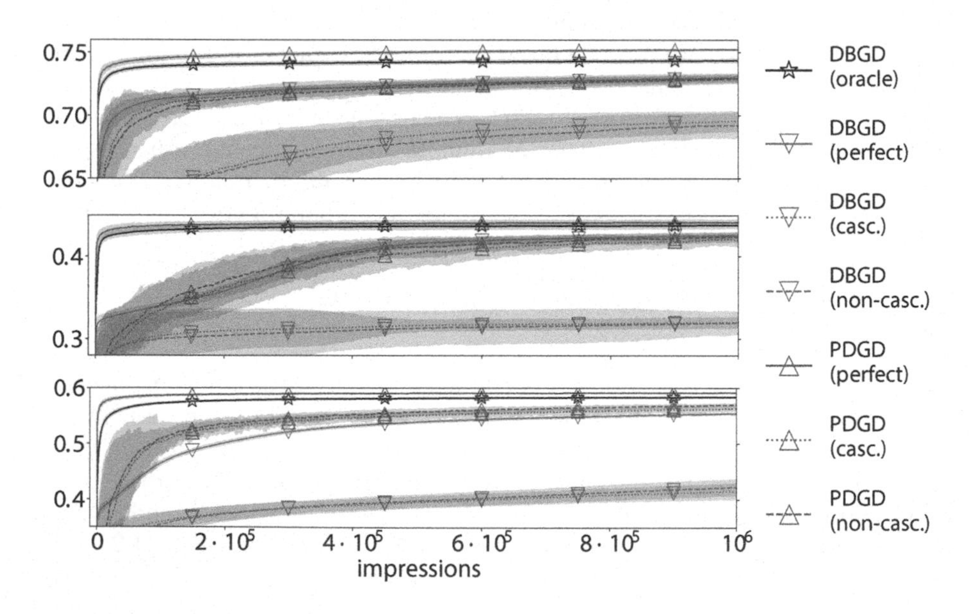

Fig. 1. Performance (NDCG@10) on held-out data from Yahoo (top), MSLR (center), Istella (bottom) datasets, under the *perfect*, and *almost random* user models: cascading (casc.) and non-cascading (non-casc.). The shaded areas display the standard deviation.

Table 2. Performance (NDCG@10) after 1,000,000 impressions for DBGD and PDGD under a *perfect* click model and two almost-random click models: *cascading* and *non-cascading*, and DBGD with an *oracle* comparator. Significant improvements and losses ($p < 0.01$) between DBGD and PDGD are indicated by ▲, ▼, and ○ (no significant difference). Indications are in order of: *oracle*, *perfect*, *cascading*, and *non-cascading*.

	Yahoo	MSLR	Istella
		Dueling Bandit Gradient Descent	
oracle	0.744 (0.001) ▼ ▲ ▲	0.438 (0.004) ▼ ▲ ▲	0.584 (0.001) ▼ ▲ ▲
perfect	0.730 (0.002) ▼ ○ ○	0.426 (0.004) ▼ ▲ ▲	0.554 (0.002) ▼ ▼ ▼
cascading	0.696 (0.008) ▼ ▼ ▼	0.320 (0.006) ▼ ▼ ▼	0.415 (0.014) ▼ ▼ ▼
non-cascading	0.692 (0.010) ▼ ▼ ▼	0.320 (0.014) ▼ ▼ ▼	0.422 (0.014) ▼ ▼ ▼
		Pairwise Differentiable Gradient Descent	
perfect	0.752 (0.001) ▲ ▲ ▲ ▲	0.442 (0.003) ▲ ▲ ▲ ▲	0.592 (0.000) ▲ ▲ ▲ ▲
cascading	0.730 (0.003) ▼ ○ ▲ ▲	0.420 (0.007) ▼ ▼ ▲ ▲	0.563 (0.003) ▼ ▲ ▲ ▲
non-cascading	0.729 (0.003) ▼ ○ ▲ ▲	0.424 (0.005) ▼ ▼ ▲ ▲	0.570 (0.003) ▼ ▲ ▲ ▲

RQ2 (whether the advantages of PDGD over DBGD of previous work generalize to extreme levels of noise and bias) and **RQ3** (whether the performance of PDGD is reproducible under non-cascading user behavior).

Our main results are presented in Table 2. Additionally, Fig. 1 displays the average performance over 1,000,000 impressions. First, we consider the performance of DBGD; there is a substantial difference between its performance under the *perfect* and *almost random* user models on all datasets. Thus, it seems that DBGD is strongly affected by noise and bias in interactions; interestingly, there is little difference between performance under the cascading and non-cascading behavior. On all datasets the *oracle* version of DBGD performs significantly better than DBGD under *perfect* user behavior. This means there is still room for improvement and hypothetical improvements in, e.g., interleaving could lead to significant increases in long-term DBGD performance.

Next, we look at the performance of PDGD; here, there is also a significant difference between performance under the *perfect* and *almost random* user models on all datasets. However, the effect of noise and bias is very limited compared to DBGD, and this difference at 1,000,000 impressions is always less than 0.03 NDCG on any dataset.

To answer **RQ2**, we compare the performance of DBGD and PDGD. Across all datasets, when comparing DBGD and PDGD under the same levels of interaction noise and bias, the performance of PDGD is significantly better in every case. Furthermore, PDGD under the *perfect* user model significantly outperforms the *oracle* run of DBGD, despite the latter being able to directly observe the NDCG of rankers on the current query. Moreover, when comparing PDGD performance under the *almost random* user model with DBGD under the *perfect* user model, we see the differences are limited and in both directions. Thus, even under ideal circumstances DBGD does not consistently outperform

PDGD under extremely difficult circumstances. As a result, we answer **RQ2** positively: our results strongly indicate that the performance of PDGD is considerably better than DBGD and that these findings generalize from ideal circumstances to settings with extreme levels of noise and bias.

Finally, to answer **RQ3**, we look at the performance under the two *almost random* user models. Surprisingly, there is no clear difference between the performance of PDGD under *cascading* and *non-cascading* user behavior. The differences are small and per dataset it differs which circumstances are slightly preferred. Therefore, we answer **RQ3** positively: the performance of PDGD is reproducible under *non-cascading* user behavior.

7 Conclusion

In this study, we have reproduced and generalized findings about the relative performance of Dueling Bandit Gradient Descent (DBGD) and Pairwise Differentiable Gradient Descent (PDGD). Our results show that the performance of PDGD is reproducible under non-cascading user behavior. Furthermore, PDGD outperforms DBGD in both *ideal* and extremely *difficult* circumstances with high levels of noise and bias. Moreover, the performance of PDGD in extremely *difficult* circumstances is comparable to that of DBGD in *ideal* circumstances. Additionally, we have shown that the regret bounds of DBGD are not applicable to the ranking problem in LTR. In summary, our results strongly confirm the previous finding that PDGD consistently outperforms DBGD, and generalizes this conclusion to circumstances with extreme levels of noise and bias.

Consequently, there appears to be no advantage to using DBGD over PDGD in either theoretical or empirical terms. In addition, a decade of OLTR work has attempted to extend DBGD in numerous ways without leading to any measurable long-term improvements. Together, this suggests that the general approach of DBGD based methods, i.e., sampling models and comparing with online evaluation, is not an optimally effective way of optimizing ranking models. Although the PDGD method considerably outperforms the DBGD approach, we currently do not have a theoretical explanation for this difference. Thus it seems plausible that a more effective OLTR method could be derived, if the theory behind the effectiveness of OLTR methods is better understood. Due to this potential and the current lack of regret bounds applicable to OLTR, we argue that a theoretical analysis of OLTR could make a very valuable future contribution to the field.

Finally, we consider the limitations of the comparison in this study. As is standard in OLTR our results are based on simulated user behavior. These simulations provide valuable insights: they enable direct control over biases and noise, and evaluation can be performed at each time step. In this paper, the generalizability of this setup was pushed the furthest by varying the conditions to the extremely difficult. It appears unlikely that more reliable conclusions can be reached from simulated behavior. Thus we argue that the most valuable future comparisons would be in experimental settings with real users. Furthermore, with the performance improvements of PDGD the time seems right for evaluating the effectiveness of OLTR in real-world applications.

Acknowledgements. This research was supported by Ahold Delhaize, the Association of Universities in the Netherlands (VSNU), the Innovation Center for Artificial Intelligence (ICAI), and the Netherlands Organization for Scientific Research (NWO) under project nr 612.001.551. All content represents the opinion of the authors, which is not necessarily shared or endorsed by their respective employers and/or sponsors.

References

1. Ai, Q., Bi, K., Luo, C., Guo, J., Croft, W.B.: Unbiased learning to rank with unbiased propensity estimation. In: The 41st International ACM SIGIR Conference on Research & Development in Information Retrieval, pp. 385–394. ACM (2018)
2. Borisov, A., Markov, I., de Rijke, M., Serdyukov, P.: A neural click model for web search. In: International World Wide Web Conferences Steering Committee, WWW, pp. 531–541 (2016)
3. Chapelle, O., Chang, Y.: Yahoo! learning to rank challenge overview. J. Mach. Learn. Res. **14**, 1–24 (2011)
4. Chuklin, A., Markov, I., de Rijke, M.: Click Models for Web Search. Morgan and Claypool Publishers, San Rafael (2015)
5. Dato, D., Lucchese, C., Nardini, F.M., Orlando, S., Perego, R., Tonellotto, N., Venturini, R.: Fast ranking with additive ensembles of oblivious and non-oblivious regression trees. ACM Trans. Inform. Syst. (TOIS), **35**(2) (2016). Article 15
6. Dumais, S.: Keynote: the web changes everything: understanding and supporting people in dynamic information environments. In: Lalmas, M., Jose, J., Rauber, A., Sebastiani, F., Frommholz, I. (eds.) ECDL 2010. LNCS, vol. 6273, pp. 1–1. Springer, Heidelberg (2010). https://doi.org/10.1007/978-3-642-15464-5_1
7. Guo, F., Liu, C., Wang, Y.M.: Efficient multiple-click models in web search. In: WSDM, pp. 124–131. ACM (2009)
8. He, J., Zhai, C., Li, X.: Evaluation of methods for relative comparison of retrieval systems based on clickthroughs. In: CIKM, pp. 2029–2032. ACM (2009)
9. Hofmann, K.: Fast and Reliable Online Learning to Rank for Information Retrieval. Ph.D. thesis, University of Amsterdam (2013)
10. Hofmann, K., Schuth, A., Whiteson, S., de Rijke, M.: Reusing historical interaction data for faster online learning to rank for information retrieval. In: WSDM, pp. 183–192. ACM (2013)
11. Hofmann, K., Whiteson, S., de Rijke, M.: Balancing exploration and exploitation in learning to rank online. In: Clough, P., Foley, C., Gurrin, C., Jones, G.J.F., Kraaij, W., Lee, H., Mudoch, V. (eds.) ECIR 2011. LNCS, vol. 6611, pp. 251–263. Springer, Heidelberg (2011). https://doi.org/10.1007/978-3-642-20161-5_25
12. Hofmann, K., Whiteson, S., de Rijke, M.: A probabilistic method for inferring preferences from clicks. In: CIKM, pp. 249–258. ACM (2011)
13. Joachims, T.: Optimizing search engines using clickthrough data. In: KDD, pp. 133–142. ACM (2002)
14. Joachims, T., Swaminathan, A., Schnabel, T.: Unbiased learning-to-rank with biased feedback. In: WSDM, pp. 781–789. ACM (2017)

15. Lefortier, D., Serdyukov, P., de Rijke, M.: Online exploration for detecting shifts in fresh intent. In: CIKM, pp. 589–598. ACM, November 2014
16. Liu, T.Y.: Learning to rank for information retrieval. Found. Trends Inform. Retrieval **3**(3), 225–331 (2009)
17. Oosterhuis, H., de Rijke, M.: Balancing speed and quality in online learning to rank for information retrieval. In: CIKM, pp. 277–286. ACM (2017)
18. Oosterhuis, H., de Rijke, M.: Sensitive and scalable online evaluation with theoretical guarantees. In: CIKM, pp. 77–86. ACM (2017)
19. Oosterhuis, H., de Rijke, M.: Differentiable unbiased online learning to rank. In: CIKM, pp. 1293–1302. ACM (2018)
20. Oosterhuis, H., Schuth, A., de Rijke, M.: Probabilistic multileave gradient descent. In: Ferro, N., Crestani, F., Moens, M.-F., Mothe, J., Silvestri, F., Di Nunzio, G.M., Hauff, C., Silvello, G. (eds.) ECIR 2016. LNCS, vol. 9626, pp. 661–668. Springer, Cham (2016). https://doi.org/10.1007/978-3-319-30671-1_50
21. Qin, T., Liu, T.Y.: Introducing letor 4.0 datasets. arXiv preprint arXiv:1306.2597 (2013)
22. Radlinski, F., Craswell, N.: Optimized interleaving for online retrieval evaluation. In: WSDM, pp. 245–254. ACM (2013)
23. Sanderson, M.: Test collection based evaluation of information retrieval systems. Found. Trends Inform. Retrieval **4**(4), 247–375 (2010)
24. Schuth, A., et al.: Probabilistic multileave for online retrieval evaluation. In: SIGIR, pp. 955–958. ACM (2015)
25. Schuth, A., Oosterhuis, H., Whiteson, S., de Rijke, M.: Multileave gradient descent for fast online learning to rank. In: WSDM, pp. 457–466. ACM (2016)
26. Schuth, A., Sietsma, F., Whiteson, S., Lefortier, D., de Rijke, M.: Multileaved comparisons for fast online evaluation. In: CIKM, pp. 71–80. ACM (2014)
27. Vakkari, P., Hakala, N.: Changes in relevance criteria and problem stages in task performance. J. Doc. **56**, 540–562 (2000)
28. Wang, H., Langley, R., Kim, S., McCord-Snook, E., Wang, H.: Efficient exploration of gradient space for online learning to rank. In: SIGIR, pp. 145–154. ACM (2018)
29. Wang, X., Bendersky, M., Metzler, D., Najork, M.: Learning to rank with selection bias in personal search. In: SIGIR, pp. 115–124. ACM (2016)
30. Yue, Y., Joachims, T.: Interactively optimizing information retrieval systems as a dueling bandits problem. In: ICML, pp. 1201–1208. ACM (2009)
31. Yue, Y., Patel, R., Roehrig, H.: Beyond position bias: Examining result attractiveness as a source of presentation bias in clickthrough data. In: WWW, pp. 1011–1018. ACM (2010)
32. Zhao, T., King, I.: Constructing reliable gradient exploration for online learning to rank. In: CIKM, pp. 1643–1652. ACM (2016)
33. Zoghi, M., Whiteson, S., de Rijke, M., Munos, R.: Relative confidence sampling for efficient on-line ranker evaluation. In: WSDM, pp. 73–82. ACM (2014)

Simple Techniques for Cross-Collection Relevance Feedback

Ruifan Yu, Yuhao Xie, and Jimmy Lin(✉)

David R. Cheriton School of Computer Science,
University of Waterloo, Ontario, Canada
jimmylin@uwaterloo.ca

Abstract. We tackle the problem of transferring relevance judgments across document collections for specific information needs by reproducing and generalizing the work of Grossman and Cormack from the TREC 2017 Common Core Track. Their approach involves training relevance classifiers using human judgments on one or more existing (source) document collections and then applying those classifiers to a new (target) document collection. Evaluation results show that their approach, based on logistic regression using word-level *tf-idf* features, is both simple and effective, with average precision scores close to human-in-the-loop runs. The original approach required inference on every document in the target collection, which we reformulated into a more efficient reranking architecture using widely-available open-source tools. Our efforts to reproduce the TREC results were successful, and additional experiments demonstrate that relevance judgments can be effectively transferred across collections in different combinations. We affirm that this approach to cross-collection relevance feedback is simple, robust, and effective.

Keywords: Relevance classifier · Logistic regression · Query expansion

1 Introduction

High-quality test collections form vital resources for guiding research in information retrieval, but they are expensive and time consuming to construct. Thus, when faced with new collections, tasks, or information needs, researchers aim to exploit existing test collections as much as possible. Learning a ranking function from one collection and applying it to another is perhaps the most obvious example, but in this paper we tackle a different use case: the transfer of relevance judgments across document collections for the *same* information need. We characterize this process as *cross-collection* relevance feedback. Suppose a user has already searched a particular document collection and the system has recorded the user's relevance judgments: Can the system then automatically take advantage of these judgments to provide a better ranking for the *same* information need on a *different* document collection? The answer is *yes*, and in this paper we reproduce and then generalize a simple yet highly effective solution using existing open-source tools.

L. Azzopardi et al. (Eds.): ECIR 2019, LNCS 11437, pp. 397–409, 2019.
https://doi.org/10.1007/978-3-030-15712-8_26

The cross-collection relevance feedback scenario can arise in a number of ways, the most common of which is when the user searches different verticals. For example, suppose the user first searches web documents and after some time, realizes that scholarly publications might better address her needs. Another case might be different sub-collections, for example, exploiting judgments on the New York Times to search the Washington Post. Yet another might be temporal segments of the same collection—for example, in a meta-analysis, a researcher might repeat the same search periodically to examine updated documents. In this paper, we focus on the case of transferring relevance judgments between sub-collections of the same genre (newswire documents), thus avoiding issues related to stylistics and genre mismatch.

Resources for studying cross-collection relevance feedback exist because various evaluation campaigns have reused topics (i.e., information needs) across different document collections at different points in time. For example, topics from the TREC 2004 Robust Track [7] were reused for the same track in TREC 2005 [8], which used a different document collection. Another more recent example is the TREC 2017 Common Core Track [2], a renewed effort to focus on the classic *ad hoc* retrieval task, which also reused topics from the TREC 2004 Robust Track. The work of Grossman and Cormack [4] achieved the highest effectiveness of all non-manual runs in the TREC 2017 Common Core Track.

The contribution of this paper is the successful reproduction of the work of Grossman and Cormack (hereafter, GC for short) for cross-collection relevance feedback. We confirm, via a reimplementation from scratch, that the simple technique proposed by GC is highly effective. Our efforts extend beyond replicability (per ACM definitions[1]), as our technical infrastructure resulted in an implementation that differed from GC in several ways. Additionally, we leverage popular open-source data science tools to provide a solid foundation for follow-on work. Finally, we generalize GC by examining different combinations of source and target collections, demonstrating that cross-collection relevance feedback reliably yields large increases in effectiveness.

2 Approach

We begin with a discussion of why GC is worth reproducing. First, the technique is extremely effective. Figure 1, reproduced from the TREC 2017 Common Core Track overview paper [2] shows the effectiveness of runs (in terms of average precision) that contributed to the judgment pools. GC is indicated by the run `WCrobust0405`, ranking third out of all submitted runs. The color coding of the figure indicates the run type: green dots (`Auto-Fdbk`) represent runs that take advantage of automatic feedback from existing judgments and blue dots (`Manual-NoFdbk`) represent manual human-in-the-loop runs. The two runs that were more effective than `WCrobust0405` involved humans who interactively searched the target collection to find relevant documents. The surprising

[1] https://www.acm.org/publications/policies/artifact-review-badging.

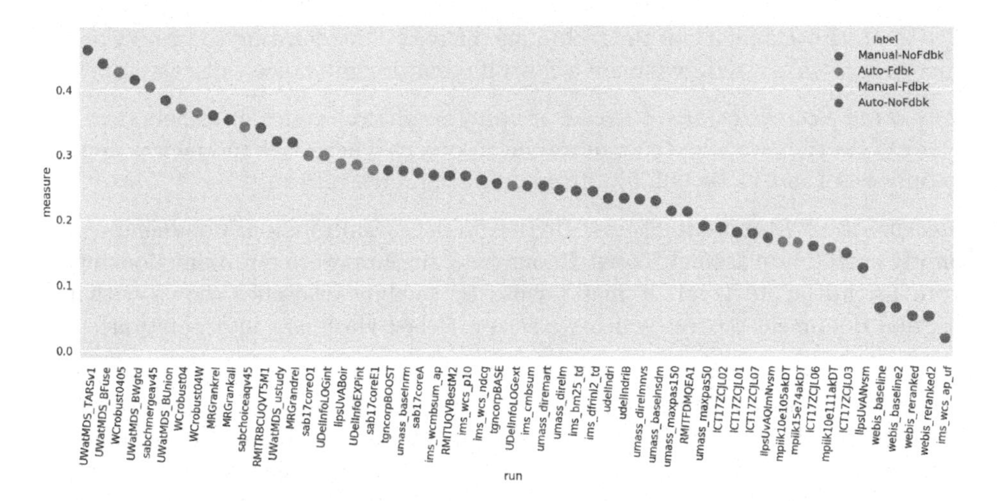

Fig. 1. Effectiveness (average precision) of runs that contributed to the pool in the TREC 2017 Common Core Track, reproduced from [2].

observation here is that relevance transfer (i.e., automatically exploiting existing judgments) approaches the effectiveness of humans manually searching the target collection. Second, the technique described by GC is very simple: the description in their overview paper is only a paragraph. This combination of effectiveness and simplicity makes GC worthy of detailed study.

At a high-level, GC trained logistic regression classifiers on the union of relevance judgments from the TREC 2004 and 2005 Robust Tracks. A separate classifier was trained for each topic, capturing notions of relevance for that specific information need. Documents were represented in terms of word-level *tf-idf* features on the union of the collections used in the 2004 and 2005 evaluations, as well as the collection used in Common Core 2017. Each logistic regression model was learned using Sofia-ML[2] and then applied to the entire Common Core collection. The top 10, 000 documents, in decreasing order of classifier score, comprised the final ranked list for each topic.

To aid in our efforts, Gordon Cormack kindly supplied us with the source code used to generate the runs. However, the source code comprised a series of complex bash scripts that were not documented; although we were able to examine the code to recover the gist of its functionality, we were not able to successfully run the code to replicate the results. During our reimplementation, we did not encounter any need to specifically ask the authors questions. However, there was one important detail critical to effectiveness that was left out of their description—we were able to glean this only by looking at the source code (more details in Sect. 3).

[2] https://github.com/glycerine/sofia-ml.

Given the simplicity of the technique, instead of attempting to *exactly* reproduce GC from scratch, we made a few different design choices, discussed below:

Reranking Search Results. Instead of applying the relevance classifiers over the *entire* collection, we adopted a reranking approach where each model was applied to only the top $k = 10,000$ hits from an initial retrieval run.

Incorporating Document Scores. In the final GC submission, documents were simply sorted by classifier scores. In our case, since we were reranking documents from an initial retrieval, it made sense to combine classifier scores with the original document scores (which we accomplished via linear interpolation).

Leveraging Widely-Used Open-Source Tools. We aimed to build an implementation to serve as the foundation of future efforts, and thus decided to leverage widely-used open-source tools: the Python machine learning package `scikit-learn` and the Anserini IR toolkit [9,10]. In particular, our use of Python meant that we could take advantage of Jupyter notebooks and other modern data science best practices for interactive data exploration and manipulation.

3 Implementation

To be precise, our initial efforts focused on reproducing the run `WCrobust0405` submitted by GC for the TREC 2017 Common Core Track (henceforth, Core17 for convenience). The run leveraged relevance judgments from the TREC 2004 and 2005 Robust Tracks (henceforth, Robust04 and Robust05, respectively). Core17 used the New York Times Annotated Corpus; Robust04 used TREC Disks 4 & 5 (minus Congressional Records) and Robust05 used the AQUAINT document collection. All 50 topics in Core17 are contained in Robust04, while Core17 and Robust05 only share 33 common topics. The run `WCrobust0405` used training data from Robust04 and Robust05 (where available); relevance judgments from Core17 served as a held-out test set.

All source code for replicating results reported in this paper is available in the Anserini code repository[3] (post v0.3.0 release, based on Lucene 7.6) at commit `9548cd6` (dated Jan. 19, 2019).

The per-topic breakdown of relevance judgments is shown in Fig. 2, which plots both the volume of judged documents as well as the proportion of relevant documents. It is immediately clear that both the volume and the proportion of relevant labels vary across topics as well as collections. Furthermore, there are usually many more non-relevant judgments than relevant judgments (and this skew is especially severe for some topics). Although this observation isn't surprising, it reminds us that we are dealing with an unbalanced classification problem, and that the prior probability of relevance varies greatly.

[3] http://anserini.io/.

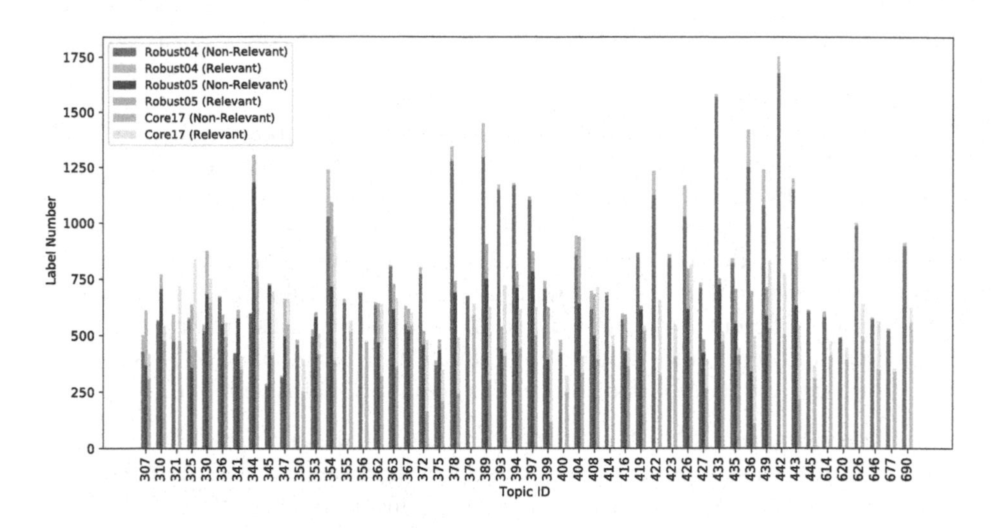

Fig. 2. Per-topic analysis of judgments from Robust04, Robust05, and Core17.

3.1 Feature Extraction and Classifier Training

We began by indexing all three collections (Robust04, Robust05, and Core17) using the Anserini IR toolkit [9,10], which is based on the popular open-source Lucene search toolkit. Anserini provides convenient tools to dump out raw *tf-idf* document vectors for arbitrary documents. Data preparation consisted of extracting these document vectors for all judgments in Robust04 and Robust05 for each topic. The features for these document vectors are comprised of stemmed terms as processed by standard Lucene analyzers. Although we extracted document vectors for each collection individually, the output is post-processed so that the final feature space is the union of vocabulary terms from the training corpora (Robust04 and Robust05 in this case). This meant that out-of-vocabulary terms may be observed at inference time on Core17 data.

Our implementation differs from GC: their brief description (in the track overview paper) suggests that their *tf-idf* document vectors are computed with respect to the union of the three collections (although this point is not explicit). In our case, eliminating *a priori* knowledge about the target collection makes our implementation more general. Furthermore, our approach leverages existing IR tools to extract document vectors, making it easier to vary source/target collections (see additional experiments later). However, we do not believe that this detail has a substantive impact on effectiveness.

As the last step, all feature vectors were converted to unit vectors by L_2 normalization. This was an important detail not mentioned in the GC description, but has a large impact on effectiveness since document lengths vary across collections. Our initial efforts did not include this normalization, and we were not able to reproduce effectiveness values anywhere close to those reported by GC. We realized this omission only after consulting the source code of the original implementation. Perhaps in retrospect, the need for normalization is obvious, but this detail provides an example of the difficulty of reproducibility, where small implementation decisions make a big difference.

The feature vectors prepared in the manner described above were then fed to the Python machine learning package `scikit-learn` [6]. Each topic was treated as an independent training dataset to learn a relevance classifier for that particular information need. One advantage of using `scikit-learn` is that we can easily explore a wide range of different models. We did in fact do so, but discovered that different models as well as variations within families of models (for example, different loss functions, regularization methods, and optimization algorithms) did not make much of a difference in terms of effectiveness. For brevity, we decided to report results with three representative models:

- Logistic regression (LR). We used the so-called "balanced" mode to automatically adjust class weights to be inversely proportional to class frequencies.
- Support vector machines (SVM). We used a linear kernel and the "balanced" mode as well.
- Gradient-boosted decision trees (GB Tree). Specifically, LightGBM [5].

In addition to evaluating each model individually, we also explored an ensemble of all models using simple score averaging.

3.2 Reranking Retrieval Results

Classifiers trained in the manner described above capture relevance with respect to an information need at the lexical level, which can then be applied to a new (target) collection to infer document relevance with respect to the same information need. GC accomplished this by applying inference on *every* document in the target collection and generating a ranked list based on the classifier scores. While this approach is feasible for newswire collections that are moderate in size, especially with an efficient classifier implementation, scaling to larger collections is potentially problematic. Classifying every document is also computationally wasteful, since most of the documents in a collection will not be relevant.

Instead, we adapted GC into a reranking architecture, where the relevance classifier is used to rescore an initial candidate list of documents generated by traditional *ad hoc* retrieval techniques. In our case, we used title queries from the topics to produce the top $k = 10,000$ results using two query expansion techniques: RM3 [1] and axiomatic semantic term matching [3] (Ax for short). In both cases, we used default parameters in the Anserini implementation. Query expansion techniques provide the classifier with a richer set of documents to work on, thus potentially enhancing recall.

We applied our relevance classifiers to these initial results to generate a final ranking in two different ways: First, by ignoring the RM3 and Ax retrieval scores and reranking solely on the classifier scores. Second, by a linear interpolation between retrieval and classifier scores as follows:

$$\text{score} = \alpha \cdot \text{score}_{\text{classifier}} + (1 - \alpha) \cdot \text{score}_{\text{retrieval}}$$

Table 1. Baseline retrieval results.

Method	Robust04		Robust05		Core17	
	AP	P10	AP	P10	AP	P10
BM25	0.1442	0.3280	0.2046	0.4818	0.1977	0.4920
BM25+RM3	0.1725	0.3500	0.2716	0.5333	0.2682	0.5560
BM25+Ax	0.1779	0.3560	0.2699	0.5121	0.2700	0.5680

The first case can be viewed as a special case of the second where $\alpha = 1$. The interpolation parameter can be learned by cross validation, but experiments show that results are not particularly sensitive to the setting.

Beyond our attempt to reproduce the `WCrobust0405` run, we also ran experiments that considered different combinations of source and target document collections to examine the generality and robustness of GC.

4 Experimental Results

We first establish baselines on Robust04, Robust05, and Core17. Effectiveness measured in terms of average precision (AP) at rank 1000 and precision at rank 10 (P10) is shown in Table 1 for title queries (in all our experiments we ignored the descriptions and narratives). The rows show effectiveness with "bag of words" BM25, BM25 combined with RM3 expansion [1], and BM25 with axiomatic semantic term matching [3]; all used default Anserini parameters. Note that for Robust04 and Core17, metrics are computed over the 50 common topics, while for Robust05, metrics are computed over the 33 common topics. Consistent with the literature, query expansion yields sizeable gains in effectiveness. We find that RM3 is slightly more effective than axiomatic semantic term matching.

Table 2 shows results from our relevance transfer experiments to reproduce `WCrobust0405`: training on Robust04 and Robust05 judgments, evaluating on Core17 judgments. The table reports results applied to the initial ranked list from RM3 (left) and axiomatic semantic term matching (right); baseline effectiveness is reported in the second row (copied from Table 1). The effectiveness of `WCrobust0405` is presented in the first row. The remaining parts of the table are organized into three blocks: The first presents results where we ignore the retrieval scores and sort by the relevance classifier scores only. The second shows results from interpolating the original retrieval scores and the classifier scores, with the optimal interpolation weight α (i.e., provided by an oracle, in tenth increments, selected separately for each metric). The third block shows interpolation results with a weight of $\alpha = 0.6$. Within each block, individual rows show the effectiveness of each model; we also show results of the ensemble using simple score averaging (denoted "All Classifiers").

Focusing on optimal α values (we examine sensitivity to the interpolation weight below), we see that our results successfully reproduce the technique of GC: We achieve comparable effectiveness and demonstrate large increases over

Table 2. Relevance transfer results: train on Robust04 and Robust05, test on Core17.

	RM3		Axiomatic	
	AP	P10	AP	P10
`WCrobust0405`	0.4278	0.7500	0.4278	0.7500
Baseline	0.2682	0.5560	0.2700	0.5680
Classification Only				
LR	0.3721	0.7420	0.3605	0.7440
SVM	0.3595	0.7440	0.3445	0.7340
GB Tree	0.3069	0.6640	0.3046	0.6660
All Classifiers	0.4011	0.7700	0.3907	0.7660
Interpolation (Optimal α)				
LR	0.4198	0.7720	0.4166	0.7840
SVM	0.4153	0.7640	0.4135	0.7780
GB Tree	0.3857	0.7320	0.3945	0.7460
All Classifiers	0.4452	0.7780	0.4472	0.7840
Interpolation ($\alpha = 0.6$)				
LR	0.4198	0.7640	0.4166	0.7700
SVM	0.4153	0.7580	0.4121	0.7740
GB Tree	0.3815	0.7320	0.3893	0.7460
All Classifiers	0.4451	0.7540	0.4472	0.7740

the baselines; the absolute values of the metrics are quite close. Based on a paired t-test (which we use throughout this paper for testing statistical significance, at the $p < 0.01$ level), we find no significant differences between any of our models and `WCrobust0405` in terms of both AP and P10.

Interestingly, a classification-only approach does not appear to be effective in our reranking implementation. For both RM3 and axiomatic semantic term matching, weighted interpolation with optimal α is significantly better than the classification-only approach in terms of average precision (across all models and the ensemble), but not significantly better in terms of P10 (except for GB Tree).

In terms of different models, we observe that logistic regression (LR) and SVM yield comparable results. None of the differences (for both metrics, for both initial rankings) are statistically significant. The tree-based model (GB Tree) performs quite a bit worse than either LR or SVM; these differences, however, are not significant with the exception of GB Tree vs. SVM in terms of AP. Finally, an ensemble using simple score averaging yields effectiveness that is higher than any individual model; these differences are significant for AP, but not P10. Comparing RM3 vs. axiomatic semantic term matching, we find that differences in both AP and P10 are not significant.

An optimal interpolation weight α assumes the existence of an oracle, which is of course unrealistic in a real-world setting. To address this issue, we performed a sensitivity analysis by varying α from zero to one in tenth increments, with the results shown in Fig. 3. As expected, the curve has a convex shape, with a peak in a fairly wide range, from 0.5 to 0.7. We further ran a five-fold cross-validation experiment: training on Robust04 and Robust05 as before, but selecting a fifth of the test topics from Core17 as a validation set to select α and evaluating on the remaining topics. In each case, the optimal weight lies in this 0.5 to 0.7 range (although the exact value varies from fold to fold). From this cross-validation analysis, we conclude that 0.6 appears to be a reasonable interpolation weight that can be adopted in the absence of validation data. In Table 2, the third block of rows report results with $\alpha = 0.6$, and we see that effectiveness is quite close to the optimal settings. None of the differences in effectiveness between optimal α and $\alpha = 0.6$ are statistically significant. We further demonstrate the robustness of this setting in experiments below.

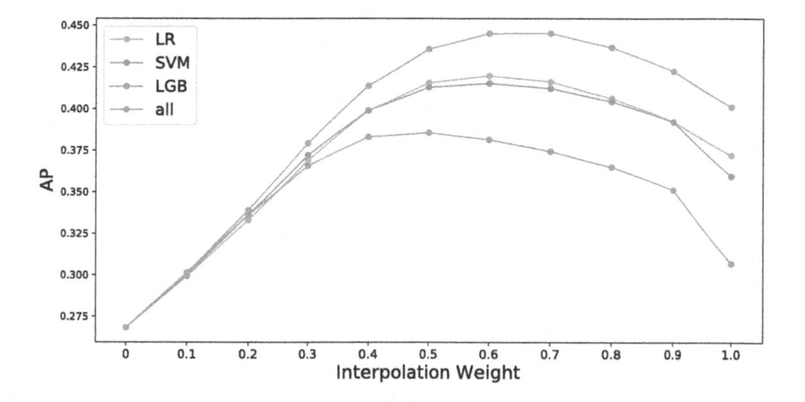

Fig. 3. AP scores with different interpolation weights.

The work of GC represents one specific instance of cross-collection relevance feedback: leveraging judgments from Robust04 and Robust05 to improve ranking effectiveness on Core17. Of course, given the available evaluation resources, it is possible to examine different combinations of source and target document collections. Such experiments allow us to examine the generality of the technique: results are reported in Table 3. Here, we treat BM25+RM3 as the baseline and the initial ranking. For simplicity, we fixed the relevance classifier to logistic regression interpolated with BM25+RM3 ($\alpha = 0.6$), denoted BM25+RM3+LR. The first column denotes the target collection used for evaluation and the second column denotes the source of the relevance judgments used for training. The first three rows are simply repeated from Table 2 for convenience. In addition to training on both Robust04 and Robust05 data together, we also tried each collection separately. Training on Robust04 alone actually corresponds to the run `WCrobust04` submitted by GC, whose effectiveness we repeat here for convenience. Note that when testing on Core17 and Robust04, the evaluation is conducted over 50 topics in all cases, and on Robust05, over 33 topics.

Table 3. Results on different combinations of source and target collections.

Test	Train	Approach	AP	P10
Core17	-	BM25+RM3	0.2682	0.5560
Core17	Robust04, Robust05	`WCrobust0405`	0.4278	0.7500
Core17	Robust04, Robust05	BM25+RM3+LR	0.4198	0.7640
Core17	Robust04	`WCrobust04`	0.3711	0.6460
Core17	Robust04	BM25+RM3+LR	0.3812	0.7360
Core17	Robust05	BM25+RM3+LR	0.3721	0.7060
Robust04	-	BM25+RM3	0.1725	0.3500
Robust04	Robust05, Core17	BM25+RM3+LR	0.3520	0.6060
Robust04	Robust05	BM25+RM3+LR	0.2802	0.5040
Robust04	Core17	BM25+RM3+LR	0.3248	0.5700
Robust05	-	BM25+RM3	0.2716	0.5333
Robust05	Robust04, Core17	BM25+RM3+LR	0.4471	0.7515
Robust05	Robust04	BM25+RM3+LR	0.3647	0.6970
Robust05	Core17	BM25+RM3+LR	0.4042	0.7242

These results generalize the classification-based relevance transfer technique of GC by demonstrating consistent and large effectiveness gains with different source and target collections. We find that the technique is both simple and robust. Moreover, results show that more relevance judgments yield higher effectiveness, even if those judgments come from different collections: training on two source collections consistently beats training on a single collection. Note that in these experiments, we used a single interpolation weight ($\alpha = 0.6$) and performed no parameter tuning. This further validates the recommendation derived from the results in Table 2.

Our final set of experiments consists of in-depth error analyses to better understand the impact of relevance transfer. Figure 4 presents per-topic analyses, comparing the effectiveness (in terms of average precision) of logistic regression interpolated with BM25+RM3 ($\alpha = 0.6$) with the BM25+RM3 baseline. Each bar represents a topic and the bars are sorted by differences in AP. We show evaluation on Core17 in the top plot, Robust04 in the middle plot, and Robust05 in the bottom plot (using all available judgments).

As is common with many retrieval techniques, relevance transfer improves many topics (some leading to spectacular improvements) but hurts some topics as well. Our implementation only decreased effectiveness for two topics on Robust05, but that test set contains fewer topics overall, so we hesitate to draw any definitive conclusions from this. Focusing on the top plot (train on Robust04 and Robust05, test on Core17), we performed manual error analysis to try and understand what went wrong. Topic 423, the rightmost bar and the worst-performing topic, is simply the named entity "Milosevic, Mirjana Markovic".

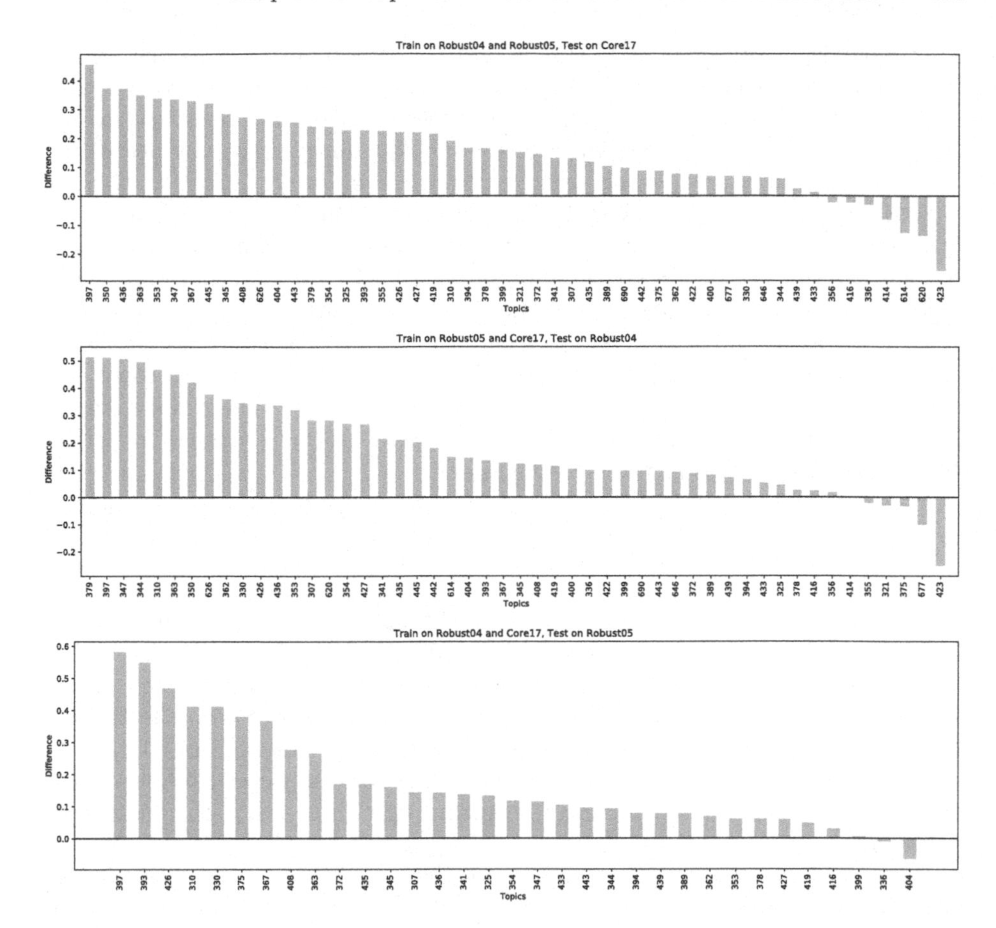

Fig. 4. Per-topic analysis, comparing interpolated logistic regression with BM25+RM3 and the BM25+RM3 baseline.

For this topic, BM25+RM3 achieves AP 0.8252; interpolated relevance classification yields AP 0.5698. Topic 620, the second worst-performing topic, is "France nuclear testing": BM25+RM3 achieves AP 0.7716, while relevance classification drops AP down to 0.6358. For this classifier, the highest-weighted feature is the term "Greenpeace", the non-profit environmental organization. This term leads the classifier astray likely because of the different time spans of the collections and specific occurrences of events. In the Robust04 and Robust05 collections, French nuclear testing was frequently associated with Greenpeace protests; in Core17, this association does not appear to be as strong. We might characterize this as an instance of relevance drift, where notions of relevance shift over time and across collections.

We further observe that both topics are relatively "easy", given the high average precision scores of the baselines. This suggests that relevance transfer has the potential to "screw up" easy topics, which is not a unique characteristic of this technique. In general, query expansion runs the risk of decreasing effectiveness on topics with already high scores, since scores can only further increase by bringing in additional relevant documents. Any bad expansion term can depress the rankings of relevant documents, thereby decreasing the overall score.

Looking at all three plots with different target collections, it is difficult to draw any firm conclusions. Comparing Core17 (top) and Robust04 (middle), we see that topic 423 performs poorly in both cases. Unfortunately, that topic is not in the overlap set with Robust05, so the result is missing from the bottom plot. However, it is *not* the case that topics perform poorly in a consistent manner—when evaluating on Core17, topic 620 is the second worst-performing topic, but when evaluating on Robust04, we observe a large effectiveness improvement. The choice of source and target collections appears to have a large impact on effectiveness differences in relevance transfer.

5 Conclusion

As a succinct summary of our results, we find that the cross-collection relevance feedback technique of GC "works as advertised". Additional experiments further demonstrate its generality across different combinations of source and target collections. We conclude that this technique is simple, robust, and effective. In addition to these experimental findings, the concrete product of our effort is an open-source computational artifact for replicating our experiments, implemented with modern tools (Lucene and `scikit-learn`) that can serve as the foundation for future work.

Although our experiments demonstrate the generality of relevance transfer with simple "bag of words" classifiers, we believe that more work is needed to better understand when an information need can benefit from existing relevance judgments on another collection. At the core, the problem formulation is one of document classification. Thus, an obvious next step is to apply the plethora of techniques involving deep learning and continuous word representations to tackle this problem. We have already begun explorations along these lines.

Acknowledgments. This research was supported in part by the Natural Sciences and Engineering Research Council (NSERC) of Canada.

References

1. Abdul-Jaleel, N., et al.: UMass at TREC 2004: novelty and HARD. In: Proceedings of the Thirteenth Text REtrieval Conference (TREC 2004). Gaithersburg, Maryland (2004)
2. Allan, J., Harman, D., Kanoulas, E., Li, D., Gysel, C.V., Voorhees, E.: TREC 2017 common core track overview. In: Proceedings of the Twenty-Sixth Text REtrieval Conference (TREC 2017). Gaithersburg, Maryland (2017)

3. Fang, H., Zhai, C.: Semantic term matching in axiomatic approaches to information retrieval. In: Proceedings of the 29th Annual International ACM SIGIR Conference on Research and Development in Information Retrieval SIGIR 2006, pp. 115–122. ACM, New York (2006). http://doi.acm.org/10.1145/1148170.1148193
4. Grossman, M.R., Cormack, G.V.: MRG_UWaterloo and WaterlooCormack participation in the TREC 2017 common core track. In: Proceedings of the Twenty-Sixth Text REtrieval Conference (TREC 2017). Gaithersburg, Maryland (2017)
5. Ke, G., et al.: LightGBM: a highly efficient gradient boosting decision tree. In: Advances in Neural Information Processing Systems, pp. 3146–3154 (2017)
6. Pedregosa, F., et al.: Scikit-learn: machine learning in Python. J. Mach. Learn. Res. **12**, 2825–2830 (2011)
7. Voorhees, E.M.: Overview of the TREC 2004 Robust Track. In: Proceedings of the Thirteenth Text REtrieval Conference (TREC 2004). Gaithersburg, Maryland (2004)
8. Voorhees, E.M.: Overview of the TREC 2005 Robust Track. In: Proceedings of the Fourteenth Text REtrieval Conference (TREC 2005). Gaithersburg, Maryland (2005)
9. Yang, P., Fang, H., Lin, J.: Anserini: enabling the use of Lucene for information retrieval research. In: Proceedings of the 40th International ACM SIGIR Conference on Research and Development in Information Retrieval SIGIR 2017, pp. 1253–1256. ACM, New York (2017). http://doi.acm.org/10.1145/3077136.3080721
10. Yang, P., Fang, H., Lin, J.: Anserini: reproducible ranking baselines using Lucene. J. Data Inf. Qual. **10**(4), Article 16 (2018)

Reproducibility (Application)

A Comparative Study of Summarization Algorithms Applied to Legal Case Judgments

Paheli Bhattacharya[1(✉)], Kaustubh Hiware[1], Subham Rajgaria[1], Nilay Pochhi[1], Kripabandhu Ghosh[2], and Saptarshi Ghosh[1]

[1] Indian Institute of Technology Kharagpur, Kharagpur, India
pahelibhattacharya@gmail.com
[2] Indian Institute of Technology Kanpur, Kanpur, India

Abstract. Summarization of legal case judgments is an important problem because the huge length and complexity of such documents make them difficult to read as a whole. Many summarization algorithms have been proposed till date, both for general text documents and a few specifically targeted to summarizing legal documents of various countries. However, to our knowledge, there has not been any systematic comparison of the performances of different algorithms in summarizing legal case documents. In this paper, we perform the first such systematic comparison of summarization algorithms applied to legal judgments. We experiment on a large set of Indian Supreme Court judgments, and a large variety of summarization algorithms including both unsupervised and supervised ones. We assess how well domain-independent summarization approaches perform on legal case judgments, and how approaches specifically designed for legal case documents of other countries (e.g., Canada, Australia) generalize to Indian Supreme Court documents. Apart from quantitatively evaluating summaries by comparing with gold standard summaries, we also give important qualitative insights on the performance of different algorithms from the perspective of a law expert.

Keywords: Summarization · Legal case judgment · Supervised · Unsupervised

1 Introduction

In countries following the *Common Law* system (e.g., UK, USA, Canada, Australia, India), there are two primary sources of law – *Statutes* (established laws) and *Precedents* (prior cases). Precedents help a lawyer understand how the Court has dealt with similar scenarios in the past, and prepare the legal reasoning

Electronic supplementary material The online version of this chapter (https://doi.org/10.1007/978-3-030-15712-8_27) contains supplementary material, which is available to authorized users.

L. Azzopardi et al. (Eds.): ECIR 2019, LNCS 11437, pp. 413–428, 2019.
https://doi.org/10.1007/978-3-030-15712-8_27

accordingly. Hence lawyers have to go through hundreds of prior cases. These cases are available as law reports/case judgments which are essentially long[1] and free-flowing with dense legal text.

Table 1. Performances of summarization approaches on legal documents of various countries, as reported in the corresponding papers.

Method	Corpus/#Documents	ROUGE-1	ROUGE-2	ROUGE-L
LetSum [15]	Federal Court of Canada/3500	0.575	0.313	0.451
CaseSummarizer [38]	Federal Court of Australia/5	0.194	0.114	0.061
Graphical Model [40]	Kerala High Court/200	0.6	0.386	–

This makes reading and comprehending the full text of a case a difficult task, even for a legal expert. In scenarios like this, summaries of the case judgments prove to be beneficial.

All popular legal retrieval systems provide summaries of case judgments. Due to the complexity of case documents, they are mostly manually summarized by legal experts. For instance, the popular Westlaw India legal system employs legal attorneys to summarize Indian legal documents [6]. Employing experts to write the summaries incurs high cost. Hence, with the advancement of the Web and large amounts of unstructured legal documents being made available everyday, there is an increasing need for *automated legal text summarization* that can work in such dynamic environments.

In this work, we explore the task of automatic summarization of Indian legal case judgments, specifically of the Supreme Court of India. It can be noted that several prior works have developed text summarization techniques on various legal text, e.g., Canadian case judgments [14,15], UK case judgments (House of Lords Judgments) [21,25], legal judgments from Indian High Courts [40], judgments from the Federal Court of Australia [38], and so on. However, *there has not been any systematic investigation of whether methodologies developed for legal text of one country, generalize well to legal text of another country.* Different countries have their own formats for law reports, and legal terminologies vary widely between different countries. Hence, a summarization approach targeted to case judgments of one country may not generalize well to case judgments from other countries.

Table 1 shows the performance of case document summarization algorithms developed for documents of different countries/courts. The performance measures (ROUGE scores) stated by the corresponding prior works are also mentioned. The datasets used in the prior works differ significantly on the nature of legal documents, size of corpus, and so on. No systematic comparisons have been performed regarding the generalizability of these methods developed for

[1] The average length of an Indian Supreme Court judgment is as high as 4,500 words. Important 'landmark' cases often span hundreds of pages, e.g. https://indiankanoon.org/doc/257876/.

documents of one country/court to documents of other countries/courts, and it is not clear which of the methods would perform well in summarizing documents for a different country/court. To bridge this gap, in this work, we apply all these (and many other) methods in a common setting, namely summarizing Indian Supreme Court case judgments. Note that Australian [1], Canadian [2] and UK [5] law reports have *section headings* and follow a certain structure, whereas Indian case law reports do not usually contain any such information and are highly unstructured, making the summarization task more challenging[2].

Additionally, there are a large number of domain-independent summarization algorithms [23,36] – including classical unsupervised algorithms as well as recent supervised neural algorithms – which can potentially be used for summarizing legal case judgements. Again, there has not been any systematic investigation of how well domain-independent summarization algorithms perform on legal text. In this scenario, we in this paper make the following contributions:

- **Generalizability and Classical Reproducibility:** We assess the performance of several domain-independent text summarization methods (both traditional unsupervised ones and recent supervised neural models) on legal case judgments. We also assess how well summarization algorithms developed for documents of one country generalize to documents of another country. Specifically, we reproduce three existing extractive text summarization algorithms specifically designed for legal texts, two for documents of other countries (Canada and Australia) and one for Indian case documents of another court. The implementations of the algorithms explored in this paper are available at https://github.com/Law-AI/summarization.
- **Comparative evaluation:** We perform an extensive evaluation of the performance of different summarization algorithms on Indian Supreme Court case judgments. The evaluation is both quantitative (in terms of comparing with gold standard summaries using ROUGE scores) and qualitative (gathering opinion from legal experts). We show that there is no one best performing summarization algorithm for legal case judgments. While one method performs quantitatively better, another can generate a qualitatively better summary.

To the best of our knowledge, this is the first work that performs a systematic comparison of the performance and generalizability of legal document summarization methods in a common setting.

2 State-of-the Art on Legal Document Summarization

We have classified the prior works into two broad categories – (i) summarization algorithms specifically for legal text, and (ii) domain-independent summarization approaches. We describe these two types of prior works in this section.

[2] https://timesofindia.indiatimes.com/india/when-even-judges-cant-understand-judgments/articleshow/58690771.cms.

2.1 Summarization Algorithms Specifically for Legal Documents

Text summarization approaches have been applied to legal texts of many countries. The survey paper [28] highlights research in this field. Existing methods for summarization of legal text can be broadly classified into (i) unsupervised, (ii) supervised and (iii) citation based approaches.

Unsupervised Approaches: These methods use linguistic and statistical signals from the text to identify important sentences for summarization. Initial attempts at summarizing legal cases was by [19,32]. A recent work on unsupervised legal text summarization infusing additional domain knowledge is *CaseSummarizer* [38]. Since we reproduce this method, details can be found in Sect. 4.2.

Supervised Approaches: Supervised approaches for legal text summarization perform a type of template-filling task. Here, the templates are the rhetorical roles (e.g., facts of the case, background, precedent and statutes, arguments, verdict of the Court, etc.). Each of these slots are filled with sentences, ranked in order of their importance. The *LetSum* project [14,15] and a method using *Graphical models* [40] have been applied for legal case document summarization. Since we reproduce these methods, details can be found in Sect. 4.2. The *Sum* project [21,22,24,25] uses several linguistic features and various machine learning techniques to classify a sentence into one of the rhetorical role labels. For summary generation, they select sentences located at the periphery of each rhetorical category.

Citation based approaches leverage other documents to summarize a target document, e.g. [17]. For a target document, they use the catchphrases of the documents cited by the target document (citphrases) and the citation sentences of documents that cite the target document (citances). Another work by the same authors [18] propose to combine these with a Knowledge Base using Ripple Down Rules that suggest different parameters based on which a sentence is to be chosen for summary.

2.2 Domain-Independent Text Summarization Algorithms

Many domain-independent text summarization algorithms have been proposed, as covered in several survey papers [7,11,23,36].

Classical Extractive Text Summarization Methods: There is a wide variety of methods, of which we describe a few. One of the earliest approaches for text summarization is *Luhn's method* [29]. There are graph-based [13,31] and matrix-based approaches [20] for summary generation. The data reconstruction approach *(DSDR)* [26] generates the summary by extracting those sentences that are more probable in reconstructing the original document. It selects sentences that minimize the reconstruction error.

Neural Network Based Summarization Algorithms: In recent years, Deep Learning has been applied to text summarization; see [12] for a survey. Supervised deep neural architectures have been proved to be extremely beneficial for generating **abstractive summaries** [10,34,39,41]. Reinforcement learning have also been applied to abstractive summarization [35,37]. Neural models have also been used for **extractive summarization** [8,9,27,33,43]. An unsupervised extractive text summarization algorithm using Restricted Boltzmann Machines (RBM) was developed in [42].

3 Data and Experimental Setup

Dataset Details: We collected 17,347 legal case documents of the Supreme Court of India from the years 1990–2018 from the website of Westlaw India (http://www.westlawindia.com). For each case judgment, Westlaw provides the full text judgment and a summary. Summaries are written by legal attorneys employed by Westlaw [6]. We use these summaries as gold standard summaries for evaluation of algorithmically generated summaries. Each document has 4,533 words and 116 sentences on average.

Training Data: We use the chronologically earlier 10,000 documents as the training set. For instance, the neural abstractive text summarization algorithm explored in this work, is trained over these 10,000 documents and their gold standard summaries (details in Sect. 4.1).

Test Data: The remaining 7,347 case documents (chronologically later ones) are used as the test set. We generate summaries and perform all quantitative evaluations on the test set. Note that we split the train-test datasets based on chronological ordering of the cases because, in practice, models trained over past cases will be applied to future cases.

Summary Length: Some algorithms require the desired length of the summary to be given as an input. For each document (in the test set), we fix the desired length of the summary to 34% of the number of words in the full text judgment of the document. This number was chosen based on the average ratio of the number of words in the gold standard WestLaw summaries and the original documents, over the entire collection.

4 Applying Summarization Algorithms to Indian Legal Case Judgments

This section describes the application of several text summarization algorithms to Indian Supreme Court legal case judgments.

4.1 Domain-Independent Text Summarization Algorithms

Traditional Unsupervised Extractive Methods: We used the publicly available implementations of LSA, LexRank (both available at [4]), Frequency Summarizer [16] and the data reconstruction method DSDR [3] for summarizing the legal documents.

Neural Network Based Extractive Summarization Method: Most of the supervised neural extractive text summarization methods require *sentence level annotations* regarding the suitability of the inclusion of the sentence in the summarization. For example, [9] uses a 0/1/2 annotation for each sentence denoting whether it should not be, may be, or should be included in the summary. Such sentence-level annotation for legal case judgments can only be done by legal experts, and the cost would be prohibitively expensive due to the large length of case judgments (116 sentences on average). Hence, we use the *unsupervised* model based on Restricted Boltzmann Machines [42] whose implementation is publicly available. We use the default parameter settings – a single hidden layer with 9 perceptrons each having learning rate of 0.1. We increased the training epochs to 25.

Neural Network Based Abstractive Summarization Method: We use the pointer generator approach for abstractive text summarization [41] that uses deep learning architectures (implementation publicly available). We trained on 10,000 documents for 18,729 epochs (over five days). The learning rate was initialized to 0.15 and it fell to 0.00001 with training. The number of decode steps are increased from 100 to 150 to incorporate more decoding words. The size of the vocabulary was increased from 50,000 to 2,00,000 since legal case documents are large. All other parameters were set to default.

4.2 Summarization Algorithms Specifically for Legal Documents

From the family of *unsupervised* algorithms (as described in Sect. 2), we reproduce the model of CaseSummarizer [38] as it is a more recent approach. The methods leveraging citations employ multiple citing and cited documents to summarize a particular document. Since all the other methods aim at summarizing a particular document using linguistic signals from that document alone (and does not use other documents), we do not consider these methods in this paper, for a fair performance evaluation. From the family of *supervised* algorithms, we reproduce the Graphical model based approach [40], because the authors have experimented on cases from Kerala High Court (Kerala is an Indian state) which would intuitively be similar to those of the Indian Supreme Court (over which we are experimenting). We also reproduce the LetSum model [15]; we choose this method over the SUM model as they provide more and understandable technical content.

Generalization Across Legal Documents of Various Countries: Note that CaseSummarizer [38] was developed for Australian legal documents and LetSum [15] was developed for Canadian legal documents. We want to understand how these algorithms generalize to documents from another country (India) and what modifications are necessary to adopt the methods.

A Challenge in Reproducing Supervised Legal Summarization Algorithms: As stated in Sect. 3, since supervised algorithms perform a slot-filling task, it is necessary to decide the rhetorical categories of sentences in a case judgment, before manual annotation. Different prior works on legal text use different rhetorical categories, as shown in Table 2. The FIRE Legal Track [30] developed a scheme of rhetorical categories for Indian Supreme Court cases (Table 2 last column), and we chose to use this annotation scheme in reproducing the methods. We also noted that, although different works use different rhetorical schemes, there is a semantic mapping between the various schemes. We use the mapping from Table 3 (developed in discussion with legal experts) while reproducing the prior works GraphicalModel [40] and LetSum [15].

Table 2. Rhetorical categories of sentences in legal case judgments, as identified in different prior works

GraphicalModel [40]	LetSum [15]	FIRE Legal Track [30]
Identifying facts, Establishing facts, Arguing, History, Arguments, Ratio, Final Decision	Introduction Context Juridical Analysis Conclusion	Fact, Issue, Argument, Ruling by lower court, Statute, Precedent, Other general standards, Ruling by the present court

We now describe how we reproduced the three summarization methods specifically for legal documents. We will make the implementations of these methods publicly available upon acceptance of the paper.

4.2.1 Unsupervised Approach: CaseSummarizer [38]

Basic Technique: Standard preprocessing techniques are done using the NLTK library. Each word is then weighted using a TF-IDF score. For each sentence, the TF-IDF values of its constituent words are summed up and normalized over the sentence length. This score is called w_{old}. A new score, w_{new}, is computed for the sentence using $w_{new} = w_{old} + \sigma\,(0.2d + 0.3e + 1.5s)$ where d is the number of 'dates' present in the sentence, e is the number of named entity mentions, s is a boolean indicating the start of the section (sentences at the start of a section are given more weightage), and σ is the standard deviation among the sentence scores.

Challenges in Reproducibility: Unlike Australian case judgments, Indian case judgments are much less structured, and do not contain section/paragraph headings. As an alternative estimate of the importance of a sentence, we used a count of the number of legal terms (identified by a legal dictionary) present in the sentence. The importance of 'dates' was not clear, and Indian case judgments have very few dates. Rather, Indian case judgments refer to Sections of particular Acts in the Indian legal system, e.g., 'section 302 of the Indian Penal Code'. Hence, for the parameter d in the formulation, we included both dates and section numbers.

Table 3. Mappings between rhetorical categories in different works

Mapping to GraphicalModel	
FIRE Legal Track [30]	**GraphicalModel [40]**
Facts	Identifying facts
Issue	Establishing facts
Precedent	Arguing
Ruling by lower court	History
Arguments+Other general standards	Arguments
Statute	Ratio
Ruling by present court	Final Decision

Mapping to LetSum	
FIRE Legal Track [30]	**LetSum [15]**
Facts + Issue	Introduction
Arguments + Ruling by lower court	Context
Statute + Precedent	Juridical Analysis
Other general standards+ Ruling by present court	Conclusion

Challenges in Applying Standard NLP Tools to Legal Texts: There is another set of challenges in applying standard NLP tools to legal texts. For instance, the authors of [38] did not clearly mention how they identified the 'entities' in the texts. So, we used the popular Stanford NER Tagger[3] for identifying named entities. We found that the tool gives many false positives. For instance, the phrase 'Life Insurance Corporation India' actually represents a single organization. But Stanford NER identifies 'Life : PERSON' , 'Insurance Corporation : ORGANIZATION', 'India : LOCATION'. Again the phrase 'Pension Rules' is identified as a PERSON. Also, we find that using the popular Python NLTK library[4] for tokenizing a legal document poses many difficulties. For instance, in legal documents, a lot of abbreviations are present. NLTK attempts to use 'fullstops' as boundaries, resulting in many incorrect parses. An example: the phrase *"issued u/s. 1(3) of the Act"* is tokenized to *'issued'*, *'u/s'*, *'.'* , *'1'*, *'('*, *'3'*, *')'*, *'of'*, *'the'*, *'Act'*.

4.2.2 Supervised Approach: LetSum [14,15]

Basic Technique: LetSum divides the text structure into five themes as mentioned in Table 2. The summary is built in four phases: (i) thematic segmentation, (ii) filtering of less important textual units including case citations, (iii) selection of candidate units, and (iv) production of the summary. Sentences are assigned a theme based on the presence of hand-engineered linguistic markers. Citation units are filtered out, which are identified by presence of numbers, certain prepositions and markers like colons, quotations, etc. A list of best candidate units for each structural level of the summary is selected, based on heuristic functions, locational features, and TF-IDF of the sentence. The final summary is produced by concatenating textual units with some manual grammatical modifications. The Introduction forms 10% of the size of summary, the context is 24%, Juridical analysis and Conclusion segments are 60% and 6% of the summary respectively.

Note that, the Production module (the last phase, which deals with manually making grammatical modifications to the selected words) was mentioned as being

[3] https://nlp.stanford.edu/software/CRF-NER.shtml.
[4] https://www.nltk.org/.

implemented in the papers [14,15], and no related future work from the authors could be found. Hence this step had to be omitted.

Challenges in Reproducibility: Indian case documents do not have a fixed structure and lack section headings, unlike Canadian documents. Also, citations to statutes are important for summaries of Indian legal case judgments; thus, we do *not* carry out the phase where citations are filtered out.

Since the writing style for judicial texts in both countries are widely different, *the linguistic markers identified for Canadian legal texts could not identify themes for sentences in Indian legal documents.* Thus, we extract cue phrases as follows. We randomly selected 25 documents from the training set (described in Sect. 3) for manual annotation by legal experts. Based on these 25 annotated documents, we rank the most frequent n-grams in a theme which are minimally present in other themes. This part heavily relies on manual annotation (which is expensive in legal domain), which we have tried to automate to a large extent in our reproduction.

Also, according to the LetSum algorithm, each sentence of the document is assigned to a theme. Within each theme, sentences are ranked based on a heuristic function. However, the *heuristic function was not specified in* [14,15], in the absence of which we ranked sentences based on their TF-IDF scores. For each theme, the maximum length in the summary is known. Hence an adequate number of textual units are correspondingly chosen. Additionally the problems of using NLTK (as stated above) are encountered here as well.

4.2.3 Supervised Approach: GraphicalModel [40]

Basic Technique: The authors identified the rhetorical roles of a sentence using Conditional Random Fields. The features identified for each sentence are presence of indicator/cue phrases, position of particular words in the sentence (beginning/end of the sentence, index) and layout features such as position of sentence in the document, capitalization, presence of digits and Part-of-Speech tags. The term distribution model (the k-mixture model) is used to assign probabilistic weights. Sentence weights are computed by summing the term probability values obtained by the model. Sentences are subsequently re-ranked twice, once based on their weights and again based on their evolved roles during CRF implementation, to generate the final summary.

Challenges in Reproducibility: Since the original paper focused on cases related to 'rent control' and the annotations available with us were not from rent-specific cases, *the cue phrases mentioned in* [40] *did not perform well in our dataset.* Hence we automate the process of identification of cue phrases as follows. For all the annotated documents, sentences were separated based on their annotations (by legal experts). Identification of cue phrases for each category was achieved by computing n-grams, along with their frequency in the specific category. An n-gram was chosen as a cue phrase for a particular role label, if its frequency across all the other categories was lower.

Default values of parameters were used for the CRF (implemented using the Python library 'pycrfsuite') since the exact parameters were not stated. The re-ranking of statements considering the identified labels was ambiguous, thus we keep appending sentences ordered by k-mixture-model as long as it does not exceed the desired length of summary. Additionally, the problems with NLTK are faced here too.

5 Results and Analysis

We applied all the summarization algorithms discussed in Sect. 4 to the 7,347 case judgments of the Indian Supreme Court (as stated in Sect. 3). All extractive text summarization approaches were executed on a LINUX 64 bit machine with 4 GB RAM and Core i5 processors. The abstractive neural network-based algorithm was executed on a Tesla K40c GPU with 12 GB RAM. The Online Resource[5] gives the summary generated by each method for a particular case judgment. We then performed two types of evaluation on the performance of the summarization algorithms - quantitative and qualitative.

5.1 Quantitative Evaluation

Following the traditional way of evaluating summaries, we compute ROUGE scores by comparing the algorithmically generated summaries with the gold standard summaries obtained from WestLaw. ROUGE scores measure the fraction of n-gram overlap with a reference summary. The ROUGE scores achieved by each method over the Indian Supreme Court case judgments are shown in Table 4. This table also reports the summary generation times of the different algorithms, for a particular case (one of the landmark cases used for the qualitative evaluation discussed below). LSA achieves the highest score in the family of classical unsupervised summarization techniques, followed by the method based on data reconstruction (DSDR). LexRank performed moderately, producing shorter sentences in the summary but its execution time was quite high. Frequency Summarizer (FreqSum) being a very naive method (only relies on frequency of words) performs poorly, though it takes the least time.

The neural network based unsupervised extractive method does not perform well from the perspective of ROUGE scores. But its execution time is lower than most of the others. The pointer-generator method for abstractive summarization perform moderately well in terms of ROUGE scores. As described in Sect. 4.1, we trained the abstractive model for 18,730 epochs (over five days). We observed that the performance gradually improves with more training; we plan to repeat the experiments with more training in future.

From the family of legal-specific summarization techniques, GraphicalModel and LetSum have comparable performance, while CaseSummarizer performs relatively poorly. This poor performance of CaseSummarizer is probably because

[5] Supplementary material, also available at https://drive.google.com/open?id=1KbcjdnvO1kHn3HNr1Jo-SI2XLbN72vD8.

Table 4. Performance of the different Text Summarization Approaches applied to Indian Supreme Court Judgments (US: unsupervised, S: supervised)

Broad class of Approaches	Methods from each Class	Type	ROUGE-1		ROUGE-2		ROUGE-L		Execution Time (sec)
			Recall	F-score	Recall	F-score	Recall	F-score	
Classical Extractive Summarization Approaches	LexRank	US	0.486	0.238	0.242	0.10	0.443	0.167	8.56
	LSA	US	**0.55**	**0.269**	**0.275**	**0.114**	**0.505**	**0.189**	2.43
	FreqSum	US	0.226	0.143	0.109	0.064	0.183	0.097	0.75
	DSDR	US	0.545	0.255	0.249	0.104	0.49	0.173	1.65
Legal Document specific Extractive Summarization	CaseSummarizer	US	0.198	0.139	0.094	0.063	0.154	0.094	7.95
	GraphicalModel	S	0.386	**0.351**	**0.171**	**0.159**	0.343	**0.297**	2.4
	LetSum	S	**0.408**	0.298	0.112	0.073	**0.371**	0.235	10.16
Neural Network based Suumarization	NeuralEx	US	0.138	0.198	0.055	0.076	0.125	0.132	1.09
	NeuralAbs	S	0.239	0.29	0.11	0.14	0.214	0.215	3.75 (GPU)

the approach is heavily dependent on the correct identification of named entities, which is difficult in case of legal documents using standard NLP tools (as discussed earlier).

5.2 Qualitative Evaluation

We gather opinion from legal experts on the quality of summaries generated by different algorithms. To this end, we chose three landmark cases well-known in Indian Law (see the Supplementary Information for details of the cases). For each case, we showed to the legal experts the gold standard summary (by Westlaw) and the summaries generated by some of the best performing algorithms according to the quantitative analysis described above. The summaries were annonymized, i.e., the experts were not told which summary was generated by which method. The experts decided to evaluate the summaries based on how well the summaries capture four important aspects of a case judgment – (1) the holding/ruling of the Court combined with the reasoning behind it, (2) the legal facts, (3) the statutes involved, and (4) precedents on which the judgments were based. They rated each summary on each aspect, on a Likert scale of 0–5 where 0 means the summary was poor and 5 means it was very good (in capturing the said aspect). The average ratings of the methods over the four aspects are shown in Fig. 1. Both our experts had high agreement on the scores given. Next, we give both the qualitative and the quantitative evaluation for understanding the trade-offs of using different algorithms.

WestLaw Gold Standard Summaries: The legal experts opined that the WestLaw summaries, though well-written, focus only on two aspects – facts and statute – while they do not cover the other two aspects well. The holding appeared at the end of the summary but the reasoning was not present. Two out of the three summaries evaluated did not contain precedents. Due to these limitations of the WestLaw summaries, some of the automatic methods we studied were given higher scores by the experts.

Latent Semantic Analysis (LSA): This algorithm achieves the maximum ROUGE recall and F-score among the classical unsupervised techniques.

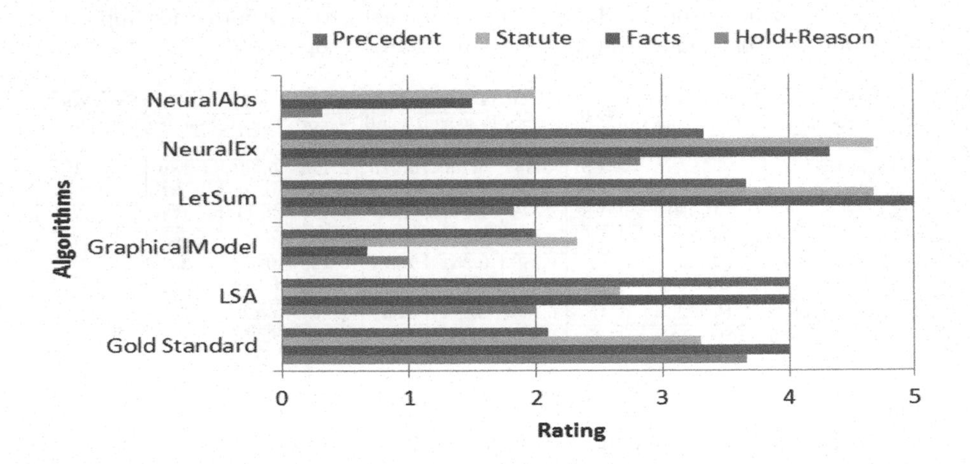

Fig. 1. Average ratings on a scale of 0–5 by legal experts of the summaries of three different cases generated by the algorithms

LSA achieves higher recall (but lesser F-score) when compared with the legal-specific summarization algorithms GraphicalModel and LetSum. The experts judged that initial parts of the summaries were nicely written and very relevant sentences were extracted. The facts of the case were presented well. Although LSA is an unsupervised approach, its inherent topic modeling features enables it to extract important sentences from each of the rhetorical categories. However, there are two limitations – (i) LSA has a tendency to pick lengthy sentences, and (ii) although the initial parts of the summary are good, the quality degrades drastically after covering around half of the document.

GraphicalModel: Graphical model achieves a comparable performance with LetSum w.r.t. ROUGE scores, while taking much lesser time. In fact, this method achieves the best F-score among all the extractive methods. However, according to the legal experts, the overall quality of the summary was not as promising as it seems from the ROUGE scores. GraphicalModel could extract well parts of the case where the statutes were quoted, and the arguments of the case. But the other two aspects were not reflected well in the summary.

LetSum: The performance of LetSum is comparable to GraphicalModel in terms of ROUGE scores. It has by far the highest execution time which is a drawback for online applications. According to the experts, the facts of the case and the parts of the case that described the statutes and precedents are covered well in the summaries. Like LSA, the initial part of the summary was good but the quality degrades gradually. The major drawback of this approach is its readability. As mentioned in their paper, LetSum extracts *textual units* and not complete sentences, which hampers the readability.

Neural Extractive (NeuralEx): This method does not perform well in terms of ROUGE score, though it performs slightly better than CaseSummarizer which uses legal knowledge. Interestingly, the legal experts felt that the quality of the summaries was much better than that of all the other techniques. The summary has a high coverage, that is, it could extract sentences from all the rhetorical categories. Another important factor is that the execution time of this method is lesser than that of most other algorithms.

Neural Abstractive (NeuralAbs): The algorithm performs moderately in terms of ROUGE scores. But its disadvantage is in the running time – the training is resource intensive and the summary generation procedure is expensive. The summaries could partially represent the facts and statutes of the case. The other aspects did not occur much in the summary, simply because the reference WestLaw summaries on which the model was trained, did not have the other two aspects. It is possible that this method will perform better if trained over better quality summaries, but it is difficult to get good quality summaries in such high numbers as is necessary for training this model.

6 Concluding Discussion and Future Directions

In this paper, we have compared several text summarization approaches on legal case judgments from the Supreme Court of India. To our knowledge, this is the first systematic comparison of summarization algorithms for legal text summarization. We make the implementations of the algorithms explored in this paper at https://github.com/Law-AI/summarization. Our analysis leads to following insights.

(1) We understand that no one method can be considered as the best. While one method can best represent the facts of the case (LetSum), another might represent the statutes and precedents cited better (GraphicalModel). Simultaneously, in an online setting, the execution time is also a very important factor.
(2) None of the methods implemented could give the holding/ruling of the case combined with reasoning. This aspect is a very important part of the summary, because based on this a lawyer will decide whether or not to include the case as an argument in his favour.
(3) ROUGE scores might not always be the best evaluation metric to measure the quality of domain-specific summaries. ROUGE measures only n-gram overlaps and does not take into account whether the sentences represent all the facets of the document (e.g., rhetorical categories for legal documents).
(4) General summarization methods that require no knowledge of the domain may perform well quantitatively (LSA) and qualitatively (NeuralEx) but not both. Legal-specific summarization methods try to achieve the best of both the worlds. But their performances are highly dependent on manual annotations/gold-standard summaries for training and correct identification of domain-specific information in the text. An important future challenge is to develop a good and sufficiently large set of gold standard summaries for training supervised methods (especially neural models).

Acknowledgment. We sincerely acknowledge Prof. Uday Shankar and Uma Jandhyala from Rajiv Gandhi School of Intellectual Property Law, Indian Institute of Technology Kharagpur, India for their valuable feedback.

References

1. Australian Case Document. http://www.judgments.fedcourt.gov.au/judgments/Judgments/fca/single/2018/2018fca1517
2. Canadian Case Document. https://www.canlii.org/en/ca/fct/doc/2018/2018fc980/2018fc980.html
3. DSDR. https://gist.github.com/satomacoto/4248449
4. LSA and LexRank. https://pypi.python.org/pypi/sumy
5. UK Case Document. https://www.supremecourt.uk/cases/docs/uksc-2016-0209-judgment.pdf
6. Westlaw. https://legal.thomsonreuters.com/en/products/westlaw
7. Allahyari, M., et al.: Text summarization techniques: a brief survey. arXiv preprint arXiv:1707.02268 (2017)
8. Cao, Z., Wei, F., Li, S., Li, W., Zhou, M., Houfeng, W.: Learning summary prior representation for extractive summarization. In: Proceedings of the 53rd Annual Meeting of the Association for Computational Linguistics and the 7th International Joint Conference on Natural Language Processing (Volume 2: Short Papers), vol. 2, pp. 829–833 (2015)
9. Cheng, J., Lapata, M.: Neural summarization by extracting sentences and words. In: Proceedings of the 54th Annual Meeting of the Association for Computational Linguistics (Volume 1: Long Papers), vol. 1, pp. 484–494 (2016)
10. Chopra, S., Auli, M., Rush, A.M.: Abstractive sentence summarization with attentive recurrent neural networks. In: Proceedings of the 2016 Conference of the North American Chapter of the Association for Computational Linguistics: Human Language Technologies, pp. 93–98 (2016)
11. Das, D., Martins, A.F.: A survey on automatic text summarization. Lit. Surv. Lang. Stat. II Course CMU **4**, 192–195 (2007)
12. Dong, Y.: A survey on neural network-based summarization methods. CoRR abs/1804.04589 (2018). http://arxiv.org/abs/1804.04589
13. Erkan, G., Radev, D.R.: Lexrank: graph-based lexical centrality as salience in text summarization. J. Artif. Int. Res. **22**(1), 457–479 (2004)
14. Farzindar, A., Lapalme, G.: Legal text summarization by exploration of the thematic structure and argumentative roles. Text Summarization Branches Out (2004)
15. Farzindar, A., Lapalme, G.: Letsum, an automatic legal text summarizing system. Legal knowledge and information systems, JURIX, pp. 11–18 (2004)
16. Text summarization with NLTK (2014). https://tinyurl.com/frequency-summarizer
17. Galgani, F., Compton, P., Hoffmann, A.: Citation based summarisation of legal texts. In: Anthony, P., Ishizuka, M., Lukose, D. (eds.) PRICAI 2012. LNCS (LNAI), vol. 7458, pp. 40–52. Springer, Heidelberg (2012). https://doi.org/10.1007/978-3-642-32695-0_6
18. Galgani, F., Compton, P., Hoffmann, A.: Combining different summarization techniques for legal text. In: Proceedings of the Workshop on Innovative Hybrid Approaches to the Processing of Textual Data, pp. 115–123. Association for Computational Linguistics (2012)

19. Gelbart, D., Smith, J.: Beyond boolean search: flexicon, a legal tex-based intelligent system. In: Proceedings of the 3rd International Conference on Artificial Intelligence and Law, pp. 225–234. ACM (1991)
20. Gong, Y., Liu, X.: Generic text summarization using relevance measure and latent semantic analysis. In: SIGIR, pp. 19–25 (2001)
21. Grover, C., Hachey, B., Hughson, I., Korycinski, C.: Automatic summarisation of legal documents. In: Proceedings of the 9th International Conference on Artificial Intelligence and Law, pp. 243–251. ACM (2003)
22. Grover, C., Hachey, B., Korycinski, C.: Summarising legal texts: sentential tense and argumentative roles. In: Proceedings of the HLT-NAACL 03 on Text Summarization Workshop, vol. 5, pp. 33–40. Association for Computational Linguistics (2003)
23. Gupta, V., Lehal, G.S.: A survey of text summarization extractive techniques. J. Emerg. Technol. Web Intell. **2**(3), 258–268 (2010)
24. Hachey, B., Grover, C.: Sentence classification experiments for legal text summarisation (2004)
25. Hachey, B., Grover, C.: Extractive summarisation of legal texts. Artif. Intell. Law **14**(4), 305–345 (2006)
26. He, Z., et al.: Document summarization based on data reconstruction. In: Proceedings of AAAI Conference on Artificial Intelligence, pp. 620–626 (2012)
27. Kågebäck, M., Mogren, O., Tahmasebi, N., Dubhashi, D.: Extractive summarization using continuous vector space models. In: Proceedings of the 2nd Workshop on Continuous Vector Space Models and their Compositionality (CVSC), pp. 31–39 (2014)
28. Kanapala, A., Pal, S., Pamula, R.: Text summarization from legal documents: a survey. Artif. Intell. Rev., 1–32 (2017). https://doi.org/10.1007/s10462-017-9566-2
29. Luhn, H.P.: The automatic creation of literature abstracts. IBM J. Res. Dev. **2**(2), 159–165 (1958)
30. Agrawal, M., Mehta, P., Ghosh, K.: Overview of information access in legal domain fire 2013 (2013). https://www.isical.ac.in/~fire/wn/LEAGAL/overview.pdf/
31. Mihalcea, R., Tarau, P.: TextRank: bringing order into texts. In: EMNLP (2004)
32. Moens, M.F., Uyttendaele, C., Dumortier, J.: Abstracting of legal cases: the salomon experience. In: Proceedings of the 6th International Conference on Artificial Intelligence and Law, pp. 114–122. ACM (1997)
33. Nallapati, R., Zhai, F., Zhou, B.: Summarunner: a recurrent neural network based sequence model for extractive summarization of documents. In: AAAI, pp. 3075–3081 (2017)
34. Nallapati, R., Zhou, B., dos Santos, C., Gulcehre, C., Xiang, B.: Abstractive text summarization using sequence-to-sequence RNNS and beyond. In: Proceedings of The 20th SIGNLL Conference on Computational Natural Language Learning, pp. 280–290 (2016)
35. Narayan, S., Cohen, S.B., Lapata, M.: Ranking sentences for extractive summarization with reinforcement learning. In: Proceedings of the 2018 Conference of the North American Chapter of the Association for Computational Linguistics: Human Language Technologies, Volume 1 (Long Papers), vol. 1, pp. 1747–1759 (2018)
36. Nenkova, A., McKeown, K.: A Survey of Text Summarization Techniques. In: Aggarwal, C., Zhai, C. (eds) Mining Text Data, pp. 43–76. Springer, Boston (2012). https://doi.org/10.1007/978-1-4614-3223-4_3
37. Paulus, R., Xiong, C., Socher, R.: A deep reinforced model for abstractive summarization. arXiv preprint arXiv:1705.04304 (2017)

38. Polsley, S., Jhunjhunwala, P., Huang, R.: Casesummarizer: a system for automated summarization of legal texts. In: Proceedings of COLING 2016, The 26th International Conference on Computational Linguistics: System Demonstrations, pp. 258–262 (2016)
39. Rush, A.M., Chopra, S., Weston, J.: A neural attention model for abstractive sentence summarization. In: Proceedings of the 2015 Conference on Empirical Methods in Natural Language Processing, pp. 379–389 (2015)
40. Saravanan, M., Ravindran, B., Raman, S.: Improving legal document summarization using graphical models. In: Proceedings of the 2006 Conference on Legal Knowledge and Information Systems: JURIX 2006: The Nineteenth Annual Conference, pp. 51–60. IOS Press (2006)
41. See, A., Liu, P.J., Manning, C.D.: Get to the point: summarization with pointer-generator networks. arXiv preprint arXiv:1704.04368 (2017)
42. Verma, S., Nidhi, V.: Extractive summarization using deep learning. arXiv preprint arXiv:1708.04439 (2017)
43. Yin, W., Pei, Y.: Optimizing sentence modeling and selection for document summarization. In: IJCAI, pp. 1383–1389 (2015)

Replicating Relevance-Ranked Synonym Discovery in a New Language and Domain

Andrew Yates[1(✉)] and Michael Unterkalmsteiner[2]

[1] Max Planck Institute for Informatics, Saarbrücken, Germany
ayates@mpi-inf.mpg.de
[2] Software Engineering Research Laboratory, Blekinge Institute of Technology, Karlskrona, Sweden
michael.unterkalmsteiner@bth.se

Abstract. Domain-specific synonyms occur in many specialized search tasks, such as when searching medical documents, legal documents, and software engineering artifacts. We replicate prior work on ranking domain-specific synonyms in the consumer health domain by applying the approach to a new language and domain: identifying Swedish language synonyms in the building construction domain. We chose this setting because identifying synonyms in this domain is helpful for downstream systems, where different users may query for documents (e.g., engineering requirements) using different terminology. We consider two new features inspired by the change in language and methodological advances since the prior work's publication. An evaluation using data from the building construction domain supports the finding from the prior work that synonym discovery is best approached as a learning to rank task in which a human editor views ranked synonym candidates in order to construct a domain-specific thesaurus. We additionally find that FastText embeddings alone provide a strong baseline, though they do not perform as well as the strongest learning to rank method. Finally, we analyze the performance of individual features and the differences in the domains.

Keywords: Synonym discovery · Thesaurus construction · Domain-specific search · Replication · Generalization

1 Introduction

The vocabulary mismatch problem [8] and its detrimental effect on recall [7] have long been recognized by the information retrieval community. In the absence of query expansion, whether by using pseudo relevance feedback or a lookup method, finding relevant information is a matter of constructing a query that expresses the user's information need using the same terms found in relevant documents. In the case of domain-specific search tasks, such as search in medical documents [12], legal documents [19], patents [27] and software engineering

L. Azzopardi et al. (Eds.): ECIR 2019, LNCS 11437, pp. 429–442, 2019.
https://doi.org/10.1007/978-3-030-15712-8_28

artifacts [11,15], domain-specific synonyms complicate information retrieval as either the searcher or the retrieval engine need to be aware of terms that may be used interchangeably. These specialized information retrieval tasks can benefit from thesauri that are crafted for their specific domain.

We chose to replicate and generalize a paper on using learning to rank for human-assisted synonym discovery, because we are interested in improving the synonyms in a classification system (ontology) from the building construction domain. While the users of the classification system are professionals, they are usually specialized in subsets of the construction business and need to coordinate with users specialized in other subsets of the domain. An improved set of synonyms would likely help the collaboration between the different parties using the classification system, such as when searching requirements documents.

Yates et al. [25] proposed a method that produces a ranked list of domain-specific synonyms using a domain-specific corpus as input. Their learning to rank approach uses a set of features that outperformed previous synonym discovery methods that relied on single statistical measures: pointwise mutual information over term co-occurrences [23] and pointwise total correlation between two terms and the syntactic context (dependency relations) in which the terms appear [10].

In this paper we replicate Yates et al.'s method on a different domain (building construction) and language (Swedish) in order to evaluate the generalizability of the approach. We used the original implementation, adapting preprocessing and features to the new setting. As such, we perform an inferential reproduction [9] where we draw similar conclusions as the original paper after evaluating it on a completely different dataset. Our contributions are *(1)* an evaluation of the method proposed by Yates et al. on a new language and different domain and *(2)* the proposal and evaluation of new features inspired by recent methodology advances and the differences introduced by the new domain and language.

2 Methodology

In this section we describe the synonym discovery task before describing our replication of the experimental setup used by Yates et al. We refer to the paper by Yates et al. as the original study and to this paper as the replication study.

2.1 Problem Formulation

As in the original study, we define *synonym discovery* as the task of identifying a target term's correct synonyms from among a set of synonym candidates. Due to the difficulty of this problem, the original study argued that it is best approached as a synonym search task in which a domain-specific corpus is coupled with a learning to rank method in order to help a user quickly identify a target term's synonyms. For example, in the original study, the user might search for the target term *alopecia* with the intent of identifying domain-specific synonyms such as *hair loss* and *missing hair*. Such synonyms would ideally be ranked highly, but because of the task's inherent difficulty, incorrect terms like *greying hair* and *headache* are also likely to be ranked highly. This ranked list can then be used by a human editor in order to manually build a list of domain-specific synonyms. For example, in the original study the method was used to augment a thesaurus mapping expert medical terms (e.g., *alopecia*) to lay synonyms often used in social media in place of the expert terms (e.g., *losing hair*).

The original study proposed to re-frame the evaluation of synonym discovery approaches from a TOEFL[1] style problem (i.e., given a term, pick the correct synonym from n candidate terms) to a ranking problem (i.e., given a term, evaluate how many true synonyms are ranked in the top X% of n candidates). This problem re-framing stems from the observation that the TOEFL style test represents the synonym discovery problem only when there is a sufficiently large number of synonym candidates. However, by increasing the number of incorrect choices, the evaluated approaches, including their own, were not able to answer the TOEFL-style question in most cases. Ranking synonym candidates according to their probability of being a true synonym of a target terms mirrors the synonym discovery problem more accurately. Ultimately, the involvement of a human editor is required to build an accurate domain-specific thesaurus.

One motivation for performing domain-specific synonym discovery is that we would like to cater for both propositional synonyms and near-synonyms. Propositional synonyms refer to terms that can be used interchangeably without affecting the truth condition of a statement. For example, the statements *He is a statesman* and *He is a politician*, referring to the same person, can both be true. This type of synonymity is concerned with identity rather than similarity of meaning, while near-synonyms refer to terms with similar meaning and are context-dependent [22]. For example, in the building construction domain, the term *kylelement (cooling panel)* and the term *förångare (evaporator)* are synonymous, and are used to describe a thermal cooling object. However, in the maritime domain, evaporators are used off-shore to produce fresh water, a different function from the building construction domain. These subtle domain-specific differences call for an approach that takes context into account and allows for the inclusion of human expertise when selecting near-synonyms from synonym candidates.

Formally, given a target term w_t and a set of candidate terms $\mathbb{C}$, a synonym discovery method ranks each candidate term $w_c \in \mathbb{C}$ with respect to its likelihood of being a synonym for the target term w_t in the target domain.

2.2 Replication Design

The objectives of this replication study are to *(1)* evaluate the generalizability of the original finding that synonym discovery is best approached as a ranking problem, *(2)* to evaluate the generalizability of the learning to rank method proposed by Yates et al. in order to determine whether it is still the best approach in a new language and domain, and *(3)* to investigate whether the approach can be improved in our new setting by incorporating methodological improvements

[1] Test Of English as a Foreign Language.

that were not considered in the original work (i.e., a language-specific feature, a contemporaneous term embedding feature, and a more sophisticated learning to rank model). We generalize over both the experimental setting (i.e., language and domain) and over time by considering both features specific to our new setting and new approaches that have become popular since the original work's publication. We evaluate the original approach, the best-performing baseline in the original work, a new embedding-based baseline, and a variant of the original approach using an improved learning to rank model on both the TOEFL task and the relevance ranking task from the original work.

2.3 Baselines

We compare the learning to rank method proposed in the original study with several baselines. We include the baseline that performed best in the original study as well two additional baselines based on embedding similarity and the similarity of dependency relation contexts.

PMI. The best-performing baseline in the original study was pointwise mutual information (PMI) calculated over term co-occurrences in sliding windows as proposed by Terra and Clarke [23]. We calculate PMI over 16-term sliding windows with the constraint that each window can only cover a single sentence. $PMI(w_c, w_t)$ can then be used as the ranking function for obtaining a ranked list of synonyms for the target term w_c.

EmbeddingSim. Word embeddings have become common since the original study's publication and are often used in information retrieval and natural language processing tasks. These methods are trained in an unsupervised manner on a large corpus to create dense word representations that encode some of the words' properties. The FastText [3], word2vec [17], and GloVe [21] methods are often used. Similar to the PMI method, word embeddings capture distributional similarity, and work has shown that much of their improvements over PMI come from the training setup used rather than from the underlying algorithm [13]. We train FastText on our corpus and rank a target term's candidates based on the cosine similarity between the term embeddings for w_t and w_c. The incorporation of this baseline is an example of "generalization over time" since this method is clearly relevant to the synonym discovery task, but it was not available at the time of the original study's publication.

LinSim. Lin's similarity measure [14] was originally proposed as a method for identifying synonyms and other related words. LinSim was used as a feature in the original study, but not as a separate baseline. This measure is similar to Hagiwara's methods [10] from the original study in that it considers pointwise mutual information over term contexts defined by dependency relations. It has the advantage of being less computationally expensive to compute on large corpora, however, so in this study we use it in place of Hagiwara's supervised and unsupervised methods.

2.4 Supervised Approaches

We evaluate two supervised learning to rank approaches on our dataset: a logistic regression as proposed in the original study, which is a pointwise method that has been used for learning to rank in other contexts [26], and LambdaMART [6], a pairwise method. Given a target term w_t and a candidate term w_c, we compute the following features for use with both supervised methods:

- `Windows`: the number of windows containing both w_t and w_c, normalized by the smaller of the two counts. With the Wikipedia and Trafikverket corpora, a window is defined as a sequence of up to 16 terms appearing in a single sentence. With the Web corpus, a window is defined as a sequence of up to 16 terms appearing in a HTML element. Let $count_{win}(x)$ be the number of windows containing the term x and $count_{win}(x, y)$ be the number of windows containing both terms x and y. This feature is then calculated as

$$Windows(w_t, w_c) = \frac{count_{win}(w_t, w_c)}{min(count_{win}(w_t), count_{win}(w_c))}$$

- `LevDist`: the Levenshtein distance (i.e., edit distance) between w_t and w_c.
- `NGram`: the probability that the target term w_t appears in a specific position in a n-gram given that the candidate term w_c has also appeared in this position. As in the original work, we consider all trigrams that appear in our corpora. Let $count_{ng}(x)$ be the number of unique n-grams a term x appears in and $count_{ng}(x, y)$ be the number of unique n-grams in which both terms x and y appear in the same position (e.g., given the trigrams *rate/of/building* and *rate/of/construction*, both *construction* and *building* appear in the same position). This feature is then calculated as

$$NGram(w_t, w_c) = \frac{count_{ng}(w_t, w_c)}{count_{ng}(w_c)}$$

- `POSNGram`: the probability that the target term w_t appears in a specific position in a part of speech n-gram given that the candidate term w_c has also appeared in this position. As in the original work, this is equivalent to `NGram` after replacing each term with its part of speech.
- `LinSim`: the similarity between w_t and w_c as computed using Lin's similarity measure [14]. This measure requires dependency parsing, which we perform with MaltParser [18]. This feature also serves as one of our baselines.
- `RISim`: the cosine distance between the vectors for w_t and w_c, as computed using random indexing. We compute these vectors use the SemanticVectors package[2] [24] with its default parameters.
- `Decompound`: the number of components shared by w_t and w_c, normalized by the minimum number of components in either term. We use the SECOS decompounder [16] to split each term into their components, which decompounds each term using several strategies. We always choose the decompounding strategy that results in the largest number of components.

[2] https://github.com/semanticvectors/semanticvectors/.

- `EmbeddingSim`: the cosine distance between word embeddings for w_t and w_c. We used FastText embeddings [3] trained on our corpus. This feature also serves as one of our baselines.

As described in the baselines section, we do not consider the two features from the original work that were based on Hagiwara's definition of contexts. The Decompound and EmbeddingSim features did not appear in the original work. We introduce the Decompound feature to account for the fact that many of our target terms are compound nouns; it is often the case that their synonyms share components with the target term. For example, the target term *apparatskåp* (*device cabinet*) shares the component *skåp* (*cabinet*) with its domain-specific synonym *elskåp* (*electrical cabinet*). We introduce the EmbeddingSim feature, on the other hand, because embedding-based similarity measures based on FastText [3], word2vec [17], and GloVe [21] have become popular alternatives to random indexing since the original work's publication.

3 Replication

3.1 Dataset

The original experiment focused on the medical side effect domain, with a corpus written in the English language. In this replication, we focus on the building construction business, in particular the provisioning and building of roads and public transportation infrastructure, and change the language to Swedish. We use the synonyms defined in CoClass [2] as the ground truth. CoClass is a hierarchical classification system, implementing ISO 12006-2:2015 [1], that is intended to facilitate the life-cycle management of construction projects. It is co-developed by the Swedish Transportation Agency (*Trafikverket*) and consultancy firms. Table 1 provides an overview of the dataset differences between the two studies.

Table 1. Comparison of the original experiment and the replication

	Original	Replication
Domain	Medical side effects	Building construction
Language	English	Swedish
Terms with/without synonyms	1,791/0	574/856
Average number of synonyms per term	2.8 ($\sigma = 1.4$)	3.8 ($\sigma = 4.4$)
Min/Max number of synonyms per term	2/11	1/46
Phrases/single term proportion	67%/33%	26%/74%
Corpus size (Number of documents)	400,000	4,241,509

While in the original study all terms were associated with at least 2 synonyms, only 574 of the terms in CoClass (40%) are associated with any synonym. Since our goal is eventually to improve the classification system with newly discovered synonyms, we did not remove synonyms that did not appear in the initially constructed corpus, which was crawled from Trafikverkets' publicly accessible document database (1,100 documents) and the Swedish Wikipedia (3,7 million articles). Only a subset of CoClass terms were found in this corpus, however. Therefore, we devised the following strategy to construct a corpus: for each term in CoClass, we searched the public internet for this term using the Bing Search API[3], contributing 540,409 documents to the total corpus. Since each API call returns at most 50 hits, our budget was limited, and some terms in CoClass were common, we used a crawling strategy focused on identifying documents containing uncommon terms. More specifically, we restricted the number of crawled websites c based on the number of search results r for each term:

$$c = \begin{cases} 2500, & \text{if } r <= 10000 \\ 1000, & \text{if } 10000 < r < 100000 \\ 500, & \text{if } r >= 100000 \end{cases} \quad (1)$$

Category c_{500} contained 494 search terms, while c_{1000} contained 708 and c_{2500} contained 2261 search terms. Within c_{2500}, 528 search terms produced no hits at all. The search results demonstrate that the terminology in CoClass is very specialized as a large amount of terms were not even found on the publicly accessible internet.

3.2 Implementation Details

We preprocess our corpus using the efselab toolkit[4] [20] to tokenize and lemmatize the input text. In the case of the Wikipedia and Trafikverket corpora, we additionally perform sentence segmentation. On the Web corpus we use textract[5] to identify text inside of HTML elements (e.g., between <p> and </p> tags) and treat these text spans as sentences. We use efselab to perform part-of-speech tagging and MaltParser [18] to perform dependency parsing for the features that require this information. Our preprocessing differs slightly from the original work due to the changes in our input language and corpus. The original work used a tokenizer based on the Natural Language Toolkit[6] (NLTK), the Porter stemmer, NLTK's part-of-speech tagging, and RASP3 [5] for dependency parsing.

The original work took advantage of large, mature domain-specific thesauri to generate synonym candidates from the target domain. Such thesauri are not available in our language and domain, so we were forced to consider every term that appeared in our Wikipedia or Web corpora as a synonym candidate. We filtered these candidates by removing terms with a low term frequency, terms

[3] https://azure.microsoft.com/en-us/services/cognitive-services/bing-web-search-api/.
[4] https://github.com/robertostling/efselab.
[5] https://textract.readthedocs.io.
[6] https://www.nltk.org/.

that did not appear much more often in our domain-specific Trafikverket corpus than in Wikipedia, and terms that were not tagged as a noun by our part-of-speech tagger. In particular, we required the candidates to have a TF of at least 300, to occur at least 30 times more often in Trafikverket than in Wikipedia, and to be tagged as a noun at least 50% of the time. These filtering steps reduced the total number of synonym candidates from approximately 867,000 to 26,000 (97% reduction) at the cost of reducing the candidates' coverage of true synonyms in CoClass by approximately 26% (i.e., 74% of the synonyms remained after filtering). This left us with 290 target terms that both appeared in our corpus and had true synonyms in our candidate list.

As in the original work, we use the logistic regression implementation from scikit-learn[7] and scale the features to unit variance. We use the LambdaMART implementation from pyltr[8] with query subsampling set to 0.5 (i.e., 50% of the queries used to train each base learner) and the other parameters at their default values. For the word embedding feature, we used FastText[9] [3] to train 100-dimensional embeddings on our corpus using the skipgram method. FastText's other parameters were kept at their default settings.

Our code, the CoClass ground truth, and the URLs of documents in our corpus are available online.[10]

3.3 Experimental Setup

We conduct two experiments in which we compare the LogReg and LambdaMART learning to rank methods against three baselines: PMI, EmbeddingSim, and LinSim. In the TOEFL-style evaluation, we confirm the original study's conclusion that synonym discovery is best approached as a ranking problem. We then evaluate the methods' ability to produce useful ranked lists of synonym candidates in the relevance ranking evaluation.

Each method receives a target term and a set of candidates as input and outputs a ranked list of the candidates. In order to mirror the original study's evaluation, each target term is associated with up to 1,000 incorrect candidates that are randomly sampled from the full set of candidates $\mathbb{C}$. The supervised methods are trained with ten-fold cross validation. We create the cross validation folds based on target terms, so each target term appears in only one fold. We describe the metrics used by the two evaluations in their respective sections.

3.4 General TOEFL-Style Evaluation

As in the original work, we first perform a TOEFL-style evaluation to illustrate the difficulty of the domain-specific synonym discovery task. In this evaluation, methods are required to identify a target term's true synonym given one

[7] http://scikit-learn.org.
[8] https://github.com/jma127/pyltr.
[9] https://github.com/facebookresearch/fastText/.
[10] https://github.com/andrewyates/ecir19-ranking-synonyms.

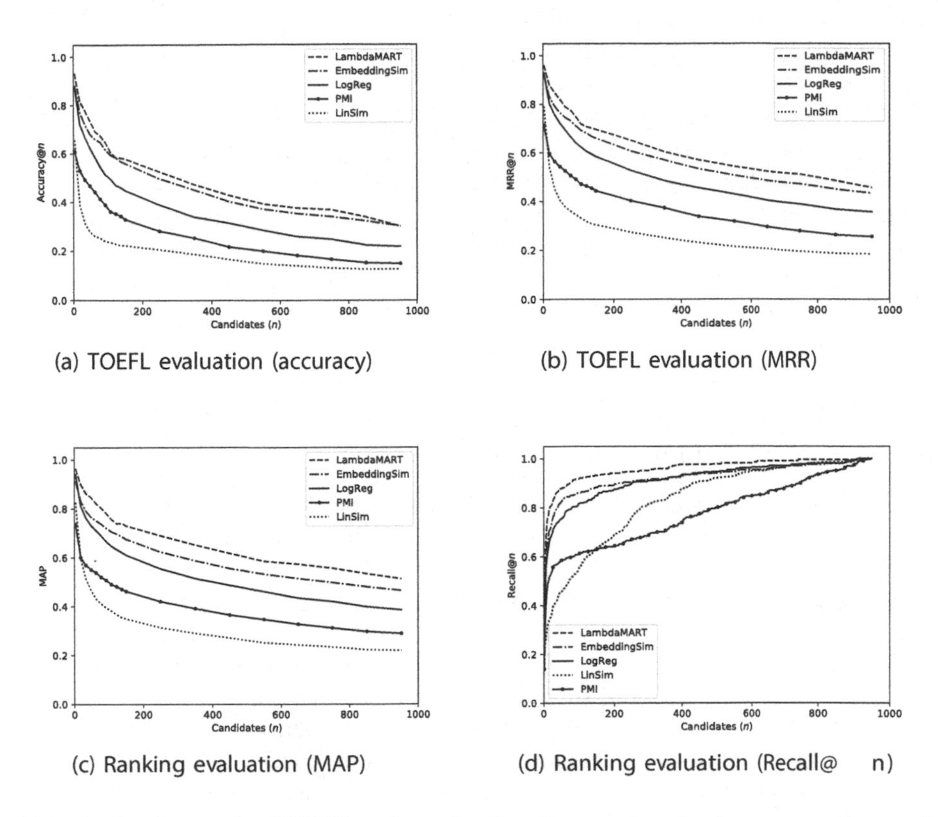

Fig. 1. Results on the TOEFL-style evaluation (top row) and relevance ranking evaluation (bottom row). While the top methods perform well in the TOEFL evaluation for low values of n, the performance of every method decreases as n is increased to more realistic values. The EmbeddingSim baseline performs well on its own, but is surpassed by the LambdaMART model that incorporates it as a feature.

correct synonym candidate and n incorrect candidates. When $n = 3$ this corresponds to the TOEFL evaluation commonly used in prior work on discovering domain-independent synonyms. The original work made the argument that this evaluation is unrealistically easy and demonstrated that, in the consumer health domain, methods are unable to accurately identify synonyms when n is increased to realistic values (e.g., $n = 1000$). In this section we repeat the general TOEFL-style evaluation in order to demonstrate that considering only $n = 3$ incorrect candidates is still unrealistically easy with our Swedish corpus focused on the building construction domain.

For each pair consisting of a target term and a correct synonym candidate, we randomly sample n incorrect candidates and feed the candidates as input to a synonym discovery method. As in the original work, we aggregate each method's predictions to calculate accuracy@n. We additionally report MRR (Mean Reciprocal Rank), which is a more informative metric because correct results in positions past rank 1 also contribute to the score. The result are shown in Fig. 1a (accuracy) and b (MRR). While LogReg, LambdaMart, and EmbeddingSim perform well at low values of n, their accuracy when approaching $n = 1000$ is less than 50%. The methods perform similarly in terms of MRR. While LogReg, the method from the original work, continues to outperform PMI and LinSim, the new EmbeddingSim baseline performs substantially better. This may be due to the fact that LogReg is a linear model and thus has difficulty weighting EmbeddingSim substantially higher than its other features. LambdaMART, the alternate learning to rank model evaluated in this work, is able to outperform both LogReg and EmbeddingSim. This illustrates that the synonym discovery task remains difficult in the new domain. For use cases where recall is important, such as ours, the task is best approached as a ranking problem.

3.5 Relevance Ranking Evaluation

In this section we evaluate the synonym discovery methods' ability to rank synonym candidates, so that they may be considered by a human editor. For each target term, we feed every synonym candidate as input to a synonym discovery method and calculate MAP (Mean Average Precision) and recall@n over the resulting rankings. In the context of this task, recall@n is the more interpretable metric: it indicates the fraction of correct synonyms that a human editor would find after reading through the top n results. The results are shown in Fig. 1c (MAP) and d (recall@n). In general the ranking of methods mirrors that from the TOEFL-style evaluation, with LambdaMART performing best. The top three methods perform similarly for different values of n, whereas PMI and LinSim perform differently. PMI performs better for low values of n, while LinSim begins to outperform PMI at roughly $n = 175$. As in the previous evaluation, the new LambdaMART and EmbeddingSim methods outperform LogReg. LambdaMART achieves 88% recall at $n = 50$, followed by EmbeddingSim with 82% recall and LogReg with 76% recall. This illustrates that the top performing methods can produce a useful ranking despite their low accuracy.

3.6 Feature Analysis

In this section we evaluate the contribution of individual features to the learning to rank models' performance. The MAP@150 achieved by each individual feature is shown in Fig. 2a. EmbeddingSim performs substantially better than the other features. Windows, NGram, and LinSim are the only other features to achieve a MAP above 0.3, with Decompound, LevDist, and POSNGram performing poorly when used in isolation. In the original work, LevDist was the best performing single feature, with Windows, RISim, and the dependency context features (LinSim and Hagiwara) performing the next best. In our new domain, Windows and LinSim continue to perform well, but LevDist and RISim perform poorly. The difference in the domains and language may account for LevDist's decreased impact, since it is only a useful feature when synonym candidates have significant character overlap with target terms. EmbeddingSim was not included

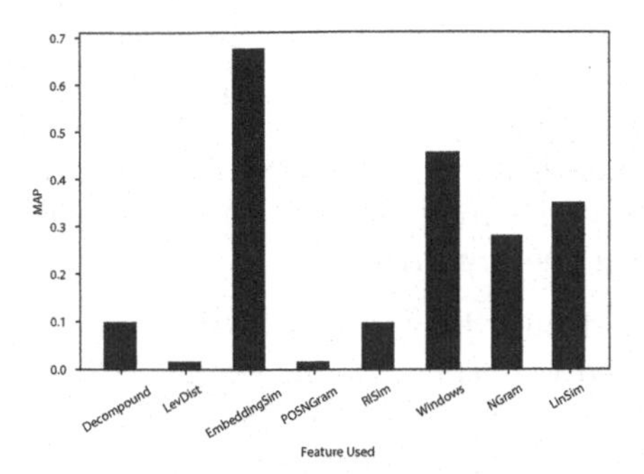

(a) MAP when single features are used.

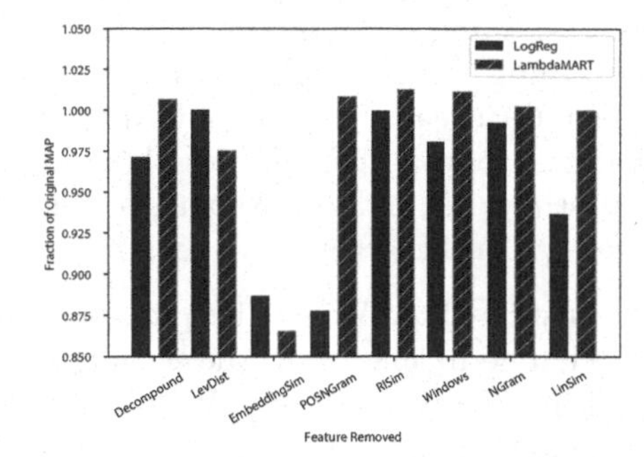

(b) MAP (fraction of max) when single features are excluded from the model.

Fig. 2. MAP@150 when only one feature is used (left) and the fraction of the entire model's MAP@150 when a single feature is excluded from the model (right). LevDist, the strongest feature in the original work, appears to have less utility in the new setting. Windows and LinSim, strong features in the original work, continue to perform well here.

in the original work, but it is similar to RISim in that both methods are intended to capture distributional semantics.

We analyze the decrease in each model's performance when a single feature is removed in Fig. 2b. The y-axis indicates each method's MAP as a fraction of the original MAP after a feature is removed. For example, removing the EmbeddingSim feature reduces the performance of both LambdaMART and LogReg to 85–90% of their MAPs when all features are used. With the exception of LevDist and EmbeddingSim, the LambdaMART model consistently achieves a smaller decrease in performance when any single feature is removed. As in the single feature analysis, EmbeddingSim is the best performing feature, and RISim does not contribute much to the models' performance. While POSNGram performed poorly in isolation, removing the feature decreases the performance of LogReg by approximately 12%, indicating that it is providing a useful signal used in conjunction with other features. Similarly, removing Decompound or LevDist decreases LogReg's or LambdaMART's performance by approximately 2.5%, respectively, despite the fact that they performed poorly as single features.

4 Generalizability

In order to better understand the generalizability of the learning to rank approach to synonym discovery, we discuss differences between this study and the original one in terms of the methodology and results.

Corpus Creation. Due to its focus on identifying synonyms of medical side effects, the original study used a corpus of 400,000 English forum posts related to health. In this study we focused on Swedish language synonyms in the building construction domain. This is a formal, specialized domain in comparison to health-related social media, which made it more difficult to identify documents containing target terms or synonym candidates. To create a corpus with sufficient term co-occurrence information, we created a Swedish language corpus that was both larger and less homogeneous than the corpus used in the original study: 4.2 million Webpages from Wikipedia, the Swedish Transportation Agency (Trafikverket), and searches against the Bing API.

Preprocessing. While the preprocessing details differed in this study due to the change in languages, the techniques used were conceptually similar to those used in the original study. We use MaltParser in place of RASP3, and efselab's tokenization and lemmatizer in place of NLTK and the Porter stemmer.

Features and Method. We introduced two new features that were not present in the original study. Motivated by prior work that showed decompounding can improve recall in the German language [4], we introduced the Decompounder feature. This feature uses the SECOS decompounder [16] to split compound nouns into their components and measures the overlap between a target term's and candidate term's components. Figure 2b suggests that this feature slightly contributes to the performance of the LogReg method, but does not positively influence the performance of LambdaMART. EmbeddingSim, which computes the similarity between FastText embeddings, is the second new feature we introduced. FastText considers character n-grams when representing terms, which may make the Decompound feature redundant. We additionally introduced experiments on a new learning to rank model, LambdaMART, in order to compare its performance with the LogReg model used in the original work. We found that LambdaMART substantially outperformed LogReg, indicating the utility of using a more advanced model. We found that methods generally performed better on our domain and corpus. For example, LogReg achieved 50% recall@50 in the original study, whereas it achieves 76% recall@50 in this work. It is difficult to attribute these performance differences to specific factors, with the language, domain, and language register (i.e., professional language in this study and casual, lay language in the original study) all differing between the two studies.

5 Conclusions

In this work we replicated the synonym discovery method proposed by Yates et al. [25] in a new language (i.e., Swedish rather than English) and in a new domain (i.e., building construction rather than medical side effects). We found that in the new domain, the proposed LogReg method outperformed the PMI baseline as before. Motivated by methodological advances and the difference in languages, we introduced two new features and an alternate learning to rank method which we found to outperform the original approach.

These results provide evidence that *(1)* synonym discovery can be effectively approached as a learning to rank problem and *(2)* the features proposed in the original work are robust to changes in both domain and language. While our replication cannot provide evidence that the new EmbeddingSim feature works well in other settings, it does provide evidence that EmbeddingSim does not make the features used in the original work obsolete.

References

1. Building construction - Organization of information about construction works - Part 2: Framework for classification. Tech. Rep. 12006–2:2015, ISO (May 2015). https://www.iso.org/standard/61753.html
2. CoClass, September 2018. https://coclass.byggtjanst.se/en/about
3. Bojanowski, P., Grave, E., Joulin, A., Mikolov, T.: Enriching word vectors with subword information. Trans. Assoc. Comput. Linguist. **5**, 135–146 (2017)
4. Braschler, M., Ripplinger, B.: How effective is stemming and decompounding for german text retrieval? Inf. Retrieval **7**(3), 291–316 (2004)
5. Briscoe, T., Carroll, J., Watson, R.: The second release of the rasp system. In: Proceedings of the COLING/ACL on Interactive Presentation Sessions. COLING-ACL 2006 (2006)
6. Burges, C.J.: From ranknet to lambdarank to lambdamart: An overview. Tech. Rep. MSR-TR-2010-82 (2010)
7. Carpineto, C., Romano, G.: A survey of automatic query expansion in information retrieval. ACM Comput. Surv. (CSUR) **44**(1), 1 (2012)
8. Furnas, G.W., Landauer, T.K., Gomez, L.M., Dumais, S.T.: The vocabulary problem in human-system communication. Commun. ACM **30**(11), 964–971 (1987)
9. Goodman, S.N., Fanelli, D., Ioannidis, J.P.: What does research reproducibility mean? Sci. Transl. Med. **8**(341), 341ps12 (2016). http://stm.sciencemag.org/content/8/341/341ps12
10. Hagiwara, M.: A supervised learning approach to automatic synonym identification based on distributional features. In: Proceedings Annual Meeting of the Association for Computational Linguistics on Human Language Technologies: Student Research Workshop, pp. 1–6. ACM (2008)
11. Haiduc, S., Bavota, G., Marcus, A., Oliveto, R., De Lucia, A., Menzies, T.: Automatic query reformulations for text retrieval in software engineering. In: Proceedings International Conference on Software Engineering (ICSE), pp. 842–851. IEEE (2013)
12. Kang, Y., Li, J., Yang, J., Wang, Q., Sun, Z.: Semantic analysis for enhanced medical retrieval. In: International Conference on Systems, Man, and Cybernetics (SMC), pp. 1121–1126. IEEE, October 2017
13. Levy, O., Goldberg, Y., Dagan, I.: Improving distributional similarity with lessons learned from word embeddings. Trans. Assoc. Comput. Linguist. **3**, 211–225 (2015)
14. Lin, D.: Automatic retrieval and clustering of similar words. In: Proceedings of the 17th International Conference on Computational Linguistics, vol. 2, pp. 768–774. Association for Computational Linguistics (1998)
15. Lucia, A.D., Fasano, F., Oliveto, R., Tortora, G.: Recovering traceability links in software artifact management systems using information retrieval methods. ACM Trans. Softw. Eng. Methodol. (TOSEM) **16**(4), 13 (2007)
16. Martin Riedl, C.B.: Unsupervised compound splitting with distributional semantics rivals supervised methods. In: Proceedings of The 15th Annual Conference of the North American Chapter of the Association for Computational Linguistics: Human Language Technologie, pp. 617–622. San Diego, CA, USA (2016)

17. Mikolov, T., Sutskever, I., Chen, K., Corrado, G.S., Dean, J.: Distributed representations of words and phrases and their compositionality. In: Advances in Neural Information Processing Systems, pp. 3111–3119 (2013)
18. Nivre, J., et al.: Maltparser: a language-independent system for data-driven dependency parsing. Nat. Lang. Eng. **13**(2), 95–135 (2007)
19. Oard, D.W., Baron, J.R., Hedin, B., Lewis, D.D., Tomlinson, S.: Evaluation of information retrieval for e-discovery. Artif. Intell. Law **18**(4), 347–386 (2010)
20. Östling, R.: Part of speech tagging: shallow or deep learning? Northern Eur. J. Lang. Technol. (NEJLT) **5**, 1–15 (2018)
21. Pennington, J., Socher, R., Manning, C.D.: Glove: global vectors for word representation. In: Empirical Methods in Natural Language Processing (EMNLP), pp. 1532–1543 (2014). http://www.aclweb.org/anthology/D14-1162
22. Stanojević, M.: Cognitive synonymy: a general overview. Facta Universitatis-Series: Linguist. Lit. **7**(2), 193–200 (2009)
23. Terra, E., Clarke, C.L.: Frequency estimates for statistical word similarity measures. In: Proceedings Conference of the North American Chapter of the Association for Computational Linguistics on Human Language Technology, pp. 165–172. AMC (2003)
24. Widdows, D., Cohen, T.: The semantic vectors package: New algorithms and public tools for distributional semantics. In: 2010 IEEE Fourth International Conference on Semantic Computing (ICSC), pp. 9–15. IEEE (2010)
25. Yates, A., Goharian, N., Frieder, O.: Relevance-ranked domain-specific synonym discovery. In: Advances in Information Retrieval - 36th European Conference on IR Research ECIR 2014 (2014)
26. Zeng, H.J., He, Q.C., Chen, Z., Ma, W.Y., Ma, J.: Learning to cluster web search results. In: Proceedings of the 27th Annual International ACM SIGIR Conference on Research and Development in Information Retrieval SIGIR 2004, pp. 210–217 (2004)
27. Zhang, L., Li, L., Li, T.: Patent mining: a survey. ACM SIGKDD Explor. Newslett. **16**(2), 1–19 (2015)

On Cross-Domain Transfer in Venue Recommendation

Jarana Manotumruksa[1(✉)], Dimitrios Rafailidis[2], Craig Macdonald[1], and Iadh Ounis[1]

[1] University of Glasgow, Glasgow, UK
j.manotumruksa.1@research.gla.ac.uk,
{craig.macdonald,iadh.ounis}@glasgow.ac.uk
[2] Maastricht University, Maastricht, Netherlands
dimitrios.rafailidis@maastrichtuniversity.nl

Abstract. Venue recommendation strategies are built upon Collaborative Filtering techniques that rely on Matrix Factorisation (MF), to model users' preferences. Various *cross-domain* strategies have been proposed to enhance the effectiveness of MF-based models on a target domain, by transferring knowledge from a source domain. Such cross-domain recommendation strategies often require user overlap, that is common users on the different domains. However, in practice, common users across different domains may not be available. To tackle this problem, recently, several cross-domains strategies without users' overlaps have been introduced. In this paper, we investigate the performance of state-of-the-art cross-domain recommendation that do not require overlap of users for the venue recommendation task on three large Location-based Social Networks (LBSN) datasets. Moreover, in the context of cross-domain recommendation we extend a state-of-the-art sequential-based deep learning model to boost the recommendation accuracy. Our experimental results demonstrate that state-of-the-art cross-domain recommendation does not clearly contribute to the improvements of venue recommendation systems, and, further we validate this result on the latest sequential deep learning-based venue recommendation approach. Finally, for reproduction purposes we make our implementations publicly available.

Keywords: Cross-domain recommendation · Venue suggestion · Transfer learning

1 Introduction

Location-Based Social Networks (LBSN) such as Foursquare and Yelp have become popular platforms that allow users to find interesting venues to visit based on their preferences, share their location to their friends using checkins, as well as leave comments on venues they have visited. Matrix Factorisation

L. Azzopardi et al. (Eds.): ECIR 2019, LNCS 11437, pp. 443–456, 2019.
https://doi.org/10.1007/978-3-030-15712-8_29

(MF) [1] is a popular collaborative filtering technique that is widely used to predict users' ratings/checkins on venues by leveraging explicit/implicit feedback. The major challenge in MF-based venue recommendation systems is the sparsity problem, as users can visit a very limited number of venues. Consequently, the checkin user-venue matrix in LBSNs is extremely sparse. The sparsity problem hinders the effectiveness of MF-based models. To alleviate the sparsity problem, various MF-based models have been proposed to exploit additional sources of information such as friendships and textual content of comments [2–5]. Moreover, previous studies have shown that the sequential properties of user's interactions, that is the sequences of checkins or clicks, play an important role in alleviating sparsity for tasks such as movie recommendation and venue recommendation [6–10]. Indeed, sequential-based recommendation is more challenging than traditional recommendation (e.g. rating prediction problem) because a user's previous interactions may have a strong influence on their current preferences. For example, users are more likely to visit a bar directly after they have visited a restaurant.

Apart from these additional sources of information and the sequential properties of users' interactions, recently, Cross-Domain Collaborative Filtering (CDCF) models have been proposed to alleviate the sparsity problem in a target domain [11–15]. CDCF aims to improve the quality of recommendations in a target domain by leveraging the knowledge extracted from source domains. Thus, CDCF is a form of *transfer learning*, for which the existing approaches can be divided into two categories: namely *overlapping* and *non-overlapping* CDCF. The *overlapping* CDCF models, such as [14–17], transfer knowledge from a source domain to a target one based on explicit links of users/items between the domains. For example, a user who has both Twitter and Foursquare accounts, *overlapping* CDCF aims to transfer user's preferences extracted from the user's tweets on Twitter (source domain) to improve the venue recommendation on Foursquare (target domain). *Non-overlapping* CDCF models like [11–13, 18] aim to transfer useful information from the source domain to the target one without any explicit links, which is the most challenging problem in cross-domain recommendation. An example of *non-overlapping* CDCF is a newly opened book e-commerce website that would like to build a recommender system. Due to the lack of user-book interactions at the outset, the effectiveness of MF-based models for the book e-commerce website would be low, due to the sparsity problem. Since the movie domain is related to the book domain in some aspects [19] – for example comedy movies correspond to humorous books – CDCF hypothesises that the rating matrix available from a popular movie rating website would alleviate the sparsity problem for the newly opened book e-commerce site, despite there being no users or items shared between the two domains.

Most of the *non-overlapping* CDCF models [12, 13, 18] are based on a Codebook Transfer technique (CBT) proposed by Li *et al.* [11]. CBT aims to transfer rating patterns from the source domain (e.g., the popular movie rating website) to the target domain (e.g., the newly opened book website). Although several studies have shown that CBT-based models can improve recommendation effectiveness, Cremonesi and Quadrana [20] demonstrated that CBT is not able to transfer knowledge between non-overlapping domains. This brings doubt into the usefulness of CBT. In this work, we are the first to explore the effectiveness of CBT-based *non-overlapping* CDCF in the context of cross-domain venue recommendation. Our assumption is that CDCF could enable knowledge from a source domain (e.g. Foursquare) to enhance the quality of recommendations for a target domain (e.g. Yelp). In particular, our contributions are summarised below:

- We investigate the performance of a of state-of-the-art *non-overlapping* CDCF framework, CrossFire [13] in the context of venue recommendation.
- Inspired by [13], we extend the state-of-the-art sequential-based deep learning venue recommendation framework of [10] using the CBT-based technique, in order to evaluate its effectiveness in the cross-domain sequential-based venue recommendation task. To the best of our knowledge, this is the first work that studies the effectiveness of cross-domain recommendation strategies by leveraging sequences of user-venue interactions.
- We conduct comprehensive experiments on three large-scale real-world datasets from Foursquare, Yelp and Brightkite to evaluate the performance of CBT-based *non-overlapping* CDCF. Our experimental results demonstrate that CrossFire is not able to transfer useful knowledge from the source domain to the target domain for the venue recommendation task, being consistent with the previous study of [20]. In particular, we find that the CBT-based technique does not clearly contribute to the improvements observed from CrossFire, compared to the traditional single-domain MF-based models. Indeed, through experiments conducted when equating the source and target domains, we show that such improvements may not be explained by transfer of knowledge between source and target domains. We postulate that, in fact, that the improvements are gained from the additional parameters introduced by CrossFire, which makes it more flexible than the traditional single-domain MF-based approaches. In addition, our experimental results on a state-of-the-art sequential-based deep learning venue recommendation framework of [10] further validate this result.

This paper is organised as follows. Section 2 provides the problem statement of cross-domain venue recommendation. Then, we describe single-domain MF-based approaches and *non-overlapping* Cross-Domain Collaborative Filtering approaches (CDCF). The experimental setup for our experiments is detailed in Sect. 3, while comprehensive experimental results comparing the effectiveness of *non-overlapping* CDCF approaches with various single-domain MF-based approaches are reported in Sect. 4. Concluding remarks follow in Sect. 5.

2 Cross-Domain Venue Recommendation Frameworks

In this section, we first formulate the problem statement of the cross-domain venue recommendation task, without overlaps between domains. Then, we briefly introduce state-of-the-art MF-based strategies for the single domain and cross-domain recommendation tasks. After presenting the original CBT model of [11], we detail the cross-domain recommendation framework of CrossFire, proposed by Shu *et al.* [13]. Note that this framework was not originally proposed for the cross-domain venue recommendation task but is sufficiently flexible to be applied to it. Next, we present the sequential-based deep learning model of [10] for venue suggestion in a single domain, and then our proposed extended model of Deep Recurrent Transfer Learning (DRTL) for the cross-domain recommendation task.

2.1 Problem Statement

The task of cross-domain venue recommendation is to exploit knowledge from a source domain to enhance the quality of venue recommendation in a target domain. The task of venue recommendation in the target domain is to generate a ranked list of venues that a user might visit given his/her historical feedback, that is the previously visited venues from the checkin data. Let $\mathcal{U}^s$, $\mathcal{V}^s$ and $\mathcal{U}^t$, $\mathcal{V}^t$ be the sets of users and venues in the source and target domains, respectively. Let $\mathcal{V}_u^s$ ($\mathcal{V}_u^t$) denote the list of venues the user u in the source (target) domain has previously visited, sorted by time. $\mathcal{S}_u^s$ ($\mathcal{S}_u^t$) denote the list of sequences of visited venues of user u in the source (target) domain, for example, if $\mathcal{V}_u^s = (v_1, v_2, v_3)$, then $\mathcal{S}_u^s = ((v_1), (v_1, v_2), (v_1, v_2, v_3))$. $s_t \in \mathcal{S}_u^s$ ($\mathcal{S}_u^t$) denotes the sequence of visited venues of user u at time t in the source (target) domain (e.g. $s_2 = (v_1, v_2)$). All checkins by all users in the source (target) domain are represented as a matrix $C_s \in \mathbb{R}^{m^s \times n^s}$ ($C_t \in \mathbb{R}^{m^t \times n^t}$) where m^s, n^s and m^t, n^t are the number of users and venues in the source and target domains, respectively. Let $c_{u,i}^s \in C_s$ ($c_{u,i}^t \in C_t$) denote a user $u \in \mathcal{U}^s$ who visited venue $i \in \mathcal{V}$ in the source (target) domain. Note that $c_{u,i}^s = 0$ means that user u has neither left a rating nor made a checkin at venue i. The goal of a cross-domain recommendation task is to generate a personalised list of venues in the target domain t, by exploiting both the checkin matrices C_s and C_t of the source and target domains, respectively.

2.2 Traditional MF-Based Models

Matrix Factorisation (MF) is a collaborative filtering technique that assumes users who share similar preferences, like visiting similar venues, are likely to influence each other [1]. The goal of MF is to reconstruct the checkin matrix $C \in \mathbb{R}^{m \times n}$ where m and n are the number of users and venues by calculating the dot product of latent factors of users $U \in \mathbb{R}^{m \times d}$ and venues $V \in \mathbb{R}^{n \times d}$, where d is the number of latent dimensions. In particular, the MF model is trained by minimising a loss function, which consists of sum-of-squared-error terms between the actual and predicted checkins, as follows:

$$\min_{U,V} \|C - UV^T\|_F^2 \tag{1}$$

where $\|.\|_F^2$ denotes the Frobenius norm.

Collective Matrix Factorisation (CMF) is a MF-based model that leverages both user-venue interactions and social information [21]. CMF aims to approximate the checkin matrix C and social link matrix[1] S, simultaneously. Built upon the MF model, the authors introduced a latent factor of friends, $F \in \mathcal{R}^{m \times d}$, which is used to capture the social relationship between users. In particular, the loss function of CMF is defined as follows:

$$\min_{U,V,F} \|C - UV^T\|_F^2 + \|S - UF^T\|_F^2 \tag{2}$$

2.3 Cross-Domain Collaborative Filtering Models

Codebook Transfer (CBT) is a cross-domain collaborative filtering technique proposed by Li *et al.* [11], which assumes that user-venue interactions across different domains might share similar checkin/rating patterns. CBT consists of two steps: extracting the patterns/knowledge of user-venue interactions from a source domain and exploiting the extracted patterns/knowledge to enhance the quality of venue recommendation in a target domain t. To extract the patterns/knowledge from the source domain, based on the collaborative filtering technique, CBT aims to approximate the checkin matrix C_s of the source domain s by finding a decomposition of C_s, i.e. a dot product of the latent factors of users $U_s \in \mathcal{R}^{m \times d}$, venues $V_s \in \mathcal{R}^{n \times d}$ and the shared checkin patterns[2] $W \in \mathcal{R}^{d \times d}$. The loss function of CBT is defined as follows:

$$\min_{U_s,W,V_s} \|C_s - U_s W V_s^T\|_F^2 \tag{3}$$

Next, CBT approximates the checkin matrix C_t of the target domain t and exploits the shared patterns W to boost the venue recommendation accuracy in target domain t as follows:

$$\min_{U_t,V_t} \|C_t - U_t W V_t^T\|_F^2 \tag{4}$$

CBT only updates the latent factors of the target domain t, that is the latent factors U_t and V_t, while keeping the checkin patterns W fixed.

CrossFire, proposed by Shu *et al.* [13], is a cross-domain recommendation framework. In applying it to venue recommendation, it aims to transfer knowledge extracted from the user-venue interactions and users' social links of the source domain to improve the quality of venue recommendation in the target domain. In particular, the objective of the CrossFire framework is to jointly approximate the checkin matrices C_s and C_t and the social link matrix S_s and S_t of the source and target domains, respectively. Built upon the CBT-based

[1] $S \in \{0, 1\}^{m \times m}$ is the adjacency matrix representing the relationship between users.
[2] The shared patterns denote similarities between the latent factors of the domains.

technique, the loss function of CrossFire is defined as follows:

$$\min_{U_s,U_t,V_s,V_t,W,Q} (\|C_s - U_sWV_s^T\|_F^2 + \|C_t - U_tWV_t^T\|_F^2) \\ + (\|S_s - U_sQU_s^T\|_F^2 + \|S_t - U_tQU_t^T\|_F^2) \tag{5}$$

where $W \in \mathcal{R}^{d\times d}$ is a matrix with the checkin patterns in the latent space, as the CBT method, and $Q \in \mathcal{R}^{d\times d}$ is the social patterns in the latent space shared across the source and target domains. Unlike the CBT-based technique, the CrossFire framework jointly updates the latent factors U_s, U_t, V_s, V_t, W, Q. More details on the optimisation strategy of CrossFire are available in [13].

2.4 Sequential-Based Venue Recommendation Frameworks

Deep Recurrent Collaborative Filtering (DRCF) proposed by Manotumruksa *et al.* [10], is a state-of-the-art sequential framework, which leverages deep learning algorithms such as Multi-Level Perceptron (MLP) and Recurrent Neural Networks (RNN) to capture users' *dynamic* preferences from their sequences of checkins. The DRCF framework consists of three components: Generalised Recurrent Matrix Factorisation (GRMF), Multi-Level Recurrent Perceptron (MLRP) and Recurrent Matrix Factorisation (RMF). In particular, given the sequence of a user's checkins $S_{i,t}$, the predicted checkin $\hat{c}_{u,i}$ is estimated as:

$$\hat{c}_{u,i} = a_{out}(h(\phi^{GRMF} \oplus \phi^{MLRP} \oplus \phi^{RMF})) \tag{6}$$

where a_{out} is the activation function, h is the hidden layer, and $\oplus$ denotes the concatenation operation. ϕ^{GRMF}, ϕ^{MLRP} and ϕ^{RMF} denote the GRMF, MLRP and RMF models, which are defined as follows:

$$\phi^{GRMF} = \left[d_{u,t} \otimes p_u \otimes q_i\right] \tag{7}$$

$$\phi^{MLRP} = \left[a_L(h_L(...a_1(h_1(d_{u,t} \oplus p_u \oplus q_i))))\right] \tag{8}$$

$$\phi^{RMF} = (d_{u,t} + p_u) \circledcirc q_i^{Gd} \tag{9}$$

where $\otimes$ is the element-wise product operation, $\circledcirc$ is the dot-product operation, $d_{u,t}$ is the user's *dynamic* preferences of user u at time t that are projected from the RNN layer, p_u and q_i are the latent factors of user u and venue i and L is the number of layers. Next, instead of training the DRCF framework to minimise the *pointwise* loss between the predicted checkin $\hat{c}_{u,i}$ and observed checkin $c_{u,i}$, the DRCF framework follows the Bayesian Personalised Ranking (BPR) strategy of [22] to learn DRCF's parameters, as follows:

$$\min_{\theta_e,\theta_r,\theta_h} \sum_{u\in\mathcal{U}} \sum_{s_t\in S_u} \sum_{j\in\mathcal{V}-s_t} \log(\sigma(\hat{c}_{u,i} - \hat{c}_{u,j}))$$

where i is the most recently visited venue in s_t and $\sigma(x)$ is the sigmoid function. θ_e, θ_r and θ_h are the parameter sets of latent factors, RNN layers and hidden layers, respectively.

Deep Recurrent Transfer Learning (DRTL) is our proposed extension of DRCF to perform cross-domain recommendation for the venue suggestion task, exploiting the user-venue interactions from the source domain based on the CrossFire framework. In particular, we extend the loss function of DRCF as:

$$\min_{\theta_e^s,\theta_r^s,\theta_h^s,\theta_e^t,\theta_r^t,\theta_h^t} \sum_{u\in\mathcal{U}^s}\sum_{s_t\in S_u}\sum_{j\in\mathcal{V}^s-s_t}\log(\sigma(\hat{c}^s_{u,i}-\hat{c}^s_{u,j})) + \sum_{u\in\mathcal{U}^t}\sum_{s_t\in S_u}\sum_{j\in\mathcal{V}^t-s_t}\log(\sigma(\hat{c}^t_{u,i}-\hat{c}^t_{u,j}))$$

where θ_e^s, θ_r^s, θ_h^s (θ_e^t, θ_r^t, θ_h^t) are the parameters of the latent factors, the RNN layers and the hidden layers for the source (target) domain. $\hat{c}^s_{u,i}$ ($\hat{c}^s_{u,i}$) is the predicted checkin for the source (target) domain, defined as follows:

$$\hat{c}^s_{u,i} = a_{out}(h^s(\phi^s_{GRMF} \oplus \phi^s_{MLRP} \oplus \phi^s_{RMF})) \tag{10}$$

$$\hat{c}^t_{u,i} = a_{out}(h^t(\phi^t_{GRMF} \oplus \phi^t_{MLRP} \oplus \phi^t_{RMF})) \tag{11}$$

Next, we extend the GRMF, MLRP and RMF models of the DRCF framework by adding the shared checkin pattern W, as follows:

$$\phi^s_{GRMF} = \left[d^s_{u,t} \otimes p^s_u \otimes q^s_i \otimes W_{GRMF}\right] \quad \phi^t_{GRMF} = \left[d^t_{u,t} \otimes p^t_u \otimes q^t_i \otimes W_{GRMF}\right] \tag{12}$$

$$\phi^s_{MLRP} = \left[a_L(h^s_L(...a_1(h^s_1(d^s_{u,t} \oplus p^s_u \oplus q^s_i \oplus W_{MLRP}))))\right] \tag{13}$$

$$\phi^t_{MLRP} = \left[a_L(h^t_L(...a_1(h^t_1(d^t_{u,t} \oplus p^t_u \oplus q^t_i \oplus W_{MLRP}))))\right] \tag{14}$$

$$\phi^s_{RMF} = (d^s_{u,t} + p^s_u) \circledcirc q^s_i \circledcirc W_{RMF} \quad \phi^t_{RMF} = (d^t_{u,t} + p^t_u) \circledcirc q^t_i \circledcirc W_{RMF} \tag{15}$$

where W_{GRMF}, W_{MLRP} and W_{RMF} are the checkin pattern parameters shared between the source and target domains. Inspired by CrossFire, we include these shared parameters into the GRMF, MLRP and RMF models, such that useful checkin patterns can be transferred from the source domain to the target domain e.g., ϕ^t_{RMF} and ϕ^s_{RMF} can share information via W_{RMF}.

3 Experimental Setup

In this section, we evaluate the effectiveness of the *non-overlapping* CDCF approach of CrossFire by comparing with single-domain MF-based approaches. In addition, we extend the Deep Recurrent Collaborative Filtering framework (DRCF), to perform cross-domain recommendation using the CBT-based technique, namely DRTL. Then, we compare the effectiveness of DRTL by comparing with the CRCF. In particular, we aim to address the following research questions:

RQ1 *Can a non-overlapping CDCF approach that relies on Codebook Transfer extract useful checkin patterns from a source domain that can enhance the quality of venue recommendation in a target domain?*

RQ2 *Can we enhance the effectiveness of a state-of-the-art sequential venue recommendation technique on a target domain by incorporating Codebook Transfer from a source domain?*

3.1 Datasets and Measures

We conduct experiments on publicly available large-scale LBSN datasets. In particular, we use two checkin datasets from Brightkite[3] and Foursquare[4], and a rating dataset from Yelp[5]. We follow the common practice from previous works [10,22,23] to remove venues with less than 10 checkins/ratings. Table 1 summarises the statistics of the filtered datasets. To evaluate the effectiveness of cross-domain venue recommendation frameworks, following previous studies [10,22–24], we adopt a *leave-one-out* evaluation methodology: for each user, we select her most recent checkin/rating as a ground truth and randomly select 100 venues that she has not visited before as the testing set, where the remaining checkins/ratings are used as the training and validation set. The venue recommendation task is thus to rank those 101 venues for each user, aiming to rank highest the recent, ground truth checkin/rating. Note that previous works [12,13,15,20] on *non-overlapping* CDCF use Mean Absolute Error (MAE) and Root Mean Square Error (RMSE) to evaluate the quality of rating prediction. In contrast, we evaluate the quality of recommendation in terms of Hit Ratio (HR)[6] and Normalised Discounted Cumulative Gain (NDCG) on the ranked lists of venues – as applied in previous studies [10,23,24]. In particular, HR considers the ranking nature of the task, by taking into account the rank(s) of the venues that each user has previously visited/rated in the produced ranking, while NDCG goes further by considering the checkin frequency/rating value of the user as the graded relevance label. Lastly, significance tests use a paired t-test.

Table 1. Statistics of the three evaluation datasets.

	Brightkite	Foursquare	Yelp
Number of users	14,374	10,766	38,945
Number of venues	5,050	10,695	34,245
Number of ratings or checkins	681,024	1,336,278	981,379
Number of social links	33,290	164,496	1,598,096
% density of user-venue matrix	0.93	1.16	0.07

3.2 Source and Target Domains

In the subsequent experiments, we use all three LBSN datasets, separated into training and testing datasets as described above. We report results both without using any cross-domain transfer (i.e. using the MF and CMF baselines), and

[3] https://snap.stanford.edu/data/.
[4] https://archive.org/details/201309_foursquare_dataset_umn.
[5] https://www.yelp.com/dataset_challenge.
[6] Hit Ratio (HR) is a simplification of Mean Reciprocal Rank (MRR), which has been commonly used in top-N evaluation for recommendation systems [24–26] when ground-truth data are extracted from the implicit feedback.

when conducting non-overlapping cross-domain transfer. In doing so, we set one LBSN dataset as the Source domain, and one as the Target domain. Finally, to determine if the cross-domain transfer from other domains brings new information, we set the source and target domains to be equal - but while retaining a fair train/test split.

3.3 Implementations and Parameter Setup

We implement all techniques using the Keras deep learning framework[7]. Following [10,13,24], we set equal the dimension of the latent factors d of MF-based approaches and cross-domain recommendation frameworks, $d = 10$, and the number of hidden layers $L = 3$ across the three datasets. Note that since the impact of the hidden layer's number L and dimension size d have been previously explored in [23,24], we omit varying the size of the hidden layers and the dimension of the latent factors in this study. Following Manotumruksa *et al.* [10], we randomly initialise all hidden, latent factors, and RNN layers' parameters for the source and target domains, $\theta_r^s, \theta_e^s, \theta_h^s, \theta_r^t, \theta_e^t, \theta_h^t$, with a Gaussian distribution, setting the mean to 0 and the standard deviation to 0.01, and then we apply the mini-batch Adam optimiser [27] to optimise those parameters. In doing so, we achieve faster convergence than stochastic gradient descent and automatically adjust the learning rate for each iteration. We initially set the learning rate to 0.001[8] and set the batch size to 256.

4 Experimental Results

Table 2 reports the effectiveness of single-domain MF-based approaches and the CrossFile (*non-overlapping* CDCF) approach in terms of HR@10 and NDCG@10 on the three evaluation datasets. The Target Domain row indicates the dataset/domain for which we generate venue recommendations. The Source Domain row indicates the dataset that is used as auxiliary information for *non-overlapping* CDCF. Note that the single-domain MF-based approaches do not leverage the auxiliary information from the source domain.

On inspection of the results of CrossFire, we observe that it consistently and significantly outperforms MF and CMF, for both HR and NDCG, across both the Brightkite and Yelp datasets. This observation is consistent with the results reported in [13] when transferring knowledge between books and movie domains. In contrast, MF is more effective than CrossFire in terms of HR and NDCG on the Foursquare dataset. These results bring doubt that CrossFire generalises across different datasets. Next, when CrossFire uses Brightkite as both the source and target domains, then CrossFire is more effective than any other setup (i.e. Foursquare or Yelp as the source domain). Similarly, for CrossFire, we observe that when CrossFire uses Yelp as both the source and target domains, it outperforms CrossFire with Brightkite or Foursquare as the source domain.

[7] https://bitbucket.org/feay1234/transferlearning.

[8] The default learning rate setting of the Adam optimiser in Keras.

Table 2. Performance in terms of HR@10 and NDCG@10 of single-domain and cross-domain MF-based approaches. The best performing result in each row is highlighted in bold and $*$ indicates significant differences in terms of paired t-test with $p < 0.01$, comparing to the best performing result.

Target domain	Brightkite				
Model	MF	CMF	CrossFire	CrossFire	CrossFire
Source domain	–	–	Brightkite	Foursquare	Yelp
HR	0.5252*	0.5931*	**0.6140**	0.5906*	0.5668*
NDCG	0.3224*	0.3444*	**0.3670**	0.3546*	0.3421*
Target domain	Foursquare				
Model	MF	CMF	CrossFire	CrossFire	CrossFire
Source domain	–	–	Brightkite	Foursquare	Yelp
HR	**0.6897**	0.6750*	0.6737*	0.6483*	0.6722*
NDCG	**0.4279**	0.3692*	0.4159*	0.3997*	0.4189*
Target domain	Yelp				
Model	MF	CMF	CrossFire	CrossFire	CrossFire
Source domain	–	–	Brightkite	Foursquare	Yelp
HR	0.3458*	0.3472*	0.4364	0.4331*	**0.4399**
NDCG	0.1782*	0.1773*	**0.2332**	0.2275*	0.2331

These experimental results are counter intuitive, because we expect no improvement from CrossFire when the source and the target domains are identical – indeed, there should be no useful checkin patterns to be transferred from the source domain to the target one, as no new information has been obtained. Note that none of previous works [12,13,15,20] on *non-overlapping* CDCF reports the effectiveness of their proposed approaches when setting the source domain equal to the target domain. In response to research question RQ1, our experimental results demonstrate that the CBT-based strategy of CrossFire does not clearly contribute to the improvements of the recommendation accuracy, compared to the traditional single-domain MF-based models. Indeed, we postulate that the observed improvements (for Brightkite and Yelp) are gained from the additional parameters introduced in CrossFire (namely, W and Q in Eq. (5)), which make the CrossFire more flexible than the traditional single-domain MF-based approaches, and not by transferring knowledge from the source domain. This new evidence, for the venue recommendation task, supports the arguments of Cresmoni and Quadrana [20], namely that a CBT-based strategy cannot effectively transfer knowledge when the source and target domains do not overlap.

Next, Table 3 reports the effectiveness of single-domain and *non-overlapping* sequential-based venue recommendation frameworks: namely the Deep Recurrent Collaborative Filtering (DRCF) and the proposed extended CBT-based strategy of Deep Recurrent Transfer Learning (DRTL), respectively. In Table 3 we

Table 3. Performance in terms of HR@10 and NDCG@10 of several sequential-based venue recommendation frameworks. The best performing result in each row is highlighted in bold and $*$ indicates significant differences in terms of paired t-test with $p < 0.01$, comparing to the best performing result.

Target domain	Brightkite			
Model	DRCF	DRTL	DRTL	DRTL
Source domain	–	Brightkite	Foursquare	Yelp
HR	0.5252*	0.6975*	**0.7083**	0.7036
NDCG	0.3224*	0.5244*	**0.5341**	0.5335*
Target domain	Foursquare			
Model	DRCF	DRTL	DRTL	DRTL
Source domain	–	Brightkite	Foursquare	Yelp
HR	**0.8595**	0.8360*	0.8444*	0.8300*
NDCG	**0.7096**	0.6700*	0.6719*	0.6632*
Target domain	Yelp			
Model	DRCF	DRTL	DRTL	DRTL
Source domain	–	Brightkite	Foursquare	Yelp
HR	0.5019*	0.5496*	**0.5577**	0.5350*
NDCG	0.2858*	0.3197	**0.3215**	0.3059*

observe similar results as reported in Table 2. For example, DRTL consistently and significantly outperforms DRCF, for HR and NDCG, across the Brightkite and Yelp datasets, while DRCF is more effective than DRTL in terms of HR and NDCG on the Foursquare dataset.

In particular, when using Brightkite or Yelp as the target domain, we found that the performances of DRTL in terms of HR@10 and NDCG@10 with Foursquare as the source domain are more effective than other setups (i.e. Brightkite or Yelp as source domains). These results imply that DRTL may be able to transfer useful checkin patterns from the Foursquare dataset to enhance the quality of venue recommendation on the Brightkite and Yelp datasets. We note that the Foursquare dataset is larger than the Brightkite dataset, and hence it is possible that the checkin patterns extracted from Foursquare are reasonably useful for improving the effectiveness of recommendation system on the Brightkite dataset. Interestingly, the checkin patterns extracted from Foursquare are also useful for the Yelp dataset, perhaps due to the higher density of the checkins in the Foursquare dataset (see Table 1).

On the other hand, as postulated above for CrossFire, a possible reason for the increased effectiveness of the DRTL model is the increased parameter space allowing more flexible learned models. To investigate this further, Fig. 1 plots the number of parameters of each approach of the nine examined approaches (MF, CMF, 3x CrossFire, DRCF, 3x DRTL) versus the resulting effectiveness

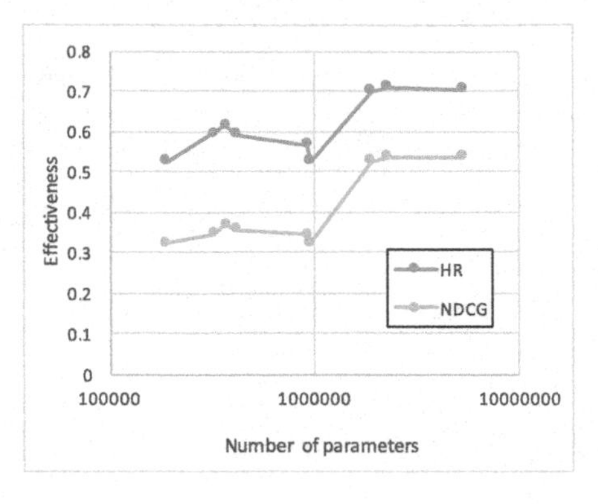

Fig. 1. Number of model parameters vs. effectiveness across all approaches, for the Brightkite dataset.

Target	HR	NDCG@10
Brightkite	0.62	0.66
Foursquare	0.53	0.67*
Yelp	0.85*	0.86*

Table 4. Spearman's ρ for $N = 9$ approaches: number of model parameters vs. effectiveness.

(HR & NDCG@10) on the Brightkite target domain. For instance, for MF, the number of parameters is defined by $m \times d + d \times n = 14374 \times 10 + 10 \times 5050 \approx 188,060$[9]. Examining the figure, some moderate correlation can easily be observed. We quantify this correlation using Spearman's ρ for each target domain and evaluation measure in Table 4. Here, we observe that in 3 out of the 6 settings, the observed correlations are significant, supporting our postulate that the increasing parameter space of the models – thereby allowing further flexibility – could explain the increasing effectiveness.

Overall, in response to research question RQ2, our experimental results demonstrate that the CBT technique of DRTL *appears to work* in the same settings that CrossFire works on. This may provide evidence that the checkin patterns extracted from the source domain that is larger than the target domain are useful for enhancing the quality of venue recommendation in sequential-based venue recommendation. However, we also provide some evidence that these improvements can be explained by the increased parameter space of the jointly-optimised transfer learning models used by the CBT technique.

5 Conclusions

In this paper, we investigate the performance of a state-of-the-art *non-overlapping* cross-domain venue recommendation framework, CrossFire, that relies on the CodeBook Transfer (CBT) technique. Moreover, we extend the state-of-the-art sequential-based venue recommendation framework to perform cross-domain venue recommendation based on CrossFire. Our comprehensive experiments on three large-scale datasets from Brightkite, Foursquare and Yelp show that the CBT-based technique does not clearly contribute to the improvements of CrossFire, compared to the traditional single-domain MF-based

[9] Recall that we remove sparse users and venues.

approaches in the context of venue recommendation. In fact, such improvements may be due to the additional parameters introduced by the CBT-based technique. Regarding sequential-based recommendation, our experiments demonstrate that the CBT-based technique can enhance the effectiveness of a state-of-the-art sequential-based venue recommendation framework, namely DRCF. In particular, the results imply that the checkin patterns extracted from the source domain that is larger than the target domain can be useful for enhancing the effectiveness of DRCF. However, we also examined the parameter space of the resulting models, which showed at least moderate correlation (significant in 3 out of 6 cases) with the resulting effectiveness, suggesting that at least some of the benefit in CBT simply arises from the increased size of the parameter spaces.

As future work, we plan to investigate a *non-overlapping* cross-domain venue recommendation strategy that can effectively transfer knowledge across two domains. We will consider users' checkin behaviours in certain regions, instead of taking into account on how users checkin on platforms in general as state-of-the-art CDCF strategies do. It is the special characteristic of the cross-domain venue recommendation task that makes the CDCF approaches less stable, since users' checkin behaviours will highly depend on the regions that the users are located. For example, we will consider users' checkin behaviour at the center of a certain city in two different platforms like Yelp and Foursquare, and then weight the transfer learning accordingly in the cross-domain venue recommendation task.

References

1. Koren, Y., Bell, R., Volinsky, C.: Matrix factorization techniques for recommender systems. Computer **8**, 30–37 (2009)
2. Ma, H., Zhou, D., Liu, C., Lyu, M.R., King, I.: Recommender systems with social regularization. In: Proceedings of WSDM (2011)
3. Manotumruksa, J., Macdonald, C., Ounis, I.: Matrix factorisation with word embeddings for rating prediction on location-based social networks. In: Jose, J.M., et al. (eds.) ECIR 2017. LNCS, vol. 10193, pp. 647–654. Springer, Cham (2017). https://doi.org/10.1007/978-3-319-56608-5_61
4. Manotumruksa, J., Macdonald, C., Ounis, I.: Regularising factorised models for venue recommendation using friends and their comments. In: Proceedings of CIKM (2016)
5. Yuan, F., Guo, G., Jose, J., Chen, L., Yu, H.: Joint geo-spatial preference and pairwise ranking for point-of-interest recommendation. In: Proceedings of ICTAI (2016)
6. Yu, F., Liu, Q., Wu, S., Wang, L., Tan, T.: A dynamic recurrent model for next basket recommendation. In: Proceedings of SIGIR (2016)
7. Tang, S., Wu, Z., Chen, K.: Movie recommendation via BLSTM. In: Amsaleg, L., Guðmundsson, G.Þ., Gurrin, C., Jónsson, B.Þ., Satoh, S. (eds.) MMM 2017. LNCS, vol. 10133, pp. 269–279. Springer, Cham (2017). https://doi.org/10.1007/978-3-319-51814-5_23
8. Zhang, Y., et al.: Sequential click prediction for sponsored search with recurrent neural networks. In: Proceedings of AAAI (2014)
9. Cheng, C., Yang, H., Lyu, M.R., King, I.: Where you like to go next: successive point-of-interest recommendation. In: Proceedings of IJCAI (2013)

10. Manotumruksa, J., Macdonald, C., Ounis, I.: A deep recurrent collaborative filtering framework for venue recommendation. In: Proceedings of CIKM (2017)
11. Li, B., Yang, Q., Xue, X.: Can movies and books collaborate? Cross-domain collaborative filtering for sparsity reduction. In: Proceedings of IJCAI (2009)
12. Zang, Y., Hu, X.: LKT-FM: a novel rating pattern transfer model for improving non-overlapping cross-domain collaborative filtering. In: Ceci, M., Hollmén, J., Todorovski, L., Vens, C., Džeroski, S. (eds.) ECML PKDD 2017. LNCS (LNAI), vol. 10535, pp. 641–656. Springer, Cham (2017). https://doi.org/10.1007/978-3-319-71246-8_39
13. Shu, K., Wang, S., Tang, J., Wang, Y., Liu, H.: Crossfire: cross media joint friend and item recommendations. In: Proceedings of WSDM (2018)
14. Farseev, A., Samborskii, I., Filchenkov, A., Chua, T.S.: Cross-domain recommendation via clustering on multi-layer graphs. In: Proceedings of SIGIR (2017)
15. Hu, L., Cao, J., Xu, G., Cao, L., Gu, Z., Zhu, C.: Personalized recommendation via cross-domain triadic factorization. In: Proceedings of WWW (2013)
16. Hu, L., Cao, L., Cao, J., Gu, Z., Xu, G., Yang, D.: Learning informative priors from heterogeneous domains to improve recommendation in cold-start user domains (2016)
17. Mirbakhsh, N., Ling, C.X.: Improving top-n recommendation for cold-start users via cross-domain information. In: Proceedings of TKDD (2015)
18. He, M., Zhang, J., Yang, P., Yao, K.: Robust transfer learning for cross-domain collaborative filtering using multiple rating patterns approximation. In: Proceedings of WSDM (2018)
19. Coyle, M., Smyth, B.: (Web search)^{shared}: social aspects of a collaborative, community-based search network. In: Nejdl, W., Kay, J., Pu, P., Herder, E. (eds.) AH 2008. LNCS, vol. 5149, pp. 103–112. Springer, Heidelberg (2008). https://doi.org/10.1007/978-3-540-70987-9_13
20. Cremonesi, P., Quadrana, M.: Cross-domain recommendations without overlapping data: myth or reality? In: Proceedings of RecSys (2014)
21. Singh, A.P., Gordon, G.J.: Relational learning via collective matrix factorization. In: Proceedings of SIGKDD (2008)
22. Rendle, S., Freudenthaler, C., Gantner, Z., Schmidt-Thieme, L.: BPR: Bayesian personalized ranking from implicit feedback. In: Proceedings of UAI (2009)
23. He, X., Zhang, H., Kan, M.Y., Chua, T.S.: Fast matrix factorization for online recommendation with implicit feedback. In: Proceedings of SIGIR (2016)
24. He, X., Liao, L., Zhang, H., Nie, L., Hu, X., Chua, T.S.: Neural collaborative filtering. In: Proceedings of WWW (2017)
25. Xiang, L., et al.: Temporal recommendation on graphs via long-and short-term preference fusion. In: Proceedings of SIGKDD, pp. 723–732. ACM (2010)
26. Lee, S., Song, S.i., Kahng, M., Lee, D., Lee, S.g.: Random walk based entity ranking on graph for multidimensional recommendation. In: Proceedings of RecSys, pp. 93–100. ACM (2011)
27. Kingma, D., Ba, J.: Adam: a method for stochastic optimization. arXiv preprint arXiv:1412.6980 (2014)

The Effect of Algorithmic Bias on Recommender Systems for Massive Open Online Courses

Ludovico Boratto[1(✉)], Gianni Fenu[2], and Mirko Marras[2]

[1] Data Science and Big Data Analytics Unit, EURECAT, C/ Bilbao 72, 08005 Barcelona, Spain
ludovico.boratto@acm.org

[2] Department of Mathematics and Computer Science, University of Cagliari, V. Ospedale 72, 09124 Cagliari, Italy
{fenu,mirko.marras}@unica.it

Abstract. Most recommender systems are evaluated on how they accurately predict user ratings. However, individuals use them for more than an anticipation of their preferences. The literature demonstrated that some recommendation algorithms achieve good prediction accuracy, but suffer from popularity bias. Other algorithms generate an item category bias due to unbalanced rating distributions across categories. These effects have been widely analyzed in the context of books, movies, music, and tourism, but contrasting conclusions have been reached so far. In this paper, we explore how recommender systems work in the context of massive open online courses, going beyond prediction accuracy. To this end, we compared existing algorithms and their recommended lists against biases related to course popularity, catalog coverage, and course category popularity. Our study remarks even more the need of better understanding how recommenders react against bias in diverse contexts.

Keywords: Recommendation · Algorithmic bias · Learning Analytics

1 Introduction

Recommender systems are reshaping online and online interactions. They learn behavioural patterns from data to support both individuals [44] and groups [22] at filtering the overwhelming alternatives our daily life offers. However, the biases in historical data might propagate in the items suggested to the users, leading to potentially undesired behavior [28]. Therefore, it is important to investigate how various biases are modelled by recommenders and affect their results [40].

Offline experiments on historical data are predominant in the field [25]. However, they often compute prediction accuracy measures that give no evidence on biased situations hidden in the recommended lists [6]. The literature is therefore going one step beyond predictive accuracy. For instance, some recommenders focus on a tiny catalog part composed by popular items, leading to popularity

L. Azzopardi et al. (Eds.): ECIR 2019, LNCS 11437, pp. 457–472, 2019.
https://doi.org/10.1007/978-3-030-15712-8_30

bias [30,39,48]. Others generate a category-wise bias because the rating distribution greatly varies across categories [26]. Historical patterns can promote social biases, such as gender discrimination in publishing [19]. In addition, prediction accuracy might not correlate to online success [14]. Recent movie and book recommenders make good rating predictions, but focus on few popular items and lack in personalization [3]. In contrast, in tourism, higher prediction accuracy corresponds to better perceived recommendations [13]. In view of these context-dependent results, inspired by [31] and recent algorithmic bias studies, assessing how recommenders manage bias in unexplored contexts becomes crucial.

Online education represents an emerging interesting field for this kind of investigation. Large-scale e-learning platforms offering Massive Open Online Courses (MOOCs) have attracted lots of participants and the interaction within them has generated a vast amount of learning-related data. Their collection, processing and analysis have promoted a significant growth of Learning Analytics [46] and have opened up new opportunities for supporting and assessing educational experiences [15,49]. The market size on this field is expected to grow from USD 2.6 billion in 2018 to USD 7.1 billion by 2023 [38]. These data-driven approaches are being viewed as a potential cure for current educational needs, such as personalization and recommendation [34]. Existing techniques mainly suggest digital educational material (e.g., slides or video-lectures) by leveraging collaborative and content-based filtering [17], while large-scale approaches for online course recommendation have been recently introduced in academia [33] and industry (e.g., Course Talk [2] and Class Central [1]). As these technologies promise to play a relevant role in personalized e-learning, the chance of introducing bias increases and any ignored bias will possibly affect a huge number of people [45]. Entirely removing any bias from algorithms is currently impracticable, but uncovering and mitigating them should be a core objective. In the e-learning recommendation context, this means putting more emphasis on the effects the algorithms have on learners rather than on prediction accuracy [20].

In this paper, we study how recommenders work in the context of MOOCs. We conducted an offline evaluation of different recommendation strategies, which took as input the ratings left by learners after attending MOOCs. We compared the courses recommended by classic and recent methods against data biases, by assessing: (*i*) how the effectiveness varies when considering algorithms that optimize the rating prediction or the items' ranking, (*ii*) how course popularity and coverage and concentration biases affect the results, and (*iii*) how popularity bias in the course categories evaluated by the learners propagates in the recommended lists. These biases might have educational implications (e.g., course popularity bias might affect knowledge diversification, while course category popularity bias might limit learner's multi-disciplinary knowledge). Our results provide evidence on the need to go beyond prediction accuracy, even in the MOOC context.

The rest of this paper is structured as follows. Section 2 presents the dataset and the recommenders. Section 3 evaluates the prediction and the ranking accuracy of the algorithms. Section 4 uncovers some biases and Sect. 5 discusses their

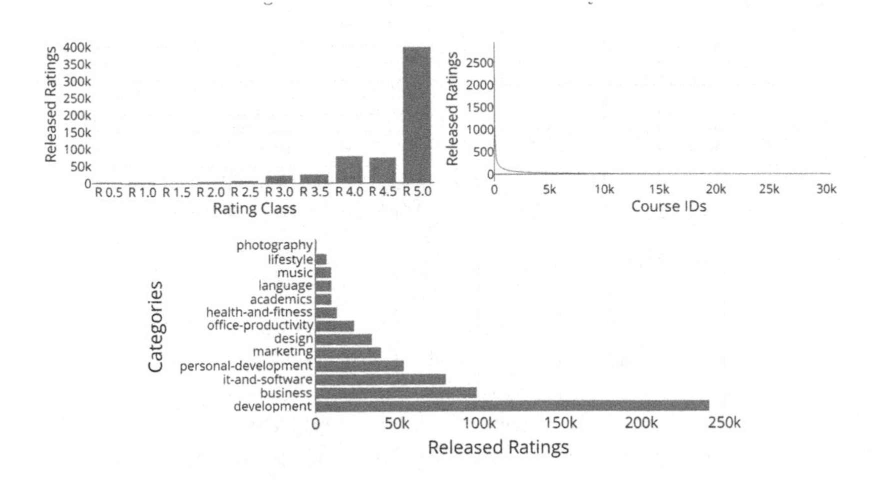

Fig. 1. Sample distributions highlighting bias on the COCO dataset. Ratings per class (top-left). Ratings per course (top-right). Ratings per category (bottom).

impact and their relation with previous studies. Finally, Sect. 6 concludes the paper. The code accompanying this paper is made publicly available[1].

2 Experimental Setup

In our experiments, we leveraged the Java recommendation framework *LibRec* [27] to evaluate several collaborative filtering algorithms on a large-scale online course dataset. Both the dataset and the algorithms are described as follows.

2.1 Dataset

To the best of our knowledge, only one dataset contains both the target MOOCs context and the data size to assess recommendation bias. COCO [16] includes information from one of the most popular course marketplaces for online learning at scale. This public dataset includes 43K courses, distributed into a taxonomy of 15 first-level categories. Over 4M learners provided 6M 5-star ratings and 2M textual reviews. To maintain the evaluation computationally tractable, we took only learners who released at least 10 ratings. The re-sampled dataset includes 37K users, who gave 600K ratings to 30K courses. Figure 1 shows in detail the biases in the dataset towards positive rating classes (common also for learning objects [21]), course popularity and course category popularity.

[1] The code accompanying this paper can be downloaded at http://bit.ly/2AEban5.

2.2 Algorithms

We focused on collaborative filtering due to its popularity also in e-learning contexts [9,17]. We ranged from K-Nearest-Neighbours (KNN) to Learning-to-Rank (LTR) approaches. It should be noted that while we can generate a ranking of the items a user has not evaluated yet by predicting missing ratings, LTR methods are optimized to maximize the ranking quality, generating diverse recommendations against prediction-based algorithms. The algorithms are described below.

Non-Personalized (NP) baselines:
- *Random*: randomly recommending items;
- *MostPop*: recommending the most frequently-consumed items;
- *ItemAvg*: recommending the items with the highest average rating;
- *UserAvg*: recommending the items with the highest user average rating.

Standard Collaborative Filtering (SCF) algorithms:
- *ItemKNN*: item-based collaborative filter (Cosine, K-NN, $k = 100$);
- *UserKNN*: user-based collaborative filter (Cosine, K-NN, $k = 100$).

Matrix Factorization (MF) methods:
- *SVD++*: gradient descent matrix factorization ($LatentFactors = 40$) [36];
- *WRMF*: weighted regular matrix factorization ($LatentFactors = 40$)[29].

Learning-To-Rank (LTR) algorithms:
- *AoBPR*: a variant of BPR manipulating uniform sampling pairs [42];
- *BPR*: bayesian personalized ranking technique for implicit feedback [43];
- *Hybrid*: hybrid integrating diversity and accuracy-focused approaches [50];
- *LDA*: a filtering approach leveraging Latent Dirichlet Allocation [24].

The algorithms were selected as a representative sample as done in other related studies [31]. In what follows, each method is identified by its short name.

3 Comparing Prediction and Ranking Effectiveness

First, we evaluate the recommendation effectiveness, considering metrics that evaluate rating prediction accuracy against those that measure the ranking quality. Like in similar studies [18], we employed a 5-fold cross validation based on a user-sampling strategy. We split the users in five test sets. Each set was the test set of a given fold. In each fold, for each user in the corresponding test set, we selected 5 ratings to be the test ratings, while the rest of their ratings and all the ratings from users not in that test set were the train ratings. Each algorithm was run in both rating prediction and top-10 item ranking mode. We chose top-10 recommendations since they probably get the most attention and 10 is a widely employed cut-off [44]. Root Mean Squared Error (RMSE) evaluated the accuracy of the rating predictions (i.e., the lower the better). Area Under the Curve (AUC), precision, recall, and Normalized Discounter Cumulative Gain (NDCG) [32] measured the recommended list accuracy (i.e., the higher the better).

Table 1. The accuracy of the algorithms on rating prediction (RMSE) and top-10 ranking (AUC, Precision, Recall, NDCG). The results are sorted by increasing RMSE.

Family	Method	RMSE	AUC	Prec@10	Rec@10	NDCG
MF	SVD++	**0.68**	0.50	0.005	0.001	0.008
NP	UserAvg	0.70	0.50	0.004	0.007	0.005
SCF	UserKNN	0.71	0.68	0.050	0.101	0.095
SCF	ItemKNN	0.76	0.69	0.051	0.102	0.092
NP	ItemAvg	0.78	0.50	0.005	0.008	0.005
NP	MostPop	1.07	0.60	0.023	0.046	0.038
LTR	BPR	2.08	0.69	0.054	0.109	0.094
LTR	AoBPR	2.34	0.69	0.054	0.108	0.094
NP	Random	2.36	0.50	0.004	0.008	0.005
LTR	LDA	4.11	0.66	0.042	0.085	0.074
LTR	Hybrid	4.11	0.55	0.018	0.037	0.029
MF	WRMF	4.12	**0.71**	**0.062**	**0.124**	**0.114**

Table 1 shows the results. The best ones are printed in bold in case they were significantly different from all others. In this paper, we used paired two-tailed Student's t-tests with a $p = 0.05$ significance level. The MF approach SVD++ significantly outperformed all the other schemes. However, the rather simple non-personalized UserAvg yielded comparable accuracy to SVD++ and was better than other computationally expensive schemes like ItemKNN and BPR. The latter was significantly better than the other LTR approaches. ItemAvg, which simply considers an item's average rating, achieved results in line with ItemKNN. The WRMF method performed, somewhat surprisingly, worse than a lot of the traditional ones. The ranking of the algorithms on RMSE is not consistent with respect to other contexts [31]. This confirms that the dataset characteristics like size, sparsity, and rating distributions can greatly affect the recommendation accuracy [5]. The results on item ranking led to a completely different algorithm ranking. BPR and AoBPR achieved the best performance together with WRMF and KNN. Except Hybrid, the LTR methods performed consistently better than MostPop. In line with the results in [35] for learning object recommendation, MostPop performed quite poorly, probably due to the wide range of categories included in the dataset. Although Item-KNN is rather simple, it performed much better than almost all the NP baselines and reached results comparable to LTR schemes. SVD++ led to mediocre results, while it was the best method in rating prediction. In contrast, WRMF achieved the highest accuracy in this setup.

While the accuracy of some algorithms is almost equal, the top-10 lists greatly varied. In view of these differences, we calculated the average overlap of courses recommended by each pair of algorithms to the same user (Fig. 2). The overlap is low, except for (WRMF, UserKNN), (UserKNN, LDA), and (AoBPR, BPR),

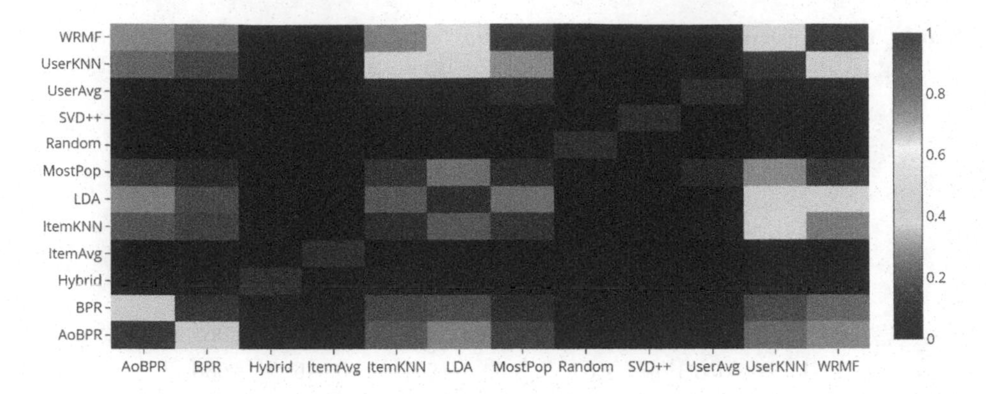

Fig. 2. The average overlap per user between the top-10 lists recommended by each pair of algorithms. The hotter the color of the rectangle the higher the overlap.

where the agreement was above 35%. Hybrid and SVD++ recommended courses which are not typically proposed by the other algorithms, for example. Since MostPop works well and has some similar recommendations with respect to other algorithms, it is possible that they also tend to recommend popular courses.

4 Uncovering Bias in Course Recommendation Rankings

This section includes the experimental comparison of the algorithms and their recommended lists against different bias causes: course popularity and the related coverage and concentration, and course category popularity.

4.1 Interacting with Course Popularity Bias

Even though it is often assumed that recommending what is popular helps high-quality content emerge, popularity can bias future success without reflecting that hidden quality. First, there could be social influence among learners and a lack of independence. Second, engagement and popularity metrics could be subjected to manipulation by fake reviews or social bots. Third, the cost of learning how to evaluate quality could lead to courses with boundless popularity irrespective of differences in quality. Fourth, long-tail courses could be often desirable for more personalized recommendations and knowledge diversification among learners, and are important for generating a better understanding of learners' preferences. Moreover, long-tail recommendation can drive markets and social good. Suffering from popularity bias could impede novel courses from rising to the top and the market could be dominated by a few large institutions or well-known teachers. With this in mind, we explored how the course popularity in data influences the algorithms. We evaluated how popular are the courses provided by an algorithm, in order to assess its capability to suggest relevant but not popular ones.

Table 2. The popularity of the recommended items based on the average rating and the average number of ratings. The algorithms are sorted by decreasing average rating.

Family	Algorithm	Avg./Std. Dev. rating	Avg./Std. Dev. number of ratings
MF	SVD++	**4.76/0.21**	134/267
NP	MostPop	4.71/0.07	**1545/588**
NP	ItemAvg	4.70/0.42	15/3
MF	WRMF	4.68/0.17	404/393
LTR	LDA	4.64/0.14	586/515
SCF	UserKNN	4.63/0.21	192/296
NP	UserAvg	4.60/0.20	341/524
LTR	AoBPR	4.58/0.25	71/152
SCF	ItemKNN	4.55/0.23	88/168
LTR	BPR	4.55/0.27	67/144
NP	Random	4.47/0.58	20/73
LTR	Hybrid	4.44/0.72	11/57

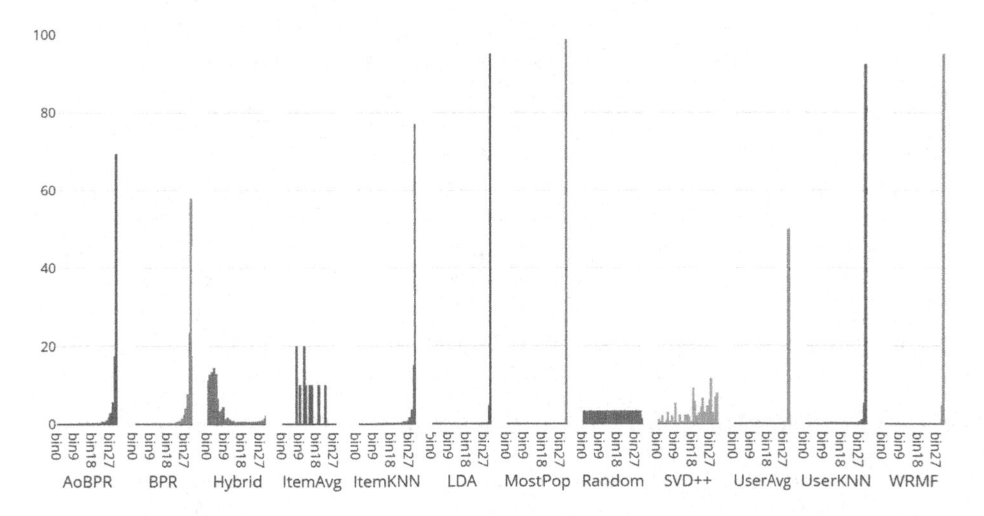

Fig. 3. The distribution of the recommended courses with respect to all the courses in the catalog grouped into 31 bins with 1000 courses each. X-axis shows the bins ranked by increasing popularity in the dataset. Y-axis shows the percentage of the recommended courses belonging to each bin.

Table 2 presents the popularity of the recommended courses as the number of ratings they received. MostPop has, by design, the highest average popularity, since it recommends best sellers. The recommended courses received about 1,500 ratings on average. LDA and WRMF also showed a popularity bias, with

586 and 404 ratings per recommended course, respectively. On the other hand, some algorithms are not biased towards course popularity. SVD++, ItemKNN, AoBPR, and BPR recommended a lot of courses from the long tail. Interestingly, only Hybrid recommended niche and unpopular courses, and its average number of ratings (11) is lower than the average number of ratings per course in the catalog (20). NP baselines achieved a good trade-off between popular and less popular courses. To obtain a detailed picture, we sorted the courses according to the number of ratings in the dataset and organized them in bins of 1000 courses (Fig. 3); the first bin contains the least rated courses, while subsequent ones consider courses of increasing popularity. Except Hybrid, Random, and SVD++, all the algorithms often recommended courses from the bin of the most popular ones ($bin30$). In BPR, course popularity seems to be directly related with the chance of being recommended. SVD++ and Hybrid seem to be good options to recommend niche courses. Interestingly, Hybrid tends to recommend more unpopular courses than popular ones. For ItemAvg, the plot is a rough indicator, since its histogram is based on a small number of recommended courses.

Receiving a lot of ratings does not imply people liked a course. The correlation between number of ratings and average rating is weak, 0.11. Therefore, we measured the average rating of a course as another popularity indicator. It does not tell if the course is really liked by a large number of people, but it can help to see if some algorithms tend to concentrate on highly-rated and probably less-known courses. Table 2 shows that a lot of algorithms recommend courses that were rated, on average, above 4.44 (the global average is 4.47). Furthermore, some algorithms (i.e., SVD++, PopRank, and WRMF) recommended a lot of courses with a high average rating, and low-rated courses are rarely recommended. LDA focuses on high-rated courses (4.64) and is significantly different from other LTR methods. For algorithms not optimized for rating prediction, the average rating is comparably low and closer to the global average. This means that they do not take the average rating into account and recommended also low-rated courses. These algorithms might recommend controversial courses. The average rating of the MostPop recommendations is 4.71, so well-known courses are also top-rated.

4.2 Exploring Bias on Catalog Coverage and Concentration

To check if the recommender system is guiding users to long-tail or niche courses, we should count how many courses in the catalog are recommended. Hence, we looked at the course space coverage and concentration effects of the algorithms.

We counted the number of different courses appearing in the lists (Table 3). The results show that the coverage can be quite different across the algorithms. Except Random, only Hybrid recommend more courses than all other techniques, almost half of the whole catalog. This is in line with the idea behind Hybrid: balancing diversity and rating prediction accuracy. However, in our context, we found it achieved good diversity, but low prediction accuracy. Other LTR approaches provided a coverage of around 20%, except LDA (1%). KNN methods showed a limited catalog coverage, confirming the results in [47] for learning

Table 3. The catalog coverage per algorithm out of 30.399 courses. GINI indexes are computed for the ratings per course distributions in the recommended lists.

Family	Algorithm	Coverage	Catalog percentage	Gini index
NP	Random	**30399**	**100.00**	**0.16**
LTR	Hybrid	12735	41.90	0.77
LTR	BPR	6514	21.43	0.85
LTR	AoBPR	5857	19.27	0.89
SCF	ItemKNN	4653	15.31	0.89
SCF	UserKNN	1183	3.89	0.89
MF	SVD++	1121	3.68	0.88
MF	WRMF	457	1.50	0.68
LTR	LDA	200	0.65	0.64
NP	MostPop	29	0.09	0.63
NP	UserAvg	14	0.04	0.17
NP	ItemAvg	12	0.04	0.28

objects. In contrast to the learning object scenario [37], the algorithms performing best on prediction accuracy are not the best ones also for the catalog coverage. These differences went unnoticed if only the accuracy was considered.

Catalog coverage does not reveal how often each course was recommended. Thus, we captured inequalities with respect to how frequently the courses appeared. For each course suggested by an algorithm, we counted how often it is contained in the lists of that algorithm. The courses are sorted in descending order, according to the times they appeared in the lists, and grouped in bins of 10 courses. $Bin1$ contains the most recommended courses. Figure 4 shows the four bins (out of 3040) with the 40 most frequently recommended courses. The Y-axis shows the percentage of recommendations the algorithm has given for the courses in the corresponding bin with respect to the total number of suggestions provided by that algorithm. While SVD++ and ItemKNN recommended a number of different courses, most of them were rarely proposed. BPR, AoBPR, and WRMF, which had a good catalog coverage, provided about 20% of the courses from the 40 most often recommended ones. In Table 3, we show the Gini index to observe the inequality with respect to how often certain courses are recommended, where 0 means equal distribution and 1 corresponds to maximal inequality [25]. Except for the NP baselines, Hybrid and BPR have the weakest concentration bias. Compared to BPR, Hybrid's Gini index is significantly lower, showing a more balanced distribution of recommendations among courses.

4.3 Exposing Course Category Popularity Bias

E-learning recommender systems are often equipped with a taxonomy that associates each course with one or more categories. This attribute does not imply the

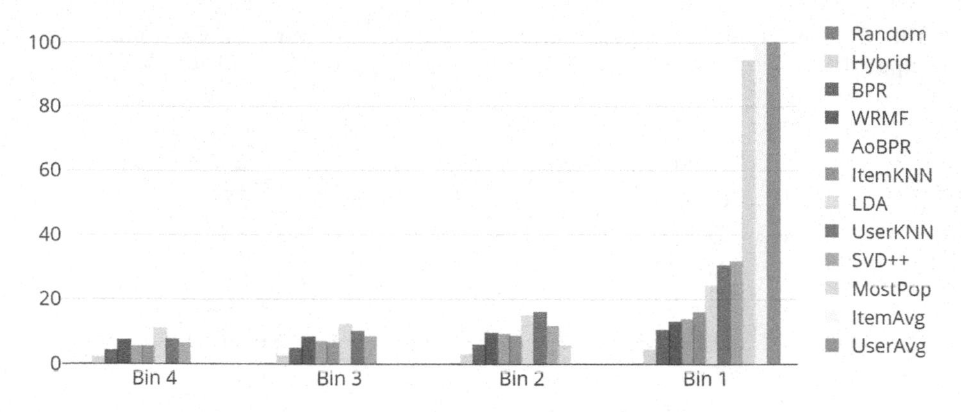

Fig. 4. The distribution of the number of recommendations for the 40 most recommended courses for each algorithm, grouped into 4 bins with 10 courses each. Each coloured column in X-axis is associated to an algorithm. For each algorithm within a bin, Y-axis shows the percentage of recommendations for the courses in the corresponding bin with respect to the total recommendations provided by that algorithm.

quality of a course, but the distribution of the number of ratings can greatly vary across categories. Nonetheless it is natural, given by the heterogeneity of users and courses, it makes aggregated ratings commonly used by algorithms incomparable across categories and thus prone to bias issues. The course category popularity bias inherits large part of the drawbacks held by global popularity bias, and could even influence how learners perceive the recommendations as useful for deepening the knowledge in a preferred category or for fostering a multi-disciplinary knowledge in unexplored categories. Therefore, we focused on the popularity of the category to which courses belong and how the popularity bias affecting course categories in data propagates in the recommended lists.

We counted how many different course categories appeared in the lists. UserAvg exhibited only 3 out of 13 different categories, while MostPop and ItemAvg recommended 5 and 8 categories, respectively. Except for LDA (10 categories), all the other algorithms provided a full coverage on categories. To obtain a clear picture, we sorted the 13 categories according to their increasing number of ratings in the dataset. *Bin12* represents the most popular category. For each algorithm, we counted how many recommendations per category were provided in the recommended lists. Figure 5 shows the distribution of the recommendations per category. BPR, ItemKNN, LDA, and WRMF showed a bias to the most popular category. More than 50% of their recommendations came from it. Hybrid and SVD++ offered a more uniform distribution across categories.

In this context, it was also important to measure how much each algorithm reinforces or reduces the bias to a given category. Figure 6 shows the bias related to course category popularity. Each rectangle shows the increment/decrement on the recommended courses per category with respect to the ratings per category in the dataset. Considering that "development" is the most popular category in COCO, when producing recommendations, MostPop reinforces its popularity by 50%.

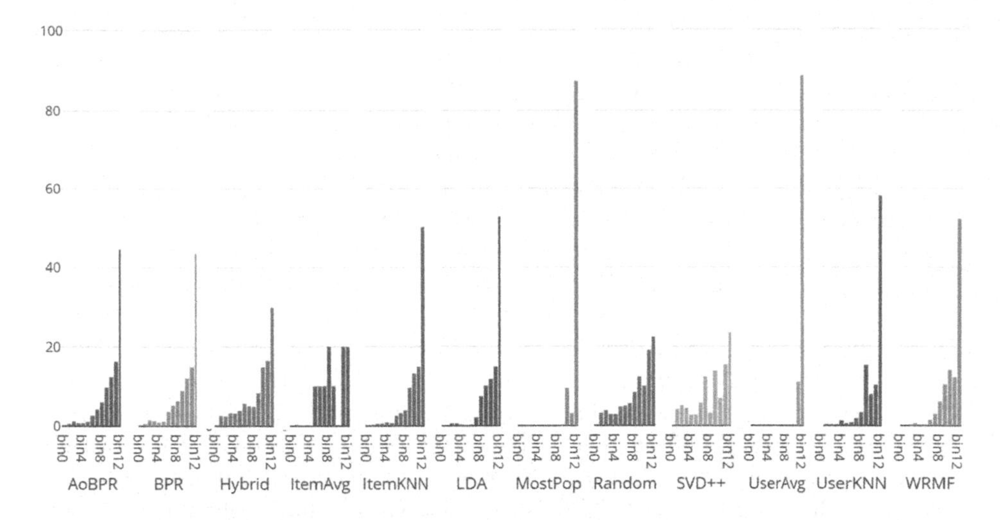

Fig. 5. The distribution of the recommended courses with respect to the course categories. X-axis shows the category bins ranked by their increasing popularity in the dataset. Y-axis shows the relative frequency of the recommended items for each bin.

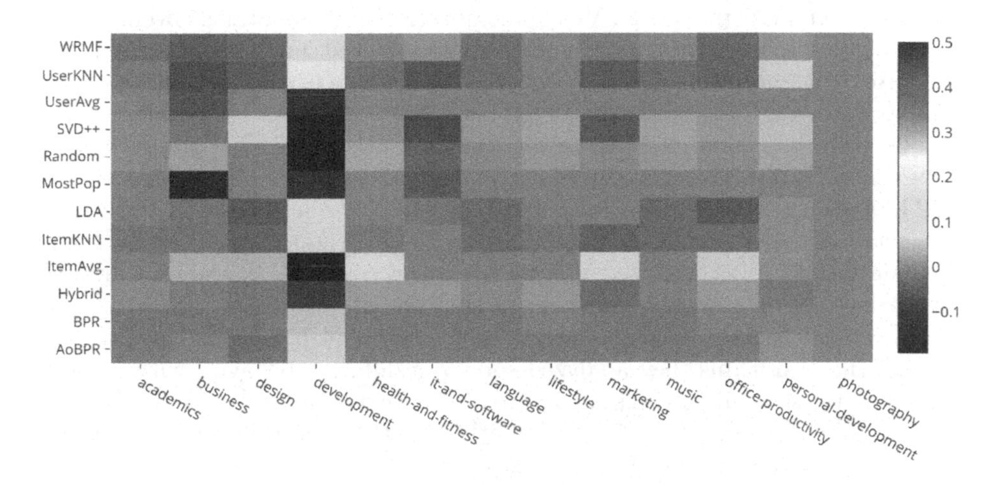

Fig. 6. The reinforcement produced by algorithms with respect to course categories. The hotter the color the higher the reinforcement of that algorithm to that category.

On the other hand, Hybrid and SVD++ caused a 10% popularity reduction in courses of this category. Hence, their recommendations can potentially meet the needs of those not interested only in "development" courses.

5 Discussion

The differences in terms of RMSE or NDCG between some algorithms are very small. For instance, the best-performing techniques, SVD++ and UserAvg, have a difference of 0.02 in RMSE and the same happens for ItemKNN and ItemAvg or LDA, Hybrid, and WRMF. The algorithm ranking based on their item ranking accuracy was quite different, going in contrast with the observations made by [31] for movies (i.e., SVD++ was the best algorithm for both prediction and ranking). However, the analysis regarding catalog coverage and concentration showed that the differences between the algorithms can be more marked and contrasting with respect to the ones reached for prediction accuracy. If the goal is to point learners to different areas of the course catalog, the choice should not be based on accuracy alone. In fact, Hybrid did not perform well on prediction accuracy, but covered half of the catalog and is less influenced by concentration bias. In contrast, SVD++ had a better prediction accuracy, but recommended only 4% of the courses. AoBPR, BPR, and ItemKNN significantly outperformed SVD++ in catalog coverage, even though they achieved a poor accuracy on rating prediction. KNN methods tended to reinforce the bias towards popular courses, as shown by [10] for music. Considering course categories, Hybrid achieved the best trade-off between catalog coverage and category distribution in the recommended lists. While SVD++ demonstrated a low catalog coverage, the recommended courses were more uniformly distributed among categories w.r.t. BPR, AoBPR, and ItemKNN. The latter suggested lots of courses from the most popular category. This went unnoticed if we only considered the catalog coverage. Overall, no algorithm was better than the others, but we observed that Hybrid and SVD++ reached the best trade-off across all the dimensions.

The recommender system community has long been interested in this social dimension of recommendation; similarly to us, representative studies that highlight algorithmic bias analyzed accuracy, catalog coverage, concentration, and popularity bias on several algorithms in the contexts of movies, music, books, social network, hotels, games, and research articles [4,12,30,31,41]. However, some of the algorithms they analyzed showed a different behavior with respect to the one the same algorithms showed in our context. Category-wise biases have been studied on movies data [26]. Differently from them, we went in-depth on the distribution of the recommended courses with respect to the course categories and highlighted the reinforcement generated by the algorithms on popular categories. Conversely, other works analyzed fairness on users' attributes, such as gender on books and gender with age on movies and music [18,19]. Popularity and diversity biases at user profile level have been recently considered [11].

6 Conclusions and Future Work

In this paper, we analyzed existing recommendation algorithms in terms of their predictive accuracy on course ratings and rankings in the context of MOOCs. Then, through a series of experiments, we demonstrated that, despite comparably minor differences with respect to accuracy, the algorithms can be quite different on which courses they recommend. Moreover, they can exhibit possibly undesired biases and consequent educational implications. Offline analysis cannot replace user studies, but our work can provide a better understanding on how generalizable state-of-the-art recommenders are to new contexts, in our case to MOOCs. Furthermore, it can foster more learner-oriented evaluations of the recommenders applied to MOOCs, going beyond classical prediction accuracy.

In next steps, we plan to investigate more algorithms, such as content-based recommenders, and exploit the semantics of content, such as course descriptions or learners' reviews [7,8,23]. Moreover, we will consider other types of bias related to demographic attributes and user profiles, as examples. Then, we will design context-specific countermeasures to the biases we have uncovered.

Acknowledgments. Mirko Marras gratefully acknowledges Sardinia Regional Government for the financial support of his PhD scholarship (P.O.R. Sardegna F.S.E. Operational Programme of the Autonomous Region of Sardinia, European Social Fund 2014–2020, Axis III "Education and Training", TG 10, PoI 10ii, SG 10.5).

This work has been partially supported by the Italian Ministry of Education, University and Research under the programme "Smart Cities and Communities and Social Innovation" during "ILEARNTV, Anytime, Anywhere" Project (DD n.1937 05.06.2014, CUP F74G14000200008 F19G14000910008), and by the Agència per a la Competivitat de l'Empresa, ACCIÓ, under "AlgoFair" Project.

References

1. Class Central. https://www.class-central.com/. Accessed 17 Jan 2019
2. Coursetalk. https://www.coursetalk.com/. Accessed 17 Jan 2019
3. Abdollahpouri, H., Burke, R., Mobasher, B.: Controlling popularity bias in learning-to-rank recommendation. In: Proceedings of the Eleventh ACM Conference on Recommender Systems, pp. 42–46. ACM (2017)
4. Adamopoulos, P., Tuzhilin, A., Mountanos, P.: Measuring the concentration reinforcement bias of recommender systems. rN (i) **1**, 2 (2015)
5. Adomavicius, G., Bockstedt, J., Curley, S., Zhang, J.: De-biasing user preference ratings in recommender systems. In: Joint Workshop on Interfaces and Human Decision Making in Recommender Systems, p. 2 (2014)
6. Bellogín, A., Castells, P., Cantador, I.: Statistical biases in information retrieval metrics for recommender systems. Inf. Retrieval J. **20**(6), 606–634 (2017)
7. Boratto, L., Carta, S., Fenu, G., Saia, R.: Using neural word embeddings to model user behavior and detect user segments. Knowl. Based Syst. **108**, 5–14 (2016)
8. Boratto, L., Carta, S., Fenu, G., Saia, R.: Semantics-aware content-based recommender systems: design and architecture guidelines. Neurocomputing **254**, 79–85 (2017)
9. Cechinel, C., Sicilia, M.Á., SáNchez-Alonso, S., GarcíA-Barriocanal, E.: Evaluating collaborative filtering recommendations inside large learning object repositories. Inf. Process. Manag. **49**(1), 34–50 (2013)
10. Celma, Ò., Cano, P.: From hits to niches? Or how popular artists can bias music recommendation and discovery. In: Proceedings of the 2nd KDD Workshop on Large-Scale Recommender Systems and the Netflix Prize Competition, p. 5. ACM (2008)

11. Channamsetty, S., Ekstrand, M.D.: Recommender response to diversity and popularity bias in user profiles. In: Proceedings of the Thirtieth International Florida Artificial Intelligence Research Society Conference, FLAIRS 2017, Marco Island, Florida, USA, 22–24 May 2017, pp. 657–660 (2017). https://aaai.org/ocs/index.php/FLAIRS/FLAIRS17/paper/view/15524
12. Collins, A., Tkaczyk, D., Aizawa, A., Beel, J.: Position bias in recommender systems for digital libraries. In: Chowdhury, G., McLeod, J., Gillet, V., Willett, P. (eds.) iConference 2018. LNCS, vol. 10766, pp. 335–344. Springer, Cham (2018). https://doi.org/10.1007/978-3-319-78105-1_37
13. Cremonesi, P., Garzotto, F., Turrin, R.: User-centric vs. system-centric evaluation of recommender systems. In: Kotzé, P., Marsden, G., Lindgaard, G., Wesson, J., Winckler, M. (eds.) INTERACT 2013. LNCS, vol. 8119, pp. 334–351. Springer, Heidelberg (2013). https://doi.org/10.1007/978-3-642-40477-1_21
14. Cremonesi, P., Koren, Y., Turrin, R.: Performance of recommender algorithms on top-n recommendation tasks. In: Proceedings of the Fourth ACM Conference on Recommender Systems, pp. 39–46. ACM (2010)
15. Dessì, D., Fenu, G., Marras, M., Recupero, D.R.: Bridging learning analytics and cognitive computing for big data classification in micro-learning video collections. Comput. Hum. Behav. **92**, 468–477 (2018)
16. Dessì, D., Fenu, G., Marras, M., Reforgiato Recupero, D.: COCO: semantic-enriched collection of online courses at scale with experimental use cases. In: Rocha, Á., Adeli, H., Reis, L.P., Costanzo, S. (eds.) WorldCIST'18 2018. AISC, vol. 746, pp. 1386–1396. Springer, Cham (2018). https://doi.org/10.1007/978-3-319-77712-2_133
17. Drachsler, H., Verbert, K., Santos, O.C., Manouselis, N.: Panorama of recommender systems to support learning. In: Ricci, F., Rokach, L., Shapira, B. (eds.) Recommender Systems Handbook, pp. 421–451. Springer, Boston (2015). https://doi.org/10.1007/978-1-4899-7637-6_12
18. Ekstrand, M.D., et al.: All the cool kids, how do they fit in? Popularity and demographic biases in recommender evaluation and effectiveness. In: Conference on Fairness, Accountability and Transparency, pp. 172–186 (2018)
19. Ekstrand, M.D., Tian, M., Kazi, M.R.I., Mehrpouyan, H., Kluver, D.: Exploring author gender in book rating and recommendation. In: Proceedings of the 12th ACM Conference on Recommender Systems, pp. 242–250. ACM (2018)
20. Erdt, M., Fernández, A., Rensing, C.: Evaluating recommender systems for technology enhanced learning: a quantitative survey. IEEE Trans. Learn. Technol. **8**(4), 326–344 (2015)
21. Farzan, R., Brusilovsky, P.: Encouraging user participation in a course recommender system: an impact on user behavior. Comput. Hum. Behav. **27**(1), 276–284 (2011)
22. Felfernig, A., Boratto, L., Stettinger, M., Tkalčič, M.: Group Recommender Systems: An Introduction. SECE. Springer, Cham (2018). https://doi.org/10.1007/978-3-319-75067-5
23. Fenu, G., Nitti, M.: Strategies to carry and forward packets in VANET. In: Cherifi, H., Zain, J.M., El-Qawasmeh, E. (eds.) DICTAP 2011. CCIS, vol. 166, pp. 662–674. Springer, Heidelberg (2011). https://doi.org/10.1007/978-3-642-21984-9_54
24. Griffiths, T.: Gibbs sampling in the generative model of latent Dirichlet allocation (2002)
25. Gunawardana, A., Shani, G.: Evaluating recommender systems. In: Ricci, F., Rokach, L., Shapira, B. (eds.) Recommender Systems Handbook, pp. 265–308. Springer, Boston (2015). https://doi.org/10.1007/978-1-4899-7637-6_8

26. Guo, F., Dunson, D.B.: Uncovering systematic bias in ratings across categories: a Bayesian approach. In: Proceedings of the 9th ACM Conference on Recommender Systems, pp. 317–320. ACM (2015)
27. Guo, G., Zhang, J., Sun, Z., Yorke-Smith, N.: LibRec: a Java library for recommender systems. In: UMAP Workshops, vol. 4 (2015)
28. Hajian, S., Bonchi, F., Castillo, C.: Algorithmic bias: from discrimination discovery to fairness-aware data mining. In: Proceedings of the 22nd ACM SIGKDD International Conference on Knowledge Discovery and Data Mining, pp. 2125–2126. ACM (2016)
29. Hu, Y., Koren, Y., Volinsky, C.: Collaborative filtering for implicit feedback datasets. In: Eighth IEEE International Conference on Data Mining, ICDM 2008, pp. 263–272. IEEE (2008)
30. Jannach, D., Kamehkhosh, I., Bonnin, G.: Biases in automated music playlist generation: a comparison of next-track recommending techniques. In: Proceedings of the 2016 Conference on User Modeling Adaptation and Personalization, pp. 281–285. ACM (2016)
31. Jannach, D., Lerche, L., Kamehkhosh, I., Jugovac, M.: What recommenders recommend: an analysis of recommendation biases and possible countermeasures. User Model. User-Adap. Inter. **25**(5), 427–491 (2015)
32. Järvelin, K., Kekäläinen, J.: Cumulated gain-based evaluation of IR techniques. ACM Trans. Inf. Syst. (TOIS) **20**(4), 422–446 (2002)
33. Jing, X., Tang, J.: Guess you like: course recommendation in MOOCs. In: Proceedings of the International Conference on Web Intelligence, pp. 783–789. ACM (2017)
34. Klašnja-Milićević, A., Vesin, B., Ivanović, M., Budimac, Z., Jain, L.C.: Recommender systems in E-learning environments. E-learning Systems. ISRL, vol. 112, pp. 51–75. Springer, Cham (2017). https://doi.org/10.1007/978-3-319-41163-7_6
35. Kopeinik, S., Kowald, D., Lex, E.: Which algorithms suit which learning environments? A comparative study of recommender systems in TEL. In: Verbert, K., Sharples, M., Klobučar, T. (eds.) EC-TEL 2016. LNCS, vol. 9891, pp. 124–138. Springer, Cham (2016). https://doi.org/10.1007/978-3-319-45153-4_10
36. Koren, Y.: Factorization meets the neighborhood: a multifaceted collaborative filtering model. In: Proceedings of the 14th ACM SIGKDD International Conference on Knowledge Discovery and Data Mining, pp. 426–434. ACM (2008)
37. Manouselis, N., Vuorikari, R., Van Assche, F.: Collaborative recommendation of E-learning resources: an experimental investigation. J. Comput. Assist. Learn. **26**(4), 227–242 (2010)
38. MarketsandMarkets: Education and learning analytics market report (2018). https://www.marketsandmarkets.com/Market-Reports/learning-analytics-market-219923528.html
39. Nagatani, K., Sato, M.: Accurate and diverse recommendation based on users' tendencies toward temporal item popularity (2017)
40. Olteanu, A., Castillo, C., Diaz, F., Kiciman, E.: Social data: biases, methodological pitfalls, and ethical boundaries (2016)
41. Pampın, H.J.C., Jerbi, H., O'Mahony, M.P.: Evaluating the relative performance of collaborative filtering recommender systems. J. Univ. Comput. Sci. **21**(13), 1849–1868 (2015)
42. Rendle, S., Freudenthaler, C.: Improving pairwise learning for item recommendation from implicit feedback. In: Proceedings of the 7th ACM International Conference on Web Search and Data Mining, pp. 273–282. ACM (2014)

43. Rendle, S., Freudenthaler, C., Gantner, Z., Schmidt-Thieme, L.: BPR: Bayesian personalized ranking from implicit feedback. In: Proceedings of the Twenty-Fifth Conference on Uncertainty in Artificial Intelligence, pp. 452–461. AUAI Press (2009)
44. Ricci, F., Rokach, L., Shapira, B.: Recommender systems: introduction and challenges. In: Ricci, F., Rokach, L., Shapira, B. (eds.) Recommender Systems Handbook, pp. 1–34. Springer, Boston (2015). https://doi.org/10.1007/978-1-4899-7637-6_1
45. Selwyn, N.: Data entry: towards the critical study of digital data and education. Learn. Media Technol. **40**(1), 64–82 (2015)
46. Siemens, G., Long, P.: Penetrating the fog: analytics in learning and education. EDUCAUSE Rev. **46**(5), 30 (2011)
47. Verbert, K., Drachsler, H., Manouselis, N., Wolpers, M., Vuorikari, R., Duval, E.: Dataset-driven research for improving recommender systems for learning. In: Proceedings of the 1st International Conference on Learning Analytics and Knowledge, pp. 44–53. ACM (2011)
48. Wasilewski, J., Hurley, N.: Are you reaching your audience? Exploring item exposure over consumer segments in recommender systems. In: Proceedings of the 26th Conference on User Modeling, Adaptation and Personalization, pp. 213–217. ACM (2018)
49. Xing, W., Chen, X., Stein, J., Marcinkowski, M.: Temporal predication of dropouts in MOOCs: reaching the low hanging fruit through stacking generalization. Comput. Hum. Behav. **58**, 119–129 (2016)
50. Zhou, T., Kuscsik, Z., Liu, J.G., Medo, M., Wakeling, J.R., Zhang, Y.C.: Solving the apparent diversity-accuracy dilemma of recommender systems. Proc. Nat. Acad. Sci. **107**(10), 4511–4515 (2010)

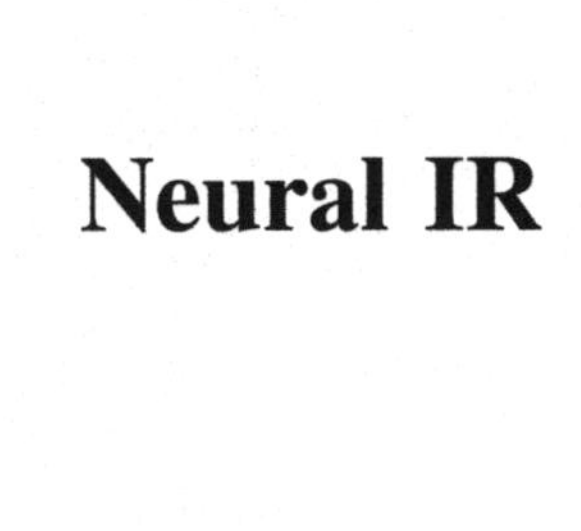

Neural IR

Domain Adaptive Neural Sentence Compression by Tree Cutting

Litton J. Kurisinkel[1(✉)], Yue Zhang[2], and Vasudeva Varma[1]

[1] International Institute of Information Technology, Hyderabad, Hyderabad 500031, India
litton.jKurisinkel@research.iiit.ac.in, vv@iiit.ac.in
[2] Westlake Institute for Advanced Study, West Lake University, Hangzhou, Zhejiang Province, China
yue.zhang@wias.org.cn

Abstract. Sentence compression has traditionally been tackled as syntactic tree pruning, where rules and statistical features are defined for pruning less relevant words. Recent years have witnessed the rise of neural models without leveraging syntax trees, learning sentence representations automatically and pruning words from such representations. We investigate syntax tree based noise pruning methods for neural sentence compression. Our method identifies the most informative regions in a syntactic dependency tree by self attention over context nodes and maximum density subtree extraction. Empirical results show that the model outperforms the state-of-the-art methods in terms of both accuracy and F1-measure. The model also yields a comparable accuracy in readability and informativeness as assessed by human evaluators.

Keywords: Sentence summarization · Sentence compression · Syntactic tree pruning

1 Introduction

Sentence simplification is the task of deriving a noise-free condensation that holds the most abstract information from a noisy complex sentence. The task is useful for solving NLP problems such as summarization. Significant amount of research has been done on the task. Extractive methods [3,8,12] remove the noisy portions of a sentence while leaving the result grammatically correct (Fig. 1). Abstractive techniques [2] construct a semantic representation for the original sentence and generates the most informative content in its own learnt writing style.

Among extractive method for sentence compression, traditional method leverage statistical m models by exploiting useful manual features to prune noisy regions of a syntactic tree [5]. Recently, neural network has been used for the task [2,6,18]. The basic idea is to leverage a recurrent neural network to automatically extract syntactic and semantic features from a sentence, before classifying

L. Azzopardi et al. (Eds.): ECIR 2019, LNCS 11437, pp. 475–488, 2019.
https://doi.org/10.1007/978-3-030-15712-8_31

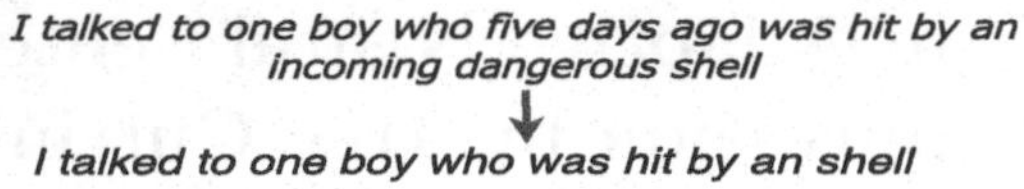

I talked to one boy who five days ago was hit by an incoming dangerous shell

↓

I talked to one boy who was hit by an shell

Fig. 1. Sentence compression

whether each word can be removed. Such methods have shown highly competitive accuracies compared with traditional statistical techniques, without the need to define manual features.

On the other hand, neural methods without syntax typically require a large amount of training data to avoid overfitting, and can generalize poorly across domains. For example, [2] trained their model on Gigaword [15], which consists of 3.8 million article-title pairs. Apparently, datasets on such scales may not be available across domains. [23] show that syntactic information can be useful for alleviating cross-domain performance loss by adding embeddings of syntactic category information to input word representations. This demonstrates that syntax as a layer of abstract structural information can be useful for reducing domain variance. The method of [23] gives the current state-of-the-art accuracies on the task. However, unlike traditional syntax-based statistical methods, their method does not explicitly leverage tree structural knowledge.

We try to combine the potential strengths of syntactic tree noise-pruning approaches and neural models to identify the informative regions in a syntactic dependency tree for domain adaptive sentence simplification. Our method computes the probability of each node to be retained for a simplification process by self attention over context. In addition, we leverage maximum density subtree extraction algorithm to trace out informative regions in a syntactic dependency tree without disturbing grammaticality.

On standard benchmark datasets, our method yields state-of-the-art result measured in terms of both F1-measure and accuracy. To the best of our knowledge, we are the first to investigate tree structures for neural sentence compression. Our code will be released.

2 Problem Definition

We solve the problem of sentence compression as a syntactic dependency tree node selection task. Formally, denote a dependency tree T as

$$T \leftarrow \{N, E\}, \tag{1}$$

where N is a set of nodes containing words $w_1, w_2, w_3, .., w_n$ of the sentence and E is the set of dependency edges of T. The compressed sentence S is represented by a set of binary labels $\{y_1, y_2, y_3, .., y_n\}$.

$$y_i = \begin{cases} 1 \text{ word at the node } w_i \text{ is retained} \\ 0 \text{ otherwise} \end{cases}$$

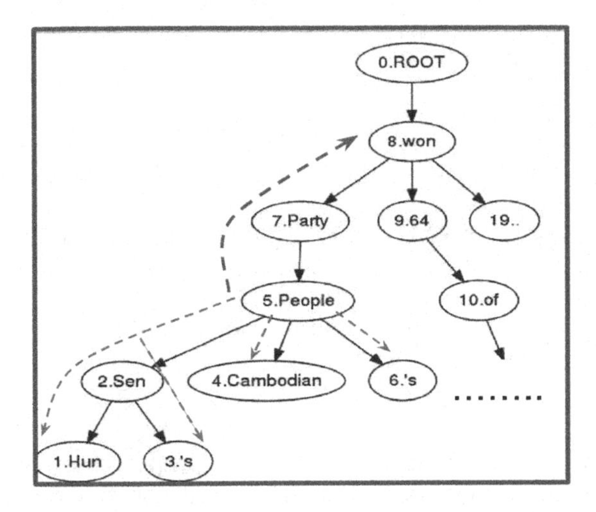

Fig. 2. Marking context nodes for the node 'People' (Color figure online)

We aim to create a model which can predict the label of a node word using the syntactic context within a dependency tree. By leveraging syntax for label prediction, we expect the model to be generalizable across domains. We train our model using an annotated dataset containing a set of parallel records consisting of dependency trees and corresponding deletion/retention labels of the node words, denoted as $D = \{(T_j, S_j)\}_{j=1}^{N}$.

3 Approach

This section discusses our tree-based neural models for node label prediction and subsequent sentence compression.

3.1 Base Model: Bi-LSTM

Inspired by Wang et al. [23], our base model takes an input sentence as a sequence of words $w_1, w_2, w_3, ..., w_n$. The model utilizes the sequential context representation of a word computed using a Bi-LSTM [9] to decide whether each word in the input sentence is to be retained.

$$x_i = \mathbf{E}_w(w_i) \tag{2}$$

$$\overrightarrow{h_i} = \mathbf{LSTM}_{\overrightarrow{\theta}}(\overrightarrow{h_{i-1}}, x_i) \tag{3}$$

$$\overleftarrow{h_i} = \mathbf{LSTM}_{\overleftarrow{\theta}}(\overleftarrow{h_{i-1}}, x_i) \tag{4}$$

$$h_i = \overrightarrow{h_i} \oplus \overleftarrow{h_i} \tag{5}$$

LSTM is computed using update equations.

$$\begin{aligned}
\mathrm{i}_t &= \tanh(W_{xi}x_t + W_{hi}h_{t-1} + b_i) \\
j_t &= \sigma(W_{xj}x_t + W_{hj}h_{t-1} + b_j) \\
f_t &= \sigma(W_{xf}x_t + W_{hf}h_{t-1} + b_f) \\
o_t &= \sigma(W_{xo}x_t + W_{ho}h_{t-1} + b_o) \\
c_t &= f_t \otimes c_{t-1} + i_t \otimes j_t \\
h_t &= \tanh(c_t) \otimes o_t
\end{aligned}$$

θ are the parameters of **LSTM** [10] and E_w is the word embedding lookup table. Using the computed representation h_j of word w_j, the label y_j is predicted as

$$P(y_j|h_j) = \mathrm{softmax}(\mathbf{W}h_j + \mathbf{b}) \tag{6}$$

3.2 Our TreeLSTM Model

We use the concept of bottom up Tree-LSTM proposed by Tai et al. [19] to compute the dense representation for an input syntactic tree node. Given a tree, let *C(j)* denote the set of children of the node j. The Child-Sum Tree-LSTM transitions are:

$$\widetilde{h_j} = \sum_{k \epsilon C(j)} h_k \tag{7}$$

$$i_j = \sigma(\mathbf{W}^{(i)}x_j + \mathbf{U}^{(i)}\widetilde{h_j} + \mathbf{b}^{(i)}) \tag{8}$$

$$f_{jk} = \sigma(\mathbf{W}^{(f)}x_j + \mathbf{U}^{(f)}h_k + \mathbf{b}^{(f)}), k\epsilon C(j) \tag{9}$$

$$o_j = \sigma(\mathbf{W}^{(o)}x_j + \mathbf{U}^{(o)}\widetilde{h_j} + \mathbf{b}^{(o)}) \tag{10}$$

$$u_j = \tanh(\mathbf{W}^{(u)}x_j + \mathbf{U}^{(u)}\widetilde{h_j} + \mathbf{b}^{(u)}) \tag{11}$$

$$c_j = i_j \odot u_j + \sum_{k \epsilon C(j)} f_{jk} \odot c_k \tag{12}$$

$$h_j = o_j \odot \tanh(c_j) \tag{13}$$

where x_j is the input at each node step, σ denotes the logistic sigmoid function and $\odot$ denotes elementwise multiplication. Input $x_j \epsilon \mathrm{R}^d$ is the embedding of word w_j at the node j.

Using the dense representation h_j of node j, the label y_j is predicted as

$$P(y_j|h_j) = \mathrm{softmax}(\mathbf{W}h_j + \mathbf{b}) \tag{14}$$

$W^{(i)}$, $U^{(i)}$, $b^{(i)}$, $W^{(f)}$, $U^{(f)}$, $b^{(f)}$, $W^{(o)}$, $U^{(o)}$, $b^{(o)}$, $W^{(u)}$, $U^{(u)}$, $b^{(u)}$, W and b are model parameters which are optimized using training dataset.

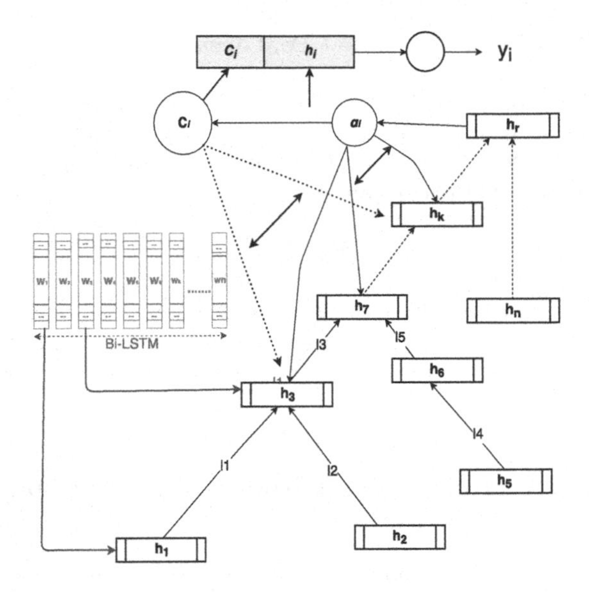

Fig. 3. Tree LSTM states are represented on the right side. Double ended arrows represents self attention based context vector computation.

3.3 Tree LSTM Model + Bi-LSTM

We further combine sequential context and tree context to compute the dense representation of a node. This is achieved by taking sequential context vector of the word as node input in TreeLSTM. Let w_1, w_2,..., w_n be the sequence of words in the sentence from which the input syntactic tree is parsed out. Also t_1, t_2,..., t_n be the corresponding POS tag sequence of words. We compute the dense representation of a word in the sequential context using Bi-LSTM [9] as

$$xs_i = \mathbf{E}_w(w_i) \oplus \mathbf{E}_t(t_i) \tag{15}$$

$$\overrightarrow{hs_i} = \mathbf{LSTM}_{\overrightarrow{\theta}}(\overrightarrow{hs_{i-1}}, xs_i) \tag{16}$$

$$\overleftarrow{hs_i} = \mathbf{LSTM}_{\overleftarrow{\theta}}(\overleftarrow{hs_{i-1}}, xs_i) \tag{17}$$

$$hs_i = \overrightarrow{hs_i} \oplus \overleftarrow{hs_i} \tag{18}$$

$\mathrm{E}_w(w_i)\epsilon\mathrm{R}^d$ and $\mathrm{E}_t(t_i)\epsilon\mathrm{R}^t$ are word and tag embeddings, respectively, $\oplus$ represents concatenation of vectors and $\overleftarrow{\theta}$ and $\overrightarrow{\theta}$ are parameters of the Bi-LSTM model. The computed sequential context representation hs_j computed using Eq. 18 for the word w_j is used as the input x_j (Eqs. 8–11) at the corresponding node j in TreeLSTM. Using the computed node representation, label y_j is predicted using Eq. 14.

3.4 Tree LSTM Model + Label Features

Tracking dependency labels in TreeLSTM is crucial for effectively defining the syntactic context of a node. Along with sequential context, the current model incorporates dependency labels to the parent in the node input. The node input in TreeLSTM (Eqs. 8–11) is computed as

$$x_j = \mathbf{E}_{dl}(dl_j) \oplus hs_j \tag{19}$$

where $\mathbf{E}_{dl}(dl_j)\epsilon \mathrm{R}^{dl}$ is the embedding for dependency label of node j to the parent node and hs_j sequential context representation of word w_j computed using Eq. 18. After computing each node representation, the label of the node is predicted using Eq. 14.

3.5 Self Attention over Context Nodes for Computing Context Vector

The representation of syntax context of each node is computed by attending over its context nodes. Formally, context nodes of node i are defined as follows

$TopContextNodes_{i,dt}$ ← All nodes within a path distance dt towards root from node i
$BottomContextNodes_{i,db}$ ← All nodes within a path distance db towards leaves from node i

In Fig. 2, for the node '*People*', the top context nodes are along the blue line and bottom context nodes are along the red line. dt and db are constants, which are optimized emprically. $ContextNodes_i$ is the union of $TopContextNodes_{i,dt}$ of $BottomContextNodes_{i,db}$.

The overall network architecture is shown in Fig. 3. Node representations are computed using the same method as described in Sect. 3.4. The context vector C_i for node i is computed by weighting the context nodes with the attention weights computed based on the representation of node i.

$$C_i \leftarrow \sum_j \beta_j h_j, j \in ContextNodes_j \tag{20}$$

$$\beta_j \leftarrow \mathrm{softmax}(\alpha_j) \tag{21}$$

$$\alpha_j \leftarrow h_j^T \mathbf{W}_a h_i, \tag{22}$$

where W_a is a model parameter, h_i is the dense representation for node i, β_j is the attention weight for context node j. The label of the node i is predicted as follows,

$$P(y_i|h_i) \leftarrow \mathrm{softmax}(\mathbf{W}(C_i \oplus h_i) + \mathbf{b}) \tag{23}$$

4 Training

The training goal is to minimize the average negative-log likelihood loss in predicting the label of each node for each tree.

$$\text{Minimize:}\frac{-\sum_i log(P(y_{i_a}))}{N}, \tag{24}$$

where y_{i_a} is the actual label of node i and N is the number of nodes in the syntactic tree.

5 Maximum Density Subtree Cut Algorithm for Sentence Simplification

For a tree T with weights set for all its nodes, density is defined as the average weight of nodes. Maximum density subtree cut algorithm extracts the subtree with maximum density value within given size limits. We define the density of syntactic dependency tree T as follows,

$$Density(T) \leftarrow \frac{\sum_{j \epsilon Nodes(T)} P(y_i = 1)}{N} \tag{25}$$

where $P(y_i = 1)$ is the probability for the word at node i to be retained, computed by the neural network as represented in Eq. 23, $Nodes(T)$ is the set of nodes in T and N is the total number of nodes in T. Consequently, for a given maximum number of nodes, maximum density subtree $T^{'}$ of a syntactic tree contains a set of nodes which are highly probable to be retained. For a syntactic dependency tree T with N nodes, the pruning process can be represented as follows.

$$T^{'} \leftarrow getMaxDenSubtree(T, c * N, q * N) \tag{26}$$

where c and q are constants which are empirically optimized. We use a greedy maximum density subtree cut algorithm ensures that a node with a word is essential to maintain grammaticality as decided by its relation with parent node is not removed without removing its parent, irrespective of their $P(y_j = 1)$ value. We list the dependency relations *nsubj, csubj, nsubjpass, xsubj, aux, xcomp, pobj, acomp, dobj, case, det, poss, possessive, auxpass, ccomp, neg, expl, cop, prt, mwe, pcomp, iobj, number, quantmod, predet, dep*, and *mark* as essential to maintain grammaticality. During each iteration, the algorithm searches for the next subtree to be pruned out within grammatical constraints while enforcing the size constraints.

6 Experiments

In order to validate the cross-domain effectiveness of our model, we used two different datasets representing two different domains for training and evaluation.

- **Google News Data Set:** The parallel dataset[1] released by Filippova et al. [6] contains 10000 sentences and corresponding compressed version.
- **BNC News:** The second dataset consists of 1500 sentences taken from British National Corpus (BNC) and the American News Text corpus before 2008 and their ground truth compressed versions. The dataset[2] is collected and released by Cohn et al. [4].

We split the Google News dataset into training (1001–9000), testing (1–1000) and validation sets (9000–10000). The offset and size of each set remain exactly same as those of Wang et al. [23]. The BNC News dataset is utilized as a cross-domain testset.

Table 1. Tuning *dt* and *db* dataset

dt	*db*	F1-Measure
1	3	0.81
3	4	0.81
4	3	**0.83**
4	5	0.82
6	3	0.81
6	6	0.80
3	4	0.82

6.1 Experimental Settings

We evaluate our approaches using F1 score and accuracy. The former is derived from precision and recall values, where precision is defined as the percentage of retained words that overlap with the ground truth and recall is defined as the percentage of words in the ground truth compressed sentences that overlap with the generated compressed sentences. The latter is defined as the percentage of tokens for which the predicted label y_i is correct.

We evaluate five different variations of our method.

- **TreeLSTM:** A bottom up TreeLSTM [19] taking word embedding as input at each node as described in Sect. 3.2.
- **TreeLSTM + Bi-LSTM:** A bottom up TreeLSTM taking sequential context representation of a word computed using a Bi-LSTM as input at each node as described in Sect. 3.3.

[1] http://storage.googleapis.com/sentencecomp/compression-data.json.
[2] http://jamesclarke.net/research/resources/.

Table 2. Comparison with base systems: F1, Acc - Accuracy, CR - Compression Ratio, GN - Google News

Method	Size	GN			BNC		
		F1	Acc	CR	F1	Acc	CR
LSTM [6]	2M	0.80	–	0.39	–	–	–
LSTM+[6]	2M	**0.82**	–	0.38	–	–	–
Abstractive seq2seq (ala [23])	3.8M	0.09	0.02	0.16	0.14	0.06	0.21
LSTM (baseline [23])	8K	0.74	0.75	0.45	0.51	0.48	0.37
LSTM+ (baseline [23])	8K	0.77	0.78	0.47	0.54	0.51	0.38
BiLSTM [23]	8K	0.75	0.76	0.43	0.52	0.50	0.34
BiLSTM+SynFeat [23]	8K	0.80	0.82	0.43	0.57	0.54	0.37
Our models							
TreeLSTM	8K	0.73	0.72	0.46	0.49	0.47	0.33
TreeLSTM + Bi-LSTM	8K	0.80	0.81	0.44	0.53	0.53	0.36
TreeLSTM + Bi-LSTM + Dep	8K	0.80	0.835	0.42	0.57	0.54	0.36
TreeLSTM + Bi-LSTM + Attention	8K	0.81	**0.845**	0.43	0.59	0.54	0.36
Syntactic constraints							
Traditional ILP [3]	N/A	0.54	0.56	0.62	0.64	0.56	0.56
BiLSTM+SynFeat+ILP [23]	8K	0.78	0.78	0.57	0.66	0.58	0.53
TreeLSTM+Attn+ MDT(CR= 0.21)	8K	0.67	0.66	0.21	0.42	0.41	0.21
TreeLSTM+Attn+ MDT(CR=0.38)	8K	0.81	0.83	0.38	0.59	0.53	0.38
TreeLSTM+Attention+ MDT	8K	0.79	0.79	0.55	**0.70**	**0.60**	0.54

Table 3. Human evaluation

	RD	**IF**
Traditional ILP	3.3	3.27
BiLSTM+SynFeat+ILP	4.21	4.1
TreeLSTM + Self Attention+ MDT	**4.30**	**4.25**

- **TreeLSTM + Bi-LSTM + Dependency Features:** A bottom up TreeLSTM taking the concatenation of sequential context representation of a word and dependency label embedding as input at each node as described in Sect. 3.4.
- **TreeLSTM + Bi-LSTM + Self Attention:** In this setting, to estimate the probability for each node to be retained, we take the weighted context of the node into account as described in Sect. 3.5. Remaining settings are same as the setting described above.

- **TreeLSTM + Self Attention+ MDT:** In this method, $P(y_i = 1)$ for each node i is computed using the method explained above. Subsequently, maximum density subtree (MDT) cut algorithm is applied as explained in Sect. 5 to decide the final label of each node.

Settings: Our model was trained using Adam [13] with the learning rate initialized at 0.001. The TreeLSTM hidden layer dimension is set to 200. The dimension of the hidden layers of bi-LSTM is 100. Word embeddings are initialized from GloVe 100-dimensional pre-trained embeddings [16]. POS and dependency embeddings are randomly initialized with 40-dimensional vectors. Word embeddings are set as updatable during training. The batch size is set as 20. Constants c, q (Eq. 26), dt, and db (Sect. 3.5) are set to 0.7, 0.3, 4 and 3 respectively for maximum accuracy in validation data using grid-search. The Table 1 list the accuracy for different values of dt and db in development dataset. We used Satndford Parser[3] for creating syntactic parse trees.

6.2 Results

Table 2 shows the performance of our approaches, the method of [23] and other baselines. The models include BiLSTM without incorporating any syntactic feature, BiLSTM+SynFeat in which they incorporate syntactic features in a BiLSTM and BiLSTM+SynFeat+ILP in which they use ILP to predict the final label y for each word in the input sentence. Their baselines are LSTM [6], LSTM+ in which syntactic features are incorporated with LSTM, Traditional ILP [3] and Abstractive seq2seq which is an abstractive sequence-to-sequence[4] model trained on 3.8 million Gigaword title-article pairs [15].

Effectiveness of Neural Tree Model. A simple bottom-up TreeLSTM with word-embedding as input does not yield good results. We observe that explicit use of syntactic features is necessary for defining the syntactic context of a node. The performance dramatically increased when sequential context representation consisting of word and POS tag information is used as input to the TreeLSTM at each node (TreeLSTM + Bi-LSTM). TreeLSTM, which jointly tracks sequential context representation of a word and dependency label while computing node representations (TreeLSTM + Bi-LSTM + Dependency Features), outperforms BiLSTM+SynFeat in terms of accuracy in Google News dataset and yields comparable results in BNC dataset. This shows that, a TreeLSTM with explicit use of syntactic features can leverage the syntactic context better than Bi-LSTM utilizing syntactic features (BiLSTM+SynFeat).

Computing context vector representation (TreeLSTM + Bi-LSTM + Self Attention) and subsequent labelling of nodes outperforms BiLSTM+SynFeat in all domains Table 2. Our observation is that computing context vector representation using self attention can effectively model the syntactic context of a node and identifies the regions of a syntactic tree, which holds abstract information relevant for sentence compression.

[3] https://nlp.stanford.edu/software/lex-parser.shtml.
[4] http://opennmt.net/.

Effectiveness Maximum Density Tree Cut. Incorporation of multi-density subtree extraction with neural model (TreeLSTM + Self Attention+ MDT) outperforms the state-of-art BiLSTM+SynFeat+ILP on both of Google News and BNC datasets, proving its efficiency in cross-domain application. This is probably because it locates regions of a syntactic dependency tree holding retainable abstract information, explicitly relying on dependency relations to maintain grammaticality. As a result, the approach enjoys more flexibility in the search for abstract content than an ILP method with hard constraints. Also, self attention based context vector computation can identify patterns of informativeness more accurately (Fig. 4).

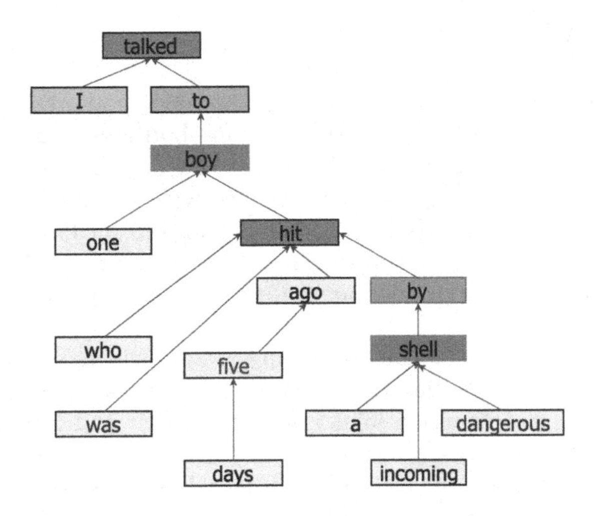

Fig. 4. Tree Node Weights: Grey scale shows the probability to be retained

Evaluating Cross Domain Effectiveness by Fixing Compression Ratio. The sentence compressions in BNC dataset is relatively larger than Gigaword title-article pairs [15]. Taking this into account, in order to have a fair comparison with Abstractive seq2seq, we have tested our MDT model also by pre-fixing the model with their compression ratio in the BNC dataset as shown in Table 2. We have also tested the MDT model by setting the compression ration of LSTM+ in the BNC datset. In both cases, MDT has a better score than the two baselines systems, which rules out the chance for higher compression ratio being the reason for better results in new domains. Approaches with syntactic constraints can ensure a fare comparison as all of them share a similar compression ratio.

7 Human Evaluation

90 randomly chosen source sentences and corresponding compressions produced by TreeLSTM + Self Attention+ MDT, BiLSTM+SynFeat+ILP and Traditional ILP are chosen for human evaluation. These three approaches exhibits almost

similar average compression ratio and don't enjoy any advantage due to higher compression ratio ensuring a level comparison. The human raters were asked to rate the informativeness (**IF**) and readability (**RD**) of the compressed outputs on a scale of 5. The compressed sentences take a random order during each rating to avoid any kind of bias. The average scores obtained are shown in the Table 3. TreeLSTM + Self Attention+ MDT yields scores which are comparable with those of BiLSTM+SynFeat+ILP. The results show that MDT with constraints on grammaticality can better preserve readability than Traditional ILP. None of the methods has an advantage due to compression ratio as all of them exhibit similar average compression ratio in test.

8 Related Work

Our work belongs to the line of extractive sentence compression approaches. A seminal graph-based sentence compression method was suggested by Mcdonald et al. [14]. They assign a score for each word pair existing in the original sentence and search for a compressed sentence with the maximum total score within a given length limit. Clarke et al. [3] uses an Integer Linear Programming (ILP) framework for sentence compression. They compute a relevance score for each word and then incorporates the scores in the ILP formulation for ranking candidates.

Filippova et al. [7] apply ILP over syntactic dependency trees to trace-out a proper subtree corresponding to a grammatically correct simplified sentence. Berg-Kirkpatrick et al. [1] use a joint model to extract and compress for multi-document summarization. Their approach weighs bi-grams using a supervised linear model. In contrast, our method uses neural network to estimate informativeness, while utilizing generalizability of syntactic tree based approaches.

There has been work in the past which tried to impose syntactic constraints via soft logic [12,25] or by hard structural rules [17,21,22,24]. However, a combination of neural methods and multi-density tree cut algorithm can more flexibly search for compressed representation of the source sentence. Also, the current work investigates domain adaptability of sentence simplification methods by an optimum combination of neural methods and syntactic tree based approaches.

There has been work which builds neural network models trained on large datasets, both for extractive [6] and abstractive [2,18] sentence compression. However, these techniques require a large amount of training data. They also tend to overfit in the domain of training data and end up with a poor performance in a cross-domain settings [23]. There are has been previous works to improve text simplification techniques such as multi- document summarization [11] and headline generation [20]. However scope and challenges of extractive sentence summarization is different from these problems.

The most recent work, which effectively merge the potential strengths of syntax based approaches and data-driven neural network model, is Wang et al. [23]. They make significant improvement in cross-domain sentence compression with a relatively smaller training set. However, they have predominantly relied on the sequential context of a word along with syntactic information to decide whether a word needs to be retained or not, despite that the syntactic context of word is better definable within a syntactic tree. Our approach decides whether to retain or to remove a word based on its context within a dependency tree. In this sense, we extend the work of Wang et al. [23] by incorporating the strength of traditional syntax tree noise-pruning methods by using a maximum density subtree extraction algorithm.

9 Conclusion

We investigated an approach for sentence compression which utilize the possibilities of neural models in syntactic tree pruning for sentence compression. Our method yields the best results in terms of F1-measure and accuracy in two different domains proving its domain adaptability. There is scope for research in future for a method which jointly learn to compute weights and extract subtrees. Parsing errors can be overcome by using the top-K parse trees generated by the parser and train the current approach using all the trees.

References

1. Berg-Kirkpatrick, T., Gillick, D., Klein, D.: Jointly learning to extract and compress. In: Proceedings of the 49th Annual Meeting of the Association for Computational Linguistics: Human Language Technologies-Volume 1, pp. 481–490. Association for Computational Linguistics (2011)
2. Chopra, S., Auli, M., Rush, A.M.: Abstractive sentence summarization with attentive recurrent neural networks. In: Proceedings of the 2016 Conference of the North American Chapter of the Association for Computational Linguistics: Human Language Technologies, pp. 93–98 (2016)
3. Clarke, J., Lapata, M.: Global inference for sentence compression: an integer linear programming approach. J. Artif. Intell. Res. **31**, 399–429 (2008)
4. Cohn, T., Lapata, M.: Large margin synchronous generation and its application to sentence compression. In: EMNLP-CoNLL, pp. 73–82 (2007)
5. Cohn, T.A., Lapata, M.: Sentence compression as tree transduction. J. Artif. Intell. Res. **34**, 637–674 (2009)
6. Filippova, K., Alfonseca, E., Colmenares, C.A., Kaiser, L., Vinyals, O.: Sentence compression by deletion with LSTMs. In: EMNLP, pp. 360–368 (2015)
7. Filippova, K., Strube, M.: Dependency tree based sentence compression. In: Proceedings of the Fifth International Natural Language Generation Conference, pp. 25–32. Association for Computational Linguistics (2008)
8. Galanis, D., Androutsopoulos, I.: An extractive supervised two-stage method for sentence compression. In: Human Language Technologies: The 2010 Annual Conference of the North American Chapter of the Association for Computational Linguistics, pp. 885–893. Association for Computational Linguistics (2010)
9. Graves, A., Jaitly, N., Mohamed, A.r.: Hybrid speech recognition with deep bidirectional LSTM. In: IEEE Workshop on Automatic Speech Recognition and Understanding (ASRU), pp. 273–278. IEEE (2013)

10. Hochreiter, S., Schmidhuber, J.: LSTM can solve hard long time lag problems. In: Advances in Neural Information Processing Systems, pp. 473–479 (1997)
11. Hong, K., Conroy, J.M., Favre, B., Kulesza, A., Lin, H., Nenkova, A.: A repository of state of the art and competitive baseline summaries for generic news summarization. In: LREC, pp. 1608–1616 (2014)
12. Huang, M., Shi, X., Jin, F., Zhu, X.: Using first-order logic to compress sentences. In: AAAI (2012)
13. Kingma, D., Ba, J.: Adam: a method for stochastic optimization. arXiv preprint arXiv:1412.6980 (2014)
14. McDonald, R.T.: Discriminative sentence compression with soft syntactic evidence. In: EACL (2006)
15. Napoles, C., Gormley, M., Van Durme, B.: Annotated gigaword. In: Proceedings of the Joint Workshop on Automatic Knowledge Base Construction and Web-scale Knowledge Extraction, pp. 95–100. Association for Computational Linguistics (2012)
16. Pennington, J., Socher, R., Manning, C.: Glove: global vectors for word representation. In: Proceedings of the 2014 Conference on Empirical Methods in Natural Language Processing (EMNLP), pp. 1532–1543 (2014)
17. Qian, X., Liu, Y.: Polynomial time joint structural inference for sentence compression. In: Proceedings of the 52nd Annual Meeting of the Association for Computational Linguistics (Volume 2: Short Papers), vol. 2, pp. 327–332 (2014)
18. Rush, A.M., Chopra, S., Weston, J.: A neural attention model for abstractive sentence summarization. arXiv preprint arXiv:1509.00685 (2015)
19. Tai, K.S., Socher, R., Manning, C.D.: Improved semantic representations from tree-structured long short-term memory networks (2015)
20. Takase, S., Suzuki, J., Okazaki, N., Hirao, T., Nagata, M.: Neural headline generation on abstract meaning representation. In: Proceedings of the 2016 Conference on Empirical Methods in Natural Language Processing, pp. 1054–1059 (2016)
21. Thadani, K.: Approximation strategies for multi-structure sentence compression. In: Proceedings of the 52nd Annual Meeting of the Association for Computational Linguistics (Volume 1: Long Papers), vol. 1, pp. 1241–1251 (2014)
22. Thadani, K., McKeown, K.: Sentence compression with joint structural inference. In: Proceedings of the Seventeenth Conference on Computational Natural Language Learning, pp. 65–74 (2013)
23. Wang, L., Jiang, J., Chieu, H.L., Ong, C.H., Song, D., Liao, L.: Can syntax help? Improving an LSTM-based sentence compression model for new domains. In: Proceedings of the 55th Annual Meeting of the Association for Computational Linguistics (Volume 1: Long Papers), vol. 1, pp. 1385–1393 (2017)
24. Yao, J.g., Wan, X.: Greedy flipping for constrained word deletion. In: AAAI, pp. 3518–3524 (2017)
25. Yoshikawa, K., Hirao, T., Iida, R., Okumura, M.: Sentence compression with semantic role constraints. In: Proceedings of the 50th Annual Meeting of the Association for Computational Linguistics: Short Papers-Volume 2, pp. 349–353. Association for Computational Linguistics (2012)

An Axiomatic Approach to Diagnosing Neural IR Models

Daniël Rennings, Felipe Moraes, and Claudia Hauff(✉)

Delft University of Technology, Delft, The Netherlands
d.j.a.rennings@student.tudelft.nl, {f.moraes,c.hauff}@tudelft.nl

Abstract. Traditional retrieval models such as BM25 or language models have been engineered based on search heuristics that later have been formalized into axioms. The axiomatic approach to information retrieval (IR) has shown that the effectiveness of a retrieval method is connected to its fulfillment of axioms. This approach enabled researchers to identify shortcomings in existing approaches and "fix" them. With the new wave of neural net based approaches to IR, a theoretical analysis of those retrieval models is no longer feasible, as they potentially contain millions of parameters. In this paper, we propose a pipeline to create *diagnostic datasets for IR*, each engineered to fulfill one axiom. We execute our pipeline on the recently released large-scale question answering dataset `WikiPassageQA` (which contains over 4000 topics) and create diagnostic datasets for four axioms. We empirically validate to what extent well-known deep IR models are able to realize the axiomatic pattern underlying the datasets. Our evaluation shows that there is indeed a positive relation between the performance of neural approaches on diagnostic datasets and their retrieval effectiveness. Based on these findings, we argue that diagnostic datasets grounded in axioms are a good approach to diagnosing neural IR models.

1 Introduction

Over the past few years, deep learning approaches have been increasingly applied to IR tasks [28]. At the same time, the IR community has identified a number of issues [7,8,28], hindering the kind of progress seen in other research areas such as natural language processing (NLP) and computer vision. Overall, our community lacks adequately large-scale public datasets for training (an exception is the recently released `WikiPassageQA` [6]), shared public code repositories of neural IR models (although some progress has been recently made [11,21]) and approaches to interpret and analyze neural IR models (here, [5] is an exception).

In this paper, we focus our attention on the last issue—the analysis of neural IR models. While many neural models have been proposed, few have turned out to outperform properly tuned BM25 or language modeling baselines. Traditional retrieval models have been engineered based on search heuristics that later have been formalized into *axioms*—formal constraints that should be fulfilled by a good model—which enable us to analytically investigate to what extent retrieval

L. Azzopardi et al. (Eds.): ECIR 2019, LNCS 11437, pp. 489–503, 2019.
https://doi.org/10.1007/978-3-030-15712-8_32

models fulfill them [13–16]. This analytical approach enabled researchers to identify shortcomings in existing retrieval models and "fix" them [1,4,12,18,27,30], in order to achieve higher retrieval effectiveness. Ideally, we employ a similar axiomatic approach to diagnose & fix neural IR models in order to reap the benefits deep learning has offered in other fields. However, as these models may contain millions of parameters [28], this is not possible.

Instead, we propose a pipeline for the creation of *diagnostic datasets for IR*, each engineered to fulfill one axiom. This approach follows the tradition in NLP and computer vision where dataset creation for diagnostic purposes is well-known—consider for instance `bAbI` [44] for automatic text understanding & reasoning, adversarial examples for `SQUAD` [22], a popular reading comprehension dataset, and `CLEVR` [23], a dataset for language & visual reasoning.

We execute our pipeline for four axioms on the answer-passage retrieval dataset `WikiPassageQA` [6] which contains more than 4,000 topics. It has been shown to be a difficult dataset for a range of neural models. We empirically validate to what extent four well-known deep IR models are able to realize the axiomatic patterns underlying the datasets. We find that, indeed, there is a positive relation between the performance of neural approaches on the diagnostic datasets and their retrieval effectiveness.

We believe these findings to be more insightful for IR researchers to improve neural models than, for instance, the probing of neural net layers via NLP tasks [5] or simply evaluating deep models with a range of metrics on standard test collections. The **main contribution** of our work is to showcase that a transformation from an analytical axiom to a diagnostic dataset is possible and offers us a new tool to diagnose retrieval models that are too complex to be analyzed theoretically.

2 Related Work

Axiomatic Approach to IR. Fang et al.'s [13] seminal work introduced six retrieval constraints (later coined *axioms* [15]) that a reasonable retrieval function should satisfy. Formalizing retrieval heuristics into constraints—e.g. *given a single-term query w and two equally long documents, the retrieval score of the document with a higher frequency of w should be ranked higher* (also known as constraint `TFC1`)—enabled the authors to *analytically* evaluate a number of existing retrieval functions. The main assumption of the work—the effectiveness of retrieval functions is connected to their fulfillment of retrieval constraints—was empirically validated. Fang et al. [14] also proposed the use of perturbed document collections to gather further insights on retrieval functions fulfilling the same set of axioms. This approach has not been followed-up upon in works other than [31].

Apart from the diagnosis of existing functions, Fang et al. derived novel retrieval functions based on their initial set of constraints [15] and later extended their list of axioms from purely term-matching to semantic-matching based constraints [12,16]. Others have contributed query term proximity [18,42], document length normalization [27] and query term discrimination [1] constraints, consistently showing that traditional models improve when slightly altered to satisfy those constraints. While most of the more than twenty existing axioms have been designed for standard retrieval models, a number of axioms have been proposed for more specialized cases, such as statistical translation models [24] and pseudo-relevance feedback [3,4,30].

Lastly, we point to two works closest to ours. Hagen et al. [18] explored the re-ranking of a given result list based on the aggregated re-ranking preferences of twenty-three axioms. Similar to our work, this application of axioms to an actual result list (instead of "hypothetical" documents containing one or two different query terms used in the analytic evaluation of retrieval functions) requires the *extension* and *relaxation* of axioms. Pang et al. [35] investigated differences in neural IR models and learning to rank approaches with hand-crafted features. Through a manual error analysis, weaknesses in deep IR models were identified and connected to retrieval constraints.

Neural IR Models. By now, deep learning has become the mainstay in a number of research fields, yielding impressive improvements on long-standing tasks. The information retrieval community has also seen a large number of proposed neural IR models, which can be categorized as *interaction-based*, *representation-based* or a *hybrid* between the two [17], based on the manner they model the query and document. While interaction-based neural approaches (e.g., DeepMatch [26], DRMM [17], MatchPyramid [36], ANMM [45]) use the local interactions between the query and document as input to the deep net, representation-based approaches (e.g., DSSM [20], C-DSSM [39], ARC-I [19]) strive to create good representations of the query and the document separately; hybrid approaches such as Duet [29], ARC-II [19] and MVLSTM [43] incorporate both an interaction- and representation-based component. Despite the motivation for representation-based approaches and the need for semantics over syntax matching, a recent comparative study [32] has shown the deep interaction-based approaches to clearly outperform the representation-based approaches in terms of retrieval effectiveness. Whereas most interaction- and representation-based approaches compute relevance at the document level, Fan et al. [10] recently proposed a hierarchical neural matching model (HiNT) which employs a local matching layer and global decision layer, to capture relevance signals at the passage and document level which compete with each other. Another recent work has achieved state-of-the-art performance by creating a neural pseudo relevance feedback framework (NPRF) that can be used with existing neural IR models as building blocks [25].

While many works have presented novel neural approaches, few works have focused on diagnosing neural IR models. While studies such as the one conducted in [32] enable us to empirically determine which type of approach performs better, they can only provide relatively coarse-grained insights (in this case: interaction-based performs better than representation-based). In contrast, Cohen et al. [5] recently proposed to *probe* neural retrieval models by training them, and then using each layer's weights as input to a classifier for different types of

NLP tasks (sentiment analysis, POS tagging, etc). The performance on those tasks provides insights into the kind of information that each layer captures. While this is indeed useful to realize, it does not provide an immediate insight into how to improve an existing neural approach. It is also quite labor-intensive as this probing is not model agnostic—in contrast to our work.

While in the IR community, the diagnosing of deep nets is in its infancy, the computer vision and NLP communities have proposed a number of different manners to open up this black box that a typical deep net is. The 20 bAbI tasks [44] were developed specifically to diagnose text understanding and reasoning systems, while Jia and Liang [22] proposed an adversarial evaluation scheme of the SQUAD dataset by inserting distracting sentences into text passages (and as a result all evaluated models dropped sharply in their accuracy). CLEVR [23], a dataset for language and visual reasoning, consists of a large number of rendered images (constructed from a limited universe of objects and relationships) and automatically generated questions.

Here, we propose to bring the approach of diagnostic dataset creation into the IR community, based on well-established axioms.

3 Creating Diagnostic Datasets

Out of the more than twenty IR axioms proposed by now, we have selected four among those in [13,14,40], and converted them for our purpose of diagnostic dataset creation. Two of the axioms (TFC1 and M-TDC) were selected as they capture a fair amount of relevance, while being present in existing datasets—including the one we work with. Combined, TFC1 and M-TDC essentially represent the TF-IDF statistic, a pervasive component in most IR models [6,9,48,49]. A third axiom (TFC2) constraints the difference in scores between pairs of documents instead of individual documents. We include TFC2 to show our methodology can handle such axioms as well. Finally, we selected the LNC2 axiom to showcase how we can generate a diagnostic dataset from an existing corpus through creating artificial data when extracting a diagnostic dataset does not yield enough data points.

We now describe the axioms we consider and propose (1) an *extension* of each axiom in order to match realistic queries and documents[1], and, (2) a *relaxation* of extended axioms such that the strictly defined query and document relations are relaxed to enable selection and generation of sufficient amounts of data. Whereas step (1) allows us to move from one- or two-term queries to arbitrary query lengths and from two- or three-document instances to any number of documents, step (2) allows us to make use of query/document pairings that *approximately* fulfill a particular relationship.

Finally, a note on notation: we refer to an original axiom as Axiom; its extended and relaxed variant is referred to as $\overline{\texttt{Axiom}}$.

[1] For completeness, we note that we did not observe a single instance of query-document pairs or triplets in our WikiPassageQA corpus that satisfies any of the four original (non-extended, non-relaxed) axioms considered here.

3.1 TFC1: Extension and Relaxation

The TFC1 axiom [13] favours documents with more occurrences of a query term and is formally defined as follows: let $\mathbf{q} = \{w\}$ be a single-term query and $\mathbf{d_1}$ and $\mathbf{d_2}$ be two documents of equal length, i.e. $|\mathbf{d_1}| = |\mathbf{d_2}|$. Further, let $c(w, \mathbf{d})$ be the count of term w in document $\mathbf{d}$ and $S(\mathbf{q}, \mathbf{d})$ be the retrieval status value a retrieval function assigns to $\mathbf{d}$, given $\mathbf{q}$. TFC1 then states that if $c(w, \mathbf{d_1}) > c(w, \mathbf{d_2})$ holds, $S(\mathbf{q}, \mathbf{d_1}) > S(\mathbf{q}, \mathbf{d_2})$ should hold as well.

We now *extend* this axiom to multiple-term queries and *relax* it to incorporate documents of approximately the same length, resulting in $\overline{\text{TFC1}}$. Formally: let $\mathbf{q} = \{w_1, w_2, ..w_{|\mathbf{q}|}\}$ be a multi-term query and $|\mathbf{d_i}| \approx |\mathbf{d_j}|$, i.e. $|\mathbf{d_i}| - |\mathbf{d_j}| \leq |\delta_{\text{TFC1}}|$. Here, δ_{TFC1} is an adjustable parameter that may be set according to the document corpus and retrieval task. Additionally, we *relax* the constraint that $\mathbf{d_i}$ must have a larger count for *every* query term than $\mathbf{d_j}$. We now require $c(w, \mathbf{d_i}) \geq c(w, \mathbf{d_j})\ \forall w \in \mathbf{q}$ and $\sum_{w \in \mathbf{q}} c(w, \mathbf{d_i}) > \sum_{w \in \mathbf{q}} c(w, \mathbf{d_j})$, i.e. there is at least one query term with a higher term count in $\mathbf{d_i}$. If this relaxed constraint is fulfilled, then $\overline{\text{TFC1}}$ states that $S(\mathbf{q}, \mathbf{d_i}) > S(\mathbf{q}, \mathbf{d_j})$.

3.2 TFC2: Extension and Relaxation

Axiom TFC2 [13] encapsulates the intuition that an increase in retrieval status value due to an increase in term count becomes smaller as the absolute term count increases. Formally, the axiom considers the case of $\mathbf{q} = \{w\}$ and $|\mathbf{d_1}| = |\mathbf{d_2}| = |\mathbf{d_3}|$. If $c(w, \mathbf{d_1}) > 0$, $c(w, \mathbf{d_2}) - c(w, \mathbf{d_1}) = 1$ and $c(w, \mathbf{d_3}) - c(w, \mathbf{d_2}) = 1$ (i.e. the absolute term count of w in $\mathbf{d_1}$ is smallest and in $\mathbf{d_3}$ is largest), then $S(\mathbf{q}, \mathbf{d_2}) - S(\mathbf{q}, \mathbf{d_1}) > S(\mathbf{q}, \mathbf{d_3}) - S(\mathbf{q}, \mathbf{d_2})$.

We define $\overline{\text{TFC2}}$ for multi-term queries and documents of approximately the same length. Formally, we consider $\mathbf{q} = \{w_1, w_2, ..w_{|\mathbf{q}|}\}$ and $|\mathbf{d_i}| \approx |\mathbf{d_j}| \approx |\mathbf{d_k}|$, i.e. $max_{\mathbf{d_a}, \mathbf{d_b} \in \{\mathbf{d_i}, \mathbf{d_j}, \mathbf{d_k}\}}(|\mathbf{d_a}| - |\mathbf{d_b}|) \leq |\delta_{\text{TFC2}}|$. Every document has to contain at least one query term and the differences in term count are no longer restricted to be exactly 1. This leads to the constraints $\sum_{w \in \mathbf{q}} c(w, \mathbf{d_k}) > \sum_{w \in \mathbf{q}} c(w, \mathbf{d_j}) > \sum_{w \in \mathbf{q}} c(w, \mathbf{d_i}) > 0$ and $c(w, \mathbf{d_j}) - c(w, \mathbf{d_i}) = c(w, \mathbf{d_k}) - c(w, \mathbf{d_j})\ \forall w \in \mathbf{q}$. The latter constraint does not mean that the difference has to be the same for every query term, instead we enforce this equality in term count difference on a term level. If these constraints hold, then according to $\overline{\text{TFC2}}$, $S(\mathbf{q}, \mathbf{d_j}) - S(\mathbf{q}, \mathbf{d_i}) > S(\mathbf{q}, \mathbf{d_k}) - S(\mathbf{q}, \mathbf{d_j})$.

3.3 M-TDC: Extension and Relaxation

The TDC axiom was originally proposed by Fang et al. [13] to favour documents with more occurrences of less popular query terms in the collection. Shi et al. modified the TDC axiom to M-TDC [40] to fix undesired behavior in some cases. Formally, M-TDC is defined as follows. Let $\mathbf{q} = \{w_1, w_2\}$ be a two-term query and assume $|\mathbf{d_1}| = |\mathbf{d_2}|$, $c(w_1, \mathbf{d_1}) = c(w_2, \mathbf{d_2})$ and $c(w_2, \mathbf{d_1}) = c(w_1, \mathbf{d_2})$. If $idf(w_1) \geq idf(w_2)$—i.e. w_1 is rarer in the corpus than w_2—and $c(w_1, \mathbf{d_1}) \geq c(w_1, \mathbf{d_2})$, then $S(q, \mathbf{d_1}) \geq S(q, \mathbf{d_2})$.

We define $\overline{\texttt{M-TDC}}$ for multi-term queries $\mathbf{q} = \{w_1, w_2, ..w_{|\mathbf{q}|}\}$ and pairs of documents $\mathbf{d_i}, \mathbf{d_j}$ that (i) differ in at least one count of a query term, (ii) have the same total count of query terms (sum of term frequencies), and, (iii) have approximately the same length, i.e. $|\mathbf{d_i}| - |\mathbf{d_j}| \leq |\delta_{\texttt{M-TDC}}|$.

For a query-doc-doc triplet to be included in our axiomatic dataset with retrieval score preference $S(\mathbf{q}, \mathbf{d_i}) \geq S(\mathbf{q}, \mathbf{d_j})$, all query terms for which $c(w, d_i) \neq c(w, d_j)$ need to appear at least once in a *valid* query term pair. We evaluate all possible query term pairs (with $w_a \neq w_b$) and consider a query term pair to be *valid* when the following conditions hold: (i) $idf(w_a) \geq idf(w_b)$, (ii) $c(w_a, \mathbf{d_i}) = c(w_b, \mathbf{d_j})$ and $c(w_b, \mathbf{d_i}) = c(w_a, \mathbf{d_j})$, (iii) $c(w_a, \mathbf{d_i}) > c(w_a, \mathbf{d_j})$, and (iv) $c(w_a, \mathbf{q}) \geq c(w_b, \mathbf{q})$.

3.4 LNC2: Extension and Relaxation

The LNC2 [14] axiom prescribes that over-penalizing long documents should be avoided: if a document is replicated k times, its retrieval status value should not be lower than that of its un-replicated variant. The axiom was defined under the assumption that redundancy is not an issue, which we also follow here. Formally the axiom is defined as follows: let $\mathbf{q} = \{w_q\}$, $c(w_q, \mathbf{d_1}) > 0$, $k > 1, k \in \mathbb{N}$, $|\mathbf{d_1}| = k \times |\mathbf{d_2}|$, and for $\forall w \in \mathbf{d_1}$, $c(w, \mathbf{d_1}) = k \times c(w, \mathbf{d_2})$. If those constraints are met, $S(\mathbf{q}, \mathbf{d_1}) \geq S(\mathbf{q}, \mathbf{d_2})$ should hold. This axiom can simply be extended to $\overline{\texttt{LNC2}}$ by defining $\mathbf{q}$ for multi-term queries and documents $\mathbf{d_i}$ and $\mathbf{d_j}$. No additional relaxation is required.

3.5 From $\overline{\texttt{Axiom}}$ to Dataset

Having defined extended and relaxed variants of our axioms, we now describe how to obtain a diagnostic dataset for each. Given a corpus with standard pre-processing applied, we determine the number of instances the (i) original axiom, (ii) relaxed axiom and (iii) relaxed & extended axiom can be found in it. As the axioms are defined over retrieval status values (instead of relevance labels), we do not require relevance judgments and thus, almost any dataset is suitable as source dataset. We can sample queries and document pairs/triplets from such a dataset at will; we keep those in our diagnostic datasets that satisfy our axioms. Due to the very restrictive nature of the original axioms, we expect few instances that fulfill their conditions to be found in most existing corpora. In the case of the extended axiom $\overline{\texttt{LNC2}}$, we expect only spam documents to satisfy the axiom. For this axiom we move beyond *extracting* instances from a given corpus and artificially *create* instances instead by appending each selected document in the dataset $k - 1$ times to itself for a set of values $k > 1$. Figure 1 shows a graphical overview of our pipeline, with examples from **WikiPassageQA**.

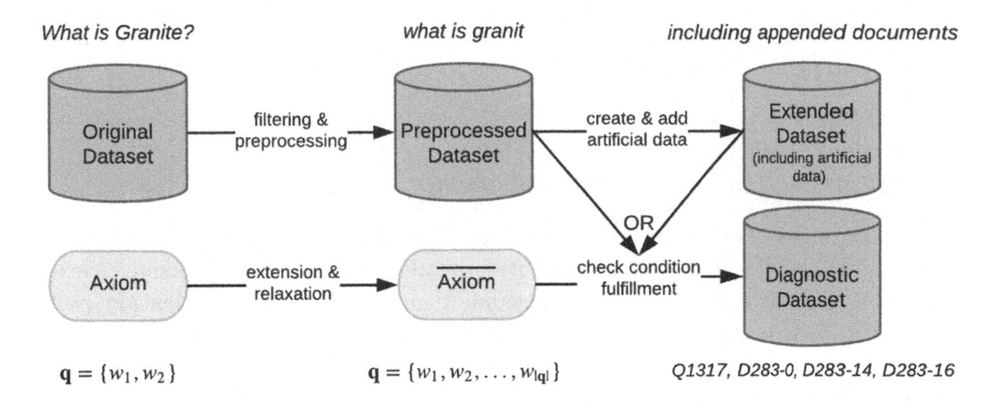

Fig. 1. Overview of the diagnostic dataset creation pipeline. In *italics*, we show an example for the $\overline{\texttt{TFC2}}$ axiom as extracted from question 1317 on passages from Wikipedia document 283 in the `WikiPassageQA` dataset, and refer to appended documents as an example of artificial data (for $\overline{\texttt{LNC2}}$).

3.6 Evaluating IR Models

The evaluation through diagnostic datasets as presented in this work is **model-agnostic**. Given a trained and tuned model, we record the fraction of diagnostic dataset instances that the model scores according to the axiomatic preferences. Thus, a model that is able to rank all instances correctly achieves an *axiomatic score* of 1.0. In line with past works on axiomatic approaches to IR, we expect there to be a positive correlation between models' retrieval effectiveness and their axiomatic scores.

4 Experiments

We now introduce the corpus we use for our study in more detail, then discuss details of the diagnostic datasets created from this corpus. Subsequently, we introduce the employed retrieval models and finally explore to what extent our traditional baselines and neural IR models satisfy the constraints encapsulated in the diagnostic datasets and how this relates to their retrieval effectiveness.

4.1 `WikiPassageQA`

We empirically validate our diagnostic dataset creation pipeline on the answer passage retrieval dataset `WikiPassageQA` [6]. As it contains thousands of topics, it is a suitably large dataset for the training of neural models.

The dataset consists of 861 Wikipedia documents, split into passages of six sentences each, yielding 50, 477 unique passages in total—each containing 135.2

words on average. Crowd-workers created a total of 4, 154 questions. For example, for the Wikipedia document on *Granite*[2] the created questions include:

- *What is the occurrence of granite?*
- *How does weathering affect granite?*
- *What are the geochemical origins of granite?*

The questions contain 9.5 terms on average (minimum 2[3], maximum 39). The binary passage-level relevance judgments were also sourced from the same crowd-workers and later validated by a subsequent mechanical turk verification poll. On average, there are 1.7 relevant passages per question[4].

The corpus has been developed for the *answer passage retrieval* task: given a query (the question) and a Wikipedia document (more concretely, all passages making up that document), rank the passages such that those containing the answer to the question are ranked on top. `WikiPassageQA` has been released with a pre-defined train/dev/test split that we maintain in our work.

As in [6], we employ mean average precision (MAP), mean reciprocal rank (MRR) and precision at k documents (P@k) to report retrieval effectiveness. As noted earlier, in terms of *axiomatic performance*, we report the fraction of precedence constraints each model satisfies.

In terms of pre-processing, we apply stemming[5] but not stopword removal, as the latter may actually remove informative terms from the text such as the question words *what* and *why*.

4.2 Diagnostic Datasets

Given the nature of our corpus, we do not need to randomly sample document pairs or triplets. Instead, we consider all possible pairs or triplets (depending on the axiom in question) of passages within a single Wikipedia document and the respective questions and keep those instances in our diagnostic datasets that satisfy our extended and relaxed axiomatic constraints—once more, keeping in place the train/dev/test split of the original corpus. Since the Wikipedia documents are already split into six-sentence passages, we do not manually set a threshold (δ) of allowed document length differences for this corpus and instead accept all passage pairs/triplets as sufficiently similar in length.

For the $\overline{\texttt{LNC2}}$ axiom, we have two options: we can either (1) add duplicated documents to the test set only (which may be considered "unfair" to the neural models as they have never seen this type of document in the training data) or (2) we add duplicated documents to all (train/dev/test) splits. Here, we report the axiomatic performance of our investigated models across both options.

---

[2] https://en.wikipedia.org/wiki/Granite.

[3] The question is *define Hydroelectricity.*

[4] The corpus statistics reported here differ slightly from those reported in [6] as we filtered out instances with empty question texts, duplicated questions and questions appearing in both the training and test set.

[5] We employed the `nltk.stem.SnowballStemmer` for the English language.

In Table 1 we report the number of extracted diagnostic instances per axiom. Let us first consider the three axioms ($\overline{\texttt{TFC1}}$, $\overline{\texttt{TFC2}}$ and $\overline{\texttt{M-TDC}}$) based on data extraction: depending on the axiom, we extract between 42 K and 3.5 M instances. One may question the need for the introduced relaxations that go beyond document length relaxation. We incorporated those as document length relaxation alone was insufficient for this corpus: as a concrete example, for $\overline{\texttt{TFC1}}$ (when extended and *only* with document length relaxation applied), we only found six instances that could be extracted from `WikiPassageQA`.

Let us now consider $\overline{\texttt{LNC2}}$, whose instances are not extracted from the corpus, but instead require data generation based on the original corpus. We created instances with $k = \{2, 3, 4\}$ times the original content and maintain the original labels (e.g. a passage that was labelled relevant in its original form is labelled relevant in its artificial form as well, as supported by the `LNC2` axiom). We only considered passages up to 240 words in eventual length, due to experimental constraints[6], leading to a total of 10 K and 100 K instances respectively for the two variants of $\overline{\texttt{LNC2}}$. Note, that for $\overline{\texttt{LNC2}}^{All}$ we train the neural models not only on the original train split of `WikiPassageQA`, but add the generated instances to the training data as well; this addition does not significantly alter the fraction of relevant to non-relevant answer passages.

Table 1. Number of instances per axiom ($\overline{\texttt{TFC1}}$, $\overline{\texttt{TFC2}}$, $\overline{\texttt{M-TDC}}$) extracted from `WikiPassageQA`. For $\overline{\texttt{LNC2}}$ we report the number of artificial diagnostic instances, in two variants: the duplication of document content restricted to the test set ($\overline{\texttt{LNC2}}^{Test}$) and across the train/dev/test sets ($\overline{\texttt{LNC2}}^{All}$).

	$\overline{\texttt{TFC1}}$	$\overline{\texttt{TFC2}}$	$\overline{\texttt{M-TDC}}$	$\overline{\texttt{LNC2}}^{Test}$	$\overline{\texttt{LNC2}}^{All}$
Parameters				$k = \{2, 3, 4\}$, $doc\_len_{max} = 240$	
Train	2,758,223	837,838	33,509	0	82,785
Dev	376,902	50,772	3,958	0	10,485
Test	353,621	183,898	4,497	10,074	10,074
Total	3,488,746	1,072,508	41,964	10,074	103,344

4.3 Retrieval Models

For our experiments we opted for the retrieval baselines BM25 and query likelihood with Dirichlet smoothing (QL) as implemented in the `Indri` toolkit [41] (version 5.11) and four neural IR models as implemented in `MatchZoo`[7] [11].

We tuned the hyper-parameters of BM25 and QL on the train and development parts of `WikiPassageQA`, optimizing for MAP, resulting in the following settings: $k1 = 0.4, b = 0.1, k3 = 1$ (BM25) and $\mu = 750$ (QL).

[6] Concretely, we use the `MatchZoo` toolkit for our neural models and ran into issues when the maximum document length was set to include longer passages, see also https://github.com/faneshion/MatchZoo/issues/264.

[7] Version https://github.com/NTMC-Community/MatchZoo/tree/e564565.

For our neural models, we employed the `MatchZoo` retrieval toolkit which has been employed in a number of prior studies, including [2,38,47]. `MatchZoo` contains architecture configurations[8] that have been optimized for the `WikiQA` dataset [46], an open-domain question answering dataset, similar in spirit to `WikiPassageQA`, though defined on the document, not passage level. Due to the computational requirements of neural model training, we limited the maximum query length and passage length to 20 and 240 terms respectively— as a result, in more than 99% of all training instances the entire question and entire passage was considered. We maintained the default `MatchZoo` configurations, including learning rates and optimizers. All neural models were trained for 400 iterations.

Initially, we considered all neural models implemented in `MatchZoo`; however, for a number of models (especially the models incorporating a representation-based component such as CDSSM and MV-LSTM) we observed a significant drop in retrieval effectiveness in the `WikiPassageQA` dataset compared to `WikiQA` when relying on the pre-configured model architectures. As a concrete example, MV-LSTM dropped in MAP from 0.62 in `WikiQA` to 0.22 in `WikiPassageQA`. This lack of model robustness to the corpus is a well-known problem for neural models. Since neural architecture search [50] is beyond the scope of our work, we here consider the four best-performing models, which are all interaction-based, in line with the findings reported in [32]. Concretely, the four models are:

- DRMM [17], an interaction-based model that employs a histogram representation of the similarity between a query and a document;
- aNMM [45], an attention based neural matching model, specifically designed for ranking short text in an interaction-based fashion;
- Duet [29], a hybrid of an interaction-based and representation-based model: it combines two separate deep neural networks, one employs a local representation, and another employs distributed representations for matching the query and the document;
- MatchPyramid [36], a hybrid model that mimics image recognition in its text matching and employs a convolutional neural network.

4.4 Retrieval Model Performance

The main results of our study are shown in Table 2 where we present the models' retrieval effectiveness on the original `WikiPassageQA` corpus as well as the fraction of axiomatic instances each model ranks correctly.

Let us first consider the retrieval effectiveness of our models. As found in several prior studies [17,34,37], and as already indicated in [6] with regard to the `WikiPassageQA` dataset, neural models struggle to outperform decades-old retrieval baselines that contain just a handful of hyper-parameters (and recall that we only report the best-performing neural models wrt. retrieval effectiveness). Only DRMM and aNMM are able to significantly outperform the traditional models, with an increase in MAP from 0.54 (QL) to 0.55 (DRMM) and

[8] https://github.com/faneshion/MatchZoo/tree/e564565/examples/wikiqa/config contains the configurations (learning rate, optimizer, etc.) per model.

0.57 (aNMM) respectively. These results are not unexpected, as DRMM is considered to be one of the most competitive neural IR models to date [33], and DRMM and aNMM are similar in the sense that they both model the interaction between query terms and document terms to build a matching matrix [33]. Furthermore, similar to [37] we find that DRMM outperforms MatchPyramid, and similar to [21] we observe that MatchPyramid in turn outperforms Duet.

Moving on to the axiomatic performance of our models, we find both BM25 and QL to satisfy the precedence constraints of the vast majority of instances across all four axioms: both models satisfy more than 90% of the $\overline{\texttt{M-TDC}}$ instances and more than 70% of the $\overline{\texttt{TFC1}}$ instances. The largest difference in percentage of satisfied axiomatic instances can be found in $\overline{\texttt{TFC2}}$ (BM25 satisfies 98% of instances, QL only 63%), which can explained by the fact that QL with Dirichlet smoothing employs a document length dependent smoothing component. Overall, the results are in line with our expectations: as QL and BM25 conditionally satisfy all axioms according to their analytical analyses [13,14] they should satisfy a large percentage of our extended and relaxed axiomatic instances as well. However, these numbers do not reflect the (un)conditional fulfillment of BM25 and QL per original axiom on a one-to-one basis, for which one possible explanation is our relaxation of the document length difference δ.

When we consider the axiomatic scores of our evaluated neural models we observe a clear gap: while for $\overline{\texttt{TFC1}}$ (i.e., documents with more query terms should have higher retrieval scores) between 69–85% instances are satisfied, for $\overline{\texttt{TFC2}}$ (i.e., the increase in retrieval score becomes smaller as the absolute term count increases) and $\overline{\texttt{M-TDC}}$ (i.e., documents with more occurrences of rare query terms are favoured) this drops to at most 76%. When considering the $\overline{\texttt{LNC2}}$ axiom, we find that only aNMM is able to learn the underlying pattern to some degree (38% of satisfied instances) without observing instances of duplicated documents in training ($\overline{\texttt{LNC2}}^{Test}$); the remaining neural models correctly rank between 0 and 19% of instances. Once we include the diagnostic dataset instances in the training regime ($\overline{\texttt{LNC2}}^{All}$) all models have learned to some degree that duplicated document content should not be penalized, but still, none of the models is able to satisfy even half of the diagnostic instances. Finally, we note that aNMM achieves a higher retrieval effectiveness than QL, while QL outperforms aNMM across all four diagnostic datasets. This is an indication that fulfillment of those four axioms alone is not a perfect indicator of retrieval effectiveness—after all, more than twenty have been proposed in the literature. We leave the evaluation of additional axioms to future work. The correlation between retrieval effectiveness in MAP and the average axiomatic score across all axioms is 0.44 ($N = 6$ retrieval models); this is a positive trend, but not a significant one due to the overall low number of models compared.

What we have gained are insights into the type of patterns our neural models have (not) learned and can use those insights to "fix" the models, just like the traditional IR models were fixed based on their axiomatic analyses. As an example, we may want to train Duet on additional triplets, $\mathbf{q}, \mathbf{d_i}, \mathbf{d_j}$, for which $S(\mathbf{q}, \mathbf{d_i}) > S(\mathbf{q}, \mathbf{d_j})$ according to $\overline{\texttt{TFC1}}$, as Duet currently performs worst on this axiom across the evaluated models.

Table 2. Overview of models' retrieval effectiveness and fraction of fulfilled axiom instances. $^{1/2/3/4}$ denote statistically significant improvements (Wilcoxon signed rank test with $p < 0.05$) in retrieval effectiveness.

	Retrieval effectiveness			Performance per axiom				
	MAP	MRR	P@5	TFC1	TFC2	M-TDC	LNC2^{Test}	LNC2^{All}
[1] BM25	$0.52^{3,4}$	$0.60^{3,4}$	0.18^{3}	0.73	**0.98**	**1.00**	**0.80**	**0.80**
[2] QL	$\mathbf{0.54}^{1,3,4}$	$\mathbf{0.62}^{1,3,4}$	$\mathbf{0.19}^{3}$	**0.87**	0.63	**0.94**	0.68	0.68
[3] Duet	0.25	0.29	0.10	0.69	0.56	0.48	0.19	0.47
[4] MatchPyramid	0.44^{3}	0.51^{3}	0.18^{3}	0.79	0.58	0.63	0.00	0.19
[5] DRMM	$0.55^{1,2,3,4}$	$0.64^{1,2,3,4}$	$0.20^{1,2,3,4}$	0.84	**0.60**	**0.76**	0.05	0.12
[6] aNMM	$\mathbf{0.57}^{1,2,3,4}$	$\mathbf{0.66}^{1,2,3,4}$	$\mathbf{0.21}^{1,2,3,4}$	**0.85**	0.56	0.69	**0.38**	**0.47**

5 Conclusions

In this paper, we have proposed a novel approach to empirically analyze retrieval models that is rooted in the axiomatic approach to IR. Today's neural models, with potentially millions of parameters are too complex for any kind of analytical evaluation; instead, we take inspirations from the NLP and computer vision communities and propose the use of model-agnostic *diagnostic datasets* in order to determine what kind of search heuristics neural models are able to learn. We have shown for four specific axioms how to extend and relax them, in order to make them match realistic datasets. We have applied our diagnostic dataset creation pipeline to the `WikiPassageQA` corpus and evaluated two traditional baselines and four neural models. As a model's axiomatic performance does not require a labelled dataset (i.e., no relevance judgments are required), we can apply our pipeline to almost any dataset containing queries and documents.

Our future work will extend this work in several directions: we will (i) investigate the impact of the adopted document length (δ) relaxation; (ii) extend and relax additional axioms to enlarge our set of diagnostic datasets; (iii) empirically evaluate a larger set of neural models and subsequently attempt to "fix" them (through training data augmentation or the adaptation of their loss function); ad (iv) evaluate a wider range of datasets in order to determine the impact of the retrieval task on the models' axiomatic performance.

Overall, we believe that the axiomatic approach to diagnosing neural IR models presented in this work is a step forward to gaining valuable insights into the black boxes that deep models are generally considered to be.

Acknowledgements. This work was funded by NWO projects LACrOSSE (612.001.605) and SearchX (639.022.722) and Deloitte NL.

References

1. Ariannezhad, M., Montazeralghaem, A., Zamani, H., Shakery, A.: Improving retrieval performance for verbose queries via axiomatic analysis of term discrimination heuristic. In: SIGIR 2017, pp. 1201–1204 (2017)
2. Chen, H., et al.: MIX: multi-channel information crossing for text matching. In: KDD 2018, pp. 110–119 (2018)
3. Clinchant, S., Gaussier, E.: Is document frequency important for PRF? In: Amati, G., Crestani, F. (eds.) ICTIR 2011. LNCS, vol. 6931, pp. 89–100. Springer, Heidelberg (2011). https://doi.org/10.1007/978-3-642-23318-0_10
4. Clinchant, S., Gaussier, E.: A theoretical analysis of pseudo-relevance feedback models. In: ICTIR 2013, pp. 6–13 (2013)
5. Cohen, D., O'Connor, B., Croft, W.B.: Understanding the representational power of neural retrieval models using NLP tasks. In: ICTIR 2018, pp. 67–74 (2018)
6. Cohen, D., Yang, L., Croft, W.B.: WikiPassageQA: a benchmark collection for research on non-factoid answer passage retrieval. In: SIGIR 2018, pp. 1165–1168 (2018)
7. Craswell, N., Croft, W.B., de Rijke, M., Guo, J., Mitra, B.: SIGIR 2017 workshop on neural information retrieval. In: SIGIR 2017, pp. 1431–1432 (2017)
8. Craswell, N., Croft, W.B., de Rijke, M., Guo, J., Mitra, B.: Report on the second SIGIR workshop on neural information retrieval. SIGIR Forum **51**(3), 152–158 (2018)
9. De Boom, C., Van Canneyt, S., Demeester, T., Dhoedt, B.: Representation learning for very short texts using weighted word embedding aggregation. Pattern Recogn. Lett. **80**, 150–156 (2016)
10. Fan, Y., Guo, J., Lan, Y., Xu, J., Zhai, C., Cheng, X.: Modeling diverse relevance patterns in Ad-hoc retrieval. In: SIGIR 2018, pp. 375–384 (2018)
11. Fan, Y., Pang, L., Hou, J., Guo, J., Lan, Y., Cheng, X.: MatchZoo: A Toolkit for Deep Text Matching. arXiv preprint arXiv:1707.07270 (2017)
12. Fang, H.: A re-examination of query expansion using lexical resources. In: ACL HLT 2008, pp. 139–147 (2008)
13. Fang, H., Tao, T., Zhai, C.: A formal study of information retrieval heuristics. In: SIGIR 2004, pp. 49–56 (2004)
14. Fang, H., Tao, T., Zhai, C.: Diagnostic evaluation of information retrieval models. ACM Trans. Inf. Syst. **29**(2), 7:1–7:42 (2011)
15. Fang, H., Zhai, C.: An exploration of axiomatic approaches to information retrieval. In: SIGIR 2005, pp. 480–487 (2005)
16. Fang, H., Zhai, C.: Semantic term matching in axiomatic approaches to information retrieval. In: SIGIR 2006. pp. 115–122 (2006)
17. Guo, J., Fan, Y., Ai, Q., Croft, W.B.: A deep relevance matching model for ad-hoc retrieval. In: CIKM 2016, pp. 55–64 (2016)
18. Hagen, M., Völske, M., Göring, S., Stein, B.: Axiomatic result re-ranking. In: CIKM 2016, pp. 721–730 (2016)
19. Hu, B., Lu, Z., Li, H., Chen, Q.: Convolutional neural network architectures for matching natural language sentences. In: NIPS 2014, pp. 2042–2050 (2014)
20. Huang, P.S., He, X., Gao, J., Deng, L., Acero, A., Heck, L.: Learning deep structured semantic models for web search using clickthrough data. In: CIKM 2013, pp. 2333–2338 (2013)
21. Hui, K., Yates, A., Berberich, K., de Melo, G.: Co-PACRR: a context-aware neural IR model for ad-hoc retrieval. In: WSDM 2018, pp. 279–287 (2018)

22. Jia, R., Liang, P.: Adversarial examples for evaluating reading comprehension systems. In: EMNLP 2017, pp. 2021–2031 (2017)
23. Johnson, J., Hariharan, B., van der Maaten, L., Fei-Fei, L., Zitnick, C.L., Girshick, R.: CLEVR: A diagnostic dataset for compositional language and elementary visual reasoning. In: CVPR 2017, pp. 1988–1997 (2017)
24. Karimzadehgan, M., Zhai, C.: Axiomatic analysis of translation language model for information retrieval. In: ECIR 2012, pp. 268–280 (2012)
25. Li, C., et al.: NPRF: a neural pseudo relevance feedback framework for ad-hoc information retrieval. In: EMNLP 2018, pp. 4482–4491 (2018)
26. Lu, Z., Li, H.: A deep architecture for matching short texts. In: NIPS 2013, pp. 1367–1375 (2013)
27. Lv, Y., Zhai, C.: Lower-bounding term frequency normalization. In: CIKM 2011, pp. 7–16 (2011)
28. Mitra, B., Craswell, N.: An introduction to neural information retrieval. Found. Trends Inf. Retrieval **13**(1), 1–126 (2018)
29. Mitra, B., Diaz, F., Craswell, N.: Learning to match using local and distributed representations of text for web search. In: WWW 2017, pp. 1291–1299 (2017)
30. Montazeralghaem, A., Zamani, H., Shakery, A.: Axiomatic analysis for improving the log-logistic feedback model. In: SIGIR 2016, pp. 765–768 (2016)
31. Na, S.H.: Two-stage document length normalization for information retrieval. TOIS 2015 **33**(2), 8:1–8:40 (2015)
32. Nie, Y., Li, Y., Nie, J.Y.: Empirical Study of Multi-level Convolution Models for IR Based on Representations and Interactions. In: ICTIR 2018, pp. 59–66 (2018)
33. Onal, K.D., Zhang, Y., Altingovde, I.S., Rahman, M.M., Karagoz, P., Braylan, A., Dang, B., Chang, H.L., Kim, H., McNamara, Q., et al.: Neural information retrieval: at the end of the early years. Inf. Retrieval J. **21**(2–3), 111–182 (2018)
34. Pang, L., Lan, Y., Guo, J., Xu, J., Cheng, X.: A Study of MatchPyramid Models on Ad-hoc Retrieval. arXiv preprint arXiv:1606.04648 (2016)
35. Pang, L., Lan, Y., Guo, J., Xu, J., Cheng, X.: A Deep Investigation of Deep IR Models. arXiv preprint arXiv:1707.07700 (2017)
36. Pang, L., Lan, Y., Guo, J., Xu, J., Wan, S., Cheng, X.: Text matching as image recognition. In: AAAI 2016, pp. 2793–2799 (2016)
37. Pang, L., Lan, Y., Guo, J., Xu, J., Xu, J., Cheng, X.: DeepRank: a new deep architecture for relevance ranking in information retrieval. In: CIKM 2017, pp. 257–266 (2017)
38. Rao, J., Yang, W., Zhang, Y., Ture, F., Lin, J.: Multi-Perspective Relevance Matching with Hierarchical ConvNets for Social Media Search. arXiv preprint arXiv:1805.08159 (2018)
39. Shen, Y., He, X., Gao, J., Deng, L., Mesnil, G.: Learning semantic representations using convolutional neural networks for web search. In: WWW 2014, pp. 373–374 (2014)
40. Shi, S., Wen, J.R., Yu, Q., Song, R., Ma, W.Y.: Gravitation-based model for information retrieval. In: SIGIR 2005, pp. 488–495 (2005)
41. Strohman, T., Metzler, D., Turtle, H., Croft, W.B.: Indri: A language model-based search engine for complex queries. In: International Conference on Intelligence Analysis, pp. 2–6 (2004)
42. Tao, T., Zhai, C.: An exploration of proximity measures in information retrieval. In: SIGIR 2007, pp. 295–302 (2007)
43. Wan, S., Lan, Y., Guo, J., Xu, J., Pang, L., Cheng, X.: A deep architecture for semantic matching with multiple positional sentence representations. In: AAAI 2016, pp. 2835–2841 (2016)

44. Weston, J., Bordes, A., Chopra, S., Rush, A.M., van Merriënboer, B., Joulin, A., Mikolov, T.: Towards AI-Complete Question Answering: A Set of Prerequisite Toy Tasks. arXiv preprint arXiv:1502.05698 (2015)
45. Yang, L., Ai, Q., Guo, J., Croft, W.B.: aNMM: ranking short answer texts with attention-based neural matching model. In: CIKM 2016, pp. 287–296 (2016)
46. Yang, Y., Yih, W.T., Meek, C.: WikiQA: a challenge dataset for open-domain question answering. In: EMNLP 2015, pp. 2013–2018 (2015)
47. Yang, Z., et al.: A deep top-k relevance matching model for ad-hoc retrieval. In: Zhang, S., Liu, T.-Y., Li, X., Guo, J., Li, C. (eds.) CCIR 2018. LNCS, vol. 11168, pp. 16–27. Springer, Cham (2018). https://doi.org/10.1007/978-3-030-01012-6_2
48. Zhai, C., Lafferty, J.: A study of smoothing methods for language models applied to ad hoc information retrieval. In: SIGIR 2001, pp. 334–342 (2001)
49. Zhang, X., Zhao, J., LeCun, Y.: Character-level convolutional networks for text classification. In: NIPS 2015, pp. 649–657 (2015)
50. Zoph, B., Le, Q.V.: Neural Architecture Search with Reinforcement Learning. arXiv preprint arXiv:1611.01578 (2016)

Cross Lingual IR

Term Selection for Query Expansion in Medical Cross-Lingual Information Retrieval

Shadi Saleh(✉) and Pavel Pecina

Institute of Formal and Applied Linguistics, Faculty of Mathematics and Physics, Charles University, Prague, Czech Republic
{saleh,pecina}@ufal.mff.cuni.cz

Abstract. We present a method for automatic query expansion for cross-lingual information retrieval in the medical domain. The method employs machine translation of source-language queries into a document language and linear regression to predict the retrieval performance for each translated query when expanded with a candidate term. Candidate terms (in the document language) come from multiple sources: query translation hypotheses obtained from the machine translation system, Wikipedia articles and PubMed abstracts. Query expansion is applied only when the model predicts a score for a candidate term that exceeds a tuned threshold which allows to expand queries with strongly related terms only. Our experiments are conducted using the CLEF eHealth 2013–2015 test collection and show significant improvements in both cross-lingual and monolingual settings.

1 Introduction

In Cross-lingual Information Retrieval (CLIR), search queries are formulated in a language which differs from the language of documents. Machine Translation (MT) of queries into the document language is a common method which reduces this task into monolingual retrieval [19]. In this work, we tackle the *vocabulary mismatch problem* which occurs when MT fails to select the most effective query translation option and subsequently, a term-matching IR system fails to retrieve relevant documents because the terms in the translated query and terms in the relevant documents do not match.

The proposed method is based on a simple linear regression model that predicts the retrieval performance for each candidate expansion term when combined with a query translated by a Statistical Machine Translation (SMT) system. The model features are obtained from the SMT system, external document sources (Wikipedia, PubMed) and information from the document collection. The model is used to score each term from a candidate pool and those scored above a (pre-trained) threshold are automatically added to the translated query. As a result, the queries are expanded with strong candidates only. If no strong candidates are available, the queries remain unchanged. This prevents performance drop caused by adding irrelevant terms to the query.

L. Azzopardi et al. (Eds.): ECIR 2019, LNCS 11437, pp. 507–522, 2019.
https://doi.org/10.1007/978-3-030-15712-8_33

The work presented in this paper is focused on cross-lingual retrieval in the domain of medicine and health. The experiments were conducted using the CLEF eHealth 2013–2015 IR collection. The method, however, is domain-independent and can be used in monolingual retrieval too (after excluding the cross-lingual features). Our results demonstrate a significant improvement over the baseline system which exploits plain query translation using a domain-adapted SMT system. In the monolingual setting, our model significantly outperforms both the monolingual baseline system (no expansion) and the standard Kullback-Leiber divergence (KLD) method for automatic query expansion.

2 Related Work

2.1 Query Expansion

Web search user queries tend to be short. The average web search query length, as reported by Gabrilovich et al. [9], is about 2.5 terms. The information represented in these terms might be too brief and/or vague. This is considered to be a challenge for IR systems that follow the term-matching approach, since they fail to find relevant documents which do not contain the terms specified in the query. Query expansion (QE) can be done automatically, or by interaction with users (e.g. selecting one or more terms to be added to the query), which is known as interactive query expansion [12]. In this study, we will focus on automatic query expansion.

Blind Relevance Feedback (BRF) is one of the most popular techniques for QE, also known as pseudo-relevance feedback [33]. First, an initial retrieval is conducted using the base query and top m ranked documents are selected as a source for term candidates. Then each term in these documents is scored using some approaches like a combination of its frequency (TF) in these documents and its inverse document frequency (IDF) in the collection. Finally, the highest scored m terms are added to the base query and a final retrieval is done. However, there is a risk when following this approach because one or more of these m documents might be irrelevant; thus, adding terms from these documents might drift the information away from the intended one. QE can have significant improvement on one of the main evaluation metrics (such as MAP (mean average precision), precision at 10 documents, or recall) and degrades the others; thus, the use of QE should consider the context of the IR application when using query expansion [13]. Pal et al. [26] employed WordNet to weight a candidate term and measure its usefulness for expansion. They leveraged the similarity score of the top retrieved documents using BRF assumption, and excluded terms from WordNet which do not appear in these documents. They calculated different similarity scores between the query term and the candidate term based on term distribution in the document collection. Then they linearly combined these scores to select the weights of the expansion terms. This approach brought an improvement over the use of base queries on multiple TREC collections. Ermakova and Mothe [8] used local context analysis by choosing terms which surround query terms from documents that are retrieved from the initial retrieval. They assumed that document terms which appear closely to query terms are more likely to be good candidates for expansion. They experimented their approach on TREC Ad-Hoc track datasets from three years (1997–1999)[1] and the WT10G dataset [5]. Cao et al. [3] showed that when QE is based only on term distribution, it can not distinguish good terms, which will improve the IR performance, and bad terms which will harm it. They presented a classification model that is integrated into a BRF method. It uses features from the collection to predict the usefulness of the expansion terms and select only the good ones.

2.2 Query Expansion in Cross-Lingual Information Retrieval

Cross-lingual Information Retrieval (CLIR) enables users to search in a collection that is different than their language. In order to conduct a retrieval that is based on term-matching, both documents and queries should be represented in one language [24]. Query-translation is the most common approach in CLIR, wherein user queries are translated into the document language and then a monolingual retrieval is conducted. Popular machine translation techniques struggle translating short queries because of the lack of linguistic information that is required to solve ambiguity, which eventually causes information loss in the translated queries [32]. Query expansion in CLIR helps to solve this issue by adding relevant terms to the translated queries. Chandra and Dwivedi [4] used Google Translate[2] to translate queries from Hindi into English in the FIRE 2008 dataset. Then, they did an initial retrieval using the translated queries and created a set of candidate terms. They applied different methods for term selection. They found that adding the term which has the lowest frequency in the top 3 ranked documents gave the best result.

2.3 Query Expansion in Medical Information Retrieval

Query expansion in the medical domain is considered to be a more difficult task. Approaches which work on the general domain might not work perfectly when applied in the medical domain. Nikoulina et al. [21] reported that simply merging the top 5 scored translation hypotheses (as a special QE approach) to create queries in the CLIR task outperformed the baseline system in the general domain data. However, the same approach did not work when tested on the medical domain [34]. Kullback-Leiber Divergence (KLD) for query expansion (explained in Sect. 3.4) failed to outperform the baseline system (using initial queries) during the CLEF 2011 medical retrieval task [16]. Choi and Choi [6] used Google Translate to translate the queries into English (from Czech, French and German) during their participation in the CLEF eHealth 2014 CLIR task [10]. Then, they annotated each query with medical concepts using MetaMap [2], and the top scored concepts were added to the original query. Finally, they weighted the original query and the expanded query with 0.9 and 0.1 respectively. Query

[1] https://trec.nist.gov.
[2] http://translate.google.com.

expansion approach outperformed their baseline system by 18% for Czech, 4% for German and 4% for French. Liu and Nie [18] participated in the monolingual task of CLEF eHealth 2015 [11], and presented a system which expanded queries with UMLS [15] concepts and terms extracted from Wikipedia articles. Authors claim that Wikipedia abstracts are similar to the way that users pose queries (more generic), while the titles of Wikipedia articles contain medical terms. However, using only Wikipedia to expand the queries did not help. Only a system that combined Wikipedia with MetaMap [2] improved the baseline system. Employing MeSH (Medical Subject Heading)[3] for QE was investigated thoroughly. Wright et al. [39] presented a simple method that expands queries with five synonyms from MeSH. Nunzio and Moldovan [23] expanded a query with one MeSH term that is related to the base query, when there was more than one MeSH candidate term, they created multiple expanded queries, then for each expanded query, they conducted retrieval and merged the retrieved documents by different approaches like averaging document scores or summing them.

Word embeddings became a well-known technique in representing terms in high dimensional vectors. This allows estimating semantic and syntactic similarities between terms. Term vectors can be generated using famous models like word2vec [20] and Glove [31]. The main idea is to expand the query with terms that are semantically related and appear in a position close to the query terms [22,40,41]. Multiple researchers confirmed that embeddings models that are trained on medical data like PubMed articles are not significantly better than those which are trained on general domain data, such as news [42].

3 Experimental Setting

3.1 Test Collection

The training and evaluation data used in our work is described in [36]. It is adopted from the IR tasks of the CLEF eHealth Lab series 2013–2015 [10,27,38]. The **document collection** is taken from the IR task in 2015 eHealth Task 2: User-Centred Health Information Retrieval [11], and consists of about 1.1 million web pages (documents) crawled from various medical-domain websites. We cleaned the documents using HTML-Strip Perl module[4]. We did not perform any preprocessing (stemming, lemmatisation) since it showed degrading in our previous experiments. The **query set** contains 166 items used during the three years of the CLEF eHealth IR tasks as test queries. The queries were originally created in English and then manually translated into seven languages (Czech, French, German, Spanish, Swedish, Polish, and Hungarian) to allow cross-lingual experiments. As proposed in [36], we used 100 queries for training the model parameters (feature weights, term selection threshold, IR model parameters) and 66 queries for testing (measuring retrieval performance). See Table 1 for query examples and [36] for additional details.

[3] https://www.nlm.nih.gov/mesh.
[4] http://search.cpan.org/dist/HTML-Strip/Strip.pm.

Table 1. Query samples from the extended CLEF eHealth test collection.

Query id	Query title
2013.02	*Facial cuts and scar tissue*
2013.41	*Right macular hemorrhage*
2013.30	*Metabolic acidosis*
2014.04	*Anoxic brain injury*
2014.21	*Renal failure*
2014.17	*Chronic duodenal ulcer*
2015.08	*Cloudy cornea and vision problem*
2015.59	*Heavy and squeaky breath*
2015.48	*Cannot stop moving my eyes medical condition*

3.2 Machine Translation of Queries

In our experiments, we employ a statistical machine translation (SMT) system for query translation. This system was developed under the Khresmoi project [7]. It is based on Moses [17], a state-of-the-art phrase-based system, trained on a combination of in-domain (EMEA, PatTR, COPPA, UMLS, etc.) and general-domain (e.g., EuroParl, JRC Acquis and News Commentary corpus) resources. The system employs several special features [28] that allow for optimal translation of medical search queries. For an input text, an SMT system produces a list of translation hypotheses ranked by their translation quality, the best one is referred to as *1-best* translation. In this research, we employ seven SMT models to translate queries from Czech, German, French, Hungarian, Polish, Spanish and Swedish into English.

3.3 Baseline Retrieval System

Our baseline CLIR system is designed as follows: The non-English queries are translated into English (using *1-best* translations produced by the SMT system described above) which reduces the CLIR task into a monolingual IR task. For indexing and retrieval, we use Terrier, an open source search engine [25], and its implementation of the language model with Bayesian smoothing with Dirichlet prior [37]. The default value of the smoothing parameter μ is set to 2500 (this has been proven to work well in our previous work [35]). For comparison purposes, we also report results of a monolingual system which employs the English (reference) translations of the queries. It sets a theoretical maximum which a CLIR system can reach when a query translation is completely correct.

Retrieval results are evaluated by the standard *trec_eval* tool[5] using two evaluation metrics: precision at top 10 documents (*P@10*) which is used as the main evaluation measure in our work, and preference-based measure *BPREF* which

[5] http://trec.nist.gov/trec_eval.

considers if the judged relevant documents are ranked above the judged irrelevant ones. Statistical significance tests are performed using the paired Wilcoxon signed-rank test [14], with α set to 0.05.

3.4 KLD Query Expansion

To compare our query expansion method with other approaches, we report the results of Kullback-Leiber Divergence (KLD) for query expansion as it is implemented in Terrier [1]. In KLD, the top n ranked documents (pseudo-relevant documents) are retrieved using the base query, then each term in these documents is scored by the equation below, where $P_r(t)$ is the probability of term t in the pseudo-relevant documents, and $P_c(t)$ is the probability of term t in the document collection c. Finally the top m scored terms are added to the base query and a final retrieval is done using the new expanded query.

$$Score(t) = P_r(t) \cdot log\left(\frac{P_r(t)}{P_c(t)}\right)$$

We set n and m to 7 and 2 respectively by grid-search tuning (using the monolingual English training queries) as shown in Fig. 3.

4 Term Selection Model

The proposed CLIR query expansion method is performed in four steps. First, a set of candidate terms (candidate pool) are collected from various sources. Second, each term from the candidate pool is assigned a vector of features describing its potential to identify relevant documents. Third, the features are combined in a regression model to score each candidate term. Finally, terms with scores exceeding a given threshold are selected to expand the query. Figure 1 shows the architecture of our presented model in detail. The following sections explain the term selection process.

4.1 Candidate Pool

Three sources of candidate terms are considered in our experiments:

Machine Translation (MT). For each source query, we collect all the terms from the 100 highest-scored translation hypotheses as produced by the SMT system. The motivation behind this is based on the fact that the translation hypotheses might contain alternative translations of query terms (synonyms/other related terms).

English Wikipedia (Wiki). The base query (*1-best* translation) is used to retrieve articles from an indexed Wikipedia collection. Only titles and abstracts are used to build the index following the same settings as in our baseline model. The titles of the top 10 ranked retrieved articles are added to the expansion pool.

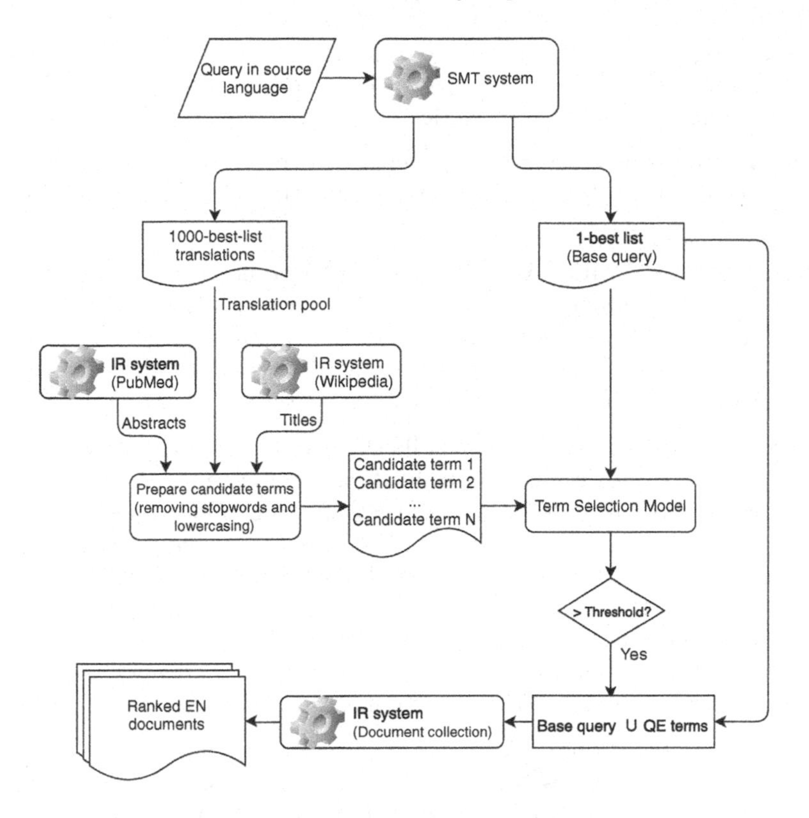

Fig. 1. System architecture overview.

The use of Wikipedia titles is to tackle the challenge when users pose a query in the medical domain using non-medical terms by describing the symptoms of a specific disease. Disease names usually appear in the title and their symptoms are described in the abstract [18].

PubMed. We also enrich the candidate pool with terms from the PubMed articles [30] following the settings as the Wikipedia articles. PubMed articles (both abstracts and titles) are indexed, then the top 10 ranked articles are retrieved using the *1-best* translation as a base query and added to the candidate pool.

4.2 Feature Set

Each term from the candidate pool is described by a set of features designed to reflect the term's usefulness for expansion:

IDF, which is calculated in the document collection.

Translation pool frequency, i.e. the frequency of the term in the 100 highest-scored translation hypotheses as produced by the SMT system. When a term appears in multiple hypotheses, this means that the probability of being a relevant translation to one of the terms in the original query is high. This feature is excluded in our monolingual QE model.

Wikipedia frequency, i.e. the frequency of the term in the top 10 Wikipedia articles retrieved from the Wikipedia index using the *1-best* translation as a base query.

Retrieval Status Value (RSV), which is the difference of the RSV value (the score of the Dirichlet retrieval model) of the highest-ranked document retrieved using the base query, and the RSV value of the highest-ranked document retrieved using the base query expanded by the candidate term. This feature tells us the contribution of the candidate term to the RSV score.

Query similarity, i.e. an average similarity between a candidate term t_m and the query term obtained using a pre-trained model of *word2vec* embeddings on 25 millions articles from PubMed[6]. First, we get the word embeddings for each term in the original query and we sum those embeddings to get a vector that represents the entire query. Then we take the embeddings for t_m, and calculate the cosine similarity between the query vector and the t_m vector. It is important to point out here that choosing terms that are similar to each term of the query caused significant drift in the information need, for example: *mother* was suggested as a similar term to *baby*, and *white* as a similar term to *black*.

Co-occurrence frequency, the co-occurrences of a candidate term t_m and the query terms $t_i \in Q$ indicates how likely t_m is related to the original query Q, we sum up the co-occurrence frequency for each term in query Q and the candidate term t_m in all documents d_j in the collection C, as shown below:

$$co(t_m, Q) = \sum_{d_j \in C, t_i \in Q} tf(d_j, t_i) tf(d_j, t_m)$$

Term frequency, first, we perform retrieval from the collection using a query that is constructed from the *1-best* translation, then we calculate the term frequency of a candidate term t_m in the top 10 ranked documents from the retrieval result.

UMLS frequency, this feature represents how many times a term appeared in the UMLS lexicon [15], as an attempt to give more weight to the medical terms.

4.3 Regression Model

The term selection model is based on linear regression. Training instances are candidate terms for the training queries after translating those queries from all seven languages into English. Each term t from a candidate pool of a given query is assigned a value computed as the difference of P@10 obtained by the baseline query (*1-best-list* translation) and P@10 obtained by the expanded baseline

[6] https://www.ncbi.nlm.nih.gov/CBBresearch/Wilbur/IRET/DATASET/.

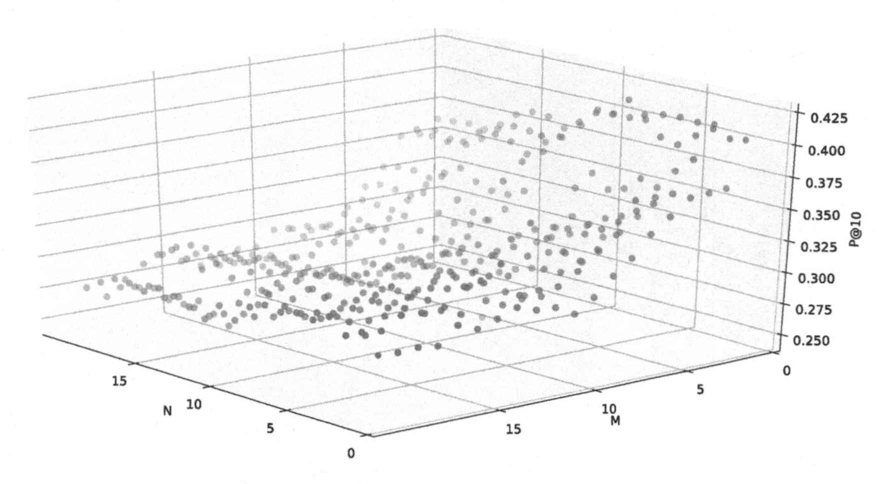

Fig. 2. Tuning KLD parameters, number of documents (N) and number of expansion terms (M) on the monolingual queries.

query with the term t. Expansion terms increasing P@10 for the given query are assigned positive values, terms decreasing P@10 are assigned negative values, and terms without any effect on the retrieval performance for that query are assigned zero. The purpose is to expand the queries with terms that can improve the performance, rather than terms that harm the performance. The feature vectors are centered and reduced. This is done independently on each feature on the training set, then we use the scaler coefficient to standardise the test set. We consider P@ difference as the objective function, and we use the proposed feature set to train the model. Linear Regression (LR) models the relationship between the dependent variable (P@10 in our case) and the regressors x (term feature values).

We use ordinary least squares linear regression as it is implemented in *scikit* package [29]. There might be one or more good candidate terms for expansion. To select these terms, we set a threshold value for the predicted score. The threshold value is tuned on the training set for all languages as shown in Fig. 3. All terms which have a score equal or higher than the threshold are added to the base query. This allows us to avoid expanding queries with irrelevant terms.

5 Experiments and Results

Results of all experiments for the seven languages are presented in percentages in Table 2 (in terms of P@10) and Table 3 (in terms of BPREF). For each language, the underlined score denotes the best result, and the scores in bold refer to results which are not significantly different (given the Wilcoxon signed-rank test) from the best (underlined) score. TS refers to the proposed QE technique based on term selection, and the text in the brackets denote the candidate term sources: machine translation (MT) hypotheses, Wikipedia titles (Wiki),

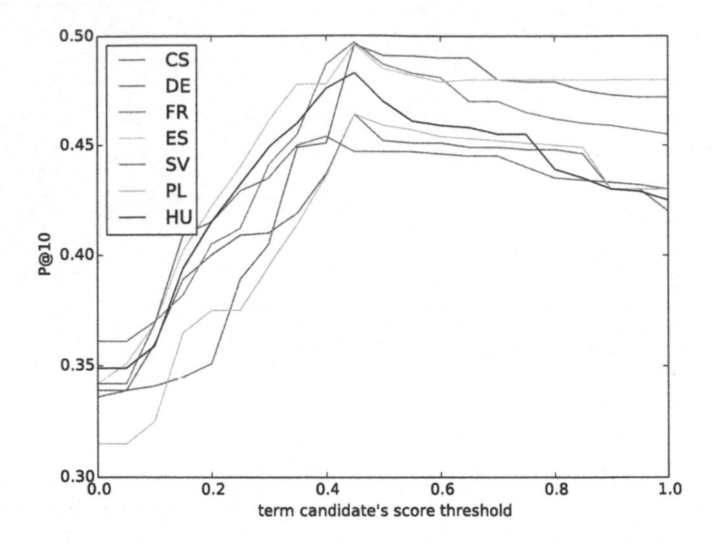

Fig. 3. Tuning threshold for term candidate selection based on their predicted scores.

and PubMed articles (PubMed). The monolingual experiment (exploiting the reference English queries) sets a theoretical upper-boundary for the results of the CLIR experiments. It is 53.03 in terms of P@10 and 39.94 in terms of BPREF. (These values hold for all the languages since the reference translations of the source queries are the same). Monolingual+KLD refers to the result of the KLD-based query expansion applied to the reference translations of the queries. In terms of P@10, the result went down substantially. This can be explained because either the indexed documents are not good enough as a source of candidate expansion terms, or because there is no criteria to prevent

Table 2. Experiment results in terms of P@10 (percentage)

System/query language	CS	FR	DE	ES	HU	PL	SV
Monolingual	53.03	53.03	53.03	53.03	53.03	53.03	53.03
+KLD	48.18	48.18	48.18	48.18	48.18	48.18	48.18
+TS(PubMed)	55.76	55.76	55.76	55.76	55.76	55.76	55.76
Baseline	47.27	48.03	44.24	46.97	45.91	42.12	40.00
+KLD	39.85	45.76	38.33	42.12	42.12	39.24	36.36
+TS(MT)	47.42	48.03	43.03	46.82	46.21	**42.42**	<u>**41.52**</u>
+TS(Wiki)	44.85	44.70	43.03	43.18	**47.12**	41.06	39.70
+TS(PubMed)	50.15	47.12	43.33	45.30	43.48	37.58	36.52
+TS(MT ∪ Wiki)	<u>**52.58**</u>	<u>**49.55**</u>	<u>**47.12**</u>	<u>**48.33**</u>	<u>**47.88**</u>	**42.42**	<u>**41.52**</u>
+TS(MT ∪ PubMed)	50.30	**48.79**	**45.45**	**48.03**	42.73	38.48	34.85
+TS(MT ∪ Wiki ∪ PubMed)	**52.12**	**48.94**	**45.45**	**47.42**	**47.58**	<u>**43.18**</u>	**41.21**

Table 3. Experiment results in terms of BPREF (percentage)

System/query language	CS	FR	DE	ES	HU	PL	SV
Monolingual	39.94	39.94	39.94	39.94	39.94	39.94	39.94
+KLD	41.22	41.22	41.22	41.22	41.22	41.22	41.22
+TS(PubMed)	41.41	41.41	41.41	41.41	41.41	41.41	41.41
Baseline	36.79	35.65	35.38	37.24	37.08	33.77	20.94
+KLD	36.21	**38.34**	34.84	**39.64**	36.59	**34.33**	32.11
+TS(MT)	36.80	35.49	35.64	37.05	37.03	**33.92**	**33.38**
+TS(Wiki)	36.82	36.10	36.09	36.17	**38.77**	**33.82**	**34.23**
+TS(PubMed)	**39.16**	**38.14**	**39.15**	**39.47**	36.87	33.51	**33.78**
+TS(MT ∪ Wiki)	**40.49**	**38.82**	**40.86**	37.93	36.95	**33.92**	**33.38**
+TS(MT ∪ PubMed)	38.90	**37.63**	36.09	**38.87**	36.57	**34.16**	**33.67**
+TS(MT ∪ Wiki ∪ PubMed)	**40.21**	**37.15**	36.02	37.93	37.70	**33.86**	32.98

Table 4. Precision (percentage) of selected terms manually checked by a medical expert (first raw) and with respect to the terms that appeared in the reference English queries (second raw)

Measure/query language	CS	FR	DE	ES	HU	PL	SV
Precision w.r.t. manual judgments	87.60	89.33	90.84	87.50	96.43	90.91	87.50
Precision w.r.t. reference translations	21.49	14.04	13.74	25.00	21.43	36.36	12.50

expanding some queries with low scored term candidates. The proposed term selection (TS) method applied to the monolingual retrieval (using PubMed only as a source of candidate terms) seems to be much more promising. The P@10 score is as high as 55.76. This system improved the results for 13 queries and degraded 4 queries. The rest of the queries did not change due to the low scores of candidate terms as predicted by the model. In terms of BPREF, both KLD and TS bring a small improvement which is not statistically significant. P@10 scores of the CLIR baseline systems (exploiting *1-best* translation) range between 40.00 and 48.03 depending on the query language. The KLD-based expansion in CLIR brings the scores even lower (36.36–45.76) which is in line with the monolingual expansion experiments. Though, for some queries (10 on average), P@10 improved, and results for more queries (20 on average) degraded. The proposed term selection experiments show consistent improvement over the baseline. The best system uses terms from MT and Wiki for expansion. Samples of queries that are improved by this system are shown in Table 5. The CLIR system improved 21 queries in Czech, 18 in French, 14 for German and 11 in Spanish, 10 queries in Hungarian, 2 queries in Polish, and 3 queries in Swedish. While it degraded 11 queries in Czech, 12 in French, 11 in German, 4 in Spanish, 5 queries in Hungarian, 2 queries in Polish, and 2 queries in Swedish. The performance of the rest of the queries did not change. The average result in Czech is very close to

Table 5. Examples of queries from different systems including Mono (*ref*), Baseline (*base*), and expansion terms to the baseline query (*QE*). The scores in parentheses refer to query P@10 scores in percentages

Query: 2015.18 (Czech)
ref: poor gait and balance with shaking (50.00)
base: bad posture and balance with tremor (60.00)
QE: poor balanced shaking (70.00)

Query: 2014.21 (French)
ref: white patchiness in mouth (10.00)
base: renal impairment (00.00)
QE: kidney disease function dysfunction failure insufficiency deficiency poor (30.00)

Query: 2013.11 (German)
ref: chest pain and liver transplantation (50.00)
base: breast pain and liver transplantation (10.00)
QE: chest hepatic graft thoracic (40.00)

Query: 2014.11 (Spanish)
ref: Diabetes type 1 and heart problems (40.00)
base: type 1 diabetes and heart problems (40.00)
QE: cardiac disease (60.00)

Table 6. Examples of queries degraded in the QE approach (*QE*) with respect to Mono (*ref*), Baseline (*base*), the scores in parentheses refer to query P@10 scores in percentages

Query: 2013.41 (Czech)
ref: right macular hemorrhage (60.00)
base: amacular bleeding right (70.00)
QE: hemorrhage haemorrhage side blood (30.00)

Query: 2013.41 (French)
ref: right macular hemorrhage (60.00)
base: macular hemorrhage right eye (80.00)
QE: eyes haemorrhage hemorrhagic bleeding (50.00)

Query: 2015.65 (German)
ref: weird brown patches on skin (10.00)
base: strange brown spots on the skin (40.00)
QE: spot patches cutaneous patch (10.00)

Query: 2014.31 (Spanish)
ref: Acute renal failure (80.00)
base: acute renal failure (80.00)
QE: kidney disease (60.00)

the monolingual performance. Table 6 shows examples of queries that degraded in the TS(MT ∪ Wiki) system.

For further analysis of the expansion quality, we report in Table 4 the percentage of relevant expansion terms calculated by the two methods. In the first method, we provided a medical doctor with query titles, their narratives (to understand the topic for each query) and the expansion terms as suggested by the TS(MT ∪ Wiki) system. We asked them to identify the expanded terms whether they are relevant to the topic or not. The second method is an automatic evaluation that is done by checking if the expansion terms exist in the reference queries. For example in the Czech system, 78.51% of the expansion terms did not appear in the reference query; however, we could not tell if they are relevant or not. In contrast, when checked by a medical doctor, it appeared that only 12.4% of them are irrelevant to the topic.

6 Conclusion

In this work, we have addressed the problem of automatic query expansion for cross-lingual information retrieval as an attempt to improve the information represented in user queries. The presented model is based on machine translation of queries from a source language to a document language and machine learning to predict relevant expansion terms from a rich source of term candidates. The feature set is based on information derived from the collection, external resources (Wikipedia and PubMed articles), and word-embeddings. Fine-tuning the threshold value of the term predicted score helps to expand queries only when there is a good candidate. This prevents expanding queries when candidate terms are irrelevant to the topic. Our evaluation has shown that our approach helps significantly improving the baseline system in cross-lingual and mono-lingual settings.

Acknowledgments. This work was supported by the Czech Science Foundation (grant n. 19-26934X).

References

1. Amati, G., Carpineto, C., Romano, G.: Query difficulty, robustness, and selective application of query expansion. In: McDonald, S., Tait, J. (eds.) ECIR 2004. LNCS, vol. 2997, pp. 127–137. Springer, Heidelberg (2004). https://doi.org/10.1007/978-3-540-24752-4_10
2. Aronson, A.R.: Effective mapping of biomedical text to the UMLS Metathesaurus: the MetaMap program. In: Proceedings of AMIA Symposium, pp. 17–21 (2001)
3. Cao, G., Nie, J.Y., Gao, J., Robertson, S.: Selecting good expansion terms for pseudo-relevance feedback. In: Proceedings of the 31st Annual International ACM SIGIR Conference on Research and Development in Information Retrieval, SIGIR 2008, pp. 243–250. ACM, New York (2008)
4. Chandra, G., Dwivedi, S.K.: Query expansion based on term selection for Hindi-English cross lingual IR. J. King Saud Univ. Comput. Inf. Sci. (2017)
5. Chiang, W.T.M., Hagenbuchner, M., Tsoi, A.C.: The wt10g dataset and the evolution of the web. In: Special Interest Tracks and Posters of the 14th International Conference on World Wide Web, WWW 2005, pp. 938–939. ACM, New York (2005)
6. Choi, S., Choi, J.: Exploring effective information retrieval technique for the medical web documents: Snumedinfo at clefehealth2014 task 3. In: Working Notes of CLEF 2015 - Conference and Labs of the Evaluation forum, vol. 1180, pp. 167–175. CEUR-WS.org, Sheffield (2014)
7. Dušek, O., Hajič, J., Hlaváčová, J., Novák, M., Pecina, P., Rosa, R., et al.: Machine translation of medical texts in the Khresmoi project. In: Proceedings of the Ninth Workshop on Statistical Machine Translation, pp. 221–228, Baltimore (2014)
8. Ermakova, L., Mothe, J.: Query expansion by local context analysis. In: Conference francophone en Recherche d'Information et Applications (CORIA 2016), pp. 235–250. CORIA-CIFED, Toulouse (2016)
9. Gabrilovich, E., Broder, A., Fontoura, M., Joshi, A., Josifovski, V., Riedel, L., Zhang, T.: Classifying search queries using the web as a source of knowledge. ACM Trans. Web **3**(2), 5 (2009)

10. Goeuriot, L., et al.: ShARe/CLEF eHealth evaluation lab 2014, Task 3: user-centred health information retrieval. In: Proceedings of CLEF 2014, pp. 43–61. CEUR-WS.org, Sheffield (2014)
11. Goeuriot, L., et al.: Overview of the CLEF eHealth evaluation lab 2015. In: Mothe, J., et al. (eds.) CLEF 2015. LNCS, vol. 9283, pp. 429–443. Springer, Cham (2015). https://doi.org/10.1007/978-3-319-24027-5_44
12. Harman, D.: Towards interactive query expansion. In: Proceedings of the 11th Annual International ACM SIGIR Conference on Research and Development in Information Retrieval, pp. 321–331. SIGIR 1988, ACM, New York (1988)
13. Harman, D.: Information retrieval. In: Relevance Feedback and Other Query Modification Techniques, pp. 241–263. Prentice-Hall Inc., Upper Saddle River (1992)
14. Hull, D.: Using statistical testing in the evaluation of retrieval experiments. In: Proceedings of the 16th Annual International ACM SIGIR Conference on Research and Development in Information Retrieval, pp. 329–338. ACM, Pittsburgh (1993)
15. Humphreys, B.L., Lindberg, D.A.B., Schoolman, H.M., Barnett, G.O.: The unified medical language system. J. Am. Med. Inform. Assoc. **5**(1), 1–11 (1998)
16. Kalpathy-Cramer, J., Muller, H., Bedrick, S., Eggel, I., De Herrera, A., Tsikrika, T.: Overview of the clef 2011 medical image classification and retrieval tasks. In: CLEF 2011 - Working Notes for CLEF 2011 Conference, vol. 1177. CEUR-WS (2011)
17. Koehn, P., Hoang, H., Birch, A., Callison-Burch, C., Federico, M., Bertoldi, N., et al.: Moses: open source toolkit for statistical machine translation. In: Proceedings of the 45th Annual Meeting of the Association for Computational Linguistics, Demo and Poster Sessions, pp. 177–180, Stroudsburg (2007)
18. Liu, X., Nie, J.: Bridging layperson's queries with medical concepts - GRIUM @CLEF2015 eHealth Task 2. In: Working Notes of CLEF 2015 Conference and Labs of the Evaluation forum, vol. 1391. CEUR-WS.org, Toulouse (2015)
19. McCarley, J.S.: Should we translate the documents or the queries in cross-language information retrieval? In: Proceedings of the 37th Annual Meeting of the Association for Computational Linguistics on Computational Linguistics, pp. 208–214, College Park (1999)
20. Mikolov, T., Sutskever, I., Chen, K., Corrado, G., Dean, J.: Distributed representations of words and phrases and their compositionality. In: Proceedings of the 26th International Conference on Neural Information Processing Systems, NIPS 2013, vol. 2, pp. 3111–3119. Curran Associates Inc., Red Hook (2013)
21. Nikoulina, V., Kovachev, B., Lagos, N., Monz, C.: Adaptation of statistical machine translation model for cross-lingual information retrieval in a service context. In: Proceedings of the 13th Conference of the European Chapter of the Association for Computational Linguistics, pp. 109–119, Stroudsburg (2012)
22. Nogueira, R., Cho, K.: Task-oriented query reformulation with reinforcement learning. In: Proceedings of the 2017 Conference on Empirical Methods in Natural Language Processing, pp. 574–583 (2017)
23. Nunzio, G.M.D., Moldovan, A.: A study on query expansion with mesh terms and elasticsearch. IMS unipd at CLEF ehealth task 3. In: Working Notes of CLEF 2018 - Conference and Labs of the Evaluation Forum, Avignon, France, 10–14 September 2018. CEUR-WS, Avignon (2018)
24. Oard, D.W.: A comparative study of query and document translation for cross-language information retrieval. In: Farwell, D., Gerber, L., Hovy, E. (eds.) AMTA 1998. LNCS (LNAI), vol. 1529, pp. 472–483. Springer, Heidelberg (1998). https://doi.org/10.1007/3-540-49478-2_42

25. Ounis, I., Amati, G., Plachouras, V., He, B., Macdonald, C., Johnson, D.: Terrier information retrieval platform. In: Losada, D.E., Fernández-Luna, J.M. (eds.) ECIR 2005. LNCS, vol. 3408, pp. 517–519. Springer, Heidelberg (2005). https://doi.org/10.1007/978-3-540-31865-1_37
26. Pal, D., Mitra, M., Datta, K.: Improving query expansion using wordnet. J. Assoc. Inf. Sci. Technol. **65**(12), 2469–2478 (2014)
27. Palotti, J.R., Zuccon, G., Goeuriot, L., Kelly, L., Hanbury, A., Jones, G.J., Lu pu, M., Pecina, P.: CLEF eHealth Evaluation Lab 2015, Task 2: Retrieving information about medical symptoms. In: CLEF (Working Notes), pp. 1–22. Springer, Heidelberg (2015)
28. Pecina, P., Dušek, O., Goeuriot, L., Hajič, J., Hlavářová, J., Jones, G.J., et al.: Adaptation of machine translation for multilingual information retrieval in the medical domain. Artif. Intell. Med. **61**(3), 165–185 (2014)
29. Pedregosa, F., et al.: Scikit-learn: machine learning in Python. J. Mach. Learn. Res. **12**, 2825–2830 (2011)
30. Peng, Y., Wei, C.H., Lu, Z.: Improving chemical disease relation extraction with rich features and weakly labeled data. J. Cheminformatics **8**(1), 53 (2016)
31. Pennington, J., Socher, R., Manning, C.: Glove: global vectors for word representation. In: Proceedings of the 2014 Conference on Empirical Methods in Natural Language Processing (EMNLP), pp. 1532–1543 (2014)
32. Pirkola, A., Hedlund, T., Keskustalo, H., Järvelin, K.: Dictionary-based cross-language information retrieval: problems, methods, and research findings. Inform. Retrieval **4**(3–4), 209–230 (2001)
33. Rocchio, J.J.: Relevance feedback in information retrieval. The SMART Retrieval Syst. Exp. Autom. Doc. Process. 313–323 (1971)
34. Saleh, S., Pecina, P.: Reranking hypotheses of machine-translated queries for cross-lingual information retrieval. In: Fuhr, N., Quaresma, P., Gonçalves, T., Larsen, B., Balog, K., Macdonald, C., Cappellato, L., Ferro, N. (eds.) CLEF 2016. LNCS, vol. 9822, pp. 54–66. Springer, Cham (2016). https://doi.org/10.1007/978-3-319-44564-9_5
35. Saleh, S., Pecina, P.: Task3 patient-centred information retrieval: Team CUNI. In: Working Notes of CLEF 2016 - Conference and Labs of the Evaluation forum. CEUR-WS.org, Evora (2016)
36. Saleh, S., Pecina, P.: An Extended CLEF eHealth Test Collection for Cross-lingual Information Retrieval in the medical domain. In: Advances in Information Retrieval - 41th European Conference on IR Research, ECIR 2019, Cologne, Germany, April 14–18, 2019, Proceedings. Lecture Notes in Computer Science, Springer (2019)
37. Smucker, M.D., Allan, J.: An investigation of Dirichlet prior smoothing's performance advantage. University of Massachusetts, Technical report (2005)
38. Suominen, H., et al.: Overview of the ShARe/CLEF eHealth evaluation lab 2013. In: Forner, P., Müller, H., Paredes, R., Rosso, P., Stein, B. (eds.) CLEF 2013. LNCS, vol. 8138, pp. 212–231. Springer, Heidelberg (2013). https://doi.org/10.1007/978-3-642-40802-1_24
39. Wright, T.B., Ball, D., Hersh, W.: Query expansion using mesh terms for dataset retrieval: OHSU at the biocaddie 2016 dataset retrieval challenge. J. Biol. Databases Curation 2017, Database (2017)
40. Zamani, H., Croft, W.B.: Embedding-based query language models. In: Proceedings of the 2016 ACM International Conference on the Theory of Information Retrieval, ICTIR 2016, pp. 147–156. ACM, New York (2016)

41. Zamani, H., Croft, W.B.: Relevance-based word embedding. In: Proceedings of the 40th International ACM SIGIR Conference on Research and Development in Information Retrieval, pp. 505–514. SIGIR 2017. ACM, New York (2017)
42. Zuccon, G., Koopman, B., Bruza, P., Azzopardi, L.: Integrating and evaluating neural word embeddings in information retrieval. In: Proceedings of the 20th Australasian Document Computing Symposium, p. 12. Stroudsburg (2015)

Zero-Shot Language Transfer for Cross-Lingual Sentence Retrieval Using Bidirectional Attention Model

Goran Glavaš[1(✉)] and Ivan Vulić[2]

[1] Data and Web Science Group, University of Mannheim, Mannheim, Germany
goran@informatik.uni-mannheim.de
[2] Language Technology Lab, University of Cambridge, Cambridge, UK
iv250@cam.ac.uk

Abstract. We present a neural architecture for cross-lingual mate sentence retrieval which encodes sentences in a joint multilingual space and learns to distinguish true translation pairs from semantically related sentences across languages. The proposed model combines a recurrent sequence encoder with a bidirectional attention layer and an intra-sentence attention mechanism. This way the final fixed-size sentence representations in each training sentence pair depend on the selection of contextualized token representations from the other sentence. The representations of both sentences are then combined using the bilinear product function to predict the relevance score. We show that, coupled with a shared multilingual word embedding space, the proposed model strongly outperforms unsupervised cross-lingual ranking functions, and that further boosts can be achieved by combining the two approaches. Most importantly, we demonstrate the model's effectiveness in zero-shot language transfer settings: our multilingual framework boosts cross-lingual sentence retrieval performance for unseen language pairs without any training examples. This enables robust cross-lingual sentence retrieval also for pairs of resource-lean languages, without any parallel data.

Keywords: Cross-lingual retrieval · Language transfer · Bidirectional attention model · Sentence retrieval

1 Introduction

Retrieving relevant content across languages (i.e., cross-lingual information retrieval, termed CLIR henceforth) requires the ability to bridge the lexical gap between languages. In general, there are three distinct approaches to CLIR. First, translating queries and/or documents "using dictionaries or full-blown machine translation (MT)" to the same language enables the use of monolingual retrieval models [17, 23, 31]. Second, the lexical chasm can be crossed by grounding queries and documents in an external multilingual knowledge source (e.g., Wikipedia or BabelNet) [10, 46]. Finally, other systems induce shared cross-lingual semantic

L. Azzopardi et al. (Eds.): ECIR 2019, LNCS 11437, pp. 523–538, 2019.
https://doi.org/10.1007/978-3-030-15712-8_34

spaces (e.g., based on bilingual word embeddings) and represent queries and documents as vectors in the shared space [9,48,49].

Each line of work comes with certain drawbacks: (1) robust MT systems require huge amounts of parallel data, while these resources are still scarce for many language pairs and domains; (2) concept coverage in multilingual knowledge bases like BabelNet [29] is still limited for resource-lean languages, and all content not present in a knowledge base is effectively ignored by a CLIR system; (3) CLIR models based on bilingual semantic spaces require parallel or comparable texts to induce such spaces [40,49].

Due to smaller quantities of text which the ranking functions can exploit, sentence retrieval is traditionally considered more challenging than standard document-level retrieval [21,23,28,32]. Cross-lingual sentence retrieval typically equals to identifying parallel sentences in large text collections: the so-called *mate retrieval* task [30,37,42], which benefits the construction of high-quality sentence-aligned data for MT model training [41]. To this end, it is crucial to distinguish exact translation pairs from sentence pairs that are only semantically related. This is why CLIR models that exploit coarse-grained representations, e.g., by inducing latent topics [9,48] or by aggregating word embeddings [49], are not suitable for modeling such subtle differences in meaning.

In this work, we propose a neural architecture for cross-lingual sentence retrieval which captures fine-grained semantic dependencies between sentences in different languages and distinguishes true sentence translations pairs from related sentences. Its high-level flow is illustrated in Fig. 1.

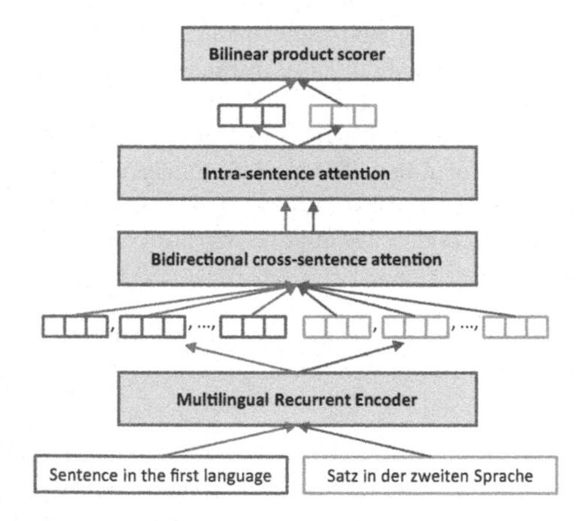

Fig. 1. High-level overview of the bidirectional cross-lingual attention (BiCLA) model.

First, a joint multilingual word embedding space is coupled with a recurrent encoder that is *shared* between the two languages: this enables contextualization of word representations for word sequences in both languages. Further, selective cross-sentence contextualization is achieved by means of a bidirectional attention mechanism stacked on top of the shared encoder. The attention layer enables the model to assign more weight to relevant information segments from the other sentence. Cross-sentence informed representations are then aggregated into a fixed-size sentence vectors via an intra-sentence attention mechanism. Finally, the model predicts the ranking score for a sentence pair by computing a bilinear product between these fixed-size sentence representations.

We evaluate the model in a mate sentence retrieval task on the Europarl data. We experiment with four languages of varying degrees of similarity, showing that our bidirectional attention model significantly outperforms state-of-the-art unsupervised CLIR models, recently proposed in [19]. Most importantly, we demonstrate the effectiveness of the model in zero-shot language transfer: a model trained on parallel data for one language pair (e.g., German and English) successfully performs CLIR for another language pair (e.g., Czech and Hungarian). Finally, we observe that the proposed bidirectional attention model complements state-of-the-art unsupervised CLIR baselines: we obtain further significant performance gains by ensembling the models.

2 Related Work

Cross-Lingual Information Retrieval. Early CLIR methods combined dictionary-based term-by-term translations of queries to the collection language [4,17,34] with monolingual term-based retrieval models [39,45]. Such models suffer from two main drawbacks: (1) inability to account for in-context meaning of query words and multi-word expressions and (2) inability to capture semantic similarity between queries and documents.[1]

One way to mitigate these issues is to represent documents and queries using concepts from external (multilingual) knowledge resources. Sorg and Cimiano [46] exploit Wikipedia as a multilingual knowledge base and represent documents as vectors where dimensions denote Wikipedia concepts. Franco-Salvador et al. [10] link the document text to concepts in BabelNet [29] and then measure document similarity by comparing BabelNet subgraphs spanned by the linked concepts. These methods, relying on external structured knowledge, are limited by the coverage of the exploited knowledge bases. Another limiting factor is their core dependence on the quality of concept linkers [26,33], required to associate the concepts from text with knowledge base entries: any piece of text that is not linked to a knowledge base concept is effectively ignored by the model.

Another class of models for cross-lingual text comparison is based on the induction of shared multilingual semantic spaces in which queries and documents in both languages are represented as vectors. These are induced using (Probabilistic) Latent Semantic Analysis (LSA) [9,35], Latent Dirichlet Allocation [25,48], or Siamese Neural Networks [51]. In contrast to directly learning bilingual document representations, Vulić and Moens [49] obtain bilingual word embeddings and then compose cross-lingual document and query representations by simply summing the embeddings of their constituent words.

[1] E.g., a German term *"Hund"* translated as *"dog"* still does not match a term *"canine"* from a relevant document.

Cross-Lingual Sentence Matching. Approaches to extracting parallel sentences have ranged from rule-based extraction from comparable documents [30,38,42], over classifiers trained on sentence-aligned parallel data [27,41] to cross-lingual sentence retrieval [37]. Supervised approaches typically exploit pretrained SMT models or their components (e.g., word alignment models) to produce features for classification [27,41]. These are often coupled with a rich set of domain-specific features computed from metadata of the bilingual data at hand [41]: all this limits the portability of such models.

Another related task is cross-lingual semantic similarity of short texts (STS) [1,7]. The best-performing cross-lingual STS models [6,14,47] all employ a similar strategy: they first translate the sentences from one language to the other (resource-rich) language (i.e., English) and then apply supervised, feature-rich and language-specific (e.g., they rely on syntactic dependencies and named entity recognizers) regression models. Their dependence on full-blown MT systems and resource-intensive and language-specific features limits their portability to arbitrary (resource-lean) language pairs.

3 Bidirectional Attention CLIR Model

First we describe the induction of a multilingual word vector space and then the components of our bidirectional cross-lingual attention model (BiCLA).

Multilingual Word Vector Space. Multiple methods have recently been proposed for inducing bilingual word vector spaces by learning linear projections from one monolingual space to another [2,8,24,43]. A multilingual vector space for N languages is then induced by simply learning $N - 1$ bilingual projections with the same target space (e.g., English). A comparative evaluation by Ruder et al. [40] indicates that all of the above models produce multilingual spaces of similar quality. Due to its large language coverage and accessible implementation, we opt for the model of Smith et al. [43]. They learn the projection by exploiting a set of (10 K or less) word translation pairs.

3.1 BiCLA: Cross-Lingual Sentence Retrieval Model

The architecture of the BiCLA model is detailed in Fig. 2. We encode the sentences from both languages with the same bidirectional long short-term memory network (Bi-LSTM): word vectors from a pre-trained bilingual space are input to the network. The sentences are then made "aware of each other": we compute the vector representations of each sentence's tokens by attending over Bi-LSTM encodings of other sentence's tokens. Next, we use an intra-sentence attention mechanism to aggregate a fixed-size encoding of each sentence from such cross-sentence contextualized vectors of its tokens. Finally, we compute the relevance score for the sentence pair as the bilinear product between fixed-size representations of the two sentences obtained through intra-sentence attention. In what follows, we describe all components of the BiCLA model.

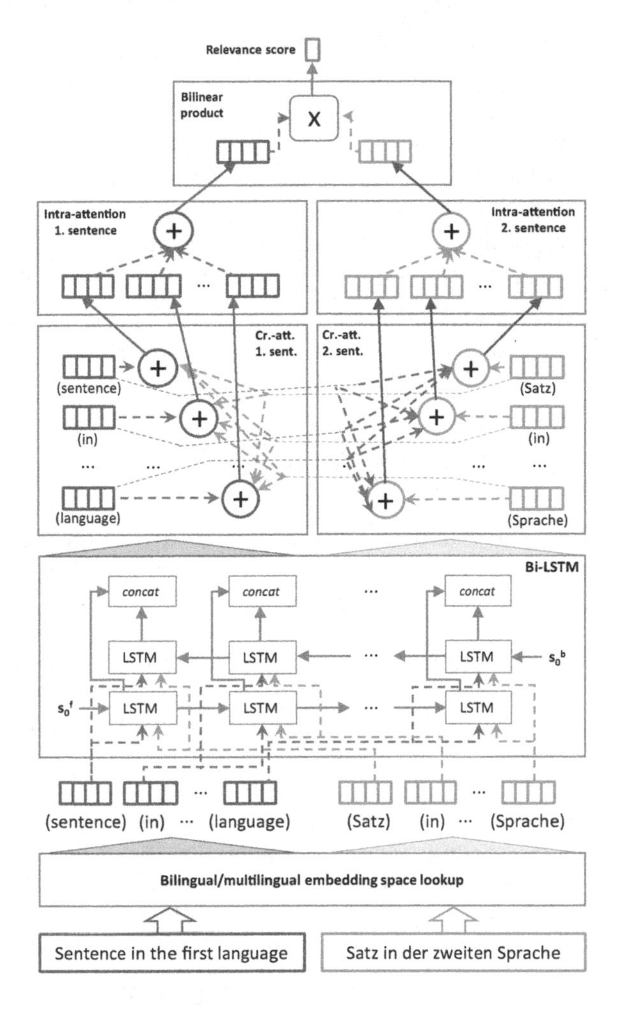

Fig. 2. Schema of the BiCLA model.

1. Recurrent Encoder. We encode both sentences using the same shared Bi-LSTM encoder [12].[2] Given an input sequence of T tokens $\{t_i\}_{i=1}^{T}$, the Bi-LSTM layer produces a sequence of T within-sentence *contextualized* token representations $\{h_i = [h_i^f, h_i^b]\}_{i=1}^{T}$, where $h_i^f \in \mathbb{R}^H$ is the i-th token vector produced by the forward-pass LSTM and $h_i^b \in \mathbb{R}^H$ is the i-th token vector produced by the backward-pass LSTM, with H as LSTMs hidden state size. Vector h_i^f contextualizes the i-th token with the meaning of its left context (i.e., preceding tokens), whereas h_i^b makes the representation of the i-th token aware of its right context (following tokens).

[2] We also experimented with two different Bi-LSTMs for encoding sentences in two languages, but this exhibited poorer performance.

2. Bidirectional Cross-Sentence Attention. In neural machine translation [3,22], the attention mechanism allows to focus more on parts of the source sentence that are most relevant for translation generation at a concrete position. In sentence retrieval, the goal is to semantically *align* two sentences and determine their semantic compatibility. To this end, we define a bidirectional attention layer that allows to represent tokens of the first sentence by focusing on representations of only relevant tokens from the second sentence, and vice versa. Let $\{h_i^{S_1}\}_{i=1}^{T_1}$ and $\{h_j^{S_2}\}_{j=1}^{T_2}$ be the sequences of token representations of the input sentences produced by Bi-LSTM. The cross-sentence representation $\overline{h}_i^{S_1}$ of the i-th token of the first sequence is then computed as the weighted average of token vectors $\{h_j^{S_2}\}_{j=1}^{T_2}$ of the second sentence, and vice versa:

$$\overline{h}_i^{S_1} = \sum_{j=1}^{T_2} \alpha_{i,j} \cdot h_j^{S_2}; \qquad \overline{h}_j^{S_2} = \sum_{i=1}^{T_1} \beta_{j,i} \cdot h_i^{S_1}. \tag{1}$$

Attention weights $\alpha_{i,j}$ and $\beta_{j,i}$ are obtained by computing the softmax functions over respective raw matching scores $m_{i,j}$ and $n_{j,i}$ which are, in turn, computed on the basis of a bilinear product of token vectors $h_i^{S_1}$ and $h_j^{S_2}$:

$$\alpha_{i,j} = \frac{e^{m_{i,j}}}{\sum_{k=1}^{T_2} e^{m_{i,k}}}; \quad m_{i,j} = \tanh\left(h_i^{S_1} W_{ca}^1 h_j^{S_2} + b_{ca}^1\right); \tag{2}$$

$$\beta_{j,i} = \frac{e^{n_{j,i}}}{\sum_{k=1}^{T_1} e^{n_{j,k}}}; \quad n_{j,i} = \tanh\left(h_j^{S_2} W_{ca}^2 h_i^{S_1} + b_{ca}^2\right). \tag{3}$$

$W_{ca}^1, W_{ca}^2 \in \mathbb{R}^{2H \times 2H}$ and $b_{ca}^1, b_{ca}^2 \in \mathbb{R}$ are attention parameters.[3] Using the cross-attention mechanism we *contextualize* one sentence in terms of the other in a *localized* manner: the vector $\overline{h}_i^{S_1}$ of a first sentence token aggregates information from the semantically most relevant parts of the other sentence, and vice-versa for $\overline{h}_j^{S_2}$.

3. Intra-Sentence Attention. Bi-LSTM contextualizes token representations within the sentence, whereas the cross-attention contextualizes them with respect to the other sentence. We finally produce the task-specific fixed-size sentence representations by aggregating their respective contextualized token vectors. Because not all parts of a sentence are equally contributing to its meaning, we learn how to aggregate the fixed-size sentence representation by means of an intra-sentence attention mechanism. Our intra-sentence attention is a simplified version of the recently introduced self-attention networks [18,20]. The sentence embeddings e_1 and e_2 are computed as weighted sums of their cross-sentence contextualized token vectors:

$$e_1 = \sum_{i=1}^{T_1} \gamma_i \overline{h}_i^{S_1}; \qquad e_2 = \sum_{j=1}^{T_2} \delta_j \overline{h}_j^{S_2}. \tag{4}$$

[3] Note that by constraining $W_{ca}^1 = W_{ca}^2$ and $b_{ca}^1 = b_{ca}^2$ we enforce a symmetric bidirectional cross-attention mechanism. However, the above asymmetric attention gave better performance.

The weights γ_i and δ_j are computed as non-linear transformations of dot products between token vectors and intra-sentence attention parameter vectors:

$$\gamma_i = \tanh\left(\overline{h}_i^{S_1} \cdot v_{ia}^1 + b_{ia}^1\right); \delta_j = \tanh\left(\overline{h}_j^{S_2} \cdot v_{ia}^2 + b_{ia}^2\right).$$

We learn the parameters $v_{ia}^1, v_{ia}^2 \in \mathbb{R}^{2H}$ and $b_{ia}^1, b_{ia}^2 \in \mathbb{R}$ during training.[4]

4. Bilinear Scoring. Finally, we can quantify a similarity (i.e., relevance) score for the cross-lingual pair of sentences from the obtained fixed-size representations e_1 and e_2. We combine the vectors e_1 and e_2 into a relevance score r with a bilinear product function, which was previously successfully applied to, e.g., relation prediction for knowledge base completion [44,50] and predicting semantic matches at a word level [11]:

$$r(S_1, S_2) = \tanh\left(e_1 W_B e_2 + b_B\right), \tag{5}$$

where $W_B \in \mathbb{R}^{2H\times 2H}$ and $b_B \in \mathbb{R}$ are the bilinear product parameters.

5. Objective and Optimization. The model has to assign higher scores $r(S_1, S_2)$ to sentence pairs where S_2 is a complete semantic match (i.e., a translation) of S_1 than to semantically related sentence pairs with only a partial semantic overlap. Therefore, BiCLA relies on a contrastive loss function that maximizes the difference in scores between positive sentence pairs and corresponding negative pairs. Let $\{(S_1^{(i)}, S_2^{(i)})\}_{i=1}^N$ be the collection of positive pairs in our training set: these are exact translations. For each source language sentence $S_1^{(i)}$ we create K negative training pairs $\{(S_1^{(i)}, S_2^{(k_j)})\}_{j=1}^K$. Half of the these K pairs are created by pairing $S_1^{(i)}$ with $K/2$ randomly selected target language sentences. The remaining pairs are created by coupling $S_1^{(i)}$ with $K/2$ semantically most similar sentences in the other language (excluding the target sentence from the positive example), according to a baseline heuristic function as follows. Let $e(t)$ retrieve the embedding of the term t from the shared bilingual embedding space. We then compute the heuristic similarity between sentences S_1 and S_2 as:

$$sim(S_1, S_2) = \cos\left(\sum_{t_1 \in S_1} e(t_1), \sum_{t_2 \in S_2} e(t_2)\right) \tag{6}$$

By taking the most similar sentences according to the above heuristic, we create – for each positive sentence pair – $K/2$ corresponding negative pairs in which there is at least some semantic overlap between the sentences. The contrastive loss objective for the given training set of translation pairs $\{(S_1^{(i)}, S_2^{(i)})\}_{i=1}^N$ is finally defined as follows:

$$J = \sum_{i=1}^{N}\sum_{j=1}^{K} \epsilon - \left(r\left(S_1^{(i)}, S_2^{(i)}\right) - r\left(S_1^{(i)}, S_2^{(k_j)}\right)\right). \tag{7}$$

[4] Again, we could enforce the symmetric intra-sentence attention for both sentences by setting $v_{ia}^1 = v_{ia}^2$ and $b_{ia}^1 = b_{ia}^2$, but doing so resulted in lower performance in our experiments.

The hyper-parameter ϵ defines the margin between scores of positive and negative pairs. The final objective function J_{MIN} augments the contrastive loss function J with the L_2 regularization of parameters $\mathbf{\Omega}$: $J_{MIN} = J + \lambda\|\mathbf{\Omega}\|_2$, with λ as regularization factor.

4 Evaluation

We first describe the important aspects of the experimental setup – datasets, baselines, and details on model training. We then report and discuss BiCLA performance in (1) standard mate retrieval and (2) zero-shot language-transfer experiments.

4.1 Experimental Setup

Data. We use the parallel Europarl corpus [16][5] in all experiments. Since one of our goals is to examine retrieval performance for languages of varying degree of similarity, we experiment with the Europarl data in English (EN), German (DE), Czech (CS), and Hungarian (HU).[6] The Europarl datasets for all six language-pair combinations (see Table 1) were preprocessed by (1) removing stopwords and (2) retaining only sentence pairs in which each sentence has at least three tokens represented in the bilingual embedding space. For each language pair, we use a set of 1000 randomly selected sentence pairs as test data. All the remaining pairs are used for the BiCLA training. The datasets' sizes, in terms of number of sentence pairs, are shown in Table 1.

Table 1. Sizes of all datasets used in experiments.

Language pair	Train size (# pairs)	Test size (# pairs)
CS-DE	506,495	1,000
CS-EN	572,889	1,000
CS-HU	543,959	1,000
DE-EN	1,584,202	1,000
HU-DE	501,128	1,000
HU-EN	556,774	1,000

Multilingual Embedding Space. We use precomputed 300-dimensional monolingual FASTTEXT word embeddings [5][7] for all four languages. We then induce a shared four-lingual embedding space using the lexicon-based projection method with pivoting from [43], outlined in Sect. 3.[8]

[5] http://opus.nlpl.eu/Europarl.php.

[6] English, German, and Czech belong to the family of Indo-European languages (EN and DE are representatives of the Germanic branch, and CS is in the Slavic branch), whereas Hungarian belongs to the Uralic language family.

[7] https://github.com/facebookresearch/fastText.

[8] https://tinyurl.com/msrmwee.

Baselines. We compare BiCLA with the standard query likelihood model [36], two state-of-the-art unsupervised CLIR models [19], and the reduced architecture without the bidirectional attention:

(1) Standard query likelihood retrieval model [36] with Jelinek-Mercer smoothing [13] (**QLM**). The model computes the relevance by multiplying source sentence terms' probabilities under the unigram language model of a target language sentence, smoothed with the their probabilities under the language model of the whole target collection:

$$rel(q, d) = \prod_{t \in q} \mu P(t|M_S) + (1 - \mu) P(t|M_{SC}). \quad (8)$$

$P(t|M_S)$ is probability of term t under the local language model of the sentence S, $P(t|M_{SC})$ is the probability of t under the global language model of the target collection SC, and $\mu = 0.95$ is the interpolation coefficient. Designed for monolingual retrieval, QLM's CLIR performance crucially depends on the amount of lexical overlap between languages. We thus employ QLM merely as a "sanity check" baseline.

(2) Aggregating word embeddings from the shared embedding space (**AGG**) [19,49]. AGG computes sentence embeddings by averaging the embeddings of their tokens, obtained from the shared multilingual embedding space. The relevance score is the cosine similarity between aggregated sentence embeddings. We used AGG also as a heuristic for creating negative instances for the contrastive loss (cf. Eq. (6) in Sect. 3.1).

(3) Term-by-term translation using shared embedding space (**TbT**). Each query token is replaced by the most similar target language token, according to the cosine similarity in the shared space. After the term-by-term translation of the query, we use the monolingual QLM to rank the target sentences. TbT and AGG have recently exhibited state-of-the-art performance on several benchmarks for document-level CLIR [19].

(4) BiCLA without the bidirectional cross-sentence attention (**InAtt**). In InAtt, the intra-sentence attention layer is stacked directly on top of representations produced by the Bi-LSTM encoder. The comparison between BiCLA and InAtt directly reveals the contribution that bidirectional cross-sentence attention has on sentence CLIR performance.

Ensemble Models. By design, BiCLA aims to capture semantic similarity stemming from semantic alignments of longer sequences (i.e., phrases, clauses), implicitly capturing semantic compositionality. In contrast, baselines AGG and TbT make simpler similarity assessments: AGG assumes the sentence meaning to be a linear combination of word meanings whereas TbT actually measures the lexical overlap, using the multilingual embedding space as the translation dictionary. Given this complementarity between BiCLA and the baselines, we also evaluate ensemble rankers: they rank target sentences by interpolating between ranks assigned by individual models.

Model Configuration. We train BiCLA in mini-batches of size $N_b = 50$ sentence pairs, each consisting of 10 micro-batches containing one positive sentence pair and $K = 4$ corresponding negative sentence pairs (two created randomly and two using AGG as unsupervised similarity heuristic). We optimize the parameters with the Adam algorithm [15], setting the initial learning rate to 10^{-4}. We tune the hyperparameters on a validation set in a fixed-split cross-validation. We found the following optimal values: BiLSTM state size $H = 100$, regularization factor $\lambda = 10^{-4}$, contrastive margin $\epsilon = 1$. The loss on the validation set was also used as the criterion for early stopping of the training.

4.2 Results and Discussion

First we show the results for the basic mate CLIR evaluation, with train and test set involving the same language pair. We then examine BiCLA's behavior in language transfer settings – a model trained on one language pair is used to perform CLIR for another language pair. In both cases, we evaluate the performance of BiCLA alone and ensembled with the unsupervised CLIR baselines, AGG and TbT.

Task 1: Base Evaluation. We first show the results for the base evaluation task where the train and test set involve the same language pair. We treat each sentence of the source language as a query and the 1,000 test sentences in the target language as a target sentence collection. Performance is reported in terms of the standard mean average precision (MAP) measure. Table 2 summarizes the MAP scores for six language pairs (first language is always the query/source language).

Table 2. Cross-lingual mate retrieval performance.

Model	DE-EN	CS-EN	HU-EN	CS-DE	HU-DE	CS-HU
QLM	.303	.121	.141	.064	.054	.083
AGG	.390	.547	.372	.374	.356	.378
TbT	.490	.563	.357	.228	.142	.124
InAtt	.506	.597	.462	.495	.422	.404
BiCLA	**.604**	**.665**	**.569**	**.562**	**.577**	**.575**

BiCLA strongly outperforms all baselines for all six language pairs.[9] The baselines are able to reduce the gap in performance only for two language pairs: DE-EN and CS-EN. The baseline scores generally tend to decrease as the languages in the pair become more distant. In contrast, BiCLA exhibits fairly stable performance across all language pairs – the performances for pairs of more distant languages (e.g., HU-DE or CS-HU) are on par with the performances for pairs of closer languages (e.g., DE-EN and CS-DE). Full BiCLA outperforms InAtt,

[9] We tested the significance over 1000 average precision scores obtained for individual queries (which are, in our case, equal to reciprocal rank scores, since there is only one relevant sentence in the other language for each query) using the two-tailed Student's t-test. BiCLA significantly outperforms all baselines with $p < 0.01$.

a model without the bidirectional attention layer, by a wide margin, confirming our intuition that fine-grained cross-sentential semantic awareness is crucial for better recognition of sentence translation pairs.

We next investigate the extent to which the supervised BiCLA model complements the state-of-the-art unsupervised CLIR baselines, AGG and TbT [19]. To this end, we ensemble the models at the level of the rankings they produce for the queries. We evaluate three different ensembles: (1) BiCLA and AGG, (2) BiCLA and TbT, and (3) BiCLA and both AGG and TbT. For each of the ensembles we show the performance with different weight configurations, i.e., different weight values w_{BiCLA}, w_{AGG}, and w_{TbT} assigned to individual models, BiCLA, AGG. and TbT, respectively. The results of ensemble methods are shown in Table 3. Almost all ensemble models (exception is BiCLA+TbT with large TbT weight for distant languages), exhibit better performance than BiCLA on its own. BiCLA+AGG and BiCLA+AGG-TbT ensembles with larger weights for BiCLA yield performance gains between 10 and 20% MAP with respect to the BiCLA model alone, suggesting that BiCLA indeed complements the best unsupervised CLIR models. Note that the ensembles in which BiCLA gets a larger weight than the unsupervised models exhibit the best performance. For example, a BiCLA+AGG-TbT ensemble with weights $w_{BiCLA} = 0.8$, $w_{AGG} = 0.1$, and $w_{TbT} = 0.1$ significantly outperforms the same ensemble with equal weights (i.e., $w_{BiCLA} = w_{AGG} = w_{TbT} = 0.\dot{3}$), for all language pairs except CS-EN. Big boosts of ensembles with small contributions from unsupervised baselines suggest that (1) in most cases, BiCLA does a much better job than unsupervised baselines and (2) in cases where BiCLA fails, the unsupervised baselines perform very well – even their small contribution significantly improves the ranking.

Task 2: Zero-Shot Language Transfer. We next investigate the predictive capability of BiCLA in transfer learning settings, that is, we test whether a model trained for one language pair may successfully, both on its own and in ensembles with unsupervised models, perform sentence CLIR for another language pair. The language transfer results are shown in Table 4: rows denote the language pair of the train set and columns the language pair of the test set. We analyze the results in view of three types of language transfer: (1) source language transfer (SLT) (same collection language in training and test); (2) target language transfer (TLT) (same query language in training and test), and (3) full language transfer (FLT) (query and collection language in test are both different than in training). BiCLA outperforms all baselines in 7/8 SLT settings, 5/8 TLT settings, and (only) 2/14 FLT experiments. It is not surprising to observe better performance in SLT and TLT settings than in FLT: BiCLA seems to be able to account for the change of one language, but not for both simultaneously. Although BiCLA alone does not outperform the unsupervised baselines in most FLT setting (e.g., DE-EN→CS-HU), we find the SLT and TLT results significant. Drops in performance in some of the SLT and TLT setups, compared to respective basic setups (no language transfer), are almost negligible (e.g., DE-EN→HU-EN has only a 3-point lower MAP compared to the basic HU-EN setup). SLT and TLT seem to work even when we switch between distant languages (e.g., when we replace the query language from DE to HU), which we find encouraging.

Table 3. Performance of ensemble models.

	Weights	DE-EN	CS-EN	HU-EN	CS-DE	HU-DE	CS-HU
BiCLA + AGG	.5; .5	.662	.802	.622	.666	.597	.675
	.7; .3	.686	.818	.653	.690	.624	.708
	.9; .1	.706	.784	.660	.691	**.647**	**.702**
BiCLA + TbT	.5; .5	.651	.827	.680	.508	.396	.300
	.7; .3	.650	.832	.685	.553	.442	.351
	.9; .1	.649	.802	.687	.618	.561	.473
BiCLA + AGG + TbT	$.\dot{3}$; $.\dot{3}$; $.\dot{3}$	.656	.846	.651	.599	.449	.401
	.6; .2; .2	.685	**.859**	.683	.647	.525	.463
	.8; .1; .1	**.708**	.845	**.726**	**.697**	.626	.568

Table 4. BiCLA performance in language transfer settings (underlined results denote base CLIR results from Table 2, without language transfer. Bold scores indicate BiCLA transfer scores that are above all baseline scores, cf. Table 2).

	DE-EN	CS-EN	HU-EN	CS-DE	HU-DE	CS-HU
DE-EN	.604	**.602**	**.537**	.231	.177	.213
CS-EN	.356	.665	**.421**	**.440**	.307	**.419**
HU-EN	.440	.484	.569	.371	**.414**	.369
CS-DE	.299	.493	.292	.562	**.374**	**.398**
HU-DE	.365	.398	**.448**	**.399**	.577	.329
CS-HU	.360	**.604**	**.432**	**.524**	**.459**	.575

Table 5. Results of language transfer ensembles. Bold scores denote ensembles that surpass the performance of the unsupervised AGG model alone.

	DE-EN	CS-EN	HU-EN	CS-DE	HU-DE	CS-HU
AGG	.390	.547	.372	.374	.356	.378
Ensemble: BiCLA (transfer) + AGG						
DE-EN	–	**.818**	**.634**	**.520**	**.447**	**.518**
CS-EN	**.564**	–	**.550**	**.630**	**.520**	**.624**
HU-EN	**.578**	**.679**	–	**.556**	**.562**	**.537**
CS-DE	**.549**	**.771**	**.529**	–	**.569**	**.594**
HU-DE	**.544**	**.667**	**.562**	**.550**	–	**.499**
CS-HU	**.544**	**.773**	**.534**	**.639**	**.584**	–

Finally, we evaluate the ensemble between BiCLA and AGG (since, on average, AGG exhibits better performance than TbT) in zero-shot language transfer. We assign equal weights to both rankers, i.e., $w_{BiCLA} = w_{AGG} = 0.5$. The results of the zero-shot language transfer ensembles are shown in Table 5. While on its own BiCLA outperforms the unsupervised models in half of the language transfer setups, when ensembled with the AGG baseline, it drastically boosts the CLIR performance on *all* test collections. This also holds for all FLT setups – where BiCLA is trained on a completely different language pair from the language pair of the test collection. For example, a BiCLA model trained on DE-EN, when ensembled with AGG, boosts the CS-HU retrieval by almost 15 MAP points. We hold this to be the most important finding of our work – it implies that we can exploit readily available large parallel corpora for major languages in order to train a model that significantly improves mate retrieval for pairs of under-resourced languages, for which we have no parallel resources.

5 Conclusion

We have presented a novel neural framework for mate sentence retrieval across languages. We introduced the bidirectional cross-lingual attention (BiCLA) model, a multi-layer architecture which learns to encode sentences in a shared cross-lingual space in such a way to recognize true semantic similarity between sentences: this means that BiCLA is able to distinguish true translation pairs from only semantically related sentences with partial semantic overlap. A series of experiments for six language pairs have verified the usefulness of the model – we have shown that BiCLA outperforms unsupervised retrieval baselines, and that further gains, due to the complementarity of the two approaches, can be achieved by combined ensemble methods. Most importantly, we have shown that the multilingual nature of the BiCLA model allows for a zero-shot language transfer for CLIR: a model trained on one pair of languages (e.g., German and English) can be used to improve CLIR for another pair of languages (e.g., Czech and Hungarian). This indicates that we can perform reliable cross-lingual sentence retrieval even for pairs of resource-lean languages, for which we have no parallel corpora.

In future work, we plan to experiment with deeper architectures and more sophisticated attention mechanisms. We will also test the usability of the framework in other related retrieval tasks and evaluate the model on more language pairs. We make the BiCLA model code along with the datasets used in our experiments publicly available at: https://github.com/codogogo/bicla-clir.

Acknowledgments. The work described in this paper has been partially supported by the "Eliteprogramm für Postdoktorandinnen und Postdoktoranden" of the Baden Württemberg Stiftung, within the scope of the AGREE (*Algebraic Reasoning over Events from Text and External Knowledge*) grant. We thank the anonymous reviewers for their useful comments.

References

1. Agirre, E., et al.: Semeval-2016 task 1: semantic textual similarity, monolingual and cross-lingual evaluation. In: SemEval, pp. 497–511. ACL (2016)
2. Artetxe, M., Labaka, G., Agirre, E.: Learning bilingual word embeddings with (almost) no bilingual data. In: Proceedings of the 55th Annual Meeting of the Association for Computational Linguistics (Volume 1: Long Papers), pp. 451–462. Association for Computational Linguistics, Vancouver, July 2017. http://aclweb.org/anthology/P17-1042

3. Bahdanau, D., Cho, K., Bengio, Y.: Neural machine translation by jointly learning to align and translate. In: International Conference on Learning Representations (2014)
4. Ballesteros, L., Croft, B.: Dictionary methods for cross-lingual information retrieval. In: Wagner, R.R., Thoma, H. (eds.) DEXA 1996. LNCS, vol. 1134, pp. 791–801. Springer, Heidelberg (1996). https://doi.org/10.1007/BFb0034731
5. Bojanowski, P., Grave, E., Joulin, A., Mikolov, T.: Enriching word vectors with subword information. Trans. ACL **5**, 135–146 (2017). http://arxiv.org/abs/1607.04606
6. Brychcín, T., Svoboda, L.: UWB at semeval-2016 task 1: semantic textual similarity using lexical, syntactic, and semantic information. In: SemEval, pp. 588–594. ACL (2016)
7. Cer, D., Diab, M., Agirre, E., Lopez-Gazpio, I., Specia, L.: Semeval-2017 task 1: semantic textual similarity multilingual and crosslingual focused evaluation. In: Proceedings of the 11th International Workshop on Semantic Evaluation (SemEval-2017), pp. 1–14. Association for Computational Linguistics, Vancouver, August 2017. http://www.aclweb.org/anthology/S17-2001
8. Conneau, A., Lample, G., Ranzato, M., Denoyer, L., Jégou, H.: Word translation without parallel data. arXiv preprint arXiv:1710.04087 (2017)
9. Dumais, S.T., Letsche, T.A., Littman, M.L., Landauer, T.K.: Automatic cross-language retrieval using latent semantic indexing. In: AAAI Spring Symposium on Cross-language Text and Speech Retrieval, vol. 15, p. 21 (1997)
10. Franco-Salvador, M., Rosso, P., Navigli, R.: A knowledge-based representation for cross-language document retrieval and categorization. In: Proceedings of the 14th Conference of the European Chapter of the Association for Computational Linguistics, pp. 414–423 (2014)
11. Glavaš, G., Ponzetto, S.P.: Dual tensor model for detecting asymmetric lexico-semantic relations. In: Proceedings of the 2017 Conference on Empirical Methods in Natural Language Processing, pp. 1757–1767. Association for Computational Linguistics, Copenhagen, September 2017. https://www.aclweb.org/anthology/D17-1185
12. Hochreiter, S., Schmidhuber, J.: Long short-term memory. Neural Comput. **9**(8), 1735–1780 (1997)
13. Jelinek, F., Mercer, R.: Interpolated estimation of markov source parameters from sparse data. In: Proceedings of Workshop on Pattern Recognition in Practice 1980, pp. 381–402 (1980)
14. Jimenez, S.: Sergiojimenez at semeval-2016 Task 1: effectively combining paraphrase database, string matching, WordNet, and word embedding for semantic textual similarity. In: SemEval, pp. 749–757. ACL (2016)
15. Kingma, D.P., Ba, J.: Adam: a method for stochastic optimization. In: Proceedings of ICLR (Conference Track) (2015). https://arxiv.org/abs/1412.6980
16. Koehn, P.: Europarl: a parallel corpus for statistical machine translation. In: Proceedings of the 10th Machine Translation Summit, pp. 79–86 (2005)
17. Levow, G.A., Oard, D.W., Resnik, P.: Dictionary-based techniques for cross-language information retrieval. Inf. Process. Manage. **41**(3), 523–547 (2005)
18. Lin, Z., et al.: A structured self-attentive sentence embedding. In: Proceedings of the International Conference on Learning Representations (2017)
19. Litschko, R., Glavaš, G., Ponzetto, S.P., Vulić, I.: Unsupervised cross-lingual information retrieval using monolingual data only. arXiv preprint arXiv:1805.00879 (2018)

20. Liu, Y., Sun, C., Lin, L., Wang, X.: Learning natural language inference using bidirectional LSTM model and inner-attention. arXiv preprint arXiv:1605.09090 (2016)
21. Losada, D.E.: Statistical query expansion for sentence retrieval and its effects on weak and strong queries. Inf. Retrieval **13**(5), 485–506 (2010)
22. Luong, T., Pham, H., Manning, C.D.: Effective approaches to attention-based neural machine translation. In: Proceedings of the 2015 Conference on Empirical Methods in Natural Language Processing, pp. 1412–1421. Association for Computational Linguistics, Lisbon, September 2015. http://aclweb.org/anthology/D15-1166
23. Martino, G.D.S., et al.: Cross-language question re-ranking. In: Proceedings of SIGIR, pp. 1145–1148 (2017)
24. Mikolov, T., Le, Q.V., Sutskever, I.: Exploiting similarities among languages for machine translation. arXiv preprint arXiv:1309.4168 (2013)
25. Mimno, D., Wallach, H.M., Naradowsky, J., Smith, D.A., McCallum, A.: Polylingual topic models. In: Proceedings of the 2009 Conference on Empirical Methods in Natural Language Processing, vol. 2, pp. 880–889. Association for Computational Linguistics (2009)
26. Moro, A., Raganato, A., Navigli, R.: Entity linking meets word sense disambiguation: a unified approach. Trans. Assoc. Comput. Linguist. **2**, 231–244 (2014)
27. Munteanu, D.S., Marcu, D.: Improving machine translation performance by exploiting non-parallel corpora. Comput. Linguist. **31**(4), 477–504 (2005)
28. Murdock, V., Croft, W.B.: A translation model for sentence retrieval. In: Proceedings of the conference on Human Language Technology and Empirical Methods in Natural Language Processing, pp. 684–691. Association for Computational Linguistics (2005)
29. Navigli, R., Ponzetto, S.P.: Babelnet: the automatic construction, evaluation and application of a wide-coverage multilingual semantic network. Artif. Intell. **193**, 217–250 (2012)
30. Nie, J.Y., Simard, M., Isabelle, P., Durand, R.: Cross-language information retrieval based on parallel texts and automatic mining of parallel texts from the web. In: Proceedings of the 22nd annual international ACM SIGIR conference on Research and development in information retrieval, pp. 74–81. ACM (1999)
31. Oard, D.W.: A comparative study of query and document translation for cross-language information retrieval. In: Farwell, D., Gerber, L., Hovy, E. (eds.) AMTA 1998. LNCS (LNAI), vol. 1529, pp. 472–483. Springer, Heidelberg (1998). https://doi.org/10.1007/3-540-49478-2_42
32. Otterbacher, J., Erkan, G., Radev, D.R.: Using random walks for question-focused sentence retrieval. In: Proceedings of the conference on Human Language Technology and Empirical Methods in Natural Language Processing, pp. 915–922. Association for Computational Linguistics (2005)
33. Pappu, A., Blanco, R., Mehdad, Y., Stent, A., Thadani, K.: Lightweight multilingual entity extraction and linking. In: Proceedings of the Tenth ACM International Conference on Web Search and Data Mining, pp. 365–374. ACM (2017)
34. Pirkola, A.: The effects of query structure and dictionary setups in dictionary-based cross-language information retrieval. In: Proceedings of the 21st Annual International ACM SIGIR Conference on Research and Development in Information Retrieval, pp. 55–63. ACM (1998)
35. Platt, J.C., Toutanova, K., Yih, W.t.: Translingual document representations from discriminative projections. In: Proceedings of the 2010 Conference on Empirical Methods in Natural Language Processing, pp. 251–261. Association for Computational Linguistics (2010)

36. Ponte, J.M., Croft, W.B.: A language modeling approach to information retrieval. In: SIGIR, pp. 275–281. ACM (1998)
37. Rauf, S.A., Schwenk, H.: Parallel sentence generation from comparable corpora for improved SMT. Mach. Transl. **25**(4), 341–375 (2011)
38. Resnik, P., Smith, N.A.: The web as a parallel corpus. Comput. Linguist. **29**(3), 349–380 (2003)
39. Robertson, S., Zaragoza, H., et al.: The probabilistic relevance framework: Bm25 and beyond. Found. Trends® Inf. Retrieval **3**(4), 333–389 (2009)
40. Ruder, S., Vulić, I., Søgaard, A.: A survey of cross-lingual word embedding models (2017). arXiv preprint arXiv:1706.04902
41. Smith, J.R., Quirk, C., Toutanova, K.: Extracting parallel sentences from comparable corpora using document level alignment. In: Human Language Technologies: The 2010 Annual Conference of the North American Chapter of the Association for Computational Linguistics, pp. 403–411. Association for Computational Linguistics (2010)
42. Smith, J.R., Saint-Amand, H., Plamada, M., Koehn, P., Callison-Burch, C., Lopez, A.: Dirt cheap web-scale parallel text from the common crawl. In: Proceedings of the 51st Annual Meeting of the Association for Computational Linguistics (Volume 1: Long Papers), vol. 1, pp. 1374–1383 (2013)
43. Smith, S.L., Turban, D.H., Hamblin, S., Hammerla, N.Y.: Offline bilingual word vectors, orthogonal transformations and the inverted softmax. In: Proceedings of International Conference on Learning Representations (ICLR 2017, Conference Track) (2017)
44. Socher, R., Chen, D., Manning, C.D., Ng, A.: Reasoning with neural tensor networks for knowledge base completion. In: Proceedings of the 2013 Annual Conference on Neural Information Processing Systems, pp. 926–934 (2013)
45. Song, F., Croft, W.B.: A general language model for information retrieval. In: Proceedings of the Eighth International Conference on Information and Knowledge Management, pp. 316–321. ACM (1999)
46. Sorg, P., Cimiano, P.: Exploiting wikipedia for cross-lingual and multilingual information retrieval. Data Knowl. Eng. **74**, 26–45 (2012)
47. Tian, J., Zhou, Z., Lan, M., Wu, Y.: Ecnu at semeval-2017 task 1: Leverage kernel-based traditional nlp features and neural networks to build a universal model for multilingual and cross-lingual semantic textual similarity. In: Proceedings of the 11th International Workshop on Semantic Evaluation (SemEval-2017), pp. 191–197 (2017)
48. Vulić, I., De Smet, W., Moens, M.F.: Cross-language information retrieval models based on latent topic models trained with document-aligned comparable corpora. Inf. Retrieval **16**(3), 331–368 (2013)
49. Vulić, I., Moens, M.F.: Monolingual and cross-lingual information retrieval models based on (bilingual) word embeddings. In: Proceedings of the 38th International ACM SIGIR Conference on Research and Development in Information Retrieval, pp. 363–372. ACM (2015)
50. Yang, B., Yih, W.t., He, X., Gao, J., Deng, L.: Embedding entities and relations for learning and inference in knowledge bases. In: Proceedings of the 2015 International Conference on Learning Representations (2015)
51. Yih, W.T., Toutanova, K., Platt, J.C., Meek, C.: Learning discriminative projections for text similarity measures. In: Proceedings of the Fifteenth Conference on Computational Natural Language Learning, pp. 247–256. Association for Computational Linguistics (2011)

QA and Conversational Search

QRFA: A Data-Driven Model of Information-Seeking Dialogues

Svitlana Vakulenko[1(✉)], Kate Revoredo[2], Claudio Di Ciccio[1], and Maarten de Rijke[3]

[1] Vienna University of Economics and Business, Vienna, Austria
{svitlana.vakulenko,claudio.di.ciccio}@wu.ac.at
[2] Graduate Program in Informatics, Federal University of Rio de Janeiro, Rio de Janeiro, Brazil
katerevoredo@ppgi.ufrj.br
[3] University of Amsterdam, Amsterdam, The Netherlands
derijke@uva.nl

Abstract. Understanding the structure of interaction processes helps us to improve information-seeking dialogue systems. Analyzing an interaction process boils down to discovering patterns in sequences of alternating utterances exchanged between a user and an agent. Process mining techniques have been successfully applied to analyze structured event logs, discovering the underlying process models or evaluating whether the observed behavior is in conformance with the known process. In this paper, we apply process mining techniques to discover patterns in conversational transcripts and extract a new model of information-seeking dialogues, QRFA, for Query, Request, Feedback, Answer. Our results are grounded in an empirical evaluation across multiple conversational datasets from different domains, which was never attempted before. We show that the QRFA model better reflects conversation flows observed in real information-seeking conversations than models proposed previously. Moreover, QRFA allows us to identify malfunctioning in dialogue system transcripts as deviations from the expected conversation flow described by the model via conformance analysis.

Keywords: Conversational search · Log analysis · Process mining

1 Introduction

Interest in information-seeking dialogue systems is growing rapidly, in information retrieval, language technology, and machine learning. There is, however, a lack of theoretical understanding of the functionality such systems should provide [24]. Different information-seeking models of dialogue systems use different terminology as well as different modeling conventions, and conversational datasets are annotated using different annotation schemes (see, e.g., [23,30]). These discrepancies hinder direct comparisons and aggregation of the results.

L. Azzopardi et al. (Eds.): ECIR 2019, LNCS 11437, pp. 541–557, 2019.
https://doi.org/10.1007/978-3-030-15712-8_35

Moreover, the evaluation of conversational datasets has largely been conducted based on manual efforts. Clearly, it is infeasible to validate models on large datasets without automated techniques.

Against this background, we create a new annotation framework that is able to generalize across conversational use cases and bridge the terminology gap between diverse theoretical models and annotation schemes of the conversational datasets collected to date. We develop and evaluate a new information-seeking model, which we name QRFA, for Query, Request, Feedback, Answer, which shows better performance in comparison with previously proposed models and helps to detect malfunctions from dialogue system transcripts. It is based on the analysis of 15,931 information-seeking dialogues and evaluated on the task of interaction success prediction in 2,118 held-out dialogues.

The QRFA model is derived and evaluated using process mining techniques [28], which makes this approach scalable. We view every conversation to be an instance of a general information-seeking process. This inclusive perspective helps us to extract and generalize conversation flows across conversations from different domains, such as bus schedules and dataset search.

To the best of our knowledge, we present the first grounded theory of information-seeking dialogues that is empirically derived from a variety of conversational datasets. Moreover, we describe the methodology we used to develop this theory that can be used to revise and further extend the proposed theory. We envision that the model and the approach that we describe in this paper will help not only to better understand the structure of information-seeking dialogues but also to inform the design of conversational search systems, their evaluation frameworks and conversational data sampling strategies. More concretely, we discovered a set of functional components for a conversational system as different interaction patterns and the distribution over the space of next possible actions.

The remainder of the paper is organized as follows. Section 2 provides a short review of discourse analysis and a gentle introduction to process mining. In Sect. 3 we discuss related work in two contexts: theoretical models of conversational search and process mining for discourse analysis. Section 4 provides details of our approach to mining processes from conversations. In Sect. 5 we report on the results of applying our conversation mining approach to several conversational datasets and we describe the model we obtained as a result. We conclude in Sect. 6.

2 Background

2.1 Discourse Analysis

A plethora of approaches have been proposed for discourse analysis, all focusing on different aspects of communication processes [18]. The main bottleneck in traditional approaches to discourse analysis, especially ones grounded in social and psychological theories (e.g., [9]), is context-dependence of the conversational semantics. Many studies rely on a handful of sampled or even artificially constructed conversations to illustrate and advocate their discourse theories, which limits their potential for generalization.

In this work we describe and demonstrate the application of a data-driven approach that can be applied to a large volume of conversational data, to identify patterns in the conversation dynamics. It is directly rooted in Conversational Analysis (CA) [17], which proposes to analyze regularities such as adjacency pairs and turn-taking in conversational structures, and Speech Act Theory (SAT) [2,19] to identify utterances with functions enabled through language (speech acts). To this end we leverage state-of-the-art techniques developed in the context of process mining [28], which has traditionally been applied in the context of operational business processes such as logistics and manufacturing, to discover and analyze patterns in sequential data.

Process mining (PM) has been designed to deal with structured data organized into a process log rather than natural language, such as conversational transcripts. However, we view a conversation as a sequence of alternating events between a user and an agent, thus a special type of process, a communication or information exchange process, that can be analyzed using PM by converting conversational transcripts into process logs. Basic concepts and techniques from PM, which we adopt in our discourse analysis approach, are described in the next subsection.

2.2 Process Mining

A process is a structure composed of events aligned between each other in time. The focus of PM is on extracting and analyzing process models from event logs. Each event in the log refers to the execution of an activity in a process instance. Additional information such as a reference to a resource (person or device) executing the activity, a time stamp of the event, or data recorded for the event, may be available.

Two major tasks in PM are *process discovery* and *conformance checking*. The former is used to extract a process model from an event log, and the latter to verify the model against the event log, i.e., whether the patterns evident from the event log correspond to the structure imposed by the model. It is possible to verify conformance against an extracted model as well as against a theoretical (independently constructed) model.

In this work, we adopt state-of-the-art PM techniques to analyze conversational transcripts by extracting process models from publicly available datasets of information-seeking dialogues, and to verify and further extend a theoretical model of information-seeking dialogues based on the empirical evidence from these corpora.

3 Related Work

3.1 Theoretical Models of Information-Seeking Dialogues

The first theoretical model of information-seeking dialogues has been proposed by Winograd and Flores [31] and further extended by Sitter and Stein [21] to the COnversational Roles (COR) model. The authors envision an implementation of a human-computer dialogue system that could support necessary functionality to provide efficient information access and illustrate it as a transition network over a set of speech acts (Fig. 1). This model describes a use case of a "conversation for action" and is mainly focused on tracking commitments rather than analyzing language variations. We use the COR model as our baseline and show in an empirical evaluation that it is not able to adequately reflect the structure of information-seeking dialogues across four publicly available datasets and propose an alternative model.

Belkin et al. [3] argue for a modular structure of an interactive information retrieval (IR) system that would be able to support various dialogue interactions. The system should be able to compose interactions using a set of scripts, which provide for various information-seeking strategies (ISSs) that can be described using the COR model. The authors introduce four dimensions to describe different ISSs and propose to collect cases for each of the ISSs to design the scripts. We closely follow their line of work by accumulating empirical evidence from publicly available conversational datasets to validate both the COR model and the ISS dimensions proposed by Sitter and Stein [21] that form the basis for accumulating the body of sample scripts describing various ISSs.

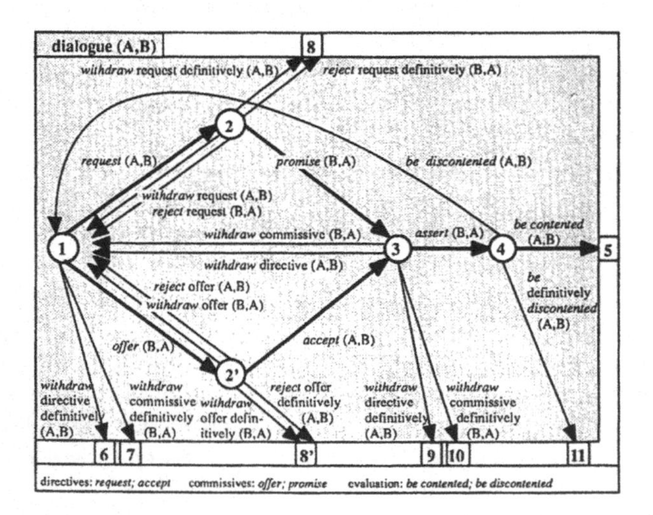

Fig. 1. COnversational Roles (COR) model of information-seeking dialogues proposed by Sitter and Stein [21].

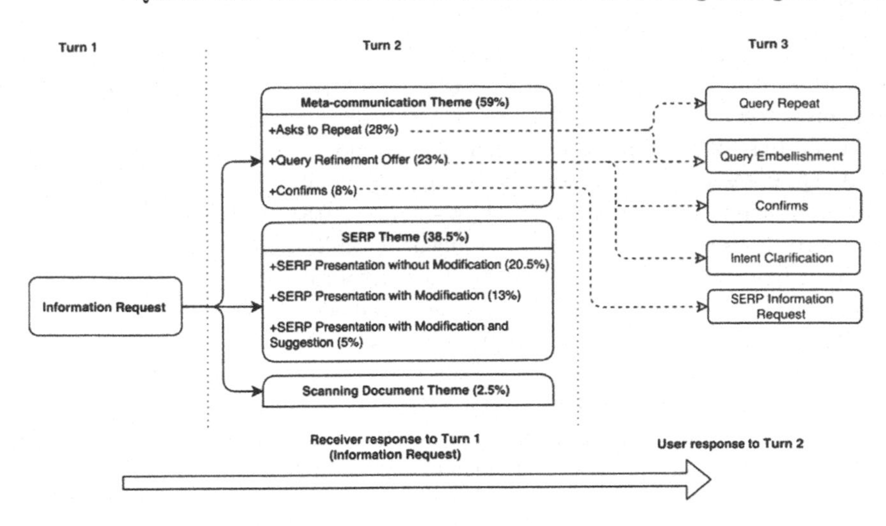

Fig. 2. Spoken Conversational Search (SCS) model by Trippas et al. [23] (SERP is short for Search Engine Results Page.)

More recently, Radlinski and Craswell [15] have also proposed a theoretical model for a conversational search system. They propose a set of five actions available to the agent and a set of five possible user responses to describe the user-system interactions. However, they do not describe in detail the conversation flow between these actions. In contrast, Trippas et al. [23] empirically derive a model of Spoken Conversational Search (SCS) based on collected conversational transcripts (Fig. 2). The SCS model describes the case of using a web search engine via a speech-only interface and is limited to sequences of three conversation turns. We show how such an empirical approach to analyzing and structuring conversation transcripts into conversation models can be performed at scale on multiple conversational datasets (including SCS) using process mining techniques.

3.2 Process Mining from Conversations

There are relatively few prior publications that demonstrate the benefit of applying process mining (PM) techniques to conversational data. Di Ciccio and Mecella [6] use a corpus of e-mail correspondence to illustrate how the structure of a complex collaborative process can be extracted from message exchanges. Wang et al. [29] analyze a sample of discussion threads from an on-line Q&A forum by applying process mining and network analysis techniques and comparing patterns discovered across different thread categories based on their outcomes (solved, helpful and unhelpful threads).

Richetti et al. [16] analyze the performance of a customer support service team by applying process mining to conversational transcripts that were previously annotated with speech acts using a gazetteer. Their results reveal similar structural patterns in the conversational flow of troubleshooting conversations with different durations, i.e., less and more complex cases, which require additional information seeking loops.

To the best of our knowledge, there is no prior work going beyond the individual use cases mining conversations from a specific domain. In contrast, we analyze multiple heterogeneous conversational datasets from various domains to be able to draw conclusions on structural similarities as well as differences stemming from variance introduced by labeling approaches and specific characteristics of the underlying communication processes.

Before applying the proposed approach, utterances have to be annotated with activity labels, such as speech acts [20]. The task of utterance classification is orthogonal to our work. Dialogue corpora to be used for process mining can be manually annotated by human annotators or automatically by using one of the classification approaches proposed earlier [4,10,11,22].

4 An Empirical Approach to Extracting a Conversation Model from Conversational Transcripts

We consider every conversation C in a transcript $\mathcal{C}$ to be an instance of the same communication process, the model of which we aim to discover. A conversation is represented as a sequence of utterances $C = \langle u_1, u_2, \ldots \rangle$. An *utterance*, in this case, is defined rather broadly as a text span within a conversation transcript attributed to one of the conversation participants and explicitly specified during the annotation process (utterance labeling step). We denote the set of all utterances as $U = \{u_1, u_2, \ldots\}$. Our approach to conversational modeling consists of three steps: (1) utterance labeling, (2) model discovery, and (3) conformance checking, which is useful for model validation and error detection in conversation transcripts.

4.1 Utterance Labeling

A utterance u_i in a conversation C can be mapped to multiple labels $l_i^1, \ldots, l_i^n$, each belonging to pre-defined label sets $L^1, \ldots, L^n$, respectively. The label sets, not necessarily disjoint, may correspond to different annotation schemes. We denote the general multi-labeling of utterances as a mapping function $\hat{\lambda} : U \rightarrow L_1 \times \cdots \times L_n$. It can stem from manual annotations of different human analysts, or categories returned by multiple machine-learned classifiers. For the sake of readability, we assume in the remainder of this section that all annotations share a single set of labels L, thus the mapping function used henceforth is reduced to $\lambda : U \rightarrow L$.

4.2 Extracting the Model of the Conversation Flow

We apply a process discovery approach to collect patterns of the conversation structure from transcripts. The goal of process discovery is to extract a model that is representative of empirically observed behavior stored in an *event log* [28].

Event logs can be abstracted as sets of sequences (*traces*), where each element in the sequence (the *event*) is labeled with an activity (*event class*) plus optional *attributes*. We reduce a conversation transcript $\mathcal{C}$ along with its labeling to an event log, as follows: a conversation C is a trace, an utterance u is an event, and the label of u, $\lambda(u) = l$, is the event class.

There is a wide variety of algorithms that can be applied for process discovery. Imperative workflow mining algorithms, such as the seminal α-algorithm [28] or the more recent Inductive Miner (IM) [13], extract procedural process models that depict the possible process executions, in the form of, e.g., a Petri net [27]. Other approaches, such as frequent episode mining [12] or declarative constraints mining [7,8], extract local patterns and aggregate relations between activities. One such relation is the *succession* between two activities, denoting that the second one occurs eventually after the first one. In our context, succession between l_i and l_j holds true in a sub-sequence $\langle u_i, \ldots, u_j \rangle$, with $i < j$, if $u_j \mapsto l_j$ and $u_i \mapsto l_i$. The frequency of such patterns observed across conversations can indicate dependencies between the utterance labels l_i and l_j. Those dependencies can be used to construct a model describing a frequent behavior (model discovery) as well as to detect outliers breaking the expected sequence (error analysis), upon the setting of thresholds for minimum frequency. The discovery algorithm described in [8] requires a linear pass through each sequence to count, for every label $l \in L$, (1) its number of occurrences per sequence, and (2) the distance at which other labels occur in the same sequence in terms of number of utterances in-between.

4.3 Conformance Analysis and Model Validation

The goal of the model validation step is conformance checking, i.e., to assess to which extent the patterns evident from the transcript fit the structure imposed by the model. We use it here to also evaluate the predictive power of the model, i.e., the ability of the model to generalize to unseen instances of the conversation process. A good model should fit the transcripts but not overfit it. The model to be validated against can be the one previously extracted from transcripts, or a theoretical model, i.e., an independently constructed one. In the former case, we employ standard cross-validation techniques by creating a test split separate from the development set that was used to construct the model.

Model quality can be estimated with respect to a conversation transcript. Likewise, the quality of the conversation can be estimated with respect to a pre-defined model. In other words, discrepancies between a conversation model and a transcript indicate either inadequacy of the model or errors (undesired behavior or recording malfunctions) in the conversation.

To compute fitness, i.e., the ability of the model to replay the event log, we consider the measure first introduced in [26], based on the concept of alignment. Alignments keep consistent the replay of the whole sequence and the state of the process by adding so-called non-synchronous moves if needed. The rationale is, the more non-synchronous moves are needed, the lower the fitness is. Thus, a penalty is applied by means of a cost function on non-synchronous moves.

Table 1. Dataset statistics.

Dataset	Dialogues	Utterances	Labels
SCS	39	101	13
ODE	26	417	20
DSTC1	15,866	732,841	37
DSTC2	2,118	40,854	21

Table 2. New functional annotation schema for information-seeking conversation utterances.

	Proactive		Reactive	
User	**Q**uery	Information	**F**eedback	Positive
		Prompt		Negative
Agent	**R**equest	Offer	**A**nswer	Results
		Understand		Backchannel
				Empty

For every sequence fitness is computed as the complement to 1 of the total cost of the optimal alignment, divided by the cost of the worst-case alignment. Log fitness is calculated by averaging the sequence fitness values over all sequences in a log.

5 QRFA Model Development and Results

In this section, we apply the approach to extraction and validation of a conversational model proposed in Sect. 4 to develop a new model (QRFA). We (1) collect *datasets* of publicly available corpora with information-seeking dialogue transcripts; (2) analyze and link their annotation schemes to each other and to the COR model (Fig. 1); (3) analyze the conversation flows in the datasets; and (4) *evaluate* QRFA and compare the results with COR as a baseline.

5.1 Conversational Datasets

We used publicly available datasets of information-seeking dialogues that are annotated with utterance-level labels (see the dataset statistics in Table 1).

Spoken Conversational Search. This dataset[1] (*SCS* [23]) contains human-human conversations collected in a controlled laboratory study with 30 participants. The task was designed to follow the setup, in which one of the conversation participants takes over the role of the information Seeker and another of the Intermediary between the Seeker and the search engine. It is the same dataset that was used to develop the SCS model illustrated in Fig. 2. All dialogues in the dataset are very short and contain at most three turns, with one label per utterance. The efficiency of the interaction and the user satisfaction from the interaction are not clear.

[1] https://github.com/JTrippas/Spoken-Conversational-Search.

Open Data Exploration. This dataset[2] (*ODE*) was collected in a laboratory study with 26 participants and a setup similar to the SCS but with the task formulated in the context of conversational browsing, in which the Seeker does not communicate an explicit information request. The goal of the Intermediary is to iteratively introduce and actively engage the Seeker with the content of the information source. All dialogues in this dataset contain one label per utterance. The majority of the conversation transcripts (92%) exhibit successful interaction behavior leading to a positive outcome, such as satisfied information need and positive user feedback (only 2 interactions were unsuccessful), and can be considered as samples of effective information-seeking strategies.

Dialog State Tracking Challenge. These datasets[3] (*DSTC1* and *DSTC2* [30]) provide annotated human-computer dialogue transcripts from an already implemented dialogue system for querying bus schedules and a restaurant database. The transcripts may contain more than one label per utterance, which is different from the previous two datasets. The efficiency of the interaction and user satisfaction from the interaction with the agent are not clear.

5.2 QRFA Model Components

Since all datasets and the theoretical model that we consider use different annotation schemes, we devise a single schema to be able to aggregate and compare conversation traces. To the best of our knowledge, no such single schema that is able to unify annotations across a diverse set of information-seeking conversation use cases has been proposed and evaluated before. Our schema is organized hierarchically into two layers of abstraction to provide a more simple and general as well as more fine-grained views on the conversation components.

First, we separate utterances into four basic classes: two for User (Query and Feedback) and two for Agent (Request and Answer). This distinction is motivated by the role an utterance plays in a conversation. Some of the utterances explicitly require a response, such as a question or a request, while others constitute a response to the previous utterance, such as an answer. Such a distinction is reminiscent of the Forward and Backward Communicative Functions that are foundational for the DAMSL annotation scheme [5]. The labels also reflect the roles partners take in a conversation. The role of the Agent is to provide Answers to User's Queries. During the conversation the Agent may Request additional information from the User and the Agent may provide Feedback to the Agent's actions.

[2] https://github.com/svakulenk0/ODExploration_data.
[3] https://www.microsoft.com/en-us/research/event/dialog-state-tracking-challenge.

The initial set of four labels (QRFA) are further subdivided to provide a more fine-grained level of detail. See Table 2 and the descriptions below:

Query provides context (or input) for *Information* search (question answering), as the default functionality provided by the agent (e.g., "Where does ECIR take place this year?"), but can also *Prompt* the agent to perform actions, such as cancel the previous query or request assistance, e.g., "What options are available?"

Request is a pro-active utterance from the agent, when there is a need for additional information (Feedback) from the user. It was the only class that caused disagreement between the annotators, when trying to subdivide it into two groups of requests: the ones that contain an *Offer*, such as an offer to help the User or presenting the options available (e.g., "I can group the datasets by organization or format"), and the ones whose main goal is to *Understand* the user need, such as requests to repeat or rephrase the Query (e.g., "Sorry I am a bit confused; please tell me again what you are looking for").

Feedback from the user can be subdivided by sentiment into *Positive*, such as accept or confirm, and *Negative*, such as reject or be discontented.

Answer corresponds to the response of the agent, which may contain one of the following: (1) *Results*, such as a search engine result page (SERP) or a link to a dataset, (2) *Backchannel* response to maintain contact with the User,

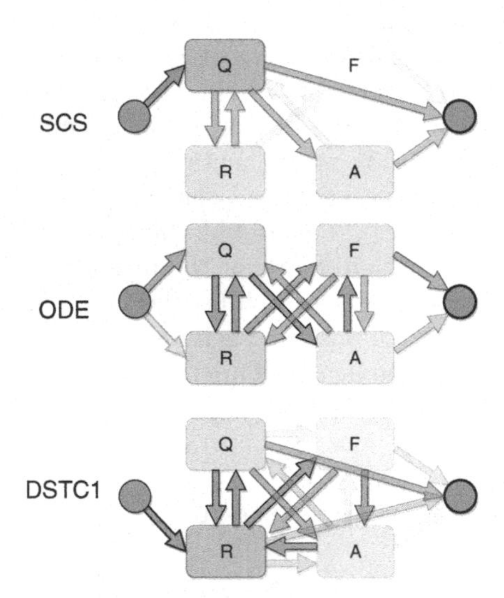

Fig. 3. Conversation flows in the SCS, ODE and DSTC1 datasets. Color intensity indicates frequency.

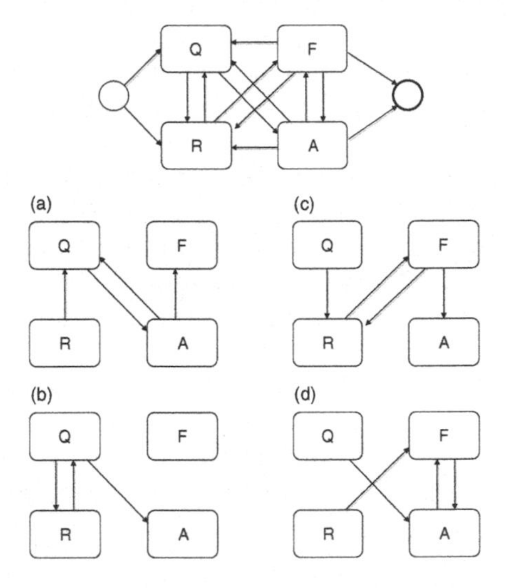

Fig. 4. The QRFA model for conversational search composed of the "four virtuous cycles of information-seeking": (a) question answering loop, (b) query refinement loop, (c) offer refinement loop, (d) answer refinement loop.

such as a promise or a confirmation (e.g., "One moment, I'll look it up.") and (3) *Empty* result set (e.g., "I am sorry but there is no other Indian restaurant in the moderate price range").

Two authors of this paper independently aligned the annotation schemes of the datasets and the COR model to match the single schema with an inter-annotator agreement of 94%. We found the first more abstract level of annotation sufficient for our experiments to make the conversation models easier to interpret. The complete table containing alignments across the schemes is made available to the community to enable reproducibility and encourage future work in this direction.[4]

5.3 QRFA Model Dynamics

We used the ProM Episode Miner plug-in [12] and a declarative process mining tool, MINERful[5] [7,8], to discover frequent sequence patterns in the conversation transcripts.

Figure 3 illustrates the conversation flows in each of the three datasets used for model discovery (one of the datasets, DSTC2, is held out for model evaluation). Color intensity (opacity) indicates the frequency of the observed sequences between the pairs of utterances within the respective dataset (the frequency counts for all transitions across all the datasets are available on-line[6]). An empirically derived information-seeking conversation model would be the sum of the models extracted from the three conversation transcripts.

However, an empirically derived model guarantees neither correctness nor optimality since the transcripts (training data) may contain errors, i.e., negative patterns. Instead of blindly relying on the empirical "as is" model, we analyze and revise it (re-sample) to formulate our theoretical model of a successful information-seeking conversation (Fig. 4). For example, many conversations in the SCS and DSTC1 datasets are terminated right after the User Query (Q→END pattern) for an unknown reason, which we consider to be undesirable behavior: the User's question is left unanswered by the Agent. Therefore, we discard this transition from our prescriptive model, which specifies how a well-structured conversation "should be" (Fig. 4). Analogously to discarding implausible transitions, the power of the theoretical modeling lies in the ability to incorporate transitions are still considered legitimate from a theoretical point of view even though they were not observed in the training examples. Incompleteness in empirically derived models may stem from assumptions already built into the systems by their designers, when analyzing dialogue system logs, or also differences in the annotation guidelines, e.g., one label per utterance constraint. In our case, we noticed that adding the FQ transition, which was completely absent from our training examples, will make the model symmetric. The symmetry along

[4] https://github.com/svakulenk0/conversation_mining/blob/master/annotations/alignments_new.xls.
[5] https://github.com/cdc08x/MINERful.
[6] https://github.com/svakulenk0/conversation_mining/tree/master/results/.

the horizontal axes reflects the distribution of the transitions between the two dialogue partners. Hence, the FQ transition mirrors the AR transition, which is already present in our transcripts, but on the User side. The semantics of an FQ transition is that the User can first give feedback to the Agent and then follow up with another question. Trippas et al. [24, Figure 1, Example 1] empirically show that utterances in information-seeking dialogues tend to contain multiple moves, i.e., can be annotated with multiple labels.

The final shape of a successful information-seeking conversation according to our model is illustrated in Fig. 4. To analyze this model in more detail, we decompose it into a set of connected components, each containing one of the cycles from the original model. We refer to them as "four virtuous cycles of information-seeking,"[7] representing the possible User-Agent exchanges (feedback loops) in the context of: (a) question answering, (b) query refinement, (c) offer refinement, and (d) answer refinement. To verify that the loops actually occur and estimate their frequencies, we mined up to 4-label sequences from the transcripts using the Episode Miner plug-in.

5.4 QRFA Model Evaluation

Our evaluation of the QRFA model is twofold. Firstly, we measure model fitness with respect to the conversational datasets including a held-out dataset (DSTC2), which was not used during model development, to demonstrate the ability of the model to fit well across all available datasets and also generalize to unseen data. Secondly, we hypothesize that deviations from the conversation flow captured in the QRFA model signal anomalies, i.e., undesired conversation turns. Therefore, we also compare the model's performance on the task of error detection in conversational transcripts with human judgments of the conversation success.

Fitness and Generalization. To analyze the model fit with respect to the actual data we applied the conformance checking technique proposed by Adriansyah et al. [1], available as a ProM plug-in under the name "Replay a Log on Petri Net for Conformance Analysis." To this end, we translated the COR and QRFA models into the Petri net notation and ran a conformance analysis for each model on every dataset. Exit and Restart activities are part of the "syntactic sugar" added for the Petri net notation and we set them to *invisible* in order to avoid counting them, when assigning the costs during analysis.

Table 3 (top) contains the fitness measures of the COR and QRFA models for all the datasets. We use the default uniform cost function that assigns a cost of 1 to every non-synchronous move. Fitness is computed separately for each sequence (dialogue) as a proportion of the correctly aligned events. We measure generalization as the fitness of a model on the sequences that are not

[7] A virtuous cycle refers to complex chains of events that reinforce themselves through a feedback loop. A virtuous circle has favorable results, while a vicious circle has detrimental results.

considered for its creation. The ability of the model to generalize to a different held-out dataset is significant (0.99 on average). This result demonstrates the out-of-sample generalizability of the model, which is a more challenging task than testing the model on the held-out (test) splits from the same datasets (label frequency distributions) used for the model development. Remarkably, the baseline COR model managed to fully fit only a single conversation across all four datasets. This comparison clearly shows the greater flexibility that the QRFA model provides, which in turn indicates the requirement for information-seeking dialogue systems to be able to operate in four different IR modes (Fig. 3) and seamlessly switch between them when appropriate.

Conversation Success and Error Detection. Since only one of the datasets, namely ODE, was annotated with a success score, we add manual annotations for the rest of the datasets (2 annotators, inter-annotator agreement: 0.85). We produced annotations for 89 dialogues in total, for the full SCS dataset and a random sample for each of the DSTC datasets.[8] Criteria for the success of a conversational interaction are defined in terms of informational outcomes, i.e., the search results were obtained and the information need was satisfied, as well as emotional outcomes, i.e., whether the interaction was pleasant and efficient.

Table 3. Evaluation results of QRFA and COR models of information-seeking dialogues on the conversational datasets in terms of model fitness/generalization (top) and error detection abilities (bottom). The gold standard (GS) column refers to the manual annotations of the conversational datasets with the conversation success score (inter-annotator agreement: 0.85).

Dataset	SCS			ODE			DSTC1			DSTC2			Average	
Metric/Model	COR	QRFA	GS	COR	QRFA	GS	COR	QRFA	GS	COR	QRFA	GS	COR	QRFA
Average/case	0.58	0.89		0.74	1		0.66	0.96		0.7	0.99		0.67	0.96
Max.	0.8	1		1	1		1	1		0.91	1		0.93	1
Min.	0.4	0.8		0.6	1		0	0		0.53	0.8		0.38	0.65
Std. Deviation	0.17	0.1		0.09	0		0.08	0.05		0.05	0.02		0.10	0.04
Cases with value 1	0	0.46	0.37	0.04	1	0.92	0	0.14	0.07	0	0.83	0.79	0.01	0.61
Error detection Precision		1			1			1			0.67			0.92
Error detection Recall		0.78			0			0.83			0.57			0.55

Results of the conversation success prediction task are summarized in Table 3 (bottom); QRFA correlates well with human judgments of conversation success based on the model fitness obtained via conformance checking (Cases with value 1). We also took a closer look at the cases annotated as unsuccessful in terms of fitness to the QRFA model and reported Precision/Recall metrics for the

[8] https://github.com/svakulenk0/conversation_mining/tree/master/annotations/dialogue_success.

conversation failure detection task. For example, the model predicted all conversations in the ODE corpus as success (100% success rate) and overlooked 8% that actually failed, hence the Recall for conversation failure detection is 0 in this case.

Table 3 shows that half of the errors affecting conversation success are due to a violation of structural requirements formulated via the QRFA model. The model overestimates the success rate of a dialog agent since only syntactic information in some cases is not enough to evaluate the overall performance, such as the quality of the answer obtained. However, it shows very promising results, clearly indicating the faulty cases, such as the situations when the user's query was left unanswered by the agent (SCS and DSTC1).

Our evaluation shows that the QRFA model reflects the patterns of successful information-seeking conversations and the deviations from its shape likely indicate flaws in the conversation flow. These results are demonstrated across four conversational datasets from different domains. In particular, then, QRFA does not overfit the errors from the datasets used for development and it generalizes to the held-out dataset.

We conclude that the QFRA model satisfies the four quality criteria for a process model defined by van der Aalst [28]: (1) fitness – it fits across four conversational datasets without overfitting, which allows it to successfully detect deviations (errors) in the information-seeking process (Table 3); (2) precision – all types of interaction described by the model are observed in the conversation transcripts; (3) generalization – the model is able to describe the structure and deviations in previously unseen conversations; and (4) simplicity – it contains a minimal number of elements necessary to describe the conversation dynamics in information-seeking dialogues.

6 Conclusion

We have proposed an annotation schema and a theoretical model of information-seeking dialogues grounded in empirical evidence from several public conversational datasets. Our annotation schema resembles the approach used in DAMSL, where utterances are classified into Forward and Backward Communicative Function, but adopts labels to our information-seeking setting, where roles are more distinct due to information asymmetry between participants. The patterns that we have discovered extend and correct the assumptions built into the COR model and also incorporate frameworks previously proposed within the information retrieval community.

Our empirical evaluation indicates that, however simple, the QRFA model still provides a better fit than the most comprehensive model proposed previously by explaining the conversational flow in the available information-seeking conversational datasets. Moreover, we have described an efficient way to provide sanity checking diagnostics of a dialogue system using process mining techniques (conformance checking) and have shown how the QRFA model helps to evaluate the performance of an existing dialogue system from its transcripts.

In future work, we plan to evaluate the QRFA model against new conversational datasets and further extend it to a finer granularity level if required. Our experiments so far have utilized hand-labeled conversation transcripts. Introducing automatically generated labels may propagate errors into the model extraction phase. Nevertheless, discovering patterns in raw conversational data that is automatically tagged with semantic labels is an exciting research direction [25]. In addition, the predictions of the QRFA model may be an informative signal for evaluating or training reinforcement learning-based dialogue systems [14].

Wide adoption of information-seeking dialogue systems will lead to a massive increase in conversational data, which can potentially be used for improving dialogue systems. We believe that QRFA and similar models will become important for informing the design of dialogue systems, motivating collection of new information-seeking conversational data, specifying the functional requirements the systems should satisfy, and providing means for their evaluation.

Acknowledgements. The work of S. Vakulenko and C. Di Ciccio has received funding from the EU H2020 program under MSCA-RISE agreement 645751 (RISE_BPM) and the Austrian Research Promotion Agency (FFG) under grant 861213 (CitySPIN). S. Vakulenko was also supported by project 855407 "Open Data for Local Communities" (CommuniData) of the Austrian Federal Ministry of Transport, Innovation and Technology (BMVIT) under the program "ICT of the Future." M. de Rijke was supported by Ahold Delhaize, the Association of Universities in the Netherlands (VSNU), and the Innovation Center for Artificial Intelligence (ICAI).

All content represents the opinion of the authors, which is not necessarily shared or endorsed by their respective employers and/or sponsors.

References

1. Adriansyah, A., van Dongen, B.F., van der Aalst, W.M.: Conformance checking using cost-based fitness analysis. In: 15th International Enterprise Distributed Object Computing Conference, pp. 55–64. IEEE Computer Society (2011)
2. Austin, J.L.: How to do Things with Words. Oxford Paperbacks, Oxford University Press, Oxford (1976)
3. Belkin, N.J., Cool, C., Stein, A., Thiel, U.: Cases, scripts, and information-seeking strategies: on the design of interactive information retrieval systems. Expert Syst. Appl. **9**(3), 379–395 (1995)
4. Cohen, W.W., Carvalho, V.R., Mitchell, T.M.: Learning to classify email into "speech acts". In: Proceedings of the 2004 Conference on Empirical Methods in Natural Language Processing, pp. 309–316 (2004)
5. Core, M.: Coding dialogs with the DAMSL annotation scheme. In: Working Notes of the AAAI Fall Symposium on Communicative Action in Humans and Machines, pp. 1–8 (1997)
6. Di Ciccio, C., Mecella, M.: Mining artful processes from knowledge workers' emails. IEEE Internet Comput. **17**(5), 10–20 (2013)
7. Di Ciccio, C., Mecella, M.: A two-step fast algorithm for the automated discovery of declarative workflows. In: Symposium on Computational Intelligence and Data Mining, pp. 135–142. IEEE (2013)

8. Di Ciccio, C., Mecella, M.: On the discovery of declarative control flows for artful processes. ACM Trans. Manag. Inf. Syst. **5**(4), 24:1–24:37 (2015)
9. Goffman, E.: Erving Goffman: Exploring the Interaction Order. Polity Press, Cambridge (1988)
10. Jeong, M., Lin, C., Lee, G.G.: Semi-supervised speech act recognition in emails and forums. In: Proceedings of the 2009 Conference on Empirical Methods in Natural Language Processing, pp. 1250–1259 (2009)
11. Jo, Y., Yoder, M., Jang, H., Rosé, C.P.: Modeling dialogue acts with content word filtering and speaker preferences. In: Proceedings of the 2017 Conference on Empirical Methods in Natural Language Processing, pp. 2179–2189 (2017)
12. Leemans, M., van der Aalst, W.M.P.: Discovery of frequent episodes in event logs. In: Ceravolo, P., Russo, B., Accorsi, R. (eds.) SIMPDA 2014. LNBIP, vol. 237, pp. 1–31. Springer, Cham (2015). https://doi.org/10.1007/978-3-319-27243-6_1
13. Leemans, S.J.J., Fahland, D., van der Aalst, W.M.P.: Process and deviation exploration with inductive visual miner. In: Proceedings of the BPM Demo Sessions 2014 Co-located with the 12th International Conference on Business Process Management, p. 46 (2014)
14. Li, Z., Kiseleva, J., de Rijke, M.: Dialogue generation: from imitation learning to inverse reinforcement learning. In: 33rd AAAI Conference on Artificial Intelligence, AAAI 2019, January 2019
15. Radlinski, F., Craswell, N.: A theoretical framework for conversational search. In: Proceedings of the 2017 Conference on Conference Human Information Interaction and Retrieval, pp. 117–126. ACM (2017)
16. Richetti, P.H.P., de A.R. Gonçalves, J.C., Baião, F.A., Santoro, F.M.: Analysis of knowledge-intensive processes focused on the communication perspective. In: Carmona, J., Engels, G., Kumar, A. (eds.) BPM 2017. LNCS, vol. 10445, pp. 269–285. Springer, Cham (2017). https://doi.org/10.1007/978-3-319-65000-5_16
17. Schegloff, E.A.: Sequencing in conversational openings. Am. Anthropologist **70**(6), 1075–1095 (1968)
18. Schiffrin, D.: Approaches to Discourse: Language as Social Interaction. Blackwell Textbooks in Linguistics. Wiley, Hoboken (1994)
19. Searle, J.R.: Speech Acts: An Essay in the Philosophy of Language. Cambridge University Press, Cambridge (1969)
20. Searle, J.R.: A classification of illocutionary acts. Lang. Soc. **5**(1), 1–23 (1976)
21. Sitter, S., Stein, A.: Modeling the illocutionary aspects of information-seeking dialogues. Inf. Process. Manage. **28**(2), 165–180 (1992)
22. Stolcke, A., et al.: Dialogue act modeling for automatic tagging and recognition of conversational speech. Comput. Linguist. **26**(3), 339–373 (2000)
23. Trippas, J.R., Spina, D., Cavedon, L., Sanderson, M.: How do people interact in conversational speech-only search tasks: a preliminary analysis. In: Proceedings of the 2017 ACM on Conference on Human Information Interaction and Retrieval, pp. 325–328. ACM (2017)
24. Trippas, J.R., Spina, D., Cavedon, L., Joho, H., Sanderson, M.: Informing the design of spoken conversational search: perspective paper. In: Proceedings of the 2018 Conference on Human Information Interaction and Retrieval, pp. 32–41. ACM (2018)
25. Vakulenko, S., de Rijke, M., Cochez, M., Savenkov, V., Polleres, A.: Measuring semantic coherence of a conversation. In: Vrandečić, D., et al. (eds.) ISWC 2018. LNCS, vol. 11136, pp. 634–651. Springer, Cham (2018). https://doi.org/10.1007/978-3-030-00671-6_37

26. van der Aalst, W.M., Adriansyah, A., van Dongen, B.F.: Replaying history on process models for conformance checking and performance analysis. Wiley Interdisciplinary Reviews: Data Min. Knowl. Discov. **2**(2), 182–192 (2012)
27. van der Aalst, W.M.P.: The application of petri nets to workflow management. J. Circuits Syst. Comput. **8**(1), 21–66 (1998). https://doi.org/10.1142/S0218126698000043
28. van der Aalst, W.M.P.: Process Mining: Data Science in Action, 2nd edn. Springer, Heidelberg (2016). https://doi.org/10.1007/978-3-662-49851-4
29. Wang, G.A., Wang, H.J., Li, J., Abrahams, A.S., Fan, W.: An analytical framework for understanding knowledge-sharing processes in online Q&A communities. ACM Trans. Manag. Inf. Syst. **5**(4), 18 (2015)
30. Williams, J., Raux, A., Henderson, M.: The dialog state tracking challenge series: a review. Dialogue Discourse **7**(3), 4–33 (2016)
31. Winograd, T., Flores, F.: Understanding Computers and Cognition: A New Foundation for Design. Intellect Books, Bristol (1986)

Iterative Relevance Feedback for Answer Passage Retrieval with Passage-Level Semantic Match

Keping Bi(✉), Qingyao Ai, and W. Bruce Croft

College of Information and Computer Sciences, University of Massachusetts Amherst, Amherst, MA, USA
{kbi,aiqy,croft}@cs.umass.edu

Abstract. Relevance feedback techniques assume that users provide relevance judgments for the top k (usually 10) documents and then re-rank using a new query model based on those judgments. Even though this is effective, there has been little research recently on this topic because requiring users to provide substantial feedback on a result list is impractical in a typical web search scenario. In new environments such as voice-based search with smart home devices, however, feedback about result quality can potentially be obtained during users' interactions with the system. Since there are severe limitations on the length and number of results that can be presented in a single interaction in this environment, the focus should move from browsing result lists to iterative retrieval and from retrieving documents to retrieving answers. In this paper, we study iterative relevance feedback techniques with a focus on retrieving answer passages. We first show that iterative feedback is more effective than the top-k approach for answer retrieval. Then we propose an iterative feedback model based on passage-level semantic match and show that it can produce significant improvements compared to both word-based iterative feedback models and those based on term-level semantic similarity.

Keywords: Iterative relevance feedback · Answer passage retrieval · Passage embeddings

1 Introduction

In typical relevance feedback (RF) techniques, users are provided with a list of top-ranked documents and asked to assess their relevance. The judged documents, together with the original query, are used to estimate a new query model using an RF model, which further acts as a basis for re-ranking. There were extensive studies of RF [1,2,4,9,13,16,25,27–29,37] based on the vector space model (VSM) [30], the probabilistic model [18] and, more recently, on the language model (LM) for Information Retrieval (IR) approach [23]. Despite the effectiveness of RF, the overhead involved in obtaining user relevance judgments has meant that it is not used in typical search scenarios.

L. Azzopardi et al. (Eds.): ECIR 2019, LNCS 11437, pp. 558–572, 2019.
https://doi.org/10.1007/978-3-030-15712-8_36

With mobile and voice-based search becoming more popular, it becomes feasible to obtain feedback about result quality during users' interactions with the system. In these scenarios, the display space or voice bandwidth leads to severe limitations on the length and number of results shown in a single interaction. Thus, instead of providing a list of results, an iterative approach to feedback may be more effective. There has been some work in the past on iterative relevance feedback (IRF) with only a few results in each interaction using the VSM [1,2,13], but this has not been looked at for a long time. In addition, the space and bandwidth limitations make the retrieval of longer documents less desirable than shorter answer passages. Motivated by these reasons, in this paper, we present a detailed study of methods for IRF focused on answer passage retrieval.

Although they could be applied to any text retrieval scenario, most existing RF algorithms use word-based models originally designed for document retrieval. Answer passages, however, are much shorter than documents, which could potentially present problems for accurate estimation of word weights in the existing word-based RF methods. Moreover, the limitations on the length and number of results in IRF mean that there is even less relevant text available at every iteration. Given these issues, introducing complementary information from semantic space may help to estimate a more accurate RF model. Dense vector representations of words and paragraphs in distributed semantic space, called embeddings, [5,8,17,21,32], have been effectively applied to many natural language processing (NLP) tasks. Embeddings have also been used in pseudo relevance feedback based on documents [24,35], but their impact in iterative and passage-based feedback is not known. Besides, these previous work use semantic similarity at the term level and does not consider semantic match at larger granularity. This had led us to incorporate passage-level semantic match in IRF for answer passage retrieval to improve upon word-based IRF and other embedding-based IRF using term-level semantic similarity.

In the paper, we first investigate whether iterative feedback based on different frameworks is effective relative to RF with a list of top k (k = 10) results on answer passage retrieval. The results indicate that IRF is significantly more effective on answer passage collections. In addition, we propose an embedding-based IRF method using passage-level similarity for answer passage retrieval. This method incorporates the similarity scores computed with different types of answer passage embeddings and fuses them with other types of IRF models. The model we propose significantly outperforms IRF baselines based on words or semantic matches between terms. Combining both term-level and passage-level semantic match information leads to additional gains in performance.

2 Related Work

In this section, we first review previous approaches to RF and IRF. We then discuss related work on embeddings of words and paragraphs applied to IR and some previous studies on answer passage retrieval.

Relevance Feedback. In general, there are mainly three types of relevance feedback methods for ad-hoc retrieval, which are based on the vector space model (VSM) [30], the probabilistic model [18] and the language model (LM) [23]. Basically, they all extract expansion terms from annotated relevant documents and re-weight the original query terms so as to estimate a more accurate query model to retrieve better results.

Rocchio [27] is generally credited as the first RF technique, developed on the VSM. It refines the vector of a user query by bringing it closer to the average vector of relevant documents and further from the average vector of non-relevant documents. In the probabilistic model, expansion terms are scored according to the probability they occur in relevant documents compared to non-relevant documents [12,25]. Salton et al. [29] studied various RF techniques based on the VSM and probabilistic model and showed that the probabilistic RF models are in general not as effective as the methods in the VSM.

More recently, feedback techniques have been investigated extensively based on LM, among which, the relevance model [16] and the mixture model [37] are two well-known examples that empirically perform well. In the third version of the relevance model (RM3) [16], the probabilities of expansion terms are estimated with occurrences of the terms in feedback documents. The mixture model [37] considers a feedback document to be generated from a mixture of a corpus language model, and a query topic model, which is estimated with the EM algorithm. Some recent work [4,9] extend the mixture model by considering additional or different language models as components of the mixture.

Iterative Relevance Feedback. In contrast to most RF systems that ask users to give relevance assessments on a batch of documents, Aalsberg et al. [1] proposed the alternative technique of incremental RF based on Rocchio. Users are asked to judge a single result shown in each interaction, then the query model can be modified iteratively through feedback. This approach showed higher retrieval quality compared with standard batch feedback. Later, Lwayama et al. [13] showed that the incremental relevance feedback used by Aalsberg et al. works better for documents with similar topics, while not as well for documents spanning several topics. In this paper, we investigate how IRF performs on retrieval of answer passages instead of documents using more recent retrieval models.

Some recent TREC tracks [10,33] have made use of iterative and passage-level feedback, but they focus on document retrieval with different objectives and require a large amount of user feedback. The Total Recall track [33] aims at high recall, where the goal is to promote all of the relevant documents before non-relevant ones. The target of the Dynamic Domain track [10] is to identify documents satisfying all the aspects of the users' information need with passage-level feedback. In contrast, we focus on iterative feedback for the task of answer passage retrieval and investigate IRF with a fixed small amount of feedback.

Word and Paragraph Embeddings for RF. Dense representations, called embeddings, of words and paragraphs, have become popular and been used [5,21,32,36] to abstract the meaning from a piece of text in semantic space. Two well-known techniques to train word and paragraph embeddings are Word2Vec [20] and Paragraph Vectors (PV) [17] respectively. The similarity of word embeddings can be used to compute the transition probabilities between words [24,35] and incorporated with the VSM or relevance model [16] to solve problems of term mismatch. Basically, these approaches use semantic match at word level and are in the form of query expansion. In contrast, our approach uses semantic match at passage level and is not based on query expansion.

3 Word-Based IRF Models

In IRF, the topic model of users' intent can be refined each iteration after a small number of results are assessed. Therefore, re-ranking is triggered earlier in IRF than in standard top-k RF methods. On the one hand, earlier re-ranking may produce better results with fewer iterations, which essentially reduces the user efforts in search interactions. On the other hand, having only a small amount of feedback information in each iteration may hurt the accuracy of model estimation and cause topic drift in the iterative process.

We convert several representative models to iterative versions and investigate the performance of the IRF models on answer passage retrieval. Since LM and VSM are the two most effective frameworks for RF, we study iterative feedback under these two frameworks. We use RM3 [16] and the Distillation (or Distill) model [4] to represent the LM framework, and Rocchio [27] for VSM. RM3 is a common baseline for pseudo RF that has also been used for RF. Distillation is one of the most recent RF methods, which is an extension of the mixture model by incorporating a query specific non-relevant topic model. Rocchio [27] is the standard feedback model in VSM. As for the retrieval models for initial ranking, we use Query Likelihood (QL) for LM, and BM25 [26] for VSM respectively.

To keep the query model from diverging to non-relevant topics, we maintain two pools for relevant and non-relevant results, which accumulate all the judgments until the ith iteration. During the ith iteration, judged relevant results and non-relevant results are added to the corresponding pool. Expanded query models are then estimated from the relevant document pool by RM3 and from both relevant and non-relevant pools by Distillation and Rocchio. Detailed introduction about the IRF models and the experiments can be found in [3].

4 Passage Embedding Based IRF Models

Word-based RF methods were initially designed for document retrieval and usually based on query expansion. In contrast to documents, answer passages do not have sufficient text to estimate the probabilities or weights of the expansion terms accurately, especially for IRF when fewer results are available in each iteration. To alleviate the problem of text insufficiency in IRF, we incorporate semantic information about paragraphs to the IRF models. Paragraph embeddings are shown to be capable of capturing the semantic meanings of passages [5,17,32], which could potentially help us build more robust IRF models by supporting semantic matching between passages.

In this section, we propose to use paragraph embeddings to improve the performance of IRF for answer passage retrieval. In contrast to existing word-based and embedding-based RF methods, this approach does not extract expansion terms to update the query model. Instead, it represents the relevance topic from feedback passages with embeddings. Similar to Rocchio, we assume a relevant passage should be near the centroid of other relevant passages in the embedding space. Also, we only focus on positive feedback as negative feedback has been shown to have little benefit for RF when positive feedback is available in previous studies [1]. Therefore, our model can be viewed as an embedding version of Rocchio with only positive feedback.

We first describe the methods we use to obtain the semantic representations for answer passages. Then we will introduce the passage embedding based iterative feedback model.

4.1 Passage Embeddings

Aggregated Word Embeddings. One common way of representing passages is to use aggregated embeddings of words in the paragraph. Word2Vec is a well-known method of training word embeddings [20,21]. It projects words to dense vector space and uses a word to predict its context or predicts a word by its context. In our experiments, we also use average word embeddings trained from Word2Vec both with and without IDF weighting as passage representations.

Paragraph Vectors. The other way of representing passages is using specially designed paragraph vectors models as in [5,17,32]. The models we use are PV-HDC [32] with or without corruption, shown in Fig. 1. PV-HDC is an extension of the initially proposed paragraph vector model [17], where a document vector is first used to predict an observed word, and afterward, the observed word is used to predict its context words. The recent work of training paragraph representation through corruption [5] shows advantages in many tasks such as sentiment analysis. It replaces the original part of paragraph representation with a corruption module, where the global context $\tilde{u}$ is generated through an unbiased dropout corruption at each update and the paragraph representation is calculated as the average embeddings of the words in $\tilde{u}$. The final representation is simply the average of the embeddings of all the words in the paragraph. We also investigate other models such as the original PV models, DM, DBOW, [17], and the Parallel Document

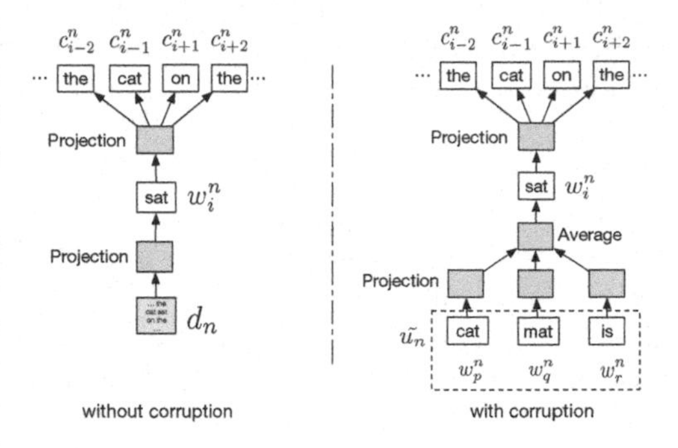

Fig. 1. HDC models used in our experiments. Red words are local context, and blue words are global context. (Color figure online)

Context Model (PDC) [32], both with and without corruption, but HDC is better in most cases. So we exclude the other models in the paper.

4.2 IRF with Passage-Level Semantic Similarity

As an alternative to query expansion based RF methods, we propose to represent the whole semantic meaning of a passage and a passage set with vectors in the embedding space and measure the similarity between them without explicitly extracting any expansion terms. Specifically, we represent the relevance topic in the ith iteration as the embedding of the relevant passage pool and fuse the similarity between a passage with the relevance topic with other RF methods. Thus the score function is shown as follows,

$$score(Q^{(i)}, d) = score_{rf}(Q^{(i)}, d) + \lambda_{sf} score_{sem}(RP^{(i)}, d) \quad (1)$$

$Q^{(i)}$ is the expanded query model estimated by iterative version of RF models such as RM3, Distillation and Rocchio; d is the candidate passage; $RP^{(i)}$ denotes the relevant passage pool in the ith iteration; $score_{rf}$ denotes the score calculated from other RF models; $score_{sem}$ is the semantic match score between passages, which is the commonly used cosine similarity in the paper; λ_{sf} is the coefficient of incorporating the passage embedding based similarity; Similar to Rocchio, we assume the topic of a passage set is the centroid of these passages and we consider a relevant passage pool can be represented by average vectors of the passages in it. Thus the similarity between a passage and the pool is

$$score_{sem}(RP^{(i)}, d) = \cos\left(\frac{1}{|RP^{(i)}|} \sum_{d_r \in RP^{(i)}} \boldsymbol{d}_r, \boldsymbol{d}\right) \quad (2)$$

where $\boldsymbol{d}_r$ and $\boldsymbol{d}$ is the vector representation of d_r and d in the embedding space.

Our method has two advantages over existing RF methods. One is that compared to expansion term based methods that only alleviate word-level mismatch, semantic similarity of larger granularity is captured in our method. The other is the flexibility of combining this semantic match signal with different types of approaches such as RM3, Distillation, Mixture, Rocchio, and other embedding-based feedback approaches.

5 Experiments of Word-Based IRF

In this section, we introduce the experimental setup and results of word-based IRF on answer passage retrieval.

5.1 Experimental Setup

Data. In our experiments, we used WebAP and PsgRobust for answer passage retrieval. Statistics of the datasets are summarized in Table 1. WebAP [34] is a web answer passage collection built on Gov2. It uses a subset of queries that are likely to have passage-level answers from Gov2 and retrieved the top 50 documents with the Sequential Dependency Model (SDM) [19]. After that, relevant documents were annotated for relevant answer passages. Overall, 3843 passages from 1200 documents are annotated as relevant. In our experiments, we split the rest of the documents into non-overlapping 2 or 3 (randomly chosen) contiguous sentences as non-relevant passages and used topic descriptions as questions.

PsgRobust[1] is a new collection we created for answer passage retrieval. It is based on the TREC Robust collection following a similar approach as WebAP but without manual annotation. In PsgRobust, we assume that *top-ranked passages in relevant documents can be considered as relevant* and *all passages in non-relevant documents are irrelevant*. We first retrieved the top 100 documents for each title query in Robust with SDM [19] and generated answer passages from them with a sliding window of random lengths (2 or 3 sentences) and no overlap. After that, we retrieved top 100 passages with SDM again and treated those from relevant documents as the relevant passages. Similar to WebAP, we used the descriptions of Robust topics as questions and have 246 queries with non-zero relevant answer passages in total. The Recall@100 in the initial retrieval process is 0.43, which means that 43% of relevant documents for all queries were included in the passage collection on average. By manually checking some randomly sampled passages marked as relevant, we found most of them are indeed relevant passages for the questions. There are 6589 relevant passages from 3544 documents for the 246 queries in total.

Table 1. Statistics of experimental datasets.

Dataset	#Psg	PsgLen	Vocab	#Query	#Psg/D	#RelPsg/Q	#RelPsg/D
WebAP	379k	45	59k	80	114.3	48.0	3.2
PsgRobust	383k	46	64k	246	17.1	26.8	1.9

We also considered other collections that have passage-level annotation such as the DIP2016Corpus [11] and the dataset from the Dynamic Domain track [33]. However, they either are trivial for RF tasks (almost all top 10 results retrieved by BM25 are relevant in DIP2016Corpus) or have few queries (only 26 and 27 for the two domains of the Dynamic Domain track). Other popular question answering datasets usually only have one relevant answer for each query and thus are not suitable for our RF task either. Therefore, we only report the results of WebAP and PsgRobust in this paper.

System Settings. All the methods we implemented are based on the Galago toolkit [7] [2]. Stopwords were removed from all collections using the standard

[1] This dataset is publicly available at https://ciir.cs.umass.edu/downloads/PsgRobust/.

[2] http://www.lemurproject.org/galago.php.

INQUERY stopword list and words were stemmed with Krovetz Stemmer [15]. To compare iterative feedback with typical top-k feedback in a fair manner, we fixed the total number of judged results as 10 and experimented with 1, 2, 5, and 10 iterations, where 10, 5, 2, 1 results were judged during each iteration respectively. Then 10Doc-1Iter is exactly the top-k feedback. We pay more attention to the settings of one or two results per iteration which are more likely to be in a real interactive search scenario considering the limitation of presenting results. True labels of results were used to simulate users' judgments.

All the parameters were set using 5-fold cross-validation over all the queries in each collection with grid search. For WebAP and PsgRobust, we tuned μ of QL in $\{30, 50, 300, 500, 1000, 1500\}$ and k of BM25 from $\{1.2, 1.4, \cdots, 2\}$, b set to 0.75 as suggested by [22]. The number of expansion terms m is from $\{10, 20, \cdots, 50\}$. The range to scan parameters for RM3, Distillation and Rocchio is similar as the corresponding original paper. They are not shown here due to space limits.

Evaluation. The evaluation should only focus on the ranking of unseen results. So we use freezing ranking [6,28], as in [1,14], to evaluate the performance of IRF. The *freezing ranking* paradigm freezes the ranks of all results presented to the user in the earlier feedback iterations and assigns the first result retrieved in the ith iteration rank $iN + 1$, where N is the number of results shown in each iteration. Note that all the previously shown results are filtered out in the following retrieval to remove duplicates and the final result list concatenates $(\#Iter - 1) * N$ freezing results with the rest candidates ranked in the last iteration, where $\#Iter$ is the total number of iterations. Then we use mean average precision at cutoff 100 (MAP) and $NDCG@20$ to measure the performance of results overall and on the top. As suggested by Smucker et al. [31], statistical significance is calculated with Fisher randomization test with threshold 0.05.

Table 2. Performance of IRF on answer passage collections. $D \times I$ stands for $Doc \times Iter$. '*' denotes significant improvements over the standard top 10 feedback model (10×1).

Dataset	Method	*MAP* of freezing rank lists					*NDCG*@20 of freezing rank lists				
	(D×I)	Initial	(10×1)	(5×2)	(2×5)	(1×10)	Initial	(10×1)	(5×2)	(2×5)	(1×10)
WebAP	RM3	0.076	0.100	0.107*	**0.113***	**0.113***	0.143	0.170	0.180*	0.185*	**0.187***
	Distill	0.076	0.099	0.104*	0.109*	**0.111***	0.143	0.166	0.177*	0.185*	**0.187***
	Rocchio	0.081	0.106	0.112*	0.118*	**0.119***	0.150	0.169	0.181*	0.190*	**0.191***
PsgRobust	RM3	0.248	0.293	0.299*	0.306*	**0.308***	0.319	0.356	0.363*	0.372*	**0.373***
	Distill	0.248	0.292	0.299*	0.311*	**0.313***	0.319	0.354	0.362*	0.375*	**0.379***
	Rocchio	0.191	0.268	0.280*	0.285*	**0.286***	0.292	0.341	0.356*	0.361*	**0.364***

5.2 Results and Discussion

The performance of the initial retrieval with QL and BM25 and the IRF experimental results are shown in Table 2. All the feedback methods are significantly better than their retrieval baselines, i.e. RM3 and Distillation compared with QL, Rocchio compared with BM25, in terms of both MAP and $NDCG@20$.[3]

[3] On PsgRobust, BM25 and Rocchio underperform QL, RM3 and Distillation respectively by a large margin. Because its labels are generated based on retrieval with SDM, this collection favors approaches in the framework of LM more than VSM.

In addition, on both WebAP and PsgRobust, the MAP and $NDCG@20$ of RM3, Distillation and Rocchio increase as the ten results are judged in more iterations. In other words, IRF is much more effective for answer passage retrieval compared with top-k feedback. Performance goes up when re-ranking is done earlier even when we have only a small number of passages, probably because answer passages are usually focused on a single topic and less likely to cause topic drift. Since MAP and $NDCG@20$ show similar trends using IRF under different settings, we only show MAP in Sect. 6.2 due to the space limitations.

6 Experiments of Passage Embedding Based IRF

We compare our method with word-based and embedding-based RF baselines in two groups of experiments. One is the same as in Sect. 5, i.e. retrieval with a different number of iterations and 10 results judged in total. The other focuses on identifying more relevant passages given only one relevant answer passage. We first describe the experimental setup and then introduce the two groups of experiments in Sects. 6.2 and 6.3.

6.1 Experimental Setup

In this part, we again use WebAP and PsgRobust for experiments. All comparisons are based on LM (RM3, Distillation, and Rocchio) and VSM (Rocchio) to see whether the complementary semantic match benefits in both frameworks. We also include the Embedding-based Relevance Model (ERM) [35] as a baseline. ERM revises $P(Q|D)$ in the original RM3 as a linear combination of $P(Q|D)$ computed from exact term match and $P(Q|w, D)$, which takes the semantic relationship between words into account. The translation probability between words is computed with the cosine similarity of their embeddings transformed with the sigmoid function.[4] Statistical significance in all the result tables is calculated with Fisher randomization test with threshold 0.05.

Embeddings Training. Four paragraph representations are tested in the four groups of experiments, where the base models (BM) can be RM3, ERM, Distillation (or Distill) and Rocchio:

$BM + W2V / BM + idfW2V$: uniformly or idf-weighted average word vectors trained with the skip-gram model [20].

$BM + PVC / BM + PV$: paragraph vectors trained with the HDC structure with or without corruption [5,32].

Embeddings of words or paragraphs were trained with each local corpus respectively. Words with the frequency less than 5 were removed. No stemming was done across the collections. 10 negative samples were used for each target word. The learning rate and batch size were 0.05 and 256. The dimension of

[4] We also tried the true RF version of BM25-PRF-GT [24], which is a generalized translation model of BM25 based on word embeddings and Rocchio. Due to its inferior performance on our dataset, we did not include the experiments here.

embedding vectors was set to 100. We also tried other hyper-parameters for training embeddings, and the results were similar under different settings. For PVC, corruption rate q [5] was set to 0.9. All the neural networks of training embeddings were implemented using TensorFlow[5].

Parameter Settings. We used the best settings of the baseline models and tuned the parameters of the semantic signals with 5-fold cross-validation for different paragraph embeddings. All the parameters of ERM are tuned in the same range as [35] suggests. λ_{sf} in Eq. 1 is selected from $\{5, 10, 15, \cdots, 40\}$ for WebAP, and $\{0.5, 1, 1.5, \cdots, 5\}$ for PsgRobust respectively.

6.2 Iterative Feedback with Embeddings

First, we conducted IRF experiments with different number of iterations and 10 results judged in total, as described in Sect. 5. We use MAP at cutoff 100 of freezing rank lists as the evaluation metric, which is described in Sect. 5.1.

Results and Discussion. We show the experimental results of using language model baselines (RM3, ERM, Distillation) in Table 3 and include Rocchio as a baseline in Fig. 2. We can see in general the four representations of paragraphs all can improve performance significantly over the word-based and embedding-based baselines under most iteration settings. ERM performs similar to RM3 on WebAP, and our method based on RM3 and ERM also perform similarly. On PsgRobust, ERM performs slightly better than RM3 and our method also performs slightly better combined with ERM than RM3.[6] This shows that incorporating passage-level semantic similarity in embedding space produces improvements to both the word-based RF models and the embedding-based RF model using semantic similarity at term level.

The conclusion that IRF shows advantages over top-k feedback still holds when we incorporate word-based RF models with passage-level semantic match. In addition, there is no one representation better than the others all the time, which implies for different datasets, with different baselines, some representations show their advantages fitting the specific property underlying the setting.

6.3 Retrieval Given One Relevant Passage

As we mentioned in Sect. 4, the small amount of text in answer passages during each iteration may not be enough to build word-based RF models. The extreme case is when we have only one short passage as positive feedback. Effective re-ranking after *the first positive feedback* will show the user a second relevant answer in fewer iterations and make users less likely to leave after several interactions. Therefore, it is particularly important to perform well given the first

[5] https://www.tensorflow.org/.

[6] The reason why ERM does not perform well will be shown in Sect. 6.3 where we discuss the performance difference of ERM on the two tasks.

Table 3. Performance of different IRF models. '$*$' and '†' denote significant improvements over word-based (RM3, Distillation) and embedding-based (ERM) baselines respectively. ($10Doc \times 1Iter$) represents standard top-k feedback.

Method	*MAP* on **WebAP**				*MAP* on **PsgRobust**			
(Doc×Iter)	(10×1)	(5×2)	(2×5)	(1×10)	(10×1)	(5×2)	(2×5)	(1×10)
RM3	0.100	0.107	0.113	0.113	0.293	0.299	0.306	0.308
ERM	0.101	0.107	0.113	0.116	0.294	0.301	0.310	0.310
RM3+W2V	**0.107***†	**0.115***†	0.117	0.116	**0.298***†	0.303*†	0.312*†	0.312*
RM3+idfW2V	0.106*†	0.113*†	0.121*†	0.119*	**0.298***†	0.303*	**0.313***†	0.313*†
RM3+PV	0.102*	0.113*†	**0.123***†	**0.123***	**0.298***†	**0.305***†	**0.313***	**0.314***
RM3+PVC	**0.107***†	0.114*†	0.120*†	0.114	0.297*†	0.303*	0.308	0.311*
ERM+W2V	**0.107***†	**0.116***†	0.119*†	0.118	**0.299***†	0.304*†	0.313*†	0.312*
ERM+idfW2V	0.106*†	0.114*†	0.121*†	0.118	**0.299***†	0.304*†	**0.314***†	**0.314***†
ERM+PV	0.103*	0.115*†	**0.122***†	**0.121***	**0.299***†	**0.307***†	**0.314***†	0.313*
ERM+PVC	**0.107***†	0.114*†	0.121*†	0.114	0.298*†	0.304*†	0.312*	0.313*†
Distillation	0.099	0.104	0.109	0.111	0.292	0.299	0.311	0.313
Distill+W2V	**0.106***	**0.114***	**0.120***	0.113	0.297*	0.304*	0.314*	0.319*
Distill+idfW2V	**0.106***	0.113*	0.116*	0.115	0.297*	**0.306***	0.316*	0.319*
Distill+PV	0.103*	0.110*	0.118*	0.116*	**0.298***	**0.306***	**0.317***	**0.320***
Distill+PVC	0.105*	0.112*	**0.120***	**0.120***	0.297*	0.304*	0.315*	0.317*

positive feedback from users. We designed the second type of experiment to be answer retrieval given one relevant passage.

For each query, we randomly assign a relevant passage to the model as positive feedback and then retrieve from the remaining results. To make the results more reliable, we randomly draw a relevant passage for each query ten times and do ten retrievals. Then we evaluate the performance of each model based on the overall rank lists from the ten retrievals. We take QL and BM25 as baseline retrieval models that do not consider feedback. Similar to the first group of experiments, we use RM3, Distillation, Rocchio as word-based RF baselines in the framework of LM and VSM, and ERM as the embedding-based RF baseline. We use $P@1$ (precision@1), MRR (mean reciprocal rank) to evaluate the ability of a model to identify a second relevant passage in the next interaction given only one positive feedback. MAP at cutoff 100 measures the ability of the model to identify all the other relevant answers.

Results and Discussion. In Table 4, feedback methods are always better than their base retrieval models, i.e. QL, BM25. In general, with the four paragraph representations, the improvements of MAP over the baselines are always significant; $P@1$, MRR can also be improved significantly in many cases. This shows that incorporating the passage semantic similarity can improve significantly over

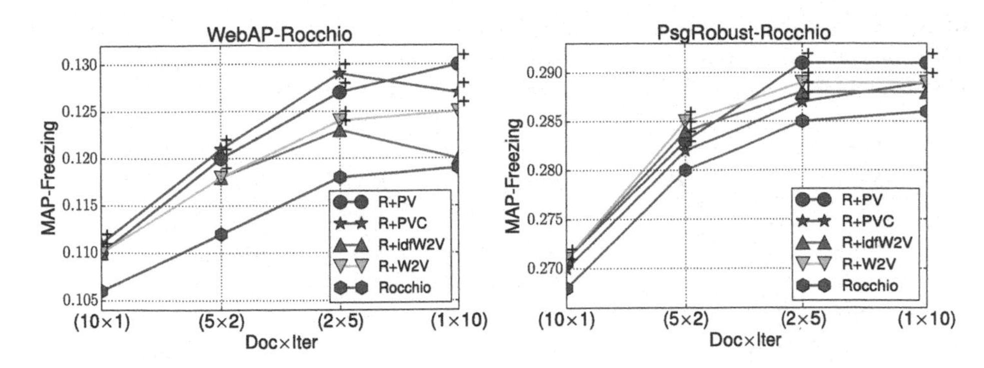

Fig. 2. Performance of our method with different paragraph representations compared with Rocchio. '+' means significant difference.

both the word-based RF baselines and the embedding-based RF baseline with only term-level semantic match information.

In contrast to the IRF experiments, ERM performs much better than RM3 in this task. The reason may be that in the IRF experiments, there are more relevant passages for RM3 to extract expansion terms and alleviate the term mismatch problem, which makes the term-level semantic match from ERM less helpful. In this task, the text for RM3 is not enough to estimate an accurate model and ERM is effective with semantic match. Since our method considers semantic match at passage level, its benefit does not overlap with that from term-level semantic match.

On WebAP, our method combined with RM3 performs similarly to ERM when using PV and PVC and worse than ERM using W2V and idfW2V. On PsgRobust, incorporating our method to RM3 performs better than ERM in terms of $P@1$ and MRR, but worse than ERM with MAP. This shows that incorporating embedding similarity to do RF at passage level or term level alone with little information are comparable to each other. When we combine these two ways of doing RF together, the performance can be further improved, which is shown from the significant improvements upon ERM when we add the passage similarity signal to ERM on both datasets. This is consistent with our claim that the semantic similarity of the passage level is complementary to the term level when combined with word-based RF models since they capture two different granularities of semantic match.

Different from the IRF experiments, the performance of paragraph vectors are better than W2V and idfW2V. This indicates that when there is little feedback information, more accurate representations lead to better performance. In addition, with scarce user feedback, PVC is more effective than PV, probably because it is less susceptible to overfitting a small dataset due to many fewer parameters, i.e. vocabulary size versus corpus size.

Table 4. Performance of different IRF methods on finding other relevant answers given one relevant answer. '$*$' and '$\dagger$' denote significant improvements over word-based (RM3, Distillation, Rocchio) or embedding-based (ERM) baselines respectively.

Dataset	WebAP			PsgRobust		
Model	*P@1*	*MRR*	*MAP*	*P@1*	*MRR*	*MAP*
QL	0.259	0.373	0.071	0.367	0.486	0.231
RM3	0.498	0.602	0.116	0.515	0.634	0.299
ERM	0.516	0.615	0.125	0.513	0.634	0.307
RM3+W2V	0.488	0.598	0.120*	0.524*†	0.643*†	**0.304***
RM3+idfW2V	0.488	0.597	0.120*	0.525*†	0.643*†	**0.304***
RM3+PV	**0.525***	0.625*	0.122*	0.521	0.641*	0.301*
RM3+PVC	0.524*	**0.635*†**	**0.123***	**0.526*†**	**0.644*†**	0.303*
ERM+W2V	0.513	0.622*	0.131*†	0.529*†	0.648*†	0.312*†
ERM+idfW2V	0.525*	0.627*	0.130*†	**0.534*†**	0.650*†	0.312*†
ERM+PV	**0.556*†**	0.648*†	0.131*†	0.531*†	0.649*†	0.311*†
ERM+PVC	**0.556*†**	**0.658*†**	**0.134*†**	**0.534*†**	**0.653*†**	**0.313*†**
Distillation	0.494	0.597	0.113	0.516	0.635	0.299
Distill+W2V	0.489	0.593	0.117*	**0.528***	**0.645***	**0.304***
Distill+idfW2V	0.489	0.595	0.117*	0.525*	0.643*	**0.304***
Distill+PV	0.519*	0.621*	0.120*	0.514	0.638	0.297
Distill+PVC	**0.534***	**0.638***	**0.123***	0.524*	0.643*	0.303*
BM25	0.298	0.399	0.072	0.35	0.479	0.176
Rocchio	0.516	0.616	0.121	0.522	0.641	0.279
Rocchio+W2V	0.531	0.640*	0.140*	0.526	0.645*	**0.282***
Rocchio+idfW2V	0.536	0.642*	0.139*	0.526	0.644	**0.282***
Rocchio+PV	**0.576***	**0.668***	0.138*	0.518	0.642	0.280*
Rocchio+PVC	0.560*	**0.668***	**0.143***	**0.528***	**0.647***	0.281*

7 Conclusion and Future Work

We first showed that IRF is effective on answer passage retrieval. Then we showed that, with passage-level semantic match the performance of iterative feedback and retrieval given one relevant passage can produce significant improvements compared with word-based RF models in the framework of both LM and VSM. The IRF experiments also show our method is better than the embedding-based baseline using term-level similarity. The retrieval experiment based on one relevance passage shows that combining the word and passage level granularities leads to the best performance.

Our method focuses more on user requests of "more like this". We know diversity is also very important to provide users more informative results and we will take it into account in our future work. In addition, we will consider IRF on answer passage retrieval with end-to-end neural models.

Acknowledgments. This work was supported in part by the Center for Intelligent Information Retrieval and in part by NSF IIS-1715095. Any opinions, findings and conclusions or recommendations expressed in this material are those of the authors and do not necessarily reflect those of the sponsor.

References

1. Aalbersberg, I.J.: Incremental relevance feedback. In: Proceedings of the 15th Annual International ACM SIGIR Conference, pp. 11–22. ACM (1992)
2. Allan, J.: Incremental relevance feedback for information filtering. In: Proceedings of the 19th Annual International ACM SIGIR Conference, pp. 270–278. ACM (1996)
3. Bi, K., Ai, Q., Croft, W.B.: Revisiting iterative relevance feedback for document and passage retrieval. arXiv preprint arXiv:1812.05731 (2018)
4. Brondwine, E., Shtok, A., Kurland, O.: Utilizing focused relevance feedback. In: Proceedings of the 39th International ACM SIGIR Conference, pp. 1061–1064. ACM (2016)
5. Chen, M.: Efficient vector representation for documents through corruption. arXiv preprint arXiv:1707.02377 (2017)
6. Cirillo, C., Chang, Y., Razon, J.: Evaluation of feedback retrieval using modified freezing, residual collection, and test and control groups. Scientific Report No. ISR-16 to the National Science Foundation (1969)
7. Croft, W.B., Metzler, D., Strohman, T.: Search Engines: Information Retrieval in Practice, vol. 283. Addison-Wesley, Reading (2010)
8. Dai, A.M., Olah, C., Le, Q.V.: Document embedding with paragraph vectors. In: NIPS Deep Learning Workshop (2015)
9. Dehghani, M., Azarbonyad, H., Kamps, J., Hiemstra, D., Marx, M.: Luhn revisited: significant words language models. In: Proceedings of the 25th ACM International on Conference on Information and Knowledge Management, pp. 1301–1310. ACM (2016)
10. Grossman, M.R., Cormack, G.V., Roegiest, A.: TREC 2016 total recall track overview. In: TREC (2016)
11. Habernal, I., et al.: New collection announcement: focused retrieval over the web. In: Proceedings of the 39th International ACM SIGIR Conference, pp. 701–704. ACM (2016)
12. Harman, D.: Relevance feedback revisited. In: Proceedings of the 15th Annual International ACM SIGIR Conference, pp. 1–10. ACM (1992)
13. Iwayama, M.: Relevance feedback with a small number of relevance judgements: incremental relevance feedback vs. document clustering. In: Proceedings of the 23rd Annual International ACM SIGIR Conference on Research and Development in Information Retrieval, pp. 10–16. ACM (2000)
14. Jones, G., Sakai, T., Kajiura, M., Sumita, K.: Incremental relevance feedback in Japanese text retrieval. Inf. Retrieval **2**(4), 361–384 (2000)
15. Krovetz, R.: Viewing morphology as an inference process. In: Proceedings of the 16th Annual International ACM SIGIR Conference, pp. 191–202. ACM (1993)
16. Lavrenko, V., Croft, W.B.: Relevance-based language models. In: ACM SIGIR Forum, vol. 51, pp. 260–267. ACM (2017)
17. Le, Q., Mikolov, T.: Distributed representations of sentences and documents. In: Proceedings of the 31st International Conference on Machine Learning (ICML-14), pp. 1188–1196 (2014)

18. Maron, M.E., Kuhns, J.L.: On relevance, probabilistic indexing and information retrieval. J. ACM (JACM) **7**(3), 216–244 (1960)
19. Metzler, D., Croft, W.B.: A Markov random field model for term dependencies. In: Proceedings of the 28th Annual International ACM SIGIR Conference, pp. 472–479. ACM (2005)
20. Mikolov, T., Chen, K., Corrado, G., Dean, J.: Efficient estimation of word representations in vector space. arXiv preprint arXiv:1301.3781 (2013)
21. Mikolov, T., Sutskever, I., Chen, K., Corrado, G.S., Dean, J.: Distributed representations of words and phrases and their compositionality. In: Advances in Neural Information Processing Systems, pp. 3111–3119 (2013)
22. Mogotsi, I.: Christopher D. Manning, Prabhakar Raghavan, and Hinrich Schütze: Introduction to Information Retrieval (2010)
23. Ponte, J.M., Croft, W.B.: A language modeling approach to information retrieval. In: Proceedings of the 21st Annual International ACM SIGIR Conference, pp. 275–281. ACM (1998)
24. Rekabsaz, N., Lupu, M., Hanbury, A., Zuccon, G.: Generalizing translation models in the probabilistic relevance framework. In: Proceedings of the 25th ACM CIKM Conference, pp. 711–720. ACM (2016)
25. Robertson, S.E., Jones, K.S.: Relevance weighting of search terms. J. Assoc. Inf. Sci. Technol. **27**(3), 129–146 (1976)
26. Robertson, S.E., Walker, S., Jones, S., Hancock-Beaulieu, M.M., Gatford, M., et al.: Okapi at TREC-3. NIST Special Publication SP **109**, 109 (1995)
27. Rocchio, J.J.: Relevance feedback in information retrieval. In: The Smart Retrieval System-experiments in Automatic Document Processing (1971)
28. Ruthven, I., Lalmas, M.: A survey on the use of relevance feedback for information access systems. Knowl. Eng. Rev. **18**(2), 95–145 (2003)
29. Salton, G., Buckley, C.: Improving retrieval performance by relevance feedback. J. Am. Soc. Inf. Sci. **41**, 288–297 (1990)
30. Salton, G., Wong, A., Yang, C.S.: A vector space model for automatic indexing. Commun. ACM **18**(11), 613–620 (1975)
31. Smucker, M.D., Allan, J., Carterette, B.: A comparison of statistical significance tests for information retrieval evaluation. In: Proceedings of the 16th ACM CIKM Conference, pp. 623–632. ACM (2007)
32. Sun, F., Guo, J., Lan, Y., Xu, J., Cheng, X.: Learning word representations by jointly modeling syntagmatic and paradigmatic relations. In: ACL, vol. 1, pp. 136–145 (2015)
33. Yang, G.H., Soboroff, I.: TREC 2016 dynamic domain track overview. In: TREC (2016)
34. Yang, L., et al.: Beyond factoid QA: effective methods for non-factoid answer sentence retrieval. In: ECIR (2016)
35. Zamani, H., Croft, W.B.: Embedding-based query language models. In: Proceedings of the 2016 ACM ICTIR, pp. 147–156. ACM (2016)
36. Zamani, H., Croft, W.B.: Relevance-based word embedding. In: Proceedings of the 40th International ACM SIGIR Conference. SIGIR 2017 (2017)
37. Zhai, C., Lafferty, J.: Model-based feedback in the language modeling approach to information retrieval. In: Proceedings of the Tenth CIKM Conference, pp. 403–410. ACM (2001)

Topic Modeling

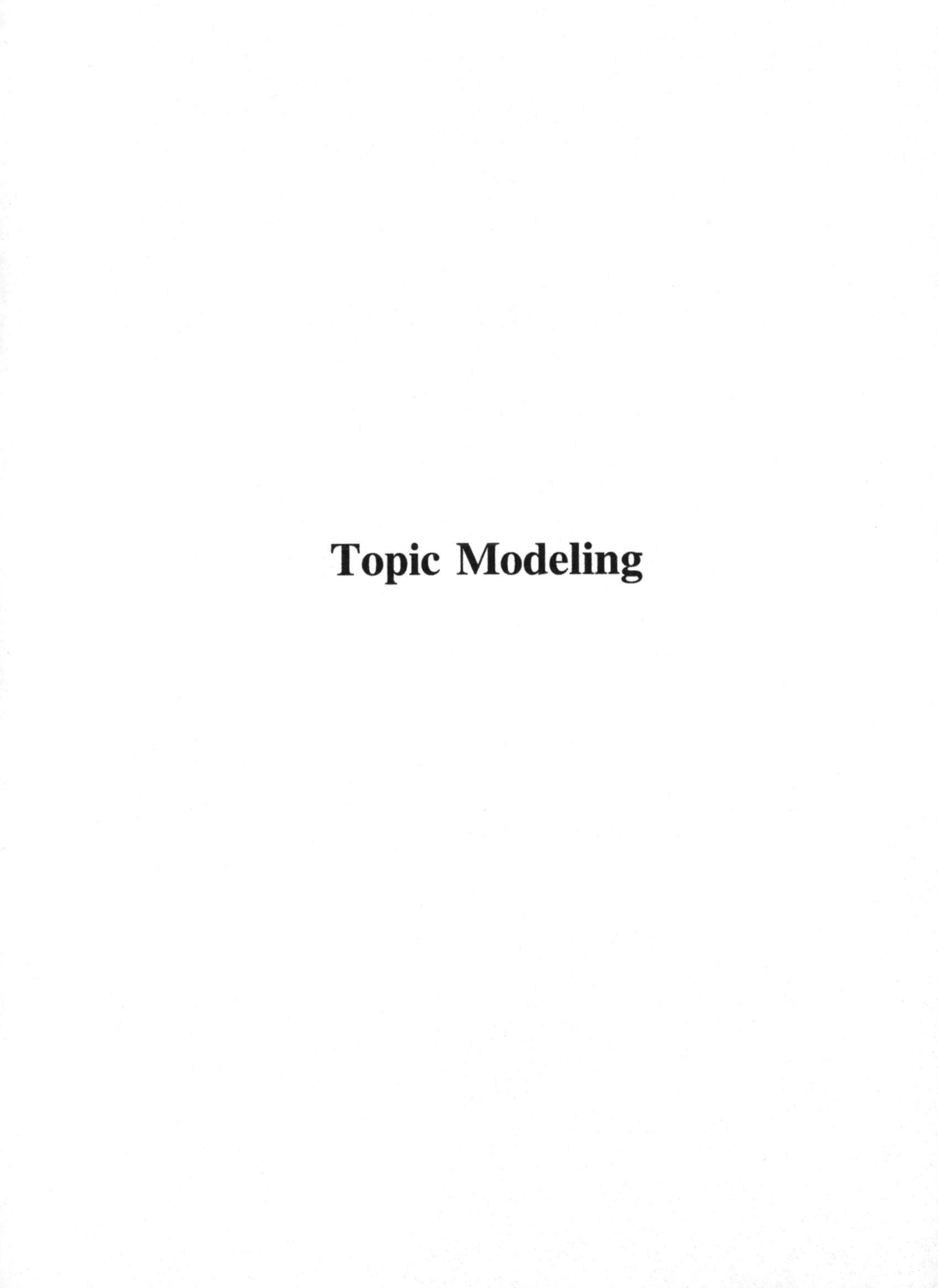

LICD: A Language-Independent Approach for Aspect Category Detection

Erfan Ghadery[1(✉)], Sajad Movahedi[1], Masoud Jalili Sabet[2], Heshaam Faili[1], and Azadeh Shakery[1]

[1] School of ECE, College of Engineering, University of Tehran, Tehran, Iran
{erfan.ghadery,s.movahedi,hfaili,shakery}@ut.ac.ir
[2] Center for Information and Language Processing (CIS) LMU Munich, Munich, Germany
masoud@cis.lmu.de

Abstract. Aspect-based sentiment analysis (ABSA) deals with processing and summarizing customer reviews and has been a topic of interest in recent years. Given a set of predefined categories, Aspect Category Detection (ACD), as a subtask of ABSA, aims to assign a subset of these categories to a given review sentence. Thanks to the existence of websites such as Yelp and TripAdvisor, there exist a huge amount of reviews in several languages, and therefore the need for language-independent methods in this task seems necessary. In this paper, we propose Language-Independent Category Detector (LICD), a supervised method based on text matching without the need for any language-specific tools and handcrafted features for identifying aspect categories. For a given sentence, our proposed method performs ACD based on two hypotheses: First, a category should be assigned to a sentence if there is a high semantic similarity between the sentence and a set of representative words of that category. Second, a category should be assigned to a sentence if sentences with high semantic and structural similarity to that sentence belong to that category. To apply the former hypothesis, we used soft cosine measure, and for the latter, word mover's distance measure is utilized. Using these two measures, for a given sentence we calculate a set of similarity scores as features for a one-vs-all logistic regression classifier per category. Experimental results on the multilingual SemEval-2016 datasets in the restaurant domain demonstrate that our approach outperforms baseline methods in English, Russian, and Dutch languages, and obtains competitive results with the strong deep neural network-based baselines in French, Turkish, and Spanish languages.

Keywords: Aspect-based sentiment analysis · Aspect category detection · Consumer reviews · Soft cosine measure · Word mover's distance

1 Introduction

No one shops alone, even someone who goes shopping alone. People usually care about other people's comments and recommendations. With the advent of web

L. Azzopardi et al. (Eds.): ECIR 2019, LNCS 11437, pp. 575–589, 2019.
https://doi.org/10.1007/978-3-030-15712-8_37

2.0 people tend to express their opinions on the web and share their experiences with each other. Therefore, sentiment analysis and opinion mining for online reviews are attracting a lot of attention [19].

One of the subtasks introduced in SemEval-2016 [21] ABSA is Aspect Category Detection (ACD). Given a review sentence and a set of pre-defined categories, ACD aims to assign a subset of these categories to the review sentence. The pre-defined categories consist of two subcategories: Attribute and Entity in the form of A#E (e.g. 'RESTAURANT#GENERAL', 'FOOD#QUALITY'). For example, the sentence 'The food here is rather good, but only if you like to wait for it.' contains a sentiment about the aspect FOOD#QUALITY while containing another sentiment about the aspect SERVICE#GENERAL. Therefore, given the above sentence, an ACD method should assign these two categories to the review sentence.

Most of the previously developed systems on this task are based on supervised machine learning techniques. Among these supervised methods, some systems use classic classification algorithms such as SVM and Logistic Regression [13,32]. Typically, these methods need a lot of effort to extract hand-crafted features which is time-consuming, and the performance of these methods are dependent on selecting an appropriate set of features. On the other hand, following the recent interest in neural network methods, some authors proposed neural network-based approaches [31,33,34]. Eventually, the performance of these neural-based methods depends on the availability of a sufficient amount of training data. In this paper, we propose a language-dependent method that does not require expensive hand-crafted features or language-dependent external resources[1]. Also, our method can perform ACD well even with a small amount of training data, which can be an advantage, especially for low-resource languages.

Our proposed method, LICD, detects categories belonging to a given sentence based on two hypotheses as follows. First, the given sentence belongs to a specific category if it has high semantic similarity to a set of key-words that represent that category. Second, the semantic and structural similarity between sentences is used. If a given sentence is close to another sentence of a specific category with regard to the aforementioned features, then, the given sentence also should belong to that category. To explore the first hypothesis, for each category, we choose a set of words using a feature selection method. These words constitute the representative set of that category. Then, we calculate the semantic similarity between the given sentence and each representative set to obtain a similarity score for each sentence category pair. In order to assess the mentioned semantic similarity, we utilize soft cosine similarity measure [27]. To inspect the second hypothesis, given a sentence, we retrieve the k-most semantically and structurally similar sentences to the sentence. Therefore, we need a similarity metric that measures both structural and semantic similarity at the same time. In this paper, we used the word mover's distance [14] to serve as the similarity metric for this purpose. The similarity score between a given sentence and a category will

[1] Note that, if we want to remove stopwords in the preprocessing step a list of stopwords is required.

be the sum of the inverse of word mover's distance values over the neighbor sentences that contain that category. Based on the two hypotheses mentioned above and using soft cosine measure and word mover's distance, we calculate a set of similarity scores for a given sentence and provide these scores as features to train a set of one-vs-all logistic regression classifier (one classifier for each category), where the output of classifiers is a probability distribution over the predefined categories. Categories that exceed a threshold are assigned to the given sentence. It is worth noting that instead of soft cosine and word mover's distance, any measure that has similar characteristics can be used.

We evaluate our language-independent approach on the multilingual SemEval-2016 datasets [21] in the restaurant domain with data available in 6 languages. Experimental results show that our method outperforms baseline methods in English, Russian, and Dutch languages while obtaining competitive results compared to baselines in French, Turkish, and Spanish.

2 Related Works

ABSA has been well studied in recent years. Schouten and Frasincar's work [26] provides a comprehensive survey on aspect level sentiment analysis specifically. The pioneering work of Hu and Liu [10] started this field. They used Association rule mining to extract frequent nouns and noun phrases assuming aspect terms to be noun or noun phrases, followed by applying a set of rules to prune redundant aspect terms. Another early work in this field is [28]. In this paper, authors detect implicit aspects using point-wise mutual information (PMI) to discriminate aspects from notional words.

In [13], authors tackle ACD using a set of one-vs-all SVM classifiers, one for each category, trained with features extracted from lexicon resources. They used n-grams, word clusters, etc. learned from the Yelp dataset as features. This system was ranked 1st in SemEval-2014 ACD sub-task. [32] has done similar work, but trained an SVM classifier for each predefined Entity and Attribute. They created a set of lexicons using train data, where lexicons are stemmed and un-stemmed n-grams with their precision, recall, and F1-scores calculated from the train data. Ganu et al. in [9] also proposed to use one-vs-all SVM classifiers for aspect category detection. They just used stemmed words as features. Unlike these methods, our method does not need expensive feature engineering and relies only on similarity measures.

In recent years, neural network- based methods, especially deep learning methods, have been proposed to address the ACD task. Khalil et al. [11] used a combination of a Convolutional Neural Network (CNN) and an SVM classifier for predicting aspect categories of a given sentence in the restaurant domain. They used term frequency (TF) weighted by inverse document frequencies (IDF) as features to train an SVM classifier. Also, for the laptop domain, they used one CNN classifier. Zhou et al. [34] proposed a method to learn the representation of words on a large set of reviews with noisy labels. After obtaining word vectors, through a neural network stacked on the word vectors, they generate deeper and hybrid features for providing to a logistic regression classifier. Deep neural-based methods rely on sufficient enough of training data in order to perform well. However, our proposed method has acceptable performance even with a small amount of training data. Therefore, in low-resource languages, our method can be an alternative to deep learning-based methods.

3 Technical Ingredients

In this section, we will describe Word Mover's Distance and Soft Cosine Similarity measures in more details.

3.1 Word Mover's Distance

Word mover's distance [14] is a text distance measure, inspired from the Earth Mover's Distance [23]. This method interprets the distance between two documents as a transformation problem. Therefore, the distance between two documents is the distance that the embedded words of the first document need to travel to become the embedded words of the second document in the embedding space.

Let D and D' be two documents and let $X \in R^{d \times n}$ be the embedded word vectors, where d and n are embedding dimension and number of words, respectively. x_i is embedding vector of word i in R^d. Let $T \in R^{n \times n}$ be a flow matrix, where $T_{ij} \geq 0$ denotes how much of word i in D travels to word j in D'. The word mover's distance between two documents is given by the following equation,

$$WMD(D_i, D_j) = min_{T \geq 0} \sum_{i,j \in [1..n]} T_{ij}.c(i,j) \tag{1}$$

subject to

$$\sum_{i,j \in [1..n]} T_{ij} = d_i, \quad \forall i \in [1..n] \quad and \quad \sum_{i,j \in [1..n]} T_{ij} = d_j, \quad \forall j \in [1..n] \tag{2}$$

where $c(i, j)$ is defined as the Euclidean distance between x_i and x_j in embedding space.

3.2 Soft Cosine Similarity

Soft cosine [27] is a method for calculating the similarity between two feature vectors, in our case two documents, even when they have no features in common. For measuring the similarity between words, soft cosine can leverage Levenshtein distance, WordNet similarity, or cosine similarity in word embedding space. As described in [6], given two documents X and Y, soft cosine similarity can be defined as Eq. 3,

$$cos_M(X, Y) = \frac{X^t.M.Y}{\sqrt{X^t.M.X}.\sqrt{Y^t.M.Y}} \tag{3}$$

$$X^t.M.Y = \sum_{i=1}^{n}\sum_{j=1}^{n} x_i m_{i,j} y_j \tag{4}$$

where M is a similarity matrix between words. The similarity between words is defined as the cosine similarity between word embedding vectors of words.

4 LICD: Language-Independent Category Detector

We describe LICD in this section. The architecture of the LICD is depicted in Fig. 1. It contains the following three main components: SCM similarity calculation, WMD similarity calculation, and classification. In order to identify aspect categories belonging to a given sentence, first of all, we calculate two similarity scores per category: SCM similarity, which is the score between the given sentence and set of representative words, and WMD similarity, which is the score between the sentence and the category calculated using word mover's distance. This gives us $2 \times c$ similarity scores for each sentence, where c is the number of categories. These scores are provided to a one-vs-all logistic regression classifier as features. In the following subsections, we will discuss each of these components in detail.

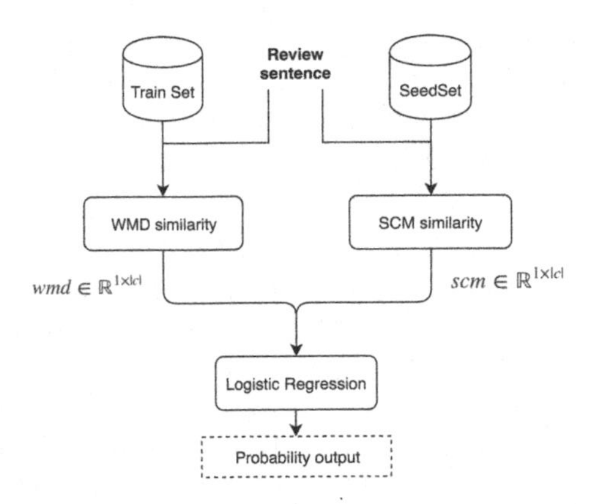

Fig. 1. The architecture of LICD.

4.1 SCM Similarity Calculation

Seed Selection. Every category has some specific words that can represent the category well. For example, the words 'food' and 'delicious' are good representative words for the 'FOOD#QUALITY' category. Each of these words has several semantically similar words (e.g 'cuisine' is a synonym for 'food'). A list of representative words plus their semantically similar words can represent a category well. To extract the set of representative words for different categories, we investigated different feature selection methods [8,30].

The best performance was achieved by the chi-square (χ^2) method. The selected words for each category constitute its seed words. After obtaining seed words, a list of top-k most similar words to each seed word is extracted from a word embedding model as semantically similar words. Table 1 illustrates an example of two extracted seed words and top-3 similar words for 'FOOD#QUALITY' and 'AMBIENCE#GENERAL' categories.

Table 1. A list of top 3 most similar words for 2 seed words of the categories 'FOOD#QUALITY' and 'AMBIENCE#GENERAL'.

Category	Seed word	Top 3 similar words
FOOD#QUALITY	Food	Cuisine - foods - grub
	Delicious	Delish - yummy - tasty
AMBIENCE#GENERAL	Atmosphere	Ambiance - ambience - environment
	Decor	Ambiance - decoration - decore

We compose a sentence using seed word followed by its semantically similar words. We can see this composed similar words as a sentence that describes a specific aspect of a category. The set of these sentences for each category constitute its SeedSet. For example, according to Table 1, the SeedSet of 'FOOD#QUALITY' category will contain the sentences 'food cuisine foods grub' and 'delicious delish yummy tasty'.

SCM Similarity. A sentence is likely to belong to a specific category if there is a high semantic similarity between sentence and the SeedSet of that category. Soft cosine similarity has an advantage compared to the more traditional method of cosine similarity on bag-of-words for computing the similarity between two sentences in that it also takes into account the semantic similarity between words. This advantage motivates us to utilize soft cosine similarity for obtaining similarity values between sentences and SeedSets of "the SeedSets" of each category. If a review contains words that are semantically related to one of the sentences in $SeedSet_i$, we expect the given review and $SeedSet_i$ to gain a high soft cosine similarity value.

For each sentence, we calculate a SCM similarity vector $\mathrm{scm} \in \mathbb{R}^{c}$, where c is the number of categories. We define $\mathrm{scm_i(x)}$ as the soft cosine similarity score between sentence x and the ith category, which is calculated by averaging soft cosine similarity values between x and each sentence in the ith category's SeedSet. Since each sentence in a SeedSet covers one of the aspects of the category. So for review sentences that belong to a category, the average of the soft cosine similarity between these review sentences and the sentences belonging to the SeedSet of that category will yield larger values compared to sentences not belonging to that category. Furthermore, it gives relatively higher value to a review sentence that covers at least one of the category's aspects compared to the sentences that do not cover any aspects of a category. $scm_i(x)$ is calculated using Eq. 5,

$$scm_i(x) = \frac{\sum_{s \in SeedSet_i} softcossim(x,\ s)}{|SeedSet_i|} \tag{5}$$

where $\mathrm{SeedSet_i}$ is the SeedSet of the ith category, $|\mathrm{SeedSet_i}|$ is the number of sentences in $\mathrm{SeedSet_i}$, and softcossim is the soft cosine similarity measure.

4.2 WMD Similarity Calculation

People tend to use similar sentence structures to express similar emotions [16]. For example, two sentences 'The food was good' and 'The cuisine was perfect' are two structurally similar sentences that express a similar opinion about the same category. Because part of the word mover's distance algorithm involves matching words, this measure can be quite efficient in measuring the similarity between a set of both structurally and semantically similar documents. Our motivation for utilizing this measure is based on the hypothesis that categories of a review sentence can be captured from it's structurally similar and semantically relevant sentences. Table 2 shows an example of the closest sentences to a given sentence retrieved by word mover's distance measure. As we see in this example, the word mover's distance measure tends to provide a minimum distance for the sentences that share a similar structure and semantically similar words.

Table 2. Example of the closest sentences to a given sentence retrieved by word mover's distance.

Given sentence: Ambiance is relaxed and stylish	
Closest sentences	Distance
The atmosphere is relaxed and casual	10.02
Atmosphere is nice and relaxed too	10.78
Zero ambiance to boot	16.38
Decor is charming	17.91
The decor is very simple but comfortable	18.00

In this part, analogous to SCM similarity, for a given sentence we calculate WMD similarity vector $\mathrm{wmd} \in \mathbb{R}^{\mathrm{c}}$, where c is the number of categories. We define $\mathrm{wmd_i(x)}$ as the word mover's distance similarity score between sentence x and the ith category. For calculating WMD similarity vector wmd, first, we retrieve the top-k closest sentences to the given sentence, where the distance measure is the word mover's distance. Then, the similarity score between the given sentence and the ith category will be the sum of similarity scores between the given sentence and all top-k closest sentences that have category i as their label, where similarity score is defined as the inverse of word mover's distance value.

We choose to sum the similarity scores to emphasize the majority categories between the top-k closest sentences and hash out the non-common categories.

Let $X = \{x_\ell, y_\ell\}_{\ell=1}^{n}$ be the train set, where x is a train sentence and y is the set of labels corresponding to x, and n is the number of train sentences. So $\mathrm{wmd_i(x)}$ can be calculated using Eq. 6,

$$wmd_i(x) = \sum_{x_k \in neighbor(x)} sim_i(x,\, x_k) \tag{6}$$

where $neighbor(x)$ is the set top-k closest neighbors to the sentence x and $\mathrm{sim_i(x,\, x_k)}$ is the similarity value between sentences x and $\mathrm{x_k}$ which is defined as follows,

$$sim_i(x,\, x_k) = \begin{cases} \frac{1}{1+wmdistance(x,\, x_k)} & \text{if}\, i \in y_k \\ 0 & \text{otherwise} \end{cases} \tag{7}$$

where $wmdistance(x,\, x_k)$ is word mover's distance between x and x_k sentences and $\mathrm{y_k}$ is the set of labels corresponding to sentence x_k.

One of the shortcomings of word mover's distance measure is its high time complexity. To speed up finding the closest neighbors using word mover's distance, we first find a large number of closest neighbors using cosine similarity between the average of the embedding vectors of words in the sentences to obtain a smaller and approximately relevant set of sentences. Then we find closest neighbors based on word mover's distance measure on the smaller set of obtained sentences.

4.3 Classification

For each category, a logistic regression classifier is trained using *scm* and *wmd* vectors obtained from the previous sections. The output of the classifiers is a probability distribution over the predefined categories. A category is assigned to a sentence if it's probability exceeds a threshold. A single optimum threshold for all the categories is found using a simple linear search.

5 Experiments

5.1 Datasets

For our experiments, we used the SemEval-2016 Task 5 datasets for English [21], French [2], Russian [15], Spanish, Dutch [7], and Turkish languages in restaurant domain. The number of training sentences is 2000, 1711, 1733, 3490, 2070, and 1104 for English, Dutch, French, Russian, Spanish, and Turkish, respectively, with 676, 575, 696, 1209, 881, and 144 test sentences. These statistics are borrowed from [21].

5.2 Experiment Setup

As the pre-processing step, tokenization and stop word removal is performed using the NLTK tool [3]. The scikit-learn package [20] is used to implement logistic regression classifier and chi-square feature selection method. For word mover's distance and soft cosine similarity and training word embeddings, we used the implementation provided by the gensim package [22]. CBOW method [17] is used for training word embeddings for the English language on the unlabeled Yelp restaurant dataset[2] with the dimension size of 300. For other languages, we used pre-trained word embeddings provided by [24]. All the parameters are optimized on the validation set. The best number of seed words for English, French, Spanish, Dutch, Turkish, and Russian are found as 1, 3, 5, 19, 15, and 25, respectively. The best number of nearest neighbors for WMD similarity for English, French, Spanish, Dutch, Turkish, and Russian languages are found as 9, 9, 15, 15, 10 and 10, respectively. For CONV and LSTM baselines we used multi-label soft margin as loss function. Both models are trained using a minibatch size of 128 and Adam optimizer [12] with a learning rate of 1×10^{-3}. Both models are trained for a maximum of 100 epochs for which early stopping is performed with patience set to 10. For the CONV baseline, the number of kernels, hidden layer size, and drop out rate is set to 300, 150, and 0.5, respectively. For the LSTM baseline, hidden size and drop out rate is set to 150 and 0.5, respectively.

5.3 Baseline Methods

Since one of the contributions of our method is being language-independent, for each language, we listed the current best model for that language. The compared methods are as follows:

- **NLANGP** [31]: This model utilizes the output of a CNN model and a set of features including word clusters, name lists, and head words to train a set of binary classifiers for each category.
- **XRCE** [4]: In this method the classification task is done in two steps. First, aspects with explicitly mentioned targets are classified using an ensemble method containing a Singular Value Decomposition and a One-vs-all Elastic Net Regression using a set of features. Then, aspects with implicit targets are classified using the same set of features but at the sentence level.
- **UFAL**: [29] This method uses a LSTM based Deep Neural Network. As features, this model only utilizes word embeddings. This is the only team that participated in all the languages in SemEval2016.
- **TGB**: [5] In this model, the classification task is done in two stages: At first, a set of one-vs-all logistic regression classifiers are trained for each attribute and entity. In the second stage, the output of the aforementioned classifiers along with other textual features (such as n-grams) are used as features for another set of one-vs-all logistic regression models for each category.

[2] https://www.yelp.com/dataset/challenge.

- **GTI**: [1] This work utilizes a set of one-vs-all SVM classifiers trained on textual features such as bi-grams, POS tags, lemmas, etc.
- **INSIGHT**: [25] This method utilizes a CNN with word embeddings as features. The word embeddings for languages other than English were initialized randomly.
- **UWB**: This method trains a Maximum Entropy Classifier using a large number of features including word embeddings, several kinds of bag-of-words features, POS tags, etc.
- **SemEval 2016 Baseline**: [21] For evaluation, SemEval 2016 provided a baseline. The baseline used an SVM with a linear kernel. For features, a set of 1000 most common unigrams was used.

The results of the baselines mentioned above are borrowed from [21]. Among several competitors in the SemEval-2016 workshop, the best results in English, French, Spanish, Russian, Dutch, and Turkish languages are achieved by NLANGP, XRCE, GTI, UFAL, TGB, and UFAL, respectively. Furthermore, we implement two methods as deep neural network-based baselines as follows:

- **CONV**: A CNN with one convolutional layer followed by a max-pooling layer and two fully-connected layers. We used tanh activation function for the convolutional layer and ReLU for the hidden layer. We used the sigmoid activation function on the output.
- **LSTM**: A bidirectional Long Short-Term Memory network with one layer followed by two fully-connected layers. We used ReLU after recurrent layer and sigmoid at the output layer.

5.4 Results and Analysis

For evaluation, we used micro F1-score of all the category labels, similar to the evaluation metric for SemEval-2016 ACD subtask. Table 3 shows the best results of the baseline methods and LICD in all languages. The best result for each language is marked in bold. LICD outperforms best systems in English, Russian, and Dutch languages. These results show the effectiveness of our approach in multilingual environments. In French, Spanish, and Turkish languages LICD achieves result competitive to the best-performed baselines. Our method is ranked 2nd among baselines in French and Turkish, and is ranked 3rd among baselines in Spanish. However, we would like to emphasize that the best systems in French and Turkish are deep neural network methods, which means their success depends on the existence of a sufficient volume of train data. However, as we will show in further experiments, LICD does not need a large volume of data, and with just about half of the training samples can achieve reasonable results. On the other hand, the best result for Spanish belongs to GTI method, which used several hand-crafted features such as n-gram, POS-tag, lemmas, and words. LICD is based on text matching and thus does not need such hand-crafted features.

5.5 Analysis of Different Components

Table 4 shows the comparison results between the different components of our method on the English dataset. When we trained the classifier with just *scm* vectors as features, the F1-score is 67.81, while F1-score obtained by *wmd* vectors as features is 70.60. Although *wmd* achieves better result compared to *scm*, the best result is obtained when we combine it with *scm* features. This shows the effectiveness of combining two different kinds of *scm* and *wmd* features. The interesting point is that results obtained by each of *scm* and *wmd* features alone outperforms most of the baseline methods.

Regarding computational complexity of our method, given n as the size of the vocabulary, the complexity of word mover's distance computation [14] is $O(n^3 log(n))$ and the complexity of computing soft-cosine similarity [18] is $O(n^3)$, respectively. In order to be able to utilize this method in the real-world applications, we need to reduce its computational complexity which we plan to address in the future works.

Table 3. The F1-score of LICD compared to baseline methods in each language.

Method	English	French	Spanish	Russian	Dutch	Turkish
CONV	71.78	60.90	67.53	68.95	57.88	61.79
LSTM	73.30	**63.27**	65.78	68.22	60.94	**64.80**
NLANGP	73.03	-	-	-	-	-
XRCE	68.70	61.20	-	-	-	-
UWB	68.20	-	61.96	-	-	-
INSIGHT	68.10	-	61.37	62.80	56	49.12
GTI	67.714	-	**70.58**	-	-	-
TGB	63.91	-	63.55	-	60.15	-
UFAL	59.3	49.92	58.81	64.82	53.87	61.02
SemEval	59.92	52.61	54.69	55.88	42.81	58.90
LICD	**73.54**	61.57	65.99	**69.76**	**61.30**	62.96

Table 4. Different component analysis results.

Components	F1(%)
scm	67.81
wmd	70.60
scm + wmd	73.54

5.6 Effect of Training Set Size and the Number of Seed Words

Fig. 2 presents the F1-score of the LICD method compared to CONV and LSTM baselines with different sizes of the training data. To perform this experiment, we split the original training set into ten equally divided portions, using stratified sampling method. Initially, we use only one part of the training set and gradually increase the number of utilized training set portions by one portion at each step. In all the experiments evaluation was performed on the full test set. Due to the limitation of space, we only show the result of the conducted experiment in three languages English, Russian, and Spanish. According to Fig. 2, in all of the three languages, our method achieves better or competitive results compared to the other baselines when there are few numbers of train samples. Figure 2(a) shows that in English, LICD achieves a stable F1-score approximately in the range of 40–70% (800–1300 sentences) of the training data volume, while the performance of other baselines increases steadily up to the end. Figure 2(b) shows the results in Spanish. We observe that deep learning-based baselines perform very poorly when the number of training samples is lower than 400 train sentences, especially the LSTM baseline. Similarly, we can see from Fig. 2(c) that in the Russian, LICD performs better than deep learning-based baselines when the training data volume is less than 50% (1745 sentences) of the original training data volume. These results indicate that LICD is not very sensitive to the volume of training data and can obtain reasonable performance compared to the results corresponding to the full-sized train data.

Fig. 3 shows the effect of the number of seed words in 3 languages. For some languages like English, with a few seeds, here with just one seed, we achieve the best result, but in others, we need more seeds to achieve the best performance of our system. This behavior may occur because of the characteristics of languages. In some languages, few words can represent a category, while in other languages we need more words to represent a category. Furthermore, we observe that increasing the number of seed words after a certain point yields no improvement in the performance of the system as the seed words start to contain general terms. These general terms are not discriminative, and therefore, do not influence the results.

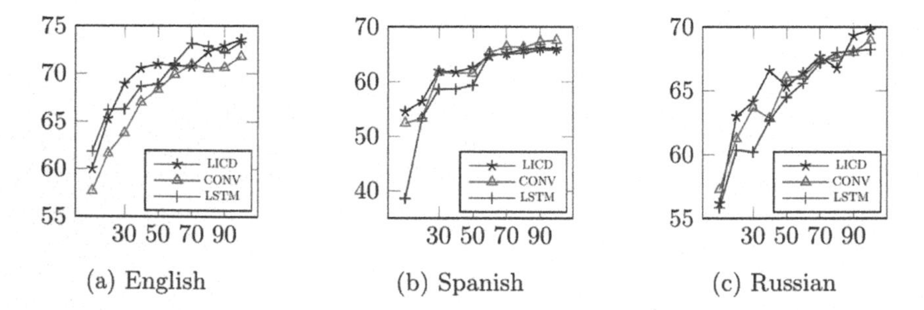

Fig. 2. Effect of different sizes of training data. The vertical axis represents F1-Score, and the horizontal axis represents data volume (in % of total).

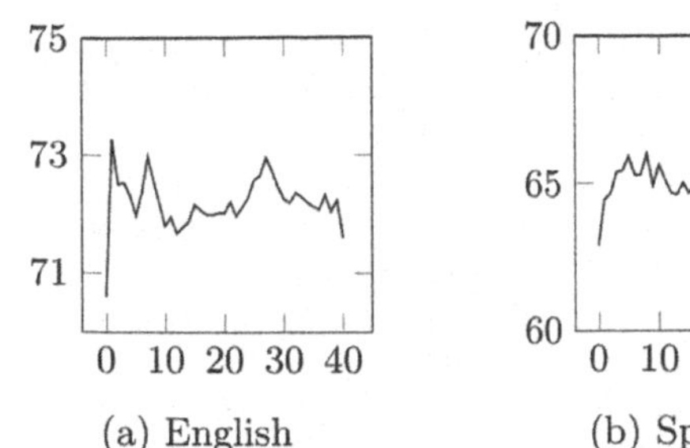

(a) English

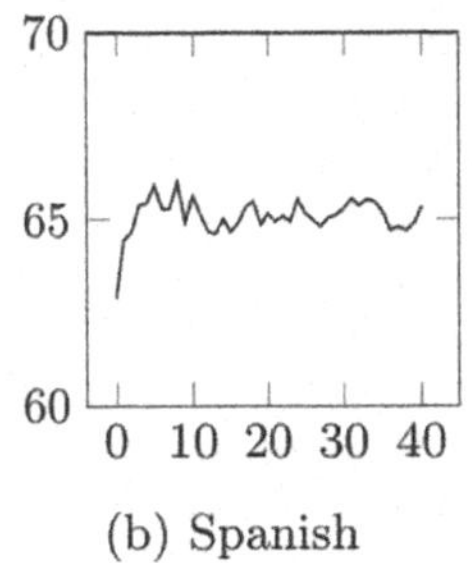

(b) Spanish

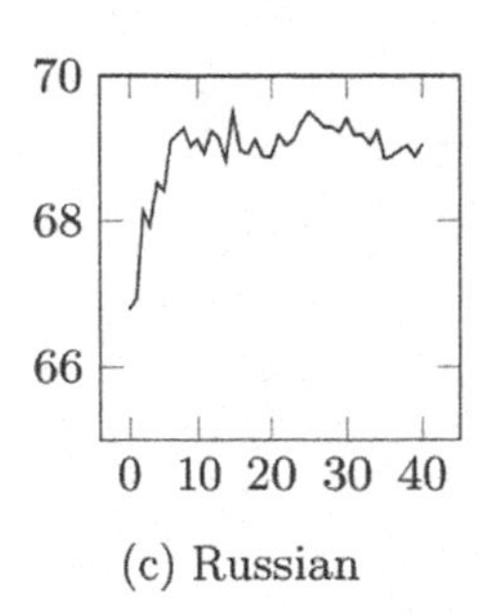

(c) Russian

Fig. 3. Effect of the number of seed words. The vertical axis represents F1-Score, and the horizontal axis represents number of seed words.

6 Conclusions

We proposed LICD, a text matching based method for aspect category detection, which does not require feature engineering or any language-specific tools. We proposed to use soft cosine and word mover's distance to assess the similarity between a given sentence and the set of seed words and find structurally and semantically similar sentences to the given sentence respectively. Experimental results on multilingual datasets demonstrate that our method outperforms baselines in several languages, and achieves competitive results in others. For future works, we plan to lower the computational complexity of our model and investigate the suitability of other document distance measures for this task.

References

1. Alvarez-López, T., Juncal-Martinez, J., Fernández-Gavilanes, M., Costa-Montenegro, E., González-Castano, F.J.: GTI at SemEval-2016 task 5: SVM and CRF for aspect detection and unsupervised aspect-based sentiment analysis. In: Proceedings of the 10th International Workshop On Semantic Evaluation (SemEval-2016), pp. 306–311 (2016)
2. Apidianaki, M., Tannier, X., Richart, C.: Datasets for aspect-based sentiment analysis in French. In: LREC (2016)
3. Bird, S., Klein, E., Loper, E.: Natural Language Processing with Python: Analyzing Text with the Natural Language Toolkit. O'Reilly Media Inc., Sebastopol (2009)
4. Brun, C., Perez, J., Roux, C.: XRCE at SemEval-2016 task 5: feedbacked ensemble modeling on syntactico-semantic knowledge for aspect based sentiment analysis. In: Proceedings of the 10th International Workshop on Semantic Evaluation (SemEval-2016), pp. 277–281 (2016)
5. Çetin, F.S., Yıldırım, E., Özbey, C., Eryiğit, G.: TGB at SemEval-2016 task 5: multi-lingual constraint system for aspect based sentiment analysis. In: Proceedings of the 10th International Workshop on Semantic Evaluation (SemEval-2016), pp. 337–341 (2016)
6. Charlet, D., Damnati, G.: SimBow at SemEval-2017 task 3: soft-cosine semantic similarity between questions for community question answering. In: Proceedings of the 11th International Workshop on Semantic Evaluation (SemEval-2017), pp. 315–319 (2017)

7. De Clercq, O., Hoste, V.: Rude waiter but mouthwatering pastries! an exploratory study into Dutch aspect-based sentiment analysis. In: Tenth International Conference on Language Resources and Evaluation (LREC2016), pp. 2910–2917. ELRA (2016)
8. Forman, G.: An extensive empirical study of feature selection metrics for text classification. J. Mach. Learn. Res. **3**(Mar), 1289–1305 (2003)
9. Ganu, G., Elhadad, N., Marian, A.: Beyond the stars: improving rating predictions using review text content. In: WebDB, vol. 9, pp. 1–6. Citeseer (2009)
10. Hu, M., Liu, B.: Mining and summarizing customer reviews. In: Proceedings of the Tenth ACM SIGKDD International Conference on Knowledge Discovery and Data Mining, pp. 168–177. ACM (2004)
11. Khalil, T., El-Beltagy, S.R.: NileTMRG at SemEval-2016 task 5: deep convolutional neural networks for aspect category and sentiment extraction. In: Proceedings of the 10th International Workshop on Semantic Evaluation (SemEval-2016), pp. 271–276 (2016)
12. Kingma, D.P., Ba, J.: Adam: A method for stochastic optimization. arXiv preprint arXiv:1412.6980 (2014)
13. Kiritchenko, S., Zhu, X., Cherry, C., Mohammad, S.: NRC-Canada-2014: detecting aspects and sentiment in customer reviews. In: Proceedings of the 8th International Workshop on Semantic Evaluation (SemEval 2014), pp. 437–442 (2014)
14. Kusner, M., Sun, Y., Kolkin, N., Weinberger, K.: From word embeddings to document distances. In: International Conference on Machine Learning, pp. 957–966 (2015)
15. Loukachevitch, N., Blinov, P., Kotelnikov, E., Rubtsova, Y., Ivanov, V., Tutubalina, E.: SentiRuEval: testing object-oriented sentiment analysis systems in Russian. In: Proceedings of International Conference Dialog, vol. 2, pp. 3–13 (2015)
16. Ma, W., Suel, T.: Structural sentence similarity estimation for short texts. In: FLAIRS Conference, pp. 232–237 (2016)
17. Mikolov, T., Chen, K., Corrado, G., Dean, J.: Efficient estimation of word representations in vector space. arXiv preprint arXiv:1301.3781 (2013)
18. Novotnỳ, V.: Implementation notes for the soft cosine measure. In: Proceedings of the 27th ACM International Conference on Information and Knowledge Management, pp. 1639–1642. ACM (2018)
19. Pang, B., Lee, L., et al.: Opinion mining and sentiment analysis. Found. Trends® Inf. Ret. **2**(1–2), 1–135 (2008)
20. Pedregosa, F., et al.: Scikit-learn: machine learning in Python. J. Mach. Learn. Res. **12**, 2825–2830 (2011)
21. Pontiki, M., et al.: SemEval-2016 task 5: aspect based sentiment analysis. In: Proceedings of the 10th International Workshop on Semantic Evaluation (SemEval-2016), pp. 19–30 (2016)
22. Řehůřek, R., Sojka, P.: Software framework for topic modelling with large corpora. In: Proceedings of the LREC 2010 Workshop on New Challenges for NLP Frameworks, pp. 45–50. ELRA, Valletta, May 2010. http://is.muni.cz/publication/884893/en
23. Rubner, Y., Tomasi, C., Guibas, L.J.: A metric for distributions with applications to image databases. In: Sixth International Conference on Computer Vision 1998, pp. 59–66. IEEE (1998)
24. Ruder, S., Ghaffari, P., Breslin, J.G.: A hierarchical model of reviews for aspect-based sentiment analysis. arXiv preprint arXiv:1609.02745 (2016)

25. Ruder, S., Ghaffari, P., Breslin, J.G.: Insight-1 at SemEval-2016 task 5: deep learning for multilingual aspect-based sentiment analysis. arXiv preprint arXiv:1609.02748 (2016)
26. Schouten, K., Frasincar, F.: Survey on aspect-level sentiment analysis. IEEE Trans. Knowl. Data Eng. **1**, 1 (2016)
27. Sidorov, G., Gelbukh, A., Gómez-Adorno, H., Pinto, D.: Soft similarity and soft cosine measure: similarity of features in vector space model. Computación y Sistemas **18**(3), 491–504 (2014)
28. Su, Q., Xiang, K., Wang, H., Sun, B., Yu, S.: Using pointwise mutual information to identify implicit features in customer reviews. In: Matsumoto, Y., Sproat, R.W., Wong, K.-F., Zhang, M. (eds.) ICCPOL 2006. LNCS (LNAI), vol. 4285, pp. 22–30. Springer, Heidelberg (2006). https://doi.org/10.1007/11940098_3
29. Tamchyna, A., Veselovská, K.: UFAL at SemEval-2016 task 5: recurrent neural networks for sentence classification. In: Proceedings of the 10th International Workshop on Semantic Evaluation (SemEval-2016), pp. 367–371 (2016)
30. Tan, S., Zhang, J.: An empirical study of sentiment analysis for chinese documents. Expert Syst. Appl. **34**(4), 2622–2629 (2008)
31. Toh, Z., Su, J.: NLANGP at SemEval-2016 task 5: improving aspect based sentiment analysis using neural network features. In: Proceedings of the 10th International Workshop on Semantic Evaluation (SemEval-2016), pp. 282–288 (2016)
32. Xenos, D., Theodorakakos, P., Pavlopoulos, J., Malakasiotis, P., Androutsopoulos, I.: AUEB-ABSA at SemEval-2016 task 5: ensembles of classifiers and embeddings for aspect based sentiment analysis. In: Proceedings of the 10th International Workshop on Semantic Evaluation (SemEval-2016), pp. 312–317 (2016)
33. Xue, W., Zhou, W., Li, T., Wang, Q.: MTNA: a neural multi-task model for aspect category classification and aspect term extraction on restaurant reviews. In: Proceedings of the Eighth International Joint Conference on Natural Language Processing (Volume 2: Short Papers), vol. 2, pp. 151–156 (2017)
34. Zhou, X., Wan, X., Xiao, J.: Representation learning for aspect category detection in online reviews. In: AAAI, pp. 417–424 (2015)

Topic Grouper: An Agglomerative Clustering Approach to Topic Modeling

Daniel Pfeifer[1(✉)] and Jochen L. Leidner[2,3(✉)]

[1] Department of Medical Informatics, Heilbronn University of Applied Sciences, Max-Planck-Str. 39, 74081 Heilbronn, Germany
daniel.pfeifer@hs-heilbronn.de

[2] Refinitiv Labs, 30 South Colonnade, London E14 5EP, UK
leidner@acm.org

[3] University of Sheffield, 211 Portobello, Sheffield S1 4DP, UK

Abstract. We introduce Topic Grouper as a complementary approach in the field of probabilistic topic modeling. Topic Grouper creates a disjunctive partitioning of the training vocabulary in a stepwise manner such that resulting partitions represent topics. Topic generation is based on a simple probabilistic model and agglomerative clustering, where clusters are formed as sets of words from the vocabulary. The resulting binary tree of topics may act as a containment hierarchy typically with more general topics towards the root of tree and more specific topics towards the leaves. As opposed to other topic modeling approaches, Topic Grouper avoids the need for hyper parameter optimizations.

As part of an evaluation, we show that Topic Grouper has reasonable predictive power but also a reasonable complexity. It can deal well with stop words and function words. Also, it can handle topic distributions, where some topics are more frequent than others. We present examples of computed topics which appear as conclusive and coherent.

Keywords: Topic modeling · Topic analysis · Clustering · Probabilistic topic models · Text collection browsing · Exploratory data analysis

1 Introduction

During the last two decades, probabilistic *topic modeling* has justly become an active sub-field of information retrieval and machine learning. Hereby, each topic is typically represented via a multinomial distribution over the collection's vocabulary. Related ideas and solutions were formed in the two seminal publications on *probabilistic Latent Semantic Indexing* (pLSI) ([12]) and *Latent Dirichlet Allocation* (LDA) ([4]).

Regarding pure document clustering, the two major machine learning directions are *Expectation Maximization* (EM) including *k*-Means on the one hand and hierarchical clustering including agglomerative clustering on the other. However, in the case of topic modeling *the opportunities of hierarchical clustering*

L. Azzopardi et al. (Eds.): ECIR 2019, LNCS 11437, pp. 590–603, 2019.
https://doi.org/10.1007/978-3-030-15712-8_38

have been overlooked. In this paper, we aim to close this gap by presenting *Topic Grouper* as *a topic modeling approach relying on agglomerative clustering*, where sets of words from the vocabulary form respective clusters.

2 Basic Concepts and Related Work

2.1 Agglomerative Clustering

Hierarchical agglomerative clustering (HAC) or simply agglomerative clustering is the process of clustering the clusters in turn iteratively, based on a similarity measure between clusters from a previous iteration. It was first described in the 1960s by Ward, Jr. [30], by Lance and Williams [16,17], and a few others. *Cluster distance* is usually the term for the inverse of a similarity measure underlying a clustering procedure. Standard cluster distances derived from the so-called Lance-Williams formula include single linkage, complete linkage and group average linkage, but many others have been proposed ([21,31]).

Cluster distances, such as the one developed here, may not necessarily meet standard mathematical distance axioms, as agglomerative clustering can do without ([30]). Moreover, our cluster distance is *model-based*, as it is governed by a simple generative model. Model-based agglomerative clustering has rarely been investigated: [14] give a model-based interpretation of some standard cluster distances and partly extend them under the same framework. [26] develop a recursive probabilistic model for a clustering tree in order to explain the data items merged at each tree node.

A common critique of agglomerative clustering is its relatively high time complexity typically amounting to $O(k^2)$ or more given the number of data items k ([31]). Also, space complexity is often in $O(k^2)$ depending on the chosen cluster distance. In the case of our contribution and additionally in the case of text, *k corresponds to the vocabulary size, which can be limited* even for large text collections, e.g. by simple filtering criteria such as high document frequency. This offers the potential for a reasonable computational overhead in the context of topic modeling.

A major asset of agglomerative clustering is the *tree structure of its clusters* often assumed to reflect containment hierarchies. Also, it is widely held that agglomerative clustering offers better and more computationally stable clusters than competing procedures such as k-Means ([13], p. 140).

2.2 Non-Hierarchical Topic Models

Here, we briefly recap non-hierarchical or *flat* probabilistic topic models in order to describe our contribution using the same terminology below. Let

- D be the set of training documents with size $|D|$,
- V be the vocabulary of D with size $|V|$,
- $f_d(w)$ be the frequency of a word $w \in V$ with regard to $d \in D$.

Given a set of topic references T with $|T| = n$, the goal of non-hierarchical topic modeling is to estimate n topic-word distributions $p(w|t)_{w \in V}$ (one for each $t \in T$) and $|D|$ document-topic distributions $p(t|d)_{t \in T}$ (one for each $d \in D$). Together, these distributions are meant to maximize $p(D) = \prod_{d \in D} p(d)$, where $p(d)$ is the probability of all word occurrences in d regardless of their order. Yet, how this is done in detail, depends on the topic modeling approach: Under pLSI ([12]) we have[1]

$$p(d) = c_d \cdot \prod_{w \in V} p(w|d)^{f_d(w)} \text{ and } p(w|d) = \sum_{t \in T} p(w|t) \cdot p(t|d).$$

The n topic-word distributions form a corresponding topic model $\phi = \{\phi_t\}$. Each $\phi_t = p(w|t)_{w \in V}$ represents the essence of a topic, where t itself is just for reference.

As a more sophisticated Bayesian approach, LDA puts all potential topic-word distributions under a Dirichlet prior β in order to determine $p(D)$ ([4]). In this case, an approximation of

$$\Phi = argmax_{\phi} \left((\prod_{t \in T} p(\phi_t)) \cdot \prod_{d \in D} p(d|\phi, \alpha) \right) \text{ with } \phi_t \sim Dirichlet(\beta)$$

may be considered a topic model ([4]). Hereby, α is an additional Dirichlet prior to determine

$$p(d|\phi, \alpha) = \int p(\theta_d) \cdot \prod_{w \in V} \left(\sum_{t \in T} \phi_t(w) \cdot \theta_d(t) \right)^{f_d(w)} d\theta_d \text{ with } \theta_d \sim Dirichlet(\alpha).$$

Alternatively to the *argmax* operator, ϕ may be integrated out leading to a corresponding point estimate for Φ ([10]).

There exist several methods and various derived algorithms readily available to approximate Φ under LDA including variational Bayes, MAP estimation and Gibbs sampling. LDA's hyper parameters α and β are usually set based on heuristics, via an EM step nesting the actual topic inference procedure or via a hyper parameter search (e.g. see [2]).

2.3 Hierarchical Topic Models

Traditional topic models create flat topics; however, it may be more appropriate to have a hierarchy comprising multiple levels of super-topics and increasingly specialized sub-topics.

One of the early attempts towards hierarchical topic models is Hofmann's Cluster Abstraction Model (CAM) [11], using an instance of EM with annealing.

[1] The factor $c_d = (\sum_{w \in V} f_d(w))! / \prod_{w \in V, f_d(w)>0} f_d(w)!$ accounts for the underlying "bag of words model" where word order is ignored. It is usually omitted in publications because if two approaches are compared, the expression turns out to be an identical factor for both of them ([6]). We therefore also set $c_d := 1$.

Blei *et al.* [3] discuss an extension of the "Chinese restaurant process" (CRP) from [1]: Using their so-called "*nested Chinese restaurant process*" (nCRP) the authors propose *Hierarchical LDA* (hLDA) in order to estimate a topic tree of a fixed depth L. Documents are thought to be generated by first choosing a path of length L along the tree and then mixing the document's topics via the chosen path where each path node represents a topic to be inferred. The corresponding document-topic distribution is subject to a Dirichlet distribution with the prior α. Under hLDA, higher level topics tend to be common across many documents, but do not necessarily form semantic generalizations of lower level topics. I.e., the model tends to push function words towards the root of tree and rather domain-specific words towards the leaves. Besides L and α, hLDA requires the further priors γ, η.

The *Pachinko Allocation Model* (PAM) by Wei and McCallum [19] is a hierarchical topic model based on multiple Dirichlet processes. The PAM requires a directed acyclic graph (DAG) as a prior, where leaf nodes correspond to words from the vocabulary, parents of leaf nodes correspond to *flat*, word-based topics and other nodes represent mixture components over their children's mixture components. The PAM has similar hyper parameters as LDA including α, β and the number of word-based topics n.

The *recursive Chinese Restaurant Process* (rCRP) from Kim *et al.* [15] is another extension of the CRP to infer hierarchical topic structures. In contrast to hLDA, the sampling of a document-topic distribution is generalized in a way that permits a document's topics to be drawn from the entire (hierarchical) topic tree, not just from a single path.

The *Nested Hierarchical Dirichlet Processes* (nHDP) from [23] are perhaps the most sophisticated approach to produce tree-structured topics: Based on Blei *et al.* [3] it uses the nCRP to produce a global topic tree. Every document obtains its specific topic tree which is derived from the global tree via a Hierarchical Dirichlet Process according to Teh *et al.* [25]. The approach mandates a hyper parameter α for its basic nCRP and β for document-level trees. The authors provide efficient inference procedures and offer impressive results on small as well as very large text data sets, where the vocabulary on the large data sets is reduced to about 8.000 words.

A commonality of all these approaches is the need for hyper parameters – usually several scalars. A developer applying a related approach may therefore struggle with its complexity and with setting hyper parameters. Although some of the above-mentioned solutions scale up to large data sets the resulting topic trees remain rather shallow. In contrast, Topic Grouper offers deep trees and requires no hyper parameters.

2.4 Evaluation Regime

Since typically, there exists no ground truth regarding topic models, a well-established *intrinsic* evaluation scheme is to compute the log probability for test documents $d \in D_{test}$ withheld from the training data. In this context, estimating (the log of) $p(d|\Phi, \alpha)$ via an LDA topic model Φ with its Dirichlet prior α is a non-trivial problem in itself. [28] introduce their so-called "left-to-right" method for this purpose and [5] presents a refined and unbiased version of the former method named "left-to-right sequential". We report results based on the latter algorithm since it acts as a gold standard (see [5]). Both methods can also be applied to topic models Φ produced by Topic Grouper. Like [4] and others we report on a model's *perplexity* as a derived measure to aggregate the predictive power of Φ over D_{test}:

$$perplexity(D_{test}) := \exp\left(-\sum_{d \in D_{test}} \log p(d|\Phi, \alpha) / \sum_{d \in D_{test}} |d|\right).$$

To generate LDA models we use Gibbs sampling according to [10]. Moreover, we adopt a commonly used heuristic from [10] for LDA's hyper parameters implying $\beta = 0.1$ and $\alpha = 50/n$.

As an alternative type of evaluation we also make use of *synthetically* generated data. *In this case the true topics S of words are known*, which allows us to consider *error rate* as an additional quality measure and to examine some basic qualities of our approach: The idea is to compare a topic model Φ against the *true* topic-word distributions used to generate a dataset.

The following definition of error rate *err* assumes that the perfect number of topics is already known, such that $|T| := |S|$ is preset for training. The order of topics in topic models is unspecified, so we try every *bijective* mapping $\pi : T \to S$ when comparing a computed model Φ with a true model $\tilde{p}(w|s)_{w \in V, s \in S}$ and favor the mapping that minimizes the error:

$$err := \min_{\pi} \frac{1}{2|T|} \sum_{t \in T} \sum_{w \in V} |\Phi_t(w) - \tilde{p}(w|\pi(t))|.$$

The measure is designed to range between 0 and 1, where 0 is perfect.

3 Topic Grouper

3.1 Generative Cluster Distance

Let $T(n) = \{t \mid t \subseteq V\}$ be a (topical) partitioning of V such that $s \cap t = \emptyset$ for any $s, t \in T(n)$, $\bigcup_{t \in T(n)} t = V$ and $|T(n)| = n$. Further, let the *topic-word assignment* $t(w)$ be the topic of a word w such that $w \in t \in T(n)$.

In the following we also make use of the variables D, V and f_d as specified in Sect. 2.2. Our principal goal is to find an *optimal* partitioning $T(n)$ for each n along

$$argmax_{T(n)} q(T(n)), \text{with}$$

$$q(T(n)) := \prod_{d \in D} \prod_{w \in V, f_d(w) > 0} (p(w|t(w)) \cdot p(t(w)|d))^{f_d(w)}.$$

The idea is that each document $d \in D$ is considered to be *generated* via a simple stochastic process where a word w in d occurs by

- first sampling a topic t according to a probability distribution $p(t|d)_{t \in T(n)}$,
- then sampling a word from t according to the topic-word distribution $p(w|t)_{w \in V}$

and so, the total probability of generating D is proportional to $q(T(n))$.

The optimal partitioning consists of n pairwise disjunctive subsets of V, whereby each subset is meant to represent a topic. By definition every word w must be in exactly one of those sets. This may help to keep topics more interpretable for humans because they do not overlap with regard to their words. On the other hand, polysemic words can only support one topic, even though it would be justified to keep them in several topics due to multiple contextual meanings. Note that the approach considers a solution for every possible number of topics n ranging between $|V|$ and one.

To further detail our approach, we set

- $f(w) := \sum_{d \in D} f_d(w) > 0$, since otherwise w would not be in the vocabulary,
- $|d| := \sum_{w \in V} f_d(w) > 0$, since otherwise the document would be empty,
- $f_d(t) := \sum_{w \in t} f_d(w)$ be the topic frequency in a document d and
- $f(t) := \sum_{w \in t} f(w) = \sum_{d \in D} f_d(t)$ be the number of times t is referenced in D via some word $w \in t$.

Concerning $q(T(n))$ we use maximum likelihood estimations for $p(t(w)|d)$ and $p(w|t(w))$ based on D:

- $p(t(w)|d) \approx f_d(t(w))/|d|$, which is > 0 if $f_d(w) > 0$,
- $p(w|t(w)) \approx f(w)/f(t(w))$, which is always > 0 since $f(w) > 0$.

Unfortunately, constructing the optimal partitionings $\{T(n) \mid n = 1 \ldots |V|\}$ is computationally hard. *We suggest a greedy algorithm that constructs suboptimal partitionings instead*, starting with $T(|V|) := \{\{w\} \mid w \in V\}$ as step $i = 0$. At every step $i = 1 \ldots |V| - 1$ the greedy algorithm joins two different topics $s, t \in T(|V| - (i-1))$ such that $q(T(|V| - i))$ is maximized while $T(|V| - i) = (T(|V| - (i-1)) - \{s, t\}) \cup \{s \cup t\}$ must hold. Essentially, this results in an *agglomerative clustering approach, where topics, not documents, form respective clusters.*

For efficient computation we first rearrange the terms of $q(T(n))$ with a focus on topics in the outer factorization:

$$q(T(n)) = \prod_{t \in T(n)} \prod_{d \in D, f_d(t) > 0} \left(p(t|d)^{f_d(t)} \cdot \prod_{w \in t} p(w|t)^{f_d(w)} \right)$$

The rearrangement relies on the fact that every word belongs to exactly one topic and enables the "change of perspective" towards topic-oriented clustering.

We maximize $\log q(T(n))$ instead of $q(T(n))$ which is equivalent with respect to the argmax-operator. This leads to

$$\log q(T(n)) = \sum_{t \in T(n)} \sum_{d \in D, f_d(t) > 0} (f_d(t) \cdot \log p(t|d) + \sum_{w \in t} f_d(w) \cdot \log p(w|t)) \approx \sum_{t \in T(n)} h(t)$$

with the maximum likelihood estimation

$$h(t) := \sum_{d \in D, f_d(t) > 0} f_d(t) \cdot (\log f_d(t) - \log |d|) + \sum_{w \in t} f(w) \cdot \log f(w) - f(t) \cdot \log f(t). \tag{1}$$

Using these formulas the best possible join of two (disjunctive) topics $s, t \in T(n)$ results in $T(n-1)$ with

$$\log q(T(n-1)) \approx \log q(T(n)) + \Delta h_n,$$
$$\Delta h_n := max_{s,t \in T(n)} \Delta h(s,t) \text{ and } \Delta h(s,t) := h(s \cup t) - h(s) - h(t). \tag{2}$$

From the perspective of clustering procedures $-\Delta h(s,t)$ is the cluster distance between s and t. Note though, that it does not adhere to standard distance axioms.

3.2 Algorithm Sketch and Complexity

Topic Grouper can be implemented via adaptations of standard agglomerative clustering algorithms: E.g. one may apply the *efficient hierarchical agglomerative clustering* (EHAC) from [20]. EHAC manages a map of priority queues in order to represent evolving clusters during the agglomeration process.

Considering the resulting algorithm, we can reuse $h(s)$ and $h(t)$ from prior computation steps in order to compute $h(s \cup t)$ efficiently: Regarding expression (1) from above, let $i(t) := \sum_{w \in t} f(w) \cdot \log f(w)$. We have $f_d(s \cup t) = f_d(s) + f_d(t)$, $f(s \cup t) = f(s) + f(t)$ and $i(s \cup t) = i(s) + i(t)$, and so

$$h(s \cup t) = \sum_{d \in D, f_d(s) + f_d(t) > 0} (f_d(s) + f_d(t)) \cdot (\log(f_d(s) + f_d(t)) - \log |d|) + \\ i(s) + i(t) - (f(s) + f(t)) \cdot \log(f(s) + f(t)). \tag{3}$$

The terms $i(u)$ and $f(u)$ with $u = s, t$ will have been computed already during the prior steps of the resulting algorithm, i.e. when t and s were generated as topics. Thus, the computation of all sums over words w can be avoided with respect to $h(s \cup t)$. This is essential for a reasonable runtime complexity.

EHAC's original time complexity is in $O(k^2 \log k)$ and its space complexity in $O(k^2)$ with k being the initial number of clusters. However, this implies that the cost of computing the distance between two clusters is in $O(1)$. In the case of Topic Grouper the latter cost is in $O(|D|)$ instead, because one must compute the value of h from Eq. 3. The factor "$\log k$" from EHAC's original time complexity accounts for access to priority queue elements – in the case of Topic Grouper this is dominated by the cost to compute h-values. Putting it together, the time complexity for Topic Grouper is on the order of $|V|^2 \cdot |D|$ and its space complexity is in $O(|V|^2)$.[2]

[2] The stated space complexity, $O(|V|^2)$, can be problematic if the vocabulary is large. We also devised an alternative clustering algorithm, MEHAC, whose space complexity is in $O(|V|)$ but its *expected* time complexity is still in $O(|V|^2 \cdot |D|)$ (not detailed here).

4 Experiments

4.1 Synthetic Data

This section provides a first evaluation by applying error rate according to Sect. 2.4 to a simple *synthetically* generated dataset. We use a data generator as introduced in [24]: It is based on $|V| = 400$ (artificial) words equally divided into 4 *disjoint* topics $S = \{s_1, \ldots, s_4\}$. The words are represented by numbers, such that $0 \ldots 99$ belongs to s_1, $100 \ldots 199$ to s_2, and so on.

Concerning the 100 words of a topic s_i, the topic-word distribution $\tilde{p}(w|s_i)_{w \in V}$ is drawn independently for each topic from a Dirichlet distribution with a symmetric prior $\tilde{\beta} = 1/100$, such that $\sum_{w=(i-1)\cdot 100}^{i\cdot 100-1} \tilde{p}(w|s_i) = 1$. A resulting dataset holds 6,000 documents with each document consisting of 30 word occurrences. A document-topic distribution $\tilde{p}(w|d)_{w \in S}$ is drawn independently for each document via a Dirichlet with the prior $\tilde{\alpha} = (5, 0.5, 0.5, 0.5)^\top$, where topic 1 with $\tilde{\alpha_1} = 5$ is meant to represent a typical "stop word topic", which is more likely than other topics.

Figure 1 shows the error rate of LDA as well as of Topic Grouper for the synthetic dataset and for $n = 4$ learned topics. Regarding LDA, the depicted values are averaged across 50 runs per data point, whereby the random seed for the related Gibbs sampler was changed for every run. LDA's hyper parameter α changes along the X axis such that $\alpha = \sum_i \tilde{\alpha}_i = 6.5$ and $\alpha_2 = \alpha_3 = \alpha_4$ always hold. The results stress the importance of hyper parameter choice for model quality under LDA with regard to α in concordance with [27]. Note that a symmetric $\alpha = 1.625$, in other words $\alpha_i = 1.625$ for $i = 1 \ldots 4$, fails to deliver low error rates. Topic Grouper delivers good error rates right away. As its results are independent of α (and β) but also deterministic, they are included as a horizontal line. We also added results for pLSI, but it attains only mediocre and volatile results, heavily depending on its random initialization values.

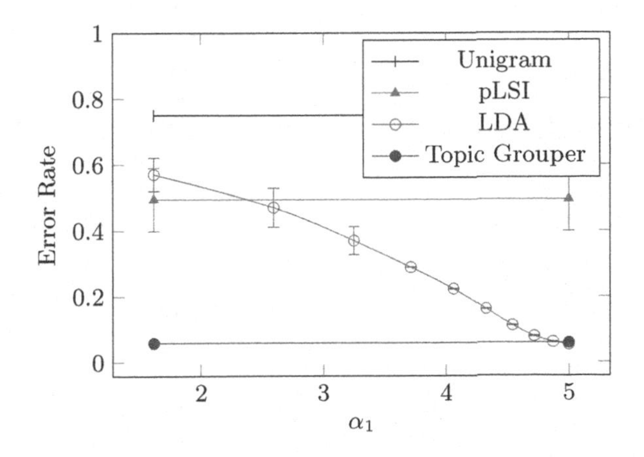

Fig. 1. Error rate depending on α for a dataset generated according to Tan *et al.* [24]

4.2 Real-World Data

Retailing. Regarding retailing, a shopping basket or an order is equivalent to a document. Articles correspond to words from a vocabulary and item quantities transfer to word occurrence frequencies in documents. In this context, topics represent groups of articles as typically bought or ordered together. Therefore, inferred topic models may be leveraged to optimize sales-driven catalog structures, to develop layouts of product assortments ([9]) or to build recommmender systems ([29]).

The "Online Retail" dataset is a "transnational dataset which contains all the transactions occurring between 01/12/2010 and 09/12/2011 for a UK-based ... online retail" obtained from the UCI Machine Learning Repository ([8]).[3] We performed data cleaning by removing erroneous and inconsistent orders. Item quantities are highly skewed with about 5% above 25, some reaching values of over 1,000. This is due to a mixed customer base including consumers and wholesalers. We therefore excluded all order items with quantities above 25 to focus on small scale (parts of) orders. We randomly split such preprocessed orders into 90% training and 10% test data, keeping only articles that were ordered at least 10 times in the training data. The resulting training dataset covers $|V| = 3,464$ articles, $|D| = 17,086$ orders and 427,150 order items. The resulting average sum of item quantities per order is about 154.

Figure 2 shows that the performance of LDA begins to degrade at 80 topics. Topic Grouper is competitive although its underlying topic model is more restrained (as each article or word, respectively, belongs to exactly *one* topic).

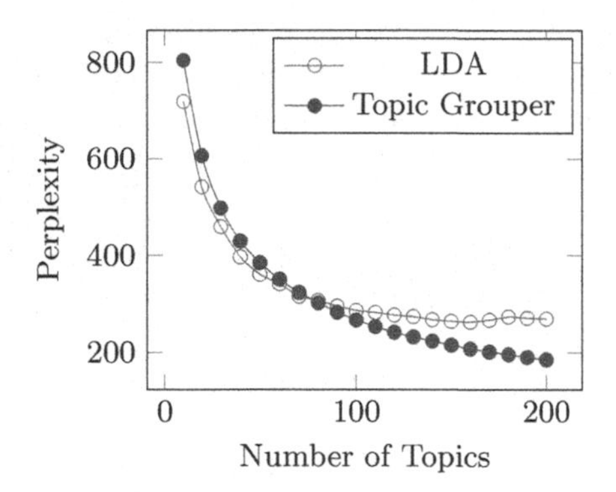

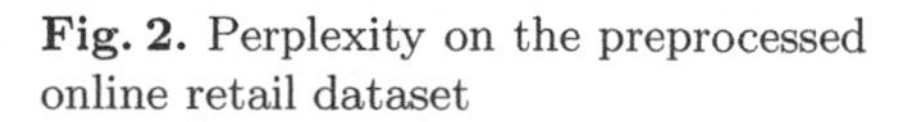
Fig. 2. Perplexity on the preprocessed online retail dataset

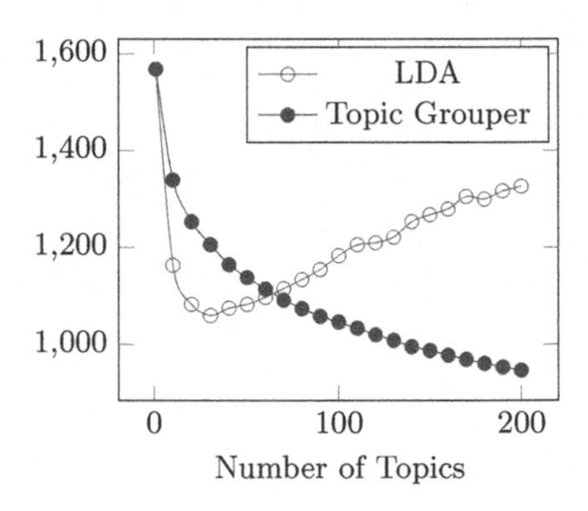

Fig. 3. Perplexity on the preprocessed NIPS dataset

[3] See https://archive.ics.uci.edu/ml/datasets/Online+Retail (cited 2018-09-10).

Text. The NIPS dataset is a collection of 1,500 research publications from the Neural Information Processing Systems Conference. We used a preprocessed version as is of the dataset from the UCI Machine Learning Repository.[4] We split the document set on a 90% to 10% basis and only kept words occurring at least five times in the training data. This way we ended up with $|V| = 8,801$ words left for training. Figure 3 shows that Topic Grouper is competitive. In this case the common heuristic hyper parameter as referred to in Sect. 2.4 affects LDA's performance.

4.3 Example Model

Topic Grouper returns hierarchical topic models by design. The hierarchy of topics may be explored interactively assuming that larger topics form a kind of semantic abstraction of contained smaller topics. Figure 4 presents a corresponding tree for the TREC AP Corpus containing 20,000 newswire articles:[5] We performed Porter stemming and kept every stem that occurs at least five times in the dataset. Moreover, we removed all tokens containing non-alphabetical characters or being shorter than three characters. This led to $|V| = 25,047$ words left for training.

All nodes below level six are collapsed in order to deal with limited presentation space. Each node contains the five most frequent words of a respective topic. More frequent topics are shaded in blue (as they tend to collect low content words and stop words), whereas less frequent word sets are shaded in red. Topics are identified by the number n under which they were generated.

The tree may be interpreted is as follows: The root forks into node (4) covering economy and weather as well as node (2) covering other topics and function words. Function words are mainly gathered along the path (1)/(2)/(3)/(6)/(11) and the sub-path (9)/(12)/(23). Node (4) forks into financial topics (14) and topics covering production and weather (17). Node (53) is on weather and potentially different weather regions. Node (46) covers agriculture and water supply whereas node (81) focuses on energy. Regarding node (14), we suspect that stock trading in (30) is separated from general banking and acquisitions in (31). Other topics in the tree seem equally coherent such as "home and family" (59), "public media" (25), "jurisdiction and law" (42), "military and defense" (50), and so forth. We find that such interpreted topics often meet the idea of being more general towards the root and more specific towards the leaves. However mixed topics also arise such as topic (21) combining "drug trafficking" in (73) with "military and defense" in (50).

Alternatively, Topic Grouper enables *flat topic views* via $T(n)$ based on the same model as a respective tree: Fig. 5 lists every second topic (for space reasons) from $T(40)$ as sorted by frequency for the AP Corpus dataset. Topics 47 and 69 gather function words and therefore have high frequency. Most other topics seem conclusive and coherent.

[4] See https://archive.ics.uci.edu/ml/datasets/Bag+of+Words (cited 2018-09-10).
[5] See https://catalog.ldc.upenn.edu/LDC93T3A (cited 2018-09-10).

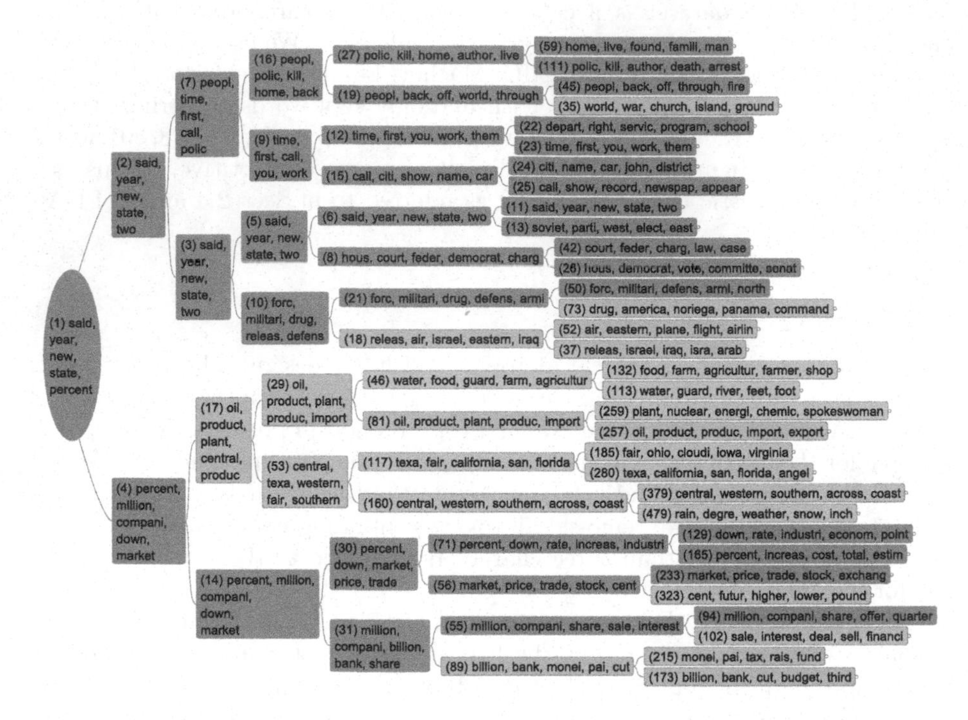

Fig. 4. Tree diagram as a result of topic grouper on the AP corpus extract

n	$f(t)$	Top Seven Words per Topic t						
47	538739	year	new	two	dai	week	three	month
40	305812	state	govern	nation	unit	american	includ	countri
69	281349	said	report	offici	sai	befor	against	told
42	176138	court	feder	charg	law	case	rule	order
71	119423	percent	down	rate	increas	industri	econom	point
51	115641	presid	bush	plan	meet	talk	administr	propos
59	112332	home	live	found	famili	man	children	life
67	96161	commun	visit	miss	travel	becam	histori	art
49	89151	call	show	newspap	appear	televis	radio	publish
74	82919	john	william	robert	richard	paul	wait	king
46	77385	water	food	guard	farm	agricultur	river	farmer
85	73131	democrat	vote	run	campaign	republican	won	dukaki
58	65857	world	war	church	mass	cathol	jewish	conflict
111	62540	polic	kill	author	death	arrest	counti	shot
41	62094	union	south	white	black	worker	job	strike
70	51630	west	east	german	germani	british	europ	northern
110	46693	parti	elect	communist	opposit	reform	conserv	seat
72	45998	island	ground	beach	princ	scale	relief	coup
81	43377	oil	product	plant	produc	import	nuclear	energi
95	34542	israel	iraq	isra	arab	palestinian	iraqi	gulf

Fig. 5. Every second topic of $T(40)$ sorted by frequency for the AP corpus dataset

5 Summary, Discussion and Future Work

We have presented Topic Grouper as a novel and complementary method in the field of probabilistic topic modeling based on agglomerative clustering: Initial clusters or topics, respectively, each consist of one word from the vocabulary of the training corpus. Clusters are joined on the basis of a simple probabilistic model assuming that each word belongs to exactly one topic. Thus, topics or clusters, respectively, form a disjunctive partitioning of the vocabulary.

Using a simple synthetic dataset, where each word belongs to just one original topic, we examined some of its basic qualities: Topic Grouper manages to recover original topics at low error rate even if their a-priori probabilities are rather unbalanced. pLSI fails under these conditions. LDA is able to recover the original topics but only if its hyper parameter α is asymmetrical, i.e. vectorial, and adjusted accordingly.

Regarding two real world datasets, Topic Grouper's predictive performance matched or surpassed LDA at larger topic numbers, where LDA was put under a common heuristic for its hyper parameters. It is noteworthy that the practical training performance of Topic Grouper on a regular PC with 8GB RAM ranged between a few minutes (e.g. for the Online Retail dataset) to a few hours (e.g. for the AP Corpus).

The tree-based model also offering flat topic views is an important asset. It allows for deep tree structures to be produced even on small-sized datasets. Another benefit is the method's simplicity and that it requires no configuration or hyper parametrization and no stop word filtering. The fact that each word is in exactly one topic is a considerable limitation and falls short for polysemic words and for words applied in multiple topical contexts. The approach seems appealing for shopping basket analysis where articles stand for themselves: Related models may then aid in forming sales-driven catalog structures or layouts of product assortments since in both cases, a clear-cut to decision on where to place an article is customary.

An important concern for further work is *model smoothing*, i.e. on how to relax the constraint of each word being in exactly one topic: Regarding flat topic views, we experimented with a combination of Topic Grouper and LDA, where LDA acts as post-processing step. Compiling related experimental results is work in progress. In the future, we also would like to substantiate model quality via *extrinsic evaluation* methods which resort to external resources: e.g. [7] describe two human experiments, one study on *word intrusion* and another one on *topic intrusion*, respectively, whereas [18] and [22] trend to more automated topic evaluation approaches.

The results of this paper can be reproduced via a prototypical Java library named "TopicGrouperJ" published on GitHub.[6] The library features implementations of the mentioned algorithms MEHAC and EHAC. The code to regenerate any result file of the above-described experiments is also available for full reproducibility.

[6] See https://github.com/pfeiferd/TopicGrouperJ.

References

1. Aldous, D.J.: Exchangeability and related topics. In: Hennequin, P.L. (ed.) École d'Été de Probabilités de Saint-Flour XIII — 1983. LNM, vol. 1117, pp. 1–198. Springer, Heidelberg (1985). https://doi.org/10.1007/BFb0099421
2. Asuncion, A., Welling, M., Smyth, P., Teh, Y.W.: On smoothing and inference for topic models. In: Proceedings of the Twenty-Fifth Conference on Uncertainty in Artificial Intelligence, UAI 2009, pp. 27–34. AUAI Press, Arlington (2009)
3. Blei, D.M., Jordan, M.I., Griffiths, T.L., Tenenbaum, J.B.: Hierarchical topic models and the nested Chinese restaurant process. In: Proceedings of the 16th International Conference on Neural Information Processing Systems, NIPS 2003, pp. 17–24. MIT Press, Cambridge (2003)
4. Blei, D.M., Ng, A.Y., Jordan, M.I.: Latent Dirichlet allocation. J. Mach. Learn. Res. **3**, 993–1022 (2003)
5. Buntine, W.: Estimating likelihoods for topic models. In: Zhou, Z.-H., Washio, T. (eds.) ACML 2009. LNCS (LNAI), vol. 5828, pp. 51–64. Springer, Heidelberg (2009). https://doi.org/10.1007/978-3-642-05224-8_6
6. Buntine, W., Jakulin, A.: Discrete component analysis. In: Saunders, C., Grobelnik, M., Gunn, S., Shawe-Taylor, J. (eds.) SLSFS 2005. LNCS, vol. 3940, pp. 1–33. Springer, Heidelberg (2006). https://doi.org/10.1007/11752790_1
7. Chang, J., Boyd-Graber, J.L., Gerrish, S., Wang, C., Blei, D.M.: Reading tea leaves: How humans interpret topic models. In: Bengio, Y., Schuurmans, D., Lafferty, J.D., Williams, C.K.I., Culotta, A. (eds.) Advances in Neural Information Processing Systems 22: 23rd Annual Conference on Neural Information Processing Systems, 7–10 December 2009, Vancouver, British Columbia, Canada, pp. 288–296. Curran Associates, Inc. (2009)
8. Chen, D., Sain, S.L., Guo, K.: Data mining for the online retail industry: a case study of RFM model-based customer segmentation using data mining. J. Database Mark. Customer Strategy Manag. **19**(3), 197–208 (2012)
9. Chen, M.C., Lin, C.P.: A data mining approach to product assortment and shelf space allocation. Expert Syst. Appl. **32**(4), 976–986 (2007)
10. Griffiths, T.L., Steyvers, M.: Finding scientific topics. Proc. Nat. Acad. Sci. **101**(Suppl. 1), 5228–5235 (2004)
11. Hofmann, T.: The cluster-abstraction model: unsupervised learning of topic hierarchies from text data. In: Proceedings of the 16th International Joint Conference on Artificial Intelligence, IJCAI 1999, vol. 2, pp. 682–687. Morgan Kaufmann, San Francisco (1999)
12. Hofmann, T.: Probabilistic latent semantic analysis. In: Proceedings of the Fifteenth Conference on Uncertainty in Artificial Intelligence, UAI 1999, pp. 289–296. Morgan Kaufmann, San Francisco (1999)
13. Jain, A.K., Dubes, R.C.: Algorithms for Clustering Data. Prentice-Hall Inc., Upper Saddle River (1988)
14. Kamvar, S.D., Klein, D., Manning, C.D.: Interpreting and extending classical agglomerative clustering algorithms using a model-based approach. In: Proceedings of the Nineteenth International Conference on Machine Learning, ICML, pp. 283–290. Morgan Kaufmann, San Francisco (2002)
15. Kim, J.H., Kim, D., Kim, S., Oh, A.: Modeling topic hierarchies with the recursive Chinese restaurant process. In: Proceedings of the 21st ACM International Conference on Information and Knowledge Management, CIKM 2012, pp. 783–792. ACM, New York (2012)

16. Lance, G., Williams, W.: A generalized sorting strategy for computer classifications. Nature **212**, 218 (1966)
17. Lance, G., Williams, W.: A general theory of classificatory sorting strategies. I. hierarchical systems. Comput. J. **9**, 373–380 (1967)
18. Lau, J.H., Newman, D., Baldwin, T.: Machine reading tea leaves: automatically evaluating topic coherence and topic model quality. In: Proceedings of the 14th Conference of the European Chapter of the Association for Computational Linguistics, EACL 2014, pp. 530–539. Association for Computational Linguistics, Gothenburg (2014)
19. Li, W., McCallum, A.: Pachinko allocation: DAG-structured mixture models of topic correlations. In: Proceedings of the 23rd International Conference on Machine Learning, Pittsburgh, ICML 2006, PA, USA. ACM, New York (2006)
20. Manning, C.D., Raghavan, P., Schütze, H.: Introduction to Information Retrieval. Cambridge University Press, New York (2008)
21. Murtagh, F.: A survey of recent advances in hierarchical clustering algorithms. Comput. J. **26**(4), 354–359 (1983)
22. Newman, D., Lau, J.H., Grieser, K., Baldwin, T.: Automatic evaluation of topic coherence. In: Human Language Technologies: The 2010 Annual Conference of the North American Chapter of the Association for Computational Linguistics, HLT, pp. 100–108. Association for Computational Linguistics, Stroudsburg (2010)
23. Paisley, J., Wang, C., Blei, D.M., Jordan, M.I.: Nested hierarchical Dirichlet processes. IEEE Trans. Pattern Anal. Mach. Intell. **37**(2), 256–270 (2015)
24. Tan, Y., Ou, Z.: Topic-weak-correlated latent Dirichlet allocation. In: 7th International Symposium on Chinese Spoken Language Processing, pp. 224–228 (2010)
25. Teh, Y.W., Jordan, M.I., Beal, M.J., Blei, D.M.: Sharing clusters among related groups: hierarchical Dirichlet processes. In: Saul, L.K., Weiss, Y., Bottou, L. (eds.) Advances in Neural Information Processing Systems 17, NIPS, pp. 1385–1392. MIT Press (2005)
26. Vaithyanathan, S., Dom, B.: Model-based hierarchical clustering. In: Proceedings of the Sixteenth Conference on Uncertainty in Artificial Intelligence, UAI, pp. 599–608. Morgan Kaufmann, San Francisco (2000)
27. Wallach, H.M., Mimno, D.M., McCallum, A.: Rethinking LDA: Why priors matter. In: Bengio, Y., Schuurmans, D., Lafferty, J.D., Williams, C.K.I., Culotta, A. (eds.) NIPS, pp. 1973–1981. Curran Associates, Inc. (2009)
28. Wallach, H.M., Murray, I., Salakhutdinov, R., Mimno, D.: Evaluation methods for topic models. In: Proceedings of the 26th Annual International Conference on Machine Learning, ICML 2009, pp. 1105–1112. ACM, New York (2009)
29. Wang, C., Blei, D.M.: Collaborative topic modeling for recommending scientific articles. In: Proceedings of the 17th ACM SIGKDD International Conference on Knowledge Discovery and Data Mining, KDD 2011, pp. 448–456. ACM, New York (2011)
30. Ward Jr., J.H.: Hierarchical grouping to optimize an objective function. J. Am. Stat. Assoc. **58**(301), 236–244 (1963)
31. Xu, R., Wunsch, D.I.: Survey of clustering algorithms. IEEE Trans. Neural Networks **16**(3), 645–678 (2005)

Metrics

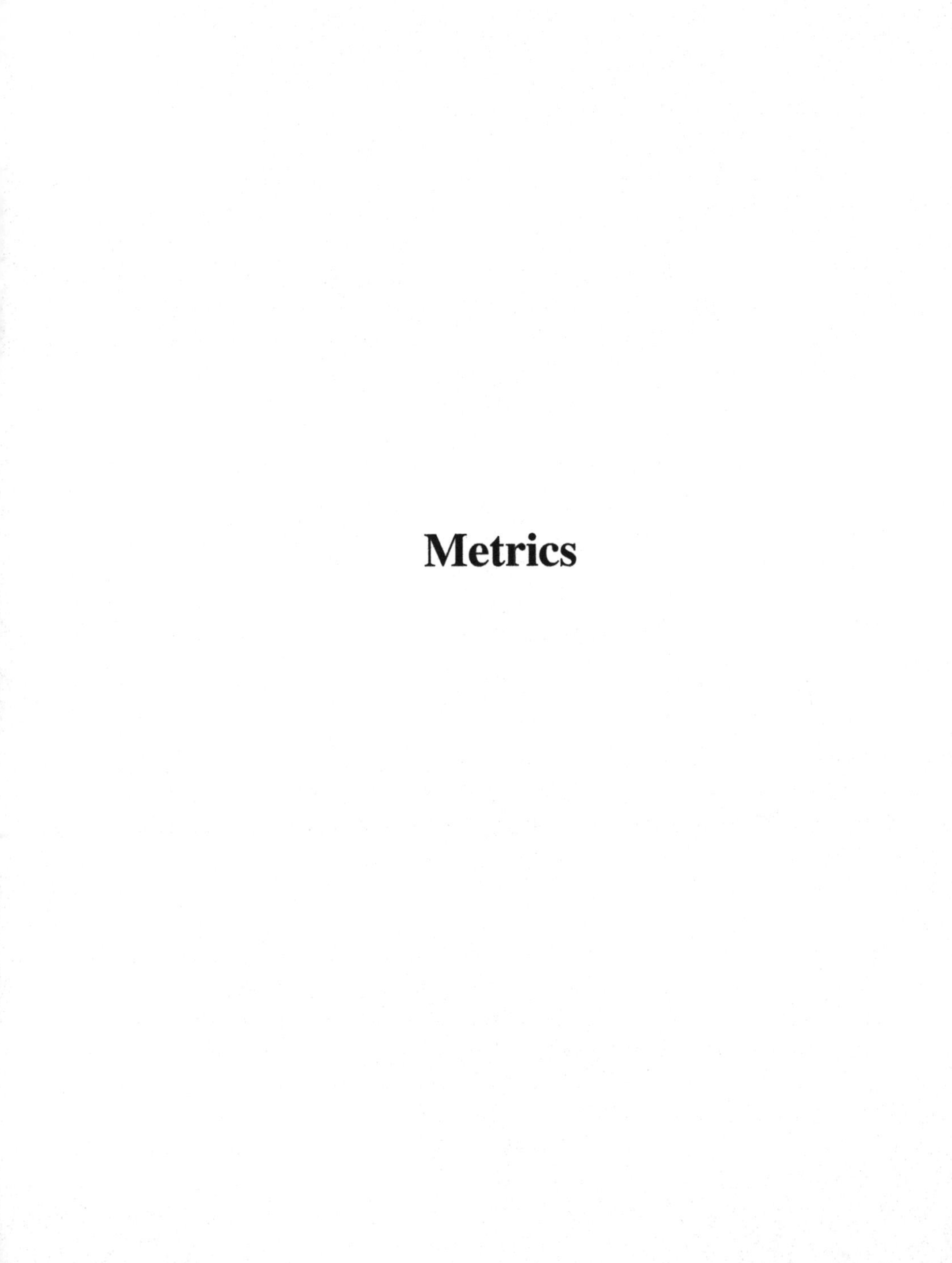

Meta-evaluation of Dynamic Search: How Do Metrics Capture Topical Relevance, Diversity and User Effort?

Ameer Albahem(✉), Damiano Spina, Falk Scholer, and Lawrence Cavedon

RMIT University, Melbourne, Australia
{ameer.albahem,damiano.spina,falk.scholer,lawrence.cavedon}@rmit.edu.au

Abstract. Complex dynamic search tasks typically involve multi-aspect information needs and repeated interactions with an information retrieval system. Various metrics have been proposed to evaluate dynamic search systems, including the Cube Test, Expected Utility, and Session Discounted Cumulative Gain. While these complex metrics attempt to measure overall system "goodness" based on a combination of dimensions – such as topical relevance, novelty, or user effort – it remains an open question how well each of the competing evaluation dimensions is reflected in the final score. To investigate this, we adapt two meta-analysis frameworks: the Intuitiveness Test and Metric Unanimity. This study is the first to apply these frameworks to the analysis of dynamic search metrics and also to study how well these two approaches agree with each other. Our analysis shows that the complex metrics differ markedly in the extent to which they reflect these dimensions, and also demonstrates that the behaviors of the metrics change as a session progresses. Finally, our investigation of the two meta-analysis frameworks demonstrates a high level of agreement between the two approaches. Our findings can help to inform the choice and design of appropriate metrics for the evaluation of dynamic search systems.

Keywords: Evaluation · Dynamic search · Intuitiveness Test · Metric Unanimity

1 Introduction

In many search scenarios, users interact with search systems multiple times to find documents relevant to a complex information need. During the search session, users might submit multiple queries [5], paginate [12] or provide fine-grained relevance feedback [23]. In recent years, *dynamic search* systems that can learn from user feedback and adapt subsequent results have been developed [5,12,23]. As a result of the complex actions and processes that may be part of dynamic search, different dimensions such as topical relevance, novelty and the amount of user effort can all play a role in the overall user satisfaction with search results [11]. To evaluate such systems, several information retrieval effectiveness metrics

L. Azzopardi et al. (Eds.): ECIR 2019, LNCS 11437, pp. 607–620, 2019.
https://doi.org/10.1007/978-3-030-15712-8_39

have been proposed that attempt to model some or all of these dimensions in their formulations. As these metrics address multiple (and sometimes competing) aspects of performance, it is unclear whether they overall behave as intended, by rewarding relevant and novel documents while minimizing user effort.

In this work, we methodically study which dynamic search metrics are better able to capture different dimensions of effectiveness. Since dynamic search includes multiple iterations of interactions between a user and a dynamic search system, we also investigate how the length of a search session impacts on the ability of complex metrics to model these dimensions.

Recently, the TREC Dynamic Domain (DD) Track [21–23] and the CLEF Dynamic Search Lab [13] adopted metrics such as the Cube Test (CT) [14] and Session Discounted Cumulative Gain (sDCG) to evaluate dynamic search in an interactive setup where systems are expected to learn from user feedback and adapt their outputs dynamically. Luo et al. [14] compared these metrics with other widely used effectiveness measures by considering their discriminative power – the ability of metrics to detect statistically significant differences in the retrieval results of different systems. Specifically, CT was compared with diversity metrics (Alpha Normalized Discounted Cumulative Gain (α-nDCG) [9] and Intent-Aware Expected Reciprocal Rank (ERR-IA) [6]) and Time-Biased Gain [18]. However, this approach does not inform researchers about the behaviors of the metrics in terms of measuring the key effectiveness dimensions.

There are different approaches to study the behavior of evaluation metrics. One of such approaches consists of analyzing how closely a metric matches the behavior or preferences of users across a set of search systems [20]. Another option is axiomatic analysis, which consists of defining formal properties that metrics may or may not satisfy [1–4, 10, 15]. A complementary proposal is the use of statistical analysis over metric scores – this is the approach followed in this work.

To study complex dynamic search metrics and how they reflect effectiveness dimensions, we apply two meta-analysis frameworks: the Intuitiveness Test proposed by Sakai [16], and the Metric Unanimity framework proposed by Amigó et al. [3]. The *Intuitiveness Test* measures the extent to which complex metrics are able to capture key properties that are important to measure in a search task evaluation. In previous work, the test was used to analyze the intuitiveness of a range of diversity metrics such as α-nDCG and ERR-IA over a range of tasks including search result diversification [16, 17] and aggregated search [24]. Chuklin et al. [7] applied the test to evaluate the intuitiveness of click models in aggregated search.

Metric Unanimity [3] relies on the intuition that, if a system is superior to another system in every key property of an evaluation, then this should be unanimously reflected by metrics that measure these properties. Amigó et al. [3] analyzed the Metric Unanimity of a wide range of relevance and diversity metrics. However, they did not study the impact of session length on metric behavior, which is one of the key dimensions in dynamic search evaluation.

Our work provides the following key contributions:

- An evaluation of effectiveness metrics for dynamic search,[1] a domain where new complex metrics have been proposed, but whose ability to reflect different dimensions is not clear;
- Proposing a new dimension – *User Effort* – which has not been investigated in prior research into metric intuitiveness and unanimity;
- Investigating the agreement between the two meta-analysis frameworks (Intuitiveness Test and Metric Unanimity) that have previously been used to study evaluation metrics separately.

The results of our analysis show that the Normalized Cube Test [19] (nCT) is generally more intuitive than other metrics when measuring all dimensions simultaneously. However, dynamic search metrics and diversity metrics can provide complementary information, and should both be reported to provide better insights into search effectiveness. The results also shed light on how metric behavior varies with search session length, where metrics tend to agree with each other more at earlier iterations. Lastly, both frameworks show a high level of agreement.

2 Methodology

2.1 Multidimensional Intuitiveness Test

The Intuitiveness Test proposed by Sakai [16] measures the ability of complex metrics to capture different dimensions of search task effectiveness evaluation. For instance, the test can be used to measure the extent to which diversity metrics such as α-nDCG or ERR-IA can intuitively capture relevance, or coverage of subtopics, two potentially competing dimensions of system effectiveness for search result diversification. In this framework, evaluation metrics are divided into *simple* and *complex* metrics: the former model a single dimension of retrieval effectiveness (for instance, Topical Relevance), while the latter incorporate two or more dimensions. In our work, we adapt the Intuitiveness Test [16] to evaluate how *intuitive* dynamic search metrics are in measuring multiple effectiveness dimensions simultaneously.

Algorithm 1 describes the process for comparing two complex metrics, M_{c1} and M_{c2}, given a set of simple metrics $M_s \in \mathcal{M}_S$ where each embodies a particular effectiveness dimension. In the algorithm, $M_x(t, r)$ denotes the final effectiveness score that the complex metric M_x assigned for the output of a run r (typically a ranked list of retrieved documents) produced by a given system in response to a search topic t. Where the complex metrics M_{c1} and M_{c2} disagree with each other regarding which run is more effective (line 12), the simple metrics are used to judge which of the complex metrics is more closely aligned with the simple metrics $\mathcal{M}_S$, and therefore more strongly embodies the effectiveness dimension that each simple metric M_s in $\mathcal{M}_S$ represents. A key assumption underlying this test is that the chosen simple metrics appropriately represent

[1] Code is available at https://github.com/aalbahem/ir-eval-meta-analysis.

```
1  Input: Complex Metrics M_c1 and M_c2; Simple Metrics 𝓜_S; Pairs of runs
     ⟨r1, r2⟩ ∈ 𝓡; Set of topics t ∈ 𝓣;
2  Output: Intuitiveness of M_c1 and M_c2;
3  Disagreements = 0; Correct1 = 0; Correct2 = 0;
4  foreach pair of runs ⟨r1, r2⟩ ∈ 𝓡 do
5      foreach topic t ∈ 𝓣 do
6          δM_c1 = M_c1(t, r1) − M_c1(t, r2);
7          δM_c2 = M_c2(t, r1) − M_c2(t, r2);
8          δ𝓜_S = {};
9          foreach M_s in 𝓜_S do
10             δ𝓜_S.add(M_s(t, r1) − M_s(t, r2));
11         end
12         if δM_c1 × δM_c2 < 0 then
13             Disagreements++;
14             if ∀δM_s ∈ δ𝓜_S, δM_c1 × δM_s > 0 then
15                 Correct1++;
16             end
17             if ∀δM_s ∈ δ𝓜_S, δM_c2 × δM_s > 0 then
18                 Correct2++;
19             end
20         end
21     end
22 end
23 Intuitiveness(M_c1|M_c2, 𝓜_S) = Correct1/Disagreements;
24 Intuitiveness(M_c2|M_1, 𝓜_S) = Correct2/Disagreements;
```

Algorithm 1. Multi-dimensional Intuitiveness Calculation, based on [16].

the effectiveness dimensions to be considered; the choice of the simple metrics is therefore crucial to the analysis. Note that the focus of the test is only on the cases where complex metrics disagree with each other, since complex metrics generally correlate with each other [8,16].

Algorithm 1 differs from the original version described by Sakai [16] in two aspects. First, we amended the original algorithm to support comparing complex metrics based on two or more simple metrics.[2] In lines 14 and 17, an agreement occurs if the complex metric agrees with *all* of the simple metrics in the set $\mathcal{M}_S$. Second, we refine the condition for agreement between a simple and a complex metric, such that this is only met when both metrics have the same preferences for the pair of runs under consideration (lines 14 and 17); for example, given a topic, both the simple and complex metrics prefer run r_1, or they both prefer run r_2. We discard the cases where the simple metric gives both runs the same scores (tie), i.e. $\delta M_s = 0$. The original paper does not discard these cases when reporting results; however, in practice they are discarded when evaluating statistical significance using a sign test. In our experiments, we found the number of ties

[2] Sakai [16] evaluated metrics considering diversity and relevance simultaneously, but the procedure was not detailed.

```
1  Input: A complex metric M_c; Set of simple Metrics 𝓜_S; Pairs of runs
   ⟨r_1, r_2⟩ ∈ 𝓡; Set of topics t ∈ 𝓣;
2  Output: Metric Unanimity (MU) of M_c with the set of metrics 𝓜_S
3  Δm_{i,j} = 0  Δ𝓜_{S i,j} = 0  Δm𝓜_{S i,j} = 0
4  foreach pair of runs ⟨r_1, r_2⟩ ∈ 𝓡 do
5      foreach topic t ∈ 𝓣 do
6          if M_c(t, r_1) == M_c(t, r_2) then
7              Δm_{i,j} += 0.5
8          end
9          else
10             Δm_{i,j} += 1
11         end
12         if (∀m ∈ 𝓜_S, m(t, r_1) ≥ m(t, r_2)) Or (∀m ∈ 𝓜_S, m(t, r_1) ≤ m(t, r_2))
           then
13             Δ𝓜_{S i,j}++;
14         end
15         if (∀m ∈ 𝓜_S ∪ {M_c}, m(t, r_1) ≥ m(t, r_2)) Or
           (∀m ∈ 𝓜_S ∪ {M_c}, m(t, r_1) ≤ m(t, r_2)) then
16             Δm𝓜_{S i,j}++;
17         end
18     end
19 end
20 MU(M_c, 𝓜_S) = PMI(Δm_{i,j}, Δ𝓜_{S i,j})
```

21 $\mathrm{PMI}(\Delta m_{i,j}, \Delta \mathcal{M}_{S\,i,j}) = \log\left(\frac{P(\Delta m_{i,j}, \Delta \mathcal{M}_{S\,i,j})}{P(\Delta m_{i,j} \times P(\Delta \mathcal{M}_{S\,i,j})}\right) = \log\left(\frac{\frac{\Delta m \mathcal{M}_S}{|\mathcal{R}|}}{\frac{\Delta m_{i,j}}{|\mathcal{R}|} \times \frac{\Delta \mathcal{M}_{S\,i,j}}{|\mathcal{R}|}}\right)$

Algorithm 2. Metric Unanimity calculation based on Amigó et al. [3]

for simple metrics are high, which could obfuscate the actual trends regarding which complex metric has greater intuitiveness.

When conducting the sign test for our experimental results, we use the number of times a complex metric agrees with the simple metrics as the number of successes, i.e., the final $Correct_1$ for M_{c1} and $Correct_2$ for M_{c2}; the number of trials is $Correct_1 + Correct_2$; and the hypothesized probability of success is 0.5.

2.2 Metric Unanimity

Amigó et al. [3] define Metric Unanimity as the Point-wise Mutual Information between improvement decisions of a metric M and improvements captured simultaneously by a set of metrics $\mathcal{M}'$. In their work, Amigó et al. [3] represent $\mathcal{M}'$ by a set of various metrics such as ad hoc and diversity metrics, which is a mixture of simple and complex metrics. In this study, which investigates to what extent complex metrics embody different effectiveness dimensions, we instead instantiate M to be a complex metric M_c, and $\mathcal{M}'$ to be a set of one or more simple metrics, as described below.

Let $\Delta m_{i,j}$ denote the event that a complex metric M_c captures improvements between a pair of systems (r_1, r_2); similarly, let $\Delta \mathcal{M}_{S i,j}$ denote the event that all metrics in $\mathcal{M}_S$ simultaneously capture the improvements between system pairs. Then we calculate Metric Unanimity (MU) of M_c using Algorithm 2. Line 21 shows how the point-wise mutual information between the two events is calculated. A higher value of MU for a complex metric implies that it more effectively measures the individual dimensions of interest.

2.3 Simple Metrics

Sakai [16] suggested two simple (gold) metrics to measure the intuitiveness of diversification metrics: (i) Precision (prec) to measure the ability of metrics to capture topical relevance; and (ii) Subtopic Recall[3] (st-rec) to measure the ability of metrics to capture diversity.

In addition to these simple metrics that measure topical relevance and diversity, we study a new dimension not previously explored: *user effort*. As a first attempt, we propose three simple metrics to define user effort in terms of the time spent by users when inspecting the ranking:

- Reciprocal Iteration (r-it), 1/number of iterations.
- Negative Iteration (neg-it), $-1 \times$ number of iterations.
- Total Time (tot-time), computed based on the user model presented as part of the Time-Biased Gain metric [18]: $\sum_{d \in D} 4.4 + r_d \times (0.018 l_d + 7.8)$, where $r_d = 0.64$ if d is a relevant document in the ranking D, otherwise it is 0.39 and l_d is the number of words in d. We use the document length statistics provided by the TREC Dynamic Domain 2016 track to compute this.

The first two approaches, r-it and neg-it, are straightforward estimates based simply on the number of iterations in which a user interacts with a dynamic search system; both assume that the amount of user effort is the same for all documents. The third approach, tot-time, uses richer information and instead estimates user effort based on parameters that include the time to scan search result snippets, and read document content. The r-it approach mirrors the CT metric effort estimation used in the TREC Dynamic Domain track [22]; we investigate neg-it to test whether using the same information differently changes metric behaviors. Note that *user effort* could also be measured by considering other factors such as cognitive effort; in this work, we primarily represent user effort by simple metrics based on the number of iterations or time spent in dynamic search tasks, and leave other factors for future research.

In our experiments, we use different combinations of the simple metrics that reflect Topical Relevance, Diversity, and User Effort, to define $\mathcal{M}_S$. The different sets of simple metrics are then used to measure the intuitiveness and unanimity of complex metrics proposed in the literature to evaluate dynamic search and diversity tasks, which are described below.

---

[3] Also known as Intent Recall [16].

2.4 Complex Metrics

In this work, we study metrics used in dynamic search evaluation campaigns such as the TREC Dynamic Domain Track [21–23] and the CLEF Dynamic Search Lab [13]. In particular, we consider the Average Cube Test (ACT) [22] and the normalized version of CT(nCT) [14], Expected Utility (nEU), and Session Discounted Cumulative Gain (nsDCG) [19], which capture three key dimensions of dynamic search: Topical Relevance, Diversity and User Effort. We also study the Rank-Biased Utility (RBU) metric, recently introduced by Amigó et al. [3], which models these three dimensions of dynamic search, and was designed by incorporating ideas from different ad hoc and diversity metrics.

Since Diversity – supporting the retrieval of documents for different subtopics of a complex task – is a key dimension of dynamic search systems, we also study two well-known metrics for search result diversification: Alpha Normalized Discounted Cumulative Gain (α-nDCG) and Intent-Aware Expected Reciprocal Rank (nERR-IA), calculated using collection-dependent normalization as described by Clarke et al. [8]. These metrics are also complex, in that they combine two dimensions: Topical Relevance and Diversity. Therefore, comparing them with the dynamic search metrics may provide additional insights in terms of the relation between dynamic search metrics and the Diversity and Topical Relevance dimensions.

2.5 TREC Dynamic Domain Collections

The TREC Dynamic Domain (DD) track ran for three years, from 2015 to 2017 [21–23]. This track models an interactive search setup, where systems receive aspect-level feedback repeatedly and need to dynamically find relevant and novel documents for the query subtopics using the least possible number of iterations.

We make use of these collections and conduct our analysis of complex metrics based on the formal runs submitted to these tracks. While the specifics of the tracks differed slightly across the years (e.g. the number of search topics were 118, 53 and 60 in 2015, 2016 and 2017, respectively) the search tasks being modelled remained consistent, and we therefore conduct our analysis of metric behaviour both for specific instances of the track, as well as aggregating across all three years.

A total of 32, 21 and 11 runs were submitted to TREC DD 2015, 2016 and 2017, respectively. The runs include different diversification algorithms, relevance feedback methods, and retrieval models. In evaluating the runs, we used the official track evaluation script. We also evaluated these runs using the standard web diversity metrics implemented by the Web Diversity evaluation script `ndeval`. As TREC DD considers passage-level relevance judgments, we generate the document-level relevance judgments by summing up the respected subtopic-passage judgments, the same approach that is followed by the official TREC DD evaluation script. For the TREC DD evaluation, the participating runs were required to return 5 documents per iteration, thus to evaluate a standard web diversity metric at the n-th iteration, we calculated the scores at a cutoff of $5 \times n$.

Table 1. Results of the multidimensional intuitiveness test for the TREC DD 2016 data. For each pair of metrics, the intuitiveness scores for (metric in row)/(metric in column) are reported; the fraction of disagreements is shown in parenthesis (the ratio of disagreements to the total number of cases). $^+$ indicates statistically significant differences according to sign test with $p < 0.05$.

Topical Relevance (Rel, prec) and Diversity (Div, st-rec) and User Effort (Eff, tot-time)						
it	nCT	nEU	nsDCG	α-nDCG	nERR-IA	RBU
1 ACT	0.0084/0.0823$^+$ (4.27%)	0.0215/0.0135 (14.65%)	0.0128/0.0171 (4.23%)	0.0044/0.0558$^+$ (6.14%)	0.0212/0.0466$^+$ (6.38%)	0.0077/0.0694$^+$ (5.84%)
1 nCT	-	0.0307$^+$/0.0028 (15.86%)	0.0684$^+$/0.0143 (5.67%)	0.0284/0.0269 (6.04%)	0.0440$^+$/0.0214 (7.18%)	0.0049/0.0245 (1.84%)
1 nEU	-	-	0.0118/0.0206 (15.34%)	0.0056/0.0302$^+$ (17.63%)	0.0105/0.0268$^+$ (17.18%)	0.0031/0.0303$^+$ (17.53%)
1 nsDCG	-	-	-	0.0042/0.0506$^+$ (6.42%)	0.0192/0.0410$^+$ (6.59%)	0.0130/0.0746$^+$ (5.56%)
1 α-nDCG	-	-	-	-	0.1126$^+$/0.0000 (1.36%)	0.0195/0.0265 (6.47%)
1 nERR-IA	-	-	-	-	-	0.0155/0.0418$^+$ (7.55%)
10 ACT	0.0019/0.0405$^+$ (14.47%)	0.0315$^+$/0.0235 (36.90%)	0.0357/0.0279 (24.23%)	0.0255/0.0330 (20.51%)	0.0473$^+$/0.0193 (18.68%)	0.0173/0.0192 (29.64%)
10 nCT	-	0.0349$^+$/0.0114 (36.40%)	0.0355$^+$/0.0121 (31.98%)	0.0259$^+$/0.0126 (30.68%)	0.0424$^+$/0.0058 (29.54%)	0.0293$^+$/0.0015 (18.16%)
10 nEU	-	-	0.0210/0.0242 (34.70%)	0.0158/0.0279$^+$ (36.54%)	0.0306/0.0244 (36.27%)	0.0258/0.0356$^+$ (35.68%)
10 nsDCG	-	-	-	0.0226/0.0533$^+$ (11.16%)	0.0560$^+$/0.0330 (14.48%)	0.0189/0.0240 (48.10%)
10 α-nDCG	-	-	-	-	0.1291$^+$/0.0017 (5.24%)	0.0195/0.0174 (46.15%)
10 nERR-IA	-	-	-	-	-	0.0146/0.0278$^+$ (43.85%)

3 Results and Analysis

3.1 Intuitiveness Test Analysis

We first report and analyze results of the Intuitiveness Test (see Algorithm 1) between different pairs of complex metrics, and at different iterations, when the three dimensions – Topical Relevance, Diversity and User Effort – are considered through the simple metrics prec, st-rec and tot-time, respectively.

Table 1 shows results for the TREC Dynamic Domain 2016 runs, for early and late iterations.[4] Consistent trends were observed in the data for the other years. The results show that the metric nCT – which was defined to cover the three dimensions – is more intuitive overall (usually significantly better, and never significantly worse) than metrics designed to only cover Topical Relevance and Diversity dimensions (e.g. α-nDCG or nERR-IA). With respect to metrics that model all dimensions (nCT and RBU), nCT is generally more intuitive at iteration 1 and statistically significantly more intuitive at iteration 10.

[4] Due to space limitations, in Table 1 we only show the results for the TREC DD 2016 runs, which is the second edition of the track and had almost as twice as many runs as the last edition.

The results also demonstrate that the behavior of metrics in terms of intuitiveness is dependent on the iteration in the dynamic search session at which they are measured. In particular, metrics tend to disagree with each other in their preferences of runs more at iteration 10 than at the first iteration.

3.2 Ranking of Metrics Based on the Intuitiveness Test

To gain a broader understanding of intuitiveness, it is desirable to aggregate the low-level results of individual intuitiveness test evaluations, such as those that were presented in Table 1. Ideally, given pairwise comparisons between complex metrics such as from the previous section, a ranking of the intuitiveness of metrics can be induced. However, statistical significance is not transitive. As a result, in Table 2, we report a ranking of metrics, based on the number of times that a complex metric obtains a significantly higher intuitiveness score against the other metrics. Representative combinations of the dimensions (Topical Relevance (Rel), Diversity (Div), and User Effort (Eff)) and iterations (1, 3, and 10) are reported.[5]

The aggregation process involves summing the number of times that one complex metric obtained a significantly higher intuitiveness score than another, across the TREC Dynamic Domain tracks from 2015, 2016 and 2017, and converting this count into a ranking, such that a rank of 1 indicates that a metric obtained the highest count of significantly higher scores, while a rank of 7 indicates the lowest count. For example, for the topical relevance dimension (column *Rel*) at iteration 1 (sub-column *1*), nERR-IA has a rank of 5, indicating that it is more intuitive than two other metrics across different years.

In terms of Topical Relevance, ACT is more intuitive in late iterations, whereas nsDCG is more intuitive than other metrics in early iterations. For Diversity, α-nDCG is more intuitive than other metrics, regardless of the iterations. Here, nCT and RBU start more intuitive than other metrics but become less intuitive in later iterations.

In terms of User Effort, complex metrics that directly model this dimension (ACT, nCT, nEU and RBU) are, as may be hoped, more intuitive than other metrics.

When considering Topical Relevance and Diversity together, α-nDCG has higher intuitiveness scores than other metrics. In addition, nCT shows good performance in early iterations. We suspect this may be due to its emphasis on time, which becomes greater in later iterations.

[5] Other combinations and iterations are not reported due to lack of space, but overall trends were consistent with these settings. We also calculated the ranking of metrics based directly on their intuitiveness test relationship (i.e. without taking statistical significance into account); overall trends were again consistent with those presented here.

Table 2. Intuitiveness Test-based ranking of complex metrics using the number of times that a complex metric obtained a statistically significantly higher Intuitiveness Test score than other metrics, across all TREC Dynamic Domain years.

Metric/Iteration	Rel				Div				Eff				Rel and Div				Rel and Div and Eff			
	1	3	10	All	1	3	10	All	1	3	10	All	1	3	10	All	1	3	10	All
ACT	4	5	1	3	4	5	4	4	4	4	3	4	5	4	3	5	4	4	2	4
nCT	3	3	1	2	2	3	6	3	2	3	2	2	2	3	5	4	1	1	1	1
nEU	6	6	6	7	6	7	5	5	1	1	3	3	6	5	4	6	5	4	6	6
nsDCG	1	1	2	1	5	4	2	2	6	6	6	7	5	2	1	2	4	4	5	5
α-nDCG	1	2	3	4	1	1	1	1	5	6	5	6	1	1	1	1	1	3	3	2
nERR-IA	5	4	5	6	3	2	3	2	3	5	4	5	4	3	2	3	3	4	6	6
RBU	2	3	4	5	2	6	7	6	2	2	1	1	3	4	6	7	2	2	4	3

Modelling User Effort. We also experimented with two other simple metrics for modeling user effort, as described in Sect. 2.3: r-it and neg-it. Compared to tot-time, they agree with nCT more than nEU, whereas tot-time agrees with nEU more than nCT. This is likely because nCT uses the number of iterations to represent time, hence iteration-based simple metrics of User Effort may have biased the analysis toward nCT. On the other hand, nEU uses document length as an estimate of user effort, and since tot-time also uses document length as one part of its calculation, which may lead to better agreement for nEU. Of the three simple metrics that we explored to represent User Effort, we recommend tot-time as the most suitable, since it more closely models user behaviour when interacting with search results, taking the relevance of answers, and the amount of time required to process both document summaries and document content, into account.

3.3 Ranking Based on Metric Unanimity

For Metric Unanimity (Algorithm 2), the MU scores between complex metrics and different sets of simple metrics (to represent the effectiveness dimensions individually, in pairs, or all together) are calculated. A ranking of metrics was then induced, using the frequency with which one complex metric showed a higher unanimity than another complex metric – i.e. $\mathrm{MU}(M_{c1}) > \mathrm{MU}(M_{c2})$ – across all pairwise comparisons.[6]

Table 3 shows the ranking of the complex metrics, for representative sets of combinations of simple metrics (columns) and iterations (sub-columns); the displayed configurations are consistent with those shown for the Intuitiveness Test analysis.

[6] The Metric Unanimity framework differs from the Intuitiveness Test framework in that there is no equivalent concept of underlying "number of successes", therefore a significance test similar to the sign test in the Intuitiveness Test framework cannot be carried out.

Table 3. Metric Unanimity-based ranking of complex metrics using the number of times that a complex metric obtained a higher Metric Unanimity score than other metrics, across all TREC Dynamic Domain years.

Metric/Iteration	Rel				Div				Eff				Rel and Div				Rel and Div and Eff			
	1	3	10	All	1	3	10	All	1	3	10	All	1	3	10	All	1	3	10	All
ACT	5	5	2	4	5	3	5	5	2	3	4	4	6	5	2	5	6	4	3	3
nCT	1	1	1	1	1	1	6	3	1	2	2	2	1	1	1	1	1	1	1	1
nEU	6	4	5	7	7	5	4	6	1	1	3	3	7	4	4	6	5	3	4	4
nsDCG	4	2	4	2	6	4	2	4	4	5	7	7	5	2	3	3	7	6	7	7
α-nDCG	2	3	3	3	2	1	1	1	3	5	6	6	2	2	1	2	3	5	5	5
nERR-IA	3	4	4	5	4	2	3	2	2	4	5	5	3	3	3	4	4	7	6	6
RBU	5	4	6	6	3	5	7	7	1	2	1	1	4	4	5	7	2	2	2	2

In general, when considering all dimensions, nCT and RBU are ranked higher than other metrics for different iterations. For the different years, nCT unanimously agreed more with the simple metrics than other metrics. This mirrors its behavior in the Intuitiveness Test analysis. However, considering other dimension combinations, metrics might have different behavior than ones observed with the Intuitiveness Test. For instance, for the topical relevance dimension (column Rel), nCT consistently ranked first across iterations, while in the Intuitiveness Test analysis, it ranked better in late iterations.

We therefore formally analyze the correlation between the Metric Unanimity ranking and the ranking induced by the Intuitiveness Test, which we describe in the next section.

3.4 Comparing Intuitiveness Test and Metric Unanimity Rankings

The Spearman rank correlations between the rankings induced by the Intuitiveness Test and the Metric Unanimity analysis, for different combinations of dimensions and iterations, are shown in Table 4.

When considering the effectiveness dimensions, the correlations are generally high. In particular, for Diversity, User Effort, and the combination of all three dimensions, all correlation coefficients are strong at 0.6 or higher. Regarding iterations, for combinations of two or more dimensions, the correlations are generally higher in early iterations than later iterations. Similar patterns were observed for the individual years (again not included due to space constraints). However, statistically significant correlations were found more frequently for the TREC DD 2015 and 2016 than for the 2017 edition. This is likely due the lower number of submitted runs (11) in 2017, while 2015 and 2016 received 32 and 21 runs, respectively.

With regard to comparing the Intuitiveness Test and Metric Unanimity frameworks, both approaches are motivated differently. Moreover, both differ in their fundamental units of comparison (as noted previously, one practical implication of this is that an additional significance test can be applied as part of the Intuitiveness Test approach). Nevertheless, high correlations are observed between both meta-evaluation approaches. This provides strong evidence to indicate that the observed metric behavior is not due to either framework being biased towards certain complex metrics, but instead is a reflection of the ability of complex metrics to capture the properties instantiated by the chosen simple metrics.

Table 4. Spearman rank-order correlation coefficient between the Intuitiveness Test ranking (Table 2) and the Metric Unanimity ranking (Table 3). * indicates a statistically significant correlation with $p < 0.05$.

Iteration	Rel	Div	Eff	Rel and Div	Rel and Div and Eff
1	0.44	0.94*	0.96*	0.92*	0.82*
3	0.71	0.89*	0.99*	0.75	0.71
10	0.80*	0.96*	0.99*	0.39	0.65
All	0.89*	0.94*	1.00*	0.79*	0.61

4 Conclusions

In this work, we investigated the ability of complex effectiveness metrics to cover three key dimensions for dynamic search tasks: Topical Relevance, Diversity, and User Effort.

Our analysis – using both the Intuitiveness Test proposed by Sakai [16] and the Metric Unanimity approach proposed by Amigó et al. [3] – showed that complex metrics can differ substantially in their ability to capture the various dimensions. Across iterations and datasets, nCT captures the key properties better than other metrics. However, the results also showed that α-nDCG can provide complementary information to nCT, and therefore we recommend that both should be reported when considering the effectiveness of dynamic search systems. In addition, the results showed that the behaviour of metrics can change as a search session progresses: metrics tend to disagree with each other more at later iterations. Thus, we also recommend reporting results at different iterations of a search session. Finally, our investigation demonstrated a high level of correlation between the Intuitiveness Test and Metric Unanimity. This provides a solid understanding of how complex effectiveness metrics agree or differ in relation to simple metrics that represent specific dimensions.

Future work includes the exploration of the impact of assigning different weights to the dimensions in the meta-analysis frameworks, which currently implicitly assume equal importance. We also intend to extend this analysis by considering other metrics for each of the dimensions, e.g. considering cognitive complexity for user effort. It will also be interesting to apply these frameworks to study the suitability of complex metrics that are used in domains that are typified by different search tasks, such as slow search, or high-recall search. Finally, we plan to study how the framework can be applied to assist in the construction of new metrics, to ensure that they sufficiently cover desired properties.

Acknowledgement. This research was partially supported by Australian Research Council (projects LP130100563 and LP150100252), and Real Thing Entertainment Pty Ltd.

References

1. Albahem, A., Spina, D., Scholer, F., Moffat, A., Cavedon, L.: Desirable properties for diversity and truncated effectiveness metrics. In: Proceedings of Australasian Document Computing Symposium, pp. 9:1–9:7 (2018)
2. Amigó, E., Gonzalo, J., Verdejo, F.: A general evaluation measure for document organization tasks. In: Proceedings of SIGIR, pp. 643–652 (2013)
3. Amigó, E., Spina, D., Carrillo-de Albornoz, J.: An axiomatic analysis of diversity evaluation metrics: introducing the rank-biased utility metric. In: Proceedings of SIGIR, pp. 625–634 (2018)
4. Busin, L., Mizzaro, S.: Axiometrics: an axiomatic approach to information retrieval effectiveness metrics. In: Proceedings of ICTIR, pp. 8:22–8:29 (2013)
5. Carterette, B., Kanoulas, E., Hall, M., Clough, P.: Overview of the TREC 2014 session track. In: Proceedings of TREC (2014)
6. Chapelle, O., Metlzer, D., Zhang, Y., Grinspan, P.: Expected reciprocal rank for graded relevance. In: Proceedings of CIKM, pp. 621–630 (2009)
7. Chuklin, A., Zhou, K., Schuth, A., Sietsma, F., de Rijke, M.: Evaluating intuitiveness of vertical-aware click models. In: Proceedings of SIGIR, pp. 1075–1078 (2014)
8. Clarke, C.L., Craswell, N., Soboroff, I., Ashkan, A.: A comparative analysis of cascade measures for novelty and diversity. In: Proceedings of WSDM, pp. 75–84 (2011)
9. Clarke, C.L., et al.: Novelty and diversity in information retrieval evaluation. In: Proceedings of SIGIR, pp. 659–666 (2008)
10. Ferrante, M., Ferro, N., Maistro, M.: Towards a formal framework for utility-oriented measurements of retrieval effectiveness. In: Proceedings of ICTIR, pp. 21–30 (2015)
11. Jiang, J., He, D., Allan, J.: Comparing in situ and multidimensional relevance judgments. In: Proceedings of SIGIR, pp. 405–414 (2017)
12. Jin, X., Sloan, M., Wang, J.: Interactive exploratory search for multi page search results. In: Proceedings of WWW, pp. 655–666 (2013)
13. Kanoulas, E., Azzopardi, L., Yang, G.H.: Overview of the CLEF dynamic search evaluation lab 2018. In: Bellot, P., et al. (eds.) CLEF 2018. LNCS, vol. 11018, pp. 362–371. Springer, Cham (2018). https://doi.org/10.1007/978-3-319-98932-7_31
14. Luo, J., Wing, C., Yang, H., Hearst, M.: The water filling model and the cube test: multi-dimensional evaluation for professional search. In: Proceedings of CIKM, pp. 709–714 (2013)
15. Moffat, A.: Seven numeric properties of effectiveness metrics. In: Banchs, R.E., Silvestri, F., Liu, T.-Y., Zhang, M., Gao, S., Lang, J. (eds.) AIRS 2013. LNCS, vol. 8281, pp. 1–12. Springer, Heidelberg (2013). https://doi.org/10.1007/978-3-642-45068-6_1
16. Sakai, T.: Evaluation with informational and navigational intents. In: Proceedings of WWW, pp. 499–508 (2012)

17. Sakai, T.: How intuitive are diversified search metrics? Concordance test results for the diversity U-Measures. In: Banchs, R.E., Silvestri, F., Liu, T.-Y., Zhang, M., Gao, S., Lang, J. (eds.) AIRS 2013. LNCS, vol. 8281, pp. 13–24. Springer, Heidelberg (2013). https://doi.org/10.1007/978-3-642-45068-6_2
18. Smucker, M.D., Clarke, C.L.: Time-based calibration of effectiveness measures. In: Proceedings of SIGIR, pp. 95–104 (2012)
19. Tang, Z., Yang, G.H.: Investigating per topic upper bound for session search evaluation. In: Proceedings of ICTIR, pp. 185–192 (2017)
20. Turpin, A., Scholer, F.: User Performance versus precision measures for simple web search tasks. In: Proceedings of SIGIR, pp. 11–18 (2006)
21. Yang, H., Frank, J., Soboroff, I.: TREC 2015 dynamic domain track overview. In: Proceedings of TREC (2015)
22. Yang, H., Soboroff, I.: TREC 2016 dynamic domain track overview. In: Proceedings of TREC (2016)
23. Yang, H., Tang, Z., Soboroff, I.: TREC 2017 dynamic domain track overview. In: Proceedings of TREC (2017)
24. Zhou, K., Lalmas, M., Sakai, T., Cummins, R., Jose, J.M.: On the reliability and intuitiveness of aggregated search metrics. In: Proceedings of CIKM, pp. 689–698 (2013)

A Markovian Approach to Evaluate Session-Based IR Systems

David van Dijk[1], Marco Ferrante[2], Nicola Ferro[2], and Evangelos Kanoulas[1](✉)

[1] University of Amsterdam, Amsterdam, The Netherlands
d.v.van.dijk@hva.nl, e.kanoulas@uva.nl
[2] University of Padua, Padua, Italy
ferrante@math.unipd.it, ferro@dei.unipd.it

Abstract. We investigate a new approach for evaluating session-based information retrieval systems, based on Markov chains. In particular, we develop a new family of evaluation measures, inspired by random walks, which account for the probability of moving to the next and previous documents in a result list, to the next query in a session, and to the end of the session. We leverage this Markov chain to substitute what in existing measures is a fixed discount linked to the rank of a document or to the position of a query in a session with a stochastic average time to reach a document and the probability of actually reaching a given query. We experimentally compare our new family of measures with existing measures – namely, session DCG, Cube Test, and Expected Utility – over the TREC Dynamic Domain track, showing the flexibility of the proposed measures and the transparency in modeling the user dynamics.

Keywords: Information retrieval · Evaluation · Sessions · Markov chains

1 Introduction

Evaluation measures are an intrinsic part of experimental evaluation. Even if a growing attention is called in the field for developing stronger theoretical foundations [1–3,9,12], they are often formulated and justified in a somewhat informal and intuitive way rather then being based on well-founded mathematical models. Carterette [4] has made a post-hoc attempt to propose a unifying framework which explains modern evaluation measures based on three components: a browsing model, a model of document utility, and a utility accumulation model. According to this framework, measures such as RBP [21], DCG [13] or ERR [5] can be defined as expectations of the utility, total utility, and effort, respectively

Electronic supplementary material The online version of this chapter (https://doi.org/10.1007/978-3-030-15712-8_40) contains supplementary material, which is available to authorized users.

L. Azzopardi et al. (Eds.): ECIR 2019, LNCS 11437, pp. 621–635, 2019.
https://doi.org/10.1007/978-3-030-15712-8_40

over a probabilistic space defined by the chance of a user to browse the next in rank document in a provided ranking.

When it comes to session search, defining an evaluation measures based on a rigorous mathematical model becomes an even more challenging task. Session search involves multiple iterations of searches in order for a user to accomplish a complex information need, with multiple queries being issued or reformulated and multiple runs of search results being returned by the search engine and examined by the user. The difficulty of defining session evaluation measure comes from the question of how to assess the value of a relevant document not only along a certain ranking but across rankings of different queries within a session, or, in other words, how to mathematically model the dynamics of a user across the entire search session. Session evaluation measures proposed in the literature, such as the session Discounted Cumulated Gain (sDCG) [14], the Expected Utility (EU) [28], the Expected Session measures (esM) [15], or the Cube Test (CT) [19], typically extend single ranking evaluation measures in an ad-hoc manner that results in a lack of a sound, clear, and extensible mathematical framework. In this paper, we focus on the following research question: *How can we mathematically model user dynamics over a multi-query session and inject them into an effectiveness measure?*

To answer this question, we represent queries in a session and documents within a ranked result list for a query as states in a Markov chain. We then define an event space of user actions when searching: (a) moving along a ranking of documents, (b) reformulating her query, and (c) abandoning the search session, and the probabilities of each one of these actions. Different instantiations of these probabilities give rise to different transition probabilities among the states of the Markov chain which allow us to model the different and perhaps complex user behaviors and paths in scanning the ranked result lists in a session.

Finally, we conduct a experimental evaluation of the Markov Session Measures (MsM) using three standard Text REtrieval Conference (TREC) collections developed by the Dynamic Domain Track (DDT) [26,27], based on which we show the flexibility of MsM in modeling a wide variety user dynamics, as well as how close MsM is to existing measures in terms of user dynamics.

2 Related Work

Järvelin et al. [14] extended the Discounted Cumulated Gain (DCG) measure to consider multi-query sessions. The measure – session Discounted Cumulated Gain (sDCG) – discounts documents that appear lower in the ranked list for a given query, as well as documents that appear in follow up query reformulations. sDCG underlies a deterministic user model with the user stepping down the ranked list until a fixed reformulation point and then moving to the next query in the session until all ranked lists in the session have been scanned. Luo et al. [19] proposed the Cube Test (CT) which is also based on a deterministic user model of browsing a ranked list up to a certain reformulation point and then continuing to browse the results of the next query. Departed from the work of Smucker and Clarke [22,23] who defined the Time-Biased Gain (TBG) measure, Luo et al. inject the time it takes users to read relevant documents as a discounting factor of the utility of a document. Differently from the aforementioned deterministic user models, both Yang and Lad [28] and Kanoulas et al. [15] took a probabilistic approach and defined a session measure as an expectation over a set of possible browsing paths. Yang and Lad introduced Expected Utility (EU) and, to define the probability of a user following a certain path, they followed the Rank-Biased Precision (RBP) approach [21], replacing RBP's stopping condition with a reformulation condition. Kanoulas et al., instead, first defined a reformulation probability that allows for an early abandonment and then, for those queries that are being realized, they introduced a stepping-down probability, similar to RBP. Our approach differs from sDCG and CT by considering a probabilistic event space of user actions across the states of a Markov chain, which represent documents in the different ranked lists and positions within them. Further, it offers a solid mathematic framework that from Yang and Lad and Kanoulas et al. by avoiding the unreasonable assumptions their approaches make, but also offering the ability to extend the framework to more advanced user dynamics.

Markov-based approaches have been previously exploited in IR, for example: Markov chains have been used to generate query models [17], for query expansion [7,20,25], and for document ranking [8], or to address the placement problem in the case of two dimensional results [6]. Ferrante et al. [10] use Markov chains to define evaluation measures over a single ranked list. This work is an extension of their work to session retrieval. However, differently from their work that depends on the computation of an invariant distribution and which makes the assumption that there is no absorbing state, our work takes a random walk approach and assumes the presence of an absorbing state.

Finally, according to a conducted laboratory user study, Liu et al. [18] have recently suggested some desirable features of a session-based evaluation measure: (1) the most useful document in a query is the most important; (2) the weighting function between queries should be normalized; (3) the primacy effect is not suitable for session evaluation; (4) the recency effect has a stronger influence on user's session satisfaction. Our MsM measure addresses some of the requirements formulated by Liu et al.: (1) because it handles graded relevance and an higher gain can be assigned to the most useful document; (2) because modelling the whole session with a single Markov chain seamlessly normalizes scores across queries; (3) and (4) because by setting appropriate transition probabilities and discount functions, it is possible to smooth the effect of the first documents (primacy) and emphasise the importance of latter queries (recency).

3 The Model

Section 3.1 introduces our Markovian model of the user dynamics over a multi-query session. Section 3.2 exploits this Markovian model to define the Markov Session Measure (MsM) which can be used to evaluate session-based IR systems.

3.1 Multi-query Session Dynamics

For a given task, a user can generate a sequence of queries, each of which originates a ranked list of documents. Given D the whole document corpus and $N \in \mathbb{N}$ the length of a run, $D_j(N) = \{(d_{1,j}, \dots, d_{N,j}) : d_{n,j} \in D, d_{n,j} \neq d_{m,j}$ for any $m \neq n\}$ is the ordered set of documents retrieved by a system run for the j-th query. The sets $D_j(N)$ for $j \in \mathbb{N}$ may not be disjoint and the same document may appear in many queries. Without loss of generality we assume that every run has the same length N.

Let $k \in \mathbb{N}$ the number of queries in a session. The whole search session is defined as a matrix of documents, where columns are the runs $D_1(N), \dots, D_k(N)$ corresponding to each query

$$\begin{bmatrix} d_{1,1} & d_{1,2} & d_{1,3} & \dots & d_{1,k} \\ d_{2,1} & d_{2,2} & d_{2,3} & \dots & d_{2,k} \\ d_{3,1} & d_{3,2} & d_{3,3} & \dots & d_{3,k} \\ \vdots & \vdots & \vdots & \ddots & \vdots \\ d_{N,1} & d_{N,2} & d_{N,3} & \dots & d_{N,k} \end{bmatrix}$$

The user moves among the documents according to some dynamics, that we assume to be Markovian, i.e. the user decides which document to visit only on the basis of the last document considered. Moreover, we assume that the user starts her search from the first document in the first query, i.e. first row and first column, as typically assumed by any evaluation measure. Then, she moves among the documents in the first column until she decides to change column, i.e. to reformulate the query, or to abandon the search session. In case of query reformulation, she passes to the next result list and, as before, she starts from the first document of the subsequent column, i.e. the next query, and so on until she ends the search.

We define the sequence of positions of the documents visited by the user as a stochastic process, $(X_n = (X_n^1, X_n^2))_{n \geq 1}$, where $X_n = (i, j)$ means that the n-th document visited by the user is $d_{i,j}$, the i-th document of the j-th column. We assume that this process is a Markov chain on the state space $S = \{1, \dots, N\}^\infty \cup \{(F, j), j \in \mathbb{N}\}$ where $X_n = (F, j)$ represents the fact that the user ends his search after visiting $n-1$ documents and formulating j queries. The transition matrix of this Markov chain

$$p_{(i_n, j_n),(i_{n+1}, j_{n+1})} = \mathbb{P}[X_{n+1} = (i_{n+1}, j_{n+1}) | X_n = (i_n, j_n)],$$

undergoes these constraints:

1. $p_{(i_n, j_n),(i_{n+1}, j_{n+1})} = 0$ if $j_{n+1} \neq j_n, j_n + 1$, i.e. the user can either move within a column of documents or pass to the next one;
2. $p_{(i_n, j_n),(i_{n+1}, j_n+1)} = 0$ if $i_{n+1} \neq 1$, i.e. when the user leaves a column, she goes to the first document of the next one;
3. $p_{(i_n, j_n),(F, j_n)} > 0$ for any $i_n \neq F$;
4. $p_{(F, j_n),(F, j_n)} = 1$ for any j_n, i.e. the states (F, j)'s are all *absorbing*.

Example 1. Let us assume that the stochastic process $(X_n)_{n\geq 1}$ takes the following values:

$$X_1 = (1,1), X_2 = (3,1), X_3 = (1,2), X_4 = (2,2), X_5 = (6,2),$$
$$X_6 = (3,2), X_7 = (1,3), X_8 = (3,3), X_9 = (F,3).$$

This means that the user performed 3 queries and considered 8 documents before stopping, as shown in the following graph

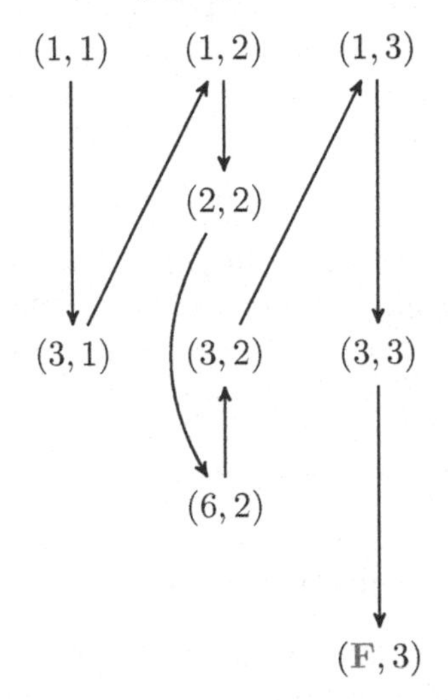

In order to determine how many queries have been issued and how long the search lasted, we define the following sequence of stopping times. Recall that the *stopping time* for a Markov chain $(X_n)_{n\geq 1}$ is a random variable T with values in $\mathbb{N} \cup \{\infty\}$ such that for any $n \in \mathbb{N}$ the event $\{T = n\}$ depends only on $\{X_m, m \leq n\}$. The stopping time

$$H = \inf\{n \geq 1 : X_n^1 = F\}$$

determines the number of steps done by the process, with the convention that $\inf \varnothing = \infty$. It allows us to define the (random) number K of queries performed during the search

$$K = X_H^2$$

K is the second component of the process $(X_n, n \in \mathbb{N})$ once absorbed in $(F, \cdot)$.

Then, we define the random times to leave any query as

$$\begin{aligned} H^1 &:= \inf\{n \geq 1 : X_n^2 = 2\} \\ H^2 &:= \inf\{n \geq 1 : X_n^2 = 3\} \\ &\vdots \\ H^{K-1} &:= \inf\{n \geq 1 : X_n^2 = K-1\} \end{aligned}$$

Thanks to these stopping times, we are able to determine how many documents of any query have been visited by a user. Indeed, defined $H^0 = 1$, the user has considered $H^1 - H^0$ documents of the first query, $H^2 - H^1$ documents of the second query, $H^3 - H^2$ documents of the third query and so on until the last query, where the number of documents visited is $H - H^{K-1}$. In the previous example, we have $H = 9$, $K = 3$, $H^1 = 3$, $H^2 = 7$, and the user has visited, respectively, 2, 4 and 2 documents in the three queries before stopping the search.

By means of these stopping times, we can define the events corresponding to the end of the search session in any given query. Indeed, if $H^1 = \infty$, it means that the user never passes to the second query and $(F, 1)$ is the unique absorbing state, $A^1 = \{\omega : H^1(\omega) = \infty\}$ corresponds to the event "the user visits just the documents in the first query". Analogously, for any $j > 1$, we can define the event $A^j = \{\omega : H^1(\omega) < \infty, \dots, H^{j-1}(\omega) < \infty, H^j(\omega) = \infty\}$ that the user ends search after considering the first j queries. The events $\{A^j, j \in \mathbb{N}\}$ form a partition of the underlying probability space.

In the following, these events are used to measure how "often" a random user visits each query during her search and to obtain, as a consequence, a weight to be assigned to any query. Moreover, to evaluate the effectiveness of the search within the queries actually visited by the user, we evaluate how far (stochastically) any state is, i.e. any document of any query, from the initial state $(1, 1)$ and discount its relevance proportionally to this "random" distance.

3.2 Evaluation of Multi-query Sessions

As previously discussed, evaluation measures typically apply a deterministic discount of the gain/utility of a document by a function of its rank position. We replace these deterministic discounts operating a two-step stochastic procedure:

- Given that the search generates k queries, we consider the probabilities that the search ends in $(F, 1)$, $(F, 2)$, $\dots$, (F, k), respectively;
- Given that the user does not end her search before the query j, i.e. she visits the documents of the query j, we compute the discount at each rank position of the j-th query according to the expected number of steps needed to reach that rank position starting from $(1, 1)$.

The user can stop her search after considering only the first run, or the first two runs, or the first three runs and so on. This is equivalent to considering that the Markov chain is absorbed in $(F, 1)$, or $(F, 2)$ and so on until (F, k). We are able to evaluate the absorption probabilities in any of these states $h = (h^1, \dots, h^k)$ starting from the probabilities of the events $A^1, \dots, A^{k-1}$ defined above. Indeed, we have $h^j = \mathbb{P}[A^j]$ for any $j < k$, and $h^k = 1 - h^1 - \dots - h^{k-1}$.

Let us define π_j as the probability that the user visits the query j before ending the search

$$\pi_j = \sum_{l=j}^{k} h^l = 1 - \sum_{l=1}^{j-1} h^l.$$

To evaluate the "expected distance" from state $(1, 1)$ for the documents in query j, we define the following family of stopping times for any $i \leq N$, since the search does not end before this query j

$$H^{(i,j)} = \inf\{n \geq 1 : X_n = (i, j)\}.$$

These stopping times allow us to evaluate how long it takes to reach the document at depth i in query j, and these values are used to perform the average inside the columns.

Thus, given that the search does not end before document (i, j), we define the weight at position (i, j) as

$$e(i, j) = \mathbb{E}_{(1,1)}[H^{(i,j)}] = \mathbb{E}[H^{(i,j)}|X_1 = (1, 1)]. \tag{1}$$

To evaluate the contribution of the j-th query to the multi-session search, we compute

$$E(j) = \sum_{i=1}^{N} \phi(e(i, j))\ GT(d_{i,j}) \tag{2}$$

where $GT(d_{i,j}) \in \mathbb{N}_0$ is the gain corresponding to document $d_{i,j}$ (0 for not relevant documents) and the discount function ϕ is a positive, monotone real function. Choosing it decreasing we discount the relevance of the documents and queries far from the top (primacy according to Liu et al. [18]), while choosing it increasing we give more weight to the relevance of those documents and queries (recency according to Liu et al.). Examples of the function ϕ are: *reciprocal linear weight*, i.e. $\phi(x) = \frac{1}{x}$; *reciprocal logarithmic weight*, i.e. $\phi(x) = \frac{1}{1+\log_{10}(x)}$; and, *logarithmic weight*, i.e. $\phi(x) = 1 + \log_{10}(x)$.

Finally, the new *Markov Session Measure* (MsM) combines the contribution of the k queries in a search session as

$$MsM = \sum_{j=1}^{k} \pi_j E(j). \tag{3}$$

Overall, MsM expresses the expectation of the stochastic time $E(j)$, i.e. number visited documents, it takes for a user to accumulate gain during the search, monotonically transformed by a weighting function ϕ which can put more emphasis either on the start or the end of the search, weighted by the probability π of actually continuing to query.

4 Experimental Setup

In this section, we evaluate the behavior of the proposed measure, answering the following research questions:

RQ1. How does MsM compare to existing session evaluation measures regarding the ranking of retrieval systems?

RQ2. Which factors of the user dynamics affect these correlations and to what extent?

Computation of the MsM Measure. We developed an efficient way of computing the MsM measure, avoiding the most general and immediate approach of using a large block-diagonal matrix, where each sub-matrix would represent a single query in the session. For space reasons the pseudo-code of the algorithm is omitted here but it is available in the electronic appendix available as Online Resource 1. Moreover, to further ease the reproducibility of experiments, the source code of the actual implementation is available at: https://github.com/ekanou/Markovian-Session-Measures.

Data Collection. To answer RQ1 and RQ2 we ran experiments on the TREC 2015, 2016, and 2017 Dynamic Domain Track (DDT) collection. The search tasks in DDT focus in domains of special interests, which usually produce complex and exploratory searches with multiple rounds of user and search engine interactions. The DDT collection consists of a set of topics, and multi-query sessions corresponding to each topic. In DDT retrieval systems were provided with the first query, they returned a ranked list of 5 documents, and based on passage annotations in these documents, a jig (user simulation) returned a follow-up query. IR systems had the chance to decide when to stop providing users with ranked lists of documents.

Session Evaluation Measures. We compare MsM to the normalized [24] versions of session DCG (sDCG) [14], Expected Utility (EU) [28], and Cube Test (CT) [19]. Since we are not dealing with diversity, we simplify them by using a gain function that ignores subtopic relevance.

Model Instantiations. As an exemplification for experimentation purposes, we consider two user's models, with two different set of assumptions. The first model, called **Random-Walk model**, assumes a user who after considering a document she decides, according to constant probabilities, to proceed to the next document (p), to the previous document (q), to stop her search (s), or to reformulate a new query (r). From the transition matrix point of view, this model is determined by the following assumptions:

$$
\begin{aligned}
p_{(i,j),(i+1,j)} &= p \text{ if } i \neq F \\
p_{(i,j),(i-1,j)} &= q \text{ if } i \neq 1 \\
p_{(i,j),(1,j+1)} &= r \text{ if } i \neq F \\
p_{(i,j),(F,j)} &= s \text{ if } i \neq F \\
p_{(F,j),(F,j)} &= 1 \text{ for any } j
\end{aligned}
$$

where $p + q + r + s = 1$ and $p > 0$, $q \geq 0$, $r > 0$ and $s > 0$.

The second one, called **Forward model**, is a special case of the first one, inspired by the RBP philosophy, where the backward probability, q, is set to 0, i.e. it assumes that the user moves only forward in the ranked list.

Experiments. To answer RQ1 we experimented with both the Forward and the Random-Walk model, using reciprocal log and linear weight, introduced earlier, while the probabilities p, r, and s were set to values on a grid in $[0,1)^3$ with a step of 0.05, under the constraint that $p+r+s=1$. Given that the results of the Forward and Random-walk model were highly correlated, and due to space limitations, we present results only of the Forward model.

To answer RQ2 we experiment with both the Forward and Random-Walk model, while we factorize user types by three characteristics: (a) *patience*, in terms of the total number of documents they are willing to examine, (b) *browsing pattern* in terms of whether they prefer to scan the ranked list or reformulate, and (c) *decisiveness* in terms of deciding whether a document is relevant once they observe it, or moving back to it and re-examining it after they have examined more documents. We control patience by setting the stopping probability, s, to three distinct values, 0.01, 0.1, and 0.3; the first type of user will on average view around 50 documents, the second around 9, and the third around 3, before they quit their search. We control the browsing pattern by setting the probabilities of walking down the ranked list, p, and reformulating, r to a set of values, such that the user either demonstrates a bigger willingness to scan the ranked list, to reformulate, or to have a balanced behaviour. Last, we control decisiveness by either not allowing the user to walk backwards, hence setting the backwards probability, q, to 0, or allowing the user to do so, by setting q to 0.1. [1]

5 Results and Analysis

5.1 RQ1 – Correlation Analysis

To answer RQ1, we conduct a correlation analysis using Kendall's τ [16] among the rankings of systems produced by the different evaluation measures. Ferro [11] has shown that, even if the absolute correlation values are different, removing or not the lower quartile runs produces the same ranking of measures in terms of correlation; similarly, it was shown that both τ and AP correlation τ_{ap} [29] produce the same ranking of measures in terms of correlation. Therefore, we focus only on Kendall's τ without removing lower quartile systems.

Figure 1 presents the average Kendall's τ correlation between different instantiations of MsM measure and sDCG, EU and CT, respectively, on rankings of systems in DDT. The x-axis in all three plots corresponds to the forward probability, p, while the y-axis to the reformulation probability, r. The stopping probability, s, can be inferred, given that $p+r+s=1$. The colorbar shows the actual correlation values.

Figure 1a corresponds to the correlation with sDCG. It can be observed that the highest correlation is achieved along the secondary diagonal, i.e. when the stopping probability is 0.05, with the maximum value obtained when p is 0.55, r is 0.40 and s is 0.05. This shows that the browsing model of sDCG penalizes

[1] Advanced user dynamics that condition probabilities on the relevance of the viewed document, similar to ERR, are also possible with MsM but are left as future work.

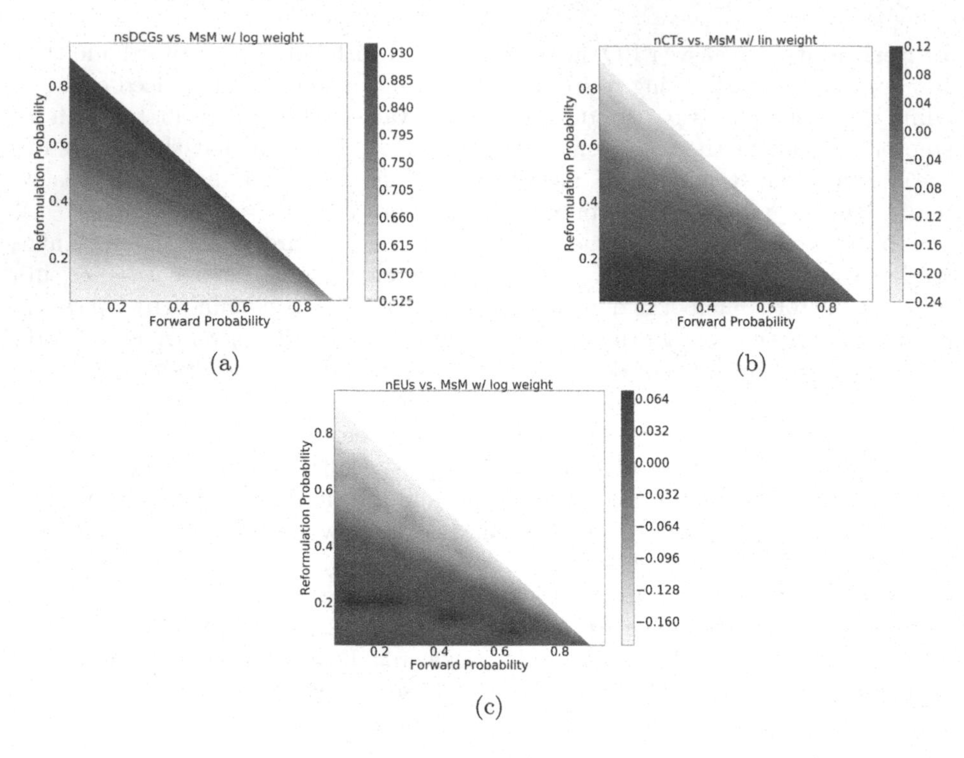

Fig. 1. Average Kendall's τ correlation.

documents both lower in the ranking and further in the session queries. Figure 1b corresponds to the correlation with CT. It can be observed that the highest correlation is achieved along the x-axis, i.e. when the reformulation probability is 0.05, with the maximum value obtained when p is 0.10, r is 0.05, and s is 0.85. As a reminder CT does not penalize documents that appear lower in the ranking; it only penalizes documents that appear further in the session, with a reciprocal linear weight of the index of the query in the session. This is captured by the plot: the high forward probability essentially dictates little penalization within a ranking, while the low reformulation probability dictates a high penalization across queries in the session. The overall low correlation (0.12) however also designates that the penalization model of CT can be hardly modeled in a probabilistic manner. Figure 1c corresponds to the correlation with EU. The highest correlation is achieved when p is 0.10, r is 0.20 and s is 0.70. The plot demonstrates a pattern of high correlations that is in between the high correlation patterns of sDCG and CT. The high correlation at low reformulation probabilities also shows that EU expects a user to move forward a ranked list and reformulate only at the end of it. In conclusion, to some extent, the MsM measure provides some insights on the implicit user models of existing measures, even if some of the assumptions made in those measures do not always allow high correlation scores.

5.2 RQ2 – Analysis of Variance

To answer RQ2, we conduct an Analysis of Variance (ANOVA) of the different factors that may influence the correlation between MsM and existing session evaluation measures. For space reasons, we report this analysis in the case of sDCG and EU, being possible to draw similar conclusions also in the case of CT.

Table 1. Analysis of the factors influencing correlation with sDCG.

Source	SS	DF	MS	F	p	$\hat{\omega}^2_{\langle fact \rangle}$
Track	0.1526	2	0.0763	5.3294	0.0083	0.1382
Patience	0.2712	2	0.1356	9.4704	0.0003	0.2388
Browsing	0.4330	2	0.2165	15.1245	<e–4	0.3435
Weight	0.0014	1	0.0014	0.1004	0.7528	–
Error	0.6582	46	0.0143			
Total	1.5167	53				

Table 2. Analysis of the factors influencing correlation with EU.

Source	SS	DF	MS	F	p	$\hat{\omega}^2_{\langle fact \rangle}$
Track	0.0084	2	0.0042	0.3962	0.6752	–
Patience	0.0255	2	0.0127	1.2022	0.3098	–
Browsing	0.9185	2	0.4593	43.3024	<e–4	0.6104
Weight	0.0057	1	0.0057	0.5333	0.4689	–
Error	0.4879	46	0.0106			
Total	1.4459	53				

Table 1 shows the three-way ANOVA for analysing the factors in the MsM measures which influence the correlation with nsDCGs. The Track factor represents the effect of one of the three tracks (DD 2015, 2016, and 2017); the Patience factor represents the effect of the patience of the user in scanning the list (impatient, balanced, patient); the Browsing factor represents the attitude to walk down the list or reformulate new queries (down, balanced, reformulate); the Weight factor represents the type of discount (linear, log). The ANOVA analysis shows that the Track, Patience, and Browsing factors are statistically significant while the Weight one is not; we also conducted an ANOVA analysis (not reported here for space reason) to test the interaction among these factors but none of them is significant. The Tukey Honestly Significant Difference (HSD) test shows that the `impatient` user is significantly different from the `balanced` and `patient` ones, which are not significantly different from each other, being the `impatient` user the lowest one in terms of correlation and the `patient` the highest one. The Tukey HSD test also shows that the `balanced` browsing pattern is significantly different from the `down` and `reformulate` ones, which are not significantly different from each other, being the `reformulate` strategy the lowest one in terms of correlation and the `balanced` the highest one. The Strength of Association (SOA) ω^2 shows that the Track factor is a medium-size effect while the Patience and Browsing factors are large-size effects, being the browsing pattern the most prominent one. Overall, this analysis suggests that the most prominent motivations of similarity between MsM measures and sDCG are a balanced browsing pattern and a balanced/patient user, which is the user model actually implemented in sDCG.

Table 2 shows the three-way ANOVA for analysing the factors in the MsM measures which influence the correlation with EU. The ANOVA analysis shows that only the Browsing factor is statistically significant while all the others are not; we also conducted an ANOVA analysis (not reported here for space reason) to test the interaction among these factors but none of them is significant. The Tukey HSD test shows that all the browsing patterns are significantly different, being the `balanced` strategy the lowest one in terms of correlation and the `down` the highest one. The SOA ω^2 shows that the Browsing factor is a large-size effect. Overall, this analysis suggests that the most prominent motivation of similarity between MsM measures and EU is a down browsing pattern.

Overall, the ANOVA analysis also highlights that the Track factor, when significant, is not the most influencing one and this supports the previous observation about a consistent behaviour of the measures across the tracks and reporting the correlation values averages across tracks.

6 Conclusions and Future Work

We considered the problem of evaluating multi-query sessions. Differently from past attempts we provided a mathematical formulation of the user dynamics on the basis of a Markov chain that allows for a strong theoretical underpinning of the deduced measure. The measure proposed provides a flexible but at the same time mathematically sound and intuitive parametrization on the basis of the expected user behavior. We experimented with different variations of the measure each making its own assumption regarding (a) the chance of the user to return to an already seen document in a ranked list; (b) the patience of the user to move down in a ranked list as opposed to reformulating her query; and, (c) the patience of the user in the overall use of the information retrieval system. We showed that the produced measures can indeed capture different user behaviors, and through a correlation analysis we attempted to provide a better understanding of existing session measures and the implicit assumptions in their user models.

What we present in this work is a rather flexible framework to construct session evaluation measures of interest. A number of future directions could be explored: (a) identifying the right parameters that will reduce the proposed MsM to existing session measures, providing a theoretical underpinning of those measures and better expandability; (b) injecting more advanced user dynamics in the MsM by e.g. modeling transition probabilities as conditional probabilities on the relevance of the visited documents; (c) learning parameters using query logs or leveraging user studies; and, (d) expanding the discrete Markov chain to a continuous-time Markov chain to naturally incorporate time in the measure.

Acknowledgements. This research was supported by the NWO Innovational Research Incentives Scheme Vidi (016.Vidi.189.039). All content represents the opinion of the authors, which is not necessarily shared or endorsed by their respective employers and/or sponsors.

References

1. Allan, J., et al.: Research frontiers in information retrieval - report from the third strategic workshop on information retrieval in Lorne (SWIRL 2018). SIGIR Forum **52**(1), 34–90 (2018)
2. Amigó, E., Fang, H., Mizzaro, S., Zhai, C.: Report on the SIGIR 2017 workshop on axiomatic thinking for information retrieval and related tasks (ATIR). SIGIR Forum **51**(3), 99–106 (2017)
3. Busin, L., Mizzaro, S.: Axiometrics: an axiomatic approach to information retrieval effectiveness metrics. In: Kurland, O., Metzler, D., Lioma, C., Larsen, B., Ingwersen, P. (eds.) Proceedings of the 4th International Conference on the Theory of Information Retrieval (ICTIR 2013), pp. 22–29. ACM Press, New York (2013)
4. Carterette, B.: System effectiveness, user models, and user utility: a conceptual framework for investigation. In: Proceedings of the 34th International ACM SIGIR Conference on Research and Development in Information Retrieval, SIGIR 2011, pp. 903–912. ACM, New York (2011). https://doi.org/10.1145/2009916.2010037
5. Chapelle, O., Metzler, D., Zhang, Y., Grinspan, P.: Expected reciprocal rank for graded relevance. In: Cheung, D.W.L., Song, I.Y., Chu, W.W., Hu, X., Lin, J.J. (eds.) Proceedings of the 18th International Conference on Information and Knowledge Management (CIKM 2009), pp. 621–630. ACM Press, New York (2009)
6. Chierichetti, F., Kumar, R., Raghavan, P.: Optimizing two-dimensional search results presentation. In: King, I., Nejdl, W., Li, H. (eds.) Proceedings of the 4th ACM International Conference on Web Searching and Data Mining (WSDM 2011), pp. 257–266. ACM Press, New York (2011)
7. Collins-Thompson, K., Callan, J.: Query expansion using random walk models. In: Herzog, O., Schek, H.J., Fuhr, N., Chowdhury, A., Teiken, W. (eds.) Proceedings of 14th International Conference on Information and Knowledge Management (CIKM 2005), pp. 704–711. ACM Press, New York (2005)
8. Daniłowicz, C., Baliński, J.: Document ranking based upon Markov chains. Inf. Process. Manag. **37**(4), 623–637 (2001)
9. Ferrante, M., Ferro, N., Pontarollo, S.: A general theory of IR evaluation measures. IEEE Trans. Knowl. Data Eng. (TKDE). **31**(3), 409–422 (2019)
10. Ferrante, M., Ferro, N., Maistro, M.: Injecting user models and time into precision via Markov chains. In: Proceedings of the 37th International ACM SIGIR Conference on Research and Development in Information Retrieval, SIGIR 2014, pp. 597–606. ACM, New York (2014). https://doi.org/10.1145/2600428.2609637

11. Ferro, N.: What does affect the correlation among evaluation measures? ACM Trans. Inf. Syst. (TOIS) **36**(2), 19:1–19:40 (2017)
12. Fuhr, N.: Salton award lecture: information retrieval as engineering science. SIGIR Forum **46**(2), 19–28 (2012)
13. Järvelin, K., Kekäläinen, J.: Cumulated gain-based evaluation of IR techniques. ACM Trans. Inf. Syst. (TOIS) **20**(4), 422–446 (2002)
14. Järvelin, K., Price, S.L., Delcambre, L.M.L., Nielsen, M.L.: Discounted cumulated gain based evaluation of multiple-query IR sessions. In: Macdonald, C., Ounis, I., Plachouras, V., Ruthven, I., White, R.W. (eds.) ECIR 2008. LNCS, vol. 4956, pp. 4–15. Springer, Heidelberg (2008). https://doi.org/10.1007/978-3-540-78646-7_4. http://dl.acm.org/citation.cfm?id=1793274.1793280
15. Kanoulas, E., Carterette, B., Clough, P.D., Sanderson, M.: Evaluating multi-query sessions. In: Proceedings of the 34th International ACM SIGIR Conference on Research and Development in Information Retrieval, SIGIR 2011, pp. 1053–1062. ACM, New York (2011). https://doi.org/10.1145/2009916.2010056
16. Kendall, M.G.: Rank Correlation Methods. Griffin, Oxford (1948)
17. Lafferty, J., Zhai, C.: Document language models, query models, and risk minimization for information retrieval. In: Kraft, D.H., Croft, W.B., Harper, D.J., Zobel, J. (eds.) Proceedings of the 24th Annual International ACM SIGIR Conference on Research and Development in Information Retrieval (SIGIR 2001), pp. 111–119. ACM Press, New York (2001)
18. Liu, M., Liu, Y., Mao, J., Luo, C., Ma, S.: Towards designing better session search evaluation metrics. In: Collins-Thompson, K., Mei, Q., Davison, B., Liu, Y., Yilmaz, E. (eds.) Proceedings of the 41th Annual International ACM SIGIR Conference on Research and Development in Information Retrieval (SIGIR 2018), pp. 1121–1124. ACM Press, New York (2018)
19. Luo, J., Wing, C., Yang, H., Hearst, M.: The water filling model and the cube test: multi-dimensional evaluation for professional search. In: Proceedings of the 22nd ACM International Conference on Information and Knowledge Management, CIKM 2013, pp. 709–714. ACM, New York (2013). https://doi.org/10.1145/2505515.2523648
20. Maxwell, K.T., Croft, W.B.: Compact query term selection using topically related text. In: Jones, G.J.F., Sheridan, P., Kelly, D., de Rijke, M., Sakai, T. (eds.) Proceedings of the 36th Annual International ACM SIGIR Conference on Research and Development in Information Retrieval (SIGIR 2013), pp. 583–592. ACM Press, New York (2013)
21. Moffat, A., Zobel, J.: Rank-biased precision for measurement of retrieval effectiveness. ACM Trans. Inf. Syst. **27**(1), 2:1–2:27 (2008). https://doi.org/10.1145/1416950.1416952
22. Smucker, M.D., Clarke, C.L.A.: Stochastic simulation of time-biased gain. In: Proceedings of the 21st ACM International Conference on Information and Knowledge Management, CIKM 2012, pp. 2040–2044. ACM, New York (2012). https://doi.org/10.1145/2396761.2398568
23. Smucker, M.D., Clarke, C.L.: Time-based calibration of effectiveness measures. In: Proceedings of the 35th International ACM SIGIR Conference on Research and Development in Information Retrieval, SIGIR 2012, pp. 95–104. ACM, New York (2012). https://doi.org/10.1145/2348283.2348300
24. Tang, Z., Yang, G.H.: Investigating per topic upper bound for session search evaluation. In: Proceedings of the ACM SIGIR International Conference on Theory of Information Retrieval, ICTIR 2017, pp. 185–192. ACM, New York (2017). https://doi.org/10.1145/3121050.3121069

25. Yan, X., Gao, G., Su, X., Wei, H., Zhang, X., Lu, Q.: Hidden Markov model for term weighting in verbose queries. In: Catarci, T., Forner, P., Hiemstra, D., Peñas, A., Santucci, G. (eds.) CLEF 2012. LNCS, vol. 7488, pp. 82–87. Springer, Heidelberg (2012). https://doi.org/10.1007/978-3-642-33247-0_10
26. Yang, H., Frank, J., Soboroff, I.: TREC 2015 dynamic domain track overview. In: The Twenty-Forth Text REtrieval Conference (TREC 2015) Proceedings, Gaithersburg, Maryland (2016)
27. Yang, G.H., Soboroff, I.: TREC 2016 dynamic domain track overview. In: Proceedings of The Twenty-Fifth Text REtrieval Conference, TREC 2016, Gaithersburg, 15–18 November 2016
28. Yang, Y., Lad, A.: Modeling expected utility of multi-session information distillation. In: Azzopardi, L., et al. (eds.) ICTIR 2009. LNCS, vol. 5766, pp. 164–175. Springer, Heidelberg (2009). https://doi.org/10.1007/978-3-642-04417-5_15
29. Yilmaz, E., Aslam, J.A., Robertson, S.E.: A new rank correlation coefficient for information retrieval. In: Chua, T.S., Leong, M.K., Oard, D.W., Sebastiani, F. (eds.) Proceedings of the 31st Annual International ACM SIGIR Conference on Research and Development in Information Retrieval (SIGIR 2008), pp. 587–594. ACM Press, New York (2008)

Correlation, Prediction and Ranking of Evaluation Metrics in Information Retrieval

Soumyajit Gupta[1], Mucahid Kutlu[2(✉)], Vivek Khetan[1], and Matthew Lease[1]

[1] University of Texas at Austin, Austin, TX, USA
{smjtgupta,vivek.khetank,ml}@utexas.edu
[2] TOBB University of Economics and Technology, Ankara, Turkey
m.kutlu@etu.edu.tr

Abstract. Given limited time and space, IR studies often report few evaluation metrics which must be carefully selected. To inform such selection, we first quantify correlation between 23 popular IR metrics on 8 TREC test collections. Next, we investigate prediction of unreported metrics: given 1–3 metrics, we assess the best predictors for 10 others. We show that accurate prediction of MAP, P@10, and RBP can be achieved using 2–3 other metrics. We further explore whether high-cost evaluation measures can be predicted using low-cost measures. We show RBP(p = 0.95) at cutoff depth 1000 can be accurately predicted given measures computed at depth 30. Lastly, we present a novel model for ranking evaluation metrics based on covariance, enabling selection of a set of metrics that are most informative and distinctive. A *greedy-forward* approach is guaranteed to yield sub-modular results, while an *iterative-backward* method is empirically found to achieve the best results.

Keywords: Evaluation · Metric · Prediction · Ranking

1 Introduction

Given the importance of assessing IR system accuracy across a range of different search scenarios and user needs, a wide variety of evaluation metrics have been proposed, each providing a different view of system effectiveness [6]. For example, while *precision@10* (P@10) and *reciprocal rank* (RR) are often used to evaluate the quality of the top search results, *mean average precision* (MAP) and *rank-biased precision* (RBP) [32] are often used to measure the quality of search results at greater depth, when recall is more important. Evaluation tools such as `trec_eval` compute many more evaluation metrics than IR researchers typically have time or space to analyze and report. Even for knowledgeable researchers with ample time, it can be challenging to decide which small subset of IR metrics

M. Kutlu—Work began while at Qatar University.

L. Azzopardi et al. (Eds.): ECIR 2019, LNCS 11437, pp. 636–651, 2019.
https://doi.org/10.1007/978-3-030-15712-8_41

should be reported to best characterize a system's performance. Since a few metrics cannot fully characterize a system's performance, information is effectively lost in publication, complicating comparisons to prior art.

To compute an unreported metric of interest, one strategy is to reproduce prior work. However, this is often difficult (and at times impossible), as the description of a method is often incomplete and even shared source code can be lost over time or difficult or impossible for others to run as libraries change. Sharing system outputs would also enable others to compute any metric of interest, but this is rarely done. While Armstrong *et al.* [2] proposed and deployed a central repository for hosting system runs, their proposal did not achieve broad participation from the IR community and was ultimately abandoned.

Our work is inspired in part by work on biomedical literature mining [8,23], where acceptance of publications as the most reliable and enduring record of findings has led to a large research community investigating automated extraction of additional insights from the published literature. Similarly, we investigate the viability of predicting unreported evaluation metrics from reported ones. We show accurate prediction of several important metrics is achievable, and we present a novel ranking method to select metrics that are informative and distinctive.

Contributions of our work include:

- We analyze correlation between 23 IR metrics, using more recent collections to complement prior studies. This includes *expected reciprocal rank* (ERR) and RBP using graded relevance; key prior work used only binary relevance.
- We show that accurate prediction of a metric can be achieved using only $2-3$ other metrics, using a simple linear regression model.
- We show accurate prediction of some high-cost metrics given only low-cost metrics (*e.g.* predicting RBP@1000 given only metrics at depth 30).
- We introduce a novel model for ranking top metrics based on their covariance. This enables us to select the best metrics from clusters with lower time and space complexity than required by prior work. We also provide a theoretical justification for metric ranking which was absent from prior work.
- We share[1] our source code, data, and figures to support further studies.

2 Related Work

Correlation between Evaluation Metrics. Tague-Sutcliffe and Blustein [45] study 7 measures on TREC-3 and find R-Prec and AP to be highly correlated. Buckley and Voorhees [10] also find strong correlation using Kendall's τ on TREC-7. Aslam *et al.* [5] investigate why RPrec and AP are strongly correlated. Webber *et al.* [51] show that reporting simple metrics such as P@10 with complex metrics such as MAP and DCG is redundant. Baccini *et al.* [7] measure correlations between 130 measures using data from the TREC-(2-8) ad hoc task,

[1] https://github.com/smjtgupta/IR-corr-pred-rank.

grouping them into 7 clusters based on correlation. They use several machine learning tools including Principal Component Analysis (PCA) and Hierarchical Clustering Analysis (HCA) and report the metrics in particular clusters.

Sakai [41] compares 14 graded-level and 10 binary level metrics using three different data sets from NTCIR. Correlation between P($^+$)-measure, O-measure, and normalized weighted RR shows that they are highly correlated [40]. Correlation between precision, recall, fallout and miss has also been studied [19]. In addition, the relationship between F-measure, break-even point, and 11-point averaged precision has been explored [26]. Another study [46] considers correlation between 5 evaluation measures using TREC Terabyte Track 2006. Jones *et al.* [28] examine disagreement between 14 evaluation metrics including ERR and RBP using TREC-(4-8) ad hoc tasks, and TREC Robust 2005–2006 tracks. However, they use only binary relevance judgments, which makes ERR identical to RR, whereas we consider graded relevance judgments. While their study considered TREC 2006 Robust and Terabyte tracks, we complement this work by considering more recent TREC test collections (*i.e.* Web Tracks 2010–2014), with some additional evaluation measures as well.

Predicting Evaluation Metrics. While Aslam *et al.* [5] propose predicting evaluation measures, they require a corresponding retrieved ranked list as well as another evaluation metric. They conclude that they can accurately infer user-oriented measures (*e.g.* P@10) from system-oriented measures (*e.g.* AP, R-Prec). In contrast, we predict each evaluation measure given only other evaluation measures, without requiring the corresponding ranked lists.

Reducing Evaluation Cost. Lu *et al.* [29] consider risks arising with fixed-depth evaluation of recall/utility-based metrics in terms of providing a fair judgment of the system. They explore the impact of evaluation depth on truncated evaluation metrics and show that for recall-based metrics, depth plays a major role in system comparison. In general, researchers have proposed many methods to reduce the cost of creating test collections: new evaluation measures and statistical methods for incomplete judgments [3,9,39,52,53], finding the best sample of documents to be judged for each topic [11,18,27,31,37], topic selection [21,24,25,30], inferring some relevance judgments [4], evaluation without any human judgments [34,44], crowdsourcing [1,20], and others. We refer readers to [33] and [42] for detailed review of prior work for low-cost IR evaluation.

Ranking Evaluation Metrics. Selection of IR evaluation metrics from clusters has been studied previously [7,41,51]. Our methods incur lower cost than these. We further provide a theoretical basis to rank the metrics using the proposed determinant of covariance criteria, which prior work omitted as an experimental procedure, or by inferring results using existing statistical tools. Our ranking work is most closely related to Sheffield [43], which introduced the idea of unsupervised ranking of features in high-dimensional data using the covariance information of the feature space. This enables selection and ranking of features that are highly informative yet less correlated with one another.

3 Experimental Data

To investigate correlation and prediction of evaluation measures, we use runs and relevance judgments from TREC 2000–2001 & 2010–2014 Web Tracks (WT) and the TREC-2004 Robust Track (RT) [48]. We consider only *ad hoc* retrieval. We calculate 9 evaluation metrics: AP, bpref [9], ERR [12], nDCG, P@K, RBP [32], *recall* (R), RR [50], and R-Prec. We use various cut-off thresholds for the metrics (e.g. P@10, R@100). Unless stated, we set the cut-off threshold to 1000. The cut-off threshold for ERR is set to 20 since this was an official measure in WT2014 [17]. RBP uses a parameter p representing the probability of a user proceeding to the next retrieved page. We test $p = \{0.5, 0.8, 0.95\}$, the values explored by Moffat and Zobel [32]. Using these metrics, we generate two datasets.

Topic-Wise (TW) Dataset: We calculate each metric above for each system for each separate topic. We use $10, 20, 100, 1000$ cut-off thresholds for AP, nDCG, P@K and R@K. In total, we calculate 23 evaluation metrics.

System-Wise (SW) Dataset: We calculate the metrics above (and GMAP as well as MAP) for each system, averaging over all topics in each collection.

4 Correlation of Measures

We begin by computing Pearson correlation between 23 popular IR metrics using 8 TREC test collections. We report correlation of measures for the more difficult TW dataset in order to model score distributions without the damping effect of averaging scores across topics. More specifically, we calculate Pearson correlation between measures across different topics. We make the following observations from the results shown in Fig. 1.

- R-Prec has high correlation with bpref, MAP and nDCG@100 [5,10,45].
- RR is strongly correlated with RBP(p = 0.5), decreasing as its p parameter increases (while RR always stops with the first relevant document, RBP becomes more of a deep-rank metric as p increases). That said, later Fig. 2 shows accurate prediction of RBP($p = 0.95$) even with low-cost metrics.
- nDCG@20, one of the official metrics of WT2014, is highly correlated with RBP(p = 0.8), connecting with Park and Zhang's [36] noting p = 0.78 is appropriate for modeling web user behavior.
- nDCG is highly correlated with MAP and R-Prec, and its correlation with R@K consistently increases as K increases.
- P@10 (ρ = 0.97) and P@20 (ρ = 0.98) are most correlated with RBP (p = 0.8) and RBP(p = 0.95), respectively.
- Sakai and Kando [38] report that RBP(0.5) essentially ignores relevant documents below rank 10. Our results are consistent: we see maximum correlation between RBP(0.5) and nDCG@K at K = 10, decreasing as K increases.
- P@1000 is the least correlated with other metrics, suggesting that it captures a different effectiveness measure of IR systems than other metrics.

While a varying degree of correlation exists between many measures, this should not be interpreted to mean that measures are redundant and trivially exchangeable. Correlated metrics can still correspond to different search scenarios and user needs, and the desire to report effectiveness across a range of potential use cases is challenged by limited time and space for reporting results. In addition, showing two metrics are uncorrelated shows only that each captures a different aspect of system performance, and not whether each aspect is equally important or even relevant to a given evaluation scenario on interest.

Test Set	Document Set	#Sys	Topics
WT2000 [22]	WT10g	105	451-500
WT2001 [49]	WT10g	97	501-550
RT2004 [48]	TREC 4&5*	110	301-450, 601-700
WT2010 [14]	ClueWeb'09	55	51-99
WT2011 [13]	ClueWeb'09	62	101-150
WT2012 [15]	ClueWeb'09	48	151-200
WT2013 [16]	ClueWeb'12	59	201-250
WT2014 [17]	ClueWeb'12	30	251-300

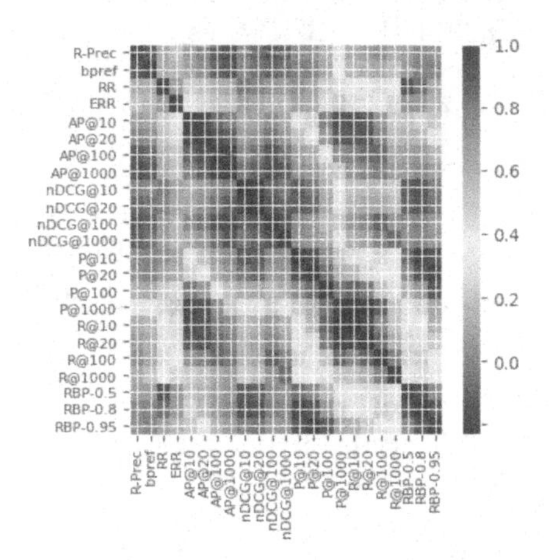

Fig. 1. Left: TREC collections used. *RT2004 excludes the congressional record. **Right:** Pearson correlation coefficients between 23 Metrics. Deep green entries indicate strong correlation, while red entries indicate low correlation. (Color figure online)

5 Prediction of Metrics

In this section, we describe our prediction model and experimental setup, and we report results of our experiments to investigate prediction of evaluation measures. Given the correlation matrix, we can identify the correlated groups of metrics. The task of predicting an independent metric m_i using some other dependent metrics m_d under a linear regression model is $m_i = \sum_{k=1}^{K} \alpha^k m_d^k$.

Because a non-linear relationship could also exist between two correlated metrics, we also tried using a radial basis function (RBF) Support Vector Machine (SVM) for the same prediction. However, the results were very similar, hence not reported. We further discuss this at the end of the section.

Model & Experimental Setup. To predict a system's missing evaluation measures using reported ones, we build our model using only the evaluation measures of systems as features. We use the SW dataset in our experiments for prediction because studies generally report their average performance over a set of topics, instead of reporting their performance for each topic. Training data combines WT2000-01, RT2004, WT2010-11. Testing is performed separately on WT2012, WT2013, and WT2014, as described below. To evaluate prediction accuracy, we report coefficient of determination R^2 and Kendall's τ correlation.

Table 1. System-wise prediction of a metric using varying number of metrics K=[1−3]. Kendall's τ scores higher than 0.9 are bolded.

Predicted metric	Independent variables			WT2012		WT2013		WT2014	
				τ	R^2	τ	R^2	τ	R^2
bpref	nDCG	-	-	0.805	−0.693	0.885	0.079	**0.915**	−1.174
	nDCG	R-Prec	-	0.872	−0.202	0.850	0.094	0.824	−0.989
	nDCG	R-Prec	R@100	**0.906**	0.284	0.844	0.645	0.866	0.390
ERR	RR	-	-	0.764	−1.874	0.734	0.293	0.704	−1.004
	RR	RBP(0.8)	-	0.790	−1.809	0.777	0.392	0.714	−0.686
	RR	RBP(0.8)	R@100	0.796	−1.728	0.741	0.478	0.704	−0.473
GMAP	bpref	-	-	0.729	−1.216	0.704	−2.982	0.739	−1.034
	nDCG	RBP(0.5)	-	0.817	0.877	0.777	0.600	0.767	0.818
	nDCG	RBP(0.95)	RR	0.817	0.882	0.748	0.514	0.794	0.854
MAP	R-Prec	-	-	0.885	0.754	0.824	0.667	**0.952**	0.819
	R-Prec	nDCG	-	**0.904**	0.894	**0.905**	0.760	**0.958**	0.897
	R-Prec	nDCG	RR	**0.924**	0.916	**0.901**	0.779	**0.947**	0.922
nDCG	bpref	-	-	0.805	−2.101	0.885	−0.217	**0.915**	−2.008
	bpref	GMAP	-	0.803	−0.079	0.809	0.574	0.872	0.024
	bpref	GMAP	RBP(0.95)	0.794	−0.113	0.801	0.556	0.850	−0.032
P@10	RBP(0.8)	-	-	0.884	0.942	0.832	0.895	0.866	0.893
	RBP(0.8)	RBP(0.5)	-	**0.941**	0.994	0.882	0.966	**0.914**	0.988
	RBP(0.8)	RBP(0.5)	RR	**0.946**	0.994	0.885	0.968	**0.914**	0.987
RBP(0.95)	R-Prec	-	-	0.824	0.346	0.651	−0.786	0.607	−2.401
	bpref	P@10	-	**0.911**	0.952	0.718	0.873	0.728	0.591
	bpref	P@10	RBP(0.8)	**0.911**	0.967	0.720	0.868	0.744	0.639
R-Prec	R@100	-	-	0.899	0.708	0.871	0.624	**0.935**	0.019
	R@100	RBP(0.95)	-	**0.909**	0.952	0.820	0.882	0.820	0.759
	R@100	RBP(0.95)	GMAP	**0.924**	0.970	0.833	0.914	0.841	0.825
RR	RBP(0.5)	-	-	0.782	0.904	0.806	0.927	0.810	0.878
	RBP(0.5)	RBP(0.8)	-	0.869	0.918	0.809	0.919	0.820	0.942
	RBP(0.5)	RBP(0.8)	ERR	0.876	0.437	0.818	0.924	**0.915**	0.824
R@100	R-Prec	-	-	0.899	0.423	0.871	0.232	**0.935**	−1.075
	R-Prec	GMAP	-	0.899	0.433	0.871	0.238	**0.940**	−1.077
	R-Prec	RR	ERR	0.881	−0.104	0.823	0.355	**0.935**	−1.187

Results (Table 1). We investigate the best predictors for 10 metrics: R-Prec, bpref, RR, ERR@20, MAP, GMAP, nDCG, P@10, R@100, RBP(0.5), RBP(0.8) and RBP(0.95). We investigate which K evaluation metric(s) are the best predictors for a particular metric, varying K from $1-3$. Specifically, in prediction of a particular metric, we try all combinations of size K using the remaining 11 evaluation measures on WT2012 and pick the one that yields the best Kendall's τ correlation. Then, this combination of metrics is used to predict the respective metric separately for WT2013 and WT2014. Kendall's τ scores higher than 0.9 are bolded (a traditionally-accepted threshold for correlation [47]).

bpref: We achieve the highest τ correlation and interestingly the worst R^2 using only nDCG on WT2014. This shows that while predicted measures are not accurate, rankings of systems based on predicted scores can be highly correlated with the actual ranking. We observe the same pattern of results in prediction of RR on WT2012 and WT2014, R-prec on WT2013 and WT2014, R@100 on WT2013, and nDCG in all three test collections.

GMAP & ERR: Both seem to be the most challenging measures to predict because we could never reach $\tau = 0.9$ correlation in any of the prediction cases of these two measures. Initially, R^2 scores for ERR consistently increase in all three test collections as we use more evaluation measures for prediction, suggesting that we can achieve higher prediction accuracy using more independent variables.

MAP: We can predict MAP with very high prediction accuracy and achieve higher than $\tau = 0.9$ correlation in all three test collections using R-Prec and nDCG as predictors. When we use RR as the third predictor, R^2 increases in all cases and τ correlation slightly increases on average (0.924 vs. 0.922).

nDCG: Interestingly, we achieve the highest τ correlations using only bpref; τ decreases as more evaluation measures are used as independent variables. Even though we reach high τ correlations for some cases (e.g. 0.915 τ on WT2014 using only bpref), nDCG seems to be one of the hardest measures to predict.

P@10: Using RBP(0.5) and RBP(0.8), which are both highly correlated measures with P@10, we are able to achieve very high τ correlation and R^2 in all three test collections ($\tau = 0.912$ and $R^2 = 0.983$ on average). We reach nearly perfect prediction accuracy ($R^2 = 0.994$) on WT2012.

RBP(0.95): Compared to RBP(0.5) and RBP(0.8), we achieve noticeably lower prediction performance, especially on WT2013 and WT2014. On WT2012, which is used as the development set in our experimental setup, we reach high prediction accuracy when we use 2–3 independent variables.

R-Prec, RR and R@100: In predicting these three measures, while we reach high prediction accuracy in many cases, there is no independent variable group yielding high prediction performance on all three test collections.

Overall, we achieve high prediction accuracy for MAP, P@10, RBP(0.5) and RBP(0.8) on all test collections. RR and RBP(0.8) are the most frequently selected independent variables (10 and 9 times, respectively). Generally, using a single measure is not sufficient to reach $\tau = 0.9$ correlation. We achieve very high prediction accuracy using only 2 measures for many scenarios.

Note R^2 is sometimes negative, whereas theoretically the value of the coefficient of determination should lie in $[0, 1]$. R^2 compares the fit of the chosen model with a horizontal straight line (the null hypothesis); if the chosen model fits worse than a horizontal line, then R^2 will be negative[2].

Although the empirical results might suggest that the relationship between metrics are linear because non-linear SVMs did not improve results much, the negative values of R^2 contradict this observation, as the linear model clearly did not fit well. Specifically, we tried out RBF SVM's using different kernel sizes of $\{0.5, 1, 2, 5\}$, without significant result changes as compared to linear regression. Additional non-linear models could be further explored in future work.

5.1 Predicting High-Cost Metrics Using Low-Cost Metrics

In some cases, one may wish to predict a "high-cost" evaluation metric (i.e., requiring relevance judging to some significant evaluation depth D) when only "low-cost" evaluation metrics have been reported. Here, we consider prediction of Precision, MAP, nDCG, and RBP [32] for high-cost $D = 100$ or $D = 1000$ given a set of low-cost metric scores ($D \in \{10, 20, ..., 50\}$): precision, bpref, ERR, infAP [52], MAP, nDCG and RBP. We include bpref and infAP given their support for evaluating systems with incomplete relevance judgments. For RBP we use $p = 0.95$. For each depth D, we calculate the powerset of the 7 measures mentioned above (excluding the empty set $\emptyset$). We then find which elements of the powerset are the best predictors of the high-cost measures on WT2012. The set of low-cost measures that yields the maximum τ score for a particular high-cost measure for WT2012 is then used for predicting the respective measure on WT2013 and WT2014. We repeat this process for each evaluation depth $D \in \{10, 20, ..., 50\}$ to assess prediction accuracy as a function of D.

Figure 2 presents results. For depth 1000 (Fig. 2a), we achieve $\tau > 0.9$ correlation and $R^2 > 0.98$ for RBP in all cases when $D \geq 30$. While we are able to reach $\tau = 0.9$ correlation for MAP on WT2012, prediction of P@1000 and nDCG@1000 measures performs poorly and never reaches a high τ correlation. As expected, the performance of prediction increases when evaluation depth of high-cost measures are decreased to 100 (Fig. 2a vs. Fig. 2b).

Overall, RBP seems the most predictable from low-cost metrics while precision is the least. Intuitively, MAP, nDCG and RBP give more weight to documents at higher ranks, which are also evaluated by the low-cost measures, while precision@D does not consider document ranks within the evaluation depth D.

[2] https://stats.stackexchange.com/questions/12900/when-is-r-squared-negative.

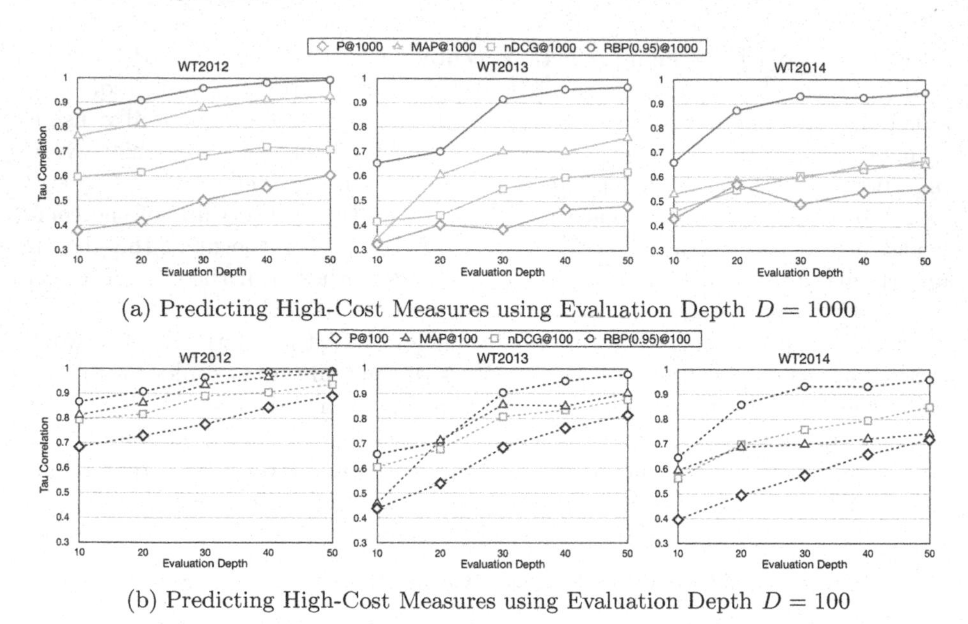

(a) Predicting High-Cost Measures using Evaluation Depth $D = 1000$

(b) Predicting High-Cost Measures using Evaluation Depth $D = 100$

Fig. 2. Linear regression prediction of high-cost metrics using low-cost metrics

6 Ranking Evaluation Metrics

Given a particular search scenario or user need envisioned, one typically selects appropriate evaluation metrics for that scenario. However, this does not necessarily consider correlation between metrics, or which metrics may interest other researchers engaged in reproducibility studies, benchmarking, or extensions. In this section, we consider how one might select the most informative and distinctive set of metrics to report in general, without consideration of specific user needs or other constraints driving selection of certain metrics.

We thus motivate a proper metric ranking criteria to efficiently compute the top L metrics to report amongst the S metrics available, i.e., a set that best captures diverse aspects of system performance with minimal correlation across metrics. Our approach is motivated by Sheffield [43], who introduced the idea of unsupervised ranking of features in high-dimensional data using the covariance information in the feature space. This method enables selection and ranking of features that are highly informative and less correlated with each other.

$$\Omega^* = \arg \max_{\Omega:|\Omega| \leq L} \det(\Sigma(\Omega)) \tag{1}$$

Here we are trying to find the subset Ω^* of cardinality L such that the covariance matrix Σ sampled from the rows of and columns of the entries of Ω^* will have the maximum determinant value, among all possible sub-determinant of size $L \times L$. The general problem is NP-Complete [35]. Sheffield provided a backward rejection scheme that throws out elements of the active subset Ω until it is left with L elements. However, this approach suffers from large cost in both time and space (Table 2), due to computing multiple determinant values over iterations.

We propose two novel methods for ranking metrics: an *iterative-backward* method (Sect. 6.1), which we find to yield the best empirical results, and a *greedy-forward* approach (Sect. 6.2) guaranteed to yield sub-modular results. Both offer lower time and space complexity vs. prior clustering work [7,41,51].

Table 2. Complexity of ranking algorithms.

Algorithm	Time complexity	Space complexity
Sheffield [43]	$O(LS^4)$	$O(S^3)$
Iterative-Backward	$O(LS^3)$	$O(S^2)$
Greedy-Forward	$O(LS^2)$	$O(S^2)$

6.1 Iterative-Backward (IB) Method

IB (Algorithm 1) starts with a full set of metrics and iteratively prunes away the less informative ones. Instead of computing all the sub-determinants of one less size at each iteration, we use the adjugate of the matrix to compute them in a single pass. This reduces the run-time by a factor of S and completely eliminates the need for additional memory. Also, since we are not interested in the actual values of the sub-determinants, but just the maximum, we can approximate $\Sigma_{adj} = \Sigma^{-1} \det(\Sigma) \approx \Sigma^{-1}$ since $\det(\Sigma)$ is a scalar multiple.

Once the adjugate Σ_{adj} is computed, we look at its diagonal entries for values of the sub-determinants of size one less. The index of the maximum entry is found in Step 7 and it is subsequently removed from the active set. Step 9 ensures that adjustments made to rest of the matrix prevents the selection of correlated features by scaling down their values appropriately. We do not have any theoretical guarantees for optimality of this IB feature elimination strategy, but our empirical experiments found that it always returns the optimal set.

Algorithm 1. Iterative-Backward Method

1: ***Input***: $\Sigma \in \mathbb{R}^{S \times S}$, L : number of channels to be retained
2: Set counter $k = S$ and $\Omega = \{1 : S\}$ as the active set
3: **while** $k > L$ **do**
4: $\quad \Sigma_{adj} \approx \Sigma^{-1}$ ▷ Approximate adjugate
5: $\quad i^* \leftarrow \arg\max_{i \in \Omega} diag(\Sigma_{adj}(i))$ ▷ Index to be removed
6: $\quad \Omega^{k+1} \leftarrow \Omega^k - \{i^*\}$ ▷ Augment the active set
7: $\quad \sigma_{ij} \leftarrow \sigma_{ij} - \sigma_{ii^*}\sigma_{i^*j}/\sigma_{i^*i^*}, \forall i, j \in \Omega$ ▷ Update covariance
8: $\quad k \leftarrow k - 1$ ▷ Decrement counter
9: ***Output***: Retained features Ω

6.2 Greedy-Forward (GF) Method

GF (Algorithm 2) iteratively selects the most informative features to add one-by-one. Instead of starting with the full set, we initialize the active set as empty, then grow the active set by greedily choosing the best feature at each iteration, with lower run-time cost than its backward counterpart. The index of the maximum entry is found in Step 6 and is subsequently added to the active set. Step 8 ensures that the adjustments made to the other entries of the matrix prevents the selection of correlated features by scaling down their values appropriately.

Algorithm 2. Greedy-Forward Method

1: ***Input***: $\Sigma \in \mathbb{R}^{S \times S}$, L : number of channels to be selected
2: Set counter $k = 0$ and $\Omega = \varnothing$ as the active set
3: **while** $k < L$ **do**
4: $\quad i^* \leftarrow \arg\max_{i \notin \Omega} \sum_{j \notin \Omega} \sigma_{ij}^2 / \sigma_{ii}$ ▷ Index to be added
5: $\quad \Omega^{k+1} \leftarrow \Omega^k \cup \{i^*\}$ ▷ Augment the active set
6: $\quad \sigma_{ij} \leftarrow \sigma_{ij} - \sigma_{ii^*}\sigma_{i^*j}/\sigma_{i^*i^*}, \forall i, j \notin \Omega$ ▷ Update covariance
7: $\quad k \leftarrow k + 1$ ▷ Increment counter
8: ***Output***: Selected features Ω

A feature of this greedy strategy is that it is guaranteed to provide sub-modular results. The solution has a constant factor approximation bound of $(1 - 1/e)$, *i.e.* even under worst case scenario, the approximated solution is no worse than 63% of the optimal solution.

Proof. For any positive definite matrix Σ and for any $i \notin \Omega$:

$$f_\Sigma(\Omega \cup \{i\}) = f_\Sigma(\Omega) + \frac{\sum_{j \notin \Omega} \sigma_{ij}^2}{\sigma_{ii}}$$

where σ_{ij} are the elements of $\Sigma(\notin \Omega)$ *i.e.* the elements of Σ not indexed by the entries of the active set Ω, and f_Σ is the determinant function $\det(\Sigma)$. Hence, we have $f_\Sigma(\Omega) \geq f_\Sigma(\Omega')$ for any $\Omega' \subseteq \Omega$. This shows that $f_\Sigma(\Omega)$ is a monotonically non-increasing and sub-modular function, so that the simple greedy selection algorithm yields an $(1 - 1/e)$-approximation. □

6.3 Results

Running the Iterative-Backward (IB) and Greedy-Forward (GF) methods on the 23 metrics shown in Fig. 1 yields the results shown in Table 3. The top six metrics are the same (in order) for both IB and GF: MAP@1000, P@1000, NDCG@1000, RBP($p - 0.95$), ERR, and R-Prec. They then diverge on whether R@1000 (IB) or bpref (GF) should be at rank 7. GF makes some constrained choices that

Table 3. Metrics are ranked by each algorithm as numbered below.

IB	1. MAP@1000	2. P@1000	3. NDCG@1000	4. RBP-0.95	5. ERR
	6. R-Prec	7. R@1000	8. bpref	9. MAP@100	10. P@100
	11. NDCG@100	12. RBP-0.8	13. R@100	14. MAP@20	15. P@20
	16. NDCG@20	17. RBP-0.5	18. R@20	19. MAP@10	20. P@10
	21. NDCG@10	22. R@10	23. RR	-	-
GF	1. MAP@1000	2. P@1000	3. NDCG@1000	4. RBP-0.95	5. ERR
	6. R-Prec	7. bpref	8. R@1000	9. MAP@100	10. P@100
	11. RBP-0.8	12. NDCG@100	13. R@100	14. MAP@20	15. P@20
	16. RBP-0.5	17. NDCG@20	18. R@20	19. P@10	20. MAP@10
	21. NDCG@10	22. R@10	23. RR	-	-

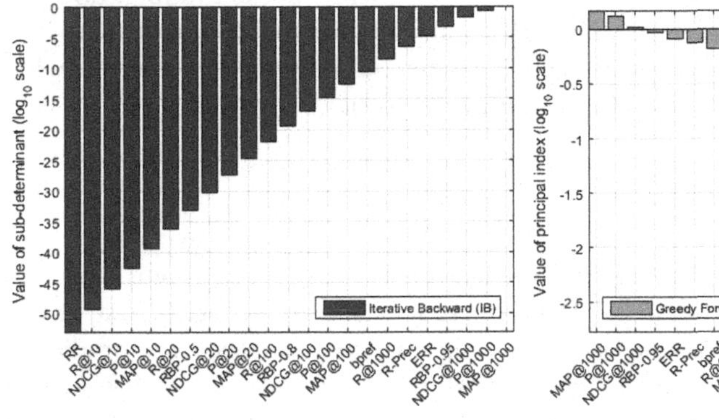

(a) Iterative Backward.
Left-to-Right: metrics discarded

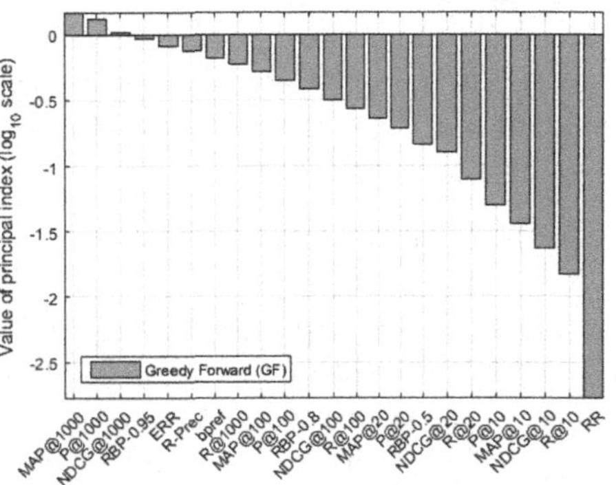

(b) Greedy Forward.
Left-to-Right: metrics included

Fig. 3. Metrics ranked by the strategies. Positive values on the GF plot shows values computed by the greedy criteria were positive for the first three selections.

lead to swapping of ranks among some metrics (bpref and R@1000, RBP-0.8 and NDCG@100, RBP-0.5 and NDCG@20, P@10 and MAP@10). However, due to the sub-modular nature of the greedy method, the approximated solution is guaranteed to incur no more than 27% error compared to the true solution. Both methods assigned lowest rankings to NDCG@10, R@10, and RR.

Figure 3a shows the metric deleted from the active set at each iteration of the IB strategy. As irrelevant metrics are removed by the maximum determinant criteria, the value of the sub-determinant increases at each iteration and is empirically maximum among all sub-determinants of that size. Figure 3b shows the metric added to the active set at each iteration by the GF strategy. Here we add a metric that maximizes the greedy selection criteria. We can see that over iterations the criteria value steadily decreases due to proper updates made.

The ranking pattern shows that the relevant, highly informative and less correlated metrics (MAP@1000, P@1000, nDCG@1000, RBP-0.95) are clearly ranked at the top. While ERR, R-Prec, bpref, and R@1000 may not be as informative as the higher ranked metrics, they still rank highly because the average information provided by other measures (*e.g.* MAP@100, nDCG@100 *etc.*) decreases even more in presence of already selected features MAP@1000, nDCG@1000 *etc.* Intuitively, even if two metrics are informative, both should not be ranked highly if there exists strong correlation between them.

Relation to Prior Work. Our findings are consistent with prior work in showing that we can select best metrics from clusters, although we report lower algorithmic (time and space) cost procedures than prior work [7,41,51]. Webber *et al.* [51] consider only the diagonal entries of the covariance; we consider the entire matrix since off-diagonal entries indicate cross-correlation. Baccini *et al.* [7] use both Hierarchical Clustering (HCA) of metrics which lacks ranking, does not scale well, and is slow, having runtime $O(S^3)$ and memory $O(S^2)$ with large constants. Their results are also somewhat subjective and subject to outliers, while our ranking is computationally effective and theoretically justified.

7 Conclusion

In this work, we explored strategies for selecting IR metrics to report. We first quantified correlation between 23 popular IR metrics on 8 TREC test collections. Next, we described metric prediction and showed that accurate prediction of MAP, P@10, and RBP can be achieved using 2–3 other metrics. We further investigated accurate prediction of some high-cost evaluation measures using low-cost measures, showing RBP(p = 0.95) at cutoff depth 1000 could be accurately predicted given other metrics computed at only depth 30. Finally, we presented a novel model for ranking evaluation metrics based on covariance, enabling selection of a set of metrics that are most informative and distinctive.

We proposed two methods for ranking metrics, both providing lower time and space complexity than prior work. Among the 23 metrics considered, we predicted MAP@1000, P@1000, nDCG@1000 and RBP(p = 0.95) as the top four metrics, consistent with prior research. Although the timing difference is negligible for 23 metrics, there is a speed-accuracy trade-off, once the problem dimension increases. Our method provides a theoretically-justified, practical approach which can be generally applied to identify informative and distinctive evaluation metrics to measure and report, and applicable to a variety of IR ranking tasks.

Acknowledgements. This work was made possible by NPRP grant# NPRP 7-1313-1-245 from the Qatar National Research Fund (a member of the Qatar Foundation). The statements made herein are solely the responsibility of the authors.

References

1. Alonso, O., Mizzaro, S.: Can we get rid of TREC assessors? using mechanical turk for relevance assessment. In: Proceedings of the SIGIR 2009 Workshop on the Future of IR Evaluation, vol. 15, p. 16 (2009)
2. Armstrong, T.G., Moffat, A., Webber, W., Zobel, J.: Improvements that don't add up: ad-hoc retrieval results since 1998. In: Proceedings of the 18th ACM Conference on Information and Knowledge Management, pp. 601–610. ACM (2009)
3. Aslam, J.A., Pavlu, V., Yilmaz, E.: A statistical method for system evaluation using incomplete judgments. In: Proceedings of the 29th Annual International ACM SIGIR Conference on Research and Development in Information Retrieval, pp. 541–548. ACM (2006)
4. Aslam, J.A., Yilmaz, E.: Inferring document relevance from incomplete information. In: Proceedings of the Sixteenth ACM Conference on Information and Knowledge Management, pp. 633–642. ACM (2007)
5. Aslam, J.A., Yilmaz, E., Pavlu, V.: A geometric interpretation of r-precision and its correlation with average precision. In: Proceedings of the 28th Annual International ACM SIGIR Conference on Research and Development in Information Retrieval, pp. 573–574. ACM (2005)
6. Aslam, J.A., Yilmaz, E., Pavlu, V.: The maximum entropy method for analyzing retrieval measures. In: Proceedings of the 28th Annual International ACM SIGIR Conference on Research and Development in Information Retrieval, pp. 27–34. ACM (2005)
7. Baccini, A., Déjean, S., Lafage, L., Mothe, J.: How many performance measures to evaluate information retrieval systems? Knowl. Inf. Syst. **30**(3), 693 (2012)
8. de Bruijn, L., Martin, J.: Literature mining in molecular biology. In: Proceedings of the EFMI Workshop on Natural Language Processing in Biomedical Applications, pp. 1–5 (2002)
9. Buckley, C., Voorhees, E.M.: Retrieval evaluation with incomplete information. In: Proceedings of the 27th Annual International ACM SIGIR Conference on Research and Development in Information Retrieval, pp. 25–32. ACM (2004)
10. Buckley, C., Voorhees, E.M.: Retrieval system evaluation. In: TREC: Experiment and Evaluation in Information Retrieval, pp. 53–75 (2005)
11. Carterette, B., Allan, J., Sitaraman, R.: Minimal test collections for retrieval evaluation. In: Proceedings of the 29th Annual International ACM SIGIR Conference on Research and Development in Information Retrieval, pp. 268–275. ACM (2006)
12. Chapelle, O., Metlzer, D., Zhang, Y., Grinspan, P.: Expected reciprocal rank for graded relevance. In: Proceedings of the 18th ACM Conference on Information and Knowledge Management, pp. 621–630. ACM (2009)
13. Clarke, C., Craswell, N.: Overview of the TREC 2011 web track. In: TREC (2011)
14. Clarke, C., Craswell, N., Soboroff, I., Cormack, G.: Overview of the TREC 2010 web track. In: TREC (2010)
15. Clarke, C., Craswell, N., Voorhees, E.M.: Overview of the TREC 2012 web track. In: TREC (2012)
16. Collins-Thompson, K., Bennett, P., Clarke, C., Voorhees, E.M.: TREC 2013 web track overview. In: TREC (2013)
17. Collins-Thompson, K., Macdonald, C., Bennett, P., Voorhees, E.M.: TREC 2014 web track overview. In: TREC (2014)
18. Cormack, G.V., Palmer, C.R., Clarke, C.L.: Efficient construction of large test collections. In: Proceedings of the 21st Annual International ACM SIGIR Conference on Research and Development in Information Retrieval, pp. 282–289. ACM (1998)

19. Egghe, L.: The measures precision, recall, fallout and miss as a function of the number of retrieved documents and their mutual interrelations. Inf. Process. Manage. **44**(2), 856–876 (2008). Evaluating Exploratory Search Systems Digital Libraries in the Context of Users Broader Activities
20. Grady, C., Lease, M.: Crowdsourcing document relevance assessment with mechanical turk. In: Proceedings of the NAACL HLT 2010 Workshop on Creating Speech and Language Data with Amazon's Mechanical Turk, pp. 172–179. Association for Computational Linguistics (2010)
21. Guiver, J., Mizzaro, S., Robertson, S.: A few good topics: experiments in topic set reduction for retrieval evaluation. ACM Trans. Inf. Syst. (TOIS) **27**(4), 21 (2009)
22. Hawking, D.: Overview of the TREC-9 web track. In: TREC (2000)
23. Hirschman, L., Park, J.C., Tsujii, J., Wong, L., Wu, C.H.: Accomplishments and challenges in literature data mining for biology. Bioinformatics **18**(12), 1553–1561 (2002)
24. Hosseini, M., Cox, I.J., Milic-Frayling, N., Shokouhi, M., Yilmaz, E.: An uncertainty-aware query selection model for evaluation of IR systems. In: Proceedings of the 35th International ACM SIGIR Conference on Research and Development in Information Retrieval, pp. 901–910. ACM (2012)
25. Hosseini, M., Cox, I.J., Milic-Frayling, N., Vinay, V., Sweeting, T.: Selecting a subset of queries for acquisition of further relevance judgements. In: Amati, G., Crestani, F. (eds.) ICTIR 2011. LNCS, vol. 6931, pp. 113–124. Springer, Heidelberg (2011). https://doi.org/10.1007/978-3-642-23318-0_12
26. Ishioka, T.: Evaluation of criteria for information retrieval. In: IEEE/WIC International Conference on Web Intelligence, 2003. WI 2003. Proceedings, pp. 425–431. IEEE (2003)
27. Jones, K.S., van Rijsbergen, C.J.: Report on the need for and provision of an "ideal" information retrieval test collection (British library research and development report no. 5266), p. 43 (1975)
28. Jones, T., Thomas, P., Scholer, F., Sanderson, M.: Features of disagreement between retrieval effectiveness measures. In: Proceedings of the 38th International ACM SIGIR Conference on Research and Development in Information Retrieval, pp. 847–850. ACM (2015)
29. Lu, X., Moffat, A., Culpepper, J.S.: The effect of pooling and evaluation depth on IR metrics. Inf. Retrieval J. **19**(4), 416–445 (2016)
30. Mizzaro, S., Robertson, S.: Hits hits TREC: exploring IR evaluation results with network analysis. In: Proceedings of the 30th Annual International ACM SIGIR Conference on Research and Development in Information Retrieval, pp. 479–486. ACM (2007)
31. Moffat, A., Webber, W., Zobel, J.: Strategic system comparisons via targeted relevance judgments. In: Proceedings of the 30th Annual International ACM SIGIR Conference on Research and Development in Information Retrieval, pp. 375–382. ACM (2007)
32. Moffat, A., Zobel, J.: Rank-biased precision for measurement of retrieval effectiveness. ACM Trans. Inf. Syst. (TOIS) **27**(1), 2 (2008)
33. Moghadasi, S.I., Ravana, S.D., Raman, S.N.: Low-cost evaluation techniques for information retrieval systems: a review. J. Informetr. **7**(2), 301–312 (2013)
34. Nuray, R., Can, F.: Automatic ranking of information retrieval systems using data fusion. Inf. Process. Manage. **42**(3), 595–614 (2006)
35. Papadimitriou, C.H.: The largest subdeterminant of a matrix. Bull. Math. Soc. Greece **15**, 96–105 (1984)

36. Park, L., Zhang, Y.: On the distribution of user persistence for rank-biased precision. In: Proceedings of the 12th Australasian Document Computing Symposium, pp. 17–24 (2007)
37. Pavlu, V., Aslam, J.: A practical sampling strategy for efficient retrieval evaluation. Technical report, College of Computer and Information Science, Northeastern University (2007)
38. Sakai, T., Kando, N.: On information retrieval metrics designed for evaluation with incomplete relevance assessments. Inf. Retrieval **11**(5), 447–470 (2008)
39. Sakai, T.: Alternatives to Bpref. In: Proceedings of the 30th Annual International ACM SIGIR Conference on Research and Development in Information Retrieval, pp. 71–78. ACM (2007)
40. Sakai, T.: On the properties of evaluation metrics for finding one highly relevant document. Inf. Media Technol. **2**(4), 1163–1180 (2007)
41. Sakai, T.: On the reliability of information retrieval metrics based on graded relevance. Inf. Process. Manage. **43**(2), 531–548 (2007)
42. Sanderson, M.: Test Collection Based Evaluation of Information Retrieval Systems. Now Publishers Inc (2010)
43. Sheffield, C.: Selecting band combinations from multispectral data. Photogramm. Eng. Remote Sens. **51**, 681–687 (1985)
44. Soboroff, I., Nicholas, C., Cahan, P.: Ranking retrieval systems without relevance judgments. In: Proceedings of the 24th Annual International ACM SIGIR Conference on Research and Development in Information Retrieval, pp. 66–73. ACM (2001)
45. Tague-Sutcliffe, J., Blustein, J.: Overview of TREC 2001. In: Proceedings of the Third Text Retrieval Conference (TREC-3), pp. 385–398 (1995)
46. Thom, J., Scholer, F.: A comparison of evaluation measures given how users perform on search tasks. In: ADCS2007 Australasian Document Computing Symposium. RMIT University, School of Computer Science and Information Technology (2007)
47. Voorhees, E.M.: Variations in relevance judgments and the measurement of retrieval effectiveness. Inf. Process. Manage. **36**(5), 697–716 (2000)
48. Voorhees, E.M.: Overview of the TREC 2004 robust track. In: TREC, vol. 4 (2004)
49. Voorhees, E.M., Harman, D.: Overview of TREC 2001. In: TREC (2001)
50. Voorhees, E.M., Tice, D.M.: The TREC-8 question answering track evaluation. In: TREC, vol. 1999, p. 82 (1999)
51. Webber, W., Moffat, A., Zobel, J., Sakai, T.: Precision-at-ten considered redundant. In: Proceedings of the 31st Annual International ACM SIGIR Conference on Research and Development in Information Retrieval, pp. 695–696. ACM (2008)
52. Yilmaz, E., Aslam, J.A.: Estimating average precision with incomplete and imperfect judgments. In: Proceedings of the 15th ACM International Conference on Information and Knowledge Management, pp. 102–111. ACM (2006)
53. Yilmaz, E., Aslam, J.A.: Estimating average precision when judgments are incomplete. Knowl. Inf. Syst. **16**(2), 173–211 (2008)

Modeling User Actions in Job Search

Alfan Farizki Wicaksono(✉), Alistair Moffat, and Justin Zobel

School of Computing and Information Systems,
The University of Melbourne, Melbourne 3010, Australia
wicaksonoa@student.unimelb.edu.au

Abstract. Users of online job search tools interact with result pages in three different ways: via *impressions*, via *clicks*, and via *applications*. We investigate the relationship between these three kinds of interaction using logs provided by Seek.com, an Australian-based job search service. Our focus is on understanding the extent to which the three interaction types can be used to predict each other. In particular we examine models for inferring impressions from clicks, thereby providing system designers with new options for evaluating search result pages.

1 Introduction

Good search engine design requires that the list of documents generated in response to a query be presented to users in a way that allows them to efficiently identify material of interest. In the context of online job search, a results page is a list of job summaries; and a user might proceed in any of a variety of ways, such as examining many summaries before clicking on any of them; or clicking on each summary in turn; or abandoning the search and issuing a new query. Modeling of these interaction patterns is a key element towards good design.

Each chronology of user decisions can be represented as an *action sequence*, the ordered series of activities performed by a particular user interacting with a ranked list of results, defined as $\mathcal{A} = \langle(a_1, r_1), (a_2, r_2), (a_3, r_3) \ldots\rangle$, where (a_t, r_t) is an *action* comprised of two elements: the type of action, a_t, and the rank position at which the action took place, $r_t \geq 1$. The interaction logs used in this work were provided by the Australasian job search site Seek.com, and contain three types of action. An impression, $a_t =$ "I", is defined by Seek.com to have occurredretrieval when a job summary is fully visible on screen for at least 0.5 s. While an impression is an imprecise measurement, the overall collection of impressions is nevertheless a valuable resource for exploring and understanding user behavior. A click, $a_t =$ "C", is recorded when the user selects a particular summary and loads the corresponding job details page. An application, $a_t =$ "A", occurs when the user clicks the "apply" button in the job details page, before they actually fill out and submit the job application (akin to the "purchase" click in online shopping). This action is a good signal for *relevance*. For example, one action sequence excerpted from the Seek.com interaction logs commences with:

$$(\text{"I"}, 1), (\text{"I"}, 2), (\text{"I"}, 3), (\text{"I"}, 4), (\text{"I"}, 5), (\text{"C"}, 4), (\text{"A"}, 4), (\text{"I"}, 5), \ldots$$

L. Azzopardi et al. (Eds.): ECIR 2019, LNCS 11437, pp. 652–664, 2019.
https://doi.org/10.1007/978-3-030-15712-8_42

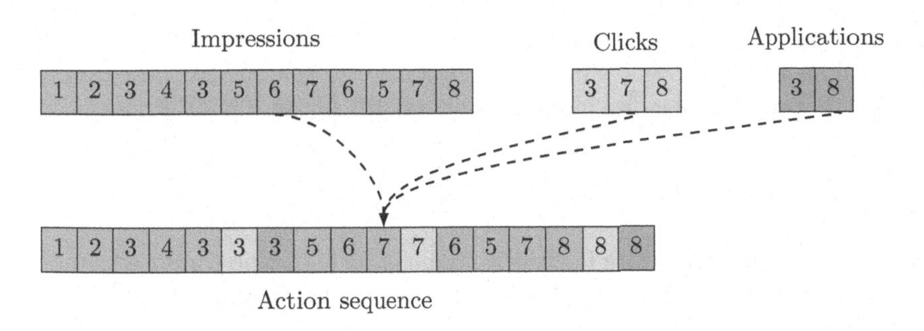

Fig. 1. An action sequence as a combination of three sub-sequences.

where the user is presumed to have examined the summaries at ranks 1 through 5, then clicked on the summary at rank 4, then started an application for that position, then viewed the summary at rank 5 again. Each action sequence can also be regarded as being the interleaving of three component sequences, as illustrated in Fig. 1.

Examination of a collection of action sequences may provide an understanding of how each of the actions relate to each other. In particular, it is desirable to be able to infer the sequence of impressions from a sequence of clicks, since the latter is almost always observable, and the former may not be. For example, some interaction logs only contain clicks, and some browsers may obscure impressions. Figure 2 illustrates this possibility. In the absence of impressions, previous analysis has typically employed clicks to infer which documents in the SERP the user has inspected [1,2,11], on the assumption that the last click represented the last impression. But that assumption lacks supporting evidence. Zhang et al. [17] propose a method to extrapolate user impressions beyond that last click, but did not have access to resources that would have allowed their method to be validated.

Here we develop models for approximating impressions using click information. We first identify predictable patterns in regard to the three elements, impression, click, and application, and use them to infer impressions from clicks. We then evaluate this approach relative to hypothesized characteristics of user behavior, including *conditional continuation probabilities* [8,9], seeking consistency between the inferred characteristics and those derived from the real impression information. Our results show that there are two key factors that have

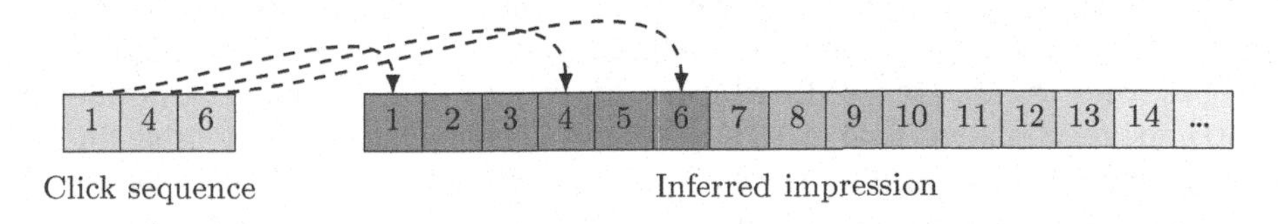

Fig. 2. Click information as a predictor of impressions.

considerable bearing: the deepest click rank; and the number of distinct clicked results. The value of the resultant methodology is that it can be employed to compute parameters for user models such as RBP [8] and INSQ [9].

2 Related Work

The most relevant prior work is that of Zhang et al. [17], who proposed a model to predict which documents in the ranking had been inspected by users. Their approach, which infers impressions based on clicks, was used to identify parameters for weighted-precision metrics that empirically matched predicted user behavior. Zhang et al.'s work used MSN query logs containing 12 million clicks, but no other information such as mouse-hover or eye-tracking records; an absence that meant it was not possible for them to validate their impression model.

Other work on search behavior is also pertinent. White and Drucker [14] study the variability of user's search behavior when interacting with results pages, using a collection of *search trails* from major commercial search engines, in which each trail represents an interaction graph that begins with a query submission and ends once the user has completed their search task. Their findings suggested two types of users: *navigators* who tend to solve problems sequentially, and *explorers* who tend to pose many queries and visit many pages at the same time [14]. Klöckner et al. [6] used eye-tracking experiments to investigate how rankings are scanned. They found that the majority of users employed a *depth-first* strategy, progressing down the ranking from top to bottom and deciding sequentially whether or not to click to open the linked document. Cutrell and Guan [3] also used eye-fixation data, but to study how the presentation of search results affects the behavior of users. One of their findings that may have implication for our work is that before the user clicked on a result, they viewed almost all results before it, and only a few results beyond it. They also confirmed the "top-to-bottom" reading behavior that previously had been investigated by Joachims et al. [5]. Thomas et al. [13] also used eye-tracking; they concluded that users follow a "two steps forward, one step back" approach to viewing result pages, with backward steps almost as common as forwards ones.

Two recent investigations have explored the behavior of online job seekers: Spina et al. [12] study interaction logs from an online job site to investigate the characteristics of job seekers in terms of click-through and query submission; and Mansouri et al. [7] select job-related queries from among millions of Web queries, and study several aspects such as query formulation and job search intensity across the week. There have also been studies on e-commerce search logs. Parikh and Sundaresan [10] analyze around 115 million eBay queries and suggest that the frequency distribution of distinct queries follows a power-law distribution. Hasan et al. [4] extended that same investigation. A notable finding is that query frequency, a measure of query popularity, positively correlates with the number of retrieved results in eBay, showing a balance between supply and demand.

3 Predictable Patterns

Our dataset is a representative sample of user interaction logs from Seek.com, a job search site servicing a large English-language market [15]. Action sequences for two distinct modalities were employed: online job search using a mobile-based Android/iOS application, in which search engine results pages have no pagination and continuous scrolling; and job search using a desktop-based web browser, in which results pages are paginated, each containing 20 results. We used a total of 20,000 action sequences in response to Android/iOS queries, and the same number for browser queries.

Impressions as a Prelude to Clicks and Applications. As an initial summary of user behavior, the left graph in Fig. 3 shows the mean number of distinct impressions below, and also beyond, each of the click actions, through until the time of that particular click. For each value of r_t the graph bars showing below and beyond are offset from the marked value of r_t matched on the vertical scale, to create a visual representation of overall viewing activities. While users typically examined almost all of the job summaries before and including rank r_t in the lead up to any click action at rank r_t, they were also consistently recorded as viewing a number of results beyond that rank prior to the click, reinforcing the "two steps forward, one step back" observation already noted. For example, prior to click actions at rank 5, on average a browser user inspected 4.9 results at ranks 1 to 5, and 2.6 results at ranks 6 and deeper. Comparing the Android/iOS users and browser users, it also seems that browser users examine a broader range of job summaries before each click than the Android/iOS users. The right graph in Fig. 3 depicts similar analysis, but now pivoting on applications instead of clicks. In most of the conditions the vertical bars are longer, suggesting that active job seekers inspected the SERP more deeply before returning to summaries they had already viewed in order to make an application. Overall, the fact that the positional distribution of impressions across all clicked ranks follows a similar pattern suggests that it should be possible to infer impressions

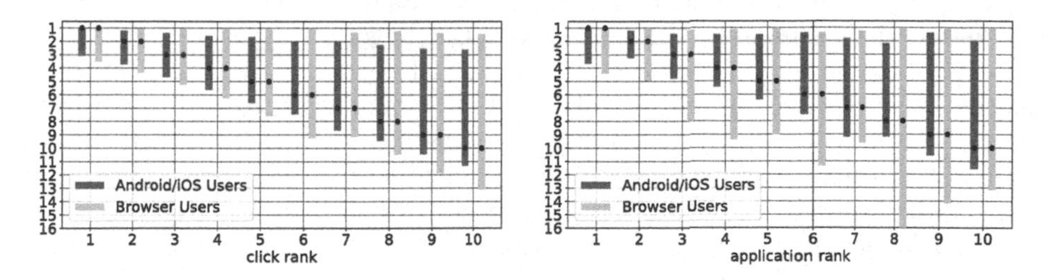

Fig. 3. Mean number of distinct ranks examined prior to and including rank r_t, and beyond rank r_t, as bars above and below the marked reference point r_t indicating the rank position of a subsequent click or application action. On the left, the data is stratified by the ranks r_t at which clicks occurred; on the right the data is stratified by the ranks r_t at which applications occurred.

Table 1. Mean number of distinct ranks clicked up to and including rank r_t, and beyond rank r_t, stratified by the ranks r_t at which applications occurred.

r_t	Android/iOS		Browser	
	Before	Beyond	Before	Beyond
1	1.00	0.10	1.00	0.13
2	1.25	0.02	1.16	0.06
3	1.47	0.17	1.10	0.24
4	1.53	0.09	1.27	0.27
5	1.76	0.02	1.52	0.17

Table 2. Mean number of distinct summaries that were examined beyond the deepest click rank, stratified by the ranking position of the deepest click.

Deepest click	1	2	3	4	5	6	7	8	9	10	11	12	13	14	15
Android/iOS	3.7	5.0	4.7	5.3	6.7	5.8	6.3	6.9	6.8	7.5	6.7	6.7	8.4	7.6	9.3
Browser	3.7	4.9	5.5	5.3	4.9	5.9	5.2	6.3	6.6	4.8	6.7	7.0	5.5	6.9	6.8

from click information; and that the additional signal provided by application actions may strengthen that relationship.

Clicks as a Prelude to Applications. Table 1 shows the mean number of distinct ranks prior to and beyond rank r_t that were clicked before a job application took place at rank r_t, adopting the same measurement methodology as used for Fig. 3. The most predictable pattern is that (unsurprisingly) an application action at some rank r_t is always preceded by a click action at depth r_t. However, a click action is not necessarily followed by a job application action in either the short or long term, and additional click actions at ranks both shallower and deeper than r_t may occur before the application at r_t is pursued.

Impressions Beyond the Deepest Click. Table 2 shows the mean number of distinct job summaries that were examined beyond the deepest click rank observed in each action sequence. Users typically examined multiple job summaries beyond even the deepest observed click; moreover, as the rank of the deepest click increases, the mean number of distinct summaries viewed beyond the deepest click rank also increases. This suggests that the deepest impression rank could be predicted using the deepest click rank; and that this correction should also be adaptive to the rank position of the deepest click [17].

Clicks Versus Impressions. In recent work Wicaksono and Moffat [16] consider three methods denoted "L", "M", and "G" for inferring an approximation for $C(i)$, the conditional continuation probability [1,9] of the user viewing the summary at rank $i + 1$, given that they have viewed the summary at rank i. They compute an empirical value $\hat{C}(i)$ based on observed interactions; we apply that same process here, and also using clicks instead of impressions, considering

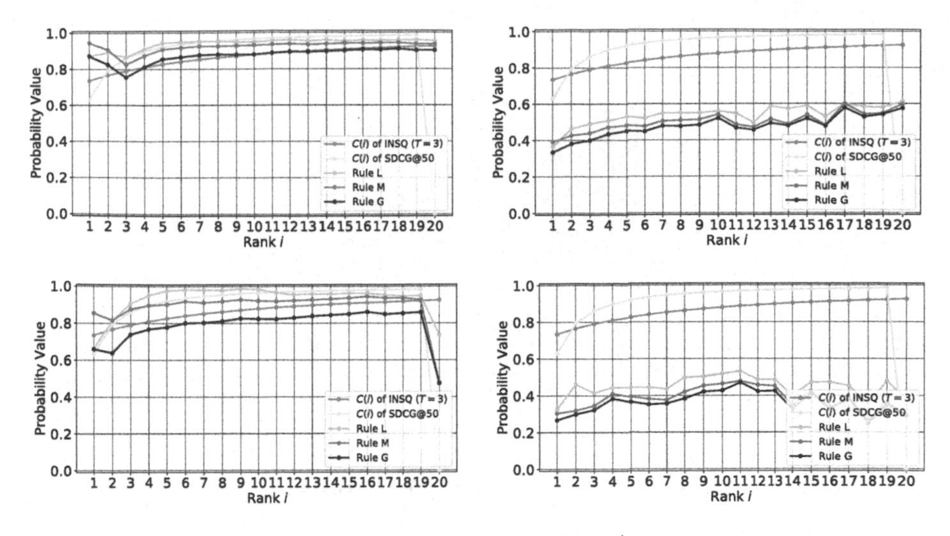

Fig. 4. Observed conditional continuation probability, $\hat{C}(i)$, across twenty results per query, estimated from impressions (left column) and clicks (right column). The top row is for Android/iOS users, and the lower row for browser users.

Table 3. Best fit parameters for RBP and INSQ across the first twenty results, computed using clicks and impressions. The click-based results suggest strongly top-weighted behavior; the impression-based parameters are more plausible.

Model	Impression		Click	
	Android/iOS	Browser	Android/iOS	Browser
RBP	$\phi = 0.85$	$\phi = 0.75$	$\phi = 0.42$	$\phi = 0.33$
INSQ	$T = 4.0$	$T = 1.9$	$T = 1.4 \times 10^{-13}$	$T = 2.3 \times 10^{-9}$

only the first twenty results (the first page for browser-based users), and using micro-averaging across contributions (the latter because clicks are top-heavy and much sparser than impressions).

Figure 4 shows the resulting empirical conditional continuation probabilities $\hat{C}(i)$, and compares them with two reference curves, those for SDCG and INSQ [9]. Continuation probabilities estimated using clicks are markedly different to those derived from impressions. Table 3 then compares the best-fit parameters for the metrics RBP and INSQ, also computed using clicks and impressions. Compared to impressions, the use of clicks results in underestimation of the persistence parameter ϕ (used in RBP) and volume of relevance parameter T (used in INSQ), suggesting that clicks are not a direct surrogate for impressions. Thomas et al. [13] make a similar observation, based on eye-tracking experiments. However, we argue that impressions can nevertheless be inferred from clicks, and develop that theme in the next section.

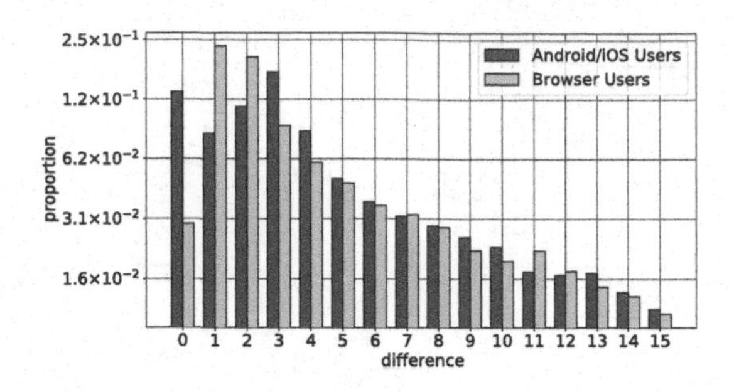

Fig. 5. Distribution of *diff*, the difference between the rank of the deepest click and the rank of the deepest impression. If *diff* = 0 the two occurred at the same rank in the SERP.

4 Predicting Impression Distributions

This section describes ways that patterns of impressions can be modeled.

Regression-based Prediction. A simple option is to build on these three assumptions:

1. the user reads summaries from top to bottom of the ranking; and
2. inspects all summaries at ranks 1 to n if the one at rank n is clicked; and
3. may also inspect summaries deeper than rank n before or after clicking at rank n.

The first of these assumptions is supported by previous experiments [3,5,15]; and the Seek.com interaction logs provide empirical support for the other two [15], including the behavior depicted in Fig. 3.

Figure 5 provides further evidence in support of the second assumption. It shows the distribution of the difference between the deepest click rank and the deepest impression rank across all queries (denoted *diff*), through to rank 15. When no clicks occurred the deepest click rank was set to zero. The difference is always greater than or equal to zero, since the set of clicks is subset of the set of impressions. Browser (Android/iOS) users inspected one or more job summaries beyond the deepest click rank 97.0% (86.3%) of the time; and the expected number of summaries inspected beyond the deepest click rank is 6.5 (8.0).

We then analyzed the contributions of two click-related characteristics to the difference *diff* between the deepest click rank and the deepest impression rank. These characteristics were the rank position of the deepest click (denoted *dc*), and the number of distinct items clicked (denoted *nc*), the two factors used by Zhang et al. [17]. Linear regression was employed to find the best coefficients (w_i) for the linear combination:

$$diff = f(dc, nc; \mathbf{w}) = w_0 + w_1 \cdot dc + w_2 \cdot nc\,.$$

Table 4 shows those best-fit values. With other factors held equal, *diff* tends to increase with *dc* ($w_1 > 0$) and decrease with *nc* ($w_2 < 0$), with generally

Table 4. Linear regression quantifying the effect that the deepest click rank and the number of clicks have on the numeric difference between the deepest click rank and the deepest impression rank.

Factor	Android/iOS		Browser	
	Coef	p	Coef	p
Intercept	$w_0 =$ 6.06	0.000	$w_0 =$ 4.70	0.000
Deepest click rank	$w_1 =$ 0.19	0.000	$w_1 =$ 0.17	0.000
Number of clicks	$w_2 = -0.88$	0.000	$w_2 = -0.15$	0.235

small p values indicating a high degree of confidence in the direction of those relationships. Note that the zero-click case (that is, $dc = 0$) was excluded in this regression analysis since $dc = 0$ indicates a different type of interaction and may be an amalgam of many behavioral patterns. The coefficients presented in Table 4 support the third of the assumptions given above.

Figure 5 already showed the distribution of $\hat{P}(\mathit{diff} = n)$ for $0 \leq n \leq 15$. Assuming that the user sequentially inspects the summaries from top to bottom, the cumulative distribution $P(\mathit{diff} \geq n)$ is the fraction of times that the user reads all summaries from rank $DC(u,q)$ to $DC(u,q) + n$, where $DC(u,q)$ is the deepest click rank observed for user u after posing query q (zero if no clicks were observed). Let $P(\mathit{imp} = i \mid u, q)$ be the probability that user u records an impression action at rank i for query q. Based on the three listed assumptions, a general framework is given by:

$$P(\mathit{imp} = i \mid u, q) = \begin{cases} 1 & i \leq DC(u,q)\,, \\ \hat{P}(\mathit{diff} \geq (i - DC(u,q)) \mid u) & \text{otherwise}\,. \end{cases} \tag{1}$$

If we also assume that all users have the same behavior in regard to *diff*, then $\hat{P}(\mathit{diff} \geq n \mid u) = \hat{P}(\mathit{diff} \geq n)$. Thus, the problem is to approximate $\hat{P}(\mathit{diff} \geq n)$.

Model 1. We build a heuristic approach by approximating $\hat{P}(\mathit{diff} \geq n)$ with a mathematical function that has a "similar behavior" to it, and then use the impression and click logs to select parameters. Figure 6 depicts the $\hat{P}(\mathit{diff} \geq n)$ observed from the Seek.com interaction logs, aggregated over all users and queries, and suggests that an exponential decay function be considered as a proxy. Hence, we define

$$\hat{P}(\mathit{diff} \geq n) = e^{-n/K}\,, \tag{2}$$

where $K > 0$ is a parameter that controls the decay rate. Computing best-fit values for the Seek.com logs then yields $K_{android} = 7.05$ and $K_{browser} = 6.27$.

Model 2. Table 4 suggests that *diff* is sensitive to two factors, the deepest click rank *dc*, and the number of distinct clicked items, *nc*. After defining K as a response to *dc* and *nc*, we obtain:

$$K_{android} = 5.30 + 0.29 \cdot dc - 1.10 \cdot nc\,, \text{ and}$$
$$K_{browser} = 3.72 + 0.19 \cdot dc + 0.54 \cdot nc\,.$$

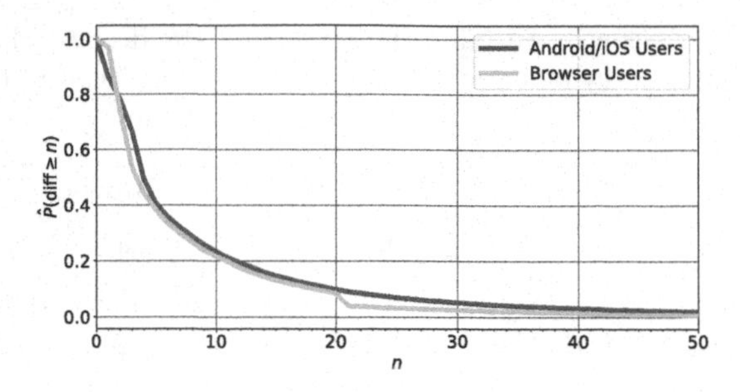

Fig. 6. $\hat{P}(\mathit{diff} \geq n)$ observed from the Seek.com interaction logs.

Impression Model 2 then defines $\hat{P}(\mathit{diff} \geq n)$ as

$$\hat{P}(\mathit{diff} \geq n) = e^{-n/g(K)},$$

where $g(x) = \ln(1 + e^x)$ is a "softplus" function that maps x to zero as x goes to $-\infty$, while approximately preserving the value of x when $x > 0$.

The ZPM Impression Model. Zhang et al. [17] propose the use of the *click gap distribution* of a user, $P(\mathit{gap} = n \mid u, q)$, that is, the probability that the user u views n consecutive documents without any click after posing query q. The ZPM impression model is then defined as:

$$P(\mathit{imp} = i \mid u, q) = \begin{cases} 1 & \text{when } i \leq DC(u, q) \\ P(\mathit{gap} \geq (i - DC(u, q)) \mid u) & \text{otherwise}, \end{cases} \quad (3)$$

where $P(\mathit{gap} \geq n \mid u)$ is determined by averaging $P(\mathit{gap} \geq n \mid u, q)$ across all queries issued by user u; and the overall impression model for a single user, $P(\mathit{imp} = i \mid u)$, is computed by averaging impression model $P(\mathit{imp} = i \mid u, q)$ across all queries. Zhang et al. also address the question of smoothing for $P(\mathit{gap} \geq n \mid u)$, needed because the clicks observed from a single user are usually sparse:

$$P(\mathit{gap} \geq n \mid u) = \alpha_u P(\mathit{gap} \geq n \mid u) + (1 - \alpha_u) P(\mathit{gap} \geq n), \quad (4)$$

where $P(\mathit{gap} \geq n)$ is the global click gap distribution across all users, and α_u is the smoothing parameter for user u, computed via:

$$\alpha_u = \frac{CT(u)}{CT(u) + \mu}. \quad (5)$$

In this expression, $CT(u)$ is the number of clicks observed from user u, and μ is an empirical constant.

Inferring $C(i)$ From Impression Models. Wicaksono and Moffat [16] propose three heuristics for computing empirical estimates of $C(i)$ given sequences of impressions, but require access to the impression sequences in order to do so. In the absence of impression information (the situation assumed here), we propose an alternative approach to estimate $C(i)$, starting with the click sequences instead, and using the impression models. Let $N(i,u,q)$ and $D(i,u,q)$ respectively be the nominal numerator and nominal denominator for user u and query q that contribute to the conditional continuation probability at rank i,

$$N(i,u,q) = P(imp = i+1 \mid u,q) \quad \text{and} \quad D(i,u,q) = P(imp = i \mid u,q)\,;$$

and assume that the overall estimate is generated by micro-averaging across users and queries:

$$\hat{C}(i) = \frac{\sum_{u \in U} \sum_{q \in Q(u)} N(i,u,q)}{\sum_{u \in U} \sum_{q \in Q(u)} D(i,u,q)}, \tag{6}$$

where U is a set of users and $Q(u)$ is a set of queries from user u in the interaction logs. For example, suppose we have three queries q_1, q_2, and q_3 recorded from an Android/iOS user; and the deepest click ranks are 1, 7, and 8, respectively. Using Model 1, the first ten values of $P(imp = i \mid u,q)$ are:

$$P(imp = i \mid u,q_1) = \langle 1.00, 0.87, 0.75, 0.65, 0.56, 0.48, 0.42, 0.36, 0.31, 0.27 \rangle\,,$$
$$P(imp = i \mid u,q_2) = \langle 1.00, 1.00, 1.00, 1.00, 1.00, 1.00, 1.00, 0.87, 0.75, 0.65 \rangle\,,$$
$$P(imp = i \mid u,q_3) = \langle 1.00, 1.00, 1.00, 1.00, 1.00, 1.00, 1.00, 1.00, 0.87, 0.75 \rangle\,.$$

Based on these, estimated values $\hat{C}(i)$ for $i = 2$ and $i = 9$ are:

$$\hat{C}(2) = \frac{0.75 + 1.00 + 1.00}{0.87 + 1.00 + 1.00} = 0.96 \quad \hat{C}(9) = \frac{0.27 + 0.65 + 0.75}{0.31 + 0.75 + 0.87} = 0.87\,.$$

Validating Impression Models. The impression models were evaluated on held-out dataset that contains 100,103 action sequences from Android/iOS-based queries. The continuous scrolling in the Android/iOS-based interface means that these action sequences do not have page boundary effects that arise when results are paginated, see the left-hand pair of graphs in Fig. 4. We use these held-out action sequences as a test set for measuring the quality of impression predictions in two different ways – via fit against estimated conditional continuation probability, $\hat{C}(i)$; and as a probability weighting vector $W(i)$ resulting from the use of $C(i)$ as a weighted-precision effectiveness metric [1,9]. The latter is, of course, why we are interested in impression distributions in the first place.

Continuation Probability. We compare the $\hat{C}(i)$ values estimated using impression models, with parameters developed using the original query sets and then applied to the clicks in the held-out action sequences, against the corresponding "true" $\hat{C}(i)$ values derived from the impressions present in the held-out action sequences. Figure 7 does this visually, showing the different estimates of $\hat{C}(i)$ for $1 \leq i \leq 50$, and Table 5 provides details by reporting weighted-by-frequency mean squared error (WMSE) differences between the "true" $\hat{C}(i)$ values and four different estimation mechanisms tuned using the original action sequences. Model 2 performs well compared to the other options and provides a superior way of inferring user behavior based of the two common factors of deepest click rank, *dc*, and the number of distinct items clicked, *nc*.

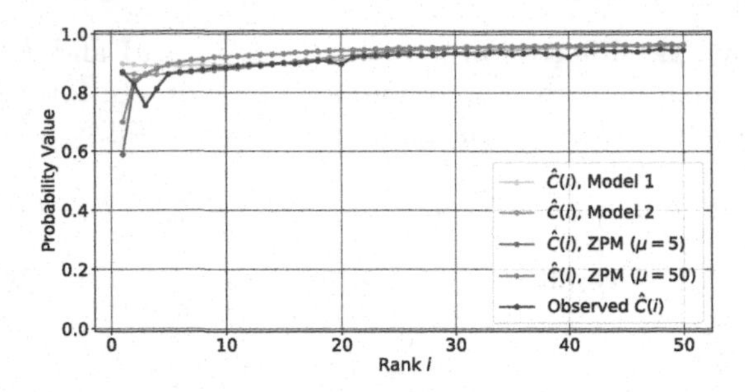

Fig. 7. Observed conditional continuation probability, $\hat{C}(i)$, across the first 50 results for each query, estimated using impression models derived from the original queries applied to the click sequences of the held-out queries; and, as a reference point, the true impression sequences of the held-out action sequences.

Table 5. Weighted-by-frequency mean squared error (WMSE) between the observed $\hat{C}(i)$ values and the $\hat{C}(i)$ values estimated using impression models, measured used the held-out action sequences. Lower numbers are better. Model 2 significantly outperformed the other models (Wilcoxon signed-rank test, $p < 0.01$).

Model	WMSE (top-10)	WMSE (top-50)
Clicks	172.0×10^{-3}	166.8×10^{-3}
ZPM ($\mu = 5$)	6.7×10^{-3}	4.2×10^{-3}
ZPM ($\mu = 50$)	4.2×10^{-3}	2.7×10^{-3}
Model 1	4.4×10^{-3}	2.6×10^{-3}
Model 2	2.1×10^{-3}	1.3×10^{-3}

Impression Distribution. Recall that $W(i)$ is the weight associated with the ith item in the SERP in terms of a weighted-precision effectiveness metric, and is a direct estimate of the probability of the user viewing the ith summary in the ranking and thereby generating an impression. In a weighted-precision metric these weights $W(i)$ are non-increasing, with $W(i) \geq W(i+1)$, which implies that the probability of viewing the documents at a deep rank is less than the probability of viewing a document at a shallow rank. We employ KL-divergence to measure the difference between pairs of probability distributions, again comparing the output of the four impression models (tuned on the original action sequences, and then applied to the held-out click sequences) with the "true" weights derived from the held-out impression sequences. Note that the operational definition of the distribution is the total observed numbers of impressions (or clicks) per rank position, normalized by the total number of impressions (or clicks) overall. Table 6 shows the resultant values when computed across the first 10 and 50 items in each SERP. Under this alternative evaluation process Model 1 outperforms Model 2, and, for the depth 50 evaluation, the ZPM methods also outperform Model 2. Nevertheless, all of the impression models yield outcomes that are better than those provided by the click distribution in all of the four evaluations. Determining which of WMSE or KL-divergence is the more useful "nearness" criteria (or whether there is a third assessment approach that might be applied) is an area for future work.

Table 6. KL-divergence scores assessing $\hat{W}(i)$ estimated using impression models P, and the reference distribution I, with $\hat{W}(i)$ estimated using held-out impression sequences across 10 and 50 items in each ranking. Lower numbers are better.

P	$KL(P \,\|\, I)$ (top-10)	$KL(P \,\|\, I)$ (top-50)
Click distribution	4.62×10^{-2}	8.87×10^{-2}
ZPM ($\mu = 5$)	0.71×10^{-2}	4.14×10^{-2}
ZPM ($\mu = 50$)	0.40×10^{-2}	3.24×10^{-2}
Model 1	0.22×10^{-2}	4.03×10^{-2}
Model 2	0.29×10^{-2}	4.54×10^{-2}

5 Conclusion

We have examined the patterns of clicks and impressions in rich interaction logs derived from a job search service, and confirmed that before making each of their clicks (say, at rank i) users have usually inspected the great majority of job summaries ranked at positions ahead of i, plus several summaries beyond rank i. Based on these findings, we developed an impression model that infers which documents in the ranking are likely to have been examined by the user, based on the observed click sequence, thereby allowing estimation of parameters for user models such as RBP and INSQ from such logs. Our study also confirmed that the deepest click rank and the number of distinct clicked items are key factors in terms of predicting how many additional job summaries the user viewed beyond their deepest click action.

Acknowledgment. This work was supported by the Australian Research Council's Linkage Projects scheme (project number LP150100252) and by Seek.com.

References

1. Azzopardi, L., Thomas, P., Craswell, N.: Measuring the utility of search engine result pages. In: Proceedings of SIGIR, pp. 605–614 (2018)
2. Carterette, B.: System effectiveness, user models, and user utility: a conceptual framework for investigation. In: Proceedings of SIGIR, pp. 903–912 (2011)
3. Cutrell, E., Guan, Z.: What are you looking for? An eye-tracking study of information usage in web search. In: Proceedings of CHI, pp. 407–416 (2007)
4. Hasan, M.A., Parikh, N., Singh, G., Sundaresan, N.: Query suggestion for e-commerce sites. In: Proceedings of WSDM, pp. 765–774 (2011)

5. Joachims, T., Granka, L., Pan, B., Hembrooke, H., Gay, G.: Accurately interpreting clickthrough data as implicit feedback. SIGIR Forum **51**(1), 4–11 (2017)
6. Klöckner, K., Wirschum, N., Jameson, A.: Depth- and breadth-first processing of search result lists. In: Proceedings of CHI, pp. 1539–1539 (2004)
7. Mansouri, B., Zahedi, M.S., Campos, R., Farhoodi, M.: Online job search: study of users' search behavior using search engine query logs. In: Proceedings of SIGIR, pp. 1185–1188 (2018)
8. Moffat, A., Zobel, J.: Rank-biased precision for measurement of retrieval effectiveness. ACM Trans. Inf. Syst. **27**(1), 2.1–2.27 (2008)
9. Moffat, A., Bailey, P., Scholer, F., Thomas, P.: Incorporating user expectations and behavior into the measurement of search effectiveness. ACM Trans. Inf. Syst. **35**(3), 24.1–24.38 (2017)
10. Parikh, N., Sundaresan, N.: Inferring semantic query relations from collective user behavior. In: Proceedings of CIKM, pp. 349–358 (2008)
11. Smucker, M.D., Clarke, C.L.A.: Time-based calibration of effectiveness measures. In: Proceedings of SIGIR, pp. 95–104 (2012)
12. Spina, D., et al.: Understanding user behavior in job and talent search: an initial investigation. In: SIGIR Workshop. eCommerce (2017)
13. Thomas, P., Scholer, F., Moffat, A.: What users do: the eyes have it. In: Banchs, R.E., Silvestri, F., Liu, T.-Y., Zhang, M., Gao, S., Lang, J. (eds.) AIRS 2013. LNCS, vol. 8281, pp. 416–427. Springer, Heidelberg (2013). https://doi.org/10.1007/978-3-642-45068-6_36
14. White, R.W., Drucker, S.M.: Investigating behavioral variability in web search. In: Proceedings of WWW, pp. 21–30 (2007)
15. Wicaksono, A.F., Moffat, A.: Exploring interaction patterns in job search. In: Proceedings of Australasian Document Computing Symposium, pp. 2.1–2.8 (2018)
16. Wicaksono, A.F., Moffat, A.: Empirical evidence for search effectiveness models. In: Proceedings of CIKM, pp. 1571–1574 (2018)
17. Zhang, Y., Park, L.A.F., Moffat, A.: Click-based evidence for decaying weight distributions in search effectiveness metrics. Inf. Retr. **13**(1), 46–69 (2010)

Image IR

Automated Semantic Annotation of Species Names in Handwritten Texts

Lise Stork[1,2(✉)], Andreas Weber[3], Jaap van den Herik[1,2], Aske Plaat[1,2], Fons Verbeek[1,2], and Katherine Wolstencroft[1,2]

[1] Leiden Institute of Advanced Computer Science, Leiden, the Netherlands
{l.stork,k.j.wolstencroft,f.j.verbeek,a.plaat}@liacs.leidenuniv.nl
[2] Leiden Centre of Data Science, Leiden, the Netherlands
h.j.vandenherik@law.leidenuniv.nl
[3] University of Twente, Enschede, the Netherlands
a.weber@utwente.nl

Abstract. In this paper, scientific species names from images of *handwritten* species observations are automatically recognised and annotated with semantic concepts, so that they can be used for document retrieval and faceted search. Until now, automated semantic annotation of such named entities was only applied to printed or digital text. We employ a two-step approach. First, word images are classified, identifying elements of scientific species names; `Genus, species, author`, using (i) visual structural features, (ii) position, and (iii) context. Second, the identified species names are semantically annotated according to the *NHC-Ontology*, an ontology that describes species observations. Internationalised Resource Identifiers (IRIs) are assigned to the elements so that they can be linked and disambiguated at a later stage by individual researchers. For the identification of scientific species names, we achieve an average *F1* score of 0.86. Moreover, we discuss how our method will function in a semi-automated annotation process, with a fruitful dialogue between system and user as the main objective.

Keywords: Deep learning · Ontologies · Taxonomy · Scientific names · Semantic annotation · Historical biodiversity research

1 Introduction

Handwritten material brought back from biodiversity expeditions is an important source of information for naturalists and historians. An abundance of these records is available for research [21]. Much of these data, however, remain computationally inaccessible and difficult to explore [4]. This presents an interesting challenge to both the field of *information extraction* and *document retrieval*. Scientific descriptions or depictions of species observations carefully employ the systematic organisation of species variations. Thus, despite the often difficult nature of the data - hard-to-read, multi-lingual, historical texts - document retrieval can exploit the systematic organisation of the document content.

L. Azzopardi et al. (Eds.): ECIR 2019, LNCS 11437, pp. 667–680, 2019.
https://doi.org/10.1007/978-3-030-15712-8_43

Since the onset of field work in biodiversity expeditions, species observation data have been manually recorded by researchers. Records are fittingly named *field books* [15]. Starting from the first part of the 18th century, Linnaean taxonomy and binomial nomenclature was generally used for the classification and naming of species [18]. Therefore, most historical field books found today in musea and other institutions adhere to Linnaeus's *Systema Naturae* [19]. Due to a common system for the classification of organisms, historical species names can potentially be referenced and compared to current ones, allowing researchers to study the changes in biodiversity over time. However, transforming raw historical biodiversity data to usable structured knowledge is still one of the main challenges of historical taxonomy research [7,22].

In this work we use state-of-the-art techniques from computer vision and semantic web technologies to (i) identify the elements of scientific species names in handwritten document images, and (ii) link and structure the elements, using an ontology for species observations. We use the MONK handwriting recognition system [23] to segment the document images into single word images. Our main contribution is the identification and semantic annotation of scientific species names from word images containing *handwritten* text. We build on previous work [27], where an ontology and software for semantic annotation of species observation records was constructed and tested with domain experts. Here, we advance these methods by automating the process of semantic annotation. Biological taxonomies, once extracted from field books, can be used by algorithms aiming to exploit query expansion techniques, while it allows users to semantically query, or browse through, field book collections. As the species names are structured via a controlled vocabulary that is well-used in the domain, extracted species names can also be federated across collections.

2 Species Classification and Nomenclature

In the binomial nomenclature, scientific names consist of minimally two and maximally four types of elements. The first type identifies the genus to which the organism belongs. The second type is called the *specific epithet*, the specific species within that genus. Commonly, the binomial is followed by the author name, and the date when the name was published in literature. It is also common for a name to have more than one author. Below in Fig. 1, an example of a scientific species name from a field note is given; it dates back to 1821.

Species names are ambiguous due to evolving taxonomical systems, nomenclature and opposing views within the science of classification [12,18]. Therefore, scientific names become valuable for scientific research when they are compared to synonyms or homonyms from alternative classifications and their respective meta-data. In the rest of this paper, we will use the term *scientific name* to refer to, minimally, a *genus* and *species* tuple or *genus*, *species* and *author* triple.

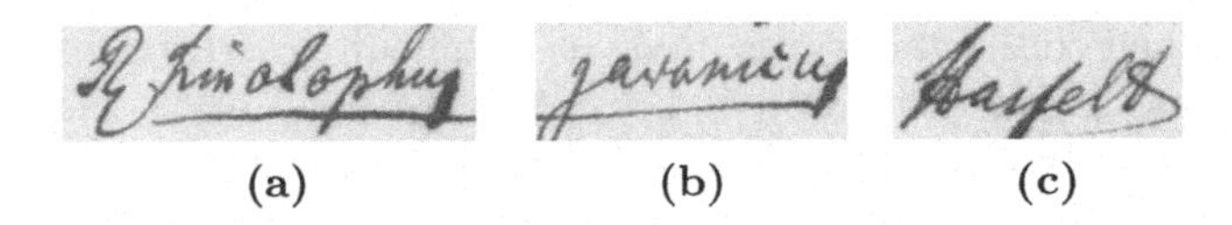

(a) (b) (c)

Fig. 1. A scientific name in binomial nomenclature: (a) *Rhinolophus* (genus) (b) *javanicus* (species) (c) *Hasselt* (author of the name: *Johan Coenraad van Hasselt*)

3 Related Work

Organisations and researchers that dedicate themselves to the preservation of natural history collections, such as *IdigBio*[1] or the *Biodiversity Heritage Library* [9], continuously develop new methods to digitise specimen collections in a cost-effective and sustainable way, in order to facilitate ongoing species research. The automatic extraction of scientific names from text is essential for the management of archival resources. Therefore, there are several examples of methods for extracting and disambiguating species names from printed texts, but extracting the same information from handwritten texts is much more of a challenge. Taxongrab [13], for example, automatically extracts species names from printed biological texts. The Biodiversity Heritage Library, that aggregates scans of biodiversity publications and field notes, indexes scientific names extracted from the publications - printed text - in their collection, to improve accessibility for taxonomists. They match the text, extracted via Optical Character Recognition (OCR), with the Taxonomic Name Server (TNS) to identify likely scientific names [9]. They are not the only ones exploiting the power of automatic text processing for the digitisation of natural history collections. Software has been developed to parse OCR output of printed text to formalised Darwin Core[2] entries for archival and retrieval purposes [10]. Drinkwater and others [8] investigate the aid of OCR in the digitisations of herbarium specimen labels, finding significant increase in time effectiveness using OCR output to sort specimens prior to database submission, and to add data to minimal database records. They explicitly note that OCR is currently only possible for typed and printed labels and not for handwritten text.

Handwritten Text Recognition (HTR) is one of the more challenging tasks within the field of Document Image Analysis and Recognition (DIAR), mainly due to the huge variety in writing styles and languages, paper degradation, overlapping words and historical handwriting. The recognition of named entities - real word objects, such as: *locations*, *persons*, *organisations* - in handwritten text can help document understanding and searchability of the text, and can potentially aid handwriting recognition [5]. Formerly, Named Entity Recognition and Classification (NERC) was a task solely used on digital text [17], but it has just recently also been applied directly to handwritten text [1,5,25,28]. Especially when few instances of words exist and a collection consists of many different

[1] https://www.idigbio.org/.
[2] http://rs.tdwg.org/dwc/.

handwritings and connected words, making it difficult to create character-based representations, the identification of key words can help make the text searchable, and potentially aid HTR. Moreover, in many cases, full-text transcriptions of entire pages of field books are not required in order to make them digitally accessible.

In this contribution, we develop a novel approach to identify domain specific named entities, elements of *scientific species names*, in historical *handwritten* document images. Rather than first transcribing the text and performing NERC afterwards on the digital text, we exploit characteristics of the document images to identify the named entities, using terms from the NHC-Ontology[3] to classify and organise them. We argue that the ability to quickly index handwritten document images based on scientific names, ranks and authors, helps users to navigate through large collections of documents in online libraries, such as the Biodiversity Heritage Library. It opens up possibilities for faceted search, semantic querying and semantic recommendations. Additionally, maintaining a link to the word image and its position in the full document image is important to allow the repetition of image processing experiments as well as to allow researchers to view the original document and therefore the extracted text in context.

4 Data

One of the main issues history of science and natural history researchers encounter is the inaccessibility of natural history archival collections. Field books, drawings and specimens are physically stored in museum collection facilities or research institutes, hidden from external researchers and policymakers interested in long-term developments of global biodiversity [7].

Table 1. Data set class count

Class	Genus	Species	Author	Other	Total
y	0	1	2	3	
n	177	167	144	17309	17797

Transcribed field books exist online, but (to the best of our knowledge) no segmented and annotated images of handwritten species observations are available online for experimental research using image processing methods. Therefore, word images from 240 field notes from a natural history collection have been segmented and semantically annotated. This has been carried out in the context of the project *Making Sense of Illustrated Handwritten Archives* [31].[4] From a field book on mammals, we selected field notes from four different writers, to account for different handwriting styles and structures, ensuring a representative

[3] https://makingsense.liacs.nl/rdf/nhc/.
[4] http://www.makingsenseproject.org.

data set to demonstrate how the automated methods perform on heterogeneous, real-world data. The segmented word images were obtained from a nichesourcing effort, with the help of a handwriting recognition system MONK and a group of domain expert labellers. The word images were subsequently manually annotated using four classes, as shown in Table 1. Two of four classes are taxonomical entities. The third class refers to the publisher of the taxonomical name, and lastly we have the class *Other*, which includes all words that do not belong to any of the previously mentioned classes. The final counts of examples per class are shown in Table 1. The process of labelling and annotating words is time-consuming and, in our case, requires expert knowledge. Therefore, limited training data is available. As machine learning methods generally require a very large number of annotated samples, methods have to be adjusted to the data set size to acquire a predictive model that generalises well. These adjustments are described in Sects. 5 and 6. This is also one of the challenges of such projects; to create an adaptive learning system with a generic method that learns from small amounts of annotated data, but adapts to new data and performs better over time when more data is annotated. The data set used in this work can be found online.[5]

5 Scientific Name Extraction Model

Below we describe the methods that were used in this work. The full pipeline is shown in Fig. 2, the blue rectangle indicating the scope of this work.

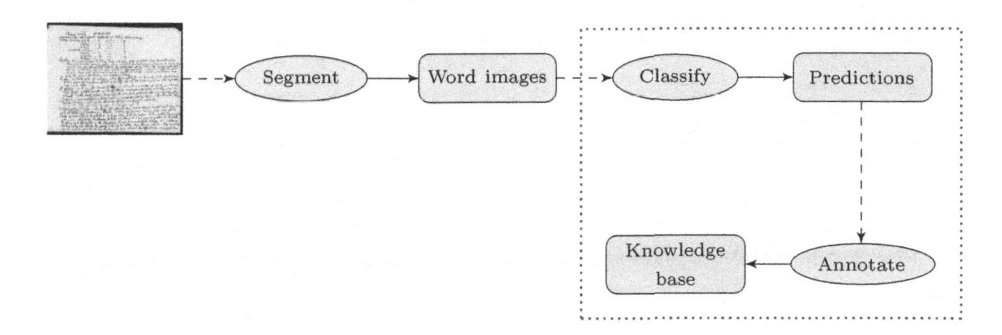

Fig. 2. The full pipeline: automated semantic annotation of scientific names

We used the MONK handwriting recognition system, developed by Schomaker, for word segmentation [3,23,29,30]. First, the system segments handwritten document images into lines and second, relative to those lines, into word zones that potentially hold words. The system allows the labelling, at the word level, of word images by domain experts. It then uses these labels for

[5] 10.5281/zenodo.2545573.

HTR. In this work, the word images were manually annotated using four semantic concepts, or classes: *genus*, *species*, *author* and *other*. The classification of each word image to its corresponding semantic class is discussed in Sect. 5.1. In Sect. 5.2, we discuss the semantic annotation of the classified word images using the NHC-Ontology[6] for species observations.

5.1 Classification of Word Images

To classify the word images to one of four classes, we use three distinct features; *visual structural features, position and context*. We chose to create one single neural architecture, built with help of Keras [16], that could be trained end-to-end, so that the classification error is only propagated once, in contrast to using predictions from multiple classifiers and combining them after training to form a single prediction. The final architecture is explained visually in Fig. 3, and will be discussed below.

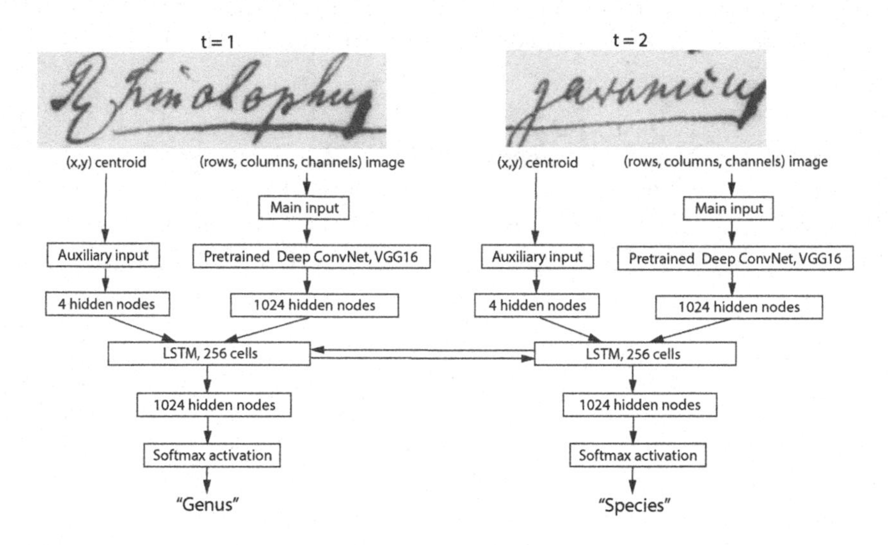

Fig. 3. The MLP-CNN-LSTM architecture, "unrolled" for both time steps t.

Visual Structural Features. The feature detector that was used in this work for the detection of visual structural features is a Convolutional Neural Network (CNN) [14]. It has been shown that CNNs outperform other neural networks on image recognition tasks [26]. The basic network used here is a deep CNN for object recognition developed and trained by Oxford's Visual Geometry Group (VGG) and called the VGG network [26]. We use their configuration, with 16 convolutional layers, and import weights from the VGG, pre-trained on the ImageNet task [6]. Previous work [20] has demonstrated that transferring image representations with CNNs overcomes the problem of training with limited training

[6] http://www.makingsense.liacs.nl/rdf/nhc/.

data, e.g., less than a few thousand training images, despite differences in image statistics between the *source* data set and *target* data set. By, for instance, training on the ImageNet task, the VGG model learns filters on various different scales, which can be used as feature extractors for other types of images. These features, extracted from handwritten documents with help of the convolutional part of the VGG network, are used for training a simple Multi-Layer Perceptron (MLP) on our task.

Position. In addition to visual features, the position of a word in a document often provides a good descriptive feature for the recognition of a named entity. The position is therefore often used as a feature in the field of NERC, however, it has been used more often in text [17] than in images [1,5,28]. In this work, we use the *relative* centroid of a word image's position in the image as input features to a simple MLP. Hence, each training example $(\mathbf{x}^{(i)}, y^{(i)})$, $\mathbf{x}^{(i)} \in \mathbb{R}^2$, where every x_i lies within the interval [0,1], is used to train a simple MLP with 4 hidden layers. To train the entire model end-to-end, we concatenated the last hidden layers of both models. The merged hidden layer therefore has a size of $1024 + 4 = 1028$.

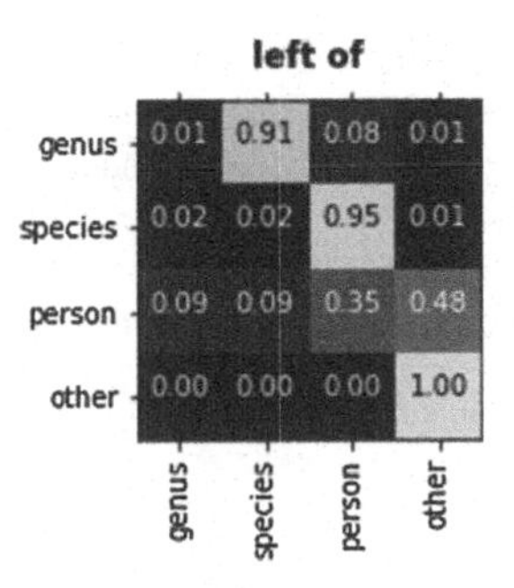

Fig. 4. Adjacency matrix that shows frequencies for word bi-grams (sequences of two adjacent words). E.g., 'genus' was **left of** 'species' 91% of the time 'genus' was encountered.

Context. As a third feature type, we introduce context: the characteristics of adjacent word images, specifically *bi-grams*. Figure 4 shows frequencies for word image bi-grams. First, horizontal pairwise alignment was calculated for each pair of word images $(w^{(i)}, w^{(j)})$, $w \in W$, where $i \neq j$. They were seen as horizontally aligned if $y_1^{(i)} < y_c^{(j)} < y_2^{(i)}$, where $y_1^{(i)}$ indicates the first y coordinate of the bounding box of $w^{(i)}$, $y_2^{(i)}$ the second, and $y_c^{(j)}$ the y coordinate of the centroid of $w^{(j)}$.

Second, the right neighbouring word of $w^{(i)}$ was retrieved by calculating all pairwise vertical distances for the horizontally aligned words: $dist_{ij} = x_c^{(i)} - x_c^{(j)}$, where $x_c^{(i)}$ and $x_c^{(j)}$ refer to the x coordinates of the centroids of $w^{(i)}$ and $w^{(j)}$.

The smallest negative distance indicated right adjacency. The adjacency matrix only takes into account instances that actually *have* an adjacent word, as it could be that a word is surrounded by white space on every side.

As expected, the different classes have strong co-occurrence dependencies. Therefore, we converted the data set to sequences of size two (bi-grams), and added a last layer to the model architecture for sequence prediction. For an adequate prediction we used a Bidirectional Long Short-Term Memory (BLSTM) neural network, a type of Recurrent Neural Network (RNN) that implements a *memory node* in order to learn long-term dependencies between features [24]. By using the bidirectional variant of the LSTM [11], dependencies can be learned in both horizontal orientations, see Fig. 3. This is beneficial for our work, as in the bi-gram *species-author*, the identification of the 'author' class largely depends on the visual characteristics of the word image left adjacent to it.

```
nc:taxon1 rdf:type dwc:Taxon
          nhc:scientificNameAuthorship nc:author1
          nhc:taxonRank nc:species

nc:author1 rdf:type foaf:Person

nc:anno1 oa:hasBody nc:taxon1
         oa:hasTarget nc#image1.jpg#xywh=x,y,h,w
         oa:hasTarget nc#image1.jpg#xywh=x,y,h,w

nc:anno2 oa:hasBody nc:author1
         oa:hasTarget nc#image1.jpg#xywh=x,y,h,w
```

Listing 1.1. Example of a semantically annotated species name

5.2 Semantic Annotation of Word Images

The NHC-Ontology[7] is an ontology for species observations, based on the Darwin Semantic Web (DSW) Ontology, and written in OWL.[8] The ontology is centered around the description of meta-data relating to the observation of an organism, and allows a researcher to describe to which various taxon groups an organism is identified by a researcher. The model uses the Web Annotation Data Model[9] to link bounding boxes of word images to their semantic labels. In the examplary fragment above, listing 1.1, two images refer to a genus and a species, which together constitute one taxonomical name `nc:taxon1`[10] of rank `nc:species`. They are linked to the publisher of the name with the `nhc:scientificNameAuthorship` predicate.

[7] http://makingsense.liacs.nl/rdf/nhc/, https://github.com/lisestork/nhc-ontology/.
[8] https://www.w3.org/OWL/.
[9] https://www.w3.org/TR/annotation-model/.
[10] *nc:* is the prefix for the http://makingsense.liacs.nl/rdf/nhc/nc# namespace.

6 Experiments and Results

To analyse the influence of the three features on the predictive performance of the model, we conducted multiple experiments where we tested the performance of the pre-trained CNN, MLP-CNN and MLP-CNN-BLSTM.

6.1 Experimental Methodology

Before training, the images were scaled by dividing them by 255 so that they would fall within the range [0–1]. All images were re-sized to the average image dimensions: $y = 74$, $x = 139$. No data augmentation was used. Based on horizontal adjacency, as explained in Subsect. 5.1, image bi-grams were constructed, sequences of $l = 2$, as input to the BLSTM.

The word images were shuffled, keeping together word images from the same page, and thereafter split into a train and test set. As one word image could occur in two bi-grams, we hereby avoid that word images from the test set were also in the training set, which would bias the classification results. However, by shuffling the pages, we still ensure that the model does not overfit to one writing style or structure. We used 80% of the word images for training and the remaining partition as test set, making sure that 20% of the scientific name elements were in the test set. As classes in the word bi-grams were highly imbalanced, we used random minority oversampling with replacement, to increase the counts of samples from minority classes in the training data. When training a CNN, oversampling is thought to be the best method to deal with imbalanced data sets with few examples in minority classes, and appears to work best if the oversampling totally eliminates the imbalance [2]. However, as we are dealing with sequences rather than singular samples, we chose to oversample sequences, e.g., *species-author*. Converted back to singular images, this would result in a *step imbalance* with a small imbalance ratio $p = \pm 1.1$ rather than a large imbalance ratio of $p = \pm 16$ [2].

The networks were all trained using the Adam classifier with a learning rate of 10^{-4} and categorical cross-entropy loss. Each network was trained using early stopping with patience 2, meaning that training was stopped when, for two epochs, the validation error was increasing. Per epoch, the weights were only stored if the predictive performance had increased compared to the previous epoch. In the testing phase, thresholding was applied to the output of the networks to compensate for oversampling the data during training, as oversampling alters prior probability distributions. One way to perform thresholding is to simply correct for these prior probabilities, by dividing the output of the network for each class, then seen as posterior probabilities, by the estimated prior probabilities. In our case, the imbalance was not completely eliminated, so the thresholds were calculated as the ratio between the original class counts and those after oversampling.

As a final step, the output of the model that performed best was used to test the whole pipeline. Word images from the test set, that were classified as scientific names, were assigned IRIs within the project's namespace, e.g., `nc:taxon1`.

The names were linked and semantically enriched using terms from the ontology and transformed to the Resource Description Framework (RDF) format. The code can be found online.[11]

6.2 Results and Discussion

Table 2 summarises the final classification results for each network. Due to a large class imbalance, precision and recall were used to assess the predictive power of the classifier. Reporting accuracies would be misleading, as they would portray the underlying distribution rather than the predictive power of the model (if the model would always predict 'Other', it would be a bad predictor for the task, but the accuracy would be 93%, as the 'Other' class accounts for 93% of the data). The table indicates that the BLSTM produced the highest average $F1$ scores for each class. The addition of the BLSTM layer specifically increases precision and recall scores for the author names. This makes sense; without context these appear similar to regular words. The input of centroid data to the network does not have an effect on the recall or precision of author names, but does increase precision for the retrieval of species names. Figure 5 shows 4 images from the test set that were misclassified. While both the CNN and MLP-CNN network misclassify most of the same word images, the output of the MLP-CNN-BLSTM is quite different. Image (a) and (b) were both misclassified by the networks without the BLSTM layer, but were correctly classified by the

Table 2. Classification results per network

Method	Class	Precision	Recall	F1-score	Support
CNN	Genus	0.80	0.78	0.79	36
	Species	0.64	0.97	0.77	33
	Author	0.78	0.78	0.78	32
	Other	1.00	0.97	0.98	525
	avg/total	0.82	0.77	0.80	626
+MLP	Genus	0.85	0.81	**0.83**	36
	Species	0.81	0.88	**0.84**	33
	Author	0.78	0.78	0.78	32
	Other	0.99	0.99	0.99	525
	avg/total	0.96	0.96	0.96	626
+BLSTM	Genus	0.86	0.89	**0.88**	36
	Species	0.94	0.91	**0.92**	33
	Author	0.78	0.88	**0.82**	32
	Other	1.00	0.99	0.99	525
	avg/total	0.98	0.97	0.98	626

[11] https://github.com/lisestork/asa-species-names.

final model. Image (a) for example, was classified as 'Species', while actually being labelled as an author name. Visually, it resembles a species name; it is underlined and appears in a similar position on the page. Without context of other words it is challenging to correctly classify such images without proper historical knowledge of the domain. Image (b) was misclassified as 'Other', but correctly identified as an author name in the BLSTM model, most likely due to the visual characteristics of the word image that is left adjacent. On the other hand, image (c) and (d) are together misclassified as a species name and its author by the BLSTM network, while they were correctly classified by the other networks. Eyeballing the images, we see that they are adjacent and visually resemble these classes (capitals, underlining).

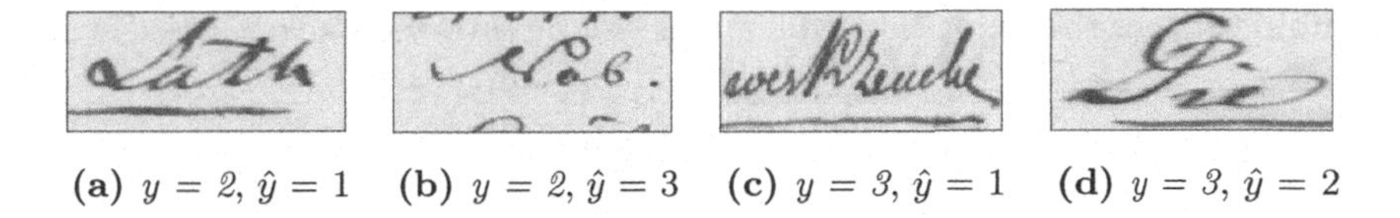

(a) $y = 2, \hat{y} = 1$ **(b)** $y = 2, \hat{y} = 3$ **(c)** $y = 3, \hat{y} = 1$ **(d)** $y = 3, \hat{y} = 2$

Fig. 5. Four misclassified examples. Classlabels relate to those discussed in Table 1

In Table 3, we present retrieval scores for the identification of complete scientific names from field book pages. A python script parsed the recognised species elements from the test set, and connected them together using the NHC-Ontology. A total of 27 out of 36 species names were retrieved, with an *F1* score of 0.86. Interestingly, there were no false-positives among the final predictions. Figure 6 shows one of the correctly classified scientific names. The final RDF data set can be queried through our online SPARQL endpoint.[12]

Table 3. Final classification results for the detection of scientific names

Method	Class	Precision	Recall	F1-score	Support	Total
+BLSTM	Scientific names	1.0	0.75	0.86	27	36

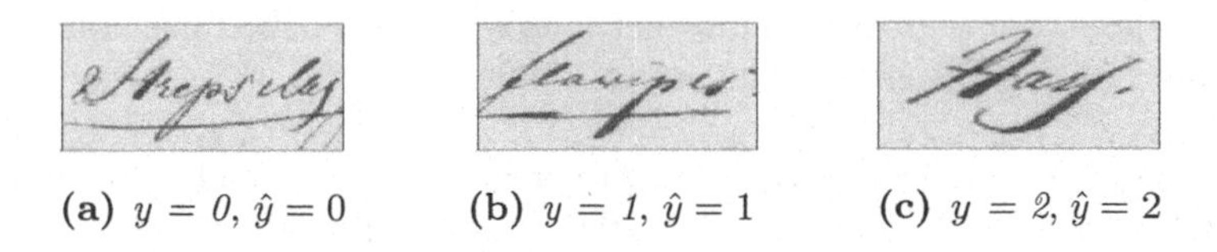

(a) $y = 0, \hat{y} = 0$ **(b)** $y = 1, \hat{y} = 1$ **(c)** $y = 2, \hat{y} = 2$

Fig. 6. A correctly classified scientific species name: (a) *Genus* (b) *Species* (c) *Person*

[12] http://makingsense.liacs.nl/rdf4j-server/repositories/SN, can be queried through a query editor such as: https://yasgui.org/.

6.3 A Semi-automated Process

This work serves as a step within the development of an adaptive system, with the MONK handwriting recognition system at its core [31], for the segmentation, recognition and semantic annotation of handwritten words, named entities and illustrations from historical biodiversity collections. Using labelling input from domain experts, representations of the document images are learned in order to generate new, machine learned, labels. Simultaneously, domain experts can provide contextual knowledge on specific biodiversity expeditions from which the annotation process can benefit. For example, named entities - such as author names - can be used to pre-populate the knowledge base so that they can be retrieved during the semantic annotation process. Moreover, domain experts can link the - validated - automatically identified scientific names to word images containing higher ranks, so that collections can be browsed using faceted search.

7 Conclusions and Future Work

In this work we show that we can accurately identify and classify components of handwritten species observation records from different features: visual structural features, position and context. We show that our methods are applicable even though the data set contains four authors with different handwriting styles and different processes of recording their species observations. A major challenge of working with handwritten text is its irregularity. Our results show that we can mitigate this challenge by building up multiple pieces of evidence for classification by learning from multiple features. Each of the different features we examine in our model adds information and improves the overall results. In addition, as the results are extracted and structured in RDF as part of the process, they are immediately available for search and comparison with other archives - historical or present day.

The entire data set used for these experiments is part of the same expedition archive. Although we represent multiple authors and styles, the next step would be to demonstrate the generic nature of our results by analysing biodiversity records from other expeditions. Once we establish that, we will extend our methods to identify other common classes from biodiversity data, for example, locations, dates and anatomical entities. In the context of the Making Sense project, we aim to integrate the new methods with established methods for automated handwriting recognition, using the MONK system.

Acknowledgements. This work is supported by the Netherlands Organisation for Scientific Research (NWO), grant 652.001.001, and Brill publishers.

References

1. Adak, C., Chaudhuri, B.B., Blumenstein, M.: Named entity recognition from unstructured handwritten document images. In: 12th IAPR Workshop on Document Analysis Systems (DAS), pp. 375–380. IEEE (2016)
2. Buda, M., Maki, A., Mazurowski, M.A.: A systematic study of the class imbalance problem in convolutional neural networks. Neural Netw. **106**, 249–259 (2018)
3. Bulacu, M., van Koert, R., Schomaker, L., van der Zant, T.: Layout analysis of handwritten historical documents for searching the archive of the cabinet of the Dutch queen. In: Proceedings of the Ninth International Conference on Document Analysis and Recognition, vol. 1, 2, pp. 357–361. IEEE (2007)
4. Canfield, M.R.: Field Notes on Science & Nature. Harvard University Press, Cambridge (2011)
5. Carbonell, M., Villegas, M., Fornés, A., Lladós, J.: Joint recognition of handwritten text and named entities with a neural end-to-end model. arXiv preprint arXiv:1803.06252 (2018)
6. Deng, J., Dong, W., Socher, R., Li, L., Li, K., Fei-Fei, L.: Imagenet: a large-scale hierarchical image database. In: Proceedings of the 2009 Computer Vision and Pattern Recognition, CVPR, pp. 248–255. IEEE (2009)
7. Drew, J.A., Moreau, C.S., Stiassny, M.L.: Digitization of museum collections holds the potential to enhance researcher diversity. Nature Ecol. Evol. **1**(12), 1789–1790 (2017)
8. Drinkwater, R.E., Cubey, R.W., Haston, E.M.: The use of optical character recognition (OCR) in the digitisation of herbarium specimen labels. PhytoKeys **38**, 15–30 (2014)
9. Gwinn, N.E., Rinaldo, C.: The biodiversity heritage library: sharing biodiversity literature with the world. IFLA J. **35**(1), 25–34 (2009)
10. Heidorn, P.B., Wei, Q.: Automatic metadata extraction from museum specimen labels. In: Proceedings of the 2008 International Conference on Dublin Core and Metadata Applications, pp. 57–68 (2008)
11. Hochreiter, S., Schmidhuber, J.: Long short-term memory. Neural Comput. **9**(8), 1735–1780 (1997)
12. Kennedy, J.B., Kukla, R., Paterson, T.: Scientific names are ambiguous as identifiers for biological taxa: their context and definition are required for accurate data integration. In: Ludäscher, B., Raschid, L. (eds.) DILS 2005. LNCS, vol. 3615, pp. 80–95. Springer, Heidelberg (2005). https://doi.org/10.1007/11530084_8
13. Koning, D., Sarkar, I.N., Moritz, T.: Taxongrab: extracting taxonomic names from text. Biodivers. Inf. **2**, 79–82 (2005)
14. LeCun, Y., Bottou, L., Bengio, Y., Haffner, P.: Gradient-based learning applied to document recognition. Proc. IEEE **86**(11), 2278–2324 (1998)
15. MacGregor, A. (ed.): Naturalists in the Field. Brill, Leiden (2018)
16. Chollet, F., et al.: Keras (2015). https://keras.io
17. McCallum, A., Li, W.: Early results for named entity recognition with conditional random fields, feature induction and web-enhanced lexicons. In: Proceedings of the Seventh Conference on Natural Language Learning, vol. 4, pp. 188–191. Association for Computational Linguistics (2003)
18. Miracle, M.E.G.: On whose authority? Temminck's debates on zoological classification and nomenclature: 1820–1850. J. Hist. Biol. **44**(3), 445–481 (2011)
19. Müller-Wille, S.: Names and numbers: "data" in classical natural history, 1758–1859. Osiris **32**(1), 109–128 (2017)

20. Oquab, M., Bottou, L., Laptev, I., Sivic, J.: Learning and transferring mid-level image representations using convolutional neural networks. In: Proceedings of the 2014 IEEE Conference on Computer Vision and Pattern Recognition, pp. 1717–1724 (2014)
21. Page, L.M., MacFadden, B.J., Fortes, J.A., Soltis, P.S., Riccardi, G.: Digitization of biodiversity collections reveals biggest data on biodiversity. BioScience **65**(9), 841–842 (2015)
22. Sarkar, I.N.: Biodiversity informatics: organizing and linking information across the spectrum of life. Briefings Bioinform. **8**(5), 347–357 (2007)
23. Schomaker, L.: Design considerations for a large-scale image-based text search engine in historical manuscript collections. It - Inf. Technol. **58**(2), 80–88 (2016)
24. Schuster, M., Paliwal, K.K.: Bidirectional recurrent neural networks. IEEE Trans. Signal Process. **45**(11), 2673–2681 (1997)
25. Shi, Z.: Datefinder: detecting date regions on handwritten document images based on positional expectancy. Master's thesis, University of Groningen, Groningen, the Netherlands (2016)
26. Simonyan, K., Zisserman, A.: Very deep convolutional networks for large-scale image recognition. CoRR, abs/1409.1556 (2014)
27. Stork, L., et al.: Semantic annotation of natural history collections. Web Semant. Sci. Serv. Agents World Wide Web (2018). https://doi.org/10.1016/j.websem.2018.06.002
28. Toledo, J.I., Sudholt, S., Fornés, A., Cucurull, J., Fink, G.A., Lladós, J.: Handwritten word image categorization with convolutional neural networks and spatial pyramid pooling. In: Robles-Kelly, A., Loog, M., Biggio, B., Escolano, F., Wilson, R. (eds.) S+SSPR 2016. LNCS, vol. 10029, pp. 543–552. Springer, Cham (2016). https://doi.org/10.1007/978-3-319-49055-7_48
29. Van der Zant, T., Schomaker, L., Haak, K.: Handwritten-word spotting using biologically inspired features. IEEE Trans. Pattern Anal. Mach. Intell. **30**(11), 1945–1957 (2008)
30. van Oosten, J.-P., Schomaker, L.: Separability versus prototypicality in handwritten word-image retrieval. Pattern Recogn. **47**(3), 1031–1038 (2014)
31. Weber, A., Ameryan, M., Wolstencroft, K., Stork, L., Heerlien, M., Schomaker, L.: Towards a digital infrastructure for illustrated handwritten archives. In: Ioannides, M. (ed.) Digital Cultural Heritage. LNCS, vol. 10605, pp. 155–166. Springer, Cham (2018). https://doi.org/10.1007/978-3-319-75826-8_13

Tangent-V: Math Formula Image Search Using Line-of-Sight Graphs

Kenny Davila[1(✉)], Ritvik Joshi[2], Srirangaraj Setlur[1], Venu Govindaraju[1], and Richard Zanibbi[2]

[1] University at Buffalo, Buffalo, NY 14260, USA
{kennydav,setlur,govind}@buffalo.edu
[2] Rochester Institute of Technology, Rochester, NY 14623, USA
{rj9336,rlaz}@cs.rit.edu

Abstract. We present a visual search engine for graphics such as math, chemical diagrams, and figures. Graphics are represented using Line-of-Sight (LOS) graphs, with symbols connected only when they can 'see' each other along an unobstructed line. Symbol identities may be provided (e.g., in PDF) or taken from Optical Character Recognition applied to images. Graphics are indexed by pairs of symbols that 'see' each other using their labels, spatial displacement, and size ratio. Retrieval has two layers: the first matches query symbol pairs in an inverted index, while the second aligns candidates with the query and scores the resulting matches using the identity and relative position of symbols. For PDFs, we also introduce a new tool that quickly extracts characters and their locations. We have applied our model to the NTCIR-12 Wikipedia Formula Browsing Task, and found that the method can locate relevant matches without unification of symbols or using a math expression grammar. In the future, one might index LOS graphs for entire pages and search for text *and* graphics. Our source code has been made publicly available.

Keywords: Graphics search · Mathematical Information Retrieval (MIR) · Image search · PDF symbol extraction

1 Introduction

Modern search engines find relevant documents for text-based queries with high efficiency. However, not all information needs are satisfied by text. Most text search engines index graphical elements using textual metadata or ignore graphical elements altogether. To address this, retrieval systems have been created for specific graphic types, but they rely heavily upon notation-specific language models. At the same time, recently developed techniques have been used to extract tables, figures and other graphics automatically from large corpora (e.g., PDFFigures [11] for SemanticScholar[1]), presenting new opportunities for search within and across graphic types.

[1] https://www.semanticscholar.org.

L. Azzopardi et al. (Eds.): ECIR 2019, LNCS 11437, pp. 681–695, 2019.
https://doi.org/10.1007/978-3-030-15712-8_44

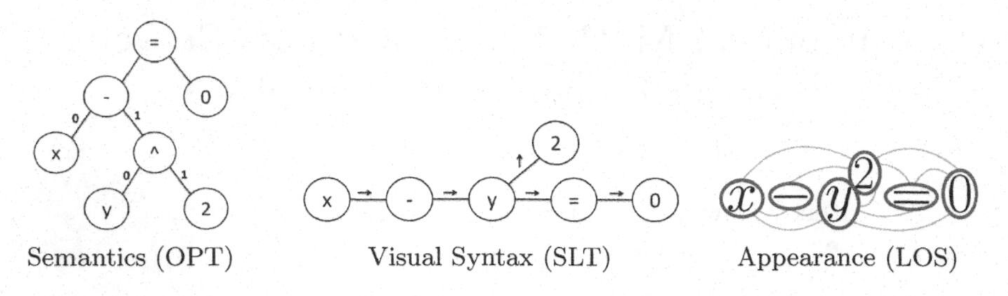

Fig. 1. Formula Structure Representations for $x - y^2 = 0$. State-of-the-art formula retrieval systems use Operator Tree (OPT) and Symbol Layout Tree (SLT) representations (e.g., MCAT [20] and Tangent-S [14]). Our visual search engine uses only domain-agnostic Line-of-Sight (LOS) graphs to represent structure.

We propose a visual graphics search engine, Tangent-V, that is applicable to vector images with known symbols (e.g., in PDF), and raster images with Optical Character Recognition (OCR) output giving recognized symbols locations and labels (e.g., for PNG images). Our method is based upon finding correspondences in Line-of-Sight graphs [7] that represent which symbols 'see' each other along an unobstructed line (see Fig. 1). The language model requires only a set of symbols, allowing it to be applied to multiple graphic types such as math, chemical diagrams, and figures using a single index. In addition, our retrieval model supports wildcards that can be matched to any symbol.

Our main concern in this work is testing the viability of this purely visual approach, and comparing this method's behavior to that of notation-specific techniques. We benchmark our system using the NTCIR-12 Wikipedia Formula Browsing Task benchmark [30]. Despite the absence of explicit formula structure or a detailed language model, our approach achieves BPref results comparable to the state-of-the-art Tangent-S [14] search engine, for *both* PDF images (symbols known) and PNG images (symbols from OCR).

2 Background

Our search engine for graphics found in PDF and PNG images is a specialized form of Content-Based Image Retrieval (CBIR). Many CBIR approaches use a Bags-of-Visual Words (BoVW) framework [28], retrieving objects based on image features ('words'). Traditionally, visual words are defined by local image descriptors (e.g SIFT [22] or SURF [6]), and an inverted index of visual words in images is used for lookup. Later CBIR models use Deep Learning techniques [15] to learn local features [25] or even complete image representations such as hashes or embeddings [4,8,16,27].

For images containing notation (e.g., math), the spatial location of a symbol is important because it affects the structure and semantics of the graphic (see Fig. 1). Some CBIR techniques consider spatial constraints, for example by locating candidates using spectral models, and then re-ranking the most promising matches using spatial verification (e.g., using RANSAC [26]). Affine transformations [3,19,21] and elastic distortions [35] are also useful for spatial validation. Other models include spatial information during indexing [36].

To successfully index and retrieve images including notation, our approach combines topology (by indexing the relative locations of symbols) with spatial verification during the retrieval process. In the following, we summarize methods for graphic representation and search, along with recent methods designed specifically for mathematical information retrieval (our application).

Graphics Representation and Search. Graphics may be represented in three ways: *Semantics*, *Visual Syntax* and *Appearance*. Semantic representations encode the domain-specific information represented in a graphic. Figure 1 shows an Operator Tree (OPT) representing the operations and arguments in the expression $x - y^2 = 0$. As another example, table semantics may be represented by an indexing relation from category labels to values [29]. Often the same indexing relation can be represented in a table, plot, or bar graph. Visual Syntax provides a graphics-type-specific representation of visual structure. For example, in Fig. 1 we see a Symbol Layout Tree (SLT) formula representation, giving the symbols on each writing line and the relative positions of writing lines. For tables, visual syntax can be defined using a two dimensional grid of cells [32]. For bar charts and scatter/line plots, visual syntax can be represented by the placement of axes, axis labels, bars/lines/points, ticks, and values [1,2,10,11]. Finally, appearance representations describe only objects/symbols and their relative positions. One example is the Line-of-Sight graph see Fig. 1 [17,18]). We want search techniques applicable across graphic types, so we use LOS graphs to capture visual structure (see Fig. 1).

Mathematical Information Retrieval (MIR). We apply our model to math formula retrieval, placing our work within the field of MIR [31]. Because math expressions are structured, traditional text-based search systems are inadequate for MIR [31]. We distinguish two math formula retrieval modalities: Image-based and Symbolic. For Image-based approaches, symbols and their relationships are initially unknown. Few methods have been proposed for MIR using images directly, and none have used standard benchmarks for evaluation. Zanibbi and Yu [34] used dynamic time warping over pixels projections to search for typeset formula images using handwritten queries. Chatbri et al. [9] use a connected component matching process to cast votes for candidate query matches in images. All of these methods avoid fully recognizing math expressions in images because it is challenging [23].

In contrast, for symbolic approaches, both the symbols and structure are known (e.g., from LaTeX or MathML). The math retrieval tasks at the NTCIR conferences [30] have produced improved symbolic MIR systems. Among other systems, this includes the MCAT [20] and Tangent-S [14] formula search engines that we use for benchmarking in our experiments. Both systems make use of Visual Syntax (SLT) and Semantic (OPT) representations for search, and retrieve formulas using paths in SLTs and OPTs, followed by finer-grained structural analysis and re-ranking.

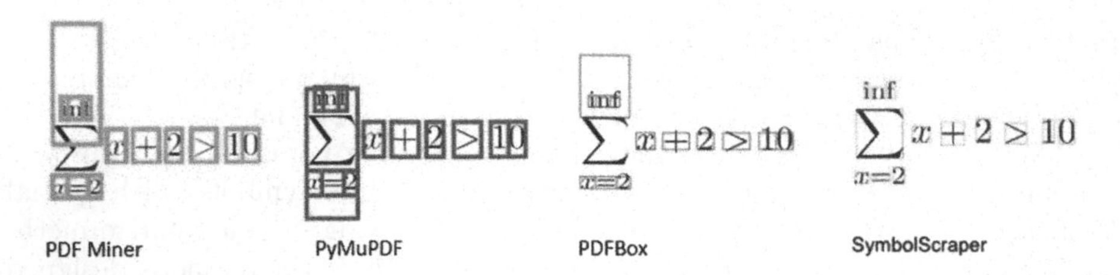

Fig. 2. Symbol Bounding Box Extraction Comparison. SymbolScraper captures the exact location of all symbols. Other tools add or omit character ascender and descender regions, and mislocate large operators (e.g., $\sum$).

Tangent-V [12] generalizes Tangent-S [14], which performs retrieval using symbol pairs in SLTs and OPTs. Tangent-V uses symbol pairs taken directly from images: for PDFs, using symbols extracted directly from the file, and for PNGs using symbols identified with our open-source OCR system [13]. Previously Tangent-V was successfully applied to retrieval of specific handwritten formulas in videos using LaTeX queries [12]. In this paper, we observe the effectiveness of Tangent-V for more general search within isolated formula images.

3 Extracting Symbols from PDF Documents

We use an extension of the Apache PDFBox Java library to extract symbol locations and codes from *born-digital* PDF files (e.g., created using Word or LaTeX). Available tools for extracting symbols provide imprecise locations (see Fig. 2), or require image processing and/or OCR [5]. Our SymbolScraper tool is open source, and available for download.[2]

In PDF, each character has a vector representation containing a character code, font attributes, and writing line position. However, specific symbol locations must be inferred from font attributes. We identify bounding boxes around symbols using font metrics and the character outlines (*glyphs*) embedded in PDF files. Glyphs are defined by a sequence of line segments, arcs, and lifts/moves of the 'pen' used to draw character outlines. Most characters have a single outline, however some symbols such as parentheses may be drawn using multiple glyphs to support smooth rendering at different scales. To capture these we assume that intersecting character outlines belong to a single symbol.

4 Line-Of-Sight Graphs

To capture spatial relationships between neighboring symbols, we identify symbols and then construct an LOS graph (see Fig. 3(a)). After constructing an LOS graph, we use the graph edges to construct an inverted index over symbol pairs for search.

[2] SymbolScraper: https://www.cs.rit.edu/~dprl/Software.html.

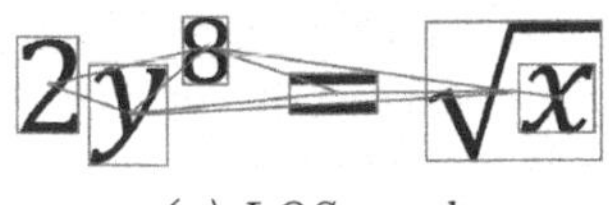

(a) LOS graph

(b) OCR results

Actual Symbol	OCR Label	Prob
2	2	0.98
y	y	0.95
8	8	0.70
	&	0.25
=	=	0.99
$\sqrt{}$	$\sqrt{}$	0.96
x	x	0.97

(c) Inverted Index Entries over Symbol Label Pairs

Labels u	Labels v	Pair ID	Probs. u	Probs. v	3D Disp. dx	3D Disp. dy	3D Disp. dz	Size Ratio	Order
2	y	1	0.98	0.95	0.93	0.36	0.00	0.90	1
2	8	2	0.98	0.70	0.98	-0.22	0.00	1.41	1
&	2	2	0.98	0.25	-0.98	0.22	0.00	0.71	-1
8	y	3	0.70	0.95	-0.77	0.64	0.00	0.64	1
&	y	3	0.25	0.95	-0.77	0.64	0.00	0.64	1
=	y	4	0.99	0.95	-0.99	0.11	0.00	0.76	1
$\sqrt{}$	y	5	0.96	0.95	-0.99	0.06	0.00	1.74	1
=	8	6	0.99	0.70	-0.95	-0.32	0.00	1.18	1
&	=	6	0.25	0.99	0.95	0.32	0.00	0.85	-1
$\sqrt{}$	8	7	0.96	0.70	-0.99	-0.13	0.00	2.71	1
&	$\sqrt{}$	7	0.25	0.96	0.99	0.13	0.00	0.37	-1
=	$\sqrt{}$	8	0.99	0.96	1.00	0.00	0.00	0.44	1
$\sqrt{}$	x	9	0.96	0.97	0.39	0.08	0.92	2.21	1

Fig. 3. Indexing an image-based LOS graph (a) using OCR results (b). The inverted index entries in (c) map lexicographically sorted symbol label pairs to LOS graph edges with their attributes. All LOS edges with the '8' require two entries, to capture the two OCR label outputs for the symbol ('8' and '&').

Symbol Nodes. Each node in an LOS graph represents a candidate symbol with its location and set of labels with confidences. For born-digital PDFs, we use SymbolScraper to obtain these directly from the file (see Sect. 3).

For binary images, we use connected black pixel regions (connected components, or CCs) as symbol candidates, and run our open-source OCR system [13] trained on 91 mathematical symbol classes (e.g. digits, operators, latin and greek letters, etc.) to obtain the most likely symbol labels. To better capture symbols comprised of multiple CCs such as 'i' and 'j,' we try merging each CC with its two closest neighboring CCs. If one of these merged symbols has a higher classification confidence than the average of the top label confidences for each individual CC, the merged symbol is kept. Note that our OCR model does not recognize all symbols found in our test collection. However, since the images are typeset, a reasonably consistent label assignment is expected for symbols belonging to the same class, allowing the proposed model to describe them well using multiple labels, even if the specific symbol is unknown to the OCR system.

LOS Edges. Once the symbols (nodes) of the LOS graph are defined, two symbols (nodes) are connected if they can "see" each other. We test visibility by drawing lines from the bounding box center of each symbol to the vertices of the convex hull of the other symbol. If one of these lines does not intersect a third symbol's convex hull, then there is a line of sight between the symbols. Starting with a fully connected graph, the LOS graph is generated by pruning edges between symbols that fail the visibility test.[3] Some graphs include large empty regions, allowing distant elements to see each other, producing a very dense graph. To avoid this, we prune LOS edges more than twice the median symbol distance apart. This substantially reduces both the index size and retrieval times.

5 Indexing Line-of-Sight Graphs

Using LOS graphs edges, we create an inverted index from symbol pairs to LOS edges connected to symbols of the given type (see Fig. 3). Figure 3 shows the LOS graph for $2y^8 = \sqrt{x}$, along with OCR symbol confidences and entries for the inverted index.

Symbol Probabilities. For PDF input, symbols are known, and the probability of each label is 1.0. For image input, OCR produces a list of class probabilities for each symbol in decreasing order. The top-n classes are selected until a cumulative probability of at least 80% is obtained, or n class labels have been selected ($1 \leq n \leq 3$). In Fig. 3(b), the top class for all symbols has a probability larger than 80% except for the symbol "8." For the"8," we also include the second-highest probability label "&," at which point the cumulative probability is greater than our threshold.

Graph Edge Identifiers. Graph edges have unique global identifiers. When symbols have multiple labels from OCR, their associated LOS edges are entered in the index using pairs of candidate labels. Figure 3(c) shows all 13 index entries for the 9 LOS graph edges. Edge #2 has two entries, in the postings for $(2, 8)$ and $(\&, 2)$. Symbol label pairs for keys are sorted in lexicographic order (i.e., by unicode value). The LOS edge identifiers allow postings for symbols with multiple OCR hypotheses to be merged during retrieval.

Displacement Vectors and Label Order. The relative position of symbol centers are represented using a 3D unit vector $\langle d_x, d_y, d_z \rangle$. The third dimension is non-zero when a symbol center lies within the bounding box of the other (see Fig. 4). We fit an enclosing sphere around the bounding boxes of each symbol, and define r_{max} as the larger radius for the two symbols. If symbol centers are at a distance smaller than r_{max}, d_z is computed as:

$$d_z = \sqrt{(r_{max})^2 - \left(\sqrt{d_x^2 + d_y^2}\right)^2} \quad (1)$$

or $d_z = 0$ otherwise. Displacement vectors are normalized and indexed as a unit vector along with their *label order*. The label order indicates whether a given symbol pair is consistent with the direction of the displacement vector (1) or if the displacement vector has been inverted (-1). In Fig. 3(c), LOS edge #2 $(2, 8)$ uses an inverted ordering for the combination of labels $(\&, 2)$.

Size Ratios (s_p). We also index the ratio of symbol sizes for an LOS edge. At retrieval time, we prune edge matches with large differences in symbol size ratios. For symbols u and v, $s_p(u, v)$ is the length of the bounding box diagonal

[3] Faster algorithms may be used [7].

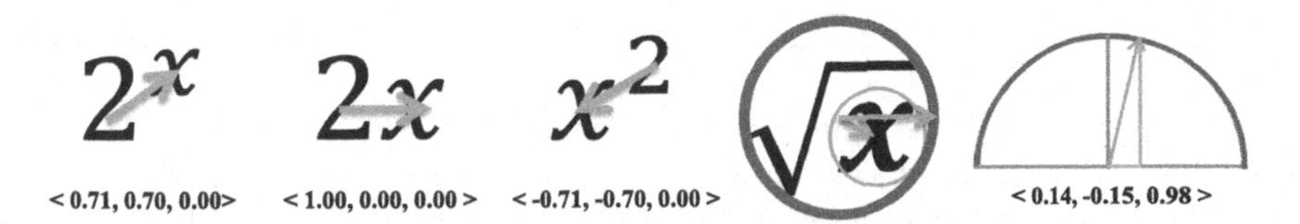

Fig. 4. 3D Unit Vectors Between Symbol Centers. Two directions are enough to represent relative positions for 2^x, $2x$ and x^2. However, the bounding boxes of $\sqrt{}$ and x are overlapping in $\sqrt{x}$, so the center of x is projected onto a sphere around $\sqrt{}$.

for u divided by the bounding box diagonal length for v. A posting with label order -1 indicates that size ratio has been inverted relative to order of symbols associated with an entry (e.g., for entry $(\&, 2)$ in Fig. 3).

Isolated Symbols. To index isolated symbols, we introduce same-symbol pairs using self-edges for symbols in LOS graphs with three or fewer symbols. This allows queries such as x to match small graphs (e.g., x^2). Single symbol 'pairs' have displacement vector $< 0, 0, 1 >$, size ratio 1, and label order 1.

6 Retrieval

Our system uses a two-layer retrieval model. The first layer (the *core engine*) finds all graphs with LOS edges matching the query. Matched graphs are ranked using an edge-based metric, after which the top-k ($k = 1000$) candidates are passed to the second layer (*re-ranking*), which revises scores using an alignment algorithm.

Notation. We define Ω_x as the set of possible symbol identities for LOS graph node x. Given a candidate graph (M), a matched LOS edge for query and candidate symbol pairs (q_1, q_2) and (c_1, c_2) is represented by (Q, C), where $Q = ((\Omega_{q_1}, q_1), (\Omega_{q_2}, q_2))$ and $C = ((\Omega_{c_1}, c_1), (\Omega_{c_2}, c_2))$. The corresponding displacement vectors between symbol pairs on candidate and query edges are unit vectors $\boldsymbol{q}$ and $\mathbf{c}$. The conditional probability of symbol class ω given visual features for query symbol q_1 is denoted by $p(\omega|q_1)$.

6.1 Layer 1: Core Engine

LOS edges matching the query are retrieved from the inverted index using pairs of lexicographically sorted symbols. For example, after applying OCR to 'x^2' we obtain one LOS edge with class label lists $\Omega_u = \langle 2 \rangle$ for the '2' and $\Omega_v = \langle x, X \rangle$ for the 'x'. We lookup postings for both $(2, x)$ and $(2, X)$ in the index, merging postings for edges that appear in both posting lists. We support matching wildcards in queries, which are mapped to single symbols on candidates. This is limited compared to domain-specific MIR retrieval models which can match wildcards to sub-graphs [30]. Given a query pair containing a wildcard, we retrieve all index entries satisfying the given pattern. For example, the pair $(X, 2)$ will match all index entries containing a 2 (e.g. $(1, 2)$, $(+, 2)$, $(2, x)$, etc). Edges containing two wildcards are ignored (these match all index entries).

Retrieved edges in posting lists are filtered, removing candidate edges with large differences in displacement angles and/or symbol size ratios relative to the query edge. First, candidate edge displacement vectors ($\boldsymbol{c}$) with an angular difference of greater than $\pm 30°$ relative to the query ($\boldsymbol{q}$) are removed. Candidate edge C is also filtered if its symbol size ratio is less than half, or more than twice the ratio for the corresponding query edge, given by $s_r(Q, C) < 0.5$, where $s_r(Q, C)$ is:

$$s_r(Q, C) = \frac{min\left(s_p\left(q_1, q_2\right), s_p\left(c_1, c_2\right)\right)}{max\left(s_p\left(q_1, q_2\right), s_p\left(c_1, c_2\right)\right)} \tag{2}$$

The initial edge-based ranking metric is an edge recall, weighted by symbol confidences and differences in displacement vectors. Our edge-based scoring function $S(M)$ for matched LOS subgraph M adds the product of symbol confidences and angular differences, summing over common symbol classes for matched symbols. For a given symbol class ω, we combine the probabilities for that class in corresponding query/candidate symbols p and c using their minimum probability: $f(\omega, q, c) = min\,(\ p(\omega|q),\ p(\omega|c)\ )$. We found this produces more stable results than using the product of the probabilities [12]. For a wildcard pair $Q_{\mathrm{w}} = ((\{X\}, q_{\mathrm{w}}), (\{\omega_j\}, q_2))$, matching a concrete pair $C = ((\{\omega_i\}, c_1), (\{\omega_j\}, c_2))$, we set $p(\omega_{\mathrm{w}}|q_{\mathrm{w}}) = p(\omega_j|q_2)$. This forces our model to prefer wildcard matches only when attached to strong concrete symbol matches.

$$s_{\Omega}(Q, C) = \sum_{\substack{\omega_i \in \Omega_{q_1} \cap \Omega_{c_1} \\ \omega_j \in \Omega_{q_2} \cap \Omega_{c_2}}} f(\omega_i, q_1, c_1)\ f(\omega_j, q_2, c_2) \tag{3}$$

$$s_{\angle}(Q, C, \theta) = \begin{cases} \frac{\boldsymbol{q}\cdot\boldsymbol{c} - cos(\theta)}{1 - cos(\theta)}, & \text{if } \boldsymbol{q}\cdot\boldsymbol{c} \geq cos(\theta) \\ 0 & \text{otherwise} \end{cases} \tag{4}$$

$$S(M) = \sum_{(Q,C)\in M} s_{\Omega}(Q, C)\ s_{\angle}(Q, C, 30°) \tag{5}$$

We keep only the top-k ($k = 1000$) matched graphs after computing an optimistic greedy estimation of the maximum $S(M)$ score candidate graphs. For each graph, matched edges are added to $S(M)$ in decreasing order of weighted recall score ($s_{\Omega}(Q, C)\ s_{\angle}(Q, C, 30°)$ in Eq. 5), while enforcing a 1-to-1 matching constraint between query and candidate edges.

6.2 Layer 2: Re-Ranking

For each candidate graph selected by the core engine, connected components from matched edges are identified. To ensure that the connected components match query graph LOS structure, we require components to preserve a one-to-one mapping from candidate to query symbols (nodes). The first row in Fig. 5

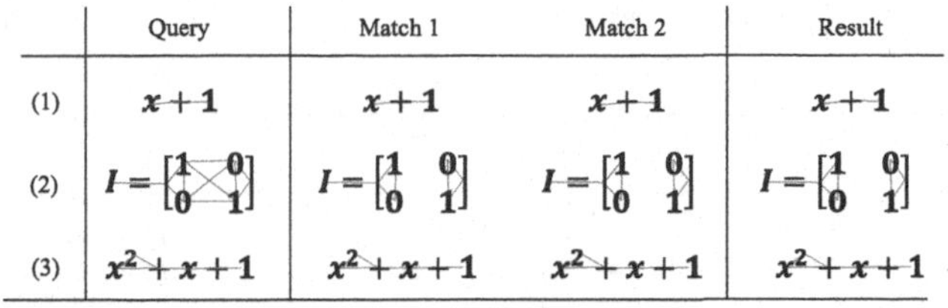

	Query	Match 1	Match 2	Result
(1)	$x+1$	$x+1$	$x+1$	$x+1$
(2)	$I=\begin{bmatrix}1 & 0\\ 0 & 1\end{bmatrix}$	$I=\begin{bmatrix}1 & 0\\ 0 & 1\end{bmatrix}$	$I=\begin{bmatrix}1 & 0\\ 0 & 1\end{bmatrix}$	$I=\begin{bmatrix}1 & 0\\ 0 & 1\end{bmatrix}$
(3)	x^2+x+1	x^2+x+1	x^2+x+1	x^2+x+1

Fig. 5. Structural Alignment Steps. Matching nodes and edges are shown in green/blue and red respectively. (1) Match growing, two connected pairs matching different portions of $x + 1$ are merged into a single larger match. (2) Joining disconnected subgraphs, two partial matches on disconnected subgraphs which are spatially consistent are merged into a single larger match. (3) Incompatible match removal, keeping only the best match.

shows two matched candidate LOS edges being merged ("$x+$" and "$+1$") to form a connected component ("$x + 1$"). After growing all connected components, the top-M ($M = 50$) components are selected for further processing, each scored using the $S(M)$ metric.

Queries often match disjoint LOS subgraphs. This occurs due to some combination of OCR errors, unmatched symbols, or pruned LOS edges. To connect subgraphs into larger matches, we greedily merge disjoint matches that preserve a one-to-one query/candidate symbol mapping and have a very low *spatial distortion cost* after merging. Using the highest scoring match as the reference match, we compute an affine transformation matrix that will translate and scale the candidate nodes into the query space. Based on the reference match, the center of the bounding box of its candidate subgraph is translated to the center of the bounding box of its query subgraph. Then, the diagonals of the same bounding boxes are used to define a scaling factor, used to re-scale candidate nodes into the query space scale. We then use this transformation matrix to project candidate nodes from the second match into the query space. Representing symbol bounding boxes as 4D vectors (x_1, y_1, x_2, y_2), we compute the average of all euclidean distances between the bounding box of each candidate node from the second match and the bounding box of their corresponding query nodes. Finally, we normalize this average euclidean distance by the average diagonal length of query node bounding boxes, and we use this as our *spatial distortion cost*. If this cost exceeds a threshold ($max_{dist} = 0.5$), matches will not be joined. An example is shown in the middle row of Fig. 5, where two components of a matrix are disconnected, but then merged.

This procedure may produce multiple candidate matches. We again apply greedy filtering, selecting the next largest match that does not contain previously selected candidate nodes. In our current evaluation, only one match is allowed per candidate graph; therefore we keep only the highest scoring match. Candidate graphs are sorted by their final scores, with ties broken by sorting by increasing number of unmatched edges.

Table 1. Statistics for different index conditions for the NTCIR-12 MathIR Wikipedia Collection.

Property	PNG Top-1	PNG Top-2	PNG Top-3	PDF All
Index Entries	3,923	4,147	4,186	36,593
Graph Edges	10,462,843	10,462,843	10,462,843	9,591,932
Pair Instances	10,462,843	33,532,368	60,232,395	9,728,374
Size on Disk (GB)	2.61	3.09	3.63	1.57
Query Times (seconds)				
Core - Avg (Std)	6.59 (4.79)	10.66 (7.35)	15.77 (12.18)	4.39 (4.09)
Full - Avg (Std)	9.93 (7.40)	14.99 (10.87)	21.84 (19.20)	6.36 (5.47)

7 Evaluation

Benchmark. For evaluation, we use the NTCIR-12 MathIR Wikipedia Formula Browsing Task [30]. The collection contains 591,608 instances of approximately 328,685 unique formulas taken from English Wikipedia. The NTCIR-12 query set has 40 topics, with 20 containing wildcards. During the competition, the top-20 hits from participating systems were pooled, with each scored by two human assessors (university students). Assessors rated hits using 0, 1, or 2 to indicate whether a hit is irrelevant, partially relevant, or relevant. The two assessor scores are then added. 'Fully Relevant' hits are those with a combined assessor score ≥ 3, while 'Partially Relevant' hits are those with a combined score ≥ 1.

Indexing and Retrieval. Using LaTeX, we render each formula in PDF and PNG formats, and then create an index for each. For PNGs, we trained our symbol classifier using classes in the CROHME 2016 dataset [24] with LaTeX-generated synthetic data: the 101 classes were grouped based on similar shapes into 91 classes. For PDFs, we used the extraction tool described in Sect. 3 to obtain precise symbol locations and classes. For PNG, we constructed three indices for when at most 1, 2, or 3 class labels are permitted per symbol (see Sect. 5). Metrics for the indices are provided in Table 1.

Both Precision@K and BPref are computed using the official competition relevance judgments for this task, and the `trec_eval` tool[4] (see Tables 2 and 3). Our experimental system had an Intel processor i7-7820X with 64 GB of RAM, and a Nvidia GTX 1080 GPU. Most operations run on a single thread except for vector operations in long posting lists executed on the GPU. Mean query execution times for all indexing and retrieval conditions are shown in Table 1.

Discussion. As one expects, the LOS-based Precision@K values are lower than those obtained by domain-specific state-of-the-art methods for formula retrieval; but this is partly because many formulas without judgments in the top-20 and treated as irrelevant. However, for BPref scores the LOS approach produces more comparable results based on human pairwise preferences. In fact, the LOS model on PDFs achieves slightly better BPref values for Partially Relevant hits for queries without wildcards than domain-specific retrieval models such as Tangent-S [14].

[4] http://trec.nist.gov/trec_eval.

Table 2. Average Precision@K values per topic for NTCIR-12 MathIR Wikipedia Formula Browsing Task.

		Relevant (%)			Partially Relevant (%)		
		P@5	P@10	P@20	P@5	P@10	P@20
MCAT [20]		**49.00**	**39.00**	**28.25**	**91.00**	**84.00**	**76.87**
Tangent-S [14]		44.00	31.50	21.62	70.00	60.75	51.12
LOS PDF							
All	Core	23.00	18.00	13.00	41.00	33.75	27.13
	Reranked	29.50	22.25	17.37	41.50	38.00	32.37
LOS PNG							
Top-1	Core	23.00	17.25	12.13	41.50	33.75	25.50
Top-2	Core	20.50	16.25	13.00	40.50	33.50	27.37
Top-3	Core	19.50	14.50	11.37	38.00	30.25	24.37
Top-1	Reranked	26.00	19.00	14.50	46.00	37.00	30.50
Top-2	Reranked	27.00	19.75	15.62	47.50	38.25	32.50
Top-3	Reranked	27.50	20.25	16.12	47.00	38.50	33.25

Table 3. Average BPref values per topic for NTCIR-12 MathIR Wikipedia Formula Browsing Task. Results shown for all queries (40) & queries without/with wildcards (20/20).

		Relevant (%)			Partially Relevant (%)		
		All	Concr.	Wild.	*All*	Concr.	Wild.
MCAT [20]		52.02	57.02	**47.02**	53.56	56.98	50.13
Tangent-S [14]		**55.30**	**63.61**	46.99	56.20	58.72	**53.68**
LOS PDF							
All	Core	39.17	48.30	30.04	55.13	60.00	50.26
	Reranked	53.05	59.85	46.26	**56.44**	60.32	52.57
LOS PNG							
Top-1	Core	36.78	49.14	24.42	46.32	55.99	36.64
Top-2	Core	42.05	50.67	33.43	50.97	59.26	42.68
Top-3	Core	40.53	50.33	30.73	51.88	58.05	45.71
Top-1	Reranked	46.04	55.93	36.15	47.51	57.01	38.00
Top-2	Reranked	49.74	58.96	40.52	52.43	**60.86**	44.00
Top-3	Reranked	50.75	59.20	42.30	53.52	59.72	47.31

As expected, PDF results are almost always better than PNG results. We consider PDFs as the better condition for our model, since they have a more accurate label assignment, producing fewer index entries (see Table 1). This means that more unique combinations of symbols pairs are being considered, with shorter postings lists overall. On the other hand, too specific labels for some variations of known symbols (e.g x vs $\hat{x}$) may prevent the system from ranking partial matches properly.

In contrast, PNG results are degraded by noise. Considering only 91 unique symbol shapes can cause problems for out-of-vocabulary symbols. This results in a smaller number of index entries with longer posting lists. Adding extra labels for each symbol causes the index to quickly multiply in size, and produces slower retrieval times (see Table 1). However, we can obtain slight improvements for both Precision@K and BPref for re-ranked results when these extra labels are indexed. This is a trade-off between retrieval time and rank quality, which may worthwhile for applications where higher recall is more important than speed.

The initial core results can be retrieved in shorter times compared to the full model (see Table 1). However, re-ranking helps in almost all conditions, and Precision@K is always increased after re-ranking. In comparison, the MCAT system takes several minutes on average when unification is used [20]. The Tangent-S system is implemented with several core engine optimizations making it faster (avg of 2.67 s) than our core engine, but it has slower re-ranking times [14,33] with greater variance in execution times than our proposed re-ranking.

We implemented our model using Python. All queries and retrieval conditions were computed using a single process except for GPU-accelerated vector operations. MCAT systems uses 50 processors for variable unification [20]. Our current prototype tests the effectiveness of the retrieval model, and future work includes various low level optimizations that will increase efficiency like the ones used in the pair-based engine of Tangent-S [33]. Overall, our model finds many relevant formulas despite a lack of domain-specific knowledge. We expect our LOS appearance-based model will also provide meaningful results for other graphic types, with little need for domain-specific fine tuning.

8 Conclusion

We have presented our Tangent-V model for visual graphics search, along with its application to retrieving mathematical formulas. Our model considers only symbol labels and their relative positions, without any facility for unification of numbers, identifiers, or variable names. Despite this simple approach, our model finds relevant results, and outperforms existing domain-specific formula search engines in terms of BPref for partially relevant matches. This confirms that appearance alone can provide meaningful formula search results, and we are interested in seeing how this generalizes to other notations (e.g., chemical diagrams and figures). We are also interested in replacing OCR in raster images (e.g., in PNG) with visual feature-based descriptors.

Previously our model was successfully applied to cross-modal search, by matching handwritten versions of formulas taken from course notes in LaTeX [12]. This work confirms that our approach is promising for not just locating specific formulas, but formulas similar to a query.

In the future, we want to explore support for unification of symbols, and modify our scoring metrics to consider the context of a match, preferring identical matches to those surrounded by extra symbols. Finally, our implementation can be optimized in a number of ways, including re-implementing the Python prototype in C/C++, and structuring posting lists to reduce the number of candidate matches considered. Source code for Tangent-V is publicly available.[5]

Acknowledgements. We are grateful to Chris Bondy for his help with designing SymbolScraper. This material is based upon work supported by the National Science Foundation (USA) under Grant Nos. HCC-1218801, III-1717997, and 1640867 (OAC/DMR).

[5] https://cs.rit.edu/~dprl/Software.html#tangent-v.

References

1. Al-Zaidy, R.A., Giles, C.L.: Automatic extraction of data from bar charts. In: Proceedings of the 8th International Conference on Knowledge Capture, K-CAP 2015, Palisades, NY, USA, 7–10 October 2015, pp. 30:1–30:4 (2015). https://doi.org/10.1145/2815833.2816956, http://doi.acm.org/10.1145/2815833.2816956
2. Al-Zaidy, R.A., Giles, C.L.: A machine learning approach for semantic structuring of scientific charts in scholarly documents. In: Proceedings of the Thirty-First AAAI Conference on Artificial Intelligence, 4–9 February 2017, San Francisco, California, USA, pp. 4644–4649 (2017). http://aaai.org/ocs/index.php/IAAI/IAAI17/paper/view/14275
3. Avrithis, Y., Tolias, G.: Hough pyramid matching: speeded-up geometry re-ranking for large scale image retrieval. Int. J. Comput. Vis. **107**(1), 1–19 (2014)
4. Babenko, A., Lempitsky, V.: Aggregating local deep features for image retrieval. In: Proceedings of the IEEE International Conference on Computer Vision, pp. 1269–1277 (2015)
5. Baker, J., Sexton, A.P., Sorge, V.: Extracting precise data on the mathematical content of PDF documents. In: Towards a Digital Mathematics Library (DML). Masaryk University Press, Birmingham, 27 July 2008. ISBN 978-80-210-4658-0
6. Bay, H., Tuytelaars, T., Van Gool, L.: SURF: speeded up robust features. In: Leonardis, A., Bischof, H., Pinz, A. (eds.) ECCV 2006. LNCS, vol. 3951, pp. 404–417. Springer, Heidelberg (2006). https://doi.org/10.1007/11744023_32
7. Berg, M., Cheong, O., Kreveld, M., Overmars, M.: Computational Geometry: Algorithms and Applications, 3rd edn. Springer, Heidelberg (2008). https://doi.org/10.1007/978-3-540-77974-2
8. Cao, Y., Long, M., Liu, B., Wang, J.: Deep cauchy hashing for hamming space retrieval. In: The IEEE Conference on Computer Vision and Pattern Recognition (CVPR), June 2018
9. Chatbri, H., Kwan, P., Kameyama, K.: An application-independent and segmentation-free approach for spotting queries in document images. In: ICPR, pp. 2891–2896. IEEE (2014)
10. Choudhury, S., et al.: Figure metadata extraction from digital documents. In: 12th International Conference on Document Analysis and Recognition, ICDAR 2013, pp. 135–139 (2013). https://doi.org/10.1109/ICDAR.2013.34
11. Clark, C., Divvala, S.K.: Pdffigures 2.0: mining figures from research papers. In: Proceedings of the 16th ACM/IEEE-CS on Joint Conference on Digital Libraries, JCDL 2016, Newark, NJ, USA, 19–23 June 2016, pp. 143–152 (2016). https://doi.org/10.1145/2910896.2910904, http://doi.acm.org/10.1145/2910896.2910904
12. Davila, K., Zanibbi, R.: Visual search engine for handwritten and typeset math in lecture videos and latex notes. In: 2018 16th International Conference on Frontiers in Handwriting Recognition (ICFHR), pp. 50–55, August 2018. https://doi.org/10.1109/ICFHR-2018.2018.00018
13. Davila, K., Ludi, S., Zanibbi, R.: Using off-line features and synthetic data for on-line handwritten math symbol recognition. In: ICFHR, pp. 323–328. IEEE (2014)
14. Davila, K., Zanibbi, R.: Layout and semantics: combining representations for mathematical formula search. In: SIGIR (2017)

15. Goodfellow, I., Bengio, Y., Courville, A., Bengio, Y.: Deep Learning, vol. 1. MIT Press, Cambridge (2016)
16. Gordo, A., Almazán, J., Revaud, J., Larlus, D.: Deep image retrieval: learning global representations for image search. In: Leibe, B., Matas, J., Sebe, N., Welling, M. (eds.) ECCV 2016. LNCS, vol. 9910, pp. 241–257. Springer, Cham (2016). https://doi.org/10.1007/978-3-319-46466-4_15
17. Hu, L., Zanibbi, R.: MST-based visual parsing of online handwritten mathematical expressions. In: Proceedings of the International Conference on Frontiers in Handwriting Recognition (ICFHR), Shenzhen, China (2016, to appear)
18. Hu, L., Zanibbi, R.: Line-of-sight stroke graphs and parzen shape context features for handwritten math formula representation and symbol segmentation. In: ICFHR, pp. 180–186. IEEE (2016)
19. Jégou, H., Douze, M., Schmid, C.: Improving bag-of-features for large scale image search. Int. J. Comput. Vis. **87**(3), 316–336 (2010)
20. Kristianto, G.Y., Topić, G., Aizawa, A.: The MCAT math retrieval system for NTCIR-12 MathIR task. In: Proceedings of the NTCIR-12, pp. 323–330 (2016)
21. Li, X., Larson, M., Hanjalic, A.: Pairwise geometric matching for large-scale object retrieval. In: CVPR, pp. 5153–5161, June 2015
22. Lowe, D.G.: Distinctive image features from scale-invariant keypoints. Int. J. Comput. Vis, **60**(2), 91–110 (2004)
23. Mouchère, H., Zanibbi, R., Garain, U., Viard-Gaudin, C.: Advancing the state-of-the-art for handwritten math recognition: the CROHME competitions, 2011–2014. Int. J. Doc. Anal. Recogn. (IJDAR) **19**(2), 173–189 (2016)
24. Mouchère, H., Viard-Gaudin, C., Zanibbi, R., Garain, U.: ICFHR 2016 CROHME: competition on recognition of online handwritten mathematical expressions. In: International Conference on Frontiers in Handwriting Recognition (ICFHR) (2016)
25. Noh, H., Araujo, A., Sim, J., Weyand, T., Han, B.: Largescale image retrieval with attentive deep local features. In: Proceedings of the IEEE International Conference on Computer Vision, pp. 3456–3465 (2017)
26. Philbin, J., Chum, O., Isard, M., Sivic, J., Zisserman, A.: Object retrieval with large vocabularies and fast spatial matching. In: CVPR, pp. 1–8. IEEE (2007)
27. Radenović, F., Iscen, A., Tolias, G., Avrithis, Y., Chum, O.: Revisiting oxford and paris: large-scale image retrieval benchmarking. In: The IEEE Conference on Computer Vision and Pattern Recognition (CVPR), June 2018
28. Sivic, J., Zisserman, A.: Video Google: a text retrieval approach to object matching in videos, In: ICCV, pp. 1470–1477. IEEE (2003)
29. Wang, X.: Tabular Abstraction, Editing and Formatting. Ph.D. thesis, University of Waterloo, Canada (1996)
30. Zanibbi, R., Aizawa, A., Kohlhase, M., Ounis, I., Topić, G., Davila, K.: NTCIR-12 MathIR task overview. In: Proceedings of the NTCIR-12, pp. 299–308 (2016)
31. Zanibbi, R., Blostein, D.: Recognition and retrieval of mathematical expressions. IJDAR **15**(4), 331–357 (2012)
32. Zanibbi, R., Blostein, D., Cordy, J.R.: A survey of table recognition: models, observations, transformations, and inferences. Int. J. Doc. Anal. Recogn. (IJDAR) **7**(1), 1–16 (2004)

33. Zanibbi, R., Davila, K., Kane, A., Tompa, F.: Multi-stage math formula search: using appearance-based similarity metrics at scale. In: SIGIR (2016)
34. Zanibbi, R., Yu, L.: Math spotting: retrieving math in technical documents using handwritten query images. In: ICDAR, pp. 446–451. IEEE (2011)
35. Zhang, W., Ngo, C.W.: Topological spatial verification for instance search. IEEE Trans. Multimedia **17**(8), 1236–1247 (2015). https://doi.org/10.1109/TMM.2015.2440997
36. Zhang, Y., Jia, Z., Chen, T.: Image retrieval with geometry-preserving visual phrases. In: CVPR, pp. 809–816. IEEE (2011)

Figure Retrieval from Collections of Research Articles

Saar Kuzi(✉) and ChengXiang Zhai

University of Illinois at Urbana-Champaign, Urbana, IL, USA
{skuzi2,czhai}@illinois.edu

Abstract. In this paper, we introduce and study a new task of figure retrieval in which the retrieval units are figures of research articles and the task is to rank figures with response to a query. As a first step toward addressing this task, we focus on textual queries and represent a figure using text extracted from its article. We suggest and study the effectiveness of several retrieval methods for the task. We build a test collection by using research articles from the ACL Anthology corpus and treating figure captions as queries. While having some limitations, using this data set we were able to obtain some interesting preliminary results on the relative effectiveness of different representations of a figure and different retrieval methods, which also shed some light regarding possible types of information need, and potential challenges in figure retrieval.

1 Introduction

Devising intelligent systems to assist researchers and improve their productivity is crucial for accelerating research and scientific discovery. Tools for literature search such as Google Scholar and many digital library systems are essential for researchers; their effectiveness directly affects the productivity of researchers. Conventional literature search systems often treat a literature article as a retrieval unit (i.e., a document) and the retrieval task is to rank articles in response to a query. In this paper, we introduce and study a novel retrieval task where we would treat a figure in a literature article as a retrieval unit and the retrieval task is to return a ranked list of figures from all the literature articles in a collection in response to a query.

An effective figure retrieval system is useful in many ways. First, major scientific research results (e.g., precision-recall curves in information retrieval research) are often summarized in figures and key ideas of technical approaches (e.g., neural networks and graphical models in machine learning research) are often illustrated with figures, making figures important "information objects" in research articles that researchers often want to locate and pay special attention to. While one can also navigate into relevant figures after finding a relevant article, it would be much more efficient if a researcher can directly retrieve relevant figures by using a figure retrieval system. Second, a figure search system may supply useful features for improving the ranking of literature articles in

L. Azzopardi et al. (Eds.): ECIR 2019, LNCS 11437, pp. 696–710, 2019.
https://doi.org/10.1007/978-3-030-15712-8_45

a conventional literature search system by rewarding an article whose figures also match well with a query. Third, a figure search system can be very useful for finding examples of illustrations of a concept, thus potentially having broad applications beyond supporting researchers to also generate benefit in education. For example, a figure search engine operating on a collection of research articles in the natural language processing domain can conveniently allow anyone to find some examples of parse trees, which would be useful for learning about a parse tree or just citing an example in a tutorial of natural language processing.

As a retrieval problem, figure retrieval is different from conventional retrieval tasks in many ways, making it an interesting new problem for research. First, the types of information need of users in figure retrieval are expected to be different than in document retrieval, thus potentially requiring the development of novel approaches to satisfy those needs. Another challenge in figure retrieval is how to effectively represent a figure in the collection. One way to represent figures is to treat them as independent units (i.e., image files). However, such a representation does not benefit from the rich context of a figure in the research article that contains the figure. For example, text in the article that explicitly describes the figure as well as other related parts of the article can be used to represent a figure. Finally, it would be important to study models for measuring the relevance between a figure and a query.

In this work, as a first step, we focus on textual queries (i.e., keywords) and represent figures using text extracted from their articles. We propose multiple ways to represent figures and study their effectiveness when using different retrieval methods. Specifically, we propose to represent a figure using multiple textual fields, generated using text in the article that explicitly mentions the figure and also other text in the article that might be related. We then use existing retrieval models, based on lexical similarity and semantic similarity, to measure the relevance between a figure field and a query. Finally, a learning-to-rank approach is used in order to combine different figure fields and retrieval models.

We perform experiments using research articles from the natural language processing domain (ACL Anthology). Since no data sets of queries for figure retrieval are publicly available, we created an initial test collection for evaluation in which figure captions are used to simulate queries (thus, the task is to retrieve a single figure using its caption). While having some limitations, using this data set we were able to obtain some interesting preliminary results. Specifically, our experimental results show that it is beneficial to use a rich textual representation for a figure and to combine different retrieval models. We also gain some initial understanding of the figure retrieval problem, including some illustration of potential types of information need and possible difficulties and challenges. We conclude the paper by suggesting a road map for future research on the task.

2 Related Work

In most retrieval tasks, the retrieval units are documents, though the retrieval of other units, notably entities (e.g., [1,6,20,21]) and passages (e.g., [11,23,26]) has also been studied. Our work adds to this line of research a new retrieval task where the retrieval units are figures in scientific research articles.

As an effective way to communicate research results, figures are especially useful in domains such as the biomedical domain. As a result, how to support biologists to search for figures has attracted a significant amount of attention, and multiple systems were developed [10,13,24]. These previous works have focused on the development of a figure search engine system from the application perspective, but none of those systems or algorithms used in those systems has been evaluated in terms of retrieval accuracy.

Some works [14,31] studied the ranking of figures within a given article based on the assumption that figures in an article have different levels of importance. These works suggested a set of features for ranking so as to measure the centrality of a figure in the article. The suggested features, however, have not been used for figure retrieval. In this paper, we analyze the performance of our approach as a function of the figure centrality in the article, which serves as a first step toward utilizing such features for figure retrieval in the future.

In another line of works, methods for extraction of text from figures in the biomedical domain were studied (e.g., [12,19,29]). Using the text inside a figure can potentially improve retrieval effectiveness by enriching the figure representation. Yet, these works focused mainly on testing the text extraction accuracy, and not the retrieval effectiveness. In our work, we focus on studying the effectiveness of general figure retrieval models, which we believe is required in order to establish a solid foundation for research in figure retrieval; naturally, the general retrieval models can be enhanced by using many additional techniques to enrich figure representation to further improve accuracy as happens in many other applications such as Web search, which we leave as an interesting future work.

Finally, our work is also related to the large body of work on image search. As an effort for improving image search, the ImageCLEF Track was established. In one task, for example, participants were asked to devise approaches for ranking images in the medical domain using visual and textual data [18]. Content-based Image Retrieval (CBIR) was also explored in some works [9,25]. In CBIR, the idea is to extract visual features from the image (e.g., color, texture, and shape) and use them for ranking with respect to an image query. Other works focused on combining visual and textual data for image representation and retrieval (e.g., [2,7,27]). Figures in research articles can also be viewed as images, but we study the problem from the perspective of textual representation of figures. An interesting future work would be to try to incorporate some of the approaches for image search in figure retrieval.

3 Figure Retrieval

In this section, we introduce and define the new problem of figure retrieval, discuss strategies for solving this problem, and present specific retrieval methods that we will later experiment with.

3.1 Problem Formulation

As a retrieval problem, figure retrieval treats each figure in a research article as a retrieval unit. As those figures do not naturally exist as well separated units, the notion of a collection in figure retrieval is defined based on a collection of research articles D, which can be used to build a collection of figures F_D as follows. For every article $d \in D$, k_d figures are extracted; each figure can be uniquely identified in its article by a number $i \in \{1, .., k_d\}$. Then, all figures, extracted from all articles in D, constitute the figure collection F_D.

The goal of the figure retrieval task is to rank figures in F_D according to their relevance to a user query q, where q can be a set of keywords (i.e., textual), an image, or a combination of the two. In general, a user may use keywords to describe what kind of figures he/she wants to find and may also (optionally) use one or multiple example images to define what kind of figures should be retrieved. As a first step in studying this problem, we only consider keyword queries, though we should note that a full treatment of the figure retrieval problem should also include matching any user-provided examples of images with the figure collection, which would be a very interesting direction for future work.

With a keyword query, the figure retrieval problem is quite challenging because it requires matching a keyword query with a figure, which does not necessarily have any readily available text description. Fortunately, we can extract relevant text information from the article with a figure to represent the figure; indeed, all figures have captions, which we can conveniently use to represent them. We can also extract any sentences discussing a figure in an article as an additional text description of the figure. This way, we would obtain a pseudo text document to represent each figure, which we refer to as a *figure document*. Thus, our figure collection contains a set of figures where each figure is associated with a figure document, and the main task for retrieval now is to match a query with those figure documents. This transformation of problem formulation allows us to leverage existing text retrieval models to solve the problem. There are two key technical challenges that we need to study in order to solve the problem effectively: (1) How to derive effective text representations of the figures. (2) How to measure the relevance between a figure and a query. We discuss each next in detail.

3.2 Figure Representation

While figures can be treated just as independent images (i.e., sets of pixels), they appear in the context of research articles, which offers opportunities to build a rich representation for them. For example, text in the article that explicitly mentions the figure can be utilized. Such text can be the figure caption or other parts of the article that describe or discuss the figure. Other text in the article may not explicitly mention the figure but can still be useful. The abstract of the article, for instance, may serve as a textual representation of the figure since both are in the topic of the article. Finally, other information can be derived from the context of the article which is not necessarily textual. The "authority" of the article (e.g., the number of citations) can serve as a prior for the figure relevance. Our approach to the computation of figure representation is to generate a set of textual fields for each figure, using text that explicitly mentions the figure, as well as other parts of the article.

Explicit Figure Mentions: We generate textual fields using text in the article that explicitly mentions the figure. The caption of the figure, for example, can be regarded as such text. Nevertheless, since figure captions serve as queries in our experiments, we were not able to use them for figure representation at this point. Thus, we only utilize text in the article that discusses or describes the figure (e.g., "The results for the experiment are depicted in Figure 1 ..."). While the general location of such text can be detected easily (since the figure number is explicitly mentioned), it might be challenging to determine its boundaries. That is, automatically detecting at what point in the text the discussion about the figure begins, and at what point the subject changes. A similar problem has been studied in the context of identifying the text that describes a cited article [8]. Yet, it was not studied, to the best of our knowledge, for figure retrieval. In this paper, we take the following approach for extracting this type of text. Given an explicit mention of a figure (i.e., the string "Figure i"), we include w words that precede the figure mention and w words that follow it; w is a free parameter. We denote these textual fields as **FigText** fields and generate three such fields for $w \in \{10, 20, 50\}$. In the case where a figure is mentioned several times in the text, we concatenate all of the text segments that correspond to the different mentions to form a single textual field for a given value of w; overlapping texts are merged so as to avoid textual redundancy.

General Article Text: Other parts of the article that do not explicitly mention the figure can also be useful for figure representation. This might be the case since a figure is usually related to some of the topics of the article, and these topics may also be discussed in some other parts of the article. Using this type of text can be potentially advantageous when the text that explicitly mentions the figure is very short or not highly informative. In such a case, other parts of the article can help to bridge the lexical gap between the query and the figure when measuring the relevance between them. We denote this type of fields **FigArticle** fields. We use the title, abstract, and introduction of the article to generate three separate fields, denoted **Title**, **Abs**, and **Intro**, respectively. By using these sections of the article we can obtain textual fields with different levels of length and generality. We do not use other parts of the article as these may be too general (e.g., using the entire text), or too narrow (e.g., using sections that describe the model). Furthermore, these three sections appear in almost every research article and are easy to detect automatically.

An alternative approach for using the text of an entire article section would be to select only parts of it that are presumably more related to the figure. Motivated by a previous work [30], we select a single sentence from the abstract to represent a figure. This sentence serves as an additional field and is denoted **Abs-sen**. We select a single sentence from the abstract in the following way. We measure the similarity between a sentence in the abstract and the figure using the cosine similarity between their $tf.idf$ representations; a figure is represented using the FigText field ($w = 50$). Then, we choose a single sentence with the highest similarity. If the scores for all abstract sentences with respect to a figure are zeros, we do not represent the figure with a sentence from the abstract. In that sense, using this field we can somehow measure the centrality of the figure in the article (i.e., if the similarity with all abstract sentences is zero then the figure is not likely to be central). The importance in considering the figure centrality was discussed in previous works [14,31].

3.3 Retrieval Models

As each figure is represented by a figure document which consists of multiple text segments, conventional retrieval models are applicable to measure relevance. Our study thus focuses on understanding how effective the basic standard retrieval models are for this new retrieval task, and what kind of representation of figures is the most effective. Specifically, we generate a set of features for each figure where each feature corresponds to a combination of a textual field and a retrieval model and use these features to learn a ranking function using a learning-to-rank (LTR) algorithm [15]. We use LTR so as to effectively combine the different retrieval models and textual fields. Furthermore, LTR offers a flexible framework for adding more features in the future that are not necessarily generated using text data.

In our experiments, we considered two retrieval models in order to measure the relevance between a query and a textual field. The first model we use is **BM25** [22]. This model can also be viewed as a model that measures the lexical similarity between the query and some text as it heavily relies on exact keyword matching. The second model that we use is based on word embeddings (e.g., Word2Vec [17]). Specifically, word embeddings can be used to measure the semantic similarity between the query and a textual field, thus this approach is expected to be complementary to BM25. We learn an embeddings model using the entire collection of research articles. Then, we represent the query and a textual field using the *idf* weighted average of their term vectors. Finally, the similarity between them is measured using the cosine function. This retrieval approach is denoted **W2V** in our analysis of experimental results.

4 Evaluation

Our main goal is to study the effectiveness of the various approaches we proposed for computing figure representation and ranking figures. Unfortunately, as figure retrieval is a new task, there does not exist any test collection that we can use for evaluation. Thus, we first need to address the challenge of creating a test collection.

4.1 Test Collection Creation

A test collection for figure retrieval generally consists of three components: (1) a collection of figures; (2) a set of queries; (3) a set of relevance judgments. We now discuss how we construct each of them and create the very first test collection for figure retrieval (available at figuredata.web.illinois.edu).

Figure Collection: To construct a figure collection, we leveraged the ACL Anthology reference corpus [3]. This is one of the very few publicly available full-text article collections. This corpus consists of 22,878 articles whose copyright belongs to ACL. Figures and their captions were extracted from all articles in the corpus using the PdfFigures toolkit [5], resulting in a collection of 42,530 figures; figures that were not mentioned in the text of the article at least one time were excluded from the collection. In order to extract the full text from the PDF files of the articles, we used the Grobid toolkit (github.com/kermitt2/grobid).

Queries Data Set and Relevance Judgments: Ideally, we should create our query set based on real queries from users. Unfortunately, there are no such queries available to us. To address this challenge, we opt to use figure captions as queries with the assumption that if a user would like to search for figures, it is conceivable that the user would use a sentence similar to a caption sentence of a figure. One additional benefit of this is that we can then assume that the figure whose caption has been taken as the query is relevant to the query and thus should be ranked on the top of other figures by an effective figure retrieval algorithm. Of course, we have to exclude the caption sentences from the representation of the figure, or otherwise, the relevant figure would be trivially ranked on the top of other figures by every ranking method. The other figures are assumed to be non-relevant. We note that this assumption is clearly invalid as some of those figures may also be relevant. However, it is still quite reasonable to assume that the figure whose caption has been used as a query should be regarded as more relevant than any other figures, thus measuring to what extent a method can rank this target figure on top of all others is still quite meaningful and can be used to make relative comparisons of different methods. To further improve the quality of the queries, we use only captions that have between 2 and 5 words (not including stopwords), resulting in 16,829 queries; 17%, 33%, 30%, and 20% of the queries in the data set are of length 2, 3, 4, and 5, respectively. The data set of queries was split at random such that one half was used for training the LTR algorithm and the other half was used for evaluation.

4.2 Implementation Details

The Lucene toolkit (lucene.apache.org) was used for experiments. Krovetz stemming and stopword removal were applied to both queries and figure fields. For our word embeddings-based retrieval model, we trained a CBOW Word2Vec model [17] with a window size of 5 and 100 dimensions (radimrehurek.com/gensim/models/word2vec). We used the LambdaMart algorithm [28] in order to learn an LTR model (sourceforge.net/p/lemur/wiki/RankLib). Using the LTR model for ranking the entire collection of figures is not practical as several features are quite expensive to compute for all figures (e.g., word embeddings). We address this issue by adopting a 2-phase retrieval paradigm as follows. We perform an initial retrieval of 100 figures using the FigText field with $w = 50$ (and the BM25 retrieval model). Then, we re-rank the result list using the LTR model with the entire set of features. We use the mean reciprocal rank (MRR@100) and the $success@k$ ($k \in \{1, 3, 5, 10\}$) as our evaluation measures. $success@k$ is the fraction of queries for which the relevant figure is among the top k results.

Table 1. Main result. Figure retrieval performance when different figure fields and different retrieval models are used. The differences in MRR between all LTR models and the initial retrieval are statistically significant (two-tailed paired t-test, $p < 1.0e - 7$).

		MRR	$success@1$	$success@3$	$success@5$	$success@10$
Initial retrieval		.443	.353	.497	.547	.607
LTR						
BM25	FigText	.478	.391	.531	.577	.639
	FigArticle	.126	.079	.142	.172	.218
	FigText+FigArticle	.483	.394	.538	.586	.648
W2V	FigText	.212	.129	.233	.291	.377
	FigArticle	.070	.026	.064	.096	.154
	FigText+FigArticle	.212	.127	.230	.289	.380
BM25+W2V	FigText+FigArticle	.487	.398	.541	.592	.649

4.3 Experimental Results

Main Result: The performance of our suggested approach for the figure retrieval task is presented in Table 1. We compare the effectiveness of the initial retrieval with that of the re-ranking approach in which LTR was used. In the case of LTR, we report the performance of using different figure fields and different retrieval models. The LTR performance when the BM25 retrieval model is used is reported in the upper block of the table. According to the results, this approach outperforms the initial retrieval by a very large margin when FigText fields are used. This result attests to the benefit of using different sizes of window for the FigText fields (recall that only a single window size of 50 was used for the initial retrieval). Using the FigArticle fields, on the other hand, results in an ineffective LTR model compared to the initial retrieval. Yet, according to the results, there is clear merit in combining FigText and FigArticle fields. When W2V is used as a retrieval model, we can see that it is not effective with respect to the initial retrieval. Furthermore, as in the case of BM25, FigText fields are more effective than FigArticle fields when W2V is used. Finally, when all figure fields and all retrieval models are combined, the highest performance is achieved for all evaluation measures. We conclude, based on Table 1, that the most useful figure fields are the FigText fields and the most effective retrieval model is BM25. The W2V retrieval model and the FigArticle fields, on the other hand, are not very effective when used alone and only improve performance when added on top of the other features.

Analysis of Individual Fields: The performance of using individual FigText fields and FigArticle fields for re-ranking the initial result list is reported in Fig. 1(a) and 1(b), respectively. In each graph, the performance (MRR) when a single field is used is reported (blue bar) as well as when a single field is used together with all the fields presented to its left (i.e., accumulative performance; orange bar); BM25 was used as a retrieval model. According to Fig. 1(a), all FigText fields are quite effective and the re-ranking performance increases with the size of the window. Moreover, there is a clear benefit in combining different sizes of the window as the accumulative performance also increases as a function of the window size. Indeed, the length of the text which describes a figure can often vary. In this paper, we address this issue by using different values for the text length. In future work, we plan to explore automatic approaches for setting this value dynamically on a per-figure basis. As for the FigArticle fields, the performance increases as a function of the average field length. That is, the lowest performance is achieved for the title and the highest performance is achieved for the introduction. As in the case of the FigText fields, we can see that there is always an added value when using multiple fields.

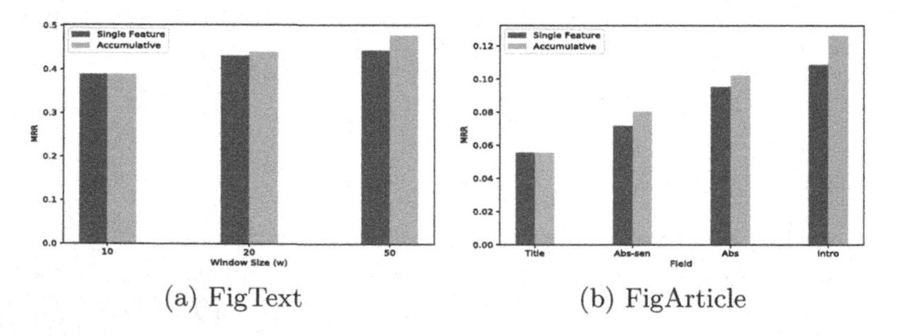

(a) FigText (b) FigArticle

Fig. 1. Performance of using individual figure fields. The performance of the FigText and FigArticle fields is depicted in Figure (a) and (b), respectively. (Color figure online)

Figure Centrality Analysis: A figure in a research article can be mentioned in the text several times. We define the number of figure mentions as the number of times the figure number was explicitly mentioned in the article (i.e., the number

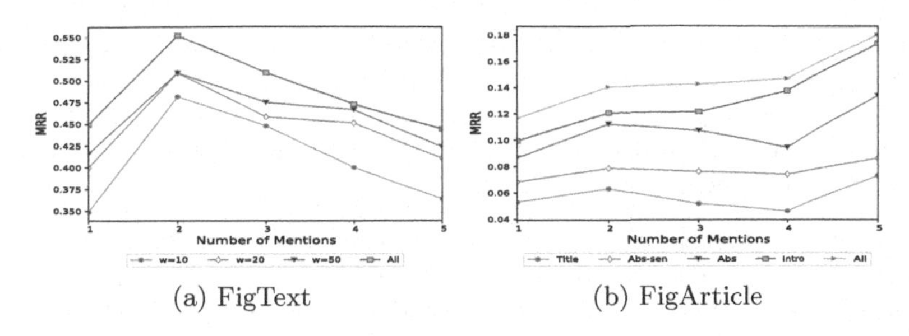

(a) FigText (b) FigArticle

Fig. 2. Performance of using different figure fields as a function of the number of mentions of the figure in the article. "All" refers to using all fields. The value of '5' in the x-axis refers to figures with *at least* five mentions. (Color figure online)

of mentions of figure i is the number of appearances of the string "Figure i" in the text). We examine the performance of using different figure fields (using BM25) for re-ranking the initial result list as a function of the number of figure mentions in Fig. 2. Figures with 1, 2, 3, 4, and 5 (or more) mentions constitute 65%, 23%, 7%, 3%, and 2% of the entire figures in the test set, respectively. The performance of using the FigText fields is depicted in Fig. 2(a). According to the graph, the poorest performance is achieved when the figure has only one mention and the highest performance is achieved for two mentions. Furthermore, increasing the number of mentions to more than two almost always results in a performance decrease. A possible explanation for that can be that when the figure is mentioned many times, there are high chances for the window of text to include irrelevant text. The results for the FigArticle fields are presented in Fig. 2(b). According to the graph, the performance almost always increases with the number of mentions for all fields. A possible explanation for that can be that once the figure is mentioned many times in the article, there are high chances that it describes a central topic in the article. Consequently, the text that does not explicitly describes the figure is expected to serve as a more reliable

Table 2. Representative queries and the rank of the relevant figure.

Query	Rank	Query	Rank
(1) Dialog strategy architecture	6	(6) word gloss algorithm	2
(2) Dependency tree english sentence	2	(7) precision recall graph query	32
(3) Performance official runs	1	(8) example graphic tree	1
(4) Full simulation naive bayes f1	9	(9) graphical model sdtm	1
(5) Hierarchical recurrent neural network	1	(10) example dependency tree	0

representation of the figure. Further exploration revealed that adding the number of mentions as an additional feature in the LTR algorithm does not result in further performance gains. An interesting future work would be to explore the effectiveness of more features that capture the centrality of a figure in an article as suggested in previous works [14,31].

Query Analysis: In Table 2, we provide ten representative examples of queries with variable performance and information needs and the corresponding rank of the relevant figure when all features are used for re-ranking the initial result list. (Rank=0 means that the relevant figure did not appear in the top 100 results.) The queries in Table 2 help to illustrate the different information needs that can be addressed by figure retrieval. For example, queries 4 and 7 describe a need for experimental results, while queries 5 and 9 describe a need for some model. Table 2 also helps to illustrate the variance in performance of different queries. For example, query 10 fails to retrieve the relevant figure presumably since this query is very general, resulting in many other figures that match those keywords. Other queries are well specified (e.g., query 4) and thus result in a much better performance. As we already mentioned, one limitation of our experiments is that only one figure is considered relevant for a query. Thus, it is plausible that in a more realistic scenario we would be able to see much better performance for these queries. Nevertheless, these examples help illustrate the potential information needs in figure retrieval and the difficulty of some queries.

We perform an analysis of the query topics in order to gain further understanding about the types of information need in figure retrieval and the effectiveness of their corresponding queries. Specifically, we learn an LDA topic model [4] using all queries in both training and test set. (We use the MeTA toolkit to learn the topic model [16].) Ten words with the highest probabilities in five topics are presented in Table 3. We also present the performance of each topic, which is calculated as follows. We first assign a topic for each query. This topic is the one with the highest probability in the multinomial distribution over topics for this query. Then, we report the average MRR of the queries in each topic. (Each topic ended up containing about 20% of the queries.) The results in Table 3 illustrate potentially five types of information need. For example, Topic 1 contains words that are frequently used in figures that describe examples in the ACL corpus (e.g., "example", "tree", and "parse"). Words that describe a model or an algorithm, on the other hand, can be seen in Topic 2. Finally, Topic 3 contains words that are related to the description of experimental results (e.g., "accuracy" and "performance"). Examining the performance of the different topics, we can see that it can be very different. For example, the worst performance is achieved for Topic 1 (potentially queries for retrieving examples), and the best performance is achieved for Topic 4 which presumably describes an information need for an experimental setup (e.g., "corpus", "annotation", and "text").

Table 3. Query topics (LDA). The average performance of the queries in each topic in terms of MRR is reported in the parenthesis.

Topic 1 (.417)	Topic 2 (.506)	Topic 3 (.501)	Topic 4 (.541)	Topic 5 (.471)
Example	Example	Result	Example	System
Tree	Algorithm	Distribution	Sample	Architecture
Sentence	Model	Accuracy	Annotation	Overview
Parse	Rule	Different	Model	Result
Structure	Learning	Set	Corpus	Process
Dependency	Word	Score	Dialogue	Question
Derive	Alignment	Data	Interface	Framework
Sample	Base	Performance	Entry	Evaluate
Graph	Process	Comparison	Structure	Flow
Rule	Graph	Training	Text	Example

5 Conclusions and Future Work

In this paper, a novel task of figure retrieval from collections of research articles is suggested and studied. According to the new task, figures of research articles are treated as retrieval units and the goal is to rank them with response to a query. We propose and study different approaches for building a representation for a figure using the article text as well as different retrieval methods. Our empirical evaluation demonstrates the benefit of using a rich textual representation for a figure and of combining different retrieval models. Furthermore, an analysis of the queries in the data set sheds some light on the potential information needs in figure retrieval and their relative difficulty.

Figure retrieval is a very promising novel retrieval task; an effective figure search engine would enable researchers to increase productivity, thus accelerating scientific discovery. Our work is only a small initial step; there are many interesting novel research directions that can be further studied in the future which we briefly discuss below.

First, as there does not exist any test collection for figure retrieval, evaluation of figure retrieval is quite challenging. Although we created a test collection, which allowed us to make some interesting relative comparisons of different methods, the test collection we constructed has two limitations: (1) captions do not necessarily represent information needs of real users; (2) captions have only one relevant figure. This data set allowed us to gain some initial understanding of the problem and study the relative effectiveness of different approaches, but those findings have to be further verified with additional experiments. Thus, a very important future work is to build a more realistic data set using a query log and verify our findings. We are currently working on collecting such data by using a figure search engine which we developed (figuresearch.web.illinois.edu).

Second, related to the challenge of constructing a test collection is a better understanding of the information needs in figure retrieval. To that end, it is necessary to conduct a user study in order to obtain some realistic queries. It would also be interesting to study what kind of queries are harder to answer. Another interesting question would be whether there are some common types of information need shared among different research disciplines. A thorough understanding of the users' information needs is also crucial for devising effective retrieval methods that are optimized with respect to user needs.

Third, in this paper, we assumed that the user query is textual. However, in the most general case, the query can involve both textual and visual information. For example, the user would describe an information need using text and also provide figure examples. This raises the question of how to create an effective representation of the user query. To that end, it would make sense to leverage ideas from the area of computer vision, creating an interesting opportunity for interdisciplinary research of information retrieval and computer vision. Furthermore, different representations of the query may also necessitate the development of new ranking models that have to combine multiple ranking criteria.

Figure representation is another subject worth exploring in future work. In this work, we used only textual information for figure representation. In the general case, however, it might be useful to combine different types of information. For example: text data, visual information, article citation information, and figure centrality information. One line of works in this direction would be to identify useful sources of information. Another direction would be to combine heterogeneous information into an effective figure representation.

Finally, devising approaches for the extraction of relevant information for representing a figure is also important. For example, devising methods for automatically identifying the text in the article that discusses a figure, and devising computer vision methods for extraction of useful information from figures to enhance retrieval accuracy are all very interesting directions for future work.

Acknowledgments. We thank the reviewers for their useful comments. This material is based upon work supported by the National Science Foundation under Grant No. 1801652.

References

1. Adafre, S.F., de Rijke, M., Sang, E.T.K.: Entity retrieval. In: Recent Advances in Natural Language Processing (RANLP 2007) (2007)
2. Ah-Pine, J., Csurka, G., Clinchant, S.: Unsupervised visual and textual information fusion in cbmir using graph-based methods. ACM Trans. Inf. Syst. (TOIS) **33**(2), 9 (2015)
3. Bird, S., et al.: The ACL anthology reference corpus: A reference dataset for bibliographic research in computational linguistics (2008)
4. Blei, D.M., Ng, A.Y., Jordan, M.I.: Latent dirichlet allocation. J. Mach. Learn. Res. **3**, 993–1022 (2003)

5. Clark, C., Divvala, S.: Looking beyond text: Extracting figures, tables, and captions from computer science papers (2015)
6. Demartini, G., Missen, M.M.S., Blanco, R., Zaragoza, H.: Entity summarization of news articles. In: Proceedings of the 33rd International ACM SIGIR Conference on Research and Development in Information Retrieval, pp. 795–796. ACM (2010)
7. Dey, S., Dutta, A., Ghosh, S.K., Valveny, E., Lladós, J., Pal, U.: Learning cross-modal deep embeddings for multi-object image retrieval using text and sketch. arXiv preprint arXiv:1804.10819 (2018)
8. Ding, Y., Zhang, G., Chambers, T., Song, M., Wang, X., Zhai, C.: Content-based citation analysis: the next generation of citation analysis. J. Assoc. Inf. Sci. Technol. **65**(9), 1820–1833 (2014)
9. Eakins, J., Graham, M.: Content-based image retrieval (1999)
10. Hearst, M.A., et al.: Biotext search engine: beyond abstract search. Bioinformatics **23**(16), 2196–2197 (2007)
11. Kaszkiel, M., Zobel, J.: Passage retrieval revisited. In: ACM SIGIR Forum, vol. 31, pp. 178–185. ACM (1997)
12. Kim, D., Yu, H.: Figure text extraction in biomedical literature. PloS one **6**(1), e15338 (2011)
13. Liu, F., Jenssen, T.K., Nygaard, V., Sack, J., Hovig, E.: Figsearch: a figure legend indexing and classification system. Bioinformatics **20**(16), 2880–2882 (2004)
14. Liu, F., Yu, H.: Learning to rank figures within a biomedical article. PloS one **9**(3), e61567 (2014)
15. Liu, T.Y.: Learning to rank for information retrieval. Found. Trends® Inf. Retr. **3**(3), 225–331 (2009)
16. Massung, S., Geigle, C., Zhai, C.: Meta: a unified toolkit for text retrieval and analysis. In: Proceedings of ACL-2016 System Demonstrations, pp. 91–96 (2016)
17. Mikolov, T., Chen, K., Corrado, G., Dean, J.: Efficient estimation of word representations in vector space. arXiv:1301.3781 (2013)
18. Müller, H., Deselaers, T., Deserno, T., Clough, P., Kim, E., Hersh, W.: Overview of the ImageCLEFmed 2006 medical retrieval and medical annotation tasks. In: Peters, C., et al. (eds.) CLEF 2006. LNCS, vol. 4730, pp. 595–608. Springer, Heidelberg (2007). https://doi.org/10.1007/978-3-540-74999-8_72
19. Murphy, R.F., Kou, Z., Hua, J., Joffe, M., Cohen, W.W.: Extracting and structuring subcellular location information from on-line journal articles: the subcellular location image finder. In: Proceedings of the IASTED International Conference on Knowledge Sharing and Collaborative Engineering, pp. 109–114 (2004)
20. Petkova, D., Croft, W.B.: Proximity-based document representation for named entity retrieval. In: Proceedings of the Sixteenth ACM Conference on Information and Knowledge Management, pp. 731–740. ACM (2007)
21. Raviv, H., Carmel, D., Kurland, O.: A ranking framework for entity oriented search using markov random fields. In: Proceedings of the 1st Joint International Workshop on Entity-Oriented and Semantic Search, p. 1. ACM (2012)
22. Robertson, S.E., Walker, S.: Some simple effective approximations to the 2-poisson model for probabilistic weighted retrieval. In: Proceedings of the 17th Annual International ACM SIGIR Conference on Research and Development in Information Retrieval, pp. 232–241. Springer-Verlag New York, Inc., London (1994)
23. Salton, G., Allan, J., Buckley, C.: Approaches to passage retrieval in full text information systems. In: Proceedings of the 16th Annual International ACM SIGIR Conference on Research and Development in Information Retrieval, pp. 49–58. ACM (1993)

24. Sheikh, A.S., et al.: Structured literature image finder: Open source software for extracting and disseminating information from text and figures in biomedical literature. Technical report, Carnegie Mellon University School of Computer Science, Pittsburgh, USA, CMU-CB-09-101 (2009)
25. Shete, D.S., Chavan, M., Kolhapur, K.: Content based image retrieval. Int. J. Emerg. Technol. Adv. Eng. **2**(9), 85–90 (2012)
26. Tellex, S., Katz, B., Lin, J., Fernandes, A., Marton, G.: Quantitative evaluation of passage retrieval algorithms for question answering. In: Proceedings of the 26th Annual International ACM SIGIR Conference on Research and Development in Informaion Retrieval, pp. 41–47. ACM (2003)
27. Vinyals, O., Toshev, A., Bengio, S., Erhan, D.: Show and tell: a neural image caption generator. In: Proceedings of the IEEE Conference on Computer Vision and Pattern Recognition, pp. 3156–3164 (2015)
28. Wu, Q., Burges, C.J., Svore, K.M., Gao, J.: Adapting boosting for information retrieval measures. Inf. Retr. **13**(3), 254–270 (2010)
29. Yin, X.C., et al.: Detext: a database for evaluating text extraction from biomedical literature figures. PLoS One **10**(5), e0126200 (2015)
30. Yu, H., Lee, M.: Accessing bioscience images from abstract sentences. Bioinformatics **22**(14), e547–e556 (2006)
31. Yu, H., Liu, F., Ramesh, B.P.: Automatic figure ranking and user interfacing for intelligent figure search. PLoS One **5**(10), e12983 (2010)

"Is This an Example Image?" – Predicting the Relative Abstractness Level of Image and Text

Christian Otto[1,2](✉), Sebastian Holzki[2], and Ralph Ewerth[1,2]

[1] Leibniz Information Centre for Science and Technology (TIB), Welfengarten 1B, 30167 Hanover, Germany
{christian.otto,ralph.ewerth}@tib.eu

[2] L3S Research Center, Leibniz Universität Hannover, Hanover, Germany
sebastian.holzki@gmx.de

Abstract. Successful multimodal search and retrieval requires the automatic understanding of semantic cross-modal relations, which, however, is still an open research problem. Previous work has suggested the metrics *cross-modal mutual information* and *semantic correlation* to model and predict cross-modal semantic relations of image and text. In this paper, we present an approach to predict the (cross-modal) relative abstractness level of a given image-text pair, that is whether the image is an abstraction of the text or vice versa. For this purpose, we introduce a new metric that captures this specific relationship between image and text at the *Abstractness Level (ABS)*. We present a deep learning approach to predict this metric, which relies on an autoencoder architecture that allows us to significantly reduce the required amount of labeled training data. A comprehensive set of publicly available scientific documents has been gathered. Experimental results on a challenging test set demonstrate the feasibility of the approach.

Keywords: Image-text relations · Multimodal embeddings · Deep learning · Visual-verbal divide

1 Introduction

In the era of big data, the proliferation of multimodal web content in online news, social networks, open educational resources, video portals, etc. is increasing drastically. Graphics and pictures in multimodal documents are a powerful communication channel to illustrate, decorate, detail, summarize, or complement textual information. This is particularly true for educational and scientific material. In this context, graphical and pictorial information can be very important to support learning scenarios as, for instance, in the recently evolved field of search as learning. To enable truly multimodal recommender systems for web search, an automatic understanding of the *multimodal* content and the

L. Azzopardi et al. (Eds.): ECIR 2019, LNCS 11437, pp. 711–725, 2019.
https://doi.org/10.1007/978-3-030-15712-8_46

inherent cross-modal relations are a prerequisite. In this respect, however, information retrieval research has not addressed yet all possible kinds of cross-modal relations between images and text in a differentiated way[1]. Typically, (multimedia) information retrieval research assumes a semantic correlation *in general* in case image and text are placed jointly on purpose. In previous work [14], we have addressed this issue and suggested two metrics (dimensions) to differentiate image-text relations by (1.) cross-modal mutual information and (2.) semantic correlation.

In this paper, we show that these two metrics do not completely cover all possible types of image-text relations, particularly when considering educational or scientific content. Therefore, we suggest an additional metric: the relative *Abstractness Level* (ABS) that measures whether an image depicts information of a related text at a more detailed or a more abstract level, or at the same level. Furthermore, we propose a deep learning approach to automatically predict the abstractness level of a given image-text pair. The system relies on an autoencoder architecture and multimodal embeddings. Since the deep learning system requires a sufficiently large amount of training data, we have gathered an appropriate dataset from a variety of Web resources. Experimental results on a demanding test set demonstrate the feasibility of the proposed system.

The paper is structured as follows. Related work is summarized in Sect. 2 from the perspectives of communication sciences and information retrieval. Section 3 motivates the new metric of abstractness level and explains the proposed deep learning system to automatically predict this metric, while Sect. 4 describes the data acquisition process for the training of the deep networks. The experimental results are presented in Sect. 5. Section 6 concludes the paper and outlines areas of future work.

2 Related Work

2.1 Image-Text Relations and the Visual/Verbal Divide

The interplay between visual and textual[2] information has been subject to research for decades in the fields of communication sciences and applied linguistics. One of the early attempts to comprehensively categorize the joint placement of images and text date back to Barthes [5], who set the groundwork for a lot of categorizations that developed later. For example, Martinec and Salway [22], Marsh and White [21], and Unsworth [32] build upon Barthes' taxonomy, which defines the *Status* relation between an image and its accompanied text. This relation describes if there is a hierarchical dependency between both modalities or if they are equally important in conveying the information intended by the author. The aforementioned taxonomies extend this distinction with different

[1] In contrast to research in communication sciences and applied linguistics where the visual/verbal divide has been researched in a very detailed way for decades.

[2] Textual information can be considered as visual information as well, of course. Here, we denote graphical and pictorial information as *visual information*.

interpretations of Halliday's [13] *logico-semantics*, which are a linguistic method to describe different types of text clauses. The application of these fine-grained distinctions of the logico-semantics to image-text pairs result in very detailed taxonomies. Figure 1 shows the latest version of Unsworth's extensions to Martinec and Salway's taxonomy [22]. While these taxonomies are comprehensive, their level of detail makes it sometimes difficult to assign an image-text pair to a particular class, as criticized by Bateman [6], for instance. Recently, we have approached this problem differently through two metrics (or dimensions) that are more general and easier to infer [14]: (1.) *Cross-modal Mutual Information (CMI)* is defined as the amount of shared entities or concepts in both modalities, ranging from 0 to 1; and (2.) *Semantic Correlation (SC)* is defined as the amount of shared meaning or context, indicating if the information contained in both modalities are aligned, uncorrelated or contradictory, i.e., ranging from −1 to 1. Furthermore, it was shown that a deep learning approach that utilizes multimodal embeddings can basically predict these interrelation metrics.

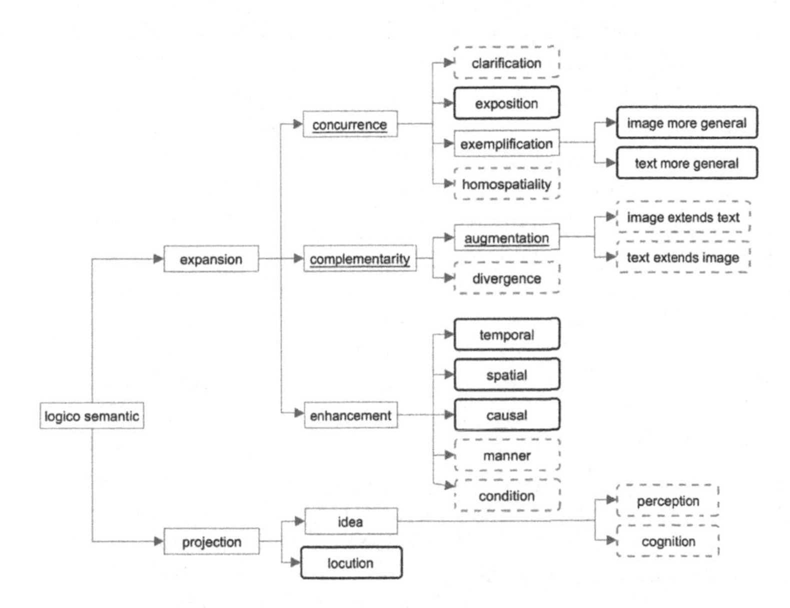

Fig. 1. The logico-semantics part of Unsworth's taxonomy [32] is shown, where blue dashed borders show extensions to Martinec and Salway [22] and underlined names were changed by the authors, but have the same meaning.

2.2 Machine Learning for Multimodal Data Retrieval

In this subsection, a brief overview of methods to encode heterogeneous modalities for machine learning and multimedia retrieval approaches is given. There are several possibilities to encode data samples consisting of distinct modalities [4].

The choice of the optimal method depends on multiple factors: the type of modality to encode, the number of training samples available, the type of classification to perform and the desired interpretability of the models. One type of algorithms utilizes *Multiple Kernel Learning*, which is an extension of kernel-based support vector machines [8,12]. They consist of a kernel specifically designed for each modality and thus allow for the fusion of heterogeneous data. Application domains are, for instance, multimodal affect recognition [15,25], event detection [36], and Alzheimer's disease classification [20]. An advantage is their flexibility in the kernel design and global optimum solutions, but they have a rather slow inference time. Deep neural networks are another technique to model multiple modalities at once. Due to their growing popularity in recent years, there is also much research on designing deep learning systems for processing multimodal data. Such research directions include approaches for audio-visual [1,23], audio-gesture [24] and textual-visual [17,27] data. The common idea is to encode each modality individually and fuse them in joint hidden layers. Especially well suited are these methods for encoding temporal information like sentences, which fits nicely to the problem addressed in this paper. For example, Cho et al. [10] use Gated Recurrent Units (GRU) to encode sentences, but it is also possible to utilize Long-Short-Term Memory (LSTM) cells [33]. More recent extensions are shown by Jia et al. [16] and Rajagopalan et al. [26], who model temporal information for textual as well as for the visual components. Neural networks are also able to learn meaningful embeddings of multimodal data in an unsupervised manner via an autoencoder architecture, which not only removes the necessity for hand-crafted features but also significantly reduces the required amount of labeled training data [14]. Cross-media and multimedia retrieval is an area of research that profits the most from techniques to bridge the semantic gap between image and text [3,18,28,34]. Fan et al. [11] implement a multi-sensory fusion network, which improves the comparability of heterogeneous media features and is therefore well suited for image-to-text and text-to-image retrieval. A self-paced cross-modal subspace matching method constructing a multimodal graph is proposed by Liang et al. [19]. It is designed to preserve the intra-modality and inter-modality similarity between the input samples. Carvalho et al. [9] proposed the *AdaMine* model, which combines instanced-based and semantic-based losses for a joint retrieval and semantic latent space learning method. This method is utilized to retrieve recipes from pictures of food and vice versa.

3 The Abstractness Metric for Image-Text Relations

In this section, we motivate and derive the new metric of relative abstractness for image-text pairs. In this respect, we analyze the existing gap in applied linguistics and communication sciences, as well as information retrieval.

3.1 Analysis

Our analysis starts with Marsh and White's [21] taxonomy, which describes different functions of images to a text. This taxonomy distinguishes three different levels according to whether the image has little or close relation to a text, or even goes beyond text. For example, two of the sub-classes in the taxonomy are *sample* and *exemplify*, which are both considered to describe a close relation of an image to text. However, it lacks a formal description that allows us to assign an illustration image to one the two classes. Similar issues can be found in other taxonomies, which is explained in the next subsection.

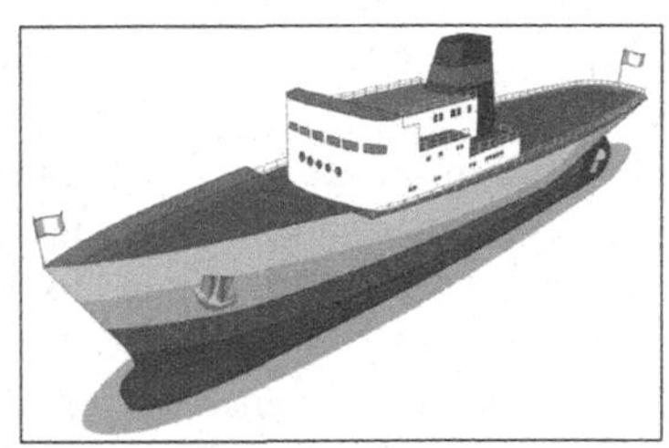

A ship is a large watercraftthat travels the world's oceans and other sufficiently deep waterways, carrying passengers or goods, or in support of specialized missions, such as defense, research and fishing.

Fig. 2. An example for the image-text classes *sample* and *exemplify* of Marsh and White. The authors implicitly use the concept of abstraction to add more depth to their categorizations.

Our example is illustrated in Fig. 2 that portrays the classes *sample* and *exemplify* by Marsh and White [21]. It shows a textual phrase and two visual representations which together in both cases create an *Illustration* example according to Barthes [5]. However, according to Marsh and White they belong to the *sample* and *exemplify* classes, where the latter one is defined as an ideal example and the first one can be any concrete instance of the described concept. Therefore, the actual distinction can be made by means of their abstractness, which is also a very important concept for scientific or educational material to improve comprehensibility.

3.2 Implications and the Abstractness Metric

We claim that the *relative difference of the Abstractness Level (ABS)* is an essential part in describing the relations between an image and text. To support this assumption, we list in Table 1 a number of image-text classes that contain a certain difference in abstractness by definition. This implies that a metric describing the relative difference in the abstractness level is indeed necessary to characterize an image-text relation. We would like to emphasize the term *relative*, since it is important that image and text are considered jointly. Therefore, a particular image can be less abstract than a text, while it is more abstract than another one. Also, in order to differentiate between abstractness levels it is necessary to have an object of reference, or a *Cross-modal Mutual Information* $CMI > 0$, as it is the case for the ship example in Fig. 2.

Table 1. Overview of image-text classes that entail a certain difference on the abstractness level between image and text. (Note: $>_a$ is read as "is more abstract than")

ABS → Reference ↓	$\mathbf{I} =_a \mathbf{T}$	$\mathbf{I} >_a \mathbf{T}$	$\mathbf{I} <_a \mathbf{T}$
Martinec & Salway [22]	Exposition, locution, idea	Image more general, enhancement by text	Text more general, enhancement by image
Unsworth [32]	Exposition, clarification, locution, perception, cognition	Text instantiates image, Enhancement by text	Image instantiates text, enhancement by image
Marsh & White [21]	Compare, contrast, concentrate, compact, model	Exemplify, isolate, contain, locate, induce perspective, emphasize, document	Sample, graph, translate, describe, define

4 Predicting the Abstractness Level of Image and Text

In this section, we present a system that automatically measures the relative abstractness level between an image and its associated text. There are no repositories and Web resources of image-text pairs that can be easily exploited to train a deep learning classifier. Consequently, we follow an autoencoder approach similar to [14], which requires much less labeled data. We collect the necessary samples from open access publications provided by Sohmen et al. [29], as explained in detail in the next subsection.

4.1 Data Acquisition

We have gathered a training dataset consisting of scientific documents since the image-text pairs therein contain different levels of relative abstraction. For the purpose of legal re-use, Sohmen et al. [29] provide illustrations of publications of the open access publisher Hindawi[3]. The majority of Hindawi articles is available under the Creative Commons Attribution License, so that they can be used for this type of research. Another advantage is that they are accessible in XML format, which makes them easier to read than files in PDF format. We crawled

[3] https://www.hindawi.com/.

Table 2. Overview of the training data used for the autoencoder.

Journal	#articles	#figures	#image-text pairs
AAI	94	1,217	3,180
ACISC	185	2,215	5,453
AM	144	2,304	6,057
MPE	8,251	106,435	273,367
Sum	8,674	112,171	288,057

Table 3. Overview of the manually labeled part of the data which is used to train and evaluate the classifier network.

Journal	T $>_a$ I	T $<_a$ I	T $=_a$ I	Sum	Percentage
AAI	322	113	145	580	19.2%
ACISC	383	173	242	798	26.5%
AM	354	169	169	858	28.4%
MPE	352	255	255	780	25.9%
Sum	1,411	710	895	3,016	-
Percentage	46.8%	23.5%	29.7%	-	-

288,057 image-text pairs from four different journals, namely *Advances in Artificial Intelligence (AAI)*, *Applied Computational Intelligence and Soft Computing (ACISC)*, *Advances in Multimedia (AM)* and *Mathematical Problems in Engineering (MPE)*. The final distribution of articles is presented in Table 2.

We manually labeled more than 3,000 image-text pairs for training and testing. The data distribution of the labeled data is presented in Table 3. Our annotation process results in a minimum number of 700 samples per image-text class which is sufficient to train a classifier that uses the embeddings of our pre-trained autoencoder network.

4.2 Representing Multimodal Data via Autoencoding

We suggest an autoencoder approach for two main reasons: First, the automatic generation of labeled training data is not possible since the available amount of annotated image-text pairs is limited. It is not reasonable to train a classifier from scratch with less than 1,000 samples per class. Second, the encoder-decoder architecture allows for adjustments that fit nicely to our scenario and also allows us to investigate if the right information is preserved by the embedding. Our design is similar to Henning and Ewerth's [14] approach, but includes some modifications that consider the nature of figures and illustrations in scientific documents as opposed to natural images. Also, we replace some components in the encoding part with more recent system components, see Fig. 3. In detail, we use a pre-trained model of the Inception-ResNet-v2 [30] without its classification

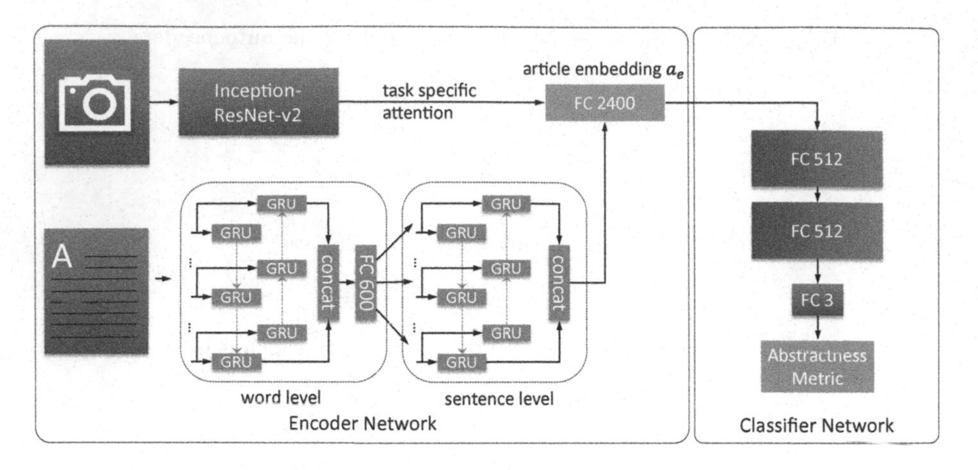

Fig. 3. Overview of the encoder and classifier network.

layers to encode the input image as well as the preprocessing pipeline suggested by Szegedy et al. [31]. The textual information is preprocessed by removing any specific XML characters and by replacing formulas with the word "formula". In addition, we truncate sentences that are longer than 50 words and paragraphs longer than 30 sentences. The resulting feature vector of image encoding is then fed into the attention mechanism of the text encoding step, where, inspired by Yang et al. [35], a bidirectional recurrent neural network (RNN) architecture is used consisting of multiple GRU cells. This way the text is encoded in a hierarchical way: first on a sentence and then on a full-text level. After concatenating the image and text embedding we receive a 2,400 dimensional *article embedding* according to [14] (Fig. 3).

4.3 Classifier

To obtain a high-quality multimodal embedding for image-text pairs, we aim at a decoder that reconstructs the image as well as text from the encoded article embedding. We compute a loss between input and output information that describes how well image and text can be reproduced from the condensed representation. A first fully-connected layer decides which parts of the embedding are important to reproduce the image and therefore generates a first 30×30 predicted reconstruction of the visual data. An alternating series of up-scaling and convolutional layers subsequently produces an image that corresponds to the size of the input image (300×300). In contrast to Henning and Ewerth [14], the convolutional layers use a kernel size of $3x3$ instead of $5x5$, which is necessary to successfully reproduce the fine lines depicted in many scientific tables or diagrams. Another difference is the size of the convolutional layers in the pipeline, where we use (128, 64, 32, 3) opposed to (32, 8, 3). The loss between input and output image is computed using mean squared error.

The textual information is reconstructed by a hierarchical unidirectional RNN consisting of LSTM cells, which proved to be more powerful than Gated Recurrent Units (GRUs) for this task. It first generates sentence features from the article embedding and uses them afterwards on a word level to estimate the original input text. Both hierarchy layers use the batch normalization technique proposed by Ba et al. [2]. An overview of the decoder network can be seen in Fig. 4. The loss between input and output is computed based on a word embedding that is based on a predefined fastText [7] vocabulary, which is reduced prior to the experiments to reduce memory usage of the model. In particular, we use the 25,000 most common words (out of about 89,000) in our dataset, which allows us to still cover 98,81% of the occurring vocabulary. All other words get the representation <*unk*>. The decoder tries to reconstruct the correct index of each word in the text from the embedding and the loss is computed using the cosine similarity between input and output feature vector. The result of the autoencoder training process is an encoder network that is able to produce highly expressive embeddings that compress visual as well as textual information to the key components, which are necessary to describe the content of the input. Based on these features we train a classifier network with our labeled (training) samples. This part of the network comprises three fully-connected layers (FC) of size (512, 512, 3), where the last one predicts one of the three different levels of *Abstractness*. The entire network architecture is displayed in Fig. 3.

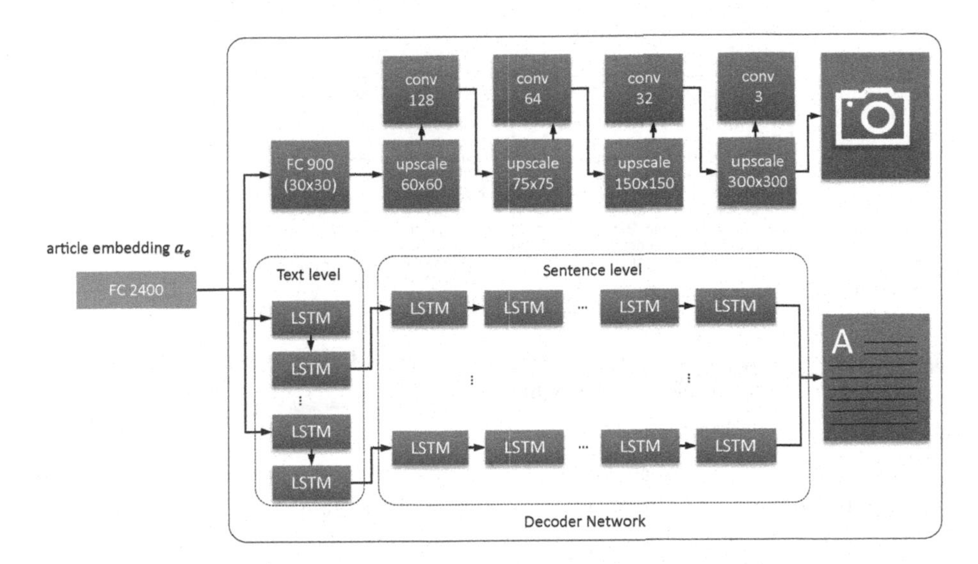

Fig. 4. Overview of the decoder network, whose input is the article embedding generated by the encoder (Fig. 3).

5 Experimental Results

This section is separated into two parts. First, we present some example results for the CNN-based autoencoder before we present experimental results for the classification of the relative *Abstractness Level* of image and text.

5.1 Autoencoder Training

The autoencoder network was trained for 360,000 iterations at a batch size of 15, which corresponds to about 19 epochs. The distribution of training samples is shown in Table 5 and an example output of the autoencoder is depicted in Fig. 5.

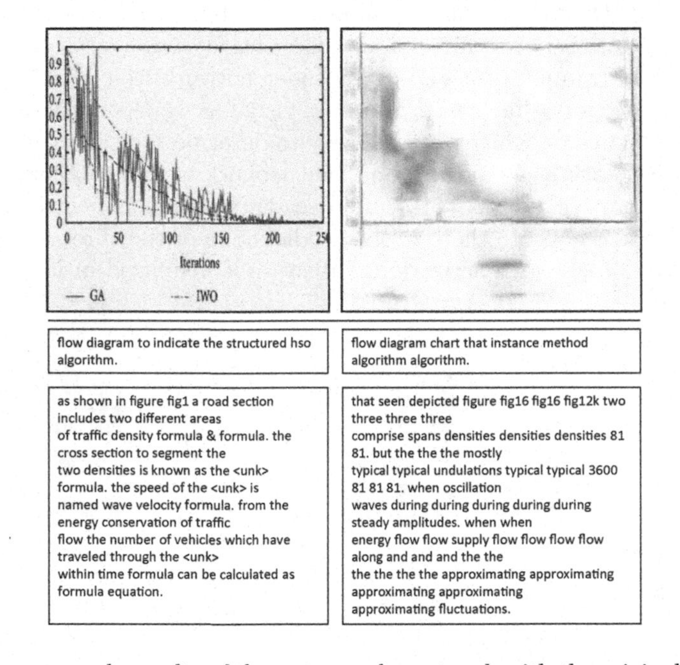

Iterations
GA
IWO
flow diagram to indicate the structured hso algorithm.
flow diagram chart that instance method algorithm algorithm.
as shown in figure fig1 a road section includes two different areas of traffic density formula & formula. the cross section to segment the two densities is known as the <unk> formula. the speed of the <unk> is named wave velocity formula. from the energy conservation of traffic flow the number of vehicles which have traveled through the <unk> within time formula can be calculated as formula equation.
that seen depicted figure fig16 fig16 fig12k two three three three comprise spans densities densities densities 81 81. but the the the mostly typical typical undulations typical typical 3600 81 81 81. when oscillation waves during during during during during steady amplitudes. when when energy flow flow supply flow flow flow flow along and and and the the the the the the approximating approximating approximating approximating approximating fluctuations.

Fig. 5. Three example results of the autoencoder network with the originals on the left and the reproduced samples on the right.

The example output of the autoencoder shows that the network is indeed able to coarsely reproduce the essential information of the visualizations, for instance, diagram borders, fine details on the axis, and the legend of the diagram. Also, the decoded text elements resemble the original text in length and to some extent even the semantic context.

5.2 Classification of the Relative Abstractness Level

As described in Sect. 4.1, we gathered a set of about 3,000 image-text pairs that we subsequently separated into a training and test set, where the latter one

Table 4. Confusion matrix for the classifier of the relative abstraction level.

Class	Image $<_a$ Text	Image $>_a$ Text	Image $=_a$ Text	Sum
I $<_a$ T	**90**	7	3	100
I $>_a$ T	14	**68**	18	100
I $=_a$ T	10	7	**83**	100
Precision	78.95%	82.93%	79.81%	-
Recall	90.00%	68.00%	83.00%	-

Table 5. Comparison of the three different classification approaches.

Classifier	$CL_{transfer}$	CL_{freeze}	$CL_{scratch}$
Accuracy	80.33%	77.33	77.00

consists of 100 random samples for each of the three classes. We have evaluated three different versions of the autoencoder and classifier networks.

1. $CL_{scratch}$: Train the classifier network as well as the encoder network from scratch, making it an end-to-end approach.
2. CL_{freeze}: Train the classifier network, but freeze the weights of the pre-trained encoder network.
3. $CL_{transfer}$: Train the classifier network and finetune the pre-trained encoder network at the same time.

Every approach was trained for about 70,000 iterations. The results are reported in Tables 4 and 5. The former shows that the classifier was able to predict the three classes successfully with a recall of 90% ($I <_a T$), 68% ($I >_a T$) and 83% ($I =_a T$). These results reflect the distribution of available labeled training data (cf. Table 3), which implies that these results can be improved by acquiring more annotated samples. Table 5 shows that a pre-trained autoencoder network outperforms a training from scratch, but is in turn outperformed by the transfer learning approach which finetunes and adapts the encoding process to the new task. This proves that a multimodal embedding is able to encode an image-text pair in a way that the *Abstractness* metric can be successfully predicted, while our autoencoder approach is able to compensate for the relatively low number of labeled training samples. Example predictions of our system are displayed in Fig. 6 (left hand side: correct, right hand side: misclassified): In the top-left image-text pair, both the text and the schematic illustrations are abstract representations, in particular for "(a)", while in the image bottom-left the image is a concretization of the text (both predicted correctly); the image in the top-right examples depicts more relevant details than the text, whereas the line chart bottom-right provides less detailed information about the experimental context than the text. As Table 4 shows, predicting $I <_a T$ was the easiest for the system, presumably because of the amount of natural images in

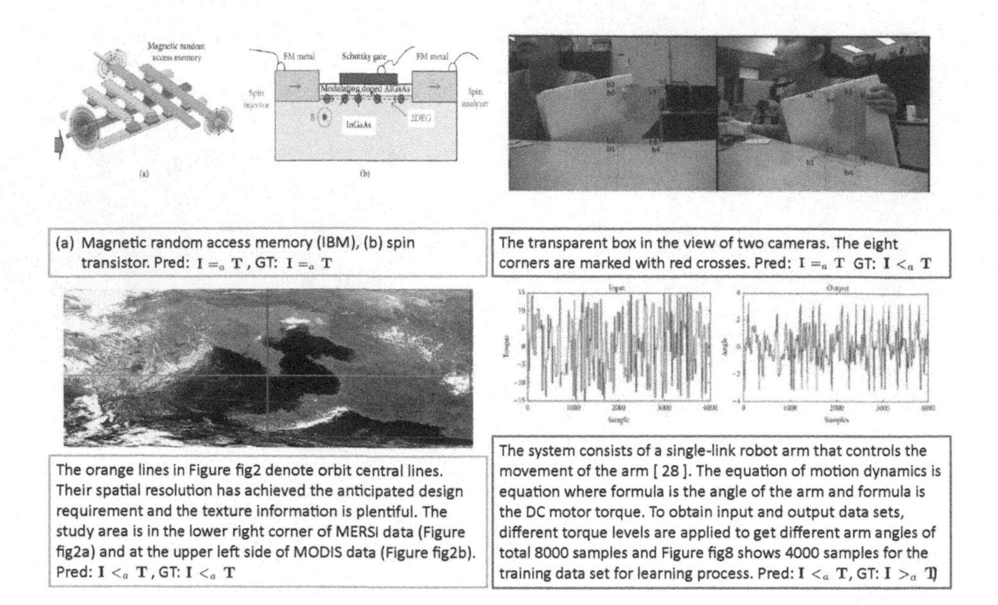
(a) Magnetic random access memory (IBM), (b) spin transistor. Pred: $\mathbf{I} =_a \mathbf{T}$, GT: $\mathbf{I} =_a \mathbf{T}$
The transparent box in the view of two cameras. The eight corners are marked with red crosses. Pred: $\mathbf{I} =_a \mathbf{T}$ GT: $\mathbf{I} <_a \mathbf{T}$
The orange lines in Figure fig2 denote orbit central lines. Their spatial resolution has achieved the anticipated design requirement and the texture information is plentiful. The study area is in the lower right corner of MERSI data (Figure fig2a) and at the upper left side of MODIS data (Figure fig2b). Pred: $\mathbf{I} <_a \mathbf{T}$, GT: $\mathbf{I} <_a \mathbf{T}$
The system consists of a single-link robot arm that controls the movement of the arm [28]. The equation of motion dynamics is equation where formula is the angle of the arm and formula is the DC motor torque. To obtain input and output data sets, different torque levels are applied to get different arm angles of total 8000 samples and Figure fig8 shows 4000 samples for the training data set for learning process. Pred: $\mathbf{I} <_a \mathbf{T}$, GT: $\mathbf{I} >_a \mathbf{T}$)

Fig. 6. Examples of correctly as well as misclassified examples from our test set, along with predicted and ground-truth labels.

this class. But, if these natural images were overlaid with additional information (top right) the system struggled to find the correct assignment.

6 Conclusions

In this paper, we have introduced a novel metric that describes the relative *Abstractness Level* between an image and associated text. We have motivated and derived the metric based on previous work on taxonomies for image-text classes in communication sciences and applied linguistics. Until now, the large variety of image-text relations had been investigated in a differentiated way mainly in these fields. We have set these taxonomies in relation to recent work in the field of multimedia retrieval, which has modeled image-text relations in a more general manner through the metrics *cross-modal mutual information* and *semantic correlation*. In this respect, our proposed metric is a contribution to the model of possible semantic (cross-modal) image-text relations in a systematic manner from an information retrieval perspective. Moreover, we have proposed a deep learning architecture to automatically predict the relative, cross-modal abstractness level of image and text. The required amount of labeled training data is minimized by the incorporation of an autoencoder network. We have evaluated three different ways of training the deep network architecture. It turned out that training the classifier network and finetune the pre-trained encoder network at the same time achieved the best results with an accuracy of 80%. In this way, an indexing method has been developed that can serve as the basis for multimodal search and retrieval. For instance, search for educational and scientific multimodal content could be improved by automatically filtering illustrations according to the classes *sample* and exemplify.

In the future, we plan to apply this indexing method to different scenarios in multimodal information retrieval, such as search as learning with multimedia data, e-learning, and recommender systems. For this purpose, we intend to build an exploration and browsing interface based on the metrics *CMI*, *SC* and *ABS*. Finally, we will evaluate the usefulness of other metrics to model cross-modal relations in a systematic way.

References

1. Afouras, T., Chung, J.S., Senior, A., Vinyals, O., Zisserman, A.: Deep audio-visual speech recognition. arXiv preprint arXiv:1809.02108 (2018)
2. Ba, J.L., Kiros, J.R., Hinton, G.E.: Layer normalization. arXiv preprint arXiv:1607.06450 (2016)
3. Balaneshin-kordan, S., Kotov, A.: Deep neural architecture for multi-modal retrieval based on joint embedding space for text and images. In: Proceedings of the Eleventh ACM International Conference on Web Search and Data Mining, pp. 28–36. ACM (2018)
4. Baltrušaitis, T., Ahuja, C., Morency, L.P.: Multimodal machine learning: a survey and taxonomy. IEEE Trans. Pattern Anal. Mach. Intell. **41**(2), 423–443 (2018)
5. Barthes, R.: Image-music-text, ed. and trans. Heath, S., Fontana 332, London (1977)
6. Bateman, J.: Text and Image: A Critical Introduction to the Visual/Verbal Divide. Routledge, Abingdon (2014)
7. Bojanowski, P., Grave, E., Joulin, A., Mikolov, T.: Enriching word vectors with subword information. arXiv preprint arXiv:1607.04606 (2016)
8. Bucak, S.S., Jin, R., Jain, A.K.: Multiple kernel learning for visual object recognition: a review. IEEE Trans. Pattern Anal. Mach. Intell. **36**(7), 1354–1369 (2014)
9. Carvalho, M., Cadène, R., Picard, D., Soulier, L., Thome, N., Cord, M.: Cross-modal retrieval in the cooking context: learning semantic text-image embeddings. arXiv preprint arXiv:1804.11146 (2018)
10. Cho, K., et al.: Learning phrase representations using RNN encoder-decoder for statistical machine translation. Association for Computational Linguistics (2014)
11. Fan, M., Wang, W., Dong, P., Han, L., Wang, R., Li, G.: Cross-media retrieval by learning rich semantic embeddings of multimedia. In: ACM Multimedia Conference, pp. 1698–1706 (2017)
12. Gönen, M., Alpaydın, E.: Multiple kernel learning algorithms. J. Mach. Learn. Res. **12**, 2211–2268 (2011)
13. Halliday, M.A.K., Matthiessen, C.M.: Halliday's Introduction to Functional Grammar. Routledge, Abingdon (2013)
14. Henning, C.A., Ewerth, R.: Estimating the information gap between textual and visual representations. In: ACM International Conference on Multimedia Retrieval (2017)
15. Jaques, N., Taylor, S., Sano, A., Picard, R.: Multi-task, multi-kernel learning for estimating individual wellbeing. In: Proceedings of the NIPS Workshop on Multimodal Machine Learning, Montreal, Quebec, vol. 898 (2015)

16. Jia, X., Gavves, E., Fernando, B., Tuytelaars, T.: Guiding the long-short term memory model for image caption generation. In: Proceedings of the IEEE International Conference on Computer Vision, pp. 2407–2415 (2015)
17. Jin, Q., Liang, J.: Video description generation using audio and visual cues. In: Proceedings of the 2016 ACM on International Conference on Multimedia Retrieval, pp. 239–242. ACM (2016)
18. Kang, C., et al.: Cross-modal similarity learning: a low rank bilinear formulation. In: ACM Conference on Information and Knowledge Management. ACM (2015)
19. Liang, J., Li, Z., Cao, D., He, R., Wang, J.: Self-paced cross-modal subspace matching. In: Proceedings of the 39th International ACM SIGIR Conference on Research and Development in Information Retrieval, pp. 569-578. ACM (2016)
20. Liu, F., Zhou, L., Shen, C., Yin, J.: Multiple kernel learning in the primal for multimodal alzheimer's disease classification. IEEE J. Biomed. Health Inform. **18**(3), 984–990 (2014)
21. Marsh, E.E., Domas White, M.: A taxonomy of relationships between images and text. J. Doc. **59**(6), 647–672 (2003)
22. Martinec, R., Salway, A.: A system for image-text relations in new (and old) media. Vis. Commun. **4**(3), 337–371 (2005)
23. Meutzner, H., Ma, N., Nickel, R., Schymura, C., Kolossa, D.: Improving audiovisual speech recognition using deep neural networks with dynamic stream reliability estimates. In: 2017 IEEE International Conference on Acoustics, Speech and Signal Processing (ICASSP), pp. 5320–5324. IEEE (2017)
24. Neverova, N., Wolf, C., Taylor, G., Nebout, F.: Moddrop: adaptive multi-modal gesture recognition. IEEE Trans. Pattern Anal. Mach. Intell. **38**(8), 1692–1706 (2016)
25. Poria, S., Cambria, E., Gelbukh, A.: Deep convolutional neural network textual features and multiple kernel learning for utterance-level multimodal sentiment analysis. In: Proceedings of the 2015 Conference on Empirical Methods in Natural Language Processing, pp. 2539–2544 (2015)
26. Rajagopalan, S.S., Morency, L.-P., Baltrušaitis, T., Goecke, R.: Extending long short-term memory for multi-view structured learning. In: Leibe, B., Matas, J., Sebe, N., Welling, M. (eds.) ECCV 2016, Part VII. LNCS, vol. 9911, pp. 338–353. Springer, Cham (2016). https://doi.org/10.1007/978-3-319-46478-7_21
27. Ramanishka, V., et al.: Multimodal video description. In: Proceedings of the 24th ACM International Conference on Multimedia, pp. 1092-1096. ACM (2016)
28. Shutova, E., Kelia, D., Maillard, J.: Black holes and white rabbits: metaphor identification with visual features. In: NAACL, pp. 160–170 (2016)
29. Sohmen, L., Charbonnier, J., Blümel, I., Wartena, C., Heller, L.: Figures in scientific open access publications. In: Méndez, E., Crestani, F., Ribeiro, C., David, G., Lopes, J.C. (eds.) TPDL 2018. LNCS, vol. 11057, pp. 220–226. Springer, Cham (2018). https://doi.org/10.1007/978-3-030-00066-0_19
30. Szegedy, C., Ioffe, S., Vanhoucke, V., Alemi, A.A.: Inception-v4, inception-resnet and the impact of residual connections on learning. In: AAAI, vol. 4, p. 12 (2017)
31. Szegedy, C., et al.: Going deeper with convolutions. In: Proceedings of the IEEE Conference on Computer Vision and Pattern Recognition, pp. 1-9. (2015)
32. Unsworth, L.: Image/text relations and intersemiosis: towards multimodal text description for multiliteracies education. In: Proceedings of the 33rd International Systemic Functional Congress, pp. 1165–1205 (2007)
33. Vinyals, O., Toshev, A., Bengio, S., Erhan, D.: Show and tell: a neural image caption generator. In: Proceedings of the IEEE Conference on Computer Vision and Pattern Recognition, pp. 3156–3164 (2015)

34. Yan, T.K., Xu, X.S., Guo, S., Huang, Z., Wang, X.L.: Supervised robust discrete multimodal hashing for cross-media retrieval. In: ACM Conference on Information and Knowledge Management, pp. 1271–1280 (2016)
35. Yang, Z., Yang, D., Dyer, C., He, X., Smola, A., Hovy, E.: Hierarchical attention networks for document classification. In: Proceedings of the 2016 Conference of the North American Chapter of the Association for Computational Linguistics: Human Language Technologies, pp. 1480–1489 (2016)
36. Yeh, Y.R., Lin, T.C., Chung, Y.Y., Wang, Y.C.F.: A novel multiple kernel learning framework for heterogeneous feature fusion and variable selection. IEEE Trans. Multimedia **14**(3), 563–574 (2012)

Short Papers

End-to-End Neural Relation Extraction Using Deep Biaffine Attention

Dat Quoc Nguyen(✉) and Karin Verspoor

The University of Melbourne, Melbourne, Australia
{dqnguyen,karin.verspoor}@unimelb.edu.au

Abstract. We propose a neural network model for joint extraction of named entities and relations between them, without any hand-crafted features. The key contribution of our model is to extend a BiLSTM-CRF-based entity recognition model with a deep biaffine attention layer to model second-order interactions between latent features for relation classification, specifically attending to the role of an entity in a directional relationship. On the benchmark "relation and entity recognition" dataset CoNLL04, experimental results show that our model outperforms previous models, producing new state-of-the-art performances.

1 Introduction

Extracting entities and their semantic relations from raw text is a key information extraction task. For example, given the sentence " David Foster is the AP 's Northwest regional reporter, based in Seattle " in the CoNLL04 dataset [27], our goal is to recognize "David Foster" as person, "AP" as organization, and "Northwest" and "Seattle" as location entities, then classify entity pairs to extract structured information: *Work_For*(David Foster, AP), *OrgBased_In*(AP, Northwest) and *OrgBased_In*(AP, Seattle). Such information is useful in many other NLP tasks. Especially in IR applications such as entity search, structured search and question answering, it helps provide end users with significantly better search experience [6,11,29].

A common relation extraction approach is to construct pipeline systems with separate sub-systems for the two tasks of named entity recognition and relation classification [2]. More recently, end-to-end systems which jointly learn to extract entities and relations have been proposed with strong potential to obtain high performance [26]. Traditional joint approaches are feature-based supervised learning methods which employ numerous syntactic and lexical features based on external NLP tools as well as knowledge base resources [12,18,20].

State-of-the-art relation extraction performance has been obtained by end-to-end models based on neural networks. Specifically, Gupta et al. (2016) [9] proposed a RNN-based model which achieved top results on the CoNLL04 dataset. Their approach relies on various manually extracted features. Other neural models employ dependency parsing-based information [19,23,31]. In particular, Miwa and Bansal (2016) [19] applied bottom-up and top-down tree-structured LSTMs

L. Azzopardi et al. (Eds.): ECIR 2019, LNCS 11437, pp. 729–738, 2019.
https://doi.org/10.1007/978-3-030-15712-8_47

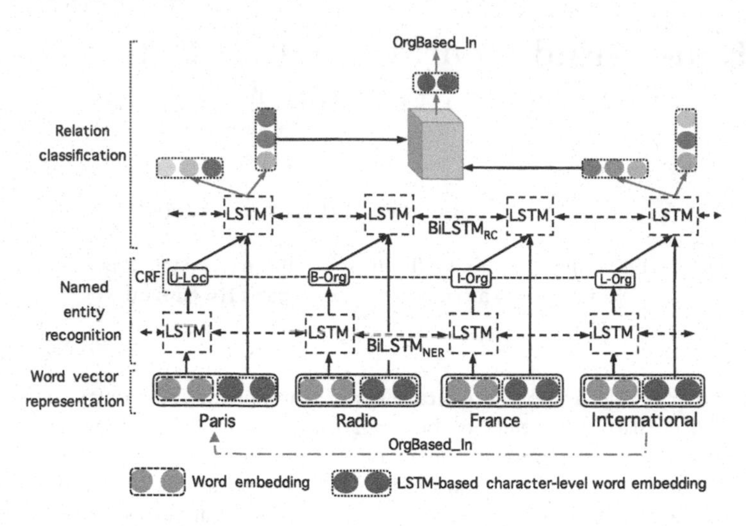

Fig. 1. Illustration of our model. Linear transformations are not shown for simplification.

to model dependency paths between entities. Zhang et al. (2017) [31] integrated implicit syntactic information by using latent feature representations extracted from a pre-trained BiLSTM-based dependency parser. Zheng et al. (2017) [32] used a softmax layer on top of a BiLSTM for entity recognition, and a CNN on top of the BiLSTM for classifying relations [22]. Adel and Schütze (2017) [1] assumed that entity boundaries are given, and trained a CNN to extract context features around the entities, and using these features for entity and relation classification. Recently, Wang et al. (2018) [30] formulated the joint entity and relation extraction problem as a directed graph and proposed a BiLSTM- and transition-based approach to generate the graph incrementally. Bekoulis et al. (2018) [4] extended the multi-head selection-based joint model [5] with adversarial training. In [5,13,33], the joint task is formulated as a sequence tagging problem, and a BiLSTM with a softmax output layer can then be used for joint prediction.

In this paper, we present a novel end-to-end neural model for joint entity and relation extraction. As illustrated in Fig. 1, our model architecture can be viewed as a mixture of a named entity recognition (NER) component and a relation classification (RC) component. Our NER component employs a BiLSTM-CRF architecture [10] to predict entities from input word tokens. Based on both the input words and the predicted NER labels, the RC component uses another BiLSTM to learn latent features relevant for relation classification. In most previous neural joint models, the relation classification part relies on a common "linear" concatenation-based mechanism over the latent features associated with entity pairs, i.e. the latent features are first concatenated into a single feature vector which is then linearly transformed before being fed into a softmax classifier. In contrast, our RC component takes into account second-order interactions over the latent features via a tensor. In particular, for relation classification we propose a novel use of the deep *biaffine* attention mechanism [7] which was first introduced in dependency parsing.

Experimental results on the benchmark "relation and entity recognition" dataset CoNLL04 [27] show that our model outperforms previous models, obtaining new state-of-the-art scores. In addition, using the biaffine attention improves the performance compared to using the linear mechanism significantly. We also provide an ablation study to investigate effects of different contributing factors in our model.

2 Our Proposed Model

This section details our end-to-end relation extraction model. Given an input sequence of n word tokens w_1, w_2, ..., w_n, we use a vector $\mathbf{v}_i$ to represent each ith word w_i by concatenating word embedding $\mathbf{e}^{(\mathrm{W})}_{w_i}$ and character-level word embedding $\mathbf{e}^{(\mathrm{C})}_{w_i}$:

$$\mathbf{v}_i = \mathbf{e}^{(\mathrm{W})}_{w_i} \circ \mathbf{e}^{(\mathrm{C})}_{w_i} \tag{1}$$

Here, for each word type w, we use a one-layer BiLSTM ($\mathrm{BiLSTM}_{\mathrm{char}}$) to learn its character-level word embedding $\mathbf{e}^{(\mathrm{C})}_{w}$ [3].

Named Entity Recognition (NER): The NER component feeds the sequence of vectors $\mathbf{v}_{1:n}$ with an additional context position index i into another BiLSTM ($\mathrm{BiLSTM}_{\mathrm{NER}}$) to learn a "latent" feature vector representing the ith word token. Then the NER component performs linear transformation of each latent feature vector by using a single-layer feed-forward network ($\mathrm{FFNN}_{\mathrm{NER}}$):

$$\mathbf{h}_i = \mathrm{FFNN}_{\mathrm{NER}}\big(\mathrm{BiLSTM}_{\mathrm{NER}}(\mathbf{v}_{1:n}, i)\big) \tag{2}$$

The output layer size of $\mathrm{FFNN}_{\mathrm{NER}}$ is the number of BIOLU-based NER labels [25]. The NER component feeds the output vectors $\mathbf{h}_{1:n}$ into a linear-chain CRF layer [16] for NER label prediction. A cross-entropy loss $\mathcal{L}_{\mathrm{NER}}$ is computed during training, while the Viterbi algorithm is used for decoding. Our NER component thus is the BiLSTM-CRF model [10] with additional LSTM-based character-level word embeddings [17].

Relation Classification (RC): Assume that $t_1, t_2, ..., t_n$ are NER labels predicted by the NER component for the input words. We represent each ith predicted label by a vector embedding $\mathbf{e}_{t_i}$. We create a sequence of vectors $\boldsymbol{x}_{1:n}$ in which each $\boldsymbol{x}_i$ is computed as:

$$\mathbf{x}_i = \mathbf{e}_{t_i} \circ \mathbf{v}_i \tag{3}$$

As for NER, the RC component also uses a BiLSTM ($\mathrm{BiLSTM}_{\mathrm{RC}}$) to learn another set of latent feature vectors, but from the sequence $\boldsymbol{x}_{1:n}$:

$$\mathbf{r}_i = \mathrm{BiLSTM}_{\mathrm{RC}}(\boldsymbol{x}_{1:n}, i) \tag{4}$$

The RC component further uses these latent vectors $\mathbf{r}_i$ for relation classification.

We propose a novel use of the deep *biaffine* attention mechanism [7] for relation classification. The biaffine attention mechanism was proposed for dependency parsing [7], helping to produce the best reported parsing performance to date [8]. First, to encode the directionality of a relation, we use two single-layer feed-forward networks to project each $\mathbf{r}_i$ into *head* and *tail* vector representations which correspond to whether the ith word serves as the head or tail argument of the relation:

$$\mathbf{h}_i^{(\text{head})} = \text{FFNN}_{\text{head}}(\boldsymbol{r}_i) \tag{5}$$

$$\mathbf{h}_i^{(\text{tail})} = \text{FFNN}_{\text{tail}}(\boldsymbol{r}_i) \tag{6}$$

Following [19], our RC component incrementally constructs relation candidates using all possible combinations of the last word tokens of predicted entities, i.e. words with L or U labels. We assign an entity pair to a negative relation class (NEG) when the pair has no relation or when the predicted entities are not correct. For example, for Fig. 1, we would have two relation candidates: *NEG*(Paris, International) and *OrgBased_In*(International, Paris). Then for each head-tail candidate pair (w_j, w_k), we apply the biaffine attention operator:

$$\mathbf{s}_{j,k} = \text{Biaffine}\left(\mathbf{h}_j^{(\text{head})}, \mathbf{h}_k^{(\text{tail})}\right) \tag{7}$$

$$\text{Biaffine}(\mathbf{y}_1, \mathbf{y}_2) = \underbrace{\mathbf{y}_1^\top \mathbf{U} \mathbf{y}_2}_{\text{Bilinear}} + \underbrace{\mathbf{W}(\mathbf{y}_1 \circ \mathbf{y}_2) + \mathbf{b}}_{\text{Linear}} \tag{8}$$

where $\mathbf{U}$, $\mathbf{W}$, $\mathbf{b}$ are a $m \times l \times m$ tensor, a $l \times (2 * m)$ matrix and a bias vector, respectively. Here, m is the size of the output layers of both $\text{FFNN}_{\text{head}}$ and $\text{FFNN}_{\text{tail}}$, while l is the number of relation classes (including NEG). Next, the RC component feeds the output vectors $\mathbf{s}_{j,k}$ of the biaffine attention layer into a softmax layer for relation prediction. Another cross-entropy loss $\mathcal{L}_{\text{RC}}$ is then computed during training.

Joint Learning: The objective loss of our joint model is the sum of the NER and RC losses: $\mathcal{L} = \mathcal{L}_{\text{NER}} + \mathcal{L}_{\text{RC}}$. Model parameters are then learned to minimize $\mathcal{L}$.

3 Experiments

3.1 Experimental Setup

Evaluation Scenarios: We evaluate our joint model on two evaluation setup scenarios: **(1)** NER&RC: A realistic scenario where entity boundaries are *not* given. **(2)** EC&RC: A less realistic scenario where the entity boundaries are given [12,20,26]. Thus the NER task which identifies both entity boundaries and classes reduces to the *entity classification* (**EC**) task. Following [20], we encode the gold entity boundaries in the BILOU scheme. Then we represent each B, I, O, L or U boundary tag as a vector embedding. As a result, the vector

$\mathbf{v}_i$ in Eq. 1 now also includes the boundary tag embedding in addition to the word embedding and character-level word embedding.

Dataset: We use the benchmark "entity and relation recognition" dataset CoNLL04 from [27]. Following [4,5], we use the 64%/16%/20% training/development/test pre-split available from Adel and Schütze (2017) [1], in which the test set was previously also used by Gupta et al. (2016) [9].

Implementation: Our model is implemented using DyNet v2.0 [21]. We optimize the objective loss using Adam [14], no mini-batches and run for 100 epochs. We compute the average of NER/EC score and RC score after each training epoch. We choose the model with the highest average score on the development set, which is then applied to the test set for the final evaluation phase. More details of the implementation as well as optimal hyper-parameters are in the Appendix. Our code is available at: https://github.com/datquocnguyen/jointRE

Metric: Similar to previous works in Table 1, we use the macro-averaged F1-score over the entity classes to score NER/EC and over the relation classes to score RC. More details of the metric are also in the Appendix. Unlike previous neural models, we report results as mean and standard deviation of the scores over 10 runs with 10 random seeds.

Table 1. Comparison with the previous state-of-the-art results on the **test** set. Recall that Setup 2 uses gold entity boundaries while Setup 1 does not. The subscript denotes the standard deviation. **(F)** refers to the use of extra feature types such as POS tag-based or dependency parsing-based features. Although using the same test set, Gupta et al. (2016) [9] reported results on a 80/0/20 training/development/test split rather than our 64/16/20 split. Results in the last two rows are just for reference, not for comparison, due to a random sampling of the test set. In particular, Miwa and Sasaki (2014) [20] used the 80/0/20 split for Setup 1 and performed 5-fold cross validation (i.e. sort of equivalent to 80/0/20) for Setup 2, while Zhang et al. (2017) [31] used a 72/8/20 split.

Model	Setup 1		Setup 2	
	NER	RC	EC	RC
Gupta et al. (2016) [9]	-	-	88.8	58.3
Gupta et al. (2016) [9] **(F)**	-	-	92.4	69.9
Adel and Schütze (2017) [1]	-	-	82.1	62.5
Bekoulis et al. (2018) [4]	83.6	62.0	93.0	68.0
Bekoulis et al. (2018) [5]	83.9	62.0	93.3	67.0
Our joint model	$\mathbf{86.2}_{0.5}$	$\mathbf{64.4}_{0.6}$	$\mathbf{93.8}_{0.4}$	$\mathbf{69.6}_{0.7}$
Miwa and Sasaki (2014) [20] **(F)**	80.7	61.0	92.3	71.0
Zhang et al. (2017) [31] **(F)**	85.6	67.8	-	-

3.2 Main Results

End-to-End Results: The first six rows in Table 1 compare our results with previous state-of-the-art published results on the same test set. In particular, our model obtains 2+% absolute higher NER and RC scores (Setup 1) than the BiLSTM-CRF-based multi-head selection model [5]. We also obtain 7+% higher EC and RC scores (Setup 2) than Adel and Schütze (2017) [1]. Note that Gupta et al. (2016) [9] use the same test set as we do, however they report final results on a 80/0/20 training/development/test split rather than our 64/16/20, i.e. Gupta et al. (2016) use a larger training set, but producing about 1.5% lower EC score and similar RC score against ours. These results show that our model performs better than previous state-of-the-art models, using the same setup.

In Table 1, the last two rows present results reported in [20] and [31] on the dataset CoNLL04. However, these results are *not* comparable due to their random sampling of the test set, i.e. using different train-test splits. Both Miwa and Sasaki (2014) [20] and Zhang et al. (2017) [31] employ additional extra features based on external NLP tools and use larger training sets than ours. Specifically, Zhang et al. (2017) integrate syntactic features by using a pre-trained BiLSTM-based dependency parser to extract BiLSTM-based latent feature representations for words in the input sentence, and then using these latent representations directly as part of the input embeddings in their model. We plan to extend our model with their syntactic integration approach to further improve our model performance in future work.

Ablation Analysis: We provide in Table 2 the results of a pipeline approach where we treat our two NER and RC components as independent networks, and train them separately. Here, the RC network uses gold NER labels when training, and uses predicted labels produced by the NER network when decoding. We find that the joint approach does slightly better than the pipeline approach in relation classification, although the differences are not significant. A similar observation is also found in [19]. Also, in preliminary experiments, we do not find any significant difference in performance of our joint model when feeding gold NER labels instead of predicted NER labels into the RC component during training. This is not surprising as the training NER score is at 99+%.

Table 2 also presents ablation tests over 5 factors of our joint model on the development set. In particular, Setup 1 performances significantly degrade by 4+% absolutely, when not using the character-level word embeddings. The performances also decrease when using a softmax classifier for NER label prediction rather than a CRF layer (here, the decrease is significant). In contrast, we do not find any significant difference in Setup 2 scores when not using either the character-level embeddings or the CRF layer, clearly showing the usefulness of the given gold entity boundaries. The 3 remaining factors, including removing NER label embeddings and not taking either the Bilinear or Linear part (in Eq. 8) into the Biaffine attention layer, do not affect the NER/EC score. However, they significantly decrease the RC score. This is reasonable because those 3 factors are part of the RC component only, thus helpful in predicting relations. More specifically, using the Biaffine attention produces about 1.5% significant

Table 2. Ablation results on the **development** set. * and ** denote the statistically significant differences against the full results at $p < 0.05$ and $p < 0.01$, respectively (using the two-tailed paired t-test). (a) Without using the character-level word embeddings. (b) Using a softmax layer for NER label prediction instead of the CRF layer. (c) Without using the NER label embeddings in our RC component, i.e. Eq. 3 would become $\mathbf{x}_i = \mathbf{v}_i$. (d) Without using the Bilinear part in Eq. 8 i.e., Biaffine would be a common Linear mechanism. (e) Without using the Linear part in Eq. 8 i.e., Biaffine reduces to Bilinear.

Model	Setup 1		Setup 2	
	NER	RC	EC	RC
Pipeline	$87.3_{0.6}$	$66.3_{0.8}$	$93.4_{0.6}$	$72.9_{0.6}$
Joint model (full)	$87.1_{0.5}$	$66.9_{0.8}$	$93.3_{0.5}$	$73.3_{0.6}$
(a) w/o Character	$82.7^{**}_{0.5}$	$63.0^{**}_{0.7}$	$93.1_{0.6}$	$73.4_{0.8}$
(b) w/o CRF	$86.4^{*}_{0.5}$	$66.0^{*}_{0.8}$	$93.5_{0.4}$	$73.2_{0.6}$
(c) w/o Entity	$87.1_{0.5}$	$64.7^{**}_{0.9}$	$93.3_{0.6}$	$72.1^{**}_{0.7}$
(d) w/o Bilinear	$86.6_{0.5}$	$65.4^{**}_{0.7}$	$93.4_{0.5}$	$72.0^{**}_{0.7}$
(e) w/o Linear	$86.8_{0.6}$	$65.9^{*}_{0.7}$	$93.3_{0.5}$	$72.6^{*}_{0.5}$

improvements to a common Linear transformation mechanism in relation classification, i.e., "w/o Bilinear" results against the full results in Table 2: 65.4% vs. 66.9% and 72.0% vs. 73.3% (although using Biaffine increases training time over using Linear by 35%, relatively).

4 Conclusion

In this paper, we have presented an end-to-end neural network-based relation extraction model. Our model employs a BiLSTM-CRF architecture for entity recognition and a biaffine attention mechanism for relation classification. On the benchmark CoNLL04 dataset, our model produces new state-of-the-art performance.

Acknowledgments. This work was supported by the ARC projects DP150101550 and LP160101469.

Appendix

Implementation Details: We apply dropout [28] with a 67% keep probability to the inputs of BiLSTMs and FFNNs. Following [15], we also use *word dropout* to learn an embedding for unknown words: we replace each word token w appearing $\#(w)$ times in the training set with a special "unk" symbol with probability $\mathsf{p}_{unk}(w) = \frac{0.25}{0.25+\#(w)}$.

Word embeddings are initialized by the 100-dimensional pre-trained GloVe word vectors [24], while character and NER label embeddings are initialized randomly. All these embeddings are then updated during training. For learning character-level word embeddings, we set the size of LSTM hidden states in $\text{BiLSTM}_{\text{char}}$ to be equal to the size of character embeddings. Here, we perform a minimal grid search of hyper-parameters for Setup 1, resulting in the Adam initial learning rate of 0.0005, the character embedding size of 25, the NER label embedding size of 100, the size of the output layers of both $\text{FFNN}_{\text{head}}$ and $\text{FFNN}_{\text{tail}}$ at 100, the number of $\text{BiLSTM}_{\text{NER}}$ and $\text{BiLSTM}_{\text{RC}}$ layers at 2 and the size of LSTM hidden states in each layer at 100. These optimal hyper-parameters for Setup 1 are then reused for Setup 2 where we additionally use the boundary tag embedding size of 100.

Metric: Similar to the previous works, when computing the macro-averaged F1 scores, we omit the entity label "Other" and the negative relation "NEG". Here, for NER an entity is predicted correctly if both the entity boundaries and the entity type are correct, while for EC a multi-token entity is considered as correct if at least one of its comprising tokens is predicted correctly. In all cases, a relation is scored as correct if both the argument entities and the relation type are correct.

References

1. Adel, H., Schütze, H.: Global normalization of convolutional neural networks for joint entity and relation classification. In: Proceedings of the 2017 Conference on Empirical Methods in Natural Language Processing, pp. 1723–1729 (2017)
2. Bach, N., Badaskar, S.: A review of relation extraction. Carnegie Mellon University, Technical Report (2007)
3. Ballesteros, M., Dyer, C., Smith, N.A.: Improved transition-based parsing by modeling characters instead of words with LSTMs. In: Proceedings of the 2015 Conference on Empirical Methods in Natural Language Processing, pp. 349–359 (2015)
4. Bekoulis, G., Deleu, J., Demeester, T., Develder, C.: Adversarial training for multi-context joint entity and relation extraction. In: Proceedings of the 2018 Conference on Empirical Methods in Natural Language Processing, pp. 2830–2836 (2018)
5. Bekoulis, G., Deleu, J., Demeester, T., Develder, C.: Joint entity recognition and relation extraction as a multi-head selection problem. Expert Syst. Appl. **114**, 34–45 (2018)
6. Blanco, R., Cambazoglu, B.B., Mika, P., Torzec, N.: Entity recommendations in web search. In: Alani, H., et al. (eds.) ISWC 2013, Part II. LNCS, vol. 8219, pp. 33–48. Springer, Heidelberg (2013). https://doi.org/10.1007/978-3-642-41338-4_3
7. Dozat, T., Manning, C.D.: Deep Biaffine attention for neural dependency parsing. In: Proceedings of the 5th International Conference on Learning Representations (2017)
8. Dozat, T., Qi, P., Manning, C.D.: Stanford's graph-based neural dependency parser at the CoNLL 2017 shared task. In: Proceedings of the CoNLL 2017 Shared Task: Multilingual Parsing from Raw Text to Universal Dependencies, pp. 20–30 (2017)

9. Gupta, P., Schütze, H., Andrassy, B.: Table filling multi-task recurrent neural network for joint entity and relation extraction. In: Proceedings of the 26th International Conference on Computational Linguistics: Technical Papers, pp. 2537–2547 (2016)
10. Huang, Z., Xu, W., Yu, K.: Bidirectional LSTM-CRF Models for Sequence Tagging. arXiv preprint arXiv:1508.01991 (2015)
11. Jiang, J.: Information extraction from text. In: Aggarwal, C.C., Zhai, C. (eds.) Mining Text Data, pp. 11–41. Springer, New York (2012). https://doi.org/10.1007/978-1-4614-3223-4
12. Kate, R.J., Mooney, R.J.: Joint entity and relation extraction using card-pyramid parsing. In: Proceedings of the Fourteenth Conference on Computational Natural Language Learning, pp. 203–212 (2010)
13. Katiyar, A., Cardie, C.: Going out on a limb: joint extraction of entity mentions and relations without dependency trees. In: Proceedings of the 55th Annual Meeting of the Association for Computational Linguistics (Volume 1: Long Papers), pp. 917–928 (2017)
14. Kingma, D.P., Ba, J.: Adam: A Method for Stochastic Optimization. CoRR abs/1412.6980 (2014)
15. Kiperwasser, E., Goldberg, Y.: Simple and accurate dependency parsing using bidirectional LSTM feature representations. Trans. ACL **4**, 313–327 (2016)
16. Lafferty, J.D., McCallum, A., Pereira, F.C.N.: Conditional random fields: probabilistic models for segmenting and labeling sequence data. In: Proceedings of the Eighteenth International Conference on Machine Learning, pp. 282–289 (2001)
17. Lample, G., Ballesteros, M., Subramanian, S., Kawakami, K., Dyer, C.: Neural architectures for named entity recognition. In: Proceedings of the 2016 Conference of the North American Chapter of the Association for Computational Linguistics: Human Language Technologies, pp. 260–270 (2016)
18. Li, Q., Ji, H.: Incremental joint extraction of entity mentions and relations. In: Proceedings of the 52nd Annual Meeting of the Association for Computational Linguistics (Volume 1: Long Papers), pp. 402–412 (2014)
19. Miwa, M., Bansal, M.: End-to-end relation extraction using LSTMs on sequences and tree structures. In: Proceedings of the 54th Annual Meeting of the Association for Computational Linguistics (Volume 1: Long Papers), pp. 1105–1116 (2016)
20. Miwa, M., Sasaki, Y.: Modeling joint entity and relation extraction with table representation. In: Proceedings of the 2014 Conference on Empirical Methods in Natural Language Processing, pp. 1858–1869 (2014)
21. Neubig, G., et al.: DyNet: The Dynamic Neural Network Toolkit. arXiv preprint arXiv:1701.03980 (2017)
22. Nguyen, T.H., Grishman, R.: Combining neural networks and log-linear models to improve relation extraction. In: Proceedings of IJCAI Workshop on Deep Learning for Artificial Intelligence (2016)
23. Pawar, S., Bhattacharyya, P., Palshikar, G.: End-to-end relation extraction using neural networks and markov logic networks. In: Proceedings of the 15th Conference of the European Chapter of the Association for Computational Linguistics: Volume 1, Long Papers, pp. 818–827 (2017)
24. Pennington, J., Socher, R., Manning, C.: Glove: global vectors for word representation. In: Proceedings of the 2014 Conference on Empirical Methods in Natural Language Processing, pp. 1532–1543 (2014)
25. Ratinov, L., Roth, D.: Design challenges and misconceptions in named entity recognition. In: Proceedings of the Thirteenth Conference on Computational Natural Language Learning, pp. 147–155 (2009)

26. Roth, D., tau Yih, W.: Global inference for entity and relation identification via a linear programming formulation. In: Introduction to Statistical Relational Learning. MIT Press, Cambridge (2007)
27. Roth, D., Yih, W.T.: A linear programming formulation for global inference in natural language tasks. In: Proceedings of the 8th Conference on Computational Natural Language Learning, pp. 1–8 (2004)
28. Srivastava, N., Hinton, G., Krizhevsky, A., Sutskever, I., Salakhutdinov, R.: Dropout: a simple way to prevent neural networks from overfitting. J. Mach. Learn. Res. **15**, 1929–1958 (2014)
29. Thomas, P., Starlinger, J., Vowinkel, A., Arzt, S., Leser, U.: GeneView: a comprehensive semantic search engine for PubMed. Nucleic Acids Res. **40**(W1), W585–W591 (2012)
30. Wang, S., Zhang, Y., Che, W., Liu, T.: Joint extraction of entities and relations based on a novel graph scheme. In: Proceedings of the Twenty-Seventh International Joint Conference on Artificial Intelligence, pp. 4461–4467 (2018)
31. Zhang, M., Zhang, Y., Fu, G.: End-to-end neural relation extraction with global optimization. In: Proceedings of the 2017 Conference on Empirical Methods in Natural Language Processing, pp. 1730–1740 (2017)
32. Zheng, S., et al.: Joint entity and relation extraction based on a hybrid neural network. Neurocomputing **257**, 59–66 (2017)
33. Zheng, S., Wang, F., Bao, H., Hao, Y., Zhou, P., Xu, B.: Joint extraction of entities and relations based on a novel tagging scheme. In: Proceedings of the 55th Annual Meeting of the Association for Computational Linguistics (Volume 1: Long Papers), pp. 1227–1236 (2017)

Social Relation Inference via Label Propagation

Yingtao Tian[1(✉)], Haochen Chen[1], Bryan Perozzi[2], Muhao Chen[3], Xiaofei Sun[1], and Steven Skiena[1]

[1] Department of Computer Science, Stony Brook University, Stony Brook, NY, USA
{yittian,haocchen,xiaofsun,skiena}@cs.stonybrook.edu
[2] Google Research, New York, NY, USA
bperozzi@acm.org
[3] Department of Computer Science, UCLA, Los Angeles, USA
muhaochen@cs.ucla.edu

Abstract. Collaboration networks are a ubiquitous way to characterize the interactions between people. In this paper, we consider the problem of inferring social relations in collaboration networks, such as the fields that researchers collaborate in, or the categories of projects that Github users work on together. Social relation inference can be formalized as a multi-label classification problem on graph edges, but many popular algorithms for semi-supervised learning on graphs only operate on the nodes of a graph. To bridge this gap, we propose a principled method which leverages the natural homophily present in collaboration networks. First, observing that the fields of collaboration for two people are usually at the intersection of their interests, we transform an edge labeling into node labels. Second, we use a label propagation algorithm to propagate node labels in the entire graph. Once the label distribution for all nodes has been obtained, we can easily infer the label distribution for all edges. Experiments on two large-scale collaboration networks demonstrate that our method outperforms the state-of-the-art methods for social relation inference by a large margin, in addition to running several orders of magnitude faster.

Keywords: Label propagation · Social relation inference · Social network

1 Introduction

In collaboration networks, edges, or social relations [12], are formed between people with shared interests. Social relations in networks are complex and nuanced, which often cannot be characterized by a single label. Consider a co-author network between researchers where the social relations between two researchers are the research areas they collaborate in. Since collaborations can occur in different research areas, the social relation between researchers is inherently multifaceted. Many applications on collaboration networks can benefit from an awareness of social relations, such as node classification [15], recommendation [11]

L. Azzopardi et al. (Eds.): ECIR 2019, LNCS 11437, pp. 739–746, 2019.
https://doi.org/10.1007/978-3-030-15712-8_48

and anomaly detection [14]. However, in many networks, such label information (social relations) is far from complete. It is thus desirable to learn to infer social relations associated with the unlabeled edges.

We formalize the task of social relation inference as a semi-supervised multi-label edge classification problem on networks. Given the network structure and a limited amount of labeled edges, our goal is to infer the labels of the rest of the edges. There are several previous studies on inferring social ties from social networks, which is similar to our definition of social relations [9,11]. However, these works assume that each edge corresponds to a single relation type, which may not be the case in collaboration networks. Moreover, they only consider first-order or second-order relationships between nodes, but fails to model higher-order relationships that play an important role in network inference tasks [2].

Another relevant area is network embeddings [4,6,8], which aim at learning low-dimensional latent representations of nodes in a network. Also, representations of larger-scale components of networks (such as edges and subgraphs) can be composed from these node representations. These representations can then be used as features for a wide range of downstream tasks on networks, including social relation inference. As a pioneering work, DeepWalk [6] generates fixed-length random walk sequences in networks and trains a skip-gram model [5] on these sequences to obtain node embeddings. While achieving state-of-the-art results on a handful of network inference tasks such as node classification and link prediction [4,6], the semantics of edges in networks are seldom exploited by network embedding models. Moreover, we find that they usually ignore the unique properties possessed by different types of networks and by different downstream tasks. Also, many of them are computationally expensive: learning network embeddings of a one-million node network can take several days on a single CPU.

In this paper, we propose a simple but effective method for social relation inference on collaboration networks. Our method is based on the observation that social relations between people in collaboration networks are determined by their shared interests. As such, the networks are highly homophilous and there is a natural connection between the (hidden) labels of the nodes, and the provided edge labels. Using this relationship, we first transform the edge labels into a node labeling. Next, to alleviate any data sparsity problem, we perform label propagation on the input network to obtain label distribution for all nodes. Label propagation [13,16] represents a class of semi-supervised learning methods which find numerous applications in graph mining. For social relation inference, we find that label propagation has several desirable properties compared to the neural methods mentioned before: it is extremely efficient and it makes good use of the high level of homophily exhibited in collaboration networks [7]. Finally, once node labels have been obtained, the label distribution of edges can be easily inferred from the label distribution of their endpoints. Experimental results on real-world networks show that our method outperforms state-of-the-art methods by a large margin.

Algorithm 1. LabelProp(G, P)

Input: graph G, initial node label distribution P, rounds of iteration k
Output: node label distribution after propagation $\hat{Y_V} \in \mathbb{R}^{|V| \times |L|}$
1: Compute the degree matrix D: $D_{ii} \leftarrow \sum_j A_{ij}$
2: Compute the transition matrix: $Q \leftarrow D^{-1}A$
3: $Y^{(0)} \leftarrow P$
4: **for** $i = 0$ to $k-1$ **do**
5: $\quad Y^{(i+1)} \leftarrow QY^{(i)}$
6: **end for**
7: $\hat{Y_V} = Y^{(k)}$
8: **return** $\hat{Y_V}$

2 Problem Definition and Notation

We hereby formalize the problem of social relation inference in collaboration networks. Let $G = (V, E)$ be an undirected graph, where V are the nodes in the graph and E represent its edges. Let A be the adjacency matrix of G. Let $L = (l_1, l_2, \cdots, l_k)$ be the set of relation types (labels). A partially labeled network is then defined as $G = (V, E_L, E_U, Y_L)$, where E_L is the set of labeled edges, E_U is the set of unlabeled edges with $E_L \cup E_U = E$. Y_L represents the relation types associated with the labeled edges in E_L, with $\forall Y_L(i) \in Y_L : Y_L(i) \subseteq L$. The objective of social relation inference is to predict the relation types Y_U of the unlabeled edges E_U: $f : G = (V, E_L, E_U, Y_L) \rightarrow Y_U$ We denote the i-th row and ij-th element of a matrix M as M_i and M_{ij}.

3 Method

3.1 Step 1: From Edge Labels to Node Labels

One challenge with social relation inference is that the labels we seek to predict are associated with edges, instead of nodes. However, most machine learning algorithms on graphs only operate on nodes. To bridge this gap, we note that collaboration networks possess a unique property: edges are typically formed between two people which have *shared interests*. Such shared interests can very well be characterized by the labels of edges. This means that we should be able to infer the latent interests of nodes based on their corresponding edge labels.

Formally, we seek to estimate the probability distribution matrix $P \in \mathbb{R}^{|V| \times |L|}$ for all nodes over the label space L. For ease of presentation, we assume that the training data is given in the form of triplets $t = (u, v, l)$, where $u, v \in V, l \in L$. In other words, if an edge has several labels, then we construct one triplet for each label. We define the set of all training triplets as T. Assume the label distribution of u and v are independent, the strength of relation l between u and v can be estimated as:

$$Pr(l|u, v) = P_{ul} \cdot P_{vl} \tag{1}$$

Our objective is to maximize the probability of observing the relations in T as given by:

$$\ell = \prod_{u \in V} \prod_{\substack{(v,l) \\ (u,v,l) \in T}} Pr(l|u,v) \tag{2}$$

Then, for a certain $u \in V$, our goal is to minimize the following objective:

$$-\log \ell_u = - \sum_{\substack{(v,l) \\ (u,v,l) \in T}} (\log P_{ul} + \log P_{vl}) \tag{3}$$

Since P is the probability distribution of labels, we have the constraint $\sum_{l \in L} P_{ul} = 1$. The Lagrangian function of Eq. (3) is:

$$\mathcal{L}(P_u, \lambda) = - \sum_{\substack{(v,l) \\ (u,v,l) \in T}} (\log P_{ul} + \log P_{vl}) + \lambda(\sum_{l \in L} P_{ul} - 1) \tag{4}$$

For all $l \in L$, we take the derivative of Eq. 4 w.r.t. P_{ul} and set it to zero:

$$-\frac{\#(u,l)}{P_{ul}} + \lambda = 0 \tag{5}$$

where $\#(u,l)$ is the number of co-occurrences of u and l in T, with v being marginalized out. It is now clear that $P_{ul} = \frac{\#(u,l)}{\lambda}$. Combined with the constraint $\sum_{l \in L} P_{ul} = 1$, we have $\lambda = \sum_{l \in L} \#(u,l)$. Finally, the closed-form estimation of P_{ul} is calculated as: $P_{ul} = \#(u,l) / \sum_{l \in L} \#(u,l)$.

Concretely, we can simply compute the relative frequency that each node co-occur with each label, which gives us the initial label distribution P of all nodes.

3.2 Step 2: Label Propagation

Labeled edges are often scarce in real-world collaboration networks. As a result, using the procedure outlined above, we may get an empty label distribution for most of the nodes (as they have no edges). To alleviate this problem, we propose using label propagation [16] on G to spread the information from labeled edges around the graph. Algorithm 1 details the process. We start from the initial label distribution obtained in Step 1 and repeatedly distribute node labels to the neighboring nodes.

3.3 Step 3: From Node Labels to Edge Labels

Once we have obtained the label distribution for all nodes, we can easily compute the label distribution for edges by reusing Eq. 1. For each edge $e = (u,v)$, the strength of relation l is $P_{ul} \cdot P_{vl}$. The ranking of relation strengths serves as our prediction of social relations.

3.4 Time Complexity Analysis

The majority of time complexity is contributed by Algorithm 1, which takes $O(k \cdot (|E| + |V| \cdot |L|))$. In our experiments, it is further shown that a small value of k is sufficient for our model to converge: empirically, we take $k = 5$ based on the performance on the validation set. We provide detailed running time comparison against baseline methods in Sect. 4.

4 Experiment

In this section, we describe the datasets for social relation inference and compare our method against a number of baselines.

4.1 Dataset

We use the processed ArnetMiner [10] datasets provided by TransNet [12]. ArnetMiner is a large-scale co-author network with over a million authors and four million collaboration relations. The social relations between researchers can be reflected by the research areas or topics they collaborate in. Concretely, for each co-author relationship, the authors of TransNet extract representative research interest phrases from the abstracts of co-authored papers as edge labels. Two collaboration networks of different scales and different amount of labels are provided in this dataset to better investigate the characteristics of different models. We use the same data split as in TransNet [12]. The statistics of the datasets are presented in Table 1.

Table 1. Statistics of the networks used in our experiments.

Dataset	# Vertices	# Edges	# Train	# Test	# Valid	# Classes
Arnet-Small	187,939	1,619,278	1,579,278	20,000	20,000	100
Arnet-Medium	268,037	2,747,386	2,147,386	300,000	300,000	500
Arnet-Large	945,589	5,056,050	3,856,050	600,000	600,000	500

4.2 Baseline Methods

The baseline methods we use are as follows: **(1) DeepWalk** [6]: This is a network embedding method that learns latent representations of nodes in a graph. **(2) LINE** [8]: This is a network embedding method that preserves both first-order and second-order proximities in networks. **(3) node2vec** [4]: This is a network embedding method that improves DeepWalk with a biased random walk phase. **(4) TransE** [1]: This is a knowledge base embedding method which simultaneously learns latent representations of nodes and relations. Since TransE models each relation separately, we split each edge with k labels into k training instances, one for each label. **(5) TransNet** [12]: This method is an extension to TransE which explicitly models edges with multiple labels. It is also the state-of-the-art method for social relation inference.

Table 2. Relation inference results on **Arnet-Small**.

Algorithm	Metrics(%)		
	hits@1	*hits*@5	*hits*@10
DeepWalk	13.88	36.80	50.57
LINE	11.30	31.70	44.51
node2vec	13.63	36.60	50.27
TransE	39.16	78.48	88.54
TransNet	47.67	86.54	92.27
Proposed	**48.89**	**90.13**	**93.90**

Table 3. Relation inference results on **Arnet-Large**.

Algorithm	Metrics(%)		
	hits@1	*hits*@5	*hits*@10
DeepWalk	5.41	16.17	23.33
LINE	4.28	13.44	19.85
node2vec	5.39	16.23	23.47
TransE	15.38	41.87	55.54
TransNet	28.85	66.15	75.55
Proposed	**29.91**	**72.32**	**80.86**

We follow the experimental setup as in TransNet [12]. For all baseline methods, we use the hyperparameter settings as described in their papers. For TransE, we use the similarity-based method to predict social relations as described in [1]. For TransNet, we follow the inference algorithm in their paper. For the three network embedding methods, we concatenate node representations as the feature vector for edges. For social relation inference, we train a one-vs-rest logistic regression model with L2 regularization implemented in LibLinear [3].

4.3 Results and Analysis

In Tables 2 and 3, we summarize the experimental results using the same data split as TransNet. Results for all baseline methods (including TransNet) are taken from the TransNet paper. We can clearly see that our simple method outperforms all baseline methods by a large margin. The performance gain over the best baseline method, TransNet, is at least 3.5% and up to 8.4% in terms of hits@5. We note that the TransNet data split uses 98%, 76% and 78% edges as training data for Arnet-Small, Arnet-Medium and Arnet-Large respectively. With such a large amount of training data, our algorithm achieves the reported performance even without performing label propagation, which proves the effectiveness of the node label inference algorithm. Moreover, our algorithm is orders of magnitude faster than all baseline methods. Using a single CPU core at 2.0 GHz, our method finishes in 5 min on Arnet-Small while all baseline methods take more than 24 h.

The only hyperparameter in our algorithm is the number of rounds of iterations k for label propagation, which is tuned on the validation set. We observe that even with only 1% of labeled edges, our label propagation algorithm converges within 5 iterations.

5 Conclusion

We study the problem of inferring social relations in collaboration networks, formulated as a semi-supervised learning problem on graphs where edges have multiple labels. Observing that edges in collaboration networks represent the shared interests of two people, we transform edge labels to node labels and perform label propagation to deal with the label sparsity problem. Experimental results on real-world collaboration networks show the superiority of our method in terms of both accuracy and efficiency.

References

1. Bordes, A., Usunier, N., Garcia-Duran, A., Weston, J., Yakhnenko, O.: Translating embeddings for modeling multi-relational data. In: Advances in neural information processing systems, pp. 2787–2795 (2013)
2. Cao, S., Lu, W., Xu, Q.: GraRep: learning graph representations with global structural information. In: Proceedings of the 24th ACM International on Conference on Information and Knowledge Management, pp. 891–900. ACM (2015)
3. Fan, R.E., Chang, K.W., Hsieh, C.J., Wang, X.R., Lin, C.J.: Liblinear: a library for large linear classification. J. Mach. Learn. Res. **9**, 1871–1874 (2008)
4. Grover, A., Leskovec, J.: node2vec: scalable feature learning for networks. In: Proceedings of the 22nd ACM SIGKDD International Conference on Knowledge Discovery and Data Mining (2016)
5. Mikolov, T., Sutskever, I., Chen, K., Corrado, G.S., Dean, J.: Distributed representations of words and phrases and their compositionality. In: Advances in Neural Information Processing Systems, pp. 3111–3119 (2013)
6. Perozzi, B., Al-Rfou, R., Skiena, S.: Deepwalk: online learning of social representations. In: Proceedings of the 20th ACM SIGKDD International Conference on Knowledge Discovery and Data Mining, pp. 701–710. ACM (2014)
7. Powell, W.W., White, D.R., Koput, K.W., Owen-Smith, J.: Network dynamics and field evolution: the growth of interorganizational collaboration in the life sciences. Am. J. Sociol. **110**(4), 1132–1205 (2005)
8. Tang, J., Qu, M., Wang, M., Zhang, M., Yan, J., Mei, Q.: Line: large-scale information network embedding. In: Proceedings of the 24th International Conference on World Wide Web, pp. 1067–1077. International World Wide Web Conferences Steering Committee (2015)
9. Tang, J., Lou, T., Kleinberg, J.: Inferring social ties across heterogenous networks. In: Proceedings of the Fifth ACM International Conference on Web Search and Data Mining, pp. 743–752. ACM (2012)
10. Tang, J., Zhang, J., Yao, L., Li, J., Zhang, L., Su, Z.: Arnetminer: extraction and mining of academic social networks. In: Proceedings of the 14th ACM SIGKDD International Conference on Knowledge Discovery and Data Mining, pp. 990–998. ACM (2008)
11. Tang, W., Zhuang, H., Tang, J.: Learning to infer social ties in large networks. In: Gunopulos, D., Hofmann, T., Malerba, D., Vazirgiannis, M. (eds.) ECML PKDD 2011. LNCS (LNAI), vol. 6913, pp. 381–397. Springer, Heidelberg (2011). https://doi.org/10.1007/978-3-642-23808-6_25
12. Tu, C., Zhang, Z., Liu, Z., Sun, M.: Transnet: translation-based network representation learning for social relation extraction. In: Proceedings of International Joint Conference on Artificial Intelligence (IJCAI), Melbourne (2017)
13. Wang, F., Zhang, C.: Label propagation through linear neighborhoods. IEEE Trans. Knowl. Data Eng. **20**(1), 55–67 (2008)

14. Xiang, B., Liu, Z., Zhou, J., Li, X.: Feature propagation on graph: a new perspective to graph representation learning. arXiv preprint arXiv:1804.06111 (2018)
15. Xu, L., Wei, X., Cao, J., Philip, S.Y.: On exploring semantic meanings of links for embedding social networks (2018)
16. Zhu, X., Ghahramani, Z.: Learning from labeled and unlabeled data with label propagation (2002)

Wikipedia Text Reuse: Within and Without

Milad Alshomary[1(✉)], Michael Völske[2(✉)], Tristan Licht[2(✉)], Henning Wachsmuth[1(✉)], Benno Stein[2(✉)], Matthias Hagen[3(✉)], and Martin Potthast[4(✉)]

[1] Paderborn University, Paderborn, Germany
{milad.alshomary,henningw}@upb.de
[2] Bauhaus-Universität Weimar, Weimar, Germany
{michael.voelske,tristan.licht,benno.stein}@uni-weimar.de
[3] Martin-Luther-Universität Halle-Wittenberg, Halle, Germany
matthias.hagen@uni-weimar.de
[4] Leipzig University, Leipzig, Germany
martin.potthast@uni-leipzig.de

Abstract. We study text reuse related to Wikipedia at scale by compiling the first corpus of text reuse cases within Wikipedia as well as without (i.e., reuse of Wikipedia text in a sample of the Common Crawl). To discover reuse beyond verbatim copy and paste, we employ state-of-the-art text reuse detection technology, scaling it for the first time to process the entire Wikipedia as part of a distributed retrieval pipeline. We further report on a pilot analysis of the 100 million reuse cases inside, and the 1.6 million reuse cases outside Wikipedia that we discovered. Text reuse inside Wikipedia gives rise to new tasks such as article template induction, fixing quality flaws, or complementing Wikipedia's ontology. Text reuse outside Wikipedia yields a tangible metric for the emerging field of quantifying Wikipedia's influence on the web. To foster future research into these tasks, and for reproducibility's sake, the Wikipedia text reuse corpus and the retrieval pipeline are made freely available.

1 Introduction

Text reuse is second nature to Wikipedia: *inside* Wikipedia, the articles grouped in a given category are often harmonized until informal templates emerge, which are then adopted for newly created articles in the same category. Moreover, passages may even be copied verbatim from one article to another when they form a hierarchical relationship. While the reuse of text inside Wikipedia has been a de facto policy for many years, neither the MediaWiki software nor tools developed by and for the Wikipedia community offer any reuse support. Unless a dedicated Wikipedia editor takes care of it, a copied passage will eventually diverge from its original, resulting in inconsistency. *Outside* Wikipedia, we distinguish reuse of Wikipedia's articles by third parties, and reuse of third-party content by Wikipedia. The former is widespread: passages of articles are manually reused

L. Azzopardi et al. (Eds.): ECIR 2019, LNCS 11437, pp. 747–754, 2019.
https://doi.org/10.1007/978-3-030-15712-8_49

in quotations and summaries, or automatically extracted to search result pages. Many sites mirror Wikipedia partially or in full; sometimes with proper attribution, other times violating Wikipedia's lenient copyrights.[1] The latter form of reuse is discouraged by Wikipedia's editing policies.[2]

With a few exceptions reviewed below, Wikipedia text reuse has not been analyzed at scale. This gap is due to the lack of open and scalable technologies capable of detecting text reuse, and the significant computational overhead required. Only recently, resulting from six consecutive shared tasks on plagiarism detection held at PAN to systematically evaluate reuse detection algorithms, new classes of algorithms emerged that specifically address the detection of various kinds of text reuse from large text corpora. To foster research into Wikipedia text reuse, we compiled the first Wikipedia text reuse corpus, obtained from comparing the entire Wikipedia to itself as well as to a 10%-sample of the Common Crawl. By scaling up the aforementioned detection algorithms, we render the computations feasible on a mid-sized cluster. A first exploratory analysis enables us to report insights on the nature of text reuse inside Wikipedia, and to quantify Wikipedia's influence on the web in terms of monetary exploitation of its content.

2 Related Work

Wikipedia's openness and success fuels tons of research about the encyclopedia[3] and how it can be exploited in different fields [8,12]. Wikipedia's influence on the web has recently become a focus of interest: for instance, posts on Stack Overflow and Reddit that link to Wikipedia have been found to outperform others in terms of interactions [20]. Other works have studied Wikipedia's role in driving research in the scientific community [19], and its importance to enrich search engines' result pages [11]. The ever increasing quality of Wikipedia drives the reuse of its content by third parties, but in a "paradox of reuse" reduces the need to visit Wikipedia itself [18], depriving the encyclopedia of potential new editors.

In general, text reuse detection is applied in many domains [2], such as the digital humanities [7], and in journalism and science (e.g., to study author perspectives [6] or to pursue copyright infringement and plagiarism [5]). Text reuse detection divides into the subtasks of *source retrieval* and *text alignment* [14,17], where the former retrieves a set of candidate reuse sources given a questioned document [9], and the latter aligns reused passages given a document pair. Approaches addressing each task have been systematically evaluated at PAN [14].

As for Wikipedia, text reuse detection has the potential to help improve the encyclopedia and to quantify its influence on the web. However, Wikipedia text reuse has only been targeted in two pioneering studies to date: Weissman et al. [21]

[1] en.wikipedia.org/wiki/Wikipedia:Copyrights.
[2] en.wikipedia.org/wiki/Wikipedia:Copyright_violations.
[3] en.wikipedia.org/wiki/Academic_studies_about_Wikipedia.

use similarity hashing to identify redundant or contradictory near-duplicate sentences within Wikipedia that may harm article quality. Similarly, Ardi and Heidemann [1] employ hashing to detect near-duplicates of complete Wikipedia articles in the Common Crawl. Both studies neglect the text alignment step, restricting the ability to perform in-depth reuse analysis. Our text reuse detection pipeline incorporates similar hashing techniques for source retrieval but further filters and refines the results through text alignment to obtain the fine-grained actual reused text passages. In this respect, our corpus better captures the author's intent of reusing a given passage of text.

3 Corpus Construction

Given two document collections D_1 and D_2, we aim to identify all cases of text reuse as pairs of sufficiently similar text spans. For within-Wikipedia detection, D_1 is the set of all English Wikipedia articles and $D_2 = D_1$, whereas otherwise D_2 is a 10%-sample of the Common Crawl (see Table 1 (left)). Our processing pipeline first carries out *source retrieval* to identify promising candidate document pairs, which are then compared in detail during *text alignment*.

3.1 Source Retrieval

In source retrieval, given a questioned document $d_1 \in D_1$, the task is to rank the documents in D_2 by decreasing likelihood of sharing reused text with d_1. An absolute cutoff rank k and/or a relative score threshold τ may be used to decide how many of the top-ranked D_2-documents become subject to the more expensive task of text alignment with d_1. The parameters are typically determined in terms of the budget of computational capacity available as well as the desired recall level. An ideal ranking function would rank all documents in D_2 that reuse text from d_1 highest; however, the typical operationalization using text similarity measures does not reach this ideal. The higher the desired recall level, the lower the precision and the higher the computational overhead.

With a goal of maximizing recall, our budget was 2 months of processing on a 130 node Apache Spark cluster (12 CPUs and 196 GB RAM each). Since Wikipedia as a whole is questioned (D_1), we generalized source retrieval toward ranking all pairs $(d_1, d_2) \in D_1 \times D_2$ based on a pruned scoring function ρ:

$$\underbrace{\exists c_i \in d_1, c_j \in d_2 : \quad h(c_i) \cap h(c_j) \neq \emptyset}_{\text{Search pruning}} \quad \rightarrow \quad \rho(d_1, d_2) = \max_{\substack{c_i \in d_1 \\ c_j \in d_2}} (\varphi(c_i, c_j)),$$

where c is a passage-length text chunk, h is a locality-preserving hash function, and φ is a text similarity measure. The idea is to view reuse as a passage-level phenomenon and to be lenient during pruning (a single hash collision suffices).

To select and fine-tune a suitable hash function h and similarity measure φ, we compiled a ground truth training set, by sampling 1000 Wikipedia articles—each at least 2000 words long—and applying our text alignment approach described

Table 1. Overview of the input dataset characteristics (left), the source retrieval performance (middle), and the retrieved text reuse cases (right).

Dataset	Count (million)
Wikipedia	
Articles	4.2
Paragraphs	11.4
Common Crawl	
Websites	1.4
Web pages	591.0
Paragraphs	187.0

Source Retrieval	Recall	Precision
Search pruning		
(1) LSH	0.32	$9.8 \cdot 10^{-6}$
(2) VDSH	**0.73**	$\mathbf{4.5 \cdot 10^{-4}}$
Ranking up to rank $k = 1000$		
(a) *tf·idf*	**0.87**	**0.007**
(b) Stop n-grams	0.74	0.007
(c) Par2vec	0.67	0.008
(d) Hybrid	0.76	0.009
VDSH + *tf·idf*	**0.66**	**0.005**

Reuse	Within	Without
Cases	110 million	1.6 million
Documents with Reuse Cases		
Articles	360,000	1 million
Pages	–	15,000
Words in Reuse Cases		
Min.	17	23
Avg.	78	252
Max.	6200	1960

below to all their pairs with all other Wikipedia articles. The source retrieval "parameters" h and φ (and thus ρ) were optimized to maximize recall of these training set text alignment results in the source retrieval phase. We considered two hashing schemes for h: (1) random projections in the form of an instantiation of the data-independent locality-sensitive hashing (LSH) family [4], and (2) variational deep semantic hashing (VDSH), a data-dependent learning-to-hash technique [3]. We further considered four text similarity measures for φ: (a) cosine similarity on a *tf·idf*-weighted word unigram representation, (b) Jaccard similarity on stop word n-grams [16], (c) cosine similarity on a simple additive paragraph vector model [13], and (d) a weighted average of (b) and (c).

Table 1 (middle) shows our evaluation results for the two components of the source retrieval pipeline. In general, the low precision values are due to the high cut-off rank (k=1000) required to collect most of the few positive cases. For search pruning, we selected VDSH with a 16-bit hash, which reduces the number of required evaluations of the ρ measure by three orders of magnitude compared to an exhaustive comparison, while retaining the majority of text reuse cases. To construct the ranking function ρ itself, we settle on cosine similarity in the *tf·idf* space as the similarity measure φ due to its superior recall compared to the other considered models.

3.2 Text Alignment

Given a candidate document pair, text alignment extracts spans of reused text—if any—through the steps of *seed generation* (identification of short exact matches), *seed extension* (clustering of short matches to form longer spans), and *post filtering*. The state of the art evaluated at PAN is determined on datasets orders of magnitude smaller than our setting, often using complex setups that turned out to be difficult to scale and to be reproduced (e.g., lacking open source implementations). We hence resorted to ideas from the literature that offer a reasonable trade-off between performance, robustness, and speed, and tuned their parameters[4] based on the standard PAN-13 training data. Our text alignment achieves a macro-averaged *plagdet* score of 0.64 (0.84 on just the unobfuscated

[4] We used word 3-gram seeds, extended via DBScan clustering ($\varepsilon = 150$, minPoints = 5), and filtered cases shorter than 200 words or with cosine similarity < 0.5.

subset) on the corresponding PAN-13 test data. In terms of raw detection performance, this is in the lower middle range of the PAN results [15].

In our pruned all-pairs search setting, an input to the text alignment step is formed by one document $d \in D_1$ and a list of all candidate documents from the other collection D_2 sorted by descending ρ-score. Text alignment is applied sequentially to this list until one of two stopping criteria is met: (1) the current candidate pair's ρ-score is below a threshold (0.025 in our implementation), or (2) the number of consecutive miss-cases (i.e., candidate pairs in which the text alignment finds no reuse) exceeds some other threshold (we use 250). Both thresholds can be configured based on the time available for text alignment; we experimentally extrapolated them from the aforementioned training set.

4 Corpus Analysis

Table 1 (right) shows basic statistics of the reuse we uncovered. Most interestingly, we find nearly 70 times more reuse cases within the Wikipedia than in the 10%-sample of the Common Crawl, but involving only one third as many articles. Based on this insight, we identify two fundamentally different kinds of text reuse within Wikipedia—the first making up for the bulk of the discrepancy. When articles use the same structure but different facts (e.g., geographical locations described in terms of their surroundings), we refer to this as *structure reuse* (Table 2, top left) and consider such cases as non-problematic (perhaps unavoidable) redundancy. On the other hand, articles may contain factually nearly-identical passages, likely after copying from one to the other. We consider such *content reuse* likely to result in inconsistency and contradiction as the articles may diverge over time (Table 2, bottom left). Ideally, such redundant sections should be replaced with a single, authoritative source. In this sense, text reuse analysis can help the Wikipedia community locate and improve articles with undesirable redundancy.

Further observations indicate that the ontological relationship between articles' topics correlates with the type of text reuse: Structure reuse occurs more frequently when articles represent concepts on the same level in the ontology tree (Table 2, top right), while two articles whose subjects are vertically aligned (e.g., "is a" or "part of" relationships) are more likely to exhibit content reuse (Table 2, bottom right). The latter association can also be envisioned as a solution to the sub-article matching task [10]: the occurrence of content reuse between articles can serve as an indicator of the ontological relationship between the concepts that they represent. However, automatically distinguishing content and structure reuse is not trivial. Our initial attempt at classifying reuse cases used a heuristic based on the ratio of reused to original text in the articles, as well as the Jaccard similarities between the sets of named entities and word 10-grams. Using two samples of 100 random structure reuse cases and 100 random content reuse cases, the heuristic achieved 100% precision for structure reuse, but only 57% for content reuse. While our heuristics identify 95.5 million (87%) of all within-Wikipedia reuse cases as structure reuse, the true number likely exceeds 100 million assuming our error estimates are accurate.

In the 10%-sample of the Common Crawl, 4,898 websites host at least one page that reuses text from a Wikipedia article for a total of 1.6 million cases.[5] We presume that Wikipedia's policy of avoiding reuse from third parties inside its articles is enforced by its editors, so that nearly all of the cases will be third parties reusing Wikipedia's articles instead. Most (94%) of the pages violate the terms of Wikipedia's license[6] by not referencing Wikipedia as a source (i.e., the term "Wikipedia" does not occur). With only a handful exceptions, such as un.org, all of the sites display advertisements, which extends to the pages containing the reuse. Furthermore, in nearly all of the cases, the reuse accounted for more than 90% of the main content, prompting usefulness questions.

Table 2. Examples of the two types of text reuse within Wikipedia—structure reuse (top) and content reuse (bottom)—and corresponding ontological article relations (right).

Title: Niedźwiedzie, Pisz County	**Title:** Zimna Woda, Zgierz County	
Niedźwiedzie is a village in the administrative district of Gmina Pisz, within Pisz County, Warmian-Masurian Voivodeship, in northern Poland. It lies approximately south-east of Pisz and east of the regional capital Olsztyn.	Zimna Woda is a village in the administrative district of Gmina Zgierz, within Zgierz County, Łódź Voivodeship, in central Poland. It lies approximately north-west of Zgierz and north-west of the regional capital Łódź.	City; Is a; Is a; Berlin; Leipzig; Structure Reuse
Title: Human tooth development	**Title:** Tooth eruption	
Tooth eruption has three stages. The first, known as deciduous dentition stage, occurs when only primary teeth are visible. Once the first permanent tooth erupts into the mouth, the teeth are in the mixed (or transitional) dentition. [...] Primary dentition stage starts on the arrival of the mandibular central incisors, typically from around six months, and lasts until the first permanent molars appear [...]	The dentition goes through three stages. The first, known as primary dentition stage, occurs when only primary teeth are visible. Once the first permanent tooth erupts into the mouth, the teeth that are visible are in the mixed (or transitional) dentition stage. [...] Primary dentition starts on the arrival of the madibular central incisors, usually at eight months, and lasts until the first permanent molars appear [...]	Opportunism; Content Reuse; Is a; Spiritual Opportunism; History of Christianity; Content Reuse; Part of; History of Christianity in Middle Ages

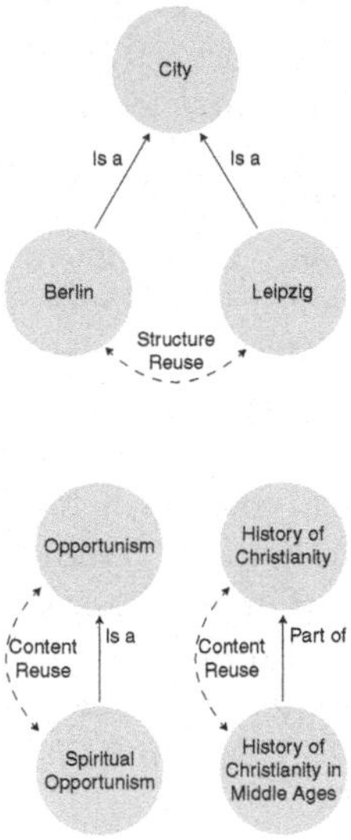

We conservatively estimate the potential advertisement revenue generated by the reused Wikipedia content. For simplicity, we assume that all reusing websites host only one ad per page and that advertisements are billed according to cost per mille (CPM), achieving a revenue per mille (RPM) of about half (1.4 USD) the average estimated CPM on the web in 2018 (2.8 USD).[7] Accounting for

[5] The top three being wikia.com (563), rediff.com (55), and un.org (28 reusing pages).
[6] en.wikipedia.org/wiki/Wikipedia:Reusing_Wikipedia_content.
[7] monetizepros.com/cpm-rate-guide/display/.

the fact that reusing pages are generally ranked lower than Wikipedia in search results, we use 10% of the monthly page view counts of reused articles (as per Wikipedia's API) as estimates for the page views of reusing pages. With these approximations, we arrive at an estimate of 45,000 USD monthly ad revenue generated by the detected 4,898 reusing sites. Extrapolated to the entire web (say, 600,000 reusing sites out of 180 million active sites as per netcraft.com), we arrive at 5.5 million USD estimated monthly ad revenue; which adds up to about 72.5% of Wikipedia's worldwide fundraising returns in the fiscal year 2016–2017.[8]

5 Conclusion

In an effort to bring text reuse analysis to very large corpora, we propose a scalable pipeline comprising the source retrieval and text alignment subtasks. We address challenges of scale primarily in the former via candidate filtering, and evaluate a set of hashing and text similarity techniques for this purpose. Our framework and the two compiled text reuse datasets—within Wikipedia and in a 10%-sample of the Common Crawl—are publicly available.[9] This way, we hope to stimulate future research targeting Wikipedia quality improvement (e.g., by template induction or automatic detection of reuse inconsistencies) and understanding Wikipedia's influence on the web at large.

References

1. Ardi, C., Heidemann, J.: Web-scale content reuse detection (extended). USC/Information Sciences Institute, Tech. Rep. ISI-TR-692 (2014)
2. Bendersky, M., Croft, W.: Finding text reuse on the web. In: Proceedings of WSDM 2009, pp. 262–271 (2009)
3. Chaidaroon, S., Fang, Y.: Variational deep semantic hashing for text documents. arXiv preprint arXiv:1708.03436 (2017)
4. Charikar, M.S.: Similarity estimation techniques from rounding algorithms. In: Proceedings of STOC 2002, pp. 380–388 (2002)
5. Citron, D.T., Ginsparg, P.: Patterns of text reuse in a scientific corpus. PNAS **112**(1), 25–30 (2015)
6. Clough, P.D., Wilks, Y.: Measuring text reuse in a journalistic domain. In: Proceedings of the CLUK Colloquium (2001)
7. Coffee, N., Koenig, J.P., Poornima, S., Forstall, C.W., Ossewaarde, R., Jacobson, S.L.: The Tesserae project: intertextual analysis of Latin poetry. Literary Linguist. Comput. **28**(2), 221–228 (2012)
8. Generous, N., Fairchild, G., Deshpande, A., Del Valle, S., Priedhorsky, R.: Global disease monitoring and forecasting with Wikipedia. PLoS Comput. Biol. **10**(11), e1003892 (2014)
9. Hagen, M., Potthast, M., Adineh, P., Fatehifar, E., Stein, B.: Source retrieval for web-scale text reuse detection. In: Proceedings of CIKM 2017, pp. 2091–2094 (2017)

[8] foundation.wikimedia.org/wiki/2016-2017_Fundraising_Report.
[9] github.com/webis-de/ECIR-19, webis.de/data/webis-wikipedia-text-reuse-18.html.

10. Lin, Y., Yu, B., Hall, A., Hecht, B.: Problematizing and addressing the article-as-concept assumption in Wikipedia. In: Proceedings of CSCW 2017, pp. 2052–2067 (2017)
11. McMahon, C., Johnson, I.L., Hecht, B.J.: The substantial interdependence of Wikipedia and Google: a case study on the relationship between peer production communities and information technologies. In: Proceedings of ICWSM 2017, pp. 142–151 (2017)
12. Mestyán, M., Yasseri, T., Kertész, J.: Early prediction of movie box office success based on Wikipedia activity big data. PLoS One **8**(8), e71226 (2013)
13. Mitchell, J., Lapata, M.: Vector-based models of semantic composition. In: Proceedings of ACL 2008, pp. 236–244 (2008)
14. Potthast, M., et al.: Overview of the 5th international competition on plagiarism detection. In: Working Notes Papers of the CLEF 2013 Evaluation Labs
15. Potthast, M., et al.: Overview of the 6th international competition on plagiarism detection. In: Working Notes Papers of the CLEF 2014 Evaluation Labs
16. Stamatatos, E.: Plagiarism detection using stopword n-grams. JASIST **62**(12), 2512–2527 (2011)
17. Stein, B., Meyer zu Eißen, S., Potthast, M.: Strategies for retrieving plagiarized documents. In: Proceedings of SIGIR 2007, pp. 825–826 (2007)
18. Taraborelli, D.: The sum of all human knowledge in the age of machines: a new research agenda for Wikimedia. In: Proceedings of the ICWSM 2015 Workshop Wikipedia, a Social Pedia: Research Challenges and Opportunities
19. Thompson, N., Hanley, D.: Science is shaped by Wikipedia: Evidence from a randomized control trial. MIT Sloan Research Paper No. 5238-17 (2018)
20. Vincent, N., Johnson, I., Hecht, B.: Examining Wikipedia with a broader lens: quantifying the value of Wikipedia's relationships with other large-scale online communities. In: Proceedings of CHI 2018, pp. 566:1–566:13 (2018)
21. Weissman, S., Ayhan, S., Bradley, J., Lin, J.: Identifying duplicate and contradictory information in Wikipedia. In: Proceedings of JCDL 2015, pp. 57–60 (2015)

Stochastic Relevance for Crowdsourcing

Marco Ferrante[1], Nicola Ferro[2(✉)], and Eleonora Losiouk[1]

[1] Department of Mathematics "Tullio Levi-Civita", University of Padua, Padua, Italy
{ferrante,elosiouk}@math.unipd.it

[2] Department of Information Engineering, University of Padua, Padua, Italy
ferro@dei.unipd.it

Abstract. It has been recently proposed to consider relevance assessment as a stochastic process where relevance judgements are modeled as binomial random variables and, consequently, evaluation measures become random evaluation measures, removing the distinction between binary and multi-graded evaluation measures.

In this paper, we adopt this stochastic view of relevance judgments and we investigate how this can be applied in the crowd-sourcing context. In particular, we show that injecting some randomness in the judgments by crowd assessors improves their correlation with the gold standard and we introduce a new merging approach, based on binomial random variables, which is competitive with respect to state-of-the-art at low numbers of merged assessors.

1 Introduction

It has been recently proposed to model relevance assessment as a stochastic process where each relevance judgement is a binomial random variable whose expectation p indicates the quantity of relevance assigned to a document [5]. This choice allowed for seamlessly modeling both binary and graded relevance judgements into a single framework and for introducing the notion of *random evaluation measures*, which are just a transformation of such binomial variables, eliminating the distinction between binary and graded evaluation measures.

In this paper, we investigate to what extent this new way of modelling relevance judgements can be applied in the context of crowdsourcing [1]. In particular, we study the following research questions:

RQ1 how the random evaluation measures can improve the robustness to variations in the assessments;

RQ2 how the proposed binomial framework can be extended to allow for merging multiple crowd-assessors.

We conduct a systematic experimentation using the TREC 2012 Crowdsourcing track [13] in order to answer the two research questions above.

The paper is organized as follows: Sect. 2 discusses some related works; Sect. 3 introduces our stochastic framework for merging crowd assessors; Sect. 4 reports the evaluation results; and, Sect. 5 draws some conclusions and outlooks possible future works.

L. Azzopardi et al. (Eds.): ECIR 2019, LNCS 11437, pp. 755–762, 2019.
https://doi.org/10.1007/978-3-030-15712-8_50

2 Related Work

2.1 Crowdsourcing

Crowdsourcing [1,9–11] has emerged as a viable option for ground-truth creation since it allows to cheaply collect multiple assessments for each document. However, it raises many questions regarding the quality of the collected assessments. Therefore, in order to obtain a ground-truth good enough to be used for evaluation purposes, the possibility of discarding the low quality assessors and/or combining them with more or less sophisticated algorithms has been considered.

State of the art crowdsourcing algorithms are *Majority Vote*, where the label with the highest number of votes, i.e. assessors, is selected, and *Expectation-Maximization* [2,6], where the Expectation-Maximization algorithm is used to iteratively select the most probable labels. More recently, AWARE [4] has been proposed as a way to compute a weighted mean of evaluation measures computed for each crowd assessor.

2.2 Binomial Relevance Framework

Ferrante et al. [5] described the relevance of a document via a **binomial random variable** $Bi(1,p)$ with parameters 1 and p, where p roughly defines the *quantity of relevance* of that document. For each topic, document pair $(t, d_i) \in T \times D$, they defined the **random ground-truth** RGT, also called random relevance, as a binomial random variable of parameters $(1, p_{t,d_i})$, where p_{t,d_i} is the quantity of relevance associated to the document d_i with respect to a topic t. In this framework, $p_{t,d_i} = 0$ corresponds to a document completely not relevant and $p_{t,d_i} = 1$ to a fully relevant document.

Thanks to the random ground-truth, they turned every evaluation measure into a random evaluation measure, by simply composing the original expression of each measure with the random relevances. To compare different systems, they needed to define an ordering among runs and, to this end, they used the expected values of the random measures defined above.

Therefore, **expected Random Rank Biased Precision (eRRBP)** is

$$\mathbb{E}\big[RBP[\hat{r}_t(\omega)]\big] = (1-\tau)\sum_{n=1}^{N} \tau^{n-1} p_{t,d_n}$$

where τ represents the persistence.

Then, **expected Random Discounted Cumulative Gain (eRDCG)** is

$$\mathbb{E}\big[DCG[\hat{r}_t(\omega)]\big] = \sum_{n=1}^{N} \frac{p_{t,d_n}}{\max\{1, \log_{10}(n)\}}$$

Finally, **expected Random Average Precision (eRAP)** is

$$\mathbb{E}\big[AP[\hat{r}_t(\omega)]\big] = \frac{1}{\widehat{RB}_t}\sum_{n=1}^{N} \frac{1}{n}\left(1 + \sum_{s=1}^{n-1} p_{t,d_s}\right) p_{t,d_n}$$

where $\widehat{RB}_t = \sum_{d \in D} \mathbb{E}\big[RGT(t,d)\big]$ is the expected recall base.

3 Random Relevance for Merging Crowd-Assessors

Let us assume that M assessors evaluate a pool of documents $\{d_1, \ldots, d_N\}$ with respect to a topic $t \in T$. According to [5], the judgment of the j-th assessor for the pair (t, d_i) is a Binomial random variable $AS_j(t, d_i)$ which models the amount of relevance of the document according to that assessor.

We assume that for any pair (t, d_i), $AS_1(t, d_i), \ldots, AS_M(t, d_i)$ are independent binomial random variables of parameters $(1, p_{t,d_i})$. Note that this i.i.d. assumption is implicitly done in all the previous works about merging crowd-assessors.

We leverage the assumption above to define the **Binomial Majority Vote (BINMV)** merging strategy, where the unknown parameter p_{t,d_i}, i.e. the merged amount of relevance of each topic/document pair, is estimated from the observed values of the random variables $AS_1(t, d_i), \ldots, AS_M(t, d_i)$ as

$$\widetilde{p}_{t,d_i} = \frac{1}{M} \sum_{j=1}^{M} AS_j(t, d_i)$$

and we define the random ground-truth $RGT(t, d_i)$ as a Binomial random variable of parameters $(1, \widetilde{p}_{t,d_i})$. As the name suggests, this strategy adopts the same logic as the Majority Vote approach but applied in the case of the random relevance.

We also define the **Quantized Binomial Majority Vote (QBINMV)** strategy which applies a sigmoid function $\frac{1}{1+\exp^{-k*(x-0.5)}}$ to the estimated parameter $\widetilde{p}_{t,d_i}$ in order to reduce the number of relevance degrees produced by the BINMV strategy and make sharper decisions towards being relevant or not relevant; in particular, we use $k = 15$.

4 Experiments

4.1 Experimental Setup

We use the TREC 21, 2012, Crowdsourcing (`T21`) [13] data set developed in the *Text Relevance Assessing Task (TRAT)*. The TRAT required participating groups to simulate the relevance assessing role of the NIST for 10 of the TREC 08, 1999, Ad-hoc topics [16], using binary relevance. In total 33 pools were submitted to TRAT; we excluded two of them (`INFLB2012` and `Orc2Stage`) because, for some topics, they did not assess any document as relevant.

Two TREC Adhoc tracks used these 10 topics over the years: the TREC 08, 1999, Ad-hoc track [16] (labeled `T08`), which contains 129 runs; and, the TREC 13, 2004, Robust track [15] (labeled `T13`), which contains 110 runs.

As in the TREC crowdsourcing track, we use correlation analysis – both Kendall's τ correlation [8] and AP correlation τ_{AP} [18] – to compare crowd assessors with respect to the gold standard pool.

We consider the following evaluation measures, to be compared against their random version: AP [3], DCG [7], and RBP [12]. We use log base 10 for DCG and gains 0 and 1 for not relevant and relevant documents, respectively; we use persistence $p = 0.8$ for RBP.

To ease the reproducibility of the experiments, the source code is available at: https://bitbucket.org/frrncl/ecir2019-ffl/.

4.2 RQ1: Robustness to Variations in the Assessments

For each crowd-assessor submitted to the T21 track, we computed the τ and τ_{AP} correlations with respect to the gold standard pool and then we averaged these scores over all the crowd-assessors. Table 1 reports the summary averages together with their confidence intervals. For each measure, we report: (i) the state-of-the-art deterministic version compared against a Deterministic Gold Standard (DGS); the random version using $p_{notrel} = 0.05$ and $p_{rel} = 0.95$ for the crowd-assessors, i.e. we allow for just a small 5% confidence on their judgements, compared against the DGS, i.e. the same used for the deterministic measures; the random version as before but compared against a Randomized Gold Standard (RGS), which is the gold standard pool but using $p_{notrel} = 0.05$ and $p_{rel} = 0.95$, i.e. we assume just a small randomness also in it.

Table 1. τ and τ_{AP} averaged over the T21 crowd-assessors. Gold standard is labelled as: DGS (Deterministic Gold Standard); RGS (Randomized Gold Standard).

		T08 Systems		T13 Systems	
		Mean τ	Mean τ_{AP}	Mean τ	Mean τ_{AP}
DGS	**AP**	0.7023 ± 0.0522	0.5802 ± 0.0655	0.7044 ± 0.0532	0.5655 ± 0.0679
DGS	**eRAP**	0.6704 ± 0.0436	0.5437 ± 0.0485	0.7033 ± 0.0355	0.5551 ± 0.0444
RGS	**eRAP**	**0.7077 ± 0.0537**	**0.5900 ± 0.0608**	**0.7471 ± 0.0469**	**0.6056 ± 0.0642**
DGS	**DCG**	0.7222 ± 0.0454	0.5998 ± 0.0482	0.7621 ± 0.0466	0.6161 ± 0.0601
DGS	**eRDCG**	0.6896 ± 0.0373	0.5737 ± 0.0422	0.7391 ± 0.0433	0.5833 ± 0.0585
RGS	**eRDCG**	**0.7858 ± 0.0356**	**0.6776 ± 0.0436**	**0.7766 ± 0.0396**	**0.6240 ± 0.0522**
DGS	**RBP**	0.6732 ± 0.0547	0.5341 ± 0.0684	0.5879 ± 0.0657	0.4534 ± 0.0706
DGS	**eRRBP**	0.6739 ± 0.0546	0.5352 ± 0.0684	0.5904 ± 0.0647	**0.4560 ± 0.0699**
RGS	**eRRBP**	**0.6749 ± 0.0545**	**0.5359 ± 0.0683**	**0.5922 ± 0.0645**	0.4555 ± 0.0698

Comparing the deterministic measures against DGS to the random ones against RGS, we can observe how the random evaluation measures substantially improve the average agreement among the gold standard and the crowd-assessors, consistently for both τ and τ_{AP} and across both tracks, T08 and T13.

Comparing the deterministic measures against DGS to the random ones against DGS, we can observe that deterministic measures tend to perform better, with the exception of RBP and eRBP whose performance are almost the same. However, it should be noted that this is by far the most unfavourable comparison for the random evaluation measures, since the DGS pool does not account for any kind of randomness and awards only the deterministic evaluation measures.

Overall, we can conclude that injecting some randomness into the evaluation measures is beneficial for compensating variations in relevance judgements in a crowdsourcing context.

Table 1 also opens an important question about what we should consider as gold standard: a deterministic or a random pool? If, for example, we consider the inter-assessor agreement issue [14,17], we should conclude that the gold standard we daily use in evaluation campaigns is far from being deterministic and, perhaps, we should move to a stochastic vision of it.

4.3 RQ2: Random Relevance for Merging Crowd-Assessors

Let $L = 31$ be the total number of available crowd-assessors and $M < L$ the number of assessors we are merging. For each of the above evaluation measures, we experimented all the $M = 2, 3, \ldots, 30$. For each value of M, there are $\binom{L}{M} = \binom{31}{M} = \frac{31!}{M!(31-M)!}$ possible ways of choosing the M assessors to be merged; we randomly sampled 10 M-tuples out of the $\binom{31}{M}$ possible ones.

Table 2 reports the average of τ_{AP} correlation over these 10 samples for both `T08` and `T13` systems; the results using Kendall's τ correlation are similar but not reported here for space reasons. As in the case of Table 1, for each measure, we report: (i) the state-of-the-art deterministic merging strategy compared against DGS; the random merging strategy compared against the DGS; the random merging strategy compared against RGS. As state-of-the-art deterministic merging strategy we considered Majority Vote (MV), Expectation-Maximization (EM), and AWARE with uniform weights.

If we compare the results of Table 2 with those of Table 1 we can note how all the merging strategies improve with respect to the performance of single crowd assessors.

When it comes to merging in the case of the deterministic state-of-the-art merging strategies, we can observe that MV is always the most effective approach for high numbers of merged assessors while AWARE is competitive for lower numbers, a more interesting case due to the less resources required. EM tends to have lower performance when using fewer assessors and they increase for more assessors but almost never reaching MV.

BINMV is especially effective with eRRBP, which always improves for low numbers of assessors with respect to MV and AWARE. However, in the case of eRAP and eRDCG deterministic state-of-the-art merging strategies tend to perform better. However, as discussed in the case of RQ1, the RGS is a more fair comparison for the random evaluation measures and, in this case, we can observe more substantial improvements for the BINMV strategy which often outperforms state-of-the-art ones.

Finally, QBINMV is typically more effective than BINMV and it often performs better than deterministic state-of-the-art merging strategies. This is probably due to the fact that it reduces the number of relevance degrees, which is almost "continuous" in the case of the BINMV strategy, and pushes towards choosing between either relevant or not relevant. This makes QBINMV closer to the deterministic evaluation measures, which use just binary relevance, and so they compete on a closer basis.

Table 2. τ_{AP} for different merging strategies and different numbers M of merged assessors using T08 and T13 systems. Gold standard is labelled as: DGS (Deterministic Gold Standard); RGS (Randomized Gold Standard).

T08 Systems

		$M = 2$	$M = 3$	$M = 4$	$M = 5$	$M = 10$	$M = 20$	$M = 30$
DGS	**MV AP**	0.5757	0.6425	0.7135	0.6920	**0.7605**	**0.7979**	**0.8103**
	EM AP	0.5722	0.6161	0.7147	0.6749	0.7445	0.7522	0.7443
	AWARE AP	**0.6797**	0.6525	0.7124	**0.6928**	0.7138	0.7034	0.7089
DGS	**BINMV eRAP**	0.5924	0.5807	0.6050	0.5978	0.5794	0.5617	0.5657
	QBINMV eRAP	0.5921	0.6299	0.7028	0.6696	0.6913	0.6848	0.6615
RGS	**BINMV eRAP**	0.6037	0.6173	0.6432	0.6259	0.6329	0.6211	0.6225
	QBINMV eRAP	0.6043	**0.6561**	**0.7226**	0.6663	0.7441	0.7445	0.7314
DGS	**MV DCG**	0.6123	0.6901	0.7116	0.6733	0.7432	**0.7868**	**0.7895**
	EM DCG	0.5441	0.6741	0.7014	0.6642	0.6770	0.6756	0.6598
	AWARE DCG	0.6190	0.6397	0.6540	0.6392	0.6461	0.6499	0.6508
DGS	**BINMV eRDCG**	0.6190	0.6397	0.6540	0.6392	0.6461	0.6499	0.6508
	QBINMV eRDCG	0.6187	0.6830	0.7096	0.6735	0.7286	0.7453	0.7544
RGS	**BINMV eRDCG**	0.6878	0.7181	0.7371	**0.7265**	0.7507	0.7502	0.7559
	QBINMV eRDCG	**0.6884**	**0.7410**	**0.7517**	0.7185	**0.7660**	0.7794	0.7892
DGS	**MV RBP**	0.6211	0.6138	0.7065	0.6661	0.7243	0.7444	0.7365
	EM RBP	0.5194	0.6087	0.6937	0.6214	0.6790	0.7053	0.6780
	AWARE RBP	0.6421	0.6374	0.7109	0.6811	0.7139	0.7152	0.7205
DGS	**BINMV eRRBP**	0.6422	0.6374	0.7109	0.6811	0.7139	0.7152	0.7205
	QBINMV eRRBP	0.6422	0.6277	**0.7241**	0.6835	**0.7453**	0.7602	**0.7716**
RGS	**BINMV eRRBP**	0.6428	**0.6381**	0.7114	0.6817	0.7147	0.7160	0.7213
	QBINMV eRRBP	**0.6429**	0.6279	0.7240	**0.6836**	0.7451	**0.7603**	**0.7716**

T13 Systems

		$M = 2$	$M = 3$	$M = 4$	$M = 5$	$M = 10$	$M = 20$	$M = 30$
DGS	**MV AP**	0.6158	0.6253	0.7167	0.7138	0.7674	**0.7995**	**0.8226**
	EM AP	0.5486	0.5575	0.7218	0.6877	0.7441	0.7833	0.7704
	AWARE AP	**0.6965**	**0.6717**	**0.7477**	**0.7221**	0.7693	0.7591	0.7600
DGS	**BINMV eRAP**	0.6315	0.6135	0.6648	0.6553	0.6485	0.6538	0.6513
	QBINMV eRAP	0.6306	0.6443	0.7106	0.7033	0.7248	0.7308	0.7248
RGS	**BINMV eRAP**	0.6381	0.6451	0.7130	0.6798	0.7271	0.7116	0.7073
	QBINMV eRAP	0.6411	0.6696	0.7320	0.6898	**0.7731**	0.7828	0.7849
DGS	**MV DCG**	0.6233	0.7039	**0.7516**	**0.7122**	**0.7923**	**0.8129**	**0.8226**
	EM DCG	0.5719	**0.7044**	0.7496	0.6972	0.7254	0.7211	0.6845
	AWARE DCG	0.6349	0.6496	0.6758	0.6542	0.6616	0.6565	0.6517
DGS	**BINMV eRDCG**	0.6349	0.6496	0.6758	0.6542	0.6616	0.6565	0.6517
	QBINMV eRDCG	0.6347	0.6919	0.7436	0.7062	0.7795	0.7815	0.7774
RGS	**BINMV eRDCG**	0.6466	0.6590	0.6990	0.6749	0.7051	0.6987	0.6975
	QBINMV eRDCG	**0.6470**	0.6938	0.7214	0.6975	0.7455	0.7541	0.7570
DGS	**MV RBP**	0.4893	0.5036	0.5880	0.5385	0.6114	0.6062	0.6187
	EM RBP	0.4106	0.4974	0.5989	0.5180	0.6270	0.6094	0.5943
	AWARE RBP	0.5614	0.5586	0.6125	**0.5717**	0.6342	**0.6307**	**0.6248**
DGS	**BINMV eRRBP**	0.5614	0.5586	0.6125	**0.5717**	0.6342	**0.6307**	**0.6248**
	QBINMV eRRBP	0.5614	0.5233	0.6038	0.5594	0.6292	0.6263	0.5988
RGS	**BINMV eRRBP**	**0.5619**	**0.5589**	**0.6137**	**0.5717**	**0.6348**	0.6304	0.6245
	QBINMV eRRBP	**0.5619**	0.5232	0.6042	0.5592	0.6297	0.6263	0.5990

5 Conclusions and Future Work

In this paper, we investigated how a stochastic approach for modelling relevance as a random binomial variable behaves in the context of crowdsourcing. We have shown how injecting some randomness in the relevance judgments of crowd-assessors improves their correlation with the gold standard (RQ1). We have also shown how the binomial relevance framework can be used to develop new merging strategies which are competitive with respect to state-of-the-art when using fewer crowd-assessors, which means reducing the required resources (RQ2). In both cases, the conducted investigation raised the issue of whether it is more appropriate to use a deterministic or a randomized gold standard.

Overall, we can appreciate the benefits of moving to a random relevance framework which is capable to unify into a single coherent vision binary to multi-graded relevance, management of incomplete information and variations in relevance judgments, and merging of crowd-assessors.

As future work, we will investigate how using a stochastic gold standard instead of a deterministic one impacts on IR evaluation. Moreover, we plan to leverage the binomial relevance framework to develop more advanced merging strategies able to also account for the quality of the assessors, instead of simply merging them in a uniform way.

References

1. Alonso, O., Mizzaro, S.: Using crowdsourcing for TREC relevance assessment. Inf. Process. Manage. **48**(6), 1053–1066 (2012)
2. Bashir, M., et al.: Northeastern university runs at the TREC12 crowdsourcing track. In: Voorhees, E.M., Buckland, L.P. (eds.) The Twenty-First Text REtrieval Conference Proceedings (TREC 2012). National Institute of Standards and Technology (NIST), Special Publication 500–298, Washington, USA (2013)
3. Buckley, C., Voorhees, E.M.: Retrieval system evaluation. In: Harman, D.K., Voorhees, E.M. (eds.) TREC. Experiment and Evaluation in Information Retrieval, pp. 53–78. MIT Press, Cambridge (2005)
4. Ferrante, M., Ferro, N., Maistro, M.: AWARE: exploiting evaluation measures to combine multiple assessors. ACM Trans. Inf. Syst. (TOIS) **36**(2), 20:1–20:38 (2017)
5. Ferrante, M., Ferro, N., Pontarollo, S.: Modelling randomness in relevance judgments and evaluation measures. In: Pasi, G., Piwowarski, B., Azzopardi, L., Hanbury, A. (eds.) ECIR 2018. LNCS, vol. 10772, pp. 197–209. Springer, Cham (2018). https://doi.org/10.1007/978-3-319-76941-7_15
6. Hosseini, M., Cox, I.J., Milić-Frayling, N., Kazai, G., Vinay, V.: On aggregating labels from multiple crowd workers to infer relevance of documents. In: Baeza-Yates, R., et al. (eds.) ECIR 2012. LNCS, vol. 7224, pp. 182–194. Springer, Heidelberg (2012). https://doi.org/10.1007/978-3-642-28997-2_16
7. Järvelin, K., Kekäläinen, J.: Cumulated gain-based evaluation of IR techniques. ACM Trans. Inf. Syst. (TOIS) **20**(4), 422–446 (2002)
8. Kendall, M.G.: Rank Correlation Methods. Griffin, Oxford (1948)
9. King, I., Chen, K.T., Alonso, O., Larson, M.: Special issue: crowd in intelligent systems. ACM Trans. Intell. Syst. Technol. (TIST) **7**(4), 77 (2016)

10. Lease, M., Yilmaz, E.: Crowdsourcing for information retrieval: introduction to the special issue. Inf. Retrieval **16**(2), 91–100 (2013)
11. Marcus, A., Parameswaran, A.: Crowdsourced data management: industry and academic perspectives. Found. Trends Databases (FnTDB) **6**(1–2), 1–161 (2015)
12. Moffat, A., Zobel, J.: Rank-biased precision for measurement of retrieval effectiveness. ACM Trans. Inf. Syst. (TOIS) **27**(1), 2:1–2:27 (2008)
13. Smucker, M.D., Kazai, G., Lease, M.: Overview of the TREC 2012 crowdsourcing track. In: Voorhees, E.M., Buckland, L.P. (eds.) The Twenty-First Text REtrieval Conference Proceedings (TREC 2012). National Institute of Standards and Technology (NIST), Special Publication 500–298, Washington, USA (2013)
14. Voorhees, E.M.: Variations in relevance judgments and the measurement of retrieval effectiveness. In: Croft, W.B., Moffat, A., van Rijsbergen, C.J., Wilkinson, R., Zobel, J. (eds.) Proceedings 21st Annual International ACM SIGIR Conference on Research and Development in Information Retrieval (SIGIR 1998), pp. 315–323. ACM Press, New York (1998)
15. Voorhees, E.M.: Overview of the TREC 2004 robust track. In: Voorhees, E.M., Buckland, L.P. (eds.) The Thirteenth Text REtrieval Conference Proceedings (TREC 2004). National Institute of Standards and Technology (NIST), Special Publication 500–261, Washington, USA (2004)
16. Voorhees, E.M., Harman, D.K.: Overview of the eighth Text Retrieval Conference (TREC-8). In: Voorhees, E.M., Harman, D.K. (eds.) The Eighth Text REtrieval Conference (TREC-8), pp. 1–24. National Institute of Standards and Technology (NIST), Special Publication 500–246, Washington, USA (1999)
17. Webber, W., Chandar, P., Carterette, B.A.: Alternative assessor disagreement and retrieval depth. In: Chen, X., Lebanon, G., Wang, H., Zaki, M.J. (eds.) Proceedings of 21st International Conference on Information and Knowledge Management (CIKM 2012), pp. 125–134. ACM Press, New York (2012)
18. Yilmaz, E., Aslam, J.A., Robertson, S.E.: A new rank correlation coefficient for information retrieval. In: Chua, T.S., Leong, M.K., Oard, D.W., Sebastiani, F. (eds.) Proceedings of 31st Annual International ACM SIGIR Conference on Research and Development in Information Retrieval (SIGIR 2008), pp. 587–594. ACM Press, New York (2008)

Exploiting a More Global Context for Event Detection Through Bootstrapping

Dorian Kodelja(✉), Romaric Besançon, and Olivier Ferret

CEA, LIST, Laboratoire Analyse Sémantique Texte et Image,
91191 Gif-sur-Yvette, France
{dorian.kodelja,romaric.besancon,olivier.ferret}@cea.fr

Abstract. Over the last few years, neural models for event extraction have obtained interesting results. However, their application is generally limited to sentences, which can be an insufficient scope for disambiguating some occurrences of events. In this article, we propose to integrate into a convolutional neural network the representation of contexts beyond the sentence level. This representation is built following a bootstrapping approach by exploiting an intra-sentential convolutional model. Within the evaluation framework of TAC 2017, we show that our global model significantly outperforms the intra-sentential model while the two models are competitive with the results obtained by TAC 2017 participants.

Keywords: Information extraction · Event detection · Global context

1 Introduction

In some domains, such as journalism, the notion of event is particularly important and can be a central dimension for guiding search among documents [7]. Detecting events from texts is a necessary step for implementing such an approach. In this article, we consider supervised event detection, which consists in identifying in texts the mentions of *a priori* known event types, *i.e.* the word or the sequence of words indicating the presence of a particular type of events. Most of the current approaches for this task are based on neural models, either convolutional [2,17], recurrent [16] or mixing the two kinds of models [4]. Moreover, the best systems of the recent evaluation campaigns for this task, such as TAC Event Nugget 2017, are based on such models. These models successfully identify a significant part of event mentions but still fail when the local context is too ambiguous for discriminating between two types of events or deciding if an event mention is actually present. For instance, in the following example:

> "[...] according to **leaked documents**. I don't trust them AT ALL [...], so I will have to read these **cables**$_{[broadcast]}$ myself."

Work partly supported by ANR under project ASRAEL (ANR-15-CE23-0018).

L. Azzopardi et al. (Eds.): ECIR 2019, LNCS 11437, pp. 763–770, 2019.
https://doi.org/10.1007/978-3-030-15712-8_51

the local sentence context is not sufficient for disambiguating the word *cables* as a trigger for a *Broadcast* event while looking at previous sentences would show that *cables* is related to the expression *leaked documents*, which is more directly linked to a *Broadcast* event. Performing such disambiguation requires exploiting contexts beyond the scope of sentences. This perspective has already been explored by [3] by adding to the input of a BiLSTM model for trigger extraction the representation of the overall document computed by the method of [10]. The underlying hypothesis is that integrating such representation accounts for the fact that a document related to the topic of war is more likely to contain *Die* or *Attack* events than *Divorce* events. However, the document representation, in that case, is general. Very recently, [20] has extended this approach in a more integrated way by exploiting a hierarchical document embedding. Similarly, our approach aims at building a document representation specifically linked to the target task but we adopt a simpler approach by relying on bootstrapping: a model focusing on a very local context is first applied to the considered document; then, its local predictions are aggregated for building a document context vector. This vector is finally exploited by a new extraction model we define. Previously, [11] introduced a global classifier to apply a second pass on the input corpus and detect ambiguous triggers missed by the local classifier. The global classifier only used the candidate word and a binary vector informing about the detection of at least one event of each event type by the local classifier. On the opposite, our system, which uses the more informative estimated distribution of the number of events for each type, is not only able to detect missed event but also to reject previously detected spurious triggers.

Our experiments on the TAC Event Nugget 2017 data show that this new model significantly outperforms our state-of-the-art local model.

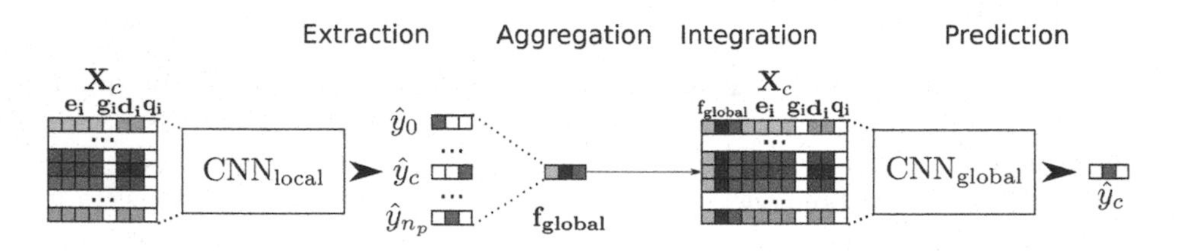

Fig. 1. Generation and integration of the global context representation

2 Method

In this article, we aim at detecting event mentions (triggers) in text and categorize them into predefined types. We consider the 38 event types defined in the DEFT Rich ERE taxonomy [1] used in the Event Nugget evaluation of the TAC campaigns [15]. Since most of the annotated triggers are single tokens [19], we only consider mono-token triggers. While this simplification does not affect performance significantly, it makes the model simpler and allows the introduction of a positional vector, which has a significant impact on results [18].

Figure 1 gives an overview of our integration of a global context in a convolutional neural network (CNN) through bootstrapping. First, a local model $\text{CNN}_{\text{local}}$ is trained to predict an event label for each word of a document. These labels are then aggregated at a specific level (in Fig. 1, labels are aggregated at the document level) and integrated into a new model. The following sections present the local and global models in more detail.

2.1 Local Event Detection Model

At the local level, our event detection model relies on a CNN based on the architecture introduced in [18]. We successively consider each token in each sentence as a candidate mention. This mention is represented by a fixed-size local context centered on the mention. We perform padding to complete the sequence when the local context goes beyond sentence boundaries. Let i_c be the index of the candidate mention and w the window size. We define $\mathbf{i}_c = [i_{c-w}, i_{c-w+1}, \ldots, i_c, \ldots, i_{c+w-1}, i_{c+w}]$ as the index vector centered on i_c. This vector is then transformed into a real-valued matrix $\mathbf{X}_c = [\mathbf{x_{c-w}}, \mathbf{x_{c-w+1}}, \ldots, \mathbf{x_c}, \ldots, \mathbf{x_{c+w-1}}, \mathbf{x_{c+w}}]$ by replacing each index i with its vector representation $\mathbf{x_i} = [\mathbf{e_i}, \mathbf{d_i}, \mathbf{g_i}, \mathbf{q_i}]$ using the concatenation of the following representations:

Word Embedding $\mathbf{e_i}$. This distributed representation of token t_i at position i is pre-trained on a large corpus to capture its semantic and syntactic properties [14].

Position Embedding $\mathbf{d_i}$. This vector encodes the relative distance from the token t_i to the candidate t_{i_c}. This embedding matrix is initialized randomly.

Dependency Vector $\mathbf{g_i}$. The size of this vector corresponds to the number of considered dependencies[1]. If a dependency of a given type is found between t_i and t_{i_c}, the corresponding value is set to 1.

Chunk Embedding $\mathbf{q_i}$. This vector encodes the type of syntactic chunk containing the token t_i, using a BIO encoding scheme: the chunks are computed by a *chunker*[2] from the syntactic tree provided by Stanford CoreNLP. This embedding matrix is initialized randomly.

A convolution layer is applied to the input matrix $\mathbf{X}_c$, made of multiple filters of different sizes. A global max-pooling is performed to get a single value for each filter. This provides a representation of the candidate in its local context, learned by the convolutional neural network. This local representation $\mathbf{f_{softmax}} = [\mathbf{f_{pooling}}]$ is then fed into a softmax layer for computing the probability distribution of the different event classes for the candidate. Finally, the highest probability class $\hat{y}_c$ is taken as prediction. To improve generalization, a dropout is applied between the embedding and the convolutional layers.

[1] We use the basic dependencies provided by par Stanford CoreNLP [13].

[2] https://github.com/mgormley/concrete-chunklink.

2.2 From Local to Global Model

As mentioned in Sect. 2, our objective is to improve the performance of our local model by integrating a representation of a more global context. Moreover, we propose to generate such global representation in connection with our target task by using bootstrapping: we first apply the $\text{CNN}_{\mathbf{local}}$ model presented above to a document. The prediction $\hat{y}_c$ for each token is then extracted and aggregated at a given level of context through *sum-pooling*, leading to a histogram of the detected event types used as the representation $\mathbf{f_{global}}$ of the global context.

Two main factors have to be defined for implementing this approach: the level of context to take into account and the place in the neural network where the representation of this context is integrated. Three levels are considered for the first factor: sentence-wide (*sentence*), a three sentence window centered on the current sentence (*wide*) or document-wide (*doc*). We use the following notation to refer to these three aggregation levels: $\mathbf{f_{global}} = \mathbf{f_{[doc/wide/sentence]}}$. Concerning the second factor, the global context representation can be integrated by concatenation either to the input matrix $\mathbf{X}_c$ by redefining $\mathbf{x_i} = [\mathbf{e_i}, \mathbf{d_i}, \mathbf{g_i}, \mathbf{q_i}, \mathbf{f_{global}}]$ or before the softmax layer: $\mathbf{f_{softmax}} = [\mathbf{f_{pooling}}, \mathbf{f_{global}}]$. Finally, 6 model configurations can be distinguished by choosing the aggregation and integration levels, with the following notation: $\text{CNN}_{\text{[doc/wide/sentence]-[input/softmax]}}$.

3 Experiments and Evaluation

Parameters and Resources. In our experiments, we use the 300 dimension word embeddings pre-trained on Google News using *word2vec* that we modify during training. The size of the chunk and position embeddings is set to 50 and the dropout probability to 0.8, based on preliminary experiments. For each window size (2, 3, 4, 5), 150 filters are used. We apply a hyperbolic tangent non-linearity to the resulting 600 filters. Following [8], our models are trained by stochastic gradient descent (SGD) using the Adadelta optimizer, a gradient clipping of the *l2* norm equal to 3 and a mini-batch size set to 50. The number of epochs is determined by early stopping on the development set. The results are averaged micro F1 scores, computed by the TAC 2017 scorer, on 10 runs.

Our training set is built by merging the DEFT_RICH_ERE_R2_V2 (LDC2015E68), DEFT_RICH_ERE_V2 (LDC2015E29) and TAC 2015 (LDC2017E02) datasets. Our development set comes from the TAC 2016 Event Nugget campaign (LDC2017E02) and we test our model on the data of the TAC 2017 Event Nugget campaign (LDC2017E02). Starting from TAC 2016, the datasets are only focused on the most difficult event types, which reduces the number of possible labels from 38 to 19. The datasets also contain few occurrences of mentions annotated with multiple distinct events types. Since most of these cases correspond to one configuration among three – *Attack/Die*, *Transfer-Money/Transfer-Ownership*, *Attack/Injure* – we introduce 3 new hybrid event types to avoid dealing with a multi-label classification task. We train our model with 42 classes (*other* class and hybrid classes included) but we skip the predictions of the removed types during validation and test. Similarly, the global

Table 1. Performance on the TAC 2016 development set depending on the aggregation level. Results are averaged over 10 runs. ‡ indicates models that are significantly better than $\text{CNN}_{\text{local}}$ ($p < 0.01$ for a bilateral t-test over the 10 runs)

methods	P	R	F
$\text{CNN}_{\text{doc-input}}$	**52.71**	47.95	**50.2** ‡
$\text{CNN}_{\text{wide-input}}$	52.00	47.6	49.69
$\text{CNN}_{\text{sentence-input}}$	49.83	49.49	49.66
$\text{CNN}_{\text{local}}$	46.42	**52.04**	49.06
$\text{CNN}_{\text{doc-input-gold}}$	**54.85**	51.02	**52.83** ‡
$\text{CNN}_{\text{sentence-input-gold}}$	54.21	47.58	50.68 ‡

Table 2. Performance on the TAC 2017 test set. † indicates ensemble models. ‡ indicates in the lower part of the table models that are significantly better than $\text{CNN}_{\text{local}}$ ($p < 0.01$ for a bilateral t-test over the 10 runs)

Methods	max			average over 10 runs		
	P	R	F	P	R	F(std)
BiLSTM CRF (Jiang) †	**56.83**	**55.57**	**56.19**	-	-	-
BiLSTM-SMO (Makarov) †	52.16	48.71	50.37	-	-	-
CNN (Kodelja)	54.23	46.59	50.14	-	-	-
$\text{CNN}_{\text{local}}$	52.21	49.55	50.84	51.90	48.92	50.36 (0.33)
$\text{CNN}_{\text{doc-input}}$	**59.13**	45.37	51.34	**58.07**	45.43	50.95 (0.41) ‡
$\text{CNN}_{\text{doc-softmax}}$	52.87	**50.35**	**51.58**	53.12	**49.61**	**51.30**(0.22) ‡
$\text{CNN}_{\text{doc-input\_softmax}}$	55.72	47.08	51.04	57.62	45.09	50.58 (0.49)
$\text{CNN}_{\text{PV-DM}}$	53.20	47.40	50.10	53.54	46.92	49.98 (0.41)

vector only aggregates the predictions from the test types. Finally, the results we present rely on the best normalization of the global context vector for each configuration, namely no normalization for the $\mathbf{f}_{[\textbf{wide/sentence}]}$ vectors while the $\mathbf{f}_{\textbf{doc}}$ was reduced and centered prior to training.

Influence of the Aggregation Level. Our first experiments concern the aggregation level used for the global representation. Aggregating the predictions at the sentence level could help to reduce intra-sentence ambiguities while a larger context could be beneficial for inter-sentence ambiguities. Table 1 compares results for different sizes while integrating this representation at the input level.

We observe that each configuration yields an improvement compared to $\text{CNN}_{\text{local}}$ but this improvement is significant only for $\text{CNN}_{\text{doc-input}}$. Since the local model used for building the global representation is not perfect, one possible interpretation of this finding is that errors tend to dilute when the local model is applied to a wider context. We ran a complementary experiment using the gold event mentions to generate the global representation (see the last two lines of Table 1) and observed that the document level aggregation is also the best choice in this configuration, confirming that this level intrinsically leads to a better global representation for the event extraction task.

Comparison to the State-of-the-Art. Our last experiments, reported in Table 2, compare the different options for the integration of the global representation (*input*/*softmax*) to the 3 best models of the event detection track of TAC 2017:

1. **BiLSTM CRF Ensemble:** [6] use an ensemble of 10 BiLSTM combined by a voting strategy. Since their neural models tend to have a good recall at the expense of precision, they combine this ensemble with a Conditional Random Field classifier to improve precision. For the BiLSTM, only word embeddings are used while the CRF use multiple features such as tokens, lemmas, roots, named entities, and POS tags.
2. **BiLSTM-SMO:** [12] introduce a BiLSTM with a softmax margin objective [5]. This objective aggressively penalizes false negatives to counterbalance the scarcity of positive samples in training data. An ensemble of 5 networks is used as well as hybrid types.
3. **CNN:** the model of [9] is similar to our local model, *i.e.* a CNN using word, position and chunk embeddings and syntactic dependencies as inputs. The main difference is the absence of hybrid types for modeling multi-type tokens.

We also compare our approach to the integration of a generic document vector in the local model, following [3] (noted $\mathbf{CNN_{PV-DM}}$). This vector of size 100 is generated using the PV-DM model [10]. Unlike our global representation, it is not specific to the task. We optimize the same integration hyperparameters as for our model, namely the level of integration and the choice of normalization. The best configuration integrates reduced and centered vectors at the softmax level.

It is difficult to compare our contributions to [6,12] since they are ensemble methods while we use a single model approach. [6] is even a rather complex ensemble method based on two different architectures combined with a specific heuristic. Furthermore, only the best score of the two models is available while average scores over several runs are more reliable [19]. However, we can note that $\text{CNN}_{\text{doc-input}}$ and $\text{CNN}_{\text{doc-softmax}}$ significantly outperform not only our local model but also the ensemble method of [12], with an advantage of *softmax* over *input* for integrating the global representation. The breakdown analysis of our gain between trigger span detection (+0.65) and trigger classification (+0.89) indicates that our representation mostly helps to filter out ambiguous non-triggers while marginally improving the classification part. Finally, we can observe that the integration of the representation proposed by [3] leads to a decrease in performance. The absence of correlation between the representation and the task is a possible explanation of this observation.

4 Conclusion and Perspectives

In this article, we propose a new representation to exploit a more global context for event extraction. This method is based on bootstrapping and more specifically on the aggregation of local predictions for building a document representation exploited by a global model. We show on the TAC 2017 evaluation data that integrating such global representation significantly increases the results of our initial state-of-the-art local model and can even outperform a BiLSTM ensemble model. We also show that a document representation linked to the target task is more effective than relying on a general document representation. While this model only exploits the output of an initial model, our work could be extended by integrating richer context representations such as internal representations produced by our initial CNN model or a document representation built on a related task trained from a large set of data following a multi-task perspective.

References

1. Bies, A., et al.: A Comparison of event representations in DEFT. in: fourth workshop on events, pp. 27–36. San Diego, June 2016. http://www.aclweb.org/anthology/W16-1004
2. Chen, Y., Xu, L., Liu, K., Zeng, D., Zhao, J.: Event extraction via dynamic multi-pooling convolutional neural networks. In: 53rd Annual Meeting of the Association for Computational Linguistics and 7th International Joint Conference on Natural Language Processing (ACL-IJCNLP 2015), Beijing, China, pp. 167–176 (2015)
3. Duan, S., He, R., Zhao, W.: Exploiting document level information to improve event detection via recurrent neural networks. In: Eighth International Joint Conference on Natural Language Processing (IJCNLP 2017), Taipei, Taiwan, pp. 352–361 (2017). https://aclanthology.coli.uni-saarland.de/papers/I17-1036/i17-1036
4. Feng, X., Huang, L., Tang, D., Ji, H., Qin, B., Liu, T.: A language-independent neural network for event detection. In: 54th Annual Meeting of the Association for Computational Linguistics (ACL 2016), Berlin, Germany, pp. 66–71 (2016). https://doi.org/10.18653/v1/P16-2011. https://aclanthology.coli.uni-saarland.de/papers/P16-2011/p16-2011
5. Gimpel, K., Smith, N.: Softmax-Margin CRFs: training log-linear models with cost functions. In: The 2010 Annual Conference of the North American Chapter of the Association for Computational Linguistics Human Language Technologies: (NAACL HLT 2010), Los Angeles, California, pp. 733–736 (2010)
6. Jiang, S., Li, Y., Qin, T., Meng, Q., Dong, B.: SRCB Entity Discovery and Linking (EDL) and event nugget systems for TAC 2017. In: Text Analysis Conference (TAC) (2017)
7. Jorge, A.M., Campos, R., Jatowt, A., Nunes, S. (eds.): First Workshop on Narrative Extraction From Text (Text2Story 2018). Grenoble, France (2018)
8. Kim, Y.: Convolutional neural networks for sentence classification. In: EMNLP (2014)
9. Kodelja, D., Besançon, R., Ferret, O., Le Borgne, H., Boros, E.: CEA LIST participation to the TAC 2017 event nugget track. In: Text Analysis Conference (TAC) (2017)
10. Le, Q., Mikolov, T.: Distributed representations of sentences and documents. In: 31st International Conference on International Conference on Machine Learning (ICML 2014), Beijing, China, pp. 1188–1196 (2014)
11. Liao, S., Grishman, R.: Using document level cross-event inference to improve event extraction. In: ACL (2010)
12. Makarov, P., Clematide, S.: UZH at TAC KBP 2017: event nugget detection via joint learning with Softmax-Margin Objective. In: Text Analysis Conference (TAC) (2017)

13. Manning, C.D., Surdeanu, M., Bauer, J., Finkel, J., Bethard, S.J., McClosky, D.: The stanford CoreNLP natural language processing Toolkit. In: 52nd Annual Meeting of the Association for Computational Linguistics (ACL 2014), system demonstrations, pp. 55–60 (2014)
14. Mikolov, T., Sutskever, I., Chen, K., Corrado, G.S., Dean, J.: Distributed representations of words and phrases and their compositionality. In: 26th International Conference on Neural Information Processing Systems (NIPS 2013), Lake Tahoe, Nevada, pp. 3111–3119 (2013)
15. Mitamura, T., Liu, Z., Hovy, E.: Events detection, coreference and sequencing: what's next? overview of the TAC KBP 2017 Event Track. In: Text Analysis Conference (TAC) (2017)
16. Nguyen, T.H., Cho, K., Grishman, R.: Joint event extraction via recurrent neural networks. In: 2016 Conference of the North American Chapter of the Association for Computational Linguistics: Human Language Technologies (NAACL HLT 2016), San Diego, California, pp. 300–309 (2016)
17. Nguyen, T.H., Grishman, R.: Event detection and domain adaptation with convolutional neural networks. In: 53rd Annual Meeting of the Association for Computational Linguistics and 7th International Joint Conference on Natural Language Processing (ACL-IJCNLP 2015), Beijing, China, pp. 365–371 (2015)
18. Nguyen, T.H., Grishman, R., Meyers, A.: New york university 2016 system for KBP event nugget: a deep learning approach. In: Text Analysis Conference (TAC) (2016)
19. Reimers, N., Gurevych, I.: Reporting score distributions makes a difference: performance study of LSTM-networks for sequence tagging. In: 2017 Conference on Empirical Methods in Natural Language Processing (EMNLP 2017), Copenhagen, Denmark, pp. 338–348 (2017)
20. Zhao, Y., Jin, X., Wang, Y., Cheng, X.: Document embedding enhanced event detection with hierarchical and supervised attention. In: 56th Annual Meeting of the Association for Computational Linguistics (ACL 2018), pp. 414–419. Association for Computational Linguistics (2018)

Faster BlockMax WAND with Longer Skipping

Antonio Mallia[1(✉)] and Elia Porciani[2(✉)]

[1] Computer Science and Engineering, New York University, New York, USA
antonio.mallia@nyu.edu
[2] Sease Ltd., London, UK
e.porciani@sease.io

Abstract. One of the major problems for modern search engines is to keep up with the tremendous growth in the size of the web and the number of queries submitted by users. The amount of data being generated today can only be processed and managed with specialized technologies.

BlockMax WAND and the more recent Variable BlockMax WAND represent the most advanced query processing algorithms that make use of dynamic pruning techniques, which allow them to retrieve the top k most relevant documents for a given query without any effectiveness degradation of its ranking. In this paper, we describe a new technique for the BlockMax WAND family of query processing algorithm, which improves block skipping in order to increase its efficiency. We show that our optimization is able to improve query processing speed on short queries by up to 37% with negligible additional space overhead.

Keywords: Top-k query processing · Inverted index · Early termination

1 Introduction

In the past two decades, the amount of data being created has skyrocketed. The key to unlock the full potential of these huge datasets is to make the most of advances in algorithms and tools capable to handle it.

Many parts of the search engine architecture, including data acquisition, data analysis, and index maintenance, are facing critical challenges. Nevertheless, query processing is still the hardest to deal with, since workload grows with both data size and query load. Although hardware is getting less expensive and more powerful every day, the size of the data and the number of searches is growing at an even faster rate. Much of the research and development in information retrieval is, indeed, aimed at improving retrieval efficiency.

While, query processing in search engines is a fairly complex process, most systems appear to process a query by first evaluating a fairly simple ranking function over an inverted index. We focus on improving this initial step, which is responsible for a significant fraction of the overall work.

L. Azzopardi et al. (Eds.): ECIR 2019, LNCS 11437, pp. 771–778, 2019.
https://doi.org/10.1007/978-3-030-15712-8_52

Traversing the index structures of all the query terms and computing the scores of all the postings is the way to evaluate exhaustively a user query. Unfortunately, the cost of each query increases linearly with the number of documents, making it very expensive for large collections. To overcome this problem, many researchers have proposed so-called early-termination techniques, for finding the top-k ranked results without computing or retrieving all posting scores.

In this work, we focus on such techniques for improving query processing efficiency without degrading effectiveness to rank K (known as *safe-to-rank*).

Our Contributions. We list here our main contributions.

1. We propose an optimization for the BlockMax WAND (BMW) family of algorithms, which exploits particular sequences of block max scores in order to perform longer skipping.
2. We embed an additional data structure that stores precomputed skips in order to overcome the run time search overhead introduced by compressing the block boundaries.

2 Background and Related Work

We first briefly explain the studied methods for both index compression and query processing. We refer to the referenced papers for more details that are omitted due to space restriction.

Index Organization. We consider a collection of documents indexed in an inverted index [14]. Each term occurring in the collection contributes a list of IDs of the documents containing it (usually along with respective document frequencies or other data used to rank documents), called a posting list. As a requirement for *Document-at-a-Time* (*DAAT*) query processing, which scans postings lists concurrently, the document IDs must be sorted in the ascending order. This allows us to compress it efficiently making it possible to keep the entire index in memory.

$\mathsf{NextGEQ_t(d)}$ is an operator which returns the smallest document ID in the inverted list of term t that is greater than or equal to d. A fast implementation of the function $\mathsf{NextGEQ_t(d)}$ is crucial for the efficiency of this process and it is strictly dependent on the compression algorithm used. One widely adopted solution to efficiently implement $\mathsf{NextGEQ_t(d)}$ operator is to divide each list into blocks that are individually encoded with the chosen encoding method.

Block-based encoding is not optimal when skipping is performed among the inverted lists, because it requires to decode an entire block to access a single element. This reflects the access pattern of early termination algorithms such as BMW, where entire segments of the posting lists are skipped. For this reason, we have chosen to use Partitioned Elias-Fano [9] as compression technique, which provides random access to compressed elements without decoding the whole sequence. Partitioned Elias-Fano has been recently proposed as an improvement of Elias-Fano, initially introduced by Vigna [12], in order to exploit the local clustering that inverted lists usually exhibit, resulting in reduced space usage.

Query Processing. Several algorithms have been proposed to accomplish exhaustive evaluation over an inverted index to find top-k documents. Broder et al. [1] introduced for the first time WAND, a solution which exploits an augmented index with maximum scores for each term of the posting lists. The algorithm maintains a top-k priority queue of the scores of the evaluated document, such that a minimum threshold needs to be met by a document to enter the top-k. The idea behind WAND is to access the posting lists with an iterator keeping the postings ordered by ID. Each iteration of the algorithm sorts the terms by the current ID of the associated iterator and adds up the maximum scores of the terms until the threshold is reached. This allows to find the minimum document ID which has to be evaluated, allowing us to safely ignore all the preceding ones. BlockMax WAND [6] further improves WAND by better estimating the upper bounds by splitting a posting list into fixed-sized blocks and storing the maximum score per block. Additionally, BlockMax WAND refines the score upper bound of a candidate ID found by WAND by using these upper bounds. This operation is fast, as it involves no block decompression. Whenever the maximum score estimation would not be sufficient to enter the top-k, we can skip all document IDs belonging to the intersection of the current blocks involved, translating to a move to the minimum document of the current block boundaries.

Variable BlockMax WAND (VBMW) [8] generalizes BMW by allowing variable lengths of blocks. More precisely, it uses a block partitioning such that the sum of differences between maximum scores and individual scores is minimized. This results in better upper bound estimation and more frequent document skipping with the downside of a computational overhead at index building time in order to compute the optimal block partitioning.

3 Our Contribution

The efficiency of early termination algorithms is closely related to the number of documents skipped during index traversing. In the case of the BlockMax WAND family, the greatest contribution to skipping happens after the block upper bound is computed, specifically when the aggregated score does not reach the threshold and a move to the next block boundary can be safely performed. Advancing the term iterator to the following minimum block boundary is driven by the intuition that no documents in the current blocks can exceed the upper bound estimation. On the other hand, this choice is not guaranteed to be optimal.

We introduce a modification of the BMW algorithm which uses a new strategy to advance the term iterator farther than the current block boundaries, which results in longer but safe document skipping. The observation behind our strategy is that when an iterator is updated, if the max score of the block after the boundary does not increase, then we can state that the sum of the blocks upper bound will still not be greater than the threshold. The next iteration will then perform unnecessary computation until a new block skipping happens again. Our solution consists of identifying the next document ID to move at, by progressively skipping entire blocks until one with greater block max score is found. Algorithm 1 depicts how the new document ID is chosen.

Our first contribution is to implement this "longer skipping" strategy in both BMW and VBMW algorithms, where we named our variations BMW-LS and VBMW-LS respectively. To the best of our knowledge, there is no evidence in literature of BMW additional data being compressed, in the way it is done for VBMW. Also, we experimented with both compressed version of the algorithms, named here C-BMW and C-VBMW (refer to [8] for the details) for the ones using the unmodified skipping strategy; C-BMW-LS and C-VBMW-LS for the ones using the new skipping strategy.

Query time search for a longer skip, although reduces the fully evaluated documents, is expensive from a computational point-of-view because of the compressed blocks information. For this reason, we introduced an alternative approach of the proposed solution which precomputes the skip size at index build time, storing the information interleaved with the blocks information. Considering that we need to store only one additional information – the distance in number of blocks to the last one that we can skip – we have chosen to encode it with a fixed number of bits. This represents a limitation for the maximum number of subsequent blocks that can be skipped, but we have experimentally observed that after a certain amount of bits per element there is no performance advantage. In our implementation and for the examined datasets, we used 3 bits per element, so that up to 7 blocks skips can be encoded. We named this implementation Precomputed Longer Skipping (PLS).

Algorithm 1. Find next doc ID

```
 1: next_docid ← MAX_DOCID
 2: for all t ∈ terms do
 3:     block ← t.block
 4:     s ← block.score
 5:     docid ← block.boundary
 6:     while block.score <= s do
 7:         docid ← block.boundary
 8:         block ← block.next
 9:     end while
10:     if docid < next_docid then
11:         next_docid ← docid
12:     end if
13: end for
```

Fig. 1. Example of Longer Skipping

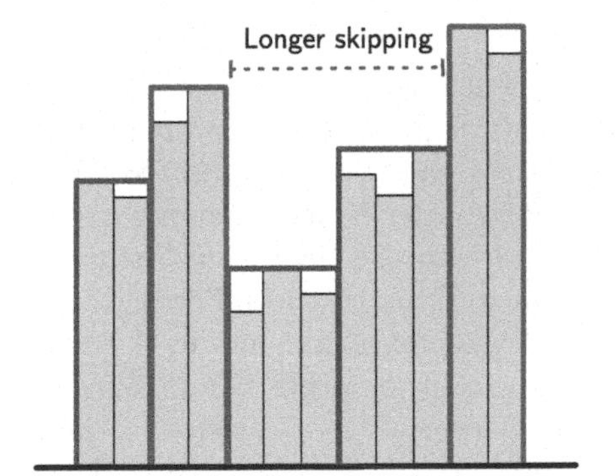

4 Experimental Results

Testing details. All the algorithms are implemented in C++14 and compiled with GCC 7.3.0 with the highest optimization settings. The tests are performed on a machine with 8 Intel Core i7-4770 Haswell cores clocked at 3.40 GHz with 32 GiB

RAM running Linux 4.15.0. The indexes are saved to disk after construction and memory-mapped to be queried so that there are no hidden space costs due to loading of additional data structures in memory. Before timing the queries we ensure that the required posting lists are fully loaded in memory. All timings are measured taking the results with minimum value of five independent runs. All times are reported in milliseconds.

The source code is available[1] for the reader interested in further implementation details or in replicating the experiments.

Datasets. We performed our experiments on the following standard datasets (Table 1).

- Gov2 is the TREC 2004 Terabyte Track test collection consisting of 25 million .gov sites crawled in early 2004; the documents are truncated to 256 kB.
- ClueWeb09 [2] is the ClueWeb 2009 TREC Category B collection consisting of 50 million English web pages crawled between January and February 2009.

For each document in the collection the body text was extracted using Apache Tika[2] and the words lowercased and stemmed using the Porter2 stemmer; no stopwords were removed. The doc IDs were assigned according to the lexicographic order of their URLs [11].

	Gov2	ClueWeb09
Documents	24622347	50131015
Terms	35636425	92094694
Postings	5742630292	15857983641

Table 1. Statistics for the test collections

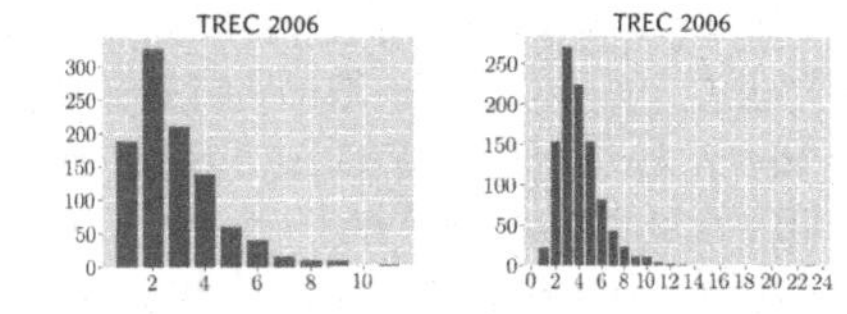

Fig. 2. Query length distribution

Queries. To evaluate the speed of query processing we use the TREC 2005 and TREC 2006 Terabyte Track Efficiency Task, drawing only queries whose terms are all in the collection dictionary. From those sets of queries, we randomly select 1000 queries for each length. Figure 2 depicts the query distribution, which clearly shows that short queries dominate.

We have used BMW and VBMW as baseline with 40 elements per block in average, in both their uncompressed and compressed form. All the results, including the query times in milliseconds for the baselines and our proposed solutions, are presented in Table 2.

In our experiments, the proposed optimization improves short queries (from 2 to 4 terms) with negligible performance degradation for longer queries. This has shown to be true for both Gov2 and ClueWeb09 datasets and without noticeable

[1] https://github.com/pisa-engine/pisa/tree/ecir19-ls.
[2] http://tika.apache.org.

Table 2. Query times (in ms) of different algorithms for several query lengths.

		Gov2					ClueWeb09				
		2	3	4	5	6+	2	3	4	5	6+
TREC 2005	BMW	1.22	3.07	4.68	7.43	16.73	4.63	11.37	16.68	25.72	55.99
	VBMW	0.99	1.91	2.69	4.21	**9.18**	3.17	6.39	8.92	14.46	32.04
	BMW-LS	0.93	2.88	4.61	7.41	17.40	2.92	10.20	16.76	26.84	60.42
	VBMW-LS	**0.78**	**1.77**	**2.63**	**4.20**	9.23	**2.18**	**5.66**	**8.57**	**14.44**	**31.95**
	C-BMW	1.33	3.39	5.13	8.27	18.26	5.19	12.78	19.09	29.19	63.32
	C-VBMW	1.10	2.08	2.93	**4.60**	**10.16**	3.53	6.97	9.86	**16.06**	36.26
	C-BMW-LS	1.38	3.42	5.32	8.26	18.74	5.36	13.11	19.42	29.93	65.08
	C-VBMW-LS	1.14	2.15	3.04	4.75	10.21	3.67	7.34	10.29	16.46	36.48
	C-BMW-PLS	1.12	3.10	5.00	8.00	18.77	3.89	11.19	18.41	29.58	65.80
	C-VBMW-PLS	**0.94**	**1.95**	**2.93**	4.71	10.17	**2.68**	**6.30**	**9.52**	16.07	**36.01**
TREC 2006	BMW	1.11	3.58	6.24	10.03	23.85	3.46	11.33	19.82	32.37	74.13
	VBMW	0.78	2.26	3.58	5.55	**12.88**	2.28	6.80	11.64	18.68	42.17
	BMW-LS	0.85	3.22	6.08	9.98	24.86	2.50	10.42	19.77	33.28	80.62
	VBMW-LS	**0.58**	**2.05**	**3.46**	**5.49**	12.92	**1.66**	**6.25**	**11.35**	**18.59**	**42.04**
	C-BMW	1.22	3.90	6.95	11.09	26.34	3.80	12.48	22.27	35.83	82.96
	C-VBMW	0.89	2.49	3.96	6.08	14.60	2.51	7.42	12.86	**20.40**	**46.87**
	C-BMW-LS	1.28	4.02	7.19	11.17	27.07	3.96	12.91	22.85	37.02	85.59
	C-VBMW-LS	0.91	2.57	4.06	6.23	14.88	2.61	7.68	13.21	20.99	47.48
	C-BMW-PLS	1.02	3.57	6.56	10.75	26.52	3.09	11.57	21.92	36.52	85.45
	C-VBMW-PLS	**0.72**	**2.34**	**3.86**	**6.07**	**14.54**	**1.98**	**6.88**	**12.51**	20.46	47.33

differences for TREC 2005 and TREC 2006. Because of its pluggability, based on the query length this optimization can be enabled at query time with the intent of maximizing query performance. Although there is a noticeable improvement for all short queries, the maximum speedup is observable using ClueWeb09 on TREC 2005 queries where BMW-LS performs 37% faster than BMW and VBMW-LS reduces by 31% the time spent to process the query. We have chosen to show the results for BMW because it could be a better choice in the case where block-based compression algorithms [13] are used and because it has a simpler and faster offline build process where compared to VBMW; our optimization is orthogonal to any further improvements built on top of BMW [3–5,7,10].

In contrast, it is interesting to notice that for the compressed version of the algorithms the run time optimization does not lead to any improvements, but actually results in a slower execution. The precomputed version of our optimization overcomes this issue and obtains almost the same gain of the run time version for the uncompressed BMW with a negligible overhead in index size (less than 1% of the total index size). PLS optimization is omitted for the uncompressed variants, considering that linear scan does not suffer the decompression overhead.

5 Conclusions

In this paper, we demonstrated the applicability of a longer skipping strategy to both BMW and VBMW, which results in marked benefits of processing time for short queries. We proposed two different variations. The former evaluates at query time the size of the possible skips and finds its best applicability with uncompressed blocks score information, while the latter precomputes and stores into the index this information which is ideal when blocks scores are compressed. Our extensive experiment analysis shows that both strategies improve on their direct competitors by up to 37%, with a negligible additional space usage in case of precomputed skips.

Finally, in the future, we also want to study the combination of our algorithm to existing and new threshold estimation techniques to study how those can be beneficial when combined with our longer skipping strategy.

Acknowledgments. Antonio Mallia's research was partially supported by NSF Grant IIS-1718680 "Index Sharding and Query Routing in Distributed Search Engines".

References

1. Broder, A.Z., Carmel, D., Herscovici, M., Soffer, A., Zien, J.: Efficient query evaluation using a two-level retrieval process. In: Proceedings of the 12th International Conference on Information and Knowledge Management, pp. 426–434 (2003)
2. Callan, J., Hoy, M., Yoo, C., Zhao, L.: Clueweb09 data set (2009). http://lemurproject.org/clueweb09/
3. Daoud, C.M., de Moura, E.S., da Costa Carvalho, A.L., da Silva, A.S., de Oliveira, D.F., Rossi, C.: Fast top-k preserving query processing using two-tier indexes. Inf. Process. Manage. **52**, 855–872 (2016)
4. Daoud, C.M., de Moura, E.S., de Oliveira, D.F., da Silva, A.S., Rossi, C., da Costa Carvalho, A.L.: Waves: a fast multi-tier top-k query processing algorithm. Inf. Retr. J. **20**, 292–316 (2017)
5. Dimopoulos, C., Nepomnyachiy, S., Suel, T.: Optimizing top-k document retrieval strategies for block-max indexes. In: Proceedings of the 6th ACM International Conference on Web Search and Data Mining, pp. 113–122 (2013)
6. Ding, S., Suel, T.: Faster top-k document retrieval using block-max indexes. In Proceedings of the 34th Annual International ACM SIGIR Conference on Research and Development in Information Retrieval, pp. 993–1002 (2011)
7. Kane, A., Tompa, F.W.: Split-lists and initial thresholds for wand-based search. In: Proceedings of the 41st Annual International ACM SIGIR Conference on Research and Development in Information Retrieval, pp. 877–880 (2018)
8. Mallia, A., Ottaviano, G., Porciani, E., Tonellotto, N., Venturini, R.: Faster block-max WAND with variable-sized blocks. In: Proceedings of the 40th Annual International ACM SIGIR Conference on Research and Development in Information Retrieval, pp. 625–634 (2017)
9. Ottaviano, G., Venturini, R.: Partitioned Elias-Fano indexes. In: Proceedings of the 37th Annual International ACM SIGIR Conference on Research and Development in Information Retrieval, pp. 273–282 (2014)
10. Rojas, O., Gil-Costa, V., Marin, M.: Efficient parallel block-max wand algorithm. In: Proceedings of the 19th International Conference on Parallel Processing, pp. 394–405 (2013)
11. Silvestri, F.: Sorting out the document identifier assignment problem. In: Proceedings of the 29th European Conference on IR Research, pp. 101–112 (2007)

12. Vigna, S.: Quasi-succinct indices. In: Proceedings of the 6th ACM International Conference on Web Search and Data Mining, pp. 83–92 (2013)
13. Yan, H., Ding, S., Suel, T.: Inverted index compression and query processing with optimized document ordering. In: Proceedings of the 18th International Conference on World Wide Web, pp. 401–410 (2009)
14. Zobel, J., Moffat, A.: Inverted files for text search engines. ACM Comput. Surv. **38**(2), 6 (2006)

A Hybrid Modeling Approach for an Automated Lyrics-Rating System for Adolescents

Jayong Kim and Mun Y. Yi(✉)

Graduate School of Knowledge Service Engineering, Korea Advanced Institute of Science and Technology, Daejeon, Republic of Korea
{kjyong,munyi}@kaist.ac.kr

Abstract. The South Korean government operates human-based lyrics-rating systems to reduce adolescents' exposure to inappropriate songs. In this study, we developed lyrics classification models for an automated lyrics-rating system for adolescents. There are two kinds of inappropriate lyrics for adolescents: (1) lyrics with inappropriate words and (2) lyrics with inappropriate content based on the semantic context. To tackle the first issue, we propose $\log CD_{\alpha}$ as a method for generating a lexicon of inappropriate words. It attained the highest performance among the lexicon-based filtering methods examined. Further, to deal with the second issue, we propose a hybrid classification model that combines $\log CD_{\alpha}$ with an RNN based model. The hybrid model composed of a 'lexicon-checking model' and a 'context-checking model' achieved the highest performance among all of the models examined, highlighting the effectiveness of combining the models to specifically target each of the two types of inappropriate lyrics.

Keywords: Lyrics classification · Offensive language detection · RNN

1 Introduction

To reduce the exposure of adolescents to lyrics that contain depictions of profanity, violence, sex, and/or substance abuse, the South Korean government operates human-based lyrics-rating systems. The Ministry of Gender Equality and Family (MOGEF) classifies lyrics as either clean or inappropriate for adolescents, and they are prohibited from accessing music records with lyrics that are considered as inappropriate.

The human-based lyrics-rating system includes the basic work of the monitoring staff, an initial review of the committee every other week, and a main review of the committee once a month [6]. Because the lyrics-rating system is a post deliberation, adolescents can still be exposed to inappropriate lyrics, particularly if the deliberation process takes time. Furthermore, a number of experts and resources are required continuously for the operation of the current lyrics-rating systems.

Automation of this rating process could be a valuable solution by saving time and resources. The purpose of this study is to develop an effective lyrics classification

L. Azzopardi et al. (Eds.): ECIR 2019, LNCS 11437, pp. 779–786, 2019.
https://doi.org/10.1007/978-3-030-15712-8_53

model for an automated lyrics-rating system. We believe that the developed models and the approaches could be easily applicable to the lyrics-rating systems of other countries.

2 Related Work

Chin et al. [3] studied an inappropriate lyrics classification model for the first time. They reported that there were two main types of inappropriate lyrics for adolescents: (1) 'lyrics that contain inappropriate words for adolescents' (Type I) and (2) 'lyrics that do not contain inappropriate words, but contain explicit content based on the context' (Type II). Although they noticed that there were two types of inappropriate lyrics, the authors did not develop a specific model that particularly dealt with them. In the present study, we developed a model that focuses on these two types of inappropriate lyrics.

It is easier to classify Type I lyrics than Type II lyrics because we only need to check for the presence of inappropriate words. The basic approach for tackling this issue is lexicon-based filtering (keyword matching) [16, 18]. However, in the absence of well-defined lexicon data, it is difficult to apply this approach. Hence, automatic generation of a lexicon of inappropriate words for adolescents is a viable, practical solution. The approach has only been studied in a limited scope for social media content [1, 12]. In the present study, we expanded it to lyrics and examined its applicability for the classification of Type I lyrics.

In order to classify Type II lyrics as inappropriate, the semantic context of the words needs to be grasped. A lexicon-based filtering approach cannot capture the context of words because it only checks for their presence in the lexicon and does not consider the other words. To understand the semantic context of words, recurrent neural network (RNN)- and convoluted neural network (CNN)-based sequential data processing models have been studied [7, 19]. Lyrics that contain even one single profanity can be classified as inappropriate according to the lyrics-rating system of MOGEF. These kinds of lyrics are Type I lyrics and RNN- or CNN-based model might not be suitable for them. These types of models can lose information on the existence of inappropriate words because they make low dimensional vectors of lyrics, not bag-of-words vectors. Therefore, to compensate for this weakness, we propose a hybrid classification model, which is composed of a 'lexicon-checking model' and a 'context-checking model' for classifying Type I and Type II inappropriate lyrics, respectively.

3 The Proposed Model

3.1 For Type I Lyrics

Automatic Lexicon Generation. Type I lyrics contain inappropriate words for adolescents, for example, *'I don't give a f***'*. This can be verified by checking whether the lyrics contain particular words in the lexicon of inappropriate words for adolescents. Filtering methods for feature selection, such as log odds ratio (LOR), correlation coefficient (CC), and supervised word weighting schemes such as relevant frequency (RF) [8] and LogCD [4] can be used for the automatic generation of a lexicon of inappropriate words. According to [15], the score that these methods give to a term t_k can be represented by the number of positive-class-documents with t_k and the number of negative-class-documents with t_k.

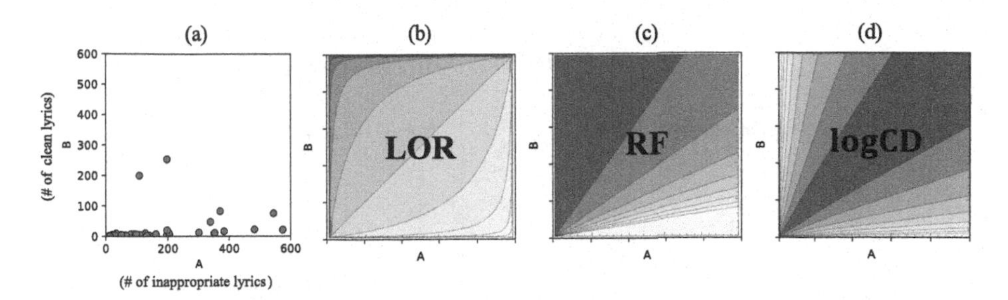

Fig. 1. (a) A scatter plot of 539 profanity words collected from [13, 14]. A axis is the number of inappropriate lyrics with a profanity word W_k, B axis is the number of clean lyrics with a profanity word W_k. (b)–(d) The scoring tendencies of the various methods according to A and B. The brighter the color in the contour plot, the higher the score.

In Fig. 1(a), the profanities in the lyrics show a clear pattern in that most of the points appear near the A axis because lyrics with even a single inappropriate word should be classified as inappropriate. Therefore, methods that give a high score to the words near the A axis are suitable for the automatic generation of a lexicon of inappropriate words for adolescents.

$$\log \mathrm{CD}_\alpha = \log\left(\frac{\frac{A+\alpha}{\#\ of\ inappropriate\ lyrics}}{\frac{B+\alpha}{\#\ of\ clean\ lyrics}}\right) \tag{1}$$

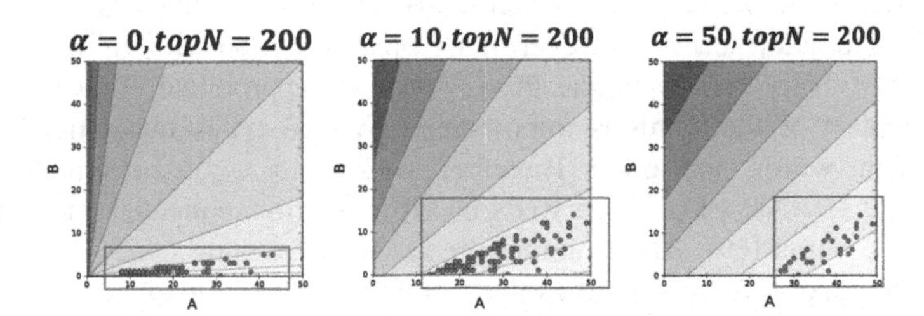

Fig. 2. Selected inappropriate words for adolescents with various α of **logCD**$_\alpha$.

We modified logCD [4] to generate a more effective lexicon (see logCD$_\alpha$ (1)). Absolute values in logCD were removed to make it assign a high score to words near the A axis but not the B axis. Further, we added α to the numerator and denominator. Figure 2 shows that as α of logCD$_\alpha$ increases, words with a low document frequency are gradually excluded from the lexicon even though they are near to the A axis. On the other hand, as α increases, words with a high document frequency are included in the lexicon of inappropriate words, even when they appeared among the clean lyrics. As words appear more frequently, the more likely they are to appear in the clean lyrics. $\log CD_\alpha$ considers this pattern and can give tolerance to the words by increasing α.

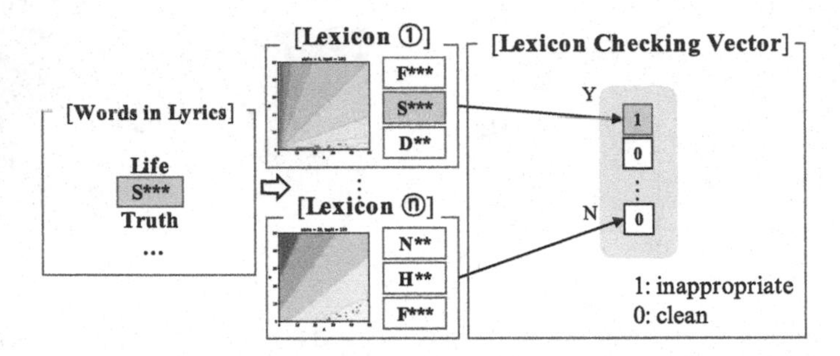

Fig. 3. The process of generating the lexicon-checking vector

The Lexicon-Checking Vector. We can create various lexicons by changing the α of $\log CD_\alpha$ and the number of words included in the lexicon. To make lexicon-checking vectors, lexicon-based filtering is carried out for each lexicon (see Fig. 3). If lyrics contain words in any one of the lexicons, the lyrics are classified as inappropriate. After conducting this process for all lexicons, predictions based on each lexicon are carried out to determine whether the lyrics are appropriate or not. We collected the top k predictions of the lexicons which performed the best when validating the data. This vector played the role of a 'lexicon-checking vector' in the hybrid model.

3.2 For Type II Lyrics

The Context-Checking Vector. Type II lyrics, for example, contain sentences like *'Take my skin off, cut out my belly'*. When we look at the overall expression, it might be inappropriate for adolescents. However, if we break the expression up into words like 'cut', 'out', or 'belly', it seems like these words are appropriate. That is to say, the semantic context of the words is important in Type II lyrics rather than the literal meaning of the words themselves. Therefore, we directly applied Hierarchical Attention Networks (HAN), which is an RNN-based model for sequential and hierarchical processing of words [19]. After training the HAN, the output value of the last layer before the Softmax layer was used as the 'context-checking vector' of the lyrics.

3.3 The Hybrid Approach for Type I and Type II Lyrics

To consider both Type I and Type II lyrics, we designed a hybrid classification model of inappropriate lyrics. The 'lexicon-checking vector' and the 'context-checking vector' of each lyric were concatenated into a single vector, after which a classifier learned these vectors.

4 Experiments and Results

The lyrics-rating results of MOGEF during 2010.1–2017.8 were collected from the MOGEF website [11]. Inappropriate lyrics from 7,468 songs and clean lyrics from 62,609 songs that did not have an 'Adults Only' tag on the music streaming sites were crawled from various lyrics databases. The class imbalance of the dataset was intended to reflect a real-world lyrics-rating system. The dataset was split into training data, validation data, and test data with a ratio of 8:1:1. All of the hyper-parameters were tuned using the validation data. α was tuned between 0 and 100 and the number of words in a lexicon (topN) was varied between 10 and 400. The lyrics were tokenized using the Part-Of-Speech taggers from the NLTK and Komoran packages. Because of the class imbalance, the F1 score and area under the precision-recall curve (PR AUC) were used as performance measures [5, 17].

4.1 Automatic Lexicon Generation

We conducted lexicon-based filtering for each lexicon generated by the diverse methods. If the lyrics contained any single token in a generated lexicon, it was classified as inappropriate. Table 1 reports that $\log CD_{\alpha}$ attained the highest performance for lexicon-based filtering as it used only 25 words. This is 0.1% of the total number of words. In addition, $\log CD_{\alpha}$ used fewer words than the existing methods that showed comparable performance. It suggests that $\log CD_{\alpha}$ is efficient for generating an effective lexicon of inappropriate words for adolescents.

Table 1. The best results of the lexicon-based filtering of each method varying topN.

Method	topN	F1
$\log CD_{\alpha=20,k=1}$	25	0.7562
Log Odds Ratio (LOR)	300	0.7368
$\log CD_{without\_absolute}$	400	0.6779
Relevant Frequency (RF) [8]	400	0.6779
Mubarak et al. [12]	21133	0.5330
Correlation Coefficient (CC)	10	0.5251
Man-made dictionary [13, 14]	539	0.4898

4.2 Hybrid Classification Model

We compared the proposed model with the bag-of-words model (TF-IDF), the document embedding model (Doc2Vec [9]), the topic modeling model (LDA [2]), and the lyrics classification model proposed by Chin et al. [3]. We tested various classifiers, namely AdaBoost, Bagging, and k-nearest neighbor (KNN), and we reported the KNN here because all of them showed similar results.

HAN produced the highest performance among the non-hybrid models (Table 2). However, the performance difference between $\log CD_{\alpha=20,k=1}$ and HAN was 1%.

Checking the presence of 25 words, $\log CD_\alpha$ achieved a comparable performance at a lower cost when compared to the deep learning model.

Table 2. The experimental results of the compared models

Model	F1	PR AUC
Doc2Vec [9] + KNN	0.5066	0.5986
TF-IDF + KNN	0.5299	0.5481
LDA [2] + KNN	0.6507	0.6720
Chin et al. [3]	0.7478	0.7774
HAN [19]	0.7665	0.8249
Hybrid(Doc2Vec + HAN) + KNN	0.7744	0.8117
Hybrid($\log CD_{\alpha=20,k=1}$ + HAN) + KNN	0.7809	0.8275
Hybrid($\log CD_{\alpha=20,k=100}$ + HAN) + KNN	**0.8049**	**0.8600**

The hybrid classification model based on HAN and $\log CD_\alpha$ showed the highest performance among all of the models compared (Table 2). It outperformed its sub-models: $\log CD_\alpha$ and HAN. The hybrid model showed a higher performance when the size of the 'lexicon-checking vector' was 100 ($\log CD_{\alpha=20,k=100}$) compared to when it was 1 ($\log CD_{\alpha=20,k=1}$), meaning that using various lexicons could improve its performance.

The improvement achieved by the hybrid model might have been simply due to combining the multiple models. To check for this, we made many hybrid models with various combinations of single models. However, even the best other combination, Hybrid(HAN + Doc2Vec), showed little improvement or less performance than its sub-models, which indicates that the proposed combination of models specifically targeting Type I and Type II lyrics was synergistic.

5 Conclusion

Automating a lyrics-rating system can save time and resource relative to the current human-based system. In this research, we developed a hybrid model of lyrics classification for an automated lyrics-rating system. Extending the extant research, we focused on two types of inappropriate lyrics for adolescents.

To classify Type I lyrics, we first found the pattern of inappropriate words for adolescents. From the pattern, we developed insight into what kinds of scoring methods might be suitable for finding inappropriate words. This approach was then applied to other text classification areas where the class of the text depended on the presence of specific words in the content, such as profanity filtering.

We proposed $\log CD_\alpha$ as an automatic generation method for lexicons of inappropriate words, which can be further used to generate domain-specific lexicons regardless of language. In addition, $\log CD_\alpha$ showed the highest performance with the fewest words for lexicon-based filtering compared to existing methods, showing that it can be applied to areas where both time and resource savings are important, such as real-time inappropriate content detection with a large amount of data.

We designed a hybrid classification model that considers both Type I and Type II lyrics by learning the 'lexicon-checking vector' and the 'context-checking vector'. This hybrid model showed the highest performance among all of the models we examined. As the hybrid modeling approach considers both the lexicon and the context together, its performance could be assessed in other document classification tasks in which both checking the lexicon and determining the context are required.

References

1. Abozinadah, E.A., Jones Jr., J.H.: A statistical learning approach to detect abusive Twitter accounts. In: Proceedings of the International Conference on Compute and Data Analysis, pp. 6–13. ACM, New York (2017)
2. Blei, D.M., Ng, A.Y., Jordan, M.I.: Latent dirichlet allocation. J. Mach. Learn. Res. **3**, 993–1022 (2003)
3. Chin, H., Kim, J., Kim, Y., Shin, J., Yi, M.Y.: Explicit content detection in music lyrics using machine learning. In: 2018 IEEE International Conference on Big Data and Smart Computing (BigComp), pp. 517–521. IEEE (2018)
4. Fattah, A.: New term weighting schemes with combination of multiple classifiers for sentiment analysis. Neurocomputing **167**, 434–442 (2015)
5. He, H.: Garci: learning from imbalanced data. IEEE Trans. Knowl. Data Eng. **21**(9), 1263–1284 (2009)
6. Hwang, S., Choi, W., Yoon, H.: Study on how to improve operating system of the commission on adolescents Protection (청소년보호위원회 운영체계 발전방안 연구).. http://www.prism.go.kr/homepage/origin/retrieveOriginDetail.do?cond_organ_id=1382000&research_id=1382000-201300016. Accessed 01 Sept 2017
7. Kim, Y.: Convolutional neural networks for sentence classification. In: Proceedings of the 2014 Conference on Empirical Methods in Natural Language Processing, pp. 1746–1751. ACL (2014)
8. Lan, M., Tan, C.L., Su, J., Lu, Y.: Supervised and traditional term weighting methods for automatic text categorization. IEEE Trans. Pattern Anal. Mach. Intell. **31**(4), 721–735 (2009)
9. Le, Q., Mikolov, T.: Distributed representations of sentences and documents. In: International Conference on Machine Learning, pp. 1188–1196, January 2014
10. Mikolov, T., Sutskever, I., Chen, K., Corrado, G.S., Dean, J.: Distributed representations of words and phrases and their compositionality. In: Proceedings of the 26th International Conference on Neural Information Processing Systems, pp. 3111–3119. Curran Associates Inc., Red Hook (2013)
11. Ministry of Gender Equality and Family: In-appropriate media for Juvenile (청소년유해매체물).. http://www.mogef.go.kr/sp/yth/sp_yth_f013.do. Accessed 01 Sept 2017
12. Mubarak, H., Darwish, K., Magdy, W.: Abusive language detection on Arabic social media. In: Proceedings of the First Workshop on Abusive Language Online, pp. 52–56. ACL (2017)
13. NamuWiki: Profanity/Korean. https://namu.wiki/w/%EC%9A%95%EC%84%A4/%ED%95%9C%EA%B5%AD%EC%96%B4. Accessed 01 Sept 2017
14. NoSwearing: List of Swear Words, Bad Words, & Curse Words. https://www.noswearing.com/dictionary. Accessed 01 Sept 2017

15. Ren, F., Sohrab, M.G.: Class-indexing-based term weighting for automatic text classification. Inf. Sci. **236**, 109–125 (2013)
16. Sood, S.O., Antin, J., Churchill, E.: Using crowdsourcing to improve profanity detection. AAAI Spring Symposium: Wisdom of the Crowd, vol. 12, p. 6 (2012)
17. Sun, Y., Wong, A.K., Kamel, M.S.: Classification of imbalanced data: a review. Int. J. Pattern Recognit. Artif. Intell. **23**(4), 687–719 (2009)
18. Xiang, G., Fan, B., Wang, L., Hong, J., Rose, C.: Detecting offensive tweets via topical feature discovery over a large scale twitter corpus. In: Proceedings of the 21st ACM international conference on Information and knowledge management, pp. 1980–1984. ACM (2012)
19. Yang, Z., Yang, D., Dyer, C., He, X., Smola, A., Hovy, E.: Hierarchical attention networks for document classification. In: Proceedings of the 2016 Conference of the North American Chapter of the Association for Computational Linguistics: Human Language Technologies, pp. 1480–1489. NAACL (2016)

Evaluating Similarity Metrics for Latent Twitter Topics

Xi Wang(✉), Anjie Fang, Iadh Ounis, and Craig Macdonald

University of Glasgow, Glasgow, UK
{x.wang.6,a.fang.1}@research.gla.ac.uk,
{iadh.ounis,craig.macdonald}@glasgow.gla.ac.uk

Abstract. Topic modelling approaches such as LDA, when applied on a tweet corpus, can often generate a topic model containing redundant topics. To evaluate the quality of a topic model in terms of redundancy, topic similarity metrics can be applied to estimate the similarity among topics in a topic model. There are various topic similarity metrics in the literature, e.g. the Jensen Shannon (JS) divergence-based metric. In this paper, we evaluate the performances of four distance/divergence-based topic similarity metrics and examine how they align with human judgements, including a newly proposed similarity metric that is based on computing word semantic similarity using word embeddings (WE). To obtain human judgements, we conduct a user study through crowdsourcing. Among various insights, our study shows that in general the cosine similarity (CS) and WE-based metrics perform better and appear to be complementary. However, we also find that the human assessors cannot easily distinguish between the distance/divergence-based and the semantic similarity-based metrics when identifying similar latent Twitter topics.

1 Introduction

Twitter has become a popular way for people to express their opinions and preferences. Researchers are often interested in examining the topics that are being discussed on such a platform [1–3]. To this end, topic modelling approaches, such as Latent Dirichlet Allocation (LDA), can be used to identify topics [2, 4]. However, redundant topics can cost researchers more time when examining their content. Therefore, it is necessary to identify the redundant topics before presenting them to the researchers. We assume that highly similar topics could be redundant and a topic similarity metric can be used to calculate the similarity among topics generated by a topic modelling approach.

We evaluate various topic similarity metrics in order to offer practical suggestions on how to effectively measure the similarities among latent topics generated from Twitter streams. A topic in a topic model is a distribution over words [4]. Commonly, the similarities of topics can be computed by using the distribution of topics over the vocabulary. Previous work has applied metrics such as the Hellinger distance (HD) [5], the Jensen Shannon (JS) divergence [6] or the

L. Azzopardi et al. (Eds.): ECIR 2019, LNCS 11437, pp. 787–794, 2019.
https://doi.org/10.1007/978-3-030-15712-8_54

cosine similarity (CS) [1,7] to measure the similarity between topics. These metrics compute the distance/divergence of topic distributions. We also propose and evaluate a new word embedding (WE)-based metric to measure the semantic similarity between topics, since word embedding has been reported to more effectively capture the semantic similarity [3,8].

We conduct a user study through crowdsourcing to examine the effectiveness of the four aforementioned similarity metrics (i.e. HD, JS, CS and WE). Our crowdsourced user study shows that the human assessors cannot easily distinguish between the distance/divergence-based and the semantic similarity-based metrics when identifying similar topics. However, we also find that, in general, the CS and WE-based metrics align the best with human judgements, as they outperform at least one other metric on our Twitter dataset. In particular, our results suggest that the CS and WE-based metrics appear to be complementary. While the CS-based metric can better assess the topic similarity when topics share the same frequent words, the WE-based metric can better capture the semantic similarity of topics. Overall, our paper contributes new insights about measuring topic similarity in Twitter, and how the topic similarity metrics perform compared to human judgements.

2 Related Work

Typically, three types of metrics can be used to capture the similarity between topics: **(1) Divergence-based metrics**. Gretarsson et al. [9] and Kim et al. [6] applied the Kullback Leibler (KL) and Jensen Shannon (JS) divergence metrics to measure the textual differences of latent topics. Kim et al. [6] concluded that the JS divergence gave the best performance when compared to the other approaches tested. **(2) Coefficient-based metrics**. The coefficient-based metrics, Jaccard's Coefficient, Kendall's τ coefficient, and discounted cumulative gain can all be used to compute the similarity between topics. However, Kim et al. [6] showed that the divergence-based metrics are better than these coefficient-based metrics, as the coefficient-based metrics require a corpus-dependent probability mass. **(3) Distance-based metrics**. Gretarsson et al. [9] estimated the similarity of latent topics by computing the L_1 distance. Later, Maiya et al. [5] adopted the Hellinger distance metric to calculate the similarity between topics. On the other hand, the most common distance metric used in the literature is the cosine similarity [6,10,11]. Indeed, the cosine similarity has been shown to provide superior performance compared to other divergence-based metrics [7].

In [12], Mikolov et al. proposed a shallow learning technique called word2vec, which represents individual words as high dimensional word embedding vectors. These word representations can be used to capture the semantic similarity between words [13]. However, it is unclear which of the aforementioned types of metrics better reflects a user's view of topic similarity on Twitter. Based on these prior studies, we choose the JS divergence, the Hellinger distance, the cosine as well as a new word embedding-based similarity for evaluation in our user study.

3 Metrics and Methodology

We introduce the used topic similarity metrics and their differences. The cosine similarity-based metric (CSM) can be applied over two distributions (i.e. two vectors) for computing the similarity of two topics. This method has been previously used in [3,14]. The JS divergence-based metric (JSM) is a symmetric form of the KL divergence. It is often used as a topic similarity metric in prior work [6,9]. The Hellinger distance-based metric (HDM) is often used to quantify the similarity between a pair of probability distributions, as in [5].

In addition, we propose a word embedding-based metric (WEM), where each topic is represented by its top n words, ranked by its words' posterior topic probabilities. We then compute the similarity of two topics by the pairwise word semantic similarity shown in Eq. (1), where W_i denotes the set of top n words for topic i, and Vec_p indicates the vector of word p in a WE model.

$$WES(\theta_i, \theta_j) = \sum_{p \in W_i} \min_{\forall q \in W_j} cosine(Vec_p, Vec_q) \tag{1}$$

Differences Among Metrics. Each of the 4 aforementioned metrics focuses on different aspects when estimating the topic similarity, providing a good representative sample of similarity metrics to compare to human judgements. The CS-based metric tends to compute the similarity using words with high frequencies. Compared to CSM, JSM and HDM alleviate the effects of high-frequency words. Moreover, while JSM tends to normalise the word probability differences, the HD-based metric applies a square root to smooth the probability differences. Unlike the CS, JS and HD-based metrics, which compute the similarity of topics using the whole topic distributions, the WE-based metric exploits instead the semantic similarity between the top-ranked words in the generated topics.

Pairwise Comparison of Metrics. We evaluate the performances of the four metrics using a pairwise approach, i.e. assessing the performances of each pair of the four metrics. This pairwise comparison method has been previously applied in the literature to compare different systems [3]. Specifically, given a topic from a topic model (we call it the **base** topic) and two metrics A & B (a metric pair), we use metric A and B to choose two **candidate** topics that are the most similar to the base topic. Two candidate topics together with their base topic are called a *topic set*. For each metric pair, we sample a number of topic sets. We conduct a user study to obtain the ground-truth from human judgements. A metric in a metric pair obtains a score of "1" if it aligns with the human judgement on a topic set, otherwise, "0". Accordingly, we use a signed-rank test on a set of generated paired scores to identify the statistically significant between each metric pair.

Twitter Dataset. We use a Twitter dataset that is related to the US 2016 election and which contains tweets posted from 01/07/2016 to 31/10/2016. This dataset has 18k sample tweets[1] collected by searching a list of keywords related

[1] This sample of tweets is in English, does not contain retweets and each tweet has at least 5 words.

to the US 2016 election (e.g. "Trump", "Hillary", "debate", "vote", "election", etc.) using the Twitter Streaming API[2]. Since the election contains numerous discussions across a range of topics, this election-related dataset allows us to obtain sufficient topics for applying a topic modelling approach such as LDA.

4 Crowdsourced User Study

We now describe how we perform the user study to obtain human judgements. The CrowdFlower[3] platform is used. Each worker is presented with multiple topic sets. Similar to [3,15], a topic is represented by the 15[4] most frequent words from its word distribution. A worker is asked to choose a topic out of the two candidate topics, which is the most similar to the base topic. If a worker cannot make a decision, they can select the option of "Either of them". To help the workers undertake the task, we provide them with guidelines that explain how to identify the most similar topic. For example, they can check whether the base and candidate topics contain words that refer to the same topic. We also provide a list of election-related hashtags (e.g. #FeelTheBern, #Wikileaks) and some commonly mentioned key players in the election (e.g. Mike Pence, Tim Kaine) with their corresponding descriptions. After the workers choose a given candidate topic, they are asked to specify how *easy* they found the question. Next, we explain how we generate topic sets and our precise used experimental setup for the user study.

Generating Topics. We apply Gibbs sampling [16], an approximate inference technique for LDA[5], to generate topics from the election Twitter data. The number of topics K is set to 90[6] and we generate 10 repeated topic models[7]. For each of the chosen topic models, we use the topic coherence metric [3], which has been shown to be particularly effective on Twitter compared to other existing metrics, to rank the 90 topics by their coherence. Then, we select the top 30 topics out of 90 from each topic model. We obtain 300 topics as the pool of base topics. For each metric pair, we randomly select 50 base topics from the base topic pool. For a given metric pair, each metric selects the most similar topic to a base topic as a candidate topic. Accordingly, we obtain 50 topic sets[8] for each metric pair (300 in total).

---

[2] https://dev.twitter.com.

[3] http://crowdflower.com.

[4] In [3,15], the top 10 words are used to estimate a given topic's coherence. However, Ramage et al. [1] argued that the top-ranked words might often be similar. Hence, we choose to use the top 15 words in this work.

[5] We use Gibbs sampling as it can still generate topics that connect well to the real topics (see [2]). We plan to study topic similarity using different LDA approaches in the future work.

[6] We found that topic models with $K = 90$ have a higher coherence according to the topic coherence metric [3] used in our experiments.

[7] Each topic model contains 90 topics.

[8] The order of topics in the topic sets is shuffled.

User Study Setup. We first limit the CrowdFlower workers to the US as the topics are related to the US election. In total, we had 60 workers who passed the test and entered the task. Among the 60 workers, 35 workers maintained the required accuracy of 70% and their judgements were retained. Each worker has to spend at least 10 s on each question and can only answer at most 20 judgements. Such a setup allows us to obtain judgements from many users. We pay a worker US$0.05 for each question. We obtain at least 3 judgements for each question. We require a minimum agreement of 60% among the 3 workers on any of their answers. Otherwise, additional workers are allocated the same question until such an agreement is reached. Among the 300 questions, 38.4% required additional workers.

Setup of the WE-Based Metric. We use tweets to train the word embedding model, since our topics are generated from tweets. First, we use the Twitter Streaming API (sample mode) to crawl a collection of random tweets posted from January to July in 2016. The size of this collection is about 200 million tweets. To obtain the embedding, we apply fastText[9] on the collected tweets. Our WE-based metric leverages this trained embedding to evaluate the similarity of the top 15 words in two topics.

Table 1. Comparison of the 6 metric pairs. Statistically significant differences are indicated by *.

	CSM vs. WEM	CSM vs. JSM	CSM vs. HDM	WEM vs. JSM	WEM vs. HDM	JSM vs. HDM
# of votes	25 vs. 25	31 vs. 19	23 vs. 27	23 vs. 24	30 vs. 19	23 vs. 23
p-Value	1.0	0.03*	0.49	0.86	0.05*	1.0

5 Results Analysis

We first report the metric preferences from our user study. Then we report a qualitative analysis of the results.

For the 300 topic sets, we obtain 900 judgements from 21 different workers. In terms of task difficulty, among the collected judgements, 22% (196) of them are labelled as "easy" and 75.6% (628) are "reasonable". Only 2.4% (66) of these judgements are "hard" for humans to make. This suggests that the task of our user study is reasonably easy for the workers. We use the method explained in Sect. 3 to calculate the *p*-value, which indicates whether two metrics perform significantly differently. The number of votes and *p*-values of the 6 metric pairs are listed in Table 1. For example, "31 vs. 19" in the CSM vs. JSM column indicates that the CSM metric (with 31 votes) significantly outperforms the JSM metric with (19 votes). Similarly, we also observe that the WEM metric is significantly better than HDM.

[9] http://fasttext.cc. The context window size is 5 and the dimension of the vector is 100.

Overall, we do not observe significant differences among the rest of 4 metric pairs. As mentioned in Sect. 3, the CS, JS and HD-based metrics consider the probabilities of all the topics' words while the WE-based metric focuses on the semantic similarity of top-ranked words. Since neither the WE-based metric nor the other 3 metrics are consistently better than the rest of metrics, this suggests that the two types of metrics align equally well with the human judgements when assessing topics similarity. On the other hand, while no metric in this study consistently beats all the others, we do observe that, in general, the CS and WE-based metrics perform the best and outperform the other 2 metrics. In addition, according to the signed-rank significance test, only CSM outperforms JSM and WEM outperforms HDM significantly. Hence, later, we further analyse the CS and WE-based metrics, their differences and why they were the preferred metrics according to human judgements.

Table 2. Topic sets of WEM vs. CSM by columns

Base topic: people #trump talking @realdonaldtrump #hrc making guy believe abt hey actually fake democratic supporter trying	**Base topic:** #wikileaks #draintheswamp #podestaemails #votetrump #voterfraud #neverhillary #trump yeah dems #alsmithdinner #corruption politics @realdonaldtrump dump readin
Candidate topic 1 (selected by CSM): #trump #putin russia putin #rednationrising talking #billclinton #tgdn morning tomorrow want iran pennsylvania gold standard	**Candidate topic 1 (selected by WEM):** #neverhillary #trumppence @realdonaldtrump polls #makeamericagreatagain watching @hillaryclinton comes watch way right nov crap #corrupthillary strong
Candidate topic 2 (selected by WEM): @realdonaldtrump @gop #hrc #america @thedemocrats say course point truth campaign support telling @reince isn moment	**Candidate topic 2 (selected by CSM):** #wikileaks emails #podestaemails @wikileaks fuck hacked according #octobersurprise trending report #assange funded @hillaryclinton staff coverage

Overall, the CS and WE-based metrics performed better than the other two metrics. However, from the signed-rank test, there is no evidence indicating that one metric is significantly better than the other. By examining the topic sets, we find that these two metrics perform differently in different scenarios. CSM is good at matching the most similar topic set when their informative words have high frequencies. The candidate topic selected by CSM is intuitively more similar to the base topic. However, CSM might fail to select the most similar one if the base topic does not share high-frequency words with any candidate topic. On the other hand, WEM can work better in this instance, as it puts more emphasis on the semantic similarity among top words. For example, "#vote" is related to "vote", "votes", "winning", etc. WEM allows to capture the semantic relationships between two topics. However, if two topics share several words with high frequencies, WEM does not outperform CSM, since CSM effectively captures the similarity. For instance, in the first column of Table 2, CSM fails to match the top words in the base topic, which results in the choice of a non-relevant candidate topic 1 (more about "putin" and "russia"), while candidate topic 2 chosen by WEM is better. On the contrary, in the second column of Table 2, when topics share the top words (e.g. "#wikileaks" and "email"), the CSM performs better than WEM. In general, we see a complementary relationship between WEM and CSM.

There are several reasons why our user study did not distinguish between 4 out of 6 metric pairs: CSM vs. WEM, CSM vs. HDM, WEM vs. JSM and JSM vs. HDM. First, two metrics can perform very similarly and thus humans cannot effectively distinguish between their chosen candidate topics. For example, for JSM vs. HDM, given a base topic, we find that 75% of the top 10 most similar topics ranked by the JSM and HDM metrics in a topic model are the same on average. Second, the number of topic set samples might not be large enough, and thus the statistical test cannot find a statistical difference between the two metrics. To conclude, our study shows that using our Twitter dataset, the CSM and WEM metrics align best with human judgements, and markedly outperform the HDM and JSM metrics in estimating the similarity of latent Twitter topics.

6 Conclusions

We studied the effectiveness of 4 commonly used similarity metrics when examining the similarity of latent topics on Twitter. We conducted a user study to ascertain which of the metrics align best with human judgements. Our study showed that, on our used Twitter dataset, the human assessors cannot distinguish between the distance/divergence-based metrics and the semantic similarity-based metric when identifying similar latent Twitter topics. However, the CS and WE-based metrics markedly outperformed the HDM and JSM metrics. In particular, we found that the CS and WE-based metrics appear to be complementary. While the CS-based metric better estimates similarity when the topics share the same high-frequency words, the WE-based metric better captures the semantic relationships among topics. Such complementarity might help to construct topic models with different requirements. As future work, we aim to conduct the same analysis on different datasets, and investigate how to seamlessly combine the CS and WE-based metrics to effectively estimate the similarity of latent topics on Twitter to further reduce redundancy in the generated topic models.

References

1. Ramage, D., Dumais, S.T., Liebling, D.J.: Characterizing microblogs with topic models. In: Proceedings of ICWSM (2010)
2. Zhao, W.X., et al.: Comparing Twitter and traditional media using topic models. In: Clough, P., et al. (eds.) ECIR 2011. LNCS, vol. 6611, pp. 338–349. Springer, Heidelberg (2011). https://doi.org/10.1007/978-3-642-20161-5_34
3. Fang, A., Macdonald, C., Ounis, I., Habel, P.: Using word embedding to evaluate the coherence of topics from Twitter data. In: Proceedings of SIGIR (2016)
4. Blei, D.M., Ng, A.Y., Jordan, M.I.: Latent Dirichlet allocation. Mach. Learn. Res. **3**, 993–1022 (2003)
5. Maiya, A.S., Rolfe, R.M.: Topic similarity networks: visual analytics for large document sets. In: Proceedings of IEEE Big Data (2014)

6. Kim, D., Oh, A.: Topic chains for understanding a news corpus. In: Gelbukh, A. (ed.) CICLing 2011. LNCS, vol. 6609, pp. 163–176. Springer, Heidelberg (2011). https://doi.org/10.1007/978-3-642-19437-5_13
7. Aletras, N., Stevenson, M.: Measuring the similarity between automatically generated topics. In: Proceedings of EACL (2014)
8. Nikolenko, S.I.: Topic quality metrics based on distributed word representations. In: Proceedings of SIGIR (2016)
9. Gretarsson, B., et al.: TopicNets: visual analysis of large text corpora with topic modeling. ACM Trans. Intell. Syst. Technol. **3**(2.23), 1–26 (2012)
10. Ramage, D., Hall, D., Nallapati, R., Manning, C.D.: Labeled LDA: a supervised topic model for credit attribution in multi-labeled corpora. In: Proceedings of EMNLP (2009)
11. Fang, A., Macdonald, C., Ounis, I., Habel, P., Yang, X.: Exploring time-sensitive variational Bayesian inference LDA for social media data. In: Jose, J.M., et al. (eds.) ECIR 2017. LNCS, vol. 10193, pp. 252–265. Springer, Cham (2017). https://doi.org/10.1007/978-3-319-56608-5_20
12. Mikolov, T., Sutskever, I., Chen, K., Corrado, G.S., Dean, J.: Distributed representations of words and phrases and their compositionality. In: Proceedings of NIPS (2013)
13. Levy, O., Goldberg, Y.: Linguistic regularities in sparse and explicit word representations. In: Proceedings of CoNLL (2014)
14. Huang, A.: Similarity measures for text document clustering. In: Proceedings of NZCSRSC (2008)
15. Fang, A., Macdonald, C., Ounis, I., Habel, P.: Topics in tweets: a user study of topic coherence metrics for Twitter data. In: Ferro, N., et al. (eds.) ECIR 2016. LNCS, vol. 9626, pp. 492–504. Springer, Cham (2016). https://doi.org/10.1007/978-3-319-30671-1_36
16. Darling, W.M.: A theoretical and practical implementation tutorial on topic modeling and Gibbs sampling. In: Proceedings of ACL HLT (2011)

On Interpretability and Feature Representations: An Analysis of the Sentiment Neuron

Jonathan Donnelly(✉) and Adam Roegiest

Kira Systems, Toronto, Canada
{jonny.donnelly,adam.roegiest}@kirasystems.com

Abstract. We are concerned with investigating the apparent effectiveness of Radford et al.'s "Sentiment Neuron," [9] which they claim encapsulates sufficient knowledge to accurately predict sentiment in reviews. In our analysis of the Sentiment Neuron, we find that the removal of the neuron only marginally affects a classifier's ability to detect and label sentiment and may even improve performance. Moreover, the effectiveness of the Sentiment Neuron can be surpassed by simply using 100 random neurons as features to the same classifier. Using adversarial examples, we show that the generated representation containing the Sentiment Neuron (i.e., the final hidden cell state in a LSTM) is particularly sensitive to the end of a processed sequence. Accordingly, we find that caution needs to be applied when interpreting neuron-based feature representations and potential flaws should be addressed for real-world applicability.

1 Introduction

Several authors [2,9,13] have investigated the idea that single neurons or groups of neurons have easily interpretable behaviour. Recent work [6] has shown evidence that interpretable neurons do not necessarily correspond to improved neural network effectiveness and that reliance on interpretable neurons may be a sign of overfitting. To this end, we focus on Radford et al.'s [9] finding that after training a large, single layer LSTM [3] language model on $\sim 86M$ Amazon Reviews [5], a single neuron emerges as a strong predictor of sentiment, which they dub the "Sentiment Neuron" ("SN"). To examine the Sentiment Neuron's predictive capabilities, we perform an ablation analysis on the language model's features to test the impact that their removal has on classification accuracy across several datasets (Sect. 3). We find that the Sentiment Neuron is not necessary to achieve effective classification and that, in some cases, it can actually decrease effectiveness. Moreover, we find that randomly choosing 100 features (neurons) from the language model more often than not produces a classifier that outperforms one based on the Sentiment Neuron alone. This indicates that the Sentiment Neuron does not contain all or most of the knowledge needed for sentiment detection.

L. Azzopardi et al. (Eds.): ECIR 2019, LNCS 11437, pp. 795–802, 2019.
https://doi.org/10.1007/978-3-030-15712-8_55

Furthermore, Radford et al.'s feature representation, where they use the final hidden cell state of the LSTM as a sequence's representation, is one of several possible valid representations. Following Howard et al. [1], we examine different methods of extracting features from an LSTM-based language model. We find that the presence of a neuron predictive of sentiment is an artefact of the network architecture regardless of how we generate features (e.g., mean-pool, final state) but its predictive power varies. In addition, a mean-pool representation appears to be more attuned to sequence length and its effects than the final state representation (Sect. 4). The main benefit to a mean-pool representation is that it is robust to adversarial examples (i.e., ending sequences with sentiment words); while the final state features are not (Sect. 5). We conclude that interpretable neurons do not guarantee success and that feature representations can and should be robust to potential adversarial cases.

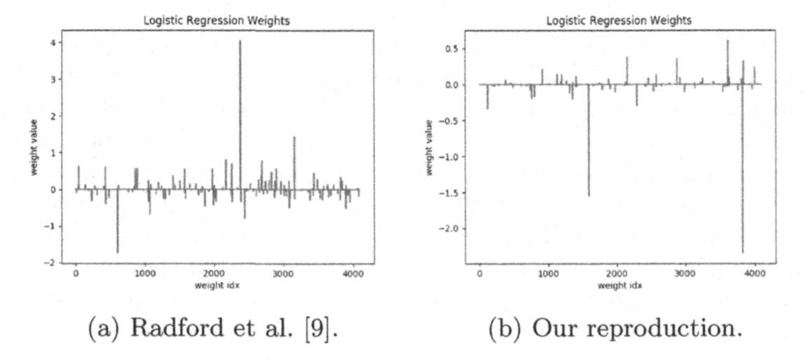

(a) Radford et al. [9]. (b) Our reproduction.

Fig. 1. Plots showing linear classifier weights trained on SST with language model features from Radford et al. [9] and our reproduction.

2 Methodology

We follow the methodology of Radford et al. [9] which trains a logistic regression classifier on top of the language model features. In the default setting, these language model features are the final hidden cell state of the LSTM over a sequence. However, we vary this in the following sections to test the effectiveness of different representations. For each dataset, we shuffle and split them into three folds: 70% training set; 10% development; and, 20% test set. The development set is used to perform a light grid search of hyper-parameters for the classifier. While full cross-validation would likely yield a superior general purpose classifier, we are only concerned with examining the effect of representations and not finding the best classifier.

We evaluate this technique using four sentiment analysis datasets: Stanford Sentiment Treebank ("SST") [10]; IMDB Large Movie Review Dataset ("IMDB") [4], Rotten Tomatoes Short Movie Reviews ("MR") [8]; and Amazon Customer Review Dataset ("CR") [12]. Wang and Manning [11] provide a useful summary of these datasets. It is worth noting that the IMDB dataset consists of full length movie reviews and an order of magnitude more examples than the short reviews in the other datasets. Code to reproduce our results, as well as train a new language model from scratch, is made publicly available.[1]

3 Neuron Ablation

We see that there is, indeed, a strongly predictive neuron when examining the weights of final state features in Fig. 1. This holds for the released weights from Radford et al. [9] and our reproduction of their model. Similar to NVidia's reproduction [7], we find that the Sentiment Neuron exhibited opposite polarity from Radford et al., which indicates that such neurons are network artefacts and not just a one-off. Although, what this neuron looks like may vary between implementations.

Table 1. Accuracy on 4 datasets using all features, SN only, and SN deleted.

Features	SST	MR	CR	IMDB
All features	91.76	87.52	91.38	92.28
SN deleted	91.87	86.96	90.72	91.77
SN only	88.52	84.52	88.33	91.46

Using Radford et al.'s weights, we can alternatively isolate and ablate the SN from the features during training. From Table 1, isolation of the SN appears to yield decreases in effectiveness across all the datasets. When we ablate the SN, there is no substantial change from using all the features. This indicates that the inclusion of the SN does not appear to add crucial information to the classifier. Indeed, inclusion of the SN appears to hinder performance on the SST dataset. In essence, their remains enough information distributed among the other neurons to still effectively train a classifier.

Based upon the ablation results, we might wonder whether any other neurons would hinder classification accuracy. Accordingly, we ablate each individual neuron and train a new classifier in turn for all neurons on the SST and IMDB datasets. Figure 2 reports the results for SST as the IMDB results are similar. It does appear to be the case that some neurons do hinder performance. The neuron which when ablated gives the highest accuracy (92.15%), yields a classifier that is only slightly better than chance when used in isolation. Thus, it appears to be the case that some features do not contribute to sentiment detection.

We can also examine the performance of classifiers trained on each neuron in isolation for SST and IMDB. Figure 3 shows that there exist neurons that rival the performance of the SN in predicting sentiment. Moreover, there are many neurons that do not appear to be good predictors of sentiment.

[1] Source code available at github.com/kirasystems/science.

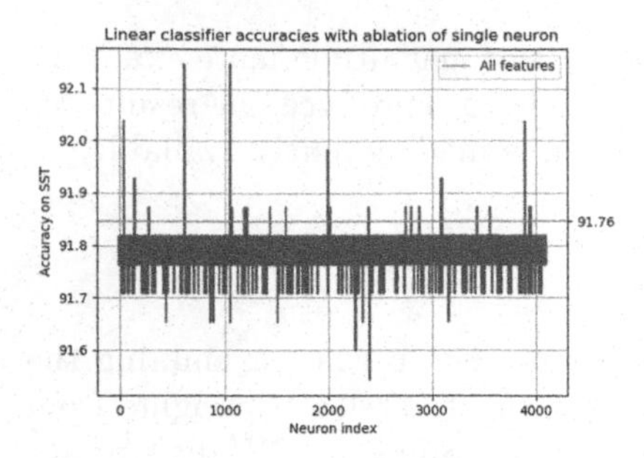

Fig. 2. Accuracy scores of classifier with each neuron ablated individually.

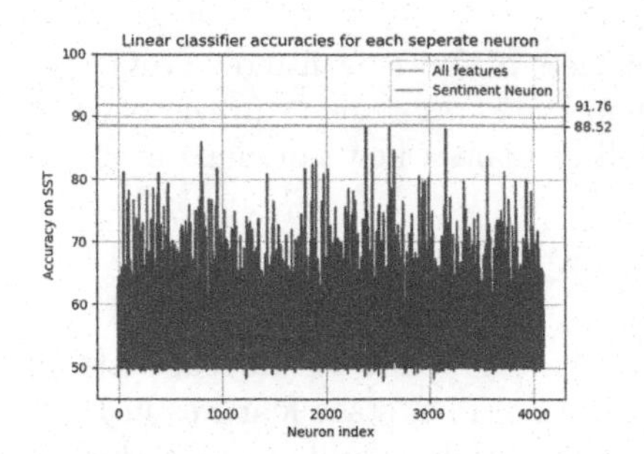

Fig. 3. Accuracy scores with linear classifier trained on each neuron individually.

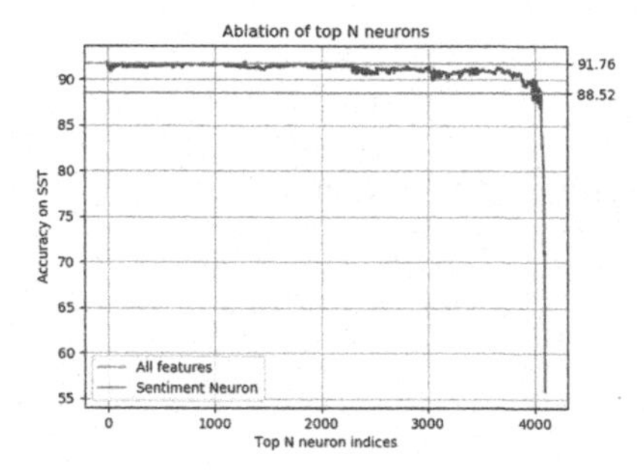

Fig. 4. Accuracy on SST as neurons are cumulatively ablated.

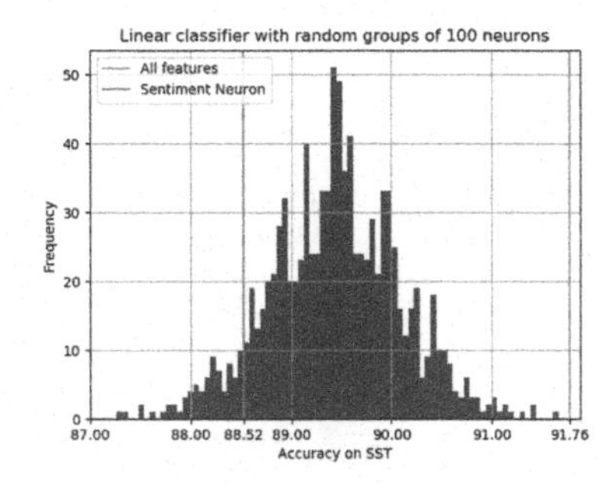

Fig. 5. Histogram showing the accuracy on 1000 random samples of 100 neurons.

Interestingly, ablating multiple of the "hindrance" neurons (i.e., those that produce a better classifier when ablated) or the less predictive neurons does not yield substantial improvements in the resulting classifier. It would appear that the full feature classifier is able to "learn around" these features.

Based upon our findings thus far, we might wonder whether or not there is a necessary "critical mass" of neurons that are needed to produce an effective classifier. In Fig. 4, we cumulatively ablate each neuron in order from highest to lowest corresponding logistic classifier weight[2] on the SST dataset. We see that with ~100 neurons remaining we achieve parity with the SN before precipitously dropping. Accordingly, if the SN is our barometer for a "good" or "sufficient" classifier then we need at least 100 neurons to rival its effectiveness. We then proceed to train 1000 different classifiers using 100 randomly selected neurons. Figure 5 shows that the resulting accuracy scores form an approximately normal distribution with the SN's score coinciding with the lower tail. Accordingly, selecting a random group of neurons will more often than not yield a superior

[2] These weights are generated from training the classifier on all available neurons.

classifier to the SN. This leads us to conclude that, contrary to Radford et al. [9], there is a meaningful amount of sentiment information stored outside of the SN.

4 Features

Up until now, we have examined the feature representation using the final hidden cell state of the LSTM but this could allow the end of a sequence to have undue influence on the feature weights. While this may matter less for short pieces of text, longer pieces of text may not be ideally represented as there would be a skew towards the end of the sequence. Accordingly, we examine different ways of integrating information across time-steps including the mean-pool, min-pool, max-pool, and absolute-pool of values. For the sake of brevity, we report the mean-pool results only as they provide the most interesting counterpoint to the final state features. Similar to Howard et al. [1], we report the concatenation of final state and mean-pool representations as such a representation may offer the advantages of both representations.

Table 2 reports the accuracy scores across our four datasets for the different possible feature representations. Generally, the concatenated features generally perform as good as one of the constituent feature representations and can sometimes outperform both. Interestingly, we do not see as much degradation of the final state features in the longer IMDB text than we had expected but this may be due to repeated use of sentiment throughout a review. The generally inferior performance of the mean-pool features is not readily apparent but we posit that the shorter reviews did not allow the mean-pool features to stabilize on a good representation. Such a hypothesis warrants further investigation but the competitive performance of the mean-pool features on IMDB provides some evidence that this length may play a role. Further, if we compare the weights of the mean-pool neurons (Fig. 6) to the final state ones (Fig. 1), we see that the mean-pool features have a greater number of influential neurons (neurons with larger corresponding weights). This increase in influential neurons may then correspond to requiring longer sequences to make accurate classifications.

In spite of these differences, there are a class of examples that neither the final state nor mean-pool features are able to classify. Across the four datasets, the

Table 2. Accuracy scores on 4 sentiment datasets using final, mean and concatenated state features.

Features	SST	MR	CR	IMDB
Final	91.76	87.52	91.38	92.28
Mean	88.03	82.97	89.52	92.82
concat(Final, Mean)	91.76	87.43	90.45	93.86

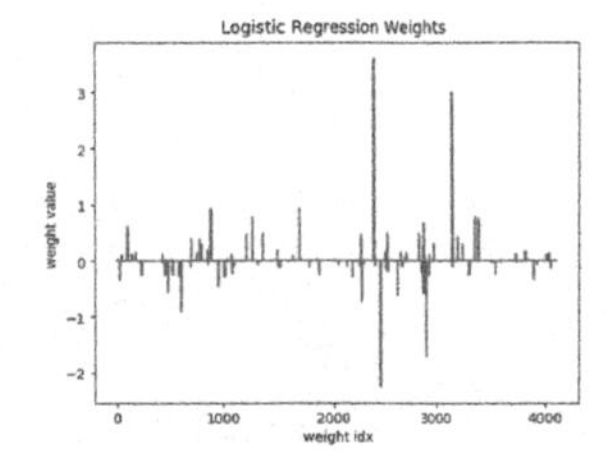

Fig. 6. Logistic regression weights for SST-model using mean-pool features.

overlap of incorrect examples ranges between 30–40% which appears to indicate that there are aspects of sentiment that neither of these feature representations capture. We note that manual examination of a selection of these incorrectly identified examples reveals that these express less obvious sentiment (e.g., *"What you would end up with if you took Orwell, Bradbury, Kafka, George Lucas and the Wachowski Brothers and threw them into a blender."*). While such examples are not unexpected, they highlight that Radford et al.'s network is unable to capture all nuances of sentiment.

5 Adversarial Examples

As suggested in the previous section, the final state features may be subject to undue influence of the ending of a sequence. We test this by adversarially adding a positive ("Wonderful") and negative ("Terrible") sentiment word to test examples and examine how this affects classification accuracy. As seen in Table 3, the effectiveness of the final state and concatenated features are substantially affected by the inclusion of a trailing sentiment word for small datasets; while the mean-pool features are robust to their inclusion. This lends credence to our hypothesis that final state features are unduly influenced by most recently processed input, whereas the mean-pool features are more capable of capturing a sequence's overall sentiment.

This trend does not hold for longer text since adding a single word does not appear to hamper the final state (or concatenated) features from achieving good scores on IMDB with only a single word added (i.e., the IMDB-1 results). However, the IMDB reviews are approximately 10 times longer than the smaller reviews, which means we may need to add a similar proportion of sentiment words to the end to achieve a similar adversarial effect. The results of the IMDB-10 setting bear this out and this indicates the choice of feature representation should take into account all possible use cases. It is not inconceivable that in the real-world such adversarial examples would occur, especially in short form communication (e.g., Twitter).

Table 3. Accuracy scores on 4 sentiment datasets when positive and negative words are appended to the test set examples. IMDB-1 denotes a single sentiment word added and IMDB-10 denotes 10 words added to the end of test examples.

Sentiment	Features	SST	MR	CR	IMDB-1	IMDB-10
Positive	Final	52.17	57.22	77.06	91.16	50.30
	Mean	89.13	82.88	90.05	92.87	89.73
	concat(Final, Mean)	53.93	61.07	79.97	93.50	61.25
Negative	Final	72.16	59.52	48.01	90.69	53.17
	Mean	89.02	83.63	90.98	92.91	92.68
	concat(Final, Mean)	73.31	57.74	42.44	93.07	81.29

6 Conclusion

Radford et al. [9] found that in a large LSTM-based language model, there is a single neuron that is predictive of sentiment. After conducting an ablation study, we find that this neuron is not necessary to achieve effective classification and that, in some cases, hinders effectiveness. Moreover, we find that the effectiveness of the Sentiment Neuron can be matched or exceeded, more often than not, by randomly selecting 100 neurons to train a classifier. This indicates that the reliance on a single "understandable" neuron may result in undue bias in the resulting classifier, which is in accord with the findings of Marcos et al. [6]. Additionally, we find that the feature representation is important and that relying on the final hidden cell state opens an end user up to exploitation by adversarial examples. Using the mean-pool of cell states across a sequence appears to be robust to this type of exploitation but is generally inferior to the final state features. Accordingly more work is necessary to find a robust but highly effective representation.

References

1. Howard, J., Ruder, S.: Universal language model fine-tuning for text classification. In: Proceedings of the 56th Annual Meeting of the Association for Computational Linguistics (Volume 1: Long Papers) (2018)
2. Karpathy, A., Johnson, J., Li, F.: Visualizing and understanding recurrent networks. In: International Conference on Learning Representations (2016)
3. Krause, B., Lu, L., Murray, I., Renals, S.: Multiplicative LSTM for sequence modelling. In: International Conference on Learning Representations (2017)
4. Maas, A.L., Daly, R.E., Pham, P.T., Huang, D., Ng, A.Y., Potts, C.: Learning word vectors for sentiment analysis. In: Proceedings of the 49th Annual Meeting of the Association for Computational Linguistics: Human Language Technologies - Volume 1, HLT 2011, pp. 142–150. Association for Computational Linguistics, Stroudsburg (2011)
5. McAuley, J.: Amazon product data. http://jmcauley.ucsd.edu/data/amazon/
6. Morcos, A.S., Barrett, D.G., Rabinowitz, N.C., Botvinick, M.: On the importance of single directions for generalization. In: International Conference on Learning Representations (2018)
7. Nvidia: Sentiment discovery. GitHub repository (2017)
8. Pang, B., Lee, L.: Seeing stars: exploiting class relationships for sentiment categorization with respect to rating scales. In: Proceedings of ACL, pp. 115–124 (2005)
9. Radford, A., Józefowicz, R., Sutskever, I.: Learning to generate reviews and discovering sentiment (2017). http://arxiv.org/abs/1704.01444
10. Socher, R., et al.: Recursive deep models for semantic compositionality over a sentiment treebank. In: Proceedings of the 2013 Conference on Empirical Methods in Natural Language Processing, pp. 1631–1642. Association for Computational Linguistics, Seattle, Washington, October 2013
11. Wang, S., Manning, C.D.: Baselines and bigrams: simple, good sentiment and topic classification. In: Proceedings of the 50th Annual Meeting of the Association for Computational Linguistics: Short Papers - Volume 2, ACL 2012, pp. 90–94 (2012)

12. Wiebe, J., Wilson, T., Cardie, C.: Annotating expressions of opinions and emotionsin language. Lang. Resour. Eval. **1** (2005)
13. Zhou, B., Khosla, A., Lapedriza, À., Oliva, A., Torralba, A.: Object detectors emerge in deep scene CNNs. In: International Conference on Learning Representations (2014)

Image Tweet Popularity Prediction with Convolutional Neural Network

Yihong Zhang(✉) and Adam Jatowt

Department of Social Informatics, Graduate School of Informatics, Kyoto University, Kyoto, Japan
yhzhang7@gmail.com, adam@dl.kuis.kyoto-u.ac.jp

Abstract. Predicting popularity of a post in microblogging services such as Twitter is an important task beneficial for both publishers and regulators. Traditionally, the prediction is done through various manually designed features extracted from post and user contexts. In recent years, deep learning models such as convolutional neural networks (CNN) have shown significant effectiveness in image processing. In this paper, we make a novel investigation of the effectiveness of deep learning models in predicting image post popularity, with the raw image as the input. In contrast to previous works that use existing model trained for object detection, we trained a CNN model targeting directly at predicting popularity. We show that a dedicated CNN is more effective than networks trained for other purposes and is comparable to text-based predictors.

Keywords: Image popularity prediction · Microblog · Deep learning

1 Introduction

Social media networks such as Twitter that has hundreds of millions monthly active users are nowadays important platforms for information sharing. On Twitter, in addition to personal users who post information about their daily lives, there are also companies who promote their products and organizations that make announcements and advertisements. It would be of great interest for these publishers to know the future popularity of their posts. In this paper, we deal with the problem of predicting tweet popularity *before* posting the tweet. This is in contrast to existing works that predict tweet popularity *after* posting the tweet, for example, based on early propagations [2]. Particularly, we aim to predict the popularity of tweets with images, as many advertisements on Twitter are based on the content of the image and some contextual information. Such before-hand prediction can have many benefits, such as allowing publisher to adjust their tweet content in order to get higher popularity.

On Twitter, there are two common measurements of tweet popularity, namely, the number of retweets and the number of likes. Retweeting is the activity to re-post someone else's tweet in the retweeting user's account, while liking is the activity to click a button on the tweet to indicate admiration, without

L. Azzopardi et al. (Eds.): ECIR 2019, LNCS 11437, pp. 803–809, 2019.
https://doi.org/10.1007/978-3-030-15712-8_56

repeating the tweet. The count of both activities received can indicate the popularity of the tweet. However, liking tends to indicate that the tweet is sentimentally admirable, and retweeting often indicates tweet containing important information, regardless of its sentiment value. In this paper, we study both measurements.

We identify two pieces of information that are critical for predicting popularity in addition to the actual content of an image, namely, the number of followers and the time elapsed in hours. On Twitter, a tweet posted by a user is usually automatically displayed in the pages of all the followers of that user, so the number of followers means the number of initial audience of the tweet, a proportion of which will then retweet or like the tweet. Also getting retweets and likes is an accumulation process, therefore the time elapsed since posting is also important. Our experimental analysis shows that, without inputting this information, one cannot get meaningful predictions. In an example application, a user would provide the image and the text message, and specify the number of followers and time elapsed to obtain a prediction of the number of retweets and likes the tweet is likely to receive once posted.

For analyzing images we will use the latest findings in deep learning. It has been shown that convolutional neural network (CNN) is particularly effective in processing images, with its ability to capture local features of the image [5]. However, existing works on predicting image popularity mostly use pre-trained network targeting object recognition. In contrast, in this paper, we propose a CNN specifically trained for predicting popularity. We extend the standard CNN with additional inputs into one of its middle layers. In the experimental analysis we will demonstrate that this dedicated network can achieve higher prediction accuracy than pre-trained networks. We also compare image-based prediction with text-based prediction.

2 Related Works

Treated either as a classification problem [6,10,12] or a regression problem [1,3,4], predicting image post popularity in existing works is mostly done through supervised learning with manually designed features. Early works using images features focus on the low-level image aspects [6,12]. Totti et al. propose a feature set that includes low-level image information such as color channel statistics, dominant colors, contrasts, and focus [12]. McParlane et al. too consider image color, while providing more advanced features such as number of faces detected and scenery information of the image, for example indoor or outdoor [6]. They also include text features such as tf-idf of image tags. Their findings show that text features are much more effective than any of the image-based features.

In recent years, neural network models for image processing have become popular, and has been adopted in image post prediction studies. Khosla et al. propose a feature set that includes output of an existing neural network trained to detect objects in the image, in addition to low-level image features such as color and texture [4]. They find that the neural network-based features achieve better results than any other features. Gelli et al. also use a neural network trained for object detection, as well as a neural network trained for sentiment detection, to generate their feature set [3]. They also tested text-based features such as BOW and recognized name entities. They found that image-based features can achieve a spearman correlation of 0.36, while for text-based features the correlation can reach 0.63. However, although the user-chosen image tags in the dataset they use contain important predictive information, such tags are not available in Twitter. Cappallo et al. propose a ranking method for predicting image popularity, also using pre-trained network for object detection [1]. The authors claim that their results compare favorably to [4] though, only some parts of features from [4] are used. In this paper, however, we compare a dedicate network with a pre-trained network and show that a dedicated network can achieve higher accuracy.

3 Hybrid CNN for Predicting Popularity

We design a deep learning model that harnesses the power of convolutional networks to comprehend local image features, and at the same time to allow an additional information input to be used for training alongside the images. The resulting model is a combination of a convolutional network and a fully-connected network. Figure 1 illustrates the structure of our proposed model.

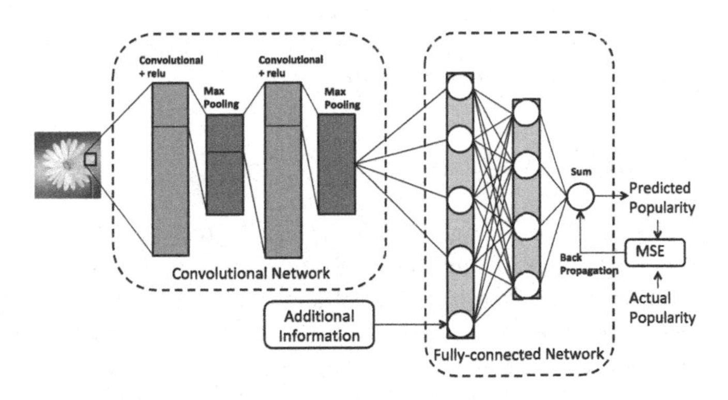

Fig. 1. Structure of the proposed hybrid CNN model. Note that only selected network nodes are shown.

3.1 Network Structure

In the convolutional network, we setup four layers. The first and third layers are convolutional layers and the second and fourth layers are max-pooling layers. For a convolutional layer, we use a number of kernels that extract local features from the image. The output of each convolutional layer is then *max-pooled* in the max-pooling layer. We use rectified linear units (relu) as the activation function for the convolution layer, because it is efficient for a large network and can avoid vanishing or exploding gradient problems in the training phase. In the fully-connected network, we setup two layers, each containing a number of fully-connected neurons. The inputs to the first layer are the outputs of the second max-pooling layer *and* additional information. As we discussed in the introduction, we use two pieces of additional information, namely, number of followers and hours elapsed. The image features and the contextual information are thus combined in a single network. Finally the outputs of the second fully-connected layer are summed in the final node, which produces the prediction value.

3.2 Network Implementation and Training

We implement the proposed network using Google Tensorflow[1]. First of all, we resize all input images to 64×64 pixels, and input them as vectors representing pixel values. After trying a number of different values, we settle on the following model parameters. For the first convolution layer, we use 32 5×5 kernels. For the second convolution layer, we use 64 5×5 kernels. For both max-pooling layers, we use pool size of 2×2 with strides of 2. For the first fully-connected layer we setup 100 neurons, and 60 neurons for the second fully-connected layer.

We use the Adam (adaptive moment estimation) optimizer implemented in Tensorflow. This optimizer is a variety of stochastic gradient descent that uses adaptive learning rates, and has proven effective in providing optimal results faster. We set the initial learning rate as 0.001. In the experiments, we run 1000 training epochs. The cost function generally converges during the training. The trained model is then applied to the test data for evaluation. We will make data and trained models available for download.

4 Experimental Analysis

We conduct experiments on image tweets dataset to test the effectiveness of our approach. Particularly, we are interested in finding out whether our dedicated hybrid CNN model can outperform an object recognition based network and text-based prediction model. In this section, we will describe our data collection process and baseline methods before discussing the evaluation results.

4.1 Data Collection

We collect a number of tweets with images using Twitter's Sample API, which returns a small random sample of all public tweets posted in realtime. For our study, we are interested in those tweets that have accumulated a certain amount of popularity over a period of time. Therefore we select from sampled tweets those that contain images and have already accumulated more than 100 retweets. We collected in this way 107,558 tweets. We also recorded the time of the collection and removed tweets with less than seven days elapsed between the posting time

[1] https://www.tensorflow.org/.

and the time of data collection. Furthermore we eliminated outliers as follows. Specifically, we removed tweets that have more than 10,000 retweets or 10,000 likes. We also removed users who have more than 100,000 followers, since it has been shown that the celebrity status of a poster can give the tweet unusually high popularity regardless of its actual content [13]. Finally, we have 33,558 tweets that satisfy all the filtering requirements. We divide the data into two equal parts as training data and testing data following the approach of [4].

4.2 Baseline Methods

We compare our method against previous popularity prediction methods based on image and text. We focus on those methods that involve deep learning.

Inception. Our image baseline method is a common approach of the previous popularity prediction works that uses vectors extracted from an object recognition network, then applies Support Vector Regression (SVR) to produce predictions. We use Inception network [11], which has 42 deep layers, and is publicly available[2]. We add short code to the Inception program to extract the output of the third pooling layer, which is a vector of 2,048 length representing the semantics of the input image. We add to this vector the two contextual information signals and run it with SVR.

BOW. Our first text based baseline is Bag-of-Words (BOW). This is a commonly used baseline for text-based analysis [9,10]. We select words that appear more than 10 times in the dataset and remove stopwords to generate a vocabulary size of 2,271. Because tweets are short, we use binary BOW for tweet representation. After adding the number of followers and count of elapsed hours, we run SVR for training and prediction.

GloVe. Our second text-based baseline uses word embedding. We use an approach similar to the method proposed in [8], which uses mean value of word embedding for each word in the tweet. Following [10], we use a pre-trained word vector representation called GloVe T100 [7], which is trained on two billion tweets. This model contains 1.2 million large vocabulary, and has a vector length of 100. We generate a vector for each tweet using the mean vector of the words in the tweet, and run SVR for training and prediction after having added the number of followers and count of elapsed hours.

4.3 Evaluation

We use two measurements as in previous studies to evaluate prediction accuracy, median error and spearman correlation [4,8]. Median error is taken as the median of absolute error between predictions and the true popularity values. We use median error because it is more stable given the large variation in tweet popularities. Spearman correlation is based on the correlation of the ranking of

[2] https://www.tensorflow.org/tutorials/image_recognition.

prediction and true values, and reflects relative popularity that is less influenced by the variance. The evaluation results are shown in Table 1. The first column lists the result of the proposed hybrid CNN method.

Table 1. Evaluation results

		Hybrid CNN	Inception	BOW	GloVe
Retweets	Median err	0.0749	0.0412	0.0904	**0.0364**
	Spearman ρ	**0.2677**	0.2060	0.2344	0.2622
Likes	Median err	0.0904	0.0579	0.0825	**0.0516**
	Spearman ρ	**0.3672**	0.3159	0.3575	0.3671

First, we compare our dedicated network to pre-trained network. For both retweets and likes, the dedicated hybrid CNN reaches higher spearman correlation value than Inception network. However, the mean absolute error is smaller for Inception network, because it is a larger network and is more stable. Then we compare dedicated network to text-based methods. We can see that hybrid CNN method reaches the same or higher spearman correlation comparing to text-based methods. In the case of retweets it also has lower median error than the BOW method. However, GloVe word embedding method achieves the lowest median error among all methods for both measures. To conclude, the dedicated network works better in predicting relative popularity than the pre-trained network and state-of-art text-based methods. However, due to it being a small network, its performance is less stable.

We also notice that the relative popularity in terms of retweet number is more difficult to predict than likes. This is reasonable because, as explained in the introduction, the act of retweeting changes the audience of the tweet. The factor of a tweet receiving many retweets, or *going viral*, is more difficult to explain and predict than a tweet receiving many likes, which is mostly based on the tweet content itself.

5 Conclusion

In this work, we study predicting image tweet popularity based on image and text contents. We propose a dedicated hybrid convolutional neural network that captures image local features with regard to popularity measurements. We compare our dedicated network to a pre-trained network built for object detection and text-based methods. Contrary to prior works that find text being better predictor than image content, we find that our dedicated network is able to make better prediction than pre-trained network, and is comparable with state-of-art text-based methods, particularly when predicting relative popularity. In future, we plan to further investigate the factors that produce popularity signals in image and text contents.

Acknowledgement. This research has been supported by JSPS KAKENHI grants (#17H01828, #18K19841) and by MIC/SCOPE (#171507010) grant.

References

1. Cappallo, S., Mensink, T., Snoek, C.G.: Latent factors of visual popularity prediction. In: Proceedings of the 5th ACM on International Conference on Multimedia Retrieval, pp. 195–202. ACM (2015)
2. Gao, S., Ma, J., Chen, Z.: Modeling and predicting retweeting dynamics on microblogging platforms. In: Proceedings of the Eighth ACM International Conference on Web Search and Data Mining, pp. 107–116. ACM (2015)
3. Gelli, F., Uricchio, T., Bertini, M., Del Bimbo, A., Chang, S.-F.: Image popularity prediction in social media using sentiment and context features. In: Proceedings of the 23rd International Conference on Multimedia, pp. 907–910. ACM (2015)
4. Khosla, A., Das Sarma, A., Hamid, R.: What makes an image popular? In: Proceedings of the 23rd International Conference on World Wide Web, pp. 867–876. ACM (2014)
5. Krizhevsky, A., Sutskever, I., Hinton, G.E.: Imagenet classification with deep convolutional neural networks. In: Advances in Neural Information Processing Systems, pp. 1097–1105 (2012)
6. McParlane, P.J., Moshfeghi, Y., Jose, J.M.: Nobody comes here anymore, it's too crowded; predicting image popularity on Flickr. In: Proceedings of International Conference on Multimedia Retrieval, p. 385. ACM (2014)
7. Pennington, J., Socher, R., Manning, C.: GloVe: global vectors for word representation. In: Proceedings of the 2014 Conference on Empirical Methods in Natural Language Processing, pp. 1532–1543 (2014)
8. Ramisa, A., Yan, F., Moreno-Noguer, F., Mikolajczyk, K.: Breakingnews: article annotation by image and text processing. arXiv preprint arXiv:1603.07141 (2016)
9. Sriram, B., Fuhry, D., Demir, E., Ferhatosmanoglu, H., Demirbas, M.: Short text classification in Twitter to improve information filtering. In: Proceedings of the 33rd International ACM SIGIR Conference on Research and Development in Information Retrieval, pp. 841–842 (2010)
10. Stokowiec, W., Trzciński, T., Wołk, K., Marasek, K., Rokita, P.: Shallow reading with deep learning: predicting popularity of online content using only its title. In: Kryszkiewicz, M., Appice, A., Ślęzak, D., Rybinski, H., Skowron, A., Raś, Z.W. (eds.) ISMIS 2017. LNCS (LNAI), vol. 10352, pp. 136–145. Springer, Cham (2017). https://doi.org/10.1007/978-3-319-60438-1_14
11. Szegedy, C., Vanhoucke, V., Ioffe, S., Shlens, J., Wojna, Z.: Rethinking the inception architecture for computer vision. In: Proceedings of the IEEE Conference on Computer Vision and Pattern Recognition, pp. 2818–2826 (2016)
12. Totti, L.C., Costa, F.A., Avila, S., Valle, E., Meira Jr., W., Almeida, V.: The impact of visual attributes on online image diffusion. In: Proceedings of the 2014 ACM Conference on Web Science, pp. 42–51. ACM (2014)
13. Wu, S., Hofman, J.M., Mason, W.A., Watts, D.J.: Who says what to whom on Twitter. In: Proceedings of the 20th International World Wide Web Conference, pp. 705–714 (2011)

Enriching Word Embeddings for Patent Retrieval with Global Context

Sebastian Hofstätter[1(✉)], Navid Rekabsaz[2], Mihai Lupu[3], Carsten Eickhoff[4], and Allan Hanbury[1,5]

[1] TU Wien, Vienna, Austria
{sebastian.hofstaetter,allan.hanbury}@tuwien.ac.at
[2] Idiap Research Institute, Martigny, Switzerland
navid.rekabsaz@idiap.ch
[3] Research Studios Austria, Vienna, Austria
mihai.lupu@researchstudio.at
[4] Brown University, Providence, USA
carsten@brown.edu
[5] Complexity Science Hub, Vienna, Austria

Abstract. The training and use of word embeddings for information retrieval has recently gained considerable attention, showing competitive performance across various domains. In this study, we explore the use of word embeddings for patent retrieval, a challenging domain, especially for methods based on distributional semantics. We hypothesize that the previously reported limited effectiveness of semantic approaches, and in particular word embeddings (word2vec Skip-gram) in this domain, is due to inherent constraints on the (short) window context that is too narrow for the model to capture the full complexity of the patent domain. To address this limitation, we jointly draw from local and global contexts for embedding learning. We do this in two ways: (1) adapting the Skip-gram model's vectors using global retrofitting (2) filtering word similarities using global context. We measure patent retrieval performance using BM25 and LM Extended Translation models and observe significant improvements over three baselines.

1 Introduction

Distributed representations of semantic and syntactic term content are surging in popularity. Several recent studies [4,7,17–22] focus on novel approaches to representing words in a vector space and show promising retrieval results in domains such as Web, news, and health search.

Prior art search (or *patent retrieval*) is a challenging retrieval domain. The nature of patent text has been shown to be a source of difficulty for retrieval models that perform very well on other domains [9]. In fact, the effectiveness of semantic resources, especially distributional semantics for patent retrieval has been disputed altogether [8].

L. Azzopardi et al. (Eds.): ECIR 2019, LNCS 11437, pp. 810–818, 2019.
https://doi.org/10.1007/978-3-030-15712-8_57

In this paper, we revisit this problem in light of recent advances in word embedding learning for document retrieval. We hypothesize that the limited effectiveness of state-of-the-art word embeddings (e.g. word2vec Skip-gram [10]) is due to their focus on local word context and that this is too narrow to capture the complexity of the patent domain language. Since fully extending contexts to the document level has also been shown not to perform well [8], we will investigate the combination of both local and global (document) contexts for embedding learning. We show that by drawing from these complementary sources of information, we can significantly improve performance in terms of recall-based measures that are central in this domain.

We use the Extended Translation variants [18] of $BM25$ and language models (LM) [14], referred to as $\widehat{BM25}$ and $\widehat{LM}$ to factor statistical semantics into the retrieval models. We examine the retrieval effectiveness using a word2vec Skip-gram embedding (based on a local window context) and observe that using $\widehat{BM25}$ and $\widehat{LM}$ with similar words from Skip-gram leads to a mild, yet statistically insignificant improvement in retrieval performance in the patent domain. The use of global context was previously suggested as an additional filter method [17] in other domains. We extend this hypothesis to the patent domain and additionally create a new vector representation based on local and global context. We employ the document-wide context of words using Latent Semantic Indexing (LSI).

To combine LSI and Skip-gram based word similarities we study two methods: (1) retrospectively adapting the Skip-gram model's vector representations based on the LSI-induced word similarities using Retrofitting [5]. (2) Inspired by the Post-Filtering method [17], we filter the Skip-gram model's result according to the LSI model similarities. In addition, motivated by previous studies [5, 12], we examine the effects of using explicitly curated semantic lexicons (e.g., WordNet). To this end, we propose two methods to combine LSI-induced similarity information and semantic lexicons.

We evaluate the methods on the CLEF-IP 2013 benchmark [13] and show a significant improvement in comparison with $BM25$ and LM as well as $\widehat{BM25}$ and $\widehat{LM}$ using Skip-gram and LSI separately.

This study fits into the larger category of research using or learning semantic resources for retrieval: some use pseudo-relevance information for training per-query word embeddings [4] or generic query embeddings [21]. Other studies follow a supervised approach to learning IR-specific word representations [19, 20] from relevance judgments. In contrast to these studies, our retrofitting approach learns a generic word embedding (no per-query overhead) and does not require industry-scale amounts of relevance judgments or sample queries.

2 Background

2.1 Retrofitting

Retrofitting [5] is an efficient post-processing method to adapt vector representations of existing word embeddings based on word-word similarities provided by a secondary resource. The method modifies the original vector representations by optimizing the following objective function:

$$\Psi(V) = \sum_{t \in T} \left[\alpha_t \left\| v_t - \widehat{v}_t \right\|^2 + \sum_{t' \in R(t)} \beta_{tt'} \left\| v_t - v_{t'} \right\|^2 \right] \tag{1}$$

where $\widehat{v}$ is the original vector and $v \in V$ denote its retrofitted vectors, T is the set of words in the embedding, and $R(t)$ is the set of similar words in the external resource. α_t represents the weight of the original vector of word t, and $\beta_{tt'}$ represents the similarity weight between the words t and t' in the external resource.

In order to minimize $\Psi(V)$, the derivative of Eq. 1 is set to zero, resulting in the following vector update formula:

$$v_t = \frac{\sum_{t' \in R(t)} \beta_{tt'} v_{t'} + \alpha_t \widehat{v}_t}{\sum_{t' \in R(t)} \beta_{tt'} + \alpha_t} \tag{2}$$

As shown in the formula, with each update v_t comes closer to the related vectors $v_{t'}$, where relatedness is defined and measured by the external resource. The retrofitting method iteratively updates the vectors with Eq. 2 until convergence.

2.2 Extended Translation Models

Rekabsaz et al. [18] introduce Extended Translation models for several probabilistic retrieval models (among which BM25 and LM) as a variant to the translation LM [3], providing a robust way of using word embeddings for document retrieval. The authors consider a form of term-term relation, based on the underlying concepts of each term, where the concepts are extracted from an embedding model. The Extended Translation models therefore, instead of counting the occurrences of a term, count the occurrences of the term's concepts in the documents. Based on this idea, they define the extended *tf* of a query term t in a document d as:

$$\widehat{tf}_{t,d} = tf_{t,d} + \sum_{t' \in R(t)} P_T(t|t') tf_d(t') \tag{3}$$

where $P_T(t|t')$ is the translation probability, and $R(t)$ is the set of similar terms, both captured from a word embedding. In addition to $\widehat{tf}$, the Extended Translation models use updated versions of other components (i.e. document length, collection and document frequency), calculated in accordance to the changes in term frequency.

3 Methodology

The focus of this paper lies on the source and measurement of global context used in the retrieval, rather than the retrieval models themselves. In this section, we propose different models to gauge the necessary word-word similarities.

SkipGram, LSI. These two baseline methods use a set of related words obtained from a word2vec Skip-gram embedding, and an LSI embedding, respectively. For each model we empirically determine a threshold on the similarity values between words by evaluating a parameter sweep over the threshold parameter.

Retro(*). This method applies retrofitting on a Skip-gram word embedding. The input resource * can be any external resource defining a similarity relation between words. Similar to [5], we set $\alpha_t = 1$ in Eq. 1, and normalize the values of β so that the sum of $\beta_{tt'}$ for word t is equal to one:

$$\beta_{tt'} = \frac{s_{tt'}}{\sum_{t'' \in R(t)} s_{tt''}} \tag{4}$$

where $s_{tt'}$ is the similarity score between the words t and t', given by the external resource. If the input resource is also a word embedding scheme (i.e., LSI), a second threshold (for selecting LSI similarities) is required in order to define $R(t)$.

PostFilter(*) This method filters the set of related words of `SkipGram` ($R(t)$), removing any words that do not also appear in the set of related words of the external resource $R^*(t)$. Hence, `PostFilter` is defined by the conjunction of both sets: $R(t) \cap R^*(t)$. In general, `PostFilter` models follow a conservative approach by considering two words related only when both the `SkipGram` and the external model agree.

ExtRetro(,*)* The Extended Retrofitting model exploits two input resources for the retrofitting procedure. The model extends eq. 2 as shown in the following:

$$v_t = \frac{\gamma \sum_{t' \in R^1(t)} \beta^1_{tt'} v_{t'} + (1-\gamma) \sum_{t' \in R^2(t)} \beta^2_{tt'} v_{t'} + \alpha_i \widehat{v}_t}{\sum_{t' \in R^1(t)} \beta^1_{tt'} + \sum_{t' \in R^2(t)} \beta^2_{tt'} + \alpha_i} \tag{5}$$

where the superscripts on $R(t)$ and β indicate the corresponding similarity models, given as input. In our experiments we set γ to 0.5 to enable both resources to have an equally strong influence.

PFRetro(,*)* The Post-Filter Retrofitting model combines the information of two external resources for the final set of related terms. It applies the `PostFilter` using the first input on the results of the `Retro` model, retrofitted by the second input resource.

4 Evaluation and Results

This section describes our experiment setup, presents and discusses the evaluation results, and finally analyzes the robustness of the methods.

4.1 Experiment Setup

Benchmark and Indexing. We conduct experiments on the CLEF-IP 2013 Claims to Passage task [13]. The collection contains approximately 2.6 million patent documents, and 50 query topics. Similar to Anderson et al. [1], we formulate the queries by selecting the top 100 words in the query documents with highest *tf idf* weights. We conduct the evaluation on the document level using the standard evaluation metrics of the task, namely MAP, PRES@1000, and RECALL@1000. For the retrieval we use Lucene and our implementation of $\widehat{BM25}$ and $\widehat{LM}$[1]. As suggested by previous studies [1,8], we do not apply stemming.

Similarity Resources. We create the Skip-gram word embedding with 300 dimensions on the complete CLEF-IP corpus using Gensim [15]. We use a window of 5 words, negative sampling of 10, down sampling of 10^{-5}, 20 epochs, and filtering words with frequency less than 100. We experiment with two types of external resources for word similarities: Document-context Latent Semantic Indexing (LSI), and semantic lexicons. The LSI word embedding is created on the CLEF-IP text corpus, following the approach in Rekabsaz et al. [17]. Similar to Faruqui et al. [5], we use four semantic lexicons: FrameNet [2], PPDB [6], only synonyms of WordNet [11] (WN.synonyms), and WordNet with synonyms, hypernyms and hyponyms (WN.synonyms+).

Baselines. We use three baselines to compare the retrieval performance of $\widehat{BM25}$ and $\widehat{LM}$, when using the word similarity methods: The first are the standard $BM25$ and LM (without adding any semantic information), which we refer to as `None`, the second are $\widehat{BM25}$ and $\widehat{LM}$ using the local-context `SkipGram` method, studied in previous work [16,18], and the third are also $\widehat{BM25}$ and $\widehat{LM}$ but using the global-context `LSI` method. We measure statistical significance of differences of the results using a two-sided paired t-test with $p < 0.05$.

Parameter Settings. The Dirichlet prior μ of the LM and also b, k_1, and k_3 for $BM25$ are shared between all method variants, hence we use the same set of values suggested by Rekabsaz et al. [18]. We explore cosine similarity threshold values to select similar words in the range of $[0.6, 1]$ with steps of 0.01. We explore LSI threshold values in the range of $[0.5, 0.9]$ with steps of 0.02. The final results are reported by applying 5-fold cross validation.

4.2 Results and Discussion

Table 1 reports retrieval performance of $\widehat{BM25}$ and $\widehat{LM}$, comparing the methods presented in Sect. 3. Contrasting the results of the baselines in the first section of the table, we observe (1) generally better performance of $\widehat{LM}$ in comparison to $\widehat{BM25}$ on the baselines across all evaluation metrics, and (2) only slight improvements of the `SkipGram` and `LSI` methods in comparison to `None`, with differences being significant mainly for MAP of the $\widehat{BM25}$ model.

[1] Our code and the Lucene extensions are available at: github.com/sebastian-hofstaetter/ir-generalized-translation-models.

The second section of Table 1 shows the effect of combining semantic lexicons with word similarities. Except for the case of PPDB on MAP of the $\widehat{LM}$ model, none of the semantic lexicon resources introduce significant improvements with respect to the baselines.

Table 1. Evaluation results of various word similarity methods on the CLEF-IP 2013 collection. Statistical significance to baselines: †: `None`, ρ: `SkipGram`, ω: `LSI`

Word similarity method	$\widehat{BM25}$			$\widehat{LM}$		
	MAP	PRES	RECALL	MAP	PRES	RECALL
`None`	0.184	0.607	0.703	0.200	0.669	0.755
`SkipGram`	0.207†	0.615†	0.679	0.200	0.665	0.758
`LSI`	0.191	0.650†ρ	0.737ρ	0.205	0.676	0.752
`Retro(FrameNet)`	0.206	0.633†	0.698	0.188	0.661	0.762
`Retro(WN.synonyms)`	0.206	0.610	0.705	0.208	0.651	0.717
`Retro(WN.synonyms+)`	0.180	0.597	0.674	0.207	0.638	0.754
`Retro(PPDB)`	0.194	0.625	0.715	0.240†$\rho\omega$	0.667	0.758
`PostFilter(LSI)`	**0.247**†$\rho\omega$	0.638†ρ	0.733ρ	0.228†$\rho\omega$	0.689	0.785
`Retro(LSI)`	0.238†ω	0.639†	0.733ρ	0.221	**0.698**	**0.812**†$\rho\omega$
`ExtRetro(LSI, PPDB)`	0.239†$\rho\omega$	0.624	0.733ρ	0.227†$\rho\omega$	0.669	0.765
`PFRetro(LSI, PPDB)`	0.246†$\rho\omega$	0.643†ρ	0.733ρ	0.218†ρ	0.686	0.788$\rho\omega$

The third section shows the results of exploiting LSI as an external resource to combine with the Skipgram word embedding. The `PostFilter(LSI)` and `Retro(LSI)` both significantly improve all baselines. Specifically, `PostFilter(LSI)` performs better on MAP (precision-based), showing significant MAP improvements to baselines in both IR models. On the other hand, `Retro(LSI)` shows stronger performance on recall-based metrics by significantly improving the baselines on RECALL using $\widehat{LM}$.

We assume that the better performance of the `PostFilter(LSI)` method on MAP is due to its conservative approach, as the method only keeps those related words which are common in both Skipgram and LSI word embeddings. The `Retro(LSI)` method, however, incorporates LSI similarity in the vector representation space, providing wider semantic similarity scopes for words (useful for recall), while still maintaining MAP results in the same range as or higher than the baselines.

Finally, the results of the methods with two resources, namely LSI and PPDB (the best performance among the semantic lexicons), are shown in the last section of the table. Neither of the methods (`Ext-Retro` and `PF-Retro`) consistently outperform `Retro(LSI)` and `PostFilter(LSI)`, suggesting that explicit semantic lexicons do not contribute to effectiveness improvements in this domain.

We continue our analysis by examining the robustness of the retrieval system when using the `Retro(LSI)` and `PostFilter(LSI)` word similarity methods. Figure 1 depicts Average Precision (AP) and RECALL per query using $\widehat{LM}$

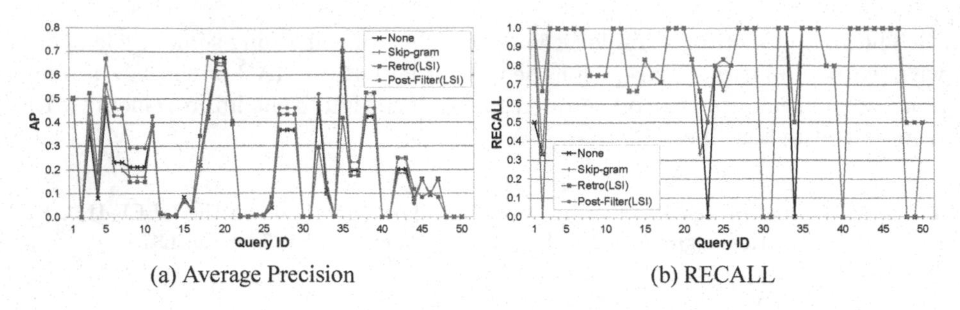

(a) Average Precision (b) RECALL

Fig. 1. Per-query retrieval performance of the $\widehat{LM}$ model on the CLEF-IP 2013 using `None`, `SkipGram`, `Retro(LSI)` and `PostFilter(LSI)`

(as it performs better in general and in particular for the baselines) with `None`, `SkipGram` (as used in [16,18]), `Retro(LSI)`, and `PostFilter(LSI)` methods. We study the robustness of the compared methods by observing the consistency of the results across queries in comparison to the results of the `None` method (no word similarity information).

Turning to `SkipGram`, we observe cases of both improved and deteriorated results in comparison to `None`, indicating a lack of robustness of the method. Tracing the reason, similar to [17] we observe several cases of topic shifting, e.g., the query term *platinum* is expanded with *palladium* and *rhodium*, causing performance losses.

In contrast, `PostFilter(LSI)` shows highly robust performance, attaining the same or a better level of performance than `None` on almost all queries on both metrics. The same characteristic applies to `Retro(LSI)` on the RECALL metric, confirming the effectiveness as well as robustness of using the `Retro(LSI)` embedding on patent retrieval for the RECALL metric.

5 Conclusion

We study the effects of enriching word embeddings for patent retrieval using a global context. Observing considerable limitations when using local-context word embeddings (word2vec Skip-gram) in patent retrieval, we suggest incorporating additional information, obtained via LSI based on global contexts. We incorporate this additional source of information via retrofitting and post-filtering methods. Using our multi-context word embeddings, we observe significant improvements over the respective retrieval baselines on the CLEF-IP 2013 task. We report early results of an ongoing line of inquiry. In the future, we intend to explore the generality of our findings by investigating retrieval domains with similar characteristics to the patent retrieval setting.

References

1. Andersson, L., Lupu, M., Palotti, J., Hanbury, A., Rauber, A.: When is the time ripe for natural language processing for patent passage retrieval? In: Proceedings of CIKM (2016)
2. Baker, C.F., Fillmore, C.J., Lowe, J.B.: The Berkeley FrameNet project. In: Proceedings of ACL (1998)
3. Berger, A., Lafferty, J.: Information retrieval as statistical translation. In: Proceedings of SIGIR (1999)
4. Diaz, F., Mitra, B., Craswell, N.: Query expansion with locally-trained word embeddings. In: Proceedings of ACL (2016)
5. Faruqui, M., Dodge, J., Jauhar, S.K., Dyer, C., Hovy, E., Smith, N.A.: Retrofitting word vectors to semantic lexicons. In: Proceedings of NAACL-HLT (2015)
6. Ganitkevitch, J., Van Durme, B., Callison-Burch, C.: PPDB: the paraphrase database. In: Proceedings of NAACL (2013)
7. Kuzi, S., Shtok, A., Kurland, O.: Query expansion using word embeddings. In: Proceedings of CIKM (2016)
8. Lupu, M.: On the usability of random indexing in patent retrieval. In: Hernandez, N., Jäschke, R., Croitoru, M. (eds.) ICCS 2014. LNCS (LNAI), vol. 8577, pp. 202–216. Springer, Cham (2014). https://doi.org/10.1007/978-3-319-08389-6_17
9. Lupu, M., Hanbury, A.: Patent retrieval. In: Foundations and Trends in Information Retrieval (2013)
10. Mikolov, T., Sutskever, I., Chen, K., Corrado, G.S., Dean, J.: Distributed representations of words and phrases and their compositionality. In: Proceedings of NIPS (2013)
11. Miller, G.A.: WordNet: a lexical database for English. Commun. ACM **38**, 39–41 (1995)
12. Nguyen, G.-H., Soulier, L., Tamine, L., Bricon-Souf, N.: DSRIM: a deep neural information retrieval model enhanced by a knowledge resource driven representation of documents. In: Proceedings of SIGIR (2017)
13. Piroi, F., Lupu, M., Hanbury, A.: Overview of CLEF-IP 2013 lab. In: Forner, P., Müller, H., Paredes, R., Rosso, P., Stein, B. (eds.) CLEF 2013. LNCS, vol. 8138, pp. 232–249. Springer, Heidelberg (2013). https://doi.org/10.1007/978-3-642-40802-1_25
14. Ponte, J.M., Croft, W.B.: A language modeling approach to information retrieval. In: Proceedings of SIGIR (1998)
15. Řehůřek, R., Sojka, P.: Software framework for topic modelling with large corpora. In: Proceedings of LREC Workshop on New Challenges for NLP Frameworks (2010)
16. Rekabsaz, N., Lupu, M., Hanbury, A.: Exploration of a threshold for similarity based on uncertainty in word embedding. In: Jose, J.M., et al. (eds.) ECIR 2017. LNCS, vol. 10193, pp. 396–409. Springer, Cham (2017). https://doi.org/10.1007/978-3-319-56608-5_31
17. Rekabsaz, N., Lupu, M., Hanbury, A., Zamani, H.: Word embedding causes topic shifting; exploit global context! In: Proceedings of SIGIR (2017)
18. Rekabsaz, N., Lupu, M., Hanbury, A., Zuccon, G.: Generalizing translation models in the probabilistic relevance framework. In: Proceedings of CIKM (2016)
19. Xiong, C., Callan, J., Liu, T.-Y.: Word-entity duet representations for document ranking. In: Proceedings of SIGIR (2017)

20. Xiong, C., Dai, Z., Callan, J., Liu, Z., Power, R.: End-to-end neural ad-hoc ranking with kernel pooling. In: Proceedings of SIGIR (2017)
21. Zamani, H., Croft, W.B.: Relevance-based word embedding. In: Proceedings of SIGIR (2017)
22. Zuccon, G., Koopman, B., Bruza, P., Azzopardi, L.: Integrating and evaluating neural word embeddings in information retrieval. In: Proceedings of Australasian Document Computing Symposium (2015)

Incorporating External Knowledge to Boost Machine Comprehension Based Question Answering

Huan Wang, Weiming Lu(✉), and Zeyun Tang

College of Computer Science and Technology, Zhejiang University, Hangzhou, China
{21621127,luwm,21721114}@zju.edu.cn

Abstract. We propose an effective knowledge representation network via a two-level attention mechanism, called KRN, to represent the background knowledge of entities in documents for boosting machine comprehension (MC). In experiments, we incorporated the KRN into several state-of-the-art MC models such as AS Reader, CAS Reader, GA Reader and BiDAF, and evaluated the performance of KRN using two datasets: WebQA and Quasar-T. Experimental results show that our KRN can improve the performance of the existing MC models.

1 Introduction

Machine Comprehension based Question Answering (MCQA) aims to answer a question with a chunk of text taken from documents [21]. Table 1 shows an example from Quasar-T [7].

A variety of machine comprehension models have been proposed recently to address this problem, such as DrQA [1], BiDAF [17], AS Reader [11], CAS Reader [5], AoA Reader [4], GA Reader [6], EpiReader [19], Reinforced mnemonic reader [9], Gated Self-Matching Network (GSMN) [20], ReasoNet [18] and FusionNet [10].

However, many existing works only rely on the information in the documents such as words, characters, part-of-speech (POS) tags and named-entity recognition (NER) tags. For example, AS Reader, CAS Reader and AoA Reader only utilized the word-level embedding in their encoder. BiDAF and GSMN utilized both word-level and character-level embedding in their encoders. MEMEN [16] and Reinforced mnemonic reader further utilized POS and NER tags embedding to enhance the capacity of the encoder. In addition, position information was proved to be helpful in [2].

Actually, human interpret texts and answer questions with respect to some background knowledge, so we think if the background knowledge can be incorporated into the MC models, the performance could be improved. Taking the example in Table 1, the tags of the entity "France" in Wikipedia are *Nationality, Capital* and *French*, and the description is *The Fifth Republic, led by Charles de Gaulle, was formed in 1958 and remains today. Algeria and nearly all the other*

L. Azzopardi et al. (Eds.): ECIR 2019, LNCS 11437, pp. 819–827, 2019.
https://doi.org/10.1007/978-3-030-15712-8_58

Table 1. An example of MCQA from Quasar-T dataset.

Question: What country 's current government was established in 1958 and is known as the Fifth Republic?
Document: The coming of the Fifth Republic In May 1958 a revolt of French settlers and army officers in Algeria against what they regarded as the effeteness of the government in Paris and its handling of the AlFifth Republic, modern France France History
Answer: France

colonies became independent in the 1960s... Obviously, they are quite related to the question, which makes the answer to be "France" more likely. However, AS Reader, CAS Reader and BiDAF, which don't resort to the background knowledge, all obtained the wrong answer "Algeria". The reason may be that the context of "Algeria" in the document is more similar to the question.

Therefore, we propose a knowledge representation network via a two-level attention mechanism, called KRN, to represent the background knowledge of entities by utilizing the tags and descriptions to boost MC. Figure 1 illustrates the MCQA framework with KRN, which shows KRN can be easily plugged into an existing MC model.

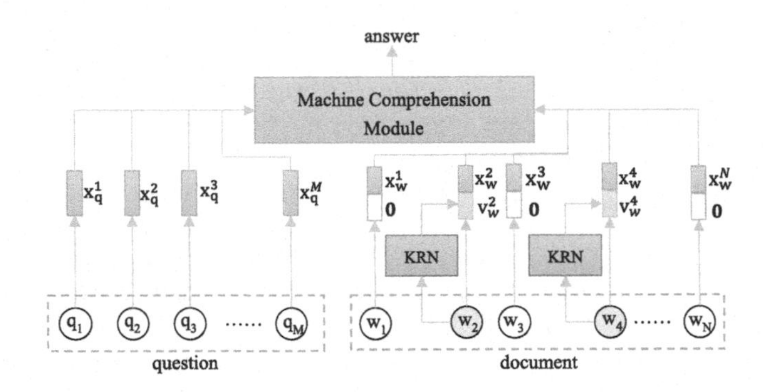

Fig. 1. QA framework with knowledge

We perform experiments on two large-scale benchmarks datasets: WebQA [13] and Quasar-T [7], and verify the KRN with a set of existing MC models. Our experiments show that our KRN can improve the performance of the most common MC models.

2 MCQA with Knowledge

In this section, we mainly focus on how KRN is incorporated into the MCQA frameworks as shown in Fig. 1.

For MCQA, given a question Q and a context C, the task is to predict an answer A, which is a chunk of text in C. Let $Q = \{q_i\}_{i=1}^{M}$ and $C = \{w_i\}_{i=1}^{L}$, which are sequences of words. For words $q_i \in Q$ and $w_i \in C$, we simply used pretrained word vectors to convert them to their word embeddings $\mathbf{x}_q^i \in \mathbb{R}^d$ and $\mathbf{x}_w^i \in \mathbb{R}^d$. We do not update the word embeddings during training. Hence, we obtain the query encoding matrix $\mathbf{Q} \in \mathbb{R}^{d \times M}$ and the context encoding matrix $\mathbf{C} \in \mathbb{R}^{d \times L}$, which are considered as the input of MC models such as AS Reader, CAS Reader, GA Reader and BiDAF.

In order to incorporate background knowledge, we proposed KRN for knowledge embedding. Specifically, we firstly used the entity link technology [8] to link mentions in C to the corresponding entities in a knowledge base. For example, a mention w_i is linked to the entity e_i, and e_i can be encoded to $\mathbf{v}_e^i \in \mathbb{R}^{2n}$ through KRN. Finally, word $w_i \in C$ is represented as the concatenation of its word-level and knowledge-level embeddings, denoted as $\mathbf{x}_i = [\mathbf{x}_w^i; \mathbf{v}_e^i] \in \mathbb{R}^{d+2n}$. For words that are not linked to the knowledge base, the knowledge-level embedding is assigned by a zero vector $\mathbf{0} \in \mathbb{R}^{2n}$.

3 Knowledge Representation Network

The goal of KRN is to generate the knowledge-level embedding $\mathbf{v}_e^i$ for an entity e_i to boost machine comprehension. In order to apply KRN in different language environments, we only rely on the description and tags of entities from encyclopedias such as Wikipedia and Baidu Baike[1] to learn the embedding, since some complicated knowledge such as knowledge graph is quite difficult to obtain in non-English environment.

Therefore, the task of KRN is formally defined as: given an entity e with its description D and tags T, it aims to learn the knowledge-level embedding v_e. Here, D is represented as a sequence of words $\{w_i\}_{i=1}^{N}$, and T is a bag of tags $\{t_i\}_{i=1}^{K}$. The architecture of KRN is shown in Fig. 2 and will be explained in the following sections.

3.1 Tags-Aware Attention Layer

Tags often represent related but more general concepts over an entity, thus we can learn a more robust representation of the entity by combining its tags via an attention mechanism.

Each tag t_i is assigned with a word embedding vector $\mathbf{x}_t^i \in \mathbb{R}^d$ through pretrained vectors, thus the representation of entity e can be formulated as a convex combination of the embeddings of itself and its tags:

$$\mathbf{u}_e = \sum_{\mathbf{x}_i \in \{\mathbf{x}_e \cup \{\mathbf{x}_t^j\}_{j=1}^{K}\}} \alpha(\mathbf{x}_i, \mathbf{x}_e) \cdot \mathbf{x}_i$$

[1] https://baike.baidu.com/.

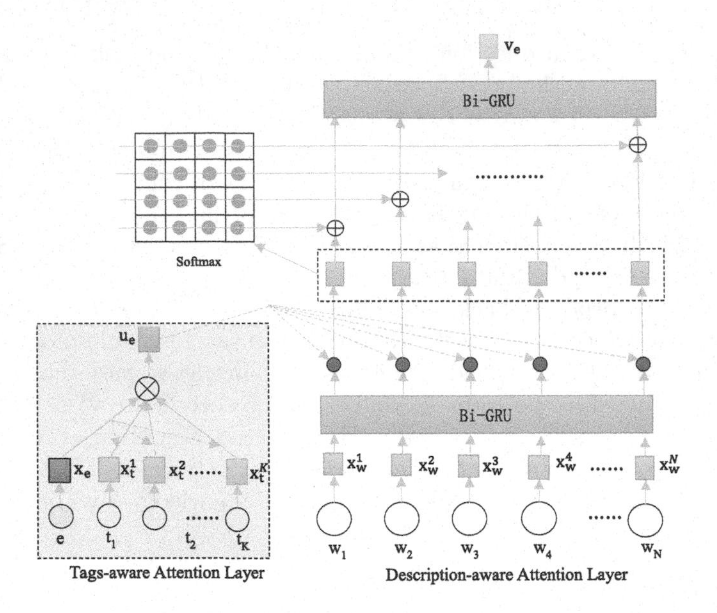

Fig. 2. Knowledge representation network

where $\mathbf{x}_e \in \mathbb{R}^d$ is the embedding of entity e, and $\alpha(\mathbf{x}_i, \mathbf{x}_e)$ is the attention weight, which is calculated by a softmax function:

$$\alpha(\mathbf{x}_i, \mathbf{x}_e) = \frac{\exp(f(\mathbf{x}_i, \mathbf{x}_e))}{\sum_{\mathbf{x}_k \in \{\mathbf{x}_e \cup \{\mathbf{x}_t^j\}_{j=1}^K\}} \exp(f(\mathbf{x}_k, \mathbf{x}_e))}$$

Here, $f(\mathbf{x}_i, \mathbf{x}_e)$ represents the compatibility between the embeddings of $\mathbf{x}_i$ and $\mathbf{x}_e$, and it is calculated by a feed-forward network with a single hidden layer (MLP):

$$f(\mathbf{x}_i, \mathbf{x}_e) = \mathbf{u}^\mathsf{T} \tanh(\mathbf{W}[\mathbf{x}_i; \mathbf{x}_e] + \mathbf{b})$$

where $\mathbf{W} \in \mathbf{R}^{l_1 \times 2d}$, $\mathbf{b} \in \mathbf{R}^{l_1}$ and $\mathbf{u} \in \mathbf{R}^{l_1}$ are trainable parameters of MLP, and l_1 is the dimension size of the hidden layer of $f(\cdot, \cdot)$.

3.2 Description-Aware Attention Layer

The description also implies the semantics of an entity. In order to distill the important parts of the description, we also used attention mechanism.

Firstly, each word w_i in the description D is assigned with a vector $\mathbf{x}_w^i \in \mathbb{R}^d$ through pretrained vectors, so the representation of the description $\mathbf{D} = [\mathbf{x}_w^i]_{i=1}^N \in \mathbb{R}^{d \times N}$. Then, attention mechanism is used to pinpoint the most relevant words by:

$$\beta = softmax(\mathbf{u}_e^\mathsf{T} \cdot \mathbf{D})$$

The representation of the description is updated by $\tilde{\mathbf{D}} = [\tilde{\mathbf{D}}_i]_{i=1}^N$, where $\tilde{\mathbf{D}}_i = \beta_i \cdot \mathbf{D}_i \in \mathbb{R}^d$ is the ith column of $\tilde{\mathbf{D}}$.

Subsequently, we used a bidirectional Gated Recurrent Unit network (Bi-GRU) [3] to produce new representation $\tilde{\mathbf{D}}'$ of the description with $\tilde{\mathbf{D}}$.

$$\tilde{\mathbf{D}}'_i = BiGRU(\tilde{\mathbf{D}}'_{i-1}, \tilde{\mathbf{D}}_i), \forall i \in [1, ..., N]$$

where $\tilde{\mathbf{D}}'_i \in \mathbb{R}^{2l_2}$ is the ith column of $\tilde{\mathbf{D}}'$, which are concatenated hidden states of BiGRU for the ith description word, and l_2 is the dimension of hidden state.

Inspired by [9], self alignment was then applied to $\tilde{\mathbf{D}}'$, which can fuse the crucial clues between words in the description into the representation. The self-coattention matrix $\mathbf{B} \in \mathbb{R}^{N \times N}$ can be calculated by:

$$\mathbf{B}_{ij} = \delta(i \neq j)\tilde{\mathbf{D}}'^{\mathsf{T}}_i \cdot \tilde{\mathbf{D}}'_j$$

where $\mathbf{B}_{ij}$ indicates the similarity between i^{th} word and j^{th} word in the description, and the diagonal of the matrix is set to zero by the δ function. Let $\mathbf{b}_i \in \mathbb{R}^N$ denote the normalized attention distribution of description for the ith word, so we can get $\mathbf{b}_i = softmax(\mathbf{B}_{:,i})$. Hence, the corresponding attended vector of the ith word is $\tilde{\mathbf{x}}'_i = \tilde{\mathbf{D}}' \cdot \mathbf{b}_i \in \mathbb{R}^{2l_2}$, and $\tilde{\mathbf{X}} = [\tilde{\mathbf{x}}'_i]_{i=1}^N \in \mathbb{R}^{2l_2 \times N}$.

Finally, $\tilde{\mathbf{D}}'$, $\tilde{\mathbf{X}}$, $\tilde{\mathbf{D}}' \circ \tilde{\mathbf{X}}$ and $\tilde{\mathbf{D}}' - \tilde{\mathbf{X}}$ are concatenated and fed into another Bi-GRU as $BiGRU(\tilde{\mathbf{D}}', \tilde{\mathbf{X}}, \tilde{\mathbf{D}}' \circ \tilde{\mathbf{X}}, \tilde{\mathbf{D}}' - \tilde{\mathbf{X}})$, where $\circ$ stands for the element-wise multiplication, $-$ means the element-wise subtraction, and n is the dimension of hidden states in BiGRU. The entity vector $\mathbf{v}_e \in \mathbb{R}^{2n}$ is obtained by combining the final forward and backward GRU hidden states.

4 Experiments

4.1 Experimental Setup

Datasets. We conduct experiments on the WebQA [13] and Quasar-T [7] datasets.

WebQA is a Chinese QA dataset for open-domain factoid QA system evaluation, where all questions are indeed asked by real-world users, and the passages are retrieved from Internet by using a search engine with questions as queries.

Quasar-T is based on a set of open-domain trivial questions, and the answer for each question is extracted from the top ranked pseudocuments from ClueWeb09 dataset.

The statistics of the datasets are shown in Table 2.

Baseline Methods. To verify the effectiveness of our KRN, we incorporated it into 4 different baseline models, including AS Reader [11], CAS Reader [5], GA Reader [6] and BiDAF [17]. We directly used open sources of AS Reader[2], BiDAF[3] and GA Reader[4], and implemented CAS Reader by ourselves.

[2] https://github.com/rkadlec/asreader.
[3] https://allenai.github.io/bi-att-flow.
[4] https://github.com/bdhingra/ga-reader.

Table 2. The statistics of the datasets, where #q represents the number of questions for training, validation, and test sets.

	#q(train)	#q(val)	#q(test)
WebQA	15480	1552	1511
Quasar-T	15838	1275	1264

Implementation Details. We used Stanford CoreNLP [14] and HanLP[5] to preprocess English and Chinese text respectively. Then, Word2vec [15] was used to train Chinese and English word embeddings with Baidu Baike and English Wikipedia, and the dimensions are both set as $d = 100$. We set the hidden size with $l_1 = 100$, $l_2 = 128$ and $n = 128$, and used Adam [12] optimizer with an initial learning rate as 0.001. In addition, we used Baidu Baike and Wikipedia for the tags and descriptions in KRN.

4.2 Main Results

The answer to each question is an entity, so the model is given a credit if its answer exactly matches the entity. We use EM (Exact Match) score, which is defined as the score of 100% accuracy in prediction, to evaluate the models. Table 3 shows the performance of baseline models and their variations which are enhanced by KRN.

From the table, we observe that KRN can achieve a better performance for MCQA on both WebQA and Quasar-T datasets by incorporating the external knowledge.

We also did the ablation experiments for the MC models with only one attention layer, and the results are also shown in Table 3. From the table, we can see that both tags-aware and description-aware attentions are helpful for KRN to boost the MCQA.

Figure 3 shows a case of the attention result of MCQA with KRN, which indicates that our two-level attention mechanism can enrich the representation of background knowledge, and then reinforce the connection between the question and answer.

[5] https://github.com/hankcs/HanLP.

Table 3. The performance of MC models and their variations, where ^{T}, ^{D} and $^\#$ denote that the models are enhanced by only tags-aware attention layer, only description-aware attention layer and KRN respectively. We restricted the numbers of documents used in Quasar-T by 20 due to memory errors for more documents.

Model	WebQA		Quasar-T	
	Val	Test	Val	Test
AS Reader	0.6475	0.6115	0.5709	0.5422
AS Reader^T	0.6488	0.6109	0.5833	0.5489
AS Reader^D	0.6572	0.6326	0.5711	0.5522
AS Reader$^\#$	**0.6707**	**0.6347**	**0.5933**	**0.5722**
CAS Reader	0.6507	0.6340	0.5846	0.5329
CAS Reader^T	0.6669	0.6347	**0.6044**	0.5578
CAS Reader^D	0.6701	0.6327	0.5850	0.5433
CAS Reader$^\#$	**0.6707**	**0.6353**	0.5878	**0.5600**
GA Reader	0.6340	0.5917	0.5641	0.5480
GA Reader^T	0.6527	0.6320	0.5656	0.5611
GA Reader^D	0.6695	0.6413	**0.5844**	0.5667
GA Reader$^\#$	**0.6772**	**0.6598**	0.5667	**0.5678**
BiDAF	0.6630	0.6155	0.5611	0.5656
BiDAF^T	0.6765	0.6565	0.5700	0.5611
BiDAF^D	**0.6843**	0.6300	**0.6000**	0.5678
BiDAF$^\#$	0.6765	**0.6585**	0.5800	**0.5756**

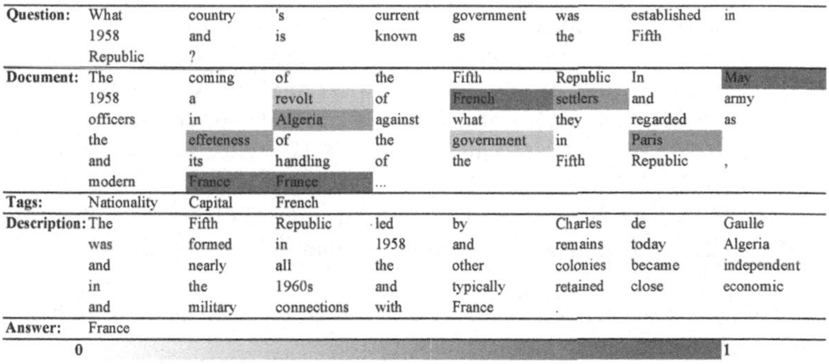

Question:	What	country	's	current	government	was	established	in
	1958	and	is	known	as	the	Fifth	
	Republic	?						
Document:	The	coming	of	the	Fifth	Republic	In	May
	1958	a	revolt	of	French	settlers	and	army
	officers	in	Algeria	against	what	they	regarded	as
	the	effeteness	of	the	government	in	Paris	
	and	its	handling	of	the	Fifth	Republic	,
	modern	France	France	...				
Tags:	Nationality	Capital	French					
Description:	The	Fifth	Republic	led	by	Charles	de	Gaulle
	was	formed	in	1958	and	remains	today	Algeria
	and	nearly	all	the	other	colonies	became	independent
	in	the	1960s	and	typically	retained	close	economic
	and	military	connections	with	France	.		
Answer:	France							

0 … 1

Fig. 3. The attention result of the example in Table 1

5 Conclusion

We propose an effective knowledge representation network via a two-level attention mechanism to address the problem of incorporating external knowledge into the existing machine comprehension models. The experiments verify the effectiveness of our model. In addition, our idea can also be extended to other NLP tasks so as to get a better depth of machine language understanding.

Acknowledgments. This work is supported by the Zhejiang Provincial Natural Science Foundation of China (No. LY17F020015), the Fundamental Research Funds for the Central Universities (No. 2017FZA5016), CKCEST, and MOE-Engineering Research Center of Digital Library.

References

1. Chen, D., Fisch, A., Weston, J., Bordes, A.: Reading Wikipedia to answer open-domain questions. In: Proceedings of the 55th Annual Meeting of the Association for Computational Linguistics (Volume 1: Long Papers), pp. 1870–1879. Association for Computational Linguistics (2017). https://doi.org/10.18653/v1/P17-1171, http://aclweb.org/anthology/P17-1171
2. Chen, Q., Hu, Q., Huang, J.X., He, L., An, W.: Enhancing recurrent neural networks with positional attention for question answering. In: Proceedings of the 40th International ACM SIGIR Conference on Research and Development in Information Retrieval, pp. 993–996. ACM (2017)
3. Cho, K., et al.: Learning phrase representations using RNN encoder-decoder for statistical machine translation. In: EMNLP (2014)
4. Cui, Y., Chen, Z., Wei, S., Wang, S., Liu, T., Hu, G.: Attention-over-attention neural networks for reading comprehension. In: ACL (2017)
5. Cui, Y., Liu, T., Chen, Z., Wang, S., Hu, G.: Consensus attention-based neural networks for Chinese reading comprehension. In: COLING (2016)
6. Dhingra, B., Liu, H., Yang, Z., Cohen, W.W., Salakhutdinov, R.: Gated-attention readers for text comprehension. In: ACL (2017)
7. Dhingra, B., Mazaitis, K., Cohen, W.W.: Quasar: datasets for question answering by search and reading. arXiv preprint arXiv:1707.03904 (2017)
8. Han, X., Sun, L., Zhao, J.: Collective entity linking in web text: a graph-based method. In: Proceedings of the 34th International ACM SIGIR Conference on Research and Development in Information Retrieval, pp. 765–774. ACM (2011)
9. Hu, M., Peng, Y., Qiu, X.: Reinforced mnemonic reader for machine comprehension. CoRR, abs/1705.02798 (2017)
10. Huang, H.Y., Zhu, C., Shen, Y., Chen, W.: Fusionnet: fusing via fully-aware attention with application to machine comprehension. arXiv preprint arXiv:1711.07341 (2017)
11. Kadlec, R., Schmid, M., Bajgar, O., Kleindienst, J.: Text understanding with the attention sum reader network. In: ACL (2016)
12. Kingma, D., Ba, J.: Adam: a method for stochastic optimization. In: ICLR (2015)
13. Li, P., et al.: Dataset and neural recurrent sequence labeling model for open-domain factoid question answering. arXiv preprint arXiv:1607.06275 (2016)
14. Manning, C.D., Surdeanu, M., Bauer, J., Finkel, J., Bethard, S.J., McClosky, D.: The Stanford CoreNLP natural language processing toolkit. In: Association for Computational Linguistics (ACL) System Demonstrations, pp. 55–60 (2014). http://www.aclweb.org/anthology/P/P14/P14-5010
15. Mikolov, T., Sutskever, I., Chen, K., Corrado, G.S., Dean, J.: Distributed representations of words and phrases and their compositionality. In: Advances in Neural Information Processing Systems, pp. 3111–3119 (2013)

16. Pan, B., Li, H., Zhao, Z., Cao, B., Cai, D., He, X.: MEMEN: multi-layer embedding with memory networks for machine comprehension. arXiv preprint arXiv:1707.09098 (2017)
17. Seo, M., Kembhavi, A., Farhadi, A., Hajishirzi, H.: Bidirectional attention flow for machine comprehension. In: ICLR (2017)
18. Shen, Y., Huang, P.S., Gao, J., Chen, W.: Reasonet: learning to stop reading in machine comprehension. In: Proceedings of the 23rd ACM SIGKDD International Conference on Knowledge Discovery and Data Mining, pp. 1047–1055. ACM (2017)
19. Trischler, A., Ye, Z., Yuan, X., Suleman, K.: Natural language comprehension with the EpiReader. In: EMNLP (2016)
20. Wang, W., Yang, N., Wei, F., Chang, B., Zhou, M.: Gated self-matching networks for reading comprehension and question answering. In: Proceedings of the 55th Annual Meeting of the Association for Computational Linguistics (Volume 1: Long Papers), vol. 1, pp. 189–198 (2017)
21. Yu, Y., Zhang, W., Hasan, K.S., Yu, M., Xiang, B., Zhou, B.: End-to-end reading comprehension with dynamic answer chunk ranking. CoRR abs/1610.09996 (2016)

Impact of Training Dataset Size on Neural Answer Selection Models

Trond Linjordet(✉) and Krisztian Balog

University of Stavanger, Stavanger, Norway
{trond.linjordet,krisztian.balog}@uis.no

Abstract. It is held as a truism that deep neural networks require large datasets to train effective models. However, large datasets, especially with high-quality labels, can be expensive to obtain. This study sets out to investigate (i) how large a dataset must be to train well-performing models, and (ii) what impact can be shown from fractional changes to the dataset size. A practical method to investigate these questions is to train a collection of deep neural answer selection models using fractional subsets of varying sizes of an initial dataset. We observe that dataset size has a conspicuous lack of effect on the training of some of these models, bringing the underlying algorithms into question.

1 Introduction

The impressive performance improvements brought by deep learning applied to certain domains—computer vision, audio speech-to-text, and natural language processing (NLP) [8,13]—has motivated a great deal of interest to apply deep learning to other domains as well, including information retrieval (IR). However, the performance improvements from deep learning relative to conventional machine learning approaches have depended on increased computational power, larger datasets to learn from, and some developments on the algorithm and architecture level. Of these three factors, large datasets may represent the least tractable challenge faced by those who would apply deep learning to new domains. Quality training data, especially for supervised learning, requires intensive effort to prepare for the actual learning process.

A category of tasks at the intersection of the fields of IR and NLP, question answering (QA) means returning a correct answer sentence in response to a grammatically well-formed, natural language question. In the present work, a specific variant of the QA task is considered, namely answer selection, the task of matching single-sentence questions with single-sentence answers. The answer selection task is simply: given a question and a set of candidate answers, select the correct answer. This task has recently been investigated as a neural IR problem [9,11].

This paper considers a practical approach to investigating the impact of training dataset size on the performance that can be achieved with various deep neural architectures for the task of answer selection. The approach taken by this

L. Azzopardi et al. (Eds.): ECIR 2019, LNCS 11437, pp. 828–835, 2019.
https://doi.org/10.1007/978-3-030-15712-8_59

paper can be summarized as follows: A pre-existing implementation of neural architectures for answer selection is investigated by truncating the training data to fractions of the original training dataset, to quantify the differences in performance by trained models given different amounts of training data from the same distribution.

One of the surprising experimental findings of this paper is that most models do not exhibit the expected behavior in terms of performance improvement in response to increased training dataset size.

2 Related Work

The impact of the size of training datasets has been investigated for convolutional neural networks (CNNs) trained on image data [2,15]. In the latter work, it was observed that model performance improves roughly logarithmically as a function of increased training data. The idea of a logarithmic relationship between performance and dataset size was further corroborated empirically by Hestness, et al. [5].

An investigation of the generalization problem in deep neural networks, i.e., the discrepancy between the performance of a trained model on training data and test data, shows that the deep neural models have a representational capacity that enables "memorization" of training data: Zhang et al. [19] show the order-of-magnitude relationship between training dataset size (sample size), input data dimensionality, and the depth of a network with sufficient parameters to fully memorize the training dataset. They report a theorem with proof such that for any finite n-sized sample of d-dimensional inputs, there exists a two-layer ReLU neural network with $2n + d$ weights that can represent any function on the sample. As a corollary, this finding extends from this hypothetical shallow and wide network to a narrow and deep network where the relationship between sample size and number of parameters is conserved. This may not be how deep neural networks learn in practice [1], but the theorem indicates the challenge that finite datasets may present to generalization in deep learning models.

3 Approach

The approach presented in this paper is practical in that dataset size was manipulated and the effects were evaluated using a pre-existing implementation of multiple neural IR models with a single original dataset. Specifically, this paper presents work on the MatchZoo project[1] [3], where a number of deep neural architectures for text matching have been implemented. Here, answer selection is considered as a form of question answering, where the question text is matched with the text of the correct answer. The original dataset used for training, validation, and testing, was the canonical WikiQA dataset [18]. The performance of the implemented models on a given dataset was characterized in terms of Mean Average Precision (MAP) over the candidate answer rankings for each question in that dataset.

[1] https://github.com/faneshion/matchzoo.

3.1 Data Preparation

The training dataset was filtered to provide the models being trained with meaningfully labelled training data. The filtering rule was simply to omit any question and its associated set of candidate answer sentences if the set of candidate answer sentences did not include both true and false candidates.

Table 1. Summary of datasets.

	Training					Valid.	Test
	10%	25%	50%	75%	100%		
#Questions	78	209	414	639	857	122	237
#QA pairs	823	2256	4321	6537	8651	1126	2341

Table 1 summarizes the datasets used in the training of the various models. Note that the same validation and test sets were used throughout, while the training dataset used was systematically varied between the original (filtered) training set (100%), and various partial training sets truncated to 10%, 25%, 50%, and 75% of the original (filtered) training set. These partial training sets were made by randomly sampling (without replacement) on the questions in the original (filtered) training set. Each selected question was then included in the respective partial training set along with all corresponding candidate answers and their labels. The percentages thus represent the probability for each question to be included in each partial dataset. However, once the random sub-sampling was accomplished, these partial training sets were fixed. Each of the models was then trained five times independently on each dataset size.

3.2 Models

A number of models were able to train and perform nominally with the code provided by the MatchZoo project [3]. The models investigated in the present paper comprised:

- **Deep Structured Semantic Model (DSSM)** [7], which extends latent semantic analysis with deep architectures; a seminal work on neural IR.
- **Convolutional Deep Structured Semantic Model (CDSSM)** [14], which uses a convolutional neural network (CNN) to extend DSSM with contextual information at the word n-gram level.
- **Architecture-I (ARC-I)** [6], an extension of CDSSM whereby siamese CNNs learn to represent two sentences, deferring matching of sentence pairs to a final multi-layer perceptron (MLP).
- **Architecture-II (ARC-II)** [6], an alternative to ARC-I where sentences interact by 1D convolution before proceeding through a 2D CNN component which is purported to learn both the representation of the individual sentences, as well as the structure of their relationship. Again, matching of the representations is determined by a final MLP.

- **Multiple positional sentence representations (MV-LSTM)** [16], follows the aforementioned models by capturing local information on multiple levels of granularity within a sentence, using bidirectional long short-term memory networks (bi-LSTMs) to represent input sentences, modeling interactions with a similarity function (tensor layer), and aggregating interactions with k-Max Pooling before a final MLP to match the obtained representations.
- **Deep relevance matching model (DRMM)** [4], distinguishes relevance matching from semantic matching, using pre-trained neural embeddings of terms and building up fixed-length matching histograms from variable-length local interactions between each query term and document. Each query term matching histogram is passed through a matching MLP, and the overall score is aggregated with a query term gate—a softmax function over all terms in that query.
- **Attention-based neural matching model (aNMM)** [17], which follows a similar structure as ARC-II, except instead of position-shared weighting, aNMM has adopted a value-shared weighting scheme "to learn the importance of different levels of matching signals," and incorporated a query term gate similar to that used in DRMM.
- **Combined local and distributed representations (DUET)** [10], which aims to combine local exact matching with embeddings of query-document pairs in semantic space. This relevance matching is enabled by both the local and distributed models, hence a "duet" of two parallel neural models. The final matching score is simply the sum of the two outputs.
- **MatchPyramid** [12], which uses a matching matrix layer to evaluate pairwise term similarity between two texts, followed by 2D convolutional and pooling layers, with a final matching MLP.
- **DRMM_TKS** [3], which is a variant of DRMM provided by the MatchZoo project, for matching short texts. The architecture is simply described by "Specifically, the matching histogram is replaced by a top-k max pooling layer and the remaining parts are fixed."

Some of these models are motivated more by ad hoc search and document retrieval, whereas others were developed specifically for answer selection and the similar task of sentence completion. However, the commonality is that all the models are designed for text matching.

4 Experiment/Results

The following experimental results show the effect of varying training set size.

4.1 Final Performance of Trained Models

Figure 1 presents the performance on the test dataset of the different models after training for 400 iterations on datasets of various sizes. These figures show that aside from the DSSM, CDSSM, and possibly MatchPyramid models, some improvement does appear to happen with greater training dataset sizes. However, by having an order of magnitude more training data (10% to 100%), only three models, CDSSM, ARC-II, and DRMM_TKS, achieve a relative improvement above 20%. Four more models, DSSM, MV-LSTM, aNMM, and DUET manage to achieve a relative improvement above 10%. For DRMM, performance even slightly decreases (by 1%). The relative improvements after having doubled (25% to 50%), tripled (25% to 75%), or quadrupled (25% to 100%) the training data size are similarly moderate for most models. Specifically, after doubling, only CDSSM and aNMM showed relative improvement above 10%, and with tripling and quadrupling, only DSSM, CDSSM, ARC-II, and aNMM showed relative improvement above 10%.

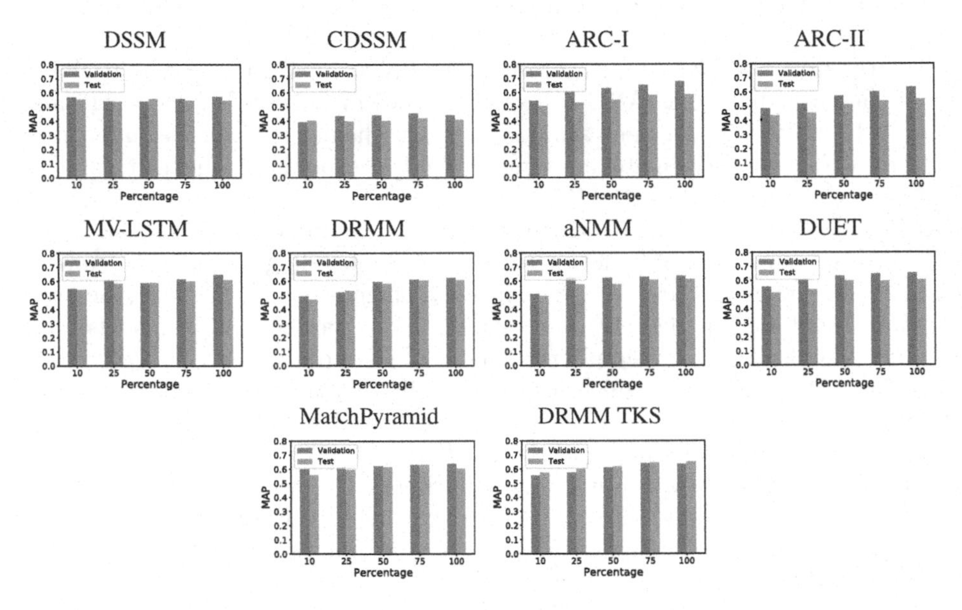

Fig. 1. Performance (as measured by mean average precision) on the validation (blue) and test (green) datasets with different training dataset sizes. (Color figure online)

4.2 Model Training Histories

Figure 2 illustrates the relationship, for each model, between the size of the training dataset and performance improvements over the course of training. We can see that most models either reach a plateau or approximately monotonically increase on the training set (shown in blue curves in Fig. 2) within the recorded training history. There are, however, a few exceptions, namely DRMM and aNMM, which do not exhibit this desired behavior. Another outlier is DRMM TKS, which improves at a drastically slow rate. It is also worth pointing out that the models DSSM and MatchPyramid overfit very quickly. This may suggest a memorization effect.

Looking at the MAP scores on the validation set (shown in blue curves in Fig. 2), we see a discrepancy from expected behavior. The desired behavior would be that these follow the same monotonically increasing trend as the red lines, with the gap between the two lines decreasing as the amount of training data increases. Most of the models, however, do not behave like that. The validation lines plateau out quickly for most models, or even degrade (DRMM, aNMM).

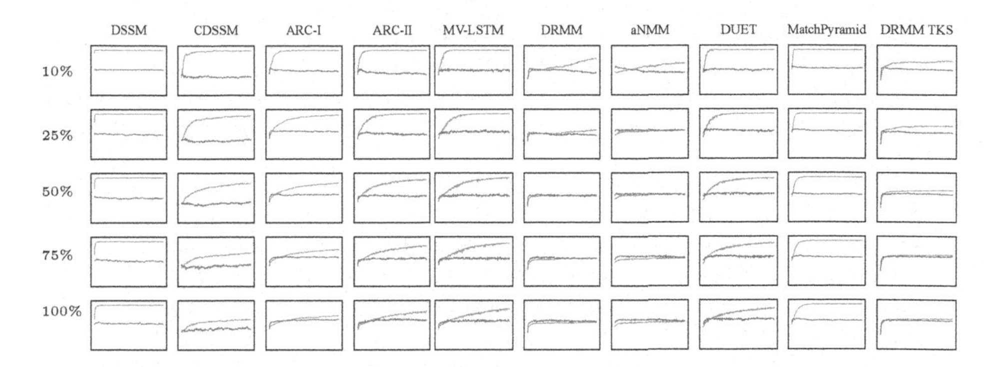

Fig. 2. Training histories for various models (columns) with varying training dataset size (rows). The red and blue lines correspond to performance for training and validation datasets, respectively. Performance is measured in terms of MAP, and indicated with respect to the y-axes, which range from 0 to 1. The x-axes indicate the number of training iterations (epochs), and range from 0 to 399. The x- and y-axes are identically scaled in each of the sub-plots. (Color figure online)

5 Conclusions

We have briefly looked at the effects of dataset size on the neural IR task of answer selection for a number of deep architectures. The consequences of reducing the available training data logarithmically (10% versus 100%) are discernible, and indicate primarily a failure to generalize. This can be seen from the discrepancy between performance improvement on training data, compared to the modest improvements on validation data. Note that these findings are based on one particular implementation, and the inner workings of the implementation were not rigorously analyzed and verified, but were assumed to correctly enact the cited algorithms. These findings show that when choosing algorithms and strategies in regard to data volume, there are factors which must be considered beyond the reported benchmarks of fully trained models. The actual performance of the models during different stages of training, relative to different scales of training data, must be considered to discover any unexpected trends. Furthermore, performance on validation sets is clearly a very important basis for comparison, to gain an intuition about how fast models generalize from different volumes of training data, and with different numbers of training epochs.

Future work may consist of a deeper investigation into the reproducibility of answer selection state-of-the-art results, as well as into quantifying the relationship between training dataset size and the impact of diverse neural models on generalizability.

References

1. Arpit, D., et al.: A closer look at memorization in deep networks. In: Proceedings of ICML 2017, pp. 233–242 (2017)
2. Cho, J., Lee, K., Shin, E., Choy, G., Do, S.: How much data is needed to train a medical image deep learning system to achieve necessary high accuracy? ArXiv e-prints (2015)
3. Fan, Y., Pang, L., Hou, J., Guo, J., Lan, Y., Cheng, X.: MatchZoo: A Toolkit for Deep Text Matching. ArXiv e-prints, July 2017
4. Guo, J., Fan, Y., Ai, Q., Croft, W.B.: A deep relevance matching model for ad-hoc retrieval. In: Proceedings of CIKM 2016, pp. 55–64 (2016)
5. Hestness, J., et al.: Deep Learning Scaling is Predictable. Empirically, ArXiv e-prints, December 2017
6. Hu, B., Lu, Z., Li, H., Chen, Q.: Convolutional neural network architectures for matching natural language sentences. In: Proceedings of NIPS 2014, pp. 2042–2050 (2014)
7. Huang, P.-S., He, X., Gao, J., Deng, L., Acero, A., Heck, L.: Learning deep structured semantic models for web search using clickthrough data. In: Proceedings of CIKM 2013, pp. 2333–2338 (2013)
8. LeCun, Y., Bengio, Y., Hinton, G.: Deep learning. Nature **521**, 436 (2015)
9. Mitra, B., Craswell, N.: Neural Models for Information Retrieval. ArXiv e-prints, May 2017
10. Mitra, B., Diaz, F., Craswell, N.: Learning to match using local and distributed representations of text for web search. In: Proceedings of WWW 2017, pp. 1291–1299 (2017)
11. Onal, K.D., et al.: Neural information retrieval: at the end of the early years. Inf. Retrieval J. **21**, 111–182 (2017)
12. Pang, L., Lan, Y., Guo, J., Xu, J., Wan, S., Cheng, X.: Text matching as image recognition. In: Proceedings of AAAI 2016, pp. 2793–2799 (2016)
13. Schmidhuber, J.: Deep learning in neural networks: an overview. Neural Networks **61**, 85–117 (2015)
14. Shen, Y., He, X., Gao, J., Deng, L., Mesnil, G.: Learning semantic representations using convolutional neural networks for web search. In: Proceedings of WWW 2014, pp. 373–374 (2014)
15. Sun, C., Shrivastava, A., Singh, S., Gupta, A.: Revisiting unreasonable effectiveness of data in deep learning era. In: Proceedings of ICCV 2017, pp. 843–852 (2017)
16. Wan, S., Lan, Y., Guo, J., Xu, J., Pang, L., Cheng, X.: A deep architecture for semantic matching with multiple positional sentence representations. In: Proceedings of AAAI 2016, pp. 2835–2841 (2016)
17. Yang, L., Ai, Q., Guo, J., Croft, W.B.: aNMM: ranking short answer texts with attention-based neural matching model. In: Proceedings of CIKM 2016, pp. 287–296 (2016)

18. Yang, Y., Yih, S.W.-T., Meek, C.: WikiQA: a challenge dataset for open-domain question answering. In: Proceedings of EMNLP 2015, pp. 2013–2018 (2015)
19. Zhang, C., Bengio, S., Hardt, M., Recht, B., Vinyals, O.: Understanding deep learning requires rethinking generalization. In: Proceedings of ICLR 2017, pp. 1–2 (2017)

Unsupervised Explainable Controversy Detection from Online News

Youngwoo Kim(✉) and James Allan

Center for Intelligent Information Retrieval, College of Information and Computer Sciences, University of Massachusetts Amherst, Amherst, USA
{youngwookim,allan}@cs.umass.edu

Abstract. Alerting users that a web page is controversial has been proposed as one method to support critical thinking about text and discourse. We propose an approach to discover controversial topics in a generic document using unsupervised training. Our approach comprises iterative training of a controversy classifier using a disagreement signal within comments and explaining the controversy of the document by generating a topic phrase describing it. Experiments show the effectiveness of our proposed training method using an EM algorithm. When controversial topic extraction is restricted to quality phrases and incorporates TextRank signals, it outperforms several baseline approaches.

Keywords: Controversy · Topic extraction · Controversy detection

1 Introduction

While search engines and social media are applauded for serving as effective information sources, they are also harshly criticized for delivering unverified and potentially harmful misinformation [14]. As an attempt to minimize such pitfalls, researchers have investigated controversy in the Web to predict misinformation and minimize the risk of it [18].

Much work on controversy has relied on certain signals available from the structure of the Web source, such as hashtags in social media which group information of similar content and thus contain inherent topic annotation [5,11,12,19–21]. However, for a generic document without implicit or explicit topic annotations, the same approaches cannot be used. Even when it can be used, it is difficult for proposed systems to identify *what* is controversial [3,7–9,13].

Previous work on controversy detection on a generic document has two limitations. First, it relies on the topics that are labeled as controversial in Wikipedia [7,8], or it relies on supervised human annotations [3], thus they may not be applicable to newly emerging topics. Second, it did not investigate how to provide an explanation for any controversy in the documents.

Our contributions are the following: First, we propose an unsupervised approach to build a controversy classifier using disagreement expressions. We aim

L. Azzopardi et al. (Eds.): ECIR 2019, LNCS 11437, pp. 836–843, 2019.
https://doi.org/10.1007/978-3-030-15712-8_60

to detect topics that are expected to generate debates with numerous disagreements, which we view as the controversy in the news media. We show that a single feature of disagreement expression in the comment is enough to build an article-content-based controversy classifier without supervised training and we propose an EM algorithm to improve the training process.

Second, we propose a method to explain which topic is controversial in the document using the content-based controversy classifier. The controversy is explained by generating the phrases that describe the controversial topic in the document. We show that the quality of generated topic phrases can be improved by quality keyphrase constraints and a keyword extraction technique.

2 Unsupervised Controversy Classification

We target online news documents that contain both the article content and users' comments about the article. We thus define a "document" to be the pair of an article's content and its comment thread, though our goal will be to train a classifier that depends only on article content. To tackle this problem, we note that if a person were asked to decide what is controversial, one way might be to observe people's reactions to the article to get a sense which topics tend to generate more controversial debates. Following Beelen et al. [3], we use the presence of disagreement expressions to recognize controversy within comments. We use a text classification approach to find such expressions (Sect. 2.2). However, because of likely errors in automatic detection we observed that disagreement expressions in a single document alone are insufficient to predict controversy. Thus, we decided to use disagreement in *comments* as a weak signal to train an article content classifier. We further improve this approach by re-training the comment text classifier using the article content classifier and iterating that process. This strategy is an instance of the EM algorithm [6,10,16].

2.1 EM Algorithm for Controversy Classifier

We build two Language Model classifiers, where one is for an article content and the second is for comments (Sect. 2.3).

Step 1. For a document x_i, the comment classifier f_c predicts whether the document is controversial, z_i, with

$$z_i = f_c(\theta_c^{(1)}, x_i) \tag{1}$$

where $\theta_c^{(1)}$ is the first set of parameters for the comment classifier. Based on the comment classifier's predictions, we assign a binary label z_i to every document in our corpus. Then the label z_i is used to get the article-content classifier's parameter set $\theta_a^{(1)}$.

$$\theta_a^{(1)} = \arg\max_{\theta} \sum_i z_i f_a(\theta, x_i) \tag{2}$$

Step 2. Again we predict each document's label using article content classifier f_a, and based on that label, get new parameters $\theta_c^{(2)}$.

$$z_i = f_a(\theta_a^1, x_i) \tag{3}$$

$$\theta_c^{(2)} = \arg\max_{\theta} \sum_i z_i f_c(\theta, x_i) \tag{4}$$

These two steps are iterated until convergence. For the controversy language model, Eqs. 2 and 4 actually update $P(w|L_C)$ and $P(w|L_{NC})$.

2.2 Initial Signal

As an initial settings of z_i, a document is assigned a pseudo-label if it has more than certain number of disagreement in its comments. To estimate the number of disagreement expressions in the comments, we trained a Convolutional Neural Network based classifier using Authority and the Alignment in Wikipedia Discussions (AAWD) corpus [4,15]. We take the first 100 comments as input and predict the number of disagreement in them. If the number of disagreement is larger than a threshold, the document is classified as controversial. We assumed the prior probability of a document having a 'controversy' label to be 0.5 and determined the threshold based on target corpus – i.e., such that half of the documents are (pseudo) controversial.

2.3 Language Model

As our primary controversy classifier model, we used the Controversy Language Model [13] which predicts a controversy ($D_C = 1$) by comparing whether the document is more likely to appear in a controversial document collection (L_C) or in a non-controversial document collection (L_{NC}).

$$\log P(D_C = 1) = \sum_{w \in D} \log P(w|L_C) - \log P(w|L_{NC}) \tag{5}$$

$P(w|L_C)$ and $P(w|L_{NC})$ are the probability of a word w in the collection of controversial documents and non-controversial documents. A document is classified controversial if $\log P(D_C) > T$ where T is set by a training corpus.

We also considered neural classifiers such as Convolutional Neural Network based text classifiers, but they turned out to be too unstable to be trained using the EM algorithm and also never outperformed the Language Model classifier.

3 Controversy Detection Explanation

When the trained classifier detects a controversial document, users will also expect an explanation of which topic in the document is actually controversial. We choose to generate topic phrase which can predict that explanation. We do this by analyzing each document token's contribution to the classification decision and generating a topic phrase based on the contribution information. First, we describe restrict candidate phrases to those meeting standard of reasonableness, so that the user can clearly understand what the output phrase implies. Then, we explain how the contribution to the classification is evaluated and how it is transformed to score each candidate phrase.

3.1 Quality Phrase as a Candidate Topic

Candidate topic phrases are restricted to be quality phrases that can be extracted from the target document. An n-gram is considered a quality phrase if (1) its document frequency exceeds a minimum, (2) it does not begin or end with stopword and (3) for the ith word w_i in the phrase, $P(w_i|w_{1:i-1}) > \lambda \cdot P(w_i)$. We used minimum document frequency = 4, $\lambda = 10$, and phrase length $n \leq 3$.

3.2 Candidates Scoring

Candidate phrases are scored based on the degree to which phrase can explain the classifier's decision. The phrase with highest score is presented as the final output. Each token in the document is assigned a **contribution score** which represents how much it contributes to the classifier decision. For the Controversy Language Model, the contribution of each word is given as $R_w = \log P(w|L_C) - \log P(w|L_{NC})$. For neural classifiers, input contribution can be evaluated using contribution analysis techniques [1,2]. A phrase's contribution score is sum of its terms' contributions, while each term's contribution is summed over all occurrences of the term in the document.

We added keyword scoring technique TextRank [17] as a **contribution independent score** for topic phrases. While the contribution to the classification decision is the most important factor in ranking explanations, we want the selected explanation to be representative and summarize other factors as well. TextRank score of the phrase is multiplied to the contribution score to achieve a final score of the candidate phrase.

4 Experiments

We present two experiments. The first experiment demonstrates the trained classifier's ability to correctly identify controversial documents. The second experiment evaluates how well the topic phrases generated from the classifier match human generated phrases. A qualitative analysis shows characteristics of topic phrases extracted from real world data.

For training, we collected unlabeled news documents from the Guardian, a British daily newspaper. We crawled the articles written in 2016 along with the related comments, which resulted in 66,763 news articles and 7,803,440 comments. We refer this collection as *Guardian16*.

Table 1. Controversy classification accuracy

Method	Accuracy
Weak signal	0.541
LM - Single iteration	0.704
LM - EM	**0.746**
LM - Supervised	0.749
Human Annotator	0.744

Differences between upper three methods are statistically significant under $p < 0.05$

4.1 Evaluation for Classification

The model is trained using the *Guardian16* corpus. Part of the articles were labeled using Amazon Mechanical Turk. 6 annotators were asked if each document is about controversial topic or not. Documents with more than 3 'controversial' annotations were assigned final controversial label. which resulted in 281 controversial and 439 non-controversial documents.

Table 1 shows the controversy classification accuracy of various methods. The 'Weak Signal' classifier is based on the number of disagreement in the comments, which was our initial label. 'LM - Single Iteration' is the controversy language model trained by 'Weak signal' without additional iteration. 'LM - EM' is our proposed method. 'LM - Supervised' is the Language Model trained in a supervised setting in which 2/3 of the 720 documents were taken as training data and remaining were regarded as test data. Three splits were made and results were averaged to get the final accuracy. 'Human Annotator' is the hypothetical classification accuracy in which one of annotator's prediction is compared against the others. Note that all 'LM' methods classify based on the article contents alone (i.e., no comment).

4.2 Evaluation for Explanation

Here, we evaluate our method's ability to predict the topic of the controversy. We collected 124 articles from iSideWith.com, which has a manually curated collection of the articles about a number of politically controversial topics. Those articles are from 13 controversial topics. Note that all of these documents are implicitly labeled as controversial. These documents are annotated with human generated topic phrases, which we adopt as ground truth for explanation and compared with output phrases from our methods.

Table 2 shows the evaluation for system generated topic phrases. As we have only one "gold" phrase, evaluation is using MRR, the reciprocal of the answer phrase's rank. Explanations generated with the conditions outperform the baseline group, which select phrases from any N-Gram ($N \leq 3$)

4.3 Quantitative Analysis

For qualitative analysis we analyzed which topics are most controversial in the collection, which we generated by accumulating individual document's controversial topic scores. For each document, the candidate topic phrases are assigned scores as explained in Sect. 3. Then the phrase scores from each document are summed to get final controversial topics at the collection level. We used 'TextRank Only' as a baseline method. Table 3 shows the top topics extracted from *Guardian16*. TextRank captures less-controversial topics such as 'women', 'people' and 'children'. In contrast, the proposed method captures clearly controversial topics as top entries.

Table 2. Explanation performance

Restriction	#	Method	P@1	MRR
N-Gram	1	LM-EM	0.10	0.19
Quality Phrases	2	Random	0.06	0.12
	3	First N Phrase	0.15^{2}	0.21^{2}
	4	LM-EM	$0.26^{1,2,3}$	$0.38^{1,2,3}$
+ TextRank	5	TextRank only	0.24	0.36
	6	LM-EM	0.33	0.41

Superscripts indicate the specified method is superior over numbered method ($p < 0.05$). Statistical significance was only measured between methods in the same group or the same model.

Table 3. Top controversial topics in the collection.

Rank	TextRank Only	Proposed
1	Trump	Trump
2	Women	EU
3	EU	Government
4	People	Labour
5	Police	Tax
6	Min	Clinton
7	Mental health	Party
8	Children	Prime minister
9	Labour	Climate change
10	Tax	Corbyn

5 Conclusions

We introduced a classifier driven approach to detect controversial topics in the news media. We showed that the EM algorithm can improve the training process and the adding quality phrase information and keyword scoring helps to generate human friendly explanation from the classifier. As future work, we expect to extend disagreement signal driven controversy detection to generic web pages outside the news media, and to generate explanations in a detail-rich format.

Acknowledgment. This work was supported in part by the Center for Intelligent Information Retrieval and in part by NSF grant #IIS-1813662. Any opinions, findings and conclusions or recommendations expressed in this material are those of the authors and do not necessarily reflect those of the sponsor. We thank Kaspar Beelen for sharing the labeled data.

References

1. Ancona, M., Ceolini, E., Oztireli, C., Gross, M.: Towards better understanding of gradient-based attribution methods for deep neural networks. In: 6th International Conference on Learning Representations (ICLR 2018) (2018)
2. Bach, S., Binder, A., Montavon, G., Klauschen, F., Müller, K.R., Samek, W.: On pixel-wise explanations for non-linear classifier decisions by layer-wise relevance propagation. PLoS ONE **10**(7), e0130140 (2015)
3. Beelen, K., Kanoulas, E., van de Velde, B.: Detecting controversies in online news media. In: Proceedings of the 40th International ACM SIGIR Conference on Research and Development in Information Retrieval, pp. 1069–1072. ACM, New York (2017). https://doi.org/10.1145/3077136.3080723
4. Bender, E.M., et al.: Annotating social acts: authority claims and alignment moves in Wikipedia talk pages. In: Proceedings of the Workshop on Languages in Social Media, pp. 48–57. Association for Computational Linguistics (2011)
5. De Clercq, O., Hertling, S., Hoste, V., Ponzetto, S.P., Paulheim, H.: Identifying Disputed Topics in the News, pp. 32–43 (2014)
6. Dempster, A.P., Laird, N.M., Rubin, D.B.: Maximum likelihood from incomplete data via the EM algorithm. J. Roy. Stat. Soc. B (Methodol.) **39**, 1–38 (1977)
7. Dori-Hacohen, S., Allan, J.: Detecting controversy on the web. In: Proceedings of the 22nd ACM International Conference on Conference on Information & Knowledge Management, pp. 1845–1848. ACM (2013)
8. Dori-Hacohen, S., Allan, J.: Automated controversy detection on the web. In: Hanbury, A., Kazai, G., Rauber, A., Fuhr, N. (eds.) ECIR 2015. LNCS, vol. 9022, pp. 423–434. Springer, Cham (2015). https://doi.org/10.1007/978-3-319-16354-3_46
9. Dori-Hacohen, S., Jensen, D., Allan, J.: Controversy detection in Wikipedia using collective classification. In: Proceedings of the 39th International ACM SIGIR Conference on Research and Development in Information Retrieval, SIGIR 2016, pp. 797–800. ACM, New York (2016). https://doi.org/10.1145/2911451.2914745
10. Fessler, J.A., Hero, A.O.: Space-alternating generalized expectation-maximization algorithm. IEEE Trans. Signal Process. **42**(10), 2664–2677 (1994). https://doi.org/10.1109/78.324732

11. Garimella, K., Morales, G.D.F., Gionis, A., Mathioudakis, M.: Quantifying controversy on social media. Trans. Soc. Comput. **1**(1), 3:1–3:27 (2018). https://doi.org/10.1145/3140565
12. Jang, M., Allan, J.: Explaining controversy on social media via stance summarization. In: The 41st International ACM SIGIR Conference on Research and Development in Information Retrieval, pp. 1221–1224. ACM, New York. https://doi.org/10.1145/3209978.3210143
13. Jang, M., Foley, J., Dori-Hacohen, S., Allan, J.: Probabilistic approaches to controversy detection. In: Proceedings of the 25th ACM International on Conference on Information and Knowledge Management, CIKM 2016, pp. 2069–2072. ACM, New York (2016). https://doi.org/10.1145/2983323.2983911
14. Kata, A.: A postmodern pandora's box: anti-vaccination misinformation on the internet. Vaccine **28**(7), 1709–1716 (2010). https://doi.org/10.1016/j.vaccine.2009.12.022
15. Kim, Y.: Convolutional neural networks for sentence classification. arXiv preprint arXiv:1408.5882 (2014)
16. Mann, G.S., McCallum, A.: Generalized expectation criteria for semi-supervised learning with weakly labeled data. J. Mach. Learn. Res. **11**(Feb), 955–984 (2010)
17. Mihalcea, R., Tarau, P.: Textrank: bringing order into texts. In: Proceedings of the 2004 Conference on Empirical Methods in Natural Language Processing, pp. 404–411. Association for Computational Linguistics, Barcelona, Spain, July 2004
18. Qazvinian, V., Rosengren, E., Radev, D.R., Mei, Q.: Rumor has it: identifying misinformation in microblogs. In: Proceedings of the Conference on Empirical Methods in Natural Language Processing, pp. 1589–1599. Association for Computational Linguistics (2011)
19. Yamamoto, Y.: Disputed sentence suggestion towards credibility-oriented web search. In: Sheng, Q.Z., Wang, G., Jensen, C.S., Xu, G. (eds.) APWeb 2012. LNCS, vol. 7235, pp. 34–45. Springer, Heidelberg (2012). https://doi.org/10.1007/978-3-642-29253-8_4
20. Yasseri, T., Sumi, R., Rung, A., Kornai, A., Kertész, J.: Dynamics of conflicts in Wikipedia. PloS ONE **7**(6), e38869 (2012)
21. Zielinski, K., Nielek, R., Wierzbicki, A., Jatowt, A.: Computing controversy: formal model and algorithms for detecting controversy on Wikipedia and in search queries. Inf. Process. Manage. **54**(1), 14–36 (2018). https://doi.org/10.1016/j.ipm.2017.08.005

Extracting Temporal Event Relations Based on Event Networks

Duc-Thuan Vo[(✉)] and Ebrahim Bagheri

Laboratory for Systems, Software and Semantics (LS3),
Ryerson University, Toronto, Canada
{thuanvd, bagheri}@ryerson.ca

Abstract. Temporal event relations specify how different events expressed within the context of a textual passage relate to each other in terms of time sequence. There have already been impactful work in the area of temporal event relation extraction; however, they are mostly supervised methods that rely on sentence-level textual, syntactic and grammatical structure patterns to identify temporal relations. In this paper, we present an *unsupervised* method that operates at the document level. More specifically, we benefit from existing Open IE systems to generate a set of triple relations that are then used to build an event network. The event network is bootstrapped by labeling the temporal disposition of events that are directly linked to each other. We then systematically traverse the event network to identify the temporal relations between indirectly connected events. We perform experiments based on the widely adopted TempEval-3 corpus and compare our work with several strong baselines. We show that our unsupervised method is able to show better performance in terms of precision and f-measure over it supervised counterparts.

1 Introduction

The extraction of temporal relationships between events that have been mentioned in a textual passage is an important task in information extraction as it enables tasks such as building event timelines and arranging plots of events. Temporal information about events can assist in understanding the evolution of news stories or the development of a narrative. There have already been many application areas such as question answering [2, 5], document summarization [5, 16] and textual entailment [12, 13] that benefit from temporal event relationships.

Several approaches to temporal relation classification use machine-learning-based classifiers [6, 7, 9, 10] that are trained based on a predefined, finite and fixed schema of relation types. The common strategy of these techniques is to generate linguistic features based on syntactic, dependency, or shallow semantic structures of the text. Based on these features, supervised learning methods are used to identify pairs of events that are related to each other, and classify them based on pre-defined relation types. However, the state-of-the-art approaches [7, 10] suffer from two key drawbacks. First, they are focused on a limited subset of features, which might not, in many cases, be present in every sentence or be sparsely available. Second, training on linguistic structures such as the output of syntactic and dependency parsers does not necessarily

L. Azzopardi et al. (Eds.): ECIR 2019, LNCS 11437, pp. 844–851, 2019.
https://doi.org/10.1007/978-3-030-15712-8_61

identify all possible types of event relations when they are presented in different sentences or different documents. For instance, consider the three sentences shown Fig. 1. In this figure, the events in both $<e_2 - e_3>$, $<e_5 - e_8>$ are related to each other by the "BEFORE" temporal relation type. Here, while e_2 and e_3 are presented in the same sentence, events e_5 and e_8 are in different sentences. As such, sentences that rely on features based on grammatical parsers can fail to identify correct relation types.

1. "... the President Bush has *approved*$_{e2}$ *duty-free treatment*$_{e3}$ for imports... "
2. "Timex had *requested*$_{e5}$ duty-free treatment... "
3. " the Philippines and Thailand would be the main *beneficiaries*$_{e8}$ of the president's *action*$_{e9}$..."

Fig. 1. Samples of direct relation events $<e_2 - e_3>$, $<e_8 - e_9>$ and indirect relation events $<e_5 - e_3>$, $<e_3 - e_9>$ in textual document.

In this paper, our objective is to address these two challenges by adopting an Open Information Extraction (Open IE) strategy [4, 8, 14, 15], which is able to extract relations and their arguments without the need to restrict the search to predefined relation types or grammatical structures. We propose a method to extract temporal event relations by using an Open IE graph-based event network, which is built based on the patterns identified and extracted by Open IE systems. Particularly, we consider and incorporate all identified Open IE patterns that consist of at least one event instance in the event network, which is then systematically traversed for identifying temporal relations. As an example in Fig. 1, both $<e_2 - e_3>$ and $<e_5 - e_3>$ relations can be extracted from two Open IE patterns, namely ("President Bush", "has approved", "duty-free treatment for imports") and ("Timex", "had requested", "duty-free treatment"), respectively. Based on the constructed event network, we employ a shortest path strategy to determine the event flow between two events.

2 The Proposed Approach

An Open IE system extracts triples in the form of (arg1, rel, arg2) representing basic propositions or assertions from text. In this context, propositions are defined as coherent and non-over-specified pieces of information. In this study, we exploit Open IE to build an event network in order to extract temporal event relations. We consider extracting temporal event relations based on both events that are directly related to each other and those that can be indirectly linked to each other through links in the network. To this end, we propose an algorithm to detect event flows in the network that will be used to identify temporal event relations. An overview of our proposed approach is illustrated in Fig. 2.

2.1 Graph-Based Event Network

Our proposed graph-based event network is built directly from triples generated by Open IE systems. Two events that are present in the same extracted pattern are considered as two event nodes in the event network that are directly connected to each other with an edge. The collection of all the extracted Open IE triple patterns are used to complete the event network. Moreover, we use reference mapping to expand all possible event relations of the event network. Reference mapping is based on the context similarity of terms in the triple patterns.

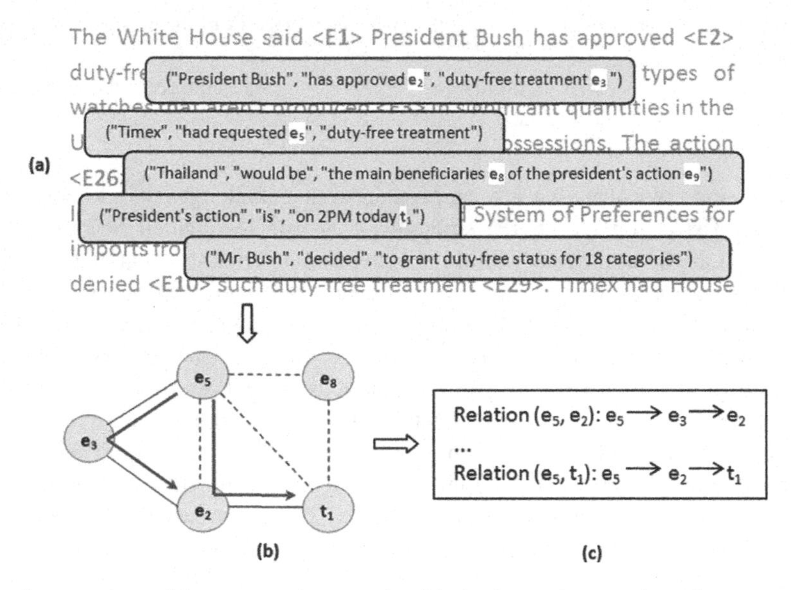

Fig. 2. An overview of the proposed approach with (a) Open IE extraction, (b) event network construction, and (c) event flow extraction.

2.2 Temporal Relation Identification

According to the TempEval-3 task description [10–12], two pairs of temporal events can be related to each other through one of four groups, namely Timex-Timex, Event-DCT, Event-Timex and Event-Event where DCT denotes *Document Creation Time* and Timex denotes *Temporal Expressions*. In this study, we present an algorithm to extract both direct and indirect temporal event relations in the event network based on event flow. Let us denote an event network as a graph $G(V, E)$ where each vertex denotes an event and each edge denotes a relation. Based on the event network, the objective is to determine the existence and type of temporal relation between two event nodes such as $\{X, Y\}$ expressed as $R\{X, Y\}$. Figure 3 presents the pseudo-code for our proposed algorithm for identifying temporal relations. The algorithm first proceeds to detect types of event relations. In case of a direct relation between two events, the *sieve* [17] implementation of the rules introduced in [3] are applied to identify relation types, and the corresponding edge in the network will be updated. In case of an indirect relation between two events, the relation will be determined based on its event flow. First, the algorithm will determine the shortest path between the two events resembling the potentially most likely temporal order of how events played out in reality.

Algorithm 1: Identifying temporal relations

```
Input:   Graph G = (V, E)
         Pair of events {X, Y}
Output:  Type of relation R(X, Y)
 1: if Direct(X, Y) ∈ G then
 2:      R(X, Y) ← Rules(E{X, Y}, G)
 3:      Update(G)
 4: else if Indirect(X, Y) ∈ G then
 5:         R(X, Y) ← Time{X,Y}⊕Tense{X,Y}
 6:         Update(G)
 7:      if R(X, Y)=NULL then
 8:            Event-flow{X,Y} ← Shortest-path({X, Y}, G))
 9:            Rules(Event-flow{X,Y}, G)
10:            R(X, Y) ←Infer(Event-flow{X,Y})
11:            Update(G)
12:      end if
13: return R(X, Y)
```

Fig. 3. Algorithm for identifying temporal relations.

Once the shortest path between events is determined, it is possible to reason over the set of temporal relations observed on the shortest path to make a determination about the type of temporal relation between the two source and target events. Temporal relations of indirect relations are inferred through transitivity of temporal relations [1] on direct relations as shown in Fig. 4. For instance, consider events e_1 and e_4 in Fig. 4a, in this example, direct relations between events e_1 and e_2 as well as e_2 and e_4 have already been identified based on *sieve* and labeled as such. Now, given the shortest path between e_1 and e_4 passes through e_2, it is possible to infer that given e_1 happened before e_2 and e_2 was before e_4 that e_1 also happened before e_4.

3 Experimentation

3.1 Experimental Results

For benchmarking our approach, we conducted experiments on TempEval-3 on Task C [7, 10, 12]. The available dataset consists of news documents separated into testing and training sets. The testing set consists of 20 documents and the training set consists of 183 documents. We built the event network based on the generated Open IE triples extracted by the LS3RyIE system [15]. Note that, reference mapping was also applied to enhance node matching in the event network. We calculated the context from the

Open IE patterns using cosine similarity then merged those nodes with a score ≥ 0.5. As a result of this process, 968 and 2,537 triples were generated by the Open IE system that were then used to build the event network for the testing and training sets, respectively. Based on the TempEval-3 task, we evaluate the approach on four categories, namely Event-Event (E-E), Event-Timex (E-T), Event-DCT (E-D), and Timex-Timex (T-T). It should be noted that unlike the state of the art baselines that are supervised temporal relation extraction methods, our work is completely unsupervised and as such we do not require separate training and testing datasets. For this reason, we report the performance of our work on the data available in both sets.

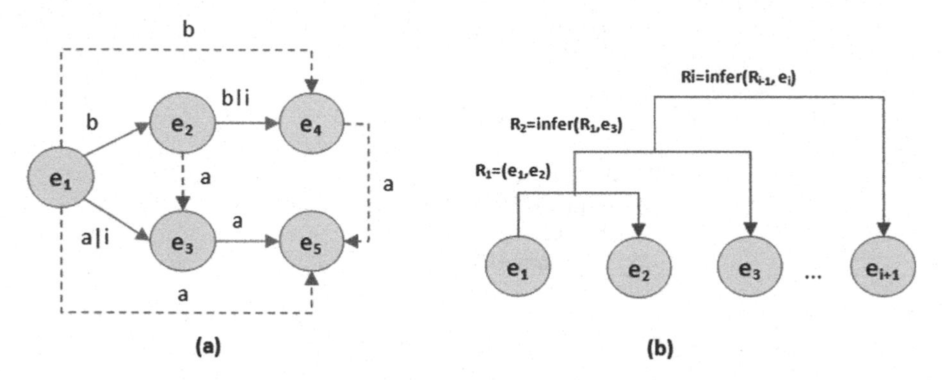

Fig. 4. Inferred relations; (a) Inferred sample relations in three nodes path ($e_1 - e_4$, $e_2 - e_3$, $e_1 - e_5$, $e_4 - e_5$) with b: before; i: includes; a: after; (b) Recursively inferred relations.

Table 1. Experimental results on four categories.

Categories	Testing set			Training set		
	Precision	Recall	F-measure	Precision	Recall	F-measure
E-E	60.63	41.81	49.49	60.51	47.21	53.04
E-T	84.82	73.21	78.58	82.37	60.06	69.47
E-D	88.24	87.50	87.86	70.63	68.75	69.68
T-T	62.50	62.50	62.50	77.77	68.62	72.91
Overall	72.92	57.54	64.32	69.58	57.18	62.77

The performance results obtained using our proposed approach on the testing and training sets are shown in Table 1. In the testing set, the system achieved F-measures of 49.49%, 78.58%, 87.86%, and 62.50% for Event-Event, Event-Time, E-DCT, and Time-Time, respectively. Regarding training set, the system obtained F-measures of 53.04%, 69.47%, 69.68% and 72.91% for Time-Time, Event-DCT, Event-Time and Event-Event, respectively. Overall, the system yielded F-measures of 64.32% and 62.77% on testing and training sets.

In Table 2, we compare our method with several strong baseline approaches designed for Task C of TempEval-3. UTTime [6] employs features based on syntactic parsing including phrase structures while Laokulrat et al. [7] extract event relations using time graphs and stacked learning. TRelPro [9] and CATENA [10] employ an SVM classifier based on event linguistic features such as POS tags, chunking, dependency paths, and others. The numbers reported in Table 2 are the results of 5-fold cross-validation evaluation strategy. The evaluation shows that our proposed method is the best performing system against the state-of-the-art baselines. It should be pointed out that our approach obtained improved performance over these baselines even though it is fully unsupervised while the baselines operate under a supervised context.

Table 2. Performance comparison.

	Precision	Recall	F-measure
UTTime [6]	55.60	57.40	56.50
TRelPro [9]	58.48	58.80	58.17
Laokulrat et al. [7]	57.60	57.90	57.80
CATENA [10]	62.60	61.30	61.90
Proposed method	70.82	57.31	63.35

3.2 Discussion

Our approach benefits from the relation patterns extracted by Open IE systems to build the initial event network and bootstraps the temporal event extraction process by determining the type of temporal relation between two directly linked events. The advantages of our proposed work are two folds: (1) it is completely unsupervised and hence does not require any hand-annotated samples by inferring indirect temporal relation types between events by systematically traversing the event network, and (2) it works at the document level and not sentence level and hence can identify temporal relations between events that have not been expressed in the same sentence. This is made possible due to the linking of different events in the network whose linking transcends individual sentences and forms a representation of events in the document.

However, our method also faces some limitations: (*i*) our proposed approach is dependent on the performance of the underlying Open IE system and hence in cases when the Open IE system cannot extract event mentions, the corresponding event nodes will not be created in the event network and hence temporal relations will be missed. The lower recall of our method, noted in Table 2, can be explained as such. (*ii*) Our method is dependent on reference mapping to identify similar event nodes in the event graph, which is currently performed through cosine similarity. However, more complex co-reference resolution methods and semantic matching of event types can improve the reference mapping process and lead to better overall performance.

4 Concluding Remarks

In this paper, we have presented an unsupervised method for extracting temporal relations between events by building an event network structure primarily based on information from Open IE systems. The event network is the basis for systematically exploring the possible temporal relations between events by considering how events can be reached from one another. We performed comparative benchmarking of our proposed method using the TempEval-3 dataset and compared our work against several strong baselines. Our experiments show that while our approach is unsupervised, it is able to outperform supervised baselines in terms of precision and f-measure. Our future work will consist of addressing the two limitations of our work, namely quantifying the impact of the performance of the Open IE systems on our work and also exploring more systematic ways for performing reference mapping.

References

1. Allen, J.F.: Maintaining knowledge about temporal intervals. Commun. ACM **26**(11), 832–843 (1983)
2. Abacha, A.B., Zweigenbaum, P.: MEANS: a medical question-answering systems combining NLP techniques and semantic web technologies. Inf. Process. Manag. **51**, 570–594 (2016)
3. Chambers, N., Cassidy, T., McDowell, B., Bethard, S.: Dense event ordering with a multi-pass architecture. Trans. Assoc. Comput. Linguist. **2**, 273–284 (2014)
4. Corro, L.D., Gemulla, R.: ClausIE: clause-based open information extraction. In: Proceedings of the 22nd international conference on World Wide Web (WWW 2013), Rio de Janeiro, Brazil, 13–17 May 2013, pp. 355–366 (2013)
5. Ji, H., Favre, B., Lin, W.P., Gillick, D., Hakkani-Tur, D., Grishman, R.: Open-domain multi-document summarization via information extraction: challenges and prospects. In: Poibeau, T., Saggion, H., Piskorski, J., Yangarber, R. (eds.) Multi-source, Multilingual Information Extraction and Summarization. Theory and Applications of Natural Language Processing. Springer, Heidelberg (2013)
6. Laokulrat, N., Miwa, M., Tsuruoka, Y.: Stacking approach to temporal relation classification with temporal inference. J. Nat. Lang. Process. **22**(3), 171–196 (2015)
7. Laokulrat, N., Miwa, M., Tsuruoka, Y., Chikayama, T.: Uttime: temporal relation classification using deep syntactic features. In: Second Joint Conference on Lexical and Computational Semantics (*SEM), Volume 2: Proceedings of the Seventh International Workshop on Semantic Evaluation (Se-mEval 2013), Atlanta, Georgia, USA, June, pp. 88–92. Association for Computational Linguistics (2013)
8. Mausam, Schmitz, M., Bart, R., Soderland, S.: Open language learning for information extraction. In: Proceedings of the 2012 conference on Empirical Methods in Natural Language Processing (EMNLP 2012), Jeju Island, Korea, 12–14 July 2012, pp. 523–534 (2012)
9. Mirza, P., Tonelli, S.: Classifying temporal relations with simple features. In: Proceedings of the 14th Conference of the European Chapter of the Association for Computational Linguistics, Gothenburg, Sweden, April, pp. 308–317. Association for Computational Linguistics (2014)

10. Mirza, P., Tonelli, S.: Catena: causal and temporal relation extraction from natural language texts. In: Proceedings of the 26th International Conference on Computational Linguistics, pp. 64–75. Association for Computational Linguistic (2016)
11. Souza, J.D., Ng, V.: Classifying temporal relations with rich linguistic knowledge. In: Proceedings of the 2013 Conference of the North American Chapter of the Association for Computational Linguistics: Human Language Technologies, Atlanta, Georgia, June, pp. 918–927. Association for Computational Linguistics (2013)
12. UzZaman, N., Lorens, H., Derczynski, L., Allen, J., Verhagen, M., Pustejovsky, J.: Semeval-2013 task 1: Tempeval-3: Evaluating time expressions, events, and temporal relations. In: Second Joint Conference on Lexical and Computational Semantics (*SEM), Volume 2: Proceedings of the Seventh International Workshop on Semantic Evaluation (SemEval 2013), Atlanta, Georgia, USA, June, pp. 1–9. Association for Computational Linguistics (2013)
13. Verhagen, M., Sauri, S., Caselli, T., Pustejovsky, J.: SemEval-2010 task 13: TempEval-2. In: Proceedings of the 5th International Workshop on Semantic Evaluation, SemEval 2010, Stroudsburg, PA, USA, pp. 57–62. Association for Computational Linguistics (2010)
14. Vo, D.T., Bagheri, E.: Open information extraction. Encycl. Seman. Comput. Robot. Intell. **1** (1) (2017). https://doi.org/10.1142/s2425038416300032
15. Vo, D.T., Bagheri, E.: Self-training on refined clause patterns for relation extraction. Inf. Process. Manag. **54**, 686–706 (2018)
16. Zhou, G., Qian, L., Fan, J.: Tree kernel based semantic relation extraction with rich syntactic and semantic information. Inf. Sci. **180**, 1313–1325 (2010)
17. Mendes, P.N., Mühleisen, H., Bizer, C.: Sieve: linked data quality assessment and fusion. In: Proceedings of the 2012 Joint EDBT/ICDT Workshops (EDBT-ICDT 2012), pp. 116–123 (2012)

Dynamic-Keyword Extraction from Social Media

David Semedo([✉]) and João Magalhães

NOVA LINCS, School of Science and Technology,
Universidade NOVA de Lisboa, Caparica, Portugal
df.semedo@campus.fct.unl.pt, jm.magalhaes@fct.unl.pt

Abstract. Traditional keyword extraction methods make the assumption that corpora is static. However, in social media, information is highly dynamic, with individual words showing a dynamic behaviour. In this paper we propose an unsupervised approach that jointly models words' temporal behaviour and keyword's semantic affinity, to address the task of dynamic-keyword extraction. Experiments show the method effectiveness and confirm the importance of exploiting keyword dynamics.

Keywords: Dynamic keyword extraction · Information extraction · Social media

1 Introduction

In social-media, topics are characterised by some keywords, whose relevance change over time. Accounting for this behaviour is essential to grasp the topic unfolding and effectively extract the most important keywords. The task of automatic keyword extraction has been widely studied, in both news articles, [4,7,8] and social media [1,10]. However, most previous work assume a static corpora, thus overlooking the temporal dynamics of keywords. Social media corpora like Twitter, are highly dynamic, with individual words possessing a dynamic behaviour. To extract *dynamic-keywords* we are interested in identifying the set of words that is relevant to a given topic on Twitter. For this task, a large and rich set of tweets covering the topic over a particular time period should be considered to perform dynamic keyword extraction.

In this paper we propose a method to extract keywords from a timeline of tweets, by exploring the dynamics of data, which are reflected on the existence of (temporal) *dynamic words*. Keywords are extracted with an unsupervised joint model of a topic-based and time-based keyword ranking, with the later being supported by a word's temporal signature model, that captures word temporal relevance and *dynamic magnitude*. The hypothesis is that words' temporal signatures will aid keyword extraction methods at better identifying words of interest whose relevance change over time. Experiments performed on corpora from two major events, have shown that our proposed method proved to be highly effective, outperforming the baseline methods.

L. Azzopardi et al. (Eds.): ECIR 2019, LNCS 11437, pp. 852–860, 2019.
https://doi.org/10.1007/978-3-030-15712-8_62

2 Related Work

Keyword extraction has been extensively addressed in the literature [4,7,9,11,14]. Graph methods are among the most popular methods [7], where PageRank [11] or topic-based PageRank [9] are the basis for keyword extraction. Recently [2] showed that using the BM25 ranking function in TextRank [11] improves keyphrase extraction results. In contrast, social media is a rich source of information where few works addressed keyword extraction. In [17], authors compared the language and coverage of topics across Twitter and traditional media, and observed that traditional media covers topics in a more comprehensive manner, but social media is a better medium to quickly spread news about a particular entity, i.e., a person, a show, an event. Thus, it is faster to get the unfolding of a topic. For instance, in [6] the authors model how a given entity is searched over time, for the task of entity ranking.

Recent work in keyword extraction from social media information were inspired by existing work [15], who applied TextRank [11] and *tf-idf* to extract keywords to annotate the user publication stream. Topic modelling techniques have also been explored to extract keyphrases from same latent-topic tweets [16]. More recently, in [10] the authors explored Brown clustering and continuous word vectors to extract keywords from individual tweets. The proposed method departs from the current literature by considering the temporal relevance of words over time.

3 Modelling Dynamic-Keywords

Given a topic's unfolding over its timespan, one expects to observe different manifestations of certain words. If this hypothesis is to be confirmed, then accounting for such *shifts* will contribute to a better candidate keyword relevance assessment.

The task of dynamic-keyword extraction from Twitter is formally defined as follows. Let $\mathcal{C}^E$ be a corpus of tweets from a major event E and str_i the topic of a published news article related to event E. These articles reflect topics of interest that were worth being covered by journalists. The article's title and publication date are denoted as s_i and t_i, respectively. Hence, given a topic $str_i = (s_i, t_i, \mathcal{C}_i^E)$, $\mathcal{C}_i^E$ is a set of representative tweets of s_i, and t_i the timestamp of when the keywords that describe str_i are extracted. Thus, given the set of words $w \in V_i$, where V_i is the vocabulary of $\mathcal{C}_i^E$, the objective is to identify a set of keywords for each topic str_i.

3.1 TempTopRank

In this section we describe the TempTopRank method: given a topic s_i, it extracts keywords w at timestamp t_i by jointly considering (i) each keyword's temporal signature under a dynamic model $\mathcal{M}_t^E$ of the event and (ii) its semantic affinity to the news topic in a given event domain embedding space $\mathcal{M}_s^E$.

Time-Based Keyword Ranking. A word's temporal behaviour on an event E is accounted through temporal word signatures. These are derived from a model $\mathcal{M}_t^E$ of the event, that is obtained from a dynamic topic modelling approach. Thus, each word temporal relevance at time t_i is given by the function $t_{rank}(w, t_i, \mathcal{M}_t^E)$. To address the temporal relevance of keywords of social media topics, we considered two factors. First, it may be the case that the news topic s_i got social media attention some days before t_i. Therefore, a temporal gap of t_{gap} days is considered. Second, we define the word relevance over each day of the topic as ϕ_w. As will be detailed later, the function ϕ_w provides a richer insight of the temporal behaviour of w over the event days than a simple word frequency count. A word's temporal relevance ranking is thus defined as:

$$t_{rank}(w, t_i, \mathcal{M}_t^E) = max(\phi_w(t)), \quad \text{s.t. } t \in [t_i - t_{gap}, t_i] \tag{1}$$

where $\phi_w(t)$ denotes the relevance estimation, ϕ_w, on day t. Thus, for each word, the highest value of ϕ_w within the temporal range $[t_i - t_{gap}, t_i]$ is chosen, corresponding to the day in which the word was more relevant.

Topic-Based Keyword Ranking. For a given event E, we learn an event domain embedding $\mathcal{M}_s^E$, for semantic affinity assessment between word pairs. Word semantic affinity can be assessed by computing a distance $d(w_1, w_2, \mathcal{M}_s^E)$ between two embeddings. Namely, semantic affinity between a candidate word w and a topic s_i, is assessed by a function $s_{rank}(w, s_i, \mathcal{M}_s^E)$ that is formulated as a distance, using the function $d(\cdot)$.

Joint Temporal-Topic Ranking. TempTopRank assesses each candidate keyword w in terms of its temporal relevance and semantic affinity to the topic, through a ranking function $ts_{rank}(w, s_i, t_i, \mathcal{M}_t^E, \mathcal{M}_s^E)$. Inspired by language models smoothing techniques [5], ts_{rank} is formulated as a mixture over t_{rank} and s_{rank} as:

$$ts_{rank}(w, s_i, t_i, \mathcal{M}_s^E, \mathcal{M}_t^E) =$$
$$t_{rank}(w, t_i, \mathcal{M}_t^E) * f_\lambda(w) + \frac{1}{s_{rank}(w, s_i, \mathcal{M}_s^E) + \beta} * (1 - f_\lambda(w)),$$

where $f_\lambda(w)$ is a dynamic smoothing factor that depends w, and dynamically controls the influence of each component. Ideally one would choose to give more importance to the temporal component only when w has a dynamic behaviour. Distances are normalised to $[0, 1]$ range and an additive constant $\beta = 1$ is added to the denominator to avoid division by zero. As s_{rank} is defined as a distance, we use its inverse.

3.2 Temporal Word Signatures

We devised a method to quantify how *dynamic* a given word is, through a temporal signature. Given the event E, and its timespan $TS_E = [t_s^E, t_f^E]$, where t_s^E

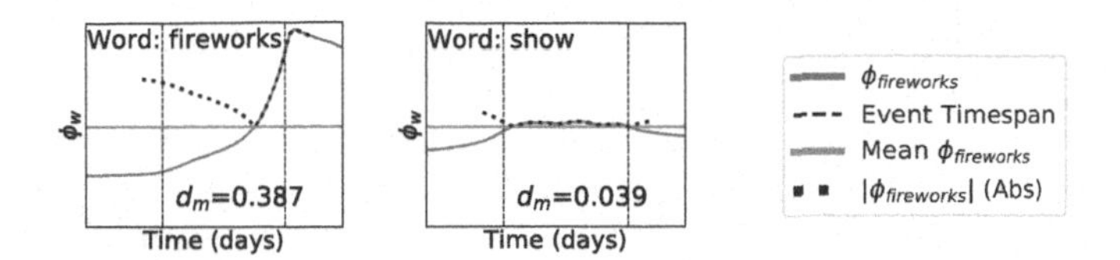

Fig. 1. Words temporal relevance. Each plot depicts the mean latent-topical temporal curve ϕ_w, over each day, on the EdFest dataset (see Sect. 4). Vertical lines mark the event timespan.

and t_f^E are the first and last days respectively, the goal is to estimate the function ϕ_w. For this purpose, we resort to Dynamic Topic Modelling (DTM) [3], to analyse the evolution of latent topics. Documents of $\mathcal{C}^E$ are arranged into a set of *time slices*, with each time slice referring to individual *days*. For each time slice, documents are modelled using a K-component topic model (LDA), where its latent topics at time slice t evolve from latent topics of slice $t-1$. We applied DTM to the each corpus $\mathcal{C}^E$, obtaining for each topic k, a temporal relevance curve ϕ_{wk}. An element-wise mean latent-topic temporal curve vector is computed as $\phi_w = \sum_{k=1}^{K} \phi_{wk}$, and normalised such that $\sum_i \phi_w(i) = 1$. Given that we average each ϕ_{wk} over the K latent-topics and that each word w reveals different behaviours on each latent-topic, we obtain a representation that encodes information about word behaviours across all topics. The mean m of the vector ϕ_w is computed, and the *dynamic magnitude* d_m of word w is defined as:

$$d_m(w) = \int_{t_s^E-\epsilon}^{t_f^E+\epsilon} |\phi_w(x) - m| \, dx, \tag{2}$$

where the integral is solved using the trapezoidal rule. $d_m(w)$ consists of the area above the mean m and below the curve ϕ_w. A small constant gap ϵ is used to capture adjacent days to TS^E. Figure 1 shows the resulting ϕ_w and d_m for sample words. It can be seen that the devised $d_m(w)$ metric does in fact provide a good quantification of how dynamic a given word is.

We resort to the word's *dynamic magnitude* $d_m(w)$ to derive the dynamic smoothing $f_\lambda(w)$. The rationale is that temporal relevance ϕ_w should only be accounted for words that vary significantly (high $d_m(w)$). Thus, we define $f_\lambda(w) = 1 - \frac{1}{exp(d_m(w))}$, where the exponential is used as a smoothing function.

3.3 Word-Topic Semantic Affinity

Word semantic affinity to a topic s_i is given by the function $s_{rank}(w, s_i|\mathcal{M}_s^E)$. To learn the model $\mathcal{M}_s^E$ of an event E, we train a continuous Skip-Gram model [12] (*word2vec*) which learns distributed vector word representations, while capturing a large number of syntactic and semantic word relationships. Namely, words with similar word-contexts lye close. $d(w_1, w_2, \mathcal{M}_s^E)$ is defined as the distance between two embeddings. The Skip-gram model was trained with negative sampling and a context window of size 4, over each event E corpus, to obtain 500-dimensional embeddings.

As each topic s_i is a sentence, we proposed a simple strategy, based on $d(w_1, w_2, \mathcal{M}_s^E)$ to implement $s_{rank}(w, s_i, \mathcal{M}_s^E)$. Embeddings are extracted for all candidate keywords $w \in V_i$ and words of s_i. Then we compute a $|V_i| \times |s_i|$ distances matrix, using ℓ_2 distance. We found that ℓ_2 performed better than *cosine*. Then, we reduce the computed matrix to a vector, by replacing each $|s_i|$ distances row by the smallest distance. Using this approach, $s_{rank}(w, s_i, \mathcal{M}_s^E)$ ranks each $w \in V_i$ by the distance of the closest word.

4 Evaluation

In this section, we empirically evaluate the effectiveness of TempTopRank, comparing its results with a set of baselines, on the task of dynamic keyword extraction.

Corpus. The corpus consists of a collection of crawled social media documents from Twitter, and has a total of 82,348 and 325,074 documents, from Edinburgh Festival 2016 (EdFest) and Tour de France 2016 (TDF), respectively. The crawling was focused around the days of each event. SPAM was discarded with content filtering techniques. Additionally, for each event, 31 news topics (each corresponding to a news article) were obtained from well reputed sources (*BBC News*, *Reuters* and *The Guardian*).

Table 1. Results on keyword relevance, using the top-10 ranked keywords for each method.

Method	EdFest 2016			TDF 2016		
	nDCG	*mAP*	*P*	*nDCG*	*mAP*	*P*
Non-temporal methods						
Random	2.32	1.13	0.65	12.73	6.85	3.87
RAKE [14]	25.57	17.26	6.45	47.25	33.97	16.77
TextRank [11]	86.16	80.13	**43.23**	93.81	91.42	55.81
Time-filtered methods						
RAKE-TF [14]	25.83	18.90	6.45	53.60	39.71	22.81
TextRank-TF [11]	82.04	77.10	32.26	94.19	92.46	**59.03**
Time-based methods						
TempRank	47.24	39.30	13.87	65.26	57.18	34.19
TempTopRank	86.94	85.81	32.26	**96.12**	**94.83**	51.94
TempTopRank+*dm*	**88.85**	**87.83**	33.55	94.45	92.59	48.39

Protocol. For each topic s_i, all methods take as input a representative topic corpus $\mathcal{C}_i^E$, consisting of the top-1000 tweets retrieved with a state-of-the-art retrieval model, using s_i as query. The topic timestamp t_i, that corresponds to the news article publication date, is only used to define the range used in Eq. 1. Stemming and expansion of trivial abbreviations (e.g. Aug→August, fest→festival) is performed to words $w \in V_i$ of $\mathcal{C}_i^E$. There will be keywords more relevant than others. Thus, we use $nDCG$@10 (normalized Discounted Cumulative Gain) as it accounts for different relevance levels, promoting ranks with highly relevant keywords at the top. We also consider mean Average Precision at 10 (mAP@10) and Precision at 10 (P@10).

Relevance Judgements. For each str^i, the top-10 ranked keywords inferred by each method were selected to be evaluated. We asked two judges to evaluate the relevance of each keyword w.r.t. to each topic s_i. The judges were both familiar with Twitter and the events considered. For each topic, we show the news title s_i, URL, and the keywords. For each str_i, judges read the news topic article and 20 representative tweets from $\mathcal{C}_i^E$. Each keyword is annotated with 0 (non-relevant), 1 (relevant) or 2 (highly relevant). Keywords used as crawling seeds and highly common words (e.g. *Edinburgh*, *Tour*, etc.) were ignored. We obtain an average Cohen's Kappa coefficient of $\kappa = 0.53 \pm 0.148$ and $\kappa = 0.49 \pm 0.095$ for *EdFest* and *TDF*, respectively, indicating *moderate* agreement.

4.1 Results and Discussion

Two state-of-the-art keyword extraction baselines are considered. Namely, we consider the TextRank [11] algorithm (Gensim [13] library implementation), and RAKE [14]. Additionally, two variants of each baseline are considered: the first one uses all tweets in $\mathcal{C}_i^E$ (referred as TextRank and RAKE); the second, referred as TextRank-TF and RAKE-TF, removes from $\mathcal{C}_i^E$ all the tweets that are outside the temporal range $[t_i - t_{gap}, t_i]$, (as Eq. 1). We evaluate TempTopRank with dynamic smoothing f_λ (TempTopRank+dm), and with static smoothing (TempTopRank), i.e. $f_\lambda(t_i) = \alpha$ (constant). The temporal ranking component (Eq. 1) is evaluated separately (TempRank). For the parameters t_{gap} and α, we experiment values in the range $[1, 10]$ and $[0, 1]$, with steps of 1 and 0.1, respectively. We report the results corresponding to the best obtained. The number of latent topics was set to $K = 10$ through grid-search.

Table 1 reports the results obtained. We can see that the two proposed TempTopRank variants outperformed all the baselines on both $nDCG$ and mAP, on the two datasets. The variant of TempTopRank+dm outperforms the two baselines, TempRank and TempTopRank on EdFest dataset. On TDF dataset it was outperformed by TempTopRank. We believe that this is due to the fact that in TDF there are mostly periodic news articles (on a per-stage basis), that are very similar semantically (e.g. different cities per stage) with most of the relevant keywords being the same. Supporting this affirmation is the fact that for TDF dataset, TempTopRank achieved the best $nDCG$ and mAP results with $\alpha = 0.3$, meaning that more importance is given to the semantic affinity component. For EdFest the best result of TempTopRank was obtained with $\alpha = 0.1$.

TextRank obtained the highest precision P score. $P@10$ assumes that any annotated keyword with score ≥1 is relevant. TextRank is a graph-based method, thus has the property that important word vertices will *promote* other less important vertices. This enables TextRank to rank up less relevant words (score = 1). Notwithstanding, given the novelty of the evaluation scenario and the adaptations made to TextRank, these results constitute a contribution, for a situation where one wishes to extract as many as possible relevant keywords, regardless of the relevance score. The results of the two created variants, RAKE-TF and TextRank-TF, allow us to confirm the adequacy of temporal range (Eq. 1) considered. Both methods are able to achieve maximum performance, each in one of the datasets, just by using documents from that range.

Table 2. Sample of 5 keywords from the top-10 retrieved keywords of TempTopRank. Annotated relevance scores are shown in parenthesis.

Topic s_i:	*Masai Graham's organ donor gag is Edinburgh fringe's funniest joke*	*Capturing the castle: Edinburgh festival's Deep Time spectacular*	*Tour de France 2016: Chris Froome completes third race victory*	*Tour de France: Chris Froome extends lead as Peter Sagan wins stage*
Extracted keywords w:	Joke (2), Masai (2), quip (1), heart (2), award (2)	deep (2), time (2), erupt (1), backdrop (2), arena (1)	Chris (2), title (1), congratulations (1), Froome (2), third (2)	stage (2), Ventoux (1), Sagan (2), retain (1), yellow (0)

All methods achieved better results on the TDF dataset. This is mainly due to two reasons: (i) as TDF is more mainstream, the corpus is larger, thus resulting in more linguistic support; and (ii) the EdFest event is wider and more complex, involving multiple types of shows (theatre, music, comedy, dance, etc.), while TDF is only based on cycling. Table 2 depicts a sample of the top-10 keywords obtained with our best method.

5 Conclusions

In this paper we proposed a dynamic-keyword extraction method from social media topics, TempTopRank, that jointly models words' temporal signatures and keyword's semantic affinity. A key novelty of TempTopRank is the introduction of a technically sound model for quantifying the temporal behaviour of words. This is achieved by estimating word's temporal densities and dynamic magnitude. Experiments showed that TempTopRank outperforms all the baselines, thus effectively exploit keyword dynamics over topics and confirming the importance of word temporal dynamics.

Acknowledgements. This work has been partially funded by the CMU Portugal research project GoLocal Ref. CMUP-ERI/TIC/0033/2014, by the H2020 ICT project COGNITUS with the grant agreement No 687605 and by the project NOVA LINCS Ref. UID/CEC/04516/2013.

References

1. Abilhoa, W.D., de Castro, L.N.: A keyword extraction method from twitter messages represented as graphs. Appl. Math. Comput. **240**, 308–325 (2014). https://doi.org/10.1016/j.amc.2014.04.090
2. Barrios, F., López, F., Argerich, L., Wachenchauzer, R.: Variations of the similarity function of textrank for automated summarization. CoRR abs/1602.03606 (2016)
3. Blei, D.M., Lafferty, J.D.: Dynamic topic models. In: Proceedings of the 23rd International Conference on Machine Learning, ICML 2006, pp. 113–120. ACM, New York (2006). https://doi.org/10.1145/1143844.1143859
4. Campos, R., Mangaravite, V., Pasquali, A., Jorge, A.M., Nunes, C., Jatowt, A.: A text feature based automatic keyword extraction method for single documents. In: Pasi, G., Piwowarski, B., Azzopardi, L., Hanbury, A. (eds.) ECIR 2018. LNCS, vol. 10772, pp. 684–691. Springer, Cham (2018). https://doi.org/10.1007/978-3-319-76941-7_63
5. Chen, S.F., Goodman, J.: An empirical study of smoothing techniques for language modeling. In: Proceedings of the 34th Annual Meeting on Association for Computational Linguistics, pp. 310–318. Association for Computational Linguistics (1996)
6. Graus, D., Tsagkias, M., Weerkamp, W., Meij, E., de Rijke, M.: Dynamic collective entity representations for entity ranking. In: Proceedings of the Ninth ACM International Conference on Web Search and Data Mining, WSDM 2016, pp. 595–604. ACM, New York (2016). https://doi.org/10.1145/2835776.2835819
7. Litvak, M., Last, M.: Graph-based keyword extraction for single-document summarization. In: Proceedings of the Workshop on Multi-source Multilingual Information Extraction and Summarization, MMIES 2008, pp. 17–24. Association for Computational Linguistics, Stroudsburg (2008). http://dl.acm.org/citation.cfm?id=1613172.1613178
8. Liu, F., Pennell, D., Liu, F., Liu, Y.: Unsupervised approaches for automatic keyword extraction using meeting transcripts. In: Proceedings of Human Language Technologies: The 2009 Annual Conference of the North American Chapter of the Association for Computational Linguistics, NAACL 2009, pp. 620–628. Association for Computational Linguistics, Stroudsburg (2009). http://dl.acm.org/citation.cfm?id=1620754.1620845
9. Liu, Z., Huang, W., Zheng, Y., Sun, M.: Automatic keyphrase extraction via topic decomposition. In: Proceedings of the 2010 Conference on Empirical Methods in Natural Language Processing, EMNLP 2010, pp. 366–376. Association for Computational Linguistics, Stroudsburg (2010). http://dl.acm.org/citation.cfm?id=1870658.1870694
10. Marujo, L., et al.: Automatic keyword extraction on twitter. In: 53rd Annual Meeting of the Association for Computational Linguistics. ACL, July 2015
11. Mihalcea, R., Tarau, P.: TextRank: bringing order into texts. In: Proceedings of EMNLP-04 and the 2004 Conference on Empirical Methods in Natural Language Processing, July 2004

12. Mikolov, T., Sutskever, I., Chen, K., Corrado, G.S., Dean, J.: Distributed representations of words and phrases and their compositionality. In: Burges, C.J.C., Bottou, L., Welling, M., Ghahramani, Z., Weinberger, K.Q. (eds.) Advances in Neural Information Processing Systems 26, pp. 3111–3119. Curran Associates Inc. (2013)
13. Řehůřek, R., Sojka, P.: Software framework for topic modelling with large corpora. In: Proceedings of the LREC 2010 Workshop on New Challenges for NLP Frameworks, pp. 45–50. ELRA, Valletta, May 2010. http://is.muni.cz/publication/884893/en
14. Rose, S., Engel, D., Cramer, N., Cowley, W.: Automatic keyword extraction from individual documents. In: Text Mining: Applications and Theory, pp. 1–20 (2010)
15. Wu, W., Zhang, B., Ostendorf, M.: Automatic generation of personalized annotation tags for twitter users. In: Human Language Technologies: The 2010 Annual Conference of the North American Chapter of the Association for Computational Linguistics, HLT 2010, pp. 689–692. Association for Computational Linguistics, Stroudsburg (2010). http://dl.acm.org/citation.cfm?id=1857999.1858100
16. Zhao, W.X., et al.: Topical keyphrase extraction from Twitter. In: Proceedings of the 49th Annual Meeting of the Association for Computational Linguistics: Human Language Technologies - Volume 1, HLT 2011, pp. 379–388. Association for Computational Linguistics, Stroudsburg (2011). http://portal.acm.org/citation.cfm?id=2002472.2002521
17. Zhao, W.X., et al.: Comparing twitter and traditional media using topic models. In: Clough, P., et al. (eds.) ECIR 2011. LNCS, vol. 6611, pp. 338–349. Springer, Heidelberg (2011). https://doi.org/10.1007/978-3-642-20161-5_34

Local Popularity and Time in top-N Recommendation

Vito Walter Anelli[1(✉)], Tommaso Di Noia[1], Eugenio Di Sciascio[1], Azzurra Ragone[2], and Joseph Trotta[1]

[1] Polytechnic University of Bari, Bari, Italy
{vitowalter.anelli,tommaso.dinoia,eugenio.disciascio, joseph.trotta}@poliba.it
[2] Bari, Italy
azzurra.ragone@gmail.com

Abstract. Items popularity is a strong signal in recommendation algorithms. It strongly affects collaborative filtering approaches and it has been proven to be a very good baseline in terms of results accuracy. Even though we miss an actual personalization, global popularity can be effectively used to recommend items to users. In this paper we introduce the idea of a *time-aware personalized popularity* in recommender systems by considering both items popularity among neighbors and how it changes over time. An experimental evaluation shows a highly competitive behavior of the proposed approach, compared to state of the art model-based collaborative approaches, in terms of results accuracy.

1 Introduction

Collaborative-Filtering (CF) [25] algorithms, more than others, have gained a key-role among recommendation approaches and have been effectively implemented in commercial systems to help users in dealing with the information overload problem. Some of them also use additional information (hybrid approaches) to build a more precise user profile in order to serve a much more personalized list of items [3,9].

However, it is well known [12] that all the algorithms based on a CF approach are affected by the so called "popularity bias" meaning that popular items tend to be recommended more frequently than those in the long tail. Initially considered as a shortcoming of collaborative filtering algorithms and then not useful to produce good recommendations [11], in some works items popularity has been intentionally penalized [17]. Very interestingly, a recommendation algorithm purely based on most popular items, has been proven to be a strong baseline [7] although it does not exploit any actual personalization. More recently, popularity has been also considered as a natural aspect of recommendation that, by measuring the user tendency to diversification, can be exploited to balance the recommender optimization goals [13]. The study of popularity in user tendencies is not completely new in the recommender systems field. Some interesting

A. Ragone—Independent Researcher.

L. Azzopardi et al. (Eds.): ECIR 2019, LNCS 11437, pp. 861–868, 2019.
https://doi.org/10.1007/978-3-030-15712-8_63

works explored these criteria for re-ranking purposes [13,17], and multiple goals optimization [11].

In the approach we present here, we introduce a more fine-grained personalized version of popularity by assuming that it is conditioned by the items that a user u already experienced *in the past.* To this extent, we look at a specific class of neighbors, that we name *Precursors*, defined as the users who already rated the same items of u in the past. This led us to the introduction of a time-aware analysis while computing a recommendation list for u. As time is considered a contextual feature, most of the works dealing with temporal aspects are considered as a sub-class of Context-Aware RS (CARS) [2]: Time-Aware RS (TARS) [1,14,24]. In TARS, the freshness of different ratings is often considered as a discriminative factor between candidate items. Usually, a time window [15] is adopted to filter out all the ratings that stand before (and/or after) a certain time relative to the user or the item. Recently, an interesting work that makes use of time windows has been proposed in [5] where the authors focus on the last common interaction between the target user and her neighbors to populate the candidate items list. In [4] social information and time are integrated dealing with the interests of the users as a series of temporal matrices. Probabilistic matrix factorization technique are adopted to learn latent factors. Regarding sequences and recommendation it is worth to mention [22], in which the authors combine an LSTM network with a low-rank matrix factorization algorithm to produce recommendation lists. A pioneer work was proposed more than a decade ago in [8] which used an exponential decay function $e^{-\lambda t}$ to penalize old ratings. An exponential decay function [14] was then used to integrate time in a latent factors model. In the last years, several Item-kNN [8,16] with a temporal decay function have been deployed. Another interesting work was proposed in [23] where three different kinds of time decay were adopted: exploiting concave, convex and linear functions.

In this paper we present `TimePop`, an algorithm that combines the notion of personalized popularity conditioned to the behavior of users' neighbors while taking into account the temporal dimension. It is worth noticing that `TimePop` works with implicit feedback to compute recommendations. Differently from some of the approaches previously described, in `TimePop` we avoid both the use of a time window and the selection of a fixed number of candidate items. Indeed, while on the one hand, a time window may severely restrict the selection of candidates, on the other hand, a fixed number of candidate items may heavily affect the algorithm results.

2 Time-Aware Local Popularity

The leading intuition behind `TimePop` is that the popularity of an item has not to be considered as a global property but it can be personalized if we consider the popularity in a neighborhood of users. We started from this observation to formulate a form of personalized popularity, and then we added the temporal dimension to strengthen this idea.

In `TimePop`, given a user u the first step is then the identification of user's neighbors who rated the same items as u but before u. We name these users

Precursors. In our intuition, Precursors represent a community of users u relies on to choose the items to enjoy. In a neighborhood of u, the same item is enjoyed by users in different time frames. This leads us to the second ingredient behind `TimePop`: personalized popularity is a function of time. The more the ratings about an item are recent, the more its popularity is relevant for the specific user. Hence, in order to exploit the temporal aspect of these ratings, the contributions of *Precursors* can be weighted depending on their freshness.

We now introduce some basic notation that will be used in the following. We use $u \in U$ and $i \in I$ to denote users and items respectively. Since we are not just interested in the items a user rated but also at when the rating happened, we have that for a user u the corresponding user profile is $P_u = \{(i_1, t_{ui_1}), \ldots, (i_n, t_{ui_n})\}$ with $P_u \subseteq I \times \Re$, being t_{ui} a timestamp representing when u rated i.

Definition 1 (Candidate Precursor and Precursor). *Given* $(i, t_{ui}) \in P_u$ *and* $(i, t_{u'i}) \in P_{u'}$*, we say that* u' *is a* ***Candidate Precursor*** *of* u *if* $t_{u'i} < t_{ui}$*. We use the set* $\hat{\P}^u$ *to denote the set of Candidate Precursors of* u*. Given two users* u' *and* u *such that* u' *is a Candidate Precursor of* u *and a value* $\tau_u \in \Re$ *we say that* u' *is a* ***Precursor*** *of* u *if the following condition holds.*

$$|\{i \mid (i, t_{ui}) \in P_u \wedge (i, t_{u'i}) \in P_{u'} \wedge t_{u'i} < t_{ui})\}| \geq \tau_u$$

We use $\P^u$ *to denote the set of* ***Precursors*** *of* u.

A user u' is a *Candidate Precursor* of u if u' rated at least one common item i before u. Although this definition catches the intuition behind the idea of *Precursors*, it is a bit weak as it considers also users u' who have only a few or even just one item in common with u and rated them before she did. Hence, we introduced a threshold taking somehow into account the number of common items in order to enforce the notion of *Precursors*. The threshold parameter τ_u in Definition 1 can be also computed automatically as:

$$\tau_u = \frac{\sum_{u' \in \hat{\P}^u} |\{i \mid (i, t_{ui}) \in P_u \wedge (i, t_{u'i}) \in P_{u'} \wedge t_{u'i} < t_{ui})\}|}{|\hat{\P}^u|} \tag{1}$$

To give an intuition on the computation of Precursors and of τ_u let us describe the simple example shown in Fig. 1.

Here, for the sake of simplicity, we suppose that there are only four users and six items and u is the user we want to provide recommendations to. Items that users share with u are highlighted in blue and items with a dashed red square are the ones that have been rated before u. We see that $\hat{\P}^u = \{u_2, u_4\}$. Indeed, although u_3 rated some of the items

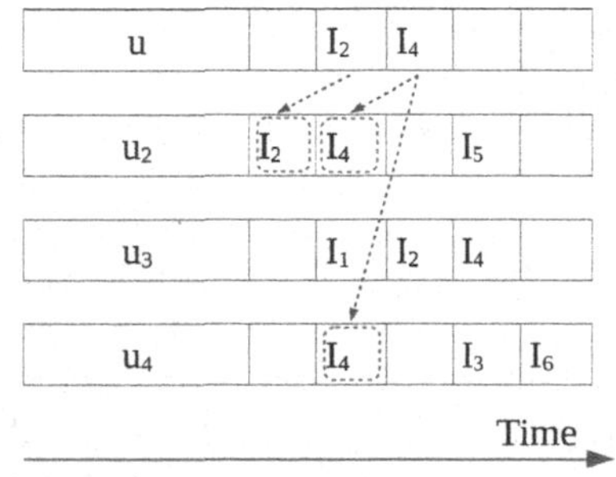

Fig. 1. Example of Precursors computation.

also rated by u they have been rated after. By Eq. (1) we have $\tau_u = \frac{3}{2} = 1.5$. Then, only u_2 results to be in $\P^u$ because she has $2 > 1.5$ shared items rated before those of u. As for u_3, it is more likely that u is a Precursor of u_3 and not vice versa.

Temporal Decay. As the definition of Precursor goes through a temporal analysis of user behaviors, we may look at the timestamp of the last rating provided by a Precursor in order to identify how active she is in the system. Intuitively, the contribution to popularity for users who have not contributed recently with a rating is lower than "active" users. On the other side, given an item in the profile of a Precursor we are interested in the freshness of its rating. As a matter of fact, old ratings should affect the popularity of an item less than newer ratings. Summing up, we may classify the two temporal dimensions as **old/recent user** and **old/recent item**. In order to quantify these two dimensions for Precursors we introduce the following timestamps: $\mathbf{t_0}$ this is the reference timestamp. It represents the "now" in our system; $\mathbf{t_{u'i}}$ is the time when u' rated i; $\mathbf{t_{u'l}}$ represents the timestamp associated to the last item l rated by the user u'. Different temporal variables are typically used [8,14], and they mainly focus on **old/recent items**. ΔT may refer to the timestamp of the items with reference to the last rating of u' [8] with $\Delta T = \mathbf{t_{u'l}} - \mathbf{t_{u'i}}$ or to the reference timestamp [14] with $\Delta T = \mathbf{t_0} - \mathbf{t_{u'i}}$. As we stated before, our approach captures the temporal behavior of both **old/recent users** and **old/recent items** at the same time. We may analyze the desired ideal behavior of ΔT depending on the three timestamps previously introduced as represented in Table 1.

Table 1. Ideal values of ΔT w.r.t. the Precursor characteristics

	recent user ($\mathbf{t_0} \approx \mathbf{t_{u'l}}$)	**old user** ($\mathbf{t_0} \gg \mathbf{t_{u'l}}$)
recent item ($\mathbf{t_{u'l}} \approx \mathbf{t_{u'i}}$)	≈ 0	$\mathbf{t_0} - \mathbf{t_{u'l}}$
old item ($\mathbf{t_{u'l}} \gg \mathbf{t_{u'i}}$)	$\mathbf{t_{u'l}} - \mathbf{t_{u'i}}$	$\mathbf{t_0} - \mathbf{t_{u'l}}$

Let us focus on each case. In the upper-left case we want ΔT to be as small as possible because both u' and the rating for i are "recent" and then highly representative for a popularity dimension. In the upper-right case, the rating is recent but the user is old. The last item has been rated very close to i but a large value of ΔT should remain because the age of u' penalizes the contribution. The lower-left case denotes a user that is active on the system but rated i a long time ago. In this case the contribution of this item is almost equal to the age of its rating. The lower-right case is related to a scenario in which both the rating and u' are old. In this scenario, the differences between the reference timestamp minus the last interaction and the reference timestamp minus the rating of i are comparable: $(\mathbf{t_0} - \mathbf{t_{u'l}}) \approx (\mathbf{t_0} - \mathbf{t_{u'i}})$. In this case, we wish the contribution of ΔT to consider the elapsed time from the last interaction (or the rating) until the reference timestamp. All the above observations lead us to define $\Delta T = |\mathbf{t_0} - 2\mathbf{t_{u'l}} + \mathbf{t_{u'i}}|$. In order to avoid different decay coefficients, in our experimental evaluation, all ΔTs are transformed in days (from milliseconds) as a common practice.

The Recommendation Algorithm. We modeled our algorithm `TimePop` to solve a *top-N* recommendation problem. Given a user u, `TimePop` computes the recommendation list by executing the following steps:

1. Compute $\P^u$;
2. For each item i such that there exists $u' \in \P^u$ with $(i, t_{u'i}) \in P_{u'}$ compute a score for i by summing the number of times it appears in $P_{u'}$ multiplied by the corresponding decay function;
3. Sort the list in decreasing order with respect to the score of each i.

For sake of completeness, in case there were no precursors for a certain user, a recommendation list based on global popularity is returned to u. Moreover, if `TimePop` is able to compute only m scores, with $m < N$, the remaining items are returned based on their value of global popularity.

3 Experimental Evaluation

In order to evaluate `TimePop` we tested our approach considering datasets related to different domains. Two of them related to the movie domain—the well-known Movielens1M dataset and Amazon[1] Movies—and a dataset referring to toys and games—Amazon Toys and Games, with 2M ratings and a sparsity of 99.99949%. "All Unrated Items" [21] protocol has been chosen to compare different algorithms where, for each user, all the items that have not yet been rated by the user all over the platform are considered. In order to evaluate time-aware recommender systems in an offline experimental setting, a typical k-folds or hold-out splitting would be ineffective and unrealistic. To be as close as possible to an online real scenario we used the fixed-timestamp splitting method [6,10], also used in [5] but with a *dataset centered* base set. The basic idea is choosing a single timestamp that represents the moment in which test users are on the platform waiting for recommendations. Their past corresponds to the training set, and the performance is evaluated with data coming from their future. In this work, we select the splitting timestamp that maximizes the number of users involved in the evaluation by setting two constraints: the training set must keep at least 15 ratings, and the test set must contain at least 5 ratings. Training set and test set for the three datasets are publicly available[2] along with the splitting code for research purposes. In order to evaluate the algorithms we measured *normalized Discount Cumulative Gain*@N (*nDCG@N*) using *Time-independent rating order condition* [6]. The metric was computed per user and then the overall mean was returned using the RankSys framework and adopting *Threshold-based relevant items* condition [6]. The threshold used to consider a test item as relevant has been set to the value of 4 w.r.t. a 1–5 scale for all the three datasets.

[1] http://jmcauley.ucsd.edu/data/amazon/.
[2] https://github.com/sisinflab/DatasetsSplits.

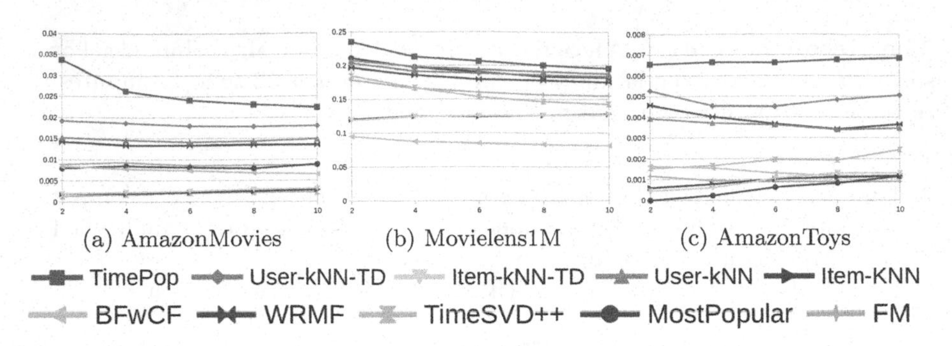

Fig. 2. nDCG @N varying N in 2..10

Baselines. We evaluated our approach w.r.t CF and time-aware techniques. **MostPopular** was included as `TimePop` is a time-aware variant of "Most Popular". From model-based collaborative filtering approaches we selected some of the best performing matrix factorization algorithms **WRMF** trained with a regularization parameter set to 0.015, α set to 1 and 15 iterations, and **FM**[3][18], computed with an ad-hoc implementation of a 2 degree factorization machine considering users and items as features, trained using Bayesian Personalized Ranking Criterion [19]. Moreover, we compared our approach against the most popular memory-based kNN algorithms, **Item-kNN**(see Footnote 3) and **User-kNN**(see Footnote 3) [20], together with their time-aware variants (**Item-kNN-TD**(see Footnote 3), **User-kNN-TD**(see Footnote 3))[8]. We included **TimeSVD++**(see Footnote 3) [14] in our comparison even though this latter has been explicitly designed for the rating prediction task. All model-based algorithms were trained using 10, 50, 100, and 200 factors; only best models are reported in the evaluation: for Movielens1M WRMF 10, FM 10; for Amazon Movies WRMF 100, FM 200; for Amazon Toys and Games WRMF 100, FM 50. Finally **BFwCF** [5] is an algorithm that takes into account interaction sequences between users and it uses the last common interaction to populate the candidate items list. In this evaluation we included the BFwCF variant that takes advantage of similarity weights per user and two time windows, left-sided and right-sided (Backward-Forward). BFwCF was trained using parameters from [5]: 100 neighbors, *indexBackWards* and *indexForwards* set to 5, normalization and combination realized respectively via *DummyNormalizer* and *SumCombiner*. Recommendations were computed with the implementation publicly provided by authors. In order to guarantee a fair evaluation, for all the time-based variants the β coefficient was set to $\frac{1}{200}$ [14]. TimeSVD++ was trained using parameters used in [14].

Results Discussion. Results of experimental evaluation are shown in Fig. 2 which illustrate nDCG (Fig. 2a, b, c) curves for increasing number of top ranked items returned to the user. Significance tests have been performed for accu-

[3] https://github.com/sisinflab/recommenders.

racy metrics using Student's t-test and p-values and they result consistently lower than 0.05. By looking at Fig. 2a we see that `TimePop` outperforms comparing algorithms in terms of accuracy on AmazonMovies dataset. We also see that algorithms exploiting a Temporal decay function perform well w.r.t. their time-unaware variants (User-kNN and Item-kNN) while matrix factorization algorithms (WRMF, TimeSVD++ and FM) perform quite bad. The low performance of MF algorithms is very likely due to the temporal splitting that makes them unable to exploit collaborative information. We may assume that the good performance of `TimePop` w.r.t. kNN algorithms are due to the adopted threshold, that emphasizes the popular items, and hence increases accuracy metrics values. Results for Amazon Toys and Games dataset are analogous to those computed for Amazon Movies. Results for Movielens additionally show that the high number of very popular items make neighborhood-based approaches perform similarly.

4 Conclusion

In this paper we presented `TimePop`, a framework that exploits local popularity of items combined with temporal information to compute top-N recommendations. The approach relies on the computation of a set of time-aware neighbors named Precursors that are considered the referring population for a user we want to serve recommendations. We compared `TimePop` against state-of-art algorithms showing its effectiveness in terms of accuracy despite its lower computational cost in computing personalized recommendations.

References

1. Adomavicius, G., Tuzhilin, A.: Multidimensional recommender systems: a data warehousing approach. In: Fiege, L., Mühl, G., Wilhelm, U. (eds.) WELCOM 2001. LNCS, vol. 2232, pp. 180–192. Springer, Heidelberg (2001). https://doi.org/10.1007/3-540-45598-1_17
2. Adomavicius, G., Tuzhilin, A.: Context-aware recommender systems. In: Ricci, F., Rokach, L., Shapira, B., Kantor, P.B. (eds.) Recommender Systems Handbook, pp. 217–253. Springer, Boston, MA (2011). https://doi.org/10.1007/978-0-387-85820-3_7
3. Anelli, V., Di Noia, T., Di Sciascio, E., Lops, P.: Feature factorization for top-n recommendation: from item rating to features relevance. In: Proceedings of RecSysKTL, pp. 16–21 (2017)
4. Bao, H., Li, Q., Liao, S.S., Song, S., Gao, H.: A new temporal and social PMF-based method to predict users' interests in micro-blogging. Decis. Support Syst. **55**(3), 698–709 (2013)
5. Bellogín, A., Sánchez, P.: Revisiting neighbourhood-based recommenders for temporal scenarios. In: Proceedings of TempRec, pp. 40–44 (2017)
6. Campos, P.G., Díez, F., Cantador, I.: Time-aware recommender systems: a comprehensive survey and analysis of existing evaluation protocols. UMAI **24**(1–2), 67–119 (2014)
7. Cremonesi, P., Koren, Y., Turrin, R.: Performance of recommender algorithms on top-n recommendation tasks. In: Proceedings of RecSys 2010, pp. 39–46 (2010)

8. Ding, Y., Li, X.: Time weight collaborative filtering. In: Proceedings of CIKM 2005, pp. 485–492. ACM (2005)
9. Fernández-Tobías, I., Braunhofer, M., Elahi, M., Ricci, F., Cantador, I.: Alleviating the new user problem in collaborative filtering by exploiting personality information. UMUAI **26**(2–3), 221–255 (2016)
10. Gunawardana, A., Shani, G.: Evaluating recommender systems. In: Ricci, F., Rokach, L., Shapira, B. (eds.) Recommender Systems Handbook, pp. 265–308. Springer, Boston, MA (2015). https://doi.org/10.1007/978-1-4899-7637-6_8
11. Jambor, T., Wang, J.: Optimizing multiple objectives in collaborative filtering. In Proceedings of RecSys 2010, pp. 55–62 (2010)
12. Jannach, D., Lerche, L., Gedikli, F., Bonnin, G.: What recommenders recommend - an analysis of accuracy, popularity, and sales diversity effects. In: Proceedings of UMAP 2013, pp. 25–37 (2013)
13. Jugovac, M., Jannach, D., Lerche, L.: Efficient optimization of multiple recommendation quality factors according to individual user tendencies. Expert Syst. Appl. **81**, 321–331 (2017)
14. Koren, Y.: Collaborative filtering with temporal dynamics. Commun. ACM **53**(4), 89–97 (2010)
15. Lathia, N., Hailes, S., Capra, L.: Temporal collaborative filtering with adaptive neighbourhoods. In: Proceedings of SIGIR 2009, pp. 796–797 (2009)
16. Liu, N.N., Zhao, M., Xiang, E., Yang, Q.: Online evolutionary collaborative filtering. In: Proceedings of RecSys 2010, pp. 95–102 (2010)
17. Oh, J., Park, S., Yu, H., Song, M., Park, S.: Novel recommendation based on personal popularity tendency. In: Proceedings of ICDM 2011, pp. 507–516 (2011)
18. Rendle, S.: Factorization machines. In: Webb, G.I., Liu, B., Zhang, C., Gunopulos, D., Wu, X. (eds.) The 10th IEEE International Conference on Data Mining, ICDM 2010, Sydney, Australia, 14–17 December 2010, pp. 995–1000. IEEE Computer Society (2010). https://doi.org/10.1109/ICDM.2010.127, http://ieeexplore.ieee.org/xpl/mostRecentIssue.jsp?punumber=5690658
19. Rendle, S., et al.: BPR: bayesian personalized ranking from implicit feedback. In: Proceedings of UAI 2009, pp. 452–461 (2009)
20. Sarwar, B., Karypis, G., Konstan, J., Riedl, J.: Analysis of recommendation algorithms for e-commerce. In: Proceedings of EC 2000, pp. 158–167 (2000)
21. Steck, H.: Evaluation of recommendations: rating-prediction and ranking. In: Proceedings of RecSys 2013, pp. 213–220 (2013)
22. Wu, C., Ahmed, A., Beutel, A., Smola, A.J., Jing, H.: Recurrent recommender networks. In: Proceedings of WSDM 2017, pp. 495–503 (2017)
23. Xia, C., Jiang, X., Liu, S., Luo, Z., Yu, Z.: Dynamic item-based recommendation algorithm with time decay. In: Proceedings of ICNC 2010, pp. 242–247 (2010)
24. Zimdars, A., Chickering, D.M., Meek, C.: Using temporal data for making recommendations. In: Proceedings of UAI 2001, pp. 580–588 (2001)
25. Rendle, S.: Using temporal data for making recommendations. In: Proceedings of ICDM 2010, pp. 995–1000 (2001)

Multiple Keyphrase Generation Model with Diversity

Shotaro Misawa(✉), Yasuhide Miura, Motoki Taniguchi, and Tomoko Ohkuma

Fuji Xerox Co., Ltd., Yokohama, Japan
{misawa.shotaro,yasuhide.miura,motoki.taniguchi,
ohkuma.tomoko}@fujixerox.co.jp

Abstract. Encoder–decoder models have achieved high performance in their application to keyphrase generation. However, keyphrases for a source text generated by these models are similar to each other because each keyphrase is independently generated. To improve the diversity, we propose a model that iteratively generates each keyphrase while considering the formerly generated keyphrase. The experimentally obtained results indicate that our model generates more diverse keyphrases with a performance that is superior or comparable to conventional models.

Keywords: Keyphrase generation · Diversity · Attention

1 Introduction

Keyphrases represent topics and summary of a source text. They are used for indexing, summarizing, and information retrieval. Automatic keyphrase assignment approaches are divided into two categories: extractive and generative. Extractive approaches assign keyphrases by selecting phrases in the source text. Generative approaches assign keyphrases by combining words from a prepared vocabulary list, including words not appeared in the source text. These generative approaches can be implemented in an encoder–decoder model: the best approach achieves higher performance in exact match than extractive models [12].

Encoder-decoder-based generative approaches tend to generate similar keyphrases for a source text. These models generate outputs by the application of beam search. However, beam search calculates the probability for each candidate independently, and the generated keyphrases have the same word having high probability [7]. Generally, because a keyphrase can be replaced with an alternative keyphrase with a similar meaning, generation of similar keyphrases is often an efficient approach to exactly match the ground-truth keyphrases for a source text. However, to generate keyphrases that broadly cover topics in a source text, the generation of diverse keyphrases becomes a more preferred approach.

For diverse keyphrase generation, we propose a model that generates each keyphrase iteratively by considering formerly generated keyphrases for a source text. Figure 1 presents an overview of our model. Our model has multiple

L. Azzopardi et al. (Eds.): ECIR 2019, LNCS 11437, pp. 869–876, 2019.
https://doi.org/10.1007/978-3-030-15712-8_64

decoders. Each decoder generates a single keyphrase with focusing on specific words in the source text by attention [1]. Additionally, each one uses different attention severally by subtracting the value derived from the attention for the former decoder.

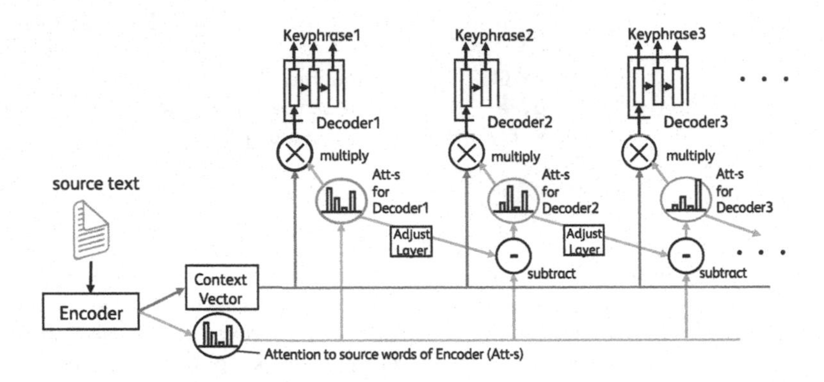

Fig. 1. Overview of our model.

The main contributions of our study are as follows: (1) Propose a model that can generate diverse keyphrases. (2) Verify the effectiveness of the model regarding diversity and performance. (3) Compare various keyphrase extraction models and generation models in terms of diversity.

2 Related Work

Keyphrase Assignment: Both unsupervised and supervised models about keyphrase extraction have been examined. Unsupervised models usually comprise two processes. The first step is extraction of candidates using heuristic rules. The second step is ranking the candidates regarding statistical features [4,11]. Supervised models solve the task as classification of the candidates or sequential labeling for the source text [15,20]. Keyphrase generation models are implemented in the encoder–decoder model: the best model is Deep Keyphrase Generation (DKG) [12]: an attentional encoder–decoder model with a copying and coverage mechanism. DKG outperforms the extractive models in terms of F1 score: however, its diversity is not compared to that of extractive models.

Diversity in Natural Language Generation: Many studies examining topics other than keyphrase generation have specifically examined diversity shortcomings in the encoder–decoder model. Methods using variational encoder–decoder have recently attracted increasing interest [2,17,21]. These models improve diversity by introducing sampling processes from the Gaussian distribution before the decoder. Another popular method to alleviate this difficulty is by adding heuristic rules to the beam search such as introducing a negative value to same words in phrases with higher probability [5,18]. Models of the abovementioned two types directly improve the diversity of outputs. Our model is designed to cover topics of a source text using attention. Our experiments compare one of these types with our model. Some studies use attention to address diversity difficulties. One of these models introduces a loss related to covering the source text by attention [16]. Another model calculates attention with consideration of the attentions of previous words in a generated sequence [13]. Such models can only consider the intra-sequence diversity within an output from a decoder, whereas our model improves the diversity among outputs.

3 Proposed Model

Simple encoder–decoder models require one-to-one alignment for training data. These models are trained on data reproduced by splitting into pairs of a source text and one of assigned keyphrases. Our model comprises one encoder and multiple decoders, which are trained on one-to-many alignments. All decoders are sequential, and each decoder considers the attention for the former decoder.

3.1 Attentional Encoder–Decoder Model with Coverage and Copying Mechanism

The base structure of our model is comprised of the encoder–decoder model. The encoder is a bidirectional Gated Recurrent Unit (GRU). Each decoder is a forward GRU. When generating each word by a decoder, the decoder specifically focuses on important words in a source text using the attention mechanism [1]. The coverage mechanism [16] is applied to improve the diversity within a single keyphrase. The copying mechanism [19] is used to generate keyphrases with out-of-vocabulary words. All decoders share the parameters: this enables the model to generate an arbitrary number of keyphrases.

3.2 Introduce Attention of Former Decoder

For each decoder to focus on a different position from that of the former decoder, attention for a decoder is calculated by subtracting attention for the former decoder. A simple subtraction has a strong constraint of focusing on the whole text, which causes the decoder to focus on unimportant parts. The introduction of a fully connected layer as an adjust layer before subtraction can degrade the constraint and alleviate the difficulty. Moreover, applying the layer can have other effects. For example, it can prevent particularly weighting of positions already focused by the former decoders by subtracting a large value for these positions. Here, the model keeps each attention value positive and regularizes the sum of the attention values as one. Equation (1) presents the attention value of ith word in a source text for decoder t.

$$\alpha_{t,i} = max(\beta_i - \gamma_i, 0) / \sum_{j \in S} max(\beta_j - \gamma_j, 0) \tag{1}$$

$$\gamma = f(\alpha_{t-1,1}, \cdots, \alpha_{t-1,N})$$

where $\alpha_{t,i}$ signifies the attention value of ith word in a source text for decoder t, S denotes a set of words in the source text, f denotes for an adjust layer, N represents a total length of the source text, γ_i denotes the ith value of $\boldsymbol{\gamma}$, and β_i expresses the attention value of the encoder to i^{th} word.

Subtraction is executed when generating only the first word in a keyphrase. The first word, which considers the formerly generated keyphrase, affects its following words. The decoder generates the following words with consideration of the former keyphrase without subtraction. If the subtraction is executed for these words, then unexpected intra-diversity would occur in the generated keyphrase.

Each decoder generates a keyphrase by beam search. However, the model does not lack in diversity because it uses the top-1 keyphrase. A decoder outputs the following phrase when the top keyphrase is same as the one by a former decoder.

4 Experiments

4.1 Experimental Settings

We conducted experiments on a dataset with titles and abstracts of scientific papers acquired from the previous work [12]. Keyphrases in the datasets are assigned by the authors. Figure 2(a) shows an example of the input and ground-truth in the dataset. Training and validation data were randomly sampled. Training and validation data have 100,000 and 10,000 papers, respectively. Test data comprise three domains: KP20k, Inspec, and Krapivin. KP20k, which has 10,000 papers, was also sampled. Inspec and Krapivin have 500 and 400 papers, respectively. Only the domain of KP20k is the same as training and validation data.

We set the beam size to 50 and cut the source text after 2,000 words. The remaining parameters are the same as those presented in an earlier report [12]. During training, the order of keyphrases follows the order decided by the authors. When a paper in the training data has more than five keyphrases, we split them into pairs of one paper and five keyphrases due to memory limitations.

We prepare extractive models; supervised models (KEA [8] and WINGNUS (WING) [14]) and unsupervised models (tfidf, Topical Page Rank (TPR) [6], and Multipartite graphs (MPG) [4]). The parameters follow the earlier work [3]. We also prepare DKG and DKG with diversity promoting beam search (DBS) [10], which is the heuristic mechanism that diversifies beam search outputs. The DBS parameter is decided based on the F1 score for validation data.

We use the F1 score based on the exact match for keyphrases while predicting top-5 and top-10 keyphrases (F@5, F@10). To evaluate the diversity, we use *distinct-1* and *disticnt-2* (*dist1, dist2*) [9], which indicate the distinct uni-gram and bi-gram rate in the top-5 generated keyphrases. For reference, we calculates *dist1* and *dist2* of ground-truth keyphrases (truth).

4.2 Experimental Results

Table 1 lists the F1 scores and diversity of all models. The table contents demonstrate that our model can generate more diverse keyphrases than other generative models. Our model outperforms other generative models in most cases in

Table 1. F@5, F@10, *dist1*, and *dist2* score of models. Underlined and bold text represent the best result in the entire and generative models for each condition. Asterisk denotes statistical significance over DKG at 5%.

	KP20k				Inspec				Krapivin			
model	F@5	F@10	*dist1*	*dist2*	F@5	F@10	*dist1*	*dist2*	F@5	F@10	*dist1*	*dist2*
tfidf	14.00	13.72	39.15	63.56	8.01	10.15	44.98	72.68	11.13	11.46	43.23	73.80
TPR	10.98	12.13	52.20	81.53	21.93	27.35	58.70	86.62	11.73	13.98	50.01	81.62
MPG	13.93	13.13	86.94	98.43	19.40	24.76	88.54	98.51	13.42	13.17	85.40	98.02
KEA	15.59	15.27	37.28	67.82	9.32	12.32	38.30	73.15	11.96	12.14	38.81	77.25
WING	15.41	15.05	41.40	73.26	10.24	14.15	46.12	80.26	11.32	12.23	45.58	83.49
DKG	26.22	**25.36**	61.67	81.57	17.19	20.06	59.18	83.19	16.93	17.50	70.66	86.82
DBS	26.42	**25.36**	61.52	81.87	17.29	20.28	59.87	83.68	16.98	17.58	70.91	86.84
ours	*26.65	25.31	*64.86	*83.29	*18.49	*21.39	*62.93	**84.78**	*19.48	*18.66	*73.09	**88.89**
truth	-	-	88.22	98.01	-	-	75.97	93.71	-	-	85.39	97.61

terms of performance. Particularly, our model is superior in F@5 score. The gap of F@10 score between DKG and our model is smaller than that of F@5 score because many papers are assigned a few keyphrases. It is difficult to train the adjust layer for generating top-10 keyphrases. The gaps of the F1 score and diversity between DKG and DBS are slight because DBS is optimized to maximize the F1 score. Generative models perform better than all extractive models in KP20k and Krapivin but perform worse in Inspec. Inspec has two advantageous characteristics for unsupervised extractive model. Firstly, it is in a different domain from training data. Secondly, most keyphrases in Inspec are extractable from the source text. These are more easily assignable than unextractable keyphrases. Krapivin also meets the aforementioned first characteristic. However, the second characteristic differs between these two datasets. The rate of extractable keyphrases is 72.85% for Inspec, whereas it is 56.14% for Krapivin.

5 Discussion

Figure 2(b) shows the top-5 outputs from DKG and our model. While the four outputs from DKG contain ‘classification’, outputs from our model does not use the same word among the outputs. Additionally, the outputs from DKG seem to focus on the title. On the contrary, our model can generate a correct keyphrase ‘scanning area’ in the middle of the abstract. The constraints about diversity in our model are not very strict. Outputs from our model mainly relate to the important positions, title and the first sentence of the abstract.

Figure 3(a) shows the distributions over the centroid positions of attention for each decoder when generating keyphrases for KP20k. This result indicates that each decoder tends to specifically focus on a different position. Additionally, the attention shifts from start to end of the source text as the model applies an adjust layer. Figure 3(b) shows the distributions over the first appearance position of the keyphrase for each order of assignation by authors. This distribution reveals that authors tend to assign keyphrases in the order of appearance. Our model reflects the characteristic. Owing to this reason, the attention shifts as shown in

Title	4D pattern structure features by three stages clustering algorithm for image analysis and classification.
Abstract	An approach for decomposition of visual image by clustering, pattern analysis and classification by structure features is considered. Hierarchical clusters such as rectangles, closed regions and integrated areas are objects of investigation. By hierarchically constructed fragments, the 4D pattern structure features are formulated. To reduce the clustering algorithm complexity, the scanning area approach is proposed. The results of pattern analysis and classification by structure features for some images are presented.
Ground-truth	visual pattern; clustering; hierarchical tree; rolling up algorithm; scanning area; rectangle; closed region; integrated area; 4d structural features; classification; indexing and retrieval

(a) Example of an input and ground-truth of the dataset.

DKG	**classification**; pattern classification; structure classification; **clustering**; image classification
Our model	**clustering**; image analysis; pattern analysis; **scanning area**; **classification**

(b) Outputs from DKG and our model.

Fig. 2. Example of datasets and outputs. Underlined words indicate the first appearance position of correct predictions. Bold keyphrases are correct predictions.

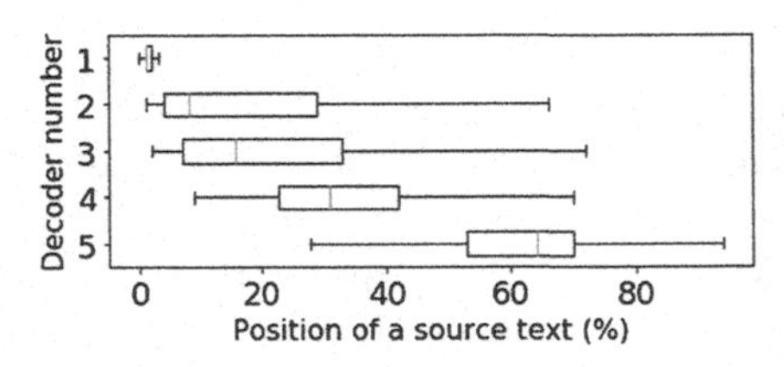

(a) Centroid of attention for each decoder.

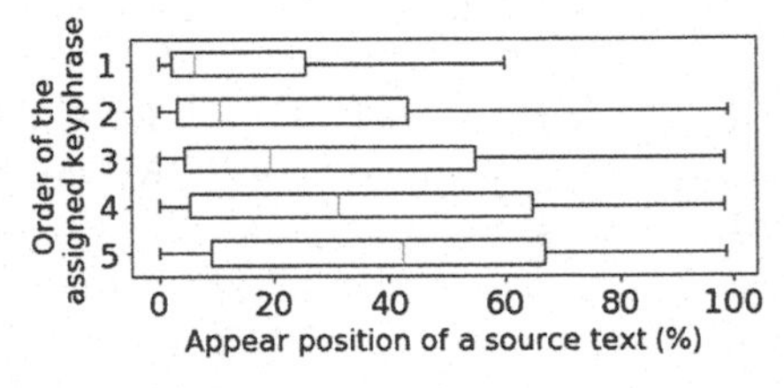

(b) Appearance of assigned keyphrases.

Fig. 3. Distribution over positions of source text.

the figure. Moreover, this figure illustrates that keyphrases are not only related to the first several sentences but also to various sentences in a source text. This indicates that the model which weakly covers sentences is suitable for this task.

A small number of keyphrases is preferable because of user interfaces and human cognitive ability. Authors of a paper tend to assign few keyphrases. Papers with more than 10 keyphrases comprise of only 3.86% in KP20k. Table 2 shows the F1 scores of two situations by splitting KP20k in terms of the number of ground-truth keyphrases: one through ten (1–10) and greater than ten (>10). Our model outperforms DBS for data with a small number of keyphrases but performs worse for many keyphrases. When authors assign many keyphrases, the amount of rephrased keyphrases is likely to increase. In such situations, a model generating similar but confident keyphrases is superior to a model generating diverse keyphrases. In fact, the *dist1* of ground-truth for papers with more than 10 keyphrases is 83.10, whereas for those with fewer than 10 keyphrases is 88.41.

Comparing generative models and extractive models in terms of the F1 score demonstrates that generative models outperform for datasets with numerous unextractable keyphrases. In terms of diversity, generative models outperform extractive models, except for MPG. MPG enhances diversity using heuristic rules related to word overlapping between candidate phrases. Table 3 demonstrates the F1 and *dist1, 2* scores for KP20k of MPG and our model with the same

Table 2. F1 scores for KP20k under each condition for the number of ground-truth.

	1–10		>10	
	F@5	F@10	F@5	F@10
DBS	26.42	24.23	**26.15**	**38.00**
ours	**26.83**	**24.32**	25.40	36.43

Table 3. Performance of MPG and our model with heuristic rules for KP20k.

	F@5	F@10	*dist1*	*dist2*
MPG	13.93	13.13	86.94	**98.43**
ours+rule	**25.28**	**23.72**	**90.39**	96.33

heuristic rules (ours+rule). Heuristics degrade the performance; however, our model achieved the same level of diversity and performed better than MPG.

6 Conclusion

We proposed a model that iteratively generates a keyphrase with consideration of formerly generated keyphrases. The experimental results demonstrate that our model can generate diverse keyphrases with superior or comparable performance to other methods. In a future study, we intend to combine the model that sorts keyphrases in a training pair to unify variations of orders for authors. Moreover, we will compare our model with other methods, such as variational encoder–decoder or a model that expands coverage mechanism from intra diversity to global diversity.

References

1. Bahdanau, D., Cho, K., Bengio, Y.: Neural machine translation by jointly learning to align and translate. arXiv preprint arXiv:1409.0473 (2014)
2. Bahuleyan, H., Mou, L., Vechtomova, O., Poupart, P.: Variational attention for sequence-to-sequence models. arXiv preprint arXiv:1712.08207 (2017)
3. Boudin, F.: pke: an open source python-based keyphrase extraction toolkit. In: Proceedings of COLING 2016, The 26th International Conference on Computational Linguistics: System Demonstrations, pp. 69–73 (2016)
4. Boudin, F.: Unsupervised keyphrase extraction with multipartite graphs. In: Proceedings of the 2018 Conference of the North American Chapter of the Association for Computational Linguistics: Human Language Technologies, vol. 2, pp. 667–672 (2018)
5. Cibils, A., Musat, C., Hossman, A., Baeriswyl, M.: Diverse beam search for increased novelty in abstractive summarization. arXiv preprint arXiv:1802.01457 (2018)
6. Florescu, C., Caragea, C.: Positionrank: an unsupervised approach to keyphrase extraction from scholarly documents. In: Proceedings of the 55th Annual Meeting of the Association for Computational Linguistics, pp. 1105–1115 (2017)
7. Gimpel, K., Batra, D., Dyer, C., Shakhnarovich, G.: A systematic exploration of diversity in machine translation. In: Proceedings of the 2013 Conference on Empirical Methods in Natural Language Processing, pp. 1100–1111 (2013)

8. Ian, H.W., Gordon, W.P., Frank, E., Gutwin, C., Craig, G.N.M.: Kea: practical automatic keyphrase extraction. In: Proceedings of the Fourth ACM Conference on Digital Libraries, pp. 254–255 (1999)
9. Li, J., Galley, M., Brockett, C., Gao, J., Dolan, B.: A diversity-promoting objective function for neural conversation models. In: Proceedings of the 2016 Conference of the North American Chapter of the Association for Computational Linguistics: Human Language Technologies, pp. 110–119 (2016)
10. Li, J., Jurafsky, D.: Mutual information and diverse decoding improve neural machine translation. arXiv preprint arXiv:1601.00372 (2016)
11. Mahata, D., Kuriakose, J., Shah, R.R., Zimmermann, R.: Key2vec: automatic ranked keyphrase extraction from scientific articles using phrase embeddings. In: Proceedings of the 2018 Conference of the North American Chapter of the Association for Computational Linguistics: Human Language Technologies, vol. 2, pp. 634–639 (2018)
12. Meng, R., Zhao, S., Han, S., He, D., Brusilovsky, P., Chi, Y.: Deep keyphrase generation. In: Proceedings of the 55th Annual Meeting of the Association for Computational Linguistics, pp. 582–592 (2017)
13. Nema, P., Khapra, M.M., Laha, A., Ravindran, B.: Diversity driven attention model for query-based abstractive summarization. In: Proceedings of the 55th Annual Meeting of the Association for Computational Linguistics, pp. 1063–1072 (2017)
14. Nguyen, T.D., Luong, M.T.: Wingnus: keyphrase extraction utilizing document logical structure. In: Proceedings of the Fifth International Workshop on Semantic Evaluation, pp. 166–169 (2010)
15. Prasad, A., Kan, M.Y.: Wing-nus at semeval-2017 task 10: Keyphrase extraction and classification as joint sequence labeling. In: Proceedings of the 11th International Workshop on Semantic Evaluation (SemEval-2017), pp. 973–977 (2017)
16. See, A., Liu, P.J., Manning, C.D.: Get to the point: summarization with pointer-generator networks. In: Proceedings of the 55th Annual Meeting of the Association for Computational Linguistics, pp. 1073–1083 (2017)
17. Shen, X., Su, H., Niu, S., Demberg, V.: Improving variational encoder-decoders in dialogue generation. arXiv preprint arXiv:1802.02032 (2018)
18. Vijayakumar, A.K., et al.: Diverse beam search: Decoding diverse solutions from neural sequence models. arXiv preprint arXiv:1610.02424 (2016)
19. Yao, T., Pan, Y., Li, Y., Mei, T.: Incorporating copying mechanism in image captioning for learning novel objects. In: IEEE Conference on Computer Vision and Pattern Recognition (CVPR), pp. 5263–5271 (2017)
20. Zhang, Y., Li, J., Song, Y., Zhang, C.: Encoding conversation context for neural keyphrase extraction from microblog posts. In: Proceedings of the 2018 Conference of the North American Chapter of the Association for Computational Linguistics: Human Language Technologies, pp. 1676–1686 (2018)
21. Zhao, T., Zhao, R., Eskenazi, M.: Learning discourse-level diversity for neural dialog models using conditional variational autoencoders. In: Proceedings of the 55th Annual Meeting of the Association for Computational Linguistics, pp. 654–664 (2017)

Author Index